24th European Cosmic Ray Symposium (ECRS 2014)

Journal of Physics: Conference Series Volume 632

Kiel, Germany
1 - 5 September 2014

ISBN: 978-1-5108-1313-7
ISSN: 1742-6588

Printed from e-media with permission by:

Curran Associates, Inc.
57 Morehouse Lane
Red Hook, NY 12571

Some format issues inherent in the e-media version may also appear in this print version.

Printed by Curran Associates, Inc. (2016)

Institute of Physics
Dirac House, Temple Back
Bristol BS1 6BE UK

Phone: 44 1 17 929 7481
Fax: 44 1 17 920 0979

techtracking@iop.org

Additional copies of this publication are available from:

Curran Associates, Inc.
57 Morehouse Lane
Red Hook, NY 12571 USA
Phone: 845-758-0400
Fax: 845-758-2634
Email: curran@proceedings.com
Web: www.proceedings.com

Table of contents

Volume 632

24th European Cosmic Ray Symposium (ECRS)
1–5 September 2014, Kiel, Germany

Accepted papers received: 29 June 2015
Published online: 13 August 2015

Papers

a. HECR I + DM + INS

b. MN + High Energy Gamma rays

c. SH + GEO

d. HECR II

The 24th European Cosmic Ray Symposium (ECRS) took place in Kiel, Germany, at the Christian-Albrechts-Universität zu Kiel from September 1 – 5, 2014. The first symposium was held in 1968 in Lodz, Poland (high energy, extensive air showers and astrophysical aspects) and in Bern (solar and heliospheric phenomena) and the two "strands" joined together in 1976 with the meeting in Leeds. The 24th ECRS covered a wide range of scientific issues divided into the following topics:

HECR-I Primary cosmic rays I (experiments)

HECR-II Primary cosmic rays II (theory)

MN Cosmic ray muons and neutrinos

GR GeV and TeV gamma astronomy

SH Energetic particles in the heliosphere (solar and anomalous CRs and GCR modulation)

GEO Cosmic rays and geophysics (energetic particles in the atmosphere and magnetosphere of the Earth)

INS Future Instrumentation

DM Dark Matter

The organizers are very grateful to the Deutsche Forschungs Gemeinschaft for supporting the symposium.

Peer review statement

All papers published in this volume of *Journal of Physics: Conference Series* have been peer reviewed through processes administered by the proceedings Editors. Reviews were conducted by expert referees to the professional and scientific standards expected of a proceedings journal published by IOP Publishing.

The knee and beyond: a review of recent results

Andrea Chiavassa

Dipartimento di Fisica dell'Università degli Studi di Torino, Via Pietro Giuria 1, 10125, Torino, Italy

E-mail: andrea.chiavassa@to.infn.it

Abstract. Primary cosmic rays of energy greater than $\sim 10^{14}eV$ must be studied by ground based experiments measuring the particles generated in the EAS (Extensive Air Shower) development in atmosphere. These experiments are mainly limited by the systematic uncertainties on an event due to their energy calibration. I will discuss the main sources of these errors: the choice of the hadronic interaction model and of the mass of the primary particle (that cannot be measured on a event by event basis).

I will then summarize some of the more recent measurements of the all particle spectrum, and I will show that, keeping into account the differences due to the energy calibration, they all agree on the spectral shape. In addition to the well known change of slope, the so called knee, two faint but significant structures have been measured by different experiments: a concavity around $10^{16}eV$ and a steepening at $\sim 10^{17}eV$.

Then I will describe the measurements of the spectra of light and heavy primaries, discussing the claimed spectral features. Using a simple calculation of the elemental spectra (based on the hypothesis that the knee energies follow a Peter's cycle) I will try to discuss if and how these results can be interpreted in a single picture.

1. Introduction

Primary cosmic rays with energy greater than $10^{14}eV$ cannot be studied by experiments operating on balloon or on satellites and their detection is only possible by means of EAS experiments. All the main characteristics of the primary particle (i.e. mass, energy and arrival direction) must therefore be derived measuring the Extensive Air Showers (EAS) generated by the interaction of the primary cosmic rays with atmospheric nuclei.

The experiments operating in the $10^{14} - 10^{18}eV$ energy range can be divided in three groups:

- Surface arrays: sampling the EAS at fixed atmospheric depth. Almost all of these arrays are able to simultaneously detect more than one EAS components: usually the electromagnetic and the muonic ones. Detecting the particle density and arrival times at different distances from the shower core these arrays derive the arrival direction of the primary cosmic ray, the number of charged particles (N_{ch}) and the number of muons (N_μ) in the EAS at observation level. Both N_{ch} and N_μ are derived as normalization of the lateral distribution of the particle density; either defined as the total number of particles at observation level or as the number of particles at a fixed distance from the shower core (distance that has to be fixed by every single experiment depending both on the detector layout and on the primary energy range studied by the array). These detectors operate with a 100% duty cycle.
- Cherenkov arrays: detecting the Cherenkov light emitted by particles during EAS development. The big advantage of these arrays is that they perform an almost calorimetric

measurement of the primary energy. The atmospheric depth of the EAS maximum (e.g. a shower parameter depending on the mass of the primary particle) can be estimated comparing the Cherenkov photon densities measured at two different distances from the shower core. These arrays operate only during clear moonless nights, thus their duty cycle is $\sim 10 - 15\%$.

- Fluorescence light telescopes: if the primary energy is greater than $\sim 5 \times 10^{16} eV$ EAS can also be detected measuring the fluorescence light emitted during their development. The major advantage of these detectors is that they allow a calorimetric determination of the primary energy and therefore their calibration does not rely on EAS simulation. The experimental observable sensible to the primary mass is the atmospheric depth of shower maximum. Like Cherenkov arrays also these detectors operate with a low duty cycle.

The main problem of these experiments is that energy, excluding fluorescence telescopes, and mass calibrations have to be performed by means of EAS simulations that are based on hadronic interaction models extrapolated from lower energy collider measurements. Recently the results coming from the LHC experiments [1] are filling this gap, nevertheless collider experiments do not cover the forward region that is the relevant one to describe the EAS development.

In spite of all the mentioned calibration problems the results of last generation experiments allowed a deeper understanding of the knee of the primary cosmic ray spectrum, the major achievements are:

- the detection, in addition to the main feature of the spectrum (i.e. the knee, at $\sim 2-4 \times 10^{15} eV$ [2]), of two faint structures, an hardening around $10^{16} eV$ and a steepening around $10^{17} eV$ [3, 4, 5].
- The measurement of the spectra of light and heavy primaries [6, 7, 8, 9]. The spectral features observed in these spectra slightly depend on the hadronic interaction model on which the energy and mass calibration of the experiments are based, while the same is not true for the measured fluxes.
- By unfolding analysis techniques the spectra of five elements (chosen as representative of five mass groups) have been measured [10, 11]. These results heavily depend on the hadronic interaction model used in the EAS simulation, therefore I will briefly mention them but I will not use them in the final discussion.
- The measurement of a large scale anisotropy at $2 \times 10^{15} eV$ [12]. The phase of the first harmonic is compatible with the ones measured at $\sim 4 \times 10^{14} eV$ [13, 14], while the amplitude is increasing.

The scenario resulting from such measurements favors an astrophysical interpretation of the knee, even if the resulting picture is not complete and, to further improve our knowledge, high precision experimental results are needed. I will show that recent results can, at least qualitatively, be described by elemental spectra having the change of slope at energies depending on Z, even if the recent ARGO-YBJ results [9] cannot be easily included in such a simple framework. Moreover in order to accurately describe the structures observed above the knee, an additional component must be introduced [15, 16].

In this review I discuss some of the more recent experimental results, while no description of the models proposed to discuss them is given. A final discussion showing the results of a simple calculation of the primary spectra and only aiming to a qualitative interpretation of the data is reported.

2. Energy and Mass Calibration of Ground Based Experiments

2.1. Energy Calibration

Surface arrays derive the energy of the primary particle either from the measured value of the number of charged particles (N_{ch}) either from the measured value of the number of muons (N_μ)

or from a combination of both.

All these shower parameters depend not only on the energy of the primary cosmic ray but also on the atmospheric depth crossed by the EAS, thus also the zenith angle θ has to be taken into account. The shower evolution in atmosphere can be treated in two different ways: a first one is based on the experimental data supposing that the arrival directions of cosmic rays are isotropically distributed, the so called "constant intensity cut" technique [17], while a second one describes the EAS atmospheric evolution by means of their simulation.

The main sources of systematic errors in the energy calibration of any EAS experiment are related to the use of a complete EAS simulation and are therefore connected to the two main hypotheses that must be done when running these codes: the mass of the primary particle generating the shower and the hadronic interaction model describing high energy interactions.

Considering the choice of the mass of the primary particle three different strategies are possible:

- the primary energy is calculated for the two extreme values of the primary mass (Hydrogen and Iron). The spectra that are obtained for these two cases represent the upper and lower limits bracketing the "true" spectrum (an example can be found in [18]).
- Starting from the primary chemical composition measured at lower energy by direct experiments and assuming its evolution with energy the mean value of the primary mass is calculated as a function of the primary energy (an example can be found in [4]). This mean value of the primary mass is then used to convert the experimental observable to primary energy.
- Combining the N_{ch} and N_{μ} values measured for each event a parameter correlated with the mass of the primary particle is evaluated. Then this parameter is used in the conversion from the experimental observable to the primary energy [3]. In this way no assumption is made about the primary chemical composition, but a further dependence on the hadronic interaction model is added (in the definition of the parameter correlated with the primary mass).

Cherenkov experiments derive the primary energy from the photon density measured at a fixed distance from the shower core, $200m$ in the case of the TUNKA-133 experiment [5]. Considering TUNKA-133 as reference experiment the correlation between $Q(200)$ and the primary energy is given by:$E_0 = CQ(200)^g$. The parameter g is evaluated by mean of the CORSIKA [19] simulation code, $g = 0.94 \pm 0.01$. The value of the constant C is obtained normalizing the integral energy spectrum of each single night of data taking to a reference one measured by the QUEST experiment[20].

The second main source of systematic error in the energy calibration of EAS experiments is the choice of the hadronic interaction model in the EAS simulation. The center of mass energy of the interaction between an high energy cosmic ray and an air nuclei is well above the ones studied at collider experiments. The reference hadronic interaction models used by current EAS experiments are: QGSJetII-02 [21], Sibyll2.1 [22] and EPOS1.99 [23]. The KASCADE-Grande experiment has evaluated the differences in the energy assignment due to the hadronic interaction model choice [24], finding that they are smaller than 20% and with a slight energy dependence. The all particle spectra derived from the KASCADE-Grande data calibrated by means of these three different hadronic interaction models are shown in figure 1.

Recently updates of two of the three previously mentioned hadronic interaction models based on the results obtained by LHC experiments have been included in the CORSIKA EAS simulation code: QGSJetII-04 [25] and EPOS-LHC [26]. First studies performed by the KASCADE-Grande collaboration [27] indicate that the differences between the energy calibrations obtained using hadronic interaction models developed before (QGSJetII-02, Sibyll2.1 and EPOS1.99) and after (QGSJetII-04 and EPOS-LHC) the LHC data are smaller than 20%.

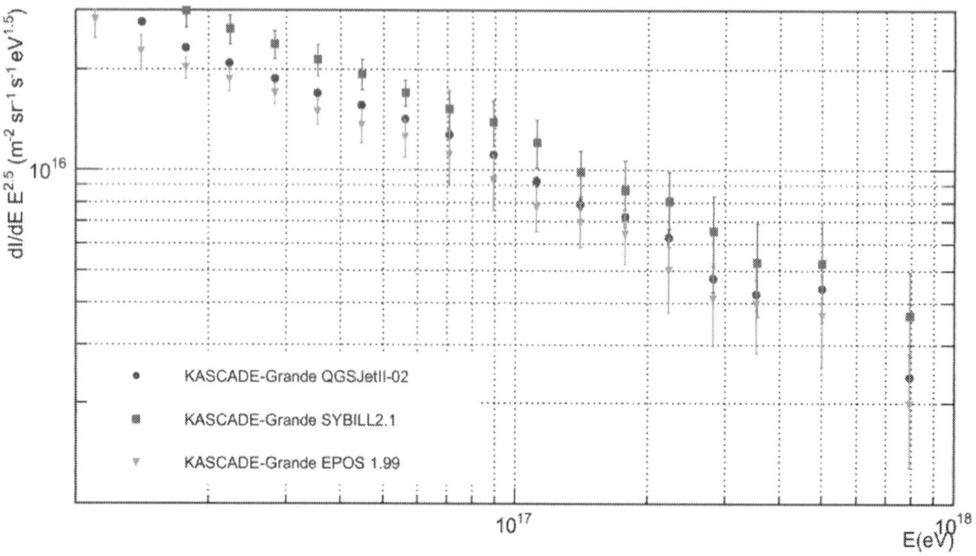

Figure 1. KASCADE-Grande measurements of the all particle energy spectra calibrated by mean of three different hadronic interaction models: QGSJetII-02, Sibyll2.1 and EPOS1.99

2.2. Mass Calibration

Due to the intrinsic fluctuations in the EAS development ground based experiments cannot access the mass of every single event. Even at shower maximum (where the fluctuations are minimized) the differences between EAS generated by primaries of similar mass (i.e. H and He, He and CNO, Si and Fe) are smaller than one standard deviation. Studies of the primary chemical composition can be performed using statistical methods like, for instance, the unfolding analysis introduced by the KASCADE experiment [10, 11], or separating events in two mass groups, i.e. light and heavy primaries.

The shower observables sizable to such purpose are:

- the measurement of the shower maximum development. With current detection techniques this information can be directly obtained only by fluorescence light telescopes, while Cherenkov detectors can infer it either from the ratio of the photon densities at two different distances from the shower core and from the measurement of the pulse width at 400 m from the core [5].
- The ratio between observables representing the charged particle and the muon numbers at observation level.
- A correlation between a parameter proportional to the number of particles at observation level and another one reflecting the shape of the particle lateral distribution.

In order to assign a value of the primary mass to any of the previously indicated experimental observables a complete EAS simulation is needed, therefore these analyses depend on the high energy hadronic interaction model.

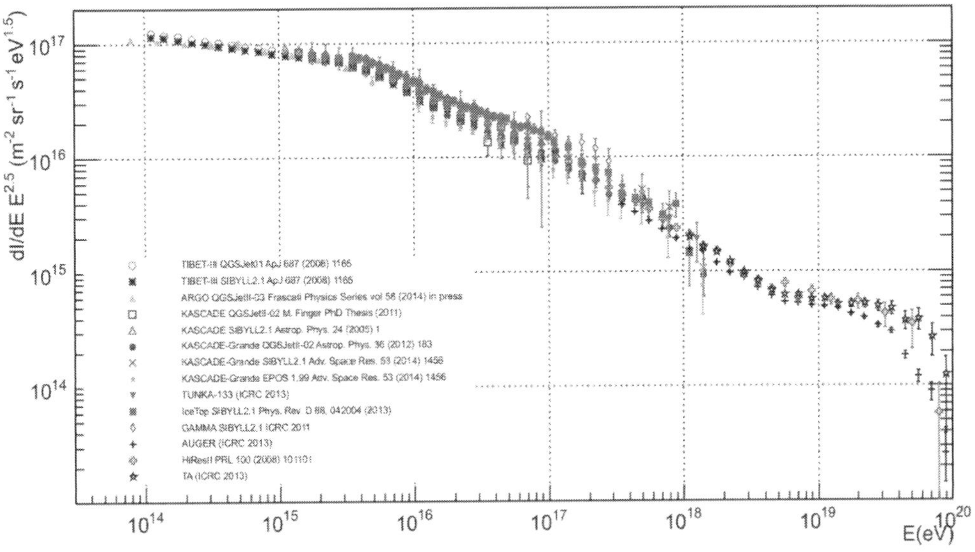

Figure 2. Compilation of all particle energy spectra measurements[28, 9, 29, 10, 3, 24, 5, 4, 30, 31, 32, 33]

3. Experimental Results

Having discussed the main sources of systematic uncertainties affecting the energy and mass measurements performed by EAS experiments, in this section I will discuss some of the more recent results obtained in the $10^{14} - 10^{18}eV$ energy range.

3.1. All Particle Energy Spectrum Measurements

Figure 2 shows a compilation of the all particle energy spectrum measured by different experiments using different experimental techniques, operating at different height above see level and calibrated by different hadronic interaction models. From this plot we can get the impression of measurements showing relevant differences between each other, but simply shifting, as already proposed in [16], the single experiment energies by an amount lower than the previously discussed systematic uncertainties (i.e. $\Delta E/E \leq 20\%$) all these results agree much better, as can be seen in figure 3. The differences shown in figure 2 can thus be attributed to the energy calibration of the experiments. There is a general agreement about the cosmic rays spectrum in the $10^{14} - 10^{17}eV$ energy range that cannot be described by a single slope power law showing three features: the well known steepening observed by Khristiansen et al. [2] more than fifty years ago at $2 - 4 \times 10^{15}eV$ (better known as the knee); a concavity at $\sim 10^{16}eV$ and a faint steepening around $\sim 10^{17}eV$ first claimed by KASCADE-Grande [3] and then confirmed by IceTop [4] and TUNKA-133 [5].

3.2. Mass Groups Energy Spectra

Ground based experiments can measure the spectra of mass groups of primary cosmic rays by two different approaches that will be described in the following.

A first analysis technique separates the events in two samples according to the measured ratio

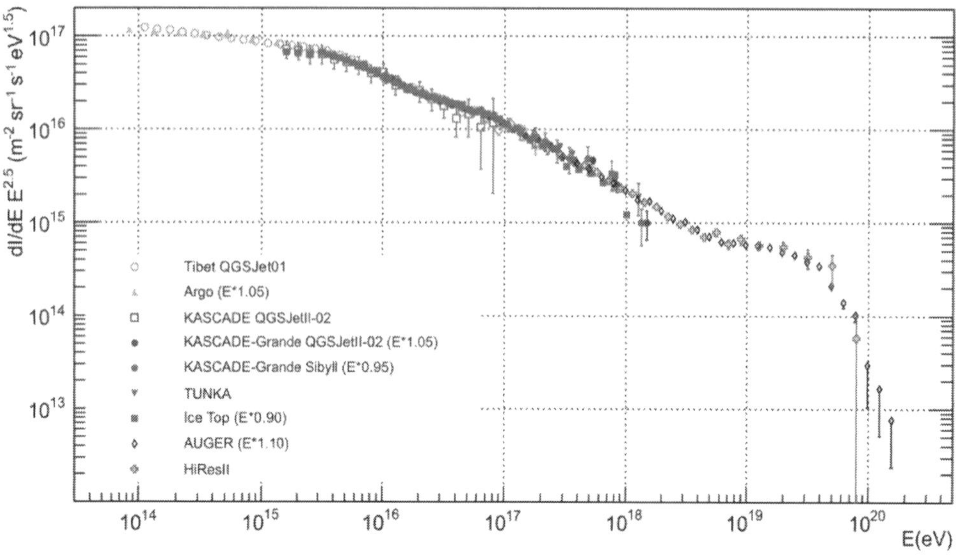

Figure 3. All particle energy spectrum measured in the $10^{14} - 10^{20} eV$ energy range. The energy calibration of the experiments are shifted by an arbitrary amount smaller than the systematic errors. References can be found in the caption of figure 2 and the energy shifts applied are indicated in the legend inside the plot.

between the muon and charged particles numbers, converted, by means of the constant intensity cut, to a reference zenith angle. The events sample with a high value of this parameter (called Y by the KASCADE collaboration who introduced this analysis technique [6]) is generated by heavy primaries and the one with low Y values by the light ones. The KASCADE-Grande experiment has then slightly modified this analysis [7, 8] calculating the ratio between muon and charged particles numbers in small intervals of zenith angles (i.e. atmospheric depth).

A similar approach has been used by the ARGO-YBJ collaboration [9] that, being a full coverage detector operating at higher altitude (i.e. almost near to shower maximum) and not having a muon detector, selected the events on the basis of the ratio between the number of particles measured within a distance of $8m$ from the shower core (Np_8) and the slope (s') of the lateral distribution function.

In both techniques a selection level must be determined to divide the two samples, this cut has to be defined according to the results of a complete EAS simulation. In the case of the KASCADE and KASCADE-Grande experiments it was shown that the level used to define the two samples affects only the absolute fluxes of the spectra, while eventual spectral features are visible independently from this choice that only enhances them. Defining the hadronic interaction model the choice of the cut value corresponds to a primary mass, while at constant primary mass defining a cut value means selecting simulations performed with a particular hadronic interaction model. Thus a spectral feature measured independently from the cut value does not depend on the hadronic interaction model used in the EAS simulation.

Figure 4 shows a summary of the, energy calibrated, KASCADE-Grande and ARGO-YBJ measurements of the light primaries spectra. In the plot also the direct measurements of the H and He spectra performed by the CREAM collaboration [38] are shown, their agreement with

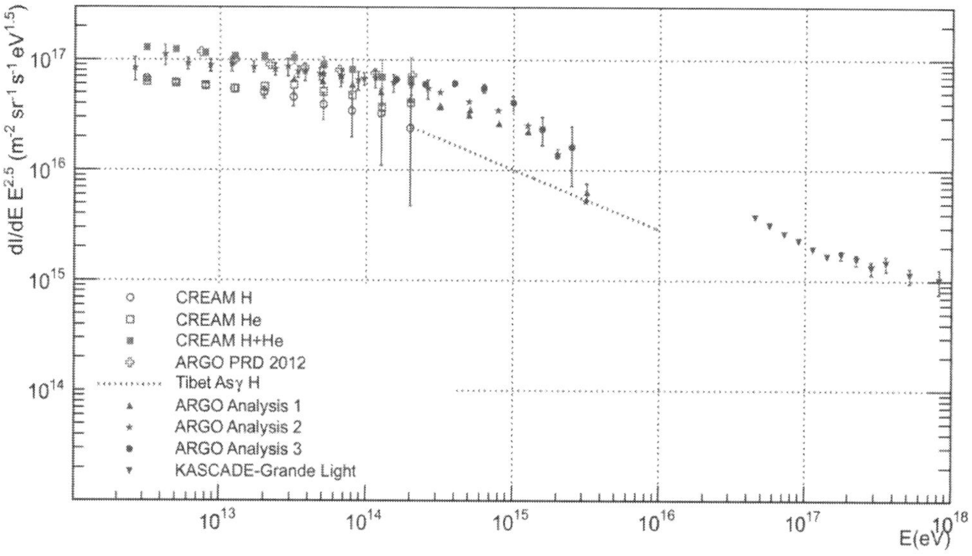

Figure 4. Spectra of the light primaries measured by ARGO-YBJ (the different analysis shown by the collaboration are described in [9]) and KASCADE-Grande. Lower energy direct measurements of the CREAM experiment are shown as a reference. The former KASCADE measurements are not shown in this plot as the light primaries spectrum was shown in a non energy calibrated way [6]

the low energies measurement of the ARGO-YBJ experiments confirms the validity of the energy calibration of this experiment and suggests also the good reliability (at least around $10^{13} eV$) of the hadronic interaction models included in the CORSIKA code.

The KASCADE experiment published [6] the spectra of the "electron rich" (i.e. light primaries) and the "electron poor" (i.e. heavy primaries) events. These spectra are not calibrated in energy (they are shown as a function of the muon density at fixed core distance), but using the integral flux above the change of slope we can identify this feature with the knee of the all particle energy spectrum.

The ARGO-YBJ collaboration has shown [9] the results of three different analysis differing between each other by an amount lower than the declared systematic errors, all the three spectra show a break well below $10^{15} eV$. But, in my opinion at the moment of writing this contribution, some details of these results are doubtful:

- different spectra have different slopes, mainly above the break.
- The spectrum of the light elements changes slope at an energy different from the one where the all particle spectrum changes its slope. It seems strange that, being the light elements a fraction near to the 50% of the cosmic rays flux, such a sharp change of slope is not visible in the all particle spectrum.

On another hand this claim is in agreement with the one of the Tibet-ASγ collaboration who measured the hydrogen spectrum above $\sim 10^{14} eV$ and found a spectrum steeper than the one detected below the knee [39]. The Tibet-ASγ collaboration concluded that the hydrogen knee is at lower energy and that the change of slope in the all particle spectrum observed at $2-4\times10^{15}eV$ is due to heavy elements.

The KASCADE-Grande experiment published the spectra of the light and heavy components in two slightly different energy ranges with selection cuts differently defined in order to enhance their spectral features:

- an hardening of the light component spectrum [8]: at $10^{17.08\pm0.08}eV$ the spectral slope changes from $\gamma = -3.25 \pm 0.05$ to $\gamma = -2.79 \pm 0.08$.
- A steepening of the heavy primaries spectrum [7]: at $log(E/eV) = 16.92 \pm 0.04$ the spectral slope γ changes from $\gamma = -2.76 \pm 0.02$ to $\gamma = -3.24 \pm 0.05$.

A different approach to derive the spectra of single elements (representative of different mass groups) is the one based on the unfolding analysis technique. The goal of this analysis, introduced by the KASCADE experiment [10], is to obtain the elemental fluxes from the bi-dimensional spectra N_e *vs* N_μ. The number of events measured in each pixel of this spectrum are compared with the one expected by a complete EAS and detector simulation. This approach allows to take into account the EAS development fluctuations and the reconstruction errors, but the results heavily depend on the hadronic interaction model used in the EAS simulation. The KASCADE collaboration reported results based on different hadronic interaction model presenting differences in the fluxes of the elemental spectra, but revealing common features: only the spectra of light elements show a change of slope and the knee energy increases with the primary atomic number.

Elemental spectra in agreement with the ones previously discussed have been obtained, at greater energies, by the KASCADE-Grande collaboration using the same unfolding algorithm [11]. The iron spectrum shows a hint of a steepening at an energy in agreement with the one where the knee of the heavy elements has been observed.

3.3. Large Scale Anisotropy

In the $10^{14}-10^{16}eV$ energy range the expected anisotropies are of the order of $10^{-3}-10^{-4}$, while the variation of the number of counts induced by atmospheric (i.e. pressure or temperature) and instrumental effects are greater by at least one order of magnitude. It is thus clear that these fake anisotropies must be removed in order to enhance the real ones. The IceCube and IceTop experiments have applied the time-scrambling algorithm [34] while KASCADE-Grande used the East-West analysis technique [35]. The experimental sensitivity in the search for large scale anisotropies of the cosmic rays arrival distribution is mainly limited by the number of events accumulated by the experiments, being the error on the amplitude proportional to $\sqrt{N}_{events}$.

Statistically significant detection of large scale anisotropies in the $10^{13}-5\times10^{16}eV$ have been obtained by EAS-TOP [14] at $100TeV$ and $400TeV$, IceCube [13] at $20TeV$ and $400TeV$ and IceTop [12] at $400TeV$ and $2PeV$. At greater energies have been published only the upper limits obtained by the KASCADE [36] and KASCADE-Grande [37] experiments. The amplitudes of the first harmonic are shown in figure 5, while the relative phases can be seen in figure 6. We can notice that there is a trend of an increasing amplitude above $10^{14}eV$ and that, at the same energy, the phase is changing (as confirmed also by KASCADE-Grande). Even if we must consider that the IceTop results cannot be well described by the sum of a dipole and a quadrupole thus an harmonic description of the anisotropy maybe a simplified one.

4. Discussion

In this section I present the results of a calculation, based on the work of T. Gaisser et al. [16], of elemental spectra following a Peter's cycle (i.e. with knee energies at constant rigidity). My goal is neither reaching a detailed interpretation of the data nor making any fit to obtain the

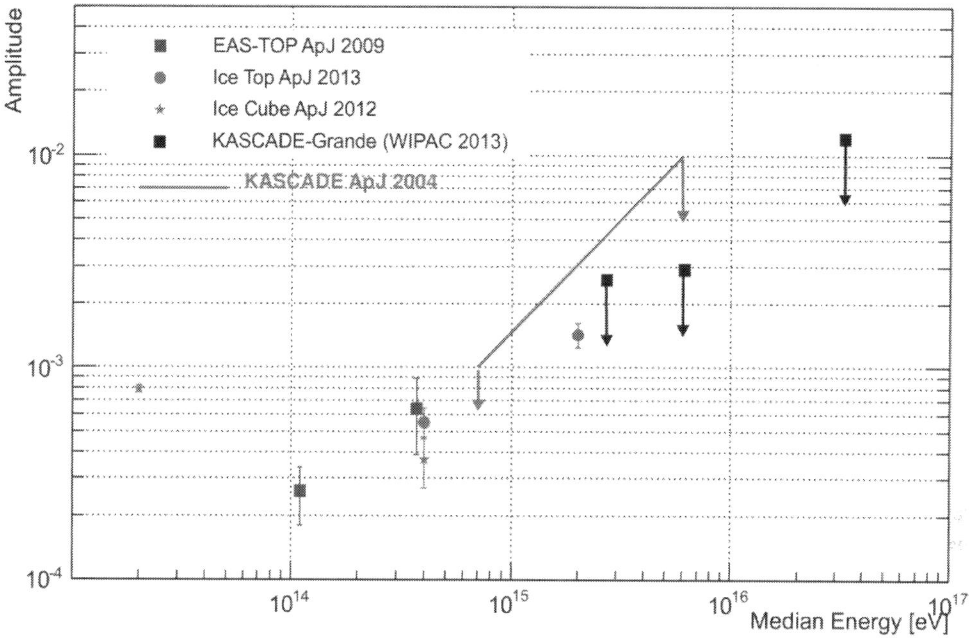

Figure 5. Amplitudes of the first harmonic measured in the $10^{13} - 5 \times 10^{16} eV$ energy range. The results of the EAS-TOP, IceTop and IceCube experiments are measurement of the first harmonic amplitude, while the KASCADE and KASCADE-Grande ones are upper limits.

elemental spectra describing them, I will only search for a qualitative agreement between the data shown in section 3 and the calculated spectra.

The elemental spectra are calculated with an expression [40] describing a power law with index changing from γ_1 to γ_2 at an energy E_{knee}.

$$\Phi(E) = KE^{\gamma_1}[1 + (\frac{E}{E_{knee}})^{\epsilon}]^{\frac{(\gamma_2-\gamma_1)}{\epsilon}} \tag{1}$$

The parameter ϵ represent the sharpness of the change of slope, and in these calculations has been kept constant ($\epsilon = 10$) corresponding to a smooth knee.

The elemental fluxes have been calculated with the following hypotheses:

1) the absolute normalization is given by the intensities measured by CREAM [38] at $10^{13}eV$.
2) The slopes of the H and He spectra before the knee are the ones measured by CREAM, the slope of the He spectrum is used also for heavier elements.
3) All the spectra suffer the same change of slope, values ranging from $\Delta\gamma = 0.5$ to $\Delta\gamma = 0.7$ have been used.
4) The knee energy has been calculated following a Peter's cycle: $E_{knee}(Z) = Z \times E_{knee}(H)$. The Hydrogen knee is fixed at two different energies: $E_{knee}(H) = 6.5 \times 10^{14}eV$ following the ARGO-YBJ results and $E_{knee}(H) = 1.5 \times 10^{15}eV$ attributing the knee of the all particle spectrum to a light element, i.e. either H or He.

Journal of Physics: Conference Series **632** (2015) 012001 doi:10.1088/1742-6596/632/1/012001

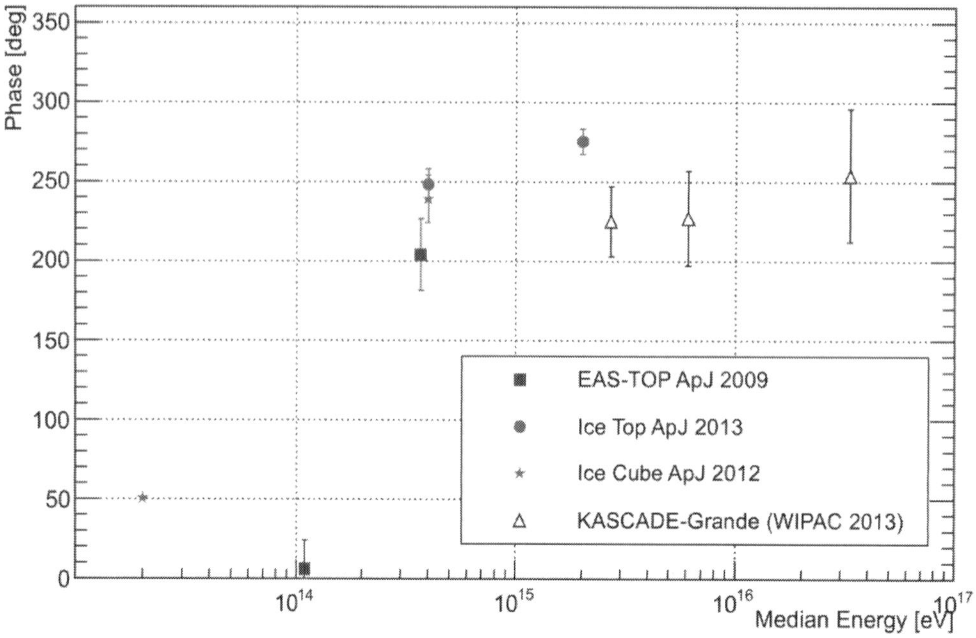

Figure 6. Phases of the first harmonic measured in the $10^{13} - 5 \times 10^{16} eV$ energy range

5) To account for the hardening of the light elements spectrum a flatter H component ($\gamma = -2.66$) becoming dominant at $\sim 10^{17} eV$ is added.

This is only one of the possible options to calculate the elemental spectra of cosmic rays, for instance one can assume different slopes for all the elements heavier than Helium or that the knee energies depends on the mass of the primaries (and not on their charge).

Comparing the results of these calculations with the mass group spectra we must always keep in mind that the intensities measured with these algorithms aim to enhance spectral feature and not to determine the absolute fluxes. To better show this qualitative agreement I multiply the calculated spectra by a factor that can be interpreted as the selection efficiency of events generated by a particular primary element. The goal of this factor is not to evaluate these efficiencies but it must be intended as an help to guide the eyes in evaluating this qualitative agreement.

In figure 7 the ARGO-YBJ light and all particle spectra are shown together with the results of the previously described calculations. We can see as the sharp knee of the light elements spectrum can be described fixing a huge value for the change of slope ($\Delta\gamma = 0.7$), but summing all elements to calculate the all particle spectrum we cannot describe the Tibet-III [28] and ARGO-YBJ measurements. This disagreement can be reduced assuming that the knee energy depends on the mass of the primaries (A) or using harder spectra for elements heavier than He, but it is in any case difficult to obtain a good description of both the light and the all particle spectra. If these results will be confirmed an additional (heavy?) component must be invoked.

Extending the elemental spectra calculations from $10^{14} eV$ to $5 \times 10^{18} eV$ we can compare the

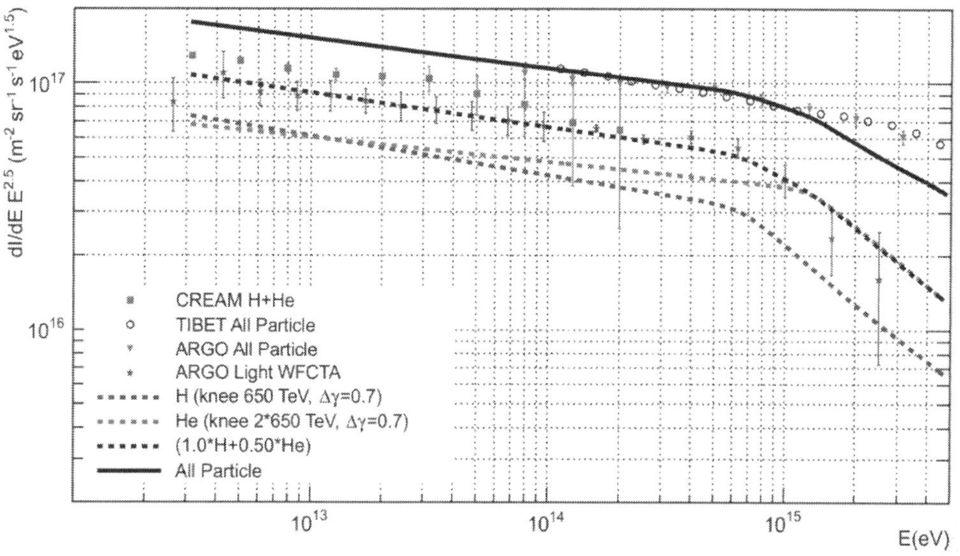

Figure 7. All particle and light elements spectra measured by the ARGO-YBJ experiment compared with the calculations described in section 4

expectations with the measurements in a wide energy range, in figure 8 I show the calculated fluxes and the experimental data (with the energy shifted by an arbitrary factor, see figure 3). Thus in addition to the simplified and schematic approach there is also an uncertainty about the absolute fluxes, therefore we can only expect a qualitative agreement between data and calculations. The shown fluxes correspond to: H (blue solid line), He (magenta dashed line), C (red dotted line), Fe (green dash-dotted line) and their sum (black thick solid line).

We can see that the knee structure is well reproduced by elemental spectra following Peter's cycle, but further details must be clarified by future experiments, such as:

- which element is dominating the spectrum at the knee?
- If all the claimed results are confirmed, is there a contradiction between measurements performed at high altitudes and at sea level?
- Which is the exact energy of the knee?
- The knee energies of different elements depend on their charge (Z) or their mass (A)?
- At energies smaller than the knee are the spectral indexes of all elements (except H) equal or not?
- Is the change of slope at the knee the same for all elements?

Figure 8 shows also that at energies above the knee such a simple calculation cannot explain the faint structures measured at $\sim 10^{16} eV$ and $\sim 10^{17} eV$, therefore (as already pointed out by T. Gaisser et al. [16]) an additional galactic component, such as the "component B" proposed by Hillas [15], is needed to explain the measured fluxes. The contribution of an additional (extra-galactic?) H flux becoming dominant at $\sim 10^{17} eV$ can, qualitatively, describe the hardening of the light elements spectrum measured by KASCADE-Grande.

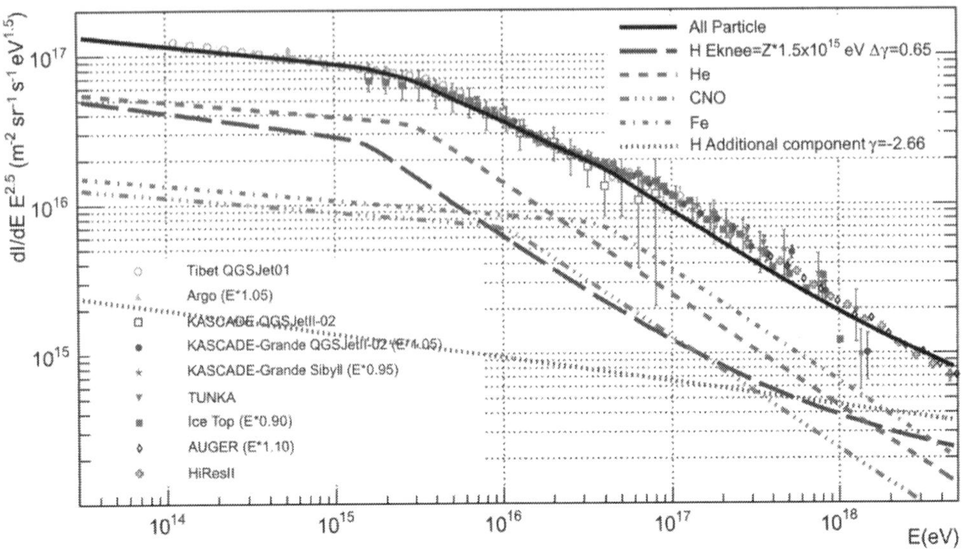

Figure 8. Summary of all particle spectrum measurements, shifted by an arbitrary amount (reported in the legend), and compared with the calculations described in section 4

Thus new, high precision measurements are required. These can be performed either by direct, high acceptance measurements or by high resolutions EAS arrays. Both techniques have their advantages and limitations:

- Direct measurements. Long duration satellite experiments, like the foreseen iss-cream [41], will be able to collect enough statistics to cover the knee energy and will have the charge resolution to separate single elements. Their main limitation will be the low energy resolution ($\sim \Delta E/E \sim 40\%$) due to the strong constrain on their mass. Calorimetric detectors will contain only a small fraction of the shower generated the primary particle interaction. Furthermore no man made sources will be available at these energies, thus the absolute energy calibration will be performed indirectly by simulations.
- EAS measurements. The main limitation of ground based experiments will always be the EAS development fluctuations. The best way to minimize them is to locate the array at high altitudes in order to detect the showers near their maximum development. This is the goal of the LHAASO experiment [42] that will be constructed in China, Sichuan province at $\sim 4400m\ a.s.l.$. This experiment will sample the electromagnetic and muonic shower components with an unprecedented active surface and will therefore reach an unprecedented resolution.

References

[1] T. Pierog, EPJ Web of Conferences **52**, 03001 (2013).
[2] G.V. Kulikov,G.B. Khristiansen, Sov. Phys. JEPT **35**, 441 (1958).
[3] W.D. Apel et al., Astropart. Phys. **36**, 183 (2012).
[4] M.G. Aartsen et al., Phys. Rev. D **88**, 042004 (2013).
[5] V.V. Prosin et al., Nucl. Instr. and Meth. A **756**, 94 (2014).

[6] T. Antoni et al., Astropart. Phys. **16**, 373 (2002).
[7] W.D. Apel et al., Phys. Rev. Lett. **107**, 171104 (2011).
[8] W.D. Apel et al., Phys. Rev. D **87**, 081101(R) (2013)
[9] G. Di Sciascio on behalf of the ARGO collaboration, Frascati Physics Series **58**, (2014) in press
[10] T. Antoni et al., Astropart. Phys. **24**, 1 (2005).
[11] W.D. Apel et al., Astropart. Phys. **47**, 54 (2013).
[12] M.G. Aartsen et al., ApJ **765**, 55 (2013).
[13] R. Abbasi et al., ApJ **746**, 33 (2012).
[14] M. Aglietta et al., ApJ **692**, L130 (2009).
[15] A.M. Hillas, J. Phys. G: Nucl. Part. Phys. **31**, (2005) R95.
[16] T.K. Gaisser et al., Front. Phys. **8(6)**, 748 (2013).
[17] J. Hersil et al., Phys. Rev. Lett., **6**, 22 (1961); D.M. Edge et al., J. Phys. A, **6**, 1612 (1973).
[18] D. Kang et al., Proc. of the 31st ICRC Lodz (Poland), #icrc1044 (2009).
[19] D. Heck et al., Report FZKA 6019, Forschungszentrum Karlsruhe (1998).
[20] E.E. Korosteleva et al., Nucl. Phys. B (Proc. Suppl.) **165**, 6837 (2007).
[21] S. Ostapchenko, Phys. Rev. D **74**, 014026 (2006).
[22] E.-J. Ahn et al., Phys. Rev. D **80**, 094003 (2009).
[23] K. Werner, F.M. Liu, T. Pierog, Phys. Rev. C **74**, 044902 (2006).
[24] W.D. Apel et al., Advances in Space Research **53**, 1456 (2014).
[25] S. Ostapchenko, Phys. Rev. D **83**, 014018 (2011).
[26] T. Pierog et al., arXiv:1306.0121, (2013).
[27] M. Bertaina et al., Nucl. Phys. B (Proc. Suppl.) **256-257**, 149 (2014).
[28] M. Amenomori et al., ApJ **678**, 1165 (2008).
[29] M. Finger, PhD Thesis, Karlsruhe University (2011)
[30] A.P. Garkaya et al., J. Phys. G: Nucl. Part. Phys. **35**, 115201 (2008).
[31] A. Aab et al., arXiv:13075059v1, (2013).
[32] R.U. Abbasi et al., Phys. Rev. Lett. **100**, 101101 (2008).
[33] T. Abu-Zayyad et al., arXiv:1305:7273, (2013).
[34] D.E. Alexandreas et al., Nucl. Instr. and Meth. A **328** , 570 (1993).
[35] R. Bonino et al., ApJ **738**, 67 (2011).
[36] T. Antoni et al., ApJ **604**, 687 (2004).
[37] A. Chiavassa et al., Jour. of Phys. Conf. Series **531**, 012001 (2014).
[38] Y.S. Yoon et al., ApJ **728**, 122 (2011).
[39] M. Amenomori et al., Physics Letters B **623**, 58 (2006).
[40] S.V. Ter-Antonyan and L. Haroyan, arXiv:hep-ex/0003006, (2000).
[41] E.S. Seo et al., Proc. of the 33rd ICRC (Rio de Janeiro, Brasil), #0629 (2013).
[42] Cao Z., Chinese Physics C **34**, 249 (2010).

The KM3NeT neutrino telescope

R. Coniglione on behalf of the KM3NeT collaboration

INFN - Laboratori Nazionali del Sud Via S. Sofia 62 - 95123 Catania - Italy

E-mail: `coniglione@lns.infn.it`

Abstract. The construction phase of an underwater high energy neutrino telescope in the Mediterranean Sea, named KM3NeT, has started. The neutrino telescope that will consist of several blocks of instrumented structures will have a size of the order of a cubic-kilometer. In this work the main elements of the detector, the status of the project and the expected performance will be briefly reported.

1. Introduction

The detection of high energy neutrinos from astrophysical sources is the main physics objective of the KM3NeT detector. The study of high energy neutrinos produced in galactic and extragalactic sources can shed light on the production mechanisms of high energy particles and, if neutrinos are detected from specific sources, can give important constraints on the acceleration models. Moreover, the detection of neutrinos will give a clear signature of the presence of hadronic processes in the production of high energy particles [1]. In a multi-messenger approach the detection of high energy neutrinos will also shed light on the still unknown origin of the observed high energy cosmic rays and high energy γ-rays. High energy γ-rays can be produced both in hadronic and leptonic mechanisms. Moreover, extragalactic γ-rays, due to their interaction with the intergalactic environment, arrive at the Earth with a reduced flux and a distorted energy spectrum. The high energy cosmic rays, that are mainly protons, are deflected by the intergalactic magnetic fields and cannot be used to point back to the sources. Neutrinos that are neutral and weakly interacting particles are neither deflected nor absorbed and can reach the Earth carrying important information on the far and violent Universe.

From the experimental point of view the detection of high energy neutrinos is a challenging task. Large detectors with size of the order of a cubic-kilometer and a very low background are necessary. The most effective technique to detect neutrinos in the energy range between 100 GeV and 100 PeV is based on the detection, by means of optical sensors, of Cherenkov photons emitted by secondary particles produced in the neutrino interactions occurring in the volume inside or surrounding the detector. To maintain the background on a reasonable level the detectors have to be shielded from the particles produced in the interaction of cosmic rays with the atmosphere. To satisfy these requirements the high energy neutrino detectors are placed under thousands of meters of ice or water. To date the IceCube detector [2], with the size of about 1 cubic-kilometer is operating at depths ranging from 1450m to 2450m in the South Pole ice. With the IceCube detector a high energy neutrino flux of astrophysical origin has been discovered [3, 4]. Even if the origin of these neutrinos is still unknown this measurement has set the starting point of neutrino astronomy.

Strengthened by the experience gained with the construction of the small detector ANTARES [5], that is taking data since 2008, the three Mediterranean collaborations (ANTARES, NEMO and NESTOR) in 2006 started the R&D for the construction of a multi cubic-kilometer detector: the KM3NeT detector. Due to its location KM3NeT is the ideal instruments to look for neutrinos from point-like Galactic sources. In fact at the latitude of the Mediterranean Sea it is possible to detect neutrinos with a low energy threshold (about 100 GeV) over a large field of view (about 3/4 of π) with an almost complete view of the galactic plane. The collaboration has defined the design and has started the construction of a research infrastructure hosting a high energy neutrino telescope and nodes for Earth and Sea sciences. The physic case was recently extended to the measurement of the Neutrino Mass Hierarchy with atmospheric neutrinos, and a feasibility study, called ORCA, was started. From this study the best sensitivity to measure the Neutrino Mass Hierarchy is obtained with a dense Mton detector that can measure atmospheric neutrinos in the GeV region using the same KM3NeT technology with a much denser optical displacement. An experimental proposal for a multi-site experiment is underway. In particular a Gigaton detector for high energy astronomy is planned at the Capo Passero Italian site while a Mton denser detector is planned in the Toulon French site.

In this paper the detector technology and the status of the project will be briefly described, focusing on the performances of the Gigaton detector for neutrino astronomy. A description of the ORCA feasibility study can be found in [6].

2. The KM3NeT technology

The KM3NeT detector consists of a three dimensional grid of optical sensors. From the measured arrival time of Cherenkov light at each optical sensor, combined with the measured spatial positions it is possible to reconstruct the trajectory of the muons producing Cherenkov light with a very good angular resolution (few tenths of a degree). In addition the amount of detected light can provide information on the energy of the particle. The optical sensors, the Digital Optical Modules (DOMs), are attached to vertical structures, called Detection Units (DUs). The DUs are flexible structures anchored to the sea floor, kept vertical by buoys and connected to shore by an electro-optical cable (Fig.1). An array of 115 DUs will constitute a detector building block. Several building blocks will constitute the detector. Simulation results show that the division into separate building blocks does not affect the expected sensitivity, while it allows for a distributed detector implementation. The choice for a distributed research infrastructure made of several building blocks with common detector technology, management, data handling and operation control was based on both technical and funding-related arguments. Moreover, the distributed infrastructure is the preferred choice of the Sea and Earth Science community that profit from KM3NeT infrastructures to construct a network of sensors in the Mediterranean Sea.

A building block of the detector devoted to high-energy neutrino astronomy is composed by 115 DUs spaced on average by 90-120m. Each DU hosts 18 optical sensors, the Digital Optical Modules (DOMs), starting 100m above the sea floor and with 36m of vertical distance between adjacent DOMs. The definition of the distances between DUs and between DOMs is the result of a detailed optimization. The optimization, performed by means of Monte Carlo simulations, was focused on the primary objective of the Gigaton experiment: the observation of neutrinos from galactic point-like sources. The DUs are supported by two parallel Dyneema© ropes. Also attached to the ropes is a single vertical electro-optical cable used to connect the DOMs to the base of the DU. It consists of a flexible, oil-filled plastic tube that is in equi-pressure with the sea water and contains 30 optical fibers for data transport and two copper wires for the provision of electrical power to the DOMs. For each DOM, a break-out box provides connection to one fibre and two wires. This configuration of a detection unit is usually referred to as a string. For deployment, a detection unit is wrapped on a spherical frame with diameter if about 2.2m

(Launcher of Optical Modules, LOM [7]) which is deposited on the seabed and then unfurls in a rotating upwards movement. After the unfurling the LOM rises to the sea surface, where it is collected for reuse.

Each DOM is a pressure-resistant glass sphere of 17-inch diameter that carries 31 3-inch photomultiplier tubes (PMTs) with low power high-voltage bases [8] and read-out electronics [9]. Each tube is surrounded by a reflector that effectively increases the collection efficiency per PMT by about 27% [10]. The lower hemisphere of each DOM contains 19 of the PMTs, which are thus downward-looking, whereas the other 12 PMTs look upwards. The front-end electronics amplify the PMT signals and transform them into digital time-over-threshold information that is fed into the readout via optical fibres. All PMT signals above an adjustable noise threshold (typically the equivalent of 0.3 photo-electrons) are sent to shore, where event candidates are selected by online filters running on a computer farm. For further details on DOM electronics and data acquisition see [11]. The PMTs are supported inside the glass sphere by a structure while the optical contact is assured by optical gel. The DOM also contains three calibration sensors: a LED nano-beacon for time calibration that illuminates the DOMs above, a compass and tiltmeter for orientation calibration and an acoustic piezo sensor glued to the inner surface for position calibration. A drawing of the DOM is shown in Fig.2.

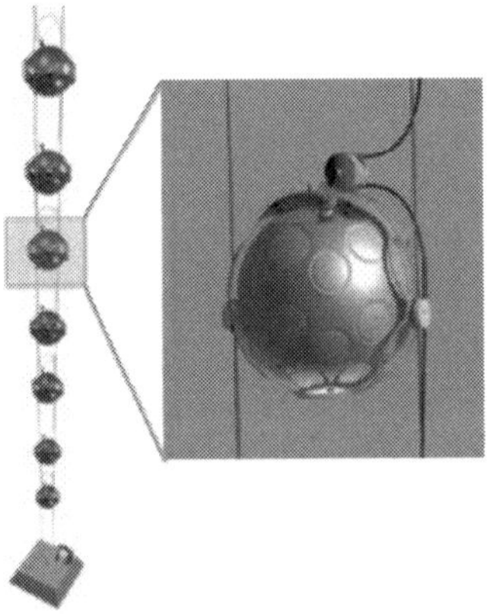

Figure 1. Drawing of DU and details of the DOM mounted into the string.

Figure 2. Drawing of the DOM.

This innovative photon sensor, which is the result of an intense R&D activity, has several advantages with respect to the conventional optical modules with one large PMT (10" diameter PMT). It houses three to four times the photo-cathode area in a single sphere and has an almost uniform angular coverage. Because the photocathode is segmented, the identification of more than one photon arriving at the DOM can be done with extremely high efficiency and purity allowing a good optical background rejection and providing a directional information on the arrival photons. These have been proved analyzing the data collected in the first DOM prototype that has been integrated in the instrumented line of the ANTARES detector [12].

3. Status of the KM3NeT project

The KM3NeT detector will be constructed with a staged approach. The KM3NeT-phase1, fully funded, is presently under construction and will reach a volume of about 0.1 km^3. A following phase, KM3NeT-phase1.5, will consist of a high-energy neutrino telescope of 2 building blocks of 115 DUs, with volume of about 1-1.6 km^3, depending on the DU spacing, and a dense detector for the measurement of the Neutrino Mass Hierarchy. The aim is to extend KM3NeT to six building blocks with a size of about 3-4.8 km^3 (full KM3NeT detector).

After the deployment of prototypes both at the Italian [13] and in the French sites [12], that proved the feasibility of the new designed detector components, the KM3NeT collaboration started the construction of KM3NeT-phase1 in the first months of 2014. In the Italian site 24 DUs based on the string concept and 8 DUs following the a flexible tower concept [14] will be deployed. A first prototype tower has been already deployed at a depth of 3500 m and took data for more than one year. From the data collected in this period the first measurement of vertical atmospheric muon flux at a depth of 3500m was measured and published in [13]. The long-term measurements of the environmental parameters confirms the low bioluminescence level at the KM3NeT-It site [15]. The foreseen detector of 24 DUs together with the 8 towers at the KM3NeT-It site, when completed, will constitute the most sensitive neutrino telescope in the Northern hemisphere and will be equivalent to about 10% of the IceCube detector. The first tower was already successfully deployed in December 2014. For the next step in the construction of KM3NeT-It, funding will be requested for a detector of two building blocks of 115 DUs each and a total instrumented volume of about 1 km^3 optimized for the neutrino astronomy.

In parallel the collaboration is preparing the deployment of 7 DUs at the French site (KM3NeT-Fr). The first string is already assembled (see Fig.3) and it is ready for the deployment that is foreseen in May 2015. The remaining six strings will be configured according the outcome of the ORCA feasibility study, with a interspacing between DUs of the order of 20m and between DOMs of about 6 m, thus providing a demonstrator for the Neutrino Mass Hierarchy experiment.

Figure 3. First string assembled.

4. Physics performances

The (cosmic) neutrinos discovered by IceCube, that correspond to a flux of about $3.6 \times 10^{-8}\,\mathrm{GeV\,cm^{-2}\,s^{-1}\,sr^{-1}}$ with a cutoff at about 3 PeV, were measured in the energy range between 10 TeV to about 1 PeV in [3]. In a more recent analysis [4] a consistent flux was measured with a lower energy threshold (about 1 TeV) but with a significant reduction in the fiducial detector volume. Many of these events are cascade events (shower of elementary particles) which are detected with a poor angular resolution (10°-15°) that prevents the possibility of source pointing. The question where the detected neutrinos came from is one still pending and many possibilities have been explored [16]. This discovery has led the KM3NeT collaboration to plan an intermediate step in the detector construction: the KM3NeT-phase1.5 that will consist of 2 blocks of 115 DUs with a volume of about 1-1.6 km^3 (depending on the DU spacing).

With its superior angular resolution, larger field of view and effective area it will allow to put some constraints on the origin of this diffuse neutrino flux. In Fig.4 the performances of the KM3NeT-phase1.5 detector is shown in terms of significance as a function of the observation time for the flux observed by IceCube. The significance for the cascade channel includes all the neutrino flavors and has been performed selecting tracks reconstructed in the full angular range (4π). The significance for the muon channel has been estimated for tracks reconstructed as up-going plus a small region of 10 degrees above the horizon. A significance of 5σ is obtained after one year in the cascade channel and after about 2.5 years in the muon channel for the flux reported by IceCube.

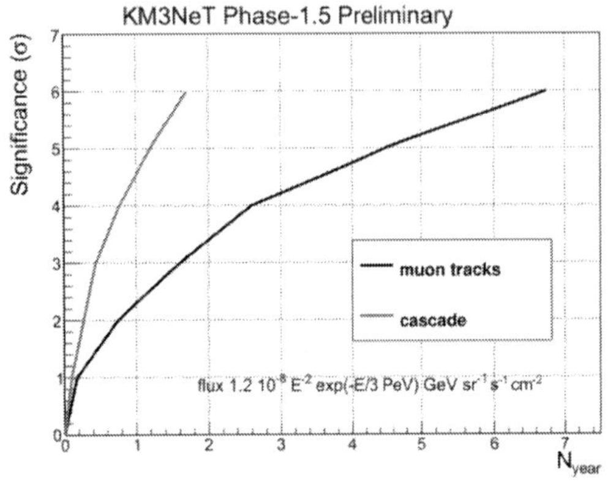

Figure 4. Significance as a function of the observation time for the detection of a neutrino diffuse flux corresponding to the signal reported by IceCube, in the up-going muon (black) and cascade (red) channels.

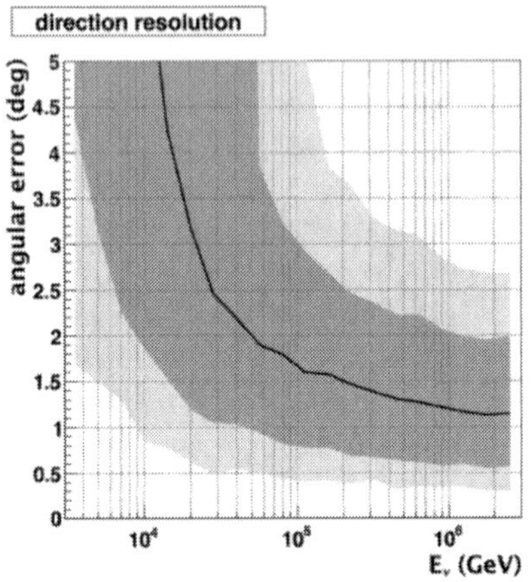

Figure 5. Angular resolution for cascade events reported as a function of the true MonteCarlo neutrino energy. Shown are the median and the 68/90% intervals of the error distribution.

Recently a more performant cascade reconstruction code has been developed showing the angular resolution reported in Fig.5. A resolution of 3 degrees is shown at 20 TeV and decreases down to about 1 degree at higher energies. No systematic uncertainties are included in this evaluation.

The Super Nova Remnants RXJ1713 and Vela Junior and the Pulsar Wind Nebula VelaX are the most intense high energy gamma-ray sources and possible neutrino sources in the Galactic plane. The expected detection capability of the KM3NeT-phase1.5 neutrino telescope is reported for the RXJ1713 and VelaX in Fig. 6. These results were estimated in the hypotheses that the sources have a neutrino spectrum derived from the high energy γ-ray spectrum following the hypotheses in [17] and in [18] [19] and that the spatial extension of the neutrino emission region is the same as for the measured γ-rays. With the KM3NeT-phase1.5 detector a significance of 3σ can be achieved in about 4-6 years of observation while with the full KM3NeT detector (6 building blocks) a significance of 5σ can be achieved in few years of observation [20]. The discovery potential at 5σ level has been evaluated also for a generic point-like source with an energy spectrum proportional to E^{-2} and is shown in Fig. 7 as a function of the source declination. The comparison with IceCube shows that KM3NeT not only complements but also overlaps the field of view of IceCube exceeding with more than one order of magnitude the IceCube sensitivity in the Southern hemisphere.

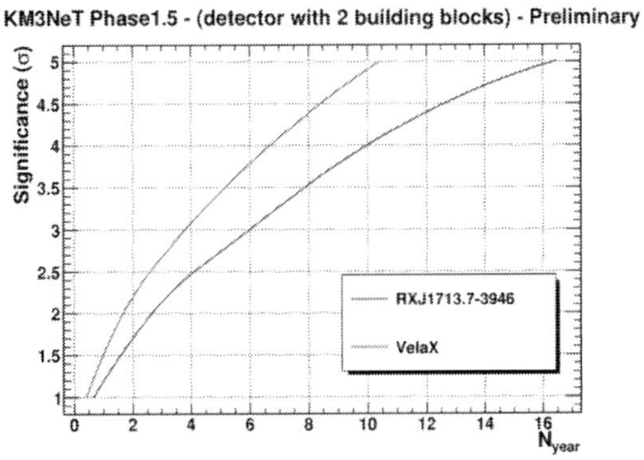

Figure 6. Significance as a function of the observation time for the detection of a neutrino diffuse from the RXJ1713 (blue) and the VelaX (red) sources.

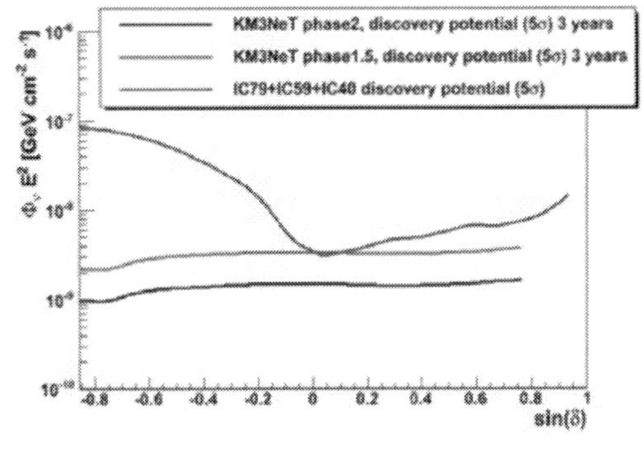

Figure 7. Discovery potential at 5σ 50% probability for point like sources with a energy spectrum proportional to E^{-2} for the KM3NeT-phase1.5 detector (red) and KM3NeT full detector (black) as a function of the source declination. The blue curve shows the discovery potential for the IceCube detector [21].

5. Summary

In this work the KM3NeT project has been briefly presented, the main technical components of the detector and performance described. The collaboration started the construction phase of the fully funded KM3NeT-phase1 that consists of 24 strings and 8 towers in the Italian site near Capo Passero (Sicily) and 7 strings in the French site close to the Toulon coasts. The first tower was already deployed and the first string is completely assembled and is ready to be deployed in the first months of 2015. Finally the detector potential for the detection of Galactic sources and the IceCube diffuse flux has been reported.

References

[1] Halzen F. and Katz U., Advances in High Energy Physics 2013 Article ID 680584 http://dx.doi.org/10.1155/2013/680584.
[2] Information about the IceCube detector can be found on `https://icecube.wisc.edu`.
[3] Aartsen M. G. et al. (IceCube Coll.), 2013 *Science* **342**1242856.
[4] Aartsen M. G. et al. (IceCube Coll.) 2015 it Phys. ReV. D **91** 022001.
[5] Ageron M. et al. 2011 *Nucl. Instrum. Meth. A* **656** 11-38, and information on `http://antares.in2p3.fr`
[6] Katz U. for the KM3NeT collaboration, Feb 2014, available on `http://arxiv.org/abs/1402.1022`.
[7] de Wolf E. et al., for the KM3NeT Coll. 2013 *Nucl. Inst. Meth.* **A725** 241.
[8] Aiello S. et al., for the KM3NeT Coll., proceedings of the VLVnT13 workshop (5-7 August 2013 - Stockholm - Sweden).
[9] Real D. for the KM3NeT Coll., proceedings of the VLVnT13 workshop (5-7 August 2013 - Stockholm - Sweden)
[10] Adrián-Martínez S. et al., KM3NeT Coll. 2013 *JINST* **8** T03006.
[11] Jansweijer P. P. M., Peek H. Z. and de Wolf E. 2013 *Nucl. Inst. Meth.* **A725** 488.
[12] Adrián-Martínez S. et al., KM3NeT Coll. 2014 *Eur. Phys. J.* **C74** 3056 DOI 10.1140/epjc/s10052-014-3056-3.
[13] Aiello S. et al., for the NEMO Coll. 2015 *Astrop. Phys.* **66** 1-7.
[14] Bagleyet P. et al., KM3NeT Coll., Technical Design Report, 2010. Available on http://www.km3net.org. ISBN 978-90-6488-033-9.
[15] Pellegriti M. G. et al. for the NEMO Coll. proceedings of the VLVnT13 workshop (5-7 August 2013 - Stockholm - Sweden).
[16] Anchordoqui L. A. et al. 2014 *Journal of High Energy Astrophysics* **1-2** 1-30.
[17] Kelner S. R., Aharonian F. A. and Bugayov V. V. 2006 *Phys. Rev. D* **74** 063007.
[18] Aharonian F. A. et al. 2012 *Astronomy and Astrophysics* **548** A38.

[19] Villante F. L and Vissani F. 2008 *Phys. Rev. D* **78** 103007; Vissani F. and Villante F. L. 2008 *Nucl. Instr. and Meth. A* **588**123; Vissani F. 2006 *Astr.Phys.* **26** 310.
[20] A. Trovato for the KM3NeT Coll., proceedings of the VLVnT13 workshop (5-7 August 2013 - Stockholm - Sweden).
[21] Aartsen M. G. et al. (IceCube Coll.) 2013 *Astroph. J.* **779** 132.

Measurement of the cosmic ray all-particle and light-component energy spectra with the ARGO-YBJ experiment

Ivan De Mitri (on behalf of the ARGO-YBJ collaboration)
Dipartimento di Matematica e Fisica "Ennio De Giorgi", Università del Salento, Lecce, Italy and Istituto Nazionale di Fisica Nucleare, INFN, Sezione di Lecce, Italy
E-mail: ivan.demitri@le.infn.it

Abstract. The ARGO-YBJ detector, located at high altitude in the Cosmic Ray Observatory of YangBaJing in Tibet (4300 m asl, about 600 g/cm^2 of atmospheric depth) provides the opportunity for the study, with unprecedented resolution, of cosmic ray physics in the primary energy region between 10^{12} and 10^{16} eV. Preliminary results of the measurements of the all-particle and light-component (i.e. protons and helium) energy spectra between approximately 5 TeV and 5 PeV are reported and discussed.

1. Introduction

There is a general consensus that galactic cosmic rays (hereafter CRs) up to the "knee" (~ 4 PeV, with 1 PeV = 10^{15} eV) mainly originate in Supernova Remnants (SNRs). Recent measurements carried out by the balloon-borne CREAM experiment [1, 2] show that the proton and helium spectra from 2.5 to 250 TeV are harder compared to lower energy measurements. As pointed out by several authors, the evolution of the proton and helium spectra and their subtle differences could be indications of the presence of different populations of CR sources contributing to the overall flux and operating in environments with different chemical compositions [3, 4]. Diffusion effects during CR propagation in the Galaxy might also play an important role.

In the knee region (and above) the measurements of the CR primary spectrum are carried out by Extensive Air Shower (EAS) arrays. In this case mass composition studies are extremely difficult and often affected by large systematic uncertainties. In the standard picture the average composition at the knee is dominated by light elements, and the knee itself is interpreted as the steepening of the p and He spectra [5].

However, different experimental results suggest an heavier composition at knee energies. For instance a hybrid measurement has been carried out by the EAS/TOP and MACRO experiments (by detecting, in coincidence, EAS Cherenkov light at 2000 m a.s.l. and underground muons below about 3000 m of water equivalent depth, respectively). The result implies a decreasing proton contribution to the primary flux well below the observed knee in the primary spectrum [6]. The same indication was previoulsy given by the analysis of the underground muon component alone by the MACRO experiment [7]. In addition, also the results of the Tibet ASγ and the BASJE experiments, located at 4300 m a.s.l and at 5200 m a.s.l. respectively, do favour a heavier composition because the proton component is no more dominant at the knee [8, 9]. Indications

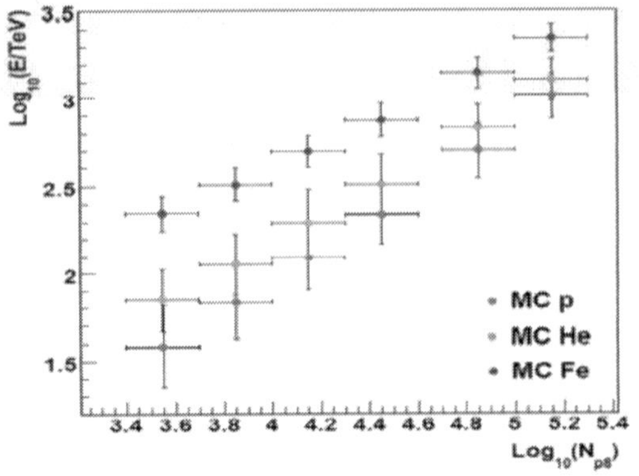

Figure 1. Primary energy as a function of the observed truncated size N_{p8} for simulated showers due to different primary nuclei.

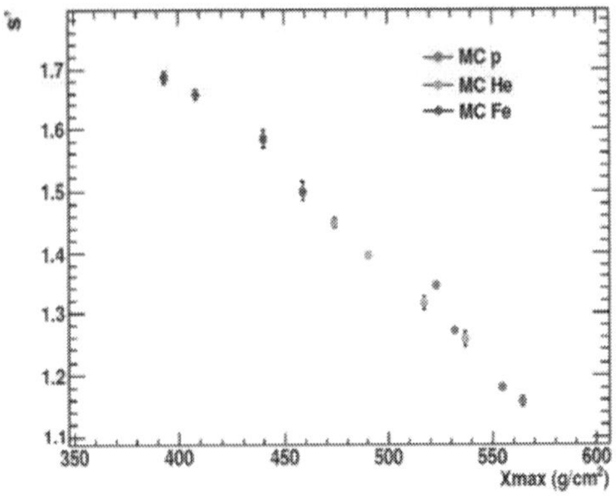

Figure 2. The age parameter s' resulting from the fit of the average LDF for simulated showers vs the corresponding X_{max} average values.

for a substantial fraction of nuclei heavier than helium at 1 PeV have also been obtained in measurements with delayed hadrons [10] as well as by the CASA-MIA collaboration [11].

A measurement of the CR primary energy spectrum (all-particle and light-component) in the energy range few TeV - 10 PeV is under way with the ARGO-YBJ experiment (for a description of the detector and a report of the latest physics results see [12]). In order to cover this wide energy range, different approaches have been followed:
- *'digital readout'*, based on the strip multiplicity, in the 5 TeV - 200 TeV range [13];
- *'analog readout'*, based on the particle density near the shower core, in the 100 TeV-10 PeV range;
- *'hybrid measurement'*, carried out by ARGO-YBJ and a wide field of view Cherenkov telescope, in the 100 TeV - PeV region [14].

Preliminary results concerning the all-particle and the light-component (i.e. p+He) spectra obtained with the analog readout are summarized in the following. More details on the results obtained with the 'hybrid measurement' are also given in [14, 15].

2. Measurement of the all-particle spectrum

The measurement of the CR energy spectrum up to 10 PeV is under way exploiting the RPC charge readout of the ARGO-YBJ detector which allows studying the structure of the particle density distribution in the shower core region up to particle densities of about $10^4/m^2$ [16].

The study of the charged particle lateral density function (LDF) at ground is expected to provide information on the longitudinal profile of the showers in the atmosphere, that is to estimate their development stage, or the so-called *age*, which is related to X_{max}, the atmospheric depth at which the cascade reaches its maximum size. This implies the possibility of selecting showers within given intervals of X_{max} or, equivalently, of X_{dm}, the grammage between the depth of the shower maximum and the detection level.

The shower development stage in the atmosphere, as observed at a fixed altitude (the detection one), depends on the energy of the interacting primary. For fixed energy, it depends on the nature of the primary: heavy primaries interact higher in the atmosphere, thus giving showers which, on average, reach their maximum at a larger distance from the detector than a lighter primary of the same energy. For this reason, the combined use of the shower energy and age estimations can ensure a sensitivity to the primary mass, thus giving the possibility of selecting a light (p+He) event sample with high efficiency.

Various observables were considered and analyzed in order to find a suitable estimator of the primary CR energy. Among them, according to MC simulations, N_{p8}, the number of particles detected within a distance of 8 m from the shower axis, resulted well correlated with energy, not biased by the finite detector size and not much affected by shower to shower fluctuations [17]. Therefore, the analysis is carried out in terms of different N_{p8} intervals to select event samples corresponding to different primary energies. Nevertheless, as shown in Fig. 1, this truncated size is a mass-dependent energy estimator parameter.

In order to have a mass-independent parameter we fitted the LDFs of individual showers (up to 10 m from the core) event-by-event, for different N_{p8} intervals and different shower initiating primaries, with a suitable function to get the shape parameter s' (see [17, 18] for details). From these studies we find that, for a given primary, the s' value decreases when N_{p8} (i.e. the energy) increases, this being due to the observation of younger (deeper) showers at larger energies. Moreover, for a given range of N_{p8}, s' increases going from proton to iron, as a consequence of older (shallower) showers. Both dependencies are in agreement with the expectations, the slope s' being correlated with the shower age, thus reflecting its development stage. This outcome has two important implications, since the measurements of s' and N_{p8} can both (i) help constraining the shower age and (ii) give information on the primary particle nature.

Concerning the first point, we show in Fig. 2 the s' values as obtained from the fit of the average LDFs, for each simulated primary type and N_{p8} interval, as a function of the corresponding X_{max} average value. As can be seen, the shape parameter s' depends only on the development stage of the shower, independently from the nature of the primary particle. That plot expresses an important universality of the LDF of detected EAS particles in terms of the lateral shower age. The LDF slope s' is a mass-independent estimator of the average X_{max}. This also implies the possibility to select most deeply penetrating showers (and quasi-constant X_{dm} intervals) at different zenith angles, an important point for correlating the exponential angular rate distribution with the interaction length of the initiating particle [19]. Obviously shower-to-shower fluctuations introduce unavoidable systematics, whose effects can be anyway quantified and taken into account.

Another implication is that s' from the LDF fit very close to the shower axis, together with the measurement of the truncated size N_{p8}, can give information on the primary particle nature, thus making possible the study of primary mass composition and the selection of a light-component data sample.

Assuming an exponential absorption after the shower maximum, we get N_{p8}^{max}, a variable linearly correlated to the size at the shower maximum, by using N_{p8} and s' measurements for each event and simply correcting with an exponential attenuation: $N_{p8}^{max} \approx N_{p8} \cdot exp[(h_0 sec\theta - X_{max}(s'))/\lambda_{abs}]$. A suitable choice of the absorption lenght λ_{abs} (=120 g/cm^2) allows to get N_{p8}^{max}, a parameter correlated with primary energy in an almost linear and mass independent way, providing an energy estimator with a Log(E/TeV) resolution of 0.10–0.15 (getting better with energy).

As described in [16], the RPC charge readout system has 8 different and overlapping gain scale settings (G0,....,G7 from smaller to larger gains) in order to explore the particle density range $\approx$20 – 10^4 particles/m^2. In this preliminary analysis the results obtained with two gain scales (so-called G1 and G4) are presented.

Selecting quasi-vertical events ($\theta < 15°$) in terms of the truncated size N_{p8} with the described procedure we reconstructed the CR all-particle energy spectrum shown in the Fig. 3 in the energy range 100 – 3000 TeV. In the plot a $\pm$14% systematic uncertainty, due to hadronic interaction models, selection criteria, unfolding algorithms, and aperture calculation, is shown by the shaded area. The statistical uncertainty is shown by the error bars. As can be seen from the figure, the

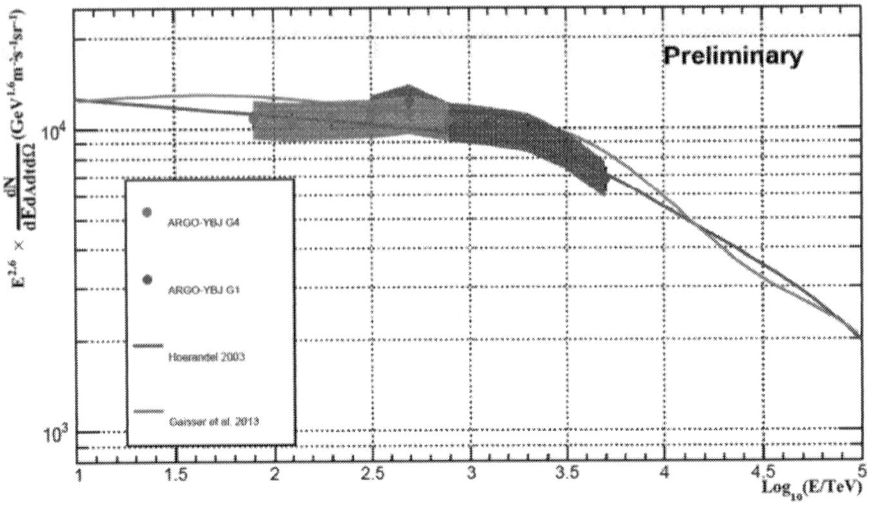

Figure 3. All-particle energy spectrum of primary CRs measured by ARGO-YBJ. Quasi-vertical events ($\theta < 15°$) recorded with two different gain scales (G1 and G4) are plotted. The systematic uncertainty is shown by the shaded area and the statistical one by the error bars. The parametrizations provided by [20] and [3] are shown for comparison.

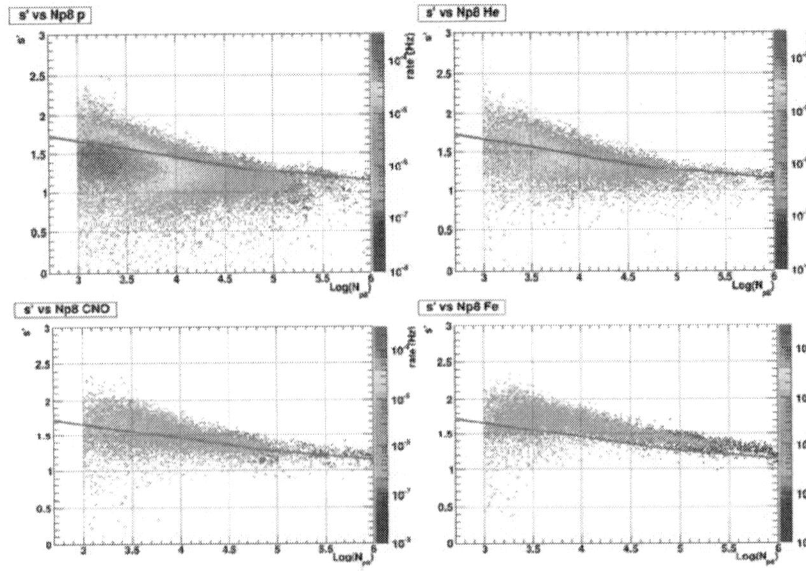

Figure 4. Relation between the LDF shape parameter s' and the truncated size N_{p8} for different nuclei. The p+He selection cut is shown by the lines.

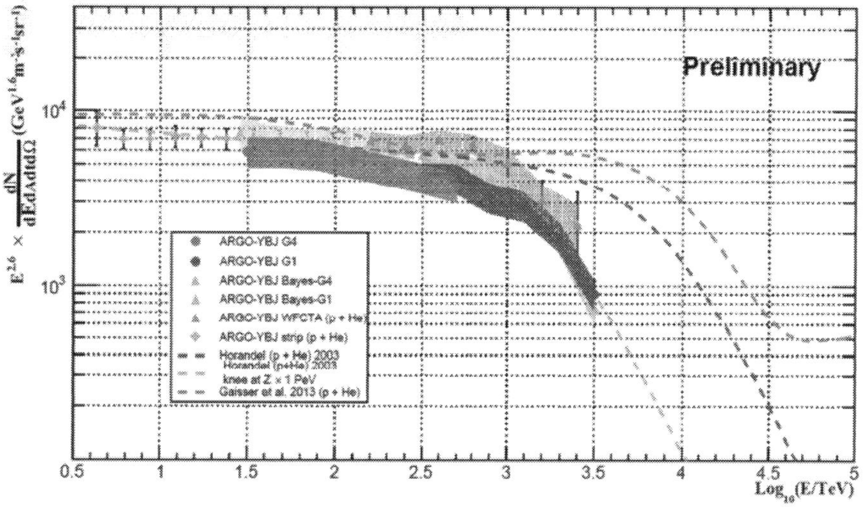

Figure 5. Light (i.e. p+He) component energy spectrum of primary CRs measured by ARGO-YBJ with four different analyses. Data recorded with two different gain scales (G1 and G4) are plotted. The systematic uncertainty is shown by the shaded area and the statistical one by the error bars. The parametrizations provided by [20] and [3] are shown for comparison. A Hörandel-like spectrum with a modified knee at Z×1 PeV is also shown.

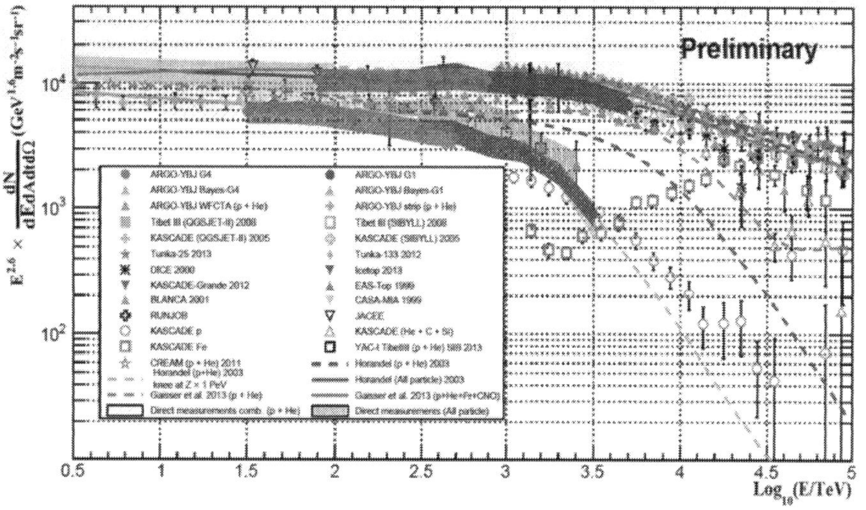

Figure 6. All particle and light (p+He) component energy spectra of primary CR measured by ARGO-YBJ and compared to different experimental results. The parametrizations provided by [20] and [3] are shown for comparison. A Hörandel-like spectrum with a modified knee at Z×1 PeV is also shown.

two gain scales overlap making us confident about the event selection and the analysis procedure. The ARGO-YBJ all-particle spectrum is in fair agreement with the parametrizations provided by [20] and [3], showing evidence of a spectral index change at an energy consistent with the position of the knee. As shown in Fig.6 this result is also consistent with previous measurements made by both direct and indirect experiments. This is also an important check on the absolute energy scale set for this analysis, whose systematic uncertainty has been anyhow conservatively estimated at the level of 10%.

3. Measurement of the light-component energy spectrum

The CR light-component energy spectrum has been measured by ARGO-YBJ from about 5 TeV to 500 TeV [13, 14, 12]. The energy range is now being extended up to the few PeV region by using the RPC charge readout information and three different approaches.

(1) A selection of events in the $s' - N_{p8}$ space allowing to get a light-component sample of showers with a contamination of heavier nuclei less than about 15% (see Fig. 4).

(2) A Bayesian unfolding technique similar to that applied to measure the light-component spectrum up to 200 TeV [13, 12].

(3) The ARGO-YBJ/WFCTA hybrid measurement [14] with a different selection procedure which increases the aperture of a factor 2.4, thus allowing the extension to larger energies (see [15] for a detailed description of the method and a discussion of the results).

The preliminary results of the three analyses (and the previous ARGO-YBJ measurement below 200 TeV) are summarized in Fig.5. The systematic uncertainty on the flux is shown by the shaded area and the statistical one by the error bars. A systematic uncertainty on the energy scale at the level of 5-10% (depending on the analysis) has also been conservatively estimated (not shown in the plots).

All three different analyses are consistent with low energy (direct) measurements and then show a clear evidence for a bending at larger energies but below 1 PeV. With respect to a single power-law with a spectral index -2.62 the deviation is observed at a level of about 6 s.d. The results obtained with the two analysis of RPC charge readout data (label 1 and 2 in the previous list) are in fair agreement. They also agree with the ARGO-YBJ/WFCTA hybrid measurement within systematic uncertainties and the possible difference in the energy scale. For comparison, the parametrizations of the light-component provided by [20] and [3] are shown by the blue and red dashed lines, respectively. A Hörandel-like spectrum with a modified knee at Z×1 PeV is also shown (a factor four lower in energy than in the original formulation).

Finally the all particle and the light-component energy spectra measured by ARGO-YBJ are compared to a compilation of different experimental results in the Fig. 6.

4. Conclusions

The CR spectrum has been studied by the ARGO-YBJ experiment in a wide energy range (TeVs→ PeVs) . This study is particularly interesting because not only it allows a better understanding of the so called 'knee' of the energy spectrum and of its origin, but also provides a powerful cross-check among very different experimental techniques.

The all-particle spectrum (measured in the energy range 100 TeV - 3 PeV) is in good agreement with both theoretical parametrizations and previous measurements, making us confident about the selection and reconstruction procedures.

The light-component (i.e. p + He) has been reconstructed with high resolution up to about 5 PeV. The ARGO-YBJ preliminary result is in agreement with direct measurements and then show a clear indication of a bending below 1 PeV. Improvements of event selection with the full statistics and a complete analysis of systematic uncertainties is currently under way.

References

[1] H.S. Ahn et al., Astrophys. J. Lett. **714**, L89 (2010).
[2] Y.S. Yoon et al., Astrophys. J. **728**, 122 (2011).
[3] T.K. Gaisser, T. Stanev, S. Tilav, Front. Phys. 8(6) (2013) 748.
[4] D. Caprioli et al., Astrop. Phys. **34**, 447 (2011).
[5] W.D. Apel et al., Astrop. Phys. **31**, 86 (2009).
[6] M. Aglietta et al., Astrop. Phys. **21**, 223 (2004).
[7] M. Ambrosio et al. Phys. Rev. D 56 (1997) 1418
[8] M. Amenomori et al., Phys. lett. B **632**, 58 (2006).
[9] H. Tokuno et al., Astrop. Phys. **29**, 453 (2008).
[10] H. T. Freudenreich et al., Phys. Rev. D **41**, 2732 (1990).
[11] M.A.K. Glasmacher et al., Astrop. Phys. **12**, 1 (1999).
[12] I. De Mitri et al., Nucl. Instr. and Meth. in Phys. Res. A742 (2014) 2-9
[13] B. Bartoli et al., Phys. Rev. D **85**, 092005 (2012).
[14] B. Bartoli et al., Chinese Phys. C **38**, 045001 (2014).
[15] Z. Cao et al., Proc. of the Vulcano workshop 2014, Frascati Physics Series, in publication
[16] S. Mastroianni et al. in Proc. 33rd ICRC, (Rio de Janeiro, 2013) **ID: 1035**
[17] I. De Mitri et al. , in Proc. 33rd ICRC, (Rio de Janeiro, 2013) **ID: 0779**
[18] A. D'Amone et al., in Proc. 33rd ICRC, (Rio de Janeiro, 2013) **ID: 0781**
[19] G. Aielli et al., Phys. Rev. D **80**, 092004 (2009).
[20] J.H. Hörandel, Astrop. Phys. **19**, 193 (2003).

Looking for dark matter on the light side

DESY 15-003

Babette Döbrich
DESY, Notkestr. 85, 22607 Hamburg, Germany
E-mail: babette.doebrich@desy.de

Abstract. Among the prominent low-mass dark matter candidates is the QCD axion but also other light and weakly interacting particles beyond the Standard Model. We review briefly the case for such dark matter and give an overview on most recent experimental efforts within laboratory searches, where we focus on experiments exploiting a potential electromagnetic coupling of such particles.

1. Three ultra-light dark matter candidates

It would be a huge break-through to find out what dark matter (DM) is made of. Whilst its constituents could be rather heavy and, e.g., leave an 'imprint' at the LHC, show up in astrophysics signatures or reveal themselves through recoil energy at direct detection setups [1], there are well-motivated candidates also on the 'light side'. Our main concern in this article are henceforth DM candidates below the eV mass-scale.

Of such ultra-light particles, the QCD axion is the most prominent dark matter candidate, see, e.g., [2] for a review. On the *theoretical side*, the axion is a pseudo-scalar pseudo-Goldstone boson that is a consequence of the Peccei-Quinn solution [3] to the strong CP problem. The strong CP problem amounts to the question why CP violation in QCD is unmeasurably small (or even absent). The effective parameter for CP violation receives contributions from the θ-angle of QCD, being essentially unconstrained a priori, and the quark mass matrix. As these parameters are unrelated from the outset, it arises the question for a natural explanation on why the CP-violating parameter is so close to zero. In essence, the axion solution to the strong CP-problem makes the parameter a dynamical variable which naturally relaxes to zero.

From an *experimental viewpoint*, it is most interesting that axions - in certain parameter regimes - constitute a perfect candidate to make up the cold dark matter (CDM) in our universe. Although the axion is very light, it can be non-thermally produced in the early universe [4, 5].

A particularly attractive feature of axion dark matter is that its viable parameter range is comparatively small and it is thus a realistic aim to confirm or exclude axions as main dark matter component with current and near-future technology. For axions as cold DM, two natural cosmological windows exist, see, e.g., [6]: In the post-inflation scenario, the spontaneous breaking of the Peccei-Quinn symmetry at a scale f, which gives rise to the axion as its pseudo Goldstone boson, takes place only after inflation. In the other scenario, the Peccei-Quinn phase transition happens before inflation. The former scenario is typically related to axions with higher masses than the latter because the decay of axionic topological defects (absent in the latter) also produces DM axions and generically the DM abundance grows with the decay constant (i.e. decreases with the axion mass).

The situation for the axion is sketched in Fig. 1 where performed experiments and the foreseen reach of some experiments that are planned for current and near future are shown. In addition, some astrophysical bounds are shown, for more insight on this matter see, e.g. [7, 8]. In Fig. 1, for the black line labeled 'axion' we have implemented $g = \frac{\alpha}{2\pi f}$ as relation of the axion-to-photon coupling and the symmetry breaking scale. For specific axion models, a $\mathcal{O}(1)$ factor enters in the relation between g and f, cf. [8]. Axions are a good cold dark matter candidate roughly between masses of 10^{-6}eV and 10^{-3}eV [9, 10] (indicated by thickening the axion line in Fig. 1. Below 10^{-6}eV, axions tend to produce too much DM and they are disfavored, although some models can justify the selection of suitable initial conditions. Note that at very low masses, there might be also effects on black hole dynamics [11]. In laboratory searches, to probe the very tiny couplings that correspond to high values of the axion symmetry breaking scale, resonant search strategies are mostly employed and will be briefly sketched in Sect. 2.1

The axion is very light and very weakly interacting. Thus it exhibits the properties of a more general class of particles dubbed *'Weakly Interacting Slim Particles' (WISPs)* [12, 13, 14]. Among the WISPs that could be cold dark matter [15] are also axion-like particles (ALPs). Such general ultralight pseudo-scalars could be pseudo-Nambu Goldstone bosons (pNGB) associated with a symmetry breaking scale different from the Peccei-Quinn scale [16, 17]. Effectively they can have a coupling to photons similarly to the QCD axion, but a 'relaxed' mass-coupling relation (i.e., they are not confined to the parameter region around the axion line in Fig. 1). Their existence is motivated in Standard Model extensions [12], and they have been also evoked to explain some astrophysical puzzles such as the observed transparency of the universe to high-energetic photons, see, e.g., [18].

Similar to axions, ALP CDM [15] can be realized through a *misalignment mechanism*: The axion mass m is strongly temperature dependent (same might apply to ALPs as pNGBs). For axions above the symmetry breaking scale (in the early universe), the axion is essentially massless. Thus at very high energies, the initial angle is not fixed (for the axion it needs not to be at its CP conserving value), i.e., it can be mis-aligned from its minimum. The equation of motion for the field in the expanding universe is that of a damped harmonic oscillator (where $3H$ quantifies the damping term and H is the Hubble expansion parameter) and initially the field is frozen when $3H \gg m$. At later times t_1, when the particle mass $m_1 = 3H$, the field starts to oscillate and behaves as cold dark matter fluid. The situation for ALP DM is reviewed in part through the dotted black lines in Fig. 1: The upper line, labeled $m_1 > 3H(T_{\rm eq})$ is an upper bound for any such ALP DM model: The mass m_1 at the which the oscillations start should be attained latest at matter-radiation equality. The lower black dotted line labeled $m_1 = m_0$ denotes 'Standard ALP DM' and is the simplest ALP DM model in which m_1 is the same as the ALP mass today (m_0). Models in which $m_1 \ll m_0$ can in principle create a sufficient DM abundance at slightly higher coupling values [15], but most model-building efforts favor ALP DM within roughly an order of magnitude above this line.

Beyond ALPs, *hidden photons* (HPs, reviewed, e.g., in [19]), which are hidden sector U(1)s coupled kinetically to the photons of the visible sector, are WISPy cold dark matter candidates [20, 15]. Such particles would - with a hidden Higgs or Stückelberg generated mass term - be manifest in photon hidden-photon oscillations similar to what is observed with neutrinos. Fig. 2 shows the viable parameter space for HP DM according to [15], in which the orange region labeled 'Xenon' denotes limits inferred from the XENON10 experiment [21] (see also [22] for novel bounds on the longitudinal HP component).

In summary: For both ALPs and hidden photons the parameter regime in which they can constitute DM is much larger than for QCD axions. Thus, in laboratory searches, also non-resonant techniques have become attractive as they allow faster scanning (albeit at reduced overall sensitivity). This is discussed in Sect 2.2. In the following we give a brief overview on laboratory searches for axion and WISP DM. We focus on searches exploiting coupling to

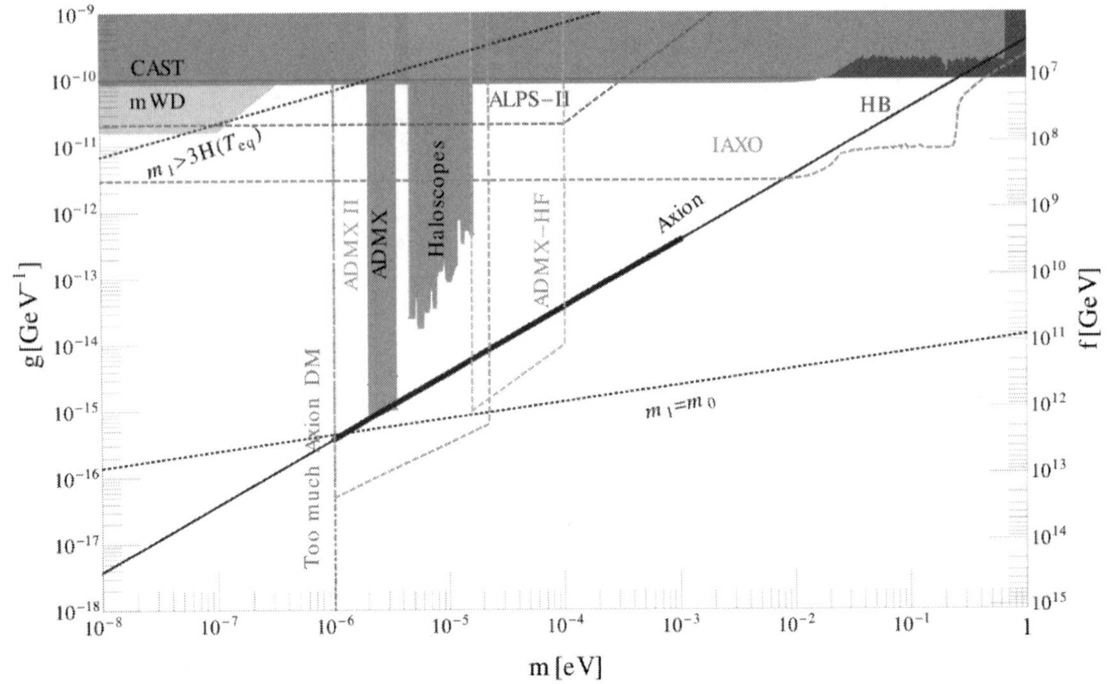

Figure 1. Parameter space of axion and axion-like particle DM (photon-coupling) in the sub-eV range. The QCD axion is predicted in an $\mathcal{O}(1)$ region around the black line labeled 'Axion'. As Dark Matter, the axion should not be much lighter than 10^{-6}eV not to overproduce DM, as indicated by the magenta dot-dashed line. On the other hand, concrete models for ALP DM have been proposed for the region round and below the dotted line labeled $m_0 = m_1$. The dotted line labeled $m_1 > 3H(T_{\rm eq})$ sets a model-building limit for ALP DM. From the observational point of view, the colored regions and and dashed lines are interesting: The colored parameter regions 'mWD', 'HB' and 'CAST', 'ADMX', 'Haloscopes' are excluded by observation and experiment, respectively. The dashed lines indicated the reach of planned experiments and experiments under construction, see text for details

light. Recent reviews on ultra-light particle dark matter with a more theory-based focus include [23, 17].

2. Laboratory setups looking for ultra-light dark matter

It might be helpful to first categorize experiments that are sensitive to light & weakly interacting particles of the above described types in the following manner:

Firstly, there are setups which could find such particles outside the range in which they likely constitute dark matter in the most immediate models and will thus not concern us in the following. This is because these setups cannot probe small enough couplings. The experimental limits of these setups are well below the sensitivity of the filled (and thereby excluded) regions in Fig. 1, but can be argued to be less model-dependent (see, e.g. [13, 14] for a comprehensive review of the WISP parameter spaces at larger coupling values).

Secondly, there are experiments which could find these particles inside the range in which they could be cold dark matter, however independent of the fact whether they actually are dark matter (indirect searches). An example is second-generation light-shining-through-a-wall LSW [24] like ALPS-II [25]. Another example are Helioscopes, that aim to convert axions or the like emitted from our sun into a detectable X-ray signal, such as with the Cern Axion Solar Telescope (CAST) ([26] for recent results, grey in Fig. 1) or the future International Axion Observatory (IAXO) [27]. In Fig. 1 the foreseen reach of ALPS-II and IAXO is shown in dashed pink and green in the upper left, respectively. The grayish region labeled 'mWD' shows an astrophysical exclusion according to [28].

Thirdly, there are experiments which are conceived such that they (probably) find only something if the above described particles are dark matter. In the following, we mention only such recent and planned setups/ideas. As motivated above, we group the searches into 'resonant' and 'non-resonant' although some setups can do both sort of scans. The former technique has an advantage of highest sensitivity, the latter of an increased scanning speed.

2.1. Resonant searches

A resonant Haloscope is the most prominent and thus far most sensitive experiment searching for ultra-light dark matter. Based on the concept suggested in [29], several such experiments have been performed, see [30, 31] for recent reviews: A narrow band microwave resonator is placed in a (typically solenoidal) strong magnetic field. If axions constitute dark matter, they should then stream in large numbers through this experimental apparatus. In the magnetic field, axions would then be converted into photons and be visible at the resonant frequency of the cavity (related to the DM particle mass). Here, the conversion probability of the axions into photons is increased by the power built-up factor of the resonator. Hence, the sensitivity of such experiments can be increased by several orders of magnitudes compared to non-resonant techniques. Due to the pseudo-scalar nature of the axion, the electric fields of the cavity-mode and the external magnetic field should have a sizable overlap. In Fig. 1, obtained Haloscope limits are shown in orange (labeled 'Haloscopes' and 'ADMX') [32, 33, 34, 35, 36, 37, 38]. ADMX has searched for KSVZ axions, over the 1.9 - 3.6 μeV mass range, whereas other performed haloscope searches were so far strictly only sensitive to ALP DM. Currently two further, more sensitive Haloscope measurements are prepared at Seattle (ADMX II) and Yale (ADMX-HF). Prospects are adapted from [39] in the yellow and green dashed areas of Fig. 1, respectively. Note that these prospects and the limits are shown under the assumption $\Omega_a = \Omega_{\rm DM}$, i.e. that axions make up all of the Dark Matter.

For Haloscope searches in the solenoid configuration, it eventually can become difficult at the high mass range to keep the cavity resonant whilst retaining a sizable cavity volume (high mass $\rightarrow$ smaller length scales) [1]. Thus, for *higher-mass DM* axion searches, additional techniques have been suggested and corresponding studies are underway. An example is the ORPHEUS setup [41] based on wire-planes in a confocal resonator. The wire-planes allow to alternate the B-field direction and keep a sizable overlap of external magnetic field and electric photon fields and thus probe higher frequencies [42]. Besides these efforts, let us also point to the emerging activities at the 'Center for Axion and Precision Physics Research' (CAPP) in Korea [43] from where novel contributions to DM axion detection are emerging currently.

A possibility for either resonant (or broadband, see below) direct dark matter searches might open also up with IAXO [27]. IAXO, although its primary mission is to act as a Helioscope like described above, could host, e.g., a cavity set-up in one of its bores. IAXO will be based on a toroidal magnet. Thus, cylindrical cavities do not suggest themselves. However, e.g., a rectangular cavity structure [44, 45] is conceivable. In principle, very large cavities could be feasible to explore a mass range below ADMX (making use of the huge magnetic volume of IAXO). This is however a limited possibility amongst else because it would likely interfere with the original purpose of IAXO as a helioscope.

If long thin cavities are realized, this opens up possibilities also for the higher mass-range

[1] & at the same time detection techniques need careful study at higher masses [40].

[46]. For testing such rectangular cavities before IAXO exists, e.g., the CAST magnet as well as a straightened[2] HERA dipole magnet could be an appropriate test-bench. In fact R&D in this direction has started on a small scale in Valencia [46]. A 1m long 1×2cm cavity (resonant at about 60 μeV) is being characterized towards this purpose. For a single HERA magnet, the magnetic length is 8.83m at 5.3T, for CAST, the magnetic length is 14.3m with 9T field. Albeit the sensitivity reach of these test setups depends on results for the cavity quality factor the noise detection level, such relatively quick R&D measurements will be crucial in determining the real DM potential of IAXO.

Another Haloscope experiment focusing (so far) on HP DM also exists at DESY, called WISPDMX [47]. It employs a former HERA proton cavity and is most sensitive in the 200-600 MHz range and can also run in 'broadband mode', see [48]. The 'broadband mode' is possible since for HPs no mode-overlap with an external field is needed.

A recent suggestion that could complement cavity searches at *very low masses* is the resonant axion DM detection through cryogenic LC-circuits in a strong external magnetic field [49]: axion CDM could induce a current in the part of an LC circuit that is immersed in an external magnetic field. Nicely, such setups are also apt to scan a large part of the HP cold DM parameter space, see [50, 51], although a careful design is likely needed that accounts for the proper shielding of the setup: As HP DM does not need the magnetic field for its conversion in a photon signal, the HP DM stream will also excite electrons in the shield, causing a parasitic contribution that needs to be accounted for in the experiment's design [51].

So far we have have described setups that exploit coupling of the light dark matter particles to photons only. Of course there are alternative techniques, focusing on the coupling of axions to other particles. Examples of recent works point out that axion DM could be found through time-varying CP-odd nuclear moments [52, 53], see also [54], atomic transitions [55] or NMR techniques [56]. As mentioned, we focus on DM-to-photon couplings in the following and turn to the non-resonant techniques now.

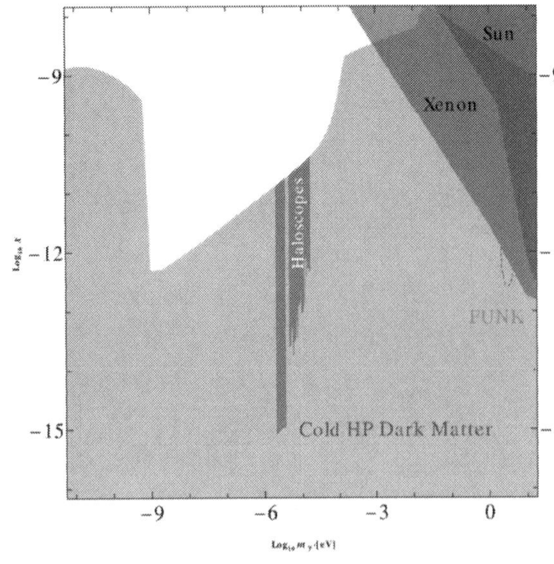

Figure 2. Parameter space (coupling $\text{Log}_{10}\,\chi$ in vs mass $\text{Log}_{10}\,m_{\gamma'}$ [eV]) of hidden photon dark matter in the sub-eV range. The blue shaded region labeled 'Cold HP Dark Matter' is the viable HP DM parameter space as discussed in [15]. The dashed green region shows the parameter reach envisaged by the non-resonant FUNK experiment in Karlsruhe in a first search in the optical (searches at other frequencies are conceivable, see [60]). Other regions see text.

[2] For HERA magnets, the cold-mass (including the beam pipe) was originally fabricated straight and then bent by its outer shell. It has been shown by the ALPS-II collaboration that the beam-pipe can be straightened to almost the full aperture [25] and thus fit appropriate test cavities (the CAST magnet has a straight beam pipe to start with). In its final stage, ALPS-II plans to use 20 HERA magnets, whose later use in DM search seems to be an intriguing possibility also.

2.2. *Broadband search*

As has become obvious from the above considerations, the resonant enhancement is a crucial factor in becoming sensitive to very small couplings and thus eventually to QCD axions. On the other hand, it can be favorable to abstain from tuning to resonance, thereby losing sensitivity in coupling, but gaining in the width of the mass-range that can be covered within a fixed amount of time[3].

Such a concept was proposed in [57, 58] as DM search with a dish antenna: axions and ALP DM (in combination with a magnetic field) as well as HP DM can excite electrons at the surface of a conductor, owing to their (albeit very weak) electromagnetic coupling to them. If this conductor is a spherical mirror, a consequence of this process would a photonic signal at the center of the mirror sphere (whereas most background photons are scattered to the focal point). Such a setup has a directional sensitivity [59] as the exact position of a signal spot will depend on the velocity distribution of the DM with respect to the dish. A global off-set of the signal can be expected due to the movement of the sun in the galactic rest frame as well as a daily modulation on the same order of magnitude (the yearly modulation is negligible due to the small velocity of the earth around the sun). Besides the signal-spot movement, a velocity distribution of the DM leads to a broadening of the signal spot. Such considerations can ultimately help to verify the dark matter nature of a signal. A dish dark matter search (Finding U(1)'s of a Novel Kind - FUNK) is currently being set up [60] at Karlsruhe. It uses a prototype mirror built for the fluorescence detector for the Pierre Auger observatory. Fig. 2 shows a plausible parameter reach for a first measurement in the optical in green.

Searching axion and ALP DM with the Dish technique is much more involved. For such an experiment, the mirror would have to be magnetized *along* the surface (for a setting $\vec{B} \parallel \vec{E}$ the conversion to pseudoscalars is maximized). For example, a magnet with reasonable field strengths $\mathcal{O}(1)$ Tesla that can host the Karlsruhe mirror would be difficult to procure and the experiment would become considerably more complicated. The parameter reach for ALP DM that could be expected in principle with the Dish technique is shown in [57], however, as we are not aware of detailed proposals for such setups, we have not included this in Fig. 1.

3. Summary

In this talk, we have briefly reviewed the case for dark matter candidates below the eV-scale, notably axions, axion-like-particles and hidden photons. We have reviewed recent experiments thriving to find them through their effective coupling to electromagnetic waves. As of today, dark matter could be light, medium mass or heavy, a single species or a mixture of particles. Thus it seems well advised to have a well-balanced variety of experimental searches until we learn more about it. For dark matter candidates in the sub-eV regime, in view of the upcoming experiments described above, we are entering very interesting times.

Acknowledgments

I would like to thank the organizers of the 24th European Cosmic Ray Symposium in Kiel for the kind invitation to present the case for light dark matter and the chance to learn about a variety of topics beyond my field.

I cordially thank A. Lindner, J. Redondo and A. Ringwald for fruitful discussion and collaboration on the above topics and helpful feed-back on this note. In addition, I would like to thank K. v. Bibber for providing some of the sensitivity data.

[3] This assumes that appropriate low-noise broad-band detectors are available which can be difficult over the entire spectral range of interest.

References

[1] See the contributions of Lars Bergstrom, Roberto Battiston and Gernot Maier to these proceedings
[2] P. Sikivie, "Axions", In 'Particle dark matter: observations, models and searches', Bertone, G. (ed.), Cambridge University Press, 2010 pp. 204-227
[3] R. D. Peccei and H. R. Quinn, Phys. Rev. D **16**, 1791 (1977). R. D. Peccei and H. R. Quinn, Phys. Rev. Lett. **38**, 1440 (1977).
[4] J. Preskill, M. B. Wise and F. Wilczek, Phys. Lett. B **120** (1983) 127; L. F. Abbott and P. Sikivie, Phys. Lett. B **120** (1983) 133; M. Dine and W. Fischler, Phys. Lett. B **120** (1983) 137.
[5] P. Sikivie, Lect. Notes Phys. **741**, 19 (2008) [astro-ph/0610440].
[6] M. P. Hertzberg, M. Tegmark and F. Wilczek, Phys. Rev. D **78**, 083507 (2008) [arXiv:0807.1726 [astro-ph]].
[7] G. G. Raffelt, Lect. Notes Phys. **741**, 51 (2008) [hep-ph/0611350].
[8] K. A. Olive *et al.* [Particle Data Group Collaboration], Chin. Phys. C **38**, 090001 (2014).
[9] M. Archidiacono, S. Hannestad, A. Mirizzi, G. Raffelt and Y. Y. Y. Wong, JCAP **1310**, 020 (2013) [arXiv:1307.0615 [astro-ph.CO]].
[10] J. E. Kim, Phys. Rev. Lett. **43**, 103 (1979);
M. A. Shifman, A. I. Vainshtein and V. I. Zakharov, Nucl. Phys. B **166**, 493 (1980).
[11] A. Arvanitaki, S. Dimopoulos, S. Dubovsky, N. Kaloper and J. March-Russell, Phys. Rev. D **81**, 123530 (2010) [arXiv:0905.4720 [hep-th]].
[12] J. Jaeckel and A. Ringwald, Ann. Rev. Nucl. Part. Sci. **60**, 405 (2010) [arXiv:1002.0329 [hep-ph]].
[13] K. Baker *et al.* Ann. Phys. (Berlin) 525, No. 6, A93-A99 (2013) [arXiv:1306.2841 [hep-ph]].
[14] R. Essig, *et al.*, arXiv:1311.0029 [hep-ph].
[15] P. Arias, D. Cadamuro, M. Goodsell, J. Jaeckel, J. Redondo and A. Ringwald, JCAP **1206**, 013 (2012) [arXiv:1201.5902 [hep-ph]].
[16] A. Ringwald, Phys. Dark Univ. **1**, 116 (2012) [arXiv:1210.5081 [hep-ph]].
[17] A. Ringwald, arXiv:1310.1256 [hep-ph].
[18] D. Horns and M. Meyer, JCAP **1202**, 033 (2012) [arXiv:1201.4711 [astro-ph.CO]].
[19] J. Jaeckel, Frascati Phys. Ser. **56**, 172 (2012) [arXiv:1303.1821 [hep-ph]].
[20] A. E. Nelson and J. Scholtz, Phys. Rev. D **84**, 103501 (2011) [arXiv:1105.2812 [hep-ph]].
[21] H. An, M. Pospelov and J. Pradler, Phys. Rev. Lett. **111**, 041302 (2013) [arXiv:1304.3461 [hep-ph]].
[22] N. Vinyoles, A. Serenelli, F. L. Villante, S. Basu, J. Redondo and J. Isern, arXiv:1501.01639 [astro-ph.SR].
[23] M. Kawasaki and K. Nakayama, Ann. Rev. Nucl. Part. Sci. **63**, 69 (2013) [arXiv:1301.1123 [hep-ph]].
[24] J. Redondo and A. Ringwald, Contemp. Phys. **52**, 211 (2011) [arXiv:1011.3741 [hep-ph]].
[25] R. Bähre, *et al.* JINST **8**, T09001 (2013) [arXiv:1302.5647 [physics.ins-det]].
[26] M. Arik *et al.* [CAST Collaboration], Phys. Rev. Lett. **112**, no. 9, 091302 (2014) [arXiv:1307.1985 [hep-ex]].
S. Aune *et al.* [CAST Collaboration], Phys. Rev. Lett. **107**, 261302 (2011) [arXiv:1106.3919 [hep-ex]].
[27] E. Armengaud *et al.*, JINST **9**, T05002 (2014) [arXiv:1401.3233 [physics.ins-det]].
[28] R. Gill and J. S. Heyl, Phys. Rev. D **84**, 085001 (2011) [arXiv:1105.2083 [astro-ph.HE]].
[29] P. Sikivie Phys. Rev. Lett. **51**, 1415 (1983) [Erratum-ibid. **52**, 695 (1984)].
[30] G. Rybka, Phys. Dark Univ. **4**, 14 (2014).
[31] I. P. Stern, AIP Conf. Proc. **1604**, 456 (2014) [arXiv:1403.5332 [physics.ins-det]].
[32] C. Hagmann, P. Sikivie, N. S. Sullivan and D. B. Tanner, Phys. Rev. D **42**, 1297 (1990).
[33] W. Wuensch, S. De Panfilis-Wuensch, Y. K. Semertzidis, J. T. Rogers, A. C. Melissinos, H. J. Halama, B. E. Moskowitz and A. G. Prodell *et al.*, Phys. Rev. D **40**, 3153 (1989).
[34] S. De Panfilis, A. C. Melissinos, B. E. Moskowitz, J. T. Rogers, Y. K. Semertzidis, W. Wuensch, H. J. Halama and A. G. Prodell *et al.*, Phys. Rev. Lett. **59**, 839 (1987).
[35] S. J. Asztalos *et al.* [ADMX Collaboration], Phys. Rev. D **69**, 011101 (2004) [astro-ph/0310042].
[36] S. J. Asztalos *et al.* [ADMX Collaboration], Astrophys. J. **571**, L27 (2002) [astro-ph/0104200].
[37] S. J. Asztalos *et al.* [ADMX Collaboration], Phys. Rev. D **64**, 092003 (2001).
[38] S. J. Asztalos *et al.* [ADMX Collaboration], Phys. Rev. Lett. **104**, 041301 (2010) [arXiv:0910.5914 [astro-ph.CO]].
[39] T. M. Shokair, J. Root, K. A. Van Bibber, B. Brubaker, Y. V. Gurevich, S. B. Cahn, S. K. Lamoreaux and M. A. Anil *et al.*, Int. J. Mod. Phys. A **29**, 1443004 (2014) [arXiv:1405.3685 [physics.ins-det]].
[40] S. K. Lamoreaux, K. A. van Bibber, K. W. Lehnert and G. Carosi, Phys. Rev. D **88**, no. 3, 035020 (2013) [arXiv:1306.3591 [physics.ins-det]].
[41] G. Rybka and A. Wagner, arXiv:1403.3121 [physics.ins-det].
[42] P. Sikivie, D. B. Tanner and Y. Wang, Phys. Rev. D **50**, 4744 (1994) [hep-ph/9305264].
[43] `http://capp.ibs.re.kr/html/capp_en/`
[44] O. K. Baker *et al.* Phys. Rev. D **85**, 035018 (2012) [arXiv:1110.2180 [physics.ins-det]]; M. Betz, F. Caspers and K. Zioutas, ATS/Note/2011/066 (2011).

[45] I. G. Irastorza and J. A. Garcia, JCAP **1210**, 022 (2012) [arXiv:1207.6129 [physics.ins-det]].
[46] J. Redondo, I. G. Irastorza, B. Gimeno, Expression of Interest to the CAST steeing committee, 2014. See also: presentation by J. Redondo at the 'IBS-MultiDark Joint focus program' in Korea, 2014 `http://wwwth.mpp.mpg.de/members/redondo/data/seminarslides/20141023.pdf`
[47] D. Horns, A. Lindner, A. Lobanov and A. Ringwald, arXiv:1309.4170 [physics.ins-det].
[48] D. Horns, A. Lindner, A. Lobanov and A. Ringwald, arXiv:1410.6302 [hep-ex].
[49] P. Sikivie, N. Sullivan and D. B. Tanner, Phys. Rev. Lett. **112**, 131301 (2014) [arXiv:1310.8545 [hep-ph]].
[50] P. Arias, A. Arza, B. Döbrich, J. Gamboa and F. Mendez, arXiv:1411.4986 [hep-ph].
[51] S. Chaudhuri, P. W. Graham, K. Irwin, J. Mardon, S. Rajendran and Y. Zhao, arXiv:1411.7382 [hep-ph].
[52] P. W. Graham and S. Rajendran, Phys. Rev. D **88**, 035023 (2013) [arXiv:1306.6088 [hep-ph]].
[53] D. Budker *et al.* Phys. Rev. X **4**, 021030 (2014) [arXiv:1306.6089 [hep-ph]].
[54] B. M. Roberts, Y. V. Stadnik, V. A. Dzuba, V. V. Flambaum, N. Leefer and D. Budker, Phys. Rev. Lett. **113**, 081601 (2014) [arXiv:1404.2723 [hep-ph]].
[55] P. Sikivie, Phys. Rev. Lett. **113**, no. 20, 201301 (2014) [arXiv:1409.2806 [hep-ph]].
[56] A. Arvanitaki and A. A. Geraci, Phys. Rev. Lett. **113**, 161801 (2014) [arXiv:1403.1290 [hep-ph]].
[57] D. Horns *et al.*, JCAP **1304**, 016 (2013) [arXiv:1212.2970].
[58] J. Jaeckel and J. Redondo, Phys. Rev. D **88**, 115002 (2013) [arXiv:1308.1103 [hep-ph]].
[59] J. Jaeckel and J. Redondo, JCAP **1311**, 016 (2013) [arXiv:1307.7181].
[60] B. Döbrich, K. Daumiller, R. Engel *et al.*, arXiv:1410.0200 [physics.ins-det]; J. Redondo and B. Döbrich, arXiv:1311.5341 [hep-ph].

Radio detection of air showers at the Pierre Auger Observatory

Qader Dorosti Hasankiadeh* **for the Pierre Auger Collaboration**†
*Institut für Kernphysik, Karlsruhe Institute of Technology (KIT), Germany
†Pierre Auger Observatory, Av. San Martín Norte 304, 5613 Malargüe, Argentina
(Full author list: `http://www.auger.org/archive/authors_2014_09.html`)

E-mail: `qader.dorosti@kit.edu`

Abstract. High-energy cosmic rays impinging onto the atmosphere of the Earth induce cascades of secondary particles: extensive air showers. Many of the particles in a shower are electrons and positrons. The electrons and positrons interact with the geomagnetic field and emit radiation. We detect such radiation at frequencies of tens to hundreds of MHz with the Auger Engineering Radio Array (AERA) at the Pierre Auger Observatory in Argentina. The objective is to investigate the properties of cosmic rays at the expected transition from Galactic to an extragalactic origin at energies around 10^{17} to 10^{18} eV. We discuss the recent progress in radio detection of ultra-high-energy cosmic rays with AERA. To this end, we elaborate on our measurements of shower properties, such as the lateral distribution of the radio emission at ground level. In addition, we present methods to reconstruct the properties of the primary particle from the radio data.

1. Introduction

The radio emission from cosmic-ray-induced air showers is dominated by two emission mechanisms, the *geomagnetic effect* [1] and the *Askaryan effect* [2]. The geomagnetic effect is caused by the transverse separation of the electrons and positrons due to the geomagnetic field. The Askaryan effect probes radio emission produced by the negative charge excess along the longitudinal development of the extensive air shower. The combination of both effects leads to radiation in a large frequency bandwidth of tens of MHz.

Radio detection of air showers benefits from a quasi-calorimetric measurement of the shower energy with a duty cycle of 100%. In addition, similar to fluorescence detectors, radio detectors are sensitive to the longitudinal air-shower development, thus can provide an estimate for the mass of the primary cosmic rays [3].

AERA [4] is located at the site of the Pierre Auger Observatory [5] in the province of Mendoza, Argentina. The current configuration of AERA, i.e. AERA-124, consists of 124 radio stations, covering an instrumented area of about 6 km^2. AERA is equipped with two types of radio stations: a) 24 radio stations equipped with Logarithmic Periodic Dipole Antennas (LPDA), which are deployed on a grid of 125 m. b) 100 radio stations using butterfly antennas (see figure 1), which are placed on grids of 250 m and 375 m. Each radio station consists of two crossed antennas aligned in the geomagnetic North-South and East-West directions, recording radio signals in a frequency range of 30 to 80 MHz. AERA operates in self trigger and external

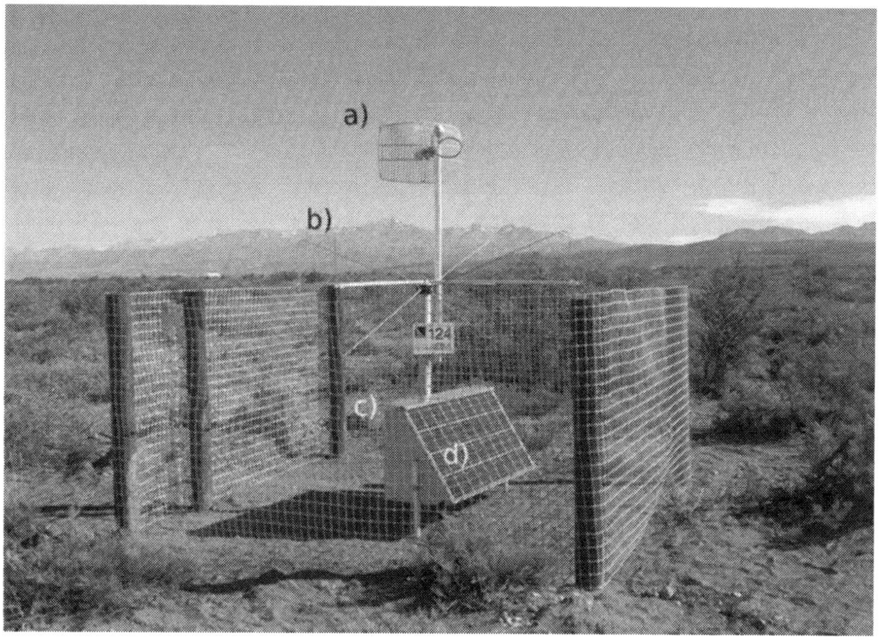

Figure 1. AERA radio station with butterfly antennas. Shown are: a) a wireless communication antenna exploited for data transfer, b) two butterfly antennas aligned in East-West and North-South directions, c) an electronics box providing the housing for electronics and a battery for power supply and d) solar panel.

trigger modes. The external trigger exploits the surface detectors (SD) or fluorescence detectors (FD) of Auger for triggering.

Exploiting the proven techniques developed in previous radio detectors, such as LOPES [6] and CODALEMA [7], AERA aims to investigate the radio emission processes taking place in high-energy extensive air showers. In addition, AERA has been calibrated to provide the measurements of the fundamental parameters of extensive air showers, such as the direction, energy and mass of the primary cosmic rays that initiated the shower. AERA covers energies around and above 10^{17} eV.

This paper presents results obtained using the first set of externally triggered events which have been detected and reconstructed with AERA.

2. Measurement of the radio emission from air showers

In order to make use of the data measured by radio detectors, we need to understand the radio emission mechanisms taking place in air showers. As mentioned before, two radio emission mechanisms, i.e. the geomagnetic and Askaryan effects, contribute mainly to the radio emission of air showers. The electric field induced by the geomagnetic effect points to the direction defined by $\vec{v} \times \vec{B}$, where $\vec{v}$ is a vector denoting the axis of an air shower and $\vec{B}$ is the geomagnetic field vector. On the other hand, the electric field induced by the Askaryan effect has a radial direction pointing towards the shower axis [8].

One can consider total electric field as the linear contribution of electric fields induced by geomagnetic $\vec{E}^G(t)$ and Askaryan $\vec{E}^A(t)$ effects:

$$\vec{E}(t) = \vec{E}^G(t) + \vec{E}^A(t), \tag{1}$$

where t is the time of the measurement of the electric field at a certain radio station. Each AERA station measures voltages in two polarisation directions, i.e. North-South and East-West. Using the information of the reconstructed direction of the event and correcting for the response pattern of the radio antennas, we transform the time traces of radio pulses measured in the two polarisation directions to the components of the electric field vector. Comparing the predicted azimuthal polarisation angle $\Phi_p(pr.)$, as defined in equation 2, with the measured azimuthal polarisation angle $\Phi_p(me.)$, one can quantify the contribution of each emission mechanism in the radiation of air showers.

$$\Phi_p(pr.) = \tan^{-1}\left(\frac{E_y^G + E_y^A}{E_x^G + E_x^A}\right), \tag{2}$$

where the subscripts x and y denote the geographic East and North directions, respectively. One can express the above equation in terms of azimuthal angles of geomagnetic Φ^G and Askaryan Φ^A contributions, in which the azimuthal angles can be calculated for a shower with a given direction (more details presented in [8]):

$$\Phi_p(pr.) = \tan^{-1}\left(\frac{\sin\Phi^G + a'\sin\Phi^A}{\cos\Phi^G + a'\cos\Phi^A}\right), \tag{3}$$

where the parameter $a' = \left|E^A\right| / \left|E^G\right|$.

We estimate the most likely values of the a' parameter, by comparing the measured and predicted azimuthal polarisation angles. Since the strength of E^G depends only on the angle α between the shower axis and the geomagnetic field, we scale E^G, dividing it by $\sin\alpha$. Thus we obtain the $a = \sin(\alpha)a'$ parameter which is independent from the shower arrival direction. Therefore, we define the a parameter as the relative strength of the electric fields induced by the Askaryan emission processes.

Figure 2 (left) presents the distribution of the measured value of the a parameter and its estimated uncertainty (black error bars) for each measurement. The distribution has a mean value of $a = (14 \pm 2)\%$ at 68% confidence level. Figure 2 (right) shows the distribution of the measured azimuthal polarisation angle as a function of the predicted azimuthal polarisation angle for the mean value of $a = 14\%$. The value of the Pearson correlation coefficient, i.e. $\rho_P = 0.93^{+0.04}_{-0.03}$ at 95% confidence level is at its maximum for the mean value of the a parameter.

A comparison between the simulation and measurement of the radio emission indicates that the general picture of the radio emission mechanisms agrees with the experimental measurements within the uncertainties (see [9]). Figure 3 shows the lateral distribution of a typical radio event measured by AERA (data points) compared with predictions from radio emission simulations (shaded areas). Two different radio emission codes, i.e. CoREAS version 1.0 [10] and ZHAireS beta v28 [11], have been applied for this comparison. There is a good agreement between the measured data and the simulation of the radio emission.

3. Event reconstruction

The measurement of the arrival time of the radio wavefront hitting the array of AERA-radio stations leads to the reconstruction of the arrival direction. When a radio event is externally triggered by the SD detector, all AERA stations equipped with buffering hardware are read out. This increases the probability of baring stations with pulses from radio frequency interferences (RFI). In order to suppress RFI pulses, we exploit the information of the event direction reconstructed by SD to impose a time window in which the radio signal is searched. The width of the signal search window is the quadratic sum of the uncertainties derived from the SD direction reconstruction and the AERA timing precision. Furthermore, we suppress the RFI

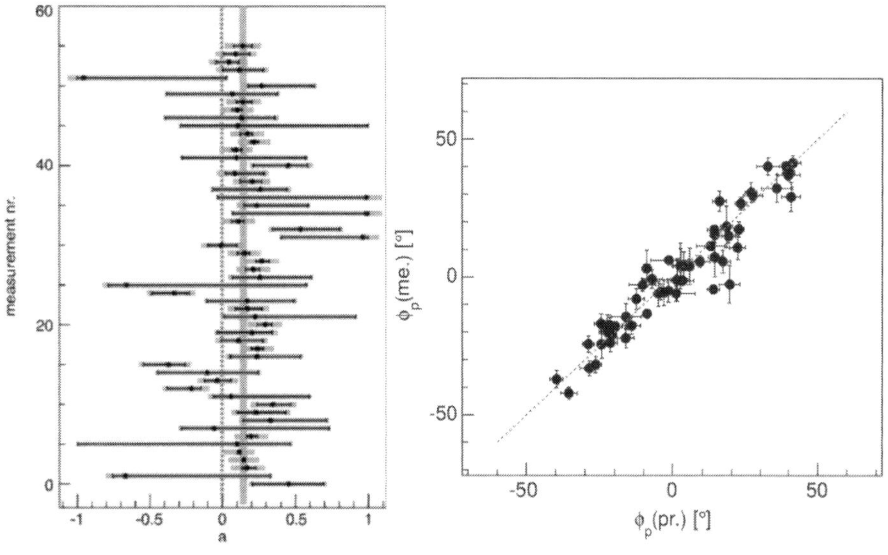

Figure 2. Left) Distribution of the a parameter and its estimated uncertainty (black error bars) per measurement, where the solid line indicates the 68% confidence belt around the mean value of the a parameter; the dashed line shows the value of $a = 0$. The orang boxes show the upper (lower) uncertainty bound when the mean value of a is larger (smaller) than the measured a. Right) Distribution of the measured azimuthal polarisation angles as a function of the predicted azimuthal polarisation angles for the optimal value of $a = (14 \pm 2)\%$. The plots are taken from [8].

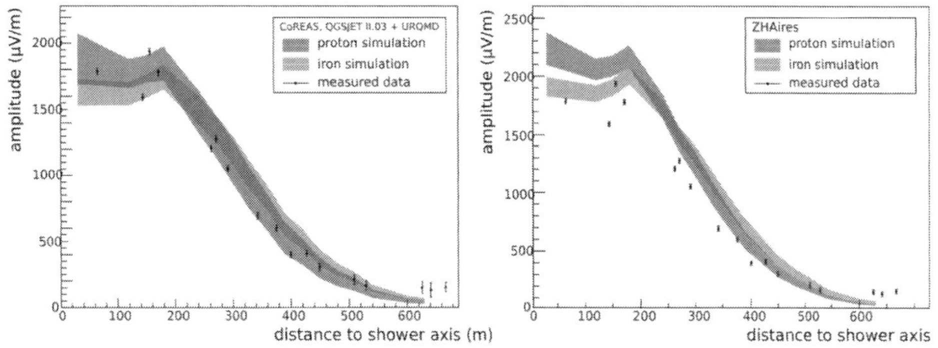

Figure 3. Lateral distribution of an event measured by AERA compared with COREAS (left) and ZHAires (right) simulation codes. Data points show the experimental data while the blue and red shaded areas show the proton and iron simulations, respectively. The smearing shown for simulations is the uncertainties of the input parameters, e.g. core location and energy, caused by shower-to-shower fluctuations. The plots are taken from [12].

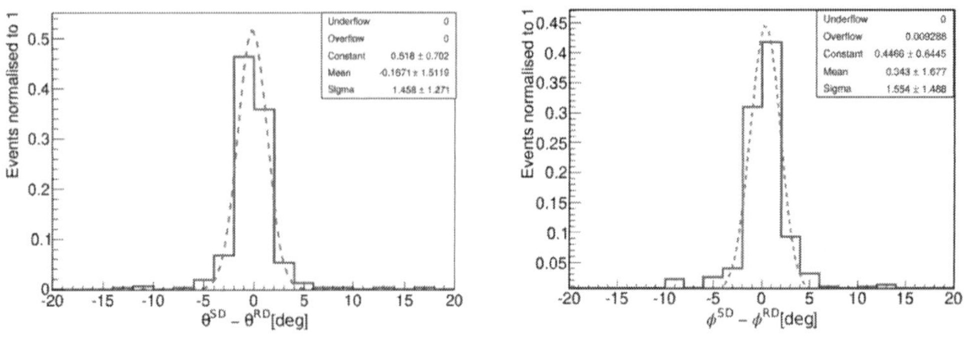

Figure 4. Performance of the direction reconstruction of AERA. Shown are the difference between the AERA and SD reconstructed zenith angle (left) and azimuth angle (right). A Gaussian function is fitted on the distributions in order to obtain the spreads of the distributions.

pulses using a cluster algorithm which selects the cluster of stations associated with the event measured with the SD detector.

Figure 4 shows the performance of the direction reconstruction of AERA, which compares the shower directions measured experimentally by SD and AERA (RD). According to the figure, the direction reconstruction of AERA agrees with that of SD within 2.1°. The events used in this analysis include low-energy events which employ only a few stations per event. The deviation is expected to decrease for high-energy events.

We exploit the amplitude information of radio pulses to reconstruct the energy of the events [13]. As explained in section 2, we transform the time traces of radio pulses in the East-West and North-South polarisation directions to the components of the electric field vector (see figure 5). Then our energy reconstruction algorithm is defined as the strength of the electric field vector at the optimal distance D_0, which is scaled with the angular dependency of the geomagnetic emission mechanism. The value of D_0 is optimised to achieve the highest energy resolution for AERA [13].

Figure 5 (left) shows a typical measured time trace of an electric field vector for East-West, North-South and vertical directions. Figure 5 (right) overimposes the measured electric field vectors and the expected electric field induced by geomagnetic and Askaryan fields, for a sub-sample of the externally triggered events. There is a good agreement between the measured and expected electric fields.

4. Conclusion

AERA is the world's largest cosmic-ray radio detector. It operates at the Pierre Auger Observatory, in Malargue, Argentina. Taking advantage of measuring cosmic rays in coincidence with Auger's surface and fluorescence detectors, AERA furnishes important cross-calibrations and comparisons with the other detection techniques employed at the Observatory. AERA primarily aims to investigate the radio emission mechanisms from air showers in an energy range between proximately $10^{16.9}$ – 10^{19} eV. In addition, operating with a duty cycle of nearly 100%, AERA intends to pursue the scientific goals of the Pierre Auger Observatory, by providing complementary information to Auger's surface and fluorescence detectors.

We have summarised results which have been obtained using the first set of externally triggered events measured by AERA. The results show that radio emission from air showers consists of the contributions of two main effects: geomagnetic and Askaryan effects. The geomagnetic emission process is dominant, while the Askaryan effect contributes with about

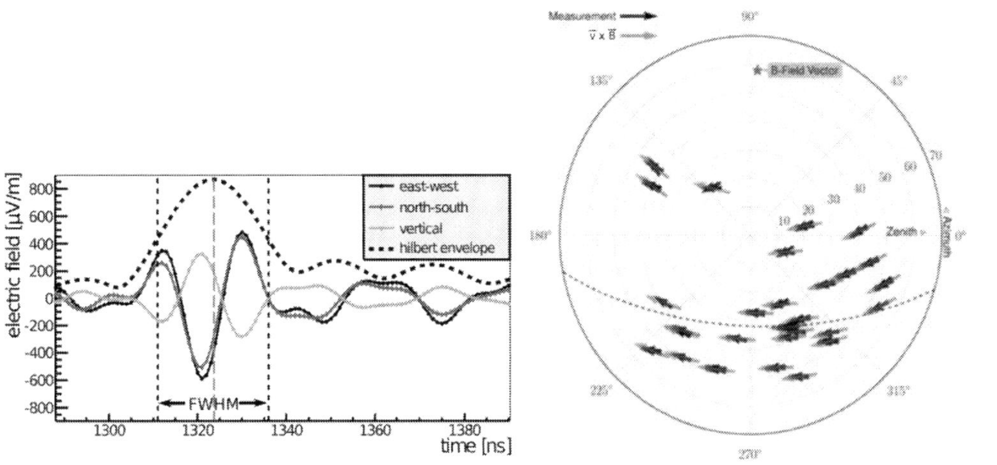

Figure 5. Left) Measured electric field vector for East-West (solid black line), North-South (solid dark grey line) and vertical (solid light grey line) polarisation components. The Hilbert transformation is applied to calculate the envelope of the time trace of the electric field vector. The dashed line shows the Hilbert envelope. The figure is adapted from [13]. Right) The measured comparison of electric field (dark black arrows) with the expected electric field (light red arrows). The figure is adapted from [14].

$(14 \pm 2)\%$ to the radio emission of air showers measured of AERA. This is in good agreement with full Monte-Carlo simulations of radio emission in air shower for the location of the Pierre Auger Observatory. Furthermore, AERA has been optimised to contribute to the measurement of fundamental air-shower parameters, i.e. direction, energy and composition. The direction of events measured by AERA scatters by an angle of 2.1° from that measured by the SD, which is expected to improve for high-energy events for which the number of employed stations per event are higher. AERA has also been calibrated to measure the event energy. The AERA-reconstructed energy correlates with the SD-reconstructed energy.

5. References

[1] Kahn F D and Lerche I 1966 *Proc. Roy. Soc. A* **289** 206–213
[2] Askaryan G A 1962 *Sov. Phys. JETP* **14** 441–443
[3] Apel W *et al.* (LOPES Collaboration) 2012 *Phys.Rev.* **D85** 071101
[4] Abreu P *et al.* (The Pierre Auger Collaboration) 2012 *JINST* **7** 10011
[5] Allekotte I *et al.* (The Pierre Auger Collaboration) 2008 *Nucl. Instrum. Meth. A* **586** 409–420
[6] Huege T *et al.* 2012 *Nucl. Instrum. Meth. A* **662** (Supplement 1) S72–S79
[7] Ravel O for the CODALEMA Collaboration 2012 *Nucl. Instrum. Meth. A* **662** (Supplement 1) S89–S94
[8] Aab A *et al.* (The Pierre Auger Collaboration) 2014 *Phys. Rev. D* **89** 052002
[9] Apel W *et al.* (LOPES Collaboration) 2013 *Astropart.Phys.* **50-52** 76–91
[10] Huege T, Ludwig M and James C 2013 *AIP Conf.Proc.* **1535** 128–132
[11] Alvarez-Muñiz J, Carvalho Washington R J and Zas E 2012 *Astropart.Phys.* **35** 325–341
[12] Schröder F G for the Pierre Auger Collaboration 2013 *33rd International Cosmic Ray Conference,* Rio de Janeiro, Brazil (*Preprint* `arXiv:1307.5059`)
[13] Glaser C for the Pierre Auger Collaboration 2013 *ARENA-2012 AIP Conference Proceedings* **1535** 68–72
[14] Weidenhaupt K 2013 *Acta Polytechnica* **53** (Supplement) 825–828

Characterization of an LSO scintillator for space applications

R Elftmann, J Tammen, S R Kulkarni, C Martin, S Böttcher, R Wimmer-Schweingruber

Institute of Experimental and Applied Physics, University of Kiel, Leibnizstr. 11-19, D-24118 Kiel, Germany

E-mail: elftmann@physik.uni-kiel.de

Abstract. Currently BGO ($Bi_4Ge_3O_{12}$) is widely used for the detection of high-energy particles in space applications because of its high stopping power, the non-hygroscopic characteristics and its ruggedness with respect to mechanical stress. The new Cerium doped LSO (Lu_2SiO_5) offers the same benefits with higher light output capabilities and a significantly shorter decay time. We investigated key characteristics of an LSO scintillator in view of its use in space missions. We characterized the intrinsic spectrum which originates from the decay of ^{176}Lu and showed that it consists of three different parts arising from different effects: the native intrinsic spectrum, chance coincidence effects and energy deposition in the readout photodiode. Furthermore we investigated the light-quenching of LSO for heavy ions with measurements performed at the Heavy Ion Medical Accelerator in Chiba (HIMAC), Japan. We found that LSO is a promising candidate for future space missions.

1. Introduction

Scintillators have been used for the detection of energetic particles since the beginning of the 20th century. Because scintillation materials differ in characteristics such as light output, ruggedness, density, or atomic number the search for the ideal scintillator is not finished. Many new materials have been discovered and developed in recent years and one of these materials was Lu_2SiO_5:Ce (LSO) [3]. Since its development it has widely replaced BGO in Positron Emission Tomography (PET) scanners because of its higher light output and shorter decay time of 42 ns compared to 300 ns of BGO. Table 1 gives a comparison between commonly used inorganic scintillation materials and LSO.

For the detection of high energy particles in space, a scintillation material with high density and high atomic number is attractive because this reduces the volume of stopping material and thus the required volume and mass of the instrument. Therefore BGO as well as LSO are both suitable candidates for these applications. While BGO is already used in various space applications LSO has not been used so far. Before LSO can be used for space applications necessary key characteristics have to be investigated like the response to heavy particles, temperature dependence of light output or the effects of the high inherent radioactivity. LSO contains a significant amount of ^{176}Lu atoms, which decay to ^{176}Hf via β^- -decay emitting three main gamma lines as shown in fig. 1. These gamma lines are almost emitted in coincidence so that all gamma lines or combinations thereof may be detected at the same time. Moreover the high amount of ^{176}Lu leads to a high intrinsic decay rate and therefore to a significant background.

	NaI(Tl)◇	BGO◇	GSO(Ce)◇	LSO(Ce)◇	CsI(Tl)†
Density (g/cm^3)	3.67	7.13	6.7	7.4	4.51
Effective atomic number	51	75	59	66	54
Radiation length (cm)	2.56	1.12	1.38	1.14	1.85
Hygroscopic	yes	no	no	no	Slight
Rugged	no	yes	no	yes	yes
Refractive index	1.85	2.15	1.85	1.82	1.80
Relative light intensity	100	15	25	75	145
Decay constant (ns)	230	300	56, 600	40	1220
Peak Wavelength (nm)	410	480	440	420	560

Table 1. Properties of common inorganic scintillation materials. ◇ taken from [4], † from [1],[2].

The higher light output of LSO compared to BGO is only determined for low ionizing particles. Heavy ionizing particles introduce an additional non-linear light loss mechanism, known as ionization quenching [7], which is different for each scintillation material and varies with the particle energy and species. This light loss mechanism is related to the high density of excited states within the scintillation detector which is produced by the heavy ionizing particles. The quenching factors for LSO have to be determined to clarify whether the light output of LSO is also superior for heavy particles and to calculate the real deposited energies of heavy particles detected with LSO.

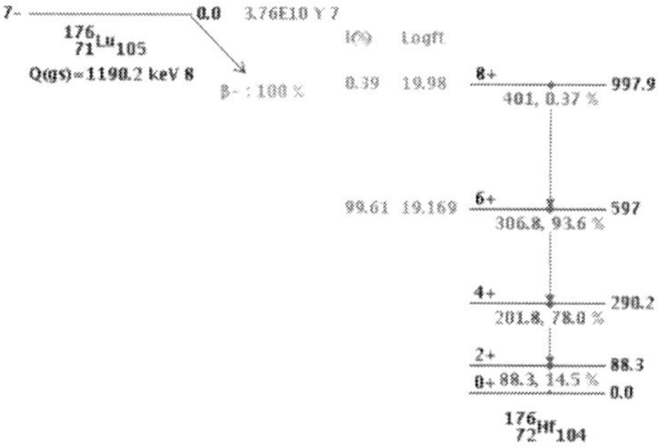

Figure 1. Decay scheme of ^{176}Lu. This isotope has an abundance of 2.59%. Taken from [5].

2. Characterization of the intrinsic spectrum of LSO:Ce

The intrinsic spectrum of LSO arises from the β^- -decay of ^{176}Lu. In order to determine the origin of different parts of the resulting spectrum, we compared a measured spectrum to a Geant4 simulation.

The LSO crystal we used for our characterization was manufactured by Scionix Holland with a Cerium doping level of 0.5%. The dimensions were 3 x 1.5 x 1 cm^3 and the crystal was initially completely polished. The readout of the scintillator was performed using a Hamamatsu Photonics PIN photodiode model S-3590-19 which was glued to one half of the 3 x 1.5 cm^2 side of the scintillator. For the glueing Dow Corning 93-500 Space-grade encapsulant was used. The S-3590-19 offers an enhanced sensitivity in the blue light region and matches the emission spectrum of LSO quite well. In order to increase optical coupling with the photodiode, the crystal surface where the photodiode was glued was roughened. The complete scintillation detector was wrapped with one layer of Millipore and two layers of PTFE tape to increase the light collection.

A picture of the scintillation detector is shown fig. 2 a). The signal of the photodiode was pre-amplified using a space grade charge-sensitive preamplifier which was designed and produced at our institute. The signal was further amplified by an Canberra amplifier model 2022 with 2 μs shaping time and the pulse height was measured with a Fast Comtec multiparameter system MPA-3. The calibration of the scintillation detector was performed with a ^{207}Bi and a ^{60}Co source.
We recorded the intrinsic spectrum of LSO for 3 hours and the result is shown as the red line in

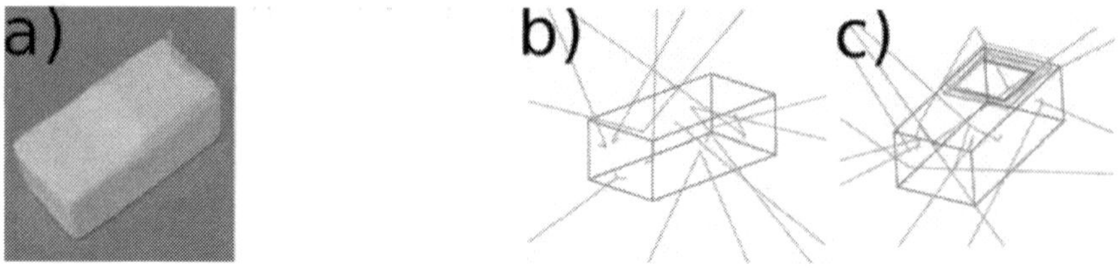

Figure 2. a): wrapped LSO scintillation detector. b): illustration of the first simulation model. c): the complete simulation model with the photodiode. The light green lines represent sample gamma trajectories.

fig. 3. With the calibrated intrinsic spectrum, shown in fig. 1, we can show that the energy range is higher than the expected Q-value of 1190.2 keV. To clarify the origin of the higher energy part of the spectrum a Geant4 simulation was performed using the Radioactive Decay module. With this module it is possible to accurately simulate the radioactive decay of unstable isotopes and to track the resulting particles.
The first approach for the simulation was to simulate the decay of ^{176}Lu and evaluate only the deposited energy in the crystal. The complete crystal was defined as a source volume so that unstable ^{176}Lu atoms were randomly distributed throughout the complete crystal. To compare the intensity of the simulation with that of the experiment, the number of decays during the time of the experiment had to be estimated. The amount of ^{176}Lu atoms in the crystal was approximated by using the mass of the crystal and the mass of a single LSO primitive cell. The decay rate can then be calculated to 1277 Bq taking into account the half life of the isotope. This rate is in good agreement with the rate of 300 $\frac{\mathrm{Bq}}{\mathrm{cm}^3}$ found in previous experiments [6]. Because the scintillation detector has a limited resolution and other line broadening effects it was necessary to introduce noise into the simulated spectrum. Therefore Gaussian noise with a σ of 30 keV was applied to the simulated spectrum. The simulated spectrum where only energy deposition in the crystal was taken into account is shown as the magenta line in fig. 3. The direct comparison between simulation and measurement shows a good agreement up to $\approx$1100 keV. Up to this energy the intensity as well as the position of the peaks match the experiment. However, while the simulated spectrum ends at 1190 keV the measured spectrum extends to energies above 4000 keV. Therefore it appears that high energy events which were measured in the experiment can not be attributed to direct energy deposition in the crystal by the intrinsic radiation.
By introducing two effects in the simulation the origin of the high energy events can be explained. The first effect is known as chance coincidence and results from the high intrinsic decay rate and the shaping time of 2 μs of the shaper. The rate r of the these events can be calculated as shown in eqn. 1 with the activity of the crystal r_{source} of 1277 Bq and the shaping time τ of 2 μs to be 3.26 s^{-1}. When we consider the chance coincidence events in the evaluation of the simulation, we can explain the first part of the high energy events as shown in blue line in fig. 3. The energy range of the simulated spectrum extends up to $\approx$2000 keV and the shape of this part matches the shape of the measured spectrum. The reason for the deviation of the spectra

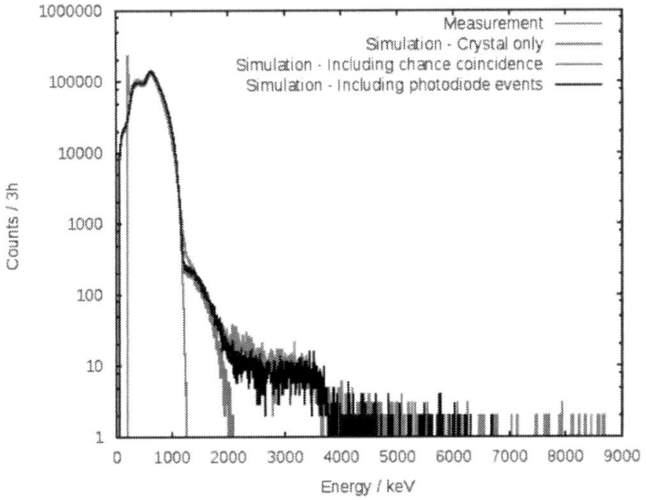

Figure 3. The red graph shows the measured intrinsic spectrum of LSO. If only energy deposition in the crystal is considered in the simulation the maximum energy is 1190 keV (magenta graph). Considering the chance coincidence counts in the simulation leads to a better match of the simulated and measured spectrum (blue graph). Photodiode hits introduce a high energy part into the intrinsic spectrum (black graph).

in the energy range between 1400 keV and 1500 keV is most probably the simple model we applied instead of the exact process of the measurement system. The electronics are sensitive to temporal separation of events within the shaping time interval and that alters the total height of the resulting peak. Thus the double events are shifted towards lower energies in the real measurement and that explains the deviation between the two spectra in that region. The consideration of these chance coincidence events is important for LSO because of its high intrinsic decay rate. Especially when large scintillator volumes are used these events will significantly influence the spectrum since the rate of double events is proportional to the single event rate squared.

$$r = (r_{source})^2 \cdot \tau \qquad (1)$$

The second effect which leads to the generation of high energy events is direct energy deposition in the readout photodiode of the scintillation detector. To study these events we implemented the photodiode and the glue which was used to glue the photodiode to the crystal in the simulation. The conversion efficiency of the photodiode is superior to that of LSO. Therefore events where the photodiode is hit by primary particles produce a higher electronic signal which is converted to higher energies when using the LSO calibration. In order to take this effect into account for the simulation the deposited energy in the photodiode had to be multiplied by a conversion factor. We determined the conversion factor with the same experimental setup as used before with a photodiode of the same model as used for the LSO readout. By comparing the two calibration functions we determined the conversion factor from photodiode to LSO to be 18.6. This factor was applied to the energy which was deposited in the photodiode in the simulation. The black line in fig. 3 shows that the last part of the intrinsic spectrum can be reconstructed using this approach. This part of the simulated spectrum matches not only the shape of the experimental spectrum but also the maximum energy. We found that the spectrum in the photodiode is almost exclusively created by electrons which pass through the glue and deposit energy in the silicon.

These events can be rejected by using a second readout photodiode. However, this results in a loss of resolution so that for high resolution applications these events should be considered.

3. Determination of quenching factors

Quenching is known as an non-linear light-loss of scintillation light [7]. To determine the quenching factors for LSO we designed and built a particle telescope consisting of the previously used LSO crystal, one 2 x 2 x 2 cm^3 BSO scintillator, an anti-coincidence photodiode between the scintillators and two tracking photodiodes. A CAD view of the instrument is shown in fig. 4. Both scintillators were wrapped in two layers of Millipore and PTFE tape. To detect the scintillation light of the scintillators each scintillator was read out by two photodiodes. Therefore events where a photodiode is directly hit by a particle can be excluded from the analysis. The telescope features two entrance windows, so that ions may enter the instrument from two sides. This makes a individual characterization of each scintillator possible. Measurements of the scintillator response to several heavy ion species were performed at the **H**eavy **I**on **M**edical **A**ccelerator in **C**hiba (HIMAC), Japan. In this work we present the quenching factors of LSO

Figure 4. The LSO crystal is shown in yellow, the BSO in blue. Circular openings at the ends allow an individual characterization of each scintillator.

for He, C, and Si. The initial energies of these ions at HIMAC are 230 MeV/nuc, 400 MeV/nuc and 800 MeV/nuc, respectively. By placing polyethylene (PE) blocks in front of the telescope we were able to decrease the ion energy. The instrument was calibrated using protons which originate from ion fragmentation processes of the primary ions inside the PE absorber and also with cosmic muons. The muons and the protons both produce only low ionization densities inside the scintillator and thus they do not exhibit quenching and can be used for calibration. Figure 5 shows the combined measurement data points for all He measurements for the various PE thicknesses. The deposited energy in the crystal is plotted against the deposited energy in the photodiode. With increasing PE thickness, the deposited energy in the LSO increases up to a certain value until at a certain thickness the particles stop in the scintillator. At this point the deposited energy decreases since the particles only deposit their remaining energy. The quenching is most severe for stopping particles because of the high dE/dx, especially at the Bragg peak. In order to determine the quenching factors for LSO we utilize Birks empirical quenching formula, equation 2, for the light output of a scintillator per path length, dL/dx, which takes into account the effect of quenching. B·dE/dx gives the ionization density of molecules along the track and with the quenching parameter k, kB gives the amount of molecules which contribute to quenching. S is the scintillation efficiency [7].

$$\frac{dL}{dx} = \frac{S\frac{dE}{dx}}{1 + kB\frac{dE}{dx}} \tag{2}$$

Using eqn.2 and the approximation for high energetic particles that dE/dx $\approx \frac{CAZ^2}{E}$ [8] with C being a proportionality constant the following equation can be calculated:

$$L(E) = C_1 \left(E - C_2 A Z^2 \ln \left| \frac{E + C_2 A Z^2}{C_2 A Z^2} \right| \right) \tag{3}$$

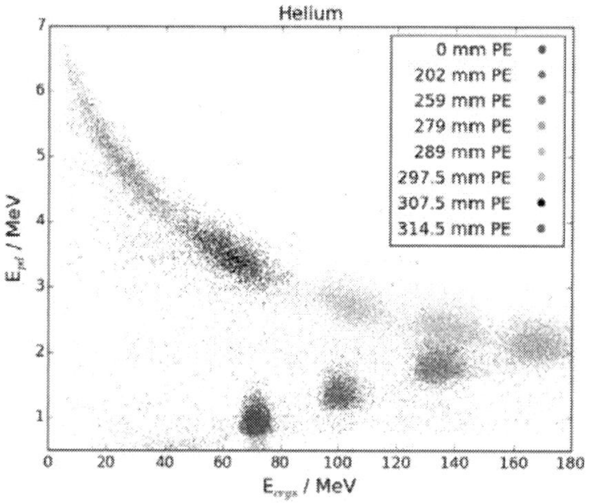

Figure 5. The measured and calibrated data points of the helium run. The data are not corrected for quenching.

The deposited energy E was calculated with a Geant4 simulation of the complete setup at HIMAC. With the energy E and the atomic mass and charge number, A and Z, we determined the parameters C_1 and C_2. For L we used the deposited energy from the measurement. Therefore the resulting quenching factors are not related to the light output of the scintillator and can be compared to other scintillators more easily. Furthermore the results can be applied to any calibrated LSO scintillation detector. For each PE thickness and ion species the most probable energy loss in the LSO and in the photodiode was determined with a fit of a 2-D Gaussian function. This was done for the simulation as well as for the experiment. The most probable energy loss in LSO of the measurement and of the simulation were then plotted against each other in fig. 6 to determine the quenching factors. The quenching parameters were determined by a least square fit of eqn. 3 to the data points of the ions which stop in the LSO crystal. For each ion species a separate set of parameters was determined. The parameter C_1 is proportional to the scintillation efficiency S and the parameter C_2 is proportional to the quenching factor kB. In our case the scintillation efficiency, parameter C_1, is particle specific and thus determined for each ion individually. The resulting curves from the least square fit are displayed in fig. 6. For illustration a dotted line is added to the plot where the measured energy E_{meas} of the experiment is the same as the simulated energy E_{sim}. This line shows where one would expect the peaks to lie of no quenching occurs. As expected the light loss due to quenching increases with the atomic mass and charge number of the ion.

Sources of experimental errors are a possible tilt in the adjustment, a non exact thickness or a density uncertainty of the PE. We considered these uncertainties by performing worst-case simulations and used the deviations from our determined values as error approximations. The uncertainty of the simulation was taken as 1% as given in [9] and [10] for the accuracy of Geant4 simulations for energy loss by ionization. The uncertainty from the determination of the most probable energy loss from the fit of the 2-D Gaussian was also taken into account for the experiment and the simulation. The resulting parameters for C_1 and C_2 are given in tab. 2. With these quenching factors it is possible to calculate the correct deposited energy for any calibrated LSO scintillation detector for the investigated ion species. While the errors for the C_1 parameters are below 4%, the errors of the C_2 parameters are quite large. The bending of the curve at low energies which is mostly described by this parameter C_2 needs more data points in order to be determined more accurately.

Ion	C_1	C_2 / MeV
He	0.712 ± 0.029	0.084 ± 0.043
C	0.529 ± 0.015	0.043 ± 0.005
Si	0.348 ± 0.014	-0.00055 ± 0.00035

Table 2. The parameters C_1 and C_2 of eqn. 3 obtained from the least square fit for each ion species.

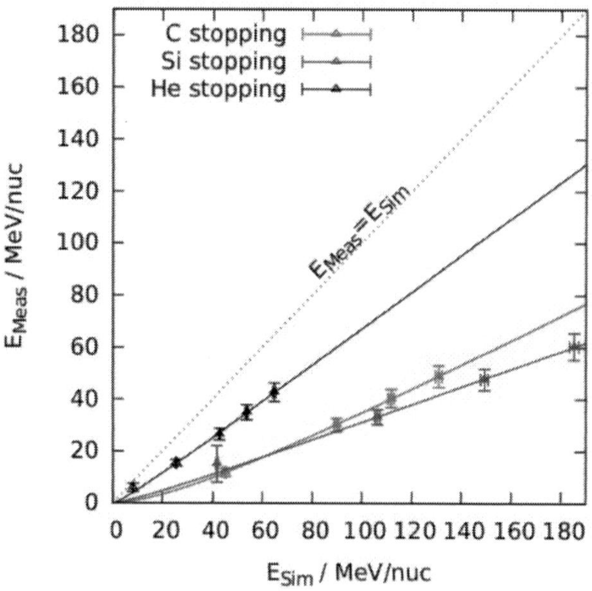

Figure 6. The energy deposited of the ions stopping in the LSO from the experiment and from the simulation is plotted against each other. For each ion species an individual fit is performed using eqn. 3.

Similar studies have been performed with the same method for BGO [11] and so we are able to compare those results with these for LSO. LSO has a five times higher light emission for light particles compared to BGO but is also known as a heavy quencher [2]. By comparing our results with BGO we can show that despite the higher quenching, the light output of LSO is superior to that of BGO for heavy ions, at least for the investigated ion species. Table 3 shows the C_1 parameter of LSO and that of BGO taken from [11]. The relative scintillation efficiencies of LSO are always lower than those of BGO. This means that a higher amount of energy is lost to quenching in LSO. However, because LSO produces five times more photons per MeV compared to BGO, LSO still emits 3.0 ± 1.2 times more light for C, 3.6 ± 1.4 times more light for He and 3.6 ± 1.2 more light for Si. Therefore LSO is indeed suited for the detection of heavy ions in future space missions and promises an increased resolution and a better signal to noise ratio. For measurements of high energy particles the intrinsic spectrum is not problematic because its maximum energy is below 4 MeV.

Ion	$S_{spec,BGO}$	$S_{spec,LSO}$	$S_{spec,LSO}/S_{spec,BGO}$	Light output
He	0.987 ± 0.202	0.712 ± 0.029	0.722 ± 0.289	3.608 ± 1.415
C	0.878 ± 0.078	0.529 ± 0.015	0.603 ± 0.239	3.013 ± 1.194
Si	0.480 ± 0.059	0.348 ± 0.014	0.726 ± 0.231	3.628 ± 1.154

Table 3. The specific scintillation efficiencies for BGO and LSO. The light output is the ratio between the specific scintillation efficiencies multiplied by a factor of 5. This factor is the difference in light output between BGO and LSO [4]. The values for BGO are taken from [11].

4. Summary

We have characterized the intrinsic spectrum of LSO:Ce resulting from the β^--decay of ^{176}Lu as measured with a photodiode glued to the LSO crystal. By comparing the intrinsic spectrum with a Geant4 simulation we could identify three parts in the spectrum originating from three different sources. The first part is the native intrinsic spectrum with an energy range up 1192 keV and the highest intensity. This part is found in all LSO scintillation detectors. The second part originates from an electronic effect known as chance coincidence and ranges up to $\approx$ 2000 keV. The intensity of this part is highly sensitive to the size of the scintillator and the shaping time of the electronics. The last part of the spectrum results from electrons which leave the crystal and deposit energy directly in the silicon of the photodiode. The crystal we used for the investigation showed a decay rate of 306 $\frac{\text{Bq}}{\text{cm}^3}$. This high intrinsic decay rate limits the use of large LSO scintillators because of the chance coincidences. Furthermore we could show that photodiode hits need to be considered for an accurate characterization of LSO.
For the second part of the characterization we build a complete particle telescope containing an LSO scintillator and performed heavy ion measurements at the HIMAC, Japan. With these we determined the quenching factors of He, C, and Si, such that the correct deposited energy of these ion species can be calculated for any calibrated LSO scintillation detector. A comparison with the currently used scintillation material BGO shows that despite the higher quenching the light emission of LSO is still at least a factor of three superior to BGO.

5. Acknowledgements

We thank the NIRS-HIMAC team for the opportunity to perform research with their accelerator, and their Accelerator Engineering Corporation (AEC) for smooth beam times.

References

[1] Saint-Gorbain Crystals 2014 http://www.crystals.saint-gobain.com/
[2] Avdeichikov V V et al 1994 Light output and energy resolution of CsI, YAG, GSO, BGO and LSO scintillators for light ions. J. Nucl. Instr. Meth. Phys. **349.1** 216-224
[3] Derenzo S E et al 2003 The quest for the ideal inorganic scintillator J. Nucl. Instr. Meth. Phys. A **505** 111-117
[4] Melcher C L and Schweitzer J S 1992 Cerium-doped lutetium oxyorthosilicate: a fast, efficient new scintillator J. IEEE Trans. Nucl. Sci. **39.4** 502-505
[5] National Nuclear Data Center 2014 http://www.nndc.bnl.gov/nudat2/
[6] Ludziejewski T et al 1995 Advantages and Limitations of LSO Scintillator in Nuclear Physics Experiments J. IEEE Trans. on Nucl. Sci. **42(4)** 328336
[7] Birks J B 1964 The theory and practice of scintillation counting
[8] Horn D et al 1992 The mass dependence of CsI (Tl) scintillation response to heavy ions J. Nucl. Instr. Meth. Phys. A **320.1** 273-276
[9] Amako K et al 2005 Comparison of Geant4 electromagnetic physics models against the NIST reference data J. IEEE Trans. Nucl. Sci. **52.4** 910-918
[10] Giani S et al 1999 Geant4 simulation of energy losses of ions INFN/AE-99/21
[11] Tammen J 2013 Scintillation efficiency of BGO Technical report Institute for experimental and applied physics, University of Kiel

Amplitude calibration with the HiSCORE-9 array

S Epimakhov[1], M Brückner[6], N Budnev[5], D Bogorodskii[7], O Chvalaev[5], A Dyachok[5], O Gress[5], D Horns[1], A Ivanova[5], S Kiruhin[5], E Konstantinov[5], E Korosteleva[3], M Kunnas[1], L Kuzmichev[3], B Lubsandorzhiev[4], N Lubsandorzhiev[3], R Mirgazov[5], R Mirzoyan[5,7], R Monkhoev[5], R Nachtigall[1], A Pakhorukov[5], V Platonov[5], V Poleschuk[5], A Porelli[1], V Prosin[3], G Rubtsov[4], M Rüger[2,6], P Satunin[4], A Saunkin[5], Yu Semeney[5], L Sveshnikova[3], V Tabolenko[5], M Tluczykont[1], D Voronin[5], R Wischnewski[2], A Zagorodnikov[5]

[1] Institut für Experimentalphysik, Universität Hamburg, Luruper Chaussee 149, 22761 Hamburg, Germany
[2] DESY, Platanenallee 6, 15738 Zeuthen, Germany
[3] Skobeltsyn Institute for Nuclear Physics, Lomonosov Moscow State University, 1 Leninskie gory, 119991 Moscow, Russia
[4] Institute for Nuclear Research of the Russian Academy of Sciences, 60th October Anniversary st., 7a, 117312, Moscow, Russia
[5] Institute of Applied Physics ISU, Irkutsk, Russia
[6] Institut für Informatik, Humboldt University, Rudower Chaussee 25, Berlin
[7] Max Planck Institut für Physik, Föhringer Ring, München, Germany

E-mail: sergey.epimakhov@desy.de

Abstract. HiSCORE is a non-imaging wide-angle Cherenkov array for the detection of extensive air showers induced by ultrahigh energy gamma-rays above 10 TeV and cosmic ray studies above 100 TeV. In October 2013 a 9-station engineering array has been deployed in Tunka valley. For HiSCORE-9, two DAQ systems are being used. The second system is a DRS4 based acquisition system with WhiteRabbit integrated time synchronization. We present the first results on the amplitude calibration from the data of this DAQ system.

1. Intoduction

HiSCORE is a new non-imaging wide-angle Cherenkov array for multi-TeV γ-ray astronomy and cosmic ray studies [1]. In October 2013 a 9-station engineering array with 150 m spacing has been deployed in Tunka valley at the site of the Tunka-133 array [2]. Currently a 34-station hybrid array in combination with an imaging telescope (TAIGA) is under construction [3].

Absolute detector calibration plays an important role in the definition of energy threshold and Monte-Carlo studies. In the past, many Cherenkov telescopes have been calibrated with the so-called excess noise factor method. The method applicable for HiSCORE optical detectors is discussed here.

2. Station and data acquisition

Each station is equipped with four 8″-PMTs. Currently ten-stage Hamamatsu R5912 PMTs are being used [4]. Each PMT has a Winston cone with 0.4 m diameter and 30° viewing angle to concentrate light onto the PMT. The station has a collection area of 0.5 m^2 and a field of view of about 0.6 sr. The signals from the 7th ("anode") and 5th ("dynode") dynodes are used. The last three dynodes and anode are shorted. Such a scheme provides a gain of $2 \cdot 10^4$ at the voltage of $1000 - 1100$ V to keep the night sky current at the accepted by the PMT level of 100 μA. In addition the signals are amplified with a factor 20 (anode) and 4 (dynode).

For HiSCORE-9, two DAQ systems are being used (see Figure 1). The first is a DRS4 based 8-channel optical station board with a custom synchronization board placed in the DAQ center [5]. The second one is a DRS4 based acquisition system with WhiteRabbit integrated time synchronization [6]. The systems serve for data quality cross checks and all necessary tests in order to achieve sub-ns synchronization and best performance. The data from the second DAQ system are used in this paper.

Anode and dynode signals from every PMT are connected to the summator. Anode signals are summed up. The sum is used for the trigger. The trigger is built with the FPGA on-board the WhiteRabbit card.

We store three anode signals to one DRS4 evaluation board and a fourth anode signal, dynode signal and sum to another DRS4 board. Additionally, every board samples the WhiteRabbit trigger signal for calibration of the time-frame of the DRS4.

Data are transferred from the RaspberryPi PC directly to the central DAQ PC via ssh connection.

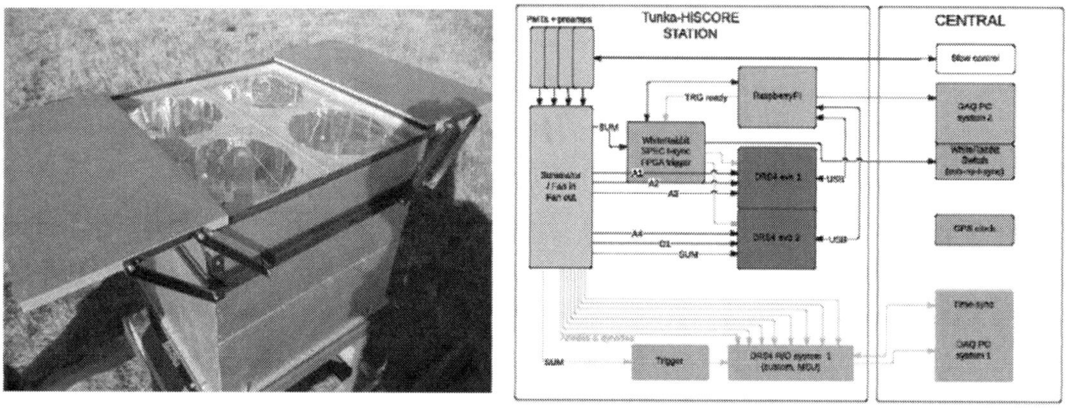

Figure 1. View of the optical station and DAQ set-up scheme.

3. Amplitude calibration

3.1. Excess noise factor method

For calibration, a fast and powerful distant LED source was used in February-March 2014 (5 runs: 22.02, 25.02, 07.03). The horizontal light source illuminated 45° reflectors at each station. Here, the amplitude calibration is discussed. Results on the time calibration are shown in [6].

By knowing the amplitude of the LED signal stored in the DRS one can extract the number of photoelectrons. Assuming that the number of photons falling on the photocathode has a Poisson variance and noise variance is negligible one can derive the conversion factor:

$$C = F\frac{\mu}{\sigma^2} \text{ [p.e./mV]}, \tag{1}$$

where F is the excess noise factor, σ is measured variance and μ is the mean of the LED amplitude distribution (see Figure 2).

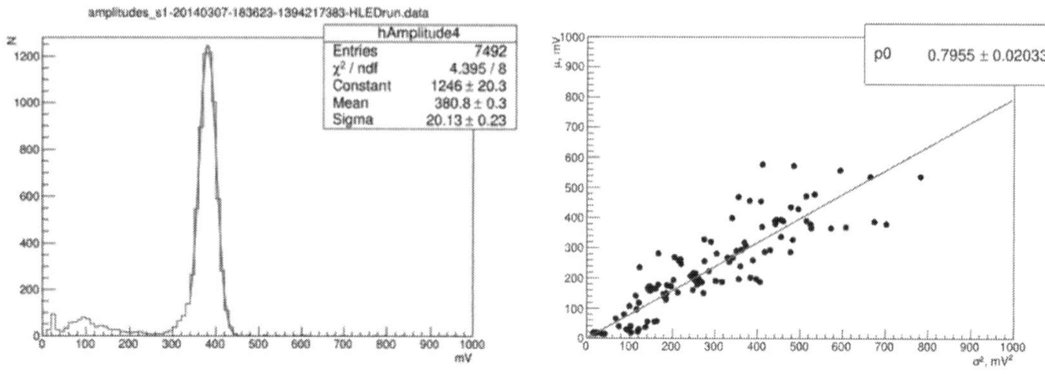

Figure 2. An example of a LED amplitude distribution and its Gaussian fit (*left*). Summary scatter plot of σ^2 and μ of anode channels for all runs ($\chi 2/ndf < 10$) (*right*).

3.2. Determination of F

The mean μ_1 and standard deviation σ_1 of the single photoelectron spectrum (with subtracted pedestal) is related to the excess noise factor of the PMT by the following definition [8]:

$$F = 1 + \frac{\sigma_1^2}{\mu_1^2} \tag{2}$$

.

However the single photoelectron spectrum is very difficult to measure by using our PMT due to its low gain. An alternative would be to use the gain to calculate the excess noise factor. The gain of each PMT is regularly measured in the laboratory before the PMT is deployed.

The excess noise factor is related to the collection efficiency of the photomultiplier CE and the secondary emission ratio δ [7]:

$$F = \frac{1}{CE}\left(1 + \frac{1}{\delta_1} + \frac{1}{\delta_1\delta_2} + ... + \frac{1}{\delta_1\delta_2...\delta_n}\right), \tag{3}$$

where δ_i is the secondary emission ratio of i-th dynode and n is total number of dynodes.

With $\delta_1, \delta_2, ..., \delta_n = \delta$, equation (3) for large n is simplified as follows:

$$F \simeq \frac{1}{CE}\frac{\delta}{\delta - 1}. \tag{4}$$

The average secondary emission ratio δ is a function of the interstage voltage of dynodes and can be derived from the gain:

$$\delta = \left(\frac{G}{CE}\right)^{1/n} = 5.4\left(\frac{G}{2 \cdot 10^4}\right)^{1/6}\left(\frac{CE}{0.8}\right)^{-1/6} \tag{5}$$

And finally one obtains the noise excess factor:

$$F = 1.5\left(\frac{CE}{0.8}\right)^{-1}. \tag{6}$$

Thus for F=1.5 the average conversion factor C is 1.2 p.e./mV.

Results on application of the calibration to data taken on 25 February 2013 are shown in Figure 3. For one chosen station, calibrated differential spectra reach their maxima at $\log_{10} A/\text{p.e.}$: 1.85, 1.79, 1.70, 1.77. Thereby the estimated discriminator threshold of the station with four PMTs:

$$A_{thr} = 10^{1.85} + 10^{1.79} + 10^{1.70} + 10^{1.77} = 241 \pm 16 \text{ p.e.} \tag{7}$$

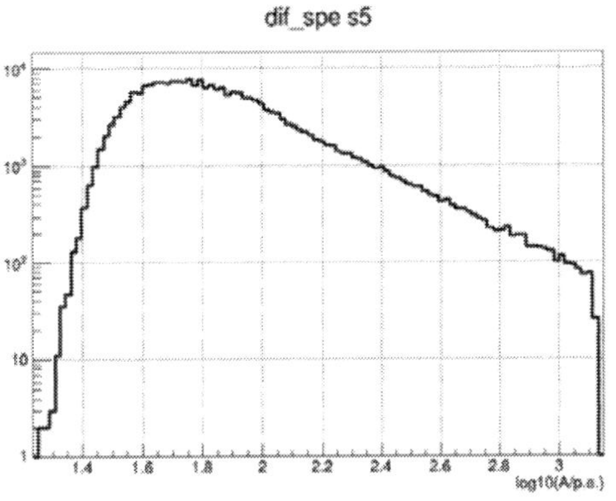

Figure 3. Calibration to photoelectrons for station five. Channel number three is shown.

4. Conclusion and outlook

The excess noise factor method was used for calibrating of many Cherenkov telescopes in the past. However classical determination of the noise factor F using single photoelectron spectrum is difficult in case of our type of PMT. An alternative way of definition of F is to use the gain of the PMT and can be used in the experiment. A dedicated calibration system will be installed in the future.

5. Acknowledgements

This work was supported by the Russian Federation Ministry of Education and Science (agreements 14.B25.31.0010, 2014/51, project 1366, zadanie 3.889.2014/K), the Russian Foundation for Basic Research (grants 13-02-00214, 15-02-10005), and by the Helmholtz Association (Grant HRJRG-303) and the Deutsche Forschungsgemeinschaft (Grant TL51-3).

References

[1] M. Tluczykont et al. Towards Gamma-Ray Astronomy with timing array, this conf.
[2] N. Budnev et al. The Tunka experiment: from cosmic ray to gamma astronomy, this conf.
[3] M. Kunnas et al. Simulation of the Tunka Area International Gamma-ray Advanced experiment (TAIGA), this conf.
[4] Hamamatsu Photonics K.K. Photomultiplier tube R5912. Retrieved from `http://neutron.physics.ucsb.edu/docs/R5912_8inch.pdf`
[5] N. Budnev et al. A prototype DAQ for the Tunka-HiSCORE: 8-channel DRS-4 board, this conf.
[6] A. Porelli et al. Timing calibration and directional reconstruction for Tunka-HiSCORE, this conf.
[7] Hamamatsu Photonics K.K. Photomultiplier tubes. Basics and applications. 2006. Retrieved from `http://psec.uchicago.edu/links/pmt_handbook_complete.pdf`
[8] Th. Schweizer et al. The Optical Calibration of the MAGIC Telescope Camera. IEEE Transactions on Nuclear Science. San Diego. 2001.

Vela as a source of cosmic rays responsible for the formation of the knee in the energy spectrum

A.D. Erlykin[1,2], **A.W. Wolfendale**[2]

[1] P.N.Lebedev Physical Institute, Leninsky Prosp. 53, Moscow 119991, Russia
[2] Department of Physics, Durham University, South Road, Durham DH1 3LE, UK

E-mail: `erlykin@sci.lebedev.ru`

Abstract. The sharpness of the knee in the cosmic ray (CR) energy spectrum observed at an energy of 3-4 PeV can be caused by the contribution of a relatively young and nearby single source. The application of the models for the acceleration and propagation of CR in the Galaxy allows us to estimate the limits for the distance to the source and its age. The Vela cluster containing supernova remnants (SNR) and pulsar is within these limits. The paper presents additional arguments based on the anisotropy and the structure of the CR energy spectrum at PeV energies and above in favour of the Vela SNR (and, possibly, the associated pulsar) being such a single source responsible for the formation of the knee.

1. Introduction

The Single Source Model (SSM) was suggested in 1997 for the explanation of the remarkable sharpness of the knee in the size spectrum of extensive air ahowers (EAS) at a primary energy of about 3-4 PeV [1]. Its basis is an obvious non-uniformity in space and time of the CR source distribution: supernova remnants (SNR), pulsars etc., as well as of the interstellar medium (ISM) properties. On the basis of accumulated statistics it is expected that the probability of a relatively close and young SN explosion at distances up to 300 pc from the Sun and of the age up to 10^5 years is rather high (~ 1).

The division of the all-particle energy spectrum into two components - that from many distant and old sources and that from a single nearby and young source is a simplification of the real situation, but such a scenario let us describe the main features of the observed spectrum and put forward a possible identification of the assumed single source.

2. Identification of the single source

2.1. The distance and the age

The identification of the source can be made by a comparison of the observed spectrum with model calculations performed for different ages and distances from the source to the observer in the solar system. Our first calculations were made assuming that such a source is a SNR and based on the ideas that CR are confined within its expanding shell until the moment when such trapping ability weakens and CR begin to leave the shell and go out into interstellar space [2,3]. In the paper [4] where we used these ideas we postulated that this escape occurs when the shock wave expanded up to the radius of 100 pc. Before that the confinement lasted up to 80 kyears and after that the accelerated particles begin to diffuse, interact with atoms of the interstellar medium during propagation and eventually escape from the Galaxy. Our calculations [5] show that the energy spectra of accelerated particles depend on time. They have a power law shape with exponents varying from 2.0 at the initial stage of SNR expansion to ~2.55 at the end of it. The overall energy spectrum of particles emergent from the SNR has also the power law shape up to the knee with the exponent of ~ 2.15. Comparing the results of our calculations with the energy spectrum of the single source derived from the experimental data we estimated the limits of the possible distances R and ages T for the SNR as $250 < R < 400$ pc and $85 < T < 115$ kyears [6]. American astronomers have found that these limits are satisfied for the SNR Monogem Ring associated with the pulsar B0656+14 [7].

In later works it was shown that there is probably no such strong confinement and high energy particles may begin to leave the shell soon after the explosion [8]. For such a model the age of the unknown SNR becomes shorter. In the extreme case of the instant release of all accelerated particles it is $5 < T < 35$ kyears, but the distance of the SNR remains the same, i.e. $250 < R < 400$ pc. These new limits are satisfied for the SNR Vela being located at a distance of 273 ± 33 pc and having an age of about 11 kyears.

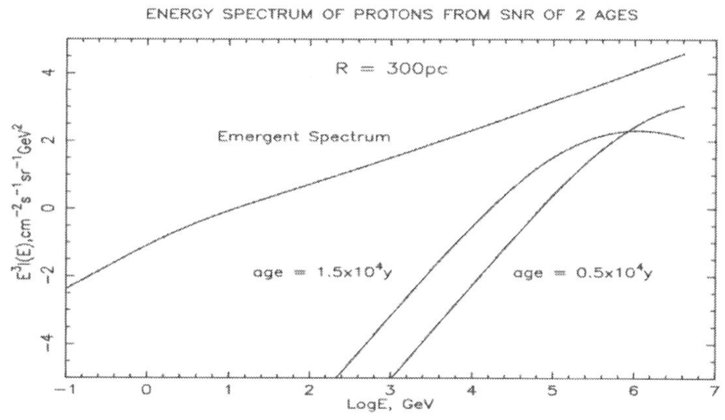

Figure 1. The energy spectrum of CR from the SNR which exploded at a distance of 300 pc from the Sun, 5 or 15 kyear ago. The Figure demonstrates that the maximum contribution to the CR flux is at PeV energies, i.e. in the region of the knee. The emergent spectrum is also shown for comparison.

Direct calculations made with an updated model of SNR acceleration show that a SN with the distance and the age similar to Vela gives the maximum contribution to the CR flux in the PeV region close to the observed knee (Fig.1)

Calculations show that such a young SNR has a relatively rapid temporal variability. With time its CR energy spectrum becomes softer, with almost no PeV particles when its age approaches 10^3 kyear (Fig.2)

2.2. Anisotropy of the CR flux

The Vela cluster is a complex and rather extended source which includes the main SNR (Vela SNR), pulsar B0833-45 with its nebula (VelaX+PWN) and one more (weak) SNR (Vela Jr). It is located in the Galactic Plane, in the third Quadrant: $\ell \sim 265°, b \sim -3°$. Due to the proximity of this source to the Solar System one may expect an excess of the CR flux close to the direction towards this source, especialy for high energy CR with their least deviation in the Galactic magnetic fields.

In our paper [9] the position of Vela on the celestial sphere was compared with the position of intensity excesses in Equatorial and Galactic coordinates for CR of different energies (Fig.3).

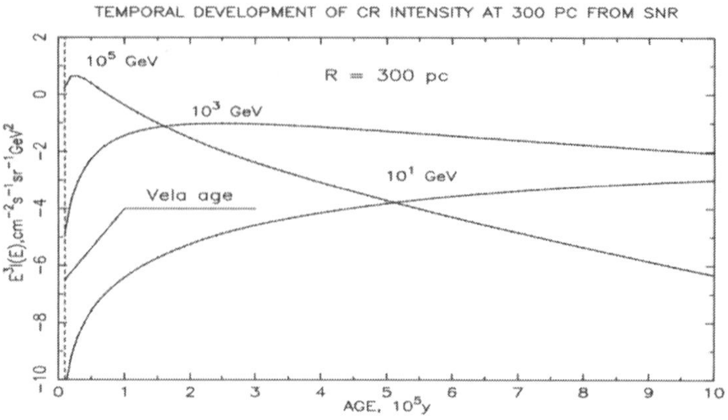

Figure 2. The temporal evolution of the CR spectrum from a SNR at 300 pc from the Sun. Differential CR intensities at 3 energies: $10^1, 10^3$ and 10^5 GeV are shown. The age of Vela is shown by the vertical dashed line at 11 kyear. The Figure demonstrates the rapid softening of the spectrum within the time range of several hundred kyears.

It is seen that the position of intensity excesses for CR of PeV energies approaching the energy of the knee tends to the direction towards Vela. The small difference of the Vela SNR and the nearest intensity excess coordinates can be related to the slightly smaller energy of the studied CR (0.4 - 2 PeV) than the energy of the knee ($\sim$ 3 PeV) or to the local magnetic field the effect of which is difficult to estimate without precise knowledge of its regular and turbulent components.

2.3. Application of the difference method for the anisotropy study

The confirmation of the possible role of the Vela SNR in the formation of the knee was found by the application of a new difference method developed for the study of anisotropy [10]. The idea of the method is that the energy spectrum of CR from a nearby and young source will be different from the background spectrum of CR particles from more distant and old sources. Accordingly, the characteristics of extensive air showers (EAS) produced by CR from such a source should be different from those from background showers. One should look for the direction in which the difference between the characteristics of showers arriving from one side of this direction and from the opposite side is a maximum. It is postulated that such a direction is the direction towards our single source.

A search of this sort has been applied to the data base of EAS detected with the GAMMA array. The array is made by an international collaboration and is located at the Aragats mountain in Armenia [10]. The data used for the analysis include about $3.38 \cdot 10^6$ EAS with the total number of charged particles $N_{ch} > 10^5$ and zenith angles $\Theta < 40°$, which correspond to the threshold primary energy of about 0.4 PeV. The parameter used for the comparison of the characteristics of EAS arriving from the direct and opposite directions was the shower age. The maximum difference was found for the direction with Galactic coordinates $\ell = 277° \pm 3°, b = -5° \pm 3°$ (Figure 4).

Taking into account the extended profile of Vela and the possible deviation of CR particles in the Galactic magnetic fields the agreement of these coordinates with the coordinates of Vela can be regarded as very satisfactory.

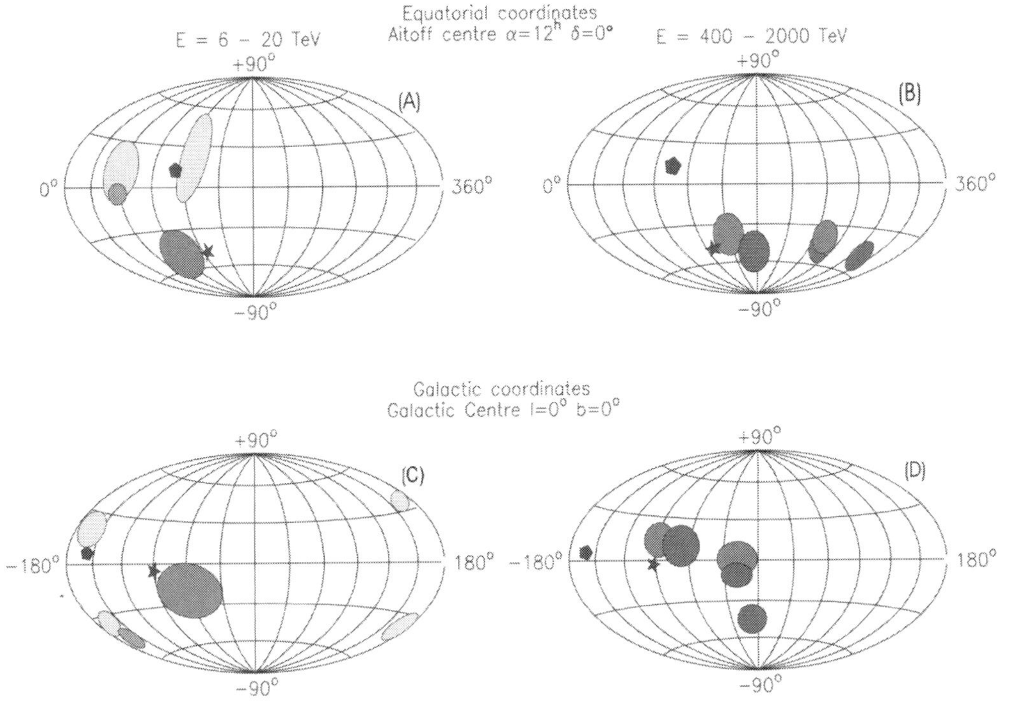

Figure 3. Schematic maps of CR excesses in Equatorial (A,B) and Galactic (C,D) coordinates for two energy domains: 6 - 20 TeV (A,C) and 400 - 2000 TeV (B,D). The excess regions observed in the MILAGRO and ARGO-YBJ experiments nearly overlap and are shown in yellow. Super-Kamiokande - green, IceCube - blue, IceTop - magenta for 400 TeV CR, red for 2000 TeV. The dark pentagon indicates the position of the Monogem Ring SNR, the dark star - the position of Vela X.

2.4. The contribution of the Vela pulsar

The pulsar B0833-45, which is a consequence of the SN exposion, can also contribute to the total CR flux from Vela. The energy spectrum of protons from this pulsar has been calculated by us in [11]. It occurs that the 'peak' of the intensity in coordinates $logE^3I$ vs $logE$ appears right at the energy of the knee at 3-4 PeV (Fig.5). If this pulsar can accelerate other nuclei as well as protons then irregularities of the spectra at higher energies above the knee might be expected. In particular, it is likely that a peak with a considerable fraction of iron nuclei might appear at energies of about 50-80 PeV. At energies above 100 PeV the contribution of the pulsar ceases and one would expect a steepening of the spectrum.

The experimental situation at energies above the knee is ambiguous. Most experiments demonstrate a flattening of the spectrum at an energy above 20 PeV, which may be caused by the end of the contribution from nuclei of the CNO group. The experiment GAMMA showed the existence of a sharp peak at an energy in the range 50-80 PeV, which can be caused by the contribution of iron nuclei [12]. The experiments Tunka-133 [13], Yakutsk [14] and Ice-Top [15] do not observe such a sharp peak, but a small increase of the intensity at these energies exists in their data. The experiment KASCADE-Grande does not show an irregularity in the spectrum at 50-80 PeV [16]. However, all experiments show a steepening at energies above 100 PeV, which can be interpreted as the end of the contribution from the single source.

Therefore, there is a high probability that both the Vela pulsar and the Vela SNR contributes to the total CR flux emitted by this source, but so far this statement cannot yet be regarded as a firm proof.

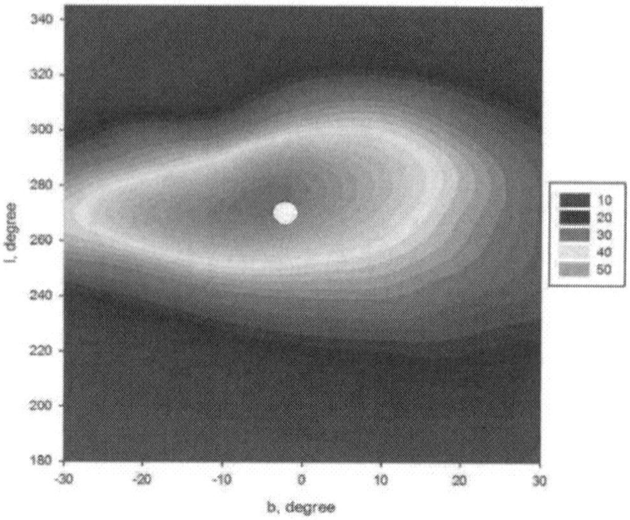

Figure 4. The contour plot for the distribution of the parameter (called χ^2/J in [10]) which describes the difference of the EAS age distributions for the direct and opposite directions, in Galactic coordinates. Contours of different colors correspond to different levels of this parameter (see inset: largest values painted red correspond to the largest difference of age distributions). The white spot at the plot center shows the position of Vela which is remarkably close to the position of the maximum difference in the EAS age distributions.

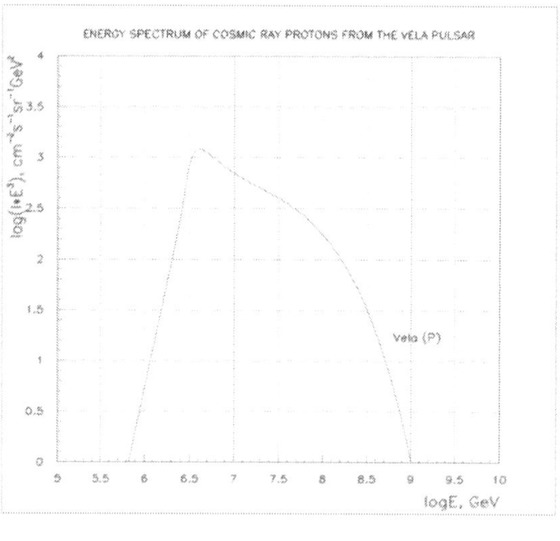

Figure 5. Energy spectrum of protons produced by the Vela pulsar [9].

3. Conclusion

Although with reservations it can be stated on the basis of:

- estimates of the distance and the age of the source,
- the data on the CR anisotropy at PeV energies,
- the application of the difference method for the study of the anisotropy,
- the possible existence of the intensity peak in the CR energy spectrum at an energy of 50-80

PeV caused by primary iron nuclei,
that it is very likely that the Vela cluster (SNR and pulsar) is the CR source responsible for the formation of the knee in the locally measured energy spectrum.

Acknowledgments
Authors thank the Kohn Foundation for the support of this work and to our GAMMA colleagues for a preview of their results.

References
[1] Erlykin A D and Wolfendale A W 1997 *J. Phys. G: Nucl. Part. Ohys.* **23** 979
[2] Völk H J 2000 astro-ph/0001331
[3] Berezhko E G and Völk H J 2000 *Astrophys.J.* **540** 923
[4] Erlykin A D and Wolfendale A W 2001 *J. Phys. G: Nucl. Part. Phys.* **27** 941
[5] Erlykin A D and Wolfendale A W 2002 *J. Phys. G: Nucl. Part. Phys.* **28** 359
[6] Erlykin A D and Wolfendale A W 2003 *J. Phys. G: Nucl. Part. Phys.* **29** 709
[7] Thorsett S E et al. 2003 *Astrophys. J.* **592** L71
[8] Ptuskin V S and Zirakashvili V N 2005 *Astron. Astrophys.* **429** 755
[9] Erlykin A D and Wolfendale A W 2013 *J. Cosmology and Astropart. Phys.* **JCAP94(2013)006**, arxiv:1303.2889
[10] Pavlyuchenko V P *et al* 2014 *Astropart. Phys. (submitted)*, arxiv:1406.0799
[11] Erlykin A D and Wolfendale A W 2004 *Astropart. Phys.* **22/1** 47
[12] Garyaka A P *et al* 2008 *J. Phys.G: Nucl. Part. Phys.* **35** 115201
[13] Berezhnev S F *et al* 2014 *Nucl. Instr. and Meth. A* **756** 94
[14] Knurenko S P and Sabourov A 2012 *23 Europ. Cosm. Ray Symp. and 32 Rus. Cosm. Ray Conf.*, paper PCR_2_466; arxiv:1112.2430
[15] Kolanoski H *et al* 2012 *23 Europ. Cosm. Ray Symp. and 32 Rus. Cosm. Ray Conf. J. Phys.: Conf. Ser.* **409** 012006
[16] Haungs A *et al* 2012 *23 Europ. Cosm. Ray Symp. and 32 Rus. Cosm. Ray Conf. J. Phys.: Conf. Ser.* **409** 012005

A template method for measuring the iron spectrum in cosmic rays with Cherenkov telescopes

Henrike Fleischhack

DESY, Platanenallee 6, 15738 Zeuthen

E-mail: henrike.fleischhack@desy.de

Abstract. Understanding the sources, acceleration mechanisms, and propagation of cosmic rays is an active area of research in astro-particle physics. Measuring the spectrum and elemental composition of cosmic rays on earth can help solve this question. IACTs, while mainly used for γ-ray astronomy and indirect searches for dark matter, can make an important contribution here. In particular, they are able to distinguish heavy nuclei in cosmic rays from protons and lighter nuclei by exploiting the direct Cherenkov light emitted by charged particles high in the atmosphere. In this paper, a method to reconstruct relevant properties of primary cosmic ray particles from the Cherenkov light emitted by the primary particles and the air showers induced by them will be presented.

1. Introduction

Cosmic rays (CRs), charged particles of extraterrestrial origin impinging on the earth's atmosphere, were discovered more than one hundred years ago. Today, we know that they mostly consist of protons and other fully ionized nuclei, and (to a lesser extent) of electrons, positrons, and anti-protons. From a few GeV to several EeV, their energy spectrum can be described by a power law with two breaks: the "knee" in the PeV range and the "ankle" in the EeV range. The flux of lower energy particles is attenuated by the Sun's and the Earth's magnetic fields and hence undergoes seasonal changes.

Many possible acceleration sites for cosmic rays have been identified, for example supernova remnants for galactic cosmic rays and gamma-ray bursts or active galactic nuclei for extra-galactic cosmic rays. However, it is still unknown by what mechanism the majority of cosmic rays are accelerated and what causes the change in spectral slope referred to as the "knee". Cosmic rays are charged particles, and are thus deflected by (inter-)galactic magnetic fields. Hence, their arrival direction does not point directly to the sources. We can only study the sources and acceleration mechanisms of cosmic rays by indirect means: Either by detecting neutral particles (neutrinos or γ-rays) which are produced at the sites of cosmic ray acceleration, or by precisely measuring the spectrum and composition of cosmic rays and comparing to theory predictions of cosmic ray acceleration and propagation. For an overview of such models and comparison of available data, see [2]. Even at energies below the knee, measurements of the flux and spectrum of different elements in cosmic rays can help constrain those models.

As the energies of cosmic rays impinging on earth span several orders of magnitude, a range of different techniques is employed to detect cosmic rays and investigate their properties. Some

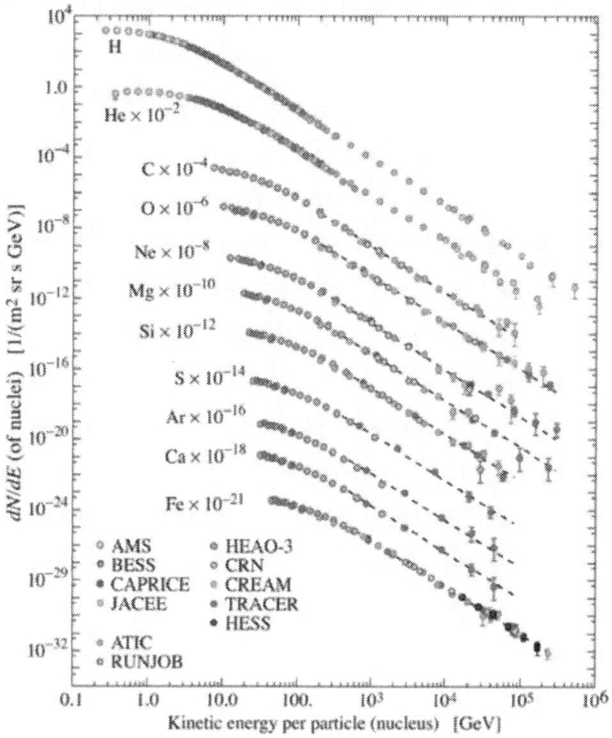

Figure 1. Energy spectra of various elements in cosmic rays, measured by different experiments over a large energy range. From [1].

of the resulting energy spectra for the different elements that make up cosmic rays are shown in figure 1.

1.1. Direct Detection Experiments

Cosmic rays in the MeV to TeV range are detected directly at the top of the atmosphere by balloon- or satellite-borne detectors. As charged particles, they will deposit energy inside the detector by ionization and bremsstrahlung. Energy and charge of the primary particle can be inferred from the deposited energy and/or the shape of the particle's track, e.g. bending in a magnetic field, or deflection due to multiple scattering off atomic nuclei. The collection area is constrained by the size/weight limits of the balloon/satellite and is usually below $1\,\mathrm{m}^2$. At energies above a few hundred GeV, the energy resolution of direct experiments tends to be limited by their size. Due to the power-law distribution of cosmic ray energies, they tend to be statistics-limited at higher energies as well. For an example, see [3].

1.2. Indirect Detection with EAS Arrays

Cosmic rays with energies in the GeV range or higher will cause a cascade of particles (composed mostly of protons, neutrons, electrons, positrons, pions, muons and neutrinos) to be produced upon interaction with the atmosphere. For primary particles of hundreds of TeV and higher, parts of this so-called "extended air shower" of particles (muons, electrons) can be detected on the Earth's surface. By measuring the density and arrival times of these particles, energy and direction of the primary particles can be inferred. Charge measurements are difficult with this type of experiment. The composition of the primary particles can, for example, be inferred on a statistical basis from measurements of the height of the shower maximum, or from the particle content of the air shower reaching the ground. Both methods rely heavily on numerical simulations of air showers.

The footprint of an air shower can be several kilometers in diameter at the surface. A large detection area is needed to make up for the low flux of cosmic rays expected at such high energies. Experiments for the detection of air showers thus consist of an array of detector stations, hundreds of meters apart. Each station detects charged particles by the energy they deposit in the detector. These arrays can have detection areas up to $10^{12}\,\mathrm{m}^2$. The surface detectors can be supported by arrays of radio antennae and/or fluorescence detectors, which measure radiation emitted by the charged component of the shower in the atmosphere. For examples, see [4, 5].

1.3. Indirect Detection with IACTs

Another way to detect extended air showers on the ground is by looking for the Cherenkov radiation emitted by the charged particles inside an air shower whenever they are faster than the (local) speed of light. Cherenkov radiation is highly beamed. The resulting light cone (footprint of the Cherenkov light on the ground) has a diameter of a few hundred meters; the size of the footprint does not depend much on the primary particle's energy.

Imaging Air Cherenkov Telescopes are a class of instruments that exploit this measurement technique. They do not rely on the charged component of the shower reaching the ground, but are able to self-trigger on the Cherenkov light. With their cameras, each made of several hundred or thousand sensitive light detectors (e.g. PMTs or silicon detectors), they can image entire air showers in Cherenkov light. From the size, shape, and position of the images in the camera, the energy and direction of the primary particles can be inferred. Arrays of several IACTs may be used together in a stereoscopic mode to improve the reconstruction.

IACTs are mostly used for γ-ray astronomy, detecting air showers produced by TeV-range photons. In that case, cosmic rays make up the main background. However, they can also be used to study properties (energy spectrum and composition) of cosmic rays.

With their effective detection area of about $10^4\,\mathrm{m}^2$, IACTs are able to cover intermediate energies (TeV range) to fill the gap between direct detection experiments and air shower arrays, cf. [6]. In contrast to air shower arrays, it is possible to measure the charge of the primary particle on an event-to-event basis.

1.4. Direct Cherenov Technique

Charged primary particles can radiate direct Cherenkov (DC) light even before starting a shower, if their velocity is high enough. This DC light can be detected with IACTs. It is very concentrated in the camera (inside one pixel in current IACTs), and located at the front of the shower image. The intensity of the DC light is proportional to the square of the charge of the primary particle, making it useful for the separation of heavy and light nuclei. At the same time, the Cherenkov light from the shower may be used to reconstruct energy and direction of the primary particle. The existence of DC light can in particular be used to select showers induced by heavy primary particles (in particular iron) in an energy range of about 10 TeV to several 100 TeV. The higher the energy, the less dense the atmosphere needs to be for the primary to emit Cherenkov light. At low energies, the particle needs to penetrate very deeply into the atmosphere to reach layers that are dense enough for Cherenkov light to be emitted. Below 10 TeV or so, the probability for the particle traveling that far in the atmosphere without interaction is very low. The higher the primary particle's energy, the higher in the atmosphere it starts emitting DC light. However, at higher energies above hundreds of TeV, the light emitted by the shower starts to drown out the DC contribution, making it impossible to infer the charge of the primary. cf. [6; 7].

Most showers detected by IACTs are induced by cosmic rays. Thus, any study of cosmic rays using IACT data can exploit a large existing data set (typically up to 1000 hours per year of operation).Showers induced by heavy nuclei (in particular iron) can be separated from light nuclei and protons due to the extra component of the DC light. In the following, iron will be used as a representative for heavy nuclei as it is the most abundant element in cosmic rays with $Z > 20$. Lighter elements such as helium, carbon and oxygen do appear more abundantly among cosmic rays. However, the intensity of the direct Cherenkov light is about ten times lower for oxygen compared to iron, making it difficult to distinguish showers induced by light elements from proton-induced showers using this technique.

Measuring the iron flux and spectrum this way is beneficial because the systematic uncertainty is complementary to other cosmic ray experiments (for IACTs, the atmosphere is the largest source of systematic uncertainty).

1.5. The VERITAS Experiment

VERITAS[1] (Very Energetic Radiation Imaging Telescope Array System) [8] is an array of four IACTs, located in southern Arizona, USA. It has been operating since 2007. Each telescope has a set of tessellated mirrors with a total mirror diameter of about 12 m^2. Each camera consists of about 500 "pixels" (PMTs), with a total field of view of about 3.5 degrees. It is sensitive to gamma-ray induced showers from 80 GeV to tens of TeV.

Over its lifetime, the telescope has seen several upgrades: one of the telescopes was moved in 2009 to increase the instrument's effective area; the PMTs were upgraded in 2012 to increase sensitivity for fainter showers. Science topics studied with VERITAS include gamma-ray astronomy, dark matter searches, and astro-particle physics.

2. Data analysis with IACTs

2.1. Standard geometrical analysis method

Cherenkov telescopes record signals from air showers. From those, properties of the primary particle have to be reconstructed. For gamma-ray astronomy, the relevant properties are energy, arrival direction, and particle class (photon or cosmic ray).

The standard geometrical reconstruction proceeds as follows. First, the signal in each pixel is integrated. The pedestal is subtracted and the signal is corrected for the gain. A cleaning is performed to reduce contributions from the night sky background.

Then, a moments analysis is performed on the surviving image pixels and the so-called 'Hillas parameters' [9] are calculated. To first order, gamma-ray induced showers develop symmetrically. Thus, the major axis of the image points back to the source of the gamma-ray. By combining information from two or more telescopes, the arrival direction of the gamma ray and the position of the shower core on the ground can be determined. Once the distance between shower core and telescope are known, lookup tables can be used to estimate the primary particle's energy from the total amount of light in the camera. Then, the second moments (width and length) are compared to the expected values for a gamma-induced shower of that energy. Showers induced by cosmic rays tend to be wider and less even than gamma-ray induced showers. Thus, the width and length of the images are a good discriminator between gamma-ray and cosmic-ray induced showers [10].

2.2. The Template Likelihood analysis method

The geometrical analysis described above works very well to reconstruct gamma-induced showers, but it does not take all available information into account. Other, more advanced techniques have been developed in the past years. One of them is the template likelihood method, which

[1] `http://veritas.sao.arizona.edu/`

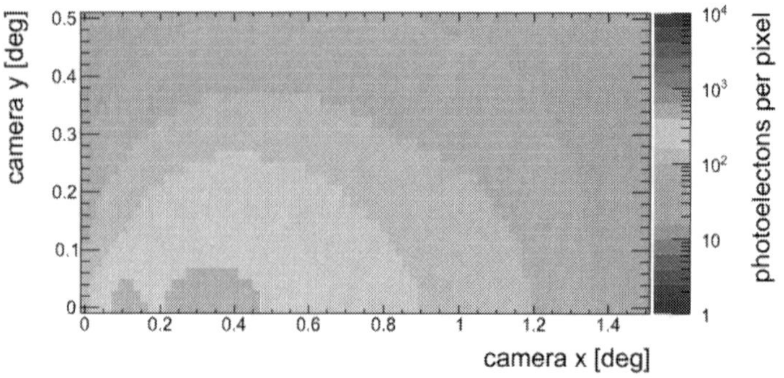

Figure 2. Average light distribution in the camera for $E = 30\,\mathrm{TeV}$, 50 m distance between the detector and shower core, $h = 33\,\mathrm{km}$. The coordinate system is centered on the primary particle's arrival direction. The x-axis is defined by the long image axis. Note the contribution from direct Cherenkov light at around 0.1 degrees from the primary's direction, clearly distinguishable from the broader contribution from the Cherenkov light emitted by the shower particles.

naturally takes the information from each pixel into account. The idea is as follows: Given a model which describes the photon distribution in the camera, depending on some properties of the primary particle, we can model the probability distribution of collected charge per pixel given those parameters as well as the detector response and its uncertainty. This gives us a likelihood function, which can be maximized to estimate the parameters for a given shower by fitting camera images to model predictions. The goodness of fit can be used to discriminate between 'signal' and 'background' events. This type of likelihood fitting has been used by similar experiments for the reconstruction of gamma-ray induced showers, see for example [11]. It is particularly useful to discriminate between iron-induced showers, which can have a contribution from direct Cherenkov light, and showers induced by protons and light elements, where the direct Cherenkov light is not detectable.

In this case, the model parameters are the primary particle's energy E, direction in the camera (X_s, Y_s), height of first interaction h and position of shower core on the ground (X_p, Y_p). The light distribution on the camera is modeled using Monte Carlo simulations of iron-induced showers at a fixed 'grid' in these parameters, interpolating between them to predict the light distribution for arbitrary parameter values. CORSIKA [12] is used to simulate air showers including Cherenkov light emission, and the grisudet package [2] for ray-tracing in the telescope.

An example for the predicted average light distribution in the camera can be seen in figure 2. In this example, the contributions from the shower light and from the direct Cherenkov light form two bumps that can be clearly separated by eye. The direct Cherenkov light is emitted at a very small angle high up in the atmosphere, and confined to a region in the camera that is less than a pixel in size. Images of showers induced by protons, light elements, and heavy elements like iron look very similar, but the direct Cherenkov contribution, which is proportional to the square of the particle's charge, can be used to distinguish between them. Hence, it is necessary to take information from all the pixels into account. The likelihood method described above does just that; an image from a proton-induced shower will not have a direct Cherenkov contribution and will thus have a worse goodness-of-fit value than an iron-induced shower when using templates generated for iron showers.

[2] `http://www.physics.utah.edu/gammaray/GrISU/`

The probability distribution per pixel is given by

$$P(q|s,\sigma_p,\sigma_e,\sigma_s) = \int \mathrm{d}\mu\; Gauss(\mu|s,\sigma_s) \cdot \sum_n Poisson(n|\mu) \cdot Gauss(q|n,\sqrt{\sigma_p^2 + n\sigma_e^2})$$
$$\approx Gauss(q|s,\sigma) \;\text{ with }\; \sigma = \sqrt{\sigma_p^2 + s\sigma_e^2 + \sigma_s^2}$$

Where

- q is the pedestal-corrected integrated charge (converted to units of photo-electrons [p.e.]).
- $s = s(E,h,X_p,Y_p,X_s,Y_s)$ and $\sigma_s = \sigma_s(E,h,X_p,Y_p,X_s,Y_s)$ are the predicted average number of photo-electrons in each pixel and its uncertainty.
- μ is the predicted number of photo-electrons in a given shower, assumed to follow a Gaussian distribution with mean s and width σ_s.
- n is the number of photo-electrons, assumed to follow a Poisson distribution with mean μ.
- σ_e^2 is the variance of the signal (integrated charge) from a single photo-electron.
- σ_p^2 is the variance of the pedestal.

The log-likelihood function to be minimized is then given by

$$-\ln L = -\sum_{\text{pixel } i} \ln L_i = \sum_{\text{pixel } i} \ln P(q_i, s_i, \sigma_i).$$

We define the goodness-of-fit function as follows:

$$G = \frac{\sum_i \ln L_i - \langle \ln P(q|s_i,\sigma_i)\rangle_q}{\sqrt{2 \cdot NDF}}.$$

It has been normalized so that it follows a distribution with a mean of zero and width of one for signal events and can thus be used to discriminate between signal and background.

3. Method Validation

To test the method, a set of templates for showers from zenith and an energy range from 10 – 100 TeV was produced. A second set of showers was produced with random core positions, with the energies following a power law spectrum. This second set of showers was passed through a VERITAS detector simulation chain, again using the grisudet package. The resulting 'fake data' files were analyzed using the likelihood fitting method described above.

As a first check, the fit was performed using the true values of the parameters as starting values. In that case, the reconstruction worked very well. However, using the results of the standard geometrical reconstruction as the starting values for the fit worsens the resolution considerably. For example, figures 3 and 4 show the resolution of the reconstructed core position and reconstructed energy, both as a function of the (known) true energy. The main problem is that the geometrical reconstruction of the core position can give very bad results, making the fit unable to reconstruct the event properly.

4. Conclusions and Outlook

Measuring the flux and spectrum of heavy nuclei in TeV cosmic rays can provide useful insights into the acceleration and propagation of cosmic rays. However, this energy range is not well covered by direct detection experiments nor by air shower arrays. IACTs are sensitive especially to heavy nuclei like iron in that energy range. They are sensitive to direct Cherenkov light

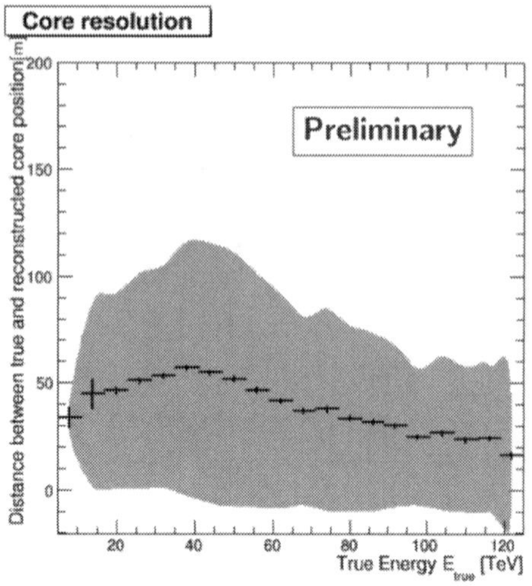

Figure 3. Distance between true and reconstructed shower core in m vs true energy. The histogram gives the mean value in each energy bin, the shaded region corresponds to the 1-σ width of the distribution.

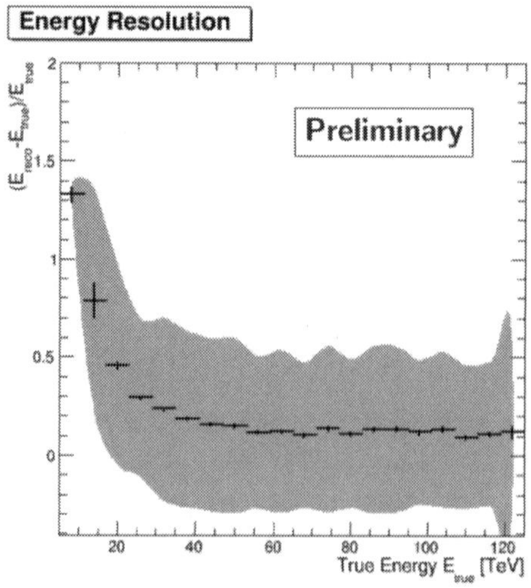

Figure 4. Relative error in reconstructed energy vs true energy. The histogram gives the mean value in each energy bin, the shaded region corresponds to the 1-σ width of the distribution.

emitted by cosmic rays high in the atmosphere. A technique for reconstructing the energy of iron-induced showers imaged by air Cherenkov telescopes, and to separate these showers from showers induced by protons and lighter elements, has been introduced. This technique uses a likelihood fitting method to compare images of cosmic ray showers to predictions based on Monte Carlo simulations. Further work is needed to either improve the starting values used for the fit, or to make the fit itself more robust against bad starting values.

Acknowledgments
This research is supported by grants from the U.S. Department of Energy Office of Science, the U.S. National Science Foundation and the Smithsonian Institution, by NSERC in Canada, by Science Foundation Ireland (SFI 10/RFP/AST2748) and by STFC in the U.K. We acknowledge the excellent work of the technical support staff at the Fred Lawrence Whipple Observatory and at the collaborating institutions in the construction and operation of the instrument. The VERITAS Collaboration is grateful to Trevor Weekes for his seminal contributions and leadership in the field of VHE gamma-ray astrophysics, which made this study possible.

References

[1] Beringer J *et al.* (Particle Data Group) 2012 *Phys. Rev. D* **86** 010001

[2] Hörandel J R 2004 *Astroparticle Physics* **21** 241–265

[3] Ahn H S *et al.* 2009 *Astrophys. J.* **707** 593–603 (*Preprint* 0911.1889)

[4] Abbasi R U *et al.* 2010 *Physical Review Letters* **104** 161101 (*Preprint* 0910.4184)

[5] Cazon L and Pierre Auger Collaboration 2012 *Journal of Physics Conference Series* **375** 052003 (*Preprint* 1201.6265)

[6] Aharonian F *et al.* 2007 *Phys. Rev. D* **75** 042004

[7] Kieda D, Swordy S and Wakely S 2001 *Astroparticle Physics* **15** 287 – 303 ISSN 0927-6505 URL http://www.sciencedirect.com/science/article/pii/S0927650500001596

[8] Holder J *et al.* 2008 *American Institute of Physics Conference Series* vol 1085 pp 657–660 (*Preprint* 0810.0474)

[9] Hillas A M 1985 *International Cosmic Ray Conference* **3** 445–448

[10] Fegan D J 1997 *Journal of Physics G Nuclear Physics* **23** 1013–1060

[11] Le Bohec S *et al.* 1998 *Nucl. Instr. Meth. Phys. Res. A* **416** 425 – 437 ISSN 0168-9002 URL http://www.sciencedirect.com/science/article/pii/S0168900298007505

[12] Heck D, Knapp J, Capdevielle J N, Schatz G and Thouw T 1998 *CORSIKA: a Monte Carlo code to simulate extensive air showers.*

TRAGALDABAS: A new high resolution detector for the regular study of cosmic rays

H. Alvarez-Pol[8], A. Blanco[3], J.J. Blanco[1], J. Collazo[9], P. Fonte[3], J.A. Garzón[9], A. Gómez [7], G. Kornakov[6], T. Kurtukian[2], L. Lopes[3], M. Morales[9,11], A. Morozova[4], M.A. Pais[4], M. Palka[5], V. Pérez Muñuzuri[10], P. Rey[7], P. Ribeiro[4], M. Seco[8], J. Taboada[10].

[1] Univ. Alcalá. Alcalá de Henares, Spain
[2] CEN-Bordeaux. Bordeaux-Gradignan, France
[3] LIP-Coimbra. Coimbra, Portugal
[4] Univ. Coimbra. Coimbra, Portugal
[5] Jagellionan Univ. Cracow, Poland
[6] TU-Darmstad. Darmstadt, Germany
[7] CESGA. S. de Compostela, Spain
[8] Depto. de Física de Partículas & IGFAE - Univ. Santiago de Compostela. S. de Compostela, Spain
[9] LabCAF - Univ. Santiago de Compostela. S. de Compostela, Spain
[10] Meteogalicia. Xunta de Galicia. S. de Compostela, Spain

E-mail: juanantonio.garzon@usc.es

Abstract. Research on cosmic rays is of big interest either for getting a better understanding about their origin and properties or because they offer very valuable information about the galactic, the solar and the Earth's environment. In order to improve our knowledge of all those fields, a high resolution cosmic ray tracking detector, TRAGALDABAS, is being commissioned at the Faculty of Physics of the Univ. of Santiago de Compostela (Spain). In this article we make overview of the main performances of the detector and we present some very preliminary results showing that the detector is taking good data, and that we are gathering a valuable sample of events, ready to be analyzed.

1. Introduction

Cosmic rays coming from the Sun, our Galaxy or other galaxies are permanently arriving to Earth. Most of them are composed by protons, all the remaining atomic nuclei, electrons and also positrons. Their energies range between almost the rest up to about 10^{20} eV for some protons or nuclei, energies that are well out reach of the existing accelerating technologies. Although cosmic rays have been studied for more than one hundred years, many unknowns remain related with their origin and with all the processes affecting them before they arrive to our atmosphere.

Figure 1 summarizes most of the methods used for the measurement of primary cosmic rays as a function of their energy. Below $\sim 10^{14}$ eV, their rate is big enough to measure them with very high accuracy hybrid detectors placed either in the ISS, in satellites or on balloons [1]. Above

[11] Present address: ICMM/CSIC-Madrid

that energy, hadrons and nuclear fragments produced in the collision of the cosmic ray with the high atmosphere, above the tropopause, start a chain of a few generations of nuclear interactions that may produce over one billion of secondary particles. All those particles constitute what is usually called an Extensive Air Shower (EAS), that spreads out randomly along the atmosphere and that may cover, when they arrive to the ground, up to several hundreds km^2. As it is not possible to cover such a big area with a dense particle detector, big and broad arrays of detectors are used for measuring the properties of a significative sample of secondary particles on the ground [2].

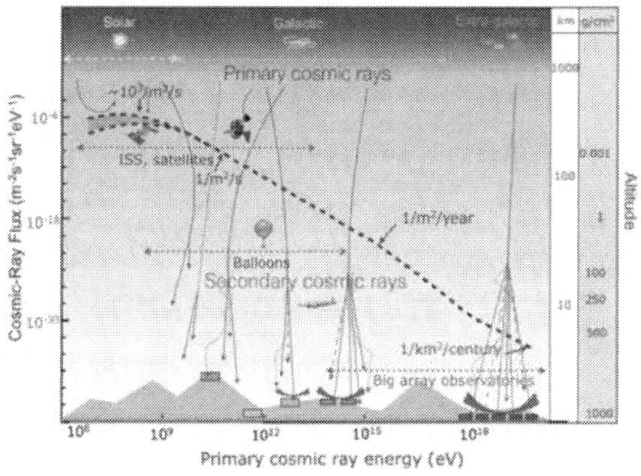

Figure 1. Energy spectrum (dark blue dashed line; left scale) of primary cosmic rays together with their corresponding estimated sources and the main detection techniques at different altitudes (right scale)

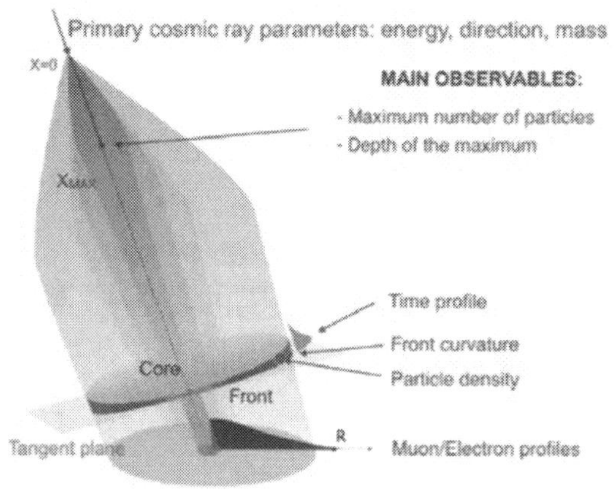

Figure 2. Simplified view of the evolution and structure of an Extensive Air Shower (EAS). Most of the secondary particles arrive within a several meters wide front.

As an alternative to the very expensive big arrays used by cosmic ray observatories to measure the high energy cosmic rays, J. Linsley [3] proposed to use mini arrays of detectors measuring a limited sample of the secondaries together with a well known parametrization relating the time width of the bundle of particles with the distance to the core of the shower. Several teams have already shown the viability of the method [4], [5], [6].

A very recent analysis of the time-space microstructure of cosmic rays showers done during the commissioning of a high resolution Time of Flight (TOF) detector designed for the nuclear physics spectrometer HADES, at GSI (Darmstadt, Germany) [7], seems to show that a high granularity relatively small RPC-based tracking detectors ($\sim 1\ m^2$) may also provide a good estimation on the arrival direction of the front of the shower and, even, some statistical separation between electrons and muons. As a consequence, it should still exist room for further improvements of the Linsley's method using detectors with more granularity and better time resolution that the ones existing when the idea was proposed.

With all those ideas in mind, a new high performance cosmic ray tracking detector based on the timing Resistive Plate Chamber (tRPC) technology, TRAGALDABAS, has been installed in the laboratory of the LabCAF group at the Univ. of Santiago de Compostela. The detector has an active area of $1.5 \times 1.2\ m^2$ and covers a vertical solid angle of $\sim 5\ sr$ offering a time resolution of $\sim 300\ ps$ and a track arriving angle resolution better than $3°$. The device is currently at the commissioning stage and it is taking test data at a rate of $\sim$7 millions of events per day. An international collaboration, composed by around 20 researchers belonging to 11 laboratories from 5 European countries, has been organized for the regular maintenance and development of tools allowing to extract all the possible information from the collected data.

In section 2 we describe some of the properties of the cosmic ray air showers. In section 3 we summarize the main features of the detector. In section 4 we show some of the preliminary results obtained during the commissioning period and, finally, in section 5 we summarize the performances of the detector and the research fields that we would like to address to.

2. The cosmic ray induced EAS (Extensive Air Showers)

EAS features have been exhaustively studied by many experiments in the last half century, and have been well parameterized for energies of the primary cosmic ray above $\sim 10^{16}$ eV. Figure 2 shows a summary of the main observables used to characterize single EAS.

A few of the most well known typical features of the EAS are the following [4] [8] [9] [10]:

- At Earth's surface, high energy showers are dominated by high energy muons and low energy electrons and gammas.
- The width of the time profile of the showers, when they arrive to the ground, is narrower near its core than on the edges.
- At a given distance of the axis of the shower, the density of particles in the shower grows with the number of particles generated in the collision region.
- The number of particles generated in the collission is related to both the energy and the mass of the primary cosmic ray.
- The ratio between electrons and muons in the shower grows with the mass of the primary cosmic ray.

As electrons and muons are the most abundant particles in EAS, both particles are the most appropriate to analyze the showers in an event by event basis. Both behave very differently not only in their amount and in their lateral distributions but also in their arrival time distribution and angular distributions [9]. As an example, Figure 3 shows the simulated arrival angle of electrons and muons respect the EAS axis as a function of the distance to the core of the shower for a $10^{16}eV$ cosmic proton. Differences in the behavior of both particles show that a detector having both good tracking and timing capability would allow to make some separation of these two components in high multiplicity events.

The analysis of the cosmic ray data of the HADES experiment showed evidences of other interesting effects:

- Most of the high multiplicity showers showed a sharp front edge having the same direction as the particles arriving in the firsts ns. This effect is compatible with the simulation shown in figure 3 .

- Using well known parameterizations of both the particle density and the arrival time width of high multiplicity events, it was possible to get an estimation of the energy of the primary cosmic ray.

- Many of the bundles of particles detected seemed to show a sparse but lumpy aspect. Perhaps those structures are due to narrow electromagnetic showers induced by the decay of high energy muons. If this is the case, such feature would allow to estimate indirectly the muon content of a shower and, as consequence, to make a guess on the mass (light, medium, heavy) of the primary cosmic ray.

Those and other features of the showers are not usually used in the identification of EAS due to the typical high cost inherent to high accuracy detectors of high energy charged particles. However, RPC technology opens the door to the development of detectors combining high granularity and high time resolution and, using several detection planes, also tracking capability at an affordable price. These features may allow to introduce in the analysis of cosmic ray showers new observables of interest, both for a better estimation of the parameters of the cosmic ray and for a deeper analysis of the relationship between the flux of cosmic rays at Earth's

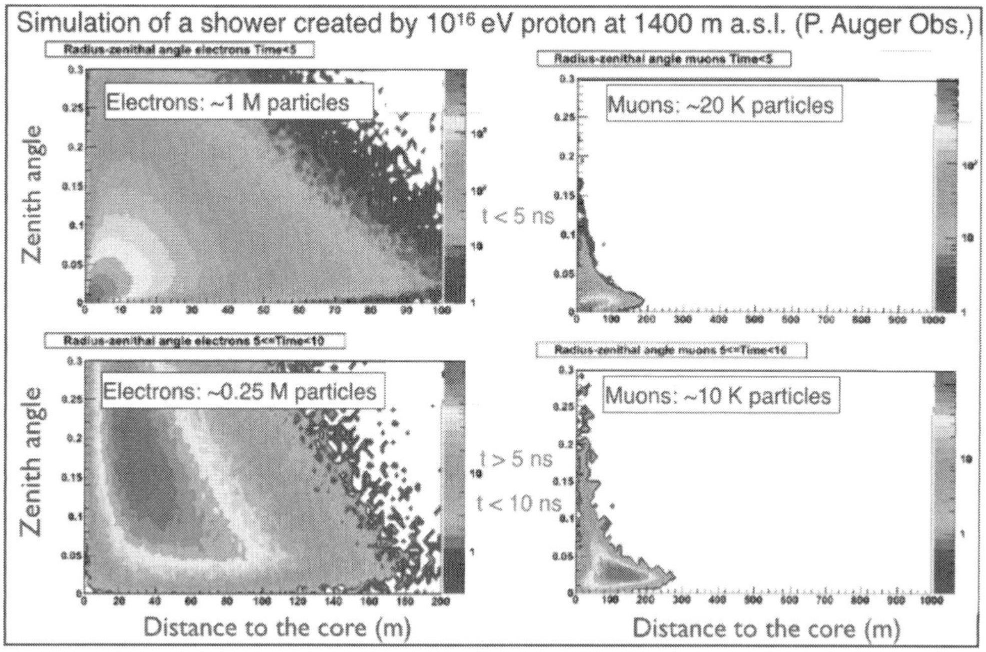

Figure 3. Shower simulation for a vertical proton with $E_0 = 10^{16}eV$, using AIRES [11]. Pictures show the zenith angle of electrons and muons as a function of the distance to the core of the shower at the height of 1400m a.s.l., both for arrival times $t < 5\ ns$ and $5\ ns < t < 10\ ns$. Note: observe the different scale on the horizontal axis and in the color code in the figures.

surface and other well known effects related with the Earth surroundings like the solar activity, the presence of solar plasma clouds, the interplanetary magnetic field, the Earth magnetic field and the Stratosphere temperature among others.

3. TRAGALDABAS

TRAGALDABAS is the acronym for "TRAsGo for the AnaLysis of the nuclear matter Decay, the Atmosphere, the earth B-Field And the Solar activity". The detector has been built following most of the features of the TRASGO philosophy [12] and most of their technical features have been already described in [13] .

Figure 4 shows a picture of the detector with its layout during the ECRS-2014 Conference.The detector was composed by two RPC planes with an active size of $1.2 \times 1.50\ m^2$, placed horizontally at a distance of $\sim 1.2\ m$ each one. It is installed on the first floor of a two-floor building, that will slightly affect the lowest energy EM component of the showers that is not relevant for most of the studies we pretend to carry out. The detector is at the geographical coordinates (N 42° 52' 34", W 8° 33' 37"), at $\sim 260\ m$ above the sea level (a.s.l.). We estimate that the corresponding geomagnetic cutoff is of the order of $\sim 6.5\ GV$.

The detector uses the RPC planes proposed as upgrade of the P. Auger observatory in Argentina [14]. Figure 5 shows the internal layout of each of the RPCs planes and how the read-out digitized pulse is made up in each pad. Each active plane is made by three slides of 2 mm-wide glasses with a 1mm-gap in between, placed inside a gas tight Methacrylate box. As ionizing gas the detector uses freon R134a at a very low flux of $\sim$ 7-8 cc/min.The external surface of the external glass plates are covered by a conductive coating where the high voltage is applied; the nominal value is $\sim 5600\ V$ with opposite sign in each side, providing an equivalent

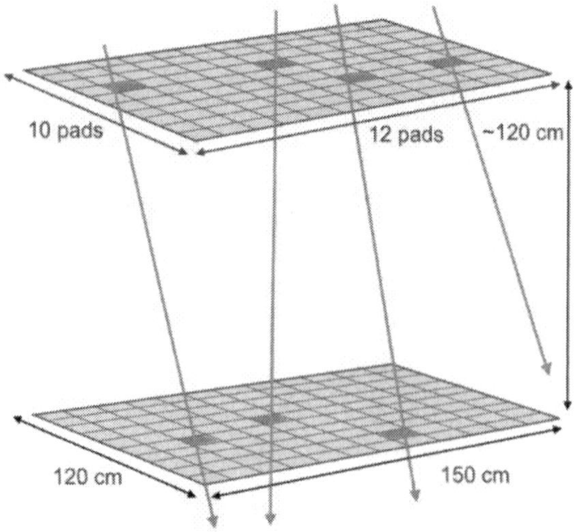

Figure 4. Left) Picture of the TRAGALDABAS detector in the laboratory. Right) RPC-planes and pad layout of the detector. The outstanding time resolution of the detector ($< 300\ ps$) will minimize the wrong assignment of pads to a track (ghost track).

voltage of $\sim 5600\ V/mm$ in each gap. The read-out is done by 120 rectangular Cu-pads placed in the outer side of the box with a size of $111 \times 116\ mm^2$ each and at a $10\ mm$ distance one of each other. Pads are separated by straight guard-electrodes 6 mm wide to prevent the crosstalk, which is almost negligible. Taking into account both the inner Methacrylate and the external Al boxes, the outer size of each RPC plane is $1650 \times 1285 \times 26\ mm^3$, its weight is of the order of $90\ kg$ and the mean radiation length is ~ 0.27.

As read-out front end electronics (FEE), the TRAGALDABAS uses the same electronics developed for the RPC TOF wall of the HADES spectrometer [15] . The FEE is based in a motherboard (MB)- daughterboard (DB) philosophy. In the 4-channel DBs, RPC signals are first amplified and integrated and then digitized in LVDS standard levels. The leading edge of the signal gives the arrival time of the particle and, in the width of the signal, the induced charge is codified by a "charge to width" (QtoW) method. Each MB houses eight DBs, giving them the low voltage and the output for both the digital read-out and the trigger signals.The four MBs collecting all the signals of a RPC plane are read-out by a TDC based read-out board (TRB.V2) [16]. The TRB.V2, developed at the GSI for the HADES experiment, is controlled by an ETRAX processor that sends the data to the external data acquisition computer via Ethernet. The joint detector-acquisition electronics chain provides a time resolution of ~280 ps sigma and an efficiency close to 100%. Every plane is equipped with a set of temperature, humidity and pressure probes, building a I2C bus, for the regular monitoring of those magnitudes.

Some of the main performances offered by the detector are:
- Granularity: 120 pads / plane; pad size: 130 cm^2.
- Angular resolution: $< 3°$
- Time resolution: ~280 ps.
- Hit efficiency: ~ 100%

4. Preliminary results

The detector is now in the commissioning phase and it is taking data with coincidence trigger between both planes, at a rate of about 7 millions of registered events per day. Every day the

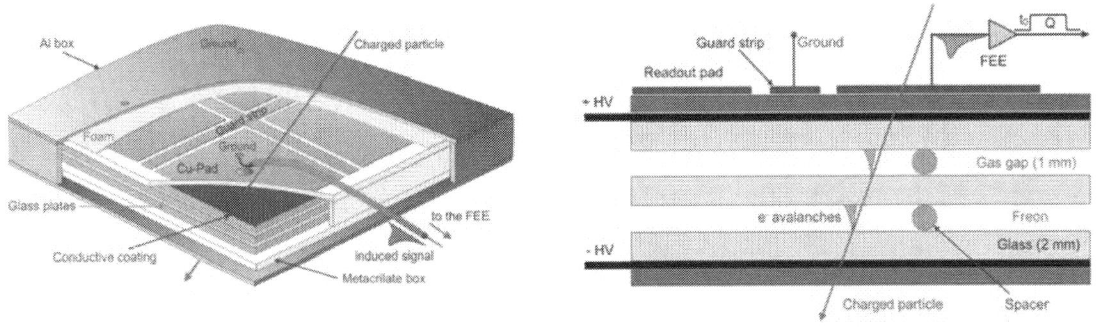

Figure 5. a) Schematic view of a detector plane showing most of each components . b) Cut of a RPC cell showing how a digitized readout signal is created. A LVDS signal codifies the arrival time t_0 of the particle in the leading edge of the signal and the induced charge Q in its width.

data is compacted and sent to a supercomputing center (CESGA) for their definitive storage and handling. Some control test have been done on single data files showing that, globally, the detector is working quite properly. The data analysis will still require, together with the well known both temperature and pressure corrections, some effort including the unpacking of the data together with the calibration, cell synchronization and track reconstruction tasks.

Figure 6 represents, on the left side, the hit multiplicity distribution of data gathered during a week. It can be seen that most of the triggers correspond to events with multiplicities $M = 1$ or $M = 2$ and more than half million of triggers every day having a multiplicity $M > 7$. On the right side, the figure shows a typical hit map of events belonging to a single file of data. The two main accumulations on the middle of the detector do correspond to the expected distribution of cosmic rays. The black region on the bottom-left side of the upper plane of the detector correspond to a couple of disabled cells. Other structures on the corners and on the edge of the detectors may be due to secondaries produced at the support structure, the NIM electronic or the power supply devices.

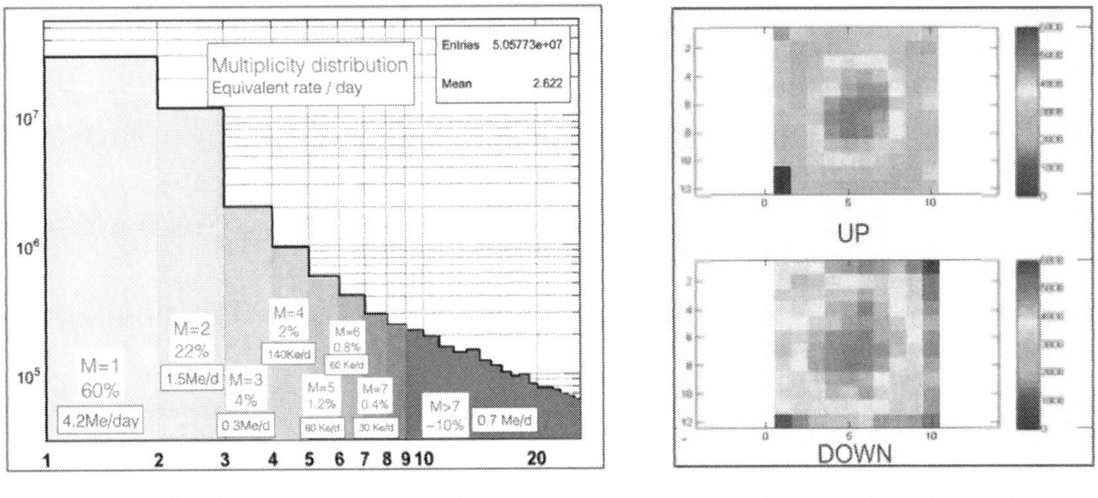

Figure 6. Left) Hit multiplicity distribution in the upper plane for a week of data collection with the corresponding equivalent rate per day. Right) Typical hit map in the upper and lower planes; the observed structures are explained in the text.

Figure 7 shows as a function of time, the evolution of the event multiplicity and the mean induced charge, together with the temperature profile of the atmosphere[12] and the position of the tropopause[13]. The upper and lower bounds given by the dotted lines correspond to the temperature interval +2° around the minimum. Data were collected between March 21st. and April 27th. 2014 when the air conditioner was not yet installed; as a consequence day by day oscillations are mainly due to the day-night changes on the room temperature.

Both quantities are slightly correlated to the altitude of the tropopause or, in other words, are anti-correlated with the ground pressure[14]. This circumstance may explains both behaviors: a higher pressure is usually associated to a higher density, and also to a higher absorption probability; as a consequence, an increase on the pressure should imply a decrease in the multiplicity of the events. Moreover, the induced charge it is expected to be correlated with the multiplicity because a high particle environment increase the chance that two or more particles arrive to the same cell increasing the collected charge.

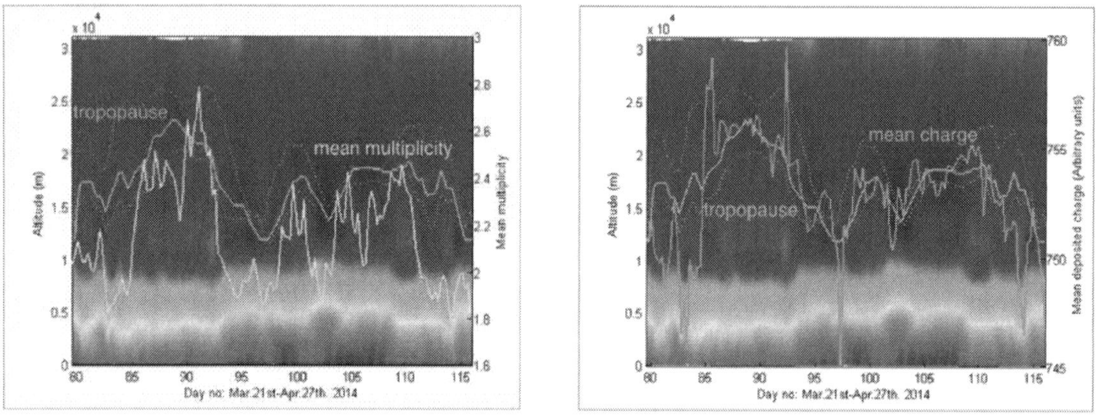

Figure 7. Time evolution of the tropopause altitude (light blue line; left axis) and several physical variables, between Mar. 21st. and Apr. 27th., 2014. The background color represents the temperature of the atmosphere: dark red for highest temperatures and dark blue for lowest temperatures. Left) Mean multiplicity (yellow line) evolution. Right) Mean charge (green line) evolution.

5. Summary

Unlike any other cosmic ray detector existing at the Earth surface, TRAGALDABAS combines both a high granularity and an outstanding time resolution with the tracking capability of charged particles. All this performances would allow to measure the multiplicity of a swarm of particles belonging to the same air shower while the good time resolution offers many advantages: on the one hand it reduces the tracking ambiguity in high multiplicity events and, on the other hand, it allows to select the first arriving particles of a cluster that are those that keep a better memory of the direction of the primary cosmic ray.

Unlike other detectors, the TRAGALDABAS doesn't make use of any intermediary absorber, being sensitive to both the muon and the electromagnetic component of the showers. According to primary results obtained in the analysis of the data taken during the commissioning of

[12] The temperature profiles have been extracted from the reanalysis NCEP/NCAR Data Base: http://www.esrl.noaa.gov/psd/data/reanalysis/reanalysis.shtml

[13] Lower temperature region between the troposphere and the stratosphere.

[14] A decrease/increase on the barometric pressure is usually associated to the ascendent/descendent air movements that push the air upwards, lifting also up the position of the tropopause

the RPC detectors of HADES [7] we expect that electromagnetic and muon components of a shower can be statistically separated, analyzing the arrival time, the angular distribution and the induced charge together with all their correlations. This possibility may open new expectations related to a better understanding of the structure of the cosmic ray air showers in a broad range of energies and also of their relationship with different phenomena related with the Earth surrounding. A few of the research fields accesible to the detector are:
- Analysis of the microstructure of cosmic ray air showers.
- Analysis of the solar activity.
- Analysis of several space weather phenomena as magnetic plasma clouds or magnetic storms.
- Analysis of the Earth's magnetic field.
- Analysis of the temperature profile of the upper atmosphere.

Despite the limited size of the detector the outstanding performances it offers may allow to overcome, in many cases, the lack of statistics problem. In any case, it is expected that it may provide a significative step forward in the understanding of the cosmic ray evolution in the atmosphere and even open the door to the discovery of new effects.

5.1. Acknowledgments

We are very grateful to all our colleagues who made possible the development of the detector and its installation in the laboratory of the LabCAF at the Univ. of Santiago de Compostela, mainly Américo Pereira and Orlando Cunha, from LIP Coimbra and Dr. Michael Traxler, from GSI. We also acknowledge the C. Desarrollo de las Ciencias, of Madrid, for their partial financial support.

References

[1] P.K.F. Grieder. *Cosmic rays at Earth., Elsevier Science* (2009).
[2] J. Blümer et al. *Cosmic rays from the knee to the highest energies., Prog. Part. & Nucl. Phys.* **63** (2009) 293338.
[3] J. Linsley. *Mini and super mini arrays for the study of highest energy cosmic rays., 19th. Int. Cosmic Ray Conf. La Jolla* **3** (1985) 434-437.
[4] T. Bezboruah et al. *A non conventional method of UHE cosmic ray detection., Astropart. Phys.* **11** (1999) 395-402.
[5] U.D. Goswami et al. *Reanalysis of GU miniarray data using CORSIKA, Astropart. Phys.* **22** (2005) 421-429.
[6] A. Iyono et al. *Linsley's EAS time structure method for the primary cosmic ray spectrum at LAAS., Astrophys. Space Sci. Trans.* **7** (2011) 335-344.
[7] D. Belver et al. *Analysis of the space-time microstructure of cosmic ray air showers using the HADES RPC TOF wall, JINST* **7** (2012) P10007.
[8] G. Agnetta et al. *Time structure of the extensive air shower front., Astropart. Phys.* **6** (1997) 301-312.
[9] W.D. Apel et al. *Time structure of the EAS electron and muon components measured by the KASCADE-Grande experiments., Astropart. Phys.* **29** (2008) 317-330.
[10] P. Sommers. *Experimental aspects fo cosmic rays, Univ. of Utah, Salt Lake City, USA* **UT84112-0830** (2008) http://cds.cern.ch/record/1017823/files/p177.pdf.
[11] S. Sciutto *AIRES User Manual*, http://www.fisica.unlp.edu.ar/auger/aires/.
[12] D. Belver et al. *TRASGO: A proposal for a timing RPCs based detector for analyzing cosmic ray air showers, Nucl. Instrum.& Meth.* **A661** (2012) S163-167.
[13] A. Blanco et al. *TRAGALDABAS: a new RPC based detector for the regular study of cosmic rays, JINST* **9** (2014) C09027.
[14] L. Lopes et al. *Resistive Plate Chambers for the Pierre Auger array upgrade, JINST* **9** (2014) C10023.
[15] D. Belver et al. *Performance of the Low-Jitter High-Gain/Bandwidth Front-End Electronics of the HADES tRPC Wall, IEEE Trans. Nucl. Sci.* **57** (2010) 2848-2856.
[16] I. Fröhlich et al. *A General Purpose Trigger and Readout Board for HADES and FAIR-Experiments, IEEE Trans. Nucl. Sci.* **55** (2008) 59-66.

KCDC – The KASCADE Cosmic-ray Data Centre

A Haungs[1], J Blümer[1,2], B Fuchs[2,a], D Kang[2], S Schoo[1], D Wochele[1], J Wochele[1], W D Apel[1], J C Arteaga-Velázquez[3], K Bekk[1], M Bertaina[4], H Bozdog[1], I M Brancus[5], E Cantoni[4,6,b], A Chiavassa[4], F Cossavella[2,c], K Daumiller[1], V de Souza[7], F Di Pierro[4], P Doll[1], R Engel[1], D Fuhrmann[8,d], A Gherghel-Lascu[5], H J Gils[1], R Glasstetter[8], C Grupen[9], D Heck[1], J R Hörandel[10], D Huber[2], T Huege[1], K H Kampert[8], H O Klages[1], K Link[2], P Łuczak[11], H J Mathes[1], H J Mayer[1], J Milke[1], B Mitrica[5], C Morello[6], J Oehlschläger[1], S Ostapchenko[12], N Palmieri[2], M Petcu[5], T Pierog[1], H Rebel[1], M Roth[1], H Schieler[1], F G Schröder[1], O Sima[13], G Toma[5], G C Trinchero[6], H Ulrich[1], A Weindl[1], J Zabierowski[11]

[1] Institut für Kernphysik, KIT - Karlsruher Institut für Technologie, Germany
[2] Institut für Experimentelle Kernphysik, KIT - Karlsruher Institut für Technologie, Germany
[3] Universidad Michoacana, Instituto de Fisica y Matemáticas, Mexico
[4] Dipartimento di Fisica, Università degli Studi di Torino, Italy
[5] National Institute of Physics and Nuclear Engineering, Bucharest, Romania
[6] Istituto di Fisica dello Spazio Interplanetario, INAF Torino, Italy
[7] Universidade São Paulo, Instituto de Física de São Carlos, Brasil
[8] Fachbereich Physik, Universität Wuppertal, Germany
[9] Fachbereich Physik, Universität Siegen, Germany
[10] Dept. of Astrophysics, Radboud University Nijmegen, The Netherlands
[11] National Centre for Nuclear Research, Department of Astrophysics, Łódź, Poland
[12] Frankfurt Institute for Advanced Studies (FIAS), Frankfurt am Main, Germany
[13] Department of Physics, University of Bucharest, Bucharest, Romania
[a] now at: DLR Stuttgart, Germany;
[b] now at: Istituto Nazionale di Ricerca Metrologia, INRIM, Torino;
[c] now at: DLR Oberpfaffenhofen, Germany;
[d] now at: University of Duisburg-Essen, Duisburg, Germany

E-mail: haungs@kit.edu

Abstract. KCDC, the 'KASCADE Cosmic-ray Data Centre', is a web portal, where data of astroparticle physics experiments will be made available for the interested public. The KASCADE experiment, financed by public money, was a large-area detector for the measurement of high-energy cosmic rays via the detection of air showers. KASCADE and its extension KASCADE-Grande stopped finally the active data acquisition of all its components including the radio EAS experiment LOPES end of 2012 after more than 20 years of data taking. In a first release, with KCDC we provide to the public the measured and reconstructed parameters of more than 160 million air showers. In addition, KCDC provides the conceptional design, how the data can be treated and processed so that they are also usable outside the community of experts in the research field. Detailed educational examples make a use also possible for high-school students and early stage researchers.

1. Introduction

The aim of the project KCDC – KASCADE Cosmic-ray Data Centre [1] is the installation and establishment of a public data centre for high-energy astroparticle physics based on the data of the KASCADE experiment (logo of KCDC, see Figure 1). KASCADE (Figure 2) was a quite successful large detector array for measuring high-energy cosmic rays via the detection of extensive air showers (EAS). KASCADE recorded data during more than 20 years on site of the KIT, Campus North, Karlsruhe, Germany (formerly Forschungszentrum Karlsruhe) at 49.1°N, 8.4°E, and 110 m a.s.l. KASCADE collected within its lifetime more than 1.7 billion events of which some 425.000.000 survived all quality cuts. Initially, about 160 million events had been made available at KCDC for public usage.

KASCADE [2], as an extensive air shower (EAS) experiment, studies the cosmic ray primary composition and the hadronic interactions in the energy range of $E_0 = 10^{14} - 10^{17}$ eV. With its extension to KASCADE-Grande in 2003 [3], the range could be extended to 10^{18} eV. EAS are generated when high-energy cosmic particles enter the Earth's atmosphere. Forward-boosted secondary particles as well as emitted light during the development of the EAS in various frequency ranges form the detectable products.

Figure 1. Logo of KCDC. The link to KCDC is: `https://kcdc.ikp.kit.edu`.

Main parts of the experiment were the Grande array spread over an area of $700 \times 700\,\mathrm{m}^2$, the original KASCADE array covering $200 \times 200\,\mathrm{m}^2$ with non-shielded and shielded detectors, a large-size hadron calorimeter, and additional muon tracking devices. The radio antenna field LOPES [4] and the microwave experiment CROME [5], as well as some smaller test experiments and monitoring equipment completed the experimental set-up of KASCADE-Grande.

One of the main results obtained by KASCADE is a picture of increasingly heavier composition above the 'knee' in the cosmic-ray energy spectrum, caused by a break in the spectrum of the light components. Conventional acceleration and propagation models predict a change of the composition towards heavier components due to the charge dependent cut-offs in the flux of the individual elements [6, 7]. The discovery of the knee in the heavy components by KASCADE-Grande, convincingly supports these theories [8].

KASCADE-Grande finally stopped the active data acquisition of all its components end of 2012 and is now decommissioned. The collaboration, however, continues the detailed analysis of nearly 20 years of high-quality air-shower data. Moreover, with KCDC, we provide to the public the edited data via a custom-made web page.

Journal of Physics: Conference Series **632** (2015) 012011 doi:10.1088/1742-6596/632/1/012011

2. KCDC in a Nutshell

The KASCADE/KASCADE-Grande experiment was a large-area detector for the measurement of cosmic ray air showers financed by taxes. The aim of KCDC is the installation and establishment of a public data centre for high-energy astroparticle physics. In the research field of astroparticle physics, such a data release is a novelty, whereas the data publication in astronomy has been established for a long time. Therefore, there are no completed concepts, how the data can be treated and processed so that they are reasonably usable outside the collaboration. The first goal of KCDC is to make the data from the KASCADE experiment available to the community.

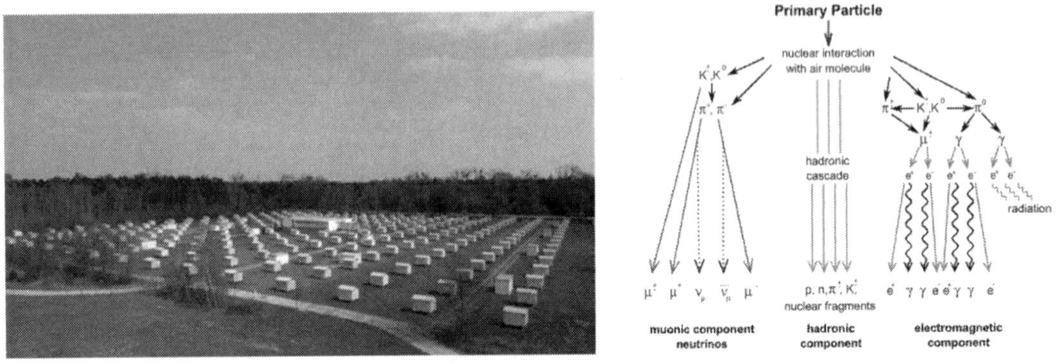

Figure 2. Photograph of the KASCADE experiment (left panel); Schematic view of an extensive air shower (EAS), where KASCADE is measuring the hadronic, muonic, and electron components (right panel).

A concept for this kind of data centre (software and hardware) is meanwhile developed, implemented, and already released as a public beta version to external users. However, the project faces thereby still open questions, e.g. how to ensure a consistent calibration, how to deal with data filtering and how to provide the data in a portable format as well as how a sustainable storage solution can be implemented. In addition, access rights and license policy play a non-negligible role and are considered in details. Readers are invited to visit KCDC under `https://kcdc.ikp.kit.edu`.

Already with the first release, KCDC provides efforts to fulfill following three basic requirements:

- **KCDC as data provider:** There is free and unlimited open access to KASCADE cosmic ray data, where a selection of fully calibrated and reconstructed quantities per individual air shower is provided. The access has to rely on a reliable data source with a guaranteed data quality.
- **KCDC as information platform:** For a meaningful usage of KCDC, a detailed experiment description as well as sufficient meta information on the provided data is needed for any kind of data analysis. This is accompanied by a reasonable description of the physics background as well as tutorials, which are focused on a level for teachers and pupils (in the present version of KCDC the tutorials are provided in German, only).
- **KCDC as long-term digital data archive:** To constitute a sustainable piece of work, KCDC serves also as archive of software and data for the collaboration as well as for the public.

3. The Web Portal

The web portal (entrance page see Figure 3) as interface between the data archive, the data centre's software and the user is one of the most important parts of KCDC. It provides the door to the open data publication, where the baseline concept follows the 'Berlin Declaration on Open Data and Open Access' [9] which explicitly requests the use of web technologies and free, unlimited access for everyone.

We declared both, the scientific and the non-scientific audience as focus of possible users. This requires extensive documentation of experiment, data, and software on a level understandable and handy for all.

Figure 3. Title page of the website of KCDC, the KASCADE Cosmic-ray Data centre.

The portal uses modern technologies, including standard internet access and interactive data selections. The selected data are provided for download via a corresponding ftp-server. Figure 4 shows schematically the basic concept of the KCDC web portal.

It is foreseen that the software behind the data centre including the web portal is also made available at a later stage. Therefore, we were anxious that KCDC provides a modern software solution justified to perform both, publishing the KASCADE data and understandable for a general audience. If KCDC is running successfully and is accepted by the community the software will be released as Open Source for free use also by other experiments. To serve as a general software solution for open access to (astroparticle) data, KCDC is build as a modular, flexible framework with a good scalability (e.g. to large computing centres). The configuration is hold to be simple and doable also via a web interface; the entire software is based solely on Open Source Software (Python, Django, HTML/Javascript, CSSdata provider, etc.)

4. Data Availability

Since November 2013, the first release of KCDC, more than 160 million events of the KASCADE experiment with 14 parameters per event are available. In the first year of operation nearly 100 users registered, where we recognized access to KCDC by IP-addresses from more than 30 countries distributed over 5 continents.

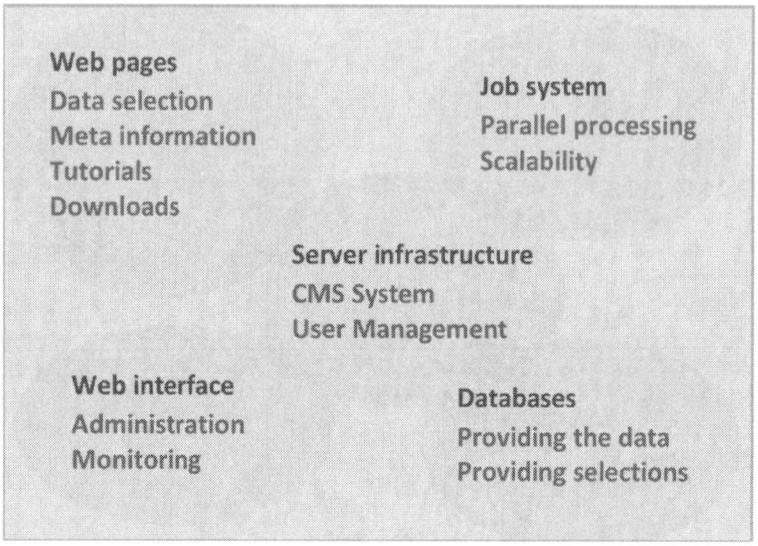

Figure 4. Scheme and key words of the KCDC web portal.

Figure 5 shows a screen shot of the data shop with the list of parameters available per event. If registered via the 'user page', the user is able to enter the data shop. A registration is necessary in order to ensure that the 'End User License Agreement' is read, i.e. the legal aspects of public data are understood (see also next chapter). At the data shop the user can select specific event samples. For each parameter a description is available in a corresponding info box appearing by a mouse-over function. After defining cuts the selection can be submitted. The user gets an Email notification when the selection has been processed and is ready for download via an ftp server. The data will be at the user's disposal in ASCII-format including a detailed header with descriptions of the selection and the data format. Also several pre-selections of KASCADE data are available directly at the data shop.

The data can be used for any analysis, presumed that the user accept the 'limited use licence' (following text is taken from the KCDC-EULA, see also next chapter):

Subject to your agreement and continuing compliance with the KCDC Terms, KIT hereby grants to you a limited, personal, nonexclusive, non-transferable, non-assignable and fully revocable license to – (a) use the webportal and (b) download and use the scientific data of the KCDC in compliance with good scientific practice – provided through the webportal or related online services for your non-commercial scientific purposes only. Commercial purposes are defined as projects for your own or third parties for which you are paid or granted values in lieu of cash for the use of the data.

There is no restriction on the kind of analysis with the provided data nor the publication of the results. However, the KCDC team would acknowledge notification on a use exceeding private education, as well as bug reports or suggestions for improvements. This can be done directly via the web portal and/or per Email to ikp-kcdc@lists.kit.edu.

5. Legal Aspects of KCDC

Opposite to software open source publications, there is no standard procedure yet available for open data publication. In cooperation with KIT and its law department we developed an own license based on the EULA (end user license agreement) model [10], adapted from that one often used for software.

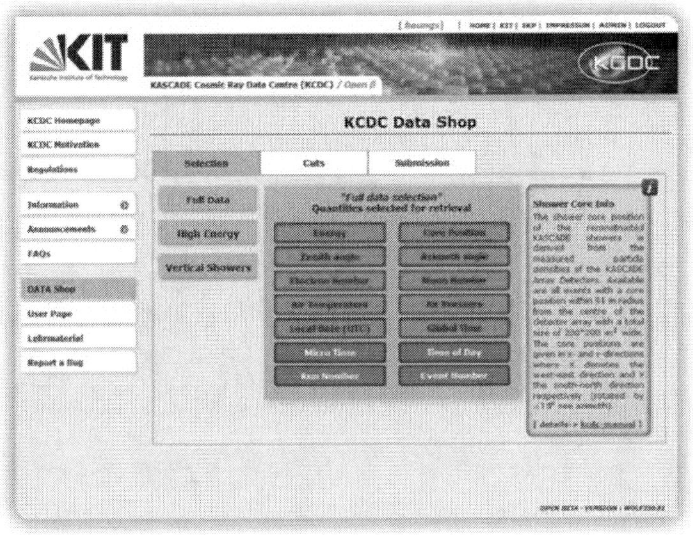

Figure 5. Screen shot of the 'data shop' of KCDC.

We had to consider a twofold issue as the license is needed for the web portal and the data. The KCDC approach is based on the EULA model, because it is flexible and adaptable to our needs, it includes the idea of requiring a good scientific practice, and it can be signed during registration and can be shipped with each data package.

In our custom-made adaption of the KCDC EULA we followed some key points from industry, like

(i) no warranty for damage by owner of web portal or data;
(ii) no guarantee for availability or uptime of the server;
(iii) in case of disputes with local laws the EULA intention is conserved;
(iv) changes are possible at any time;
(v) the termination of EULA is at our digression, only,

as well as obvious requirements from the open data idea, like

(i) free access to the data and the web portal;
(ii) good scientific practice for the work with the data;
(iii) commercial usage of the data is not prohibited[1];
(iv) the citation of collaboration, KIT, and the web portal is mandatory;
(v) free redistribution of data 'as is'.

[1] Please note, that the apparent contradiction to the statement above concerning the non-commercial use of the data, is solved by following statement in the EULA: 'As an alternative to this EULA, KIT offers a license to use the DATA commercially as well, on the basis of a commercial license agreement. If YOU are interested in such a license please contact KIT, Institute for Nuclear Physics (IKP), the KCDC Group and the contact person provided on the WEBPORTAL'.

6. Tutorials

The goal of having detailed tutorials, i.e. an 'education portal', is to provide the data also to a general public in the sense of a visible outreach of astroparticle physics. The first tutorials are already available, but presently in German, only. English and also other languages can be added without any problem and will be done in the near future.

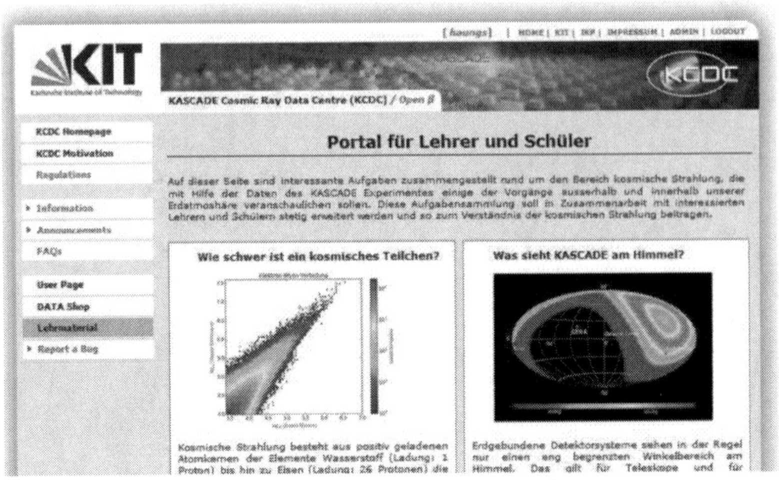

Figure 6. Impression of the tutorial section of KCDC.

The target of the tutorials are teacher and pupils in high schools. This means a tutorial has to provide

(i) a basic knowledge on KASCADE, astrophysics and related topics;
(ii) the required software and KCDC data (preferably as a pre-selection);
(iii) a step by step explanation of a simple data analysis;
(iv) a modern programming language code example;
(v) the interpretation and discussion of the outcome.

This part we develop together with local teachers and pupils.

7. Future Steps

Several tasks and ideas are on the working list for further development of KCDC (some of them are agreed on, others are under discussion):

- Extending the educational portal with more examples and in various languages
- Improving the data selection process to be more comfortable and faster
- Improving the capabilities of the data-base and corresponding software
- Adding more data of KASCADE per individual event
- Adding more events from KASCADE
- Adding simulated data
- Adding data of KASCADE-Grande and LOPES
- Adding data of other cosmic ray experiments
- Publication of the software behind KCDC
- Inclusion of KCDC in long-term data archive networks, e.g. Re3data [11],
- ....

Notice

Please note that since this article has been written, KCDC was further developed and improved. The description here is based on the *Open Beta Version: Wolf359.01*. Please visit the KCDC page and click the button 'Developer News' for more information.

Acknowledgments

The KCDC team acknowledges the continuous support of the project by the 'Helmholtz Alliance for Astroparticle Physics HAP' funded by the Initiative and Networking Fund of the Helmholtz Association. In addition we acknowledge the fruitful cooperation with Hartmut Aichert, teacher at the local Thomas-Mann-Gymnasium, Blankenloch and his pupils.

References

[1] see `https://kcdc.ikp.kit.edu`, January 2015
[2] Antoni T et al. - KASCADE, 2003 *NIM A* **513** 490
[3] Apel W D et al. - KASCADE-Grande, 2010 *NIM A* **620** 202
[4] Falcke H et al. - LOPES, 2005 *Nature* **435** 313
[5] Smida R et al. - CROME, 2013 *EPJ Web Conf.* **53** 08010
[6] Antoni T et al. - KASCADE, 2005 *Astropart. Phys.* **24** 1
[7] Apel W D et al. - KASCADE-Grande, 2011 *PRL* **107** 171104
[8] Schoo S et al. - KASCADE-Grande, 2015 *Proc. ECRS 2014, Kiel*; these proceedings
[9] see `http://en.wikipedia.org/wiki/Berlin_Declaration_on_Open_Access_to_Knowledge_in_the_Sciences_and_Humanities`, January 2015
[10] see `http://en.wikipedia.org/wiki/End-user_license_agreement`, January 2015
[11] see `http://www.re3data.org/`, January 2015

Amplitude Calibration of the Tunka Radio Extension (Tunka-Rex)

R Hiller[1], P A Bezyazeekov[2], N M Budnev[2], O A Gress[2], A Haungs[1], T Huege[1], Y Kazarina[2], M Kleifges[3], E N Konstantinov[2], E E Korosteleva[4], D Kostunin[1], O Krömer[3], L A Kuzmichev[4], R R Mirgazov[2], L Pankov[2], V V Prosin[4], G I Rubtsov[5], C Rühle[3], V Savinov[2], F G Schröder[1], R Wischnewski[6] and A Zagorodnikov[2] (Tunka-Rex Collaboration)

[1] Institut für Kernphysik, Karlsruhe Institute of Technology (KIT), Germany
[2] Institute of Applied Physics ISU, Irkutsk, Russia
[3] Institut für Prozessdatenverarbeitung und Elektronik, Karlsruhe Institute of Technology (KIT), Germany
[4] Skobeltsyn Institute of Nuclear Physics MSU, Moscow, Russia
[5] Institute for Nuclear Research of the Russian Academy of Sciences, Moscow, Russia
[6] DESY, Zeuthen, Germany

E-mail: roman.hiller@kit.edu

Abstract. Tunka-Rex is an experiment for the radio detection of cosmic-ray air showers in Siberia. It consists of 25 radio antennas, distributed over an area of 1 km^2. It is co-located with Tunka-133, an air-Cherenkov detector for cosmic-ray air showers. Triggered by Tunka-133, Tunka-Rex records the radio signal, emitted by air showers with energies above 10^{17} eV. Its goal is to probe the capabilities of a radio detector, especially for the determination of the energy and elemental composition of cosmic ray primaries. To compare the measurements of Tunka-Rex to other radio detectors or to models describing the radio emission, the radio signal in each station has to be reconstructed in terms of physical units. Therefore, all hardware components have to be calibrated. We show how the calibration is performed and compare it to simulations.

1. Introduction

Cosmic rays are a unique window in the close universe, possibly providing a look at the most violent processes in the universe in the high energy range. Because of the low flux of high-energy cosmic rays, huge devices are necessary to detect air showers in the Earth's atmosphere, with areas of up to thousands of km^2. Additionally, the different types of detectors suffer from systematics, e.g. due to the extrapolation of physics to high energies. Therefore, hybrid detectors are used, comprised by different systems, to allow for cross-checks and exploit the advantages of the different devices. Thus, there is a need for new types of detectors, which provide a new view on the measurement.

One possible candidate is the radio technique. The corresponding detector consists of several antennas, distributed in the detector field. Radio pulses were already measured and to some extent understood in the 1960s [1, 2]. Mainly due to the deflection of charged particles in the Earths atmosphere, the air shower produces a radio pulse of several 10 ns width. A second order

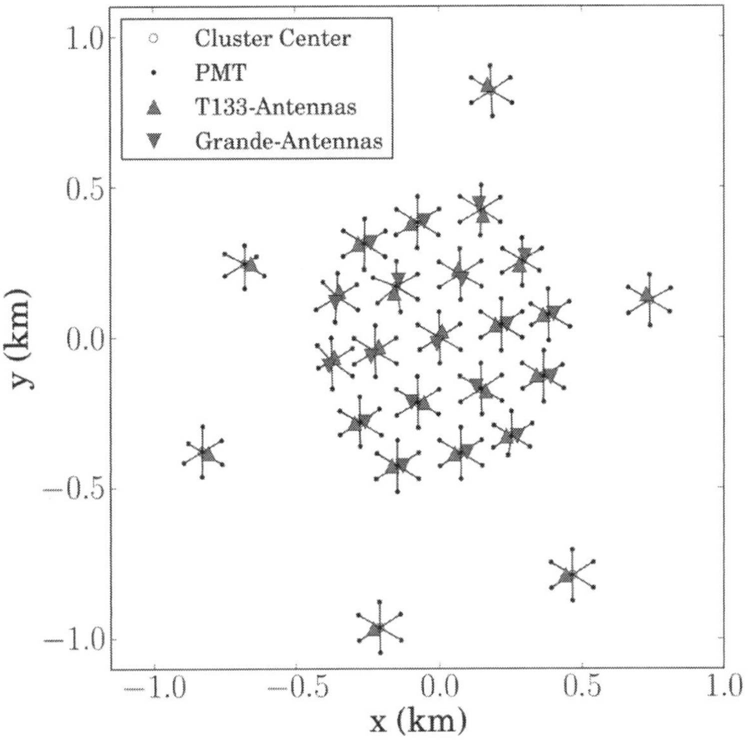

Figure 1. Map of the Tunka-Rex Layout. There is one antenna next to each cluster center of the Tunka-133 PMT stations.

contribution to the radio signal, arising from the Askaryan-effect [3], due to a time-varying net charge in the shower, was measured only recently [4]. Above around 10^{17} eV the power of the pulse exceeds the galactic background radiation several 100 m around the shower core, and thus can be detected. The main advantages of the radio detector are its almost full duty cycle and relatively low cost.

After the proof-of-principle in the 1960s, the radio technique vanished and returned only during the last decade due to improvements in electronics and digital signal processing. The feasibility of interferometry was shown and sensitivity to the energy and the depth of the shower maximum was proven [5]. Now, the capabilities of radio detectors have to be probed in more detail to prove its applicability in a contemporary detector for air showers.

With this goal the Tunka Radio Extension (Tunka-Rex) was deployed in 2012. It is a radio detector in Siberia, close to Lake Baikal and is collocated with Tunka-133, an air-Cherenkov detector for air showers above 10^{16} eV [6]. Tunka-133 triggers Tunka-Rex, therefore, Tunka-Rex can operate only during moonless nights in winter, when Tunka-133 is running. With a precision of 15% on energy and 28 g/cm^2 on X_{max} Tunka-133 also provides independent reconstruction information on the shower. Tunka-Rex and Tunka-133 can be cross-calibrated and the precision of Tunka-Rex can be determined by comparing its reconstruction results to the ones of Tunka-133 with its known precision.

2. The detector

Tunka-Rex was deployed in 2012 and its first stage consists of 25 antennas, one connected to each PMT cluster of Tunka-133 (see Fig. 1). This results in an antenna spacing of 200 m in the

Figure 2. A Tunka-Rex antenna station with its two perpendicularly aligned SALLA antennas.

central detector, covering 1 km^2 and an antenna spacing of 500 m for the outer region, covering 3 km^2. Data of the antennas are recorded together with the Tunka-133 data for the 7 PMT of a cluster after a coincidental threshold trigger on cluster level. After data acquisition, events with multiple clusters are formed, if the single cluster triggers coincide within 7 µs.
In 2014 another 19 antennas were connected to the new scintillator detector upgrade of Tunka-133, which can operate also during day and significantly extend our measurement time.
The antenna stations consist of 2 channels, each equipped with a SALLA [7] antenna, perpendicular to each other and 45° rotated relative to the north-south and east-west axis, respectively. In Fig. 2 a station is shown. The SALLA is an economic, rugged antenna with little dependence on ground conditions, but also relatively low gain. The two channels, together with the incoming direction of the signal, enable the reconstruction of the polarization.

The SALLAs are directly connected to a low-noise amplifier (LNA) with 24 dB gain at their base. To avoid noise from other electronics, e.g., of the Tunka-133 trigger hardware, the antenna station is located at a distance of at least 20 m from the cluster centers with the electronics. A possible interfering signal is attenuated and interference from the trigger can be separated from the shower signal due to its delay. 30 m of RG213 cable connect the LNA to another filter-amplifier with about 32 dB amplification in the 30-80 MHz band and 60 dB attenuation outside. It is located in the electronics box in the cluster center and its temperature is regulated to about 18 °C during operation. On each trigger of the cluster a 5 µs trace around the trigger is recorded at a sampling rate of 200 MHz with 12 bit depth.

3. Calibration

For a comparison to model calculations and other experiments, one needs to reconstruct the electrical field strength from the measurements with the antennas. The transmission of the radio signal through a linear hardware chain can formally be described by the convolution of the signal with the response function of the system. Under usual circumstances, the convolution can be inverted if the response function is known. The response function can be factorized in its single components. To determine it, we measured the transmission parameters of the analog

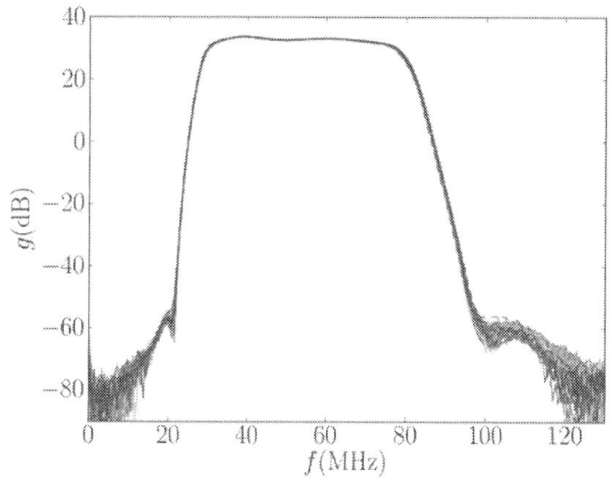

Figure 3. The gain of the filter amplifiers for all 25 filters in different colors.

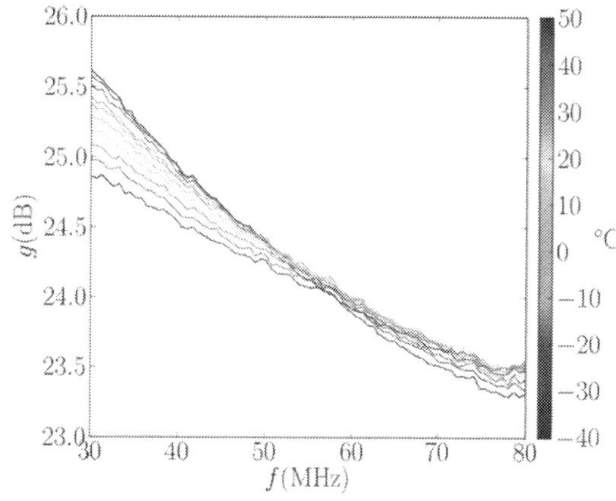

Figure 4. The gain of a LNA at different temperatures. During measurement, i.e., during winter nights we expect temperatures between -35 °C and 0 °C.

hardware chain and obtained the vector effective length (VEL) of our antenna from a simulation, combined with an amplitude calibration.

3.1. Electronics calibration

The electronics chain consists of the LNA, cables and the filter amplifier. They were calibrated using a network analyzer, which directly measures the forward transmission, i.e. the amplitude and phase shift for each frequency. This can be interpreted as the response function in the frequency domain. In Fig. 3, the amplitude gain of the filter amplifiers is shown.

The response of the full electronics chain can be retrieved by frequency wise multiplication of the single responses.

While the filter amplifiers are heated in the cluster centers, the LNAs are located outside, prone to temperature fluctuations. Therefore, we also measured the temperature dependence of the responses, by repeating the response measurement at different temperatures in a temperature chamber. In Fig. 4, the amplitude gain of the LNA at different temperatures is shown. The variation for the expected temperatures between -35 °C and 0 °C are about 0.3 dB.

3.2. Antenna calibration

The response function of the antenna is called VEL. Several other radio experiments have performed antenna calibrations to determine the VEL [7, 8, 9]. It is measured by observing a reference source with known emitted power or amplitude. One popular choice are sources in space, or for non-directional antennas, galactic noise. Alternatively, one can use a man-made reference source. Since the electronics noise for Tunka-Rex was estimated to be in the order of magnitude of the received galactic noise, and due to availability, we decided to rely on the latter option.

The VEL depends on the incoming direction of the signal and is very cumbersome to measure. Therefore, we decided to use an antenna simulation to obtain the directivity of the antenna. However, the amplitude is determined with an antenna calibration from a certain direction to have the right scale. A similar approach, using the very same calibration source, is applied by the LOPES collaboration [8], and lately at the LOFAR experiment [9]. For the simulation of the antenna the NEC2 simulation code [10] was used.

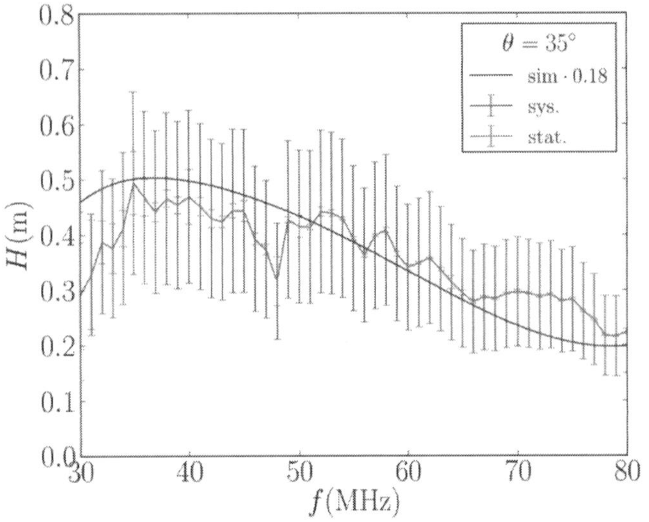

Figure 5. The vector effective length of the SALLA, obtained from the calibration of a Tunka-Rex antenna station and from simulation data. The latter is scaled down for easier comparison.

The absolute amplitude calibration was performed with the commercial calibrated source TESEQ VSQ 1000. It is a biconical antenna with known amplitude at 10 m distance in the main direction. We arranged the source with a crane at 10 m height above the antenna. The exact distance was measured with about 10 cm precision by a differential GPS. Then, we aligned it with strings from ground and recorded traces. By comparing the recorded trace with the calibration data, we obtained a total response function, from which we then removed the electronics chain to get the effective antenna height.
In Fig. 5, the resulting vector effective length of the antenna is shown and compared to the simulation. Although the frequency dependence roughly fits, the absolute scale seems to be overestimated by the simulation. Thus, in our reconstruction, we use the simulated antenna pattern and scale it by a constant factor. This result is currently cross-checked to see if it is a technical detail, or if it has physical implications, since the SALLA design was partially motivated by simulations.

The calibration was performed at 5 different zenith angles relative to the antenna, 0°, 20°, 35°,50°, and 68°. The correction factors for the simulated antenna model are for all measured zenith angles roughly the same, with no significant systematic shift. The mean standard deviation for the different zenith angles over the relevant frequency range is about 15%. We chose to normalize the simulated antenna model to the measurement at 35°, since most of our high quality events are close to this angle.

4. Summary and Outlook

The radio signal of cosmic-ray air showers can be reconstructed with measurements from Tunka-Rex by deconvolution of the signal with the hardware response. The required hardware response contains the transmission function of the hardware and the vector effective length of the antenna. The transmission parameters of the analog hardware are measured with a network analyzer. For the vector effective length of the antenna we use an antenna simulation and normalize the amplitude according to an absolutely calibrated source.
We found a significant deviation in the absolute scale of the predicted amplitudes of the simula-

tion and the one found with the calibration. However, the directivity is still well aproximated. With the calibrated detector it will be possible to directly compare the calibration curve of Tunka-Rex [11] to other calibrated radio experiments, operating in the same frequency band, and to test model predictions for the absolute amplitudes of the radio signal emitted by an air shower.

Acknowledgments
Tunka-Rex is funded by the German Helmholtz association (grant HRJRG-303) and supported by the Helmholtz Alliance for Astroparticle Physics (HAP). This work was supported by the Russian Federation Ministry of Education and Science (agreement 14.B25.31.0010, zadanie 3.889.2014/K), the Russian Foundation for Basic Research (Grants 12-02-91323, 13-02-00214, 13-02-12095, 14002-10002) and the President of the Russian Federation (grant MK-1170.2013.2).

References

[1] Allan H R 1971 *Prog. in Elem. Part. and Cosmic Ray Phys.* **10** 171.
[2] Kahn F D and Lerche I 1966 *Proc. Phys. Soc. Sect. A* **289** 206.
[3] G. A. Askaryan, Sov. Phys. JETP 14, 441 (1961).
[4] P. Abreu, M. Aglietta, E.J. Ahn *et al.*, Phys. Rev. D 89, 052002 (2014).
[5] Apel W D, Arteaga J C, Bähren L *et al.* (LOPES Collaboration) 2014 *Phys. Rev. D* **90**, 062001.
[6] Prosin V V, Berezhnev S F, Budnev N M *et al.* 2014 *Nucl.Instr. and Meth. A* **756** 94-101.
[7] Abreu P, Aglietta M, Ahlers M *et al.* 2012 *JINST* **7** P10011.
[8] Nehls S *et al.* 2008 *Nucl.Instr. and Meth. A* **589** 350-361.
[9] Nelles A *et al.* 2015 *JINST* In preparation.
[10] Burke G and Poggio A 1977 *technical report.*
[11] Kostunin D *et al.* (Tunka-Rex Collaboration) 2014 *these proceedings.*

A limit on the diffuse gamma-rays measured with KASCADE-Grande

D Kang[1], W D Apel[2], J C Arteaga-Velázquez[3], K Bekk[2], M Bertaina[4], J Blümer[1,2], H Bozdog[2], I M Brancus[5], E Cantoni[4,6,a], A Chiavassa[4], F Cossavella[1,b], K Daumiller[2], V de Souza[7], F Di Pierro[4], P Doll[2], R Engel[2], Z Feng[8], D Fuhrmann[9,c], A Gherghel-Lascu[5], H J Gils[2], R Glasstetter[9], C Grupen[10], A Haungs[2], D Heck[2], J R Hörandel[11], D Huber[1], T Huege[2], K H Kampert[9], H O Klages[2], K Link[1], P Łuczak[12], H J Mathes[2], H J Mayer[2], J Milke[2], B Mitrica[5], C Morello[6], J Oehlschläger[2], S Ostapchenko[13], N Palmieri[1], M Petcu[5], T Pierog[2], H Rebel[2], M Roth[2], H Schieler[2], S Schoo[2], F Schröder[2], O Sima[14], G Toma[5], G C Trinchero[6], H Ulrich[2], A Weindl[2], J Wochele[2], J Zabierowski[12]

[1] Institut für Experimentelle Kernphysik, KIT - Karlsruher Institut für Technologie, Germany
[2] Institut für Kernphysik, KIT - Karlsruher Institut für Technologie, Germany
[3] Universidad Michoacana, Instituto de Fisica y Matemáticas, Mexico
[4] Dipartimento di Fisica, Università degli Studi di Torino, Italy
[5] National Institute of Physics and Nuclear Engineering, Bucharest, Romania
[6] Istituto di Fisica dello Spazio Interplanetario, INAF Torino, Italy
[7] Universidade São Paulo, Instituto de Física de São Carlos, Brasil
[8] Institute of High Energy Physics, Beijing, China
[9] Fachbereich Physik, Universität Wuppertal, Germany
[10] Fachbereich Physik, Universität Siegen, Germany
[11] Dept. of Astrophysics, Radboud University Nijmegen, The Netherlands
[12] National Centre for Nuclear Research, Department of Astrophysics, Lodz, Poland
[13] Frankfurt Institute for Advanced Studies (FIAS), Frankfurt am Main, Germany
[14] Department of Physics, University of Bucharest, Bucharest, Romania
[a] now at: Istituto Nazionale di Ricerca Metrologia, INRIM, Torino;
[b] now at: DLR Oberpfaffenhofen, Germany;
[c] now at: University of Duisburg-Essen, Duisburg, Germany;

E-mail: `donghwa.kang@kit.edu`

Abstract. Using data measured by the KASCADE-Grande air shower array, an upper limit to the flux of ultra-high energy gamma-rays in the primary cosmic-ray flux is determined. KASCADE-Grande measures the electromagnetic and muonic components for individual air showers in the energy range from 10 PeV up to 1 EeV. The analysis is performed by selecting air showers with low muon contents. A preliminary result on the 90% C.L. upper limit to the relative intensity of gamma-ray with respect to cosmic ray primaries is presented and compared with limits reported by other measurements.

1. Introduction

The flux of diffuse gamma rays and its spectrum is a very interesting topic in astrophysics. This study gives information about processes and developments at extragalactic distances. Therefore, we can understand the origin and the propagation of galactic cosmic rays from investigations of the galactic diffuse emission. Since direct measurements of ultra-high energy photons by satellite or balloon-borne experiments are difficult due to the low fluxes. This detection can, however, be accomplished by large area ground-based detectors using the air shower array technique.

Extensive air showers are mainly characterized by the total electron number and the total muon number. In general, muons are produced by the decay of charged kaons or pions, which in hadronic showers are produced in nucleus-nucleus interactions, whereas in photon showers only in photoproduction processes. The ratio between the cross sections of photoproduction and nucleus-nucleus interaction processes is very small, in the order of $\sim 10^{-3}$. Therefore, the usual strategy for searches for primary gamma-rays in extensive air showers is to discriminate gamma-ray primaries from the hadronic background by identifying muon-poor extensive air showers.

There are two main plausible sources of diffuse flux of ultra-high energy gamma radiation: The first source is due to decays of neutral pions produced by the collision of cosmic-rays with interstellar gas and dust in the disk of the Galaxy. In this case, the predicted integral intensity is concentrated in the galactic plane. The second one is due to electromagnetic cascades resulting from the interactions of extremely high-energy cosmic-rays with the 2.7 K cosmological microwave background radiation or topological defects in early Universe. In this case, this results in an isotropic flux of secondary photons uniformly are distributed over cosmological distances. Therefore, the measurements of the diffuse isotropic gamma-ray flux might provide information on the ultra-high energy cosmic-ray components. In addition, this flux would represent a background for experiments searching for the gamma-ray enhancement from the direction of the galactic disk.

This contribution presents limits on the relative intensity of the gamma-ray component of the cosmic rays measured by the KASCADE-Grande experiment.

2. The KASCADE-Grande experiment

KASCADE-Grande was an extensive air shower experiment located at Karlsruhe Institute of Technology (49.1° north, 8.4° east, 110 m above sea level). It operated until the end of 2012 and all components were by now dismantled.

The KASCADE-Grande [1] array covering an area of 700×700 m^2 was optimized to measure extensive air showers up to primary energies of 1 EeV. It comprised 37 scintillation detector stations located on a hexagonal grid with an average spacing of 137 m for the measurements of the electromagnetic and muonic shower components. Each of the detector stations was equipped with plastic scintillators covering a total area of 10 m^2. The angular resolution of KASCADE-Grande is better than 0.5 ° over the whole energy range, so that it will allow us to perform also gamma-ray searches.

2.1. Selection of data

For this analysis full data sets taken from 2003 to 2012 were used, where only successfully reconstructed and precisely measured events were selected. The core positions of the showers are inside a circular area of 152,214 m^2 around the center of the array to avoid large reconstruction errors at the edges of the detector field. The zenith angle is smaller than 40° to ensure full efficiency. After applying all quality cuts, we obtained in total ca. 1.7×10^7 events for a measurement time of ca. 1865 days.

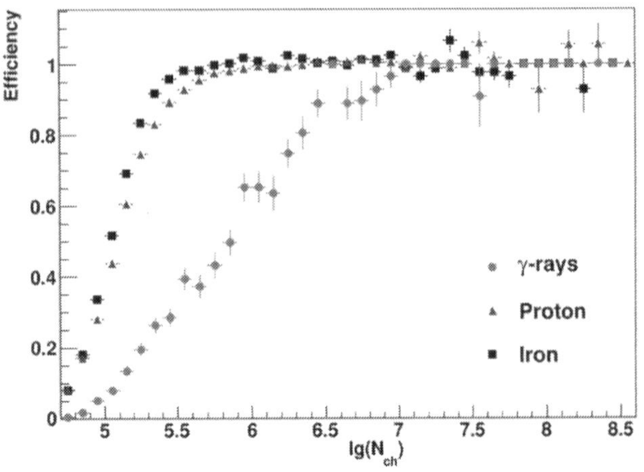

Figure 1. Trigger and reconstruction efficiency as a function of the number of charged particles for air showers induced by photons, protons and iron primaries.

3. Gamma and cosmic ray simulations

An essentail part of the present analysis is the Monte Carlo simulation of the experiment.

For the simulation of the physical processes in the air shower development the CORSIKA [2] program has been used, applying hadronic interaction models. To determine the signals in the individual detectors, all secondary particles at the ground level are passed through a complete detector simulation program using the GEANT package. The predicted observables at ground level, such as e.g. the number of electrons, muons and photons are then compared to the measurements.

The FLUKA [3] ($E < 200$ GeV) model has been used for hadronic interactions at low energies. High-energy interactions were treated with different models QGSJETII-2 [4]. Showers initiated by primary photons, protons, He, CNO, Si and iron nuclei have been simulated. The simulations covered the energy range of 10^{14} - 10^{18} eV with zenith angles in the interval 0° - 60°. The spectral index in the simulations was -2 and for the analysis it is weighted to a slope of -3. For each particle type the same number of showers was generated.

3.1. Efficiency

The trigger and reconstruction efficiency of KASCADE-Grande as a function of the shower size, i.e. the number of charged particles, is demonstrated in Fig. 1. The efficiency is defined by the ratio of the number of events in the reconstructed area to the number of events in the true area, in which the binomial statistical errors are used. Efficiencies above one are due to cuts on the reconstructed core positions, i.e. events falling outside the selected area can be reconstructed inside. Full efficiency is reached at the number of charged particles of around 10^6 for air showers induced by protons and iron primary particles, which approximately corresponds to a primary energy of 10^{16} eV. However, for showers induced by photon primaries the full efficiency is higher due to the missing muon trigger at large distances. The limit at high energies is due to the restricted area of the Grande array.

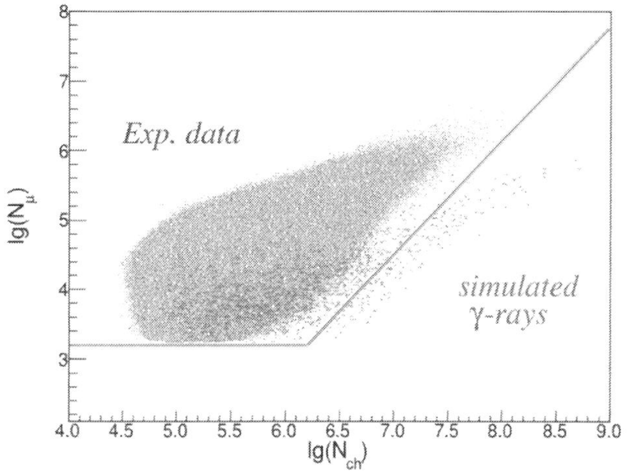

Figure 2. Scattered distribution of the measured number of muons $\lg(N_\mu)$ and number of charged particles $\lg(N_{ch})$ superimposed with simulated gamma-ray showers. The green line indicates the choice of the cut for the selection of the muon-poor showers.

4. Gamma hadron discrimination

Figure 2 shows a scatter diagram of the observed number of muons $\lg(N_\mu)$ and charged particles $\lg(N_{ch})$. The simulated gamma-ray showers are superimposed as well. The blue data points indicate the whole experimental data set, whereas the red one illustrate the simulated gamma-ray events. The shower size is corrected to a zenith angle of 20 ° using the method of constant intensity [5].

Since gamma-ray induced air showers are notable for their lack of muons, compared to hadronic showers, we select a data sample enriched in photon showers by rejecting showers containing muons, i.e., we therefore take a more conservative way in which the expected background is not subtracted from the event number below the cut line.

The selection of muon-poor showers is indicated by the straight line in Fig. 2. The events below the straight line were taken into consideration for further analysis. They amount to 1060 out of a total of 17 millions events. In the region below this line the events are expected to be mainly due to primary photons because air showers induced by heavy nuclei show a larger muon to electron ratio.

5. The analysis and results

There is no possible excess of events consistent with a gamma-ray signal seen in the data. Hence, we assume that all events below the selection line are background and set upper limits on the gamma-ray fraction of the cosmic rays. We estimate the 90% C.L. upper limit on the number of observed events, applying standard statistical methods [6] and the efficiency for gamma-ray detection.

Assuming that all events passing the gamma-ray selections are indeed gamma-rays, the 90% upper limit on the fraction of the gamma-ray integral flux relative to the cosmic-ray integral flux, I_γ/I_{CR}, is estimated by [6]

$$\frac{I_\gamma}{I_{CR}} < \frac{N_{90}}{N_{tot}\varepsilon_\gamma}\left(\frac{E_{CR}}{E_\gamma}\right)^{-\beta+1}, \tag{1}$$

Table 1. Results of the search for diffuse ultra-high energy gamma-rays at different threshold values of N_{ch}. The mean cosmic-ray energy, E_{CR}, and the mean gamma-ray energy, E_γ, are given in the unit of TeV. I_γ/I_{CR} is the 90% C.L. upper limit on the integral gamma-ray fraction, and I_γ is the 90% C.L. upper limit on the integral gamma-ray flux in the unit of photons $cm^{-2}s^{-1}sr^{-1}$.

$\lg(N_{ch})$	N_{tot}	N_{90}	ε_γ	E_{CR}	E_γ	I_γ/I_{CR}	I_γ ($\times10^{-17}$)
>6.5	6.19×10^6	358	0.61	3.21×10^4	1.64×10^4	$<2.98\times10^{-5}$	<1.88
>7	85537	351	0.88	8.72×10^4	6.29×10^4	$<2.37\times10^{-3}$	<9.83
>7.5	9640	214	~1.0	2.21×10^5	1.32×10^5	$<7.29\times10^{-3}$	<5.89
>8	1239	122	~1.0	5.31×10^5	2.75×10^5	$<2.11\times10^{-2}$	<3.30

where N_{90} is the 90% C.L. upper limit on the number of detected events, ε_γ is the efficiency for gamma-ray detection, and β is the integral cosmic-ray spectral index ($\beta = 3.0$, $E_\gamma > 4$ PeV). E_{CR} and E_γ are the mean energies of cosmic-ray and gamma-ray, respectively, which produce the same shower size (N_{ch}). Therefore, they are calculated for a fixed N_{ch} bin using the resulting Gaussian function.

Table 1 shows the results of the search for diffuse ultra-high energy gamma-rays for different threshold values of N_{ch}. To determine upper limits to the integral flux of gamma-rays at fixed gamma-ray energies, we use measurements of the all-particle primary energy spectrum of Ref. [7]. The flux limits are also given in Table 1.

Figure 3 displays the measurements on the gamma-ray fraction as a function of energy, including this work, for the energy range of 200 TeV up to 300 PeV. The limits presented here represent the first results in the energy range above 5×10^{16} eV.

These results are compared with other previous experiments [10, 11] and with theoretical curves by a specific IceCube excess model [9] in Fig. 4. The lines are IceCube excess model coming from different distances of sources of neutrinos in the galaxy, where these ν's are also responsible for primary gamma-rays. The KASCADE results [8] are compatible with the theoretical prediction of the IceCube excess model coming from < 20 kpc in the galactic plane. It might hence set some constrains on the distance of sources for the IceCube Excess. The result of the KASCADE experiment will be updated with more data. Details will be described elsewhere in a forthcoming publication.

It should be noted that all values in Fig. 4 are limits, except that one from MSU [12]; This reported positive signal seems to be disproved by the presented results.

6. Conclusion

Using full data sets measured by the KASCADE-Grande experiment, the 90% C.L. upper limits to the diffuse flux of ultra-high energy gamma-rays for the energy range of 200 TeV to 300 PeV are determined by selecting showers with low muon contents.

The best upper limit to the fraction of the gamma-ray to the cosmic-ray flux is obtained: $I_\gamma/I_{CR} < 2.98 \times 10^{-5}$ for 16.4 PeV. These stringent limits are the first results in the energy range above 10 PeV, and might constrain a limit to the background rate of muon-poor showers in the search for the galactic disk enhancement of cosmic rays.

Comparison of these experimental upper limits with the theoretical predictions is still required, since no theoretical predictions exist so far at energies around 10 PeV. The statistics on the gamma-ray simulation is currently limited, so that these preliminary results will be improved by more gamma-ray simulations and also by a new muon reconstruction.

In addition, the results of the original KASCADE experiment will be updated by including

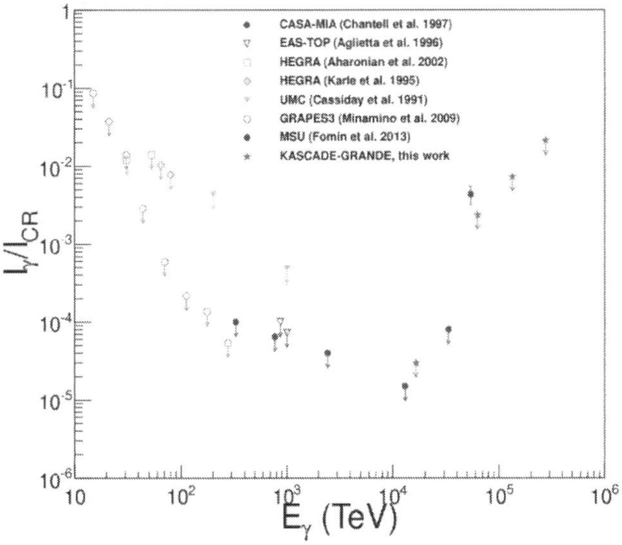

Figure 3. Fraction of gamma-rays relative to the cosmic rays. The red stars represent the results from KASCADE-Grande.

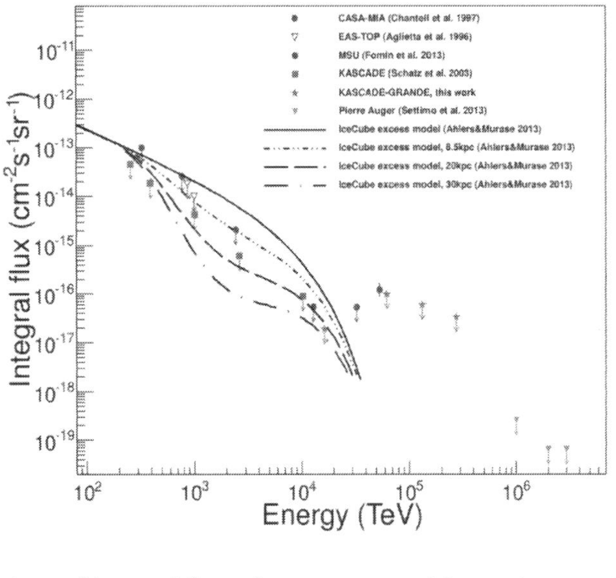

Figure 4. Comparison of integral flux of gamma-rays with previous results and with theoretical curves by an IceCube excess model [9].

eight years of more data since the publication in Ref. [8]. It should be mentioned that this might set some constraints on the distance of sources for the IceCube excess.

Acknowledgments

The authors would like to thank the members of the engineering and technical staff of the KASCADE-Grande collaboration, who contributed to the success of the experiment. The KASCADE-Grande experiment is supported in Germany by the BMBF and by the Helmholtz Alliance for Astroparticle Physics - HAP funded by the Initiative and Networking Fund of the Helmholtz Association, by the MIUR and INAF of Italy, the Polish Ministry of Science and Higher Education, and the Romanian Authority for Scientific Research UEFISCDI (PNII-IDEI grants 271/2011 and 17/2011). J.C.A.V. acknowledges the partial support of CONACyT.

References

[1] W.D. Apel et al. KASCADE-Grande Collaboration, *Nucl. Instr. Meth.* **A 620** (2010) 202
[2] D. Heck et al., Rep. FZKA 6019, Forschungszentrum Karlsruhe (1998)
[3] A. Fassò et al., CERN-2005-10, INFN/TC-05/11, SLAC-R-773 (2005)
[4] S.S. Ostapchenko, *Phys. Rev.* **D 74** (2006) 014026
[5] W.D. Apel et al. KASCADE-Grande Collaboration, *Astropart. Phys.* **36** (2012) 183
[6] O. Helene, *Nucl. Instr. Meth.* **212** (1983) 319
[7] W.D. Apel et al. KASCADE-Grande Collaboration, *Phys. Rev. Lett.* **107** 171104 (2011)171104
[8] G. Schatz et al. KASCADE Collaboration, Proc. 28th Int. Cosmic Ray Conf., Tsukuba, Japan (2003)
[9] M. Ahlers and K. Murase, arXiv:1309.4077v3 (2014)
[10] M. Aglietta et al., Astopart. Phys. **6** (1996) 71
[11] M.C.Chantell et al., *Phys. Rev. Lett.* **79** (1997) 1805
[12] Yu. A. Fomin, arXiv:1307.4988v2 (2013)

Measurement of electron-positron spectrum in high-energy cosmic rays in the PAMELA experiment

A V Karelin[12], O Adriani[1,2], G C Barbarino[3,4], G A Bazilevskaya[5], R Bellotti[6,7], M Boezio[8], E A Bogomolov[9], M Bongi[2], V Bonvicini[8], S Bottai[2], A Bruno[6,7], F Cafagna[7], D Campana[4], R Carbone[4,11], P Carlson[13], M Casolino[10,18], G Castellini[14], C De Donato[10,11], C De Santis[10,11], N De Simone[10,11], V Di Felice[10], V. Formato[8, 15], A M Galper[12], S V Koldashov[12], S A Koldobskiy[12], S Y Krutkov[9], A N Kvashnin[5], A A Leonov[12], A G Mayorov[12], V V Malakhov[12], L Marcelli[10], M. Martucci[10,17], W Menn[16], M. Merge'[10,11], V V Mikhailov[12], E Mocchiutti[8], A Monaco[7], N Mori[1,2], R. Munini[8,15], G Osteria[4], F Palma[10,11],B. Panico[4], P Papini[2], M Pearce[13], P Picozza[10,11], M Ricci[17], S B Ricciarini[2], R Sarkar[8], V. Scotti[3,4], L Rossetto[13], M Simon[16], R Sparvoli[10,11], P Spillantini[1,2], A Vacchi[8], E Vannuccini[2], G I Vasilyev[9], S A Voronov[12], Y T Yurkin[12], G Zampa[8], N Zampa[8]

[1]University of Florence, Department of Physics, I-50019 Sesto Fiorentino, Florence, Italy
[2] INFN, Sezione di Florence, I-50019 Sesto Fiorentino, Florence, Italy
[3] University of Naples "Federico II", Department of Physics, I-80126 Naples, Italy
[4] INFN, Sezione di Naples, I-80126 Naples, Italy
[5] Lebedev Physical Institute, RU-119991 Moscow, Russia
[6] University of Bari, Department of Physics, I-70126 Bari, Italy
[7] INFN, Sezione di Bari, I-70126 Bari, Italy
[8] INFN, Sezione di Trieste, I-34149 Trieste, Italy
[9] Ioffe Physical Technical Institute, RU-194021 St. Petersburg, Russia
[10] INFN, Sezione di Rome Tor Vergata, I-00133 Rome, Italy
[11] University of Rome Tor Vergata, Department of Physics, I-00133 Rome, Italy
[12] National research nuclear university MEPhI, RU-115409, Moscow, Russia
[13] KTH, Department of Physics, and the Oskar Klein Centre for Cosmoparticle Physics, AlbaNova University Centre, SE-10691 Stockholm, Sweden
[14] IFAC, I-50019 Sesto Fiorentino, Florence, Italy
[15] University of Trieste, Department of Physics, I-34147 Trieste, Italy
[16] Universitat Siegen, Department of Physics, D-57068 Siegen, Germany
[17] INFN, Laboratori Nazionali di Frascati, Via Enrico Fermi 40, I-00044 Frascati, Italy

E-mail: karelin@hotbox.ru

Abstract. At present the existing data on the cosmic ray electron energy spectra in the high energy range are fragmented, and the situation is exacerbated by their small number. In the satellite PAMELA experiment measurements at high energies are carried out by the calorimeter. The experimental data accumulated for more than 8 years of measurements, with the information of the calorimeter, the neutron detector and the scintillation counters made it possible to obtain the total spectrum of high-energy electrons and positrons in energy range 0.3-3 TeV.

1. Introduction

Study of the cosmic ray energy spectrum, obtained in direct measurements on board satellites, is one of the priority tasks in modern astrophysics.

Fluxes of cosmic electrons with energies greater than 10 GeV contain important information about the sources of their generation and the processes that occurred during their propagation inside the Galaxy. Due to significant loss in energy during propagation, unlike with protons and helium nuclei, electrons cannot propagate far from their sources. To date, measurement of electron and positron fluxes are carried out only in three satellite experiments. They are PAMELA [1], FERMI [2], the AMS-02 [3]. In previous years, in ground-based experiments measuring electron energy spectra a dramatic change in the spectrum at energies above a few hundreds of GeV was detected. It was associated with a significant decrease of electron-positron number in this energy range of the spectrum [4, 5]. However, the results obtained in the FERMI [6] and AMS-02 [7] satellite experiments, do not indicate a significant spectrum break at these energies.

Thus, the use of the calorimeter in the PAMELA experiment, providing the capability to achieve a higher energy limit, helps to clarify the current situation with the previous and current observations.

2. The PAMELA experiment

The PAMELA magnetic spectrometer is designed for high-precision measurements of the particle spectra in a wide energy range. The apparatus consists of the following detectors: scintillation time-of-flight system, TOF, magnetic spectrometer, anticoincidence system, electromagnetic calorimeter, shower scintillation detector C4 and neutron detector. The geometric factor of the PAMELA apparatus defined by the magnetic spectrometer aperture is 21.6 cm^2sr. A detailed description of both the entire PAMELA apparatus as a whole and of its separate detectors and measurement conditions is given in [1].

To increase the geometric factor and improve the statistical reliability of the results a combination of the calorimeter and the C4 detector were used. Let us consider the design of the calorimeter in more detail.

Calorimeter [8] consists of 44 (x, y) layers of silicon strip detectors, between which 22 tungsten layers ($Z = 74$, $A = 183.85$, $\rho = 18.1$ g / cm 3, $X0 = 0.3735$ cm) each 2.6 mm thick are located. Each detector layer includes a set of 96 strips with a pitch of 2.2 mm. It allows measuring particle coordinates in two planes and thus obtaining spatial distribution of secondary particle cascade developed in the calorimeter. The total thickness of the calorimeter is 0.6 of nuclear interaction length or 17 radiation lengths.

3. Measurement of electron-positron spectrum

A set of data stored in the memory while registering the particles at any trigger action of the PAMELA apparatus is called an event. Calorimeter and C4 triggers run when only a certain value of the energy is released inside the calorimeter or C4, which corresponds to the occurrence of a secondary particle cascade in the calorimeter by particles interacting within the apparatus, mainly that of the calorimeter.

The dominant background consideration for charge one particles are protons. Given that protons vastly outnumber electrons in cosmic rays, the task is to adequately separate electrons from protons, and even more challenging positrons from protons.

The contamination by nuclei in the selecting events was suppressed using scintillation detectors, when events with a particle charge more than one were selected by ionization losses. The contamination by these particles appeared to be less than 1%.

All selection conditions described below and their efficiencies were tested and optimized by Monte-Carlo simulation, beam-test data and by inter-comparing with the magnetic spectrometer data where it was possible.

The initial event selection criterion for the particle spectrum construction was exceeding of some value of the total energy release in the calorimeter that is higher than a trigger threshold. This criterion made possible the separation the high-energy particles (more than tens of GeV).

Next event selection criterion was a choice of the events with well measured direction of the primary particle along the shower axis in the calorimeter. Direction measurement in the calorimeter is possible for particles coming at an angle of not more than 75° to the vertical. To determine the shower axis direction an iterative approach was used, which is based on the least squares fit method, and yet for calculating the shower axis coordinates only those strips are used that are the closest to the shower axis, that considerably improves angular resolution [9].

Events that do not cross the calorimeter at angles greater than 15 degrees to the vertical were selected.

One of the parameters used for the selection of electrons associated with the transverse shower development is the "RMS root mean square deviation" parameter. It is an expression of the standard deviation of energy release at a certain distance from the shower axis from the energy value on the shower axis.

Besides information about the transverse shower development in the PAMELA calorimeter, information about its longitudinal profile was also used for stronger proton suppression.

Additionally, one of the criteria, that made it possible to highlight the electron component, was energy release in the region along the shower core. So the information only from strips nearest to shower axis was used while for the RMS transverse selection the information from all strips in whole plane was used.

However, events selected as a result of using the above criteria, contained not only the electrons and positrons, but also high fraction of background protons because of their, as it was mentioned above, vast number in the cosmic rays.

In order to restore the primary spectrum of electrons and positrons, it was necessary to estimate the fraction of background protons in each energy interval, which the range of measurement was divided into. This was done on the basis of cosmic-ray proton spectrum [10] experimentally measured with the PAMELA calorimeter and data on the proton fraction remaining after the selection, obtained in the simulation.

Thus, the number of protons Np in each energy interval, which the range of the electrons and positrons spectrum measurement was divided into with the PAMELA calorimeter, was calculated according to the following expression:

$$N_p(E_1 \div E_2) = \int_{E_1}^{E_2} F(E) G \varepsilon(E) \Delta t dE$$

where E1, E2 are boundary values of the proton energy intervals at which proton contribution into the given energy range of electrons spectrum cannot be considered negligible (calculated from the simulation), F (E) is a differential energy proton spectrum, taken from [10], G is geometric factor that at high energies of the particles does not depend on the energy, ε (E) is relative proton number left after the selection estimated in the simulation and Δt is time of measurement.

The value of ε (E) obtained from simulation is verified by coincidence of the proton calorimeter spectrum, that had been obtained be means of simulation, and the results of proton spectrum measurements in the similar energy range by other experiments (see for example [11, 12]).

In the last highest energy intervals proton fraction exceeded 75%, and to increase reliability of the results obtained from electrons and positrons spectrum additional electron selection with the help of neutron detector was carried out.

The threshold cut for neutrons equals to 8 helped eliminate the proton background. If the number of neutrons less than 8 it is electron or proton with some probability, but if number of neutrons is large than 8 it is with high probability proton. The neutron detector rejection was applied for last three energy bins and it reduced proton contamination there to the level of 40% while it was not applied for two first energy bins where the level of proton contamination was 15 and 40 %, respectively.

Geometric factor and events selection efficiency were calculated by simulation. Energy was determined by measuring calorimeter total energy release. For the energy range 1500 - 3000 GeV, events satisfying to selection criteria were not found, and only upper limit for the total electron and positron flux was set.

Figure 1 and 2 show results obtained for electron-positron spectrum in the energy range of 300-3000 GeV in comparison with the experimental results of other authors (ATIC[13], AMS-02 [7], FERMI [6], HESS [4], Kobayashi [5]) and with the PAMELA magnetic spectrometer measurement [14]. Errors are related to the statistical uncertainty only.

Taking into account the uncertainty in the energy measurement amounting to 15% in the HESS experiment, it can be concluded that the results of the PAMELA calorimeter are closer to the ground based experiment results of HESS and Kobayashi [5] than to the satellite ones: FERMI [6] and AMS-02 [7]. Thus the new PAMELA results show a clear spectrum drop near 300-400 GeV, while the other two satellite experiments did not indicate the existence of a sharp spectrum change.

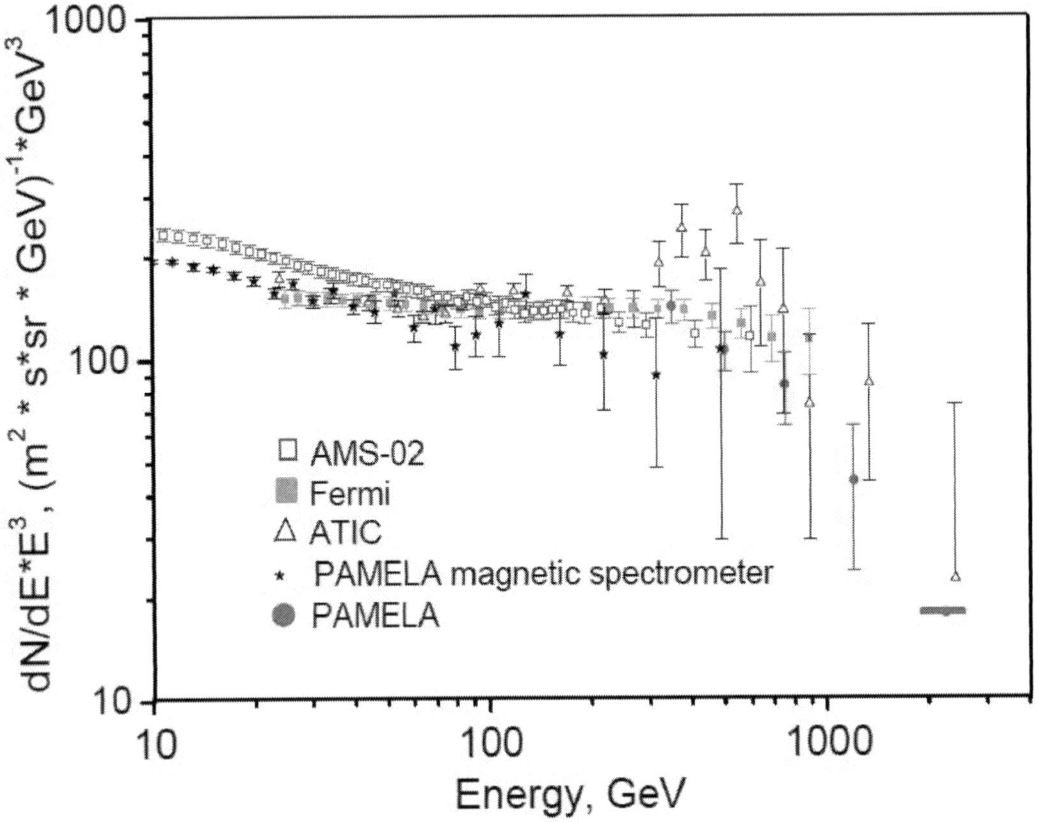

Fig.1 The obtained electron-positron spectrum in compare with direct measurements.

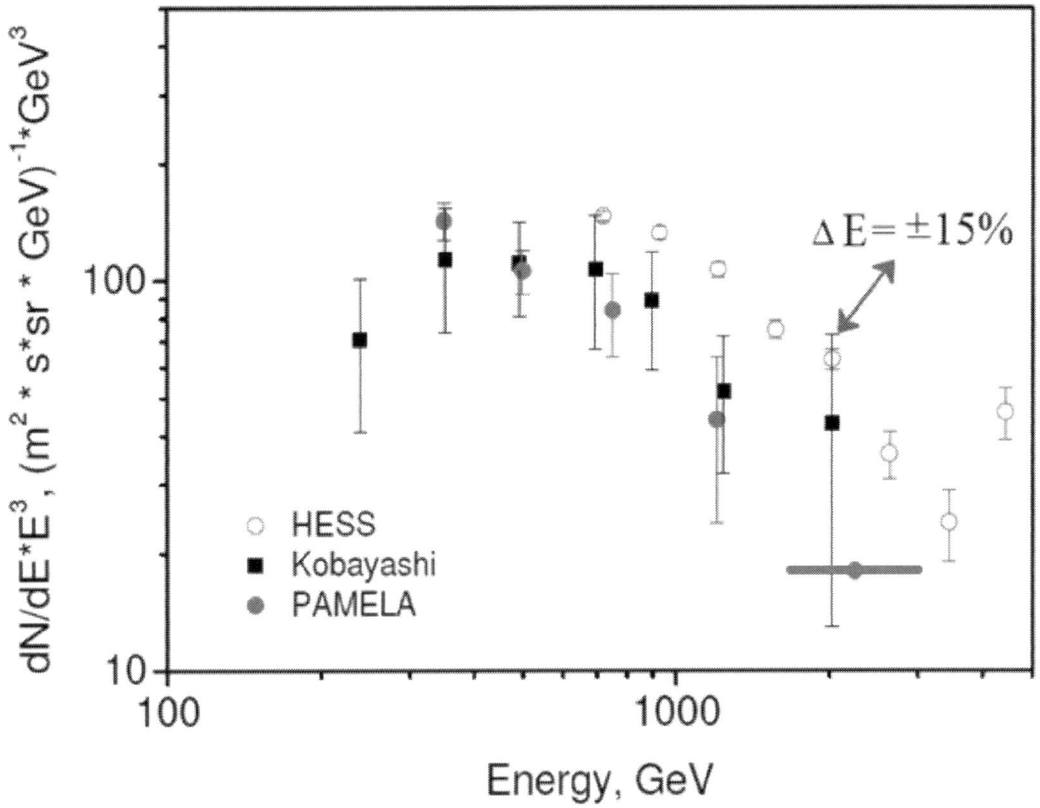

Fig.2 The obtained electron-positron spectrum in compare with ground-based measurements. The 15% for HESS measurements represents an upper limit for possible systematic energy shift.

4. Conclusion

New data obtained by the PAMELA calorimeter allowed comparison of results of the positron-electron spectra measurements with other ones in the high energy range. For the total positron-electron spectrum a sharp cutoff mentioned by HESS and Kobayashi is verified.

5. Acknowledgments

We are grateful to the Russian Science Foundation (grant no. 14-12-00373) and the President of the Russian Federation (grant no. MK-4599.2014.2) for support.

References

[1] Picozza P. et al. 2007 *Astroparticle Physics* **27** 296
[2] Abdo A. A. et al. 2009 *Phys. Rev. Lett.* **102** 181101
[3] Palmonari F. et al. 2011 *Journal of Physics: Conference Series* **335** 012 066
[4] Aharonian F. et al. 2008 Phys. Rev. Lett. **101** 261104
[5] Kobayashi et al. 1999, Proc. 26th Int. Cosmic-Ray Conf. (Salt Lake City), 61
[6] Ackermann, M. et al. 2010 *Phys. Rev. D* **82** 092004
[7] Bertucci B 2013 Precision measurement of the electron plus positron spectrum with AMS, Proc. of the 33rd Int. Cosmic Ray Conf.
[8] Boezio M. et al. 2002 *Nucl. Instrum. Meth. A.* **487** 407
[9] Borisov S.V., Voronov S.A., Galper A.M., Karelin A.V. 2013 *IET* **1** 5

[10] Adriani O. et al.2013 *Adv. Sp. Res.* 2013 **51** 219
[11] Panov A.D. et al. 2009 *Bulletin of the Russian Academy of Sciences: Physics,* **73** 564
[12] Ahn H.S. et al. 2010 *ApJ Lett* **714** 89
[13] Chang J. et al.2008 *Nature* **456** 362
[14] Adriani O. et al. 2011 *Phys. Rev. Lett.* **106** 201101–01–05

Measuring module of the Cherenkov water detector NEVOD

V V Kindin, M B Amelchakov, N S Barbashina, A G Bogdanov, V D Burtsev, D V Chernov, S S Khokhlov, V A Khomyakov, R P Kokoulin, K G Kompaniets, E A Kovylyaeva, V S Kruglikova, V V Ovchinnikov, A A Petrukhin, I A Shulzhenko, V V Shutenko, I I Yashin, E A Zadeba

National Research Nuclear University MEPhI
(Moscow Engineering Physics Institute), 115409 Moscow, Russia

E-mail: VVKindin@mephi.ru

Abstract. Quasispherical Module (QSM) of Cherenkov water detector NEVOD represents six low-noise FEU-200 photomultipliers with flat photocathodes (15 cm in diameter), oriented along the axes of orthogonal coordinate system. Such configuration allows to register Cherenkov radiation arriving from any direction with almost equal efficiency. The results of measurements of QSM characteristics in the sensitive volume of the NEVOD detector during the registration of Cherenkov radiation of single muons at different distances and angles are discussed.

Introduction

Cherenkov water detectors (CWD) are one of the most perspective instruments for ultrahigh energy cosmic rays study. For their creation, water reservoirs of natural or artificial origin are used. Inside the water reservoir the spatial lattice of optical modules (OM) is located. In optical modules for the detection of Cherenkov radiation, the photomultipliers with large area photocathode are used. Usually OM consists of a glass sphere, one photomultiplier tube with hemispherical photocathode and internal electronics. The spatial lattice is formed of vertical strings in which OMs are located at distances of several tens of meters and have a preferred direction: upwards and downwards as in the NT-200+ [1] setup or just downwards as in ANTARES [2] and IceCube [3]. Sometimes OMs are combined in clusters of two or three. Disadvantage of such optical modules is that they have a dead zone from which the Cherenkov radiation is not recorded. The solution to this problem could be the creation of a spherical photocathode photomultiplier which response is independent on the angle of incidence of Cherenkov radiation. But creation of such PMT is related with solution of serious technical problems. Therefore, PMTs with hemispherical photocathode are widely used. Particle direction in such detectors is determined by the Cherenkov radiation arrival time. Amplitude of the OM response signals is used only as additional information. The presence of selectivity in choosing the OM registration direction does not allow to create 4π-detectors that register Cherenkov radiation from any direction with the same efficiency.

Another approach to construction of OM for 4π-detectors which is based on the usage of flat photocathode photomultipliers arranged symmetrically was proposed for the first time on the 16th ICRC in 1979 [4]. The amplitude response of such PMT for relativistic charged particle depends on the angle of incidence of the Cherenkov light to the photocathode and can be calculated using the following formula:

$$A_1(R,\alpha) = \frac{S_{PMT} \cdot \cos\alpha}{2\pi(R + r_{PMT}) \cdot \sin\theta_C} \int_{\lambda_{min}}^{\lambda_{max}} \frac{dN}{d\lambda} \eta(\lambda) \cdot \exp(-\frac{R}{L(\lambda) \cdot \sin\theta_C}) d\lambda\,, \tag{1}$$

where S_{PMT} and r_{PMT} are photocathode area and radius respectively; cosα is the cosine of the angle of incidence of the Cherenkov light on the photocathode; R is a distance from the PMT photocathode center to the particle track; $dN/d\lambda$ is distribution of Cherenkov photons in wavelengths; θ_C - Cherenkov radiation angle; $\eta(\lambda)$ is PMT quantum efficiency; $L(\lambda)$ is absorption length of light in water, λ_{min} and λ_{max} are boundaries of the PMT sensitivity range.

Simple OM, which has a response of a spherical PMT can be made of six photomultipliers arranged symmetrically along the axes of orthogonal coordinate system (in the cube faces). According to the formula (1), the response of such OM is determined by the approximate expression:

$$B = \sqrt{\sum_{i=1}^{3} A_i^2} = \frac{C}{R} \times \exp(-R / l \times \sin\theta_C) \times \sqrt{\sum_{i=1}^{3} \cos^2\alpha_i}\,, \tag{2}$$

where C is a constant; l is some effective absorption length; cosα is the cosine of the angle between the normal to the plane of the photocathode and the arrival direction of the Cherenkov radiation; R is a distance from the center of the module to the particle track. Since for the cube $\sum_{i=1}^{3} \cos^2\alpha_i = 1$, the OM response will not depend on direction of Cherenkov light for distances more than about two sizes of the module. Therefore, these OM are called quasispherical (QSM).

An important advantage of quasispherical modules in comparison with spherical PMTs is that this module allows to determine the arrival direction of Cherenkov radiation using the ratio of the amplitudes of triggered photomultipliers. This principle of registration was implemented into quasispherical module (QSM) of the NEVOD detector.

1. Quasispherical module of the CWD NEVOD

For the registration of main components of cosmic rays on the Earth's surface, including neutrinos from the lower hemisphere, a multi-purpose Cherenkov water detector NEVOD [5] with a volume of 2000 m^3 was created in MEPhI. Inside the detector, the spatial lattice of quasispherical measuring modules is located. QSM consists of a waterproof metal housing, internal electronics and six FEU-200 photomultipliers with flat photocathodes (15 cm in diameter) with plastic protective illuminators. Six photomultipliers are oriented along the axes of orthogonal coordinate system (figure 1).

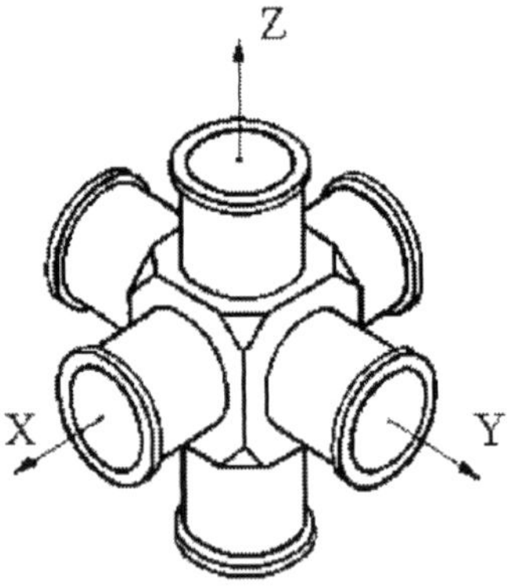

Figure 1. Quasispherical module.

The spatial lattice of such measuring modules has the property of 4π-detector and allows to reconstruct the parameters of the particle tracks on the basis of the triggered QSM amplitudes analysis only. In QSM, a low-noise FEU-200 photomultiplier (production of "JSC Ekran-optical systems") is used. Photomultiplier has a translucent antimony-potassium-cesium photocathode with maximum spectral characteristics in the range of 410-430 nm, electrostatic electron focusing system and 12-dynode louver type multiplier system. To obtain a large dynamic range of the recorded signals from 1 to 10^5 photoelectrons, the readout of signals from the 12th and 9th dynodes is used. In this case spectrometric channel of the 12th dynode allows to to measure signals in the range of 1 – 1000 ph.e. and the channel of the 9th dynode in the range of 10^2-10^5 ph.e. Internal electronics of the module provides detection of Cherenkov radiation by the photomultipliers in the regular measurement mode and the procedure of operation monitoring. Quasispherical modules are combined into strings (clusters) of 3-4 in each. Three strings of 3 QSMs or 4 strings of 4 QSMs form the plane. The current configuration of CWD NEVOD includes 4 planes of 16 QSMs and 3 planes of 9 QSMs (91 QSM in total). For the calibration of CWD spectrometric channels during long experimental series, the System of Calibration Telescopes (SCT) and the coordinate detector DECOR [6] are used. SCT consists of 80 scintillation detectors (figure 2). Forty detectors are located on the top and forty on bottom of the pool (upper and lower planes). Any pair of detectors (one detector from the upper and one from the lower planes) forms a muon telescope. Telescopes allocate single muons in the range of zenith angles from 0° to 45° with an accuracy of ~ 2°.

DECOR (figure 2) is the world's first large-scale coordinate detector specially designed for the study of cosmic radiation on the Earth's surface at large zenith angles up to the horizon. Detector consists of eight vertically arranged supermodules with total area of ~ 70 m^2, which ensure spatial and angular accuracy of the muon track reconstruction better than 1 cm and about 0.7°, respectively.

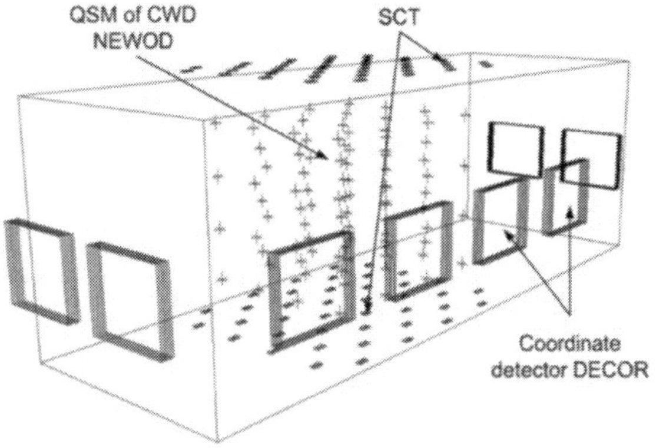

Figure 2. The layout of SCT system and DECOR detector around the CWD NEVOD.

2. Response of the QSM to the single muons

The first characteristic of the QSM is the PMT amplitude response at single-particle events in the sensitive volume of the detector. The selection of events for the PMT calibration is performed by SCT telescopes. High positioning accuracy of the allocated events allows to obtain the dependence of the PMT amplitude response on the distance to the particle tracks. An example of the event is shown in figure 3.

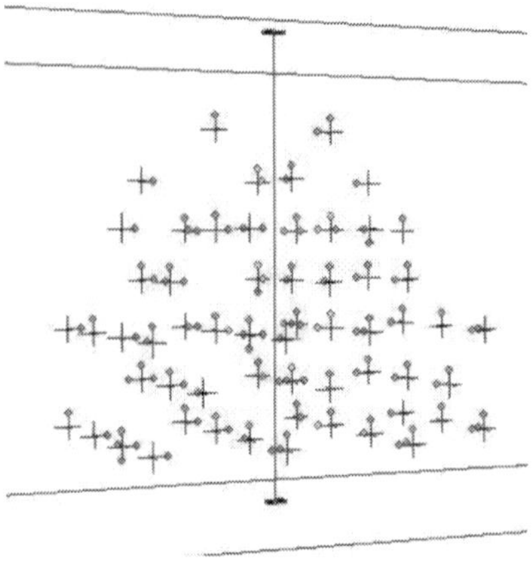

Figure 3. An example of CWD response to a vertical muon. 59 triggered QSMs, total amplitude ~ 355 ph.e., the number of upward oriented PMTs is 45, the number of downward oriented PMTs is 5, PMT registration threshold is 0.25 ph.e.

The responses of two PMTs (amplitude spectra) to the vertical muons selected by the SCT at a fixed distances R = 0.72 m and 1.0 m, and angles of incidence of the Cherenkov light 48.8° and 41.2° from the side PMT photocathode and the top PMT photocathode are shown in figures 4 and 5.

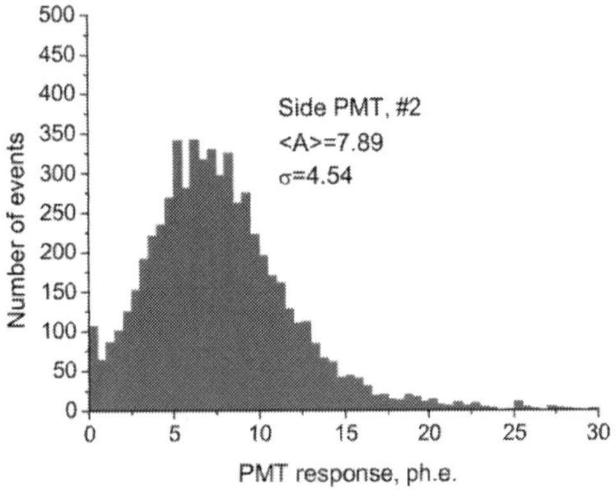

Figure 4. Amplitude responses of the side PMT of the central QSM to single muons selected by the SCT telescope.

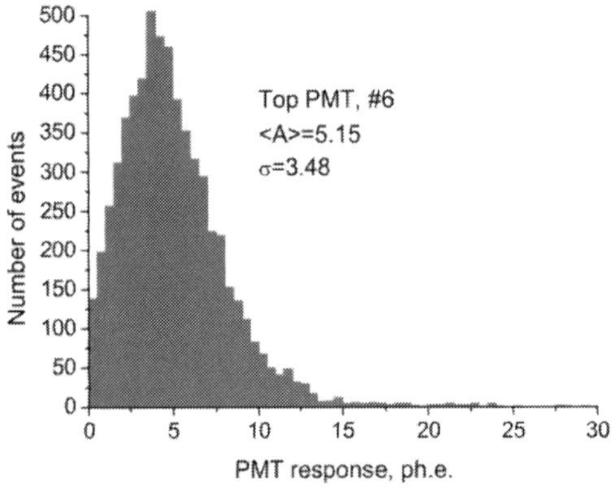

Figure 5. Amplitude responses of the top PMT of the central QSM to single muons selected by the SCT telescope.

The second characteristics of the QSM is the total QSM responses as a function of on the distance to the particle track. To obtain the dependence of the QSM response on the distance, the events with single near-horizontal tracks allocated by the DECOR supermodules were selected. An example of the event with the near-horizontal muon allocated by the DECOR supermodules is shown in figure 6.

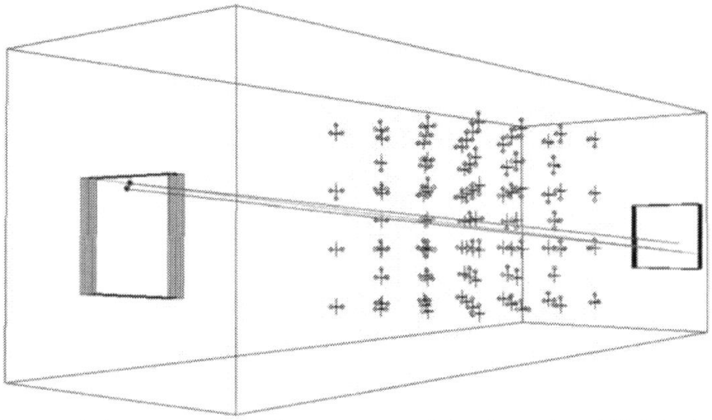

Figure 6. An example of the response of the detecting QSM lattice for the near-horizontal muon. 91 triggered QSM, total amplitude ~ 1260 ph.e.

Figure 7 shows the dependence of the average response $B=\sqrt{\sum_{i=1}^{3} A_i^2}$ of three PMTs (photomultipliers that register direct Cherenkov light) on the distance from the center of QSM to the particle track.

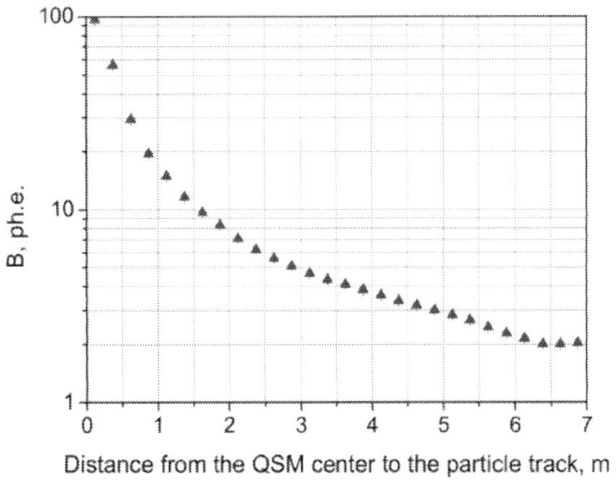

Figure 7. The dependence of the average amplitude response B on the distance from the QSM center to the particle track, registration threshold 0.25 ph.e.

For evaluation of accuracy of muon track reconstruction particles of near horizontal flux were used. The developed technique of such reconstruction is described in paper [7]. The average accuracy is better than 6.5°. Obtained results of PMTs and QSMs calibrations can be used at reconstruction of real events detected by NEVOD-DECOR complex. For evaluation of accuracy of cascade shower reconstruction the events generated by near-horizontal muons were used, too. Method of reconstruction of the cascade curves [8] in the CWD is based on the recalculation of the response of each PMT, which "sees" the track, into the number of emitting relativistic particles on the track. Estimates of the direction and geometrical position of the cascade axis based on the analysis response amplitudes of the CWD PMTs give the following results: better than 16° for the direction and about 30 cm for the axis position.

Conclusion

The main advantage of quasispherical modules compared to hemispherical PMTs is a possibility to measure Cherenkov light direction by the amplitude analysis only. This opens a way to the use in Cherenkov water detectors of PMTs with flat photocathodes, inexpensive compared to hemispherical ones.

Acknowledgments

The research has been performed in Scientific and Educational Centre NEVOD with the financial support of the Russian Ministry of Education and Science (government task and project RFMEFI59114X0002) and grant of the Leading Scientific School NSh-4930.2014.2.

References

[1] Aynutdinov V M *et al* 2006 *Nucl. Instr. and Meth.* vol 567 p 433
[2] Ageron M *et al* 2011 arXiv:1104.1607v1 [astro-ph.IM]
[3] Halzen F and Klein S 2010 *Rev. Sci. Instrum.* **81**, 081101
[4] Borog V V *et al* 1979 *Proc. 16th ICRC (Kyoto)* vol 10 p 380
[5] Aynutdinov V M *et al* 1997 *Astrophysics and Space Science* **258** 105
[6] Barbashina N S *et al* 2000 *Instrum. Exp. Tech.* **43** 743
[7] Kovylyaeva E A *et al* 2013 *J. Phys.: Conf. Ser.* **409** 012132
[8] Khokhlov S S *et al* 2014 *Proc 24th ECRS (Kiel)* Submitted to *J. Phys.: Conf. Series*

GEANT 4 simulation of the Helios cosmic ray telescope E6: Feasibility of chemical composition studies

J Marquardt, B Heber, M Hörlöck, P Kühl, R F Wimmer-Schweingruber
Christian-Albrechts-Universität zu Kiel

Abstract. In October 2011, ESA announced the selection of Solar Orbiter as one of the Cosmic Vision M missions. It's launch is foreseen in 2018. Therefore it is worthwhile to revisit the energetic particle measurements together with the solar wind and magnetic field data from Helios in light of better theoretical understanding and advanced analysis and modelling techniques developed during the past 20 years. In this contribution we will present a GEANT 4 simulation of the response function of the Helios E6 experiment, that measured electrons in the energy range from a few 100 keV to above 10 MeV, ions from 1 MeV/nucleon to above 50 MeV/nucleon, and its application to the data analysis with respect to the chemical composition of the Galactic Cosmic Rays.

1. Introduction

The elemental composition of the cosmic rays in the heliosphere has been measured a number of times during different phases in the solar cycle [1; 2] and different radial distances from the Sun beyond the Earth orbit[e.g. 3; 4].

One of these measurements was carried out by the Experiment 6 (E6) on Helios. Helios 1 and Helios 2 were launched on December 10, 1974 and January 15, 1976, respectively. The two almost identical space probes were sent into ecliptic orbits around the Sun. The orbital period around the Sun was 190 days for Helios 1 and 185 days for Helios 2. Their perihelia were 0.3095 AU and 0.290 AU, respectively.

The E6 particle telescope relies on the $\frac{dE}{dx} - E$-method [see e.g. 5]. In order to interpret the measured data a detailed understanding of the instrument is needed and can be obtained by modelling the physical processes inside the detector taking into account the instrument geometry as well as environment. [e.g. 6].

The instrument consists of a stack of 5 semiconductor, a sapphire and a plastic scintillation detector (see Fig. 1), which measures the number of particles hitting each detector and their energy losses. A rough particle identification (species, energy and incoming direction) can be obtained by analyzing the count rate data that are obtained in (anti)-coincidence of a set of active detectors [see 7]. For a statistical ensemble not only the (anti-)coincidence conditions but also the energy loss in the last three detectors are known. The channel that is analyzed here in detail counts particles in the energy range from ~13 MeV/nucleon to ~27 MeV/nucleon protons and helium.

In order to determine the chemical composition of cosmic rays, the detailed instrument response for different particle species is calculated utilizing a newly developed model of the E6 instrument. The calculations are based on the program library Geant4 that is a toolkit for the simulation of the passage of particles through matter [8]. The purpose of this contribution is to show that the Helios E6-Experiment is capable of determining the chemical composition of particles that are stopped in the detector sensor up to

Table 1. Relative Elemental Abundances at 160 MeV/Nucleon from ACE/CRIS [2; 9]

Element	Solar Minimum 1997/1998 [2]	Solar Minimum 2009-2010 [9] at 160 MeV/nucleon	Solar Maximum 1998-2001 [2]
B	1803.8 ± 10.4	1725.7 ± 19.4	1986.4 ± 11.3
C	7337.0 ± 18.4	7235.4 ± 45	6780.2 ± 18.4
N	1713.7 ± 8.4	1678.9 ± 12.3	1836.1 ± 9
O	7082.6 ± 16	7137.0 ± 42.7	6520.6 ± 15.6
Ne	998.7 ± 5.6	998.9 ± 8.4	1050 ± 5.8
Mg	1368 ± 6.1	1375.3 ± 10.3	1367.3 ± 6
Si	1000 ± 5	1000 ± 7.8	1000 ± 4.8

silicon. The chemical abundances of some light elements at 1 AU known from literature are summarized in Tab. 1.

2. Instrumentation

The Kiel experiment E6 for studying energetic cosmic rays was built as an universal detector. It was supposed to measure most effects of the cosmic radiation which occur in interplanetary space. The energy range extends from 1.3 MeV/n to > 1000 MeV/n for nuclei and from approximately 0.3 to 8 MeV for electrons. The on-board data processing system evaluates the measured pulses created during the particle transition.

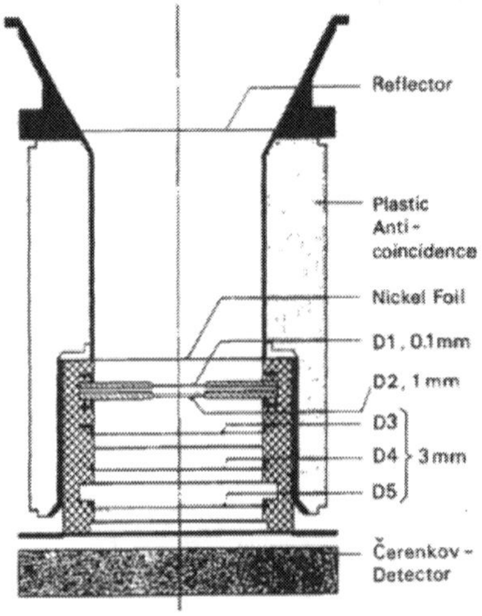

Figure 1. Schematic of the Helios E6

The detector system consists of five semiconductor-detectors of increasing thickness that are shown in Fig. 1. The figure also indicates the thicknesses of the detectors. While detectors D1(A) and D2(B) are silicon-surface-barrier-detectors with a thickness of 100 and 1000 μm, the other three are lithium drifted detectors with a thickness of 3000 μm. The first two (D1 and D2) are used for the determination of the lowest energy channels, for the definition of the geometry factor for stopping particles (energy ranges below 51 MeV/N) as well as for the discrimination between electrons and nuclei. The first detector does not respond to relativistic electrons (above 300 keV), the discrimination between electrons and nuclei is that the first detector is not triggered for electrons. To avoid false identifications by these discrimination conditions, the first and second detectors have been placed on top of each other, as close as possible. Charged particles, which penetrate the fifth detector and the aluminum absorber beneath, are detected in the Sapphire-Cerenkov-detector. The Cerenkov-threshold for this material (n=1.8) is E_s = 210 MeV/nucleon. Because the Sapphire-Cerenkov-detector also delivers scintillation light, particles in the energy range above 51 MeV/nucleon are counted in an integral channel.

3. Data Analysis

In order to determine the chemical and the isotopic abundance together with the energy spectra of each species it is common to design a detector telescope based on a stack of detectors that are surrounded by an anticoincidence detector. The characteristics for stopping particles can then be determined by the $\frac{\Delta E}{\Delta x} - E$-method [e.g., McDonald and Ludwig 5]. The principle of this method shall be explained by means of the left panel of Fig. 2. A charged particle with the energy E_0 penetrates the thin first detector

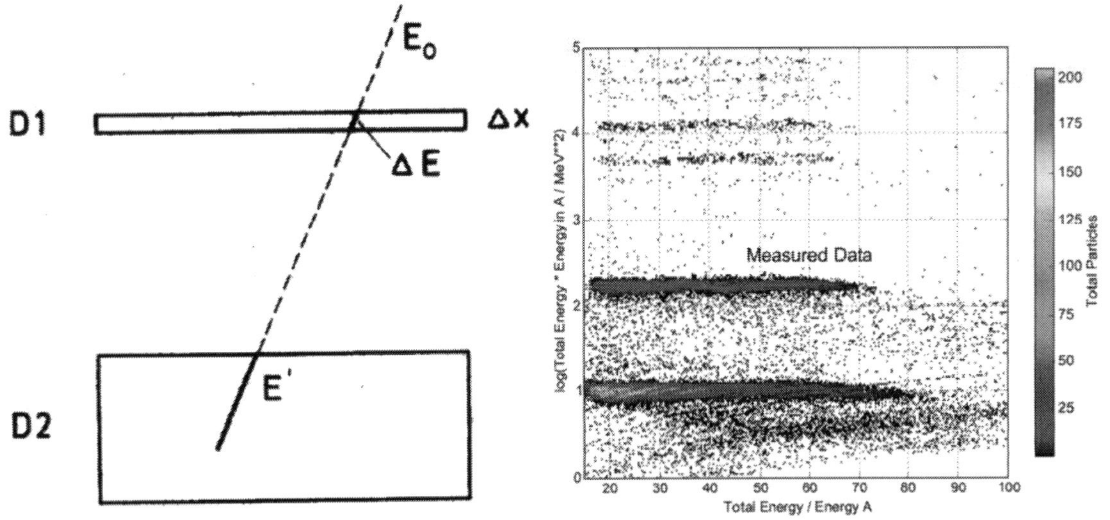

Figure 2. Left: Sample particle trajectory in a dE/dx-E-telescope. Right: $\Delta E \cdot E_0$ versus $E_0/\Delta E$-distribution for (from the bottom to the top) protons, helium, C, N, O, Ne and Si using the energy loss ΔE of the first detector and the sum of the measured energy losses in the first three detectors as E_0. The entries below the curves are caused by the non uniformity of first detector (for details see text).

D1 and stops in the thick detector D2. In the first detector, with a thickness of Δx, the particle loses the energy ΔE by ionization. The residual energy E' is deposited in the second detector. The entire kinetic energy E_0 results from the sum of ΔE and E', the energy losses in both detectors. From the Bethe-Bloch relation [5], the energy loss in both detectors can be determined for a known detector material and for particles of a known species. If $\Delta E = dE/dx \cdot \delta x \ll E'$ and for non-relativistic particles of mass M with $E = \frac{1}{2}Mv^2$ we obtain the following relation:

$$\Delta E \cdot E_0 \propto Z^2 \cdot M = const. \tag{1}$$

This constitutes a typical quantity for a particle-species. Since Z and M can only take discrete values, the relation offers a possibility to separate various particles, i.e., in addition to their energy, to determine their mass and nuclear charge. However, note that ΔE may vary due to the statistical variation of the ionization and because of the different path-lengths, Δx, due to the opening angle of the telescope. Thus in order to separate different isotopes from each other the path-length variations have to be minimized.

The concept of the $\Delta E \cdot E_0$ is best realized for particles that stop in the third detector. The ratio of the thin to thick detector is smaller than 1:10. Due to the limited telemetry, only the energy loss in the last three detectors was transmitted. Thus, for particles stopping deeper in the instrument, the corresponding ratios are 1:3 and 1:1 for particles that stop in detectors four and five, respectively. In what follows we concentrate on the measurements for ions that stop in the third detector.

Using in-flight measurements it has been become common to display the quantity $\Delta E \cdot E_0$ versus $E_0/\Delta E$. The total measured energy and the energy loss in the first detector increase with the charge number and mass. The right panel of Fig. 2 displays the quantity $\Delta E \cdot E_0$ versus $E_0/\Delta E$ using the energy loss ΔE in the first detector D1 and the total measured energy E from detector D1 to detector D3. Indeed the traces of the different elements are visible ranging from hydrogen to the CNO group and above. The entries below the curves are mainly caused by the fact that the charge collection of the first detector is non uniform from a certain radius outward - decreasing with increasing radius.

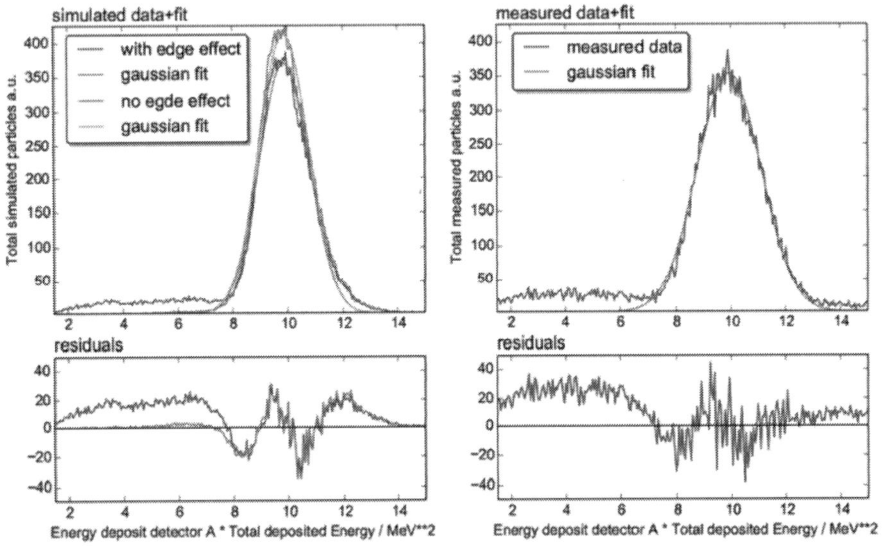

Figure 3. Measured and simulated $\Delta E \cdot E_0$ distribution of the proton peak in detector D1, demonstrating the measured and modelled edge effect.

In order to investigate this effect in more detail, the right hand panel of Fig 3 shows the measured $\Delta E \cdot E_0$-distribution close to the proton peak. The residuals to the left of the proton peak result from the so called "edge effect". Particles penetrating the detectors beyond their active region produce fewer charge carrier pairs because the electrical field does not drop immediately to zero. To confirm this, a GEANT4 (version 10.0) simulation utilizing the Bertini Cascade and the Quark Gluon string model was set up. The edge effect was modelled by multiplying a radius dependent efficiency to the signal. The corresponding results are displayed in the left panel of Fig. 3. From that figure it is evident that the model is capable to describe the edge effect in the first detector. Electronic effects like the amplification switch between high-gain and low-gain, responsible for the particle population between protons and helium, as well as noise were ignored in the simulation.

The chemical composition of galactic cosmic rays shows that a high signal to noise ratio is needed to determine the chemical composition for the CNO-group. Investigations from in-flight measurements of the second detector show an "edge effect" with a significant smaller impact than that of the first detector.

Fig. 4 displays in the upper panel on the left and on the right the measured and corrected $\Delta E \cdot E$ versus $E/\Delta E$ distribution utilizing the energy loss ΔE in the second detector. Corrections include excluding rear electrons and rear particles by comparing energy deposits in the first two detectors as well as temperature corrections. In contrast to Fig. 2 we find a deviation from a straight line at low values of $\frac{E}{\Delta E}$. This is due

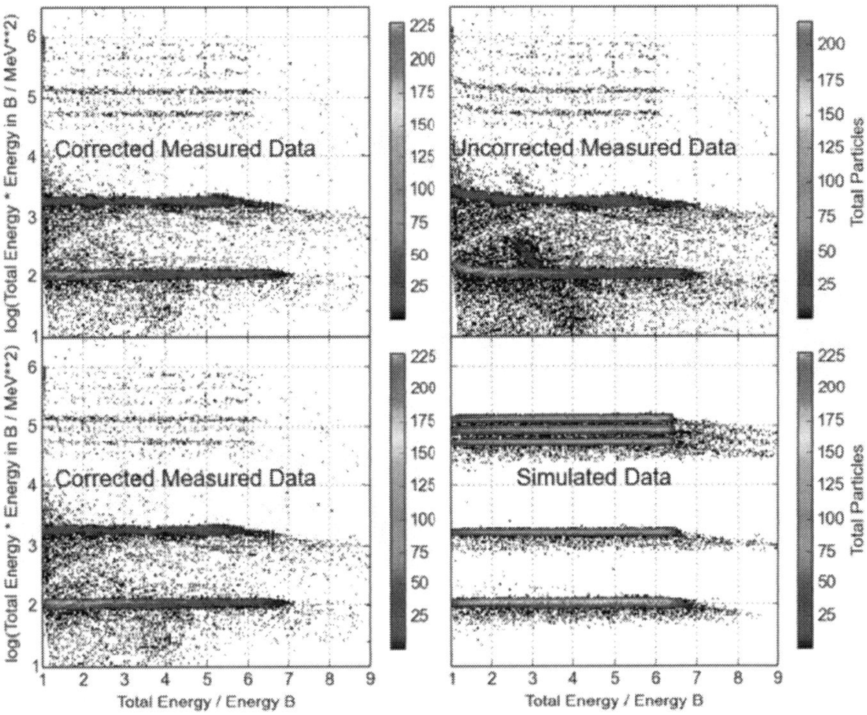

Figure 4. Upper panel uncorrected and corrected $\Delta E \cdot E$ vs. $\frac{E}{\delta E}$-measurements for particles that stop in the third detector utilizing the energy loss ΔE in the second detector. The lower panel displays the corresponding corrected simulated and measured matrix. Note that details can not be compared due to the different intensity energy spectra. For details see text.

to the fact that the energy loss is comparable to the total energy. However, that effect may be corrected by fitting a second order polynomial to the proton track. Multiplying the data by the corresponding inverse function leads to lines that are parallel to the $\frac{E}{\Delta E}$-axis. The lower row in Fig. 4 compares the simulation using an adapted model for the edge effects in the second detector and a flat energy spectrum with the measured and corrected distribution. The simulation was performed with a large number of particles in order to clearly show the tracks of protons, helium, carbon, nitrogen and oxygen. Comparing the simulated and measured corrected distributions we find a good agreement if we take into account that neither energy spectra, abundance, electrons nor and like ^{3}He have been taken into account.

4. Results

Energy spectra for galactic and if present, anomalous, protons, helium, carbon, nitrogen, and oxygen in the energy range from a few MeV/nucleon to several 100 MeV/nucleon were determined by Christian 1989, [10]. Fig. 5 displays 12 hour averaged count rates of ~13 to ~ 27 MeV/nucleon protons (black curve) and ions (red curve) from Helios 1 and Helios 2 from launch to September 1977, respectively. For our analysis only data with a proton count rate below 0.002 counts/second was used. The lower panel of Fig. 6 shows the simulated $\Delta E \cdot E$ distribution for helium and above. From that figure we find peaks that correspond to boron, carbon, nitrogen, oxygen, and neon. In order to determine the abundances, each peak was fitted with a corresponding Gaussian function. The corresponding lower panel shows the fit residuals which demonstrate the goodness of the fit. We find a typical width σ of 0.03 that allows us to separate between C, N , O, and Ne by more than 3σ. Note the small impact of the edge effect on the data between the fits compared to the strong edge effect on protons in detector A (Fig. 3). The

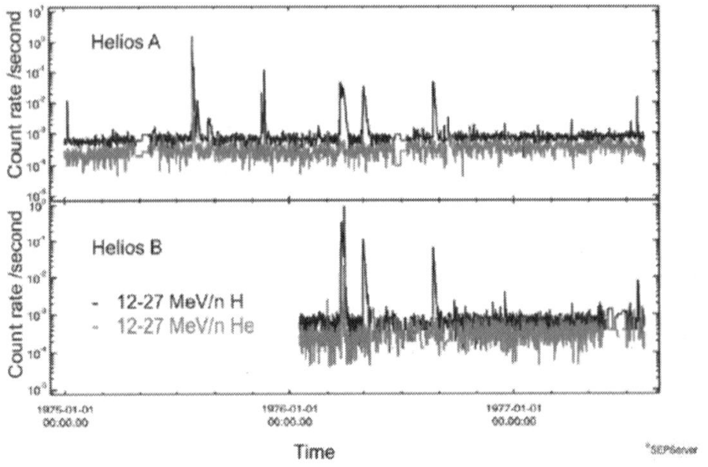

Figure 5. The upper and lower panel display the 12 hour averaged count rates of ~13 to ~ 27 MeV/nucleon protons (black curve) and ions (red curve) from Helios 1 and Helios 2 from launch to September 1977, respectively. For details see text.

Element	Position	σ	Total number	relative abundance (Si)
helium	3.27	0.03	11849	95500
boron	4.52	0.03	71	570
carbon	4.73	0.03	348	2790
nitrogen	4.94	0.03	102	820
oxygen	5.11	0.03	460	3700
neon	5.42	0.03	89	720
magnesium	5.66	0.03	119	960
silicon	5.86	0.03	124	1000

Table 2. Results of the fit using a Gaussion to the data in Fig.6. The position and σ give the position of the maximum and the width of the Gaussian in $\log(E_B \cdot E_{tot}/MeV^2)$. The total number is calculated by integration of the Gaussian function. The relative abundance is given with respect to silicon.

same was done with the measured data, leading to the results summarized in Tab. 2. In order to estimate the elemental separation of the instrument the figure displays the Gaussian fits for B, C, N, O, Ne, Mg, and Si. From this the background distribution can be analyzed further and the edge effect and possible further nuclei like ^{3}He can be investigated in more detail. The total number of counts varies between 70 for boron and 12000 for helium. Since abundances are given relative to silicon, the statistical accuracy cannot be better than 10% for helium and increases to more than 15% for boron. However, the relative abundances differ significantly from the ones given in Tab. 1. This can be explained mainly by the fact that in contrast to the work by [George et al. 2] the values are not normalized to a reference energy. While the channel analyzed here covers an energy range from 13 to 27 MeV/nucleon for helium, it measures heavier elements at much higher energies. Thus an extended analysis is needed that includes other energy channels from the Helios instrument in order to define a single energy range that is covered by all species.

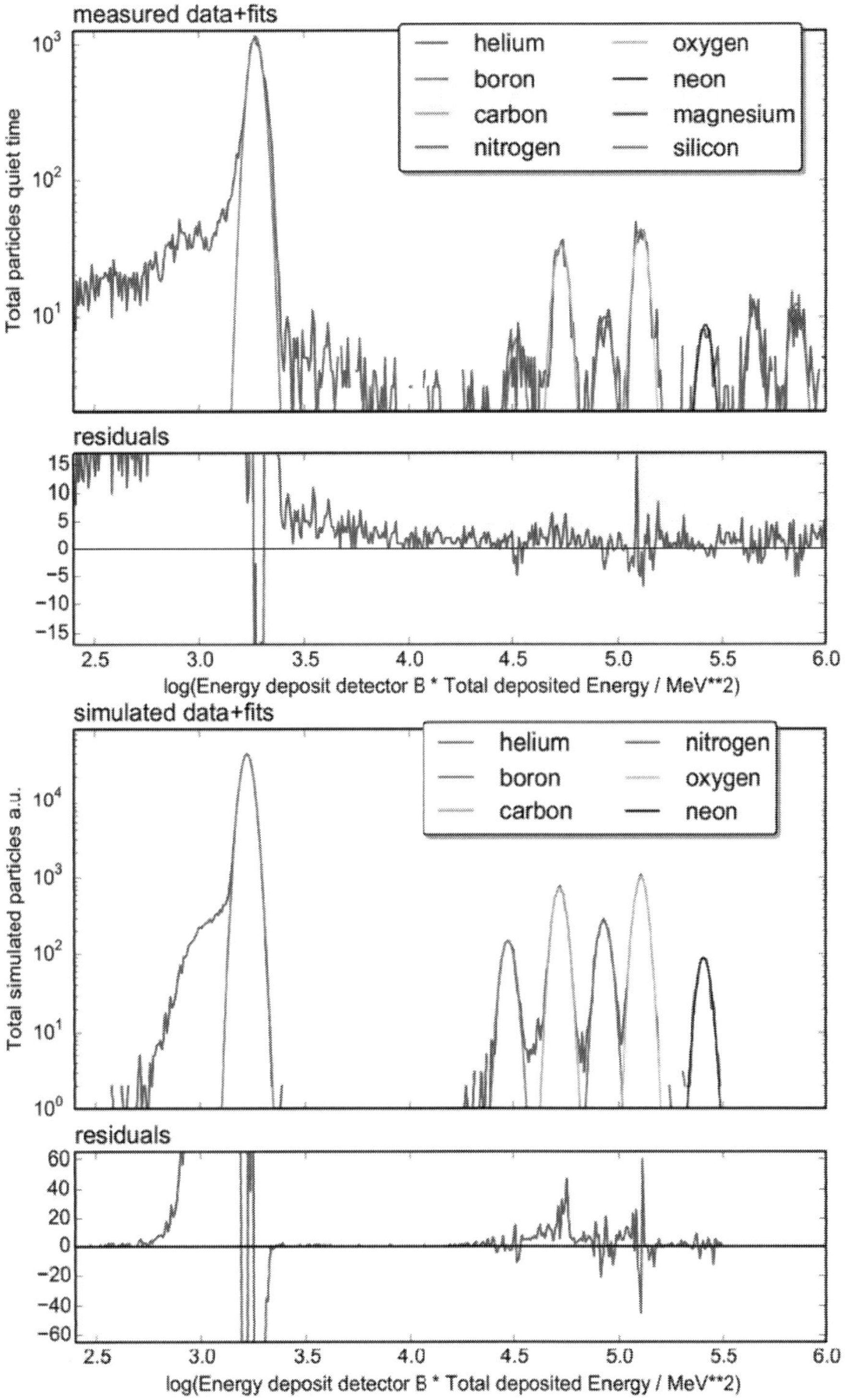

Figure 6. The blue curves show the left projection (1-dimensional representation) of the different measured elements and simulated distribution. The other colored curves are the result of fits by a Gaussian to the data. The bottom panels display the residuals of the fit, indicating the goodness of the fit. Note that the y-axis are on logarithmic and linear scale on the top and bottom, respectively.

5. Summary and future work

Here we analyzed in detail a channel that counts particles in the energy range from ~13 MeV/nucleon to ~27 MeV/nucleon for protons and helium. Using a modified $dE/dx - E$ method we could show that the instrument is capable of separating all major chemical elements from hydrogen to silicon. In order to determine their chemical composition a detailed instrument simulation for different particle species has been performed utilizing a newly developed model of the E6 instrument. This model includes the edge effect. Using a total quiet time span of four years measurements the instrument counted more than 10000 helium, about 350 carbon and 460 oxygen nuclei. This allows to investigate radial structures within the inner heliosphere. In order to determine the chemical composition at single energy intervals, other coincidence channels need to be evaluated and the corresponding response function calculated. The result of this analysis will allow us to investigate the chemical composition during the solar minimum from 1974 to 1977.

Acknowledgements

We acknowledge the use of the SEPServer data base: The SEPServer project has received funding from the European Community's Seventh Framework Programme (FP7/2007-2013) under grant agreement nr 262773.

Further we acknowledge the International Space Science Institute (ISSI) and the ISSI Team "Exploration of the inner Heliosphere what we have learned from Helios and what we want to study with Solar Orbiter".

References

[1] Engelmann J J, Ferrando P, Soutoul A, Goret P and Juliusson E 1990 *Astronomy and Astrophysics* **233** 96–111 1

[2] George J S, Lave K A, Wiedenbeck M E, Binns W R, Cummings A C, Davis A J, de Nolfo G A, Hink P L, Israel M H, Leske R A, Mewaldt R A, Scott L M, Stone E C, von Rosenvinge T T and Yanasak N E 2009 *Astrophysical Journal* **698** 1666–1681 1, 1, 4

[3] Ferrando P, Lal N, McDonald F B and Webber W R 1991 *Astronomy and Astrophysics* **247** 163–172 1

[4] Duvernois M A and Thayer M R 1996 *Astrophysical Journal* **465** 982 1

[5] McDonald F B and Ludwig G H 1964 *Physical Review Letters* **13** 783–785 1, 3, 3

[6] Heber B, Kopp A, Fichtner H and Ferreira S E S 2005 *Advances in Space Research* **35** 605–610 1

[7] Kunow H, Wibberenz G, Green G, Müller-Mellin R and Kallenrode M B 1991 *Physics of the Inner Heliosphere II. Series: Physics and Chemistry in Space* **21** 243–342 1

[8] Agostinelli S e 2003 *Nuclear Instruments and Methods in Physics Research Section A: Accelerators, Spectrometers, Detectors and Associated Equipment* **506** 250–303 1

[9] Lave K A, Wiedenbeck M E, Binns W R, Christian E R, Cummings A C, Davis A J, de Nolfo G A, Israel M H, Leske R A, Mewaldt R A, Stone E C and von Rosenvinge T T 2013 *Astrophysical Journal* **770** 117 1

[10] Christian E R 1989 *Evidence for anomalous cosmic ray hydrogen* Ph.D. thesis California Institute of Technology, Pasadena. 4

PAMELA measurements of the boron and carbon spectra

N Mori[1,2], O Adriani[1,2], G C Barbarino[3,4], G A Bazilevskaya[5], R Bellotti[6,7], M Boezio[8], E A Bogomolov[9], M Bongi[1,2], V Bonvicini[8], S Bottai[2], A Bruno[6,7], F Cafagna[7], D Campana[4], R Carbone[8], P Carlson[13], M Casolino[10,14], G Castellini[15], C De Donato[10,11], C De Santis[10,11], N De Simone[10], V Di Felice[10,18], V Formato[8,16], A M Galper[12], A V Karelin[12], S V Koldashov[12], S Koldobskiy[12], S Y Krutkov[9], A N Kvashnin[5], A Leonov[12], V Malakhov[12], L Marcelli[10,11], M Martucci[11,19], A G Mayorov[12], W Menn[17], M Mergé[10,11], V V Mikhailov[12], E Mocchiutti[8], A Monaco[6,7], R Munini[8,16], G Osteria[4], F Palma[10,11], B Panico[4], P Papini[2], M Pearce[13], P Picozza[10,11], M Ricci[19], S B Ricciarini[2,15], R Sarkar[20,21], V Scotti[3,4], M Simon[17], R Sparvoli[10,11], P Spillantini[1,2], Y I Stozhkov[5], A Vacchi[8], E Vannuccini[2], G I Vasilyev[9], S A Voronov[12], Y T Yurkin[12], G Zampa[8] and N Zampa[8]

[1]University of Florence, Department of Physics and Astronomy, I-50019 Sesto Fiorentino, Florence, Italy
[2]INFN, Sezione di Florence, I-50019 Sesto Fiorentino, Florence, Italy
[3]University of Naples "Federico II", Department of Physics, I-80126 Naples, Italy
[4]INFN, Sezione di Naples, I-80126 Naples, Italy
[5]Lebedev Physical Institute, RU-119991, Moscow, Russia
[6]University of Bari, Department of Physics, I-70126 Bari, Italy
[7]INFN, Sezione di Bari, I-70126 Bari, Italy
[8]INFN, Sezione di Trieste, I-34149 Trieste, Italy
[9]Ioffe Physical Technical Institute, RU-194021 St Petersburg, Russia
[10]INFN, Sezione di Rome "Tor Vergata", I-00133 Rome, Italy
[11]University of Rome "Tor Vergata", Department of Physics, I-00133 Rome, Italy
[12]National Research Nuclear University MEPhI, RU-115409 Moscow
[13]KTH, Department of Physics, and the Oskar Klein Centre for Cosmoparticle Physics, AlbaNova University Centre, SE-10691 Stockholm, Sweden
[14]RIKEN, Advanced Science Institute, Wako-shi, Saitama, Japan
[15]IFAC, I-50019 Sesto Fiorentino, Florence, Italy
[16]University of Trieste, Department of Physics, I-34147 Trieste, Italy
[17]Universität Siegen, Department of Physics, D-57068 Siegen, Germany
[18]Agenzia Spaziale Italiana (ASI) Science Data Center, Via del Politecnico snc I-00133 Rome, Italy
[19]INFN, Laboratori Nazionali di Frascati, Via Enrico Fermi 40, I-00044 Frascati, Italy
[20]Indian Centre for Space Physics, 43, Chalantika, Garia Station Road, Kolkata 700 084, West Bengal, India
[21]Previously at INFN, Sezione di Trieste, I-34149 Trieste, Italy

E-mail: mori@fi.infn.it

Abstract. The satellite-borne PAMELA experiment is aimed at precision measurements of the charged light component of the cosmic-ray spectrum, with a particular focus on antimatter. It consists of a magnetic spectrometer, a time-of-flight system, an electromagnetic calorimeter with a tail catcher scintillating layer, an anticoincidence system and a neutron detector. PAMELA has measured the absolute fluxes of boron and carbon and the B/C ratio, which plays a central role in galactic propagation studies in order to derive the injection spectra at sources from measurements at Earth. In this paper, the data analysis techniques and the final results are presented.

1. Introduction

During the propagation in the interstellar medium (ISM) from the acceleration sites to Earth, galactic cosmic rays undergo various physical processes which shape the injection spectra and chemical compositions down to the measured ones at Earth. Knowledge of these processes is required in order to interpret the measurements in terms of the parameters of the acceleration sources, or to estimate the backgrounds when searching for new and possibly faint contributions to the spectra.

Currently, there is still uncertainty in determining which are the most relevant physical processes [1]. The usual modelization of propagation is given in terms of a diffusive transport equation with source terms [2]. The equation accounts for diffusion, convection due to the interaction with the galactic wind, re-acceleration, spallation and radioactive decay. Direct measurements of astrophysical quantities (e.g. the hydrogen density obtained from 21-cm surveys) and fits of the distributions obtained from numerical propagation models to cosmic-ray measurements provide the parameters of the equation.

The spectra of particles which are produced exclusively by interactions of primary particles with ISM provide good constraints to propagation parameters. In particular, boron nuclei in cosmic rays are produced mainly by spallation of carbon with contributions from other elements. The boron-to-carbon ratio (B/C) has been measured by many experiments [3] and extensively studied in order to tune and benchmark propagation models (see e.g. [4] and references therein).

The PAMELA experiment [5] has recently measured the spectra of boron and carbon as well as the B/C ratio in the kinetic energy range 0.44 - 129 GeV/n [6]. In this paper the data analysis techniques and the final results for the data acquired during the period July 2006 - March 2008 are presented. Section 2 describes the detector while the data analysis procedures are detailed in section 3. The results are presented in section 4.

2. The PAMELA detector

The PAMELA main detector is a magnetic spectrometer made by tracking system placed inside the magnetic field generated by a permanent magnet. The tracking system consists of six double-sided silicon microstrip tracking layers; the readout pitches for the X (bending) and Y views are 51 μm and 66.5 μm, respectively. The intensity of the magnetic field at the center of the magnetic cavity is 0.46 T. The main aim of the tracking system is to measure the magnetic rigidity $\rho = pc/(Ze)$ of the impinging primary particle, from whose sign antimatter can be discerned from matter. The instrument trigger is provided by the Time-Of-Flight (TOF) system. It is composed by six layers of plastic scintillating pads arranged in three X-Y layers, two of them (named S1 and S2) above the tracking system and the third (S3) below it. The TOF also measures the particle velocity with a resolution of 250 ps for $|Z|=1$ particles and 70 ps for carbon [7] and the particle electric charge via multiple ionization measurements. The tracking system and the TOF are shielded by an anticoincidence system made by plastic scintillators, which allows interacting events producing secondary particles to be rejected during the offline analysis. A silicon-tungsten sampling electromagnetic calorimeter is placed below S3. It is made by 22 modules, each composed by two single-sided silicon strip detectors encompassing a tungsten converter layer. The readout pitch of the strips is 2.44 mm and the total depth is 16.3 X_0. The calorimeter provides a direct measurement of the energy of electrons and positrons. Due to its lateral segmentation a lepton/hadron rejection power of about 10^5 by means of topological shower analysis can be achieved. The rejection power is further improved by a tail-catcher scintillator and a neutron detector, both placed below the calorimeter.

For a detailed description of the PAMELA instrument and its performance see [5].

3. Data analysis

The PAMELA design and data analysis procedures have been optimized for $|Z| \sim 1$ particles, given its main focus on antimatter measurements (mainly positrons and antiprotons). In particular, an effect of this optimization is that the front-end readout electronics of the tracking system saturates when a high-Z particle like a boron or carbon nuclei crosses a silicon layer of the spectrometer releasing a high quantity of energy via ionization. This saturation produce systematic effects like a displacement of the tracking residuals when processing data with the algorithms conceived for $|Z| \sim 1$ particles. In order to get rid of these systematic effects, a position finding algorithm optimized for boron and carbon has been developed. This algorithm emulates a digital readout of saturated strip signals, which are then clustered before computing the impact point as the geometric center of the cluster ignoring all non-saturated strips. The spatial resolution obtained with this algorithm is 14 μm on the X (bending) view and 19 μm on the Y view, resulting in a Maximum Detectable Rigidity (MDR) of about 250 GV.

Before applying the tracking algorithm, clusters of strips with a total energy release less than 5 MIP have been discarded (an energy release of 1 MIP is defined as the energy released by a $|Z|=1$ minimum-ionizing particle). This selection removes the clusters produced by backscattered particles coming from the calorimeter or by delta rays, resulting in a higher tracking efficiency.

Events have been selected by requiring a single, well-fitted track with at least 4 associated hits on the X view and 3 on the Y view, in order to have a reliable measurement of the rigidity of the incoming particle. The track is required to lie inside the magnetic cavity of the spectrometer and to stay at least 1.5 mm away from the walls of the magnet, to reject particles crossing the magnet itself. This fiducial volume corresponds to a geometric acceptance of 19.9 cm^2 sr; this value has been used for computing the fluxes. Particles have also been requested to be down-going according to the TOF system. Galactic particles have been selected by imposing the requirement that the measured rigidity ρ exceeds the critical rigidity ρ_C computed in the Störmer vertical approximation by a factor of 1.3.

Boron and carbon events have been selected by means of their ionization energy releases in the TOF system, which grows as Z^2 at relativistic energies. Fig. 1 shows the mean (indicated with $\langle S2 \rangle$) of the energy releases detected by each of the two layers of scintillating pads constituting S2 as a function of the rigidity ρ. The figure also shows the selection criteria used to identify boron and carbon based on these measurement. Similar criteria are built also for S12 (the lower of the two layers constituting S1) and $\langle S3 \rangle$ (mean of ionization releases on the two layers constituting S3). Particles have been tagged as boron or carbon if they satisfy all the three charge selection criteria on S12, $\langle S2 \rangle$ and $\langle S3 \rangle$.

The efficiencies of the selection cuts described above have been measured directly from flight data. Tracking efficiency has been evaluated on a sample of non-interacting boron and carbon events selected with the TOF system whose track fitted in the calorimeter and extrapolated back to the top of payload lies entirely inside the fiducial acceptance. The energy dependence of the efficiency has been obtained at low energies by means of the velocity β measured by the TOF. However, at high energies there is no way to evaluate the energy dependence from flight data. Moreover, requiring a fitted track in the calorimeter makes the selected sample non-isotropic due to the additinal request that the primary exits from the bottom side of the calorimeter itself:

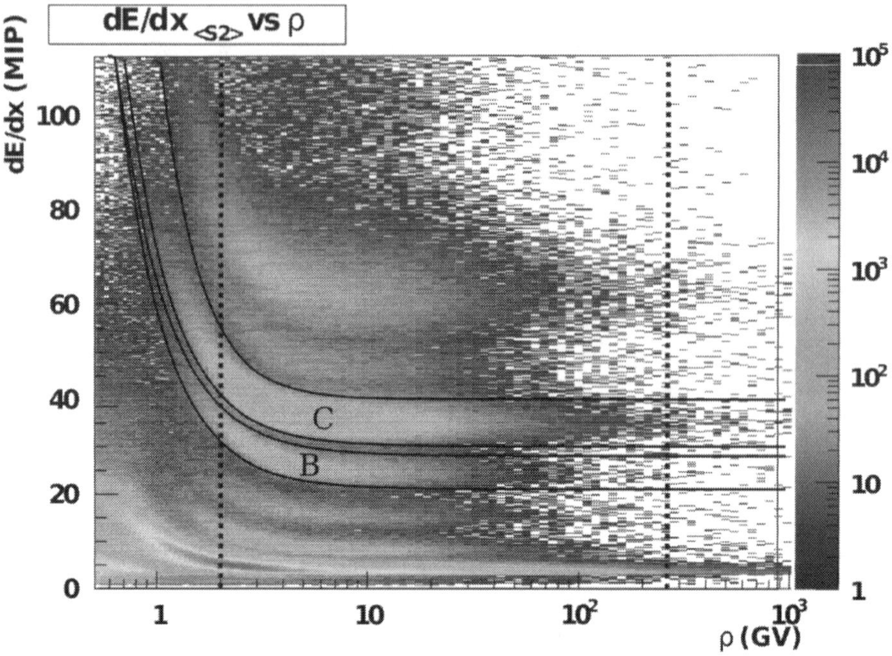

Figure 1. Mean ionization release and charge selection bands for ⟨S2⟩. The vertical dashed lines indicate the limits in rigidity of this analysis.

this ensures a good lever arm for the fit but rejects those inclined tracks which lie inside the acceptance of the instrument without crossing the last layer of the calorimeter. In order to cope with these two issues, a rigidity-dependent efficiency has been evaluated also on an isotropic sample obtained from Monte Carlo simulations. This efficiency has been subsequently divided by a Monte Carlo efficiency obtained with the same procedure used for flight data, resulting in a rigidity-dependent correction factor for the experimental efficiency. This factor is almost constant at a value of 0.97.

The efficiency of charge selection has been measured on a sample of pure boron or carbon selected by means of ionization measurements in S11 (the upper layer of S1) and in the first silicon layer of the calorimeter. These two detectors encompass the whole instrument from top to bottom, thus this selection rejects those events where the primary fragments in the apparatus due to nuclear interactions. The resulting efficiency peaks at about 75% at 3 GV for both species and then decreases with increasing energy, reaching an almost constant value of about 50% for boron and 60% for carbon above 30 GV. The presence of a possible residual contamination in the sample used to measure the efficiency has been investigated by measuring the variation of the selection efficiency of S12 when also ⟨S2⟩ and ⟨S3⟩ are used to pre-select the sample for efficiency measurement. No statistically significant variation have been observed, also when repeating the same procedure for the efficiencies of ⟨S2⟩ and ⟨S3⟩.

Monte Carlo simulations have been used to compute correction factors for various instrumental effects that cannot be investigated with flight data. Nuclear interactions in the aluminum dome on top of the instrument which destroy primaries or sweep them away from the acceptance of the detector have been investigated with a Fluka simulation [8]. The correction amounts to 14% for boron and 15% for carbon above 10 GV, where the dominant contribution is from inelastic scattering which destroys the primary. The value of the correction factor increases for both species at low energies where elastic scattering can significantly modify the trajectory of the primary and energy losses in the apparatus can significantly lower the energy of the particle, which is then swept away by the magnetic field of the instrument. Fluka simulations have been used also to quantify the production of secondaries by fragmentation of heavier species in the dome. This effect is negligible for carbon, while for boron it results in a correction factor of about 5% at some GV rising to about 20% at 200 GV, given mainly by spallation of carbon. Folding effects due to the finite rigidity resolution of the magnetic spectrometer and energy losses in the instrument (which can significantly affect the measured rigidity of particles at very low energies) have been corrected by means of a Bayesian unfolding procedure [9], using a Geant4 simulation [10] in order to obtain the smearing matrix. Since the selection of galactic particles has been done before correcting for slow down effects, some events may have been incorrectly rejected due to their rigidity being reduced by energy losses in the apparatus; a correction factor for this effect has been computed with Geant4 simulations. It amounts to 0.97 at 2 GV and it rise with energy reaching the value 1 (i.e. no correction) at 3 GV and above.

Since the events are binned according to their measured rigidity it is necessary to take into account the isotopic composition of boron when computing the boron flux as a function of kinetic energy per nucleon. Both ^{10}B and ^{11}B are present in cosmic rays, and their relative abundance is a poorly-known parameter. At relativistic energies no direct measurement is available. Estimations can be made using galactic propagation models, and the current consensus value for the ^{10}B fraction (defined as $^{10}\mathrm{B}/(^{10}\mathrm{B}+^{11}\mathrm{B})$) is 0.35 ± 0.15 and have a weak dependence on energy. But since different isotopes of the same energy have different rigidities, then the fraction is not constant when it is expressed as a function of rigidity. The procedure used to compute the rigidity-dependent ^{10}B fraction starting from the energy-independent ^{10}B fraction and the consequent boron event count as a function of kinetic energy per nucleon is described in details in appendix A of [6].

An accurate study of the systematic uncertainties of the measurement has been done considering various effects: the finite sizes of the samples used to measure the selection efficiencies and to compute correction factors, the effect of the residual contamination in the samples used to measure the charge selection efficiencies and the fluxes, the residual misalignment of the spectrometer, the requirement of fiducial containment, the disagreement between tracking efficiencies computed with flight data and with Monte Carlo data, the uncertainty on the unfolding procedure (assessed with a fold/unfold test) and the uncertainty in the isotopic composition of boron. The quadratic sum of all these factors give a total systematic uncertainty of about 6% at low energy for both species, rising to about 7% at high energy for carbon and to 10% for boron.

The live time of the instrument has been obtained from on-board clocks, and has been computed as a function of the cutoff rigidity to account for the different amount of time spent by PAMELA at different latitudes. The live time spent in regions where the cutoff value is above 2 GV is 1×10^7 s, increasing up to 3.14×10^7 s for regions with cutoff above 20 GV. The associated error is about 0.2% and has been neglected.

4. Results

The boron and carbon fluxes and the boron-over-carbon (B/C) ratio measured by PAMELA as functions of kinetic energy per nucleon are shown in fig. 2.

A quite good agreement with previous measurements can be observed, except at low energy where the effects of solar modulation are significant and must be carefully taken in to account when comparing fluxes measured in different solar periods. The fitted spectral indexes above 20 GeV/n are 3.01 ± 0.13 for boron and 2.72 ± 0.06 for carbon. A very simple fit of PAMELA data based on the GALPROP transport code [11] gives a slope of the diffusion coefficient of 0.397 ± 0.007, thus not allowing to discriminate between Kolmogorov and a Kraichnan diffusion types. However, this figure has been obtained with a specific transport code/model, and may vary when using different codes/models. Tighter constraints on propagation models and

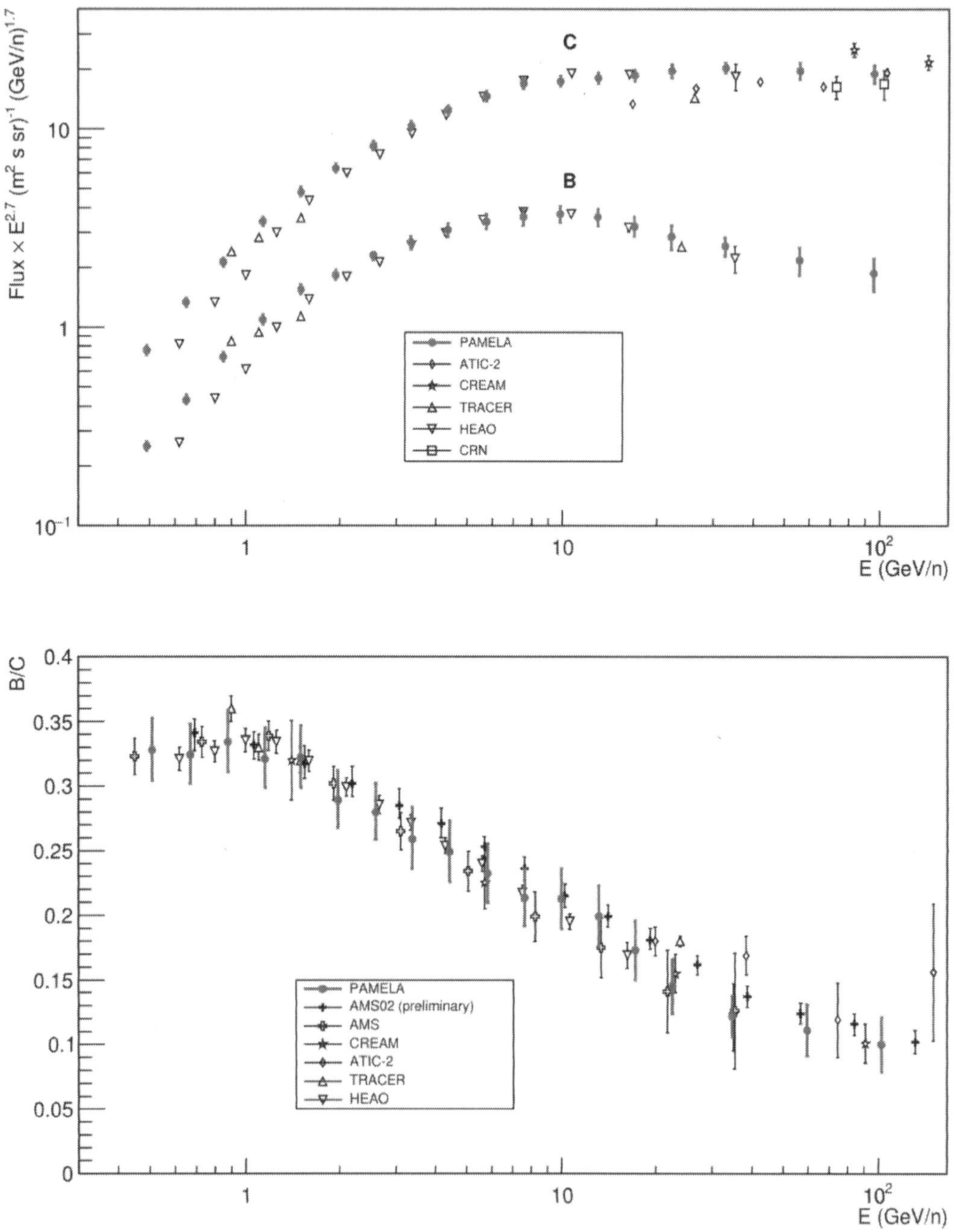

Figure 2. Boron and carbon fluxes multiplied by $E^{2.7}$ (upper panel) and B/C flux ratio (lower panel) as measured by PAMELA, together with results from other experiments [3]. The error bars for PAMELA are the quadratic sum of the statistical and the systematical contributions.

parameters could be obtained including also other quantities like the isotopic ratios of the lightest nuclei (i.e. the ^{2}H/^{1}H and ^{3}He/^{4}He ratios [12]) in the fit.

Acknowledgments
We acknowledge support from The Italian Space Agency (ASI), Deutsches Zentrum für Luft- und Raumfahrt (DLR), The Swedish National Space Board, The Swedish Research Council, The Russian Space Agency (Roscosmos) and The Russian Science Foundation.

References
[1] Maurin D, Putze A and Derome L 2010 *Astron. & Astroph.* **516** A67
[2] Ptuskin V 2012 *Astropart. Phys.* **39-40** 44
[3] Oliva A *et al* (AMS collaboration) *33rd International Cosmic Ray Conference (ICRC 2013)* http://www.cbpf.br/~icrc2013/proceedings.html
Ahn H S, Allison P S, Bagliesi M G *et al* 2008 *Astropart. Phys.* **30** 133
Obermeier A, Ave M, Boyle P *et al* 2011 *Astroph. Journ.* **742** 14
Panov D, Sokolskaya N V, Adams J H Jr *et al* 2007 *30th International Cosmic Ray Conference (ICRC 2007)* 3, ed R Caballero, J C D'Olivo *et al*
Engelmann J J, Ferrando P, Soutoul A *et al* 1990 *Astron. & Astroph.* **233** 96
Aguilar M, Alcaraz J, Allaby J *et al* 2010 *Astroph. Journ.* **724** 329
Swordy S P, Müller D, Meyer P *et al* 1990 *Astroph. Journ.* **349**,625
[4] Obermeier A, Boyle P, Hörandel J and Müller D 2012 *Astroph. Journ.* **752** 69
[5] Picozza P, Galper A M, Castellini G *et al* 2007 *Astropart. Phys.* **27** 296
[6] Adriani A, Barbarino G C, Bazilevskaya G A *et al* 2014 *Astroph. Journ.* **791** 93
[7] Campana D, Carbone R, De Rosa G *et al* 2009 *Nucl. Inst. and Meth. in Phys.* A **598** 696
[8] Battistoni G, Muraro S, Sala P R *et al* 2007 *Hadronic Simulation Workshop 2006*, ed M Albrow & R Raja, 31
[9] D'Agostini G 1995 *Nucl. Inst. and Meth. in Phys.* A **362** 487
[10] Agostinelli S, Allison J, Amako K *et al* 2003 *Nucl. Inst. and Meth. in Phys.* A **506** 250
[11] Strong A and Moskalenko I V 1998 *Astroph. Journ.* **509** 212
[12] Adriani A, Barbarino G C, Bazilevskaya G A *et al* 2013 *Astroph. Journ.* **770** 2

Journal of Physics: Conference Series **632** (2015) 012018 doi:10.1088/1742-6596/632/1/012018

A new way of air shower detection: measuring the properties of cosmic rays with LOFAR

A. Nelles[1], S. Buitink[1], A. Corstanje[1], J.E. Enriquez[1], H. Falcke[1,2,3,4], J.R. Hörandel[1,2], J.P. Rachen[1], P. Schellart[1], O. Scholten[5], S. ter Veen[1], S. Thoudam[1], and T.N.G. Trinh[5]

[1] Department of Astrophysics/IMAPP, Radboud Unversity, P.O. Box 9010, 6500 GL Njmegen, The Netherlands

[2] Nikhef, Science Park Amsterdam, 1098 XG Amsterdam, The Netherlands

[3] Netherlands Institute for Radio Astronomy (ASTRON), Postbus 2, 7990 AA Dwingeloo, The Netherlands

[4] Max-Planck-Institut für Radioastronomie, Auf dem Hügel 69, 53121 Bonn, Germany

[5] KVI-CART, University of Groningen, P.O. Box 72, 9700 AB Groningen, The Netherlands

E-mail: a.nelles@astro.ru.nl

Abstract. High-energy cosmic rays impinging onto the atmosphere of the Earth initiate cascades of secondary particles: extensive air showers. Many of the particles in a shower are electrons and positrons. During the development of the air shower and by interacting with the geomagnetic field, the electromagnetic cascade creates radiation, which we detect at frequencies of tens of MHz with the LOFAR radio telescope in the Netherlands. After many years of struggling to understand the emission mechanisms, the radio community has achieved the breakthrough. We are now able to determine direction, energy, and type of the shower-inducing primary particle from the radio measurements. The large number of antennas at LOFAR allows us to have a high precision and very detailed measurements. We will elaborate on the shower reconstruction, a precise description of the intensity of the radio signal at ground level (at frequencies from 10 to 240 MHz), a precise measurement of the shape of the radio wavefront, and on the reconstruction of the shower energy.

1. Introduction

After three years of measuring the radio emission of air showers with LOFAR (the LOw-Frequency ARay), a number of key findings have been obtained. The vast number of antennas available at the core of LOFAR has allowed us to measure the signal distribution, polarization pattern and times of arrival with unprecedented quality.

These measurements have strengthened the case that detecting the radio emission of air showers has matured as a technique. The emission mechanisms are understood as a combination of the interaction of the shower with the magnetic field of the Earth [1, 2, 3] and the accumulation of a negative charge in the shower front [4, 5, 6]. Both effects have to be considered in a medium with changing index of refraction [7, 8], which also determines the regions and frequencies in which the signals are added up coherently.

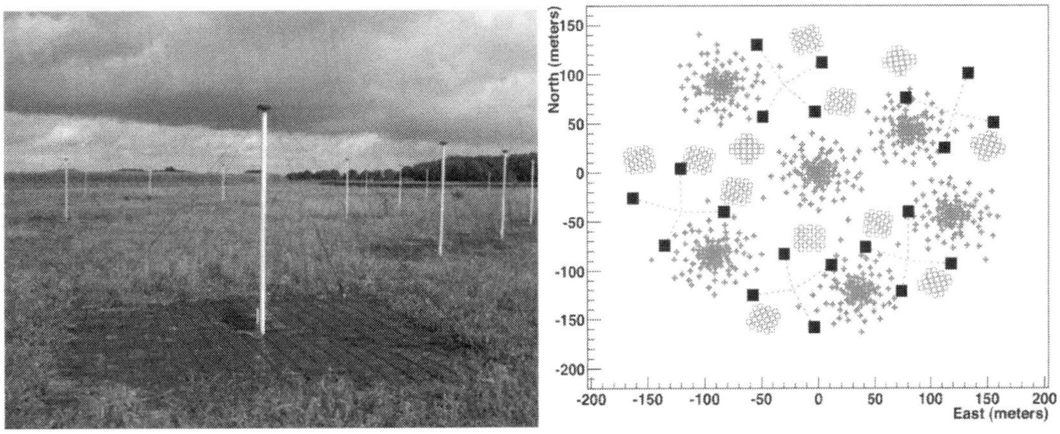

Figure 1. Left: Low-band antennas at the core of LOFAR. The amplifier is located in the black top, above the stabilizing pole. In the background, the black tiles of the high-band antennas are visible. Right: Schematic view of the superterp. Open squares are the high-band antennas, crossed depict the low-band antennas and the full squares are the LORA detectors. Dashed lines indicate the grouping of LORA detectors for electronic and triggering purposes.

2. LOFAR – The LOw-Frequency ARray

LOFAR is currently the largest distributed radio interferometer for the lowest radio frequencies [9]. Two types of antennas are employed at LOFAR. The low-band antennas (LBA, 10 – 90 MHz) are slanted dipole antennas and used as standard tool for cosmic-ray operation. The high-band antennas (HBA, 110 – 240 MHz) are fat dipole antennas, optimized for astronomical interferometric observations.

The layout of LOFAR is based on stations. About one-hundred antennas of each type are grouped on an irregular grid into a cluster, called station. Stations are positioned all over Europe with increasing density towards the core of LOFAR in the north of the Netherlands. In the center, six stations are grouped on a circular elevated area, called the superterp. This area has also been instrumented with particle detectors [10], providing triggers for measurements of air showers. The LOFAR Radboud Air Shower Array (LORA) consists of 20 detector units, (a unit contains two 0.45^2 m scintillators, type: NE 114), which are distributed between the antennas on an irregular grid. Whenever operation conditions allow and an air shower is detected in more than 13 detector units, a read-out of the transient buffer boards (TBBs) of the antennas is initialized. The antenna set and observation mode are always set by the primary astronomical observation, which also determines the data-stream that is buffered by the TBBs. Due to the multi-purpose capabilities of LOFAR, cosmic ray measurements are possible in the background of the primary observation. The TBBs store up to 5 s of data of a selected antenna type (LBA, HBA, inner or outer ring per station). Triggers can be received from inside and outside of LOFAR.

3. Data reconstruction

For every triggered data-set, the data from all selected LOFAR antennas, as well as the particle signal are available. Currently, 2.1 ms of radio data are stored following a trigger from the particle detectors. The data are processed automatically every night, delivering relatively calibrated pulse powers per antennas for air showers detected above the noises threshold [11]. An absolute calibration is currently underway [12].

The raw voltage traces from all antennas per station are first cleaned from narrow-band interference. Furthermore, relative differences between antennas are corrected for, by normalizing the average received power to a value corresponding to the expected galactic power as function of local sidereal time [11]. Using the arrival time and arrival direction reconstructed from the particle signal, the cleaned antenna signals are beamformed in order to identify the timing of a detectable air shower signal. This allows us to narrow the search window and to determine the air shower signal in absolute timing in an additional pulse-finding step. Using this time, the signals from the two dipoles are combined and corrected for the antenna response, while refitting the arrival direction iteratively. Once the final arrival direction is reached, the pulse quantities such as power, polarization, timing and corresponding uncertainties are calculated. If a pulse is successfully identified in more than one station, the results are combined and further analysis steps like a non-planar wavefront-fit and fit of the signal pattern are performed.

All results are accessible from a PostgreSQL database. Additional quality cuts, in order to exclude periods with thunderstorms or contamination with broad-band interference are applied at a later stage, allowing us to also review these contaminated measurements. The details of the reconstruction framework can be found in [11].

4. Polarization and emission mechanisms

The two main emission mechanisms can be identified by their polarization signature. The geo-magnetic effect, the interaction of the shower with the magnetic field of the Earth, will induce electric fields that are polarized in the $\vec{v} \times \vec{B}$-direction, where $\vec{v}$ is the direction of propagation of the shower and $\vec{B}$ the direction of the magnetic field at the detector. This direction of polarization will be independent of the location of the observing antenna around the shower axis. The charge excess, however, being an effect that is symmetric around the shower axis, will induce an electric field that is dependent on the position of the observer. It will therefore be changing from parallel to anti-parallel to the geo-magnetically induced electric field, when moving around the shower axis. LOFAR is highly suited to observe such changes on an event-by-event level or as function of distance to the shower axis [13].

Especially, the calculation of the fraction of charge excess a has been proven instructive. The air showers measured with LOFAR contain signals that show a degree of nearly 99% of polarization. As the geomagnetic and the charge-excess emission are linearly polarized in different directions, their relative contributions can be determined. Following [6] the expected electric field at any given time t can be written as

$$\vec{E}(t) = \vec{E}_{\mathrm{G}}(t) + \vec{E}_{\mathrm{C}}(t) \tag{1}$$

$$= (|\vec{E}_{\mathrm{G}}(t)| + |\vec{E}_{\mathrm{C}}(t)| \cos\phi')\hat{\mathbf{e}}_{\vec{v}\times\vec{B}} + \tag{2}$$

$$(|\vec{E}_{\mathrm{C}}(t)| \sin\phi')\hat{\mathbf{e}}_{\vec{v}\times\vec{v}\times\vec{B}}. \tag{3}$$

Here $\vec{E}_{\mathrm{G}}(t)$ is the electric field produced by the geomagnetic contribution that is purely polarized along the direction of the Lorentz force, and $\vec{E}_{\mathrm{C}}(t)$ the radial electric field, produced by charge-excess. The charge-excess fraction is then defined as

$$a \equiv \sin\alpha \frac{|E_{\mathrm{C}}|}{|E_{\mathrm{G}}|}, \tag{4}$$

where $|E_{\mathrm{C}}|$ is the amplitude of the electric field produced by charge-excess when the total electric field vector amplitude reaches its maximum value.

The polarization signal of one air shower measured with LOFAR is shown in figure 2. The dominant polarization in the $\vec{v} \times \vec{B}$-direction is clearly visible. The additional charge excess fraction needed is small, however, measurable. The average contribution is shown on the right of figure 2. The charge excess fraction is a function of distance to the shower axis and zenith angle varying between close to zero and 20%. This indicates that both emission mechanisms

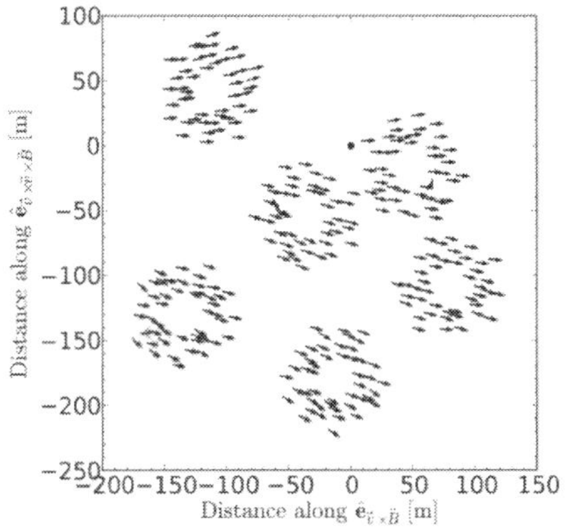

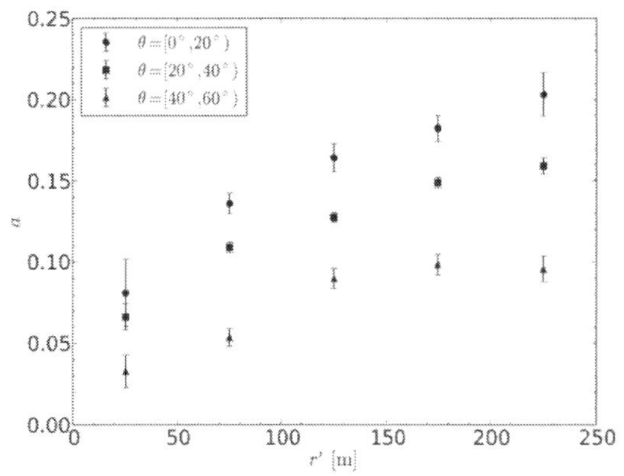

Figure 2. Left: Direction of the measured electric field vector for an air shower as detected with LOFAR. The vectors are shown in the shower plane together with their (small) uncertainties. The main direction is defined by the cross-product of propagation direction of the shower $\vec{v}$ and the geo-magnetic field $\vec{B}$. Right: The parameter a, as fractional deviation from a purely geo-magnetic polarization, as function of distance to the shower axis. The relation is shown for different zenith angle bins [13].

show a different behavior with the zenith angle, possibly the shower development and the lateral extent of the shower. Details of this analysis can be found in [13].

5. Wavefront and arrival directions

The arriving front of radio signals can be approximated to reasonable resolution with a planar wavefront. At a more closer look, it however becomes apparent that there is significant curvature in the wavefront [14, 15]. With a timing resolution of better than nanoseconds, and several hundreds of antennas, LOFAR is the ideal tool to study the details of the wavefront.

From basic assumptions, three models of a wavefront can be motivated. A sphere, a cone and a hyperbola. In order to avoid being dominated by changing conditions between air showers, these three shapes are compared on an event-by-event basis.

The inspection of a single air shower is already intuitive. Figures 3 and 4 show the three wavefront shapes in their best-fitting version for one air shower. The core position was used as a free parameter in order to allow for a fair comparison. This explains the different distribution of measurements. It is clearly visible that using a spherical fit results in too much curvature, which causes a problem at the antennas of larger distances to the shower axis. The conical fit, however, cannot be used to explain all curvature and therefore residual curvature can be seen in the bottom half of the figure. Only with the hyperbola the right balance of curvature can be obtained.

Since one needs one additional free parameter for the hyperbola, all three models cannot be compared based directly on their χ^2. Furthermore, the hyperbola can turn into a cone or a sphere for the appropriate choice of parameters. Therefore, a likelihood ratio test is performed, given the fact that both cone and sphere are a subset of the hyperbola. This results in a probability of effectively zero of either cone or sphere matching the dataset better than the hyperbola. More details of this analysis can be found in [14].

The right image in figure 4 shows the advantage of using a curved wavefront in order to fit the arrival direction. For the reconstructed arrival direction, it is not very important which form you chose. All parameterization result in the same arrival direction with a resolution of

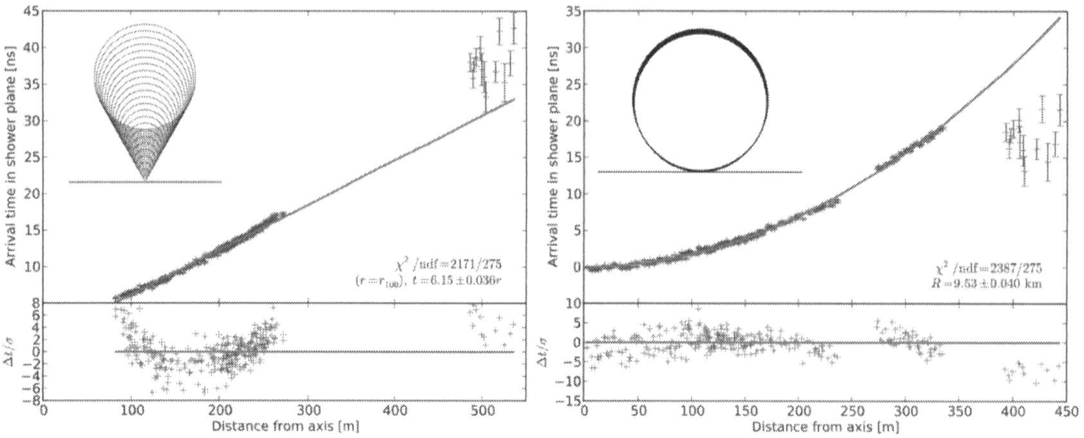

Figure 3. Two best fitting solutions of a conical (left) and spherical (right) wavefront. Shown are the arrival times in the shower plane as function of distance to the axis. Also shown are the residuals of the fit and data in units of the uncertainties [14].

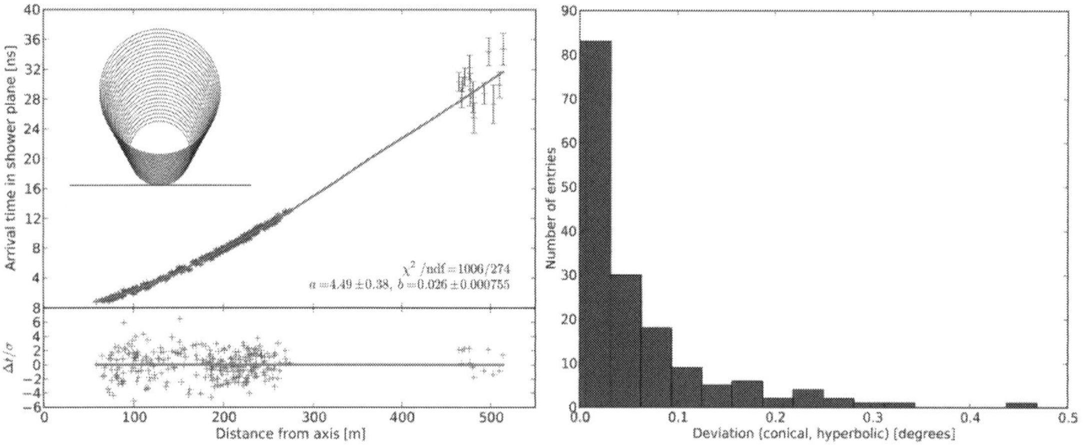

Figure 4. Left: Best fitting solutions of a hyperbolic wavefront to the arrival times of an air shower measured with LOFAR. Right: The difference in reconstruction of the arrival direction between the fit of an hyperbola and a cone. The comparison to a spherical fit looks comparable [14].

better than 0.1°. When comparing this arrival direction, however, to the results of a planar wavefront, the spread increases to about 1°. The same holds true for the comparison to the arrival directions as obtained with the particle detectors. As those already have an independent resolution of about 1°, the resolution of the curved wavefront needs to be significantly better, which is essential once arrival directions, anisotropies and possible sources are studied.

6. Energy and the depth of the shower maximum

If radio is to be used as standard detection technique, it will need to provide an independent measure of the energy of the cosmic ray and of the depth of the shower maximum, X_{max}. The data taken with LOFAR have been used to show that this is indeed possible with an excellent resolution. Two approaches have been tested. A direct comparison to simulations, as will be

Journal of Physics: Conference Series **632** (2015) 012018 doi:10.1088/1742-6596/632/1/012018

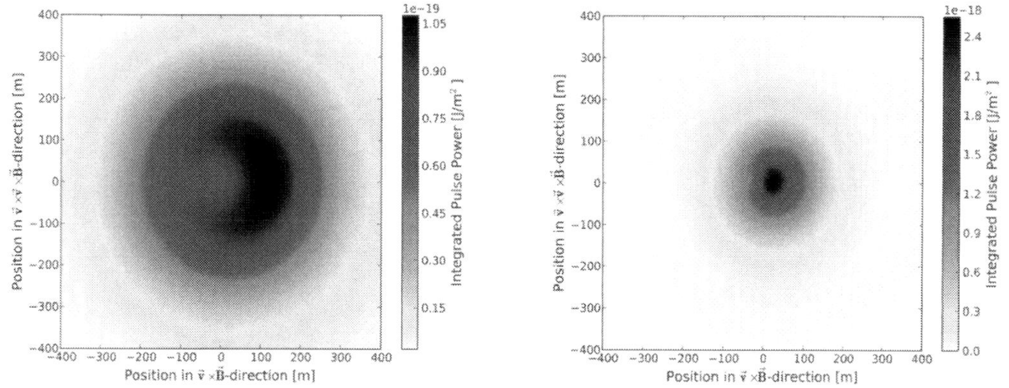

Figure 5. Two simulated signal pattern of an air shower. The left pattern was derived for a shower that is observed at a large distance to X_{max}, the right pattern for a small distance [19].

explained in section 6.1 and a parametric approach that will be discussed in section 6.2.

They both utilize the following quantities: The integrated pulse power as measured in the low-band antennas, the particle densities as measured with the LORA scintillators and derived quantities.

6.1. Direct comparison to air shower simulations

Air shower simulations predict a strong dependence of the shape of the signal distribution on the distance to the shower maximum. Given the large number of antennas at LOFAR, it is possible to test this prediction without a parameterization step in-between.

CORSIKA [16] and its radio extension CoREAS [17] are used to generate emission maps for several depths of the shower maximum, X_{max}, for a given air shower using its arrival direction and energy estimate from the particle detectors as input. The maps are generated using 160 ground positions at which the radio emission is evaluated. These ground positions are located on a star-pattern with eight arms, which is symmetric in the shower plane and centered around the shower axis. Furthermore, two arms are aligned with the $\vec{v} \times \vec{B}$-direction and two are aligned with the $\vec{v} \times (\vec{v} \times \vec{B})$-direction. This star-shaped pattern allows us to sample the shower at the relevant positions. In the $\vec{v} \times \vec{B}$-direction the largest signal is expected due to the positive interference of geomagnetic emission and charge-excess. Starting from these individual simulations, full patterns can be extrapolated as shown in figure 5. Using a robust interpolation routine delivers the total power at every given position with an uncertainty of about 2.5% with respect to dedicated simulations for a given position [18].

For every measured shower, about 40 of these patterns have to be generated in order to sample the ranges of X_{max} sufficiently. It can now be tested, which pattern best describes the data by letting the shower position vary. CORSIKA also delivers the particles generated by the air shower. The particle content can therefore also be matched to what was measured with the LORA detectors. The following fit procedure is used:

$$\chi^2 = \sum_{\text{antennas}} \left(\frac{P_{\text{ant}} - f_r^2 P_{\text{sim}}(x_{\text{ant}} - x_0, y_{\text{ant}} - y_0)}{\sigma_{\text{ant}}} \right)^2 + \sum_{\substack{\text{particle} \\ \text{detectors}}} \left(\frac{d_{\text{det}} - f_p d_{\text{sim}}(x_{\text{det}} - x_0, y_{\text{det}} - y_0)}{\sigma_{\text{det}}} \right)^2, \quad (5)$$

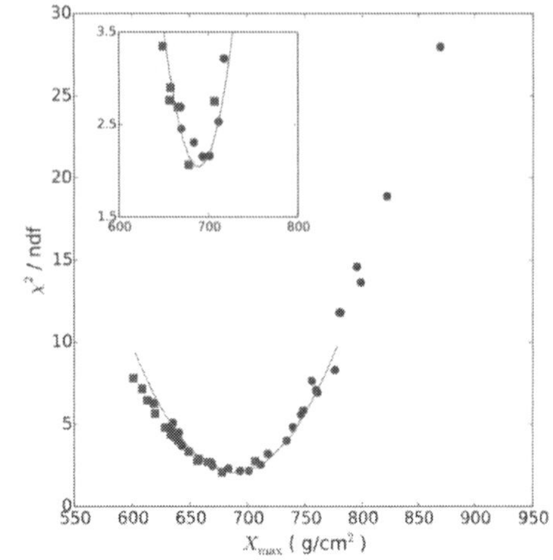

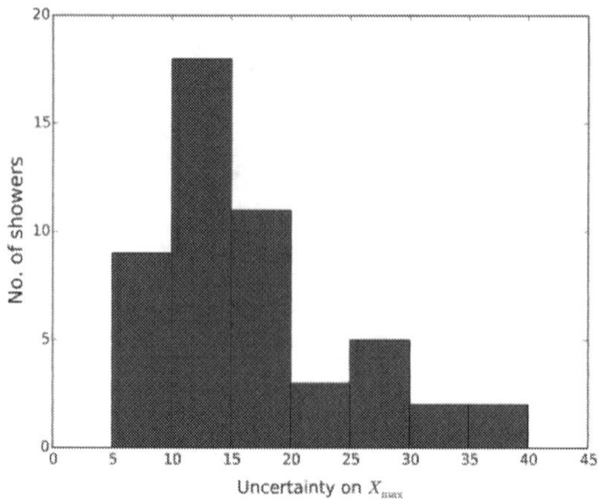

Figure 6. Left: The goodness of fit parameter χ^2/ndf for 40 air shower simulations (squares iron, circles proton primaries), all differing by the depth of shower maximum, X_{max}, of the air shower. The inset shows a zoom of the minimum of the distribution. Right: Distribution of uncertainties on X_{max} as obtained for the first 50 high-quality air showers measured with LOFAR [18].

where P_{ant} is the measured power integrated over a 55 ns window at an antenna at location (x_{ant}, y_{ant}) with noise level σ_{ant}. P_{sim} is the simulated power, d_{det} is the deposited energy as measured by a LORA detector at location (x_{det}, y_{det}) with noise σ_{det}, and d_{sim} is the simulated deposited energy. The scaling factors f_r and f_p are needed due to the missing absolute calibration and potential uncertainties in the particle reconstruction.

As shown in figure 6 the distribution of χ^2-values shows a clear preference for a range of X_{max} for a given shower, which allows us to determine the X_{max} of every shower with a very accurate estimate. The obtainable uncertainty stems from the method itself, the atmospheric correction and the uncertainty in the arrival direction [18]. The choice of hadronic interaction model seems not to be very relevant for this kind of analysis. The largest offset observed between, for example, QGSJETII-04 and EPOS-LHC was $4.3\,\mathrm{g/cm^2}$. The same holds true for the choice of simulation code, for example when using ZHAires [20] instead of CoREAS. For the first fifty air showers measured at LOFAR, the distribution of uncertainties is shown on the right in figure 6. With a mean uncertainty of $17\,\mathrm{g/cm^2}$ the results are highly competitive with respect to for example fluorescence detectors and encouraging to develop the radio technique further.

6.2. Parameterization of the signal distribution

The only drawback of the method described in section 6.1 is the large computational effort. In the current set-up, several weeks of CPU time are needed to generate a sufficient number of patterns for an air shower. The effort could be limited, if the energy was precisely known for every shower (the particle detectors are small and subject to large systematic uncertainties due to the missing sensitivity to shower-to-shower fluctuations) and if the range of X_{max}-values could be limited.

In [19] an analytic parameterization was suggested that can be used for a fast reconstruction of the shower axis. Furthermore, two fit parameters of the function are predicted to show a strong correlation with the energy of the shower and the distance to the shower maximum.

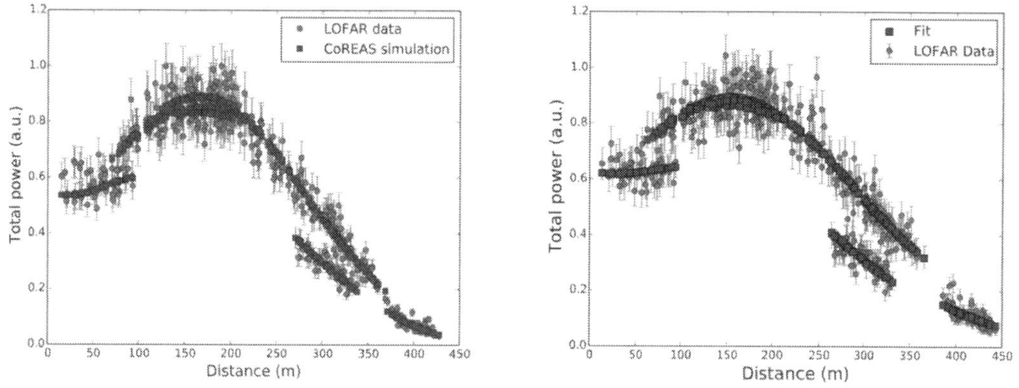

Figure 7. Comparison of the full Monte-Carlo approach (left) and the parameterized signal distribution (right). The data on the left is the originally measured pulse power in the antenna arms. On the right the pulse power corrected for the antenna response is shown. Both are in red. The best fitting simulations and the parameterization are shown in blue, respectively.

Based on [19] the following function has been tested on LOFAR data:

$$\begin{aligned} P(x', y') &= A_+ \cdot \exp\left(\frac{-[(x'-X_c)^2 + (y'-Y_c)^2]}{\sigma_+^2}\right) \\ &\quad - C_0 \cdot A_+ \cdot \exp\left(\frac{-[(x'-(X_c+x_-))^2 + (y'-Y_c)^2]}{(C_3 \cdot e^{C_1 + C_2 \cdot \sigma_+})^2}\right) \end{aligned} \quad (6)$$

The function is given in coordinates (x', y') in the shower plane, where the coordinates are aligned with the $\vec{v} \times \vec{B}$- and the $\vec{v} \times (\vec{v} \times \vec{B})$-axis. The parameters C_1, C_2 and C_3 are fixed from studies of simulations. All other parameters are used for the fit.

More than 85% of all showers recorded in three or more LOFAR stations can be reconstructed with the parameterization with good quality of fit. Showers that cannot be reconstructed are mostly showers with low signal-to-noise ratios and those where the shower axis is far away from the closest radio antennas. The shower axis that is reconstructed with the parameterization is in very good agreement with the axis obtained with the method of the direct comparison, as evidenced in figure 7. Here, the shape of the signal distribution as function of the distance to the shower axis would change immediately, if the axis was shifted. For this air shower, the axis position differs by less than three meters. The average spread between the two methods is less than 15 m. The small differences in signals and uncertainties are due to the fact that one method is applied on the instrumental signal (no correction for the antenna model on data, but application of antenna model on the simulation) and the other on the data (including its noise) corrected for the influence of the antenna model.

On data, it is also confirmed that the parameter A_+ shows a strong correlation with the energy of the primary particle. It scales quadratically with the energy, as it is a power quantity and the amplitude scales linearly with the energy due to the coherence criterion. Compared to the energy estimate given by the LORA particle detectors, the A_+-parameter is usable for more showers. First of all, the radio antennas cover a larger portion of the surface with provides a larger fiducial area. Also, the radio signal pattern can still be reconstructed if the shower axis was outside the instrumented area, which is due to the asymmetry in the signal. For the whole data-set only about half of the air showers detected in radio can reliably be reconstructed with the particle signal. Another advantage of the radio detection is that it almost delivers a calorimetric

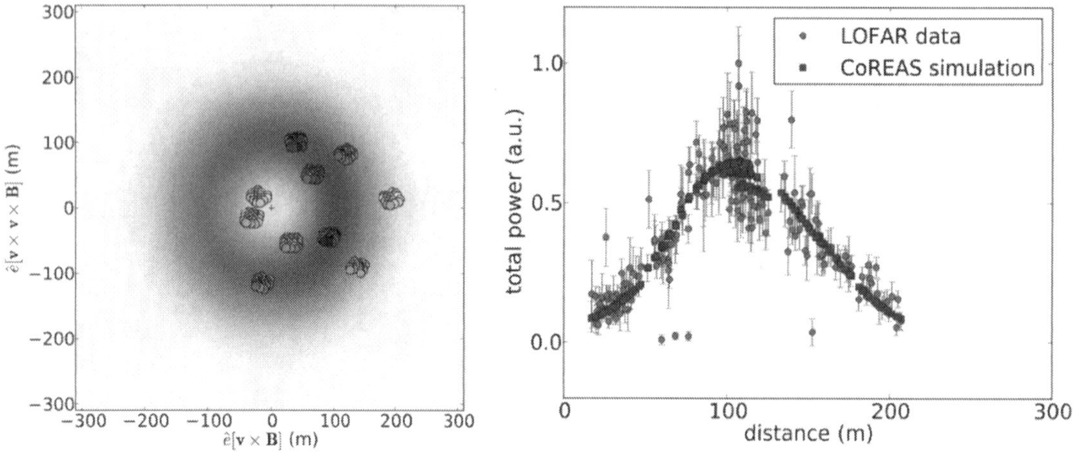

Figure 8. Air shower as measured with the high-band antennas (110 – 190 MHz) of LOFAR. Left: Signal distribution in the shower plane. The power in arbitrary units in encoded in color (from light to dark). The measured signals (circles) are combined with the best fitting simulation (background map). Right: Total power as function of the distance to the shower axis. The red circles show the measured data and the blue squares the best fitting CoREAS simulation [22].

signal. The geometric component of the shower is represented in the σ_+-parameter. Therefore, the A_+-parameter is expected to have small systematic uncertainties ($< 10\%$) with respect to the primary composition. In order to provide a good energy estimate for the simulation input, this missing uncertainty will help to significantly reduce the simulation effort, while the larger data-set allows for the retaining of more showers, which improves composition studies. Details on the application of the parameterization to the LOFAR data can be found in [21].

7. Signal distribution in different frequency ranges

Most air showers are measured at LOFAR with the low-band antennas covering the frequency range of 10 – 90 MHz. There are two reasons for this choice. First of all, the emission footprint is expected to be stronger and spread wider at lower frequencies and the first observations have also been conducted in this range. The second reason is more technical and LOFAR-specific. The high-band antennas 110 – 240 MHz have been optimized for astronomical observations. In this frequency range, the antennas are system noise dominated, which calls for a direct enhancement of the astronomical signals to ensure a good sensitivity. Thus, the signals from 16 antennas are combined directly in hardware before being forwarded to the TBBs. In this combination step a time delay is introduced making the combined set of antennas more sensitive to a fraction of the sky that corresponds to the induced delays. Signals from other directions will be distorted. This consequently means that air showers can only be measured with best sensitivity, if their arrival direction happens to coincide with the 10° field of view of the observation. This essentially limits the usefulness of the high-band antennas for studies of air showers.

However, as the primary observation determines the antenna set, there is air shower data available in the higher frequency band. Also, a small number of signals have been recorded without distortion and can be compared to simulations with the same approach as described above.

Figure 8 shows an air shower as measured in the frequency range of 110 – 190 MHz. In this frequency range, a clear ring-like structure is visible. This can be explained by the increasing importance of the relativistic time-compression for higher frequencies. At higher frequencies, the coherence is no longer given at all observer positions around the shower, which decreases the signal strength measured. At the Cherenkov angle, however, an enhancement is visible that translates into a ring of strong emission. LOFAR is the first experiment to measure this ring on an event-by-event basis and to confirm the theoretical predictions.

This ring size has been postulated to correlate with the depth of the shower maximum [23]. Indications for such a correlation are indeed found via the direct comparison to simulations. However, a much larger set is needed to confirm this relation. Furthermore, the ring size would have to be estimated with a high precision to obtain depths of the shower maximum that are well enough constrained for composition studies. Even with the large number of antennas at LOFAR, this precision is difficult to obtain. Details of this analysis can be found in [22].

8. Conclusion

The measurement and the reconstruction of the radio emission of air showers have made significant progress over the last years. The radio-telescope LOFAR with its high antenna density has been highly suitable to measure the subtle features of the emission and confirm theoretical predictions on a single-event basis. The emission shows a predominant polarization in the direction defined by the cross-product of shower axis and magnetic field, which can be attributed to the dominant emission process. A second emission process is needed to explain the secondary radial component in the signal polarization. This charge excess fraction is not constant, but is a function of the distance to the shower axis and the zenith angle of the shower. Also, we have shown that the wavefront of the radio signal is best represented by a hyperbola. Conical or spherical shapes introduce too little or too much curvature and cannot reproduce the data convincingly. A milestone in the understanding of the emission was the direct comparison of air shower simulations with the LOFAR data. Not only were all asymmetric shapes very well reproduced, but the strong dependence on the depth of the shower maximum was proven. With LOFAR data the shower maximum can be resolved with an average uncertainty of $17\,\mathrm{g/cm^2}$, which makes the radio detection of air showers a strong competitor for composition studies. Based on these simulations an analytical parameterization was developed that can provide a fast reconstruction of the air showers. This parameterization will help to speed up the detailed energy and composition studies. In addition, LOFAR data also shows that the emission mechanisms are well understood also beyond the standard frequency range of below 100 MHz. At higher frequencies the emission is clearly dominated by Cherenkov-like effects and ring structures of enhanced emission appear.

Acknowledgments

The LOFAR cosmic ray key science project very much acknowledges the scientific and technical support from ASTRON. Furthermore, we acknowledge financial support from the Netherlands Research School for Astronomy (NOVA), the Samenwerkingsverband Noord-Nederland (SNN), the Foundation for Fundamental Research on Matter (FOM) and the Netherlands Organization for Scientific Research (NWO), VENI grant 639-041-130. We acknowledge funding from an Advanced Grant of the European Research Council under the European Union's Seventh Framework Program (FP/2007-2013) / ERC Grant Agreement n. 227610.

LOFAR, the Low Frequency Array designed and constructed by ASTRON, has facilities in several countries, that are owned by various parties (each with their own funding sources), and that are collectively operated by the International LOFAR Telescope (ILT) foundation under a joint scientific policy.

References

[1] F. D. Kahn and I. Lerche, *Radiation from Cosmic Ray Air Showers*, *Proceedings of the Royal Society London A* **289** (1966) 206–213.

[2] O. Scholten, K. Werner, and F. Rusydi, *A macroscopic description of coherent geo-magnetic radiation from cosmic-ray air showers*, *Astroparticle Physics* **29** (2008) 94–103.

[3] D. Ardouin et al., *Geomagnetic origin of the radio emission from cosmic ray induced air showers observed by CODALEMA*, *Astroparticle Physics* **31** (2009) 192–200.

[4] G. Askaryan, *Excess Negative Charge of an Electron-Photon Shower And Its Coherent Radio Emission*, *Soviet Physics JETP* **14** (1962), no. 2 441–443.

[5] D. Saltzberg, P. Gorham, D. Walz, C. Field, R. Iverson, A. Odian, G. Resch, P. Schoessow, and D. Williams, *Observation of the Askaryan Effect: Coherent Microwave Cherenkov Emission from Charge Asymmetry in High-Energy Particle Cascades*, *Physical Review Letters* **86** (2001), no. 13 2802.

[6] A. Aab et al., *Probing the radio emission from air showers with polarization measurements*, *Physical Review D* **89** (Mar., 2014) 052002.

[7] K. Werner, K. D. de Vries, and O. Scholten, *A realistic treatment of geomagnetic Cherenkov radiation from cosmic ray air showers*, *Astroparticle Physics* **37** (Sept., 2012) 5–16.

[8] C. W. James, H. Falcke, T. Huege, and M. Ludwig, *General description of electromagnetic radiation processes based on instantaneous charge acceleration in "endpoints"*, *Physical Review E* **84** (Nov., 2011) 056602, [arXiv:1007.4146].

[9] M. P. van Haarlem et al., *LOFAR: The LOw-Frequency ARray*, *Astronomy and Astrophysics* **556** (Aug., 2013) A2, [arXiv:1305.3550].

[10] S. Thoudam et al., *LORA: A scintillator array for LOFAR to measure extensive air showers*, *Nuclear Instruments and Methods in Physics Research A* **767** (2014) 339–346.

[11] P. Schellart, A. Nelles, et al., *Detecting cosmic rays with the LOFAR radio telescope*, *Astronomy & Astrophysics* **560** (Dec., 2013) A98, [arXiv:1311.1399].

[12] S. Buitink et al., *Calibrating the low-band antennas of the Low-Frequency Array for cosmic ray measurements*, *to be submitted to JINST* (2014).

[13] P. Schellart et al., *Polarized radio emission from cosmic ray air showers measured with LOFAR*, *Journal of Cosmology and Astrophysics* **10** (2014) 014.

[14] A. Corstanje et al., *The shape of the radio wavefront of extensive air showers as measured with LOFAR*, *Astroparticle Physics* **61** (2015) 22–31.

[15] W. D. Apel et al., *The wavefront of the radio signal emitted by cosmic ray air showers*, *Journal of Cosmology and Astroparticle Physics* **9** (Sept., 2014) 25.

[16] D. Heck, J. Knapp, J. N. Capdevielle, G. Schatz, and T. Thouw, *CORSIKA: a Monte Carlo code to simulate extensive air showers*. Feb., 1998.

[17] T. Huege, M. Ludwig, and C. W. James, *Simulating radio emission from air showers with CoREAS*, in *American Institute of Physics Conference Series* (R. Lahmann, T. Eberl, K. Graf, C. James, T. Huege, T. Karg, and R. Nahnhauer, eds.), vol. 1535 of *American Institute of Physics Conference Series*, pp. 128–132, May, 2013.

[18] S. Buitink et al., *A method for high precision reconstruction of air shower X_{max} using two-dimensional radio intensity profiles*, *Physical Review D* **90** (2014) 082003, [arXiv:1408.7001].

[19] A. Nelles, S. Buitink, H. Falcke, J. R. Hörandel, T. Huege, and P. Schellart, *A parameterization for the radio emission of air showers as predicted by CoREAS simulations and applied to LOFAR measurements*, *Astroparticle Physics* **60** (Jan., 2015) 13–24, [arXiv:1402.2872].

[20] J. Alvarez-Muniz, W.R.Carvalho Jr., and E. Zas, *Monte Carlo simulations of radio pulses in atmospheric showers using ZHAireS*, *Astroparticle Physics* **35** (2011), no. 6 325–341.

[21] A. Nelles et al., *The radio emission pattern of air showers as measured with LOFAR - a tool for the reconstruction of the energy and the shower maximum*, *Submitted to JCAP* (2014) [arXiv:1411.7868].

[22] A. Nelles, P. Schellart, et al., *Measuring a Cherenkov ring in the radio emission from air showers at 110-190 MHz with LOFAR*, *Astroparticle Physics* **65** (2015) 11–21, [arXiv:1411.6865].

[23] K. D. de Vries, O. Scholten, and K. Werner, *The air shower maximum probed by Cherenkov effects from radio emission*, *Astroparticle Physics* **45** (May, 2013) 23–27, [arXiv:1304.1321].

Galactic Propagation of Cosmic Rays from Individual Supernova Remnants

Nils Nierstenhöfer, Philipp Graeser, Florian Schuppan and Julia Tjus
Ruhr Universität Bochum, Universitätsstraße 150, D-44780 Bochum
E-mail: `nils@tp4.rub.de`

Abstract. It is widely believed that supernova remnants are the best candidate sources for the observed cosmic ray flux up to the knee, i.e. up to ~PeV energies. Indeed, the gamma-ray spectra of some supernova remnants can be well explained by assuming the decay of neutral pions which are created in hadronic interactions. Therefore, fitting the corresponding gamma spectra allows us to derive the spectra of cosmic rays at the source which are locally injected into our Galaxy. Using these spectra as a starting point, we propagate the cosmic rays through the Galaxy using the publicly available GALPROP code. Here, we will present first results on the contribution of those SNRs to the total cosmic ray flux and discuss implications.

1. Introduction

Many popular theories for the origin of *Galactic* cosmic rays (CRs) up to the knee-region (~PeV) are based on the idea of shock acceleration of charged particles in supernova remnants (SNRs) [1–3]. That SNRs indeed provide the required energy budget to be the birthplace of the Galactic cosmic ray population has firstly been pointed out in [4].
Recent indirect observational techniques as used by the instruments H.E.S.S., MAGIC, VERITAS and MILAGRO allow for the observations of very high energy (VHE) gamma-rays from SNRs up to ~100 TeV. Including radio and X-ray observations, it is possible to test and distinguish models of leptonic or hadronic origin of those gamma-rays as has been done systematically for 24 well-identified SNRs in [5][1]. In particular, this kind of analysis provides a parametrization $j_{\rm p}(T)$ of a potential population of CR protons inside the SNRs. This indirect approach to derive $j_{\rm p}(T)$ is useful as the CRs trajectories, masses and energies are altered during their Galactic propagation - thus crucial *direct* information is lost. This is different for the VHE gamma-rays which propagate rectilinear with no significant energy losses through the Galaxy.
For the same reason, the effects of Galactic CR propagation have to be taken into account if one aims on testing whether the source spectra $j_{\rm p}(T)$ of SNRs can explain the CR flux as measured on Earth. To do so, an altered version of the publicly available GALPROP framework [6] for Galactic propagation is used, in which CR nuclei are injected according to the source spectra derived in [5]. It needs to be taken into account that these spectra represent currently observed SNRs, and thus only a short temporal phase of cosmic ray injection: while a single SNR is only active for about 10^4 years, cosmic rays diffuse in the Galaxy for about $10^6 - 10^7$ years. In

[1] While this is a first step towards modeling CR sources, an unambiguous identification is still not possible at this point, as leptonic scenarios in some cases describe the sources equally well. Here, the most optimistic scenario is chosen, where all possible sources are interpreted as hadronic accelerators.

order to have a time-averaged sample of SNRs, we assume that the CR spectra as derived from the currently observed SNRs are a representative subsample of SNR spectra and luminosities. We then perform multiple simulations with SNR locations randomly drawn from the spatial distribution of [7], which represents a model for the distribution of SNRs in the Galaxies, with an enhanced density near the Galactic Center.

The main goal of this paper is to describe our current simulation approach and to present an exemplary simulation which is focusing on the investigation of the shape of the predicted CR proton flux at Earth. In particular, we emphasize the fact that we do *not* perform any optimization of model parameters here and that we present the method of our approach with only preliminary physics results in this paper. In particular, we use the standard setting for GALPROP concerning the parameters. All details of the physics questions, including a parameter study will be discussed elsewhere in the future. One of the primary, final goals of developing this method is the evaluation of the overall flux normalization.

Finally, in this paper, we present improvements of the simulation approach and their impact on the interpretation of the results are outlined.

2. Source Spectra

The modeling of the electromagnetic spectrum of 24 SNRs has been performed in [5]. Out of those, a significant contribution of gamma-rays from hadronic interactions with a subsequent π_0-decay are found for 21 SNRs (see table 3 in [5]). These 21 source spectra $j_{\rm p}(T)$ are the input for the analysis which is presented in what follows.

More precisely, the authors of [5] apply a one zone model for the SNR and provide a parametrization of a simple power law for the proton spectra in momentum $p = \sqrt{mc^2 - (T + mc^2)^2}$, which translates into an energy spectrum of

$$j_{\rm p}(T) = a_p \left(\sqrt{\frac{T^2 + 2Tmc^2}{T_0^2 + 2T_0mc^2}}\right)^p \frac{T + 2Tmc^2}{\sqrt{T^2 + 2Tmc^2}} \tanh\left(\frac{T}{T_{\rm min}}\right) \exp\left(-\frac{T}{T_{\rm max}}\right). \qquad (1)$$

Here, T, a, p and m denote the kinetic energy, the normalization constant, the spectral index and the proton mass, respectively. A smooth low energy cut off at $T_{\rm min} = 10$ MeV and its high energy counterpart at $T_{\rm max}$ is included through the factors four and five in equation (1). In this paper, the focus is on the CR energy range from ~GeV to ~PeV. Hence, the low energy cut off is not applied.

It should be noted that for example 274 SNRs are listed in the well known Greens catalogue [8] while only ~20 SNRs have been detected in gamma-rays. Correspondingly, the prediction of a too low CR-flux is expected here.

Only SNRs that are observed in MeV-GeV or GeV-TeV are used in the analysis of [5] and are included in our analysis. Therefore, the way of selecting sources for the observation with e.g. Cherenkov telescopes, may induce a selection bias (e.g. favouring flat, luminous SNRs). Therefore, the set of 21 SNRs from [5] is not statistically complete and, consequently, the presented results need to be interpreted with adequate care. Future data taken with the instruments CTA and HAWC will provide a more complete picture of gamma-ray emitting SNRs in the Galaxy and this study can be updated accordingly.

3. Propagation

The effects of Galactic propagation on the source spectra $j_{\rm p}(T)$ was simulated using version 54 of the open source C++-framework GALPROP [6, 9, 10]. The code was extended to allow for particle injection according to $j_{\rm p}(T)$ as parametrized in equation (1). Additionally, a procedure was implemented that allows to randomly select a position for a set of SNRs from a predefined distribution of matter within the Galaxy.

3.1. Normalization

First of all it should be noted that the CR normalization in *standard* GALPROP applications is usually fixed by globally scaling the calculated cosmic ray density such that the observed CR flux at Earth is met. In the approach presented in this paper, the idea is to fix the normalization by the rate of particle injection of the individual SNRs into the Galaxy[2]. Hence, from the technical viewpoint a key ingredient was to rewrite the source spectrum $j_{\mathrm{p}}(T)$ in terms of the *internal* GALPROP units. A pragmatic approach to determine the needed conversion factor is to compare the calculation of the luminosity L in GALPROP with the corresponding integral expression on the basis of $j_{\mathrm{p}}(T)$

$$L = \alpha R_{\mathrm{SNR}}^2 c \int_{10\ \mathrm{MeV}}^{1\mathrm{e}9\ \mathrm{MeV}} dT\ \beta\ T\ j_{\mathrm{p}}(T). \tag{2}$$

Here, R_{SNR} is the radius of the SNRs. Note that $\alpha = 1$ corresponds to the case where the CR particles stream out of the SNR in radial direction. In contrast to that $\alpha = 1/2$ would arises from an averaging assuming that the CRs move in random directions inside the SNRs. The aforementioned comparison leads to the following relation between the initial source function $q_1(p(T))$ as implemented in GALPROP[3] and the spectrum parametrization $j_{\mathrm{p}}(T)$

$$q_1(p(T)) = \alpha \frac{c^2 R_{\mathrm{SNR}}^2 \beta^2}{4\pi V_{\mathrm{grid}}}\ j_{\mathrm{p}}(T). \tag{3}$$

Alternatively, the luminosity can be derived from the total energy E_{tot} of protons in the SNR via $L = E_{\mathrm{tot}}/\tau$ with a time-scale τ, representing the distribution of the total energy over the total lifetime of the remnant,

$$L = \frac{1}{\tau}\frac{4\pi}{3} R_{\mathrm{SNR}}^3 \int_{10\ \mathrm{MeV}}^{1\mathrm{e}9\ \mathrm{MeV}} dT\ T\ j_{\mathrm{p}}(T). \tag{4}$$

Using this expression for the luminosity, one finds an alternative expressions[4] for equation (3)

$$q_1'(p(T)) = \frac{\beta c R_{\mathrm{SNR}}^3}{3 V_{\mathrm{grid}} \tau}\ j_{\mathrm{p}}(T). \tag{5}$$

Assuming that E_{tot} is the energy of the SNR converted into protons and τ is the life time of the SNR, $L = E_{\mathrm{tot}}/\tau$ can be interpreted as the average luminosity in CRs.

It is a typical approach in gamma-ray astronomy to derive the total cosmic ray energy budget in order to estimate the SNRs possible contribution to the total cosmic ray budget. A back-of the envelope calculation predicts that the observed cosmic ray luminosity of the Galaxy (about $3{\cdot}10^{40}$ erg/s) can be reproduced if on average, 10^{50} erg are going into a single SNR at a Supernova rate of 1/(100 years). Here, it is assumed that cosmic rays are injected into the interaction region continuously at a constant rate during the lifetime τ with the total energy going into cosmic rays conserved over time. It is clear that this is a simplifying assumption, as it is known that at least the energy spectrum is changing with time, in particular concerning the reduction of maximum energy, see e.g. [11]. For energy spectra steeper than E^{-2}, the total energy budget is

[2] In the exemplary simulation discussed later in this paper the predicted CR flux is normalized to experimental data. The more extensive subject of modeling the CR normalization as a direct consequence of the source spectra $j_{\mathrm{p}}(T)$ will be discussed in a future paper.

[3] Further details can be found in the GALPROP documentation [9, 10] or by directly checking the source code file cr_luminosity.cc.

[4] Equation (3) has been applied in this paper.

dominated by the lower integration threshold, so effects from this temporal development should be relatively small. It is also specified in [11] that the total energy of the SNR is decreasing with time due to cooling effects. This would mean that, if we assume a constant fraction of the SNR energy going into cosmic rays at a given time, the actual average luminosity of cosmic rays would be underestimated in particular for old remnants. It is not clear, however, if the fraction of energy going into cosmic rays is actually constant over time or if it actually decreases with the available energy budget. Thus, we judge that in first order approximation, it seems reasonable to estimate the total energy of the remnant from the given value, keeping in mind the above discussed uncertainties.

In this paper, we refrain from discussing the *global* normalization and normalize the final spectrum to the observed flux. The normalizations of the individual SNRs relative to each other are fixed using equation 3.

3.2. Including CR Nuclei

Although the source spectra taken from [5] are only provided for CR protons, it is possible to include CR nuclei in GALPROP simulations (see chapter 5.5 in [9, 10]). To do so, the initial source function of nuclei $q_A(p_A)$ with mass number A and momentum p_A is related to $q_1(p_1)$ by the relative abundance X according to

$$X = \frac{A q_A(p_A)}{q_1(p_1)}. \tag{6}$$

In this context two remarks have to be made:

(i) Due to the high energy cut off in equation (1) X is not independent of the energy in what follows.

(ii) Including CR nuclei injection, the total energy in hadrons of the SNRs is artificially increased. Note, one can approximately correct the energy budget derived in [5] by down-scaling the proton normalization a_p in equation (1) appropriately.

Here, simulations are performed for all nuclei, but the resulting energy spectra are only discussed for protons. In the future, the heavy nuclei spectra will be discussed as well in order to investigate other questions connected to cosmic ray observations, sources and transport. Photohadronic effects as well as photo spallation is typically negligible at the given length scales and electromagnetic fields in the Galaxy.

3.3. Random Positions, Mapping to Closest Gridpoint and Averaging

The idea behind GALPROP is to numerically solve the transport equations that governs the propagation of CRs in the Galaxy [6]. This requires the definition of a spatial simulation grid on which the particle densities are calculated. Here, we select random position for the SNRs and map them to the closest grid point. As the grid size is larger than the extensions of the SNRs, this implies that currently the SNRs are treated as point like CR sources.

We finally calculate and present average values - e.g. the average proton flux $< dF/dT >$ or the Boron to Carbon ratio $< B/C >$ - from N sets of 21 SNRs. It is this averaging which allows one to derive a realistic prediction for CR observables.

Note that the variance of CR observables e.g. Var(dF/dT) or related quantities cannot be conclusively calculated within our current simulation procedure. The reasons for that and and a method to overcome this issue in the future is outlined in section 5.

3.4. GALPROP Settings

As a starting point we use the *standard* settings as suggested by the GALPROP web-run application web page [9, 10]. One change is applied to the standard settings: The size of

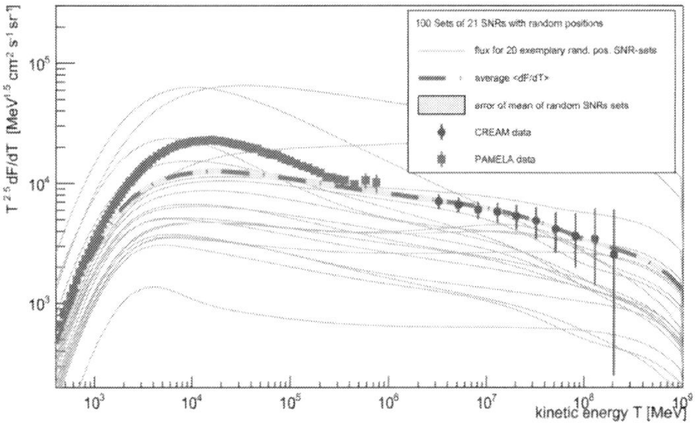

Figure 1. 20 out of 100 proton spectra dF/dT of random-position sets of 21 SNRs are exemplarily presented in grey. Mean values $< dF/dT >$ (red) and their errors (yellow band) are displayed, representing the standard deviation from averaging over 100 different configurations . Blue dots show measured data by CREAM [12], the magenta curve presents the observations by PAMELA [13]. Note that that $< dF/dT >$ is normalized to fit the CREAM data point at lowest energy.

the galaxy has been reduced in our example simulation to -10 kpc< x, y <10 kpc - to allow faster simulations. In particular we do not aim for optimizing free model parameters to improve the agreement with experimental data at this point. Here, we mainly intend to outline our computational approach and to show exemplary results - further results (including more detailed parameters studies) will be reported elsewhere in the future.

4. Exemplary Simulation

In this section an exemplary simulation is presented. The basic parameters for the underlying GALPROP runs are mainly the ones suggested by the GALPROP web-run web page [9, 10] (cf. section 3). In this example N=100 sets of 21 SNRs with random positions according to the SNR-distribution of [7] have been simulated. In each set, these SNRs are active continuously and placed randomly at different positions weighted with the SNR distributino function for each set. In this approach, we use gamma-ray data to determine the individual energy spectra, statistically distributed via the continuous distribution. This way, only a fraction of all sources will reach energies up to the knee naturally. Different models predict that there might only be a very small number of PeVatrons in the Galaxy contributing to the high-energy part of the cosmic ray energy spectrum. In our sample, many sources still have the potential to reach PeV-energies, as gamma-ray data at the highest energies do not reveal a cutoff yet. But already now, it becomes clear that not all sources can reach these energies and certainly not with a flat spectral behavior as usually assumed in the standard approach. In the future, with HAWC and CTA measurements, this question will be solved and reduce the systematic uncertainties at the highest energies.

In what follows, the bracket-notation ($< ... >$) indicates that the predicted observables that can be compared to the observations are calculated as an average of N=100 simulated sets of SNRs with random positions (cf. section 3 for details).
The predicted flux $< dF/dT >$ of cosmic ray protons is compared to experimental data from

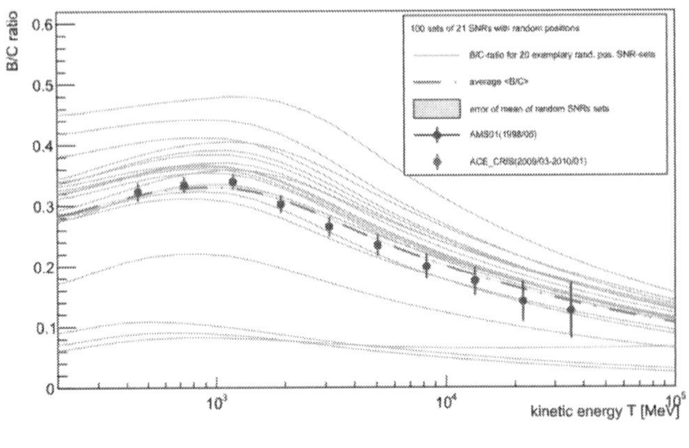

Figure 2. 20 out of 100 B/C-ratios of random-position sets of 21 SNRs are exemplarily presented in grey. Mean values $< B/C >$ (red) and their errors (yellow band) are displayed. Blue dots show measured data from AMS-01 [15].

CREAM [12] and PAMELA [13] in Fig. 1. Here, the statistical and systematic experimental uncertainties are displayed. Additionally, $< dF/dT >$ has been normalized to CREAM data at 3.25 TeV to allow for a comparison of the spectral behavior. Note that the proper solar modulation potential of 550 MV linked with the PAMELA measurements [13] is included in Fig. 1 according to [14].
We find that the spectral behavior of the measured proton spectrum is well described in the energy range of CREAM, but a discrepancy with the PAMELA data which is most obvious at 10 GeV is apparent. As we derive the individual proton spectra from gamma-ray data, we expext that best agreement should be present in the statistically well-measured range, which is around $1 - 100$ GeV for Fermi and 100 GeV-10 TeV for IACTs. Assuming that about 10% of the cosmic ray energy is going into photons, this would mean a good description between 10 GeV and 100 TeV. The disagreement is therefore not arising from any extrapolation effect.

As a cross-check, we also include the simulation of heavy nuclei and find that the simulated $< B/C >$-ratio is in good agreement with data measured by AMS-01 as can be seen in Fig. 2.

5. Discussion

In general the good agreement between $< B/C >$ and experimental data indicates that the used simulation setup is well suited to model Galactic CR propagation. The observed proton spectrum is well reproduced for energies larger than $\sim$1 TeV, but further studies are necessary to confirm this agreement.

These first results show that there is a *possible* discrepancy between $< dF/dT >$ and the observations by PAMELA which is most apparent at kinetic energies of $T \sim 10$ GeV, using GALPROP standard parameters. In the following, we will discuss both statistical and physics reasons for the disagreement.

Statistical reasons may contribute to the disagreement: our current simulation approach predicts the average CR observable, e.g. $< dF/dT >$, but not the proper corresponding variance, e.g. $Var(dF/dT)$ (see the discussion later on). However, the variance or alternative statistical measures are needed to fully quantify whether the aforementioned discrepancy of $< dF/dT >$ is statistically significant. To quantify the agreement between the predicted CR observables statistically, a measure of the *variance* is needed. The variance of the CR observables could be calculated and presented here along with the mean value. However, it should be noted that this variance has a limited statistical interpretation in the current simulation approach. It merely measures the spread of the simulated CR observable which is induced by randomly selecting positions for the 21 SNR in the Galaxy. In particular, this variance would decrease if the number of SNRs would be increased. As more than 21 SNRs are expected to contribute to the CR flux the variance calculate here would presumably be an upper limit for the true variance. In addition, the calculation of the true variance would require to include temporal aspects of the SNRs - such as their production rate and lifetime - into our simulation approach. This could be done as a generalization of our procedure to include spatial aspects via the selection random positions and mapping of SNRs to the closest grid point on the spatial simulation grid (cf. subsection 3.3). As similar issues are already addressed in the GALPROP manual [9, 10] (see chapter 6), an implementation including the temporal aspects of SNRs should be feasible. A nearby source contributing to the measured CR flux is another possible explanation for the disagreement in the proton spectra. Extending the method to provide a meaningful measure of the variance induced by spatial and temporal aspects of the SNRs will help to estimated the theoretical probability that such a constellation is realized in nature.

Possible *physics* reasons which can contribute to this potential disagreement are listed below, and are also discussed in the literature, see e.g. [16] and references therein:

(i) By using the standard GALPROP settings, we use a diffusion scalar with Kolmogorov-like diffusion, leading to a general steepening of the energy spectrum of a distant source by a power of $\sim 1/3$. Measuerments like the one of the Boron-to-Carbon ratio show that the scalar approximation would even allow for a larger steepening [17]. This would presumably give a better fit at low energies, while a discrepancy would show towards higher energies instead.

(ii) The latter behavior could be explained by missing, flat SNRs, which are not seen yet due to missing sensitivity at the highest energies: The SNR sample extracted from [5] is not statistically complete (compare section 2). It is expected to have a more complete sample of SNRs in the future, with available CTA and HAWC data. Also, future results from IceCube, KM3NeT and IceCubeGen2 will presumably help to clarify which part of the gamma-ray signal actually is of hadronic nature. Correlation studies between gamma-rays and ionization signatures can also contribute to identifying those sources that are hadronically dominated, see e.g. [18–20].

(iii) Another possibility would be a diffusion scalar with a power law behavior broken at around 100 GeV, as already discussed in [9].

(iv) Finally, the full treatment of the diffusion process with a tensor might change the picutre. Here, the difficulty is that the structure of the tensor itself is not very well known. First investigations of tensor diffusion in the Galaxy are presented inplemented in the DRAGON code ([21], so far optimized to the propagation of electrons [22]) and the PICARD code ([23], published with first test examples).

We already stressed that the parameters which define the framework for the Galactic propagation have not been optimized in our example application. Especially, using a broken or steeper power law for the diffusion coefficient as function of energy may provide a handle to reproduce the shape of the experimental proton spectrum at $T \sim$GeV.

6. Conclusion

Based on the publicly available GALPROP code for Galactic CR propagation, a procedure is introduced to predict CR observables for a given set of hadronic source spectra from SNRs. This method is based on calculating the average over multiple GALPROP run using sets of SNR with positions randomly distributed according to a given spatial distribution. In this paper, we illustrate the current procedure in which we use the proton spectra of 21 SNRs inferred from the corresponding gamma-ray measurements as presented in [5]. While the $< B/C >$ ratio is well reproduced in this simulation we find a discrepancy for the predicted proton flux $< dF/dT >$. As we only present an exemplary study with no parameter study at this point, we do not draw any conclusions from these findings yet. The aim of this paper is the presentation of the method itself. In the future, we will present a full parameter study, including the discussion of the normalization of the spectrum as well. The latter aspect will contribute to answering the question if those SNRs observed in gamma-rays are representative to explain the cosmic ray energy budget as observed at Earth. We particularily plan to extend our procedure to include (I) temporal aspects of SNR evolution, (II) additional SNRs and (III) the implemention on an inhomogeneous simulation grid.

On longer terms we intend to perform Monte-Carlo (MC) simulations of the propagation of Galactic CRs. As a basis, we suggest to use the publicly available CRPropa MC-framework to study the propagation of UHECRs in extragalactic environments [24]. Especially the redesigned object orientated structure of the upcoming version 3.0 of CRPropa seems to allow for an easy extensions for the propagation of galactic CRs [25]. With today's computer technologies, a Monte Carlo treatment of Galactic CRs down to T$\sim$10-100 TeV seems possible with reasonable run times. For lower energies it may be sufficient to switch to a diffusive approximation to avoid the time intensive numerical solution of the equation of motion in the Galactic magnetic field.

Acknowledgments
This work was supported by the DFG research unit FOR1048: *Instabilities, Turbulence and Transport in Cosmic Magnetic Fields* and by the research department on plasmas with complex interactions (Bochum). NN would like to thank A.W. Strong and C. Evoli for insights about the GALPROP code. Furthermore, we acknowledge helpful discussions with B. Eichmann.

References
[1] Fermi E 1949 *Phys. Rev.* **75**(8) 1169
[2] Blandford R D and Ostriker J P 1978 *ApJL* **221** L29
[3] Hillas A M 2005 *Journal of Phys. G Nucl. Phys.* **31** 95
[4] Baade W and Zwicky F 1934 *Proc. Natl. Acad. Sci* **567** 259
[5] Mandelartz M and Becker Tjus J 2015 *Astrop. Phys.* **65** 80
[6] Strong A W and Moskalenko I V 1998 *ApJ* **509** 212
[7] Case G L and Bhattacharya D *ApJ* **504** 761
[8] Green D A 2009 URL `http://arxiv.org/abs/0905.3699`
[9] 2014 Galprop v54 URL `http://galprop.stanford.edu/`
[10] Strong A W *et al.* 2011 GALPROP version 54: Explanatory supplement URL `http://130.79.128.5/ftp/0/aa/534/A54/galprop_v54.pdf`
[11] Cox D P 1972 *ApJ* **178** 159
[12] Yoon Y S *et al.* 2011 *ApJ* **728** 122 ISSN 0004-637X, 1538-4357
[13] Adriani O *et al.* 2011 *Science* **332** 69
[14] Gleeson L J and Axford W I 1968 *ApJ* **154** 1011
[15] Aguilar M *et al.* 2010 *ApJ* **724** 329
[16] Biermann P L *et al.* 2010 **725** 184
[17] Obermeier A *et al.* 2012 *ApJ* **752** 69 (*Preprint* 1204.6188)
[18] Becker J K *et al.* 2011 *ApJL* **739** L43 (*Preprint* 1106.4740)
[19] Schuppan F *et al.* 2012 *A&A* **541** A126 (*Preprint* 1201.4674)
[20] Schuppan F, Röken C and Becker Tjus J 2014 *A&A* **567** A50 (*Preprint* 1402.2521)
[21] Evoli C *et al.* 2008 *JCAP* **10** 018 (*Preprint* 0807.4730)
[22] Gaggero D *et al.* 2013 *Phys. Rev. Lett.* **111** 021102 (*Preprint* 1304.6718)
[23] Kissmann R 2014 *Astrop. Phys.* **55** 37 (*Preprint* 1401.4035)
[24] Kampert K H *et al.* 2013 *Astrop. Phys.* **42** 41 (*Preprint* 1206.3132)
[25] Alves Batista R *et al.* 2013 *ArXiv e-prints* (*Preprint* 1307.2643)

Close Cathode Chamber technology for cosmic particle tracking

László Oláh[1,2], Gergely Gábor Barnaföldi[1], Gergő Hamar[1], Hunor Gergely Melegh, Gergely Surányi[2,3] and Dezső Varga[1]

[1] Wigner Research Centre for Physics of HAS, PO Box 49, H-1525, Budapest, Hungary
[2] Eötvös Loránd University, 1/A Pázmány P. s., H-1117, Budapest, Hungary
[3] MTA-ELTE Geological Geophysical and Space Science Research Group, 1/A Pázmány P. s., H-1117, Budapest, Hungary

E-mail: olah.laszlo@wigner.mta.hu

Abstract. The Close Cathode Chamber (CCC) technology has been developed and found useful in a portable tracking system under harsh and varying environmental conditions due to their mechanical and operational stability. The muon flux have been measured on ground and at shallow depths underground (< 70 m.r.e.) which provides a good reference for other experiments. The multiple scattering in rock and the soft contamination of the track sample have been investigated experimentally and by GEANT4. The applicability of the sensor to detect underground rock density inhomogeneities has been demonstrated via reconstruction of an underground tunnel system. A modified muon tomograph has been built with sensitive area of 0.25 m^2 and angular resolution of 15 mrad which is useful for material discrimination via the measurement of multiple scattering and absorption of muons. The reliable tracking performance, low power consumption and fair angular resolution make the CCC technology useful also in large area tracking detectors.

1. Introduction

It has been long established both soft and hard components of cosmic rays reach the surface of Earth. The soft (or electromagnetic) component is largely absorbed after 10 radiation lengths of material. The hard component consists mostly of muons which are highly penetrating due to their small interaction cross-section and long lifetime. The flux of muons decreases by the transversed material. Muography (short for muon radiography) deduces the density length (integral of density as a function of length in a given direction) in the interior of the investigated object, such as a mountain or volcano, by the measurement of the muon absorption along different paths through the object. If path lengths (average densities) are measured, the average density (path length) can be deduced along the muon paths.

The first muography measurement has been performed by E. P. George, to determine the rock density above an Australian mine [1]. L. W. Alwarez et al. studied the inner structure of the Kephren pyramid by spark chambers to find hidden caverns [2]. Speleological application has also been demonstrated by measurements [3]. Muography has successfully been applied to the investigation of volcanoes from the reconstruction of their shape [4] to visualize and analyse the magma movements and modification patterns [5] by the measurement of the flux of near horizontal muons. The measurement of absorption and multiple scattering of muons allows to make a 3-dimensional tomographic image of the investigated object, e.g. detect hidden smuggled nuclear materials [6, 7] or monitoring the fuel level in nuclear reactors [8, 9].

Since 2010, the REGARD group [10] is also focusing on the reconstruction of hidden caverns and rock densifications by cosmic muon tracking. We found that the cost efficient and lightweight gaseous detectors, the so called Close Cathode Chambers [11, 12] are applicable for underground measurements due to their mechanical and operational stability [13, 14, 15]. In present paper, we briefly present the Close Cathode Chamber technology and its application in tracking detectors. We will quantify the required detector parameters for muography and our measurements in an underground tunnel system. Furthermore, we present encouraging preliminary results on material discrimination by muon tomography.

2. Close Cathode Chamber technology

The Close Cathode Chamber (CCC) is an asymmetric multi-wire proportional chamber (MWPC) which does not involve heavy frames to hold the wire tension. The CCC is lightweight, easy to construct and tolerant to mechanical effects such as vibration, external tension or bulging by pressure. The scheme of the CCC is shown in the *left panel* of Fig. 1. The thickness of the CCC in our application is about 12 mm. There are 20-25 μm thick sense wires and 100 μm thick field wires 1.5 mm above the baseplate. The gas gain of the CCC is around 10^4 in Ar + CO_2 80:20 gas mixture with the field wire potential and the sense wire potentials are around -600 V and +1000 V respectively. Its material budget is less than 1 % radiation length, whereas its energy cut-off is a few MeV. The CCC concept, the key operational parameters and issues of large size construction (see *right panel* of Fig. 1) have also been presented in Refs. [11] and [12].

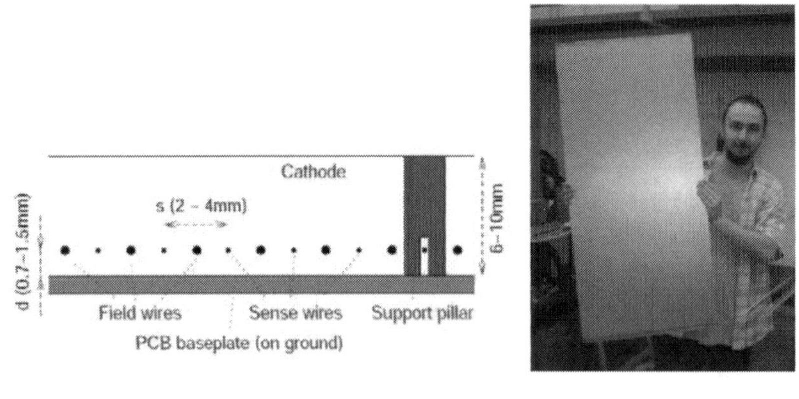

Figure 1. *Left panel*: the outline of the CCC detector. *Right panel*: CCC detector with the size of 0.5 m × 1 m (1 m long wires).

3. Portable tracking detectors for outdoor measurements

3.1. Detector structure

The surface of cosmic ray detectors needs to be as large as possible to acquire the largest statistics during the data taking. However, over-sized detectors are not appropriate to perform measurements inside unexplored underground caverns or artificial tunnels which may be difficult to access. These places require human handling of the equipments, which limits size and weight of a detector. In our case, the size of the detectors were designed to fit into caverns, described in Ref. [13], [14] and [15]. The mobile detector should be safely handled manually by a single person: the external size of the tracking system is 50 × 46 × 32 cm^3, and its weight is about

15 kg without additional power supplies and gas bottle. The *left panel* of Fig. 2 shows an installed station. The detector consists of a stack of five CCC layers. All these, including the DAQ and control, is housed inside a plexi-glass box which absorbs mechanical shocks during the deployment and isolates the detector from the humid environment. The CCC layers require low level continuous gas flow during the measurement of typically 0.5-1 L/h. A gas bottle with volume of 10 L and pressure of 150 bar is sufficient for 2 months.

Figure 2. *Left panel:* a portable detector with five tracking layers in a plexi-glass box and the additional power supply and gas system. *Right panel:* a the elements of the Raspberry Pi-based DAQ system for mobile detectors.

3.2. Data Acquisition (DAQ) systems for mobile detectors

Our first portable detector was equipped with a PIC32 micro-controller based DAQ system [13, 14, 15]. It consists of three main modules: a processor board controls the data acquisition and includes a low-voltage power supply; a high-voltage module operates the chambers; and finally a Human-Machine Interface module is responsible for maintenance and data storage. The key advantage of this DAQ is the low total power consumption of 5 W, whereas its dead time is 9.8 ms.

An upgraded DAQ, based on Raspberry Pi computer [16] and a custom designed board, called MtRD (*right panel* of Fig. 2) allows a high level remote control, data management and analysis. The MtRD is responsible for processing of data from the CCC layers. It includes a coincidence unit with up to 10 trigger lines. The actual trigger inputs and the coincidence settings can be selected manually on the MtRD. The individual trigger signals and the number of missed triggers are also recorded. The duration of the data readout (dead time) is 100 μs, considerably less than for the previous PIC32 based one. The DAQ software runs on the Raspberry Pi exploiting its fast GPIO pins. For standard operation, a convenient graphical user interface has been developed, running on any remote computer with internet connection to the Raspberry Pi (common internet link, or one of them as AP host; wired or wireless).

The electrical power for the detector is supplied through a single line nominally at +12 V DC. An efficient power converter ($\sim$ 80 %) supplies +5 V DC to the Raspberry Pi and the DAQ board. The power consumption of a complete detector system is 6 W at + 12 V DC, thus it can operate for about one week with a standard 50 Ah battery.

4. Muography experiments at shallow depths underground

4.1. Data taking time estimation as a function of zenith angle

The muon flux is maximal from the zenith, however measurements at different zenith angle are necessary under general conditions. It is then relevant to quantify the required time of data taking at a given depth and a given object – a cavern in this example. The muon flux $f(H)$ reduces as a function the density-length H, therefore the change (derivative) of $f(H)$ versus H allows one to detect the underground rock inhomogeneities. The relative error of the detection, directly related to the confidence level, for a $2R$ diameter cavern at the depth of H-h (see *left panel* of Fig. 3) is proportional to $\frac{f(H-2R)-f(H)}{\sqrt{f(H-2R)}} \approx 2R\frac{d}{dH}\left[\sqrt{f(H)}\right] = 2\,R\,f^*(H)$, where we assume $H \gg 2R$. The ratio of the required times of data taking to detect a cavern with the same confidence level is expressed by the following equation:

$$\frac{t(\theta)}{t(0)} = \frac{f^*(H)}{f^*(\frac{H}{\cos(\theta)})} \times \cos^{-2}(\theta) \times \cos^{-2}(\theta) \times A^{-1}(\theta) \qquad (1)$$

on the right side, the second factor is a reasonably good approximation of the angular distribution of cosmic muons, the third term is from the reduction of the angle of view if the cavern is located under the zenith angle of θ. Depending on detector geometry, an additional factor $A(\theta)$ determines detector acceptance relative to $\theta = 0$.

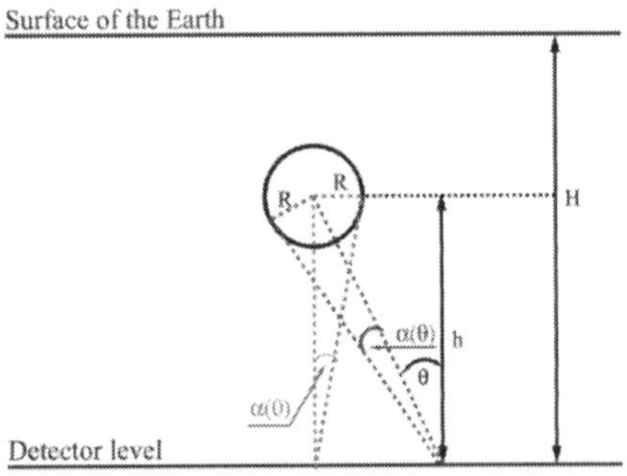

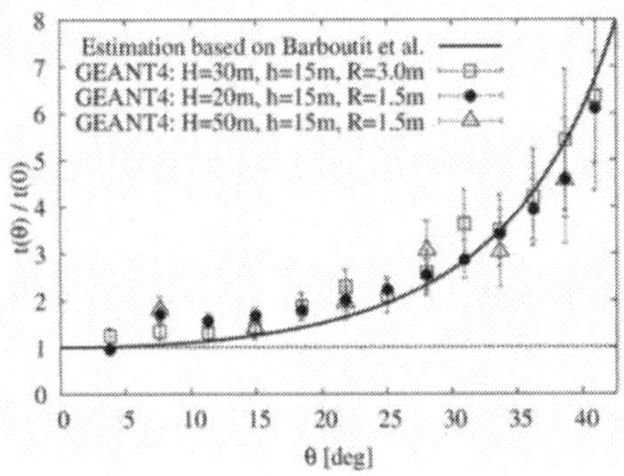

Figure 3. *Left panel:* The 2-dimensional view of the geometry of the simulations. *Right panel:* The relative duration of the data taking which is required to detect a cavern with level of significance of 3 sigma at the zenith angle of θ, with a horizontal detector.

The detector angle of view dependence has also been quantified by GEANT4 [18]. The simulation includes the penetration of atmospheric muons across a rock box with and without cavern (see in the *left panel* of Fig. 3). In our case, the CRY [19] cosmic ray generator provides the particle showers ($\mu^{\pm}$ and $e^{\pm}$) as the input for the detector simulations. Muon and electron showers have been generated within a 100 m × 100 m area at sea level. Standard rock (Z=11, A=22, and ρ = 2.65 g cm^{-3}) has been implemented with different thickness (H) and with different cavern diameters (R) as well as without any cavern. For simplicity, a "flat" detector model has been used with $A(\theta) = \cos(\theta)$; however this term can be set to $A(\theta) = 1$ if the detector is turned towards the object of interest. The time ratio of data taking to detect the cavern with 3 sigma confidence level has been calculated. Figure 3 shows the results of the GEANT4 simulation with different rock thickness (H) and cavern diameters (R) as well as the above detailed estimation which is used the muon flux versus density-length curve, measured by Barboutit *et al.* [17]. Both of the estimation and simulation results show the same trend. One can conclude that detectors with the angle of view greater than $\pm 45°$ for reconstruction of underground caverns require increased data taking time by an order of magnitude to collect sufficient statistics.

4.2. Angular resolution for cavern detection

Atmospheric muons suffer multiple scattering from the nuclei of the traversed material. The distribution of the scattering angles is approximately Gaussian with the following standard deviation:

$$\sigma = \frac{13.6\,\mathrm{MeV}}{\beta pc}\sqrt{\frac{X}{X_0}}\left[1 + 0.038\ln\left(\frac{X}{X_0}\right)\right] \tag{2}$$

where p is the momentum of the muon, X is the rock thickness, and $X_0 = 10$ cm is the radiation length of standard rock. Equation 2 is only true in the approximation that the muon momentum p does not change significantly between the entry and exit point, which is not fulfilled during muography applications: multiple scattering will increase towards the end of the muon's trajectory. Hence it is also not correct to conclude, that deeper detector will suffer from less multiple scattering.

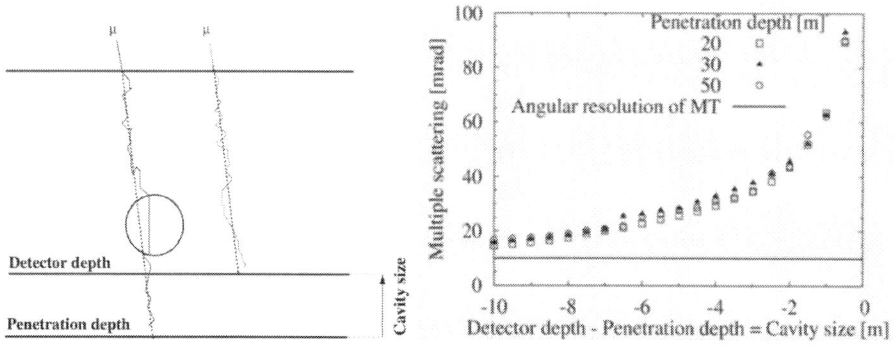

Figure 4. *Left panel:* The scheme of the muon multiple scattering of muons in rock. *Right panel:* The evolution of multiple scattering of cosmic particles which are stopped at the depth of 20 m (*empty rectangles*), 30 m (*filled triangles*) and 50 m (*empty circles*).

To quantify the effect, CRY and GEANT4 have been applied to generate atmospheric muons and simulate their penetration across the rock. Particle showers have been generated within a 300 m × 300 m area at sea level. Standard rock has been implemented with the thickness of 100 m. The angular scattering has been calculated at each half meter for those muons which could penetrate a certain thickness (penetration depth).

An interesting result arises if we measure the mean scattering angle as a function of the *difference* between the detector depth and the real penetration depth (see the *left panel of Fig. 4*). This actually corresponds to the observed cavity size: in case of a cavity (void), the increase of the muon flux happens *because of those muons* which penetrate more by the cavity size. The *right panel* of Fig. 4 shows the evolution of multiple scattering for those particles which are stopped at the depth of 20 m, 30 m and 50 m. The values fall on a universal curve. The angular resolution of the detector should be compatible with the smearing effect of multiple scattering to adequately detect rock inhomogeneities: 40 mrad smearing corresponds to 2 meter in difference between detector depth and penetration depth; in other words, a 2 m diameter cavity will appear with this smearing (because of multiple scattering) on the cosmic muon flux map *independently of its depth.* This is counter-intuitive: the multiple scattering smearing is indeed the same for larger depths at a given cavity (or other object) size.

4.3. Muon flux measurements in underground tunnel systems
The Jánossy Pit in the campus of the Wigner RCP is an underground tunnel system with three parallel tunnels under each other at 10 m, 20 m and 30 m depth with the size of $11 \times 2.5 \times 2.6\,\mathrm{m}^3$, $9.5 \times 2.5 \times 2.6\,\mathrm{m}^3$ and $23 \times 2.5 \times 2.6\,\mathrm{m}^3$ respectively. Furthermore, there are two additional tunnels at the level of 30 meter with the orientation of $\pm$ 60 deg from main tunnel. The soil density is $2.2 \pm 0.2\,\mathrm{g\,cm}^3$ and tunnels are built from concrete with the density of $2.4\,\mathrm{g\,cm}^3$. It is an excellent place to test the capabilities of the presented detector on muography at shallow depth.

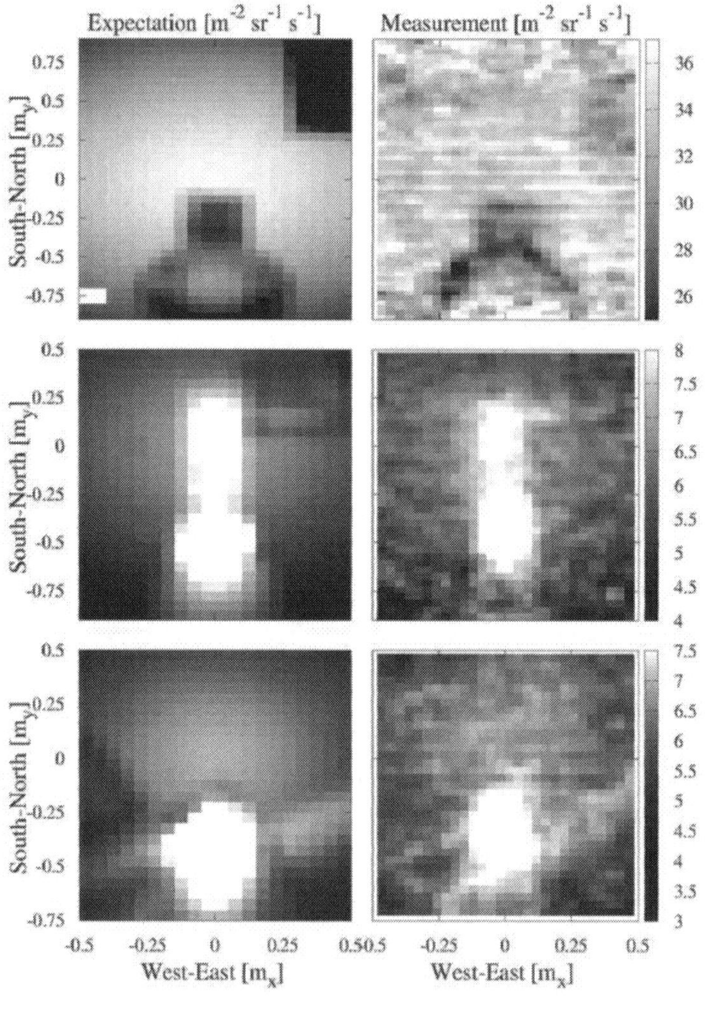

Figure 5. Muon flux measurements (*right panels*) have been compared with the GEANT4 simulation (*left panels*): the tunnels and the building structures are well reproduced by the measurements [20]. *Grayscale* shows the flux values divided with $cos^2(\theta)$.

The geometry of the Jánossy Pit has been implemented into GEANT4 with the above parameters. The expected angular dependent muon flux have been calculated from the empirical formulae of vertical muon flux versus depth [17] and the combination of the angular distribution of cosmic muons.

Figure 5 shows the expected (*left panels*) and measured (*right panels*) flux values divided with $cos^2(\theta)$. Specifically, the *upper panels* show the pit entrance dome of the tunnel system towards South and the building edge at the North-East corner from 10 meter depth. In the *middle panels*, two parallel tunnels are observed in the North-South orientation from 30 meter depth. In the lower panels, a tunnel at 20 meter depth has been observed in the North-East - South-West orientation from 30 meter depth.

5. CCC technology for material discrimination

As shown by Eq. (2), the high-Z materials (e. g. U, Pb) are particularly effective scatterers. Tracking devices placed around an object of interest (even vehicle or cargo container) can be used to obtain tomographic images about the high density regions inside the object [6, 7]. To demonstrate the applicability of CCC technology for such transmission muon tomography, an adequate tracking station has been built. The detector (see in the *left panel* of Fig. 6) consists of six tracking layers with the sensitive area of 0.5 m by 0.5 m and angular resolution of about 10 mrad. The *right panel* of Fig. 3 shows a 2-dimensional image of the scattering powers for different materials (Pb, Fe and Al). The data taking time to make an image with this precision was few hours without applicable advanced analysis methods. These presented results are encouraging in the direction that the CCC technology is useful for tomographic imaging.

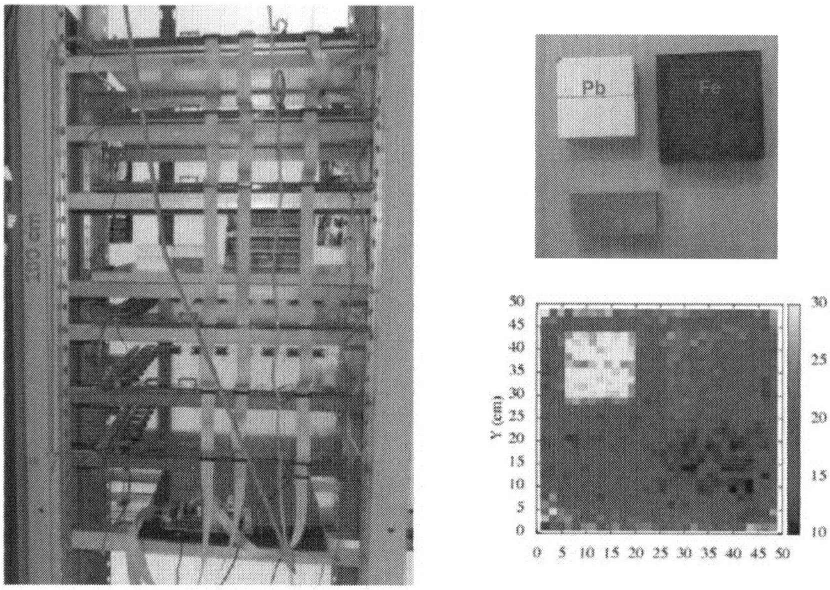

Figure 6. *Left panel: Left:* A cosmic setup for muon tomography with the investigated objects inside the middle CCC layers. *Right panel:* A photo and a 2-dimensional image of scattering strength in arbitrary units for Pb, Fe and Al absorbers.

6. Summary

The Close Cathode Chamber technology and its application in portable tracking detectors as well as their new data acquisition system have been presented. The detector parameters have been optimized for muography at shallow depth and also verified by GEANT4 simulation. The smearing effect in the angular distribution of muons due to their multiple scattering has been quantified, showing that the smearing only depends on the size of the object in study, and not on the detector depth.The angular resolution should be compatible with the smearing effect, e.g. the angular resolution of 40 mrad (2.3°) is sufficient to detect a cavern with the diameter of 2 meters. Muon flux measurements are ongoing in underground tunnels ($<$ 100 meter-rock-equivalent), showing good agreement with the expected muon flux. The results demonstrate the applicability of CCC-based portable tracking detectors to perform muography measurements. Furthermore, encouraging ground level studies have been performed as well: material discrimination is achievable by the measurement of multiple scattering of cosmic particles on the investigated materials, especially those of high Z. The reliable tracking performance, low power consumption, fair spatial and angular resolution and cost-efficiency make the CCC technology competitive also in large area tracking detectors for cosmic ray research.

Acknowledgement

Technical support by Gy. Bencédi, G. Kiss and the REGARD group is gratefully acknowledged. Authors (GGB,DV) also thank the János Bolyai Research Scholarship of the HAS. This work has been supported by the Lendület Programme of the Hungarian Academy of Sciences (LP2013-60) and the OTKA NK-106119 grant.

References

[1] E. P. George: 1955 *Commonwealth Engineer* 455-457
[2] L. W. Alvarez *et al.*: *Science* **167** 832-839
[3] E. Caffau, F. Coren, G. Giannini: 1997 *Nucl. Instrum. Meth.* A **385** 480-488
[4] K. Nagamine *et al.*: 1995 *Nucl. Instrum. Meth.* A **356** 585-595
[5] H. Tanaka *et al.*: 2014 *Nat. Commun. 5:3381 doi: 10.1038/ncomms4381*
[6] K. N. Borozdin *et al.*: 2003 *Nature* **422** 277
[7] L. J. Schultz *et al.*: 2004 *Nucl. Instrum. Meth.* A **519** 687-694
[8] H. Fujii *et al.*: 2013 *Prog. Theor. Exp. Phys.* **073C01** 20
[9] H. Miyadera *et al.*: 2013 *AIP ADVANCES 3* **052133**
[10] regard.kfki.hu
[11] D. Varga *et al.*: 2011 *Nucl. Inst. and Meth.* A **648** 163-167
[12] D. Varga *et al.*: 2013 *Nucl. Inst. and Meth.* A **698** 11-18
[13] G. G. Barnaföldi *et al.*: 2012 *Nucl. Inst. and Meth.* A **689** 60-69
[14] L. Oláh *et al.*: 2012 Geosci. Instrum. Method. Data Syst. Vol **1** 229-234
[15] L. Oláh *et al.*: 2013 *Advances in High Energy Physics* Vol **2013** Article ID 560192, 7
[16] http://www.raspberrypi.org/
[17] A. I. Barboutit and B. C. Rastin: 1983 *J. Phys. G: Nucl. Phys.* **9** 1577-1595
[18] S. Agostinelli *et al.*: 2003 *Nucl. Instrum. Meth.* A **506** 250-303
[19] http://nuclear.llnl.gov/simulation/main.html
[20] L. Oláh *et al.*: 2014 *Astroparticle, Particle, Space Physics, Radiation Interaction, Detectors and Medical Physics Applications* Vol **8** 280-284

Are the primary cosmic ray and EAS spectra the same or not?

A.A. Petrukhin

National Research Nuclear University MEPhI
(Moscow Engineering Physics Institute), Moscow 115409, Russia

E-mail: AAPetrukhin@mephi.ru

Abstract. Usually it is believed that the energy of extensive air showers (EAS) is equal or is proportional to the energy of primary particle. Of course, taking into account fluctuations in hadron interactions and in the EAS development, some difference between the energies of the EAS and PCR appears but the slopes of the spectra must be the same with a good accuracy. In this talk, an alternative approach, in which the model of PCR interaction at the knee energy is drastically changed, is considered. The consequences for interpretation of results on primary energy spectrum and composition in frame of this model are analyzed. The results agree with experimental data. Some possible experiments to check the predictions of this approach are discussed.

1. Introduction

The main characteristics of primary cosmic rays (PCR) are energy spectrum and composition. The only method of their investigation above 10^{15} eV is the study of EAS in the atmosphere and/or at the Earth's surface, and all so-called "experimental data" about energy spectrum and composition of PCR are results of interpretation of EAS parameter measurements.

In principle, two approaches to interpretation of results of EAS parameter measurements are possible. In cosmophysical approach, it is supposed that all serious changes in EAS parameters with increasing of energy are results of PCR energy spectrum and composition change. In nuclear-physical approach, corresponding changes can be connected with interaction model change. But for that really "new physics" must be included into hadron interaction model. Such approach is developed by the author since 1999 [1] and its subsequent development is described elsewhere [2, 3].

2. Cosmophysical approach

The most part of cosmic ray community supports the cosmophysical approach, and all observed changes in EAS parameter behavior are considered from this point of view. The following assumptions are taken into account:

– Energy of primary particles is equal to EAS energy which is evaluated from measured parameters in the frame of one of the known interaction models.

– All changes of EAS characteristics in dependence on energy are results of PCR energy spectrum and composition changes only.

– PCR have galactic origin with limiting energies of their acceleration and keeping in Galaxy, which are determined by particle charge Z.

Results of interpretation of various EAS measurements and their explanation in frame of the cosmophysical approach are given in figure 1, in which the main peculiarities of the energy spectrum are shown: the knee, the second knee, the ankle and GZK-cutoff.

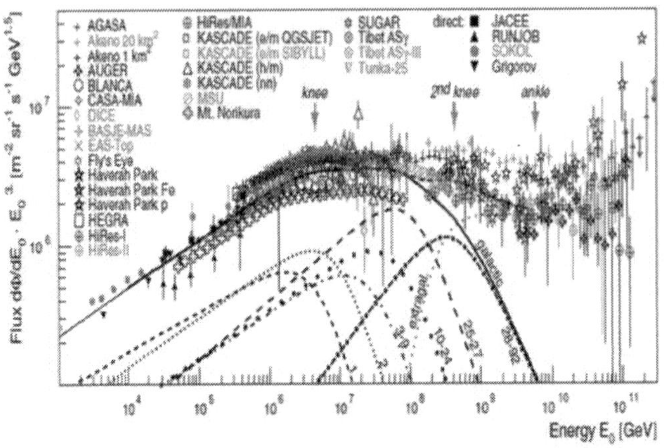

Figure 1. All-particle energy spectrum of primary cosmic rays and its explanation in frame of cosmophysical approach [4].

As it is seen from figure 1, the both knees are connected with change of PCR composition. But it follows from this figure that the composition must change in favor of heavy nuclei in the beginning slowly, than more quickly up to iron nuclei. But the results of average logarithm of mass number <ln$A$> evaluations based of N_μ/N_e – ratio measurements show another picture (figure 2).

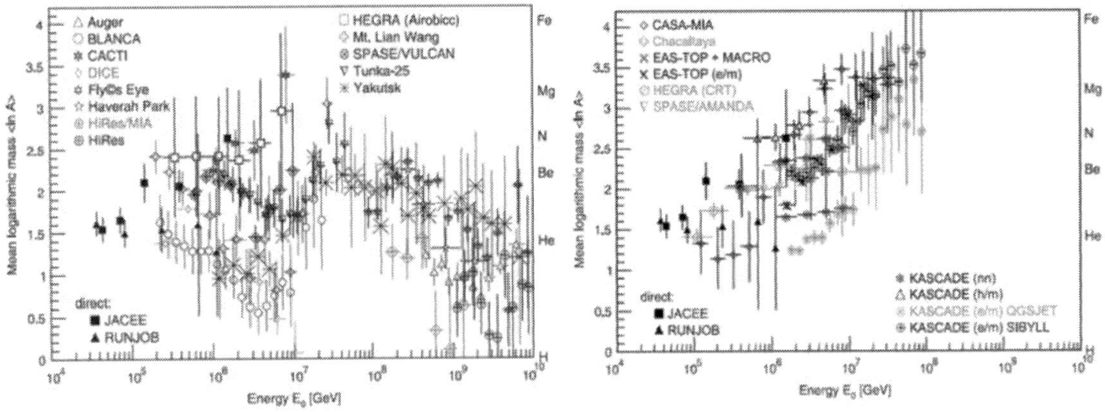

Fig. 2. Mean logarithmic mass of cosmic rays derived from the average depth of the shower maximum (left), from the measurements of electrons, muons, and hadrons at ground level (right) [5].

Above the knee <ln$A$> in the beginning is increased relatively sharply and then more slowly is decreased, and does not reach value <ln$A$> = 4.2 corresponding to pure iron composition. The same situation is in X_{max} measurements (figure 3).

Figure 3 also illustrates the fast change of composition to heavy nuclei, then slowly moving back to proton composition. These results do not agree with picture in figure 1 either.

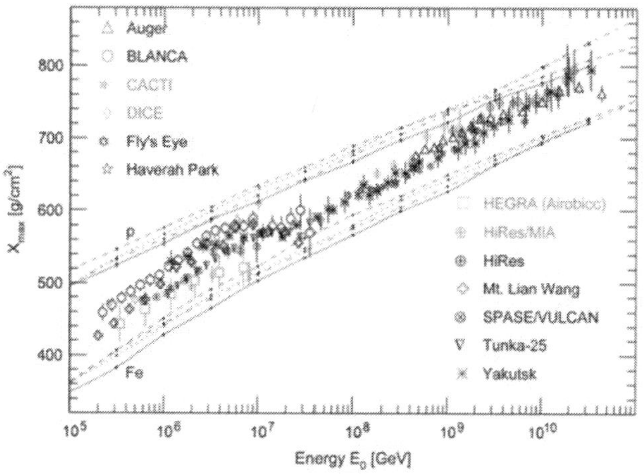

Figure 3. Average depth of the shower maximum X_{max} as a function of primary energy [5].

3. Nuclear-physical approach

A necessity of this approach consideration is connected not only with mentioned drawbacks of cosmophysical approach, but with other results obtained in investigations of cosmic rays with energies more than 10^{15} eV. In many experiments some unusual results were obtained. These results were discussed in various papers (see for example [6, 7]), therefore here the list of these events is presented only.

– In hadron experiments: halos, unusual families, alignment, penetrating cascades, long-flying component, large transferred momenta, Centauros and Anti-Centauros.

– In muon experiments: excess of VHE muons (> 100 TeV) and excess of muon bundles at large zenith angles – so-called "muon puzzle" [8].

– In EAS measurements: the mentioned above change of EAS energy spectrum (the knee), changes in the behavior of N_μ/N_e -ratio and X_{max}, and also excess of young and old showers – all these phenomena must be considered in frame of possible changes of interaction model as unusual.

For explanation of a whole set of unusual events and phenomena, a new model of hadron interaction must satisfy the following requirements:

1. Threshold behavior (unusual events appear at several PeV only).
2. Large cross-section (to be observed in cosmic ray experiments).
3. Large yield of leptons ("muon puzzle", missing energy, penetrating cascades).
4. Large orbital momentum (alignment).
5. Change of EAS development (appearance of missing energy to explain the knee, increasing of N_μ/N_e – ratio, decreasing of X_{max}, production of young and old showers).

Among possible variants of new interaction model (new type of interaction, new massive particle etc.) the most realistic is production of blobs of quark-gluon plasma. Of course, it is better to speak about quark-gluon matter (QGM), which can be considered in frame of known physical laws and rules.

4. Blobs of quark- gluon matter in cosmic rays

Production of QGM blobs provides fulfillment of two main conditions: threshold behavior, since for that a certain temperature (energy) is required; large cross section since in this case the transition from quark-quark interaction to some collective interaction of many quarks and gluons occurs, and geometrical cross section will be changed from $\sigma = \pi\lambda\!\!\!^{-2}$ to $\sigma = \pi R^2$, where R is a size of interacting quark-gluon matter blob.

For fulfillment of other conditions and explanation of corresponding unusual events, a large value of orbital angular momentum is required. The solution of this problem was proposed by two Chinese physicists [9]. They showed that orbital angular momentum L in non-central ion-ion collisions is produced (figure 4), and its value is increasing with collision energy in the center-of-mass system as $L \sim \sqrt{s}$.

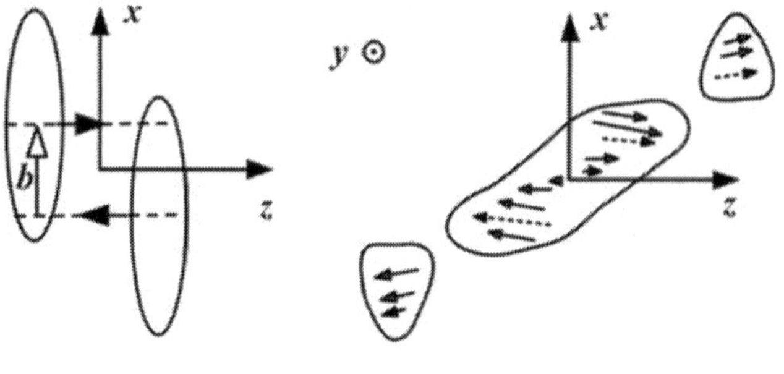

Figure 4. Production of orbital angular momentum in non-central ion-ion collisions in the center-of-mass system [9].

Further investigation of such collisions showed that the value of the orbital momentum can be very large. In figure 5, the results of L calculations for Au-Au collisions at the RHIC energy as a function of the impact parameter b in units of the nucleus radius R_A are shown [10].

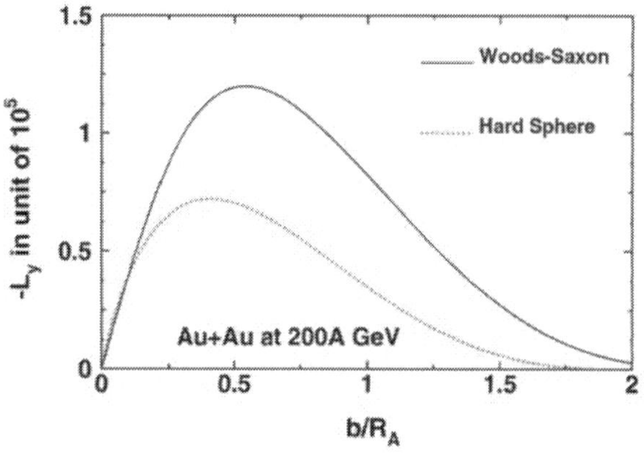

Figure 5. Total orbital angular momentum of the overlapping system in Au+Au collisions at the RHIC energy as a function of the impact parameter b [10].

The appearance of a large orbital momentum leads to production of a centrifugal barrier, which correspondingly will be also increased with grows of $\sqrt{s}$. In the center-of-mass system, the value of the centrifugal barrier for a particle with mass m will be

$$V(L) = L^2 / 2mR^2 .$$

This barrier will be large for light particles (u, d, s, c, b-quarks) but small for heavy particles (t-quarks). And though top-quarks are absent in usual hadron matter, the suppression of decay into light quarks gives time for production of even such heavy quarks as top.

Production of t-quarks with a relatively large probability drastically changes the hadron interaction model and allows explain many unusual results observed in cosmic rays. Live time of top-quarks is of the order ~ 10^{-25} s and their decay modes are

$$t(\bar{t}) \rightarrow W^+(W^-) + b(\bar{b})$$

with a consecutive chain of decays $b(\bar{b} \rightarrow)c(\bar{c} \rightarrow)s(\bar{s} \rightarrow)u(\bar{u})$. Of course b-quark can interact and produce a jet. W – bosons decay into leptons (≈ 30%) and hadrons (≈ 70%). Lepton decay mode allows explain missing energy and excess of VHE muons. W can decay into $e\nu_e$, $\mu\nu_\mu$, $\tau\nu_\tau$ with energy of each lepton about 40 GeV in the center-of-mass system. But taking into account the very large primary particle energy their energy in laboratory system will be more than 100 TeV and will be increased with growth of primary energy. This process gives an excess of VHE muons. Since energy of neutrino is not measured they give missing energy. Hadron mode (mainly pions, on average ~ 20 particles) provides a change of EAS development (by making it more quick) that allows explain changes in the behavior of N_μ/N_e – ratio and X_{max}.

5. EAS energy spectrum in new approach

In frame of the considered model, the following changes in description of interaction of cosmic rays with nuclei of the atmosphere must appear:

1. Transition from quark-quark interaction to collective interaction of many quarks and gluons must change the formula for energy in the center-of-mass system

$$\text{from} \quad \sqrt{s} = \sqrt{2m_N E_1} \quad \text{to} \quad \sqrt{s} = \sqrt{2m_c E_1},$$

where m_N is the mass of nucleon (p, n), m_c is the total mass of many quarks of the target (a part of nuclei of nitrogen or oxygen). In the simplest case, it is possible to believe, that

$$m_c = nm_N,$$

where n can be in the interval $1 \div A$. Such leap in value of $\sqrt{s}$ explains sharp inclusion of new process since the angular orbital momentum $L \sim \sqrt{s}$ and centrifugal barrier $V(L) \sim s$ will be also sharply changed.

2. Production of $t\bar{t}$-pair must decrease $\sqrt{s}$ at least by value $2m_t$ and in a general case by some value ε_t which will depend on primary particle energy and its mass. Dependence on primary particle mass A is connected with threshold energy for QGM blob production, which depends on A (see formulas below on page 7). Only the remaining part of energy in the center of mass system $\sqrt{s} - \varepsilon_t$ will be converted into the energy of usual processes of EAS development. And standard measurements and evaluation of EAS energy will give the value $E_2 = \left(\sqrt{2m_c E_1} - \varepsilon_t\right)^2 / 2m_c$ which is less than primary particle energy E_1. As a result, we will obtain steepening of the measured spectrum though real spectrum does not change its slope.

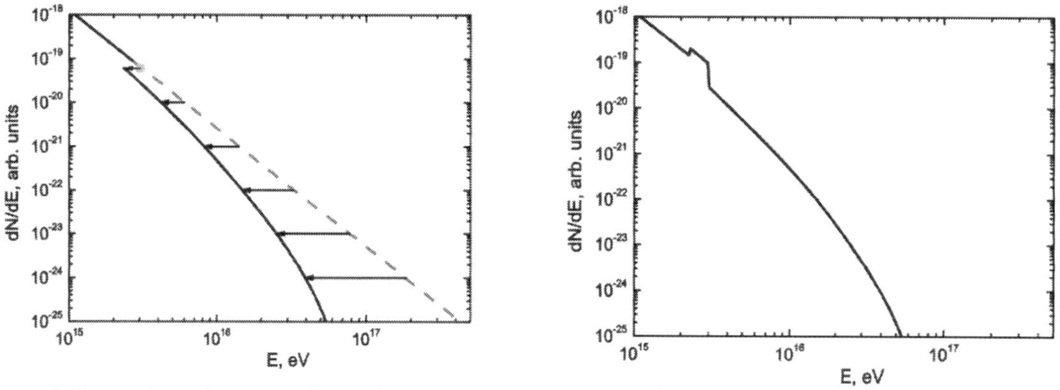

Figure 6. Formation of measured cosmic ray energy spectrum in frame of nuclear-physical approach.

Figure 6 (left) illustrates a transition from the primary spectrum with a single slope to the measured energy spectrum with a variable slope. If not to take into account this underestimation of energy, the energy spectrum with a bump can be obtained (figure 6, right). Taking into consideration the measured energy straggling, the knee will be imitated.

3. Since for the production of quark-gluon matter not only high temperature (energy) but also high density is required, it is reasonable to assume that firstly QGM will be produced in nucleus – nucleus interactions but not in proton – nucleus interactions. This means that in cosmic ray interactions the first component (at the same energy per nucleus) which will interact with the production of QGM will be iron nuclei (or more heavy), then more light nuclei and the last – protons. Figure 7 illustrates the situation.

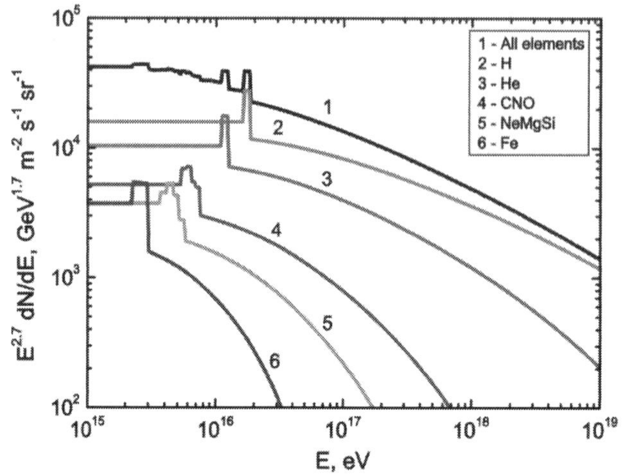

Figure 7. Changes of various nuclei spectra in the frame of the considered model.

As a result, the picture of the measured energy spectrum and composition above the knee is drastically changed (figure 8).

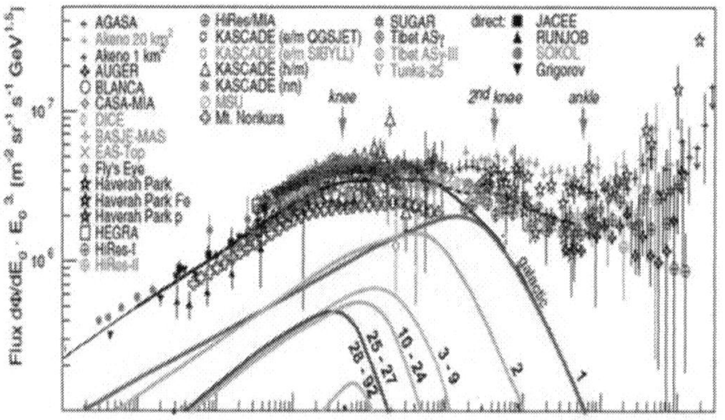

Figure 8. Measured changes of contributions of various cosmic ray components to primary spectrum.

Figure 8 is made by an analogy with figure 1. The curves are not results of calculations. They illustrate the principal change of the approach. In figure 1, *the changes* of primary spectra of different cosmic ray components are shown. In figure 8, *the results of measurements* of different cosmic ray components are indicated, which really do not change their slopes. So, in this approach the knee is the result of our interpretation of data measurements.

For calculation of all-particle energy spectrum, the simplest model of QGM blobs was used:

$$\varepsilon_t = 4m_t \sqrt{\frac{E_1}{E_{\text{th}}}} \ln\left(1.72 + \frac{E_1}{E_{\text{th}}}\right).$$

Journal of Physics: Conference Series **632** (2015) 012021 doi:10.1088/1742-6596/632/1/012021

Threshold energy E_{th} of QGM blob production is determined by the mass number of primary particle A and was taken into account as

$$E_{\text{th}} = E_{\text{knee}} \left(56 / A\right)^{0.5}.$$

Dependence of the compound mass on the mass number of primary particle was taken in the form

$$m_c = 2m_N A^{0.25}.$$

Primary spectra of various nuclei were taken as extrapolation from [11]. The 15% straggling of measured energies was assumed. The results of calculations and comparison with experimental data on the all-particle energy spectrum are given in figure 9. Very good agreement is observed.

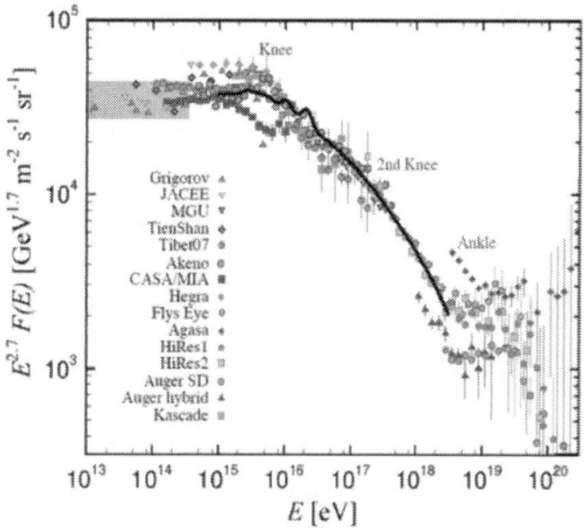

Figure 9. Calculated and experimental data on all-particle spectrum.

6. How to check new approach?

There are several possibilities to check the new approach as in LHC experiments so in cosmic ray investigations.

Of course, the most convincing results can be obtained in LHC experiments, since the considered approach has a precise signature: sharp increasing of top-quark production. Other predicted peculiarities: the increase of missing energy, excess of muons from decays of W-bosons and b, c-quarks can be observed in LHC experiments, too.

However these results cannot be obtained in p-p-interactions even at full LHC energy 14 TeV, which corresponds to 10^{17} eV in cosmic ray experiments (for p-p-interaction). If the considered approach is correct then for production of quark-gluon matter with described features at these energies the collisions of sufficiently heavy nuclei are required. But the experiments for searching of t-quark in nucleus-nucleus interactions at LHC are not even planned yet. However, the evidences in favour of the considered approach have been obtained in Pb+Pb collisions at LHC: more sharp increasing of secondary particle multiplicity and detection of imbalance events. The analysis of these events in the frame of the considered model was done in paper [12].

In cosmic rays, there are two possibilities to check the considered approach. The first of them are direct measurements in space of primary energy spectra for different nuclei to see if their energy spectra change. But such experiments are not possible in the nearest future since for that very large detectors are required.

The second and more realistic possibility are measurements of characteristics of muon flux in the atmosphere: inclusive muon energy spectrum above 100 TeV and energy deposit of EAS muon component below and above the knee.

Since for VHE (above 100 TeV) muon production the decays of heavy or/and short-lived particles are required, so the indisputable proof of existence of the excess of such muons will uniquely evidence in favor of the realization of the considered scenario. The existence of VHE muon excess in two experiments (Baksan underground scintillation telescope [13] and IceCube [14]) was observed but poor statistics [13] and some methodical uncertainties [14] require further investigations which can be done in IceCube experiment.

Energy spectrum of muon bundles is the only unexplored parameter of EAS and its direct measurements are practically impossible. But evaluation of the energy deposit of muon bundles at various EAS energies and comparison of results of the measurements below and above the knee can give an evidence for interaction model change at energies above the knee. Such experiment is now in progress at the experimental complex NEVOD-DECOR and the first preliminary results do not exclude such possibility [15].

7. Conclusion

Considered approach to interpretation of results of EAS parameters measurements allows solving problems connected with PCR energy spectrum and composition. And apparently for the first time for many ten years the experiments in cosmic rays have serious advantages compared to accelerator ones in searching of a new state of matter.

Acknowledgments

Author thanks Alexey Bogdanov and Rostislav Kokoulin for the help in preparation of this paper. The research has been performed in the Scientific and Educational Centre NEVOD with the State support from the Ministry of Education and Science of the Russian Federation (government task and project RFMEI59114X0002) and the grant of the Leading Scientific School NSh-4930.2014.2.

References

[1] Petrukhin A A 1999, *in Proceedings of the XIth Rencontres de Bloois "Frontiers of Matter"* J. Tran Thanh Van (ed), The Gioi Publishers, p. 401
[2] Petrukhin A A 2006 *Nucl. Phys. B (Proc. Suppl.)* **57** 151
[3] Petrukhin A A 2011 *Nucl. Phys. B (Proc. Suppl.)* **212** 235
[4] Hoerandel J 2007. *Mod. Phys. Lett. A* **22** 1533
[5] Bluemer J 2009 Progr. Part. & Nucl. Phys. **63** 293
[6] Slavatinsky S A 2003 *Nucl. Phys. B (Proc. Suppl.)* **3** 122
[7] Petrukhin A A 2004 *in Proceeding of the Vulcano Workshop Frontier Objects in Astrophysics and Particle Physics* F. Giovanelli & G. Mannocchi (eds), p. 489.
[8] Petrukhin A A 2014 *Nucl. Instrum. Meth. A* **742** 228
[9] Zuo-Tang Liang and Xin-Nian Wang 2005 *Phys. Rev. Lett.* **94** 102301
[10] Jian-Hua Gao *et al* 2008 *Phys. Rev. C* **77** 044902
[11] Amsler C *et al* 2008 *Phys. Lett. B* **667** 1
[12] Petrukhin A A 2013 *Acta Polytechnica (Suppl.)* **53** 707
[13] Bogdanov A G *et al* 2012 *Astroparticle Physics* **36** 224
[14] Berghaus P, Xu C 2011 *Presented at 32nd International Cosmic Ray Conference*, Beijing
[15] Kovylyaeva E A *et al* 2014 *Bulletin of the Lebedev Physics Institute* **41** 218

Interpretation of Voyager 1 data on low energy cosmic rays in galactic wind model

V S Ptuskin[1], **E S Seo**[2] **and V N Zirakashvili**[1]
[1] IZMIRAN, Troitsk, Moscow 142190, Russia
[2] University of Maryland, College Park, MD 20742, USA

E-mail: vptuskin@izmiran.ru

Abstract. The local interstellar energy spectra of galactic cosmic rays down to a few MeV/nucleon were directly measured in the experiment on the board of the Voyager 1 spacecraft. We suggest interpretation of these data based on our models of cosmic ray acceleration in supernova remnants and the diffusion in galactic wind where diffusion coefficient is determined by the cosmic ray streaming instability. The dependence of wind velocity on distance above the Galactic disk is determined.

1. Introduction

Launched in 1977 and expected to operate till 2020-2025, Voyager 1 is 130 AU from the Sun and Voyager 2 is at 107 AU. Voyager 1 is currently in "interstellar space" and Voyager 2 is in the "heliosheath" – the outermost layer of the heliosphere where the solar wind is slowed by the pressure of interstellar gas. This experiment extends our knowledge of the local interstellar cosmic ray spectrum and composition down to MeV energies [1]. It is assumed that the Voyager 1 spacecraft is measuring cosmic ray intensity not affected by the solar wind modulation. The spectra of protons, nuclei and electrons look as a natural continuation of spectra at energies above a few hundred MeV and no any additional low energy component was revealed. The intensity of the so called Anomalous Cosmic Rays (ACR) is diminished to zero and it confirms that these particles are accelerated somewhere in the inner part of the heliosheath (probably, at the hot spots of the solar wind termination shock). The flat shape of proton and nuclei spectra at energies less than 100 MeV/nucleon is likely explained by the combined effect of the leakage of these particles from the Galaxy and the heavy ionisation energy losses. The corresponding calculations were made in [2] based on the Voyager data in the outer heliosphere shortly before the spacecraft entered the interstellar medium. The interpretation of the final data [1] in the frameworks of the diffusion Galactic model with the cosmic ray diffusion coefficient which has minimum at about 1 GeV/nucleon and increases to smaller energies was given in [3]. Qualitatively, such energy dependence of diffusion is expected in a model with damping of the interstellar turbulence on cosmic rays [4]. The data on cosmic ray electrons and Galactic synchrotron radioemission were analysed in the same model in [5]. Other versions of the diffusion model for this problem were considered in [6] (the model with distributed cosmic ray reacceleration) and [7] (the model with diffusion and adiabatic energy losses in galactic wind).

In the present work, we analyse the Voyager data [1] by developing the model of cosmic ray transport in the Galaxy with galactic wind proposed in [8] for the interpretation of available at that time data on the ratio of fluxes of Boron-to-Carbon nuclei at low energies. The specific feature of this model is the deceleration of the Galactic wind flow at distances up to about 1 kpc above the galactic disk.

2. Transport of low energy cosmic rays in galactic wind model

The steady-state one-dimensional (perpendicular to the galactic plane) transport equation used in the present paper for protons and stable cosmic ray nuclei in the interstellar medium reads as follows (see [9] for detail):

$$-\frac{\partial}{\partial z}D\frac{\partial f}{\partial z}+u\frac{\partial f}{\partial z}-\frac{p}{3}\frac{du}{dz}\frac{\partial f}{\partial p}-\frac{1}{p^2}\frac{\partial}{\partial p}\left(p^2 b_{ion}\frac{\partial f}{\partial p}\right)+nv\sigma f=q, \tag{1}$$

where $f(p,z)$ is the isotropic part of cosmic ray density in momentum space; p is the particle momentum; z-axis is directed perpendicular to the galactic midplane, which has the coordinate $z=0$. $D(p)=vF(p/Z)$ is the spatial diffusion coefficient, v is the particle velocity, the function F is determined by the spectrum of interstellar MHD turbulence, Z is the particle charge. $u(z)$ is the convection velocity, the velocity of galactic wind, directed symmetrically outward of galactic plane, $u(z)=-u(-z)$. The term with $\frac{du}{dz}$ represents the adiabatic momentum gain or loss in the nonuniform gas flow with a frozen-in magnetic field whose inhomogeneities scatter cosmic rays. The term $q(p,z)$ is the source term including primary and spallation contributions (the primaries are produced by cosmic ray sources). n is the interstellar gas density, σ is the total spallation cross section. The ionization energy loss rate is $(\frac{dp}{dt})_{ion}=-b_{ion}$.

We assume in the following that the cosmic ray sources and the interstellar gas, where nuclear spallation and ionization energy losses occur, are confined in a thin galactic disk and the following approximations by the delta functions in eq. (1) are applied:

$$q=s\delta(z),\ n=\eta\delta(z),\ b_{ion}=b\delta(z). \tag{2}$$

The source spectrum is assumed in the form

$$s=Qs_0,\ s_0=R^{-2.28}[1+0.07log(R/0.7)log(R/100)],\ Q=const, \tag{3}$$

where $R=pc/Z$ is the particle magnetic rigidity in GV, c is the speed of light. Eqs. (3) are based on our results of modelling of cosmic ray shock acceleration in supernova remnants [10].

It is assumed that the diffusion coefficient D is selfconsistently determined by the streaming MHD instability of cosmic rays leaving the Galaxy [11, 12]. The corresponding equations for the diffusion coefficient are the following:

$$D=\kappa D_0,\ D_0=\beta(R\times s_0(R))^{-1},\ \kappa=const. \tag{4}$$

Here $\beta=v/c$. The expression for D_0 is replaced by $D_0=\beta(R^2\times s_0(R))^{-1}$ at the smallest energies where the ionization energy loss dominates over the leakage from the Galaxy.

We continue to find the approximate analytical solution of eq. (1), see also [12].

The distinctive feature of the diffusion-convection transport is the existence of a boundary layer of the size $z_c(p)$ adjacent to the galactic plane and defined by the approximate equation

$$z_c(p)=D(p)/u(z_c). \tag{5}$$

Here and below we omit the numerical coefficients of the order of 1 in the approximate equations.

The diffusive propagation with the diffusion coefficient $D(p)$ dominates in the cosmic ray transport at distances $0<z<z_c$ (the diffusion region) whereas the convective transport with velocity $u(z)$ dominates at $z>z_c$ (the advection region). This propagation pattern is valid if the wind velocity increases with distance, remains constant, or decreases not faster than $u \sim 1/z$. Accelerated in the sources (the supernova remnants) located in a relatively thin galactic disk, the relativistic particles fill an extensive cosmic ray halo and have low probability to return back to galactic disk if they enter the advection region where cosmic rays are carried out to the intergalactic space. Thus, for an observer in the galactic disk, the surface $z = z_c$ can be considered as an absorbing boundary.

We solve eq. (1) separately in the region $0 < |z| < z_c$, where the terms with the wind velocity $u(z)$ are neglected, and in the region $z_c < |z| \leq H$, where the diffusion term is neglected (H is the height of cosmic ray halo). The conditions of continuity for the function f and for the diffusion - convection flux $j = -D\frac{\partial f}{\partial z} - \frac{u}{3}p\frac{f}{p}$ are used to sew solutions together at $|z| = z_c$. Finally, the following equation for the cosmic ray intensity in the Galactic disk $J_0(E)$, where E is the energy per nucleon, can be derived:

$$\frac{J_0}{X} + \frac{d}{dE}\left[\left(\frac{dE}{dx}\right)_{ion} J_0\right] + \frac{\sigma}{m_{ism}} J_0 = \frac{S(E)}{\mu}, \tag{6}$$

where

$$X = X_e \varphi_{nl},\ X_e = \frac{\mu v}{2u(z_c)},\ \varphi_{nl} = \left[1 - 3\int_p \frac{dp_1}{p_1}\left(\frac{p}{p_1}\right)^5 \frac{J_0(E(p_1))}{J_0(E(p))}\right]^{-1}. \tag{7}$$

Here the source term is $S(E) = \frac{Ap^2}{v} s(p)$, m_{ism} is the average mass of interstellar atom, $\mu = m_{ism}\eta$ is the surface mass density of interstellar gas in the Galactic disk ($\mu \approx 10^{-3} g/cm^{-2}$).

Eq. (6) generalizes the well known "leaky-box" approximation [9] to the case of diffusion-convection transport equation with the dependent on distance convection velocity $u(z)$. Eq. (6) for J_0 is nonlinear since the escape length X depends on J_0 through function φ_{nl} because of the adiabatic energy loss of cosmic ray particles in the galactic wind with a nonconstant velocity.

3. Results of calculations

The determination of the model parameters that fit the observational data is performed by selecting the proper escape length X and the coefficient κ. The results of our calculations of $J_0(E)$ and the wind velocity $u(z)$ are shown in Figures 1 and 2 for

$$X_e = \frac{34\beta}{D_0(R)^{0.5} + (D_0(R)^{0.5}/1.3)^{-0.71}}\,\mathrm{g/cm^2},\ \kappa = 2 \times 10^{27}\,\mathrm{cm^2/s},\ u(z_c) = \mu v/2X. \tag{8}$$

The calculated spectra of protons and Helium are shown by solid lines. The Voyager 1 observations [1] at low energies are labeled by symbols (circles and squares). The high energy PAMELA data [13] are approximated by dotted lines. The calculations are in reasonably good agreement with the Voyager 1 data at low energies and the PAMELA data at high energies. The break in the helium spectrum at energies where transition from the calculated to the measured by PAMELA spectrum occurs in Figure 1 can reflect the difference in the source spectra of protons and Helium that was not taken into consideration in the present calculations.

The delta-function approximation (2) for the interstellar gas and the cosmic-ray source distributions in the galactic disc may affect the calculated energy spectra at the lowest energies. The account for the realistic distributions of gas and sources are beyond the scope of the analytical approach used in the present calculations. The $1D$ approximation used in our calculations is justified because the cosmic ray density in the galactic disk is determined by the transport coefficients in the vicinity z_c of the point of observation and the value of z_c is less than about 2 kpc at energies below 1 GeV/nucleon.

The derived galactic wind velocity is minimal at the distance about 2 kpc above the galactic disk and is rising linearly at larger distances where it merges with the galactic wind solution [11]

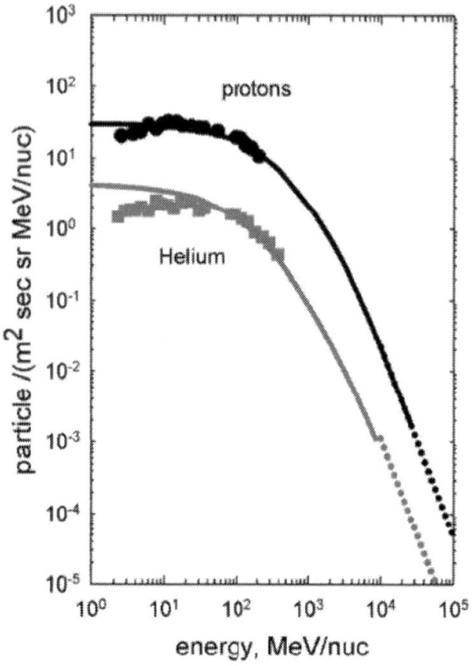

Figure 1. Calculated fluxes of protons and Helium as function of energy are shown by solid lines. Voyager 1 observations [1] at low energies are labeled by symbols (circles and squares). High energy PAMELA data [13] are approximated by dotted lines.

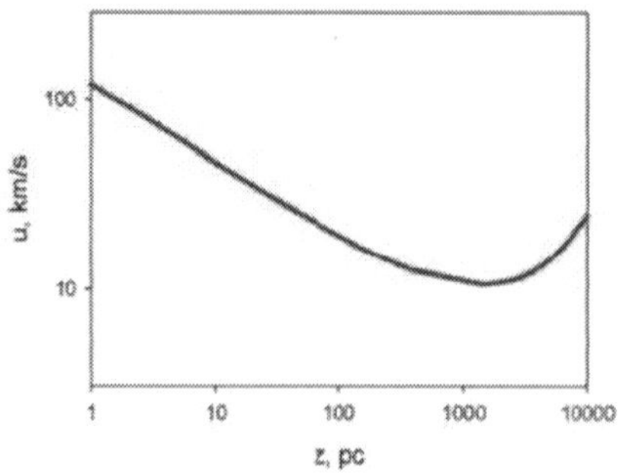

Figure 2. Calculated dependence of galactic wind velocity on distance from galactic disk.

obtained in the model of galactic wind driven by cosmic ray pressure. The decrease of velocity u with distance at $|z| < 2$ kpc may reflect the pattern of the gas flow produced by the supernova explosions in the Galactic disk.

Acknowledgments
The work was supported by the Russian Foundation for Basic Research grant 13-02-00056 and by the Russian Federation Ministry of Science and Education contract 14.518.11.7046. Part of this work was fulfilled during VSP visit to the University of Maryland, USA, where it was

supported by the NASA grant NNX13AC46G.

References

[1] Stone E C, Cummings A C, McDonald F B, Heikkila B C, Lala N and Webber W R 2013 *Science*, **341** 150
[2] Webber W R and Higbie P R 2009 *J. Geophys. Res.*, **114** 02103
[3] Webber W R, Higbie P R and McDonald F B 2013 *A comparison of the galactic cosmic ray H, He and C/O nuclei spectra measured between 5 and 500 MeV/nuc beyond 122 AU at Voyager 1 with the predictions of a diffusion model for propagation in the Galaxy, eprint arXiv*: 1308.4426
[4] Ptuskin V S, Moskalenko I V, Jones F C, Strong A W, Zirakashvili V N 2006 *Dissipation of magnetohydrodynamic waves on energetic particles: impact on interstellar turbulence and cosmic-ray transport, Astrophys. J.* **642** 892
[5] Webber W R and Higbie P R 2013 *Calculations of the propagated LIS electron spectrum which describe the cosmic ray electron spectrum below 100 MeV measured beyond 122 AU at Voyager 1 and its relationship to the PAMELA eelectron spectrum above* 200 *MeV, eprint arXiv*: 1308.6598
[6] Lave K A, Weidenbeck M E, Binns W R, Christian E R, Cummings A C, Davis A J, deNolfo G A, Israel M H, Leske R A, Mewaldt R A, Stone E C and von Rosenvinge T T 2013 *Astrophys. J.* **770** 117
[7] Schlickeiser R, Webber W R and Kempf A 2014 *Astrophys. J.* **787** 35
[8] Soutoul A and Ptuskin V S 1999 26*th Intern. Cosmic Ray Conf.* **4** 184
[9] Berezinsky V S, Bulanov S V, Dogiel V A, Ginzburg V L and Ptuskin V S 1990 *Astrophysics of Cosmic Rays* ed V L Ginzburg (Amsterdam: NorthHolland)
[10] Ptuskin V S, Zirakashvili V N and Seo E S 2010 *Astrophys. J.* **718** 31
[11] Zirakashvili V N, Breitschwerdt D, Ptuskin V S and Voelk H J 1996 *Astron. Astrophys.* **311** 113
[12] Ptuskin V S, Voelk H, Zirakashvili V N and Breitschwerdt D 1997 *Astron. Astrophys.* **321** 434
[13] Sparvoli R 2012 *Nucl. Instrum. Methods Phys. Res. A.* **692** 29

Journal of Physics: Conference Series 632 (2015) 012023 doi:10.1088/1742-6596/632/1/012023

The CALorimetric Electron Telescope (CALET) for high-energy astroparticle physics on the International Space Station

O Adriani[26], Y Akaike[7], K Asano[24], Y Asaoka[31], M G Bagliesi[30], G Bigongiari[30], W R Binns[32], S Bonechi[30], M Bongi[26], J H Buckley[32], G Castellini[26], M L Cherry[12], G Collazuol[27], K Ebisawa[9], V Di Felice[29], H Fuke[9], T G Guzik[12], T Hams[3], M Hareyama[22], N Hasebe[31], K Hibino[10], M Ichimura[4], K Ioka[11], M H Israel[32], A Javaid[12], E Kamioka[20], K Kasahara[31], J Kataoka[31], R Kataoka[15], Y Katayose[33], N Kawanaka[23], H Kitamura[16], T Kotani[31], H S Krawczynski[32], J F Krizmanic[2], A Kubota[20], S Kuramata[4], T Lomtadze[28], P Maestro[30], L Marcelli[29], P S Marrocchesi[30], J W Mitchell[14], S Miyake[5], K Mizutani[19], A A Moiseev[3], K Mori[9,31], M Mori[18], N Mori[26], H M Motz[31], K Munakata[21], H Murakami[31], Y E Nakagawa[9], S Nakahira[9], J Nishimura[9], S Okuno[10], J F Ormes[25], S Ozawa[31], F Palma[29], P Papini[26], B F Rauch[32], S B Ricciarini[26,34], T Sakamoto[1], M Sasaki[3], M Shibata[33], Y Shimizu[8], A Shiomi[17], R Sparvoli[29], P Spillantini[26], I Takahashi[1], M Takayanagi[9], M Takita[7], T Tamura[8,10], N Tateyama[10], T Terasawa[7], H Tomida[9], S Torii[8,31], Y Tunesada[24], Y Uchihori[16], S Ueno[9], E Vannuccini[26], J P Wefel[12], K Yamaoka[13], S Yanagita[6], A Yoshida[1], K Yoshida[20] and T Yuda[7]

[1] Aoyama Gakuin University, Japan
[2] CRESST/NASA/GSFC and Universities Space Research. Association, USA
[3] CRESST/NASA/GSFC and University of Maryland, USA
[4] Hirosaki University, Japan
[5] Ibaraki National College of Technology, Japan
[6] Ibaraki University, Japan
[7] ICRR, University of Tokyo, Japan
[8] JAXA/SEUC, Japan
[9] JAXA/ISAS, Japan
[10] Kanagawa University, Japan
[11] KEK, Japan
[12] Louisiana State University, USA
[13] Nagoya University, Japan
[14] NASA/GSFC, USA
[15] National Institute of Polar Research, Japan
[16] National Institute of Radiological Sciences, Japan
[17] Nihon University, Japan
[18] Ritsumeikan University, Japan
[19] Saitama University, Japan

[34] Corresponding author. Istitute of Applied Physics "Nello Carrara" (IFAC - CNR), via Madonna del Piano 10, 50019 Sesto Fiorentino (FI), Italy.

[20] Shibaura Institute of Technology, Japan
[21] Shinshu University, Japan
[22] St. Marianna University School of Medicine, Japan
[23] The University of Tokyo, Japan
[24] Tokyo Institute of Technology, Japan
[25] University of Denver, USA
[26] University of Florence, IFAC (CNR) and INFN, Italy
[27] University of Padova and INFN, Italy
[28] University of Pisa and INFN, Italy
[29] University of Rome Tor Vergata and INFN, Italy
[30] University of Siena and INFN, Italy
[31] Waseda University, Japan
[32] Washington University - St. Louis, USA
[33] Yokohama National University, Japan

E-mail: s.ricciarini@ifac.cnr.it

Abstract. The CALorimetric Electron Telescope (CALET) is a space experiment, currently under development by Japan in collaboration with Italy and the United States, which will measure the flux of cosmic-ray electrons (and positrons) up to 20 TeV energy, of gamma rays up to 10 TeV, of nuclei with Z from 1 to 40 up to 1 PeV energy, and will detect gamma-ray bursts in the 7 keV to 20 MeV energy range during a 5 year mission. These measurements are essential to investigate possible nearby astrophysical sources of high energy electrons, study the details of galactic particle propagation and search for dark matter signatures. The main detector of CALET, the Calorimeter, consists of a module to identify the particle charge, followed by a thin imaging calorimeter (3 radiation lengths) with tungsten plates interleaving scintillating fibre planes, and a thick energy measuring calorimeter (27 radiation lengths) composed of lead tungstate logs. The Calorimeter has the depth, imaging capabilities and energy resolution necessary for excellent separation between hadrons, electrons and gamma rays. The instrument is currently being prepared for launch (expected in 2015) to the International Space Station ISS, for installation on the Japanese Experiment Module - Exposure Facility (JEM-EF).

1. Introduction

The CALorimetric Electron Telescope CALET [1] [2] will be installed on the International Space Station (ISS) in 2015, with the aim of performing accurate and long-duration observations of high-energy cosmic radiation, including charged particles and photons.

An overview of the CALET mission and apparatus is given in Sect. 2, while Sect. 3 contains a selection of major science items which will be addressed by CALET.

2. Overview of CALET experiment

The CALET apparatus is currently in advanced integration and test phase and scheduled to be launched in space in 2015 with the Japanese rocket HTV-5, for installation on the Japanese Experiment Module - Exposure Facility (JEM-EF) on the ISS and subsequent operation for a target duration of 5 years.

The CALET payload (see figure 1) includes scientific and auxiliary equipment for a total mass of 650 kg, dimensions of $1.9 \times 0.8 \times 1.0$ m^3, nominal power consumption of 650 W. The main scientific instruments are the electromagnetic calorimeter (CALET-CAL, see Sect. 2.1) and the gamma-ray burst monitor (CGBM, see Sect. 2.2).

2.1. CALET-CAL instrument

The CALET calorimeter (see figure 2) is composed of a charge detector (CHD), a pre-shower imaging calorimeter (IMC) and a total absorption calorimeter (TASC). It is optimized for

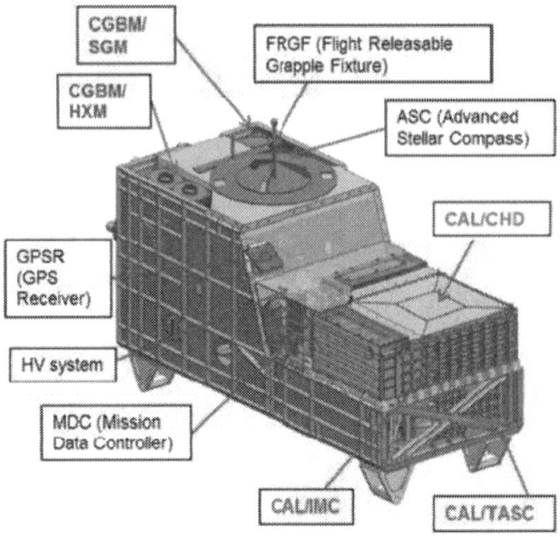

Figure 1. Drawing of the CALET payload. On the ISS the payload will be installed in such a way to have open sky on top, Earth on bottom.

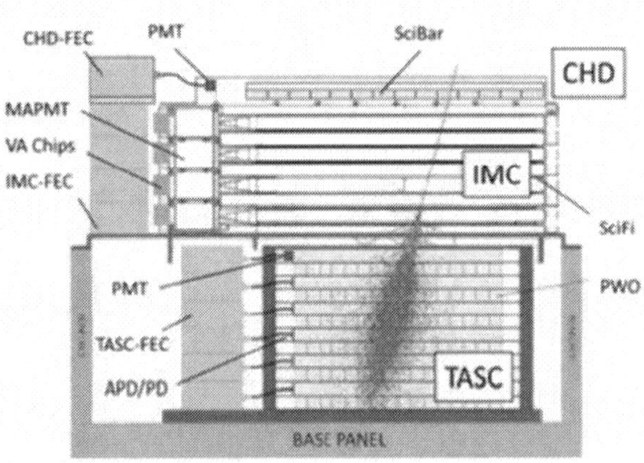

Figure 2. Drawing of the CALET-CAL instrument, with a typical shower of secondary particles generated by a 1 TeV electron incident from top.

particle identification and energy measurement of several cosmic-ray species:

- electrons and positrons in the 1 GeV - 20 TeV energy range (with no charge sign discrimination);
- photons from few GeV to $\sim$ 10 TeV;
- nuclei up to the Fe region, with energy from tens of GeV to $\sim$ 1 PeV;
- ultra-heavy (Z > 28) nuclei with E > 600 MeV/nucleon (in this case, with no energy measurement).

The CHD, positioned on top of the IMC structure, is composed of two layers of plastic scintillator with mutually orthogonal segmentation in 14 bars (SciBars), each of dimensions 3.2 $\times$ 1.0 $\times$ 44.8 cm^3 and read by a photo-multiplier tube (PMT, Hamamatsu Photonics R7400U-06) and front-end circuit (FEC) with charge sensitive amplifier, for a total of 28 channels. The

CHD determines the charge absolute value |Z| of the incoming charged particle, through the Z^2 dependence of the specific ionization loss. The very low uncertainty in the Z measurement (0.1 for light nuclei up to B, 0.3 in the Fe region) allows for resolving individual chemical elements with Z from 1 to 40 [3].

The IMC is a finely segmented sampling calorimeter, with surface area of 45×45 cm^2 and total thickness of 3 radiation lengths X_0; internally, 8 double layers of scintillating fibres (SciFi, 1 mm^2 cross-section) are interleaved with a sequence of 7 tungsten plates: 5 of thickness 0.2 X_0 and 2 of thickness 1.0 X_0. The fibres of each double layer are mutually orthogonal and arranged in belts, each read by a 64-channel multi-anode PMT (MAPMT, Hamamatsu Photonics R7600-M64) and VA front-end ASIC circuit, for a total of 7168 channels. The IMC fine granularity allows for precise determination of the incoming particle trajectory, localization of the starting point of the secondary shower possibly generated, discrimination of the primary incident particle against possible backscattering from the shower developing in the IMC and underlying TASC.

The TASC is a homogeneus calorimeter made of 192 lead tungstate (PWO) logs with dimensions $20 \times 19 \times 320$ mm^3 and arranged in 12 layers, oriented along alternatively orthogonal directions. Each log in the top layer is read by a PMT for use in the trigger system, together with CHD and IMC signals. A dual photodiode / avalanche photodiode system (APD/PD, Hamamatsu Photonics S8664-1010/S1227-33BR) is used for read-out of the other layers. The TASC is specifically designed to measure the energy of the incident particle with excellent resolution: 2% (3%) for electrons (gamma-rays) with energy $E > 10$ GeV, better than 35% for nuclei up to $\sim$ 100 TeV energy.

Moreover (see e.g. [4]), by exploiting the TASC and IMC shower imaging capabilities, a proton rejection power of $\sim 1 \cdot 10^5$ can be achieved in the electron sample, with a selection efficiency better than 80%, sufficient to keep the proton contamination below a few percent in the observation of cosmic-ray electrons in the multi-TeV region.

The total thickness of the CALET-CAL instrument is equivalent to 30 X_0 and 1.5 nuclear interaction lengths λ_I. Its effective geometrical factor is ~ 0.12 m^{2}sr for high-energy electrons and nuclei. The angular resolution is better than 0.3° for electrons and gamma-rays in the 10 GeV - 1 TeV energy range.

Prototypes of the CHD, IMC and TASC detectors were extensively tested at CERN SPS with beams of muons at 150 and 180 GeV energy, electrons with energies from 10 to 290 GeV, protons from 30 to 400 GeV, ion fragments at 13 and 30 GeV/nucleon. Results of the data analysis [5] demonstrate that the measured detector performances meet the design specifications.

2.2. CGBM instrument

The CGBM instrument [6] is dedicated to the observation of gamma-ray bursts (GRB) and other X-/gamma-ray transient phenomena, with a broad energy coverage from hard X-rays to soft gamma-rays, specifically in the range from 7 keV to 20 MeV; these data can be correlated with simultaneous gamma-ray observations of CALET-CAL instrument in the region from few GeV to several TeV.

The CGBM consists of two hard X-ray monitor (HXM) units and one soft gamma-ray monitor (SGM) unit, a support optical sensor given by the Advanced Sky Camera (ASC) and a GPS receiver (GPSR).

Each HXM unit contains a $LaBr_3$(Ce) scintillator crystal read by a PMT (Hamamatsu Photonics R6232-05) and shaped as two superposed cylinders with diameter of 6.6 and 7.9 cm respectively and thickness 0.6 cm each. The $LaBr_3$(Ce) scintillator is here used in space for the first time for celestial gamma-ray observations. The field of view is limited by a collimator within $\approx 58°$ from the vertical axis, to reduce possible contamination from cosmic X-ray background and bright X-ray astrophysical sources.

The SGM unit contains a BGO scintillator read a by PMT (Hamamatsu Photonics R6233-20) and shaped as a cylinder with 10 cm diameter and 7.6 cm thickness.

The energy resolution of HXM and SGM is $\sim$ 3% and $\sim$ 15%, respectively, at 662 keV.

3. CALET science

In the present section the main CALET science objectives are illustrated, as outlined below.

- Search for signatures of nearby astrophysical sources of cosmic rays (CR) in the multi-TeV cumulative e^-e^+ energy spectrum; accurate and high-statistics spectral measurement in the 1 GeV - 20 TeV region, searching for dark matter signatures and fine spectral features. See Sect. 3.1.
- Search for dark matter signatures in the 10 GeV - 10 TeV gamma-ray spectrum; accurate spectral measurement of gamma-ray galactic/extragalactic background radiation and bright astrophysical sources. See Sect. 3.2.
- Study of CR acceleration and propagation mechanisms, with the measurement of nuclei spectra from H to the Fe region for energies from tens of GeV up to several TeV, with the measurement of the B/C flux ratio up to several TeV/nucleon and with the determination of abundances of ultra-heavy nuclei up to Z = 40. See Sect. 3.3.

Other important science items, not specifically treated here, are listed below.

- Detection and study of X-/gamma-ray transients in the energy region 7 keV - 20 MeV with the CGBM instrument.
- Study of solar modulation of the e^-e^+ flux below $\sim$ 10 GeV.

3.1. Cosmic-ray electrons and positrons

Multi-TeV electrons originated in SuperNova Remnant (SNR) standard sources can be observed at Earth only if the SNR is younger than 10^5 years and within 1 kpc from the solar system, because of radiative synchrotron and inverse Compton energy losses, proportional to the squared energy, during propagation in the Galaxy [7]. Since the number of such nearby SNR's is very limited (Vela, Monogem, Cygnus Loop and few others), the electron energy spectrum around and above 1 TeV could exhibit spectral features and be characterized by a significant anisotropy[35] (order of 10%) in the arrival direction, related to the localization of the sources with respect to the galactic plane.

Among its contemporary experiments, CALET has unique capabilities for identifying such features in the multi-TeV region of cumulative e^-e^+ flux, namely energy resolution, proton rejection power and angular resolution, together with a long exposure in space: for example, $\sim$ 300 events/yr from the Vela SNR component with E > 1 TeV are expected to be collected and characterized after applying necessary data selection cuts, thus possibly giving a result as shown in figure 3. Besides, the CALET measurement will allow for tuning models of SNR sources, since the shape of CR electron flux in the multi-TeV range is critically affected (see e.g. [7]) by several model parameters which are still not precisely known, such as exponential cut-off energy of the accelerated spectrum, acceleration time interval, value of the diffusion coefficient.

Also in the sub-TeV energy region CALET accuracy and exposure will make it possible to significantly improve the knowledge of the detailed spectral shape and angular distribution of cumulative e^-e^+ flux. This will help in identifying the additional unknown source, either an astrophysical object (e.g. a nearby pulsar) or annihilation/decay of dark matter particles, which has been suggested to explain recent measurements showing a clear change of slope, namely an

[35] The anisotropy Δ is here defined by taking into account the maximum and minimum observed electron intensity with respect to the galactic longitude, I_{max} and I_{min}: $\Delta = (I_{max} - I_{min})/(I_{max} + I_{min})$.

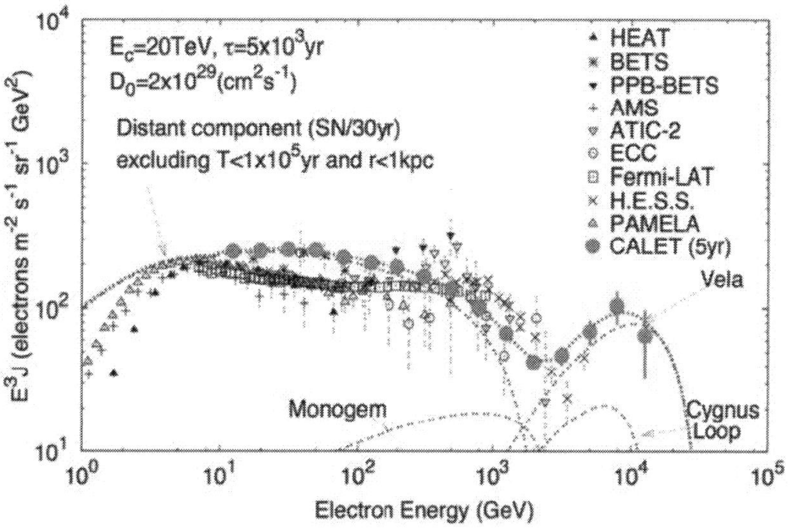

Figure 3. Expected cumulative e^-e^+ energy spectrum measured with CALET after 5 years of observation, according to a possible SNR model as discussed in [7]; a distinctive signature of nearby SNR sources can be identified by CALET in the multi-TeV region. A selection of previous measurements is also reported.

e^-e^+ flux enhancement, in the range 200 GeV - 1 TeV, not expected for standard SNR sources (see e.g. [8] and references therein).[36]

3.2. Gamma-rays

According to theories of annihilation or decay of dark-matter particles in the galactic halo, these phenomena should produce sharp gamma-ray lines in the sub-TeV to TeV energy region, superimposed on the diffuse spectrum. CALET [11] [12] will be capable of investigating such a distinctive signature, as shown e.g. in figure 4, thanks to the gamma-ray energy resolution of 3% above 100 GeV, which can be improved to 1% with a reduced (75%) on-axis effective area, obtained by selecting only events with the primary gamma-ray crossing all detector planes and at least 2 cm inside TASC layers (to assure complete lateral containment of the secondary shower).

CALET excellent energy resolution and good angular resolution (better than 0.4°, including pointing uncertainty) will allow for accurate measurements of diffuse gamma-ray emission and more than 100 bright sources at high latitude from the Fermi-LAT catalogue [13]. Given the on-axis effective area of $\approx$ 600 cm^2 for energies above 10 GeV (reduced by $\sim$ 50% at 4 GeV) and field of view of $\approx$ 45° from the vertical direction, CALET is expected to detect $\sim$ 25000 ($\sim$ 7000) photons from the galactic (extra-galactic) background with E > 4 GeV and $\sim$ 300 photons from the Vela pulsar with E > 5 GeV.

3.3. Cosmic-ray protons and nuclei

It is still not clear to what extent the energy spectra of cosmic-ray nuclei in the 10 GeV - 1 PeV region are well described by a single power law $E^{-\gamma}$ with $\gamma \approx 2.7$: in particular, the PAMELA collaboration [14] reported accurate measurements clearly indicating small but

[36] The presence of an additional source is also required to explain the now established rise of positron fraction in the 10 GeV - 500 GeV energy range (see e.g. [9], [10]).

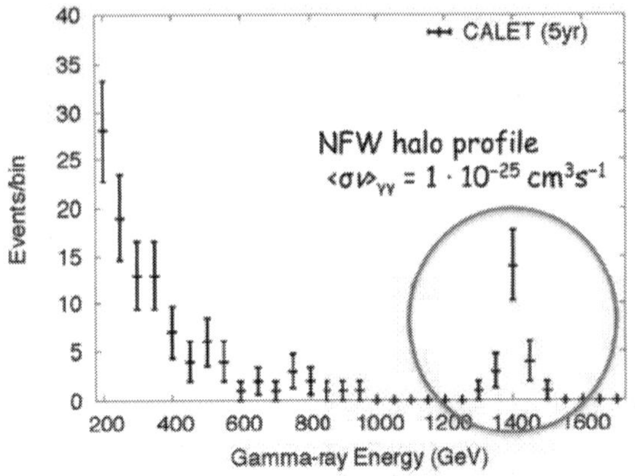

Figure 4. Expected CALET 5-years measurement of a possible 1.4 TeV gamma-ray line from dark matter in the region of galactic centre, also including galactic diffuse background, according to [11].

significant deviations from single power law spectra for both H and He in the sub-TeV region, which still need to be compared with independent high-accuracy measurements.[37] The same measurements also show a clear difference of $\approx$ 3% between the spectral indexes of H and He: it is therefore reasonable to investigate similar differences for other nuclei species. Finally, it is still an open question whether there is a spectral cut-off ("knee") below 1 PeV for the various nuclei species. These open points reflect in uncertainties in the understanding of the cosmic-ray shock acceleration mechanism taking place in SNR sources. More accurate measurements of the individual energy spectra of CR nuclei, especially approaching the PeV energy region, are therefore needed.

CALET will be able to identify CR nuclei with individual element resolution and measure their energies in the range from few tens of GeV to several hundreds of TeV. Its thickness (30 X_0, 1.5 λ_I) allows for highly efficient nuclear interaction of the incoming nucleus and subsequent good confinement of the electromagnetic core of the secondary shower, thus achieving an energy resolution better than 35% for nuclei up to $\sim$ 100 TeV. In 5 years of data taking on the ISS, CALET is expected to measure the proton energy spectrum up to $\sim$ 900 TeV and the He spectrum up to $\sim$ 400 TeV/nucleon (see figure 5) and to determine the fluxes of the more abundant heavy nuclei with good statistical precision, up to $\sim$ 20 TeV/nucleon for C and O and $\sim$ 10 TeV/nucleon for Ne, Mg, Si and Fe.

Important information on the CR propagation in the galaxy is obtained from the secondary-to-primary flux ratio for CR elements, in particular the B/C ratio [17], which is known to follow a power law in energy $E^{-\delta}$ for energies below $\sim$ 100 GeV/nucleon; an accurate measurement of δ and the determination whether the B/C ratio deviates from a simple power law above $\sim$ 100 GeV/nucleon are crucial for discriminating between different models of CR propagation. CALET will provide new data to improve the accuracy of the present measurements above 100 GeV/nucleon and extend them above 1 TeV/nucleon: in 5 years, CALET can determine δ with an absolute uncertainty of $\pm$ 0.05.

Moreover [18], by exploiting the CHD particle identification capability, CALET will measure the abundances of ultra-heavy CR nuclei at few GeV/nucleon for Z up to 40, with an expected statistics in 5 years from 2 to 4 times larger than collected by the TIGER experiment.

[37] Preliminary measurements of H and He spectra have been published by AMS-02 collaboration [15] [16], without quantitative comparisons with PAMELA results.

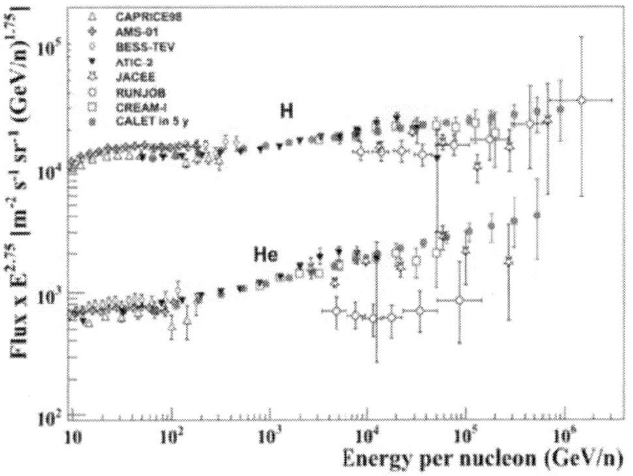

Figure 5. Expected CALET measurement of the energy spectra of H and He after 5 years of observation, compared with a selection of previous direct measurements.

References

[1] Torii S *et al* 2011 *Nucl. Instr. And Meth.* A **630** 55-7
[2] Torii S *et al* 2013 *Proc. of 33rd ICRC* 245
[3] Marrocchesi P S *et al* 2011 *Nucl. Instr. And Meth.* A **659** 477-83
[4] Akaike Y *et al* 2010 *Adv. Space Res.* **45** 690-7
[5] Akaike Y *et al* 2013 *Proc. of 33rd ICRC* 726
[6] Yamaoka K *et al* 2013 *Proc. of 33rd ICRC* 1007
[7] Kobayashi T *et al* 2004 *Astrophys. J.* **601** 340-51
[8] Ackermann M *et al* 2010 *Phys. Rev.* D **82** 092004
[9] Adriani O *et al* 2009 *Nature* **458** 607-9
[10] Accardo L *et al* 2014 *Phys. Rev. Lett.* **113** 121101
[11] Yoshida K *et al* 2013 *Proc. of 33rd ICRC* 735
[12] Mori M *et al* 2013 *Proc. of 33rd ICRC* 248
[13] Moiseev A A *et al* 2013 *Proc. of 33rd ICRC* 627
[14] Adriani O *et al* 2011 *Science* **332** 6025 69-72
[15] Haino S *et al* 2013 *Proc. of 33rd ICRC* 1265
[16] Choutko V *et al* 2013 *Proc. of 33rd ICRC* 1262
[17] Obermeier A *et al* 2012 *Astrophys. J.* **752** 69 1-7
[18] Rauch B F *et al* 2013 *Proc. of 33rd ICRC* 819

Neutrinos from SN 1987a A Puzzle Revisited

Gerd Schatz
Universität Heidelberg *and* Karlsruhe Institute of Technology
E-mail: bgschatz@t-online.de

Abstract. The smallest of the four detectors which claim to have observed neutrinos from SN 1987a registered the events more than 4 h earlier than the other three ones. This claim is not usually accepted because it is difficult to understand that the other (and larger) detectors did not register any events at the same time. It is shown that microlensing of the neutrinos by a star in-between the supernova (SN) and Earth can enhance the neutrino intensity at the position of one detector by more than an order of magnitude with respect to the other detectors. Such a configuration is improbable but not impossible. Essential for this enhancement is the small source diameter, of order 100 km. So if two bursts of neutrinos were emitted by SN 1987a at a separation of about 4 h it could be explained easily that the smallest detector observed the first burst while the other ones missed it and vice versa.

1. Introduction

Supernovae of type II, also called core collaps SN, convert of order one solar mass of protons into neutrons. They therefore produce a neutrino burst of several tens of seconds duration. This was observed once so far when the famous SN 1987a flared up in the Large Magellanic Cloud on February 23, 1987. One of the most conspicuous features of these observations is the fact that the smallest detector registered 5 events more than 4 h earlier than the other three. The following table gives some characteristics of the four detectors and summarizes the observational results (data from [1] [2] [3] [4] [9]). The claim of the LSD collaboration that the 5 events observed originate from SN1987a was soon called *unbelievable* [5] since it appeared unexplainable that the other three (and larger) detectors did not see any events at the same time. It is the purpose of this contribution to demonstrate that a mechanism exists which could explain this discrepancy in a natural way. it is not our aim to discuss the contested question of the SN 1987a neutrinos in any comprehensive way but only to draw attention to a special aspect which apparently has been overlooked in the past but is of relevance for the problem.

2. Microlensing of neutrinos

Gravitational deflection of light by a star can lead to a transient increase of the observed intensity of a background object. This effect is called microlensing and has been observed frequently in the optical range. The signal has a well defined time dependence and is strictly achromatic. Fig. 1 illustrates the situation schematically. Since all extremely relativistic particles move in

Journal of Physics: Conference Series **632** (2015) 012024 doi:10.1088/1742-6596/632/1/012024

Detector	Size [t]	Threshold [MeV]	Number of events	Time of first event [U.T.]
IMB	5000	15	8	$7^h\ 35^m\ 41^s$
Kamiokande	2140	7.5	12	$\approx\ 7^h\ 35^m$
Baksan	200	10	5	$7^h\ 36^m\ 12^s$
LSD	290	5	5	$2^h\ 53^m 11^s$

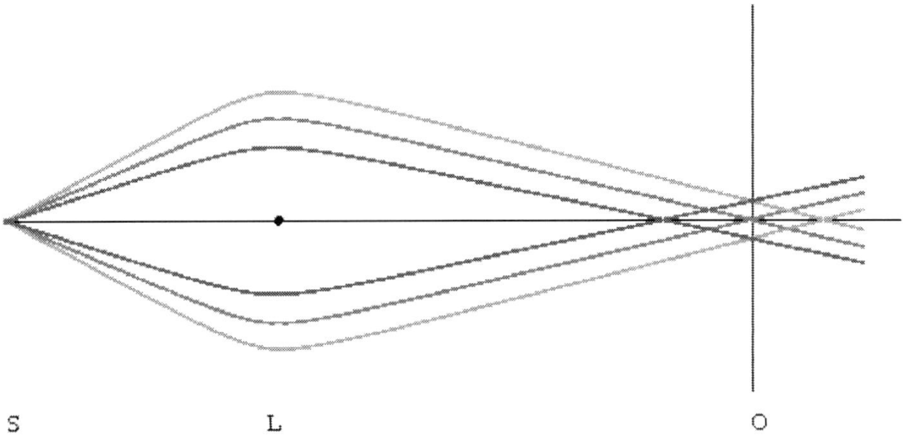

Figure 1. Schematic representation of microlensing with source S, lens L and observer O.

gravitational fields on orbits identical to those of photons the same should hold for neutrinos from SNe.

The effect of microlensing is an increase of the (light or neutrino) intensity in the plane perpendicular to the axis defined by source and lens at the distance of the observer. The time dependence results from the movement of the observer across the radiation field (stationary in the case of light). Since neutrinos and photons have identical orbits we can use all formulae derived by astronomers also for neutrinos. The radial dependence of the intensity in this plane can be calculated analytically for point sources [6] [7]. The result is shown in fig. 2. It tends to a constant value, undisturbed by the lens, at large distances from the axis and increases inversely proportionate to the distance r from the axis near the center.

The ratio of the intensity to its undisturbed value is called amplification and given by

$$A(r) \;=\; \frac{I(r)}{I(\infty)} \;=\; \frac{1}{2\ a\ r}\ \frac{R_E^2\ +\ 2\ a^2\ r^2}{\sqrt{\ R_E^2\ +\ a^2\ r^2}}$$

for point sources. Here r is the distance of the observer from the axis, $R_E = \sqrt{2R_S D_{SL} D_{LO}/D_{SO}}$

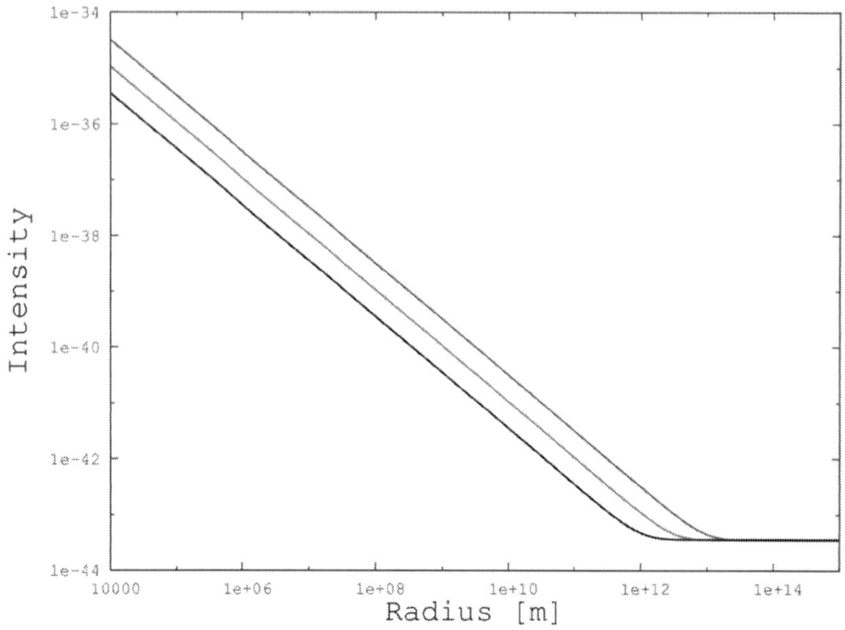

Figure 2. Dependence of intensity on radius of the observer for a point source. Different colours represent different lens positions at 10 %, 50 % and 90 % of the source-to-observer distance.

is called the Einstein radius, R_S is the Schwarzschild radius of the lens, the Ds are the distances between source, lens and observer as indicated by the indices, and $a = D_{SL}/2D_{SO}$. For large radii $r \gg R_E/a$ the amplification tends to 1 and for small ones it is inversely proportionate to r.

The singularity on the axis disappears when this formula is folded with the finite size of the source. The result is an effective truncation of intensity and amplification with a maximum on the axis of $A(0) = R_E D_{SO}/(r_s D_{LO})$. Here r_s is the source radius. For $r \gtrsim r_s D_{LO}/D_{SL}$ the point size expression holds to a good approximation. The intensities after folding are shown in fig. 3 for sources of the sizes of a SN and the Sun. They are approximately constant near the axis and at large distances and decrease inversely proportionate to r in-between.

3. Application to SN 1987a

At the moment of observation of SN 1987a the separations between the LSD detector and the other ones, projected on a plane perpendicular to the line connecting the SN with the Earth, amounted to approximately 2800 [km] (Baksan), 6100 [km] (IMB,) and 8700 [km] (Kamiokande). If the LSD detector is situated in the central plateau of the intensity distribution and another detector D in the decreasing part the ratio of amplifications is given by $2r_D D_{LO}/(r_s D_{SL})$ where r_D is the distance of the second detector from the axis. If both detectors are placed in the sloping part the ratio is simply $(r_D + r_{LSD})/r_D$. In either case it is easy to find configurations which enhance the LSD signal with respect to the other ones by an order of magnitude or more. This would be sufficient to explain that the LSD detector saw the first neutrino burst while the other ones failed to register any events not the least in view of the small event numbers involved and the correspondingly large statistical errors. Since the observer is expected to move with respect to source and lens there is a high probability that the LSD detector has moved to outside the

Journal of Physics: Conference Series **632** (2015) 012024 doi:10.1088/1742-6596/632/1/012024

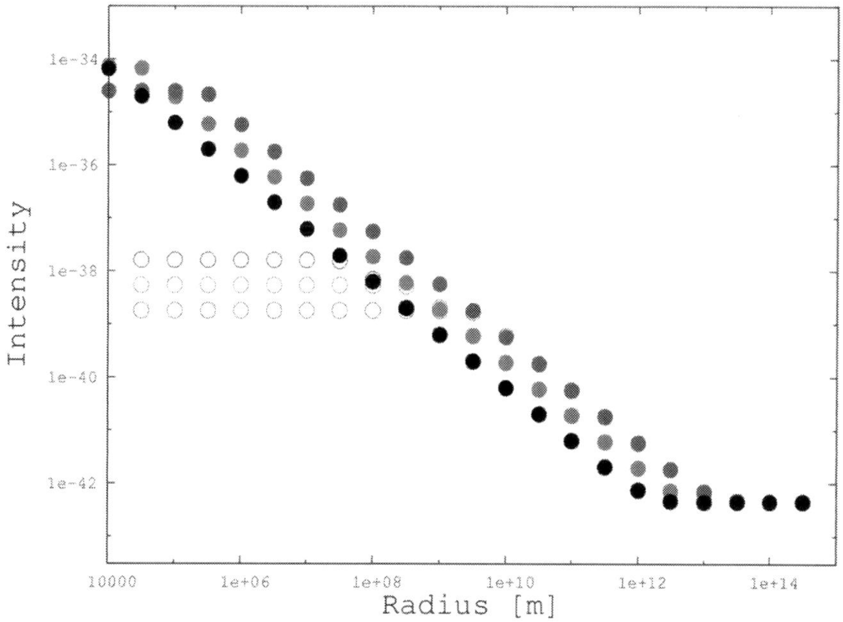

Figure 3. Neutrino intensities in the observer plane for sources of the size of a SN (full symbols) and the Sun (open symbols). The colours have the same meaning as in fig. 3.

cone of enhancement 4 h later. This would result in a high probability for the LSD detector of missing the second burst.

It is worthwhile mentioning that the described mechanism does not require exceptionally large or heavy stars as gravitational lenses. Even a planet of the size of Jupiter would do if placed at more than $\approx 10^3$ [A.U.] from the source. One should also be aware that the probability of the precurser star having companions ranging from planets to a neutron star is very high.

4. Discussion

There is no doubt that the probability of such a configuration occurring at the moment of SN explosion is very small. We refrain from trying to estimate this probability quantitatively since this would require assumptions on the spatial and mass distributions of possible lenses which are to a considerable extent arbitrary. It would in any case be much smaller than the probability of the LSD events being a statistical fluctuation which is given by $4 \cdot 10^{-4}$ [4]. But it is also clear that the required configurations are not impossible. **Hence one has to conclude that the argument against the LSD claim is not as compelling as has hitherto been assumed.** In view of the small probability of the required configurations occurring it is of course still up to the reader whether to consider the claim *unbelievable* [5] or not.

An unresolved problem which remains is the cause for two neutrino bursts originating from the SN. Hillebrandt et al. [8] proposed a mechanism which relies on the fact that all stars rotate to some extent. This will cause an acceleration of the rotational movement (pirouette effect) with a corresponding increase of centrifugal forces which might halt the collaps for a certain period of time until some angular momentum has been lost by interaction with the environment.

This might then result in a two-step collaps. An alternative mechanism invoking fragmentation of the collapsing star has been discussed more recently [9]. These explanations appear to have lost appeal, though, because the SN precursor apparently loses a considerable fraction of its angular momentum during the giant phase immediately preceding the collaps [10]. It is not unfair, though, to add that the theory of type II SNe is hardly in a definite final state so revision of this argument does not appear impossible.

Acknowledgements
I thank M. Bartelmann (Heidelberg) for patient instructions concerning general relativity, T. Huege (Karlsruhe) for equally patient assistence during my fight with *gnuplot*, and D. Huber (Karlsruhe) for his indispensible help in producing the poster displayed at the conference.

References

[1] R. M. Bionta et al., Phys. Rev. Lett. **58** (1987) 1494
[2] K. Hirata et al., Phys. Rev. Lett. **58** (1987) 1490
[3] E. N. Alexeyev et al., Phys. Lett. B **205** (1988) 209
[4] M. Aglietta et al., Europhys. Lett. **3** (1987) 1315
[5] A. W. Wolfendale, Proc. 20th ICRC, Moscow 1987, Vol. 9, p. 74
[6] P. Schneider, J. Ehlers, and E. E. Falco, Gravitational lensing, Springer-Verlag 1992
[7] B. Paczynski, ARAA **34** (1996) 419
[8] W. Hillebrandt et al., Astron. Astrophys. **180** (1987) L20
[9] V. S. Imshennik and O. G. Ryazhskaya, Astronom. Lett. **30** (2004) 14
[10] Th. Janka, private communication

Journal of Physics: Conference Series **632** (2015) 012025 doi:10.1088/1742-6596/632/1/012025

On a coherent investigation of the spectrum of cosmic rays in the energy range of 10^{14} - 10^{18} eV with KASCADE and KASCADE-Grande

S Schoo[1], W D Apel[1], J C Arteaga-Velázquez[2], K Beck[1], M Bertaina[3], J Blümer[1,4], H Bozdog[1], I M Brancus[5], E Cantoni[3,6,a], A Chiavassa[3], F Cossavella[4,b], K Daumiller[1], V de Souza[7], F Di Pierro[3], P Doll[1], R Engel[1], D Fuhrmann[8,c], A Gherghel-Lascu[5], H J Gils[1], R Glasstetter[8], C Grupen[9], A Haungs[1], D Heck[1], J R Hörandel[10], D Huber[4], T Huege[1], K H Kampert[8], D Kang[4], H O Klages[1], K Link[4], P Łuczak[11], H J Mathes[1], H J Mayer[1], J Milke[1], B Mitrica[5], C Morello[6], J Oehlschläger[1], S Ostapchenko[12], N Palmieri[4], M Petcu[5], T Pierog[1], H Rebel[1], M Roth[1], H Schieler[1], F G Schröder[1], O Sima[13], G Toma[5], G C Trinchero[6], H Ulrich[1], A Weindl[1], J Wochele[1], J Zabierowski[11]

[1] Institut für Kernphysik, KIT - Karlsruher Institut für Technologie, Germany
[2] Universidad Michoacana, Instituto de Física y Matemáticas, Morelia, Mexico
[3] Dipartimento di Fisica, Università degli Studi di Torino, Italy
[4] Institut für Experimentelle Kernphysik, KIT - Karlsruher Institut für Technologie, Germany
[5] Horia Hulubei National Institute of Physics and Nuclear Engineering, Bucharest, Romania
[6] Osservatorio Astrofisico di Torino, INAF Torino, Italy
[7] Universidade São Paulo, Instituto de Física de São Carlos, Brasil
[8] Fachbereich Physik, Universität Wuppertal, Germany
[9] Department of Physics, Siegen University, Germany
[10] Dept. of Astrophysics, Radboud University Nijmegen, The Netherlands
[11] National Centre for Nuclear Research, Department of Astrophysics, Lodz, Poland
[12] Frankfurt Institute for Advanced Studies (FIAS), Frankfurt am Main, Germany
[13] Department of Physics, University of Bucharest, Bucharest, Romania
[a] now at: Istituto Nazionale di Ricerca Metrologia, INRIM, Torino
[b] now at: DLR Oberpfaffenhofen, Germany
[c] now at: University of Duisburg-Essen, Duisburg, Germany

E-mail: sven.schoo@kit.edu

Abstract. The KASCADE experiment and its extension KASCADE-Grande have significantly contributed to the current knowledge about the energy spectrum and composition of cosmic rays (CRs) with energies between the knee and the ankle. However, the data of both experiments were analysed separately, although Grande used the muon information of the KASCADE-array. A coherent analysis based on the combined data of both arrays is expected to profit from reconstructed shower observables with even higher accuracy compared to the stand-alone analyses. In addition, a significantly larger fiducial area is available.

The aim of this analysis is to obtain the spectrum and composition of CRs in the range from 10^{14} to 10^{18} eV with a larger number of events and further reduced uncertainties using one unique reconstruction procedure for the entire energy range. This contribution will describe the motivation, the concept, and the current status of the combined analysis.

1. Introduction

Up to now the measurements of the KASCADE/-Grande ([1],[2]) arrays have been analyzed independently of each other, though the KASCADE-Grande analyses used the shielded scintillation detectors of the KASCADE array for the reconstruction of the total number of muons (N_{μ}). There are several good reasons for combining the measurements of both arrays prior to the reconstruction of the shower observables. One is the expected increase in accuracy of the reconstructed shower sizes due to the additional information from the respective other detector setup and the increased distance range for the lateral distribution of charged particles (N_{ch}). Another reason is, that a combined analysis should remove most of the differences in absolute flux observed between the KASCADE and KASCADE-Grande all-particle spectra and should result in one consistent composition measurement over the whole, combined energy range from around $10^{15} - 10^{18}$ eV and possibly even down towards 10^{14} eV. The layout of the KASCADE array and an outline of its stand-alone reconstruction of shower observables and energy will be shown in Section 2. The same information will be given for KASCADE-Grande in Section 3. Before presenting a short conclusion (Sec. 5) at the end of this article, the current status of the combined analysis will be given in Section 4.

2. KASCADE

2.1. The Experiment

The KASCADE experiment was located in Karlsruhe (Lon.: 8.4°, Lat.: 49.1°), Germany at an altitude of approximately 110 m a.s.l. The layout of the experiment with the array, the muon tracking detector and the central detector is shown on the left side of Fig. 1. The array consisted of 252 stations organized in 16 clusters. While the 192 stations of the outer 12 clusters were equipped with shielded and non-shielded scintillation detectors, the stations in the inner four clusters were built without the shielded scintillators, but with twice the number of e/γ detectors. As described in the next subsection, using this setup, the simultaneous reconstruction of N_{e} and N_{μ} is possible.

2.2. The Shower-Reconstruction

As a first step of the reconstruction the center of gravity of the signals of the non-shielded scintillation detectors (e/γ detector) is used to get a first estimate of the position of the core of the shower. The direction of the shower axis is estimated using a plane shower front. After weighting the signals of the e/γ and muon detectors with geometrical weights, the sum of these weighted deposits is used as the starting values for N_{e} and N_{μ}.

In a second step the arrival direction is improved by fitting a conically shaped shower front to the arrival times measured with the e/γ detectors. The core position, N_{e} and the shower-age are obtained by fitting a NKG-like lateral density function (LDF) [4][5] to the particle densities measured at the e/γ detectors. Prior to that, the energy deposits have been transformed into particle densities using a lateral energy correction function (LECF). The LECF describes the average energy deposited by a single charged particle depending on the distance of the station to the shower core and the size of the shower. The LECF used for the electromagnetic component is shown in Fig. 1 for three different shower sizes. The number of particles is calculated by dividing the total energy deposit by the effective energy deposited per single charged particle (f_{LECF}) taking into account also the energy deposited by photons and the e/γ ratio. The shape of the LECF is explained in detail in [3].

At this stage the result of the fit corresponds to the number of charged particles at observation level since the energy deposited by muons has not been subtracted at this point.

The reconstruction of the number of muons is done simultaneously, therefore, in the third and last step the muon LDF is known and this information is included in the fitting procedure for the e/γ detectors resulting in the reconstruction of N_{e}.

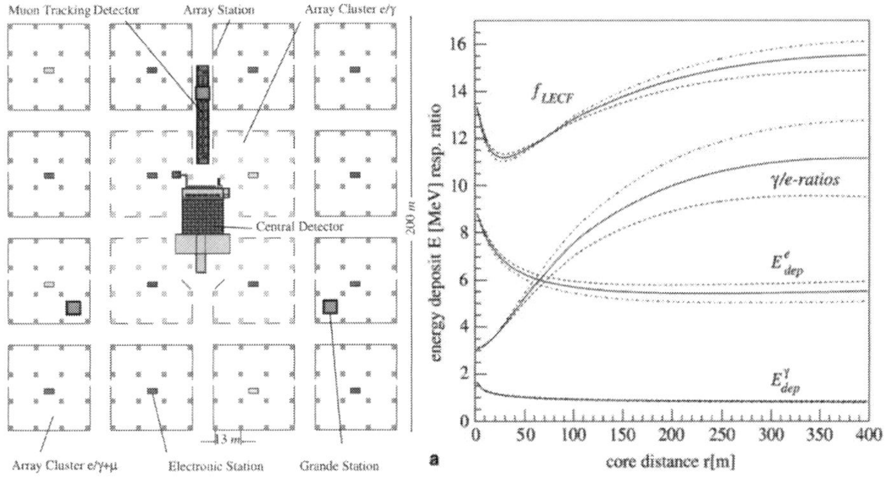

Figure 1. The layout of the KASCADE experiment is shown in the left picture. Each station was equipped with shielded and non-shielded scintillators, except for the stations of the inner four clusters, which were installed without the shielded scintillators (See [1]). In the right plot, the energy deposits per electron ($E^{\rm e}_{\rm dep}$) and γ ($E^{\gamma}_{\rm dep}$) are shown together with the γ/e ratio and the effective deposit per charged particle ($f_{\rm LECF}$) for proton primaries and $\log_{10}N_{\rm e} = 5$ (dashed lines), $\log_{10}N_{\rm e} = 6$ (solid lines), $\log_{10}N_{\rm e} = 7$ (dotted dashed lines)(See [3]).

The reconstruction of the number of muons is very similar to the one for the number of electrons. The transformation of energy deposits to particle densities takes into account the $N_{\rm e}$-dependent probability of electrons, photons or hadrons passing the shielding especially near the shower core. Since this faked muon deposit gets dominant below a distance of 40 m, stations within this distance to the shower core are excluded from the analysis. Further more, the NKG-like LDF is known to deviate from the true lateral muon distribution towards large core distances, therefore, the muon LDF is integrated only in the range from 40 – 200 m, where the KASCADE detectors provide sampling points for the fit. The result is the truncated number of muons ($N_{\mu}^{\rm tr.}$). Because of the low muon densities, the shape parameter is fixed to a value derived using simulated showers taking into account also the dependence on the shower size. The only remaining free parameter is therefore the number of muons itself.

For a more detailed description of the reconstruction procedure see [3].

2.3. The Energy Spectrum

The reconstruction of the energy spectra for all particles together and separately for five masses is based on the two-dimensional shower-size spectrum shown in Fig. 2. In principle, every nucleus with a certain mass and energy should contribute only to specific cells depending on the intrinsic shower-to-shower fluctuations. The extent of these fluctuations depends on the mass and energy of the primary particle. The 'naturally' available region in the two-dimensional spectrum gets smeared out further by inaccuracies in the reconstruction of $N_{\rm e}$ and N_{μ} and the trigger- and reconstruction probabilities for showers of a certain size.

The problem can be expressed as

$$\vec{Y} = \mathbf{R}.\vec{X} \tag{1}$$

with $\vec{Y}$ containing the cell contents in Fig. 2, $\vec{X}$ being the corresponding mass and energy and $\mathbf{R}$ being the *response matrix*. The unfolding technique used to solve this equation is described in [6], the resulting spectra for all particles, protons and helium using QGSJet01 [7] as the

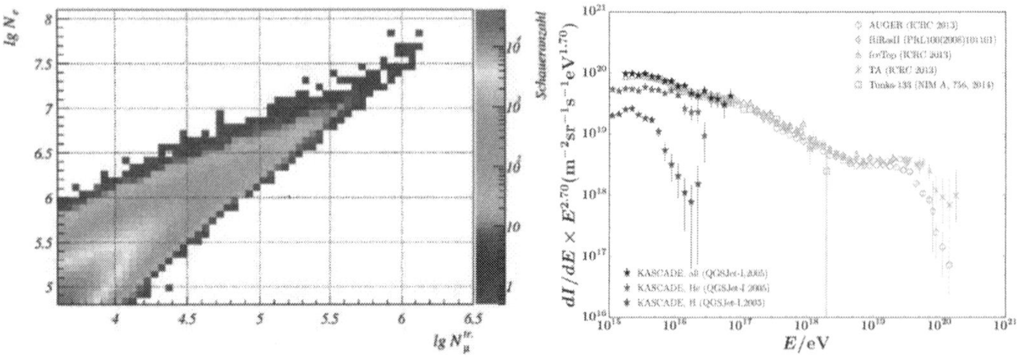

Figure 2. Left: The tow-dimensional shower-size spectrum measured with KASCADE. (See [6]); Right: The resulting, unfolded spectra for proton and helium together with the all-particle spectrum for the same analysis and other experiments.

hadronic interaction model are shown in Fig. 2. For a comparison with the results using the hadronic interaction model SIBYLL [8] and for the unfolded spectra for carbon, silicon and iron primaries, see Ref. [6].

An interesting discovery was that the knee-like structure at around 6 PeV in the all-particle spectrum might be caused primarily by a knee-like structure in the flux of helium instead of protons. However, one should not forget the dependence on the hadronic interaction model used in the simulations. A model that predicts much more muons, for example, would result in a significantly lighter composition. If the "proton-knee" is indeed located at an energy of around $2-3$ PeV, the helium knee would be expected to be at around $4-6$ PeV (in case of a charge dependence) or around $8-12$ PeV (in case of a mass dependence). Unfortunately the helium spectrum does not allow to distinguish between these two scenarios.

For iron primaries, the knee is expected to be just below 100 PeV (in case of a charge dependence), which is the upper limit of the energy range accessible with KASCADE. Its extension, KASCADE-Grande is capable of reconstructing events with energies up to 1 EeV and will be discussed in the next section.

3. KASCADE-Grande

KASCADE-Grande was located next to the KASCADE array with an overlap as it is shown in Fig. 3. Covering an energy range from 10 PeV up to 1 EeV, there is also an overlap between the energy spectra of both arrays.

The KASCADE-Grande stations were equipped with non-shielded scintillation detectors ($10\,\mathrm{m}^2$ per station) only, therefore, the reconstruction of the number of muons is only possible using the shielded detectors of the KASCADE array. The reconstruction procedure is explained in the next subsection.

3.1. The Shower-Reconstruction

The reconstruction of the shower-observables is again split into several steps. These steps will be briefly summarized in this section. A detailed description can be found in [2].

The first step is identical to the first step of the KASCADE reconstruction resulting in an estimate for the core position, arrival direction, number of charged particles (N_{ch}) and muons.

The LDF used for the electromagnetic component is again a modified NKG-function. The reconstruction of the number of muons, however, is based on a function described by Lagutin

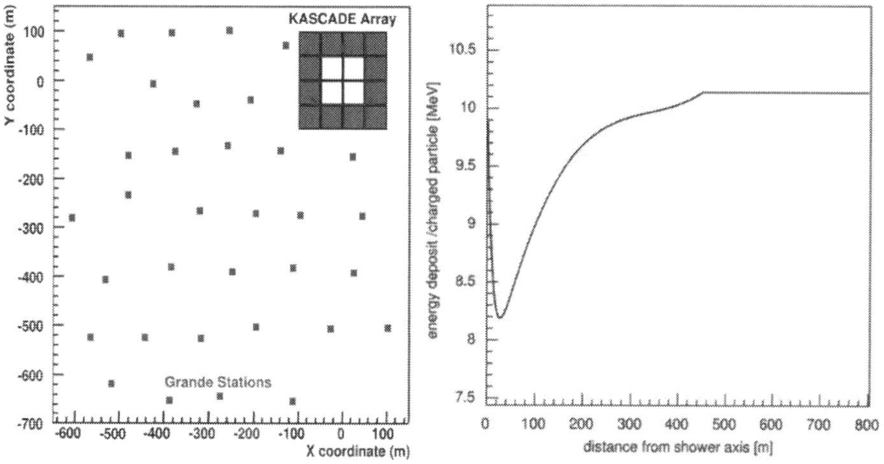

Figure 3. The KASCADE-Grande stations (rectangles) are shown relative to the KASCADE array. The LECF of the Grande stand-alone analysis for the electromagnetic component is shown on the right. Starting at $\sim$ 300 m the contribution from muons starts to get dominant. Above $\sim$ 400 m only muons contribute to the energy deposit and above 450 m the LECF corresponds to the energy deposited by a single, vertically incident muon. (See [2])

and Raikin [9].

The energy deposits are again transformed into particle densities using LECFs. The LECF for the electromagnetic component is shown in Fig. 3. The shape is mainly governed by the distance-dependent γ/e ratio and the energy distribution of electrons and photons. Unlike for the KASCADE LECF, a dependence on the shower size has been found to be negligible and does not contribute to the LECF.

In step two of the reconstruction procedure, the core position found in step one is moved on a 7×7 grid with a spacing of 8 m. Each position is used as a fixed parameter in the fit of the LDF to the charged particle densities leaving the shape parameter and N_{ch} as the only free parameters. The position resulting in the smallest χ^2 is kept as the fixed core position for the next two steps.

Next, the arrival direction is reconstructed more precisely by fitting the arrival times using a theoretical shower front derived from simulations.

After that the shape parameter and N_{ch} are fitted using again the core position obtained in step two.

The final location of the shower core is reconstructed by fitting the LDF again. This time the core position can vary freely, but N_{ch} and the slope are fixed.

Using this position, the final values for arrival direction, N_{ch} and the slope parameter are derived by repeating the steps three and four.

The number of muons can now be reconstructed by using the core position and arrival direction reconstructed for the Grande detectors and fitting the muon LDF with N_{μ} being again the only free parameter.

3.2. The Energy Spectrum

The analysis is again based on the two-dimensional shower size spectrum, which is shown on the left side in Fig. 4. Due to low statistics and larger uncertainties in N_{ch} and N_{μ} being a problem for the unfolding analysis used for the KASCADE analysis, the first Grande analyses use a different approach to be able to really reach up to 1 EeV.

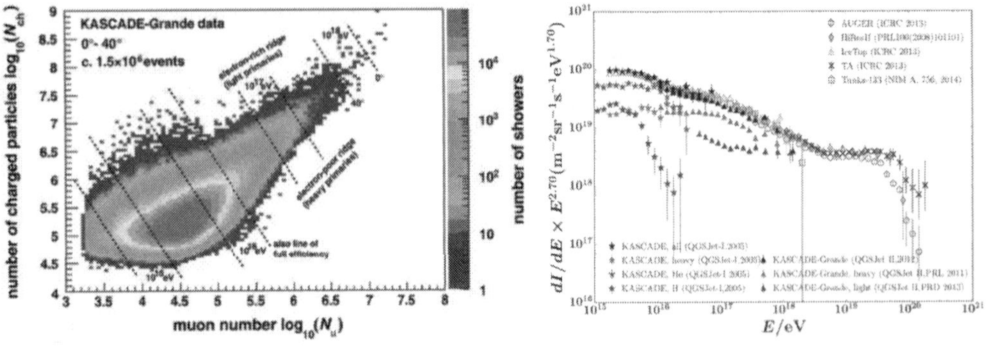

Figure 4. Left: The two-dimensional shower size spectrum derived from Grande data (See [10]). Right: All-particle spectra of several experiments are shown in addition to composition measurements of the KASCADE and KASCADE-Grande arrays.

The reconstruction of the energy is based primarily on N_{ch}, however, the ratio of N_{ch} to N_{μ} as a function of N_{ch} is used to take the mass-dependence of the energy corresponding to a certain number of charged particles into account. As explained in [11], the following formulas are used:

$$\log_{10}(E) = [a_{\mathrm{H}} + (a_{\mathrm{Fe}} - a_{\mathrm{H}}) \cdot k] \cdot \log_{10}(N_{\mathrm{ch}}) + b_{\mathrm{H}} + (b_{\mathrm{Fe}} - b_{\mathrm{H}}) \cdot k \tag{2}$$

with

$$k = \frac{\log_{10}(N_{\mathrm{ch}}/N_{\mu}) - \log_{10}(N_{\mathrm{ch}}/N_{\mu})_{\mathrm{H}}}{\log_{10}(N_{\mathrm{ch}}/N_{\mu})_{\mathrm{Fe}} - \log_{10}(N_{\mathrm{ch}}/N_{\mu})_{\mathrm{H}}} \tag{3}$$

and

$$\log_{10}(N_{\mathrm{ch}}/N_{\mu})_{\mathrm{H,Fe}} = c_{\mathrm{H,Fe}} \cdot \log_{10}(N_{\mathrm{ch}}) + d_{\mathrm{H,Fe}} \tag{4}$$

The coefficients are derived using simulated proton and iron showers. Since the k parameter is mass sensitive, it can be used to separate the events in a light and a heavy mass group. This is done by comparing the measured value of k with the mean, energy-dependent values of k for simulations of showers induced by particles of five different masses. See [10][12] for detailed descriptions of this procedure.

The resulting all-particle spectrum [11] and the spectra for the light [12] and heavy [10] mass groups are shown on the right side of Fig. 4.

The spectrum of heavy primaries shows a knee-like structure at around 80 PeV. This is the energy 26 times higher than the energy where the proton knee was found to be. This indicates a charge dependent position of the knees, however, the true composition of the heavy mass group is not entirely known. We only state that it is heavy, not that it is iron only. Iron dominating the heavy component does make sense from a theoretical point of view, though, and is indicated by the results of an unfolding analysis based on Grande data [13].

The spectrum of light primaries exhibits an ankle-like structure at about 126 PeV. The slope of the spectrum at energies below this feature is almost identical to the slope of the spectrum of heavy primaries above its knee-like structure. Assuming that the heavy galactic component reaches an end, the light component above the ankle-like feature could be considered to be of extragalactic origin. According to the QGSJetII-02 [14] based simulations, the light component should consist mainly of H + He, however, this depends on the hadronic interaction model that is used (See e.g. [15] for a comparison). If one interprets, for example, EPOS [16] simulated data in the light of QGSJetII-02 calibrations, the light component could be a pure proton component. The existence of both features does not depend on the hadronic interaction model used, only

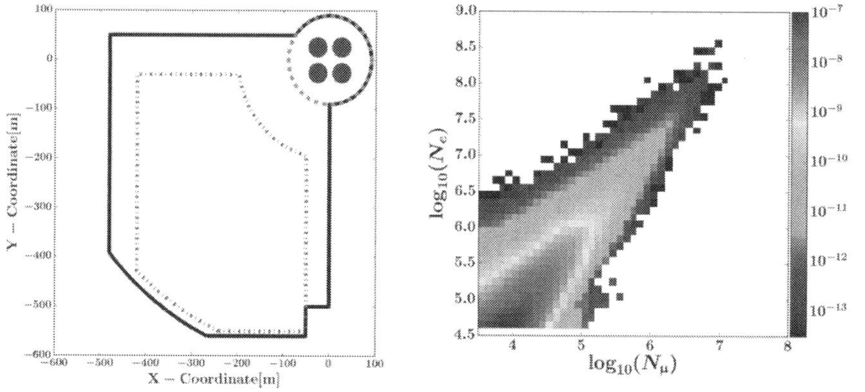

Figure 5. Left: The effective areas for the combined analysis (solid), KASCADE-Grande area (dot-dashed) and KASCADE (dashed) are shown. In addition a possible sub-KASCADE selection for reaching lower energies is shown as full circles. Right: The current two-dimensional shower spectrum of the combined analysis.

the relative fluxes of each component and the exact location of the two features change when using different models.

To improve the knowledge about the composition and to counter the differences between the results for KASCADE and KASCADE-Grande, a combined analysis using the same interaction models and reconstruction procedures is being performed. This new analysis is briefly described in the next section.

4. Combined Analysis

The combined analysis treats the KASCADE and KASCADE-Grande arrays as one single detector. The reconstruction of the shower observables follows the one described in section 2.2 (KASCADE), however, the Grande detectors are now included during the fitting processes. The energy deposits in the detectors are still translated into particle densities using the LECFs described in sections 2.2 and 3.1. Independent LECFs are needed because of the different scintillation detector configurations (thickness of the scintillators, surrounding material etc.).

Especially the events with a core position inside the Grande array and near the KASCADE detectors profit from the additional information provided by the latter, but also the reconstruction of the KASCADE events improve because of the larger distance range that gets available by adding the Grande detectors.

One of the improvements is the increase in the number of events. For KASCADE the increase compared to the stand-alone analysis shown in section 2.3 is based on a larger zenith angle range and, for part of its accessible energy range, a larger fiducial area. For the Grande energy range this is achieved mainly by using a larger fiducial area (left side of Fig. 5). Unfortunately, spanning such a large energy range, as is planned for this analysis, is not possible if the large area is to be used for the entire combined spectrum. Large showers inside or very near to the KASCADE array will result in an overestimation of the number of muons due to high energetic particles passing the shielding. Small showers farther away from the muon detectors will also result in a unreliable reconstruction of the total number of muons. Therefore, the chosen area has to depend on the size of the shower, using the KASCADE array for small showers, the total area for showers of intermediate size and excluding the KASCADE area for large showers. Although there are in total four different areas involved, we do not see significant transition effects in the two-dimensional shower-size plot, shown on the right side of Fig. 5. This is a good indication that the reconstruction works as expected. There are also plans to use a sub-KASCADE selection, shown as four full circles in Fig. 5, to try to get further down in energy towards 10^{14} eV, but this possibility is still under investigation.

5. Conclusion

The reconstruction procedure of the KASCADE array has been fully adapted to the combined use of both, the KASCADE and KASCADE-Grande detectors. The current state of the analysis is already a good basis for the final reconstruction of the energy spectrum and the determination of the mass composition in the energy range from 10^{15} (10^{14}) to 10^{18} eV. However, a lot of cross-checks on the area (energy) dependent reconstruction uncertainties are still to be done and an optimization for a reconstruction of lower energetic showers using sub-KASCADE area selections, as indicated by the full circles in Fig. 5, is also foreseen.

Acknowledgment

The authors would like to thank the members of the engineering and technical staff of the KASCADE-Grande collaboration, who contributed to the success of the experiment. The KASCADE-Grande experiment is supported in Germany by the BMBF and by the Helmholtz Alliance for Astroparticle Physics - HAP funded by the Initiative and Networking Fund of the Helmholtz Association, by the MIUR and INAF of Italy, the Polish Ministry of Science and Higher Education, and the Romanian Authority for Scientific Research UEFISCDI (PNII-IDEI grants 271/2011 and 17/2011).

References

[1] Antoni T *et al.* 2003 *Nucl. Instrum. Methods Phys. Res., Sect. A* **513** 490–510 ISSN 0168-9002

[2] Apel W *et al.* 2010 *Nucl. Instrum. Methods Phys. Res., Sect. A* **620** 202–216 ISSN 0168-9002

[3] Apel W *et al.* 2006 *Astropart. Phys.* **24** 467–483

[4] Kamata K and Nishimura J 1958 *Prog. Theor. Phys. Suppl.* **6** 93–155

[5] Greisen K 1960 *Ann. Rev. Nucl. Sci.* **10** 63–108

[6] Antoni T *et al.* 2005 *Astropart. Phys.* **24** 1–25 ISSN 0927-6505

[7] Kalmykov NN O S and AI P 1997 *Nucl. Phys. B (Proc. Suppl.)* **52** 17–28

[8] Engel R 1999 *International Cosmic Ray Conference* (*International Cosmic Ray Conference* vol 1) p 415

[9] AA L and RI R 2001 *Nucl. Phys. Proc. Suppl.* **97** 274

[10] Apel W D *et al.* (KASCADE-Grande Collaboration) 2011 *Phys. Rev. Lett.* **107**(17) 171104

[11] Apel W *et al.* 2012 *Astropart. Phys.* **36** 183 – 194 ISSN 0927-6505

[12] Apel W *et al.* 2013 *Phys. Rev. D* **87**

[13] Apel W *et al.* 2013 *Astropart. Phys.* **47** 54–66

[14] Ostapchenko S 2006 *Nucl. Phys. B, Proc. Suppl.* **151** 143 – 146 ISSN 0920-5632

[15] Apel W *et al.* 2014 *Adv. Space Res.* **53** 1456–1469

[16] Pierog T and Werner K 2009 *Nucl. Phys. B, Proc. Suppl.* **196** 102 – 105 ISSN 0920-5632 10.1016/j.nuclphysbps.2009.09.017

Anomaly in the cosmic-ray energy spectrum at GeV-TeV energies

Satyendra Thoudam[1,*] **and Jörg R. Hörandel**[1,2]

[1]Department of Astrophysics, IMAPP, Radboud University Nijmegen, 6500 GL Nijmegen, The Netherlands

[2]Nikhef, Science Park Amsterdam, 1098 XG Amsterdam, The Netherlands

E-mail: *s.thoudam@astro.ru.nl

Abstract. Recent measurements of cosmic rays by various experiments have found that the energy spectrum of cosmic rays is harder in the TeV region than at GeV energies. The origin of the spectral hardening is not clearly understood. In this paper, we discuss the possibility that the spectral hardening might be due to the effect of re-acceleration of cosmic rays by weak shocks associated with old supernova remnants in the Galaxy.

1. Introduction

The recent measurements of cosmic rays by the ATIC [1], CREAM [2], and PAMELA [3] experiments have revealed an anomaly in the cosmic-ray energy spectrum at GeV-TeV energies. The measured spectrum is found to be harder in the TeV region than at GeV energies. The spectral anomaly is difficult to explain using standard models of acceleration of cosmic rays and the nature of cosmic-ray propagation in the Galaxy, which predict a single power-law spectrum over the whole energy range.

Various explanations for the spectral anomaly have been proposed. These include hardening in the cosmic-ray source spectrum at high energies [4, 5, 6, 7], changes in the cosmic-ray propagation properties in the Galaxy [8, 9], and the effect of nearby sources [10, 11, 12, 13]. In this paper, we discuss the possibility that the anomaly can be due to the effect of re-acceleration of cosmic rays by weak shocks during their propagation through the Galaxy [14, 15]. The model assumes that cosmic rays are accelerated by strong supernova remnant shock waves. After acceleration, cosmic rays escape from the remnants and undergo diffusive propagation in the Galaxy. During the propagation, some fraction of the cosmic rays may again encounter expanding supernova remnant shock waves and get re-accelerated [16, 17]. This encounter is expected to be more often with older remnants as they occupy a larger volume in the Galaxy as compared to the younger remnants. Therefore, this process of re-acceleration is expected to be produced mainly by weaker shocks, and as weaker shocks accelerate particles with a softer spectrum, the re-accelerated particles will have an energy spectrum which is steeper than the spectrum originally produced by the strong shocks. The re-accelerated component can dominate the GeV energy region, while the non-reaccelerated component (hereafter "normal component") dominates at higher energies. In the following, we will show that this effect can explain the observed spectral anomaly.

Re-acceleration can also be produced by the same magnetic turbulence responsible for the scattering and spatial diffusion of cosmic rays in the Galaxy. This process, commonly known as distributed re-acceleration, have been predicted to produce strong features on some of the observed properties of cosmic rays at low energies, such as the peak in the secondary-to-primary ratios at ~ 1 GeV/nucleon. However, earlier studies have shown that this kind of re-acceleration is not sufficient enough to produce any noticeable features in the cosmic-ray spectrum at high energies [18]. The efficiency of distributed re-acceleration decreases with energy, and its effect becomes negligible at energies above ~ 20 GeV/nucleon. On the other hand, for the re-acceleration by old supernova remnants shock waves discussed in this paper, the efficiency does not depend strongly on energy. It depends primarily on the rate of supernova explosions in the Galaxy and the volume occupied by the supernova remnants. Its effect can be larger at higher energies as compared to the effect of distributed re-acceleration.

2. Cosmic-ray transport equation with re-acceleration

The re-acceleration of cosmic rays by weak shocks in the Galaxy is incorporated in the cosmic-ray transport equation as an additional source term following a power-law momentum spectrum [19]. The cosmic-ray transport equation including diffusion, re-acceleration and interaction losses then follows,

$$\nabla \cdot (D\nabla N) - \left[(\bar{n}v\sigma + \xi)\, N + \xi s p^{-s} \int_{p_0}^{p} du\, N(u) u^{s-1}\right] \delta(z) = -Q\delta(z) \tag{1}$$

where a cylindrical spatial coordinate (r, z) system is adopted with the origin located at the center of the Galaxy. $N(\mathbf{r}, p)$ represents the differential number density of nuclei with momentum/nucleon between p and $p+dp$, $D(p)$ is the diffusion coefficient, and $Q(r, p)$ represents the source term. The first term on the left-hand side of Eq. (1) represents diffusion. The second and third terms represent losses due to the inelastic interaction of cosmic rayss with the interstellar matter and due to the re-acceleration to higher energies respectively, where $\bar{n}$ represents the averaged surface density of interstellar atoms, $v(p)$ the particle velocity, $\sigma(p)$ the inelastic collision cross-section, and ξ corresponds to the rate of re-acceleration. The fourth term containing an integral represents the population of particles produced by the re-acceleration of low-energy particles, assuming that particles are instantaneously re-accelerated to generate a power-law distribution of index s. The effects of ionization losses and convection due to the Galactic wind are neglected as these effects are mostly important at energies below ~ 1 GeV/nucleon. Our calculation concentrates only at energies above 1 GeV/nucleon.

The propagation region is assumed to be bounded at $z = \pm H$, and unbounded in the r direction. The matter and the sources are assumed to follow a uniform distribution in the Galactic disk with a radial size R. For cosmic-ray primaries, the source term in Eq. (1) is taken as $Q(r, p) = \bar{\nu}\mathrm{H}[R - r]\mathrm{H}[p - p_0]Q(p)$, where $\bar{\nu}$ represents the supernova explosion (SNe) rate per unit surface area of the disk, $\mathrm{H}(m) = 1(0)$ for $m > 0(< 0)$ is the Heaviside step function, and p_0 is the lower cut-off in momentum/nucleon that has been introduced to account for the losses due to the ionization process. The source spectrum is assumed to follow a power-law in total momentum with an exponential cut-off at high energy. It can be written as function of momentum/nucleon as,

$$Q(p) = AQ_0(Ap)^{-q} \exp\left(-\frac{Ap}{Zp_c}\right) \tag{2}$$

where A and Z respectively represents the mass and charge numbers of the nuclei, Q_0 is a constant corresponds to the fraction f of kinetic energy of a supernova explosion that is injected into a given cosmic-ray species, q represents the source spectral index, and p_c is the cut-off momentum for protons at high energies. We have assumed that the maximum total momentum for a cosmic-ray nuclei is Z times that of the protons. The cosmic-ray diffusion coefficient is assumed to follow $D(\rho) = D_0\beta(\rho/\rho_0)^a$, where $\rho = Apc/Ze$ is the particle rigidity with e representing the charge of an electron and c the velocity of light, D_0 is the diffusion constant, a is the diffusion index, $\beta = v/c$, and ρ_0 is a constant. In Eq. (1), the re-acceleration parameter is written as $\xi = \eta V\bar{\nu}$, where $V = 4\pi\Re^3/3$ is the volume occupied by a supernova remnant of radius $\Re$, and η is a parameter introduced to take care of the unknown actual volume of supernova remnants that re-accelerate cosmic rays. We take $\Re = 100$ pc which is approximately the size of a supernova remnant of age 10^5 yr expanding in the interstellar medium with an initial shock velocity of 10^9 cm s^{-1}.

Eq. (1) is solved using the standard Green's function technique. The solution at $r = 0$ is given by,

$$N(z,p) = \bar{\nu}R\int_0^\infty dk\,\frac{\sinh[k(H-z)]}{\sinh(kH)} \times \frac{J_1(kR)}{L(p)} \times F(p) \tag{3}$$

where J_1 is a Bessel function of order 1,

$$L(p) = 2D(p)k\coth(kH) + \bar{n}v(p)\sigma(p) + \xi, \tag{4}$$

$$F(p) = Q(p) + \xi sp^{-s}\int_{p_0}^{p} u^s du\, Q(u)A(u) \times \exp\left(\xi s\int_u^p A(w)dw\right) \tag{5}$$

and, the function A is given by

$$A(x) = \frac{1}{xL(x)} \tag{6}$$

The local cosmic-ray density is obtained by taking $z = 0$ in Eq. (3). The first term on the right-hand side of Eq. (5) represents the normal cosmic-ray component, and the second term represents the re-accelerated component. Re-acceleration takes away particles from the low-energy part of the spectrum and puts them into the higher energy region. Therefore, for re-acceleration produced by weak shocks where $s > q$, the re-accelerated component might produce visible signatures, such as a bump or enhancement, in the energy spectrum. For re-acceleration by strong shocks which produces a harder particle spectrum, for instance $s = q$, the re-acceleration effect might not become visible since both the components follow the same spectra [19].

The spectrum for cosmic-ray secondaries in the Galaxy $N_2(\mathbf{r}, p)$ is obtained by solving a similar transport equation to their primaries with the source term given by,

$$Q_2(\mathbf{r},p) = \bar{n}v_1(p)\sigma_{12}(p)H[R-r]H[p-p_0]N_1(\mathbf{r},p)\delta(z) \tag{7}$$

where v_1 represents the velocity of the primary nuclei, σ_{12} represents the total fragmentation cross-section of the primary to the secondary, and N_1 is the density of the primary nuclei given by Eq. (3). The subscripts 1 and 2 denote the primary and secondary nuclei respectively.

The secondary-to-primary ratio is calculated by taking the ratio N_2/N_1. For no re-acceleration, i.e. $\xi = 0$, it can be checked that Eq. (3) reduces to the standard solution of pure-diffusion equation, and the secondary-to-primary ratio becomes inversely proportional to the diffusion coefficient at high energies (see e.g., [20]). It can be mentioned that the energy spectrum of the secondary nuclei is more sensitive to the effect of re-acceleration than the spectrum of the primary nuclei [15]. Consequently, the secondary-to-primary ratios will also be more sensitive to the re-acceleration effect than the primary energy spectrum. For a steeper re-acceleration index $s > q$, the ratio is expected to exhibit an enhancement at lower energies, and for a harder index $s = q$, the ratio will have a flattening at high energies [17, 19]. For the present calculation, since we are considering re-acceleration mainly by old supernova remnants, we will only consider the case of $s > q$ with $s \gtrsim 4$, which corresponds to a shock Mach number of ~ 1.7.

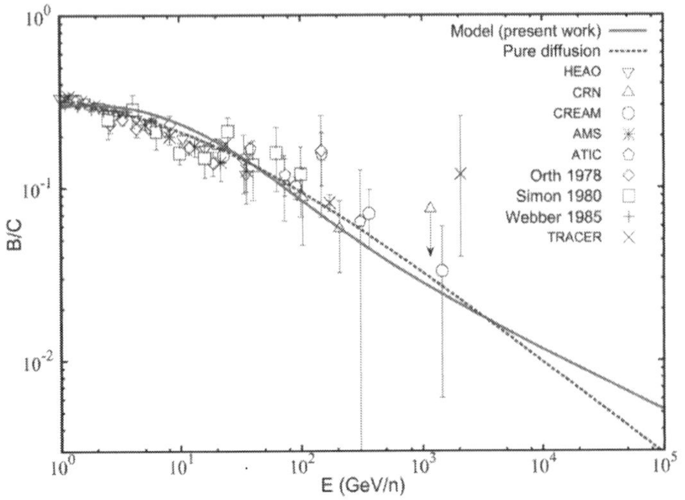

Figure 1. Boron-to-Carbon (B/C) ratio. *Solid line*: Present work including re-acceleration. *Dashed line*: Pure diffusion model without re-acceleration [11].

3. Results and discussions

We take $H = 5$ kpc, and the averaged surface density of the atomic hydrogen as $\bar{n} = 7.24 \times 10^{20}$ atoms cm^{-2} [11]. We assume that the interstellar medium consists of 10% helium. The inelastic interaction cross-sections are taken to be the same used in the calculations in Ref. [11]. We take the size of the source distribution $R = 20$ kpc, and the proton low and high-momentum cut-offs as $p_0 = 100$ MeV/c and $p_c = 1$ PeV/c respectively. The supernova explosion rate is taken as $\bar{\nu} = 25$ SNe Myr^{-1} kpc^{-1}, which corresponds to ~ 3 SNe per century in the Galaxy. The followings are taken as model parameters: the cosmic-ray propagation parameters (D_0, ρ_0, a), the re-acceleration parameters (η, s) and the source parameters (q, f).

The values of $(D_0, \rho_0, a, \eta, s)$ are determined based on the measurements of boron-to-carbon ratio, and the measured energy spectra of carbon, oxygen, and boron nuclei. The values that gives the over-all best agreement between the model calculation and the measurements are found to be $D_0 = 9 \times 10^{28}$ cm^2 s^{-1}, $\rho = 3$ GV, $a = 0.33$, $\eta = 1.02$, $s = 4.5$. These values correspond to the maximum amount of re-acceleration permitted by the available boron-to-carbon data. The result on the boron-to-carbon is shown in Figure 1 (solid line), where we have also shown the result for the case of pure diffusion (dashed line) with no re-acceleration ($\eta = 0$) taken from Ref. [11]. The best-fit carbon and oxygen source parameters are found to be $q_C = 2.24$, $f_C = 0.024\%$, and $q_O = 2.26$, $f_O = 0.025\%$ respectively, where the f's are given in units of 10^{51} ergs. The calculation assumes a force-field solar modulation parameter of $\phi = 450$ MV.

The spectra for the protons and helium nuclei calculated using the best-fit values of $(D_0, \rho_0, a, \eta, s)$ are shown in Figure 2. The top panel represents proton and the bottom panel represents helium nuclei. The data are the same as used in Ref. [11]. The source parameters required to explain the measurements are found to be $q_p = 2.21$, $f_p = 6.95\%$ for protons, and $q_{He} = 2.18$, $f_{He} = 0.79\%$ for helium. It can be seen that the model reproduces the measured data quite well, explaining the observed spectral anomaly between the GeV and TeV energy regions.

In Figure 2, the effect of re-acceleration is stronger for protons than for helium nuclei. This is partly due to the steeper source spectrum for protons, as compared to the helium nuclei, required to reproduce the overall observed spectrum. For a steeper source spectrum, a larger fraction of low-energy particles becomes available for re-acceleleration leading into an increase

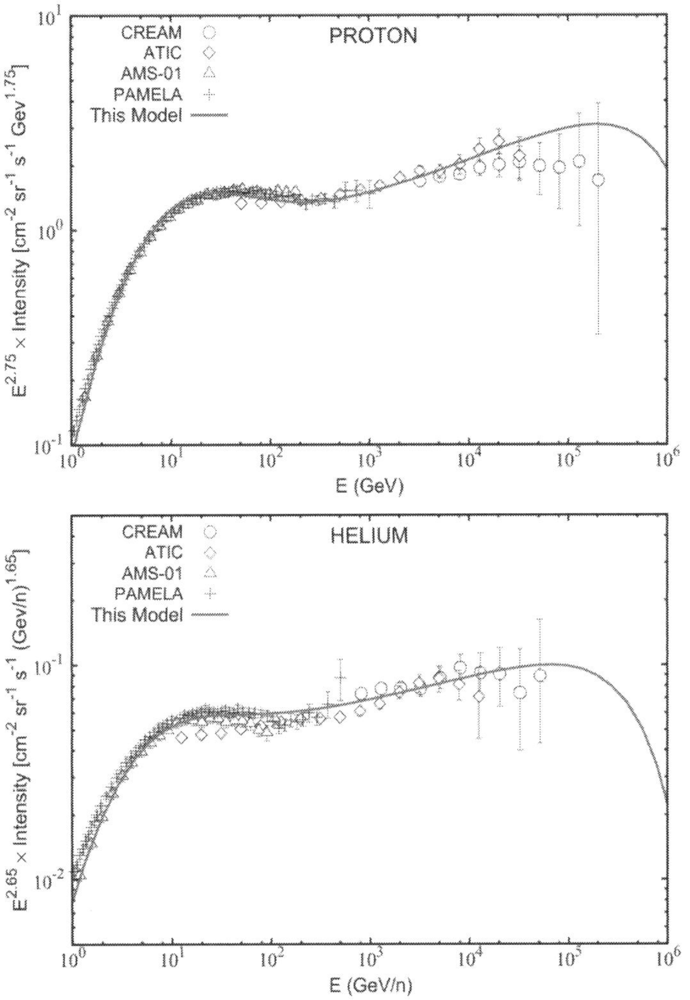

Figure 2. *Top*: Proton spectrum. *Bottom*: Helium spectrum. The lines represent our results. For the data, see the experiments listed in Ref. [11].

in the re-accelerated component relative to the normal cosmic-ray component. But, the main reason for the stronger re-acceleration effect in the case of protons is due to their smaller inelastic collision cross-sections as compared to the helium nuclei. This means that low-energy protons can be efficiently re-accelerated to higher energies before they undergo inelastic collision with the interstellar matter during their residence time in the Galaxy. For helium nuclei, the loss due to the inelastic collision becomes important relative to the re-acceleration. For heavier nuclei such as iron for which the inelastic collision cross-sections are even larger, the collision losses becomes more important and the re-acceleration effect is expected to be negligible. This is shown in Figure 3 where we have compared our model prediction for the iron nuclei with the measurements. The calculation assumes $q_{Fe} = 2.28$, and $f_{Fe} = 4.9 \times 10^{-3}\%$. As expected, the re-acceleration effect is hard to notice in Figure 3, and the model prediction above ~ 20 GeV/n follows approximately a single power-law unlike the proton and helium spectra.

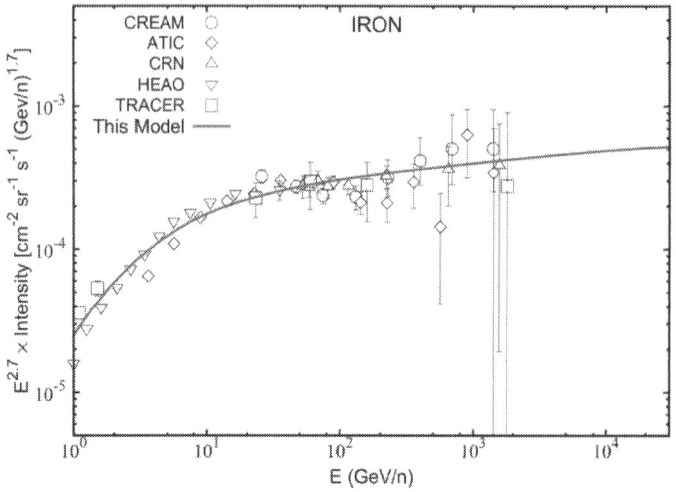

Figure 3. Iron spectrum. The line represents our result. For the data, see the experiments listed in Ref. [11].

4. Conclusions

We have presented a model in which cosmic rays, after acceleration by strong supernova remnant shock waves, are re-accelerated by weak shocks associated with old supernova remnants during propagation in the Galaxy. The re-acceleration produces a cosmic-ray component that dominates the low energy part of the spectrum, while keeping the high-energy part dominated by the population that have not suffered re-acceleration. The model is found to explain the cosmic-ray spectral anomaly at GeV-TeV energies recently observed for the protons and helium nuclei. The study shows that the re-acceleration effect is important only for light nuclei, and is negligible for heavy nuclei like iron. Our prediction can be tested by sensitive measurements of heavy cosmic-ray species in future.

References

[1] Panov, A. D. et al. 2007, *Bull. Russ. Acad. Sci.*, Vol. **71**, No. 4, pp. 494
[2] Yoon, Y. S. et al. 2011, *ApJ*, **728**, 122
[3] Adriani, O., et al. 2011, *Science*, **332**, 69
[4] Biermann P. L., et al. 2010, *ApJ*, **725**, 184
[5] Ohira, Y., Murase, K., & Yamazaki, R., 2011, *MNRAS*, **410**, 1577
[6] Yuan, Q., Zhang, B., & Bi, X. -J, 2011, *Phys. Rev. D*, **84**, 043002
[7] Ptuskin, V. S., Zirakashvili, V., & Seo, E.-S., 2013, *ApJ*, **763**, 47
[8] Tomassetti, N. 2012, *ApJ*, **752**, L13
[9] Blasi, P., Amato, E., & Serpico, P. D., 2012, *PRL*, **109**, 061101
[10] Thoudam, S., & Hörandel, J. R. 2012, *MNRAS*, **421**, 1209
[11] Thoudam, S., & Hörandel, J. R. 2013, *MNRAS*, **435**, 2532
[12] Erlykin, A. D. & Wolfendale, A. W., 2012, *APh*, **35**, 449
[13] Zatsepin, V. I., Panov, A. D. & Sokolskaya, N. V. 2013, *J. Phys.: Conf. Ser.*, **409**, 012028
[14] Ptuskin, V., Zirakashvili, V., &Seo, E.-S. 2011, *Proc. 32nd ICRC*, **6**, 234
[15] Thoudam, S., & Hörandel, J. R. 2014, *A&A*, **567**, A33
[16] Wandel, A., 1988, *A&A*, **200**, 279
[17] Berezhko, E. G., et al. 2003, *A&A*, **410**, 189
[18] Seo, E. S., & Ptuskin, V. S., 1994, *ApJ*, **431**, 705
[19] Wandel, A., Eichler, D. S., Letaw, J. R., Silberberg, R., & Tsao, C. H., 1987, *ApJ*, **316**, 676
[20] Thoudam, S., 2008, *MNRAS*, **388**, 335

Spectrum and fraction of cosmic ray positrons: results of the anomalous diffusion approach

Nikolay Volkov, Anatoly Lagutin, Alexander Tyumentsev

Altai State University, Radiophysics and Theoretical Physics Department, Barnaul, Russia, 656049, Lenin ave., 61

E-mail: volkov@theory.asu.ru, lagutin@theory.asu.ru, tyumentsev@theory.asu.ru

Abstract. We present the results of new calculations of the energy spectra of cosmic ray electrons, positrons and also positron fraction under assumption that both electrons and positrons are generated by the same Galactic sources, which accelerate particles with same power-law injection spectral index p. The value of the injection index $p \approx 2.85$ retrieved in our works from the analysis of the observed CR spectra has been used. The propagation of particles through the highly non-homogeneous interstellar medium has been described by the anomalous diffusion model.

We show that proposed approach allows the self-consistent description of the recent AMS-02 data. In contrast to the conclusion made by AMS-02 collaboration about the different origin of the high-energy positrons and electrons, our results demonstrate that the differing behavior of spectral indices of electrons and positrons with energy can be described under the assumption that both positrons and electrons have common Galactic sources.

Our prediction (Lagutin et al. Bull. Russian Academy of Sciences. Physics, 2013, Vol. 77, No. 11, p. 1312) that the e^+ to total $e^- + e^+$ ratio reaches a constant value for E $\gg$ 200 GeV is confirmed.

1. Introduction

According to standard cosmic ray (CR) electron/positron origin scenario [1–3], electron sources are continuously distributed in Galactic disk and positrons are produced only by collisions of primary CR nuclides with Galactic interstellar medium. In this CR origin picture the ratio of the secondary e^+ to total $e^- + e^+$ flux decreases with increasing energy. However, new experimental results obtained in the last decade by PAMELA [4], Fermi-LAT [5] and AMS-02 [6] collaborations demonstrate steadily increasing of the positron fraction (PF) from 10 GeV to $\sim$ 300 GeV. This is not consistent with only secondary mechanism of positron production and probably with the standard paradigm of the Galactic CR origin. The presence of new primary positron component seems unavoidable.

Models, ranging from the annihilation/decay of dark matter (DM) to the new astrophysical sources, have been proposed in order to explain the rise in the PF (see a review [7] and the references therein; see also [8–16]). Leading proposals put forth to explain reported PF results involve $e^{\pm}$ pair emission by nearby pulsars. These new charge symmetric sources of electrons and positrons are introduced in order to account for the "excess" of the positrons relative to the background fluxes, associated to galactic CR sources of the standard electron/positron origin scenario.

In this paper, the so-called alternative explanation (see [8–11]) of the rise in the PF which involves the positrons production in the astrophysical sources of the CR is investigated. We assume that both positrons and electrons have the common Galactic sources accelerated the particles with the same power-law injection spectral index p (see [9]). The value of the injection index $p \approx 2.85$ retrieved in our works [17, 18] from analysis of the observed CR spectra is used. We also suppose that the propagation of particles through the highly non-homogeneous interstellar medium is described by the anomalous diffusion model [17–19]. The anomaly in this model results from large free paths (Lévy flights) of particles between galactic inhomogeneities in the fractal-like Galactic medium. First results related to the PF behavior in this approach were discussed in our papers [20, 21].

The main goal of this paper is to validate the key assumptions of proposed approach using new results of the AMS-02 collaboration.

2. Anomalous diffusion model

In the standard scenario, the propagation of galactic cosmic ray electrons and positrons through the interstellar medium (ISM) is described by Syrovatskii's diffusion equation for density of particles $N(\mathbf{r}, t, E)$ of energy E [22]

$$\frac{\partial N(\mathbf{r}, t, E)}{\partial t} = D(E)\triangle N(\mathbf{r}, t, E) + \frac{\partial (b(E)N(\mathbf{r}, t, E))}{\partial E} + S(\mathbf{r}, t, E). \tag{1}$$

In this equation $D(E)$ is the diffusion coefficient, $b(E)$ – the rate of change of the particles energy during their propagation through the interstellar medium and $S(\mathbf{r}, t, E)$ – the source function of electrons/positrons.

This equation determines propagation process of particles (electrons/positrons) under the approximation that magnetic field inhomogeneities of the ISM are small-scale and can be considered as a homogeneous Poisson ensemble. However, during a few last decades many evidences of the existence of multiscale structures in the Galaxy have been found. Filaments, shells, clouds are entities widely spread in the ISM. A rich variety of structures that are created in the ISM can be related to the fundamental property of turbulence called intermittency. In such a fractal-like ISM we certainly do not expect that Syrovatskii's diffusion equation remains accurate enough.

Generalization of the equation (1) leads to anomalous diffusion model [17–19]. In this paper, instead of Syrovatskii's diffusion equation, the superdiffusion equation, proposed in our works [20, 21, 23] is used:

$$\frac{\partial N(\mathbf{r}, t, E)}{\partial t} = -D(E, \alpha)(-\triangle)^{\alpha/2} N(\mathbf{r}, t, E) + \frac{\partial (b(E)N(\mathbf{r}, t, E))}{\partial E} + S(\mathbf{r}, t, E). \tag{2}$$

Here $D(E, \alpha) = D_0(\alpha)E^{\delta}$ is the anomalous diffusivity. Energy losses $b(E)$ are presented in the form

$$b(E) = b_0 + b_1 E + b_2 E^2 \approx b_2(E + E_1)(E + E_2), \tag{3}$$

where $b_0 = 3.06 \cdot 10^{-16} n$ (GeV s^{-1}) is rate of the ionization losses, $b_1 = 10^{-15} n$ (s^{-1}) determines the bremsstrahlung rate, and $b_2 = 1.38 \cdot 10^{-16}$ (GeV$\cdot$ s)$^{-1}$ is the factor to take into account the inverse Compton losses and synchrotron radiation (for the intensity of the interstellar magnetic field $B = 5$ μG and background photon density $\omega = 1$ eV/cm^3). In (3) we have introduced the notation $E_1 = b_0/b_1$ and $E_2 = b_1/b_2$.

The fractional Laplacian (Riesz operator) $(-\triangle)^{\alpha/2}$ [24] describes the large free paths (Lévy flights) of particles between Galactic inhomogeneities in the fractal-like Galactic medium.

We note that in case of $\alpha = 2$ equation (2) coincides with Syrovatskii's equation (1).

Green's function $G(\mathbf{r}, t, E; E_0)$, corresponding to (2), can be found from an equation

$$\frac{\partial G(\mathbf{r}, t, E; E_0)}{\partial t} = -D(E,\alpha)(-\Delta)^{\alpha/2}G(\mathbf{r}, t, E; E_0) + \frac{\partial b(E)G(\mathbf{r}, t, E; E_0)}{\partial E} + \delta(\mathbf{r})\delta(t)\delta(E - E_0). \quad (4)$$

With the use of the standard Syrovatskii substitutions [22]

$$t' = t - \tau, \quad \tau(E, E_0) = \int\limits_E^{E_0} \frac{dE'}{b(E')}, \quad \lambda(E, E_0) = \int\limits_E^{E_0} \frac{D(E',\alpha)}{b(E')} dE',$$

and Fourier transform, it is easy to find

$$G(\mathbf{r}, t, E; E_0) = \frac{g_3^{(\alpha)}(|\mathbf{r}|\lambda^{-1/\alpha})}{\lambda^{3/\alpha}(1 - b_2t(E + E_2))^2}\delta\left(E_0 - \left\{\frac{E + E_1}{1 - b_1t(E + E_2)/(E_2 - E_1)} - E_1\right\}\right) \times \times H(1 - b_2t(E + E_2))H(t). \quad (5)$$

Here

$$E_0(t) = \frac{E + E_1}{1 - b_1t(E + E_2)/(E_2 - E_1)} - E_1$$

and $g_3^{(\alpha)}(r)$ is the probability density of three-dimentional sphericaly-symmetrical stable distribution [25]. The probability density of $g_3^{(\alpha)}(r)$ and the asymptotic behavior for $r \to \infty$ is shown in figure 1.

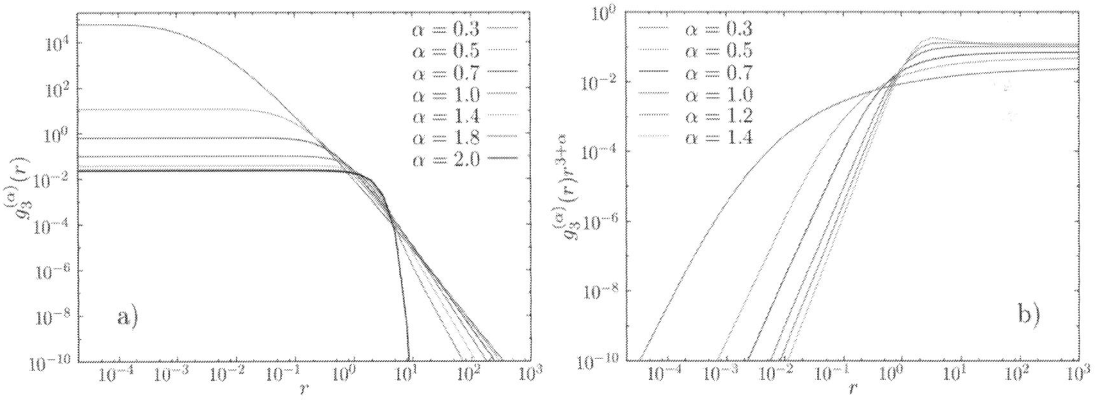

Figure 1. a) the probability density of three-dimentional sphericaly-symmetrical stable distribution $g_3^{(\alpha)}(r)$, b) the asymptotic behavior $r^{-3-\alpha}$ for $r \to \infty$

3. Solution of the anomalous diffusion equation

Using the Green's function (5), we can write a solution of the superdiffusion equation (2) for particles generated by sources, defined by the source function $S(\mathbf{r}, t, E)$

$$N(\mathbf{r}, t, E) = \int\limits_{\mathrm{R}^3} d\mathbf{r}_0 \int\limits_E^{\infty} dE_0 \int\limits_{-\infty}^{t} dt_0 G(\mathbf{r}, t, E; \mathbf{r}_0, t_0, E_0)S(\mathbf{r}_0, t_0, E_0).$$

3.1. Point pulse source
For a point pulse source with a power-law energy spectrum $S(\mathbf{r},t,E) = S_i E^{-p}\delta(\mathbf{r})\Theta(T-t)\Theta(t)$, which simulates generation of particles in supernovae, the following solution takes place.

$$N(\mathbf{r},t,E) = S_i \int\limits_{\max[0,t-T]}^{\min[t,1/b_2(E+E_2)]} dt' E_0(t')^{-p} \lambda(t',E)^{-3/\alpha} \times$$
$$\times \left(1 - b_2 t'(E+E_2)\right)^{-2} g_3^{(\alpha)}\left(|\mathbf{r}|\lambda(t',E)^{-1/\alpha}\right). \quad (6)$$

3.2. Point steady source
For a point steady source $S(\mathbf{r},E) = S_c E^{-p}\delta(\mathbf{r})$ we find

$$N(\mathbf{r},E) = S_c \int\limits_{0}^{1/b_2(E+E_2)} dt' E_0(t')^{-p} \lambda(t',E)^{-3/\alpha} \times$$
$$\times \left(1 - b_2 t'(E+E_2)\right)^{-2} g_3^{(\alpha)}\left(|\mathbf{r}|\lambda(t',E)^{-1/\alpha}\right). \quad (7)$$

4. Energy spectrum of electrons and positrons
The particle intensity $J^{\pm}(\mathbf{r},t,E) = (v/4\pi)N^{\pm}(\mathbf{r},t,E)$ was expressed as

$$J^{\pm}(\mathbf{r},t,E) = J_L^{\pm}(\mathbf{r},t,E) + J_G^{\pm}(\mathbf{r},E) + J_{\mathrm{sec}}^{\pm} \quad (8)$$

where $J_L^{\pm}$ is the local component, i.e. the contribution from nearby ($r < 1$ kpc) young ($t < 10^6$ yr) sources, $J_G^{\pm}$ is the global spectrum component determined by the multiple old ($t \geq 10^6$ yr) distant ($r \geq 1$ kpc) sources and $J_{\mathrm{sec}}^{\pm}$ is the secondary electron and positron fluxes originated from interactions of primary cosmic rays with the interstellar medium.

To describe the distribution of nearby young sources simulation of the Poisson ensemble of sources was carried out. The Poisson distribution parameter (average number of sources in the local region) was chosen ~ 10. This estimation corresponds to number of the well-known nearest supernova remnants and pulsars with $t < 10^6$ yr [26, 27]. Coordinates and times of birth of the sources were generated randomly and uniformly in the space region $r < 10^3$ pc and in the time interval $10^4 \leq t < 10^6$ yr. The duration of particle generation by local sources was assumed to be $T \approx 10^4$ yr.

Distribution of multiple old ($t \geq 10^6$ yr) distant ($r \geq 1$ kpc) sources was considered according to standard leptonic scenario (i.e. as the system of steady-state sources). To accommodate the contribution of secondary electrons and positrons produced in the collision of cosmic rays nuclei with the interstellar medium, the GALPROP code [28] was used with the parameters stated in our model.

At energies $E \lesssim 10$ GeV, the fluxes of both electrons and positrons observed in the Solar system are influenced by modulation effects. To take into account the solar modulation we used the model [29] with the potential $\Phi = 600$ MV.

5. Estimation of the model parameters
Estimations of key model parameters (injection spectral index p and exponent δ describing the energy dependence of the anomalous diffusivity) were obtained in [18] from the analysis of observed CR spectra. It was shown in [18] that with the use of representation $N = N_0 E^{-\eta}$ for observed CR spectra and the values of the exponents below and above the knee energy E_{knee},

$$\eta|_{E \ll E_{\mathrm{knee}}} = p - \delta,$$

$$\eta|_{E\gg E_{\rm knee}} = p + \delta,$$

the variation of spectral exponent $\Delta\eta = \eta|_{E\gg E_{\rm knee}} - \eta|_{E\ll E_{\rm knee}}$ can be found as $\Delta\eta = 2\delta$. Since $\eta|_{E\ll E_{\rm knee}} \sim 2.62$ and $\eta|_{E\gg E_{\rm knee}} \sim 3.16$, from the equalities above it is easy to retrieve the injection spectral index p and exponent δ of anomalous diffusivity:

$$p \approx 2.85, \quad \delta \approx 0.27.$$

We note that the estimation of $\delta \approx 0.27$ almost coincides with the value of $1/3$, accepted for the Kolmogorov turbulence [1]. Adopted value $p \approx 2.85$ for the CR sources is supported by the recent results on SNRs: for the SNR RX J1713.7-3946 the exponent $p \approx 3.0$ was obtained [30]; the Fermi-LAT experiment reported $p \approx 2.87$ for $E > 69$ GeV for the SNR IC 443 [31, 32] and $p \approx 3.3$ for the SNR W44 [33].

To assess parameter α, which is specific for our model, the results of studies of particles diffusion in cosmic and laboratory plasma have been used. It is known that for the superdiffusion regime of propagation ($1 < \alpha < 2$) the diffusion packet spreads with time as $\Delta x^2 \sim t^{2/\alpha}$. Interpretation of the data for magnetosphere leads to conclusion that the width of the diffusion packet increases with the rate $\Delta x^2 \sim t^{1.4}$ [34]. Investigation of the particles propagation before the shock wave in the circumterrestrial plasma gives the dependence of $\Delta x^2 \sim t^{1.19}$ [35]. According to these results we obtain the estimation as $\alpha \approx 1.4$.

Finally, for the anomalous diffusivity $D(E, \alpha) = D_0(\alpha)E^\delta$ we get the value of $D_0(\alpha) \approx 2 \cdot 10^{-4}$ pc$^{1.4}$/year [18].

Results of our calculations of the energy spectra of cosmic ray electrons and positrons as well as positron fraction obtained in the framework of the proposed approach are shown in figures 2–5.

6. Results and conclusions

We present the results of new calculations of the energy spectra of cosmic ray electrons, positrons and also positron fraction under assumption that both positrons and electrons are generated by the same Galactic sources, which accelerate particles with same power-law injection spectral index p. The value of the injection index $p \approx 2.85$ retrieved in our works [17, 18] from the analysis of the observed CR spectra has been used. The propagation of particles through the highly non-homogeneous interstellar medium has been described by the anomalous diffusion model [17–19]. The anomaly in this model results from large free paths (Lévy flights) of particles between galactic inhomogeneities in a fractal-like Galactic medium.

Figures 2–5 clearly show that proposed approach allows the self-consistent description of the recent AMS-02 data on electrons, positrons and positron fraction.

In contrast to the conclusion made by AMS-02 collaboration in [6] about the different origin of the high-energy positrons and electrons, our results demonstrate that the differing behavior of spectral indices of electrons and positrons with energy can be described under the assumption that both positrons and electrons have common Galactic sources. We found that the difference in spectral indices is due to the fact that in the energy region between 20 and 200 GeV in addition to the yield from the primary positron sources there is also a significant contribution of secondary positrons (see figure 5).

The positron fraction obtained in this paper (figure 3) confirms our prediction [45] that the e^+ to total $e^- + e^+$ ratio reaches a constant value for E $\gg$ 200 GeV.

Acknowledgement

This work was supported in part by the Russian Foundation for Basic Research, project no. 14-02-31524. We also acknowledge helpful comments and suggestions from an anonymous referee.

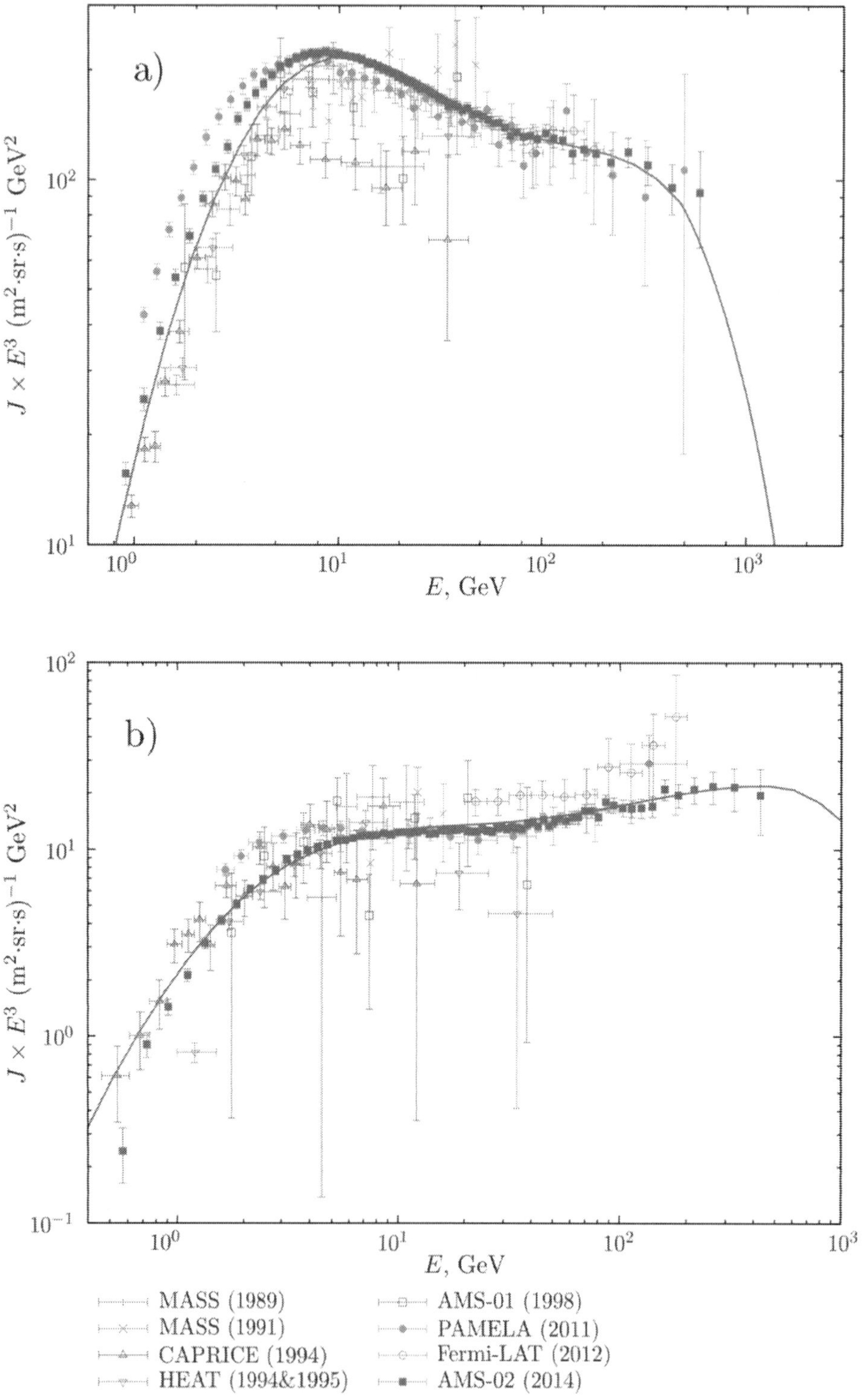

Figure 2. Spectrum (times E^3) of electrons a) and positrons b) in the proposed model. Experimental data from [4–6, 36–40]

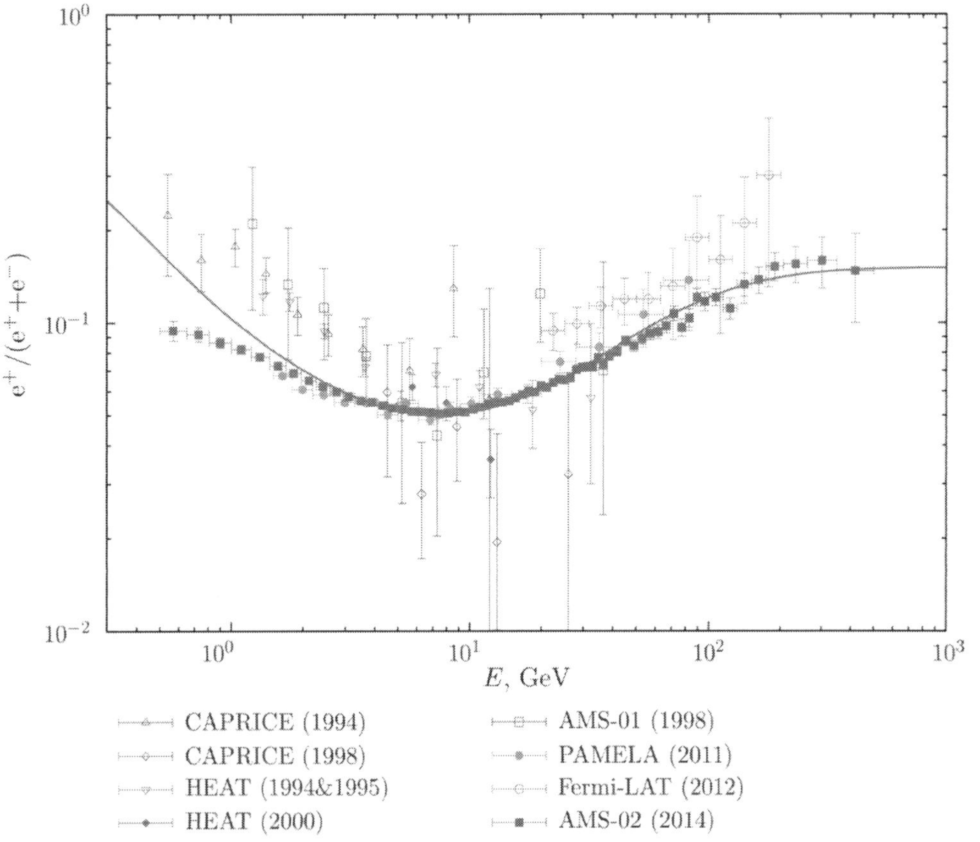

Figure 3. Positrons fraction in the proposed model. Experimental data from [4–6, 36–44]

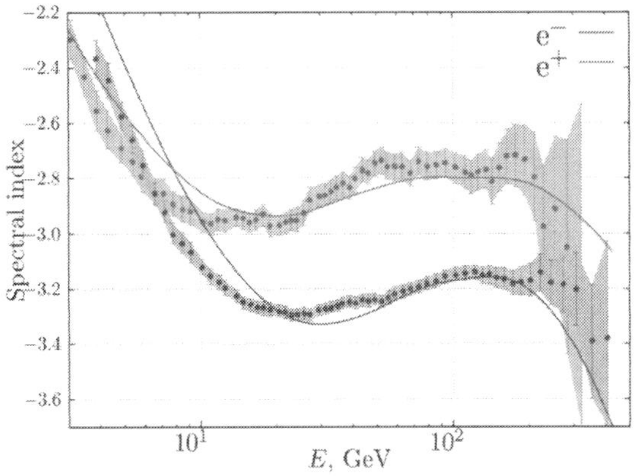

Figure 4. Spectral indices of the fluxes of electrons and positrons in the framework of the proposed model (color lines). Points are the results of AMS-02 collaboration [6]

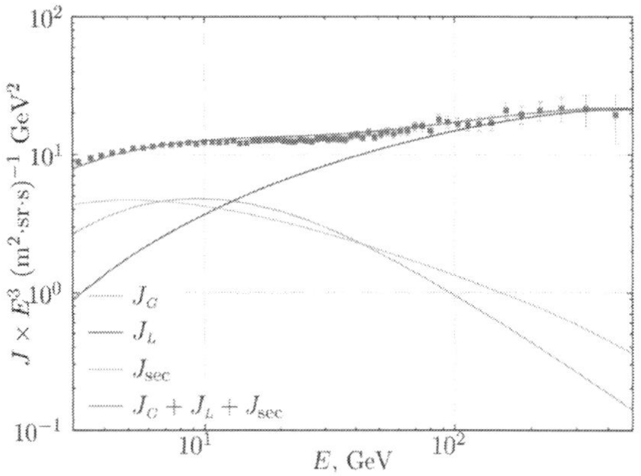

Figure 5. Positrons spectrum (times E^3) in the proposed model. Contribution from global sources (green), local group sources (blue) and secondary positrons (cyan) also presented

References

[1] Berezinskii V S, Bulanov S V, Dogiel V A and et al 1990 *Astrophysics of cosmic rays* (Amsterdam: North Holland) p 546
[2] Strong A W, Moskalenko I V and Ptuskin V S 2007 *Annu. Rev. Nucl. Part. Sci.* **57** 285
[3] Grasso D, Profumo S, Strong A W and et al 2011 *NIMPA* **630**(1) 48
[4] Adriani O, Barbarino G C, Bazilevskaya G A and (PAMELA Collaboration) 2011 *Phys. Rev. Lett.* **106**(20) 201101
[5] Ackermann M, Ajello M, Allafort A and (Fermi-LAT Collaboration) 2012 *Phys. Rev. Lett.* **108**(1) 011103
[6] Aguilar M, Aisa D, Alvino A and (AMS-02 Collaboration) 2014 *Phys. Rev. Lett.* **113**(12) 121102
[7] Serpico P 2012 *Astropart. Phys.* **39-40** 2
[8] Blasi P 2009 *Phys. Rev. Lett.* **103** 051104
[9] Kachelrieß M, Ostapchenko S and Tomas R 2011 *ApJ* **733** 119
[10] Berezhko E G and Ksenofontov L T 2013 *Journal of Phys. Conf. Ser.* **409** 012025
[11] Erlykin A D and Wolfendale A W 2013 *Astropart. Phys.* **49** 23
[12] Gaggero D and Maccione L 2013 *Phys. Rev. Lett.* **111** 021102
[13] Cholis I and Hooper D 2013 *Phys. Rev. D* **88** 023013
[14] Di Mauro M, Donato F, Fornengo N and et al 2014 *JCAP* **04** 006
[15] Cowsik R, Burch B and Madziwa-Nussinov T 2014 *ApJ* **786** 124
[16] Mertsch P and Sarkar S 2014 *Phys. Rev. D* **90** 061301
[17] Lagutin A A, Nikulin Y A and Uchaikin V V 2001 *Nucl. Phys. B. Proc. Suppl.* **97** 267
[18] Lagutin A A and Tyumentsev A G 2004 *Izv. Altai. Gos. Univ.* **35** 4 (in Russian)
[19] Lagutin A A and Uchaikin V V 2003 *NIMPB* **201** 212
[20] Volkov N, Lagutin A, Tyumentsev A and Yushkov A 2011 *Proc. of 32nd ICRC (China)* vol 6 (Beijing: Institute of High Energy Physics) p 264
[21] Volkov N and Lagutin A 2013 *J. Phys. Conf. Ser.* **409** 012042
[22] Syrovatskii S I 1959 *Astron. Zh.* **36** 17 (in Russian)
[23] Lagutin A, Volkov N, Kuzmin A and Tyumentsev A 2009 *Bull. of RAS. Phys.* **73** 581
[24] Samco S G, Kilbas A A and Marichev O I 1993 *Fractional integrals and derivatives — Theory and Applications* (New York: Gordon and Breach) p 976
[25] Uchaikin V V and Zolotarev V M 1999 *Chance and Stability* (Netherlands, Utrecht: VSP) p 594
[26] Hooper D, Blasi P and Serpico P D 2009 *JCAP* **2** 025
[27] Gendelev L, Profumo S and Dormody M 2010 *JCAP* **2** 16
[28] Moskalenko I V and Strong A W 1998 *ApJ* **493** 694
[29] Gleeson L J and Axford W I 1968 *ApJ* **154** 1011
[30] Tanaka T, Uchiyama Y, Aharonian F A and et al 2008 *ApJ* **685** 988
[31] Abdo A A, Ackermann M, Ajello M and (Fermi-LAT Collaboration) 2010 *ApJ* **712** 459
[32] Tavani M, Giuliani A, Chen A W and et al 2010 *Astrophys. Jour. Let.* **710** L151
[33] Abdo A A, Ackermann M, Ajello M and (Fermi-LAT Collaboration) 2010 *Science* **327** 1103
[34] Greco A, Taktakishvili A L, Zimbardo G and et al 2003 *J. Geophys. Res.* **108** 1
[35] Perri S and Zimbardo G 2008 *J. Geophys. Res.* **113**
[36] Golden R L, Grimani C, Kimbell B L and et al 1994 *ApJ* **436** 769
[37] Grimani C, Stephens S A, Cafagna F S and et al 2002 *A&A* **392** 287
[38] Boezio M, Carlson P, Francke T and et al 2000 *ApJ* **532** 653
[39] DuVernois M A, Barwick S W, Beatty J J and et al 2001 *ApJ* **559** 296
[40] Alcaraz J, Alpat B, Ambrosi G and (AMS Collaboration) 2007 *Phys. Lett. B* **646**(4) 145
[41] Boezio M, Barbiellini G, Bonvicini V and et al 2001 *Adv. Space Res.* **27**(4) 669
[42] Beatty J J, Bhattacharyya A, Bower C and et al 2004 *Phys. Rev. Lett.* **93**(24) 241102
[43] Adriani O, Barbarino G C, Bazilevskaya G A and (PAMELA Collaboration) 2013 *Phys. Rev. Lett.* **111**(8) 081102
[44] Accardo L, Aguilar M, Aisa D and (AMS-02 Collaboration) 2014 *Phys. Rev. Lett.* **113**(12) 121101
[45] Lagutin A A, Volkov N V and Tyumentsev A G 2013 *Bull. Russian Academy of Sciences* **77**(11) 1312

Spectra of protons and nuclei in the energy range of $10^{10} \div 10^{20}$ eV in the framework of Galactic cosmic ray origin

Nikolay Volkov, Anatoly Lagutin, Alexander Tyumentsev, Roman Raikin

Altai State University, Radiophysics and Theoretical Physics Department. 61 Lenin ave., Barnaul, 656049, Russia.

E-mail: volkov@theory.asu.ru, lagutin@theory.asu.ru, tyumentsev@theory.asu.ru, raikin@theory.asu.ru

Abstract. We consider the problem of the cosmic ray spectrum formation assuming that cosmic rays are produced by Galactic sources. The anomalous diffusion equation proposed in our recent papers is used to describe cosmic ray propagation in the interstellar medium. We show that in the framework of this approach and with generation spectrum exponent $\gamma = 2.85$ it is possible to reproduce locally observed basic features of cosmic rays in the energy region of $10^{10} \div 10^{20}$ eV: difference between spectral exponents of protons and other nuclei, mass composition variation, "knee" problem, flattening of the primary spectrum at $E \geq 10^{18} \div 10^{19}$ eV. The crucial model predictions for the mass composition behaviour in the ultra-high energy region are discussed.

1. Introduction

The problem of ultra-high energy cosmic ray (UHECR) origin remains unsolved, despite significant progress made in recent years. The suppression of the cosmic ray flux at energy of $(4 \div 6) \times 10^{19}$ eV being established unambiguously can not be reliably identified as GZK-cutoff due to the apparent differences between Telescope Array/HiRes and Auger mass composition results [1,2]. As the Auger data indicated a nuclear UHECR composition becoming heavier with energy in the region of $4 \div 40$ EeV, different models assuming that the UHECR spectra cutoff is not an extragalactic GZK feature are considered [3–5].

In this paper we examine the possibility for self-consistent description of all the basic features of the observed cosmic ray spectrum in the energy range of $10^{10} \div 10^{20}$ eV within the Galactic origin scenario. We assume the existence of Galactic sources that accelerate particles up to $\sim 10^{20}$ eV and take into account highly inhomogeneous (fractal-like) distribution of matter and magnetic fields in the Galaxy that leads to extremely large free paths of particles ("Levy flights"), along with an overwhelming contribution to the cosmic ray fluxes observed above $\sim 10^{18}$ eV from particles reaching the Solar System without scattering. The crucial model predictions for the mass composition behaviour in the ultra-high energy region are presented.

2. Anomalous diffusion model

In our recent papers an anomalous diffusion (AD) model for solution of the "knee" problem in primary cosmic-rays spectrum was proposed [6–9]. The "anomaly" in this model results from the presence of extremely large free paths (Levy flights) of particles between Galactic inhomogeneities and also the long-lasting trapping of particles in very strong magnetic field regions (Levy traps). Thus the following asymptotic behaviour of the probabilities of free paths ($p(x, E)$) and time, during which particles are caught in magnetic traps ($q(\tau, E)$), takes place:

$$\int\limits_{|x|>r} p(x,E)dx \propto A(E,\alpha)r^{-\alpha}, \quad 0<\alpha<2; \quad \int\limits_{t}^{\infty} q(\tau,E)d\tau \propto B(E,\beta)t^{-\beta}, \quad 0<\beta<1. \tag{1}$$

The cosmic rays propagation in the fractal-like interstellar Galactic medium without energy losses and nuclear interactions can be described by the anomalous diffusion equation. The equation for the concentration of particles with an energy E, generated in a fractal medium by Galactic sources with a distribution density $S(\mathbf{r}, t, E)$ can be written as

$$\frac{\partial N}{\partial t} = -D(E,\alpha,\beta)\mathrm{D}_{0+}^{1-\beta}(-\Delta)^{\alpha/2}N(\mathbf{r},t,E) + S(\mathbf{r},t,E). \tag{2}$$

Here D_{0+}^{μ} denotes the Riemann-Liouville fractional derivative [10] and $(-\Delta)^{\alpha/2}$ is the fractional Laplacian ('Riesz operator') [10]. The anomalous diffusivity $D(E,\alpha,\beta) \sim A(E,\alpha)/B(E,\beta) = D_0(\alpha,\beta)E^{\delta}$.

The Green's function $G(\mathbf{r}, t, E; E_0)$, corresponding to (2), can be found from the equation

$$\frac{\partial G(\mathbf{r},t,E;E_0)}{\partial t} = -D(E,\alpha,\beta)\mathrm{D}_{0+}^{1-\beta}(-\Delta)^{\alpha/2}G(\mathbf{r},t,E;E_0) + \delta(\mathbf{r})\delta(t)\delta(E-E_0). \tag{3}$$

With the use of the Laplace-Fourier transformations and formulae [10]

$$\int\limits_0^{\infty} e^{-\lambda t}\mathrm{D}_{0+}^{\beta}G(\mathbf{r},t,E;E_0)dt = \lambda^{\beta}\int\limits_0^{\infty} e^{-\lambda t}G(\mathbf{r},t,E;E_0)dt = \lambda^{\beta}\tilde{G}(\mathbf{r},\lambda,E;E_0),$$

$$\int\limits_{\mathrm{R}^3} \mathrm{e}^{ik\mathbf{r}}(-\Delta)^{\alpha/2}G(\mathbf{r},t,E;E_0)d\mathbf{r} = |k|^{\alpha}\int\limits_{\mathrm{R}^3} \mathrm{e}^{ik\mathbf{r}}G(\mathbf{r},t,E;E_0)d\mathbf{r} = |\mathbf{k}|^{\alpha}\tilde{G}(\mathbf{k},t,E;E_0)$$

we obtain

$$\tilde{G}(\mathbf{k},\lambda,E;E_0) = \delta(E-E_0)\lambda^{\beta-1}\int\limits_0^{\infty}\exp\left(-\left[\lambda^{\beta} + D(E,\alpha,\beta)|\mathbf{k}|^{\alpha}\right]y\right)dy.$$

Then the inverse transform yields

$$G(\mathbf{r},t,E;E_0) = \delta(E-E_0)\left(D(E_0,\alpha,\beta)t^{\beta}\right)^{-3/\alpha}\Psi_3^{(\alpha,\beta)}\left(|\mathbf{r}|\left(D(E_0,\alpha,\beta)t^{\beta}\right)^{-1/\alpha}\right), \tag{4}$$

where

$$\Psi_3^{(\alpha,\beta)}(r) = \int\limits_0^{\infty} g_3^{(\alpha)}\left(r\tau^{\beta/\alpha}\right)g_1^{(\beta,1)}(\tau)\tau^{3\beta/\alpha}d\tau$$

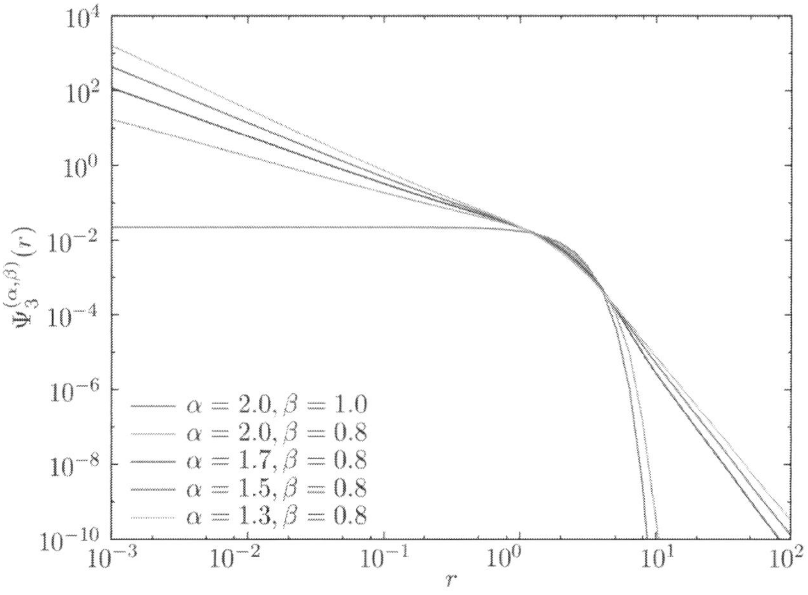

Figure 1. Fractional stable distributions densities $\Psi_3^{(\alpha,\beta)}(r)$

is the density of fractional stable distribution [11], $g_3^{(\alpha)}(r)$ ($\alpha \leq 2$) the three-dimensional spherically-symmetrical stable distribution and $g_1^{(\beta,1)}(t)$ is the one-sided stable distribution with characteristic exponent $\beta \leq 1$ [11]. The probability densities of $\Psi_3^{(\alpha,\beta)}(r)$ are shown in figure 1.

Using the Green's function (4), we can write a solution of the anomalous diffusion equation for a point pulse source $S(\mathbf{r},t,E) = S_{\text{im}}E^{-\gamma}\exp(-E/E_0(z))\delta(\mathbf{r})\Theta(T-t)\Theta(t)$ with emission time T and an exponential cutoff at energy E_0

$$N(\mathbf{r},t,E) = \frac{S_{\text{im}}E^{-\gamma}}{D(E,\alpha,\beta)^{3/\alpha}}\times$$
$$\times\left[\int\limits_{\max[0,t-T]}^{t} d\tau\tau^{-3\beta/\alpha}\Psi_3^{(\alpha,\beta)}\left(|\mathbf{r}|(D(E,\alpha,\beta)\tau^{\beta})^{-1/\alpha}\right)\right]\exp\left(-\frac{E}{E_0(z)}\right).$$

3. Energy spectrum of cosmic rays

Types of sources and their distribution in space suggest their separation into three parts as follows:

$$J(\mathbf{r},t,E) = J_G(\mathbf{r},E) + J_L(\mathbf{r},t,E) + J_{NS}(\mathbf{r},E).$$

Here

- J_G is the global spectrum component determined by the multiple old ($t \geq 10^6$ yr) distant ($r \geq 1$ kpc) sources;
- J_L is the local component, i.e. the contribution nearby ($r < 1$ kpc) young ($t < 10^6$ yr) sources;
- J_{NS} is the flux of non-scattered particles.

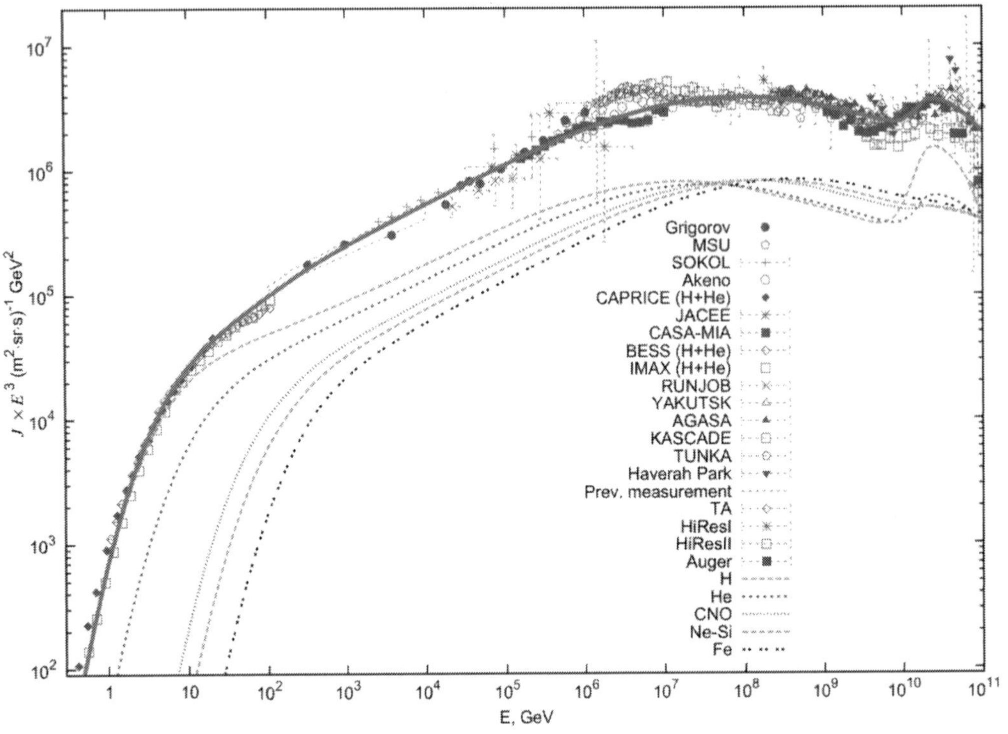

Figure 2. Comparison of our calculations of primary cosmic ray spectra with experimental data in wide energy range of $10^{10} \div 10^{20}$ eV. Curves marked 'H', 'He', 'CNO', 'Ne-Si', 'Fe' and 'Total' represent our results.

The flux of non-scattered particles J_{NS} is determined by the injected flux

$$S_0 E^{-\gamma} \exp(-E/E_0(z))$$

and the Levy flight probability $p(E, > r)$. Taking into account that for a particle with energy E the probability $p(E, > r) \sim A(E, \alpha) r^{-\alpha}$, $A(E, \alpha) \sim E^{\delta_L}$, we have

$$J_{NS}(r, E) = S_{NS} E^{-\gamma+\delta_L} \exp(-E/E_0(z)) r^{-\alpha}.$$

Since $J_{NS} \sim r^{-\alpha}, \alpha > 1$, we assume that this component is also formed by nearby ($r < 1$ kpc) sources, defining the spectrum in ultrahigh energy region, and provides the observed flattening of the spectrum at $E \geq 10^{18}$ eV.

Finally, for the global stationary component formed by old distant sources and the local component from the nearest sources the expression obtained by the solution of anomalous

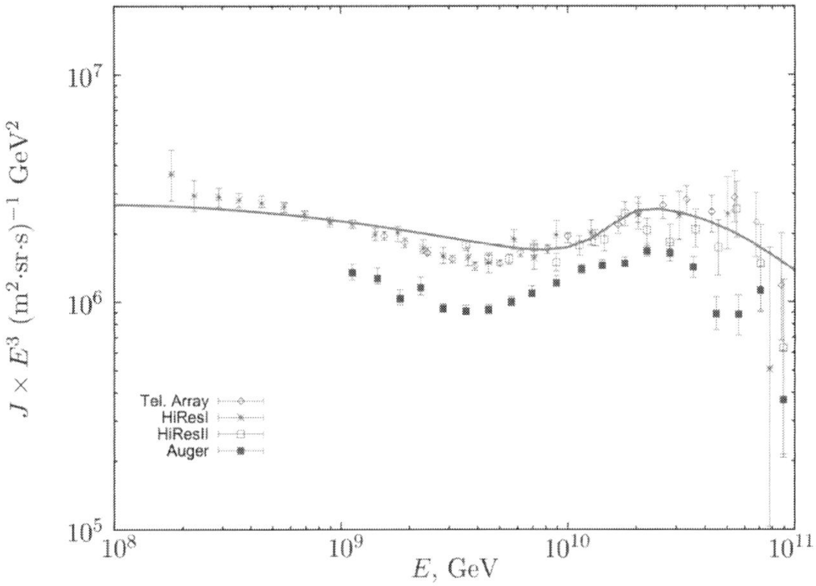

Figure 3. Comparison of our calculations of primary cosmic ray spectra with experimental data of Telescope Array/HiRes [12] and Auger [13] at ultrahigh energies.

diffusion equation is as follows:

$$J(\mathbf{r},t,E) = \frac{v}{4\pi}\Bigg[S_s E^{-\gamma-\delta/\beta} + \frac{S_{\rm im}E^{-\gamma}}{D(E,\alpha,\beta)^{3/\alpha}}\times \\ \times \sum_{\substack{r_j<1\ \rm kpc \\ t_j<10^6\ \rm yr}} \int\limits_{\max[0,t_j-T]}^{t_j} d\tau \tau^{-3\beta/\alpha}\Psi_3^{(\alpha,\beta)}\left(|\mathbf{r}_j|(D(E,\alpha,\beta)\tau^{\beta})^{-1/\alpha}\right) + \\ + S_{NS}\sum_{r_j<1\ \rm kpc} E^{-\gamma+\delta_L}|\mathbf{r}_j|^{-\alpha}\Bigg]\exp\left(-\frac{E}{E_0(z)}\right). \quad (5)$$

In (5) the first term is the contribution from the multiple old distant sources, the second term is the contribution from nearby young sources and the third term is the flux of non-scattered particles. To calculate the energy spectrum from nearby young sources, simulation of the Poisson ensemble of sources was carried out [14]. The Poisson distribution parameter (average number of sources in the local region) was chosen ~ 10. This estimation corresponds to number of the well-known nearest supernova remnants and pulsars with $t < 10^6$ yr. Coordinates and times of birth of the sources were generated randomly and uniformly in the space region $r < 10^3$ pc and in the time interval $10^4 \leq t < 10^6$ yr.

The parameters of the model are $\gamma = 2.85$, $\delta = 0.27$ as it was established earlier [6], $\delta_L = \delta/2$. The cut-off energy $E_0(z) = 4\times 10^{10} z$ GeV was fixed on the basis of comparisons with experimental data. To assess parameters $\alpha = 1.4$, $\beta = 0.8$, which are specific for our model, the results of studies of particles diffusion in cosmic and laboratory plasma have been used [15, 16]. Finally, for the anomalous diffusivity we get the value of $D_0 \approx 10^{-4}{\rm pc}^{1.4}/{\rm yr}^{0.8}$ [9]. The emission time $T = 10^4$ yr.

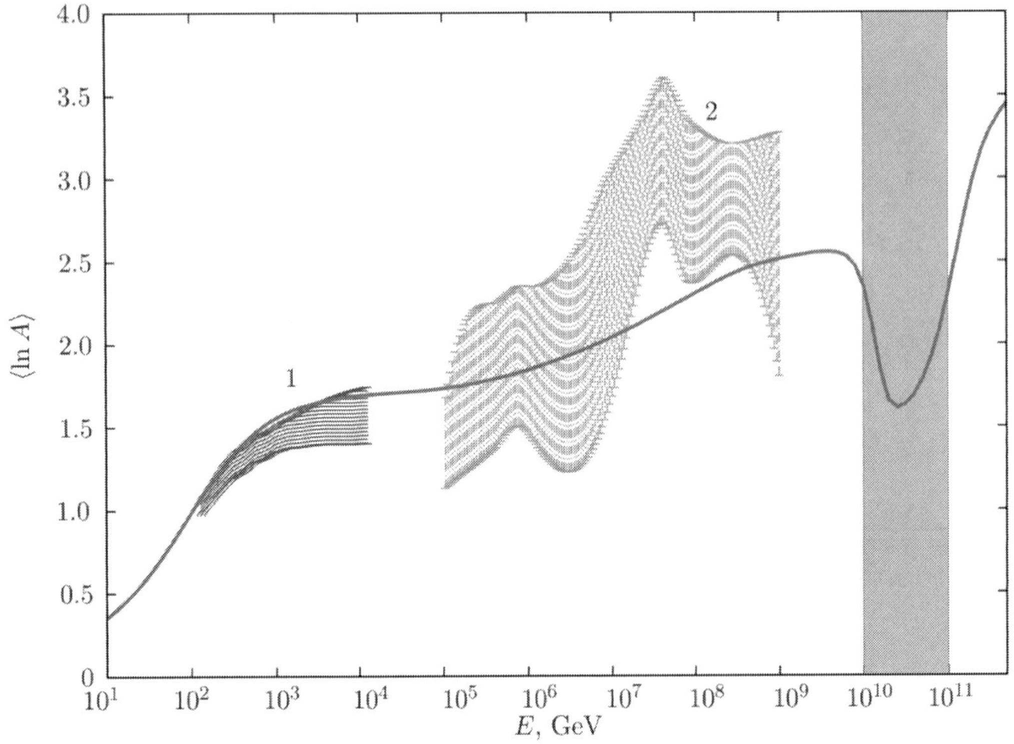

Figure 4. Mean logarithmic cosmic ray mass vs. primary particle energy. Dashed areas — experimental data [19], [20].

Our results of calculations of energy spectra of cosmic rays are shown in figures 2, 3. Note, that the "fine structure" of spectrum around the knee can be described satisfactorily in the framework of the model with two types of sources (see [17]).

An average mass number variation with energy according to our model is shown in figure 4. It was established that in the region of $(2 \div 5) \times 10^{19}$ eV the fraction of protons is $\sim 40\%$, while the fraction of iron is $\sim 15\%$. A key feature of the model is the essential dip in mean logarithm of the mass number in the region of $(4 \div 25) \times 10^{18}$ eV (from 2.6 to 1.6). The weighting of the mass composition is predicted at $E > 2.5 \times 10^{19}$ eV and $\langle \ln A \rangle$ reaches ~ 3 at $E = 2 \cdot 10^{20}$ eV. The presence of such a dip according to the experimental data could be construed in favour of our model. Note, that recent results of Auger Collaboration exhibit similar feature in $\langle \ln A \rangle$ variation with energy, although at an order of magnitude lower energies [18].

4. Conclusions

Assuming the Galactic origin of cosmic rays we propose a mechanism for the formation of ultra-high energy protons and nuclei spectra observed in the Solar System, that reproduces all the basic features of cosmic ray spectrum observed experimentally.

It is shown that taking into account both the diffusion contribution from nearby young and old distant sources and also the contribution of non-scattered radiation, due to the presence of large free paths (Levy flights) caused by anomalous diffusion, a description of the experimental data in the energy range $10^{10} \div 10^{20}$ eV can be obtained if we assume that the particle injection spectrum in the Galactic sources is

$$S = S_0 E^{-2.85} \exp(-E/E_0(z)), \quad E_0(z) = 4 \times 10^{10} z \text{ GeV}.$$

The following basic model predictions are expressed:

- the steepening of all particles spectrum in the range $E > 5 \times 10^{19}$ eV is due to the proton flux cut-off caused by energy limitation of the Galactic particle accelerators;
- at the energies $2 \div 5 \times 10^{19}$ eV the fraction of protons is ~ 40 %, and the iron fraction is about 15 % (normal composition);
- in the range $4 \div 25 \times 10^{18}$ eV mean logarithmic cosmic ray mass $\langle \ln A \rangle$ decreases rapidly from 2.6 to $\langle \ln A \rangle_{\min} \sim 1.6$;
- considerable weighting of the mass composition is observed at $E > 2.5 \cdot 10^{19}$ eV; for $E = 2 \cdot 10^{20}$ eV mean logarithmic cosmic ray mass is $\langle \ln A \rangle \sim 3$.

Acknowledgement
This work was suppored by the Russian Foundation for Basic Research grant No. 14-02-31524. The authors are greatly indebted to anonymous referee for valuable comments and suggestions that helped improve the manuscript.

References

[1] Abbasi R U et al 2008 *Phys. Rev. Lett.* **100** 101101
[2] Abraham J et al 2008 *Phys. Rev. Lett.* **101** 061101
[3] Aloisio R, Berezinsky V, Gazizov A 2011 *Astropart. Phys.* **34** 620
[4] Calvez A, Kusenko A, Nagataki S 2010 *Phys. Rev. Lett.* **105** 091101
[5] Fang K, Kotera K, Olinto A V 2013 *JCAP* **03** 010.
[6] Lagutin A A, Nikulin Yu A, Uchaikin V V 2001 *Nucl. Phys. B* **97** 267
[7] Lagutin A A, Uchaikin V V 2003 *NIM B* **B201** 212
[8] Erlykin A D, Lagutin A A, Wolfendale A W 2003 *Astropart. Phys.* **19** 351
[9] Lagutin A A and Tyumentsev A G 2004 *Izv. Altai. Gos. Univ.* **35** 4 (in Russian)
[10] Samco S G, Kilbas A A and Marichev O I 1993 *Fractional integrals and derivatives — Theory and Applications* (New York: Gordon and Breach) 976 p.
[11] Uchaikin V V and Zolotarev V M 1999 *Chance and Stability* (Netherlands, Utrecht: VSP) 594 p.
[12] Abu-Zayyad T, Aida R, Allen M (Telescope Array Experiment) 2013 *Astrophys. Journ. Lett.* **768:L1** 1
[13] Abraham J, Abreu P, Aglietta M (Pierre Auger Collaboration) 2010 *Phys. Lett. B* **685** 239
[14] Lagutin A A, Volkov N V and Tyumentsev A G 2013 *Bulletin of the Russian Academy of Sciences: Physics* **77** 1312
[15] Greco A, Taktakishvili A L, Zimbardo G and et al 2003 *J. Geophys. Res.* **108** 1
[16] Perri S and Zimbardo G 2008 *J. Geophys. Res.* **113**
[17] Lagutin A A, Tyumentsev A G, Yushkov A V 2008 *Nucl. Phys. B (Proc. Suppl.)* **175-176** 555
[18] Aab A, Abreu P, Aglietta M (Pierre Auger Collaboration) 2014 *Phys. Rev. D.* **90** 122005
[19] Shibata T 1999 *Nucl. Phys. B.* **75A** 22
[20] Hörandel J R 2003 *Astropart. Phys.* **19** 193

EAS array of the NEVOD Experimental Complex

I I Yashin[1], M B Amelchakov[1], N V Ampilogov[1], N S Barbashina[1], A G Bogdanov[1], A Chiavassa[2,3], S V Fomenko[1], N N Kamlev[1], S S Khokhlov[1], V V Kindin[1], R P Kokoulin[1], K G Kompaniets[1], V Yu Kutovoy[1], O I Likiy[1], G Mannocchi[4], V V Ovchinnikov[1], A A Petrukhin[1], O Saavedra[2], G Trinchero[4], V V Shestakov[1], I A Shulzhenko[1] and V V Shutenko[1]

[1] National Research Nuclear University MEPhI (Moscow Engineering Physics Institute), 115409 Moscow, Russia
[2] Dipartimento di Fisica dell' Università degli Studi di Torino, 10125 Torino, Italy
[3] Sezione di Torino dell' Istituto Nazionale di Fisica Nucleare – INFN, 10125 Torino, Italy
[4] Istituto di Fisica dello Spazio Interplanetario – INAF, 10133 Torino, Italy

E-mail: IIYashin@mephi.ru

Abstract. A new setup for registration of the electromagnetic component of the EAS at the "knee" region of the energy spectrum of primary cosmic rays (PCR) is now under construction on the basis of the experimental complex NEVOD-DECOR (Moscow, Russia). The EAS array detecting system has a cluster organization. Clusters are located in the MEPhI campus. The specific features of the array registering system that provides particle detection, data acquisition, cluster synchronization and events selection are discussed. The results of counter characteristics study are also presented.

1. Introduction

In 2002-2009, a method of EAS registration based on a new variable (local muon density spectra (LMDS) at large zenith angles) was developed [4] at the experimental complex NEVOD-DECOR [1-3] (Moscow). The obtained results analysis of has demonstrated some anomalous behavior of LMDS at energies of 10^{15}-10^{18} eV. It had been noticed that the intensity of muon bundles at high primary particle energies exceeds their estimated amount, even for heavy PCR composition (iron nuclei). However, the primary energy estimation based on the analysis of local density spectra of muons registered at different zenith angles is of very low accuracy ($\sigma(\lg E_0) \sim 0.4$). This is due to the contribution of the EAS to events with a fixed local muon density. This EAS have different energies and are registered at various (random) distances from their axis.

The construction of the NEVOD-EAS setup around the NEVOD-DECOR complex will provide for registering and determining of characteristics of extensive air showers with energies of 10^{15}-10^{17} eV using a traditional scintillation counter technique. The introduction of the new setup in the structure of the NEVOD-DECOR complex will enable to reduce the range of PCR energies in which the change in the muon component behavior occurs.

2. NEVOD-EAS shower array

The NEVOD-EAS setup for registration of extensive air showers is an array of scintillation counters placed on the roofs of laboratory buildings in the MEPhI campus. The measuring system of the setup is organized according to a cluster principle (figure 1). Each cluster includes 16 counters combined in 4 detector stations (DS) and is served by the Local Post (LP) of preliminary data acquisition. The cluster dimensions are ~ 20×20 m^2.The characteristic distances between the clusters are ~ 50 m. Creation of the setup will be carried out in 3 stages: central part (5 clusters, area ~ 10^4 m^2, 2014-2015), 1st step of the peripheral part (7 clusters, area ~ 2×10^4 m^2, 2016-2017), 2nd step of the peripheral part (area ~ 2×10^5 m^2, 2018-2020).

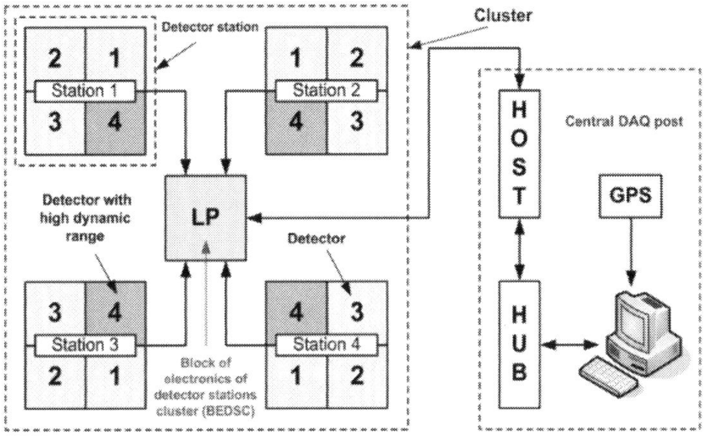

Figure 1. Structure of the NEVOD-EAS cluster.

2.1. Scintillation counter and detector station of the NEVOD-EAS shower array

Particles of the EAS electron-photon component are registered with scintillation counters [5] (figure 2a) which were previously used in the EAS-TOP [6] and KASCADE-Grande experiments [7]. The counter consists of the NE102A plastic scintillator with dimensions of 800 × 40 × 800 mm^3 and the Philips XP3462 PMT enclosed in a light-tight pyramidal housing made from stainless steel.

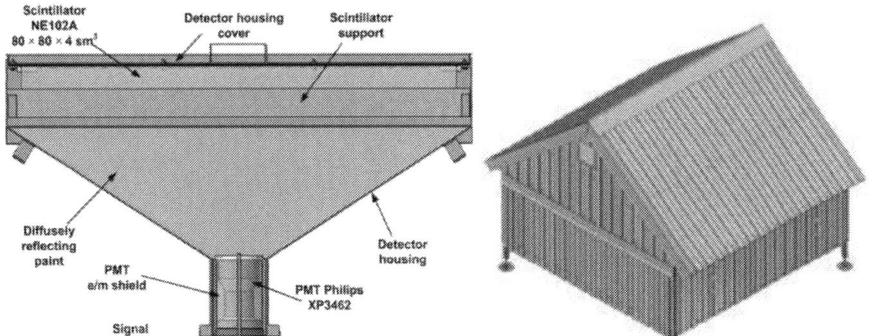

Figure 2. Counter (a) and detector station (b) of the NEVOD-EAS setup.

The inner surface of the counter housing is painted with a special diffuse-reflective coating to improve the collection of light from the flashes in the scintillator caused by the passage of particles through its sensitive volume. The scintillator is located at a distance of 30 cm above the PMT photocathode. The construction of the scintillation counter housing enables to install an additional Philips XP3462 PMT.

Detector stations are formed of 4 scintillation counters, three of them being equipped with only a "standard" photomultiplier, while the fourth one has both a "standard" and an "additional" PMT. Standard photomultipliers are selected and configured so that the most probable values of charge distributions of the responses of all four counters were almost the same. These photomultipliers are used to measure the particle density and arrival time of the detected EAS. The additional photomultiplier has a dynode system gain of approximately 90 times smaller than the standard PMT. This PMT provides a wide linearity range of the measured signals at high particle densities and allows to achieve a dynamic range of the DS ~ 10000 particles/m^2. Counters of the DS are enclosed in special housings that provide protection from moisture (figure 2b). Inside the DS housing, the DC/DC converter providing high voltage supply of the PMT and the temperature sensor measuring the DS temperature during the setup operation monitoring are installed.

2.2. Cluster of the NEVOD-EAS shower array

The NEVOD EAS cluster is formed of 4 DS placed in the vertices of a rectangle with characteristic dimensions of 20 × 20 m^2. Analog signals from the DS PMTs come via cables to the cluster LP. Local Post provides selection of events according to the intracluster trigger conditions, digitizes the amplitude information and transmits data to the Central DAQ Post via optical link. LP includes 2 boards of amplitude analysis (BAA), summator-multiplexer (SM), BAA controller (C), SM controller and power supply (+5 V and ±12 V). Figure 3 shows the scheme of the cluster Local Post.

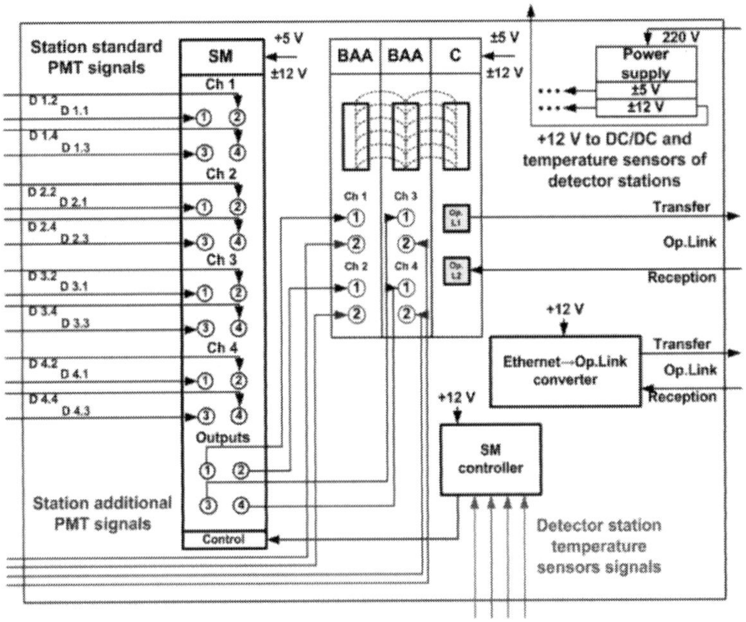

Figure 3. Scheme of the NEVOD-EAS Local Post.

BAA includes 2 channels with 2 inputs: active and passive. The signals on the inputs are digitized using 12-bit FADC with a sampling frequency of 0.2 GHz. For the active input, a 12-bit DAC is used to set the registration threshold.

The BAA controller selects events according to the specified triggering multiplicity of DS (from 1 to 4) within a time gate from 10 ns to 2 ms and transmits the data to the storage system. Each event is time-stamped with an accuracy up to 10 ns.

SM is intended for summing of analog signals from 4 standard PMTs of the DS and has 4 channels on 4 inputs. The summed signal from 4 standard PMTs of the DS is fed to the active input of the corresponding channel of BAA. The signal from the additional PMT is fed to the passive input. The SM also allows to turn on a defined input of each channel for registration of spectra from the selected detector of the DS during the setup operation monitoring. Control of the SM is performed using the SM controller. SM controller also receives and transmits data from the DS temperature sensors.

2.3. Central DAQ post of the NEVOD-EAS shower array
The main functions of the Central Post of the data acquisition are: cluster synchronization; LP control; reception and storage of experimental and monitoring data. Central Post includes: server, GPS/GLONASS sensor, communication unit and module of LP control and synchronization.

The communication unit is intended for connection of server with clusters of the setup and includes hub and media converters (Ethernet→Opt.Link), which transmit control commands to the SM controllers of Local Posts and receive information about temperature inside the detector stations.

The module of LP control and synchronization consists of several units - Hosts. Each host has 4 optical channels for control and receiving of the experimental data from the LP. Each Host serves 4 LP. The Host has an Ethernet port for connection to the communication unit.

Time synchronization of clusters is performed using a GPS/GLONASS sensor connected to the server. The sensor generates a PPS signal with a frequency of 1 Hz. When receiving the PPS signal, the server sends a synchronization signal to the main Host of the group. The main Host synchronizes other Hosts using the clock generator. Hosts send synchronization pulses to the BAA controllers of the clusters that in their turn synchronize BAA. Thus, the NEVOD-EAS registering system operates in a synchronous mode. Synchronization accuracy of the cluster is 10 ns.

3. Characteristics of the NEVOD-EAS scintillation counter
At the preliminary stage of the NEVOD-EAS creation, the characteristics of photomultiplier tubes and scintillators have been studied.

3.1. Characteristics of the PMTs and scintillators
During the Philips XP3462 PMT testing the following characteristics have been studied: the high voltage which provides a gain of the photomultiplier dynode system ~ 2×10^6; the spread of the PMT triggering delays (jitter) during its operation in the single photoelectron mode; the linearity range; the relative sensitivity (the response of the PMT to the photocathode illumination with a fixed intensity); the dependence of dark noise counting rate on the registration threshold.

At the initial stage of each PMT testing the value of high voltage providing the gain of the PMT ~ 2×10^6 was determined using the technique of single photoelectron illumination of the PMT photocathode. The other characteristics were investigated at the high voltage defined at the initial stage. Figures 4-8 show distributions of 240 tested PMTs on the high voltage, single photoelectron jitter, upper limit of linearity range, relative (to the complete sample average) sensitivity of the photocathode and dark noise counting rate at the fixed threshold.

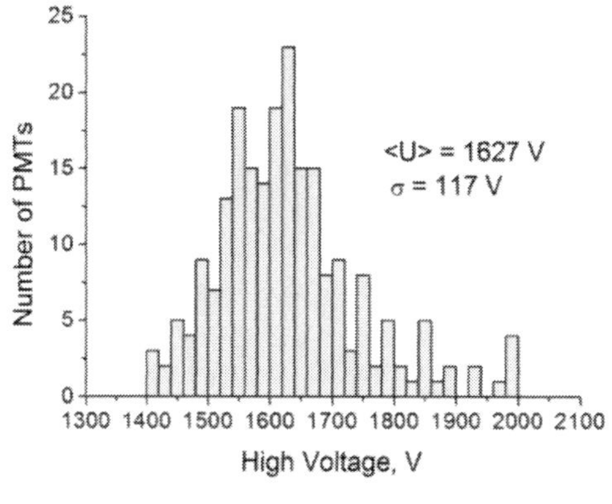

Figure 4. Distribution of the PMTs high voltage providing the PMT gain ~ 2×10^6.

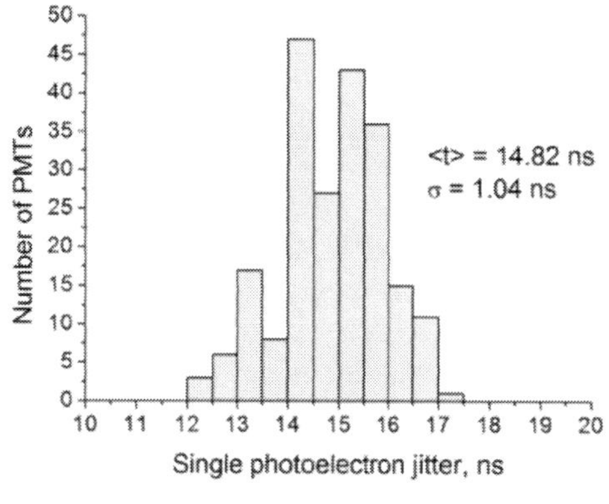

Figure 5. Distribution of the PMTs single photoelectron time jitter.

During the scintillator plate testing, the charge spectrum of counter response to the passage of near-vertical muon located by the muon telescope has been measured. The muon telescope consists of two scintillation counters with dimensions 10 × 10 cm^2 placed in the central area of the studied scintillator. Figure 9 shows the distribution of the most probable values of the measured charge spectra for 192 scintillators divided by the complete sample average charge of the muon peak (distribution of the relative light yield of the scintillators).

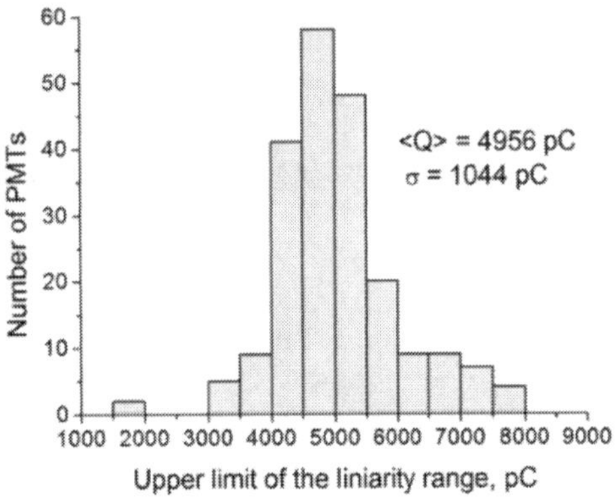

Figure 6. Distribution of the PMTs upper limit of linearity range.

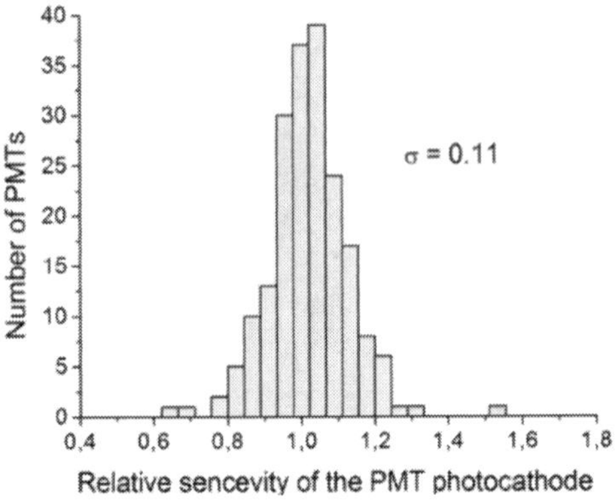

Figure 7. Distribution of the PMTs relative photocathode sensitivity.

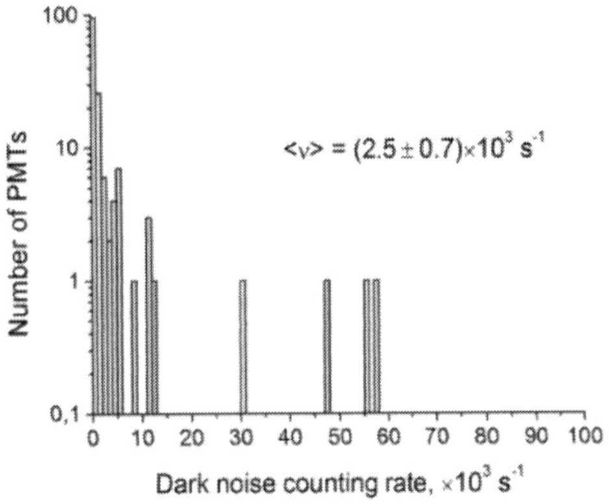

Figure 8. Distribution of the PMTs dark noise counting rate.

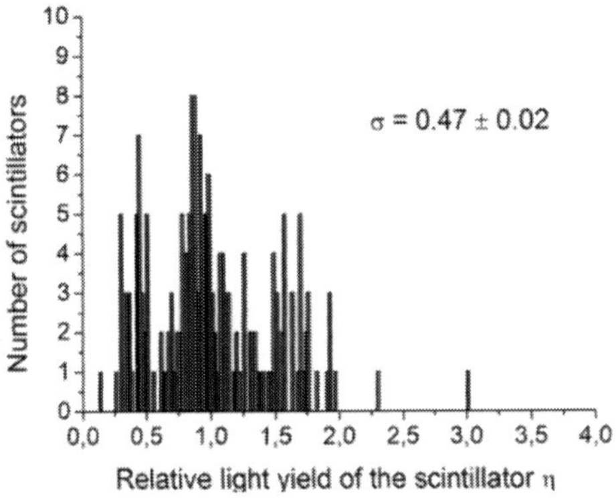

Figure 9. Distribution of the relative light yield of the scintillators.

3.2. Counter light collection non-uniformity

One of the most important tasks of the NEVOD-EAS array is to estimate the size of registered EAS which is associated with the energy deposit of the EAS fragment, fallen onto a single counter of the setup. For the accurate estimation of the energy deposit in the counter it is important that its response was independent on the place where the detected particle had passed through the scintillator. To determine the degree of the difference of the NEVOD-EAS counter response to the passage of particles through various areas of the sensitive volume of the scintillator, the light collection non-uniformity was studied. This study was performed using the muon hodoscope URAGAN [8].

Good spatial resolution of the hodoscope (1 cm for the coordinates and about 1° for the angle) enables to investigate the response of various detectors in the muon flux with known track parameters and to study in detail the structure of the scintillation counter by placing it on the surface of the hodoscope. Figure 10 shows the layout of the setup for the NEVOD-EAS counter light collection non-uniformity measurement with the coordinate detector URAGAN.

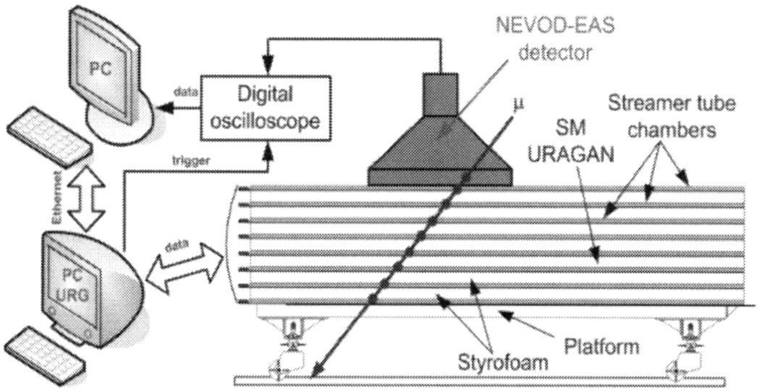

Figure 10. Layout of the setup for counter light collection non-uniformity measurement.

The main elements of this setup are: the URAGAN coordinate detector, the investigated counter, a digital oscilloscope and two computers. At the passage of the particle through the allocated triggering area, URAGAN generates a trigger signal which goes to the trigger input of the oscilloscope. On this trigger, the oscilloscope digitizes the signal of the NEVOD-EAS counter response. Then the oscilloscope buffer that stores the sweep of the counter signal is transferred to the PC. Also information about the track of the particle registered by the hodoscope is fed to the PC via Ethernet. So the event with amplitude and coordinate information is formed.

For the analysis of light collection non-uniformity of the NEVOD-EAS counter only these events in which muon has passed through the scintillator under a zenith angle less than 15° were selected.

Figure 11 shows the matrix of mean values of the charge of the counter response to the passage of single muons obtained using the URAGAN. Axes on the left bottom block of the matrix represent the geometric dimensions of the triggering area in cm. The tone shows the average charge of the detector response in pC. The top and right side blocks show the changes of the average charge of the counter response in two orthogonal sections. On the matrix this sections are marked with light lines.

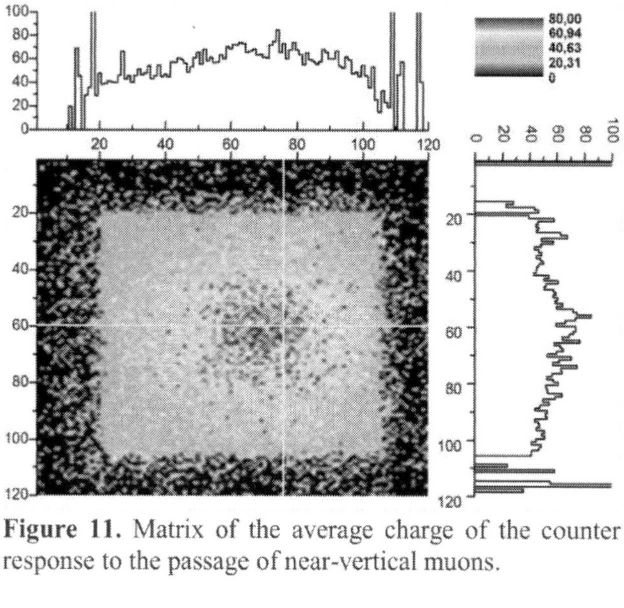

Figure 11. Matrix of the average charge of the counter response to the passage of near-vertical muons.

The increase of the counter response to particles passing through the area of the scintillator under the photomultiplier (1.5 - 2 times in comparison with the remaining area) is seen. According to this matrix, the r.m.s. light collection non-uniformity is ~ 20%.

4. Assembling and adjustment of the NEVOD-EAS counter

The assembling of the NEVOD-EAS is performed using the following principle: lower light yield of the scintillator (relative to the complete sample) should be compensated by higher sensitivity of the PMT (relative to the complete sample average), and lower PMT sensitivity should be compensated by higher scintillator light yield. The selection of scintillators and photomultipliers is based on the distributions shown in figures 7 and 9. After the assembling, the most probable response of the counter to the passage of a single muon is adjusted to the value of $\sim 13 \pm 1$ pC by varying the PMT supply voltage. Such response value provides the dynamic range of the counter of up to ~ 100 detected particles (vertical equivalent muon). Figure 12 shows the charge spectra of the counter response to the passage of charged particles measured in the self-triggering mode before (a) and after (b) the adjustment. During the adjustment of this counter the supply voltage has been reduced by 105 V.

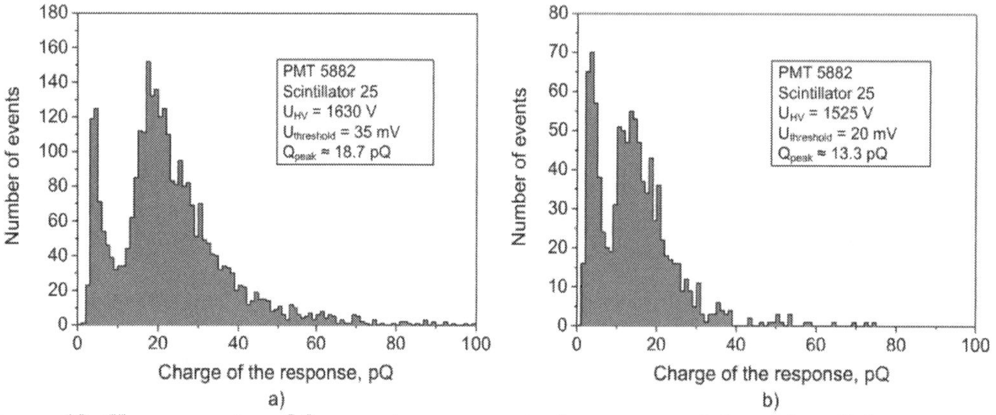

Figure 12. Charge spectra of the counter response to the passage of charged particles measured in self-triggering mode before (a) and after (b) the adjustment.

5. Conclusion

The NEVOD-EAS setup for registration of extensive air showers being created on the basis of the experimental complex NEVOD-DECOR will allow determination of the size, position of the axis and the arrival direction of EAS. The setup will be operated in conjunction with other installations of the experimental complex NEVOD and will enable to register EAS with energies of 10^{15}-10^{17} eV, as well as to verify the technique of muon bundle registration using the DECOR detector. New data obtained with NEVOD-EAS setup will allow to reduce the energy range of PCR particles responsible for generation of muon bundles with certain multiplicity arriving at different zenith angles.

Acknowledgements

This work was performed at the NEVOD Scientific and Educational Center with the financial support from the State provided by Russian Ministry of Education and Science (government task and project No. RFMEFI59114X0002) and the grant of the Leading Scientific School NSh-4930.2014.2.

References

[1] Aynutdinov V M *et al* 1997 *Astrophysics and Space Science* **258** 105
[2] Khokhlov S S *et al* 2011 *Astrophys. Space Sci. Trans.* **7** 271
[3] Barbashina N S *et al* 2000 *Instrum. Exp. Tech.* **43** 743
[4] Bogdanov A G *et al* 2010 *Phys. Atom. Nucl.* **73** 1862
[5] Shulzhenko I A *et al* 2013 *J. Phys.: Conf. Ser.* **409** 012098
[6] Aglietta M *et al* 1990 *Nucl. Phys.* B (Proc. Suppl.) **16** 493.
[7] Chiavassa A *et al* 2003 *Proc. 28th ICRC (Tsukuba)* vol 2 p 992
[8] Barbashina N S *et al* 2008 *Instrum. Exp. Tech.* **51** 180

Future instrumentation in cosmic ray research

I I Yashin

National Research Nuclear University MEPhI
(Moscow Engineering Physics Institute), 115409, Moscow, Russia

E-mail: iiyashin@mephi.ru

Abstract. This paper is based on a rapporteur talk given at the 24th European Cosmic Ray Symposium (Kiel, Germany, September 1 - 5, 2014). The object of the talk and paper is a summary based on oral talks and posters presented in the frame of the session "Future instrumentation in cosmic ray research (INS)".

1. Introduction

The program of the session "Future instrumentation in cosmic ray research (INS)" included three highlight talks, 11 contributed oral talks and 9 posters. Apart from the INS session, 12 talks and posters related to the topic of the session were given. Four themes dominated the presentations: future space missions (two highlight and two contributed talks), the TAIGA project (one highlight and two oral talks, one poster), radio detection of EAS (two oral talks and two posters) and the experimental complex NEVOD (three posters). Other talks and posters were devoted to different aspects of the development of experimental techniques and new apparatus for cosmic ray research.

2. Orbital and space missions

2.1. Fluorescence orbital detectors

There are three projects of space-born apparatus towards the realization of an idea proposed three decades ago [1], the idea to observe ultra-high energy air showers via fluorescence light in a huge atmospheric volume from space: the TUS (Tracking Ultraviolet Setup) detector on board of the "Mikhail Lomonosov" satellite to be launched in 2015 [2]; the JEM-EUSO project [3] which represents a large space-borne fluorescence detector proposed to be installed on the ISS, and the KLYPVE project recently started in SINP MSU and to be installed on board of the Russian segment of the ISS. A vital feature of future space measurements is the variable noise intensity of background UV light of the atmosphere below the satellite. First estimates of the light background have been obtained on board of Universitetsky-Tatiana-2 satellite by a single pixel of TUS camera design, at about the same orbit as that planned for the Lomonosov satellite [4]. Temporal profiles of transients in UV and red wavelengths in a FOV of 300 km diameter were measured in the atmosphere. Comparing the intensities of UHECR events and the intensity of transients from Tatiana-2 data one finds that transient events exceed the number of expected UHECR events by about two order of magnitude. Above the equatorial regions of continents, the transient rate in the TUS detector will be as high as 1 per minute. Comparison of the space-time structure of transients and UHECR showers will likely allow distinguishing both types of events, but the high trigger rate from transients challenges storage of the data before they are transmitted to Earth.

The TUS detector (see Figure 1) is developed in SINP MSU (Moscow) in cooperation with several Russian and foreign institutions and consists of a 1.8 m^2 Fresnel mirror-concentrator read out by 256 PMT pixels which will be able to monitor an area of 6400 km^2 from the orbit height of 500 km. The energy threshold for UHECR particles is about 7×10^{19} eV. The presently estimated event rates at the lowest energies seem to be rather optimistic. They will be checked in a final energy calibration during space operation. However, it is evident that three years of TUS operation will not yield decisive statistics of UHECR events beyond the GZK limit. The main goal of the TUS experiment is providing a proof of principle and acting as a "pathfinder" detector for the next large-scale space experiments (KLYPVE, JEM-EUSO).

Figure 1. The TUS detector on board of the Lomonosov satellite[1].

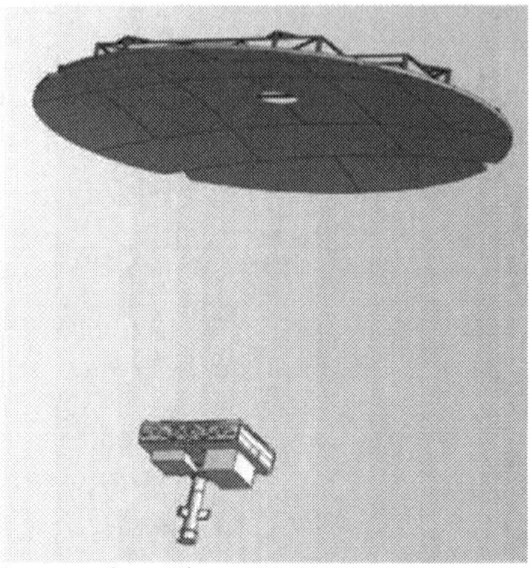

Figure 2. The KLYPVE segmented mirror-concentrator and photodetector.

Considerably improving the TUS characteristics, SINP MSU finished the preliminary design of the KLYPVE apparatus with a reflector type telescope [5]. It will be located on the outer side of the Russian segment of ISS. The optical system of this detector contains a larger primary mirror (10 m^2), which allows to considerably increase the FOV of the detector in comparison with TUS and to decrease the energy threshold (see Figure 2). The optical system has a focal distance of 3 m and a total effective FOV of 15° diameter. The angular resolution is the same as in the TUS detector, i.e., 5 mrad (pixel size 15 mm×15 mm). The annual exposure of KLYPVE will exceed the exposure of existing ground-based experiments.

In contrast to the projects TUS and KLYPVE which use reflecting mirrors, the JEM-EUSO project uses high-transmittance Fresnel lenses. The Extreme Universe Space Observatory (EUSO) project aims observingcosmic rays between 10^{20} and 10^{22} eV and is the most ambitious and sophisticated of the three projects. JEM-EUSO is alsodesigned for the ISS. It comprises a UV telescope and an atmospheric monitoring system. The telescope consists of three Fresnel lenses with an optical aperture of 4.5 m^2, and a focal surface detector formed by 137 photodetector modules composed of ~ 5000 multi-anode PMTs in total. The focal surface detector thus includes ~ $3\cdot10^5$ channels providing a spatial resolution ~ 0.074°, equivalent to ~ 0.5 km at ground. The 60° opening angle provides an observational area of ~ $1.4\cdot10^5$ km^2 in nadir mode. The annual exposure of JEM-EUSO for UHECR above 100 EeV is estimated to be an order of magnitude larger than that of Auger [6]. At the ISS orbit (400 km), JEM-EUSO is able to survey the entire celestial sphere almost uniformly. Detailed simulations show that due to these factors JEM-EUSO will be able to detect significant anisotropies of UHECR practically in all feasible astrophysical scenarios [7] thus providing a huge step towards finding sources of UHECR. In December 2013, the Japanese Space Agency (JAXA) refused to deploy the instrument on board of the JEM. The development of the full instrument of the JEM-EUSO project is currently facing certain problems, but two successful test experiments (EUSO-Balloon [8] and TA-EUSO [8]) demonstrated the capability of all sub-systems of the EUSO instrument to observe the fluorescence background from the edge of the atmosphere. The flight of the EUSO-Balloon on 24 August, 2014 was used to test a number of components of the telescope and the atmospheric monitoring system and represents a strong basis for the successful development of JEM-EUSO.

Quite recently, a new project, K-EUSO, was initiated. Its aim is to develop a new optical system with an increased FOV for KLYPVE which would improve the spatial and angular resolution and the overall performance of the instrument [8]. This work started in a close cooperation with the JEM-EUSO Collaboration. The new detector will combine the capabilities of optical systems based on mirror-concentrator and on Fresnel lens-corrector, respectively. The diameters of the reflector and the lens-corrector are 3.4 m and 1.7 m, respectively. The total length of the system is 4 m, the distance from the lens to the focal surface 70 cm. For such a system it is possible to expand the FOV up to $\pm 14^{o}$. The diameter of the image from a point-like light source (point spread function) is not larger than 6 mm in the entire FOV. The expected angular resolution of the system is $\sim 0.057^{o}$, which is equivalent to ~ 0.4 km at ground [8].

Another approach proposed in SINP MSU is the construction a Multi-Eye Telescope System (METS) consisting of several wide-field large aperture and K-EUSO optics systems (see Figure 3).

Figure 3. Two possible ways to build METS telescope.

The main idea of METS is to divide a wide FOV into several FOVs of identical smaller telescopes, with the following advantages: smaller dimensions of individual telescopes with narrower FOVs, lower cost and easier testing, calibration and launch to orbit. Table 1 comparesspot size d, effective area $S_{\rm eff}$, FOV Ω and throughput $\Omega S_{\rm eff}$ of the five projects [5].

Table 1.
Comparison of characteristic parameters of space-based fluorescence telescopes

	$S_{\rm eff}$, m^2	d, mm	Ω, sr	$\Omega\ S_{\rm eff}$, m^{2}sr
TUS	1.2	20	0.025	0.03
KLYPVE	5	20	0.05	0.25
JEM-EUSO	2	3	0.8	1.6
K-EUSO	3.5	3.5	0.2	0.7
METS	2.1	3	0.3	0.63

2.2. The CALET mission

The CALET (CALorimetric Electron Telescope) space experiment [9] will measure the flux of cosmic-ray electrons up to 20 TeV, of gamma rays up to 10 TeV and of nuclei up to 1000 TeV over five years. These measurements are essential to investigate possible nearby astrophysical sources of high energy electrons, to study the details of galactic particle propagation and to search for dark matter signatures. CALET consist of a charge detector, a 3 radiation length (r.l.) deep imaging

Tungsten/SciFi (scintillating fiber) calorimeter, and a 27-r.l. deep "total absorption calorimeter" made of segmented lead tungstate crystals (see Figure 4). CALET performances have been verified at accelerator beam tests. Flight hardware is currently under final integration and test, approved for launch by the HTV-5 Japanese carrier within end of March 2015 and to be installed on the Japanese Experiment Module (Kibo) on-board of ISS. CALET is much smaller than the Fermi-LAT apparatus but is optimized to measure high-energy electrons, mainly through its calorimeter of 30 radiation lengths depth (vs. 10.1 r.l. for LAT). A deeper calorimeter provides better energy resolution (~2% vs. 7–10% above 100 GeV), and improved hadron rejection [9]. The charge detector on CALET is designed to have excellent charge resolution for light elements. Last but not least, the addition of a veto counter makes CALET also an effective gamma-ray measuring instrument..

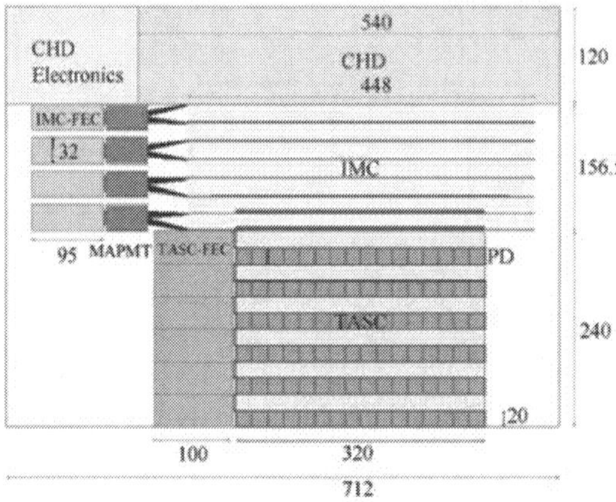

Figure 4. Layout of the CALET payload. The main components are a charge detector (CHD), a thin imaging calorimeter (IMC) and a total absorption calorimeter (TASC).

2.3. The new superconducting magnet facility

A new idea proposed in [10] suggests to equip an innovative calorimeter with an intense super conductivity magnet. In past, there were several attempts to put a superconducting (SC) magnet in space for measuring the momentum of cosmic rays. But for various reasons, the development of SC magnets ended at the design stage and only balloon experiments with prototypes were carried out. All these projects were based on the same technology: the coils are realized in NbTi SC cable, and are cooled to about 2 K by liquid super fluid He, delivered from a tank launched to orbit together with the magnet.

Nowadays, due to the development of cryogenic equipment and materials (new SC cables based in new materials such as the MgB_2 cables that can be operated up to 20 K in a few Tesla magnetic field; available commercial reliable new cryo-cooler systems), the SC magnetic system can be substantially simplified, and more complex coil system can be designed for the new generation of an 'innovative' 3D calorimeter for identifying and measuring energy and momentum of the particles in orbital experiments. Such calorimeter can have a cubic shape, allowing to increase the aperture by an order of magnitude, up to several m^2sr. The facility consists of a cubic, homogeneous, isotropic deep calorimeter equipped with a magnetic field at 4 of its faces. The complex magnet system named the "magnetic bubble" represents 4 coils at 4 sides of the cube and constitutes a null dipole magnetic torus wrapping the cube. Four detector telescopes can be lodged between each couple of coils, the fifth side of the cube is left free for a non-magnetic telescope and the sixth side is used for the mechanical support, the electronics and services. To take full advantage of the cubic shape of the calorimeter, the facility should be operated in a high orbit, possibly outside the terrestrial magnetic field. With the cubic calorimeter the facility could have an aperture of about 5 m^2sr and measure spectra of positrons and antiprotons up to several TeV, of electrons up to >10 TeV, of protons up to 3 PeV, and ion fluxes up to the actinides.

2.4. The JUICE mission

A group from the European Space Agency presented the JUpiter ICy Moons Explorer (JUICE) mission that will fly by and observe the Galilean satellites Europe, Ganymede and Callisto, characterize the Jovian system in a lengthy Jupiter-orbit phase, and ultimately orbit Ganymede for in-depth studies of habitability, evolution and the local environment [11]. The main focus of the mission is the investigation of the moon Ganymede. Other targets of the JUICE mission will be the investigation of the Jovian magnetosphere, combined with two close flybys at Europa and 8 flybys at Callisto. For the first time the high latitude magnetosphere up to 30 degrees will be studied in detail and for an extended period of time. Ganymede's magnetosphere and its boundaries will be investigated in an elliptic orbital phase while its surface and its underneath subsurface ocean will be studied in circular orbital phases at different altitudes. The science payload consists of in-situ and remote sensing instrumentation with the newest technologies onboard. Launch of JUICE is currently planned in 2022. After Jupiter orbit insertion in 2030, the spacecraft will perform a 2.5 year tour of the Jovian system focusing on observations of the atmosphere and magnetosphere of Jupiter itself. Then, gravity assists at Callisto will shape the trajectory to perform two targeted Europa flybys and raise the orbit inclination up to 30 degrees. The Europa flybys are currently planned to observe regions of likely recent activity and possible shallow liquid water. The mission will culminate in a dedicated 8 month orbital phase around Ganymede in 2032. The Ganymede phase will include high (5000 km), medium (500 km), and low (200 km) circular orbits that will have different observation conditions optimized for particular science investigations. The total mission duration is about 11 years.

3. Radio detection of air showers

A large number of contributions at this conference were devoted to the detection of cosmic rays using the MHz emission from air showers due to either geomagnetic synchrotron radiation or the Askaryan effect. The principal feasibility of this technique in terms of the measurement of the arrival direction, energy estimate and self-triggering is already established. The presentations focused on the better understanding of the signal and the way towards physics measurements with radio arrays.

3.1. LOPES radio array

The LOPES project (LOar PrototypE Station) [12] was initiated in 2001, as a prototype station of LOFAR (The LOw-Frequency Array), with the aim to detect radio emission from EAS in coincidence with the KASCADE-Grande particle detector experiment. Although the experiment was stopped together with the KASCADE experiment, the data analysis still continues. The objectives of the LOPES project were to understand the principles for radio detection of cosmic rays with modern interferometric methods, to study and calibrate the radio emission in the energy regime up to $\sim 10^{18}$ eV, and to develop and optimize the radio technique for large scale application at ultra-high energies. The LOPES array consists of 30 east-west aligned dipole antennas installed in co-location with KASCADE-Grande at the KIT, Germany (Figure 5). In [13], results of LOPES data analysis were presented.

The KASCADE-Grande array detects EAS from a high-energy cosmic ray event (10^{15} - 10^{18} eV) and triggers LOPES which then digitally records the radio signal in the frequency band from 40 to 80 MHz. 316 measured LOPES events with $E > 10^{17}$ eV and $\theta < 45°$ were selected [14]. For each measured event two corresponding CoREAS [14] simulations were performed, one for a proton as primary particle, and one for an iron nucleus. The data analysis uses interferometric methods, based on beamforming (digital shift of all traces in time by distance to wave-front) and analysis of relative timing of triggered antennas with 1 ns precision. The angle reconstructed by LOPES was compared with that obtained from the KASCADE-Grande and KASCADE reconstructions of the same events (Figure 6).

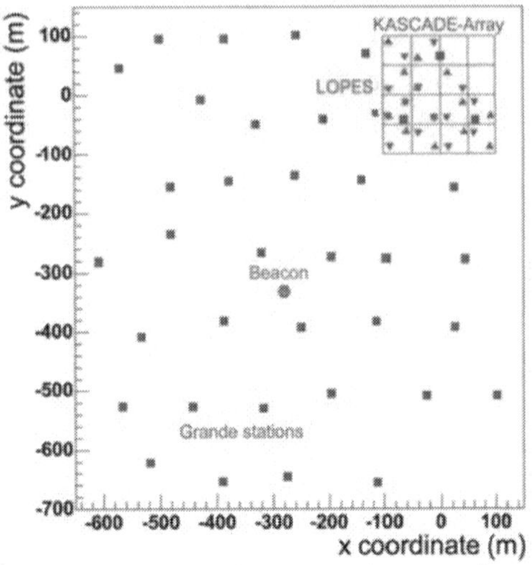

Figure 5. Layout of LOPES. Stars mark two antennas with east-west and north-south alignment at the same position, triangles antennas with either one alignment.

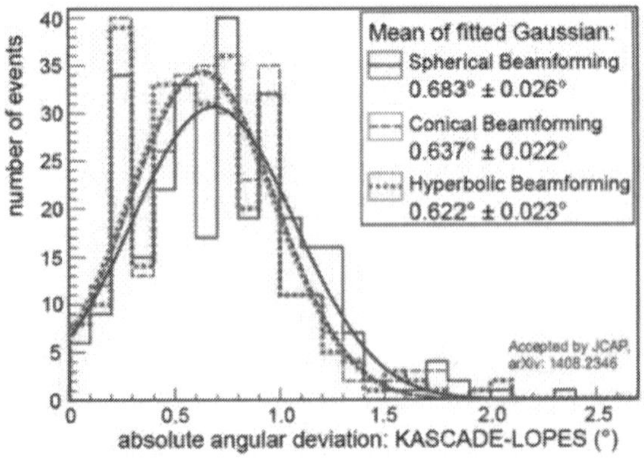

Figure 6. Comparison of the reconstructed KASCADE-Grande and LOPES directions. All stated numbers are determined with Gaussian fits [13].

The best fit indicates that the radio wave front is sufficiently well described by a hyperbolic form. At larger distances it can be approximated by a simple cone. According to the simulations with CoREAS, the cone angle is clearly correlated with the shower maximum. Also the reconstructed X_{max} is correlated with the slope of the lateral distribution. The experimentally achieved LOPES precision for the shower maximum is limited by measurement uncertainties to ~140 g/cm^2. But CoREAS simulations indicate that under better conditions the method might give an accuracy of the shower maximum, X_{max}, better than 30 g/cm^2.

The analysis of the simulated lateral distribution of the LOPES events reveals that the radio pulse amplitude at specific distances depends linearly on the primary energy E. The theoretical estimate of the energy reconstruction accuracy was obtained as about 9%. Practically, the LOPES energy reconstruction is at least as precise as that from KASCADE-Grande particle detectors, namely ~ 20%. Finally, the preliminary analysis of data obtained during 10 years showed that LOPES was a successful prototype for LOFAR and radio extension of KASCADE EAS array, but was not a precision experiment due to the high background. The precision for direction (< 0.7°) and for energy (< 20 %) is competitive with other EAS detection techniques, that for X_{max} (~ 90 g/cm^2) is worse. Radio arrays are useful extensions to particle detectors, if they are dense and are deployed in a radio-quiet environment.

In future, the LOPES data will be accessible through the public data centre KCDC (KASCADE Cosmic Ray Data Centre) which is mainly based on the data of the KASCADE experiment [15]. The developed approaches and experimental results and techniques are used for development of a new generation of EAS radio emission detectors as for instance LOFAR [16], AERA [17] and Tunka-REX [18] which were also presented in the INS session.

3.2. LOFAR

LOFAR is a novel distributed radio telescope for radio astronomy in the low frequency range. It is under construction in the northern part of the Netherlands and the north-eastern part of Germany, with baselines up to 300 km. LOFAR has the capability to measure radio emission from cosmic-ray induced air showers. In the core of the array about 2300 antennas are installed within about 4 km^2 (see Figure 7). There are two distinct antenna types: the Low Band Antenna (LBA) which operates between 10 and 90 MHz, and the High Band Antenna (HBA) between 110 and 250 MHz. For the purpose of cosmic-ray measurements a trigger is currently generated by a deployed particle detector array - the LOFAR Radboud Airshower Array (LORA) [16]. It consists of 5 stations with 4 particle scintillator detectors from the KASCADE experiment, spaced by 50–100 m distances. A trigger is sent to LOFAR when an air shower with energy $E \geq 2 \cdot 10^{16}$ eV is detected. The corresponding radio emission from showers of this energy corresponds to the LOFAR detection threshold. The basic approach for detection of radio emission from EAS was elaborated during LOPES, the LOFAR experiment is targeted to solve the following tasks: emission mechanisms; air shower simulations; shape of the shower front; arrival directions; polarization; High-Band measurements; measuring Cherenkov rings; optimal parametrization of the shower front.

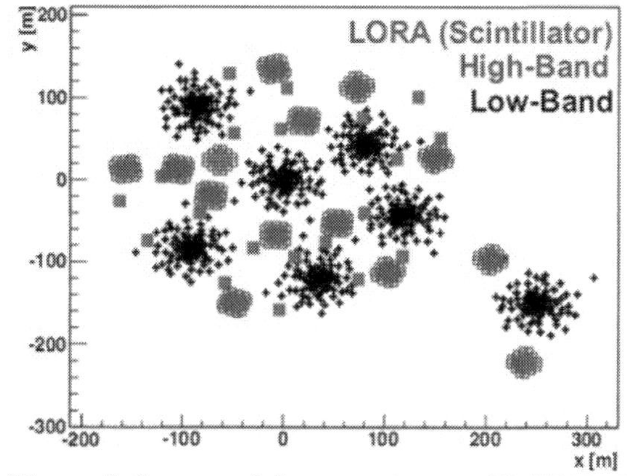

Figure 7. Layout of the central part of LOFAR. Left: the crosses indicate the LBA inner and outer antenna sets, respectively. The open squares show the positions of the HBA tiles, which are split into two groups per station. The filled squares indicate the positions of the LORA particle detectors. Right: the photo of the central part.

The measured emission is strongly polarized. The geomagnetic emission mechanism clearly dominates the polarization pattern. However, a sub-dominant charge-excess component can also be seen, and it increases with increasing observer distance from the EAS axis. For a clear understanding of the radio emission pattern of air showers detected by LOFAR, a CoREAS simulation was used. The air shower simulations had to follow all individual particles and to include a realistic atmosphere, the asymmetry through interference of emission mechanisms with different directions of electric fields (geomagnetic and Askaryan effects) and the additional relativistic time-compression (arising of Cherenkov effects for suitable geometries at frequencies above 100 MHz).

The analysis of the wave front with LOFAR confirms the LOPES result, that the shape of the wave front is best parameterized as a hyperbolic curved one near the shower axis and approximately conical further out [19].

The distribution of arrival directions of the cosmic rays detected with the LBAs radio antennas demonstrated a clear north-south asymmetry; this means that the pulse power is not a simple function of the distance to the shower axis.

One of the most important goals for radio detection is to reliably reconstruct the energy and mass of primary cosmic rays. Figure 8 illustrates two important aspects that such analyses have to take into account. The CoREAS-simulated footprint of the radio signal total field strength exhibits significant asymmetries related to the superposition of the dominant geomagnetic and sub-leading charge excess components of the radiation. The significant differences between the LDF of radio signals emitted by proton-induced and iron-induced air showers are clearly seen.

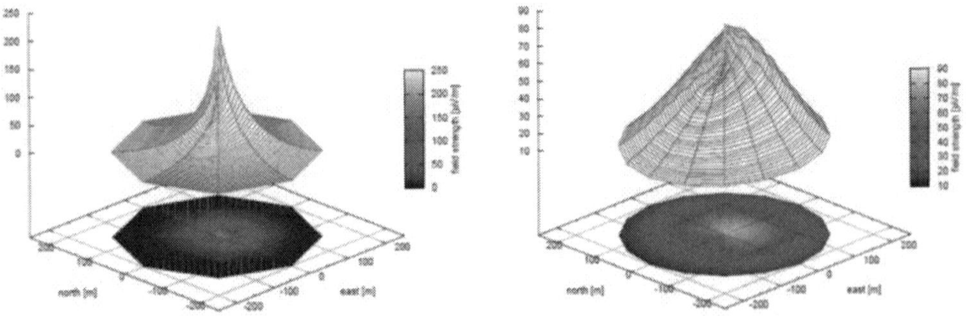

Figure 8. Footprints of the 40-80 MHz total field strength for vertical 10^{17} eV air showers induced at the LOFAR site by a proton (left) and an iron (right) primary.

LOFAR is the only dedicated experiment with high band antennas 110-250 MHz. Signals are expected to be more affected by Cherenkov enhancement and are concentrated on a ring of emission. These visible effects are related with the influence of the non-constant refractive index of the atmosphere which leads to relativistic time-compression at higher frequencies and amplification of emission for a specific angle with respect to the shower axis. The diameter of the ring thus provides information on the depth of the shower maximum. Finally, as it follows from the analysis, measuring the same air shower with both types of antennas is very promising.

This pattern is further influenced by the non-constant and non-unity refractive index of the atmosphere, which leads to visible effects of relativistic time compression of the signal. Time compression of the pulse also allows shower emission to be detectable at GHz frequencies where coherence would otherwise be lost.

3.3. AERA

The Auger Engineering Radio Array (AERA) [17] is the radio extension of the Pierre Auger Observatory located inside the infill low-energy EAS detector AMIGA [20] near the fluorescent detector site Coihueco. It allows a radio multi-hybrid measurement of EAS with the fluorescence telescopes, the water-Cherenkov and the muon detectors. Based on experience of LOPES and CODALEMA [21], AERA is deployed over 6 km^2 with 124 radio stations and constitutes two sub-arrays (see Figure 9). The first one, AERA-24, was deployed in 2011 and constitutes a dense core of 24 stations with 144 m spacing. The second part of additional 100 antennas spaced by 250 or 375 m was installed in 2013. The range of detected radio signals with AREA is 30 - 80 MHz.

For AERA-24, the Log-Periodic Dipole Antennas (LPDA) antennas are used. Each of them is composed of two independent planes 4 m×4 m, placed at 3.4 m from the ground. The new stations use Butterfly type antennas similar to those designed and developed for the CODALEMA [21] and Tunka-REX experiments [18]. The AERA Butterfly antenna consists of two 2 m×2 m active bowtie antennas. AREA can be operated in a self-trigger mode in 100% duty cycle, a part of the radio stations can get an external trigger from the water Cherenkov detectors or from AMIGA scintillators.

One of the main physics tasks of AERA is the investigation of the radio emission process of air showers in geographic locations and geomagnetic fields different to those of LOPES and Tunka-REX. The distribution of reconstructed shower arrival directions for AERA events triggered by the Auger SD array reflects a north-south asymmetry due to the geomagnetic emission mechanism [17]. The comparison of the polarization angle measured at the AREA radio stations with predictions based on a model of the components (the geomagnetic separation and the Askaryan effect) reveals a clear evidence for a radial component in the radio emission process amounting 14±2% of the geomagnetic component [17]. A preliminary test on 100 externally triggered events shows that the energy estimator can describe the Auger surface detector energy measurements with a deviation of less than 30%. Current studies are focused on evaluating the precision of AERA for estimating the arrival direction, energy and composition of air showers in the energy range from 10^{17} to 10^{19} eV.

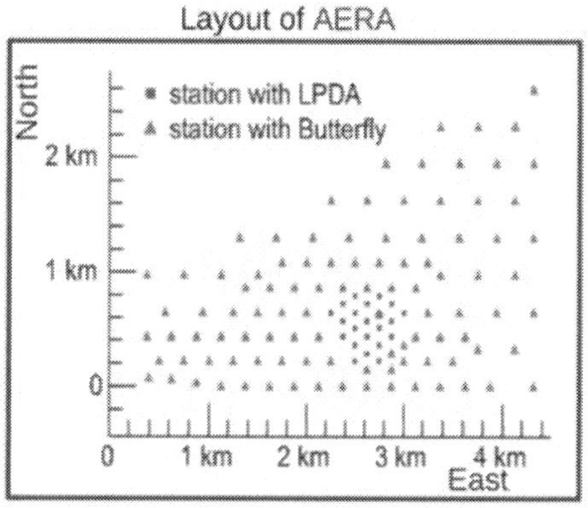

Figure 9. Layout of the Auger Engineering Radio Array in front of Coihueco. AERA-24 stations are marked with blue squares, AERA-124 stations by purple triangles.

3.4. Tunka-REX

Tunka-REX (Tunka Radio Extension) enhances the Tunka-133 air-Cherenkov array placed near the southern tip of Lake Baikal [18]. Tunka-133 detects EAS induced by cosmic rays from initial particles with energies of 10^{16} – 10^{18} eV. The radio array Tunka-REX consists of 25 antenna stations connected to the DAQ of Tunka-133 and spread over an area of 1 km^2. This combination provides the possibility of hybrid measurements and cross-calibration between the air-Cherenkov and radio techniques. At each antenna position there are two orthogonally aligned short aperiodic loaded loop antennas (SALLA) with 120 cm diameter (Figure 10).

SALLA is a cheap, simple and stable antenna with a frequency range 30 – 80 MHz. One of the main goals of Tunka-REX is demonstrating, on the basis of a cross-calibration of the radio and the Cherenkov signal emitted by air showers, that the radio detection technique is competitive to traditional EAS detection techniques of ultra-high energy cosmic rays. The analysis of results of the 2012/2013 season has shown that Tunka-REX effectively detects the radio emission from EAS [22]. After quality cuts, a strong correlation between amplitude and energy was demonstrated. The study of the sensitivity to the shower maximum is continued. Tunka-REX has a high sensitivity to inclined air-showers. The deployment of one new antenna for each of the 19 scintillator stations of Tunka-Grande and 3 antennas close to HiSCORE stations is planned [22]. A very important step towards the Tunka-REX technique was made by direct calibration of all hardware components at the KASCADE array site in KIT [23]. For these purposes, a crane lifted a commercial reference radiation source to a position 10 m above the SALLA antenna, with a frequency range from 30 MHz to 1 GHz. Calibration results reveal that the behavior of the measured frequency dependence is in agreement with simulations of antenna, however the antenna simulation predicted a gain differing by a factor 11 [23].

Figure 10. The SALLA antenna.

4. TAIGA gamma-observatory

TAIGA (Tunka Advanced Instrument for cosmic ray physics and Gamma Astronomy) is a new hybrid detector system for ground-based gamma-ray astronomy for energies from a few TeV to several PeV, and for cosmic-ray studies from 100 TeV to several 100s of PeV. TAIGA will also be located in the Tunka valley, where since 2009 the Tunka-133 Cherenkov EAS detector is in operation [24, 25]. The huge progress in high energy gamma-ray astronomy has been obtained mainly from Imaging Atmospheric Cherenkov Telescopes (IACT). The sensitivity level of modern IACTs is optimized for the energy range of 100 GeV – 20 TeV. Gamma radiation with energies above 100 GeV was detected from more than 150 sources of different types, but gamma rays with energies higher than 10 TeV are detected only from about 10 sources. However, to access the energy range from 10 TeV up to several 100 TeV, a detection area of the order of 10 km^2 or more is required. Alternatively to imaging gamma telescopes, the shower-front sampling technique allows to install large effective areas and also naturally provides large viewing angles of the instrument. Its operating principle is based on the sampling of the density and timing (arrival-time and spread) of the air shower-front with distributed arrays of detector stations [26]. The TAIGA observatory includes HiSCORE – an array of wide-angle integrating air Cherenkov stations, furthermore an array of IACTs, an array of particle detectors, both on the surface and underground, and the Tunka-133 EAS detector. Tunka-133 is a wide-FOV (~2 sr) integrating air Cherenkov detector with 185 stations spread over an area of ~3 km^2. The individual stations consist of single large PMTs (diameter 20 cm or 37 cm).

Tunka-HiSCORE will be an array of wide-FOV (~0.6 sr) integrating air Cherenkov detector stations, placed at distances of 150-200m from each other, covering an area of initially ~1 km^2 and ~100 km^2 at a later phase of the experiment. Each station consists of 4 neighbored PMTs of size either 20 cm or 25 cm in diameter. The PMTs have Winston-cone shape light-guides, which increase their light collection area by factor of four (see Figure 11). DAQ and time synchronization system consists of 8-channel optical station board (OSB) for digitization of signals from anodes and dynodes of 4 PMTs of the optical station and synchronization boards (SB) placed in the DAQ center [28]. All boards are designed on the basis of DRS-4 chip and FPGA Xilinx Spartan-6. OSBs and SB are connected via single-mode optical fibers. To synchronize all stations of the array to sub-ns precision a hybrid approach, combining a custom-made synchronization technique (100 MHz clocks distributed over separate fibers from the array center), and the new White Rabbit Ethernet-based are used [29]. The accuracy of time synchronization is <1 ns. The dead time of OBS is smaller than 0.5 ms.

Figure 11. The layout of the Tunka-HiSCORE optical box.

Methods for gamma-ray and cosmic-ray data reconstruction for HiSCORE are similar to the ones used for Tunka-133. The PMT outputs are summed up. The detector stations measure the light amplitudes and its arrival time differences over a distance of few hundred meters. This approach allows one to reconstruct with a high precision the arrival direction and the axis position of the EAS as well energy E_0 and shower maximum height X_{max}. The expected accuracy of the core location measurement is 5–6 m, that of the arrival direction about 0.1 degree, for E_0 ~ 10% and for X_{max} ~ 15-

20% [27]. The data can be used to reconstruct in detail the cosmic ray spectrum and its composition and will also improve gamma/hadron separation.

Since 2013, Tunka-HiSCORE operated as a 9-station prototype array. The array was arranged on a regular grid of 3×3 stations, with a side length of 300 m [27] (see Figure 12). The Tunka-HiSCORE stations (blue squares) are embedded in the Tunka-133 array (black circles) and cover a total area of roughly 0.1 km^2.

A test with a bright/wide-angle LED light source ~200 m outside the array has been performed to check single station performances and full array event reconstruction quality [29]. For the calibration experiment, 45°-inclined mirrors were installed on top of the stations. The event reconstruction precision estimate is obtained by fitting the relative trigger times of the 9 stations: RMS < 0.5 ns. The EAS reconstruction with the HiSCORE-9 prototype with EAS-model fit gave RMS ~ 0.6 ns.

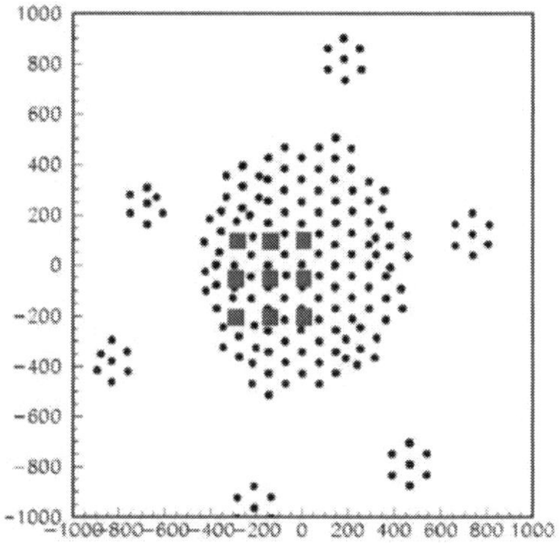

Figure 12. Layout of the Tunka-HiSCORE 9-station prototype array. Black dots mark the Tunka-133 stations.

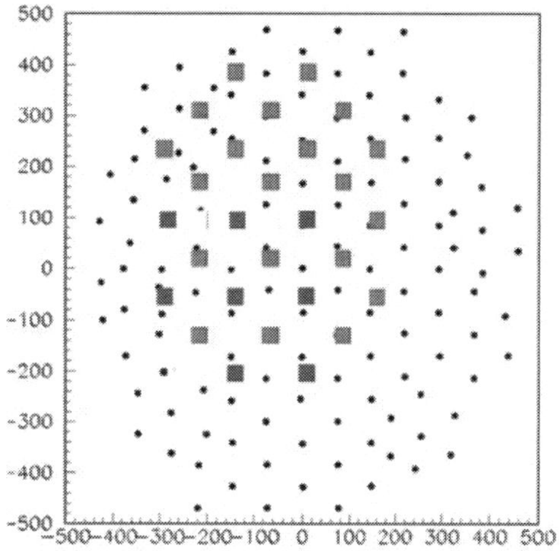

Figure 13. Layout of the Tunka-HiSCORE array: blue squares are stations deployed in 2013, red – 2014.

The configuration of the Tunka-HiSCORE stations as of 2014 is shown in Figure 13 [27]. It comprises 28 detectors at 100 m spacing forming a super-cell structure. The total area of the setup is 0.25 km^2. To increase the observation time for the Crab Nebula, all optical boxes are tilted towards South by 25 degrees. The number of gamma-ray events from the Crab Nebula at energy > 40 TeV is estimated as 15-30 for one season of observations (100 hours).

The timing array Tunka-HiSCORE alone has a poor gamma-hadron separation at low energies (10–100 TeV), with a quality factor of the order of unity below 100 TeV and only reaching 2 above few hundreds TeV. To improve the selection efficiency, an underground muon detector is foreseen to determine the type of a primary particle. The possibility to select EAS initiated by a gamma-ray is based on the fact that the number of muons in a hadron-induced EAS is on average 30 times higher than in gamma-ray events. For clear separation of muon signature at energies above 100 TeV, the total area of muon detectors should be at least a few percent of the area of the whole installation. Therefore, the overall area of shielded muon detectors should be 2000-3000 m^2 of the total area of the Tunka-HiSCORE EAS detector. For the first stage of the muon array, scintillation detectors formerly operated as part of the EAS-TOP and the KASCADE-Grande arrays are used [27]. This so-called Tunka-GRANDE detector consists of 19 scintillation stations, each of them with a surface and an underground part. The stations are located at distances about 20 m from the centres of the Tunka-133 clusters. Each station includes 12 scintillation counters with a size 80×80×4 cm^3 and is equipped with underground muon detectors with 8 identical counters in each.

The IACT array consists of up to 16 imaging Cherenkov telescopes with a reflector area of 10 m^2 equipped with imaging cameras of 400 PMT-based pixels. Single pixels will have an aperture of 0.36°.

FOV of the single telescope will be 8°×8°. The inter-telescope distances will be optimized in the range of 300-600 m.

Tunka-REX (see above) may be also included in TAIGA facility in future.

The key advantages of the gamma-observatory TAIGA is the joint operation of wide-angle and narrow-angle detectors of Tunka-HiSCORE and Tunka-IACT. By operating the telescopes in mono-scopic mode with distances of the order of 600 m between the telescopes, the total area covered per telescope is larger than the area that could be covered using the same number of telescopes as a stereoscopic system (requiring distances of roughly 300 m in the 10–100 TeV energy regime).

In Figure 14, the point-source survey sensitivity is shown for a 100 km^2 array after 1000 h of observation time (HiSCORE alone – beige, hybrid mode - red) [26]. For comparison, the point-source survey sensitivities of CTA [30], of a search for neutrinos by IceCube [31] (Milagro source stacking), and LHAASO [32, 33] (adapted to a minimum of 50 gamma-rays) are shown. For a reference, also the 50-hour pointed-observation sensitivity of CTA [34] is given.

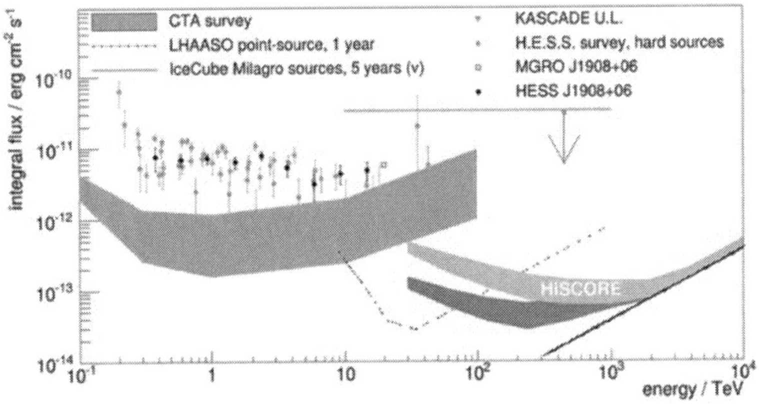

Figure 14. Point-source survey sensitivity of a 100 km^2 HiSCORE array after 1000 h of observation time [26].

5. Experimental complex NEVOD

The NEVOD-DECOR group started significant upgrades of the existing facility with new additional detection systems extending the complex research capabilities. The layout of NEVOD-DECOR complex is shown in Figure 15.

The basis of the complex is the Cherenkov water detector (CWD) with a tank of 2 kt purified water, 9×9×26 m^3. CWD NEVOD is a multipurpose detector designed for detection of all main components of CR on the Earth's surface. Therefore, the NEVOD measuring system must have 4π-sensitivity. To provide this, a quasi-spherical measuring module (QSM) was designed [35]. The QSM comprises six PMTs FEU-200 (EKRAN company, Russia) with flat 15 cm diameter cathodes directed along three orthogonal axes and placed in an aluminum housing (see Figure 16). Such a cluster of PMTs has practically isotropic sensitivity to the direction of incident Cherenkov light, since the sum of the squared amplitudes of triggered PMT signals is roughly independent of the arrival direction of the light. The detection system of CWD NEVOD is formed by a spatial lattice of QSM arranged in strings with a spacing 2.5 m along the water tank and 2.0 m across it and over the depth. Around CWD NEVOD, two systems of coordinate-tracking detectors – DECOR [36] and URAGAN [37] – are deployed. DECOR represents 8 supermodules (SM, in total 70 m^2) vertically arranged around the NEVOD water tank. Each SM consists of 8 layers of streamer tube chambers with external strip readout system. The angular and spatial resolution is ~1° and 1 cm respectively. The main task of the NEVOD-DECOR setup is the study of muon bundles over a wide range of multiplicities and zenith angles (up to the horizon).

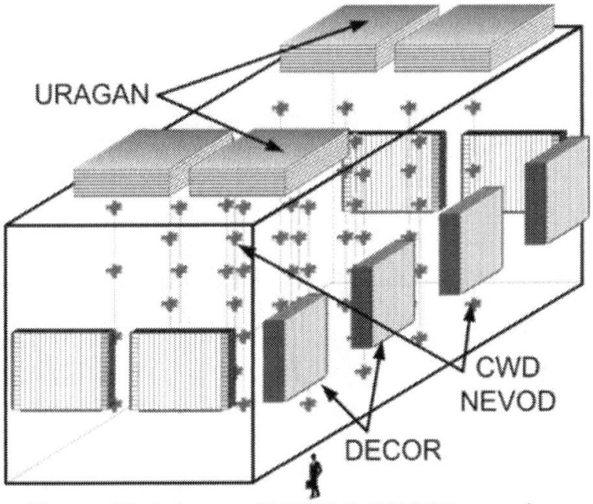

Figure 15. Scheme of NEVOD-DECOR complex.

Figure 16. Quasi-spherical measuring module of CWD NEVOD.

During the 2005 – 2010 experimental series, a new approach to primary cosmic ray study based on a new EAS observable – the local muon density spectra (LMDS) was elaborated [38, 39]. The analysis of obtained LMDS reveals a significant excess of multi-muon events generated by PCR at energies 10^{15}-10^{18} eV in comparison with simulations performed within the framework of commonly used hadron interaction models (even under assumption of heavy primaries – iron nuclei). This problem is known now as "muon puzzle".

To solve this problem, the NEVOD-DECOR group proposed the measurement of energy deposit of muon bundles in the CWD NEVOD [40]. However, DECOR SMs do not cover the entire aperture of the Cherenkov water detector and do not exclude the possibility of passing of part of muons between the super-modules, also the size of their cells limits two-track resolution by 3 cm. The new detector TREK [41] will completely cover the side aperture of the Cherenkov water detector NEVOD and significantly improve the resolution of close tracks. The TREK setup represents two vertical planes of drift chambers (IHEP, Protvino) with *X*-*Y* orientation, mounted on the outer wall of the CWD building (see Figure 17).

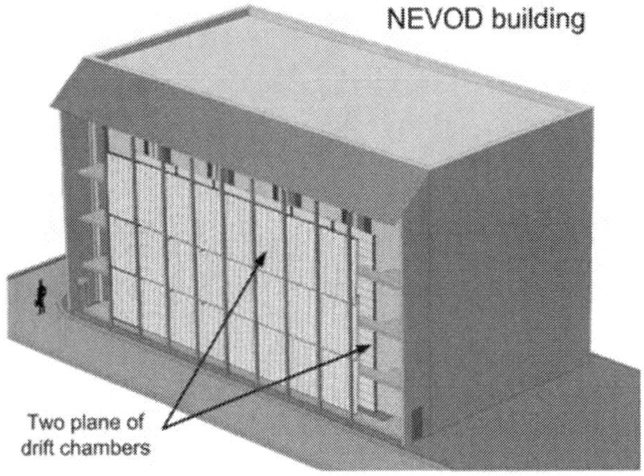

Figure 17. Layout of the TREK setup.

Each drift chamber has a large effective area (1.85 m^2), good coordinate and angular resolution with a small number of measuring channels. The detector will be operated as a part of the experimental complex NEVOD, in particular, jointly with Cherenkov water detector. There are 132 chambers in each plane. The effective area of the detector will be approximately 270 m^2.

Another planned detector is a traditional EAS array which will be installed on the roofs of the laboratory buildings around NEVOD in the MEPhI University campus [42]. The reason for the deployment of a traditional installation is the fairly poor PCR energy resolution of the LMDS method: $\sigma_{\lg E} \approx 0.4$. The measuring system of the NEVOD-EAS setup is organized on a cluster principle. Each cluster includes 16 counters combined in 4 detector stations and is served by the Local Post (LP) of preliminary data acquisition. As a detector of the EAS electromagnetic component, the scintillation counters from the former KASCADE-Grande facility are used. Figures 18 and 19 show pictures of the scintillation detector of KASCADE-Grande and a NEVOD-EAS detector station housing.

The NEVOD-EAS extension of the NEVOD-DECOR complex will allow determination of the size, position of the axis and the arrival direction of EAS. It will enable registration of EAS with energies 10^{15}-10^{17} eV, as well as the verification of the technique of muon bundle registration using the DECOR detector. New data obtained with the NEVOD-EAS setup will allow narrowing the energy range of PCR particles responsible for the generation of muon bundles with certain multiplicity arriving at different zenith angles. The NEVOD group plans to study energy deposit of muon bundles generated by PCR with energies up to 10^{18} eV in the next upgrade phase of the NEVOD complex [40].

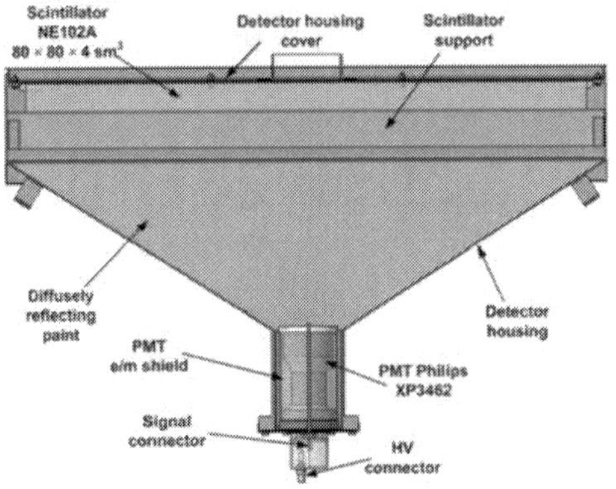

Figure 18. NEVOD-EAS counter.

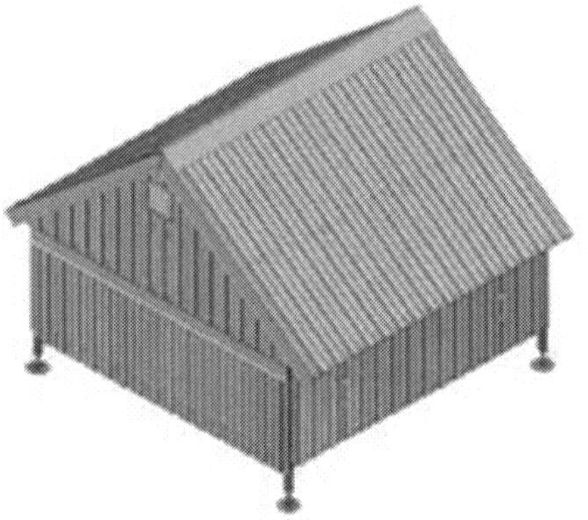

Figure 19. Detector station.

6. Underwater neutrino projects

6.1. KM3NeT

The KM3NeT Collaboration reported about the start in 2014 of the first phase of construction of a next generation high-energy neutrino telescope [43]. The main goal of KM3NeT is the discovery of high-energy neutrino sources. Its location is optimal to look directly at the Galactic centre, and it will complement the IceCube with respect to the field of view and exceed IceCube in sensitivity. The excellent optical properties of water (scattering and absorption lengths) allow the reconstruction of muon tracks with an excellent angular resolution for track events ~ 0.1° and for shower events < 2° above 100 TeV [43]. The digital optical module (DOM) used in KM3NeT is a 17-inch glass sphere, resistant to the high pressure at the sea bottom, housing 31 3-inch photomultiplier tubes and the active bases for power and the readout electronics. The total cathode area corresponds to the area of about three 10-inch PMTs. The simple requirement of a coincidence among 6-7 PMTs allows to effectively suppress the hits due to the decay of ^{40}K dissolved in the sea water. The KM3NeT detector will have a modular structure with six building blocks, each consisting of about one hundred detection units (DU). Each DU will be equipped with 18 DOMs with ~ 36 m between the DOM. A single block has 115 DUs located at a distance of 90-120 m from each other (0.5 - 0.8 km^3) (see Figure 20). The total volume of the detector will be ~ 3 - 5 km^3. The first phase of construction has been started, and shore and deep-sea infrastructures hosting the future KM3NeT detector are being prepared offshore Toulon, France and offshore Capo Passero on Sicily, Italy.

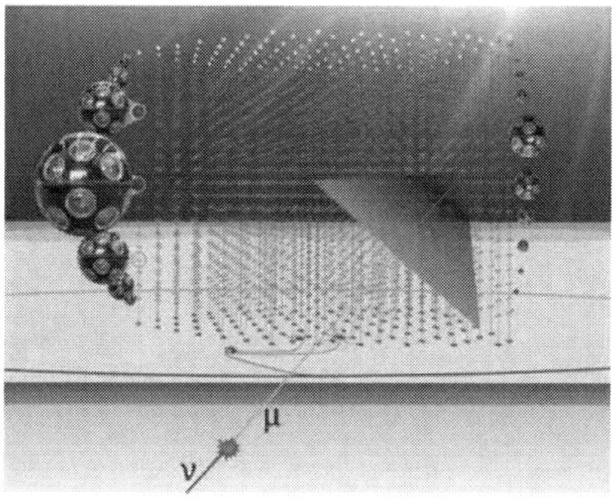

Figure 20. Layout of one block of KM3NeT neutrino telescope.

The started phase 1 of KM3NeT, which is completely funded, foresees the construction and arrangement of 31 DUs, 7 at the Toulon site and 24 at Capo Passero. During the KM3NeT Phase-1.5 (2017 - 2020) two detector blocks will be installed. It will allow detect a flux of high energy cosmic neutrinos reported by IceCube on this conference already after 1 year operation (cascade channel) [43].

6.2. GVD

Significant progress was reported for the Gigaton Volume Detector (GVD) [44]. GVD will be a second Northern kilometer-scale neutrino observatory [45]. It is the successor of the first neutrino telescope NT200+, which registered in 1996 the first high-energy atmospheric neutrinos underwater. The location of the GVD facility is Lake Baikal (104°25' *E*; 51°46' *N*), about 3 km from shore at 1360 m depth [46]. The advantage of Baikal site is the possibility to deploy strings from ice in the period of March/April.

The scientific goals are essentially the same as for KM3NeT: exploration of the neutrino sky at energies above 10 TeV from the Northern hemisphere; indirect search for dark matter; search for exotic particles like magnetic monopoles; limnology and other environmental research. GVD will consist of strings of optical modules that will be grouped in clusters (see Figure 21). Basic elements of GVD, new optical modules, FADC readout units, and underwater communication systems, were investigated and tested in Lake Baikal with prototype strings in 2008–2010.

The project is divided into two phases. The first phase includes the creation of 12 clusters (0.4 km^3, 24 strings, 2302 OMs) and is planned to be fully installed in 2019/2020. The completion of phase 2 (27 clusters, 1.5 km^3) is scheduled for mid-2020s. An important step was made in spring, 2015 with the deployment of the missing last 3 strings of the first cluster (see Figure 22).

7. New instruments and techniques

A new facility for investigation of various processes of atmospheric and extra-terrestrial origin that cause modulations of cosmic rays detected at ground level was presented by the Tragaldabas collaboration [47]. TRAGALDABAS (TRAsGo for the AnaLysis of the nuclear matter Decay, the Atmosphere, the earth's B-field And the Solar activity) is a new RPC-based cosmic ray detector which has been recently installed at the Univ. of Santiago de Compostela, Spain. The detector consists of two 1.8 m^2 planes of two 1 mm-gap glass RPCs. Each plane is readout with 120 pads. The main performances of the detectors are: an arrival time resolution of about 300 ps, tracking angular resolution below 3°, detection efficiency close to 100%, and solid angle acceptance of 5 sr. Another two planes of RPC detectors will be added in future in order to improve both the resolutions and the acceptance.

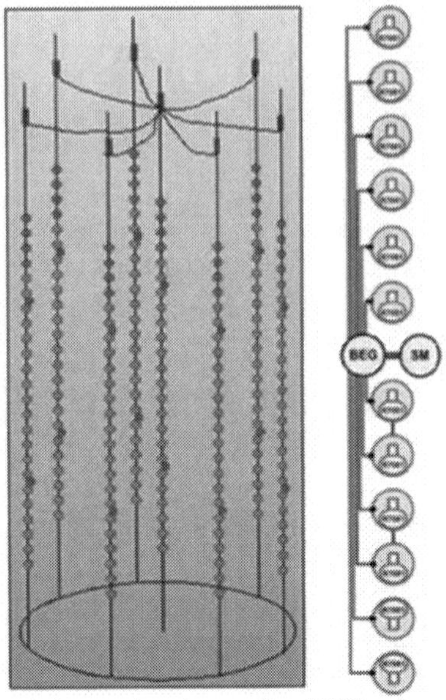

Figure 21. Schematic view of a GVD cluster (8 strings with 2 sections each), and a section of OMs (12 OMs with PMT R7081HQE).

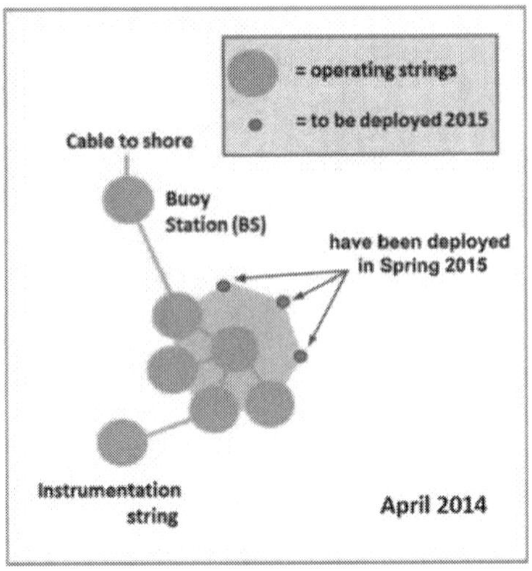

Figure 22. Top view of the first GVD cluster. It has been completed in Spring 2015.

A Hungarian group [48] presented the close cathode chamber (CCC) technology developed and proposed for the use as a portable tracking system operated under harsh and varying environmental conditions. The detector technique is based on an asymmetric multi-wire proportional chamber without outer support frames. A muon tomograph has been built with a sensitive area of 0.25 m^2 and an angular resolution of few mrad. The applicability of the sensor to detect underground rock density inhomogeneities has been demonstrated via reconstruction of an underground tunnel system. It is useful for material discrimination by measuring multiple scattering and absorption of muons.

A report of the group from Kiel was devoted to the research of material characteristics of the new Cerium doped LSO (Lu_2SiO_5) scintillator [49] that will replace BGO ($Bi_4Ge_3O_{12}$), widely used for the detection of high-energy particles in space applications. The LSO scintillator offers the same benefits with higher light output capabilities and a significantly shorter decay time. Characteristics of the LSO scintillator were investigated from the point of view of its use in space missions. Experiments with ^{176}Lu source and the Heavy Ion Medical Accelerator in Chiba (HIMAC), Japan have shown that LSO is a promising candidate for future space missions.

8. Conclusions

The reports presented in the section INS demonstrate a wide frontier of new approaches and instruments in cosmic ray research, both for ground-based and orbital experiments. We note a significant progress of EAS radio detection techniques and detectors, as well as the pre-launch readiness of orbital experiments TUS and CALET. Another progress was achieved by the TAIGA Collaboration towards creating a new observatory for gamma-ray astronomy of ultrahigh energies in Tunka valley. The observatory is targeted to the search of gamma-ray sources with PeV energies and the study of the flux of UHE primary cosmic rays on the basis of a new approach. The collaboration NEVOD-DECOR offers a new multi-component approach to UHE EAS detection. The method is based on the precise study of spatial and energy characteristics of EAS muon bundles in a wide range of zenith angles and multiplicities. The creation of the first cluster of the new underwater detector GVD in the Lake Baikal demonstrates a considerable progress in the deployment of a kilometer-scale neutrino observatory in the Northern hemisphere. The development of a new generation of space born and ground based detectors will allow, in the near future, to make a significant step in understanding the nature of cosmic ray origin.

9. Acknowledgments
I would like to thanks the organizers of the 24th European Cosmic Ray Symposium in Kiel for the invitation to give this rapporteur talk. I am very grateful to Rostislav Kokoulin and Christian Spiering for critically reading this manuscript.

10. References

[1] Benson R and Linsley J 1981 *Proc. 17th Int. Cosmic Ray Conf.* (Paris) vol 8 p 145
[2] Panasyuk M et al. 2012 *J. of Cosmology* **18** 7964-77
[3] Takahashi Y The JEM-EUSO Collaboration 2009 *New J. Phys.* **11** 065009
[4] Vedenkin N N et al 2011 *Soviet J. of Experimental and Theoretical Physics* **113** 781-90
[5] Panasyuk M I et al 2015 Air shower observations from space with TUS/JEM-EUSO/KLYPVE ECRS Kiel Germany (these proceedings)
[6] Adams J H et al 2013 *Astropart. Phys.* **44** 76-90
[7] Rouille d'Orfeuil B et al 2014 *Astron. & Astrophys.* **567** A81
[8] Haungs A for the JEM-EUSO Collaboration 2015 Physics goals and status of JEM-EUSO and its test experiments ECRS Kiel Germany (these proceedings)
[9] Ricciarini S for the CALET Collaboration 2015 The CALorimetric Electron Telescope (CALET) for high-energy astroparticle physics on the International Space Station ECRS Kiel Germany (these proceedings)
[10] Papini P and Spillantini P 2014 A multi-telescope magnetic facility in space with large acceptance and high MDR ECRS Kiel Germany (these proceedings)
[11] Krupp N 2015 JUICE: a European mission to the Jovian system ECRS Kiel Germany (these proceedings)
[12] Schröder F G et al 2013 *Journal of Physics: Conference Series* **409** 012075
[13] Schröder F G. for the LOPES Collaboration 2015 LOPES results: Radio Detection of Air Showers (these proceedings)
[14] Huege T Ludwig M James C 2013 *Proc. of the 5th ARENA* (Erlangen, Germany) vol 1535 p. 128
[15] Haungs A et al 2015 The KASCADE Cosmic-ray Data Centre KCDC ECRS Kiel Germany (these proceedings)
[16] Anna Nelles 2015 A new way of air shower detection: measuring the properties of cosmic rays with LOFAR ECRS Kiel Germany (these proceedings)
[17] Dorosti Hasankiadeh Q for the Pierre Auger Collaboration 2015 Radio detection of air showers at the Pierre Auger Observatory ECRS Kiel Germany (these proceedings)
[18] Hiller Roman for the Tunka-Rex Collaboration 2015 The Amplitude Calibration of the Tunka Radio Extension (Tunka-Rex) ECRS Kiel Germany (these proceedings)
[19] Corstanje A for the LOFAR Collaboration 2014 *Astropart. Phys.* **61** 22-31
[20] Wundheiler et al 2011 *ICRC* vol 3 p 84
[21] Ardouin D et al 2006 *Astropart. Phys.* **26** 341
[22] Kostunin D for the Tunka-Rex Collaboration 2015The Tunka Radio Extension: latest analysis results ECRS Kiel Germany (these proceedings)
[23] Hiller R for theTunka-Rex Collaboration 2015 The Amplitude Calibration of the Tunka Radio Extension (Tunka-Rex) ECRS Kiel Germany (these proceedings)
[24] Antokhonov B A et al 2011 *Nucl. Instrum.Meth. A* **639** 42–45
[25] Prosin V Berezhnev S Budnev N et al 2014 *Nucl.Instrum. Meth.* **A** **756** 94–101
[26] Tluczykont M for the TAIGA 2015 Collaboration Towards gamma-ray astronomy with timing arrays ECRS Kiel Germany (these proceedings)

[27] Budnev N M 2015 The Tunka detector complex: from cosmic-ray to gamma-ray astronomy ECRS Kiel Germany (these proceedings)
[28] Kozhin V Skurikhin A Korosteleva E 2015 DAQ and synchronization system for Tunka-HiSCORE array on the base of DRS-4 ECRS Kiel Germany (these proceedings
[29] Porelli Andrea and Wischnewski Ralf 2015 Timing calibration and directional reconstruction for Tunka-HiSCORE ECRS Kiel Germany (these proceedings)
[30] Gonzalez-Garcia M C Halzen F and Mohapatra S 2009 *Astropart. Phys* **31** 437
[31] Sinnis G 2005 *Heidelberg Gamma-Ray Symposium* (American Institute of Physics) vol 745 (Aharonian F A V"olk H J and Horns D) p 234
[32] Cao Z 2010 *Chinese Physics C* **34** 249
[33] Cui S, Liu Y, Liu Y and Ma X 2014 *Astropart. Phys*. **54** 86 – 92
[34] Actis M Agnetta G Aharonian F Akhperjanian A et al 2011 *Experimental Astronomy* **32** 193
[35] Borog V V et al 1979 *Proc. 16th ICRC* (Kyoto) vol 10 p 380
[36] Amelchakov M B et al 2002 *Izv. Akad. Nauk Ser. Fizich.* **66** 1611-1613
[37] Barbashina N S et al 2008 *Instrum. Experim. Techn.* **51** 701.
[38] Aynutdinov V M 1999 *Nucl. Phys. B –Proc. Suppl.* **75A** 318-320
[39] Barbashina N S et al 2007 *Nucl. Phys. B –Proc. Suppl.* **165** 317-323
[40] Kokoulin R 2015 Measurements of the energy deposit of inclined muon bundles in CWD NEVOD ECRS Kiel Germany (these proceedings)
[41] Zadeba E et al 2015 Large-scale drift chamber detector for registration of near-horizontal muons ECRS Kiel Germany (these proceedings)
[42] Yashin I I 2015 EAS array of the NEVOD Experimental Complex ECRS, Kiel, Germany (these proceedings)
[43] Coniglione R 2015 The KM3NeT neutrino telescope ECRS, Kiel, Germany (these proceedings)
[44] Ch. Spiering for GVD Collaboration 2014 The GVD Neutrino Project (these proceedings).
[45] Avrorin A et al 2011 *Nucl. Instrum.Meth A* **639** 30–32
[46] Belolaptikov I et al 1997*Astropart. Phys* **7** 263
[47] Juan A. Garzón for the Tragaldabas Collaboration 2015 Tragaldabas: a new high resolution detector for the regular study of Cosmic Rays ECRS, Kiel, Germany (these proceedings)
[48] László Oláh 2015 Close Cathode Chamber Technology for Cosmic Particle Tracking ECRS, Kiel, Germany (these proceedings)
[49] Elftmann R 2015 Characterization of an LSO scintillator for space applications ECRS, Kiel, Germany (these proceedings)

Status of a development of the large scale coordinate-tracking setup based on the drift chambers

E A Zadeba[1], N V Ampilogov[1], N S Barbashina[1], A G Bogdanov[1], A A Borisov[1, 2], D V Chernov[1], L I Dushkin[1], R M Fakhrutdinov[1, 2], S S Khokhlov[1], R P Kokoulin[1], A S Kozhin[1, 2], K G Kompaniets[1], V V Ovchinnikov[1], A A Petrukhin[1], V A Selyakov[1], V V Shutenko[1] and I I Yashin[1]

[1] National Research Nuclear University MEPhI (Moscow Engineering Physics Institute), 115409, Moscow, Kashirskoye sh., 31, Russia
[2] RF SSC Institute of High Energy Physics, Protvino, Russia

E-mail: EAZadeba@mephi.ru

Abstract. The large-scale coordinate detector for registration of near-horizontal muon flux of ultrahigh energy cosmic rays is being developed in MEPhI. Detector is based on the drift chambers from the neutrino experiment at the IHEP accelerator U-70, their key advantages are the large effective area (1.85 m^2), good coordinate and angular resolution with a small number of measuring channels. Detector will be operated as a part of the experimental complex NEVOD, in particular, jointly with Cherenkov water detector with volume of 2000 m^3. The current status of the project and results of studies of drift chamber characteristics are discussed.

1. Introduction

The large-scale coordinate detector for registration of near-horizontal muon flux of ultrahigh energy cosmic rays is being developed in MEPhI [1]. The detector is based on the drift chambers from the neutrino experiment at the IHEP accelerator U-70 [2], their key advantages are the large effective area (1.85 m^2), good coordinate and angular resolution with small number of measuring channels.

The project aims at solving the problem of the excess of muon bundles that increases with the energy of the primary cosmic rays [3, 4], what can be caused by both astro- or nuclear-physical reasons [5]. The only characteristic that responds differently to changes in the composition of cosmic rays and the inclusion of new physical processes, but has not been investigated to the present time, is the energy of the muon component of extensive air showers [6]. Such studies are performed at the experimental complex NEVOD-DECOR [7, 8]; however, the coordinate detector DECOR does not cover the entire aperture of the water Cherenkov detector and does not exclude the possibility of passing of part of muons between the individual supermodules of the detector; besides the size of its cells limits the possibility of separating two or more particles at small distances (less than 3 cm). The new detector TREK will completely cover the side aperture of the Cherenkov water detector NEVOD and significantly improve the resolution of close tracks.

The detector is based on the multiwire drift chambers. The chamber represents an aluminum box (figure 1) with ends limited by plexiglas plugs, to which wires, gas inlets and high voltage connectors are mounted. A uniform electric field is formed by field-forming wires with a pitch of 5 mm, the potential on them varies linearly from 0 to 12 kV. The drift chamber is filled with gas mixture of 94% Ar and 6% CO_2; at field strength of 480 V/cm electron drift velocity is constant (~ 41 mm/μs for the gas mixture) and allows to use linear time/coordinate relation. The drift chamber has four signal wires alternately shifted by ± 0.75 mm in the drift direction relative to the center of the chamber. Gas amplification is generated by potential difference of 2.2 kV between the signal and ten cathode wires arranged symmetrically at a distance of 3 mm from the center of the chamber. Two guard wires eliminate the edge effects.

With the passage of a charged particle through the working volume of the chamber drift times of electrons to the signal wires are registered; taking into account the drift velocity of electrons, these times allow to reconstruct the projection of the track on the plane perpendicular to the wires. Thus, for the reconstruction of tracks in space at least two non-parallel drift chambers are required. Spatial resolution of the chamber is 1 mm, the angular resolution is about 0.05 rad, two close tracks can be separated at a distance of 3 mm. Active area is 3.7x0.5 m^2 it constitutes 91% of the total surface.

The signals from the wires are processed by on-board amplifier-shaper that forms 50 to 100 ns signals (depending on the pulse amplitude) in the ECL levels.

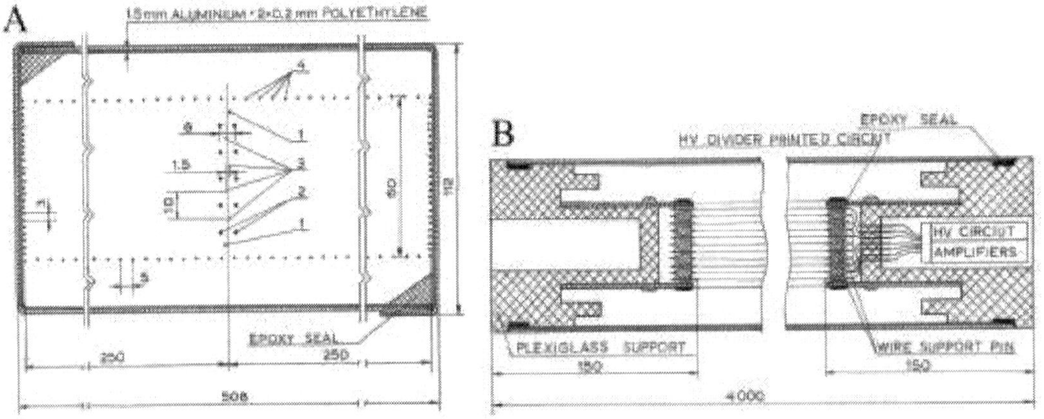

Figure 1. Schematic view of drift chamber cross section: (a) in plane perpendicular to sense wires, 1 - guard wires; 2 - cathode wires; 3 sense wires; 4 - field shaping wires; (b) in the plane parallel to sense wires.

The detector TREK consists of a two coordinate planes of 132 drift chambers installed on the external wall of the building of the EC NEVOD in a special protective box. Planes differ by the orientation of the chambers: in one plane they are placed vertically, in the other horizontally. The area of the coordinate planes is about 260 m^2, the effective area is about 94% what is provided by the installation of chambers with overlapping of the insensitive areas at the ends. The passages with hanged floors are provided between the planes, allowing installation and maintenance of the chambers. The protective box will be equipped with its own entrance and a climate control system. High- and low-voltage power supply systems and readout electronics will be located inside the NEVOD building.

Currently the project of the protective box and the drift chambers deployment is under development. Figure 2 shows the appearance of the protective box.

2. Study of drift chamber characteristics

Drift chambers are made in the mid 80-ies, for more than a decade they were used in the neutrino experiment at IHEP and for the same time they were stored. So tests for their performance are crucial. For the chambers testing and certification before installation to the TREK, a universal stand was developed, it allows to study temporal, spatial and noise characteristics of the drift chambers.

The base of the stand is an adjustable frame construction with three horizontal levels. Two of them have the compensation of the horizontal deflection of the beams under the weight of the chambers that is extremely necessary to improve the measurement accuracy. At each level, 7+7 drift chambers can be placed. On the levels with the compensation chambers are placed in two layers, differing in the orientation and forming a full XY coordinate plane.

Figure 2. Appearance of the TREK protective box on the outer wall of the NEVOD building.

The timestamp for drift chambers is provided by a pair of scintillation counters, one of which is installed at the upper frame level, and the second at its base. The area of each counter is 1 m^2, it is sufficient to work simultaneously with 8 drift chambers mounted on two adjustable levels (their width equals to 508 mm). The thickness of the scintillator is 1 cm, the light collection to PMT is performed using fibers. The signals from the counters are transferred to the coincidence circuit made on the basis of two shapers and logical "AND" in the VECTOR system.

The stand has its own system of gas supply, based on digital Bronkhorst flowmeters, managed from the computer. Chambers are connected to the gas flow pipes in 7 line rows, purging is performed about once a month, the pressure excess inside the chambers is kept to prevent ingress of oxygen (efficiency significantly decreases when the oxygen content is more than 0.5%).

Processing of the signals from the drift chambers is carried out by the 128 channel time-to-digital converter CAEN VME V1190A, which can be simultaneously connected to 32 drift chambers. Management of the TDC is performed from a computer connected via fiber to the crate controller CAEN VME V2718. A software was developed to work with the TDC, it allows to implement both manual and automatic measurements of the drift chamber characteristics, as well as performing the on-line visualization of the results.

Since the trigger signal from the coincidence circuit is produced earlier than the TDC receives signals from drift chambers it is delayed by a dual digital timer.

A four-channel oscilloscope Tektronix, triggered by the signal from the coincidence circuit, is used as a debug mode for reception of signals from separate chambers. Waveforms are transmitted via USB to the computer, where they are further processed. The computer control of the high voltage source was implemented to study drift chambers noise characteristics. The source has the control analog input to which a voltage is applied from the DAC controller. The controller operates via I2C bus emulated in the stand software. The first batch of 36 drift chambers has been checked for their performance on the stand to the date.

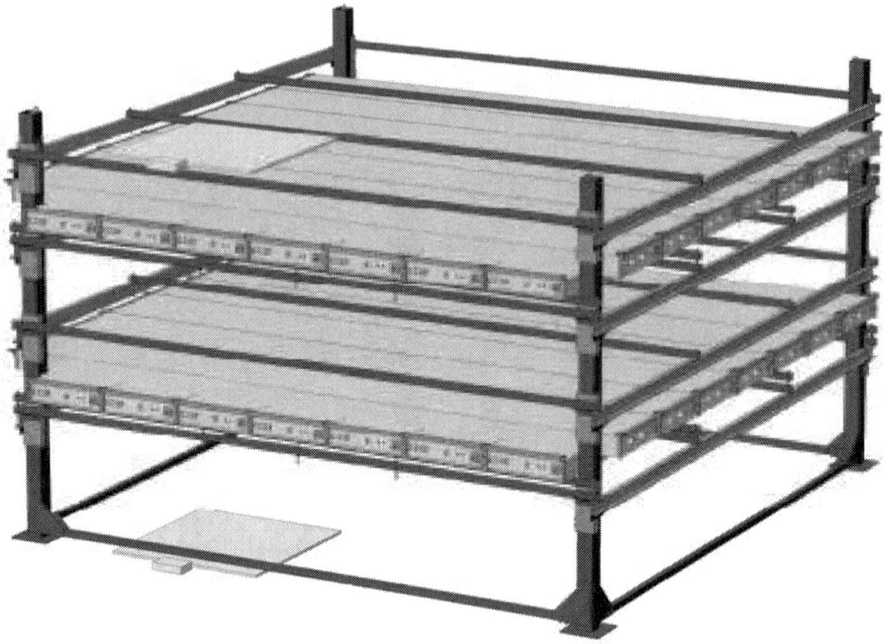

Figure 3. Deployment of drift chambers and scintillation counters in the universal test bench.

2.1. The temporal characteristics of the drift chamber

Among the temporal characteristics of the drift chamber, the main are the dependence of the number of registered signals on the drift time and the distributions of combinatorial parameters K_1 and K_2, the description of them is given below.

Figure 4 shows the distribution of events over drift time obtained during one experimental run. The maximum drift time lies in the range from 6 to 6.5 μs; dividing 250 mm wide drift gap on this value we can obtain the drift velocity of ~40 mm/μs. A large number of events near zero is caused by the noise, the fall after 5.5 μs corresponds to the decrease of the detection efficiency for such events. If the maximum drift time does not reach 5.5 μs or distribution falls from the outset, in most cases, this indicates a violation of the composition of the gas mixture, and the chamber requires additional sealing or a new gas mixture flush.

For any direct track, crossing the drift gap of the chambers, at 100% efficiency, we have 4 measures of the drift time: $t_i = x_i \cdot v^{-1}$, i = 1, 2, 3, 4 – wire order numbers, v – drift velocity of electrons. Using these periods of time for tracks that do not cross the signal wire zone in the center of the chamber, it is possible to construct linear combinations:

$$K_1 = t_1 - t_2 - t_3 + t_4 = 0, \tag{1}$$

$$K_2 = t_1 - 3 \cdot t_2 + 3 \cdot t_3 - t_4 = \pm 8 \cdot \delta \cdot v^{-1}, \tag{2}$$

where δ = 0.75 mm is the offset of the wires relative to the center of the chamber. The distribution of these parameters is built for all measurements made by means of drift chambers, it clearly shows their performance.

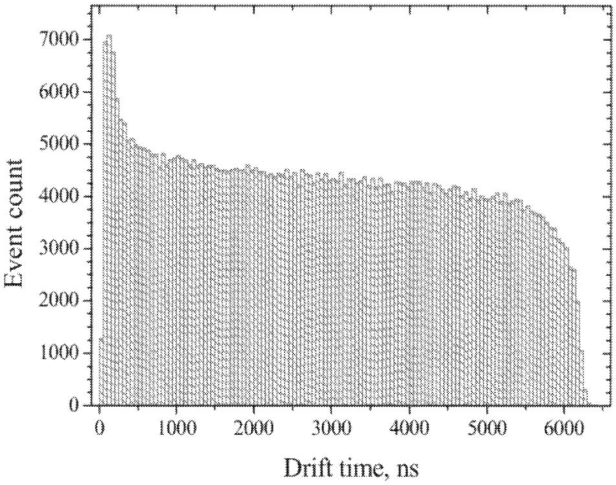

Figure 4. Distribution of events by the drift time of electrons.

In an ideal case the parameter K_1 is equal to zero, but due to Coulomb scattering of the electrons, the electronics jitter, etc., its actual value is different from zero. A typical distribution for the parameter K_1 is shown in figure 5 on the left. Assuming that the signal wires equivalent, temporal resolution (and taking into account the drift velocity of electrons and spatial resolution) of the signal wires can be obtained through the variance of K_1:

$$\sigma(t) = \sigma(K_1) \cdot 0.5 . \tag{3}$$

The distribution in figure 5 corresponds to the temporal resolution of 28 ns, or the spatial resolution of 1.2 mm. Mean value and width of the distribution are controlled during drift chamber tests, it allows to reject a chambers with poor resolution.

The parameter K_2 corresponds to the drift time of electrons at a distance of 6 mm, but the most important is the sign of this parameter, it shows at which side from the center of a chamber was a particle track is. Figure 5 (right) shows a typical distribution for this parameter. The distributions at left and right sides from the zero are not equal in area; it means that the different number of particles has passed at different sides. This is not a defect of the chamber, usually it is connected with uneven installation of scintillation counters relative to the chamber, which is inevitable for simultaneous tests of 8 chambers.

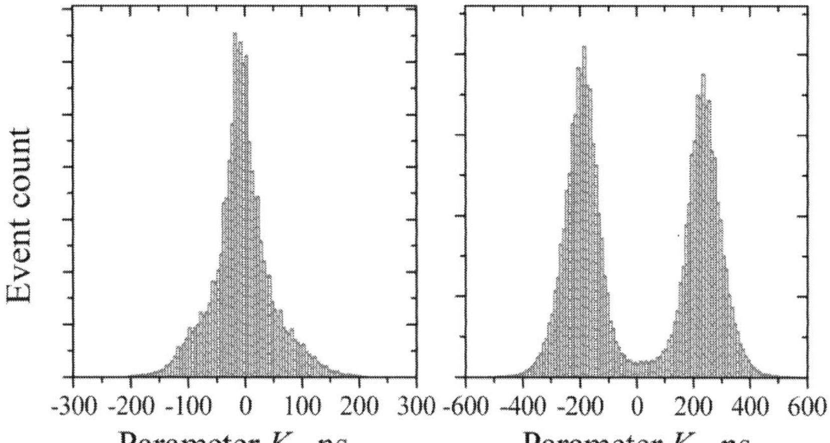

Figure 5. The distributions of events in values of parameters K_1 and K_2.

2.2. Noise characteristics

The noise characteristics of the drift chamber are the main criteria of chamber performance in addition to requirements for necessary spatial and angular resolution. The noise characteristics are the dependence of the counting rate on the voltage on signal wires. The gas amplification is not sufficint at low voltage, so efficiency significantly decreases with increasing distance from the point of ionization to the signal wires; on the other side, too high voltage causes electron emission from the cathode wires that are orders of magnitude larger than the number of signals from real events.

Figure 6 shows typical noise characteristics, the presence of a plateau in the area of 2200-2250 V is the sign of a good chamber performance. The given chamber has a nice wide plateau from 2050 to 2250 V. The gas mixture degrades over time so the width of the plateau decreases.

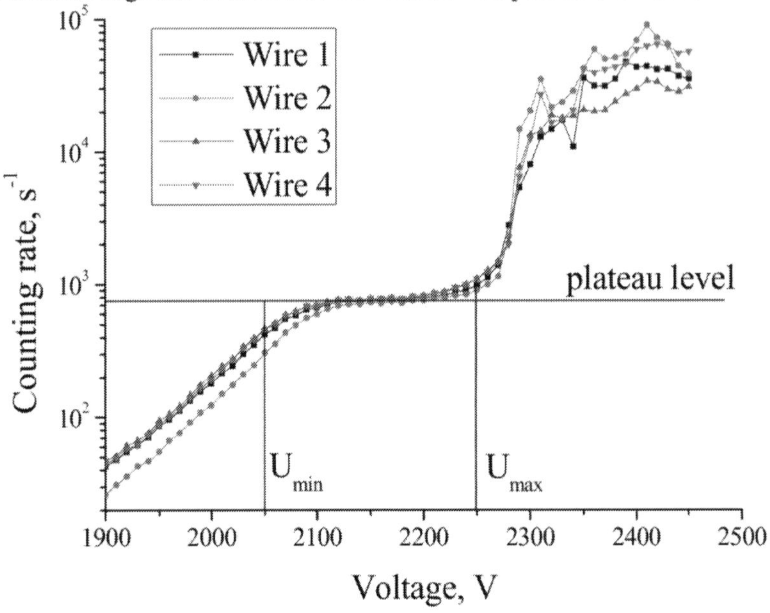

Figure 6. Noise characteristics of the drift chamber.

This characteristics should be measured not only when checking drift chambers on the stand, but it should be included in the procedure of monitoring of the TREK operation, allowing to adjust the voltage of the signal wires and to increase the gas mixture flow with increasing noise or performance drop.

3. CTUDC setup

For the improvement of the capabilities of joint operation of the NEVOD-DECOR experimental complex and the drift chamber setup, a new CTUDC setup is developed as a prototype. The CTUDC setup absolutely replicates the drift chamber deployment, data acquisition electronics and the method of data connection with NEVOD-DECOR events from the TREK. All the occurred deficiencies will be considered in the TREK detector project.

The coordinate planes of the setup are placed around the CWD (figure 7), such location allows the registration of near horizontal tracks as by own CTUDC data so by joint data with DECOR, that will significantly increase the range of muon track zenith angles (from 6° to 12° from horizon).

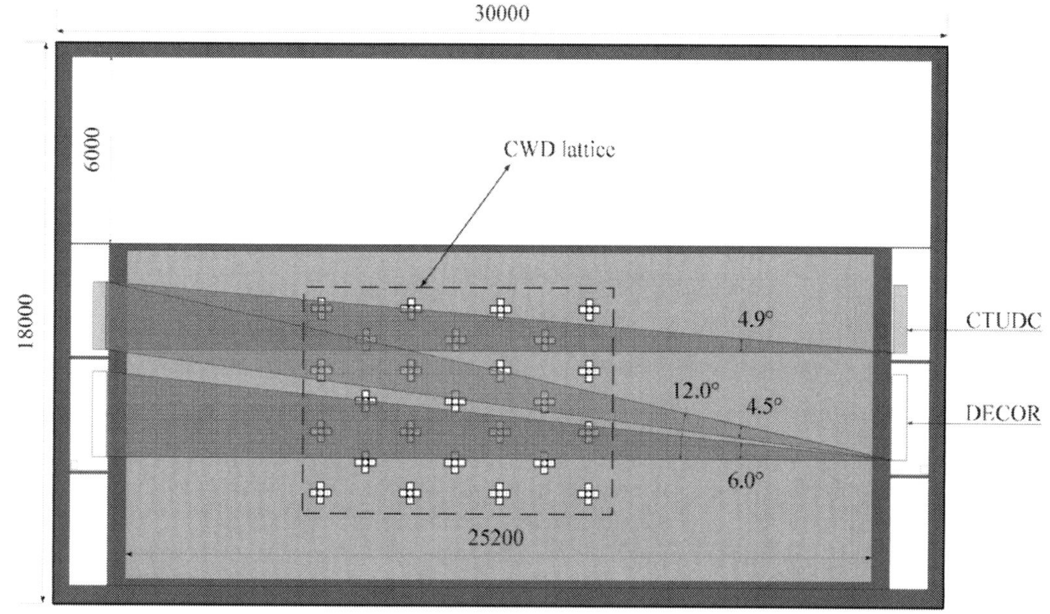

Figure 7. Location of the CTUDC coordinate planes relative to CWD and DECOR.

At present, the first coordinate plane is installed, its appearance is shown in a figure 8. The plane consists of 8 drift chambers installed in two rows, overlapped by 30 cm to exclude dead zone in the chamber ends. The effective area of the plane is 14.8 m^2. A special frame which allows a fine alignment of drift chamber mount in all coordinates is developed. It is also possible to vary a chamber overlap and an angle between chambers by regulating a distance to the floor and the walls.

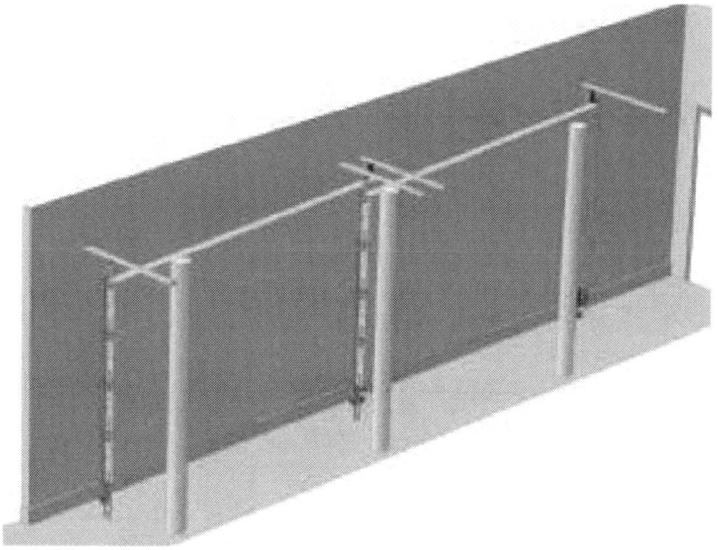

Figure 8. The coordinate plane of the CTUDC in a NEVOD building gallery.

Figure 9 shows a connection of the setup to the NEVOD-DECOR trigger system that sends a trigger signal to CTUDC time-to-digital converter. After signal comes, the CTUDC software puts an event to RAM and waits for a net package from the NEVOD server. The package keeps information on the event number, event time and some CWD and DECOR event characteristics.

The setup uses the 128-channel time-to-digital converter EM-4 developed in IHEP for E-MISS bus. The converter has 8 ns LSD and memory for 1000 32-bit words that is sufficient for the assigned task.

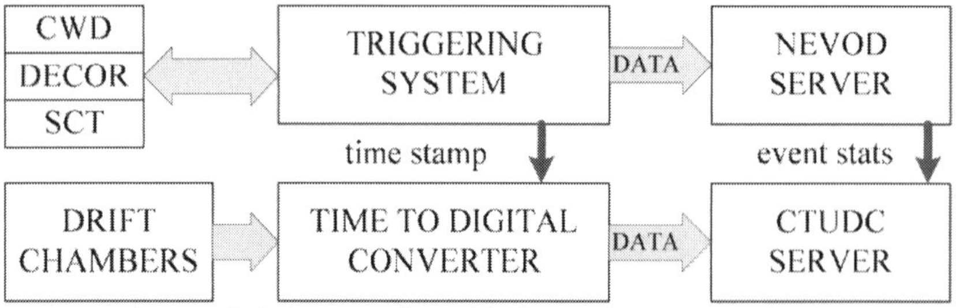

Figure 9. A scheme of a joint operation of drift chamber setup and NEVOD-DECOR.

4. Conclusion

The large-scale coordinate-tracking detector based on the drift chambers from the neutrino experiment at the IHEP accelerator U-70 is being developed in NRNU MEPhI. Detector is aimed for registration of near-horizontal muon flux as a part of experimental complex NEVOD-DECOR.

Currently, a study of the characteristics of the first batch of drift chambers obtained from IHEP after a long-term storage is being performed. For this purpose, the test stand has been developed, it will allow flow testing of the chambers before their installation into the detector.

To study the detector TREK performance, its prototype CTUDC is developed, it has the same chambers alignment and method of connection to experimental complex NEVOD-DECOR.

This work is supported by the Ministry of Education and Science of the Russian Federation (government task and project RFMEFI59114X0002), grants of the Leading Scientific School NSh-4930.2014.2 and Russian Foundation for Basic Research (project 13-02-12207-ofi-m-2013).

References

[1] Zadeba E A *et al* 2014 *Journal of Instrumentation* **9** C08018
[2] Bozhko N I *et al* 1986 *Nucl. Instr. Meth. Phys. Res. A* **243** 388
[3] Barbashina N S *et al* 2007 *Nucl. Phys. B* **165** 317
[4] Bogdanov A G *et al* 2010 *Physics of Atomic Nuclei* **73** 1852
[5] Petrukhin A A 2011 *Nucl. Phys. B (Proc. Suppl.)* **212-213** 235
[6] Kovylyaeva E A *et al* 2014 *Bull. Lebedev Phys. Inst.* **41** 218
[7] Aynutdinov V M *et al* 1998 *Astrophys. Space Sci.* **258** 105
[8] Barbashina N S *et al* 2000 *Instrum. and Experim. Tech.* **43** 743

All-sky observations with HAWC: latest results

J.C. Arteaga-Velázquez[†] **for the HAWC Collaboration**
[†]Instituto de Física y Matemáticas, Universidad Michoacana, Morelia, Mexico
E-mail: arteaga@ifm.umich.mx

Abstract. The High Altitude Water Cherenkov (HAWC) observatory is a ground-based air-shower detector designed to study cosmic rays and gamma rays with energies from 100 GeV up to 100 TeV. HAWC simultaneously surveys 2 sr of the northern sky with a high duty cycle $> 90\%$ in search for photons from point and extended sources, diffuse emission, transient events and other astrophysical phenomena at multi-TeV scales against the background of cosmic rays. In fact, the study of this background will open also the possibility of doing cosmic ray physics in the GeV – TeV regime and even to perform solar studies at HAWC. The observatory will consist of a densely packed array of 300 water Cherenkov tanks (4.5 m tall and 7.3 m diameter with 4 photomultipliers each) distributed on a 22 000 m^2 surface. Deployment started in March 2012 on a plateau situated on the Sierra Negra Volcano in the state of Puebla, Mexico, at an altitude of 4100 m. Construction is expected to be finished by the first months of 2015. In the mean time, HAWC has been taking data with a partial array and preliminary results have been already obtained. In this contribution, the results from the latest HAWC observations will be presented.

1. Introduction

Gamma-ray astronomy has received a tremendous boost with the advent of the current generation of detectors based on space-based instruments (such as AGILE [1], Fermi-LAT [2]) and ground-based observatories (e.g. HESS [3], VERITAS [4], MAGIC [5], CANGAROO [6], MILAGRO [7], Tibet-ASγ [8] and Argo-YBJ [9]) equipped with different experimental techniques that complement each other. Together they have explored the gamma-ray window starting at energies of a few MeV up to several TeV. For example, direct detection techniques on board of the Fermi-LAT space telescope have covered the low energy interval from 20 MeV to 300 MeV [2], while indirect detection techniques employed in Earth-bound detectors have extended these measurements up to the 100 TeV regime. The latter has been made possible by the commissioning of a new generation of Imaging Air Cherenkov Telescopes (IACTs), in particular HESS, VERITAS, MAGIC and CANGAROO, and air-shower particle detectors, such as MILAGRO, Tibet-ASγ and ARGO-YBJ, for the 10 GeV – 100 TeV [3, 4, 5, 6] and 100 GeV – 100 TeV [7, 8, 9] energy bands, respectively. However, despite the enormous progress that has been achieved in the last decades in this field, there are still several fundamental questions that are unsolved in gamma-ray astronomy, for instance regarding the production mechanism of the observed gamma rays in astrophysical environments, the possible existence of 100 TeV photon sources and the proposed connection between gamma-ray emitters and the origin of the cosmic rays of the highest energies.

In order to get some answers to the aforementioned enigmas, detailed observations of the 10 – 100 TeV sky with improved sensitivity and large collecting areas are needed. To accomplish this task, a whole new generation of detectors has been already devised based on existing detection techniques. Among the experimental efforts that are currently planned to lead this field in the next decades are the following: the HAWC (High Altitude Water Cherenkov) observatory [10, 11, 12], LHAASO (Large High Altitude Air Shower Observatory) [13], the Cherenkov Telescope Array (CTA) [14], the Tunka-HiScore (Hundred Square-km Cosmic Origin Explorer) detector [15] and the TAIGA (Tunka Advanced Instrument for cosmic ray physics and Gamma Astronomy) hybrid array [16]. Some of them are being designed to even increase the sensitivity of the gamma-ray measurements beyond 100 TeV [13, 14, 15, 16]. The first of these instruments of new generation to detect gamma rays is the HAWC observatory, which is the successor of the former MILAGRO detector [7].

MILAGRO pioneered the water Cherenkov technique for the detection of air showers induced by TeV photons from space [7]. The instrument consisted of a 21-million-liter pond dedicated to survey the sky at energies between 1 and 100 TeV. The operation of the detector showed that the water Cherenkov technique can be used to study the sky with wide field of views (FOV), ~ 2 sr, and duty cycles of over 90%, although with limited angular and energy resolutions, in contrast to IACT techniques.

Both, IACTs and water Cherenkov arrays use the same principle to detect air showers: the Cherenkov light emitted by relativistic charged particles of the shower in a transparent medium. However, the former detect the Cherenkov light produced by the shower along its development in the atmosphere, while the latter are limited only to the Cherenkov light emitted inside the detectors of the array during the passage of the shower.

HAWC is an extensive air shower (EAS) observatory composed of a ground-based Cherenkov array [10, 12]. The instrument is designed to study the γ and cosmic-ray sky at energies from 100 GeV to 100 TeV with a large instantaneous field of view (~ 2 sr), as well as a high duty factor ($> 95\%$). HAWC will have an improved sensitivity and angular resolution with respect to MILAGRO [17]. In addition, it will measure with a better energy resolution, a larger effective area and an increased gamma-hadron rejection [17]. These features will make HAWC an ideal detector to carry out extensive cosmic- and gamma-ray research of different phenomena and astrophysical objects. In the field of gamma-ray astronomy, with HAWC it will be possible, for instance [17, 18, 19]: to monitor the emission from extended sources, the diffuse extragalactic background and transient events; to study both point emitters and the morphology of extended sources; and to improve the spectral photon measurements into the 100 TeV regime (which will help to identify the ultimate origin of the high-energy cosmic rays). Regarding cosmic-ray research, HAWC will permit the measurement and the study of the flux of cosmic rays from our Sun and the Milky Way, and in the case of particle physics, HAWC observations will serve to put constraints on scenarios of new physics and exotic phenomena. Furthermore, there will be monitoring programs for the prompt detection of transient phenomena [17], which, in addition, will trigger monitoring campaigns of such events at complementary wavelengths. In this regard, HAWC will be part of coordinated joint efforts (e.g. GCN [20], AMON [21]) with other international observatories to carry out multi-messenger and multi-wavelength observations of a rich variety of astrophysical objects and phenomena in the universe besides transient events. In summary, the scientific potential of HAWC is vast. In this contribution, a brief review about the HAWC detector will be presented. The main objectives of the instrument, the characteristics and the status of the observatory will be described. By the end, some of the latest results obtained with HAWC, which were presented at the ECRS 2014, will be shown.

2. Science objectives

The HAWC collaboration pursues several scientific objectives in key areas of astronomy, astrophysics and particle physics. Some of the main objectives are [11, 17, 18, 19]:

- to expand the catalog of TeV gamma-ray sources;

- to extend the gamma-ray observations up to 100 TeV;
- to perform analyses of galactic and extragalactic sources;
- to study transient events;
- to carry out solar and galactic cosmic ray studies;
- to constrain scenarios for new physics;
- to initiate and participate in campaigns of multimessenger-observations.

HAWC will allow to extend the gamma-ray observations up to 100 TeV over a large fraction of the sky. This will permit to constrain and study the shape of the TeV spectra of several known GeV sources and even new emitters that might be discovered with the detector. In addition, those measurements will serve to search for high-energy cut-offs in the spectra, whose absence or presence can help to identify the mechanism of high-energy photon production, whether hadronic or leptonic, in several astrophysical gamma-ray objects.

A big variety of sources are planned to be the target for gamma-ray observations with HAWC. In our galaxy, HAWC will monitor point sources, such as the Crab, pulsar wind nebulae (PWN), supernova remnants (SNR) and binary systems; extended sources, e.g. the Cygnus Region, the galactic plane, superbubbles, young star clusters, TeV regions discovered by MILAGRO (MGRO J2019+37 and J2031+41) and the Fermi bubbles; and diffuse emission, for example, from molecular clouds and star clusters [17].

On the other hand, beyond our own galaxy, the HAWC collaboration will target the TeV photon emission from active galactic nuclei (AGNs), gamma-ray bursts (GRBs), the extragalactic background light and previously unknown gamma-ray sources [17].

The study of TeV transient phenomena in galactic and extragalactic environments, e.g. GRBs [22, 23], PWNs and AGNs, will be another priority of HAWC [17]. In this regard, HAWC observations will give an important insight into the mechanisms behind such events and the physical conditions of their sources, which are still among the unanswered questions in modern astrophysics.

In the field of cosmic ray physics, the goals of the HAWC collaboration include the search for the sources of very high-energy cosmic rays and a better understanding of the particle acceleration mechanisms in the cosmos using the TeV gamma-ray observations. In this case, gamma rays are a valuable tool, as they travel to us directly from the source, in contrast to cosmic rays, and are easier to detect than neutrinos. Another important objective for HAWC in this field is to study the distribution and propagation of galactic cosmic rays by using the measurements of the diffuse gamma-ray emission in our galaxy. On the other hand, also of interest for HAWC are the large/small-scale anisotropy measurements of high-energy cosmic rays [19, 24]. The corresponding maps contain information about the interstellar magnetic fields and possible nearby cosmic-ray sources to the Earth. Meanwhile, at lower energies, solar physics studies and the influence of the coronal mass ejections on the flux of galactic cosmic rays (Forbush effect) are also part of the cosmic-ray research agenda of the HAWC collaboration [19].

Science goals of HAWC also include other topics of investigation beyond the gamma- and cosmic-ray physics. By looking at the gamma-ray signals from the center of the Milky Way, distant galaxies and transient objects, HAWC will provide valuable information to constrain the intergalactic magnetic field [18] and several new physics models (for instance, Lorentz invariance violation (LIV), dark matter (DM) and primordial black holes (BH) scenarios [18, 25, 26]) that are deeply linked with fundamental questions in particle physics and cosmology. Even more, HAWC will look also for exotic non-relativistic DM particles, such as Q-balls, which could interact occasionally with the water of the HAWC detector producing an observable signal. For this purpose, the data acquisition (DAQ) system of the HAWC observatory has a dedicated trigger that will search for these events [18, 27].

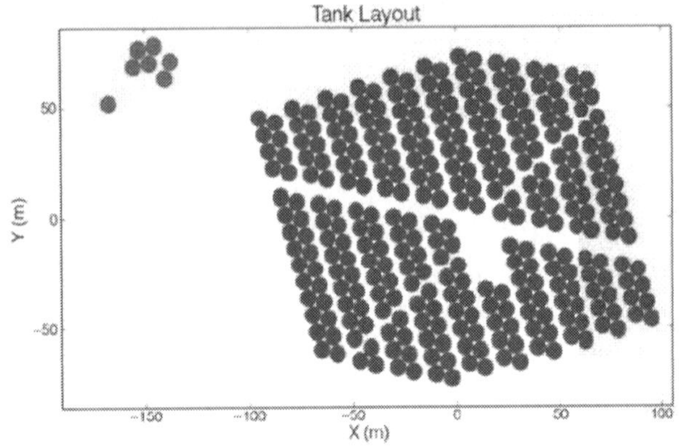

Figure 1. Left: Layout of the HAWC observatory (blue circles) and the VAMOS array with seven tanks (green circles, upper-left corner), 6 of which were instrumented. VAMOS was operated during 2011. Right: Design of of the water Cherenkov detectors of the HAWC observatory.

3. The High Altitude Water Cherenkov detector

3.1. Layout

The HAWC observatory will be a second generation gamma-ray detector using the Water Cherenkov technique that will complement observations with modern IACTs and space gamma-ray telescopes. The apparatus is currently under construction at the central part of Mexico on the northern slope of the volcano Sierra Negra (18° 59′ 41″ N, 97° 18′ 28″ W) in the state of Puebla [10, 11]. The site is located at 4100 m *a.s.l.*, which corresponds to an atmospheric depth of 640 g/cm^2. At this altitude even secondary showers with energies as low as 10 GeV can be detected with the instrument on the ground [12].

At the HAWC site the average temperature during the year is 4.3° C, and temperatures below zero degrees are observed only 5% of the time. Under these conditions, the water of the Cherenkov detectors has no risk of freezing and thus presents a homogeneous refractive index during the whole year [28]. The HAWC site is located in a seismically active zone. However, the observatory has been designed to resist earthquakes and other environmental conditions as high winds [28]. Regarding volcano risks, Sierra Negra is an extinct volcano. However, at 6 km to the Northeast of the site, there is an active volcano called Pico de Orizaba [28]. In any case, its last major activity was registered in 1545 CE and according to geophysical studies [29] the probability of a minor explosive event is 0.013 per year.

HAWC (see fig. 1, left) will consist of an array of 300 water Cherenkov detectors (WCDs) densely packed over a flat surface of 22 000 m^2 ($\approx$ 150 m $\times$ 150 m) and instrumented with 1200 upward-facing photomultipliers (PMTs) distributed at the bottom of the tanks. The large area of the detector, as compared to the deep and shallow layers of MILAGRO (which covered only 2140 m^2 and 3528 m^2, respectively), will improve the gamma-hadron separation, the acceptance for induced gamma-ray showers and the determination of the core location of EAS, which is important to achieve a better angular resolution.

For HAWC, a modular design with tanks was preferred over the single pool one of MILAGRO due to several reasons [30]. First, tanks are less expensive and present a lower cost risk/contingency than a pond. Furthermore, Cherenkov detectors with tanks are a proven technology and offers the possibility of having a flexible and expandable array. Besides, with tanks, servicing can be done without shutting down the whole detector and, even more importantly, construction and data taking can be performed simultaneously. In addition, water filling is simplified and can be done at a reduced rate. Finally, PMTs in different tanks can be optically isolated. The latter offers additional advantages for the tank design, namely, it reduces the noise rate from single muon events, as the single particles can at most illuminate a few PMTs. On the other hand, it improves the background rejection capabilities of the instrument. Moreover, it reduces redundant measurements from the same shower particles thus improving the angular resolution. In particular, scattered light that produces late tails in the PMT photon timing distributions and contributes to the noise rate in PMTs is efficiently absorbed by the tanks. And even more, by optically isolating the PMTs, it can be used a single layer of PMTs at the detector, instead of two as in MILAGRO [7], for triggering, angular reconstruction and calorimetry measurements.

In Milagro, a shallow layer of PMTs at ~ 1.5 m depth was used for triggering, shower plane reconstruction and reduction of the noise rate due to single muons. However, this shallow layer was a poor calorimeter, as it was close to the surface of the water [31]. For gamma/hadron separation, it is important to have a good calorimetric response, therefore in the Milagro design, an additional deep layer at ~ 6 m was required for calorimetry. In HAWC, the two layers are combined into a single deep one. This is possible because by using optical isolated PMTs, the problem of single muons is reduced and therefore a shallow layer is not needed.

Triggering and data acquisition is controlled from the Counting House, located at the center of the array. The building hosts, among other instruments, a laser system, which is employed for timing and charge calibration of the individual PMT channels, and an electronics facility that provides High voltage power supply for the PMTs. It also hosts the electronics and computing resources used for data taking as well as the on-line reconstruction and on-line analysis [12]. At the East edge of the array an additional facility has been installed. It is called the HAWC utility building (HUB) and contains an external platform, where the plastic liners of the WCDs are tested, and a room with a water filtration plant. In the latter, the water that is going to be used for the WCDs is first treated with a set of filters of 10 μm, charcoal, 5 μm and 1 μm to remove impurities and contaminants. In addition, the water is sterilized by exposition to a UV light source. [28]. The purified water is finally stored in a series of tanks located next to the HUB, from which it is later distributed to the WCDs. Water quality has been checked by measuring the attenuation length at 405 nm from 1 m water samples taken from 26 tanks. The results varied between 5 m and 16 m, with an average over all tanks of 10.4 m [32].

Each HAWC water Cherenkov detector is composed of a galvanized steel tank (7.3 m diameter and 5 m height) lined with a black PVC bag or bladder filled to ~ 4.5 m height (188 000 lt) with pure water [28]. The depth of the WCD was selected as a compromise between timing resolution and gamma-hadron separation [33]. In addition, four upward facing PMTs are deployed at the bottom of the tank, three 8-inch Hamamatsu PMTs on a triangle, 6 ft from the center of the tank; and a 10-inch Hamamatsu high quantum efficiency PMT at the center, which improves the detection efficiency of the observatory to low energy air-shower events [12]. Finally, on the top of the tank there is a dome cover (5 ft height) that protects the WCD from the weather and environment [28]. In regard to the steel structures of the WCDs, it is worth mentioning that they are buried 60 cm in the ground to provide a natural anchor for stability and earthquake certification [12]. Once installed, the WCDs will cover 62% of the total physical area [11]. Gaps are being left among the WCDs. They are being reserved to carry out individual maintenance tasks [12].

The DAQ of the HAWC detector is composed of two systems [22]: The main DAQ, which registers the deposited charge and relative arrival times (with ns precision) of the PMT signals generated by the passage of the EAS, and the scaler DAQ, which is used to determine the count rate (inside a window time of 10 ms) of each PMT. The data of the main DAQ are used to

Figure 2. Picture of the HAWC array. The image was taken on December 15th, 2014 after the installation of the steel frame for the 300th WCD of the observatory.

reconstruct the various air-showers observables, including the core location, the arrival direction, the energy and other useful parameters that are employed to discriminate between signal and background events [12]. The data of the scaler DAQ, on the other hand, is complementary to that of the main DAQ, for it provides count rate information on the PMTs, which is useful to monitor the good performance of the detector and is sensitive to transient events – such as gamma-ray bursts (GRBs) and solar events accompanied by the emission of energetic particles. These signals will be observed as statistically significant changes in the scaler rate [22, 23].

3.2. Status

In May 2010, the installation of HAWC's engineering array, called VAMOS (Verification And Measuring of Observatory System) began. It was built about 170 m Northwest from the center of the current HAWC array [28].

VAMOS consisted of an array of 6 WCDs, instrumented with a total of 36 PMTs, and distributed inside a $25\,\mathrm{m} \times 25\,\mathrm{m}$ surface (c.f. figure 1, left). It was fully deployed in June 2011 and continuous DAQ was carried out from October 2011 to May 2012 ($\sim 30\%$ duty cycle). VAMOS data was useful to check the behavior of the HAWC PMTs under different environmental conditions and to perform some first γ- and cosmic-ray analyses planned for the full HAWC detector (in particular, searches from GRB signals and Forbush decrease studies) [17, 19, 28]. In addition, the experience gained by building and operating VAMOS served to verify and improve the design of the HAWC array and the logistic of the HAWC construction [28].

The building of the HAWC observatory started in March 2012. Its modular design has allowed to take air-shower data while in construction. In summer of 2012, the first 30 full WCDs were installed. HAWC-30, as it was called, started operations by August 2012 and results were presented in [12, 17, 19]. HAWC-95 began data-taking in June 2013, while the full operation phase officially started in September 2013 with HAWC-111. By the end of November 2014, HAWC-250 was completely installed, and in mid-December of 2014, there were already 300 tanks deployed on the field (see fig. 2), with 251 WCDs participating in the DAQ. The full HAWC-300 array will be completed at the beginning of 2015. The observatory is planned to be inaugurated in March of that year. The construction and operation of HAWC has been in charge of the HAWC collaboration, which is composed by several scientists, engineers, postdocs, technicians and students from around 29 institutions in Mexico and the US. Before finishing this section, it should be mentioned that the HAWC results presented in this conference were based

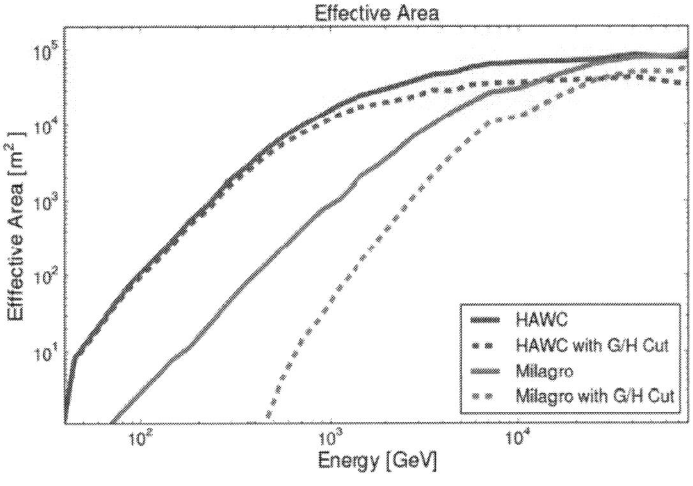

Figure 3. Expected effective area of the HAWC-300 detector to gamma rays as a function of energy. The curves were calculated with and without gamma-hadron (G/H) cuts to remove cosmic ray events. The graphs for the corresponding plots of the MILAGRO detector are also shown for comparison [10, 17].

on analyses performed with the data from the HAWC-30, HAWC-95 and HAWC-111 partial detectors.

3.3. Effective area, resolution and gamma-hadron separation

The sensitivity and the performance of the HAWC observatory have been studied with Monte Carlo (MC) simulations. The production and evolution of the cosmic- and gamma-ray induced EAS have been simulated using the CORSIKA v6990 program [34] along with the low- and high-energy hadronic interaction models FLUKA 2011 [35] and QGSJET-II [36]. Meanwhile, the detector and its response to the passage of the particle showers have been modeled using GEANT4 [37].

In fig. 3, the effective area of the full HAWC instrument to gamma rays is presented for different primary energies. Below 1 TeV, it rises with the energy, and at higher energies the effective area becomes practically independent of the energy ($\sim 10^5\,\mathrm{m}^2$). The former, is the result of the increasing probability of a primary particle of producing a detectable shower at the HAWC detector, while the latter, is a consequence of the limited accuracy of the reconstruction of EAS events falling outside the geometrical area of the HAWC array [17].

At energies below 1 TeV, the HAWC instrument will have a bigger effective area than its predecessor, the MILAGRO detector, in particular, at lower energies, i.e. $\mathcal{O}(10)$ GeV, such area will increase the statistics and sensitivity of the HAWC instrument, lowering its energy threshold below that achieved by the MILAGRO detector (~ 250 GeV) [38]. The larger effective area observed in HAWC at low energies results, in the first place, because the altitude of the site is higher than that for the MILAGRO location ($\sim 750\,\mathrm{g/cm}^2$), i.e. HAWC is closer to the EAS maximum than MILAGRO, and, in the second place, due to the extra 10-in high-gain PMTs installed in the HAWC detector, which improve the low-energy response of the instrument.

HAWC will have a much better energy resolution than MILAGRO [17], which will allow to get precise TeV spectra of gamma-ray sources to study the origin of their high-energy emission. At low energies, the energy resolution will be 100%, but it will improve at high energies, reaching values below 50% above $E \geq 10$ TeV (c.f. fig. 4). This energy resolution will imply a significant

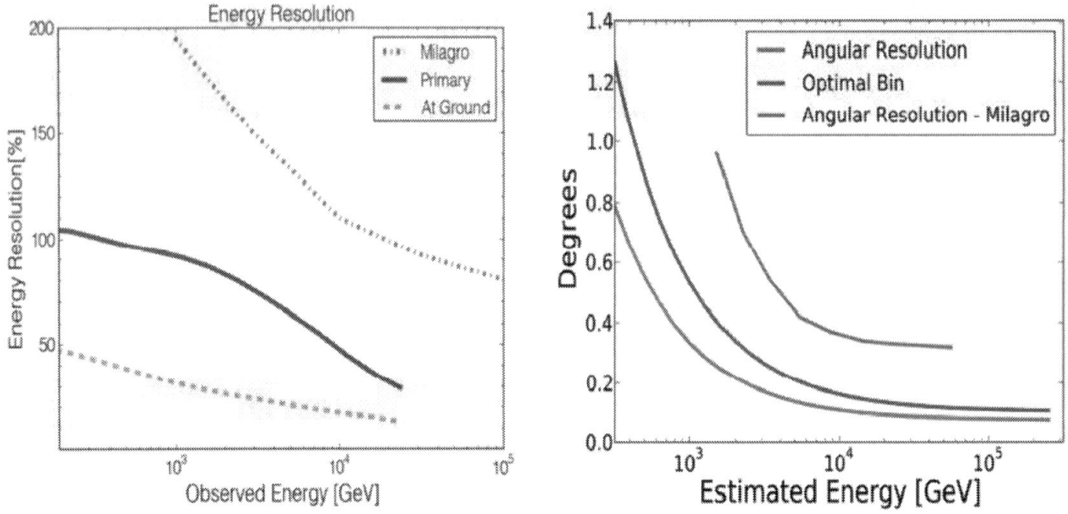

Figure 4. HAWC and MILAGRO energy (left) and angular (right) resolutions as a function of the primary energy [17]. On the right figure, the size of the optimal angular bin that maximizes the significance for HAWC-300 of a hypothetical point source at different energies is also presented. On the left figure, the continuous blue line indicates the energy resolution of HAWC including EAS longitudinal fluctuations, while the segmented blue line represents the corresponding energy resolution taking into account only detector effects.

improvement over that of first-generation Cherenkov detectors.

The angular resolution ranges from about a degree at the low energy threshold to $\sim 0.1°$ above 10 TeV. The improved angular resolution will allow more precise studies of the gamma-ray emission of point sources and the morphology of extended ones. The MC predictions for the angular resolution of the HAWC detector have been verified experimentally at least for a partial array, in particular for HAWC-30, using the measured width of the cosmic-ray Moon shadow [19]. In this case, at median energies of 2.9 TeV, it was found that the angular resolution of the detector is 1.2°, which is in good agreement with MC simulations. This value was confirmed by another experimental test performed with the same instrument by comparing independent estimations for the arrival angle of EAS from two equal-area sub-arrays composed of different PMTs from the array (in this case, the angular resolution was $1.3° \pm 0.2°$) [19].

In order to separate the gamma-ray signals from the continuous and dominant background of cosmic ray events, cuts must be applied at HAWC. In particular, rejection of cosmic ray events relies on a parameter called the compactness [39], which measures the clumpiness of the shower signals deposited at the WCDs. This compactness is a muon-content dependent quantity defined as the ratio between the total number of PMTs activated by the shower and the number of photoelectrons recorded in the largest hit channel outside a radius of 40 m from the EAS core. In general, gamma-ray induced showers are composed mostly by electromagnetic particles (photons and electrons), which produce a smooth and compact distribution of deposited charge at ground level, while hadronic EAS created by cosmic ray primaries have also a muonic component that creates a clumpy pattern of signals at the array with large charge depositions at large distances from the shower core (see fig. 5). Therefore, gamma-ray showers will be characterized by a larger compactness value than cosmic ray EAS. HAWC exploits this difference, as well as other variables, to separate the background of cosmic rays from the gamma-ray signals [10, 17, 22]. The expected efficiencies of HAWC-300 (see [10]) to photon and cosmic-ray events after applying

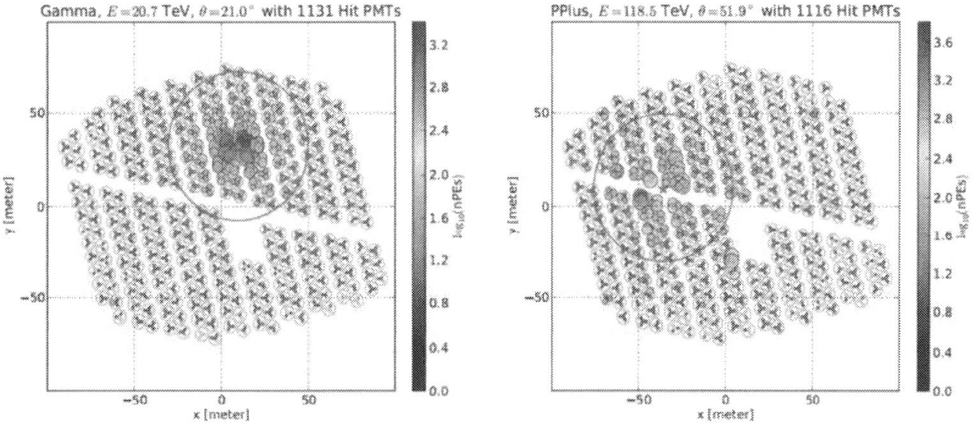

Figure 5. Deposited charge distributions at the WCDs of HAWC-300 for two simulated EAS events. Left: A gamma ray with a primary energy of 2.7 TeV and arrival zenith angle $\theta = 21°$ was used as a primary particle. Right: The initial particle is a cosmic ray proton of energy $E = 118.5$ TeV and $\theta = 51.9°$. Note that the photon-induced EAS has a more compact energy deposition pattern and a less clumpy signal structure at ground than the proton-induced shower.

the compactness cut and a cut on the angular distance of events to a Crab-like point source are shown in fig. 6 (left). From this figure, it is seen that the hadron rejection improves with increasing energy. In particular, at $10^{2.5}$ GeV, if events are required to have a compactness greater than 3.1, one retains $\sim 80\%$ of the gamma ray events and only $\sim 50\%$ of the hadron events, yielding a quality factor, Q, of 1.13. Meanwhile, at 10 TeV, if all events with compactness ≤ 14.4 are removed, 60% of the gamma ray events and only $\sim 0.45\%$ of the hadron events are retained. This results in a Q value of ~ 8.9. Milagro was only able to achieve Q factors around 1.5−2.0 [7]. Above 10 TeV, the gamma-hadron separation in HAWC is efficient enough to remove all the simulated hadronic events, so the curve presented here is limited by MC statistics.

3.4. Sensitivity to steady sources

The sensitivity of the full HAWC detector to point sources with power-law spectra of the form

$$dN/dE = \Phi_0 (E/\mathrm{TeV})^{-\alpha} e^{-(E/E_{cut})} \quad (1)$$

has been studied using MC simulations [10]. The results have shown that the HAWC-300 observatory will be 15 times more sensitive than MILAGRO, i.e. HAWC will survey the MILAGRO sky 225 times faster at the same significance. Above photon energies of 2 TeV, HAWC will have a sensitivity, at a $5\,\sigma$ level, down to $5 \times 10^{-13}\,\mathrm{cm}^{-2}\,\mathrm{s}^{-1}$($\sim 50$ mCrab) to point sources in a year of data taking over 5 sr (40%) of the sky [10]. In fig. 7, the expected sensitivity of the full HAWC instrument is shown as a function of the energy and in comparison with other gamma-ray detectors. As it can be seen from the abovementioned plot, at 1 TeV, the sensitivity of HAWC in one year of data taking will reach a value equivalent to 50 hr of observation with modern IACTs, while at energies between 10 and 100 TeV, the HAWC observatory will achieve an unprecedented sensitivity in comparison with existing space-telescopes and IACTs. The latter as a result of the altitude, large area and high duty cycle of the HAWC observatory.

The sensitivity of the HAWC detector depends also on the declination of the source and due to the geometrical latitude of the observatory, it is better (< 50 mCrab) for declination angles between $-5°$ and $45°$ [10] (see fig. 6, right). The expected sensitivity of the HAWC observatory

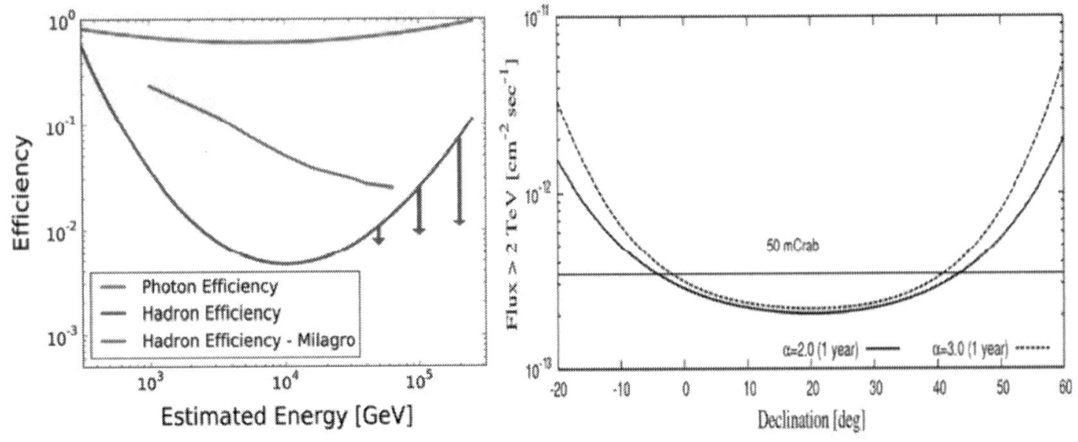

Figure 6. Left: Efficiency of HAWC to photon and cosmic-ray induced showers after applying the compactness cut [10]. The MILAGRO efficiency to cosmic ray events is also shown for comparison. Above 10 TeV, the gamma-hadron separation it is so efficient that removes all the simulated hadronic events. Right: Sensitivity of HAWC to a point source with a power-law spectrum like in eq. 1 presented as a function of declination for two different spectral indexes. The flux presented is the one required to have a $5\,\sigma$ signal in one year [10].

to a punctual source with a power-law spectrum of the form E^{-2} is presented in fig. 8. HAWC, after one year of data taking, will be sensitive to integral power-law spectra as low as 50 mCrab above 2 TeV over 5 sr ($\sim 40\%$) of the whole sky. For HAWC FOV, most celestial observations are limited to objects within 45° of the local zenith angle, however, measurements at larger zenith angle are possible, for example, of the galactic center ($\theta = 46°$), although with less sensitivity.

Observations of TeV gamma rays with the HESS telescope have shown that most Galactic sources in the TeV range are extended [40]. This could be due to the interactions of cosmic rays with material in the extended vicinity of their acceleration sites. Unfortunately, while IACT facilities, such as HESS, are quite sensitive, it is difficult for such narrow-field instruments to observe the full extent of large sources. In fact, when the source size is much larger than the point spread function of the IACTs, the sensitivity of the telescopes worsens because the background increases. Current observations of extended sources with HESS are clustered around the limit of their sensitivity, suggesting the presence of even more extended sources than currently claimed. In particular, nearby sources may have a larger angular extent than currently reported by IACTs. HAWC will be better suited than IACTs to find out this question due to its large field of view. This feature will allow, in turn, to perform also spectral studies of extended sources with HAWC. According to MC simulations [10], the expected sensitivity (integral flux > 2 TeV) of HAWC to an extended source with a pure power-law spectrum ($\alpha = 2, 3$) and a radius of $R_s = 0.5°$ will be around $5 \times 10^{-13}\,\mathrm{cm^{-2}\,s^{-1}}$, and for sources with $R_s = 3°$, it is predicted to be of the order of $2 \times 10^{-12}\,\mathrm{cm^{-2}\,s^{-1}}$.

Subtraction of the gamma-ray emission due to point and extended sources from the total flux of the galactic plane will allow to constrain better the truly diffuse emission at the galaxy [17]. The 100 GeV – 100 TeV diffuse emission measurements with HAWC will be an excellent tool to probe acceleration, propagation and density distribution of cosmic rays at different locations within our Galaxy. The HAWC site is close to the equator and the field of view covers the inner Galaxy all the way to the Galactic center. Using observations from HAWC, nearby regions such as Cygnus at $1-2$ kpc as well as the more distant inner galaxy (about 10 kpc) will be studied. The Cygnus region could be dominated by very few cosmic-ray accelerators whereas the cosmic

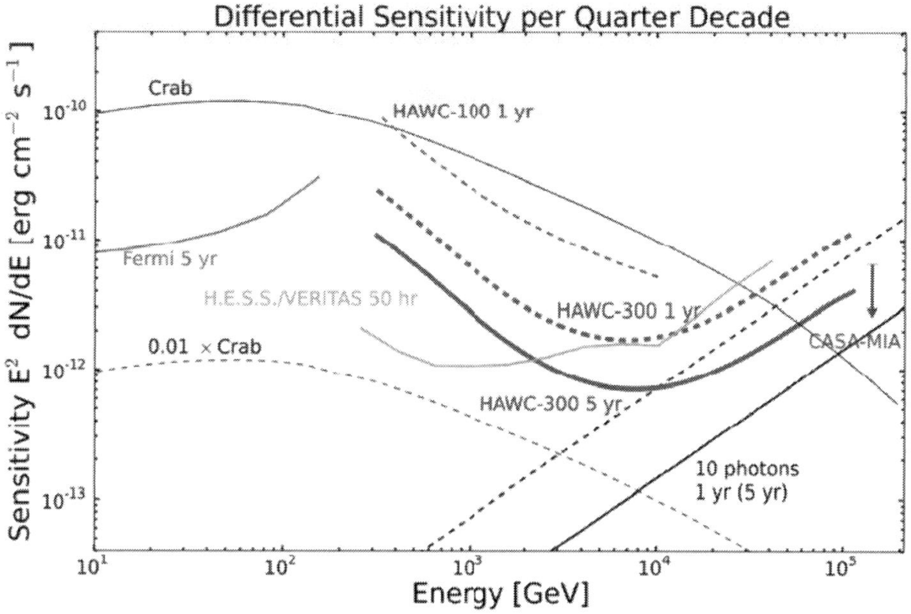

Figure 7. Differential sensitivities per quarter decade of HAWC-100 (1 year) and HAWC-300 (1 and 5 years) [10]. The corresponding sensitivities of Fermi-LAT [41], for 5-years, and two IACTs [42, 43, 44], for 50 hours, are shown for comparison. An upper limit on the Crab flux at 141 TeV derived from the CASA-MIA detector [45] is also displayed together with the measured Crab flux from [46].

rays from the inner Galaxy are from a large collection of sources and will reflect the cosmic-ray spectrum after propagation farther from their origins. These regions are hundreds of square degrees and require the large field of view of HAWC. With its wide FOV, HAWC will be able to study not only the diffuse emission from large areas along the galactic plane, but also from other structures such as the Fermi bubbles [17].

The wide field of view and high duty-cycle of HAWC constitute major advantages for the instrument during the search and monitoring of the evolution of transient events across the sky. With HAWC low-energy threshold, transient phenomena with a minimum flux of $E^2dN/dE \sim 4\times10^{-5}\,\mathrm{erg\,cm^{-2}\,s^{-1}}$ at 10 GeV, power-law spectra proportional to $E^{-1.6}$, viewing zenith angle of 20°, and duration of 10^{-2} s will be detectable at the $5\,\sigma$ level up to a redshift of $z \sim 1$ by the scaler and main DAQ systems [22]. Transients with longer duration and closer events will be detectable at HAWC at much lower fluxes. Specific details on the sensitivity of HAWC to GRBs and AGN flares can be found in [22, 23] and [47], respectively. Here, it will be only mentioned that, in case of GRBs, calculations performed in [23], and based on observations made by Fermi-LAT and other observatories, show that HAWC will be more sensitive to the short-hard variety of GRBs than to the long-soft ones. In addition, the same estimations indicate that 90% of the short-hard GRBs visible at $5\,\sigma$ significance to HAWC main (scaler) DAQ system will have a fluence above $3.5\times10^{-7}\,\mathrm{erg\,cm^{-2}}(1.7\times10^{-6}\,\mathrm{erg\,cm^{-2}})$. On the other hand, regarding transient phenomena in AGNs, it was shown in [47] that HAWC will be able to detect highly flaring events (e.g. $\times10$ quiescent state) within a fraction of a day at a $5\,\sigma$ level from Mrk421, Mrk501 and M87, which are within the FOV of the observatory. In particular, an intense outburst with an integral flux of 4 Crab above 1 TeV from these AGNs will be detected within

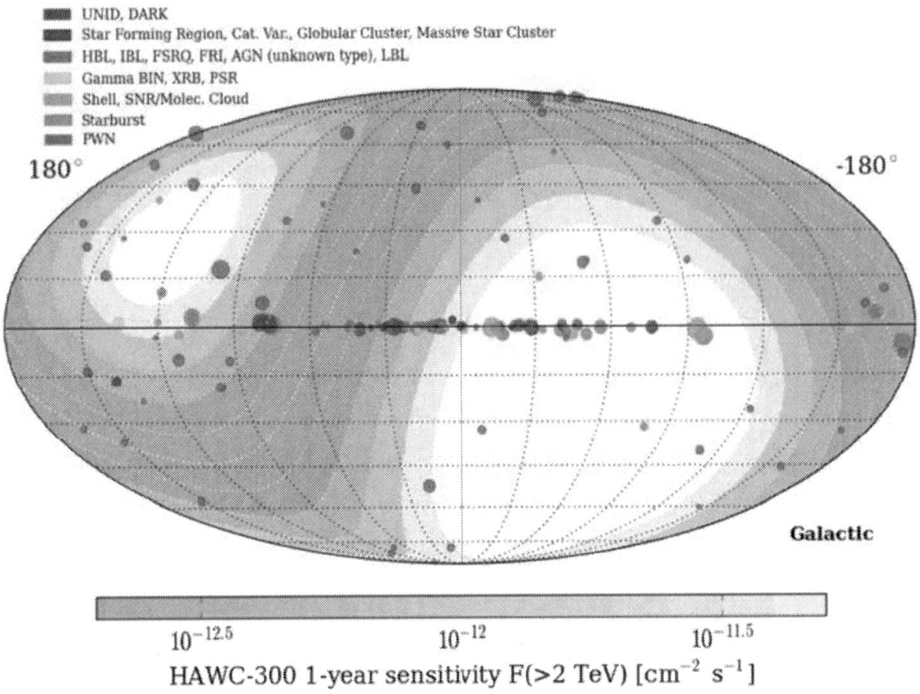

Figure 8. Sky map with the sensitivity of HAWC-300 to a point source characterized by a differential E^{-2} spectrum. The map is shown in galactic coordinates [10]. Superimposed are known sources of very high-energy from the TeV catalog.

hours.

3.5. Sensitivity to exotic physics

The sensitivity of HAWC to new physics has been also investigated. In [25], the potential of HAWC to detect high-energy photon signals due to DM self-annihilation processes in the galactic center and extragalactic systems (i.e. the Virgo cluster, the M31 galaxy and the dwarf galaxies Draco, Coma Berenices and Segue 1) was studied by means of MC simulations. A variety of weakly interacting massive particles (WIMPs) annihilation channels (into $b\bar{b}, t\bar{t}, \mu^+\mu^-, \tau^+\tau^-, W^+W^-$) with different DM density profiles were analysed. For the galactic center, both the Einasto [48] and the Navarro-Frenk-White (NFW) [49] models were employed, whereas for the extragalactic systems only one model was assumed (NFW for most of them, but for Segue 1, where Einasto was used). 95% expected limits on the $\langle\sigma v\rangle$ as a function of the dark matter mass (M_χ) were obtained for HAWC-300 assuming that no DM signal is observed above the background in five years of observation time. In fig. 9, as an example, some predicted DM limits are presented for the channel $\chi\chi \rightarrow \tau^+\tau^-$. For M_χ in the interval from $\mathcal{O}(1)$ TeV up to 10^3 TeV, the analyses performed predict that HAWC will be sensitive to cross sections as low as $\langle\sigma v\rangle \approx 3\times10^{-24}\,\mathrm{cm}^3\,\mathrm{s}^{-1}$, depending on the dark matter mass and annihilation channel.

From the dwarf galaxies studied, HAWC will have a better sensitivity to Segue 1, thanks to its strong expected DM annihilation flux and a favorable declination [25]. Other good candidates, in comparison with Segue 1, are the Virgo Cluster under the presence of dark matter substructures in the outlying regions of the halo (which could boost the gamma-ray flux at the source) and M31 with/without boost due to DM sub-halos [25]. The limits for the galactic center, on the other

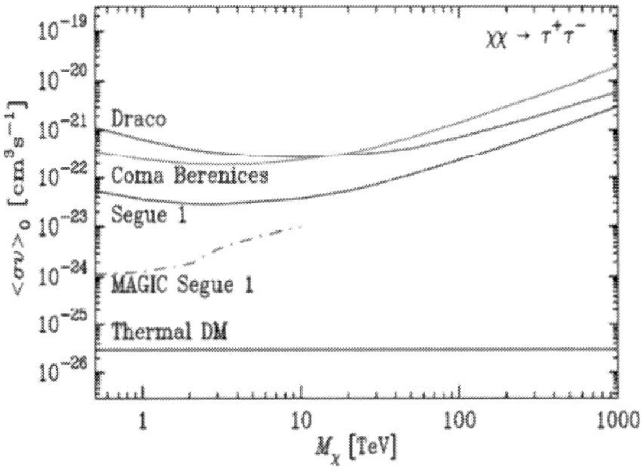

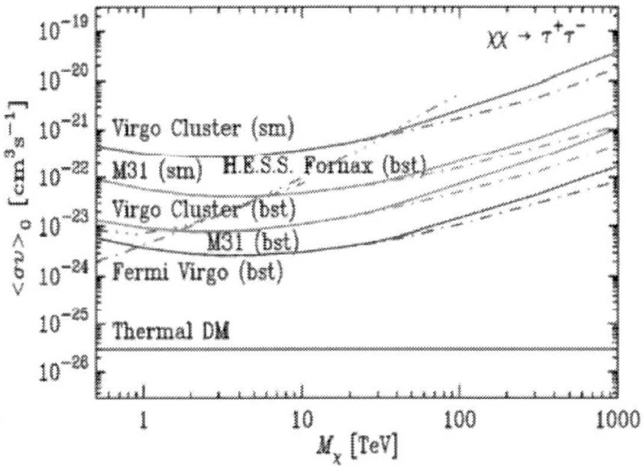

Figure 9. Expected dark matter limits (95% C.L.) from the observation of dwarf galaxies (left), the Virgo cluster and M31 (right) with HAWC-300 after five years of data taking for the $\tau^+\tau^-$ annihilation channel [25]. The limits were estimated from simulations. Solid lines were calculated for prompt emission (photon radiation from secondary particle decays and charged particles), while the dot-dashed lines include a conservative inverse-Compton component as well (from the scattering of secondary charged particles off low energy background photons). In case of M31 and the Virgo cluster, calculations were carry out assuming a smooth density profile (sm) and a boosted signal from DM substructure in their respective halos (bst). A limit from the MAGIC telescope for Segue 1 [50] (purple dot-dashed curve) is presented in the left figure. On the other hand, two limits from HESS [51] (triple-dot-dashed purple line) and Fermi-LAT [52] (dot-dashed purple line) for the Fornax and Virgo clusters are plotted, respectively, on the right figure. The cross section value for a relic thermal DM particle is also shown and is represented by the lower horizontal line.

hand, are expected to be better than Segue 1 due to a higher photon flux and its proximity to the Earth. However, because of the larger zenith angle at which the galactic center is observed from the HAWC site, those limits will be less competitive than other instruments, such as HESS, except at high M_χ (above 10 TeV, HAWC should be sensitive to DM cross sections down to $\langle\sigma v\rangle \approx 5\times10^{-24}\,\mathrm{cm^3\,s^{-1}}$) [25].

The sensitivity of HAWC-300 to other new physics scenarios, for example Q-balls and primordial black holes, are studied in references [18, 27] and [26], respectively.

4. Latest results with HAWC

4.1. Sky map

A preliminary significance sky map in gamma rays was obtained with 154 days of live time using data from the HAWC-95/111 array, whose layout is displayed in fig. 10 (left). The sky map is shown in fig. 12. To generate the map, a subset of the whole HAWC-95/111 data was reconstructed online during data taking using preliminary calibration and pointing uncertainties. The direct integration technique [53], with $\Delta t \sim 2\,\mathrm{hr}$ of integration time, was used to create the significance map. The analysis was partly optimized on the signal of the Crab Nebula, which is observed at $> 10\,\sigma$ (c.f fig. 11, left). Other signals are seen at more than $5\,\sigma$ in coincident positions with the galactic plane and extragalactic objects. In particular, multiple regions are detected along the inner region of the galactic plane. This region of the Milky Way contains different kinds of astrophysical sources, such as molecular clouds, stellar clusters, pulsars, SNRs, etc., which may be responsible for the observed emission. At the moment, ongoing analyses are

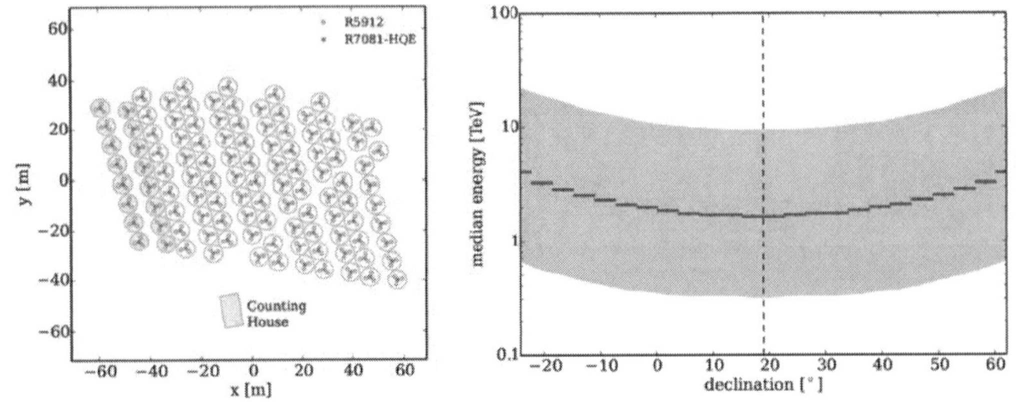

Figure 10. Left: Layouts of the HAWC-95 array (represented by the big white circles) and the HAWC-111 detector (indicated by the white and gray circles) [54]. The PMTs of low- and high-gain are represented by the open circles and the filled circles, respectively. Right: Estimated energy resolution as a function of declination for the HAWC-95/111 data used in the study of the Moon shadow and the small-scale anisotropies of cosmic rays [54].

underway to try to elucidate this question.

On the other hand, $> 7\sigma$ and $\sim 6\sigma$ excess emissions are detected from the directions of Markarian 421 (Mrk421) and Markarian 501 (Mrk501), respectively. These two objects are BL Lac galaxies with proven gamma-ray emission (see, for example, [55, 56] and references therein). Mrk421 is one of the closest blazars to Earth ($\sim$ 130 Mpc) and was the first extragalactic object detected at very high energies (VHE) [57]. The second one to be discovered at VHE

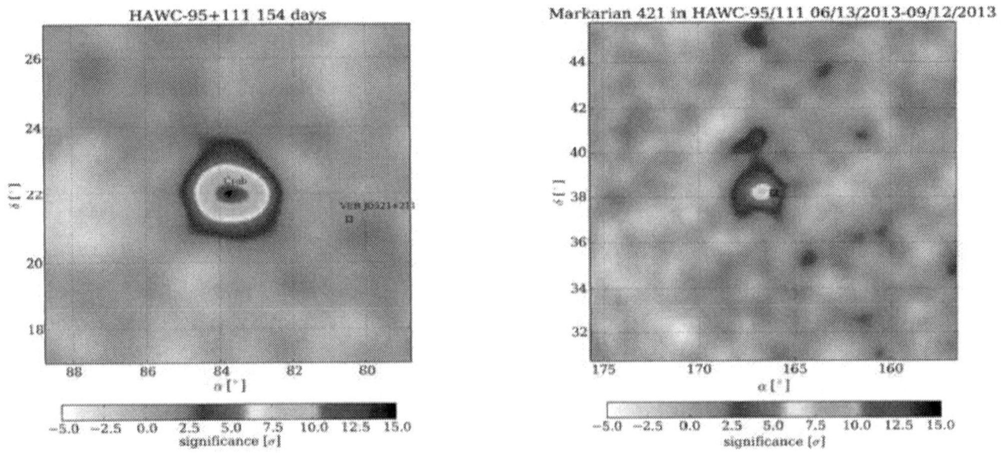

Figure 11. Left: Preliminary significance map in the direction of the Crab nebula calculated from the gamma-ray data of HAWC-95/111 with 154 days of live time. Right: Preliminary significance map obtained on Markarian 421 with HAWC-95/111 and data from the period June 13, 2013 to September 12, 2013.

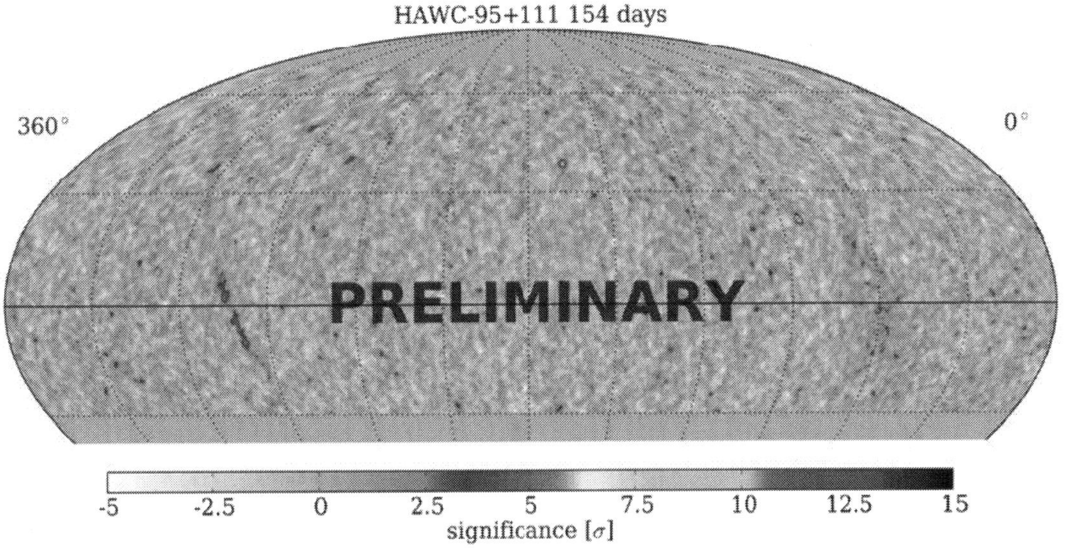

Figure 12. Preliminary sky map (in equatorial coordinates) of 154 days of data with $\sim 1/3$ of the HAWC array. The analysis is optimized on the Crab Nebula. The significance of the Crab Nebula is $> 10\,\sigma$. The galactic plane is clearly visible. An extended excess near right ascension $60°$ is due to cosmic rays (region A in fig. 14). The analysis is still under development to improve the gamma-hadron separation [59].

was Mrk501 [58], which is one of the brightest TeV sources in the sky. In fig. 11, right, a preliminary significance map from the region of Mrk421 as obtained with HAWC-95/111 is displayed. Currently Mrk421 and Mrk501, along with other extragalactic TeV sources, are the target of TeV flare searches with the HAWC array.

Finally, an extended excess near right ascension $60°$ is also observed in the sky map. This is produced by cosmic rays (region A in fig. 14). The main goal of the HAWC observatory is the investigation of VHE photons; however, cosmic ray events are also interesting for several aspects, for example, to study the cosmic ray anisotropies or to validate the detector characteristics, in terms of pointing accuracy and angular resolution, as will be seen below.

4.2. Cosmic ray studies

4.2.1. The shadows of the Moon and the Sun The Moon and the Sun cause a local deficit on the isotropic flux of cosmic rays observed at Earth, what are called the Moon and the Sun shadows, respectively. With cosmic ray data collected with a partial array of HAWC, both the shadows of the Sun and the Moon have been observed. In particular, the Moon shadow has been seen with a significance of $-23.8\,\sigma$ in HAWC-95/111 [54]. The shadow of the Moon is observed on the left side of fig. 13 for 113 sidereal days of live time. The plot is presented in moon equatorial coordinates and it is centered at the position of the Moon. A fit to the deficit region with a two-dimensional Gaussian reveals that the Moon shadow has an offset of $-1.05° \pm 0.05°$ in right ascension and of $-0.02° \pm 0.06°$ in declination, and a width of $1.26° \pm 0.05°$. The shift is the result of the deviation of charged cosmic rays due to the Earth's magnetic field and it is

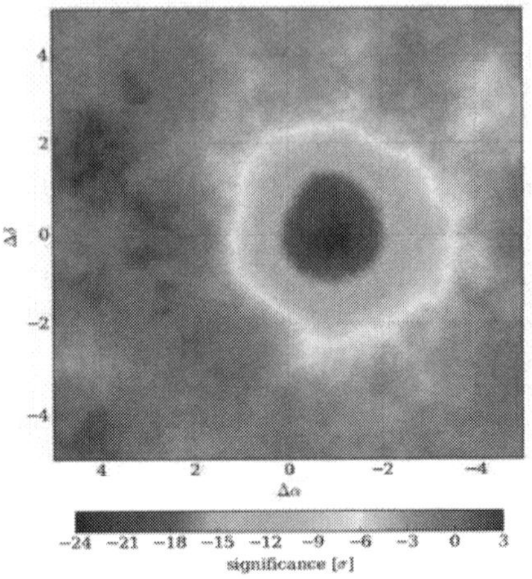

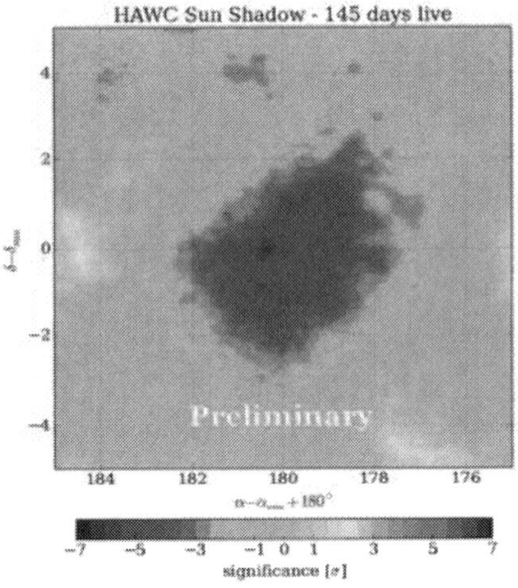

Figure 13. Left: The Moon shadow in a significance sky map of the cosmic ray flux for 113 days of data with HAWC-95/111 [54] for an integration time of $\Delta t \sim 2\,\text{hr}$. $\Delta\alpha$ and $\Delta\delta$ represent the right ascension and declination of cosmic rays with respect to the corresponding coordinates of the Moon, respectively. Right: A preliminary sky map with the shadow of the Sun obtained with 145 days of live time using HAWC-95/111 (time-integration window $\Delta t \sim 2\,\text{hr}$). The map is centered at the position of the Sun.

consistent with expectations from MC simulations. From the observed shift of the Moon shadow and simulations, a median energy of 1.6 TeV is estimated for the measured data sample [54]. The data set consisted of 48×10^9 selected events and was registered in the period from June 16, 2013 to February 27, 2014.

A preliminary sky map with the Sun shadow is displayed on the right side of fig. 13 for a live time of 145 days. The plot was obtained with 8.5×10^9 selected events collected with HAWC-95/111. The data corresponds to the DAQ period from October 22, 2012 to May 24, 2013. In this map, data is shown in equatorial coordinates with the origin at the position of the Sun. The sun shadow is observed at the sky map with a significance of $-6.4\,\sigma$, although with a shifted position relative to that of the Sun. In this case, in addition to the shift caused by the effect of the geomagnetic field on the trajectory of the charged cosmic rays, there is also a contribution from the coronal magnetic field of the Sun [60].

4.2.2. The small-scale anisotropies Using data from the HAWC-95/111 detector, small-scale anisotropies at the level of $\sim 10^{-4}$ on angular scales of $\sim 10°$ have been observed in the distribution of TeV cosmic rays [54]. The analysis was based on 49×10^9 well reconstructed events taken from June 16 of 2013, to February 27 of 2014. According to MC simulations, the whole data set has a median energy of 2 TeV. It should be mentioned that, in general, this quantity depends on the declination, as it can be seen in the right graph of figure 10, which was calculated with MC simulations [54].

HAWC's sky map for the relative intensity of cosmic rays is presented in the upper panel of figure 14 in equatorial coordinates. The map was obtained using the direct integration method [53], with a smoothing radius of 10° and integration time of 24 hr, and by removing large-scale structures greater than 60°. In the lower part of figure 14, the respective map for the significance of the deviations of the cosmic ray flux from isotropy is also shown. In those maps,

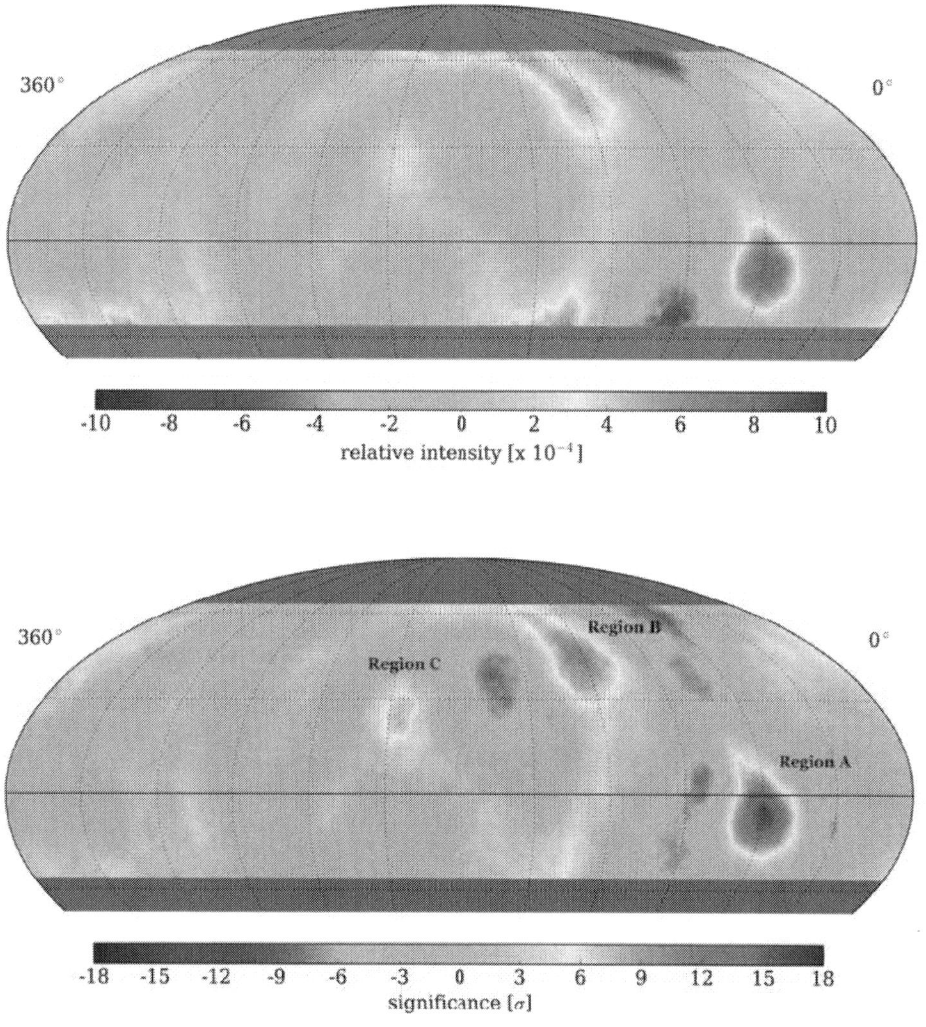

Figure 14. Top: Relative intensity (top) and significance (bottom) sky maps of the small scale cosmic ray anisotropy measured with HAWC-95/111; using an integration time of $\Delta t = 24\,\mathrm{hr}$ with a smoothing radius of $10°$ [54].

three prominent features are observed [54]. They are called regions A, B and C, respectively. The strongest excess comes from region A, in particular, from the coordinates $\alpha = 57.7°$ and $\delta = -6.3°$. This excess has a significance of $17\,\sigma$ and a relative intensity of $(8.5\pm1.0)\times10^{-4}$ [54]. According to a two-dimensional Gaussian fit, the center of Region A is located at $(\alpha, \delta) = (60.0° \pm 0.7°, -7.1° \pm 0.8°)$ and has a width of $(\Delta\alpha, \Delta\delta) = (7.1° \pm 1.3°, 7.8° \pm 1.3°)$. This region has already been observed by ARGO-YBJ [61] and MILAGRO [62], although, in the latter case, the magnitudes and locations of the peaks differ from the ones measured at HAWC, presumably due to differences in the median energies of the respective observatories.

The region B in the sky map of fig. 14 exhibits an elongated shape, which extends around $\alpha = 120°$ along all the declination range covered by HAWC, as it can be observed also in the relative intensity map in fig. 14. It has been also observed by ARGO-YBJ [61] and MILAGRO [24]. This structure seems to continue into southern celestial regions, where it could be connected with an excess observed by ICECUBE [63], however, at a slightly different right ascension ($\alpha = 122.4°$). The significance of region B has a maximum ($11.2\,\sigma$) from the direction of $(\alpha, \delta) = (122.1°, 43.8°)$, where the relative intensity has a value of $(5.2 \pm 0.9) \times 10^{-4}$.

Finally, the region C has a significance of $8.2\,\sigma$ and a relative intensity of $(2.9 \pm 0.6) \times 10^{-4}$. It is centered at $\alpha = 205.7°$ and $\delta = 22.5°$. This structure was observed by the ARGO-YBJ Collaboration, who first reported it [61]. On the contrary, at this position, MILAGRO did not observed any significant excess.

The origin of the small-scale anisotropies is not clear yet. Explanations range from conventional astrophysical explanations, such as turbulence in the interstellar magnetic field [64] and effects of the heliosphere [65], up to exotic scenarios, for instance, dark matter self-annihilations in galactic sub-halos close to the Sun [66].

4.3. GRBs limits

On April 27, 2013, during the operation of the HAWC-30 array, Fermi-LAT observed an exceptionally bright GRB [67]. GRB 130427A, as it was named, had the longest gamma-ray duration (20 hr) at $E > 100$ MeV and the highest energetic photon ever detected, with an energy of 95 GeV [67]. The burst occurred at $T_0 = $ 7:47:06:42 UTC. Unfortunately for HAWC-30, at that time two unfavorable circumstances happened, the first one is that due to maintenance the main DAQ system was turned off and only the scaler DAQ system was taking data, and the second one, that the GRB was at an unfavorable zenith angle of 57° in local coordinates, for which the effective area of the scaler DAQ system of HAWC is reduced by more than 2 orders of magnitude with respect to the zenith (c.f. fig. 15, left). This chain of events reduced any chance of detection of the GRB 130427A with HAWC-30, however, upper limits on the integral gamma-ray flux of the event were established for the interval 0.5 GeV – 1 TeV. Two of these limits are shown in the right plot of fig. 15. They were derived from an analysis of the data inside a time window of $T_0 + 11.5$ s and $T_0 + 33$ s assuming two spectral shapes for the source [32]. One of them was obtained by extrapolating a fit of the Fermi-LAT data to TeV energies under the assumption of attenuation of the flux on the extragalactic background light. The other shape consisted only of a E^{-2} power-law spectra for three different energy intervals. Expected limits were also estimated for the two DAQ systems of HAWC-300, for an overhead GRB with the same flux as the one discussed above. The limits are shown in fig. 15, right. Clearly, for such hypothetical event, the HAWC-300 detector would provide a clear signal of the GRB [32]. After completion of the observatory, HAWC will be in operations for ten years. In this period of time, better limits and larger chances of detection of GRBs are expected.

5. Conclusions

The HAWC detector is now under construction at the Volcano Sierra Negra, Mexico. Upon completion, HAWC will comprise a 22 000 m^2 array of 300 Water Cherenkov detectors dedicated to continuously monitor the northern sky in search for air showers from γ and cosmic-rays in the (100 GeV – 100 TeV) energy band. Due to its modular design, HAWC has been taking data during construction. The correct functioning of the detector as well as the analysis pipeline has been demonstrated, for instance with the observation of various sources like the galactic plane, the Crab nebula, Mrk 421 and Mrk 501. In addition, the shadows of the Moon and the Sun have been observed in cosmic rays at TeV energies with the partial array, validating the pointing accuracy and the angular resolution of the detector. The small-scale cosmic-ray anisotropy has also been observed in agreement with previously reported observations by other EAS detectors. Finally, a search for emission from the very bright GRB 130427A, though at an unfavorable zenith angle for HAWC, has also been performed. As no excess was observed, limits were set.

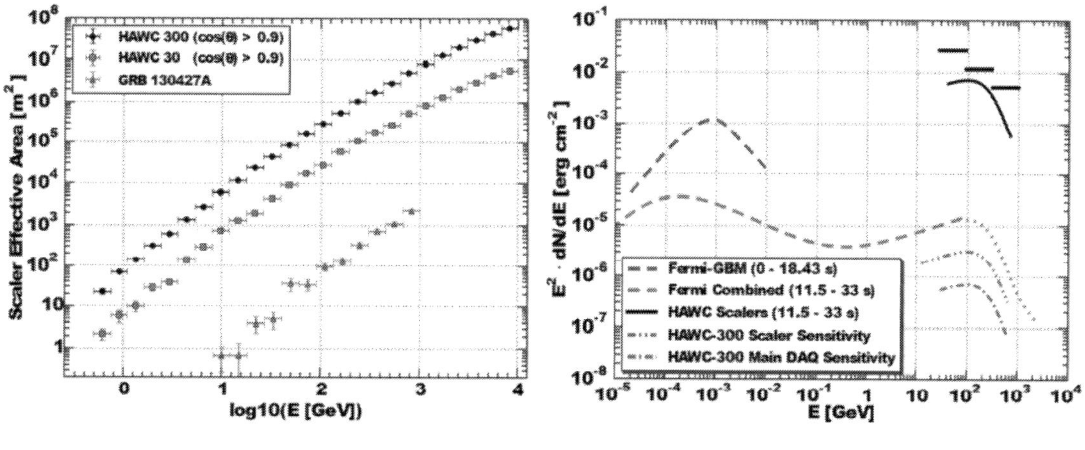

Figure 15. Left: The effective area for the scaler DAQ system HAWC-30 and HAWC-300 for events with zenith angles θ such that $\cos\theta > 0.9$. The corresponding effective area for HAWC-30 in the direction of the GRB 130427A is also shown [32]. Right: Upper limit on the integral gamma-ray flux of the GRB 130427A from the scaler system of HAWC-30. Two spectral shapes were assumed, one from a fit to the Fermi-LAT data and assuming attenuation of the flux with the extragalactic background light (black line), and another one, using a E^{-2} power-law spectrum for three different energy intervals (black horizontal lines). The expected limits from HAWC-300 for an overhead burst similar to the GRB 130427A, using data from the main DAQ and scaler system, are shown for comparison (green dashed-dotted lines).

With the HAWC detector close to being completed, more data being accumulated and more analyses underway, the HAWC observatory will provide a wealth of novel information about the very high energy universe.

Acknowledgments

We acknowledge the support from: US National Science Foundation (NSF); US Department of Energy Office of High-Energy Physics; The Laboratory Directed Research and Development (LDRD) program of Los Alamos National Laboratory; Consejo Nacional de Ciencia y Tecnología (CONACyT), México; Red de Física de Altas Energías, México; DGAPA-UNAM, México; and the University of Wisconsin Alumni Research Foundation.

References

[1] M. Tavani, et al., The AGILE mission, A&A 502 (3), 995-1013 (2009).
[2] W. B. Atwood, et al., ApJ 697, 1071-1102 (2009).
[3] J.A. Hinton et al., New Astron.Rev. 48, 331-337 (2004).
[4] R. Mukherjee et al., EPJ Web of Conferences 61, 04011 (2013).
[5] The MAGIC Telescope, Design Report, MPI-PhE/98-5; B. De Lotto et al., J. of Phys.: Conf. Ser. 375, 052021 (2012).
[6] Y. Yukawa et al., J. of Phys.: Conf. Ser. 120, 062018 (2008).
[7] R. Atkins et al., ApJ 608, 680 (2004); A.J. Smith, Proc of the 29th ICRC 10, 227 (2005).
[8] M. Amenomori, et al., ApJ 633 (2), 1005 (2005).
[9] B. Bartoli et al., ApJ 779, 27 (2013).
[10] A.U. Abeysekara et al., Astrop. Phys. 50-52, 26-32 (2013).
[11] M. Mostafa et al., Brazilian Journal of Physics, Proc. of the 33rd ICRC, arXiv:1310.7237, (2013).
[12] A.U. Abeysekara et al., arXiv:1310.0074, (2013).
[13] Cao Zhen, Chin. Phys. C 34(2), 249-252 (2010); HE Hui-hai et al., Proc. of the 31st ICRC, ID 0654, (2009).
[14] J. Hinton, S. Sarkar, D. Torres and J. Knapp (edit.), Astrop. Phys. 4, 1-356 (2013); the CTA Consortium, arxiv: 1008.3703.

[15] M. Tluczykont et al., Astrop. Phys. 56, 42 (2014).
[16] N. M. Budnev et al., JINST 9, C09021 (2014).
[17] A.U. Abeysekara et al., arXiv:1310.0071, (2013).
[18] A.U. Abeysekara et al., arXiv:1310.0073, (2013).
[19] A.U. Abeysekara et al., arXiv:1310.0072, (2013).
[20] GCN: The Gamma-ray Coordinates Network, http://gcn.gsfc.nasa.gov/
[21] AMON: Astrophysical Multimessenger Observatory Network. http://amon.gravity.psu.edu/
[22] A.U. Abeysekara et al., Astrop. Phys. 35, 641-650 (2012).
[23] I. Taboada et al., NIMA 742, 276-277 (2014).
[24] A. U. Abeysekara et al.,ApJ 796, 108 (2014).
[25] A. U. Abeysekara et al., Phys. Rev. D 90, 122002 (2014).
[26] A.A. Abdo et al., Astrop. Phys. 64, 4 (2015).
[27] P. Karn, A Search for Supersymmetric Q-balls with the High Altitude Water Cherenkov Observatory, Ph.D. thesis, University of California-Irvine (2014).
[28] A. U. Abeysekara et al., Astrop. Phys. 62, 125-133 (2015).
[29] De la Cruz-Reyna and G. Carrasco-Nunez, Journal of Volcanology and Geothermal Research 113, 307 (2002).
[30] M.M. González et al., Proc. of the 31st ICRC, (2009).
[31] G. Sinnis, A. Smith, J.E. McEnery, astro-ph/0403096 (2004).
[32] A.U. Abeysekara et al., ApJ. 800, 78 (2015), arXiv:1410.1536.
[33] M. Mostafá et al., Proc. of the 32nd ICRC, (2011).
[34] D. Heck et al., Report FZKA 6019, Forschungszentrum Karlsruhe (1998). CORSIKA: www-ik.fzk.de/ corsika
[35] A. Fassò et al., Report CERN-2005-10, INFN/TC-05/11, SLAC-R-773 (2005).
[36] S. Ostapchenko, Nucl.Phys.B (Proc. Suppl.) 151 (2006) 143; S. Ostapchenko, Phys.Rev.D 74, 014026 (2006).
[37] R. Brun, F. Carminati, GEANT-detector description and simulation tool, CERN Program Library Long Writeup, 1993. Geant4: geant4.cern.ch.
[38] G. Sinnis et al., Nuclear Physics B, Proc. Suppl. 43, 141-144 (1995).
[39] R. Atkins et al., ApJ 595, 803 (2003).
[40] S. Carrigan et al., arXiv:1307.4868, (2013); S.Carrigan et al., arXiv:1307.4690, (2013).
[41] M. Ackermann et al., ApJs 203, 4 (2012).
[42] F. Aharonian et al., A&A 457, 899 (2006).
[43] J. Aleksić et al., Astrop. Phys. 35, 435 (2012).
[44] J. Holder. private, communication (2013).
[45] A. Borione et al., ApJ 481, 313 (1997).
[46] F. Aharonian et al., ApJ 614, 897 (2004).
[47] Asif Imran, Proc. of the 32nd ICRC, vol. 8, 56 (2011).
[48] J. Stadel et al., Mon. Not. Roy. Astron. Soc. 398, L21 (2009); J. F. Navarro et al., arXiv:0810.1522; J. Einasto, Trudy Inst. Astrophys., Trudy Inst. Astrofiz. Alma-Ata 5, 87 (1965).
[49] J. F. Navarro, C. S. Frenk, and S. D. White, Astrophys.J. 490, 493508 (1997).
[50] J. Aleksi et al., JCAP 1402, 008 (2014).
[51] A. Abramowski et al., ApJ 750, 123 (2012).
[52] J. Han et al., Mon. Not. Roy. Astron. Soc. 427, 1651 (2012).
[53] R. W. Atkins et al., ApJ 595, 803811 (2003).
[54] A. U. Abeysekara et al., ApJ 796, 108 (2014).
[55] A. A. Abdo, ApJ 736, 131 (2011).
[56] A. A. Abdo et al., ApJ 727, 129 (2011).
[57] M. Punch et al., Nature 358, 477 (1992).
[58] J. Quinn et al., ApJ 456, L83 (1996); S. M. Bradbury et al., A&A 320, L5 (1997).
[59] C. M. Hui et al., Proc. of the 26th Recontres de Blois, to be published.
[60] M. Amenomori, et al., Phys.Rev.Lett. 111 (1), 011101 (2013).
[61] B. Bartoli et al., Phys.Rev. D88, 082001 (2013).
[62] A. A. Abdo et al., Phys. Rev. Lett. 101, 221101 (2008).
[63] R. Abbasi et al., ApJ 740, 16 (2011).
[64] G. Giacinti and G. Sigl, Phys. Rev. Lett. 109, 071101 (2012); M. Ahlers, Phys. Rev. Lett. 112, 021101 (2014).
[65] L. Drury, Proc. 33rd ICRC, (2013); P. Desiati and A. Lazarian, ASTRA Proc. 1, 65-71 (2014).
[66] J. P. Harding, arXiv:1307.6537, (2013).
[67] M. Ackermann et al., Science 343, 42 (2014).

Searches for point-like sources using 2007-2012 data from the ANTARES neutrino telescope

Javier Barrios-Martí, on behalf of the ANTARES Collaboration
Instituto de Física Corpuscular (CSIC - UV), C/Catedrático José Beltrán, 2, E-46980 Paterna, Spain

E-mail: `javier.barrios@ific.uv.es`

Abstract. A search for point-like sources of cosmic neutrinos has been performed using the 2007 to 2012 data collected by the ANTARES neutrino telescope. The search is performed by looking for accumulations of muon neutrinos over the expected atmospheric background. No clear signal has been found. The cluster of events with the highest significance (2.2σ) was found at equatorial coordinates (RA, DEC) = ($-46.8°$, $-64.9°$). Furthermore, 90% confidence level upper-limits have been set for 50 pre-selected astrophysical neutrino source candidates for an E^{-2} flux normalization. In addition, due to the accumulation of 7 high energy neutrinos close to the Galactic Centre recently reported by the IceCube collaboration, a specific search around 20° from the Galactic Centre has been performed. No evidence of a signal has been found in the ANTARES data within this region.

1. Introduction

The aim of neutrino telescopes is to observe the universe with the use of neutrinos, which have advantageous properties compared to other stable particles. Cosmic rays may be deflected by magnetic fields and therefore lose the information about their origin, while the mean free path of high energy photons heavily depends on their energy. Neutrinos are neutral and weakly interacting particles, so they point directly back to their origin. A first signal of cosmic neutrinos has been reported by IceCube [1, 2], which includes a cluster of events close to the Galactic Centre. Even if this cluster is not statistically significant, some authors have suggested they may be due to a single point source [3]. Due to its location in the Mediterranean Sea, the better muon rejection efficiency of the Southern Hemisphere provides a better sensitivity to Galactic sources with energies < 100 TeV. This fact is especially important for the interpretation of the observed cluster of events close to the Galactic Centre by IceCube.

2. The ANTARES neutrino telescope

The ANTARES neutrino telescope is the first neutrino telescope which operates in the Mediterranean Sea. It is located 40 km to the South of Toulon (France), at the coordinates (42° 48' N, 6° 10' E) [4]. It consists of 12 lines of 350 m with a buoy on the top to keep them vertical. Each line contains 25 storeys with 3 Optical Modules (OMs). An Optical Module constists of a photomultiplier tube (PMT) of 10" inside a glass sphere of 17". These OMs are facing downwards at an angle of 45° in order to optimize the detection of upgoing muon-neutrinos.

The detection principle is based on the indirect detection of high energy neutrinos which interact via charged current (CC) interaction with a nucleus (N) inside or nearby the detector volume:

$$\nu_l + N \rightarrow l^- + X;\ \bar{\nu}_l + N \rightarrow l^+ + X, \quad (1)$$

where a charged lepton and an hadronic shower are produced. In case this lepton is a muon, it can travel a long distance before decaying. Since this muon is travelling at relativistic speeds, it will emit Cherenkov radiation. Some of the emitted photons are detected by the PMTs, producing a signal ("hit") with the corresponding charge and time information. The information collected by the hits of a muon is used to reconstruct the direction of the muon.

3. Data selection

The data considered for this analysis was collected between January 29, 2007 to December 31, 2012. The total measurement period comprises a livetime of 1338 days, which is 70% larger compared to the previous ANTARES point-source analysis [5].

Using the hit and time position information of hits, triggered events are reconstructed. This reconstruction is done by means of a maximum likelihood (ML) method [6]. The procedure is based on a multi-step procedure to fit the direction of the reconstructed muon by maximising the ML-parameter Λ, which indicates the goodness of the reconstruction. Furthermore, the estimated uncertainty of the track direction, β, is computed from the covariance matrix.

The KM3 package [7] is used to simulate the muon propagation. Additionaly, the simulation of neutrinos and atmospheric muons is performed with the GENHEN [7] and MUPAGE [8, 9] packages, respectively. The Bartol flux [10] is used for the atmospheric neutrino simulation. In order to test our simulations, a comparison of the Λ distribution in the data and simulated samples is performed. Figure 1 shows to the comparison for zenith angles θ with $\cos\theta < 0.1$, where $0° < \theta < 90°$ correspond to down-going tracks, and $90° < \theta < 180°$ to up-going tracks.

Before performing the analysis on data, events are selected by following a blind procedure on pseudo-experiments. The final sample is chosen so that it minimises the neutrino flux needed for a 5σ discovery in 50% of the experiments for a E^{-2} source spectrum. The final cuts correspond to $\Lambda > -5.2$, $\beta < 1°$ and $\cos\theta < 0.1$, which leads to a total number of 5516 events. The estimated contamination of mis-reconstructed amospheric muons is 10%.

4. Performance of the detector

The cumulative distribution of the angular resolution can be calculated to show the pointing accuracy of the detector. The angular resolution of each event, Ψ, is defined as the difference between the true simulated direction and its reconstructed track. Figure 2 shows both the distribution for the current (blue) and previous (dashed red line) analysis [5]. Due to an improvement on the modelling of the PMT transit-time distribution the calculation of the median angular resolution is 0.38°, which corresponds to a 15% improvement.

The "acceptance" of the detector is calculated in order to obtain the number of signal events for a given source assumption. Multiplied by a given flux normalisation, Φ_0, it returns the number of cosmic neutrinos from a source. It depends on the detector effective area for a given selection cut, A_{eff}; the source energy spectrum and declination. The acceptance for a given declination, δ, can be described as

$$A(\delta) = \Phi_0^{-1} \int dt \int dE_\nu A_{eff}(E_\nu, \delta) \frac{d\Phi}{dE_\nu} . \quad (2)$$

For this analysis, the time integration covers the whole livetime of 1338 days, and the assumed energy spectrum is $\frac{d\Phi}{dE_\nu} = \Phi_0 E_\nu^{-2}$. The acceptance as a function of the declination δ is calculated and shown in Figure 3.

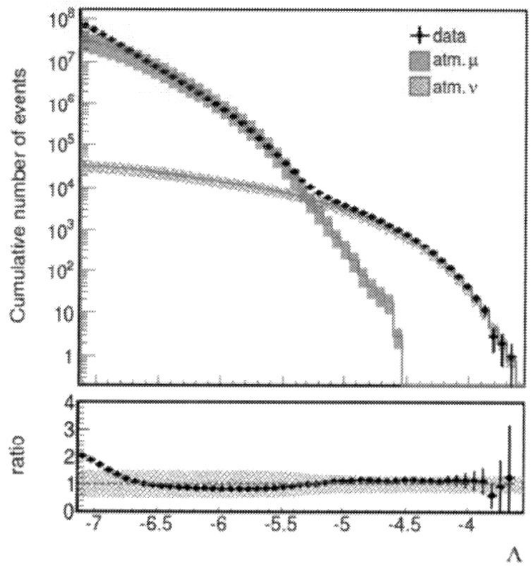

Figure 1. Cumulative distribution of the track reconstruction quality parameter, Λ, for tracks with an angular error estimate $\beta < 1°$ and coming from a direction $\cos\theta < 0.1$ (mostly upgoing). A better track reconstruction is given by larger values of Λ. Since no upgoing atmospheric muons are expected, muon-neutrinos are expected to dominate the region with larger Λ values. The ratio between data and simulation is given in the bottom panel. Data is indicated by black dots, whose uncertainties are calculated from statistical errors. The red (green) distribution indicates the simulated atmospheric neutrinos (muons), where a 30% (50%) relative error was assigned [11, 12].

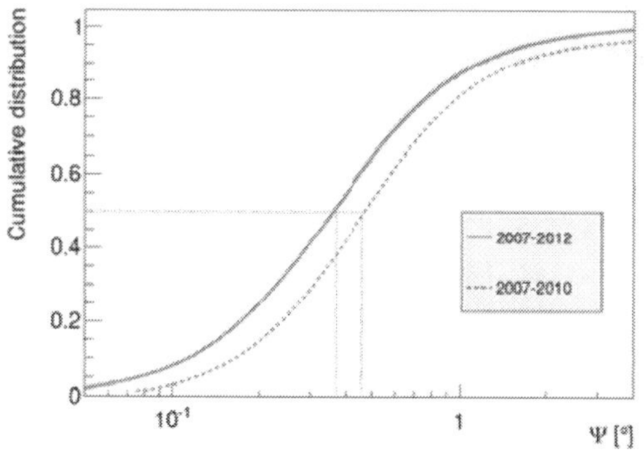

Figure 2. Neutrino angular resolution determined as the median of the cumulative distribution of the reconstruction angle, Ψ, for the present data (solid blue line) compared to the 2007-2010 analysis (dashed red line). The black-dotted line indicates the median value.

5. Search for cosmic neutrino sources

Atmospheric neutrino events are distributed isotropically, whereas cosmic neutrinos are expected to accumulate in clusters. A maximum-likelihood estimation is done in order to search for signal events. This likelihood describes the data in terms of background and signal probability density functions (PDFs):

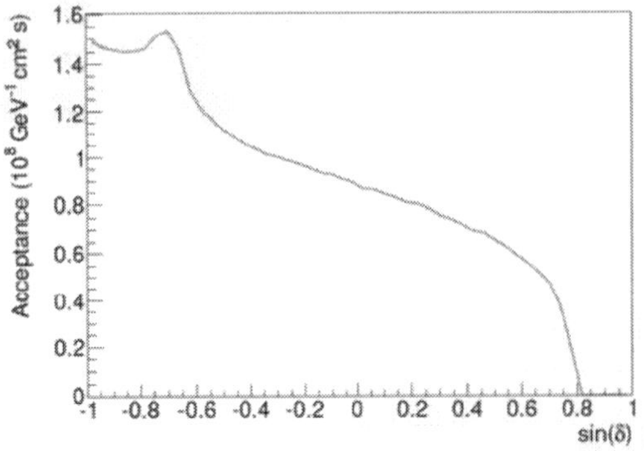

Figure 3. Acceptance (defined in Equation 2) for events passing the final selection cuts. An E^{-2} source spectrum has been assumed for the calculation.

$$\log L_{s+b} = \sum_i \log \left[\frac{n_s}{N} S_i + \left(1 - \frac{n_s}{N}\right) B_i \right], \tag{3}$$

where B_i and S_i denote the background and signal PDFs for the ith event, respectively, N indicates the events in the sample, and n_s represents the fitted number of signal source events. The signal PDF is defined as

$$S_i = \frac{1}{2\pi\beta_i^2} \exp\left(-\frac{\psi_i(\vec{x}_s)^2}{2\beta_i^2}\right) P_s(\mathcal{N}_i^{hits}, \beta_i), \tag{4}$$

where $\psi_i(\vec{x}_s)^2$ denotes the angular distance of an event i to the source direction, $\vec{x}_s = (\alpha_s, \delta_s)$, and $P_s(\mathcal{N}_i^{hits}, \beta_i)$ gives the probability for a signal event to be reconstructed with a number of hits $\mathcal{N}_i^{hits}$ and an angular error estimate of β_i.

The background PDF is defined as

$$B_i = \frac{B(\delta_i)}{2\pi} P_b(\mathcal{N}_i^{hits}, \beta_i), \tag{5}$$

where $P_b(\mathcal{N}_i^{hits}, \beta_i)$ indicates the probability for a background event to be reconstructed with an angular error estimate of β_i and a number of hits $\mathcal{N}_i^{hits}$, and $B(\delta_i)$ is the probability for an event to be at a declination δ_i.

In order to determine the significance of any cluster, the test statistic, TS, is defined as TS = $\log L_{s+b} - \log L_b$. L_b indicates the value of the likelihood for the case of no signal events (n_s = 0). Larger values of the TS designate a smaller probability of the cluster to be generated from only atmospheric events.

6. Results of the full-sky and candidate list searches

The full-sky search consists of an exploration for a significant cluster of signal events anywhere in the ANTARES visible sky. First, all clusters with at least 4 events in a half-opening angle cone of 3° are pre-selected. The likelihood, L_{s+b}, is maximised for each of them by adjusting the free parameters $\vec{x}s$ and n_s. The most significant accumulation of events of the full-sky search is located at $(\alpha, \delta) = (-46.8°, -64.9°)$. This location is compatible with the direction of the most significant accumulation found in the 2007-2010 analysis. This cluster has a post-trial p-value of 2.7% (significance of 2.2σ in the two-sided convention). The number of fitted events is $n_s = 6.2$.

6 (14) events are found in a cone of 1° (3°) from the fitted centre of the cluster. No significant excess was found. The upper limits at the 90% confidence level (C.L.) from point-sources are presented by the light blue-dashed line in Figure 5, where each dot indicates the highest upper-limit in declination bands of 1°. Figure 4 shows the p-values of the whole visible sky in steps of 0.2° in right ascension and declination.

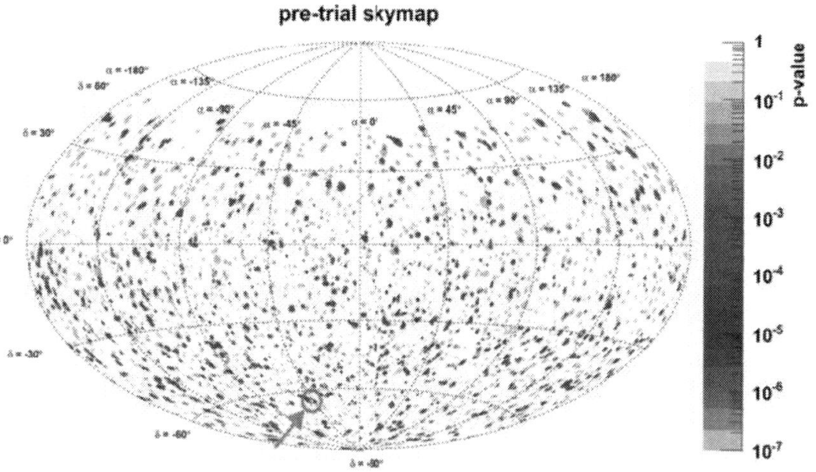

Figure 4. Pre-trial p-values of the ANTARES visible sky.

In the candidate list search, the TS is evaluated at the position of 50 neutrino candidate sources. The most significant accumulation of this search was found at the position of HESS J0632+057. The post-trial p-value is 6.1% (significance of 1.9σ) for a fitted number of source events of $n_s = 1.6$. The complete list of these sources with their corresponding flux upper limits and pre-trial p-values is shown in Table 1. Furthermore, the upper-limits and the overall fixed-source sensitivity are also represented in Figure 5, where the upper limits and sensitivities are calculated by using the Neyman method [14] at a 90% C.L.

7. Search for cosmic sources around the Galactic Centre

After the recent evidence of an excess of high energy neutrinos by IceCube [1], the presence of a point source at equatorial coordinates $\alpha = -79°$, $\delta = -23°$ has been proposed [3] in order to justify the accumulation of seven events close to the Galactic Centre. A flux normalisation of $\Phi_0 = 6 \times 10^{-8}$ GeVcm^{-2}s^{-1} would therefore be expected for this hypothetical source.

Due to the large uncertainty of the angular error estimates of these IceCube events, this source may be located at a different point in the sky. To avoid this problem, the full sky algorithm is restricted to a region of 20° around the proposed location. The likelihood used was the one presented in Ref. [5]. Because of the narrower size compared to the full sky search, the trial factor of this analysis is smaller. Apart from the point-source hypothesis, three Gaussian-like source extensions are considered (0.5°, 1° and 3°).

No significant cluster has been found. 90% C.L. upper limits for the four considered source extensions as a function of the declination of the source are represented in Figure 6. The possibility of a point source with a flux normalisation of $6{\times}10^{-8}$ GeVcm^{-2}s^{-1} is discarded in the considered region with a CL higher than 90%. In this sense, the excess of events found by IceCube around the Galactic Centre would not be caused by a sole point-source with an E^{-2} source spectrum. In addition, a source width of 0.5° for declinations below −11° is also

Table 1. Pre-trial p-values, p, fitted number of source events, n_s, and 90% C.L. flux limits, Φ_ν^{90CL}, obtained for the 50 candidate sources. The fluxes are in units of 10^{-8} GeV cm^{-2}s^{-1}.

Name	α (°)	δ (°)	n_s	p	ϕ_ν^{90CL}	Name	α (°)	δ (°)	n_s	p	ϕ_ν^{90CL}
HESSJ0632+057	98.24	5.81	1.60	0.0012	4.40	HESSJ1912+101	-71.79	10.15	0.00	1.00	2.31
HESSJ1741-302	-94.75	-30.20	0.99	0.003	3.23	PKS0426-380	67.17	-37.93	0.00	1.00	1.59
3C279	-165.95	-5.79	1.11	0.01	3.45	W28	-89.57	-23.34	0.00	1.00	1.89
HESSJ1023-575	155.83	-57.76	1.98	0.03	2.01	MSH15-52	-131.47	-59.16	0.00	1.00	1.41
ESO139-G12	-95.59	-59.94	0.79	0.06	1.82	RGBJ0152+017	28.17	1.79	0.00	1.00	2.19
CirX-1	-129.83	-57.17	0.96	0.11	1.62	W51C	-69.25	14.19	0.00	1.00	2.32
PKS0548-322	87.67	-32.27	0.68	0.10	2.00	PKS1502+106	-133.90	10.52	0.00	1.00	2.31
GX339-4	-104.30	-48.79	0.50	0.14	1.50	HESSJ1632-478	-111.96	-47.82	0.00	1.00	1.33
VERJ0648+152	102.20	15.27	0.59	0.11	2.45	HESSJ1356-645	-151.00	-64.50	0.00	1.00	1.42
PKS0537-441	84.71	-44.08	0.24	0.16	1.37	1ES1101-232	165.91	-23.49	0.00	1.00	1.92
MGROJ1908+06	-73.01	6.27	0.21	0.14	2.32	HESSJ1507-622	-133.28	-62.34	0.00	1.00	1.41
Crab	83.63	22.01	0.00	1.00	2.46	RXJ0852.0-4622	133.00	-46.37	0.00	1.00	1.33
HESSJ1614-518	-116.42	-51.82	0.00	1.00	1.39	RCW86	-139.32	-62.48	0.00	1.00	1.41
HESSJ1837-069	-80.59	-6.95	0.00	1.00	2.09	RXJ1713.7-3946	-101.75	-39.75	0.00	1.00	1.59
PKS0235+164	39.66	16.61	0.00	1.00	2.39	SS433	-72.04	4.98	0.00	1.00	2.32
Geminga	98.31	17.01	0.00	1.00	2.39	1ES0347-121	57.35	-11.99	0.00	1.00	2.01
PKS0727-11	112.58	-11.70	0.00	1.00	2.01	VelaX	128.75	-45.60	0.00	1.00	1.33
PKS2005-489	-57.63	-48.82	0.00	1.00	1.39	HESSJ1303-631	-164.23	-63.20	0.00	1.00	1.43
PSRB1259-63	-164.30	-63.83	0.00	1.00	1.41	LS5039	-83.44	-14.83	0.00	1.00	1.96
HESSJ1503-582	-133.54	-58.74	0.00	1.00	1.41	PKS2155-304	-30.28	-30.22	0.00	1.00	1.79
PKS0454-234	74.27	-23.43	0.00	1.00	1.92	Galactic Centre	-93.58	-29.01	0.00	1.00	1.85
PKS1454-354	-135.64	-35.67	0.00	1.00	1.70	CentaurusA	-158.64	-43.02	0.00	1.00	1.36
HESSJ1834-087	-81.31	-8.76	0.00	1.00	2.06	W44	-75.96	1.38	0.00	1.00	2.23
HESSJ1616-508	-116.03	-50.97	0.00	1.00	1.39	IC443	94.21	22.51	0.00	1.00	2.50
H2356-309	-0.22	-30.63	0.00	1.00	2.35	3C454.3	-16.50	16.15	0.00	1.00	2.39

excluded. Since neutrinos with $E > 2\ PeV$ coming from an E^{-2} spectrum only contribute a 7% to the flux, these limits are barely affected by a cutoff on the order of the PeV.

8. Conclusions

The latest results of the ANTARES point-source search have been described using two complementary analyses. Firstly, a scan over the whole ANTARES visible sky. Secondly, a search over a list of 50 pre-selected neutrino candidate sources. The results found are compatible with a pure background hypothesis. In the first search, the post trial p-value of the most significant cluster is 2.7% (significance of 2.2σ). The largest excess found in the candidate source list study corresponds to HESS J0632+057 with a post-trial p-value of 6.1% (1.9σ). The computed flux upper limits are the most restrictive in a significant part of the Southern Sky. The possibility that the accumulation of 7 events reported by IceCube near the Galactic Centre is produced by a single point source has been excluded.

Acknowledgements

We gratefully acknowledge the financial support of the Spanish Ministerio de Economía y Competitividad (MINECO) grants FPA2012-37528-C02-01 and Consolider MultiDark CSD2009-00064, of the Generalitat Valenciana, PROMETEOII/2014/079, and of Universitat de València, Atracció de Talent.

References

[1] Aartsen M. G., Abbasi R., Abdou Y. et al. (IceCube Collaboration) 2013a, *Phys. Rev. Lett.*, **111**, 021103.
[2] Aartsen M. G., Abbasi R., Abdou Y. et al. (IceCube Collaboration) 2013c, *Science*, **342**, 6161, 1242856.

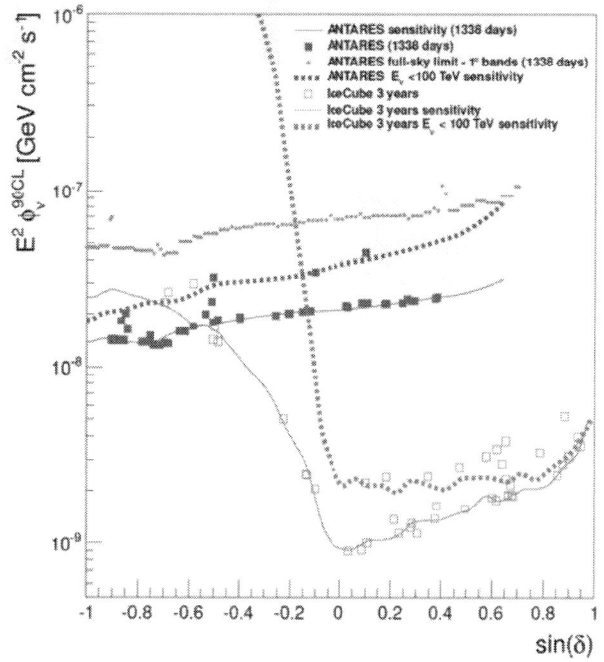

Figure 5. 90 % C.L. flux upper limits and sensitivities on the muon neutrino flux for six years of ANTARES data. IceCube results are also shown for comparison. The light-blue markers show the upper limit for any point source located in the ANTARES visible sky in declination bands of 1°. The solid blue (red) line indicates the ANTARES (IceCube) sensitivity for a point-source with an E^{-2} spectrum as a function of the declination. The blue (red) squares represent the upper limits for the ANTARES (IceCube) candidate sources. Finally, the dashed dark blue (red) line indicates the ANTARES (IceCube) sensitivity for a point-source and for neutrino energies lower than 100 TeV, which shows that the IceCube sensitivity for sources in the Southern hemisphere is mostly due to events of higher energy. The IceCube results were derived from [13].

[3] González-García M. C., Halzen F. & Niro V. 2014, *Astropart.Phys.*, **57-58**, 39.
[4] Ageron M., Aguilar J. A., Samarai I. Al et al. (ANTARES Collaboration) 2011, *Nucl. Instrum. Meth. A*, **656**, 11.
[5] Adrian-Martinez S., Samarai I.Al, Albert A. et al. (ANTARES Collaboration) 2012, *ApJ*, **760**, 53.
[6] Adrian-Martinez S., Samarai I.Al, Albert A. et al. (ANTARES Collaboration) 2013, *JCAP*, **1303**, 006.
[7] Brunner J. 2003, in VLVnT Workshop (Amsterdam), ANTARES simulation tools, ed. E. de Wolf (Amsterdam: NIKHEF), http://www.vlvnt.nl/proceedings.pdf
[8] Carminati G., Bazzotti M., Margiotta A. & Spurio M., *Comput. Phys. Commun.* (2008), **179**, 915.
[9] Bazzotti M., Carminati G., Margiotta A. & Spurio M., *Comput. Phys. Commun.* , **181**, 835.
[10] Agrawal V., Gaisser T. K., Lipari P. & Stanev T., *Phys. Rev. D*, **53**, 1314.
[11] Aguilar J. A., Albert A., Anton G. et al. (ANTARES Collaboration) 2010, *Astropart. Phys.*, **34**, 179.
[12] Barr G. D., Gaisser T. K., Robbins S. & Stanev T., *Phys. Rev. D*, **74**, 094009.
[13] Aartsen M. G., Abbasi R., Abdou Y. et al. (IceCube Collaboration) 2013b, *ApJ*, **779**, 132.
[14] Neyman J. 1937, *Phil. Trans. Royal Soc. London A*, **236**, 333.

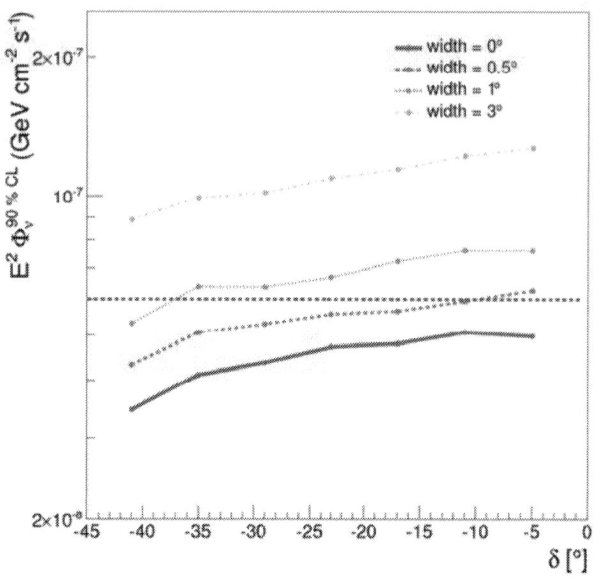

Figure 6. 90 % C.L. upper limits obtained for different source widths as a function of the declination. The blue horizontal dashed line corresponds to the signal flux given by [3].

The Tunka detector complex: from cosmic-ray to gamma-ray astronomy

**N Budnev[b], I Astapov[i], N Barbashina[i], A Bogdanov[i], D Bogorodskii[b], V Boreyko[j], M Büker[f], M Brückner[k], A Chiavassa[d], O Chvalaev[b], O Gress[b], T Gress[b], A Dyachok[b], S Epimakhov[f],
A Gafatov[b], N Gorbunov[j], V Grebenyuk[j], A Grinuk[j], A Haungs[l], R Hiller[l], D Horns[f], T Huege[l], A Ivanova[b], A Kalinin[j], N Karpov[a], N Kalmykov[a], Y Kazarina[b], V Kindin[i], N Kirichkov[b],
S Kiryuhin[b], M Kleifges[m], R Kokoulin[i], K Komponiest[i], A Konstantinov[a], E Konstantinov[b],
A Korobchenko[b], E Korosteleva[a], D Kostunin[l], V Kozhin[a], O Krömer[m], M Kunnas[f], L Kuzmichev[a], V Lenok[b], B Lubsandorzhiev[c], N Lubsandorzhiev[a], R Mirgazov[b], R Mirzoyan[e,b], R Monkhoev[b], R Nachtigall[f], A Pakhorukov[b], M Panasyuk[a], L Pankov[b], A Petrukhin[i],
V Platonov[b], V Poleschuk[b], E Popova[a], A Porelli[h], V Prosin[a], V Ptuskin[g], G Rubtsov[c], C Rühle[m], V Samoliga[b], P Satunin[g], V Savinov[b], A Saunkin[b], F Schröder[l], Yu Semeney[b],
B Shaibonov (junior)[c], A Silaev[a], A Silaev (junior)[a], A Skurikhin[a], V Slucka[j], C Spiering[h],
L Sveshnikova[a], V Tabolenko[b], A Tkachenko[j], L Tkachev[j], M Tluczykont[f], D Voronin[b],
R Wischnewski[h], A Zagorodnikov[b], V Zurbanov[b], I Yashin[i].**

[a] Skobeltsyn Institute of Nuclear Physics MSU, Moscow, Russia
[b] Institute of Applied Physics, ISU, Irkutsk, Russia
[c] Institute for Nuclear Research of RAN, Moscow, Russia
[d] Dipartimento di Fisica Generale Universiteta di Torino and INFN, Torino, Italy
[e] Max-Planck-Institute for Physics, Munich, Germany
[f] Institut für Experimentalphysik, University of Hamburg, Germany
[g] IZMIRAN, Troitsk, Moscow Region, Russia
[h] DESY, Zeuthen, Germany
[i] National Research Nuclear University (Moscow Engineering Physics Institute),
[j] Moscow, Russia JINR, Dubna, Russia
[k] Institute for Computer Science, Humboldt-University, Berlin, Germany
[l] Institut für Kernphysik, KIT, Karlsruhe, Germany
[m] Institut für Prozessdatenverarbeitung und Elektronik, KIT, Karlsruhe, Germany

E-mail: nbudnev@api.isu.ru

Abstract. TAIGA stands for "Tunka Advanced Instrument for cosmic ray physics and Gamma Astronomy" and is a project to build a complex, hybrid detector system for ground-based gamma-ray astronomy from a few TeV to several PeV, and for cosmic-ray studies from 100 TeV to 1 EeV. TAIGA will search for "PeVatrons" (ultra-high energy gamma-ray sources) and measure the composition and spectrum of cosmic rays in the knee region (100 TeV – 10 PeV) with good energy resolution and high statistics. TAIGA will include Tunka-HiSCORE (an array of wide-angle air Cherenkov stations), an array of Imaging Atmospheric Cherenkov Telescopes, an array of particle detectors, both on the surface and underground, and the TUNKA-133 air Cherenkov array.

1. Introduction

In recent years ground-based very high energy (VHE) gamma-ray astronomy became the most dynamically developing field in high energy astrophysics. Advantages of gamma rays as information carriers from powerful Galactic and extragalactic sources are due to the fact that in contrary to charged cosmic rays, they preserve their emission direction from sources, and unlike neutrinos, they are easy to detect. Most of the knowledge about sources of high energy gamma rays has been obtained from arrays of Imaging Atmospheric Cherenkov Telescopes (IACT), most prominently HEGRA [1], H.E.S.S. [2], MAGIC [3] and VERITAS [4]. The method of image shape analysis proposed by A.M. Hillas in 1985 [5] allows distinguishing, with high efficiency, Extensive Air Showers (EAS) produced by gamma rays from those initiated by charged cosmic rays. The sensitivity level of presently existing gamma-ray telescopes is optimized for the energy range of 100 GeV - 20 TeV. Gamma-radiation at ≥100 GeV was detected from more than 150 sources of different types, but radiation with energies ≥10TeV was recorded from about 10 sources only. No sources with spectra extending beyond 100 TeV have been detected so far. The so-called 4th-generation IACT array CTA [6] will have substantially higher sensitivity than the existing telescopes in the energy range of up to 100 TeV. For extension towards higher gamma-ray energies it is necessary to construct arrays with lots of telescopes, distributed over an area of a few square kilometers at least. The estimated cost of such an IACT array can be more than 100 million US dollars per square kilometer. So, it is necessary to find more cost-effective experimental solutions which would allow one to have an array with an area of the order of 10 - 100 km^2 for studying PeVatrons.

To solve many crucial tasks of high energy gamma-ray astronomy as well as of cosmic ray physics we propose to design and to construct the observatory TAIGA (Tunka Advanced Instrument for cosmic ray physics and Gamma Astronomy) - a complex hybrid detector system. TAIGA will be located in the Tunka valley, about 50 km from Lake Baikal in Siberia, Russia, were since 2009 the full-size Tunka-133 Cherenkov EAS detector is in operation [7-9]. The experience from Tunka-133, it's results on energy spectrum and mass composition of cosmic rays, and the existing infrastructure have been important factors for selecting this location. Common operation of an array of IACTs and a wide-angle array will allow abstaining from the simultaneous measurement of EAS by several closely spaced wide-angle Cherenkov detectors and instead of that increasing the distance between IACTs up to 600 m. The relatively low cost of such an array of largely spaced IACTs, combined with a wide-angle Cherenkov array (cost of a single station of the wide-angle array is only ~1/30 of the cost of an imaging telescope) allows for the necessary huge collection area and for preparing gamma-ray measurements at ~100 TeV and beyond. TAIGA will combine five arrays with different types of detectors:

- The Tunka-133 wide-field (FoV ~2 sr) integrating air Cherenkov array [8]. It consists of 175 detectors including a PMT with a hemispherical-shape photocathode of 20 cm diameter (EMI 9359 or Hamamatsu R1408).
- The low-threshold Tunka-HiSCORE wide-field (FoV ~ 0.6 sr) integrating air Cherenkov array, consisting of detector stations spaced at distances of 75-200m from each other. This array will cover an area of initially ~1 km^2, and >10 km^2 at a later phase. The detector stations consist of four PMTs of size either 20 cm or 25 cm in diameter, located close to each other. The PMTs have Winston-cone light-guides, which increase their light collection area by factor of four.
- The Tunka-IACT array – up to 16 IACTs with reflector areas about 10 m^2 and imaging cameras of 400 PMTs. The FoV of a single telescope will be 8°×8°. The inter-telescope distances will be optimized in the range of 300-600 m.
- The Tunka-Grande array – a net of surface and underground scintillation stations to measure the charged component of air showers. On order to achieve a good rejection of hadron showers at energies ≥ 100 TeV, it is desirable to cover an area of a few 1000 m^2 with underground muon detectors.
- The Tunka-Rex (Tunka Radio extension) array consists of 47 antenna stations with a spacing of 200 m, spread over about 3 km^2.

2. TheTunka-133 array

Presently, the Tunka-133 array consists of 175 wide-angle Cherenkov detectors distributed over 3 km^2 area [8]. The energy spectrum of cosmic rays has been reconstructed using data from 5 winter data taking seasons. The resulting differential energy spectrum is shown in figure 1 (left), together with the preliminary spectrum of Tunka-HiSCORE [10]. At energies about 20 PeV, the power-law index changes from $\gamma = 3.23 \pm 0.01$ to $\gamma = 2.98 \pm 0.01$. Above 300 PeV, the spectrum becomes steeper, with $\gamma = 3.35 \pm 0.11$ (the second "knee"). There is a possibility of a more complex description of the second "knee" with a small intermediate increase of the absolute value of the index to $\gamma = 3.06 \pm 0.03$ around 80 PeV. Figure 1 (right) compares the Tunka-133 data to results from other experiments.

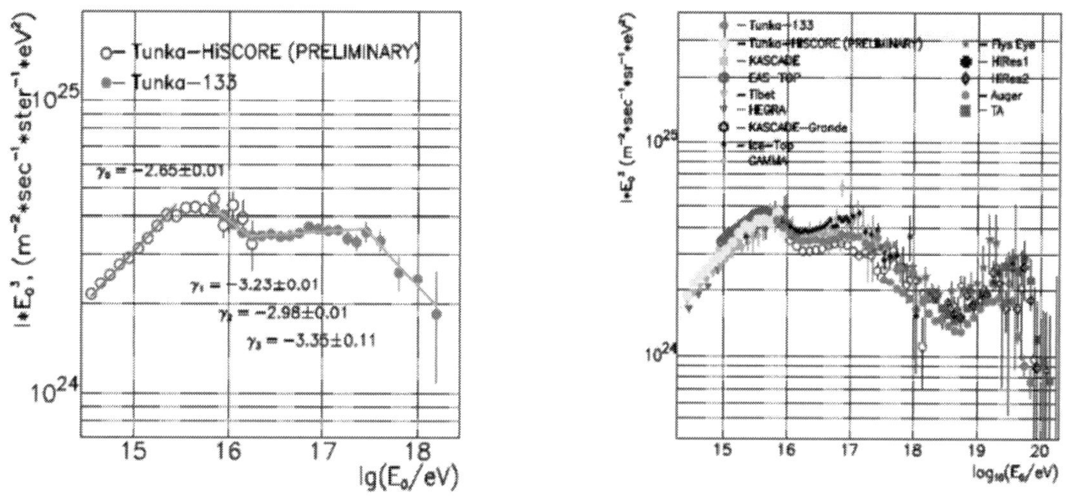

Figure 1. All-particle energy spectrum, left: Tunka data, right: comparison with other experiments.

The methods of EAS maximum depth X_{max} measurement are described in [11]. The experimental dependence of mean X_{max} and mass composition vs. primary energy E_0 are presented in figure 2. The experimental points are compared to the data of the HiRes-MIA experiment [12] and the fluorescent light detectors of Pierre Auger Observatory [13]. One sees agreement of the Cherenkov light results from the Tunka experiment with direct fluorescent light observations. The primary mass composition becomes heavier in the energy range 10 –30 PeV and lighter again in the range 100 – 1000 PeV.

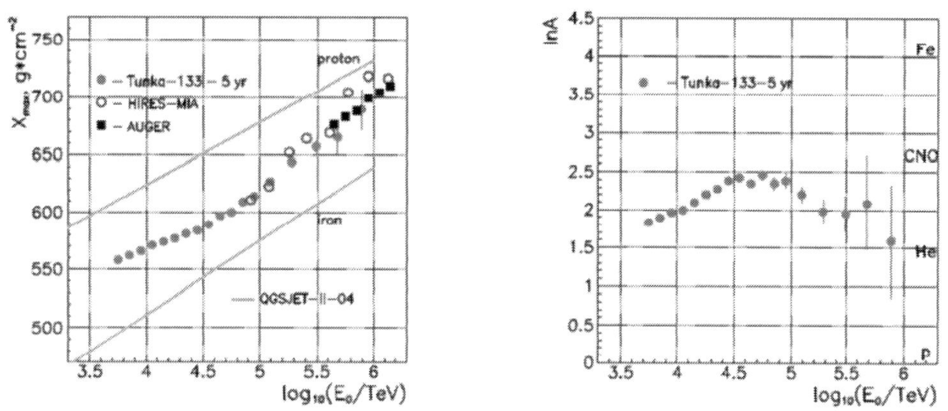

Figure 2. Left: mean depth of EAS maximum; Right: mean cosmic rays logarithmic mass vs. energy.

3. The Tunka-HiSCORE array

The principle of the Tunka-HiSCORE array follows the idea outlined in [14,21]. It is rather similar to the one used for Tunka-133. Again, one samples the Cherenkov light front of air showers. The detector stations measure the light amplitudes and its arrival time differences over a distance of few hundred meters. This approach allows one to reconstruct with high precision the arrival direction and the axis position of the EAS, as well as energy E_0 and height of the shower maximum, X_{max}. The expected accuracy of the core location measurement is 5–6 m, that of the arrival direction about 0.1°, for E_0 it is 10% and for X_{max} – 15-20%. The data can be used to reconstruct in detail the cosmic ray spectrum and its composition, but also provide some gamma/hadron separation. Since the signals from all the four large, high-sensitive PMTs in a HiSCORE station are summed-up, the array has a substantially lower energy threshold than Tunka-133. It should be possible to detect bright gamma-ray sources (as for example the Crab Nebula) at energies ≥ 20 TeV, and charged cosmic rays with energies ≥ 40 TeV.

Tunka-HiSCORE will consist of an array of wide-angle light-sensitive detector stations, distributed with a spacing of 75 – 200 m over an area of a square kilometre in a first phase, and over several tens of square kilometres in the future. A HiSCORE detector stations consist of two boxes. The first type is the optical box (figure 3) with four PMTs, each equipped with a light-collecting Winston cone of 30° half-opening angle pointing to the zenith. A plexiglas plate on top of the cones protects the PMT and the light collector from dust and humidity. Heating prevents the formation of white frost on the plate during the cold season, and a lid opening/closing mechanism is used to protect the PMT from Sun light. The optical boxes house slow-control electronics, high-voltage (HV) systems and a read-out system. For improving the dynamic range of detector, signals are read both from the anode and from one of the dynodes of the PMT.

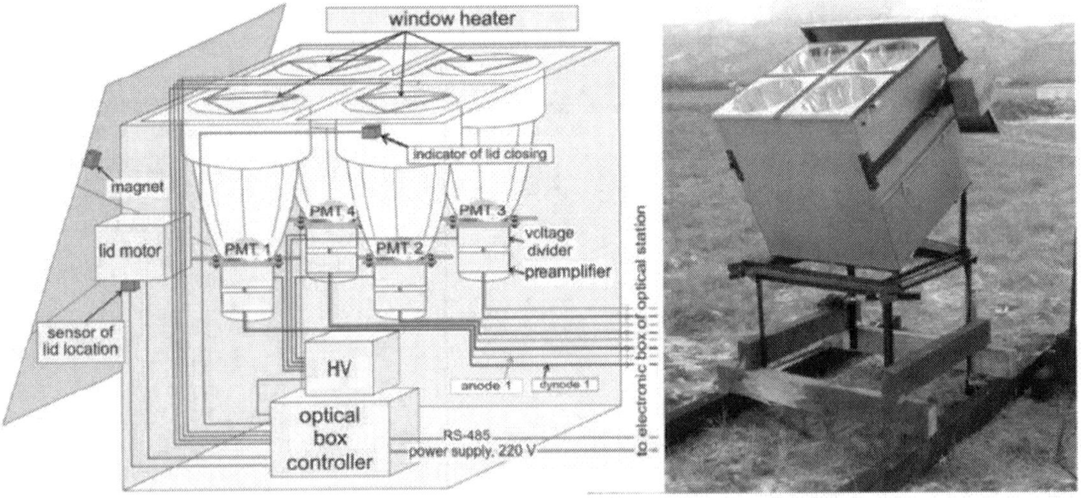

Figure 3. Schematic view (left) and outer view (right) of the Tunka-HiSCORE optical box.

The data acquisition system (DAQ), the slow control for the PMT-high voltage, the environmental control and some auxiliary electronics are placed in a special temperature-controlled box. A fast signal read-out and digitization in the GHz regime is provided by a read-out board with up to 8 channels based on the Digital Ring Sampler DRS 4 chip [15], with a depth of 1024 cells at a resolution of up to 0.2 ns per cell (5 GHz). In order to retain a sufficiently wide read-out window, a sampling frequency of 2 GHz is used, corresponding to a read-out window of 0.5 μs. To acquire the shower data, the high-gain anode signals of the PMTs are amplified by a custom-made fast preamplifier and then summed-up. To synchronize all stations in the array to sub-nsec precision we use at present a hybrid approach, combining a custom-made synchronization technique (100 MHz clocks distributed over separate fibers from the array center), and the new White Rabbit Ethernet-based technology (with synchronization and Gbit data traffic over a single fiber [16]). The main DAQ-board is designed on the basis of the DRS-4 chip and the FPGA Xilinx Spartan-6. All detector stations and the array center building are connected by single-mode optical fibers. The accuracy of time synchronization is better than 1 ns. The time step of digitization can be changed from 0.2 to 1 ns. The dead-time of the detector station electronics is smaller than 0.5 ms. For the first 9 HiSCORE stations, a fully independent DAQ has been installed, based on two 4-channel DRS-4-evaluation boards, directly synchronized and triggered by White Rabbit [20].

The configuration of the Tunka-HiSCORE stations as of 2014 is shown in figure 5. It comprises 28 detectors at 100 m spacing, forming a super-cell structure. The total area of the setup is 0.25 km^2. To increase the observation time for the gamma-ray source in the Crab Nebula, all optical boxes are tilted towards South by 25 degrees.

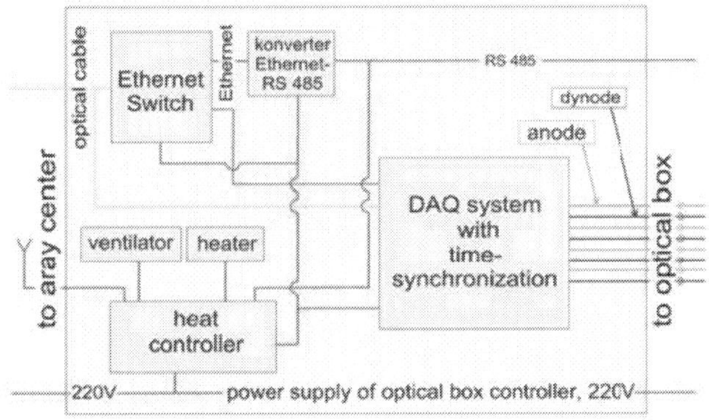

Figure 4. Layout of the Tunka-HiSCORE electronics box

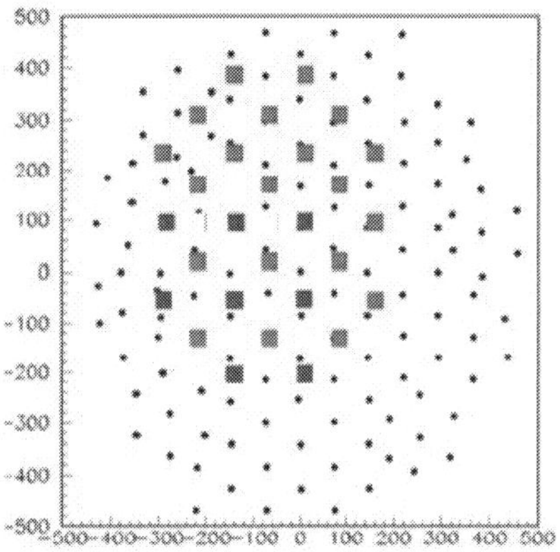

Figure 5. Layout of Tunka-HiSCORE (blue squares: stations deployed in 2013, red: in 2014) and the inner Tunka-133 stations (black dots).

The preliminary primary energy spectrum in the range 200 TeV - 20 PeV derived from the 2013 Tunka-HiSCORE data is presented in figure 1. The 2013 configuration consisted of 9 detector stations on a rectangular grid and covered an area of 0.09 km^2.

4. Tunka-Grande: a net of scintillation particle detectors

The main motivation to implement a scintillation array – Tunka-Grande – into TAIGA is to better distinguish the electron-photon component and the muon component of EAS. We intend to use surface scintillation detectors for cross-calibration of the different methods for high energy cosmic-ray studies with the sub-arrays Tunka-133, Tunka-Rex, Tunka-HiSCORE and Tunka-Grande. The underground muon detector will be helpful to determine the type of a primary particle. The possibility to select EAS initiated by gamma rays is based on the fact that the number of muons in EAS induced by charged cosmic rays is on average 30 times higher than in gamma-ray events. Moreover, the muon number in cosmic-ray induced EAS depends on the primary nuclear charge. The distinction of samples enriched with nuclei of certain masses will work well for energies above 100 TeV. However, the overall area of muon detectors S_μ should be at least 0.2-0.3% of the total array area; for Tunka-Grande that means S_μ ~2000-3000 m^2. This large area requires a strict cost reduction per unit of muon detector area.

We have deployed 19 scintillation stations, each with a surface and an underground part (figure 6). The stations are located at distances of ~20 m from the centres of the Tunka-133 clusters. In December 2014 all 19 surface detectors were put into operation. Each of them includes 12 scintillation counters with a size of 80×80×4 cm³ which have been formerly operated as part of EAS-TOP and KASCADE-Grande. At the moment seven of the scintillation stations are equipped with underground muon detectors, each segmented into eight smaller counters. DAQ, synchronization and control systems are the same as for Tunka-133.

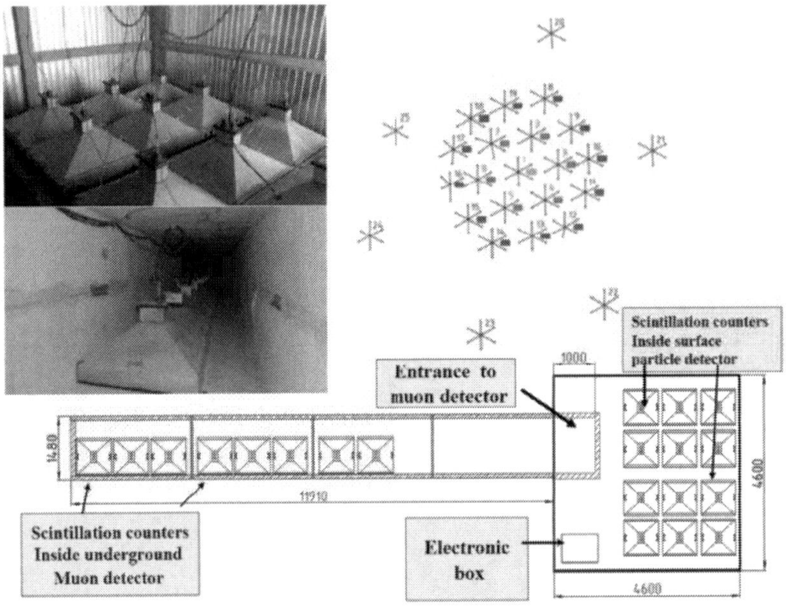

Figure 6. Tunka-Grande, 2014 configuration. Left top: inside a surface detector; Left middle: inside an underground muon detector; Right top: layout of the 19 stations (green and red rectangles - with muon detectors, blue – at present only surface detectors) embedded into the Tunka-133 array; Bottom: layout of a scintillation station.

For the next stage of Tunka-Grande we look for an optimum, cheap design for the muon detectors. The following options will be tested: 1) scintillation detectors of various types; 2) water Cherenkov detectors on the basis of up to 10 m long pipes; 3) Cherenkov detectors in the form of large area water tanks. It is supposed that muon detectors will be bundled into clusters of 30-50 m^2. Each cluster will be connected to the data acquisition center with a fiber-optics cable. The data acquisition system, the cluster synchronization with its accuracy of 1 ns and the digital trigger system are under design. The possibility to reconstruct the energy and the EAS maximum by Tunka-133 and Tunka-HiSCORE data and to estimate the number of muons will allow us to start searching of point-like and diffuse gamma radiation with energies higher than 1 PeV. The combination of Cherenkov and muon detectors of large area will help us clarifying the change of the mass composition around the knee (at 3 PeV).

5. The Tunka-IACT array

To increase the sensitivity and to decrease the threshold, to organize monitoring of certain sources and – first of all – to separate more reliably gamma-ray initiated showers from the background of charged cosmic-ray initiated showers, it is essential to add an array of Imaging Atmospheric

Cherenkov Telescopes to TAIGA. We have designed and started constructing the first Tunka-IACT, a prototype for the full array. We plan to implement a net of IACTs into TAIGA. The reflectors of these telescopes will have an area of ~10 m^2, and a focus of 4.75 m. We are designing imaging camera matrices consisting of ~400 PMTs, with a field of view of 8×8 degrees and a pixel angular size of 0.36°. Preliminary simulation results show that the joint operation of the wide-angle Cherenkov array HISCORE and a system of IACTs can be a very effective, inexpensive and a fast way to extend gamma-ray astrophysics into the yet unexplored super-high energy domain. Moreover, the IACTs improve the sensitivity at the lower part of the HiSCORE target region. The basic idea is that the two arrays (Tunka-HiSCORE and Tunka-IACT) will work in a coincidence mode and will complement each other: by using the timing and the imaging methods, showers with high precision can be measured. The results of reconstruction of core position and arrival direction of coincident EAS events from the HiSCORE array (expected accuracy of core location ~5-6m and arrival direction ~0.1 degree) will help an IACT in rejecting the high energy background events on large impact parameters of up to 300-500 m. On its turn, the IACT, observing a concrete strong source in the sky, for example the Crab Nebula, can with high efficiency tag the coincident high energy gamma event candidates, which can be used by the Tunka-HiSCORE array for training selection algorithms and improving its gamma/hadron discrimination. It is particularly important that the distance between the IACTs can be made considerably larger than for H.E.S.S., MAGIC, VERITAS or CTA which are not combined with timing arrays. Actually, for an array of 1 km^2 area, only 4-9 IACTs (the number will be specified in the course of further simulations) will be needed. Due to a dedicated, optimised design of the TAIGA configuration, the total cost of such a hybrid array can be kept an order of magnitude lower than the cost of traditional IACT configurations covering the same area but built exclusively with classical, narrow-angle IACTs.

6. The Tunka-Rex array

Tunka-Rex is a radio extension of the Tunka-133 air-Cherenkov array and the Tunka-Grande scintillation array. Every cluster of Tunka-133 includes one antenna station. Recently, for each scintillation station one additional antenna station has been deployed. These antennas measure the radio emission generated by the electromagnetic air-shower component. One major advantage of the radio detection is that it is not restricted to clear nights, as the air-Cherenkov detection. However, the threshold for radio detection is higher. The efficiency becomes comparable to the other detectors only around 10^{18} eV, exactly at those energies for which Tunka-133 is limited by low statistics. Consequently, when triggered in future by Tunka-Grande around the clock, the total statistics at the highest energies could be increased by an order of magnitude.

Each antenna station consists of two orthogonally aligned SALLA antennas, an economic and robust antenna type [17]. The antennas are connected in slave mode to the DAQ of Tunka-133 and Tunka-Grande, respectively. This means that Tunka-Rex is triggered externally, and the radio signal is read out simultaneously with the other detectors. In this way automatically hybrid events are obtained. The first two years of data from 2012 to 2014 are used for cross-calibration: The radio reconstruction of the energy and the shower maximum is compared to the air-Cherenkov reconstruction to check the precision of Tunka-Rex. Preliminary results indicate that at least the energy precision is competitive with Tunka-133 [18].

As a next step, the radio measurements can be combined with the particle detectors to increase the total accuracy for the air-shower parameters: While antenna arrays provide a calorimetric measurement of the electromagnetic component and good sensitivity to the longitudinal shower development [19], measurements of the secondary electrons and muons at ground give a complementary access to the energy and mass composition of the primary particles. In future, this concept can easily be applied to the larger muon array of TAIGA by installing additional antennas.

7. Conclusions

In 2014, the TAIGA collaboration has continued the construction of an array complex of hybrid detector systems. The goal is to search for new local Galactic sources of gamma rays with energies higher than 20-30 TeV. We will also study signals from the nearby extragalactic sources Mrk421 and Mrk-501 in order to investigate the gamma-ray absorption by intergalactic background radiation and to search for axion-photon transitions. The study of gamma radiation in the high-energy range is of interest not only for astrophysics, but also for testing theories predicting a violation of Lorenz invariance and for searching for super-heavy dark matter. Joint operation of the first Cherenkov telescopes of the IACT and the HiSCORE arrays in the energy range of 30-100 TeV will yield sensitivities of the order of 10^{-12} erg cm^{-2} s^{-1} (requiring detection of 50 events in 500 hours of observation). That would give us a good opportunity to measure the energy spectrum of gamma rays from the Tycho SNR, a main PeVatron candidate. For TAIGA – placed at 53° N/L – this source may be observed during more than 200 hours per year, given 50% of good weather condition. This sensitivity level would allow us searching for signals from the sources observed by IceCube as neutrinos, if these sources have Galactic origin, and would allow a survey for new PeVatrons.

Acknowledgments

This work was supported by the Russian Federation Ministry of Education and Science (14.B25.31.0010, №2014/51, project 1366, task № 3.889.2014/K), the Russian Foundation for Basic Research (Grants 13-02-00214, 13-02-12095, 15-02-05769), grant ISU (091-14-211, 091-14-222), by the Helmholtz Association, the RFBR (Grant HRJRG-303) and by the Deutsche Forschungsgemeinschaft (Grant TL51-3).

References

[1] Aharonian F et al. 2006 *Astronomy & Astrophysics* **454** 775–779
[2] Aharonian F et al. 2011 *Astropart. Phys.* **34** 738–747
[3] Albert J et al. 2006 *Astrophys.J.* **639** 761–765
[4] Acciari V et al. 2011 *Ap.J Letters* **730** L20
[5] Hillas M 1985 *La Jolla. Proc. 19th ICRC* **3** 445
[6] Actis M et al. 2011 *Experimental Astronomy* 121
[7] Antokhonov B A et al. 2011 *Nucl.Instrum.Meth.* A **639** 42–45
[8] Berezhnev S F et al. 2012 *Nucl.Instrum.Meth.* A **692** 98–105
[9] Prosin V, Berezhnev S, Budnev N et al. 2014 *Nucl.Instrum.Meth.* A **756** 94–101
[10] Prosin V, Berezhnev S, Budnev N. et al. 2014 *Noto, Sicily, Italy.PICAP 2014*
[11] Prosin V et al. 2014 *Nucl. Instr. and Methods in Physics Research* A **756** 94
[12] Sokolsky P 2011 *Nuclear Physics* B **74** 212–213
[13] Schulz A 2013 *Rio de Janeiro, Brazil. Proc. 33rd ICRC2013* ID=769
[14] Tluczykont M, Hampf D, Horns D et al. 2011 *Advances in Space Research* **48** 1935–1941
[15] Ritt S, Dinapoli R, Hartmann U 2010 *Nucl. Instrum. Methods Phys. Res.* A **623** 486
[16] Brückner M, Wischnewski R et al 2013 Rio de Janeiro, Brazil. Proc. 33rd ICRC2013 1158
[17] The Pierre Auger Collaboration, 2012 JINST **7** 10011
[18] Kostunin D et al. (Tunka-Rex Collaboration), 2014 Nulc. Instr. and Meth. **A742** 89
[19] Apel A et al. (LOPES Collaboration), 2012 Physical Review **D 85** 071101(R).
[20] Porelli A et al. (Tunka-HiSCORE collaboration), these proceedings
[21] Tluczykont M et al. (TAIGA collaboration), these proceedings

On conservative models of "the pair-production anomaly" in blazar spectra at Very High Energies

T.A. Dzhatdoev

Skobeltsyn Institute of Nuclear Physics, Lomonosov Moscow State University, Leninskie gory 1-2, 119991 Moscow, Russia

E-mail: timur1606@gmail.com

Abstract. For some blazars, the gamma-ray absorption features due to pair-production on the Extragalactic Background Light (EBL) are fainter than expected. The present work reviews the main models that could explain this paradox, with emphasis on conservative ones, that do not include any new physics. The models that are intrinsic to the source, do allow a very hard primary spectrum, but fail to explain a regular redshift dependence of the anomaly starting energy. The model that includes a contribution from secondary photons produced by cosmic rays (CR) near the Earth seems to require a well collimated CR beam, what is hard to achieve. Finally, the model with secondary photons produced in electromagnetic (EM) cascades initiated by primary gamma-rays is considered. In principle, it allows to decrease the statistical significance of the anomaly and, while requiring quite low EGMF strength B, does not contradict to most contemporary constraints on the B value. Additionally, it is shown that the recently observed correlation between directions to hard gamma-ray sources and voids in the Large Scale Structure is a natural feature of the EM cascade model.

1. Introduction

The number of detected extragalactic gamma-ray sources has greatly increased during the last decade. Blazars — gamma-ray loud active galactic nuclei (AGN)— constitute the great majority of these sources. By the start of 2015, ~50 blazars were discovered by ground-based detectors [1] (these instruments typically work at very high energy (VHE) range $E >100$ GeV), and $\sim 10^3$ blazars were observed by the Fermi LAT gamma-ray space telescope at $E= 100$ MeV–100 GeV [2]. Besides investigation of AGN intrinsic properties, these observations allow to study extragalactic gamma-ray propagation as well. High energy photons are subject to the $\gamma\gamma \rightarrow e^+e^-$ process on the Extragalactic Background Light (EBL) photons [3]–[4] with subsequent formation of electromagnetic (EM) cascades. Thus, primary gamma-rays with $E > E_{abs}$ ($E_{abs} = E(\tau_{\gamma\gamma} = 1)$; $\tau_{\gamma\gamma}$ is the optical depth of the pair-production process) are effectively "absorbed", and their energy is transferred to electrons, positrons (in what follows called simply "electrons") and photons with comparatively low energy. Absorption features of the $\gamma\gamma \rightarrow e^+e^-$ process were already observed by the Fermi LAT instrument [5], as well as the H. E. S. S. Cherenkov telescope [6] with high statistical significance: ~6σ and 8.8σ, respectively.

However, it appears that for some blazars modification of the gamma-ray spectrum in the optically thick regime has a somewhat anomalous character [7]. By extrapolating the spectrum from the optically thin regime, [7] found that the distribution of the flux points scatter around the predicted intensity is different for the 1< $\tau_{\gamma\gamma}$ <2 and $\tau_{\gamma\gamma}$ >2 regions, and a strong indication (significance $Z_a = 4.2\sigma$) that the observed absorption is smaller than the assumed one was obtained. This effect, observed at VHE range, was called "the pair-production anomaly" by [7].

In fact, an anomaly of such kind is a long-standing problem of VHE gamma-astronomy [8]–[9]; a number of solutions was proposed, including quite exotic ones, such as violation of Lorentz invariance (LIV) [10]–[11],[9], and oscillations of photons to Axion Like Particles (ALPs) (e.g., [12]–[14]). A search for new physical phenomena of such kind would require that all possible background processes are well known and appropriately accounted for. One such process is emission of secondary photons in EM cascades. On a subsample of spectra analysed in [7], it was recently shown that inclusion of this secondary component to the spectral fitting procedure indeed allows to considerably decrease the statistical significance of the anomaly [15]. In the present work the cascade model of the VHE anomaly in blazar spectra is compared with other models of the anomaly. The paper is organized as follows: section 2 contains a brief overview of possible processes that might relax the anomaly; the EM cascade (EMC) model is described in section 3; section 4 reviews some contemporary constraints on the Extragalactic Magnetic Field (EGMF) strength. Finally, some conclusions are presented in section 5.

2. The models of the VHE anomaly

2.1. EBL intensity models

The simplest assumption that could relax the anomaly would be that the EBL energy density is overestimated. If this is the case, the intrinsic spectrum would experience weaker modification than is believed, and for sufficiently low EBL density the VHE anomaly would be relaxed (note, however, that the dependence of the anomaly significance on the EBL intensity may have a non-trivial behaviour [16]).

We are mostly interested in the wavelength region λ=1–10 μm that roughly corresponds to 1–10 TeV minimal energy of high-energy photon that is subject to the $\gamma\gamma \rightarrow e^+e^-$ process. Many EBL models exist, including [17]–[24]; in the range of λ=1–10 μm the model of [17], that was utilized in [7], predicts one of the lowest EBL intensity among [17]–[24]. The [17] model is, in fact, a sort of lower limit on the EBL intensity. A well justified conclusion that such a limit must be revised would likely require a large amount of dedicated observational work, therefore we move to other possible explanations of the anomaly.

2.2. A list of effects that could relax the anomaly

Besides the EBL properties, a number of factors exist that may have some relation to the nature of the VHE anomaly. A non-exhaustive list of them is given below.

I. Effects, connected to the source.

I.1. Intrinsic spectrum may have a pile-up resulting from inverse Compton (IC) scattering on an ultrarelativistic outflow, thus increasing observed intensity at high energies [25].

I.2. Internal absorption of primary photons on radiation field near the source, under certain assumptions, again, may produce a very hard intrinsic spectrum [26] with a pile-up.

I.3. Another known mechanism of a hard spectrum formation is synchrotron radiation of electrons produced by ultrahigh energy (UHE) protons near the source [27].

I.4. It is believed that blazars have well collimated jets pointed towards the observer (the scheme of a typical blazar geometry taken from [28] is drawn in figure 1). Magnetic field in a typical jet likely has some turbulent component, that may scatter charged energetic particles (electrons and hadrons). This scattering angle is usually larger for charged particles with comparatively low energy, that emit predominantly low-energy photons. Thus, the beaming factor for high-energy photons would be larger, again increasing observed intensity at high energies. To the author's knowledge, this effect was never discussed in connection to the VHE anomaly yet.

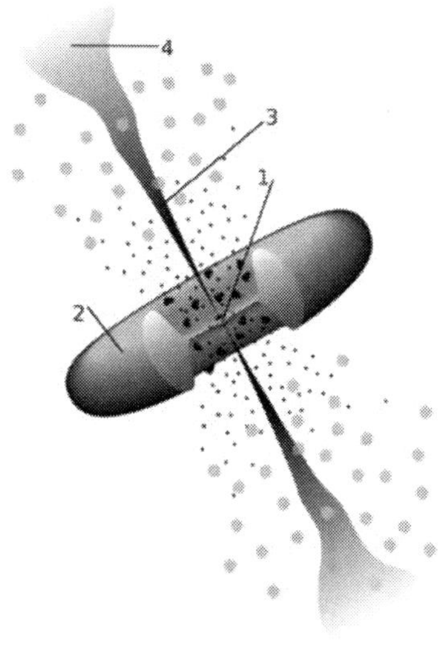

Figure 1. A sketch of a typical AGN geometry [28]. 1 — central compact object (black hole with accretion disk), 2 — dusty torus around the central object, 3 — well collimated relativistic jet with regular magnetic field $\vec{B}_{Jet}$ that can have some turbulent component $\delta\vec{B}_{Jet}$, 4 — a "lobe" at the end of the jet that likely contains a highly turbulent magnetic field region.

However, it was observed that the energy at which the anomaly shows up, exhibits strong, regular dependence on redshift (this energy is defined by the relation $E_a \approx E(\tau_{\gamma\gamma} = 2)$ [7]). The effects, intrinsic to the source, are not expected to display such a dependence, thus they can not constitute the only physical reason of the anomaly.
II. Propagation effects.
II.1. LIV effects, in principle, may suppress the $\gamma\gamma \rightarrow e^+e^-$ process [10]. This mechanism is expected to operate above a certain fixed energy, and so it is disfavoured for the same reason as the source-induced effects (see discussion in [7]).
II.2. γ-ALP oscillations (and vice versa) in magnetic field is another mechanism that makes the primary gamma-rays able to avoid strong absorption [13]. Even if we assume the existence of ALP and that the γ-ALP oscillation process is indeed allowed, there is another potential difficulty: the efficiency of direct ($\gamma \rightarrow ALP$) and reverse ($ALP \rightarrow \gamma$) conversion is required to be high enough for many sources, independently of direction to a source, and, therefore, of the Galactic magnetic field parameters in this direction, as well as of the source' magnetic field. The recent data didn't allow the authors of [29] to put any constraints on the ALP mechanism, but the same work shows that such constraints are obtainable in the future.
II.3. If blazars accelerate hadrons as well as electrons, secondary gamma-rays produced by the former may contribute to observed spectrum and thus relax the anomaly [30]–[31]. This model is discussed below.
II.4 As was already mentioned, the most basic process that may influence the shape of observed spectrum is emission of secondary photons in electromagnetic cascades [8], [25], [32], [15]. This process is discussed in section 3.

2.3. The CR beam model
Secondary photons produced by cosmic ray beam (by means of pair-production and photohadronic processes with subsequent development of electromagnetic cascades) relatively near to the Earth may enhance the intensity in the optically thick regime. Secondary gamma-ray spectra, typical for this mechanism, were studied in [33]–[36]. The case of a very distant source (with redshift $z \sim 1$) was considered in [37]. Time structure of VHE gamma-ray signal was discussed in [38]. CR beam may also produce an observable amount of neutrinos [33]–[34],[39].

This model of the VHE anomaly has many attractive features: it allows to explain the observed spectrum without invoking any exotic processes such as γ-ALP oscillations. Moreover, the shape of the secondary photon component, assuming sufficiently low strength of the EGMF $B < 10^{-14}$ G, depends mainly on the overall normalization of the primary CR spectrum. However, the model is not out of difficulties. As an example, let us consider the 1ES 1101-232 blazar spectrum (redshift z=0.186). Figure 2 shows a spectral energy distribution (SED) of the source measured by the H. E. S. S. Cherenkov telescope (red circles) together with uncertainties (dashed red lines); the assumed intrinsic spectrum is shown by black curve, the fit to the observed spectrum is shown by green line (for detals of calculations, that were performed with the publicly-available code ELMAG 2.02 [40] and the EBL model of [17], see [15]). The intrinsic spectrum curve here is normalized to the absorbed one at E= 200 GeV, where absorption effects are already small. The fit here doesn't include any secondary photon component.

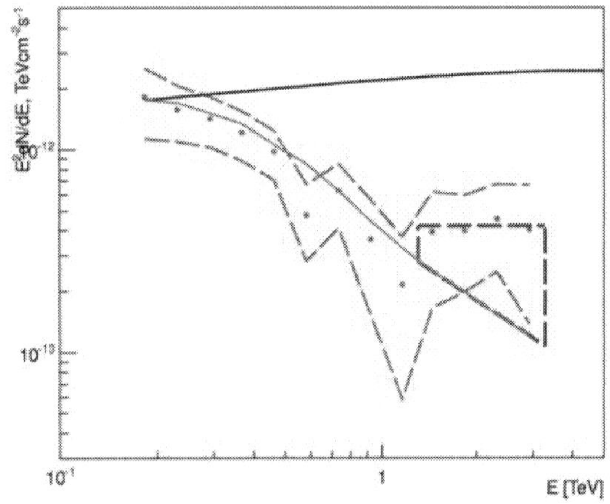

Figure 2. The fit to the 1ES 1101-232 SED [41] without account of any secondary emission [15].

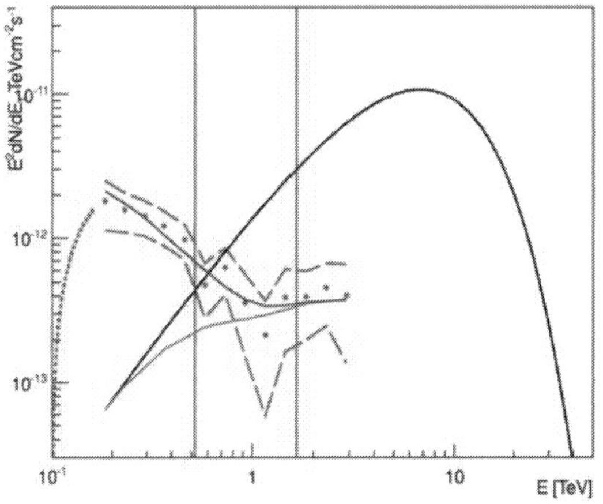

Figure 3. The fit to the 1ES 1101-232 SED including cascade component (blue curve). Figure from [15].

To explain the data in the framework of the CR beam model, an additional component from secondary photons is needed (the corresponding region of the spectrum is denoted by thick blue dashed lines). Now let us put a simple lower limit to the ratio of total power of accelerated CR W_{CR} to the same quantity for VHE electrons $W_{VHE-e} = W_{VHE-\gamma}/f_{IC}$, where f_{IC} is the fraction of energy of VHE electrons transferred to VHE photons by means of IC scattering. An important note is that gamma-rays produced by electrons deep inside blazar jets are expected to be highly beamed. On the other hand, a region of highly turbulent magnetic field (a "lobe") is expected to exist at the end of a jet, and CR particles are likely to be isotropized before they could produce secondary photons close to the Earth. Indeed, such "lobes" are observed in a number of radiogalaxies, and are expected to be able to scatter (or even confine) protons up to 10^{20} eV [42]. Therefore, if the observer is located near to the jet's axis, the observed flux from CR is effectively deamplified to a factor of $f_{beam} \sim 10^3$–10^4, assuming the angular radius of the jet θ_{Jet}=1–3°. Additionally, only a small fraction $f_{int} \sim 10^{-3}$ [31] of CR energy is converted to photons near the observer.

Assuming the intrinsic VHE gamma-ray spectrum $dN/dE \sim E^{-2}$ (what is close to the shape of black curve in Figure 2) from 100 GeV to 100 TeV, one may estimate that the secondary photons constitute a fraction $f_{sec} \sim 10^{-2}$ of energy of intrinsic VHE photons. Therefore, assuming $f_{IC} = 0.1$, we find that $W_{CR}/W_{VHE-e} \sim f_{sec} \cdot f_{beam} \cdot (f_{IC}/f_{int})$= 10^3–10^4, depending of the f_{beam} value, i.e. acceleration of hadrons is very effective, while for VHE electrons it is not. This challenges conventional models of gamma-ray blazars. Additionally, assuming

$W_{VHE-e} = 10^{44}$–$10^{45} erg/s$, we find W_{CR}= 10^{47}– $10^{49} erg/s$, a very large value. The last values are similar to those obtained by [43].

3. The EMC model

3.1. General considerations

The primary photon absorption process $\gamma\gamma \rightarrow e^+e^-$ has threshold energy $E_{\gamma 0} = m_e^2/\epsilon$, where m_e is electron mass, and ϵ is EBL photon energy (we assume c=1 here). Most of pair-production acts occur not far from the threshold; in this case each produced electron receives energy $E_e \sim E_{\gamma 0}/2$. Electrons, in their turn, produce secondary (cascade) photons by IC scattering. This process, in contrast to pair-production, doesn't have kinematical threshold, and occurs mainly on dense, low-energy cosmic microwave background (CMB) photon field; the energy of secondary photons $E_{\gamma-casc} \sim 3 \cdot (E_e/1\ TeV)^2\ GeV$ [40]. In this work we consider the case 100 GeV < $E_{\gamma 0}$ < 100 TeV, thus $E_{\gamma-casc} < 7.5\ TeV << E_{\gamma 0}$; for the case of $E_{\gamma 0} = 10\ TeV$ $E_{\gamma-casc} = 75\ GeV$, again $<< E_{\gamma 0}$. Therefore, the cascade process occurs in a peculiar, "degenerate" regime, and a typical number of cascade generations is small, n_{Int} =1–2.

In absence of EGMF the angular radius of cascade $\theta_{Int} \sim n_{Int}^{0.5}/\Gamma$, where $\Gamma = E_e/m_e > 10^5$ — Lorentz factor of a cascade electron with E_e >100 GeV= 10^5 MeV. As we have seen, $\theta_{Jet} \sim 1° \sim 10^{-2}$ rad; then $\theta_{Int} < 10^{-5} << \theta_{Jet}$, therefore the cascade process weakly modifies the angular distribution of photons from the source and 1D assumption that we shall use is well justified. Recently it was argued that plasma beam instabilities could dominate the energy losses for intergalactic cascade electrons [44]. However, this mechanism is under debate: it was noted that the relativistic length contraction effect "compresses the electric and associated magnetic fields" around charged particles, and so, in fact, they do not constitute a plasma [45]. For this reason, we do not include any electron energy loss process, except IC, in our calculations.

3.2. The signatures of cascade emission

Now let us illustrate the basic features of the EMC model of the VHE anomaly (figure 3). A fit to the measured SED was performed [15], including the secondary component from electromagnetic cascades. The notations are the same as in figure 2, but the total model SED is now denoted by blue curve; the model SED without account of the cascade component is drawn by green curve. The impact of cascades to the SED is the difference between the blue and green curves. Blue vertical line denotes the "gamma-ray horizon" (the energy, where $\tau_{\gamma\gamma}$=1); red vertical line — the energy, defined by condition $\tau_{\gamma\gamma}$=2. As in figure 2, the intrinsic SED in figure 3 was normalized to the absorbed one at E= 200 GeV.

The above-discussed rule $E_{\gamma-casc} << E_{\gamma 0}$ is clearly visible in figure 3, and the cascade component contributes mainly in the optically thin regime. However, to achieve a good fit, we were compelled to take a very hard primary spectrum, much harder, than in figure 2 for the same object neglecting the secondary component. Therefore, the model intensity in the optically thick regime in figure 3 (with account of the cascade component) is much higher, than in figure 2 (without this component), and for $\tau_{\gamma\gamma}$ >2 the fit is in much better agreement with observations. It is quite remarkable that an exceptionally good fit of the shape of the spectrum in figure 3 (13 energy bins) was achieved with a very simple parametrization of the primary spectrum $dN/dE \propto E_0^{-\gamma} \cdot exp(-E_0/E_c)$ with only two parameters γ and E_c. As well, the significance of the anomaly for $\tau_{\gamma\gamma}$ >2 drops from 2.1 σ for the fit shown in figure 2 (without secondary component) to only 0.5 σ for the new fit that includes the cascade component (for the significance evaluation method see [15]).

A possible influence of EGMF to the observed spectrum is schematically shown in Figure 3 by red dashed curve. In order to not spoil the fit, the EGMF strength B should be sufficiently small, below some value B_{Max}. Recently A. Neronov et al. [46] performed calculations of this "magnetic cutoff" shape for the case of blazar Mkn 501 (z= 0.034), and found that for $B = 10^{-16}$

G the cutoff starts to be detectable below $E \sim 100$ GeV. Given that, as was noted above, the number of generations in cascade is small, and that the last low-energy point of the observed spectrum, located at $E \sim 200$ GeV, doesn't show a prominent magnetic cutoff, we can estimate $B_{Max} \sim (1-2) \cdot 10^{-16}$ G. Plausibility of such B values will be discussed below in section 4.

A spectrum of blazar in the considered version of the EMC model has three characteristic features, or signatures: a) low-energy magnetic cutoff b) the "dip", or a kind of an "ankle" that is usually located in the energy region where $1 < \tau_{\gamma\gamma} < 2$ c) high-energy cutoff (not shown in figure 3). The third feature, the high-energy cutoff, was already studied in [35], [47]. The detailed properties of this signature are not known, but the general conclusion that, in principle, allows to discriminate between the CR beam model and the EMC model, is the following: for large $\tau_{\gamma\gamma}$ (say, >2-3) the high-energy cutoff is much more marked for the case of EMC model [35], [47]. The dip or ankle feature, as well, exists in both the CR beam (see, e.g., [34]) and the EMC model. However, the low-energy magnetic cutoff is much more marked in the EMC model when compared to the CR beam model, unless the secondary component produced by CR dominates the entire VHE spectrum. We will see in section 4 that existing hints at intergalactic cascades tend to favour the EMC model.

4. Some constraints on the EGMF strength

As we have seen, by now there exist only two conservarive models that could relax the VHE anomaly in blazar spectra, not to mention the option that the EBL intensity models must be revised: the CR beam model and the EMC model. Both require sufficiently low EGMF strength in voids to be plausible: $B < 10^{-14}$ G for the former [38], and even lower values (usually around $B < 10^{-16}$ G) for the latter.

Any conclusive measurement of EGMF strength is still absent; some constraints on this quantity are presented in figure 4 (usually they are obtained for the EGMF coherence length 1 Mpc [48]). Upper bounds on the B value are scarce and rather weak; [49] obtained a constraint $B < 10^{-9}$ G. $B = 2 \cdot 10^{-12}$ G was found to be sufficient to explain magnetic fields in galaxy clusters [50]. As well, we present a number of results obtained from non-observation of cascade component in blazar spectra at comparatively low energies, $E <100$ GeV. These constraints are highly model-dependent; [51] even found that the zero EGMF hypothesis cannot be firmly rejected. Anyway, we present the cascade constraints obtained by [52]–[54] in figure 4; these lower limits are in the range $B > 10^{-18}$–10^{-17} G. By analysing the angular distribution of arriving photons, [55] performed a search of a magnetically broadened cascade pattern and was able to exclude the range of $B = 3 \cdot 10^{-16}$–10^{-14} G at 99 % C.L. Finally, by thick red horizontal line we show the $B = 10^{-14}$ G value, above which both the CR beam model and the EMC model are ruled out [38]. Future instruments with high sensitivity (e.g. CTA [56]), or with good angular resolution, such as emulsion gamma-ray telescope [57] or the GAMMA-400 detector [58], would be able to employ the same method as [55] to put more tight constraints on the EGMF strength. Other EGMF searches based on stacking analysis of sources' angular distribution [59] or diffuse gamma-ray sky studies [60]–[61] do exist.

Finally, let us mention some recent results that actually give hints at the EM cascade presence in blazar spectra. Analysing the Mkn 501 spectrum, [46] found that it could be explained by the intergalactic cascade hypothesis with $B = 10^{-17}$–10^{-16} G. Very recently, it was shown that gamma-ray sources with hard spectra are predominantly located in directions to voids [62]. Moreover, it was found by [62] that the EBL intensity fluctuations are not sufficient to explain the observed effect. Therefore, as first suggested in [15], the most likely explanation is that [62] actually observed the low-energy magnetic cutoff of secondary component, produced by EM cascade. This feature of the EMC model is schematically shown by dashed red line in figure 3. For the case of other objects, that do not point to voids, the mean B value is likely much greater, and the cascade component may be greatly suppressed. For this reason, these latter

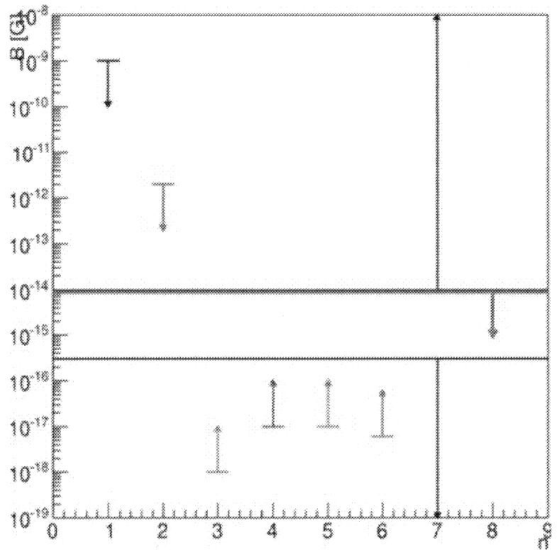

Figure 4. Some constraints on the EGMF strength in voids. n denotes the number of corresponding work. n=1 — [49]; n=2 — [50]; n=3 — [52]; n=4 — [53]; n=5 — [54], the case of the EBL model [20]; n=6 — [54], the case of the EBL lower limit from direct source counts; n=7 — [55] n=8 — [38].

objects do not show the spectral hardening observed by [62].

In principle, two other explanations of this effect are possible. The first is that the electromagnetic cascade upscatters the secondary photons to high enough energy to produce observable hardening. However, this is highly unlikely, given that a typical energy of secondary photon is much smaller than of the primary one, as discussed in subsection 3.1. The other explanation, that the effect observed by [62] is caused by the CR-initiated cascade, is unlikely as well, unless the secondary component from these cascades dominates the VHE spectrum.

5. Conclusions

In the present work a brief review of conservative models that could explain the apparent faintness of the pair-production features in blazar spectra, was given. No existing model was found to be completely free of difficulties. The source-intrinsic models, while they are able to form a very hard intrinsic spectrum, do not explain other features of the anomaly (subsection 2.2). The model with production of secondaries by CR near to the Earth seems to require either a very high W_{CR}/W_{VHE-e} ratio, or extremely good collimation of accelerated nuclei (subsection 2.3). Finally, the EM cascade (EMC) model was considered (section 3), that naturally explains some spectra with only two free parameters, doesn't contradict to contemporary constraints on the EGMF strength, and predicts the magnetic cutoff feature that may be already observed [62]. Future observations will help to test this model.

Acknowledgments

The work was supported by the Russian President grant LSS-3110.2014.2. The author is grateful to Prof. A. Kusenko for helpful discussions, to the anonymous referee for comments, and to Dr. V.V. Kalegaev, V.O. Barinova, M.D. Nguen, D.A. Parunakyan for permission to use the SINP MSU space monitoring data center computer cluster.

References

[1] http://tevcat.uchicago.edu/
[2] Ackermann M et al. (Fermi LAT) 2011 *ApJ* **743** 171
[3] Nikishov A I 1962 *Sov. Phys. JETP* **14** 393
[4] Gould R J and Shreder G 1967 *Phys. Rev.* **155** 1408
[5] Ackermann M et al. (Fermi LAT) 2012 *Science* **338** 1190
[6] Abramowski A et al. (H.E.S.S.) 2013 *A&A* **550** A4
[7] Horns D and Meyer M 2012 *JCAP* 02 033

[8] Aharonian F A et al. (HEGRA) 1999 *A&A* **349** 11
[9] Protheroe R J and Meyer H 2000 *Phys. Lett.* B **493** 1
[10] Kifune T 1999 *ApJ* **518** L21
[11] Amelino-Camelia G and Piran T 2001 *Phys.Lett.* B **497** 265
[12] Sikivie P 1983 *Phys. Rev. Lett.* **51** 1415
[13] Mirizzi A et al. 2007 *Phys. Rev.* D **76** 023001
[14] Sanchez-Conde M A et al. 2009 *Phys. Rev.* D **79** 123511
[15] Dzhatdoev T 2015 *Izvestiya Rossiiskoi Akademii Nauk* **79** 363 (In print) (*Preprint* astro-ph/1501.00259)
[16] Meyer M et al. 2012 Revisiting the Indication for a low opacity Universe for very high energy gamma-rays *Preprint* astro-ph/1211.6405
[17] Kneiske T M and Dole H 2010 *A&A* **515** A19
[18] Kneiske T M et al. 2004 *A&A* **413** 807
[19] Stecker F W et al. 2006 *ApJ* **648** 774
[20] Franceschini A et al. 2008 *A&A* **487** 837
[21] Primack J R et al. 2008 *AIP Conference Proceedings* **1085** 71
[22] Dominguez A et al. 2011 *MNRAS* **410** 2556
[23] Finke J D et al. 2010 *ApJ* **712** 238
[24] Gilmore R C et al. 2012 *MNRAS* **422** 3189
[25] Aharonian F A et al. 2002 *A&A* **384** 834
[26] Aharonian F A et al. 2008 *MNRAS* **387** 1206
[27] Oikonomou F et al. 2014 *A&A* **568** A110
[28] Padovani P 1997 *Very High Energy Phenomena in the Universe; Morion Workshop. Eds. Y. Giraud-Heraud and J. Tran Thanh Van, 1997* 7
[29] Wouters D and Brun P 2014 *JCAP* 01 016
[30] Uryson A 1998 *JETP* **86** 213
[31] Essey W and Kusenko A 2010 *APh* **33** 81
[32] d'Avezac P et al. 2007 *A&A* **469** 857
[33] Essey W et al. 2010 *Phys. Rev. Lett.* **104** 141102
[34] Essey W et al. 2011 *ApJ* **731** 51
[35] Murase K et al. 2012 *ApJ* **749** 63
[36] Essey W and Kusenko A 2014 *APh* **57** 30
[37] Aharonian F A et al. 2013 *Phys. Rev.* D **87** 063002
[38] Prosekin A et al. 2012 *ApJ* **757** 183
[39] Kalashev O E et al. 2013 *Phys. Rev. Lett.* **111** 041103
[40] Kachelriess M et al. 2012 *Comp. Phys. Comm.* **183** 1036
[41] Aharonian F A et al. (H. E. S. S.) 2006 *Nature* **440** 1018
[42] Hillas A M 1984 *ARA&A* **22** 425
[43] Razzaque S et al. 2012 *ApJ* **745** 196
[44] Broderick A E et al. *ApJ* **752** 22
[45] Venters T M and Pavlidou V 2013 *MNRAS* **432** 3485
[46] Neronov A et al. 2012 *A&A* **541** A31
[47] Takami H et al. 2013 *ApJ Lett.* **771** L32
[48] Akahori T and Ryu D 2010 *ApJ* **723** 476
[49] Blasi P et al. 1999 *ApJ* **514** L79
[50] Dolag K et al. 2005 *JCAP* 01 009
[51] Arlen T C et al. 2014 *ApJ* **796** 18
[52] Dermer C D et al. 2011 *ApJ Lett.* **733** L21
[53] Taylor A M et al. 2011 *A&A* **529** A144
[54] Vovk Ie et al. 2012 *ApJ Lett.* **747** L14
[55] Abramowski A et al. (H.E.S.S.) 2014 *A&A* **562** A145
[56] Acharya B S et al. (CTA) 2013 *APh* **43** 3
[57] Aoki S et al. 2012 Balloon-borne gamma-ray telescope with nuclear emulsion: overview and status *Preprint* astro-ph/1202.2529
[58] Galper A M et al. (GAMMA-400) 2013 *AIPCP* **1516** 288
[59] Chen W et al. 2014 Evidence for GeV Pair Halos around Low Redshift Blazars *Preprint* astro-ph/1410.7717
[60] Tashiro H et al. 2014 *MNRAS Lett.* **445** L41
[61] Chen W et al. 2014 Intergalactic magnetic field spectra from diffuse gamma rays *Preprint* astro-ph/1412.3171
[62] Furniss A et al. 2015 *MNRAS* **446** 2267

Supernova remnants and pulsar wind nebulae with Imaging Atmospheric Cherenkov Telescopes (IACTs)

Peter Eger

Max-Planck-Institut für Kernphysik, P.O. Box 103980, D-69029 Heidelberg, Germany

E-mail: peter.eger@mpi-hd.mpg.de

Abstract. The observation of very-high-energy (VHE, E>100 GeV) gamma rays is an excellent tool to study the most energetic and violent environments in the Galaxy. This energy range is only accessible with ground-based instruments such as Imaging Atmospheric Cherenkov Telescopes (IACTs) that reconstruct the energy and direction of the primary gamma ray by observing the Cherenkov light from the induced extended air showers in Earths atmosphere. The main goals of Galactic VHE gamma-ray science are the identification of individual sources of cosmic rays (CRs), such as supernova remnants (SNRs), and the study of other extreme astrophysical objects at the highest energies, such as gamma-ray binaries and pulsar wind nebulae (PWNe).

One of the main challenges is the discrimination between leptonic and hadronic gamma-ray production channels. To that end, the gamma-ray signal from each individual source needs to be brought into context with the multi-wavelength environment of the astrophysical object in question, particularly with observations tracing the density of the surrounding interstellar medium, or synchrotron radiation from relativistic electrons.

In this review presented at the European Cosmic Ray Symposium 2014 (ECRS2014), the most recent developments in the field of Galactic VHE gamma-ray science are highlighted, with particular emphasis on SNRs and PWNe.

1. Introduction

Galactic gamma-ray astronomy at very high energies (VHE, E>100GeV) has experienced a dramatic increase in activity over the past 10 years, thanks to the advent of the current generation of ground-based imaging atmospheric Cherenkov telescopes (IACTs), such as H.E.S.S. [1], MAGIC [2] and VERITAS [3]. This energy range is currently only accessible with ground-based instruments, featuring large effective areas, due to the comparatively low photon fluxes emitted from sources at these energies. Up to now, more than 70 individual Galactic sources in this energy regime have been detected, belonging to a large variety of source classes, ranging from young supernova remnants (SNRs) and pulsar wind nebulae (PWNe) to gamma-ray binaries and massive stellar clusters (see e.g. TeVCat[1]). For a comprehensive review on the subject, see [4].

The main quests of Galactic VHE gamma-ray science are 1) the study of the most extreme non-thermal astrophysical environments in the Milky Way at the highest energies, and 2) the search for the most powerful sources of Galactic cosmic rays that accelerate particles up to PeV energies ("PeVatrons"). VHE gamma rays are produced in the interaction of ultra relativistic

[1] http://tevcat.uchicago.edu

electrons or protons with the ambient interstellar matter or radiation fields via π^0 production and subsequent decay (protons) or inverse Compton scattering (electrons).

Ground-based gamma-ray astronomy with IACTs covers a very interesting energy range (100 GeV – 100 TeV) to study the origin of Galactic cosmic rays. One of the most popular explanations for the "knee" in the cosmic ray energy spectrum at a few PeV is a transition from Galactic to extra-galactic sources of cosmic rays [see, e.g., 5; 6], implying that the most powerful Galactic accelerators feature spectra with steep cut-offs at these energies. One way to prove this interpretation is the detection of individual Galactic sources of secondary gamma rays from the hadronic channel with spectra extending up to 100 TeV [see, 7, and references therein], an energy which lies at the extreme end of the capabilities of current IACTs, due to the very low fluxes at these energies.

The best candidates for accelerators of Galactic cosmic rays up to PeV energies are the expanding shocks of young (~1000 yr) SNRs which are believed to accelerate particles through the Fermi-type process [8; 9]. Indeed, VHE gamma-ray emission has been detected from several Galactic SNRs, predominantly by H.E.S.S. in a systematic scan of the Galactic Plane as well as dedicated follow-up observations [10].

The most abundant Galactic VHE gamma-ray source class, however, are pulsar wind nebulae (PWNe) which are characterized by large-scale non-thermal emission throughout the electromagnetic spectrum surrounding young and powerful pulsars. PWNe are believed to be caused by highly relativistic electrons accelerated at the termination shock of a stream of energetic particles originating from the pulsars' magnetosphere. These shock-accelerated electrons diffuse into the surrounding interstellar medium (ISM) and cool via synchrotron (SR) and inverse-Compton (IC) radiation through the interaction with ambient magnetic (SR) or photon (IC) fields [for a detailed review see, 11]. VHE gamma-ray observations offer a unique window into the acceleration processes and particle propagation present in PWNe through the observation of IC radiation, as opposed to SR which dominates at lower energies. The latter is influenced not only by the characteristics of the underlying population of relativistic electrons, but also by the magnetic field distribution which generally is an unknown quantity. In contrast, IC radiation in gamma-rays is only governed by the target photon field energy density (usually dominated by the well-known cosmic microwave background) which allows to directly measure the spectrum and spatial distribution of the relativistic electrons.

The discrimination between the two gamma-ray production channels, namely π^0 decay and IC scattering, is one of the main challenges of VHE gamma-ray astronomy. Here it is necessary to compare the VHE gamma-ray signal to emission seen in other wavelength regimes, such as radio continuum, X-rays and molecular line transitions. For instance, a correlation of the VHE gamma-ray signal with non-thermal X-ray and/or radio emission may point towards relativistic electrons which radiate through IC and SR emission in the respective energy bands. On the other hand, if the VHE gamma-ray emission corresponds well with the spatial distribution of dense molecular gas, measured through the observation of molecular line transitions, relativistic protons are the most likely explanation, due to production and subsequent decay of neutral pions in dense target environments.

2. Imaging Atmospheric Cherenkov Telescopes (IACTs)

IACTs detect gamma-rays through the observation of Cherenkov light emitted by secondary charged particles in the extensive air showers caused by the primary gamma ray (see Fig. 1). IACTs consist of large (12 - 28 m) optical reflecting mirrors coupled with very fast cameras and readout electronics to facilitate the imaging of the very short (~10 ns) flashes of Cherenkov light. The cameras of the three major IACTs H.E.S.S., VERITAS, and MAGIC, consist of ~1000 - 2000 pixels of photomultiplier tubes (PMTs) and feature a field of view of up to 5 deg.

To improve the rejection of background caused by charged primaries (mainly protons) all

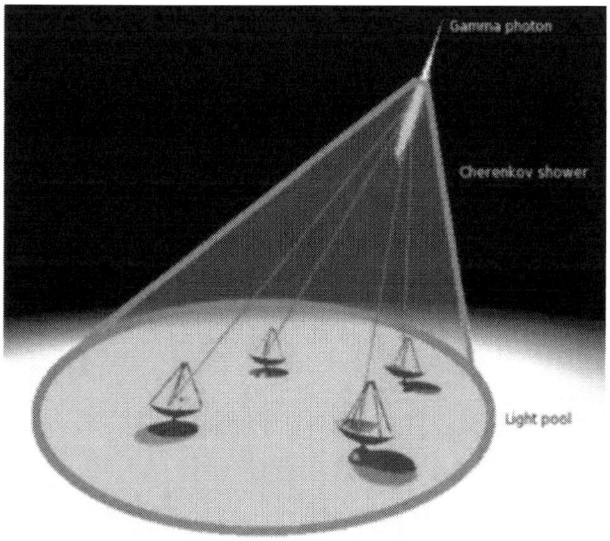

Figure 1. Illustration of the VHE gamma-ray detection technique for IACTs. The shower maximum is typically at an altitude of ~10km. The size of the Cherenkov light pool on the ground is about 200 m in diameter.

current IACTs work in an array configuration of up to five individual telescopes. Through this stereoscopic approach where the same shower is seen from several angles the Cherenkov images of the electromagnetic showers from gamma-ray primaries can be well discriminated from the predominantly hadronic showers caused by cosmic rays. Also, the reconstruction of the direction and energy of gamma rays is significantly improved in stereoscopic observations, yielding angular resolutions better than 0.1 deg and systematic uncertainties of energy reconstruction below 20% for modern IACTs.

All three major IACTs currently in operation have recently undergone large hardware upgrades. MAGIC, situated on the Canary island of La Palma, was upgraded through the addition of a second telescope of the same size enabling stereoscopic observations increasing background rejection and overall sensitivity. The cameras of the VERITAS 4-telescope array, located in Arizona (USA), were upgraded with high quantum efficiency PMTs, significantly lowering the energy threshold and sensitivity of the system [12]. The H.E.S.S. array, located in the Khomas Highland of Namibia, received a major upgrade by the addition of a fifth, larger 28 m telescope (see Fig. 2) lowering the energy threshold from 100 to about 30 GeV, as well as improving the overall effective area by about a factor of two.

Figure 2. The upgraded H.E.S.S. II array with the completed fifths, large telescope in the center.

3. Pulsar wind nebulae

A detailed description of our current understanding of the structure and evolution of PWNe is beyond the scope of this paper, and the reader is referred to [11] for a detailed review of the subject. However, I would like to highlight one very interesting aspect seen in the current sample of well observed PWNe that highlights the importance of VHE gamma-ray observations in particular. Figure 3 shows the ratio between the observed gamma-ray and X-ray luminosities of PWNe as a function of their age. Apparently, young PWNe are comparatively strong X-ray emitters whereas older systems predominantly radiate in gamma-rays. Efficient SR in X-rays requires very energetic electrons of up to 1 PeV (as present,e.g., in the Crab nebula, see [13]) coupled with strong magnetic field of several 10s to 100 μG; conditions that are only present in young and still compact PWNe.

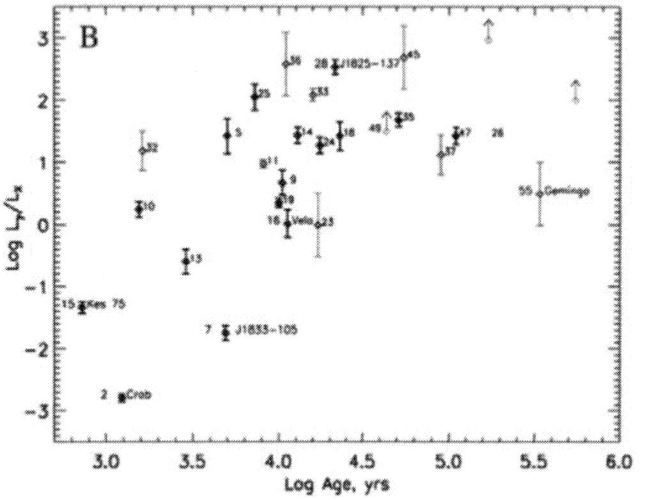

Figure 3. Ratio between the observed gamma-ray and X-ray luminosities of PWNe vs. their age [14].

During their evolution, PWNe expand in size by more than an order of magnitude which is accompanied by a corresponding decrease in the average magnetic field strength owing to the conservation of magnetic flux. Therefore, SR becomes a less and less efficient cooling mechanism as the system evolves, and at some point the B-field-independent IC emission takes over as the dominant radiation process. For a more detailed discussion of this aspect, please see [14].

VHE gamma-ray observations hence provide a new and unique window to understand PWN evolution over much longer time periods than previously possible through observations of SR. In fact, several evolved PWNe, such as HESS J1825−137 [15] and HESS J1303−631 [16], were first discovered in VHE gamma rays, while their comparatively faint X-ray counterparts were only confirmed later through deep follow-up observations.

3.1. New VERITAS observations of MGRO 2019+37

The very bright VHE gamma-ray source MGRO 2019+37 was first detected by the Milagro water Cherenkov detector with a flux of 80% of the Crab nebula but could not be clearly identified with any known astrophysical object [17]. It is located in the direction of Cygnus X, one of the richest star forming regions in the Galaxy. Recently, VERITAS observed this region and was able to decompose MGRO 2019+37 into two separate VHE gamma-ray sources, namely the pointlike source VER J2016+371 coincident with the center-filled SNR CTB 87 and the extended ($\sim 1^\circ$) source VER J2019+378 spatially consistent with the powerful pulsar PSR J2021+3651 as well as the star formation region Sh 2104.

These two sources are also very well separated through their energy spectra (see Fig. 4 and 5). The spectra of both sources are well represented by a power law, however, with largely different indices: 2.4±0.4 for VER J2019+378 and 1.75±0.3 for VER J2016+371, the latter being among the hardest spectra ever measured for Galactic VHE gamma-ray sources.

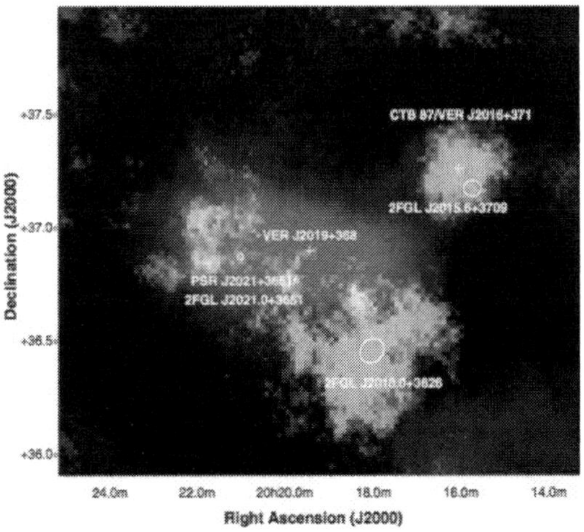

Figure 4. The region around MGRO 2019+37 as seen with VERITAS. The image shows two different energy bands: 600 GeV - 1 TeV (green) and >1 TeV (red). Image from [18].

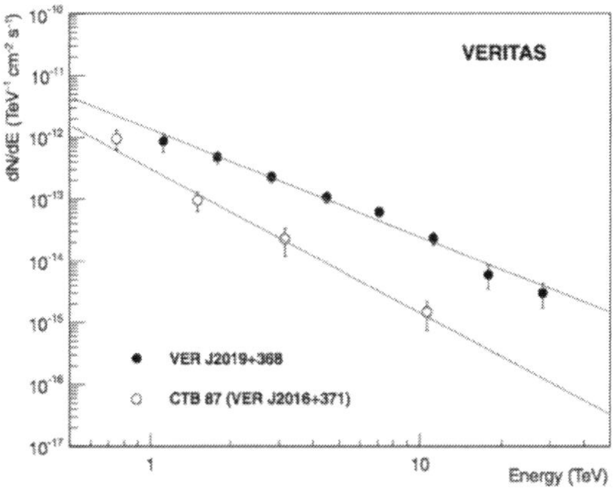

Figure 5. Differential energy spectra of VER J2016+371/CTB 87 and VER J2019+368 as measured by VERITAS. Image from [18].

VER J2016+371 most likely can be identified as a PWN due to its spatial consistency with CTB 87, and the ratio between the gamma-ray and X-ray flux being consistent with expectations for a typical magnetic field strength of 5μG. The offset between the X-ray nebula and the center of the SNR may point towards an age older than 5-10 kyr, indicating significant movement of the pulsar since its birth [18].

The case is more complicated for VER J2019+378, mainly due to its extended morphology that overlaps with several plausible counterparts. The most commonly discussed counterpart to this source is the powerful pulsar PSR J2021+3651. With a spin-down power of 3.4×10^{36} erg s^{1} it may easily account for the whole flux observed from VER J2019+378. Also, the presence of a non-thermal X-ray counterpart further strengthens a PWN interpretation. However, the tail of the X-ray emission as well as part of the VHE gamma-ray source overlap with the starforming region Sh 2104. Such regions of active star formation were discussed as potential sites of particle acceleration (see, e.g. [19]) and the partial overlap of non-thermal emission may indicate that this may contribute to some degree to the overall flux seen from VER J2019+378.

3.2. VERITAS observations of the unidentified gamma-ray source TeV J2032+4130

The gamma-ray source TeV J2032+4130 was serendipitously discovered by the HEGRA IACT array [20] and was the first source in this energy regime that could not be identified with any know object at lower wavelengths. Despite extensive efforts to uncover the nature of this source through multi-wavelength follow-up observations, no clear emission scenario could be established in the following years.

Things changed when this region was recently observed by VERITAS which was able to more tightly constrain the morphology and spectral shape of the VHE gamma-ray emission [21]. Through detailed comparisons of the VHE gamma-ray source with the distribution of the ambient ISM, a PWN scenario could be established as the most likely explanation for the observed emission. Figure 6 shows the new VERITAS gamma-ray map along with images tracing the interstellar gas distribution in this region. It seems that the VHE gamma-ray source

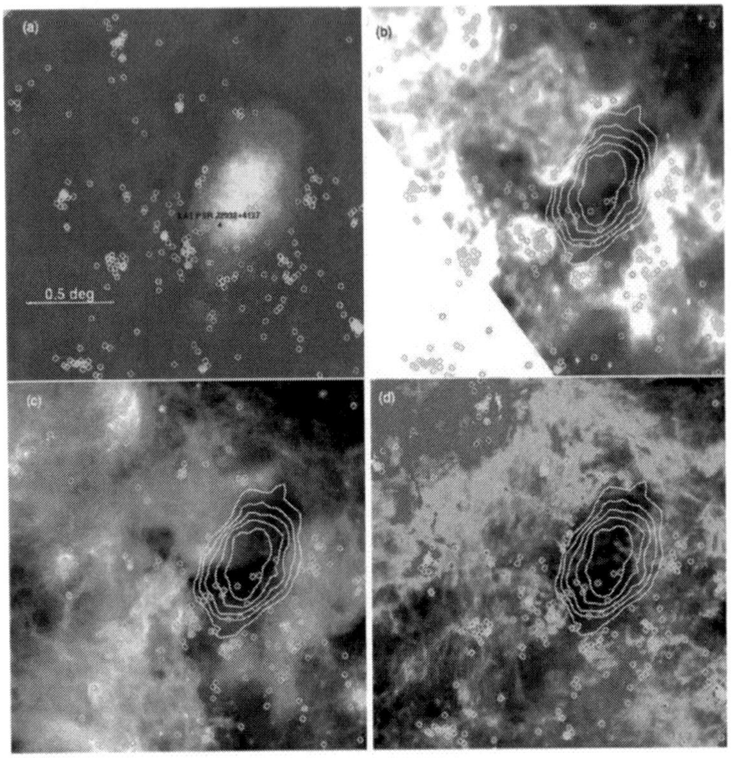

Figure 6. The four panels show the same region in different wavelength regimes: a) VERITAS significance map, b) 1.4 GHz image from the Canadian Galactic Plane Survey, c) Spitzer MIPS 24 μm image d) Spitzer GLIMPSE 8 μm image. The contours (white) are from the VERITAS source. See [21], and references therein.

completely fills a void in the ISM which is surrounded by higher density gas. The void could have been created either by a supernova explosion or by the collective wind from massive stars located in its center. In case of an SNR origin, the remnant must be quite old (>30 kyr) because younger remnants would probably still be detectable through non-thermal radio emission. Here, the VHE gamma-ray source could be the relic of an ancient PWN, which would be very faint in X-rays and thus eluding detection. This scenario is supported by the presence of an old pulsar (PSR J2032+4127), surrounded by a faint X-ray nebula, inside the void but slightly offset from the VHE source centroid.

To conclude, the recent VERITAS observations shed new light on an old mystery by discovering potentially one of the oldest PWNe ever seen in VHE gamma rays.

4. Composite SNRs

So-called composite SNRs are a critical phase in the evolution of the remnants of core-collapse supernovae where the expanding shock front of the SNR as well as a PWN in its center are efficient emitters of non-thermal radiation. On the one hand, these systems must be young enough such that the SNR itself has not yet fully dissipated into the ambient ISM. On the other hand, they must be old enough such that a sufficient amount of relativistic electrons have been injected by the central pulsar and diffused to give rise to an extended nebula of non-thermal radiation in the center of the SNR. The study of the co-evolution of SNR and PWN is a very active topic in high-energy astrophysics, particularly the interaction of the expanding PWN with the reverse-shock which moves inwards towards the SNR center. This interaction may give rise to alternating expansion and compression phases of the PWN and is probably also tightly coupled with certain features in the spectra of non-thermal radiation. Please see [22] for a detailed model describing the interesting dynamics of such systems.

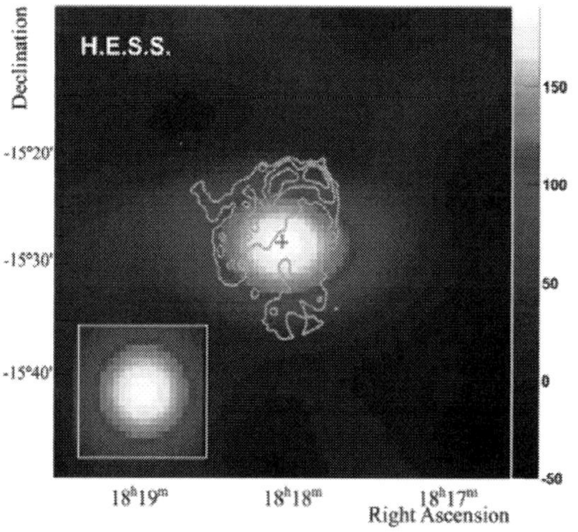

Figure 7. H.E.S.S. gamma-ray skymap of HESS J1818−154. The radio contours of SNR G15.4+0.1 are shown in green. Image from [23].

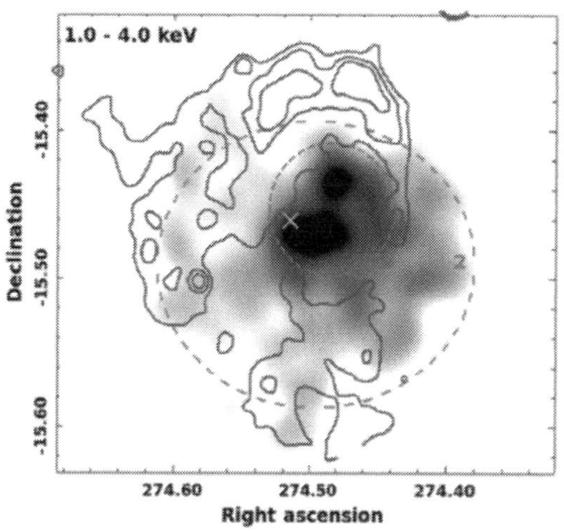

Figure 8. Map of diffuse X-ray emission detected with XMM-Newton in a follow-up observation. Image from [23].

4.1. HESS J1818−154, a new composite supernova remnant discovered in VHE gamma rays and X-rays

Recently, the HESS collaboration reported the detection of a new VHE gamma-ray source located in the center of the shell-type radio SNR G15.4+0.1 [23]. Interestingly, the VHE gamma-ray source appears to be point-like and statistically incompatible with originating from the shell of the SNR itself (see Fig. 7). The most plausible explanation therefore was the presence of a yet undetected PWN in the center of SNR G15.4+0.1, originating from a powerful but unknown pulsar.

A high-sensitivity follow-up X-ray observation with the space telescope *XMM-Newton* indeed revealed an extended diffuse non-thermal X-ray nebula compatible with the position of HESS J1818−154 (see Fig 8). At low energies (<4 keV) the X-ray source nearly covers the whole inside of the SNR, indicating that the PWN may already be in direct interaction with the reverse shock. At higher energies (>4 keV) the X-ray nebula is much more compact and probably centered on the yet undetected pulsar [23]. The energy-dependent size of the X-ray nebula probably indicates the presence of strong synchrotron cooling close to the pulsar which allows only lower energy electrons to propagate as far out as the boundaries of the SNR. The X-ray and VHE gamma-ray spectra are compatible with a standard PWN model with a moderate magnetic field strength (1.5-4.8 μG) and a total energy in electrons $<3\times10^{49}$ erg.

A detection of a powerful pulsar coincident with the newly discovered PWN would be the last missing piece to fully characterize the evolutionary state, total energy budget and potentially the distance of HESS J1818−154.

4.2. A complex gamma-ray source coincident with the SNR W41

The VHE gamma-ray source HESS J1834−087, coincident with the SNR W41, was originally detected in the first H.E.S.S. Galactic Plane survey [24] and later confirmed by dedicated observations with the MAGIC telescopes [25]. At first, due to limited statistics and resolution, it was unclear whether the VHE gamma-ray emission originates from the SNR shell or from an

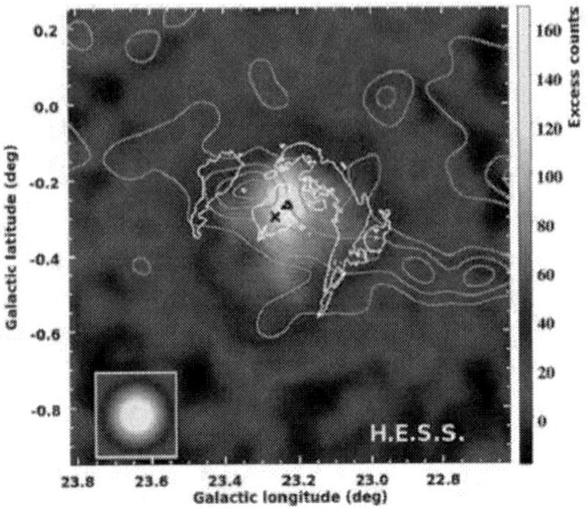

Figure 9. H.E.S.S. gamma-ray skymap of HESS J1834−087. The radio contours of the SNR W41 are indicated in white. Image from [26].

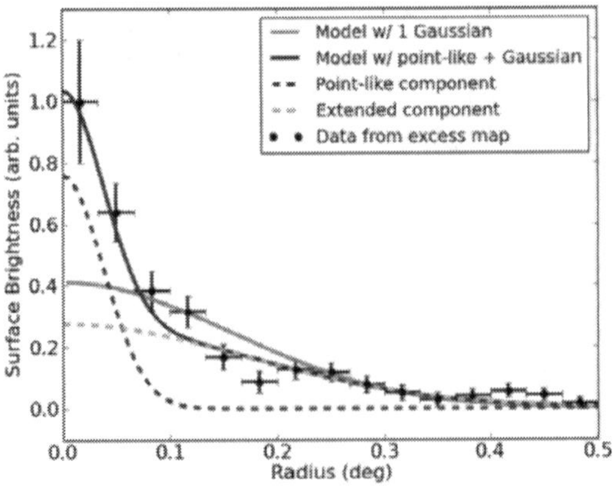

Figure 10. Radial profile of the VHE gamma-ray emission in comparison to several models, as indicated in the figure. Image from [26].

extended source in its center.

Recent dedicated deep follow-up observations with H.E.S.S., featuring an exposure of 61 h and an analysis with improved angular resolution, provide now a much more detailed view of the spectral and morphological characteristics. Fig. 9 shows the new H.E.S.S. skymap of the area with overlaid radio contours of SNR W41. From this image, and from the radial profile of excess counts (Fig. 10) it becomes clear that the morphology is quite complex. The shape of the VHE gamma-ray excess is best fit with a two-component model consisting of a point-like source in the center of the SNR and an extended Gaussian component covering the whole SNR.

Based on this new information there are now three scenarios for the origin of the gamma-ray emission: A) The whole flux is produced by a PWN. However, because the total flux from HESS J1834−087 is quite large, the pulsar would need to be very powerful with a spin-down luminosity of the order of $\dot{E} \approx 10^{37}\,\mathrm{erg\,s^{-1}}$. B) Alternatively, the whole source may arise from the interaction of hadronic cosmic rays with dense molecular clouds detected in that region. This scenario is strengthened by the fact that recent radio observations revealed the existence of OH masers evidencing a direct interaction of the SNR shock with the surrounding dense molecular gas. C) A third option would be that HESS J1834−087 is powered by scenarios A) and B) at the same time. Such an interpretation is strengthened by the favored two-component morphology of the VHE gamma-ray emission. Also, here the energy requirements for the pulsar powering the PWN would be lower, because only part of the total flux may be attributed to this object.

5. Supernova remnants

Currently, several tens of GeV and VHE gamma-ray sources are known that are most likely connected to Galactic SNRs. The sample of well-studied SNRs in this energy regime is quite heterogeneous concerning the close environments of the SNR shocks as well as the plausible radiation mechanism producing the observed gamma-ray flux. They can be roughly divided into two groups (see Fig. 11):

1) Young SNRs (<5-10 kyr) are efficient particle accelerators, often have a non-thermal X-ray/synchrotron counterpart, and mostly expand into a relatively low density medium. For these sources, both leptonic and hadronic emission scenarios are discussed in the literature.

2) Middle-aged (10 – 30 yr) interacting SNRs are characterized by dense interstellar medium in the form of molecular clouds in direct contact with the SNR shock. These SNRs are often one to two orders of magnitude brighter in gamma-rays than young non-interacting SNRs of similar size (see Fig. 11). This enhancement in flux can be likely explained by the presence of a much denser target environment for escaping relativistic protons to interact with. Therefore, the bulk of the gamma-ray emission from these SNRs is attributed to hadronic processes.

In the following subsections a few interesting and very recent examples of SNRs from both of these categories are highlighted.

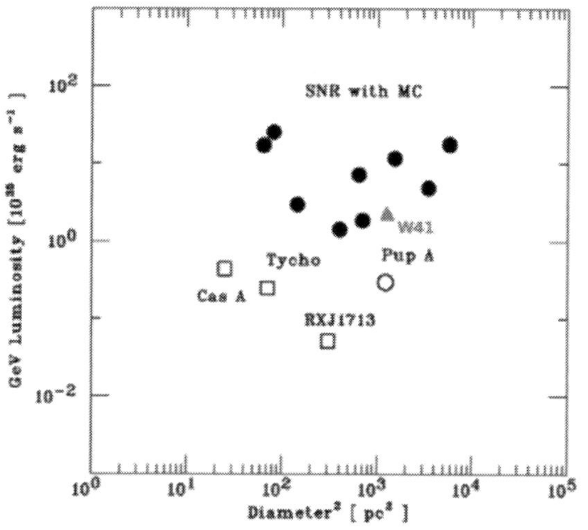

Figure 11. Physical size vs. gamma-ray luminosity for young, shell-type SNRs (open symbols) and middle-aged interacting SNRs (filled symbols). Image from [26].

5.1. VERITAS observations of Tycho's SNR

Tycho's SNR is the remnant of a Type 1a supernova observed by astronomers 1572 AD. It is an ideal laboratory to study SNR evolution coupled with cosmic-ray acceleration because its age is precisely known, and because it evolves into a very clean and low-density environment. It has been detected in gamma rays with the Fermi-LAT observatory [27] and, at higher energies by VERITAS [28], see Fig 12. The absence of an IC peak in the spectrum between the GeV and TeV energy bands, which would be typical for leptonic scenarios, argue in favor of hadronic processes (see Fig. 13). Furthermore, high-resolution X-ray observations with *Chandra* provided evidence that hadronic cosmic-ray acceleration is taking place at the forward shock of Tycho's SNR, based on detailed analyses of the morphology and spectrum of the emission in the shock region [29].

The original VERITAS spectrum had a rather high energy threshold >1 TeV, leaving a large gap between the Fermi-LAT and VHE gamma-ray coverage. As can be seen from the various models shown in Fig. 13, additional coverage in the sub-TeV regime may greatly help to distinguish between the various emission scenarios. Recent deep (105 h) follow-up observations with VERITAS [30] lowered the energy threshold significantly, due to an upgrade of the telescope cameras as well as improvements in the analysis software. Preliminary results show a significantly decreased energy threshold of a few 100 GeV. A future publication with the final spectrum analysis will certainly shed light on the details of hadronic particle acceleration and interaction in this incredibly well-studied SNR.

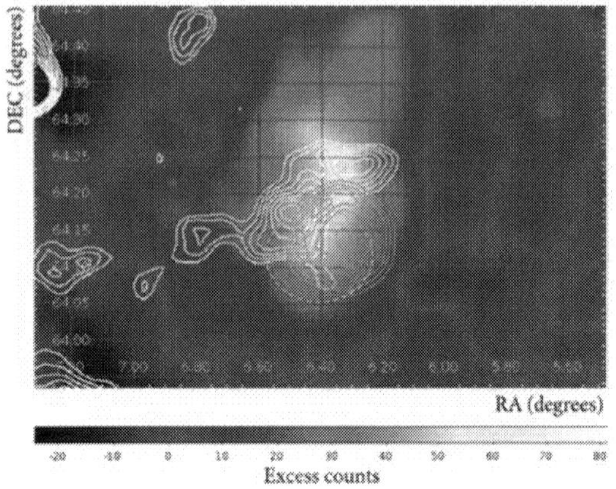

Figure 12. VERITAS map of excess events from the region around Tycho's SNR. The white contours depict ^{12}CO molecular line emission, the green contours show 1420 MHz radio continuum emission. See [30], and references therein.

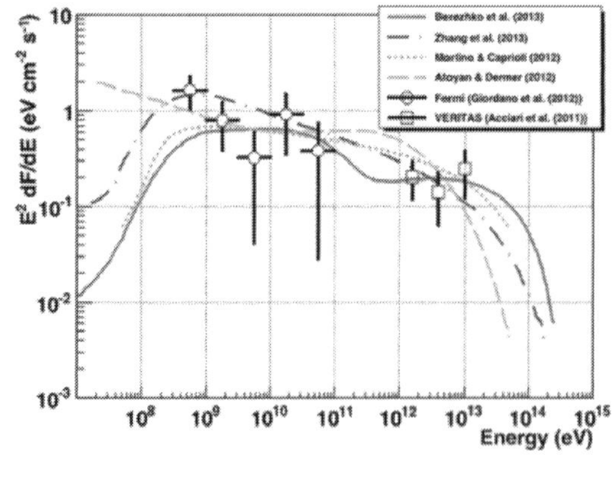

Figure 13. Gamma-ray spectrum measured with the Fermi-LAT observatory (circles, [27]) and from the original VERITAS detection (squares, [28]). Over-plotted are various model predictions (see legend). Image from [30].

5.2. HESS J1640−465, the most luminous Galactic VHE gamma-ray source

One of last years highlights in Galactic VHE gamma-ray astronomy were new observational results for the very bright and prominent source HESS J1640−465. Due to its comparatively large flux (∼10% of the Crab) emission from this object was discovered early during the H.E.S.S. Galactic plane scan [24]. Its location is consistent with the SNR G338.30.0, and follow-up observations with *XMM-Newton* revealed an extended, highly absorbed X-ray source in the center of this SNR [31]. Because of this detection, HESS J1640−465 originally was mainly discussed in the context of a PWN scenario [see e.g. 32].

A detailed study of the HI absorption features of SNR G338.30.0 revealed that this object is probably located at a distance between 8 and 13 kpc [36]. Such a large distance made HESS J1640−465 the most luminous known Galactic VHE gamma-ray emitter, and it became clear that this is potentially one of the most extreme objects of its kind, no matter what the emission scenario might be.

A recent detailed follow-up study of this source by H.E.S.S. based on a dataset more than four times larger, and featuring an analysis with improved angular resolution, shed new light on the potential origin of the gamma-ray emission [34]. According to the new data, the intrinsic size of the VHE gamma-ray source shows significant overlap with the northern part of the SNR shell (see Fig 14). This is challenging to explain in PWN scenarios where one would expect the VHE gamma-ray emission region to be contained within the SNR shell. In fact, the new H.E.S.S. study shows that the VHE gamma-ray morphology would be compatible with originating solely from the northern SNR shell. In an hadronic scenario, the fact that only the northern part of the SNR shell is visible in VHE gamma rays could be explained by the large difference in ISM density between the North and the South. The SNR is located at the southern edge of a dense HII region, and only those protons accelerated by the shock that enter this particular area may find sufficient target material to produce a detectable gamma-ray signal [34].

Recently, a new Fermi-LAT study of the counterpart of HESS J1640−465 in the 100 MeV - 100 GeV energy range presented an updated spectrum extending the gamma-ray coverage to lower energies ([35], see also Fig. 15). These authors also favor an hadronic interpretation of

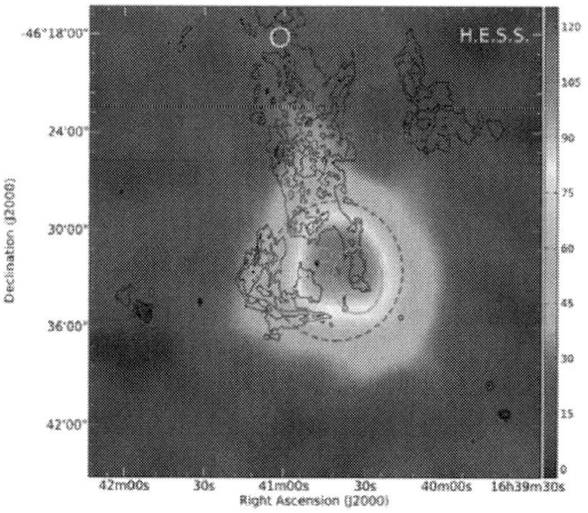

Figure 14. New H.E.S.S. skymap of HESS J1640−465. Also shown are the 610 MHz radio contours (solid, black) of the SNR G338.30.0 [33] and the intrinsic size of the VHE gamma-ray source (dashed, blue). Image from [34].

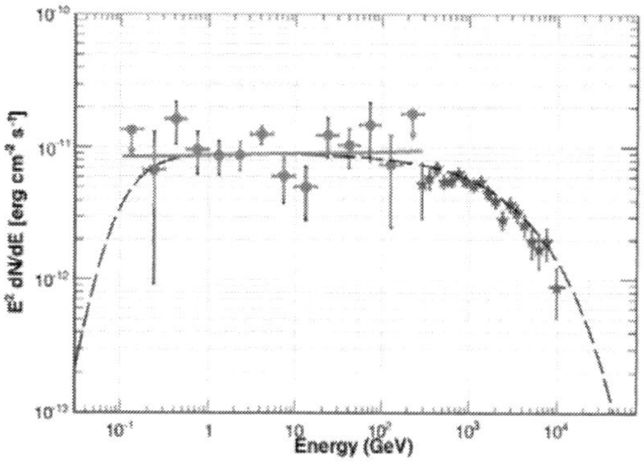

Figure 15. Combined Fermi-LAT (circles, red) [35] and H.E.S.S. spectrum (stars, blue) [34] of HESS J1640−465. The dashed line shows a hadronic model. Image from [35].

the gamma-ray signal, but cannot fully rule out a contribution from the PWN. Interestingly, a new pulsar located in the X-ray PWN was recently detected by NuStar that may be powerful enough to give rise to a significant fraction of the observed gamma-ray flux [37]. However, the spatial overlap of the VHE gamma-ray source with the SNR shell still points towards radiation mechanisms connected to the SNR itself, independent of the PWN. Despite the recent progress and new results for this very luminous object, it remains unclear if the emission is caused only by one process, or if we are observing a superposition of several distinct particle acceleration and radiation mechanisms at the same time. Only future gamma-ray observations featuring even higher angular resolution, like the Cherenkov Telescope Array (CTA) [38], may be able to provide the final answer.

5.3. A new treasure in the same region: HESS J1641−463

The detailed study of HESS J1640−465 (see Sect. 5.2) did not only provide new insights into the physics of this very prominent source, but also led to a new discovery [39]. In the tail of the point-spread function of the signal from HESS J1640−465 lies a second source, HESS J1641−463, which is an order of magnitude fainter than its bright neighbor (see Fig. 16).

Interestingly, the separation between these two sources becomes stronger with increasing energy (see Fig. 17). This is partly due to an improvement of angular resolution with increasing energy, due to the more accurate reconstruction of air showers for more energetic primary gamma rays. The main reason, however, is the significantly harder spectrum of HESS J1641−463 compared to HESS J1640−465. While the second source is barely visible at lower energies, both objects feature a comparable brightness for energies >5 TeV. A spectral analysis yields a power law photon index of $\Gamma = 2.07 \pm 0.11_{\rm stat} \pm 0.20_{\rm sys}$, which ranks among the hardest spectra ever measured from Galactic VHE gamma-ray sources. Also, the data show that the spectrum extends to an energy of at least 20 TeV without any sign of a cut-off.

Various gamma-ray production scenarios are discussed for this object, including hadronic processes within the dense target environment surrounding this source [39]. In such a case,

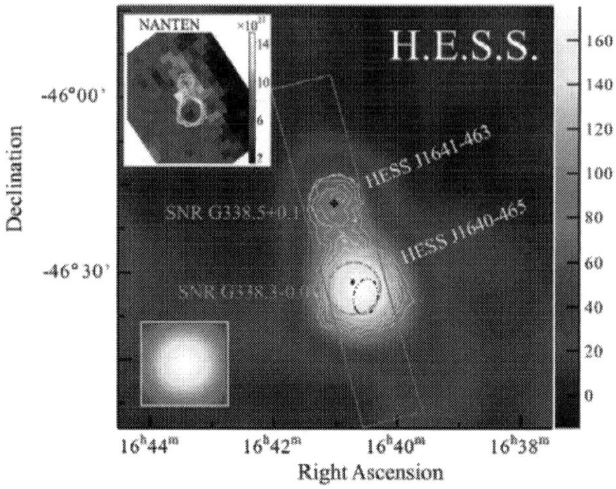

Figure 16. H.E.S.S. skymap of excess events showing HESS J1640−465 and HESS J1641−463. The box for the profiles in Fig. 17 is shown in red. Image from [39].

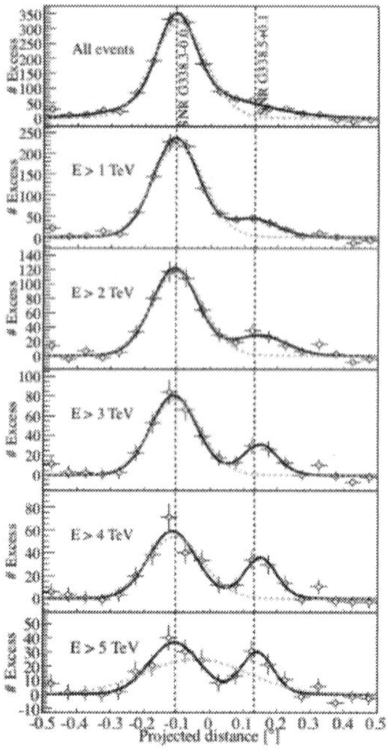

Figure 17. Profiles of excess counts along the box shown in Fig. 16 in different energy bands, as indicated on the panels. Image from [39].

owing to the hard spectrum showing no clear cut-off, the energies of the underlying relativistic protons could even extent to several hundreds of TeV to 1 PeV. Hence, HESS J1641−463 may be one of the first known representatives of a population of cosmic-ray accelerators that provide energies up to the knee in the local cosmic-ray spectrum. In one scenario, the protons originate from SNR G338.30.0, which is associated with HESS J1640−465. The reason why the latter does not look like a PeVatron today might be that the most energetic protons already escaped the source and have propagated into the dense molecular environment towards the North-East. There, they give rise to the very hard gamma-ray emission, seen from HESS J1641−463, through production and subsequent decay of neutron pions.

Alternative gamma-ray production scenarios, such as IC emission from relativistic electrons, are disfavored due to the non-detection of a spectral brake in the multi-TeV regime. However, given the statistical quality of the current dataset, such scenarios cannot be fully ruled out yet.

6. Summary

Over the last decade, the field of ground-based VHE gamma-ray astronomy with IACTs evolved into a rich and diverse new research discipline, bridging the gap between modern astronomy and astroparticle physics. The field now entered a phase of precision measurements to investigate details of the particle acceleration and subsequent radiation processes. Specifically in Galactic VHE gamma-ray science the focus lies on the discrimination between leptonic and hadronic gamma-ray production channels in the emission observed from SNRs, and the study of the evolution of PWNe generated by strong rotation-powered pulsars. Such studies do not only require a detailed analysis of the gamma-ray signal, but also the development of plausible radiation scenarios involving the whole multi-wavelength environment of the sources. Ground-based VHE gamma-ray astronomy is also an important field to explore the origin of Galactic cosmic rays. This is achieved by identifying potential acceleration sites through secondary gamma-ray production from the decay of neutral pions produced in the interaction of hadronic cosmic rays with the interstellar medium.

CTA, a next generation IACT, is currently in development and will consist of 100 individual telescopes of different sizes [38]. This new observatory will feature an increase in sensitivity by up to a factor of 10 compared to current IACTs and will provide the community with a wealth of new high-precision data to tackle the remaining open questions in an inter-disciplinary context.

References

[1] Hofmann W 2009 Proceedings of the 27th International Cosmic Ray Conference, Hamburg, Germany, **7** 2785

[2] Lorenz E (MAGIC Collaboration) 2004 New Astronomy Reviews, **48** 339–344

[3] Holder J, Acciari, V A, Aliu E, et al. (VERITAS Collaboration) 2008 American Institute of Physics Conference Series, **1085** 657-660

[4] Hinton J A and Hofmann W 2009 *ARAA* **47** 523–565 (*Preprint* 1006.5210)

[5] Stanev T, Biermann P L and Gaisser T K 1993 *A&A* **274** 902 (*Preprint* astro-ph/9303006)

[6] Hillas A M 2005 *Journal of Physics G Nuclear Physics* **31** 95

[7] Dermer C D 2012 *Physical Review Letters* **109** 091101 (*Preprint* 1206.2899)

[8] Bell A R 1978 *MNRAS* **182** 147–156

[9] Blandford R D and Ostriker J P 1978 *ApJL* **221** L29–L32

[10] Carrigan S, Brun F, Chaves R C G, Deil C, et al. for the H.E.S.S. collaboration 2013, Proceedings of the 48th Rencontres de Moriond *ArXiv e-prints* (*Preprint* 1307.4868)

[11] Gaensler B M and Slane P O 2006 *ARAA* **44** 17–47 (*Preprint* astro-ph/0601081)

[12] Kieda D B, for the VERITAS Collaboration 2013, Proceedings of the 33rd International Cosmic Ray Conference, Rio de Janiero, Brazil *ArXiv e-prints* (*Preprint* 1308.4849)

[13] de Jager O C, Harding A K, Michelson P F, et al. 1996 *ApJ* **457** 253

[14] Kargaltsev O and Pavlov G G 2010, proceedings of the *X-ray Astronomy 2009 conference; Present Status, Multi-Wavelength Approach and Future Perspectives*, **1248** 25–28 (*Preprint* 1002.0885)

[15] Aharonian F A, Akhperjanian A G, Bazer-Bachi A R, et al. (H.E.S.S. Collaboration) 2005 *A&A* **442** L25–L29 (*Preprint* astro-ph/0510394)

[16] Aharonian F, Akhperjanian A G, Aye K M, et al. (H.E.S.S. Collaboration) 2005 *A&A* **439** 1013–1021 (*Preprint* astro-ph/0505219)

[17] Abdo A A, Allen B, Berley D, et al. (MILAGRO Collaboration) 2007 *ApJL* **664** L91–L94 (*Preprint* 0705.0707)

[18] Aliu E, Aune T, Behera B, et al. (VERITAS Collaboration) 2014 *ApJ* **788** 78 (*Preprint* 1404.1841)

[19] Torres D F, Domingo-Santamaría E and Romero G E 2004 *ApJL* **601** L75–L78 (*Preprint* astro-ph/0312128)

[20] Aharonian F, Akhperjanian A, Beilicke M, et al. (HEGRA Collaboration) 2002 *A&A* **393** L37–L40 (*Preprint* astro-ph/0207528)

[21] Aliu E, Aune T, Behera B, et al. (VERITAS Collaboration) 2014 *ApJ* **783** 16 (*Preprint* 1401.2828)

[22] Gelfand J D, Slane P O and Zhang W 2009 *ApJ* **703** 2051–2067 (*Preprint* 0904.4053)

[23] Abramowski A, Aharonian F, Ait Benkhali F, et al. (H.E.S.S. Collaboration) 2014 *A&A* **562** A40 (*Preprint* 1310.6956)

[24] Aharonian F, Akhperjanian A G, Bazer-Bachi A R, et al. (H.E.S.S. Collaboration) 2006 *ApJ* **636** 777–797 (*Preprint* astro-ph/0510397)

[25] Albert J, Aliu E, Anderhub H, et al. (MAGIC Collaboration) 2006 *ApJL* **643** L53–L56 (*Preprint* astro-ph/0604197)

[26] Abramowski A, Aharonian F, Ait Benkhali F, et al. (H.E.S.S. Collaboration) 2015 *A&A* **574** A27

[27] Giordano F, Naumann-Godo M, Ballet J, et al. 2012 *ApJL* **744** L2 (*Preprint* 1108.0265)

[28] Acciari V A, Aliu E, Arlen T, et al. (VERITAS Collaboration) 2011 *ApJL* **730** L20 (*Preprint* 1102.3871)

[29] Warren J. S., Hughes J. P., Badenes C., et al. 2005 *ApJ* **634** 376–389

[30] Park N for the VERITAS Collaboration 2013, Proceedings of the Internation Cosmic Ray Conference (ICRC), Rio de Janeiro, Brazil (*Preprint* 1308.0610)

[31] Funk S, Hinton J A, Pühlhofer G, et al. 2007 *ApJ* **662** 517–524 (*Preprint* astro-ph/0701166)

[32] Slane P, Castro D, Funk S, et al. 2010 *ApJ* **720** 266–271 (*Preprint* 1004.2936)

[33] Castelletti G, Giacani E, Dubner G, et al. 2011 *A&A* **536** A98 (*Preprint* 1110.5301)

[34] Abramowski A, Aharonian F, Benkhali F A, et al. (H.E.S.S. Collaboration) 2014 *MNRAS* **439** 2828–2836 (*Preprint* 1401.4388)

[35] Lemoine-Goumard M, Grondin M H, Acero F, et al. 2014 *ApJL* **794** L16 (*Preprint* 1409.4994)

[36] Lemiere A, Slane P, Gaensler B M et al. 2009 *ApJ* **706** 1269–1276 (*Preprint* 0910.2652)

[37] Gotthelf E V, Tomsick J A, Halpern J P, et al. 2014 *ApJ* **788** 155 (*Preprint* 1405.0465)

[38] Actis M, Agnetta G, Aharonian F, et al. (CTA Consortium) 2011 Experimental Astronomy **32** 193316 (*Preprint* 1008.3703)

[39] Abramowski, A., Aharonian, F., Ait Benkhali, F., et al. (H.E.S.S. Collaboration) 2014 *ApJL* **794** L1 (*Preprint* 1408.5280)

Interpretations of the IceCube neutrino escess

M. Kachelrieß
Institutt for fysikk, NTNU, Trondheim, Norway

Abstract. The IceCube Collaboration announced 2012 evidence for the first detection of extraterrestrial neutrinos. Meanwhile, the discovery of a extraterrestrial neutrino flux of surprisingly large magnitude has been established. I discuss a selection of possible sources for these neutrinos and their signatures, concentrating on the neutrino yield from collisions of cosmic ray nuclei with gas and the possibility that Galactic sources can explain the IceCube excess. I review also the cascade bound on extragalactic neutrinos and its consequences as well as the possibility that decays of PeV dark matter are the origin of the IceCube excess.

1. Introduction

High energy (HE) neutrinos are produced by cosmic rays (CR) interacting with gas or photons in their sources, and with cosmic microwave and other background photons during propagation. Alternatively, HE neutrinos may be produced in annihilations or decays of dark matter particles. Any process involving hadronization leads mainly to the production of pions, and the ratio of charged and neutral pions produced is fixed by isospin symmetry. The production of neutrinos is thus intimately tied to the one of photons, and both depend in turn on the flux of primary CRs. This connection allowes one to estimate the minimal size of detector able to detect HE neutrinos as $\sim 1\,\mathrm{km}^3$.

The IceCube neutrino observatory has been the first detector reaching this size, being thus able to test this paradigm. In 2013, the IceCube collaboration announced evidence for the first detection of extraterrestrial neutrinos at the 4σ confidence level [1]. This announcement followed the observation of two PeV neutrino cascades [2]. The 3-year data set consists of 37 events with deposited energy in the range between 30 TeV and 2 PeV, while 15.0 background events are expected from atmospheric muons and neutrinos. In the most recent analysis [3], the energy threshold applied was reduced to 1 TeV. The determined excess of events (denoted "IceCube excess" in the following) is consistent with a diffuse intensity (summed over flavors) at the level of

$$\Phi_\nu(E) = 20.6^{+4}_{-3}\left(E_\nu/10^{14}\,\mathrm{eV}\right)^{-2.46\pm0.12}\mathrm{eV\,cm^{-2}\,sr^{-1}\,s^{-1}}\,.$$

The basic information contained in these events can be summarised by the energy spectrum, the arrival directions and the flavor composition of these events:

Energy spectrum: The energy threshold used in different analyses has been continuously reduced, reaching 1 TeV in the most recent analysis [3]. At the same time, the slope of the neutrino energy spectrum assuming an unbroken power-law $dN/dE \propto E^{-\alpha}$ in the whole energy range has become steeper, and deviates now with $\alpha = -2.46 \pm 0.12$ clearly from $\alpha = -2$. Since the intensity at $\gtrsim 10^{14}\,\mathrm{eV}$ remained unchanged, the intensity at lower energies increased, putting pressure on extragalactic neutrino models.

However, this change of the spectral index is driven by the addition of the low-energy neutrinos: In particular, if the deposited-energy threshold is raised to 60 TeV (corresponding to a sensitivity for $E_\nu > 100\,\mathrm{TeV}$), then the analysis [3] finds that the spectral index hardens to 2.26 ± 0.35, compatible with the previous high-energy result. This may indicate that above $E_\nu \sim 100\,\mathrm{TeV}$ a different source population dominates the neutrino flux. Since the maximal energy of neutrinos produced via pion production in pp interactions is approximately 10% of the energy of the proton primary, the corresponding break in the CR primary energy would correspond to $E_{\mathrm{CR}} \sim 1\,\mathrm{PeV}$. Thus the hardening of the neutrino spectrum may be associated with the Galactic CR knee and a transition to extragalactic neutrinos.

Arrival directions: No statistically significant clustering of the arrival directions has been observed, although an overdensity of events close to the Galactic center and, less pronounced, towards the Galactic plane exists.

Flavor composition: After the discovery of large ν_μ–ν_τ mixing, it was often taken for granted that the small oscillation length (of order 10 pc for $E_\nu \sim 10^{15}$ eV) together with $\nu_e{:}\nu_\mu{:}\nu_\tau = 1{:}2{:}0$ at generation leads to flavor equipartition at observation, $\nu_e{:}\nu_\mu{:}\nu_\tau = 1{:}1{:}1$. There exist however several examples of neutrino sources where at least in some energy range significant deviations from this canonical flavor ratio from pion decay can be expected [4, 5]. Specific examples are the cases where the neutrino spectrum is dominated by neutron [5] or kaon decays [6] or influenced by muon damping [7]. In the diffuse flux, one should however expect that these deviations are washed out. For a detailed discussion of the various possibilities see [8, 9].

The analysis [10] found that the ratio of track/shower events in the IceCube's 3-year data favors $\nu_e : \nu_\mu : \nu_\tau = 1{:}0{:}0$ at generation at the 92% C.L. However, the recent analysis [11] performed by the IceCube collaboration finds that the track/shower ratio is consistent with flavor equipartition at observation. The discrepancy between these two results was explained by a partial classification of CC events as showers and the not accounted systematic uncertainty of the muon background in [10].

2. Galactic sources

2.1. *Neutrino yield in CR interactions with gas*

The Galactic neutrino flux contains a guaranteed component which is produced by CRs interacting with gas during their confinement in a CR halo. This minimal Galactic neutrino flux has been discussed since the late 1970s [12].

The observed 2 PeV neutrino event requires proton energies above 20 PeV. However, already at 10^{15} eV protons represent only a subdominant fraction of the primary CR flux compared to helium and heavier nuclei [13]. Moreover, the composition of the CR flux becomes increasingly heavier in the energy range between the knee at $E_{\mathrm{k}} \approx 4\,\mathrm{PeV}$ and 10^{17} eV. Since the maximal neutrino energy in nucleus-proton collisions is a factor A lower than in pp interactions (A being the nuclear mass number), the required minimal CR energy to explain the IceCube events increases compared to pp processes. This implies in turn that the number of potential scattering events, and thus secondary fluxes, is drastically reduced, because the CR spectrum is steeply falling, $I(E) \propto E^{-3.1}$ above the knee [13]. Therefore it is essential to account correctly for the elemental composition of the Galactic CR flux, if one aims at relating the neutrino intensity required to explain the IceCube excess to the primary CR intensity.

Reference [14] quantified the neutrino yield from nucleus-proton collisions using up-to-date simulation tools for the relevant hadron production processes and including information on the elemental composition of the CR flux around and above the knee region. The simulations were based on the event generator QGSJET-II-04 [15] which includes relevant results from run I of LHC [16]. In this section, we summarise some results of this study.

Explanations for the origin of the knee fall in two main categories. First, the knee may correspond to the maximum rigidity to which CRs can be accelerated by the dominant population of Galactic CR sources [17, 18]. Second, the knee energy may correspond to the rigidity at which the CR Larmor radius r_L is of the order of the coherence length l_c of the turbulent magnetic field in the Galactic disk. As a result, a transition from large-angle to small-angle scattering or Hall diffusion is expected, the energy dependence of the confinement time changes which in turn induces a steepening of the CR spectrum [19, 20, 21]. Both possibilities lead to a rigidity-dependent sequence of knees at ZE_k, a behavior first suggested by Peters [22]. In contrast to models in category 1, those of category 2 predict both the position of the knee and the rigidity dependent suppression of the different CR components for a given model of the Galactic magnetic field [21].

Various models which describe the elemental composition of the total CR flux have been developed. The poly-gonato model [23] is a fit of rigidity-dependent knees at $E = ZE_k$ to measurements of the total CR intensity. Below and above $E = ZE_k$, the fluxes of individual CR nuclei are assumed to follow power-laws $\phi_A(E) = K_{A,1}E^{-\gamma_{A,1}}$ and $\phi_A(E) = K_{A,2}E^{-\gamma_{A,2}}$ which are smoothly interpolated. The fluxes are assumed to steepen by a common amount, $\gamma_{A,2} = \gamma_{A,1} + \delta$ with $\delta \simeq 2.1$, for more details see [23]. The steepening of the CR intensity around the knee is very pronounced in this model, $\delta \sim 2$, and as a result the composition is heavier than suggested by KASCADE-Grande data. The resulting neutrino intensity $I_\nu(E)$ is shown in Fig. 1, where we assumed that the CR nuclei cross the grammage $X = 30\,\text{g/cm}^2$; this value corresponds for primary protons with energy 1 PeV to the interaction depth $\tau_{pp} = 1$. Above few hundred TeV, the intensity is dominated by the proton contribution and is strongly decreasing with energy as $I_\nu(E) \sim E^{-4.7}$. Therefore the expected neutrino energy distribution disagrees with the neutrino spectrum suggested by the IceCube excess, in particular with the two PeV events.

The Hillas model [18] and its variants belong to the category 1, associating the knee with the maximal rigidity achievable in the dominant population of Galactic CR sources. Moreover, the Hillas model assumes that the ankle signals the transition from Galactic to extragalactic CRs. Therefore, an additional population of Galactic CR sources must exist ("the component B" of Ref. [18]) which fills the gap between the knee and the ankle. Thus the Hillas model contains two Galactic components. Each population is assumed to contain five elemental groups and cuts off at a characteristic rigidity. We present here results for the new parametrisation H3a given in Tab. 3 of Ref. [24]. The intensity of Galactic neutrinos in this model is shown in Fig. 1, which is again dominated by the proton contribution; its shape is very similar to the neutrino intensity of the poly-gonato model.

Finally, Ref. [14] considered a parametrisation of the CR flux motivated by the recent results of Ref. [21]: There, the escape of CRs from our Galaxy was studied calculating trajectories of individual CRs in models of the regular and turbulent Galactic magnetic field. For a coherence length $l_c \simeq (2-5)$ pc of the turbulent field and a reduced turbulent magnetic field, a knee-like structure at $E/Z = \text{few} \times 10^{15}$ eV was found, which is sufficiently strong to explain the knee and the energy spectra of different groups of CR nuclei as determined by KASCADE and KASCADE-Grande. The resulting neutrino intensity for $X = 30\,\text{g/cm}^2$ is shown in Fig. 2 together with the individual contribution from five elemental groups. The suppression of the neutrino intensity above the knee is less pronounced as in the previous models, since the decrease of the CR escape time $\tau_{esc}(E)$ slows down around $E/Z \simeq 10^{16}$ eV for a weak turbulent field. In the energy range between 10^{13} eV and 10^{16} eV, the neutrino intensity scales as $I_\nu(E) \propto E^{-3.2}$.

In summary, [14] found that the neutrino intensity below 10^{14} eV reflects the slope of protons and agrees therefore in all three models. In contrast, the exact position of the "neutrino knee" and the slope of neutrino intensity above this break depends on the nuclear composition and is therefore model dependent: A comparison of the neutrino intensity in the three models is shown in Fig. 1. Finally, one may yet ask which of the three composition models considered is the more realistic one in the light of recent CR data. Comparing e.g. the CR spectra predicted by the poly-gonato model to the intensities of individual groups of CR nuclei up to 10^{17} eV, measured by the KASCADE and KASCADE-Grande experiments, already an inspection by eye indicates that this model predicts a too heavy composition above the knee. The Hillas model of Ref. [24] describes well the average composition (represented e.g. by $\ln(A)$) but fails to reproduce the up-turn of the light component around 10^{17} eV observed by KASCADE-Grande. By contrast, such an up-turn around $E/Z \simeq 10^{16}$ eV is the characteristique feature of the escape model [21]. As a result, the intensity of individual groups of CR nuclei measured by KASCADE and KASCADE-Grande is well reproduced in this model [21].

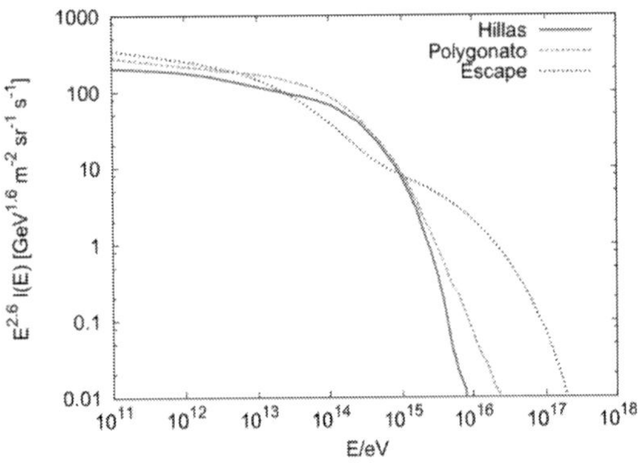

Figure 1. A comparison of the total neutrino spectra $E^{2.6}I(E)$ predicted for $X = 30\,\mathrm{g/cm^2}$ by the three CR models.

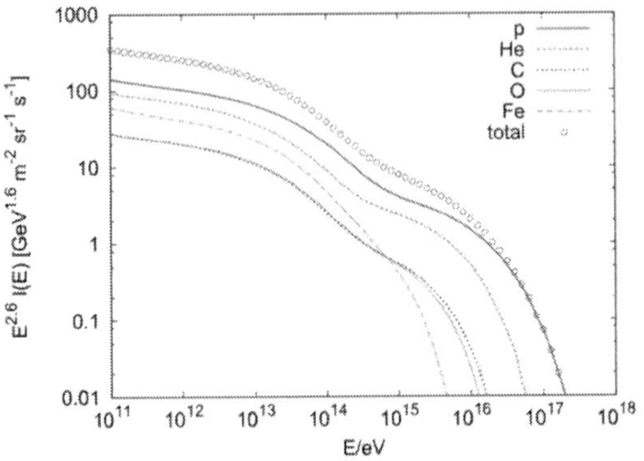

Figure 2. Total neutrino spectrum $E^{2.6}I(E)$ and individual contributions of five elemental groups for $X = 30\,\mathrm{g/cm^2}$ in the escape model.

2.2. Neutrinos from Galactic sea CRs

The slight over-density of the neutrino events in a region close to the Galactic center and towards the Galactic plane invites to speculations about a (partly) Galactic origin of these events. We consider first the guaranteed contribution from Galactic sea CRs.

The gas distribution in the Galactic disk can be modeled as $n(z) = N\exp(-(z/z_{1/2})^2)$ with $N \simeq 0.3\,\mathrm{cm^{-3}}$ at $R_\odot$ (increasing to $N \simeq 10\,\mathrm{cm^{-3}}$ at the GC) and $z_{1/2} = 0.21$ kpc. Integrating the interaction probability, $\int \mathrm{d}l\, n_{\rm CR}(E)\sigma^{pp}_{\rm inel}(E)$, with $\sigma^{pp}_{\rm inel}(E) \simeq 60$ mb results in the maximal interaction depth of $\tau \sim 0.005$ towards the GC. At the reference energy $E_* = 1$ PeV, all three parametrisations predict a neutrino intensity around $\tau_\nu E_*^{2.6}\, I_\nu(E_*) \sim 10\,\mathrm{GeV^{1.6}\,m^{-2}\,s^{-1}\,sr^{-1}}$, corresponding to $E_\nu^2 I_\nu \sim 0.1 \times \mathrm{eV\,cm^{-2}s^{-1}\,sr^{-1}}$. Thus even in the direction of the largest expected intensity, the predicted neutrino intensity due to diffuse Galactic CR interactions is about two orders of magnitude too small compared to the IceCube excess. Moreover, the neutrino events should be concentrated within $|b| \leq 1°$ [25], reflecting the very slim Galactic plane, which is much narrower than the latitude distribution of the IceCube events.

2.3. Neutrinos from Galactic CR sources

We consider next the neutrino flux produced close to recent CR sources. The propagation of CRs on distances $l >$ few $\times\, l_{\rm coh}$ can be approximated by diffusion [26]. For $l_{\rm coh} \sim 10$ pc as found in Ref. [27] for the Galactic disk, the diffusion approach is marginally justified for the time scales, 10^3 to 10^4 yr, we consider.

Galactic accelerators able to produce CRs with energies 10^{16} eV have typically only short life-times: For instance, the highest energy particles produced by a supernova remnant (SNR) are thought to escape at the end of the Sedov phase after few 100 yrs. Approximating therefore the accelerator as a bursting source, the number density $n_{\rm CR}(E,r) = dN/(dE\,dV)$ of CRs at the distance r is given by

$$n_{\rm CR}(E,r) = \frac{Q(E)}{\pi^{3/2}\, r_{\rm diff}^3} \exp\left[-r^2/r_{\rm diff}^2\right], \tag{1}$$

with $r_{\rm diff}^2 = 4Dt$ assuming no energy losses. We use as diffusion coefficient $D(E_*) = 3 \times 10^{28}\,{\rm cm}^2/{\rm s}$ at our reference energy $E_* = 1\,{\rm PeV}$ and assume that the source injects instantaneously CRs with the total energy $E_{\rm p} = 10^{50}\,{\rm erg}$ in protons with an injection spectrum $Q(E) = Q_0(E/E_0)^{-\alpha}$ between the minimal energy $E_0 = 1\,{\rm GeV}$ and a maximal energy $E_{\rm max} = 10\,{\rm PeV}$. We choose $\alpha = 2.0$ suggested by shock acceleration. Then PeV CRs are concentrated within $r_{\rm diff} \sim 35\,{\rm pc}$ after 3000 yr.

The brightest spots in the Galactic neutrino sky are likely giant molecular clouds (GMC) immersed into the CR overdensities close to recent CR sources. The neutrino flux from a point source at the distance d is given by

$$\phi_\nu(E) = \tilde{\varepsilon}_{\rm M} \frac{c\,\sigma_{\rm inel}}{4\pi d^2} \frac{M_{\rm cl}}{m_p}\, n_{\rm CR}(E)\, Y_\nu(E), \tag{2}$$

where $Y_\nu(E)$ denotes the neutrino yield $Y_\nu(E) = \phi_\nu(E)/[\tau(E)\,\phi(E)]$. Assuming as cloud mass $M_{\rm cl} = 10^5 M_\odot$ and as distance $d = 1\,{\rm kpc}$ results in the neutrino flux $E^2\phi_\nu(E) \simeq 140\ {\rm eV\,cm^{-2}\,sr^{-1}}$. Sources of this kind would be clearly visible on the neutrino sky as seen by IceCube. Choosing as source rate $\dot{N} \simeq 1/(30\,{\rm yr})$, which coincides with the Galactic SN rate, implies that the average number N_s of such sources present in the Galaxy equals $N_s \simeq 100$. Using as volume of the Galactic disk $V = \pi R^2 h \simeq 140\,{\rm kpc}^3$, there are on average 0.5 sources within one kpc distance to an observer. We note also that the presence of GMCs close to SNRs is not unnatural, since they are born most likely in OB associations.

Any source of high-energy neutrinos produces also γ-rays. In Ref. [28], the 1–10 TeV γ-ray flux from known sources in the direction towards the GC was compared with the IceCube excess. Extrapolating the γ-ray flux of these sources to higher energies implies a neutrino flux which is an order of magnitude smaller than the one required to explain the IceCube excess. This discrepancy could be explained by the slower diffusion of low-energy CRs which have not yet reached GMCs and by the additional contribution from extragalactic neutrinos.

Limits on the fraction of photons in the CR flux as e.g. those of CASA-MIA, KASCADE and IceCube can be used to constrain Galactic neutrino sources [29]. However, PeV γ-rays can be absorbed both in sources and during propagation by pair production on star light and CMB photons. Moreover, these gamma-ray limits are biased towards the northern hemisphere. As a result, they do not exclude the case that (a fraction of) the IceCube excess has a Galactic origin.

3. Extragalactic sources

3.1. Source types

The idea that neutrinos and photons are produced efficiently as secondaries in CR interactions close to the core of AGN dates back to the 1970's [30]. Gamma-ray bursts (GRB) were suggested as another promising neutrino source [31]. The negative results from searches for HE neutrinos from stacked GRBs disfavors however GRBs as the main source of the IceCube excess.

Starburst galaxies have magnetic fields which are two orders of magnitude higher than the Galactic magnetic field [32]. As a consequence, in the model of Ref. [21] where the knee is caused by the escape of CRs, its position and thus the one of the neutrino knee is shifted by up too two orders of magnitude. If moreover CRs are re-accelerated in super-bubbles, the maximal energy of CRs in starburst galaxies should be increased as well. Thus normal and starburst galaxies are a promising candidate for the neutrino events in IceCube.

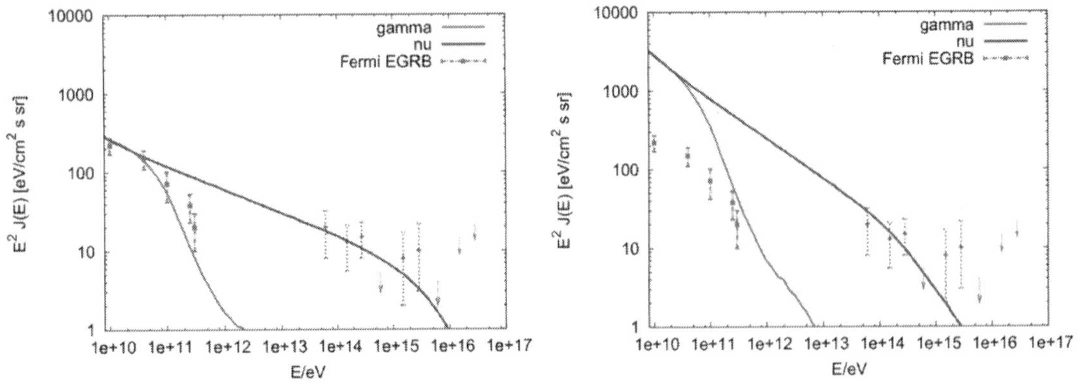

Figure 3. Intensity of the diffuse EGRB and neutrinos arising from pp collisions, left $\alpha = 2.3$ and right $\alpha = 2.5$.

3.2. Limits from the diffuse γ-ray background

A general upper bound on the cosmogenic neutrino flux can be derived from the observation of the extragalactic diffuse gamma-ray background (EGRB). Since the Universe acts as a calorimeter for electromagnetic radiation, accumulating it in the MeV–TeV range, the measured EGRB limits all processes during the history of the Universe that inject electromagnetic energy above the pair creation threshold. In [33], the measurement [34] of the EGRB by Fermi-LAT was used to constrain UHECR models and to limit the HE neutrino flux as $E^2\Phi_\nu(E) \lesssim 40\,\mathrm{eV}/(\mathrm{cm}^2\mathrm{s\,sr})$. Note that the best-fit value for the neutrino flux from the 2-year IceCube data saturated this bound.

Meanwhile, the Fermi collaboration presented results for 44 months of data taking, extending their measurement of the EGRB to higher energies [35]. As a result, the limits on the allowed cascade radiation and the limit on the cosmogenic neutrino flux drop by a factor ~ 3, i.e. to $\omega_{\rm casc} \lesssim 2 \times 10^{-7}\mathrm{eV/cm}^3$. Moreover, the contribution from unresolved sources is estimated to be as large as 50–80% [35]. In Fig. 3, we show the intensity of the extragalactic diffuse gamma-ray background (EGRB) together with diffuse neutrino intensity arising from pp collisions in extragalactic sources for two different values of the injection power-law $dN/dE \propto E^{-\alpha}$ of CRs with an exponential cutoff. Here, we normalised the diffuse neutrino intensity to the IceCube events and assumed as redshift evolution $\propto (1+z)^3$ until $z = 1$.

For $\alpha = 2.3$, the resulting EGRB saturates the EGRB measurements by Fermi-LAT [35], while the predicted EGRB overshots the data for $\alpha = 2.5$. Note also that a large fraction of the EGRB may be associated to sources like mis-aligned blazars or cascades from ultrahigh energy cosmic rays which do not contribute to the IceCube neutrino excess. Thus a spectral index of neutrino spectrum as small as -2.45 is problematic for models which rely on neutrino production via pp collisions. Similar conclusions were presented in Ref. [36]. It is clear that the EGRB is an important constraint on any proposed extragalactic model for neutrino sources.

4. Neutrinos from top-down models

Top–down model is a generic name for all proposals in which the observed CR primaries are produced as decay products of some heavy particles X with mass m_X much larger than the weak scale, $m_X \gg m_W$. These X particles can be either metastable or be emitted by topological defects at the present epoch.

The suggestion [37] of superheavy metastable relic particles as CR sources was originally motivated by the AGASA excess, although it was soon realized that stable or metastable particles

Journal of Physics: Conference Series **632** (2015) 012037 doi:10.1088/1742-6596/632/1/012037

in this mass range are generically interesting DM candidates. The idea that PeV dark matter is responsable for the IceCube excess [38] is a natural variation of this suggestion.

What are the particle candidates for PeV DM? A well-suited candidate for PeV DM is the lightest neutralino within the scenario of superheavy supersymmetry [39]. This is a unique case where the PeV DM has weak interactions and respects perturbative unitarity despite the large mass hierarchy $m_X \gg m_Z$. Since the annihilation cross section of a (point) particle is bounded by unitarity, $\sigma_{\rm ann} \propto 1/m_X^2$, the resulting flux of HE neutrinos for stable PeV DM is too small to be observable without an additional enhancement of the annihilation rate. The flux of HE neutrinos can be observable either if the PeV DM particle is meta-stable with suitable life-time or the DM is much more clumpier than predicted by the simplest inflationary scenario [40].

4.1. PeV Dark Matter

Dark matter particles with masses in the PeV range constitute (part of) the CDM and, consequently, their abundance in the galactic halo is enhanced by a factor $\sim 5 \times 10^4$ above their extragalactic abundance. Therefore, the secondary fluxes from their decays are dominated by the halo component. The quotient $r_X = \Omega_X(t_0/\tau_X)$ of relic abundance Ω_X and lifetime τ_X of the X particle is fixed by the HE neutrino flux. The value of r_X is not predicted, but calculable as soon as a specific particle physics and cosmological model is fixed. PeV dark matter has several clear signatures:

Spectral shape: The fragmentation spectra of superheavy particles calculated by different methods and different groups agree well. This allows to consider the spectral shape as a signature of models with decays or annihilations of PeV DM particles. The predicted fragmentation spectrum, $dN/dE \propto E^{-1.9}$, at $x = E/M \lesssim 0.1$ is potentially modified by virtual electroweak Bremsstrahlung and direct neutrino production at larger x, cf. Fig. 4. However, it is clear that it cannot fit the observed neutrino spectrum at low energies.

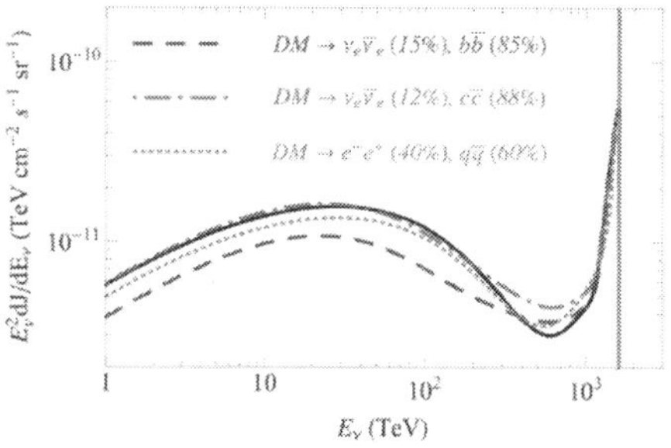

Figure 4. Neutrino spectra from DM decays for three decay channels, each chosen as sum of hard plus a soft component from fragmentation [38].

Composition : Since at the end of the QCD cascade quarks combine more easily to mesons than to baryons, the main component of the HE flux are neutrinos and photons from pion decay. Therefore, a robust prediction of this model is the appeareance of large photon fluxes, which are limited by experimental bounds.

Galactic anisotropy: The HE neutrino flux from PeV DM should show a galactic anisotropy [41], because the Sun is not in the center of the Galaxy. The degree of this anisotropy depends on how strong the CDM is concentrated near the Galactic center.

For a combined analysis of the arrival directions and the energy spectrum in the PeV DM scenario see e.g. [42]

5. Summary

The discovery of HE astrophysical neutrinos with energies $E > 10\,\text{TeV}$ by the IceCube experiment opened a new window to the Universe–and we are waiting for a more detailed view.

Acknowledgments
I am grateful to Gwenael Giacinti, Sergey Ostapchenko and Dima Semikoz for pleasant collaborations and discussions.

References

[1] M. G. Aartsen *et al.* [IceCube Collaboration], Science **342**, 1242856 (2013).
[2] M. G. Aartsen *et al.* [IceCube Collaboration], Phys. Rev. Lett. **111**, 021103 (2013).
[3] M. G. Aartsen *et al.* [IceCube Collaboration], Phys. Rev. D **91**, 022001 (2015).
[4] P. Bhattacharjee and N. Gupta, hep-ph/0501191.
[5] P. D. Serpico and M. Kachelrieß, Phys. Rev. Lett. **94**, 211102 (2005).
[6] S. Ando and J. F. Beacom, Phys. Rev. Lett. 95, 061103 (2005); K. Asano and S. Nagataki, Astrophys. J. 640, L9 (2006).
[7] T. Kashti and E. Waxman, Phys. Rev. Lett. 95, 181101 (2005).
[8] M. Kachelrieß and R. Tomas, Phys. Rev. D **74**, 063009 (2006).
[9] M. Kachelrieß, S. Ostapchenko and R. Tomas, Phys. Rev. D **77**, 023007 (2008).
[10] O. Mena, S. Palomares-Ruiz and A. C. Vincent, Phys. Rev. Lett. **113**, no. 9, 091103 (2014).
[11] M. G. Aartsen *et al.* [IceCube Collaboration], arXiv:1502.03376 [astro-ph.HE].
[12] R. Silberberg and M.M. Shapiro, Proc. 15th ICRC, v. **6**, p. 237 (1977); F. W. Stecker, Astrophys. J. **228**, 919 (1979).
[13] T. Antoni *et al.* [KASCADE Collaboration], Astropart. Phys. **24**, 1 (2005).
[14] M. Kachelrieß and S. Ostapchenko, Phys. Rev. D **90**, no. 8, 083002 (2014).
[15] S. Ostapchenko, Phys. Rev. D **83**, 014018 (2011).
[16] S. Ostapchenko, Progr. Theor. Phys. Suppl. **193**, 204 (2012); EPJ Web Conf. **52** (2013) 02001.
[17] T. Stanev, P. L. Biermann and T. K. Gaisser, Astron. Astrophys. **274**, 902 (1993).
[18] A. M. Hillas, J. Phys. G **31**, R95 (2005).
[19] V. S. Ptuskin *et al.*, Astron. Astrophys. **268**, 726 (1993).
[20] J. Candia, E. Roulet and L. N. Epele, JHEP **0212**, 033 (2002); JCAP **0305**, 003 (2003).
[21] G. Giacinti, M. Kachelrieß and D. V. Semikoz, Phys. Rev. D **90**, R041302 (2014); arXiv:1502.01608.
[22] B. Peters, Nuovo Cim. **22**, 800 (1961).
[23] J. R. Hörandel, Astropart. Phys. **19**, 193 (2003).
[24] T. K. Gaisser, T. Stanev and S. Tilav, Front. Phys. China **8**, 748 (2013) [arXiv:1303.3565 [astro-ph.HE]].
[25] V. S. Berezinsky, T. K. Gaisser, F. Halzen and T. Stanev, Astropart. Phys. **1**, 281 (1993); C. Evoli, D. Grasso and L. Maccione, JCAP **0706**, 003 (2007).
[26] G. Giacinti, M. Kachelrieß and D. V. Semikoz, Phys. Rev. Lett. **108**, 261101 (2012); Phys. Rev. D **88**, 023010 (2013).
[27] M. Iacobelli *et al.*, Astron. Astrophys. **558**, A72 (2013).
[28] A. Neronov, D. V. Semikoz and C. Tchernin, Phys. Rev. D **89**, 103002 (2014).
[29] M. Ahlers and K. Murase, Phys. Rev. D **90**, 023010 (2014).
[30] V. Berezinsky, Proc. of Neutrino 1977, Elbros; D. Eichler, Astrophys. J. **232**, 106 (1979).
[31] M. Vietri, Astrophys. J. **453**, 883 (1995); E. Waxman and J. N. Bahcall, Phys. Rev. Lett. **78**, 2292 (1997).
[32] B. C. Lacki and R. Beck, Mon. Not. R. Astron. Soc. **430**, 3171 (2013) [arXiv:1301.5391 [astro-ph.CO]].
[33] V. Berezinsky, A. Gazizov, M. Kachelrieß and S. Ostapchenko, Phys. Lett. B **695**, 13 (2011).
[34] A. A. Abdo *et al.* [Fermi-LAT Collaboration], Phys. Rev. Lett. **104** (2010) 101101.
[35] M. Ackermann, talk at 4.th Fermi symposium, Monterey 2012, `http://fermi.gsfc.nasa.gov/science/mtgs/symposia/2012/program.html`
[36] K. Murase, M. Ahlers and B. C. Lacki, Phys. Rev. D **88**, no. 12, 121301 (2013).
[37] V. Berezinsky, M. Kachelrieß and A. Vilenkin, Phys. Rev. Lett. **79**, 4302 (1997); V. A. Kuzmin and V. A. Rubakov, Phys. Atom. Nucl. **61**, 1028 (1998).
[38] A. Esmaili and P. D. Serpico, JCAP **1311**, 054 (2013) and references therein.
[39] V. Berezinsky, M. Kachelrieß and M. A. Solberg, Phys. Rev. D **78**, 123535 (2008).
[40] V. Berezinsky et al., Phys. Rev. D **81**, 103529 (2010).
[41] S. L. Dubovsky and P. G. Tinyakov, JETP Lett. **68**, 107 (1998).
[42] Y. Bai, R. Lu and J. Salvado, arXiv:1311.5864 [hep-ph].

Investigation of cascade showers in the Cherenkov water detector NEVOD

S S Khokhlov[1], M B Amelchakov[1], N S Barbashina[1], A G Bogdanov[1], V D Burtsev[1], D V Chernov[1], A N Dmitrieva[1], L I Dushkin[1], V A Khomyakov[1], V V Kindin[1], R P Kokoulin[1], K G Kompaniets[1], E A Kovylyaeva[1], V S Kruglikova[1], G Mannocchi[2], V V Ovchinnikov[1], A A Petrukhin[1], O Saavedra[3], I A Shulzhenko[1], V V Shutenko[1], G Trinchero[2], I I Yashin[1] and E A Zadeba[1]

[1]National Research Nuclear University MEPhI (Moscow Engineering Physics Institute), 115409 Moscow, Russia
[2]Istituto di Fisica dello Spazio Interplanetario, INAF, 10133 Torino, Italy
[3]Dipartimento di Fisica dell' Universita di Torino, 10125 Torino, Italy

E-mail: SSKhokhlov@mephi.ru

Abstract. A technique fot the reconstruction of cascade profiles by means of Cherenkov radiation in the water of the NEVOD detector is discussed. NEVOD is equipped with a dense spatial lattice of optical modules. The analyzed cascades have been generated either along near-horizontal muons (zenith angles between 85 and 90°), which's tracks are reconstructed by means of the tracking detector DÉCOR, or by muons with unknown tracks over a wider zenith angle range of 50-90°. Mean cascade profiles and energy spectra of cascades measured during the experimental series of about 7950 hours of 'live time' are presented.

1. Introduction

In recent years the interest in studying ultrahigh-energy muons and neutrinos of cosmic rays has grown strongly. The reasons are the so-called 'muon puzzle' [1] (the excess of both muon bundles [2, 3] and UHE muons [4, 5] with the energy increase) and the discovery of the excess of UHE neutrinos that apparently are indicative of their extraterrestrial origin [6]. In both cases, one of the most important tasks is to measure the energy of cascade showers generated by muons and neutrinos, in particular in Cherenkov water detectors (CWD). However in large-scale setups such as NT200+, ANTARES or IceCube the cascade energy is estimated by Cherenkov radiation from the cascades using a point approximation because of the large distances between the optical modules (tens of meters). The cascade energy reconstruction is based on simulation which can result in systematical uncertainties. However, the direct measurement of the cascade profile in a CWD with a dense lattice of optical modules allows to verify the models used to describe cascade profiles in Cherenkov radiation.

2. The NEVOD-DECOR complex and experimental runs

The NEVOD-DECOR experimental complex (figure 1) is a multipurpose facility created for effective detection of all cosmic ray components at ground level [7, 8, 9]. The setup is located in the campus of

MEPhI in a special four-story building and is equipped with the necessary technical and technological systems to provide the detector operation. The basis of the complex is the Cherenkov water detector NEVOD with a volume of 9×9×26 m^3. The detecting system is formed by a spatial lattice of quasi-spherical modules (QSM), each of them including six PMTs with flat cathodes directed along the coordinate axes. The QSM lattice allows to detect Cherenkov radiation from any direction with practically the same efficiency. The lattice is formed by a set of vertical strings containing 3 or 4 modules each. The distances between the modules are 2.5 m along the detector, and 2.0 m both across it and over the depth. The current detecting system consists of 91 QSM (546 PMT in total) [10,11].

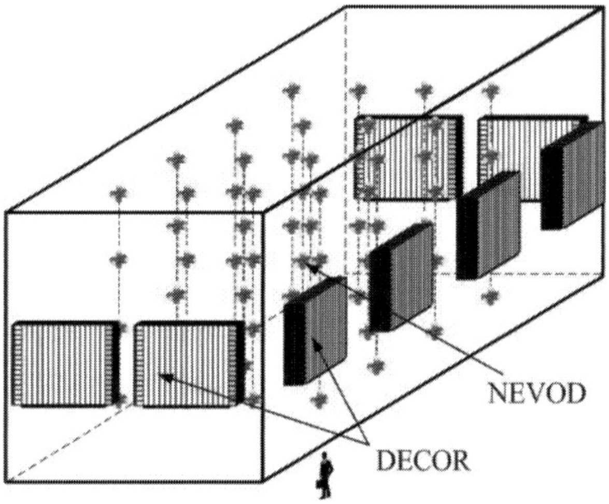

Figure 1. The NEVOD-DECOR experimental complex.

To improve the event reconstruction accuracy the coordinate detector DECOR [8,9] was constructed around the Cherenkov water calorimeter NEVOD (see figure 1). DECOR is a modular multi-layer system of plastic streamer tube chambers with resistive cathode coating. DECOR includes eight vertically suspended eight-layer assemblies (supermodules, SMs) of chambers with a total sensitive area of 70 m^2. The chamber planes are equipped with a two-coordinate external strip readout system that allows to localize charged particle tracks with about 1 cm accuracy in both coordinates (X, Y). The angular reconstruction accuracy for muon tracks crossing the single SM is better than 0.7° and 0.8° for projected zenith and azimuth angles respectively.

During the period from December 23, 2011 to March 20, 2013 a series of runs with 7945 hours of ‘live time’ was performed and 375 million events were registered.

Cascade search was conducted in two types of registered events:

• events with a single near-horizontal muon which’s track was reconstructed according to the data of the DECOR detector;

• events with large energy deposits in the volume of NEVOD.

A condition the single near-horizontal muon selection was the triggering of two DECOR supermodules located along the opposite short sides of the water tank. If track angles reconstructed on the basis of individual supermodule responses agreed within less than 5°, then it was supposed that the track segments within each SM belong to the same particle. The line connecting midpoints of track segments in each SM is taken as the track of the muon. For such geometry of the experiment, muons with zenith angles larger than 85° are selected. The threshold energy of such muons is about 7 GeV, the mean energy about 100 GeV.

An example of a cascade generated by a near-horizontal muon is shown in figure 2. Hereafter in similar figures, crosses and circles represent triggered QSMs and PMTs. The colors of circles correspond to different gradations of amplitudes: green denotes 1 ÷ 7 photoelectrons (ph.e.); yellow 8 ÷ 25 ph.e.; violet 26 ÷ 75 ph.e. and red amplitudes larger than 75 ph.e.

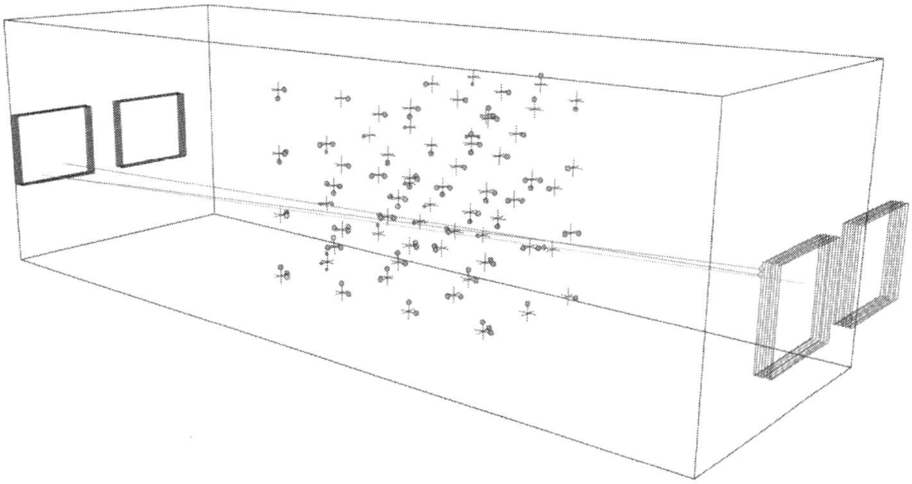

Figure 2. An example of a cascade generated by near-horizontal muon. The track of the muon is reconstructed by means of the DECOR detector.

Events with large energy deposits were selected by the trigger '60c'. A condition for this trigger was that more than 59 QSMs were hit within a time gate of 250 ns. Moreover at least two neighboring PMTs in each of these modules should be hit. This trigger was used from April 3, 2012 (5900 hours of 'live time').

For the events with single near-horizontal muons selected by means of DECOR we know the cascade axis which practically coincides with the muon track. However the statistics of such events is limited by the narrow solid angle, which is determined by the detector geometry. Events selected by trigger '60c' are registered over a wider range of angles, but for them the cascade axis must be reconstructed. In addition, the events selected by this trigger are not only cascades generated by muons but also other multi-particles events (EAS, muon bundles, hadron showers etc.) which have to be separated in the analysis.

3. The technique of cascade profile reconstruction

At distances larger than the photocathode radius, the PMT response to Cherenkov radiation of a single charged particle can be calculated by the equation:

$$A_1(R,\alpha) = \frac{S_{\mathrm{PMT}} \cdot \cos\alpha}{2\pi(R + r_{\mathrm{PMT}}) \cdot \sin\theta_C} \int_{\lambda\,\mathrm{min}}^{\lambda\,\mathrm{max}} \frac{dN}{d\lambda} \eta(\lambda) \cdot \exp(-\frac{R}{L(\lambda) \cdot \sin\theta_C}) d\lambda, \quad (1)$$

where S_{PMT} and r_{PMT} are correspondingly the area and the radius of the PMT photocathode; $dN/d\lambda$ is the wavelength distribution of photons generated per track length unit; α is the angle of Cherenkov radiation incidence on the PMT photocathode; θ_C is the angle of Cherenkov radiation in water; $\eta(\lambda)$ is the PMT quantum efficiency; $L(\lambda)$ is the absorption length of light in water (figure 3); λ_{min} and λ_{max} are the boundaries of the PMT sensitivity range. In figure 4, triangles show the dependence calculated by equation (1) with $\cos\alpha = 1$, squares represent the experimentally measured PMT response to near-horizontal muon selected by means of DECOR. Inside the sensitive volume of the detector, the muon generates secondary particles that initiate additional Cherenkov photons and increase the PMT response compared with the calculated response for a single particle.

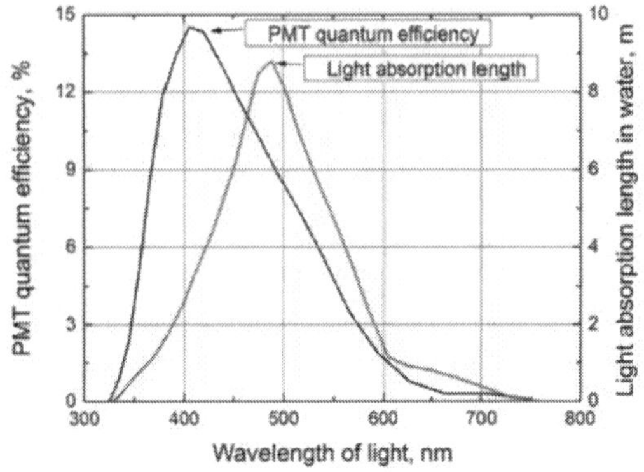

Figure 3. Dependences of the quantum efficiency of PMT and the absorption length on the wavelength of light.

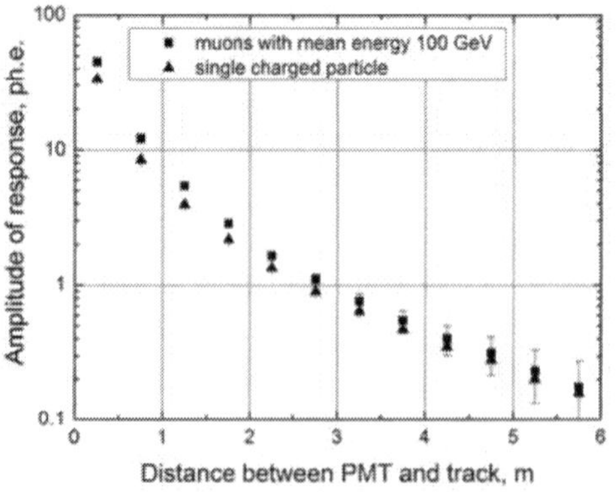

Figure 4. Dependences of the PMT response on the distance to the track, for a single charged particle and for muons with the mean energy 100 GeV.

To reconstruct the cascade profile in NEVOD we determine the number of light-emitting relativistic particles based on the PMT amplitude (figure 5) [12, 13]. This technique is used for the PMT directly illuminated by Cherenkov radiation from the cascade. We assume that the directions of the shower particles are close to the shower axis, and that this axis coincides with the track of the muon. The track is divided into bins equal to one radiation length (for water, 36.1 g/cm^2).

The number of light-emitting charged particles on a bin of the track, N_i, is calculated as the ratio between the PMT response to the event, A_{PMT}, and the PMT response to a single charged particle with the same geometric parameters of the track, A_1:

$$N_i = \frac{A_{\text{PMT}}}{A_1(R,\alpha)}. \tag{2}$$

Then the estimated numbers of particles in a bin are averaged for all PMT that "see" this segment of the track.

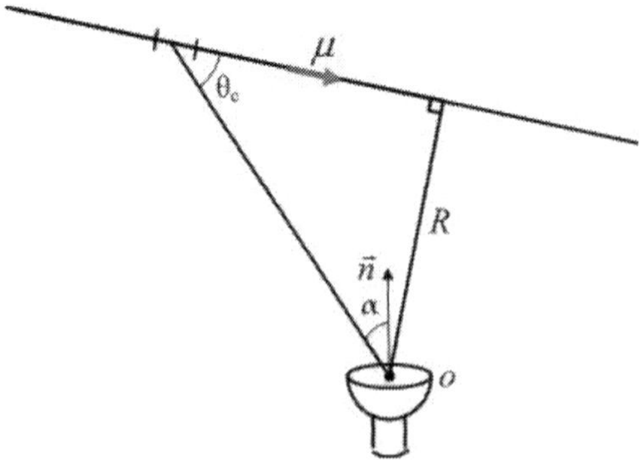

Figure 5. Geometry of the event for calculation of number of charged particles.

The resulting dependence of the number of light-emitting particles on the depth is fitted by a function constructed on the basis of the cascade profile in the one-dimensional approximation [14]:

$$N(y_0,x_0,x) = \begin{cases} 1.35, \text{ if } x < x_0 \\ \left(0.32\big/\sqrt{y_0}\right)\cdot \exp\left((x-x_0)\cdot(1-1.5\ln s)\right) + 1.35, \text{ if } x \geq x_0 \end{cases}, \tag{3}$$

where $y_0 = \ln(\varepsilon_0/\beta)$, β = 78.3 MeV is the critical energy of electrons in water, $s = 3(x-x_0)/\ (x-x_0+2y_0)$ is the shower age, ε_0 is the shower energy, and x_0 is the point of the shower generation.

To verify the technique, cascade events with fixed energy were simulated. Muons with energy E_μ = 20 GeV and gamma rays with energy 10 GeV, 31.6 GeV, 100 GeV, 316 GeV, 1 TeV and 3.16 TeV (with increments of 0.5 in the decimal logarithm of the energy) were thrown through the detector. We have generated 5000 Monte-Carlo events for cascades with energies $\varepsilon_0 = 10 \div 316$ GeV and 2700 events for cascades with energies 1 TeV and 3.16 TeV.

To account for the systematic shift, the dependence of the reconstructed parameter y_0 on the cascade energy was fitted by a linear function (figure 6):

$$y_0^{rec} = a \cdot y_0^{sim} + b\,, \tag{4}$$

where $a = 1.023 \pm 0.004$, $b = -\ 0.24 \pm 0.03$. Further these coefficients were used as correction factor.

In figure 7, the curve represents the theoretical approximation for a cascade with energy ε_0 = 316 GeV calculated by equation (3), while the squares show the average cascade profiles according to the simulation (black squares are for all shower electrons, open squares are for electrons with energy above the threshold of Cherenkov radiation in water E_{Cher}= 260 keV), triangles show the average cascade profile reconstructed by the developed technique.

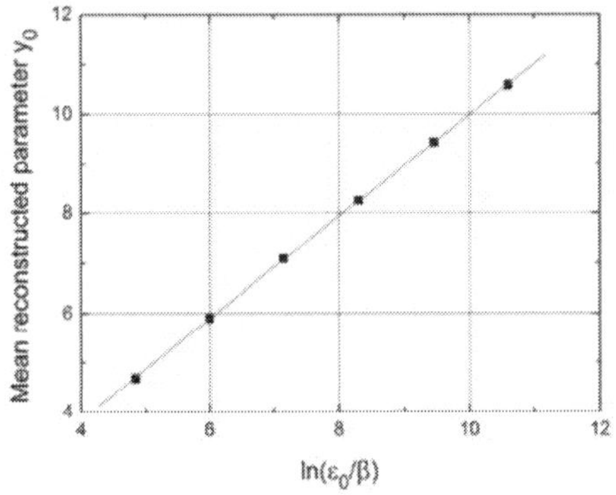

Figure 6. Dependence of reconstructed parameter y_0 on cascade energy.

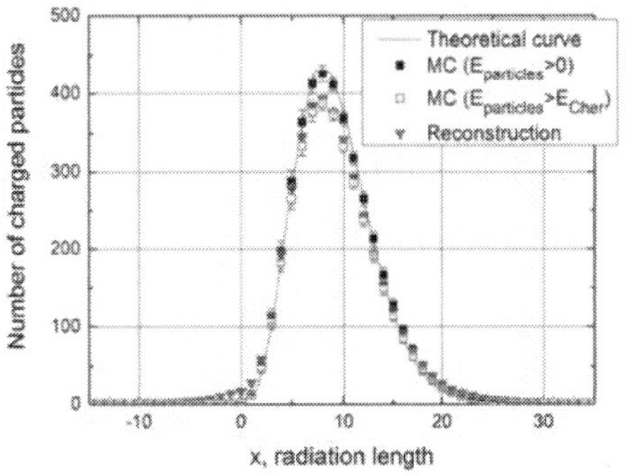

Figure 7. Theoretical, mean simulated and reconstructed cascade profiles for cascades with energy 316 GeV.

As can be seen, the average reconstructed profile follows the shape of the average simulated profile for particles with energies above the threshold. A small difference in the tails of the profile is due to the electron scattering in water which is not taken into account in the one-dimensional model of the cascade.

4. Results

4.1. Cascades generated by single near-horizontal muon with known track

1 696 530 events with single, near-horizontal muon selected by means of DECOR were registered. In 123 364 events, cascades with energy from 1 to 3000 GeV were reconstructed. In some events, double cascades were observed. An example of the cascade profile of such an event is shown in figure 8. Reconstructed energies of the cascades are 140 GeV and 42 GeV.

The cascades were sorted into groups according to the decimal logarithm of the energy, and cascade profiles were averaged within each group. The results are shown in figure 9. The mean cascade profile for showers with energy ε = 1 TeV is based on only 30 events, so this cascade profile is not completely smooth.

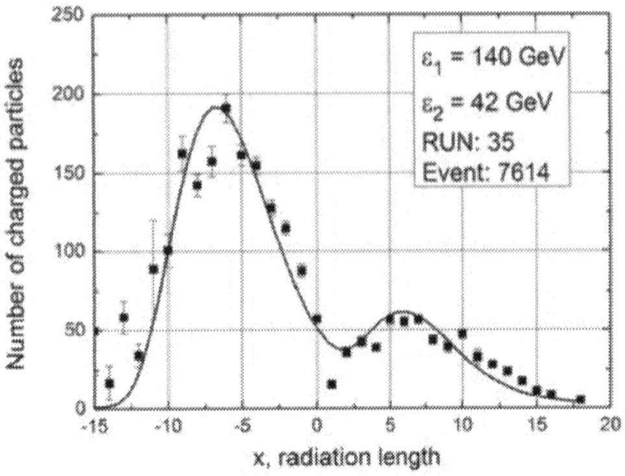

Figure 8. An example of a double cascade.

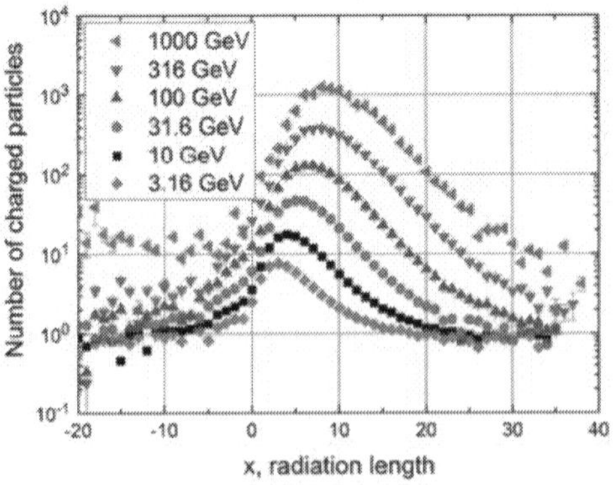

Figure 9. Mean cascade profiles.

4.2. Cascades generated by a muon with unknown track

Selection and reconstruction of the parameters of cascade events from the sample selected by the trigger '60c' are based on the compactness of location of QSMs with the highest amplitudes. Techniques were tested on events with known axes selected by means of DECOR.

In NEVOD lattice, optical modules with the highest responses in the cascade events form a compact clusters. The geometry of the cluster can be used as the cascade signature (figure 10). It was found that the rms radius of the cluster of 20 QSM with the highest response does not exceed 2.35 m in 95% of events with cascade showers with energy above 100 GeV.

The cascade axis is drawn through the center of mass of the selected QSMs (weighted with the PMT amplitudes). To reconstruct the direction of the axis in the selected cascade, the sum of the PMT surface normals weighted with their amplitudes is used. As the QSM closest to the cascade axis, with the highest response, can introduce a significant error in the determination of the resulting vector, it is excluded from the direction reconstruction. This technique was applied to the analysis of cascades generated by near-horizontal muons selected by means of DECOR. It was found that the mean error of the axis direction is about 16° and the mean error of the axis location about 30 cm.

To ensure correct cascade energy reconstruction, the events in which the cascades were developed inside the spatial lattice of NEVOD were selected. The center of mass of the QSM cluster should be at least 1.5 m away from the boundaries of the lattice. To exclude hadronic cascades, only the events with the estimated zenith angle in the range from 50° to 90° were selected.

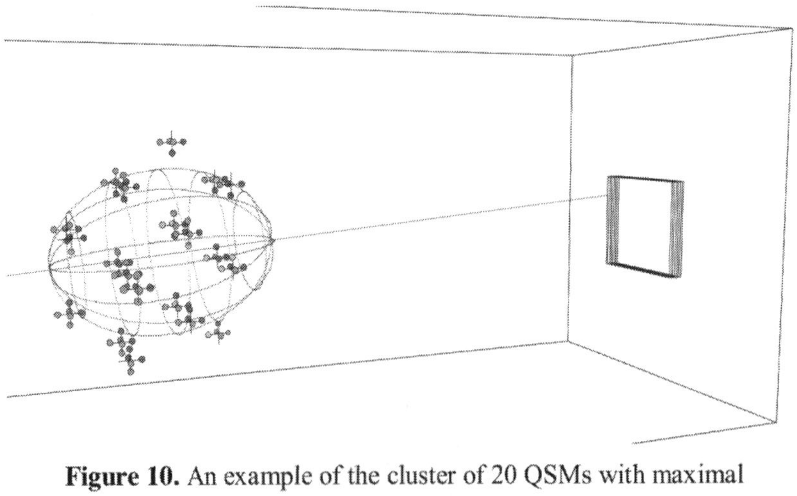

Figure 10. An example of the cluster of 20 QSMs with maximal responses.

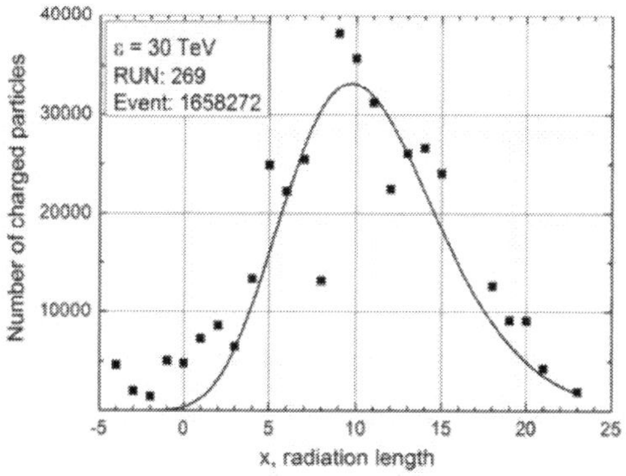

Figure 11. An example of reconstruction of the cascade profile for experimental shower.

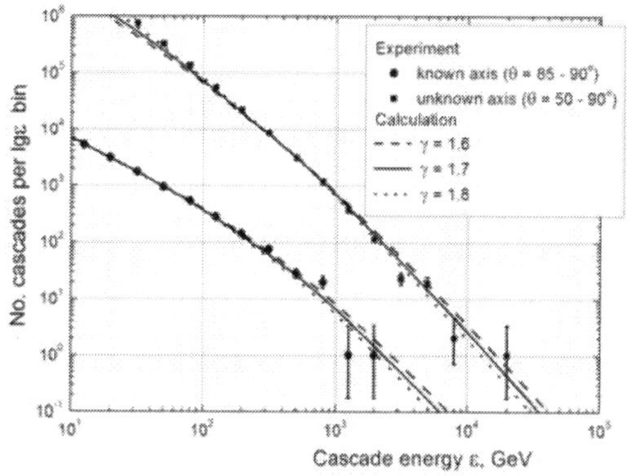

Figure 12. Energy spectra of cascade showers.

During the experimental series, about 80 million events with large energy deposits selected by trigger '60c' were registered. Using the developed criteria, 1 million cascade events with energies above 30 GeV were reconstructed. The energy of the most powerful cascade is about 30 TeV and its zenith angle is about 80°. Figure 11 shows the cascade profile of this cascade.

The differential spectra of cascades generated by muons in water in the range of zenith angles from 50° to 90° were plotted on the basis of estimated cascade energy. In figure 12 the circles show the spectrum of cascades generated by near-horizontal muons with tracks reconstructed by means of the DECOR detector; squares represent the spectrum of cascades registered by trigger '60c'; curves show the results of calculations for different values of the index γ of pion and kaon integral generation spectrum in the atmosphere. As seen from the figure, reconstruction of cascades according to the NEVOD data only increases the statistical reliability of the spectrum at high energies by nearly 2 orders of magnitude.

5. Conclusion

For the first time, cascade profiles (expressed as the number of cascade particles along the shower) were reconstructed from Cherenkov light in water, using the high-density optical module lattice of the NEVOD-DECOR experimental complex. As a result, the experimental cascade profiles in the energy range $\varepsilon = 3 \div 1000$ GeV have been obtained. These profiles can be used to verify the techniques of cascade energy reconstruction in the operating large-scale Cherenkov detectors.

Acknowledgments

This work was performed at the NEVOD Scientific and Educational Center with the financial support from the Russian Ministry of Education and Science (government task and project No. RFMEFI59114X0002), grant of the Leading Scientific School NSh-4930.2014.2, and RFBR grant No. 15-02-07763.

References

[1] Petrukhin A A 2014 *Nucl. Instr. Meth. in Phys. Res. A* **724** 228
[2] Barbashina N S *et al* 2007 *Nucl. Phys. B (Proc. Suppl.)* **165** 317
[3] Rodrigues G *(Pierre Auger Collaboration)* 2013 *EPJ Web Conf.* **53** 07003
[4] Bogdanov A G *et al* 2012 *Astropart. Phys.* **36** 224
[5] Berghaus P and Xu Chen 2011 *Proc 32nd ICRC (Beijing)*
[6] Aartsen M G *et al (IceCube Collaboration)* 2014 *Phys. Rev. Lett.* **113** 101101

[7] Aynutdinov V M *et al* 1997 *Astrophysics and Space Science* **258** 105
[8] Barbashina N S *et al* 2000 *Instrum. Exp. Tech.* **43** 743
[9] Amelchakov M B *et al* 2002 *Izv. Akad. Nauk: Ser. Fiz.* **66** 1611
[10] Khokhlov S S *et al* 2011 *Astrophys. Space Sci. Trans.* **7** 271
[11] Kindin V V *et al* 2014 *Proc 24rd ECRS (Kiel)* Submitted to *J. Phys.: Conf. Series*
[12] Khokhlov S S *et al* 2013 *J. Phys.: Conf. Ser.* **409** 012134
[13] Khokhlov S S *et al* 2014 *Bull. Leb. Phys. Inst.* **41** 292
[14] Hayakawa S 1969 Cosmic Ray Physics (NY: J. Wiley & Sons)

Status of High-Energy Neutrino Astronomy

Marek Kowalski

Deutsches Elektronen-Synchrotron (DESY), Platanenallee 6, D-15738, Zeuthen;
Humboldt-Universität zu Berlin, Institut für Physik, Newtonstrasse 15, D-12589, Berlin

E-mail: marek.kowalski@desy.de

Abstract. With the recent discovery of high-energy neutrinos of extra-terrestrial origin by the IceCube neutrino observatory, neutrino-astronomy is entering a new era. This review will cover currently operating open water/ice neutrino telescopes, the latest evidence for a flux of extra-terrestrial neutrinos and current efforts in the search for steady and transient neutrino point sources. Generalised constraints on potential astrophysical sources are presented, allowing to focus the hunt for the sources of the observed high-energy neutrinos.

1. Introduction

During the last decades, considerable efforts were made to establish high-energy neutrinos as cosmic messenger. The motivation derives from the special properties of neutrinos. They interact only weakly and hence can escape energetic and dense astrophysical environments otherwise opaque to electro-magnetic radiation. It is for that reason that neutrinos promise to provide unique insights into a number of extreme astrophysical phenomena, ranging from stellar explosions to particle acceleration in the center of active galaxies. Furthermore, the observation of high-energy neutrinos can help to decipher the more than 100 year old mystery of cosmic rays. Charged cosmic rays are expected to produce high-energy neutrinos through the interaction with ambient matter or radiation fields. However, while charged cosmic rays are deflected by magnetic fields on the way from the source to the Earth, neutrinos always point back to their origin, unaffected by matter or radiation fields encountered on their path. Hence high-energy neutrinos can provide direct information about the accelerators producing the highest-energy cosmic rays. Accordingly, with the recent discovery of high-energy neutrinos of extra-terrestrial origin by the IceCube neutrino observatory [1, 2], neutrino-astronomy has made a long-awaited step.

In this contribution I review the arguments why an astrophysical flux of neutrinos was expected to be observed (chapter 2), the currently operating open water/ice neutrino telescopes (chapter 3), the latest evidence for a flux of extra-terrestrial neutrinos and current efforts in the search for steady and transient neutrino point sources (chapter 4). Finally, generalised constraints on potential astrophysical sources are being discussed, allowing to focus the hunt for the sources of the observed high-energy neutrinos (chapter 5).

2. Expectations

The existence of high-energy astrophysical neutrinos has been motivated by the observations of high-energy hadronic cosmic rays. At their acceleration sites, but also during propagation through space, cosmic rays produce neutrinos in the collision with ambient matter and radiation fields. The resonant interaction of a high-energy proton (p) with a target photon (γ),

$$p + \gamma \rightarrow \Delta^+ \rightarrow n + \pi^+, \tag{1}$$

results in a charged pion, that decays further to produce neutrinos. From the pion decay a flux ratio of $\phi_{\nu_e} : \phi_{\nu_\mu} : \phi_{\nu_\tau} = 1 : 2 : 0$ is expected at the source, which during propagation from the source to Earth changes to $\phi_{\nu_e} : \phi_{\nu_\mu} : \phi_{\nu_\tau} = 1 : 1 : 1$ due to neutrino oscillations (small deviations are expected in case of strong magnetic fields inside the sources [3]).

Detailed model predictions exist for various source classes, ranging from Gamma-Ray Bursts to Active Galactic Nuclei (see e.g. [4] for a review of sources). To provide a rough estimate of the neutrino flux that can be expected, it is sufficient to review an early argument by Waxman & Bahcall [7]. One starts by considering the energy injection rate in protons per unit volume, $\dot{\epsilon}$, that can be constrained by the observed flux of cosmic rays on Earth. Assuming that the cosmic-rays above 10^{19} eV consist primarily of protons and are of extra galactic origin, one finds $\dot{\epsilon} = (0.5 \pm 0.015) \times 10^{44}$ergs Mpc^{-3}yr^{-1} [6]. The essential argument for formulating an upper bound on the neutrino flux — i.e. the Waxman-Bahcall bound [7] — is that within the sources, the observed cosmic ray protons are expected to have undergone the reaction of Eq. 1 at most once. This is because the reaction transforms the proton to a neutron, that being neutral, can not be confined anymore by the source magnetic fields. The neutron hence escapes the acceleration region before it eventually decays back to a proton (note that the Lorenz boost of a 10^{18} eV neutron provides for a decay length of the the size of a typical galaxy). Through Eq. 1, the cosmic ray energy density is kinematically tied to to the neutrino energy density and hence to the expected neutrino flux of high-energy neutrinos, leading to an upper bound on the flux of astrophysical neutrinos of all flavors [6]:

$$E_\nu^{-2}\phi_{\rm WB} \approx 3 \times 10^{-8}\text{GeVcm}^{-2}\text{s}^{-1}\text{sr}^{-1}. \tag{2}$$

The bound only weakly depends on the redshift evolution of a given population, such as the observed AGN luminosity function or GRB rate evolution. A flux of neutrinos at the level of the Waxman & Bahcall bound is detectable within a few years of a cubic kilometre sized detector and served as an important benchmark in evaluating the sensitivity of neutrino telescopes.

3. Neutrino Detectors

The concept of open water/ice neutrino telescopes has been successfully demonstrated in several places world-wide by now. The detection principle for these detectors is similar for all detectors: Cherenkov-light produced by charged particles — either background muons produced in air showers above the detector, or particles produced in a neutrino interaction — is recorded by a three-dimensional array of photomultipler tubes (PMTs) contained in appropriate pressure-resisting glass housings. The arrival time allows reconstructing the direction of the particles, while the total number of photons recorded is used to reconstruct its energy. Arrival direction, energy and topology allows to distinguish background from neutrino-induced events. A more refined analysis is then needed to distinguish astrophysical neutrinos from atmospheric neutrinos. In the following, an overview of the currently operating neutrino detectors is provided. Future projects are covered in a separate contribution to this conference [8].

3.1. Mediterranean Projects

The Mediterranean sea is offering deep sites (up to 4.5 km) with very transparent water. Consequently, several efforts where started to exploit the different sites: the NESTOR project deployed off the coast of the Peloponnesus [9], the NEMO project [10] deployed off the coast of Sicily as well as the ANTARES project [11] deployed off the Provence cost in France.

While technologically all projects have had significant impact on the field, only the ANTARES collaboration has been operating a sizeable detector over several years. Construction of the ANTARES detector was completed in 2008 and it now consists of 12 strings each carrying 25 storeys consisting of three PMTs each. The instrumented volume of ANTARES comprises 0.01 km^3.

3.2. Lake Baikal
In the waters of Lake Baikal, the NT200 detector was installed using the frozen lake surface during the winter as a platform for deployment. NT200 was completed in 1998, consisting of 8 strings with 192 PMTs in total. Three more strings carrying 12 PMTs each where added in 2005, resulting in the NT200+ detector [12]. The detector reaches a depth of 1100 m and, similar to ANTARES, is still in operation.

3.3. South Pole
The South Pole is home to the currently largest operating neutrino detector. The IceCube detector [13], which was deployed between 2005 and 2010, consists of 86 strings with 5160 PMTs in total. The instrumented volume comprises a cubic kilometre and it is deployed between 1450 and 2450 m depth. Unlike open-water detectors, the deployment of IceCube strings required melting holes into the antarctic ice. Once the instrumented holes are frozen again, the ice provides a very stable environment to operate a detector. The deep-ice detector is complemented by a km^2-sized air shower array at the surface.

4. Recent Results
Large neutrino detectors have significantly gained in sensitivity during the last few years, allowing to address a range of scientific topics, from neutrino oscillations and dark matter to neutrino astronomy. In light of the recent discovery of an astrophysical flux of high-energy neutrinos, the focus of this contribution is on the observation of astrophysical neutrinos. Furthermore, this review does not aim at giving a comprehensive summary of all results, but rather an overview of some of the recent results that shape our understanding of the newly discovered signal.

4.1. The discovery of astrophysical neutrinos
First hints for an excess of high-energy neutrinos was seen in data from the incomplete IceCube data from 2008 and 2009, both in the track channel (dominated by ν_μ) [14], as well as the cascade channel (dominated by ν_e and ν_τ) [15]. The subsequent discovery of two neutrino events above PeV energies in two years of data (between 2010 and 2012) provided more evidence that there is a new unaccounted component in the flux of high-energy neutrinos [16]. None of the independent observations of an excess by itself quite reached a 3 sigma significance (though combined they would have). More importantly, however, the excess was seen only in the energy spectrum and hence could be explained in principle also by a harder atmospheric neutrino component, such as expected from the decay of charmed mesons in the atmosphere. This changed for a refined analysis of the data, which focused on events starting in the detector [1]. By requiring that the neutrino interaction vertex lies within the detector (the outer detector layers provide a veto) and by reconstructing the directions of the events using the full available PMT information, one has a new handle on separating astrophysical from atmospheric neutrinos through a self-vetoing technique [17]: An atmospheric neutrino would inevitable be produced in coincidence with atmospheric muons from the same air shower. Hence, a downward going neutrino that can be identified without a coincident trough-going (atmospheric) muon track would be rather strong evidence for being of extra-terrestrial origin. The refined analysis using the self-vetoing

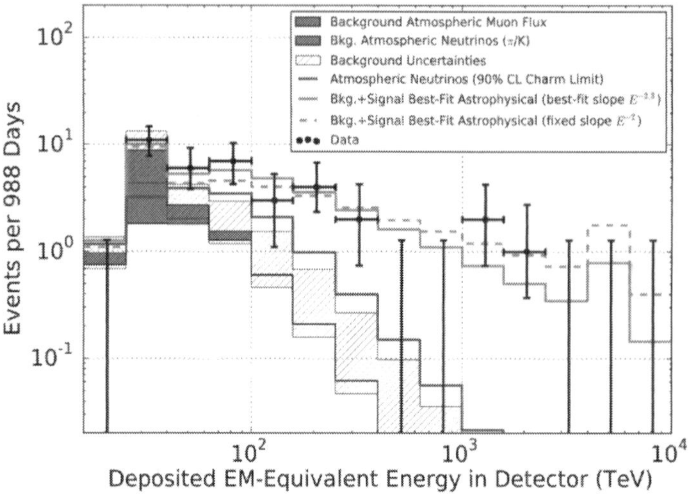

Figure 1. The visible deposited energy from the three-year search of starting events in IceCube [18].

technique has shown that the majority of the 28 starting neutrino events identified, including the two highest energy events, are consistent with being of extra-terrestrial origin [1]. The analysis was extended from 2 to 3 years of data, resulting in 37 events in total and allowing to reject the atmospheric-only hypothesis with 5.7 σ [18]. The energy spectrum of these events is shown in Fig. 1. Assuming a diffuse flux with power-law spectrum and a 1:1:1 flavor ratio, the best fit flux per neutrino flavor is

$$E^2\phi = 1.5 \times 10^{-8}(E/100\mathrm{TeV})^{-0.3}\mathrm{GeVcm^{-2}s^{-1}sr^{-1}}. \tag{3}$$

The choice to use 100 TeV as the pivot point in the parameterisation minimises the correlation between the spectral index and the normalisation. It is noteworthy that the resulting normalisation coincides with the upper bound predicted by Waxman & Bahcall. However, because the observed neutrino energies are reaching only up to a few PeV, the connection to the extragalactic cosmic rays observed with 3 to 4 orders of magnitude higher energies is not obvious.

4.2. Search for steady point sources

The analysis described in the previous section [1, 18] produces a sample of events that is dominated by astrophysical neutrinos. Most events are cascade-like and have an median angular resolution of 15°, while the track-like events are reconstructed with $\leq$ 1° resolution. The distribution of events in the sky is shown in Fig. 2. So far, the neutrino sky map appears consistent with being isotropic (if corrected for declination dependent sensitivity) which is indicative of an extragalactic population of sources. Nevertheless, a sub-dominant galactic component is possible. The apparent clustering in Fig. 2 near the galactic center is the most striking feature of the map. This location in the sky has been studied by the ANTARES collaboration, which because of the favourable field of view of ANTARES can extend the search to lower neutrino energies [19]. No excess was observed from the direction of the cluster by ANTARES, which severely constrains the possibility that it is of astrophysical origin.

The more conventional searches for neutrino point sources use all of the neutrino candidates to search for neutrino clustering or neutrinos associated with catalogs of likely sources. So

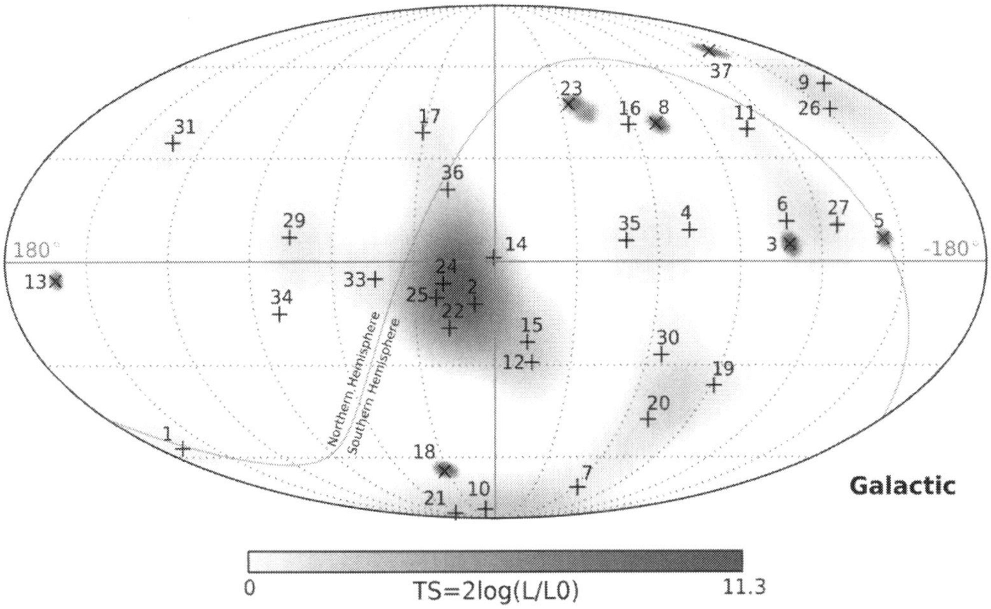

Figure 2. Sky map of 37 events from the search for starting events in IceCube [18]. Most events are cascade-like (marked with +) and have a median angular resolution of $\sim 10°$, while the track-like events (marked with $\times$) are reconstructed with $\leq 1°$ resolution.

far, no point source was identified by any of the experiments. Figure 3 shows the latest constraints by IceCube [20] and by ANTARES [21]. The IceCube sensitivity was extended to the Southern hemisphere by application of an energy cut to suppress the large background of atmospheric muons. Hence, for the Southern hemisphere the energy threshold of the search is two orders of magnitudes higher than for the search in the Norther hemisphere and the search by ANTARES. (A direct comparison of sensitivities is therefore model dependent.) IceCube records about 5 atmospheric neutrino events per year per square-degree Northern sky, above which an astrophysical source needs to produce an excess to be detected. Accordingly, the sensitivity to point sources only increases with $\sqrt{T}$ as a function of time, unless further measures are taken to improve the signal-to-noise (such as improving the angular resolution or using energy as an extra discriminator for signal-like events). ANTARES, due to its smaller size, records about 4 atmospheric neutrinos per day over one hemisphere, hence identification of a point source requires fewer events [21]. However, it also needs a much stronger source to get 5 events in ANTARES compared to 5 – and even 20 – events in IceCube.

4.3. Search for transient sources

The most prominent example of a transient source class are Gamma-Ray Bursts (GRB), which outshine the entire Universe in γ-rays during the $\sim 1 - 100$ seconds burst duration. During such a short time the background of atmospheric neutrinos is small ($\ll 1$), therefore even a single coincidence of a neutrino with a GRB would be of significant interest. No coincidence with a muon-neutrino event has been reported by any experiment so far. Testing for neutrino emission from 117 GRBs in the Northern hemisphere with IceCube [22] allowed to constrain the aggregate flux from all GRBs to below the Waxman & Bahcall bound. It also constrains the

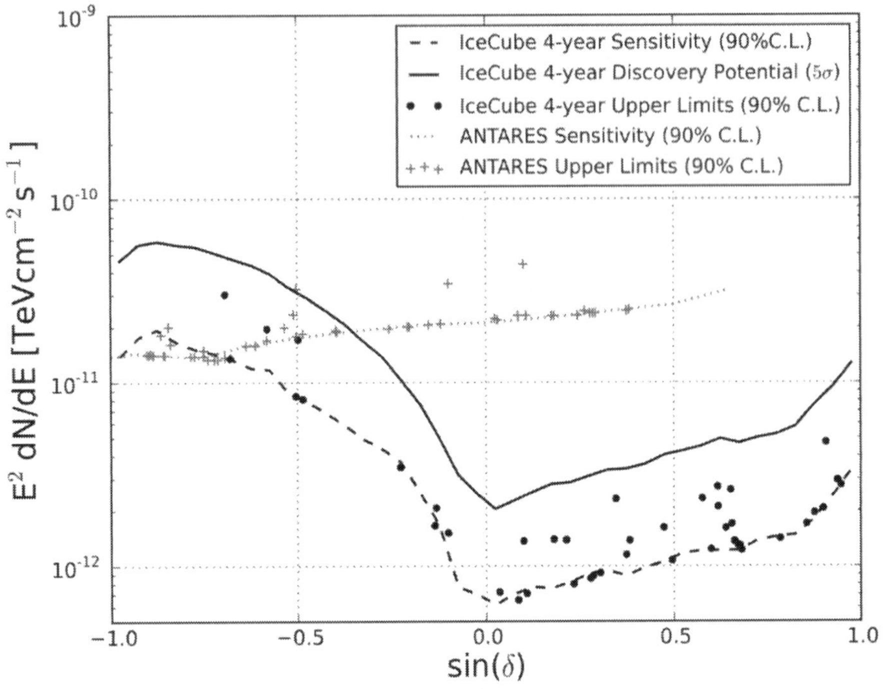

Figure 3. Sensitivity to a E^{-2} neutrino flux from individual point sources as a function of the declination [20]. The IceCube constraints are derived from the non-observation of clustering in data dominated by atmospheric muon neutrinos (Northern hemisphere) and atmospheric muons (Southern hemisphere), with correspondingly different energy thresholds [20] (see text for details). The constraints from ANTARES are derived from the non-observation of clustering in a sample dominated by atmospheric neutrinos [21]. The symbols represent flux upper limits for individual source candidates.

possibility that GRBs are the sources of cosmic rays [23], though more data as well as detailed modelling is required to rule out all scenarios (see e.g. [24]).

Other transient source classes could be off-axis or failed GRBs, where the jet does not escape the progenitor star or core collapse SNe with mildly relativistic jets, for which neutrino emission would be expected on time scales similar to GRBs. In such cases, no bright γ-ray burst would mark the stellar explosion. To nevertheless obtain sensitivity to a wide range of transient phenomena, neutrino telescopes search for event clusters both in direction and time. These searches are being performed in near realtime, allowing to trigger subsequent follow-up observations. Both ANTARES and IceCube send target-of-opportunity alerts when neutrino multiplets — two or more neutrinos coincident in time and direction — are observed [25, 26]. (ANTARES also sends alerts for the most energetic neutrino events). Follow-up observations are performed with the ROTSE and TAROT network of small robotic telescopes, with the Samuel Oschin telescope of PTF, as well as the SWIFT X-ray satellite.

5. The larger picture: Generic constraints on possible source classes

One can use the observation of a diffuse (extragalactic) flux of neutrinos, combined with a non-observation of resolved sources, to constrain the density of a generic population of sources (see

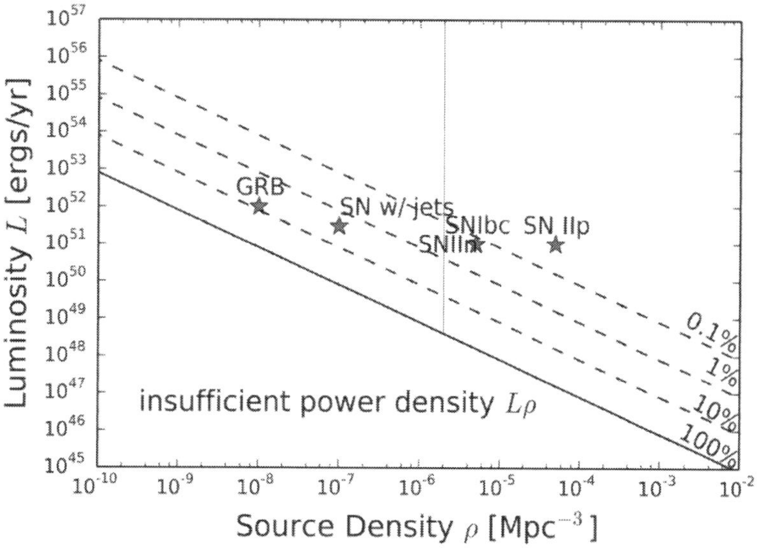

Figure 4. Constraints on potential transient sources (burst duration 100 seconds or less), responsible for the diffuse astrophysical flux detected by IceCube. The individual source populations are represented by a density[1] and typical luminosity[1] in a characteristic wavelength band. From left to right: Gamma-Ray Bursts, Supernovae with jets, Supernovae Ib/c, Supernovae IIN and Supernovae IIP. The (full) diagonal line marks the neutrino power density boundary, where the luminosity per volume, $L \cdot \rho$, saturates the observed diffuse flux; sources above the line release sufficiently abundant energy in electromagnetic radiation to explain the diffuse neutrino flux (dashed lines indicated fractional energy release in neutrinos). Sources in the vertical left (green) region can be excluded as the dominant source of the diffuse flux, since they would result in a signal within dedicated searches of IceCube (see text for details).

[27] for an early version of the argumentation). In a simplified picture, the bound on the source density can be understood as follows: The product of luminosity per source and source density, $L \cdot \rho$, corresponds to the total emission per volume and is constrained by the observed diffuse flux of neutrinos. For a distance scale to the nearest source $d = (4\pi\rho)^{-1/3}$, along with the power density $L \cdot \rho$, one can estimate the neutrino flux of the nearest object within a population $\phi = L/d^2 \propto \rho^{-1/3}$. The expected flux increases with decreasing source density and the fact that point source searches (including the search for stacked or transient sources[1] have only resulted in upper limits allows us to place constraints on possible populations and their associated source density. In the following, I discuss some current constraints derived from IceCube data using a signal simulation of generic source populations developed by Nora Linn Strothjohann and Andreas Homeier [28].

5.1. Transient sources

To start with, the simulation of a transient population of sources was assuming number density evolution and luminosity function following that observed for GRBs [5]. Assuming that the neutrino emission follows the same luminosity function and source population, one can model

[1] For transient sources the density refers to the number density of transients in a given year, while the luminosity refers to the energy released by the transient within one year. This choice of terminology allows to generalise the discussion to steady sources.

the neutrino flux expectation. For this, the spectral shape of the individual sources is chosen to match the observed diffuse flux, and the neutrino flux emitted by the total population is normalized to the observed diffuse flux. The simulation allows to predict the number of neutrino muliplets expected to be seen by IceCube (including effective areas and Poisson fluctuations due to low count rates). By requiring consistency with the observed number of neutrino multiplets, i.e. no triplet of neutrinos within 100 seconds [28], one obtains a lower bound on the source density. For a diffuse power-law flux matching the IceCube flux of Eq. 3, the bound on the source density derived from the non-observation of neutrino triplets or higher multiplicities in three years of IceCube data corresponds to $\rho \gtrsim 2 \cdot 10^{-6}\mathrm{Mpc}^{-3}\mathrm{yr}^{-1}$ [28]. The bound is broadly consistent with a recent sensitivity estimate [29] and depends only weakly on the evolution of the GRB density as a function of redshift and on the luminosity function. However, it does depend on the assumed spectral shape, which has been assumed to follow a power-law over the full energy range that IceCube is sensitive to. Fig. 4 illustrates the current constraints on transient sources. The bound from the non-observation of multiplets is shown as a vertical line. As one can see, GRBs, being very rare, are excluded as the dominant sources of the observed diffuse neutrino flux.

On the other hand, core collapse supernovae are still plausible candidates. They possess all the right ingredients for being extraordinary neutrino factories: a) they have been shown to produce ejecta with 10^{50} ergs kinetic energy, capable of efficiently accelerating CRs and b) they provide abundant amounts of target material for neutrino production, e.g. the stellar envelope or the circumstellar medium (CSM). The shock acceleration can happen at non-relativistic shock fronts [30, 31] or within (mildly) relativistic jets [32]. Other scenarios consider the spin down of rapidly rotating newborn pulsars producing large electromagnetic fields, a model that can explain also the highest energy CR [33]. The fact that the most promising SNe types (e.g. IIn, Ib/c or hypernovae) are rare and transient by nature puts them within reach of IceCube.

5.2. Steady sources

The previous considerations can be generalised to steady sources by dropping the time constraint of the multiplet search while increasing the required neutrino multiplicity to suppress the background of atmospheric neutrinos. With an angular resolution of $\sim 1^\circ$ for muon neutrinos and within 3 years of observations with IceCube, the detection of ~20 signal neutrinos from one source would be significant at the 3 σ level. Stacked searches increase the sensitivity by a factor of a few, allowing to lower the number of signal neutrinos per source. The limit derived from the non-detection of $\gtrsim 10$ signal neutrinos from the same source yields a lower bound of $\rho \gtrsim 2 \cdot 10^{-7}\mathrm{Mpc}^{-3}$ as indicated in Fig. 5. The bound agrees within a factor of two with the results reported in [34].

Figure 5 summarises the various astrophysical source populations that can potentially be responsible for the observed neutrinos, now also including steady sources. The sources populations are arranged according to their source density and the typical luminosity in electromagnetic radiation. The rarest and most luminous sources (in gamma rays) are Blazars (FSRQ and BL Lac), Fanaroff-Riley II (FR II) galaxies as well as GRBs. While all of these sources are candidates to produce the observed high-energy neutrinos, their number density is sufficiently low that one would have expected to see individual sources shining bright in the neutrino sky. The brightest sources within the sample would violate limits on (stacked) neutrino point sources, as well as limits on transient sources unless the population's density is larger than $\sim 10^{-7}\mathrm{Mpc}^{-3}$. More frequent sources, such as Fanaroff-Riley I galaxies (FRI) or Starburst galaxies are not yet restricted by point source limits, since the total flux is produced by sufficiently many sources.

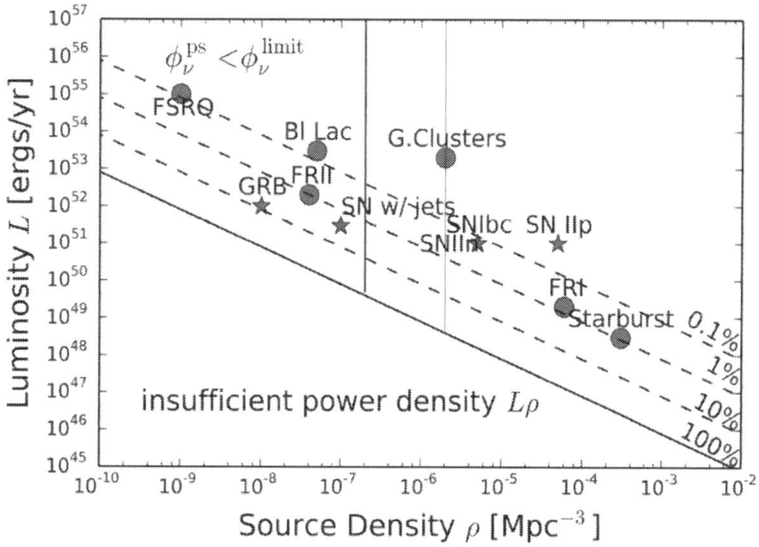

Figure 5. Figure 4 with additional constraints on steady point sources added. The figure shows the additional sources: Flat Spectrum Radio Quasars, Bl Lacs, Fanaroff-Riley II galaxies, Galaxy Clusters, Fanaroff-Riley I and Starburst galaxies (see caption of Fig. 4 for more explanation).

6. Summary and Outlook

The evidence for high-energy astrophysical neutrinos has increased significantly since their discovery in 2013. Due to their distribution in the sky, the observed neutrino events appear consistent with being of extragalactic origin. While individual sources have not yet been identified, even the absence of a steady point source and transient source signal leads to severe constraints on the possible nature of the source. The source density should be $\lesssim 10^{-7}\text{Mpc}^{-3}$ for steady sources and $\lesssim 10^{-6}\text{Mpc}^{-3}\text{yr}^{-1}$ for short transient sources (burst duration < 100 s), or individual bright sources would have been observed already.

Currently, studies are ongoing to better constrain the spectrum and flavour composition. About 1/3 of the astrophysical neutrino flux should consist of ν_τ and their identification should eventually be feasible based on their distinct "double-bang" signature [35] (the distance between hadronic interaction vertex and tau decay is ~ 50 m $\times$ (E_τ/PeV) and hence might be resolvable for the highest energy events). Soon, also other experiments hopefully reach the sensitivity to confirm IceCube's observation of astrophysical neutrinos.

Resolving the sources of high-energy neutrinos will be of particular focus for future studies. With existing neutrino detectors alone, the anticipated gain in sensitivity appears not sufficient to firmly resolve the sources. However, through combining neutrino information with observations from radio, optical, X-rays to γ−rays, there is justified hope to significantly narrow down the possibilities, and eventually identify the source.

References

[1] M. G. Aartsen *et al.* [IceCube Collaboration], Science **342** 6161, 1242856 (2013)
[2] M. G. Aartsen *et al.* [IceCube Collaboration], Physical Review Letters **113** (2014) 101101
[3] T. Kashti and E. Waxman, Phys. Rev. Lett. **95**, 181101 (2005)
[4] J. K. Becker, Phys. Rept. **458**, 173 (2008)
[5] D. Wanderman, T. Piran, MNRAS **406** (2010)
[6] E. Waxman, 2013, Proceedings of *Rencontres du Vietnam: Windows on the Universe* (2013), arXiv:1312.0558.

[7] E. Waxman and J. Bahcall, Phys. Rev. D 59, 2 (1999)
[8] C.Spiering, these proceedings.
[9] G. Aggouras, et al. [NESTOR Collaboration], Nucl. Inst. Meth. A 552, 420 (2005)
[10] M. Taiuti et al. [NEMO Collaboration], Nucl. Inst. Meth. A 626, S25 (2011)
[11] M. Ageron, et al. [ANTARES Collaboration], Nucl. Inst. Meth. A656, 11 (2011)
[12] V. Aynutdinov *et al.* [BAIKAL Collaboration], Nucl. Instrum. Meth. A **602**, 14 (2009)
[13] R. Abbasi *et al.* [IceCube Collaboration], Nucl. Instrum. Meth. A **601**, 294 (2009)
[14] M. G. Aartsen *et al.* [IceCube Collaboration], Phys. Rev. D **89**, 062007 (2014)
[15] M. G. Aartsen *et al.* [IceCube Collaboration], Phys. Rev. D **89**, 102001 (2014)
[16] M. G. Aartsen *et al.* [IceCube Collaboration], Phys. Rev. Lett. **111**, 021103 (2013)
[17] S. Schonert, T. K. Gaisser, E. Resconi and O. Schulz, Phys. Rev. D **79** 043009 (2009)
[18] M. G. Aartsen *et al.* [IceCube Collaboration], Phys. Rev. Lett. **113**, 101101 (2014)
[19] S. Adrian-Martinez *et al.* [ANTARES Collaboration], Astrophys. J. **786**, L5 (2014)
[20] M. G. Aartsen *et al.* [IceCube Collaboration], arXiv:1406.6757 [astro-ph.HE].
[21] S. Adrian-Martinez *et al.* [ANTARES Collaboration], JCAP **1405**, 001 (2014)
[22] R. Abbasi *et al.* [IceCube Collaboration], Nature **484** 351 (2012)
[23] E. Waxman and J. N. Bahcall, Phys. Rev. Lett. **78**, 2292 (1997)
[24] P. Baerwald, M. Bustamante and W. Winter, Astropart. Phys. **62**, 66 (2014)
[25] R. Abbasi *et al.* [IceCube and ROTSE Collaborations], Astron. Astrophys. **539**, A60 (2012)
[26] M. Ageron *et al.* [ANTARES Collaboration], Astropart. Phys. **35**, 530 (2012)
[27] P. Lipari, Phys. Rev. D **78** (2008) 083011
[28] N. Strothjohann, talk presented at JSI workshop, Multimessenger Astronomy in the Era of PeV Neutrinos, 2014; N. Strothjohann, Master thesis, 2014, University Bonn; A. Homeier, PhD thesis, University Bonn (in preparation)
[29] M. Ahlers and F. Halzen, arXiv:1406.2160 [astro-ph.HE].
[30] K. Murase, T. A. Thompson, B. C. Lacki and J. F. Beacom, Phys. Rev. D **84** 043003 (2011)
[31] B. Katz, N. Sapir and E. Waxman, arXiv:1106.1898 [astro-ph.HE].
[32] S. Ando and J. F. Beacom, Phys. Rev. Lett. **95** 061103 (2005)
[33] K. Fang, K. Kotera, K. Murase and A. V. Olinto, arXiv:1311.2044 [astro-ph.HE].
[34] J. Feintzeig, Ph. D. thesis, University of Wisconsin, Madison (2014)
[35] J. G. Learned and S. Pakvasa, Astropart. Phys. **3**, 267 (1995)

Simulation of the hybrid Tunka Advanced International Gamma-ray and Cosmic ray Astrophysics (TAIGA)

M Kunnas[5], I Astapov[8], N Barbashina[8], S Beregnev[11], A Bogdanov[4], D Bogorodskii[1], V Boreyko[9], M Brückner[2], N Budnev[1], A Chiavassa[6], O Chvalaev[1], A Dyachok[1], S Epimakhov[5], T Eremin[11], A Gafarov[1], N Gorbunov[9], V Grebenyuk[9], O Gress[1], T Gress[1], A Grinyuk[9], O Grishin[1], D Horns[5], A Ivanova[1], N Karpov[11], N Kalmykov[11], Y Kazarina[1], V Kindin[4], N Kirichkov[1], S Kiryuhin[1], R Kokoulin[4], K Kompaniets[4], E Konstantinov[1], A Korobchenko[1], E Korosteleva[11], V Kozhin[11], L Kuzmichev[11], V Lenok[1], B Lubsandorzhiev[10], N Lubsandorzhiev[11], R Mirgazov[1], R Mirzoyan[7,1], R Monkhoev[1], R Nachtigall[5], A Pakhorukov[1], M Panasyuk[11], L Pankov[1], A Perevalov[1], A Petrukhin[4], V Platonov[1], V Poleschuk[1], M Popescu[12], E Popova[11], A Porelli[3], S Porokhovoy[9], V Prosin[11], V Ptuskin[8], V Romanov[9], G I Rubtsov[10], Müger[23], E Rybov[1], V Samoliga[1], P Satunin[10], A Saunkin[1], V Savinov[1], Yu Semeney[1], B Shaibonov(junior)[10], A Silaev[11], A Silaev (junior)[11], A Skurikhin[11], M Slunecka[9], C Spiering[3], L Sveshnikova[11], V Tabolenko[1], A Tkachenko[9], L Tkachev[9], M Tluczykont[5], A Veslopopov[1], E Veslopopova[1], D Voronov[1], R Wischnewski[3], I Yashin[4], K Yurin[4], A Zagorodnikov[1], V Zirakashvili[8], V Zurbanov[1],

[1]Institute of Applied Physics, Irkutsk State University, Irkutsk, Russia
[2]Institute for Computer Science, Humboldt-University Berlin, Rudower Chaussee 25, 12489 Berlin, Germany
[3]DESY, Platanenallee 6, 15738 Zeuthen, Germany
[4]MEPhI (Moscow Engineering Physics Insitute), National Research Nuclear University, Moscow, Russia
[5]Institut für Experimentalphysik, Luruper Chaussee 149, 22761 Hamburg, Germany
[6]Dipartimento di Fisica Generale Universiteta di Torino and INFN, Torino, Italy
[7]Werner Heisenberg Institut, Föhringer Ring 6, 80805, München, Germany
[8]IZMIRAN, Troitsk, Moscow Region, Russia
[9]Joint Institute for Nuclear Research, Joliot-Curie 6, 141980 Dubna, Moscow region, Russia
[10]Institute for Nuclear Research of the Russian Academy of Sciences 60th October Anniversary st., 7a, 117312, Moscow, Russia
[11]Skobeltsyn institute for Nuclear Physics, Lomonosov Moscow State University, 1 Leninskie gory, 119991 Moscow, Russia
[12]Institute of Space Science, Bucharest, Romania

E-mail: `maike.kunnas@physik.uni-hamburg.de`

Abstract.
Up to several 10s of TeV, Imaging Air Cherenkov Telescopes (IACTs) have proven to be the instruments of choice for GeV/TeV gamma-ray astronomy due to their good reconstrucion quality and gamma-hadron separation power. However, sensitive observations at and above 100 TeV require very large effective areas (10 km^2 and more), which is difficult and expensive to achieve.

The alternative to IACTs are shower front sampling arrays (non-imaging technique or timing-arrays) with a large area and a wide field of view. Such experiments provide good core position, energy and angular resolution, but only poor gamma-hadron separation. Combining both experimental approaches, using the strengths of both techniques, could optimize the sensitivity to the highest energies.

The TAIGA project plans to combine the non-imaging HiSCORE [8] array with small ($\sim$10m^2) imaging telescopes. This paper covers simulation results of this hybrid approach.

1. A hybrid system

A pure IACT system uses two or more telescopes with (most often tesselated) mirrors and multi-channel cameras to take images of extensive air showers (EAS). These images give a good estimation of the nature of the primary particle, but incident angle and core position can only be precisely reconstructed if more than one image of a shower is taken. This sets an upper limit to the spacing of the telescopes of about 300 m since the telescopes need to be inside the Cherenkov light cone of the shower. But for high energy measurements, one requires huge effective areas due to the low flux, which means a huge area needs to be covered with detectors, and that is too expensive to do with telescopes.

On the other hand, shower front sampling array stations are (comparatively) cheap and give a good reconstruction of core position and incident angle, but the gamma hadron separation power is only weak at the threshold energies ($\sim$10 TeV)[2].

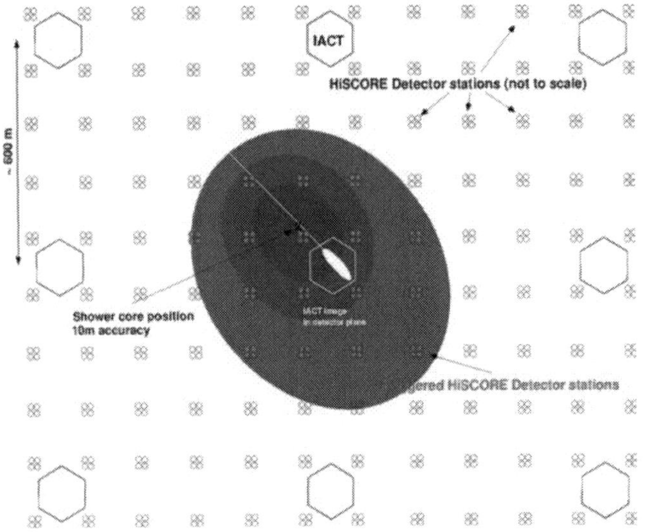

Figure 1. Principle for the combination of IACT and shower front sampling array observation.

To cover a large area with detectors, we want to increase the spacing of our telescopes to about 600 m. This way, the showers are seen by only one telescope at a time, but the reconstruction of these monoscopic images can benefit from the HISCORE timing array, combining both techniques' strengths and canceling out their weaknesses: Taking the core position and incident angle from the shower front sampling array removes the IACT's need for stereoscopy, thus enables an increase of telescope spacing without loss of gamma hadron separation power and therefore makes it possible to cover large areas with fewer telescopes (also see Tluczykont et al, these proceedings).

TAIGA itself will be a combination of small IACT telescopes (in development), the HiSCORE array (Hundred * i Square-km Cosmic Origin Explorer, shower front sampling array, currently in deployment)[10] and the Tunka-Grande scintillation array (photon-electron and muon detection, planned) at Tunka valley, Russia (51° 48' 35" N, 103° 04' 02" E, 675 m a.s.l.). The muon detectors are envisaged for gamma-hadron separation at energies beyond 100 TeV. In the future, a radio extension is planned succeeding Tunka-Rex [12]. This paper will focus on the combination of the shower front sampling array and the IACTs.

The intended telescope parameters are:

- 4.3 m tesselated mirror dish, Davies-Cotton design
- Mirror segment diameter 60 cm, 34 segments
- 4.75 m focal length
- 8° telescope Field of View (FoV)
- 397 camera pixels with about 0.38° FoV each
- 600 m spacing

The HiSCORE sampling array consists of photomultiplier tubes (PMTs) looking directly into the sky through Winston Cones with 4 PMTs per station and about 120 m spacing betweeen stations. Since 10/2013 a 9 station engineering array is running and a 28 station array is installed since 2014, allowing to start observation of the most bright gamma-sources. In the future, a large array of the order of 10 km^2 is envisaged.

2. Simulation

To evaluate the telescope design and gamma hadron separation power of the system, a Monte-Carlo (MC) simulation of the joint operation of the Tunka-HiSCORE wide-angle array and a Cherenkov telescope with image analysis is carried out. The efficiency of hadron background suppression (Q-factor) under gamma-quanta events separation is evaluated. This simulation is done in two steps:

1. Raw MC air shower data are generated with the CORSIKA code [3] including the package for EAS Cherenkov light data (IACT option, hadronic interaction models QGSJET [4] and Gheisha [5]). Here the nature of the primary and its properties (energy range, zenith angle, number of showers) are chosen by the user as well as the detector layout, i.e. the telescope/station positions.

2a. From the CORSIKA output the IACT response simulation with sim_telarray [6] is run. sim_telarray executes a raytracing routine with an automatic reconstruction of the MC images (see fig. 2, though manual reconstruction is possible) and includes all the important details of the telescope design, for example:

- Mirror dish configuration
- Mirror details
- Camera configuration
- Camera details (PMT response, funnels etc.)
- Electronic response
- Mast and camera shadowing

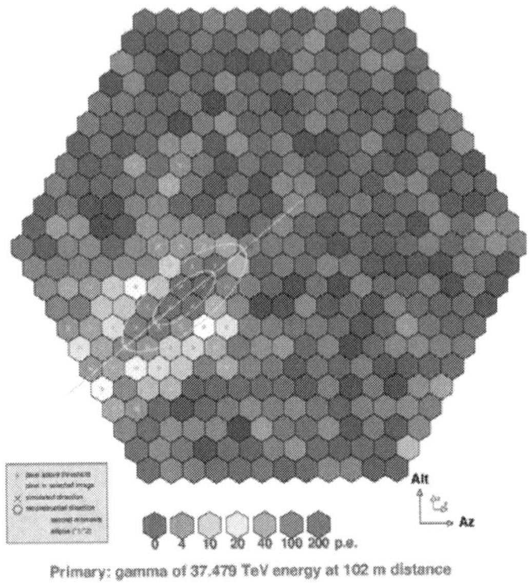

Figure 2. Example for a sim_telarray-generated EAS image of a 37.5 TeV gamma primary at 107 m core distance.

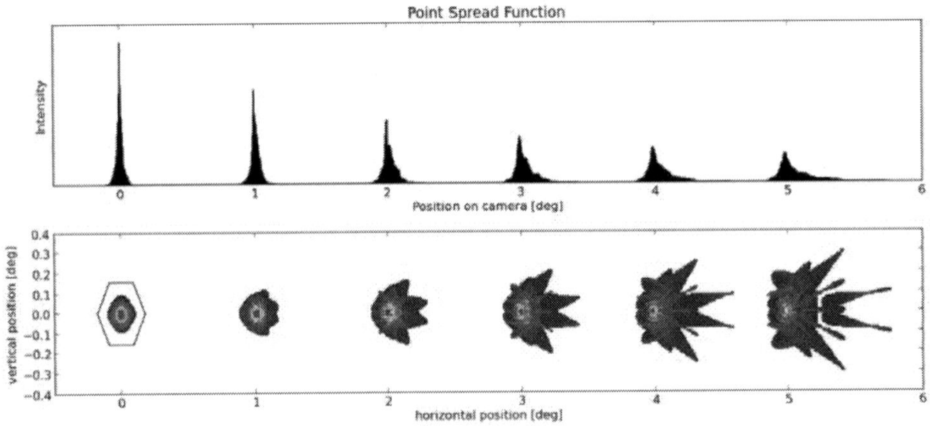

Figure 3. IACT Point Spread Function (PSF) for angles of 0-5°from the optical axis. Minimal PSF without mirror misalignment options. The structures are an effect of the outermost mirror segments.

Right now, we have over 80 million events simulated, statistics still rising.

2b. Also, from the same CORSIKA output, the HiSCORE detector response is simulated with sim_score [7], a custom simulation software based on the IACT package, to determine the core position/resolution and shower direction/angular resolution of the pure array and to evaluate different array layouts and electronic systems.

3. Simulation results

3.1. Point spread function (PSF)

One MC simulation study is the Point Spread Function (PSF) of our telescope. A good PSF is essential for good image quality and thus for good reconstruction. The mirror dish of our

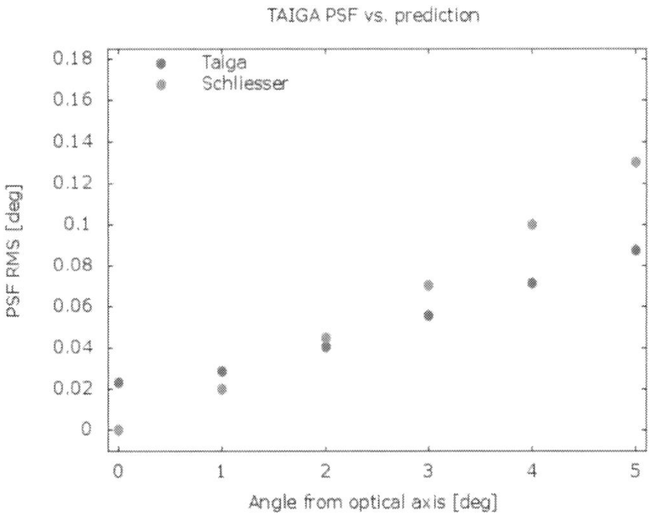

Figure 4. Comparison with expected values for the PSF for our F/D = 1.1

telescope is composed of spherical mirror tiles smaller than the overall dish size, as usual for a Davies-Cotton design. This introduces another telescope parameter: The tesselation ratio α. The tesselation ratio is the relation between individual mirror segment diameter d_{tile} and the main dish diameter D_{dish}:

$$\alpha = \frac{d_{tile}}{D_{dish}} \tag{1}$$

Therefore, the bigger the tesselation ratio, the bigger the influence of the individual mirror tiles. With the high field of view (FoV) of about 8°and a tesselation ratio of 0.13, the aberration effects of the mirror structure come into play. Therefore their impact needs to be evaluated.

Compared to the size of a single pixel, the psf shows that even for an angle of 5°to the optical axis, the PSF is still significantly smaller than a pixel (compare Fig. 3). Our PSF is not as good as the predictions made bei Schliesser and Mirzoyan [11] imply (see Fig. 4), but this is expected since our tesselation ratio is larger, increasing the effect of aberrations of the spherical mirror tiles.

3.2. Gamma-hadron separation

In a shower sampling array, the steepness of the air shower's light density function is used for primary particle identification [9]. In the low energy range (<10 TeV) however this method works only poorly, especially for gammas and protons [7].

In this range, IACT image analysis is the far more powerful tool to identify the primary's nature.

The parameter that is used for separating the different kinds of primaries with an IACT is the width of the camera image [13], derived from the second moment of the intensity distribution after image cleaning. This width depends mainly on primary energy (E), impact parameter (d) (e.g. distance between telescope and shower core) and the type of the primary. With the first two parameters measurable by the shower front sampling array, the parameter left is the type, which determines the distribution's shape. Proton showers have a broader width distribution due to their hadronic shower components. To seperate gammas from hadrons, a width cut can be applied. The parameter Q shows the quality of this cut:

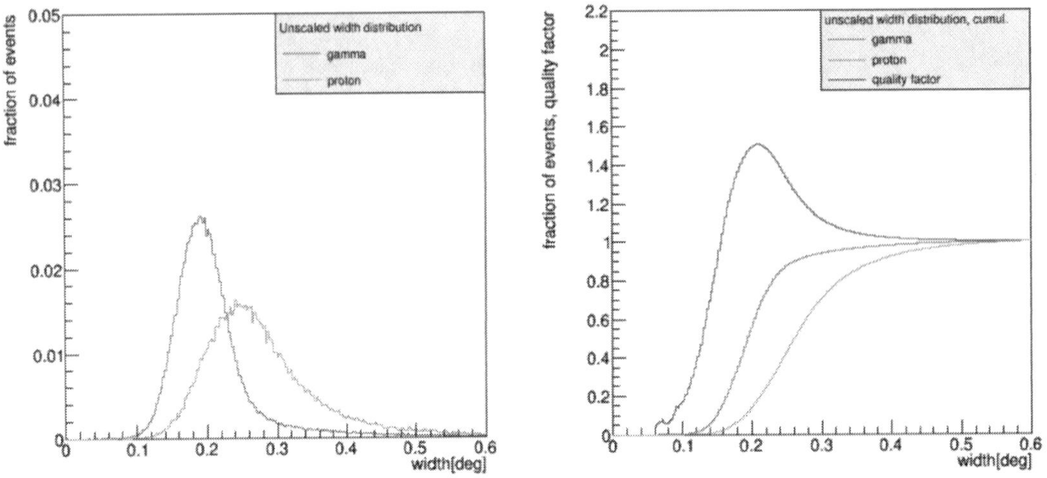

Figure 5. Raw width distribution for showers from the zenith in the energy range of 0.5-50 TeV. Red: Gammas, Green: Protons. Left: Differential distribution. It can be seen that the proton distribution is broader than the gamma distribution and a bit shifted towards bigger widths, but the two peaks still overlap much. Right: Cumulative distribution and quality factor Q for the width cut at. If a cut is made on the raw data, the maximum quality factor is less than 1.5

$$Q = \frac{\epsilon_\gamma}{\sqrt{\epsilon_p}} \tag{2}$$

with

$$\epsilon_i = \frac{n_i(< w)}{n_i} \tag{3}$$

However, the distributions for gammas and protons overlap considerably due to the varying core distances, so that a cut on this parameter is not satisfying (see fig. 5). To improve it, the measured widths are rescaled to the expected value for a gamma-ray induced shower. For this, a three-dimensional lookup-table is generated from MC data, with the parameters core distance, image size and expected width. Then, each event's width is divided by the respective lookup-table entry (speak: scaled), which makes the gamma and the proton distribution peaks move apart (see fig. 6). This is the point where the combination takes place: To find the right lookup-table entry, core distance and zenith angle are reconstructed from the sampling array, and we get the Hybrid Scaled Width (HSCW):

$$HSCW = \frac{w_{tel}}{w^{\gamma}_{MC}(d_{array}, size_{tel})} \tag{4}$$

Scaling the width increases the quality factor from about 1.4 to almost 2.2, which is significantly better than the quality for pure HiSCORE reconstruction, which achieves only a Q-factor of 1 at the threshold and only approaches 2 at several 100s of TeV. [7]

The results shown in fig. 5 and 6 are the width values. For now, we took the core position for finding the correct lookup-table entry for scaling directly from the MC data. For a first estimate on TAIGA's combined reconstruction quality, we'll randomize the MC core position by

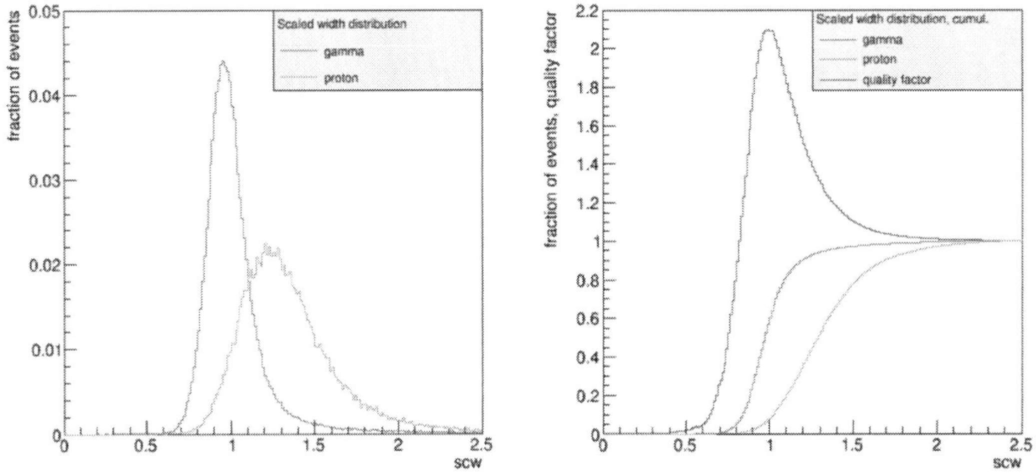

Figure 6. Hybrid scaled width (HSCW) distribution for showers from the zenith in the energy range of 0.5-50 TeV. Red: Gammas, Green: Protons. Left: Differential distribution. Compared to figure 5, the gamma and the proton distribution have moved apart significantly. Right: Cumulative distribution and quality factor Q for the width cut. If a cut is made on the HSCW, the maximum quality factor is almost 2.2

the HiSCORE core resolution and use this randomized core position in the width scaling (toy core simulation). We expect the quality to decrease only little, down to about 2.0.

As a next step, a full hybrid simulation is planned, including a combination of reconstruction algorithms. The monoscopic direction information from the IACT image will be taken into account in the reconstruction of core position from the array, lowering the uncertainties on this parameter, and the gamma-hadron separation by shower front profiling of the array will be combined with the image reconstruction. We expect to improve the Q-factor at low energies to 3.

Typical stereoscopic systems of Cherenkov telescopes achieve a gamma-hadron separation Q-factor using the mean scaled width cut of about 6. The often stated factor of 10^4 results from combining the width and directional cut (separating the isotropic hadrons by a tight angular cut around the point source of interest). This cut can be done with the hybrid array on a similar quality level, since the angular resolution of the timing array alone is comparable to stereoscopic IACT systems. Here however, we only state the width cut quality.

4. Conclusion and outlook

The TAIGA hybrid system is the first to combine the strengths of Cherenkov timing arrays and IACT telescopes. Our telescope design concept gives a good imaging quality as shown by the PSF. The combination of IACTs with the HiSCORE timing array will result in an improved gamma-hadron separation and a maximisation of the effective area.

The next step is a toy core analysis, then a full hybrid simulation both for array and IACT.

Acknowledgements

We acknowledge the support of the Russian Federation Ministry of Education and Science (agreements N 14.B25.31.0010,N2014/51, project 1366, zadanie N 3.889.2014/K), the Russian

Foundation for Basic research (grants 13-02-00214, 15-02-10005, 13-02-12095), the Helmholtz association (grant HRJRG-303), and the Deutsche Forschungsgemeinschaft (grant TL 51-3)

References

[1] Tluczykont M *et al* 2014 The HiSCORE concept for gamma-ray and cosmic-ray astrophysics beyond 10 TeV *Astropart. Phys.* 42-53
[2] Hinton J A and Hofmann W 2009 Teraelectronvolt Astronomy *Ann.Rev.Astron.Astrophys.***47** 523-65
[3] Heck D *et al* 1998 Report *FZKA 6019*, available from http://www-ik.fzk.de/corsika/physics_description/corsika_phys.html
[4] Kalmykov N N, Ostapchenko S S, and Pavlov A I 1997 *Nucl. Phys. B (Proc. Suppl.)* **52B** 17
[5] Fesefeldt H 1985 Report **PITHA-85/02**
[6] Bernlöhr K 2008 Simulation of imaging atmospheric Cherenkov telescopes with CORSIKA and sim_telarray *Astropart. Phys.* 149-58
[7] Hampf D, Tluczykont M and Horns D 2012/13 Event reconstruction techniques for the wide-angle air Cherenkov detector HiSCORE *Nucl. Inst. Meth. in Phys. Res. A* 137-46
[8] Tluczykont M *et al* 2011 *Adv. Sp. Res.* **48** 1935-41S.
[9] Budnev N *et al* 2015 The Tunka experiment: from cosmic ray to gamma-ray astronomy *these proceedings*
[10] Tluczykont M *et al* 2015 Towards gamma-ray astronomy with timing arrays *these proceedings*
[11] Schliesser A and Mirzoyan R 2005 Wide-field prime-focus imaging atmospheric Cherenkov telescopes: A systematic study *Astrop. Phys.***24** 382-90
[12] Schröder F G *et al* 2012 *AIP Conf. Proc* **1535** 111
[13] Hillas A M 1985 Cerenkov light images of EAS produced by primary gamma *Int. Cosm. R. Conf.* **3** 4458.

Timing calibration and directional reconstruction for Tunka-HiSCORE

A.Porelli[2], D.Bogorodskii[5], M.Brückner[6], N.Budnev[5], O.Chvalaev[5], A.Dyachok[5], S.Epimakhov[1], T.Eremin[3], O.Gress[5], T.Gress[5], D.Horns[1], A.Ivanova[5], S.Kiruhin[5], E.Konstantinov[5], E.Korosteleva[3], M.Kunnas[1], L.Kuzmichev[3,5], B.Lubsandorzhiev[4], N.Lubsandorzhiev[3], R.Mirgazov[5], R.Mirzoyan[5,7], R.Monkhoev[5], R.Nachtigall[1], A.Pakhorukov[5], V.Platonov[5], V.Poleschuk[5], V.Prosin[3], G.Rubtsov[4], M.Rüger[2,6], V.Samoliga[5], A.Saunkin[5], P.Satunin[4], D.Spitschan[1], Yu.Semeney[5], L.Sveshnikova[3], V.Tabolenko[5], M.Tluczykont[1], D.Voronin[5], R.Wischnewski[2], A.Zagorodnikov[5], V.Zurbanov[5]

1 : Institute for Experimental physics, University of Hamburg, Luruper Chaussee 149, 22761 Hamburg, Germany
2 : DESY, Platanenallee 6, 15738 Zeuthen, Germany
3 : Skobeltsyn institute for Nuclear Physics, Lomonosov Moscow State University, 1 Leninskie gory, 119991 Moscow, Russia
4 : Institute for Nuclear Research of the Russian Academy of Sciences 60th October Anniversary st., 7a, 117312, Moscow, Russia
5 : Institute of Applied Physics ISU, Irkutsk, Russia
6 : Institute for Computer Science, Humboldt-University Berlin, Rudower Chaussee 25, 12489 Berlin, Germany
7 : Max Planck Institut fur Physik, Fohringer Ring, Munchen, Germany

E-mail: `andrea.porelli@desy.de`

Abstract. The Tunka-HiSCORE detector follows the concept of a non-imaging wide-angle EAS Cherenkov array, designed to search for γ-ray sources above 10 TeV and to investigate the spectrum and composition of cosmic-rays above 100 TeV. A prototype array with 9 stations has been deployed in October 2013 at the site of the Tunka experiment in Russia. We describe design and performance of the array data acquisition system DAQ-2, focusing on its timing system based on the White Rabbit technology for sub-nsec time-synchronization over ethernet. First results of EAS arrival direction reconstruction, compared with MC simulations, and tests with artifical light sources verify an excellent performance of the system.

1. Introduction

The Tunka-HiSCORE detector is a ground-based air shower Cherenkov array that samples the air shower light front using the non-imaging technique for γ-ray astronomy and cosmic-rays studies [1, 2]. The detector is located in the Tunka Valley, Siberia, at the Tunka-133 array site [3]. Currently, a total area of $0.25\,\text{km}^2$ is instrumented with 28 Tunka-HiSCORE stations. A

hybrid detector system, TAIGA, is under construction, combining as main components a km^2-scale HiSCORE array with a net of imaging air Cherenkov telescopes [1, 3].
For γ-ray astronomy, the HiSCORE array must achieve a good reconstruction of the air shower arrival direction. For a pointing resolution of $\sim 0.1°$, all array stations need to be synchronized with sub-ns time precision [4, 5]. This work presents results from a timing system based on the White Rabbit technology, that was installed in 2012, and operates since 2013/14 with the first Tunka-HiSCORE prototype array. In this paper we use data from dedicated time-calibration runs as well as from normal EAS showers to verify this new technology, which has been used for the first time in a field installation for an astroparticle physics experiment.

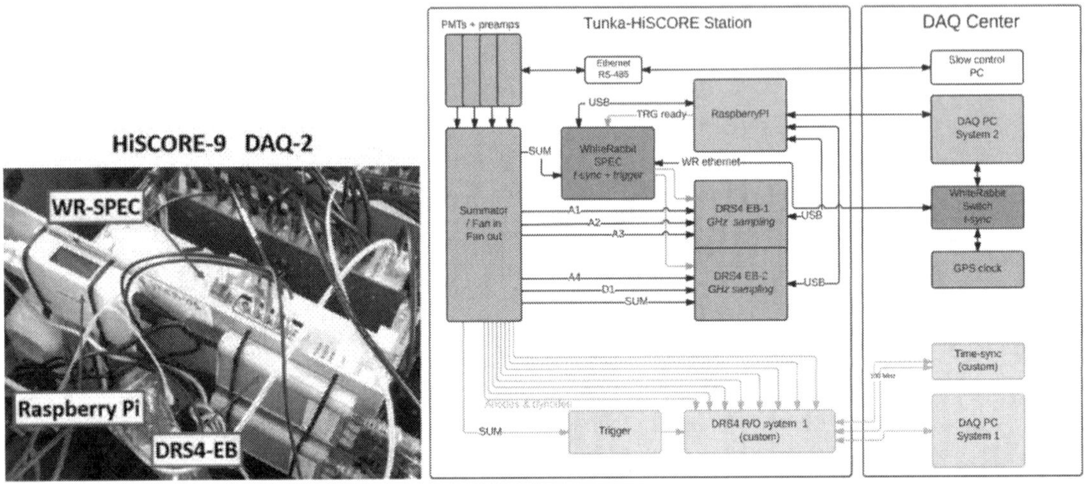

(a) (b)

Figure 1: The HiSCORE-9 DAQ-2 system, with DRS4-Evaluation Board and White-Rabbit timing system. (a) Picture of the station setup (with DRS4-EB, WR-SPEC and MiniPC); (b) Schematics of the setup in Station and Center for DAQ-2 (DAQ-1 in grey).

2. The HiSCORE-9 array

A first HiSCORE array with 9 stations has been deployed in October 2013, arranged on a regular grid of 3 x 3 stations with 150 m distance and an instrumented area of 0.09 km^2 (HiS-9 array, see section 3 and [1, 3, 6]). Each HiSCORE station is equipped with 4 Photomultiplier Tubes (PMTs) of 8 inch diameter, and with 4 light collectors (Winstone cones), yielding a collection area of 0.50 m^2 and a field of view of 0.6 sr. This first Tunka-HiSCORE array is operated by two independent data acquisition systems (DAQ-1 and DAQ-2), both based on GHz waveform-sampling by the DRS4-chip. Array time-synchronization is done by a custom system in DAQ-1 (with 100MHz clocks sent over separate fibers) and by the ethernet-based White-Rabbit system (WR) in DAQ-2. WR [7] is a new standard for time- and clock-transfer using the extended PTP standard (IEEE 1588) and commercially available basic components; it allows for scalable, flexible and cost-effective sub-nsec synchronization built from large-scale next generation experiments, like CTA, KM3NeT, IceCube-Upgrade [8].
The main purpose of the DAQ-2 system, which is discussed in this paper, is to evaluate this new timing system in a robust DAQ, built from off-the-shelf components. A direct timing performance comparison with DAQ-1 is possible for the 19 new HiSCORE stations installed in October 2014, which combine both systems on a hybrid DAQ-board. Systematic tests of

HiSCORE WR-boards operating since 2012 are given in [9]. Results from the DAQ-1 running since fall 2013 are mentioned in [3].

2.1. The HiSCORE-9 DAQ-2 system

DAQ-2 is an autonomous, distributed data taking system, with a central control server that concurrently runs a predefined and exchangable set of tasks per station. It supports remote operation over WAN from multiple locations and incorporates monitoring and self-recovery on hardware error detection. Its main hardware components per station are shown in Fig.1:

(i) **DRS4 evaluation boards**: to sample PMT pulses at up to 5 GS/s (two boards with 2x4 channels) [10]; for the chosen trace length of 1 μs sampling rate is 1 GS/s,

(ii) **White Rabbit (WR) timing system**: one WR-Node - a SPEC (*Simple PCIe Card*) - per station executes, with custom firmware, the nsec-time stamping and analog pulse triggering [8]. Each SPEC is connected by optical fiber (for synchronization and data transfer) to the main WR-switch in the DAQ center, that provides time synchronization of all the stations and interfaces to a GPS clock. The WR system provides local clocks with a relative phase stability of ~200 ps and an absolute clock precision of ~1 ns [9],

(iii) **Raspberry PI**: a mini computer for readout of DRS4 boards, data transfer to the DAQ center and to generate a "readout ready" signal (GPIB) to enable the next SPEC trigger; optionally serves for SPEC monitoring and firmware upgrades.

WR-triggering and time stamping is as follows: The sum of the four PMT anode signals, built by an analog summator board, is sent to a comparator on the WR-SPEC card, which generates a trigger if the signal stays above adjustable threshold for $t > 9$ ns [9]. Upon a trigger, the WR-clock value is latched with nsec resolution, and sent as trigger time message over the WR-fiber. Synchroneous SPEC trigger strobes are sent to both DRS4-EBs, initiating their readout. The intrinsic 1-nsec granularity of the WR-time stamps can be improved, using sub-nsec resolution of anode signals recorded by the DRS4 relative to the WR-strobes. The analysis presented here is based on the WR-time stamps only; with pulse shape analysis an additional improvement is expected.

The eight DRS4 board channels are used to sample four anode signals, their sum, and one dynode as well as the WR strobes (one per DRS4-EB). The trigger thresholds are set to ~ 10 Hz station trigger rates.

2.2. Array event building

All HiS-9 stations trigger independently, thus array events are offline selected by requesting coincidence of $\geq$ N stations within 2 μs. Since all WR time-stamps are available in real-time, an array trigger can be easily formed online. This will allow serious data reduction by e.g. station-multiplicity or topology based online array triggers.

3. Time calibration

The calibration of the relative arrival times between the stations in a non-imaging Cherenkov detector is critical to achieve precise shower pointing. To investigate the time resolution of the HiS-9 array, a series of calibration runs has been performed - to check single station and full array performance as well as long-term stability.

A bright wide-angle LED light source was positioned outside the array perimeter (~ 200 m), emitting light pulses with ~6 Hz towards all the nine stations. The LED calibration setup is depicted in Fig.2. On top of each station a 45° inclined reflecting screen was added, to redirect

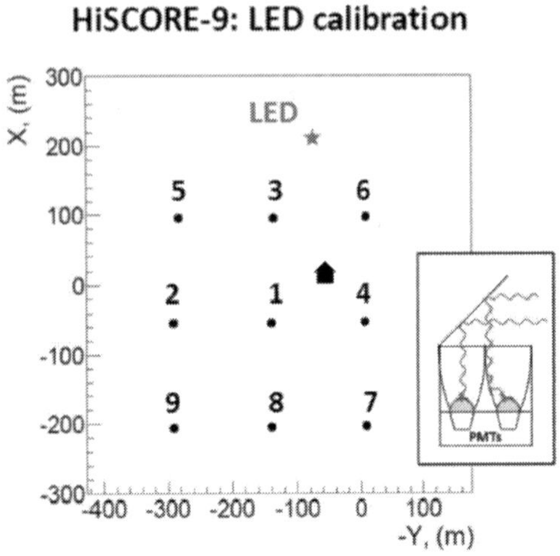

Figure 2: HiS-9 array layout: black house: DAQ center; red star: LED source position during calibration runs; in the small box a scheme of the station setup for the calibration runs.

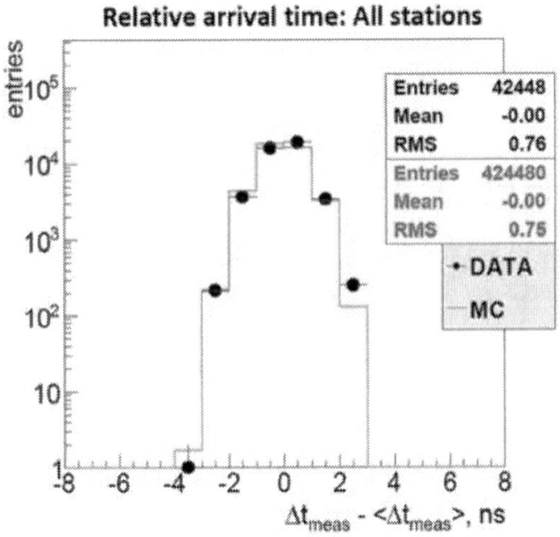

Figure 3: Measured time delay distribution, Δt_{i3} (shifted to zero, $\Delta t_{i3} - < \Delta t_{i3} >$, to plot all the 9 stations together): dots: Data; red line: simulated events.

the LED light into the stations.
The data sample for this analysis consists of two calibration runs of 15 minutes each, with the second one taken 10 days after the first, from the same position. Only events with all the nine stations triggered are used for the analysis.

3.1. Toy Monte-Carlo
We use a toy Monte-Carlo to simulate the response of the array and the measurement system, and to compare to the calibration data. A spherical model is applied for light propagation from the source to the stations; thus the arrival time is given by $t_i = R_i/c$, where t_i is the arrival time of the light to the station i, R_i is the distance between the station and the light source, and c is the speed of light. To simulate the fluctuation of the timing system (WR), a gaussian time jitter with σ_{WR} is added to the arrival time. The WR time stamps are obtained by rounding the times to 1 ns precision (ns-granularity). We chose $\sigma_{WR} = 0.45$ ns, as obtained from the raw data analysis, see section 3.2.

3.2. Data analysis
The data analysis for the time calibration consists of three steps:

1. Stations time synchronization stability check
2. Determination of the calibration constants Δt_i^{cal}
3. Reconstruction of source position coordinates

To check the stability of the time synchronization between the stations, we analyse the raw time differences distributions Δt_{ij} between station i and a reference station j (in this case $j = 3$, the closest to the LED). The standard deviation of Δt_{ij} is given by (assuming identical stations):

$$\sigma(\Delta t_{ij}) \sim \sqrt{\sigma(t_i)^2 + \sigma(t_j)^2} = \sqrt{2} \cdot \sigma(t_i) \quad (1)$$

$$\sigma(t_i) = \sqrt{(1/\sqrt{12})^2 + \sigma_{WR}^2 + \sigma_X^2} \quad (2)$$

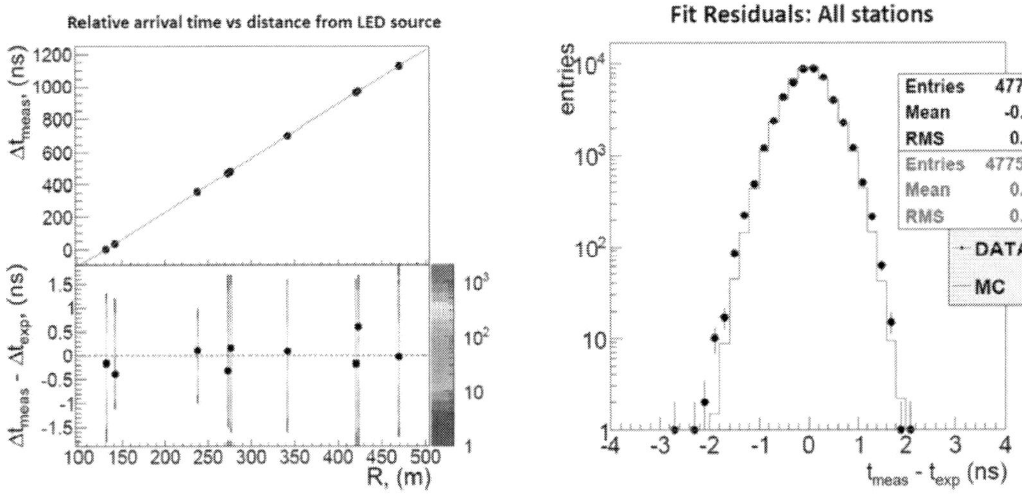

(a) (b)

Figure 4: (a) Upper panel: station relative arrival times Δt_{meas} as function of distance R from the LED light source. The red line gives $t_i = R_i/c$. Lower panel: fit residual distributions per station. Black dots show an example of a single triggered event. (b) Distribution of fit residuals (all nine stations superimposed). Black dots: data; Red line: simulated events.

where $\sigma(t_i)$ is the standard deviation of the measured time at station i, $1/\sqrt{12} = 0.29$ ns is due to the 1 ns time granularity, σ_{WR} is the time jitter of the timing system and σ_X is the time jitter due to apparatus effects (i.e. PMT, electronics, etc.).
From the Δt_{ij}-distribution in Fig.3, we find $\sigma(t_{ij}) = \sigma(\Delta t_{ij})/\sqrt{2} = 0.54$ ns, which according to eq.2 gives $\sqrt{\sigma_{WR}^2 + \sigma_X^2} = 0.45$ ns. Thus, we find as an upper limit of the WR-induced time jitter $\sigma_{WR} \sim 0.45$ ns.
In the second step we compare the distributions of the measured time differences, $\Delta t_{ij,data}$ with simulation. For each station i, we calculate the differences between the average values of the two distributions, $\Delta t_i^{cal} = \overline{\Delta t}_{ij,data} - \overline{\Delta t}_{ij,MC}$. We consider this Δt_i^{cal} an additional, station-specific time delay w.r.t. the photon arrival time, introduced by station components (analog DAQ components, PMT transient times, cables, WCones, etc.). These values are used as offset calibration constants to correct the arrival time of each station.
The third step of the analysis is the reconstruction of the LED position on an event by event basis, using all nine stations simultaneously. After correcting the arrival times by their calibration constants Δt_i^{cal}, we fit a spherical propagation model for the time t_i it takes to reach the i-th station from the source position. An example of a fitted event and the station-wise residual plots is given in Fig.4(a). The standard deviation of the fit residuals, see Fig.4(b), is consistent with the value obtained from the raw data analysis (step 1); and the reconstructed LED coordinates agree with the real ones. This confirms the consistency of the procedure, up to the array level.
To summarize, from the LED calibration runs we obtained the station calibration constants Δt_i^{cal}, and derived a conservative upper limit for the HiS-9 WhiteRabbit time jitter of $\sigma_{WR} \leq 0.45$ ns. An improved estimation of σ_{WR} will be obtained from the DRS4 recorded pulses (for sampling at ≥ 2 GS/s), and is expected to reach the $\sim$0.2 ns range, as estimated for DAQ-1 [3] and measured in previous WR-field setups [9].

4. EAS shower reconstruction

We reconstruct the first HiS-9 DAQ-2 data, and compare with Monte-Carlo simulations. The selected sample consists of 12 runs for a total of $\sim 40\,\text{h}$ data taken during clear moonless nights in February/March 2014. Array events are built according to sect.2.2; for this analysis only events with 9 triggered stations are taken into account. A preliminary amplitude calibration is used to convert PMT signals to photo-electrons.

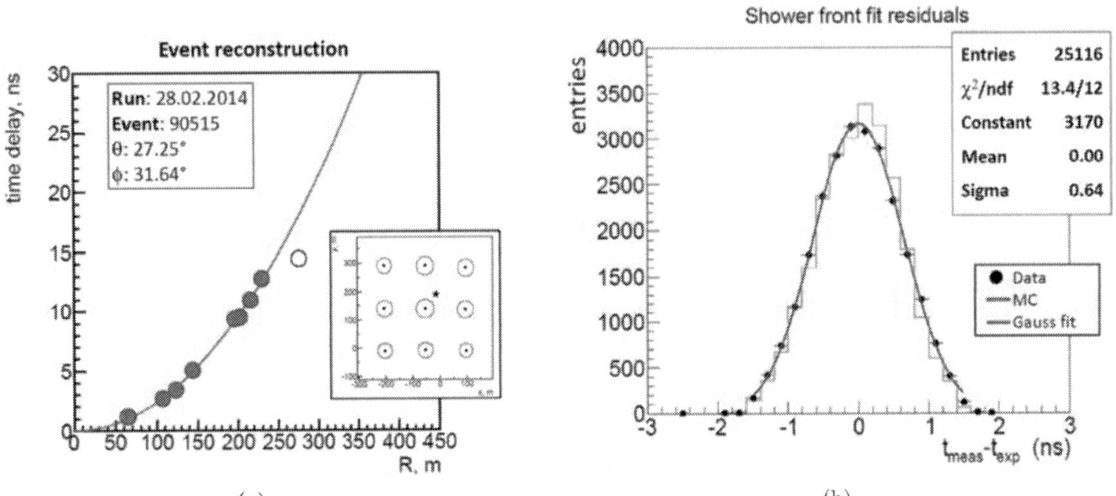

Figure 5: **EAS shower reconstruction**. (a) Arrival time delay vs distance R from the shower axis. Red/white dots: stations retained / excluded in the final fit; red line: reconstructed shower profile. Small panel: Reconstructed core position (black star), the area of the circles is proportional to $logA$, with A the station signal amplitude. (b) Distribution of fit residuals after shower reconstruction. Black dots: data; Red line: simulated events; Blue line: gaussian data fit.

4.1. Monte-Carlo simulation

To study the detector response and reconstruction performance, a dedicated Monte-Carlo simulation was done for the HiS-9 array. Air showers have been generated according to the primary proton spectrum ($dN/dE \propto E^{-2.7}$ and $10 \leq E \leq 10^3\,\text{TeV}$) using Corsika v699 [11]; the detector response is simulated with the standard HiSCORE simulation software *sim_score* [5, 2]. The shower reconstruction is done with *reco_score*, the HiSCORE reconstruction package [4]. A time jitter with a gaussian distribution of $\sigma = 0.5\,\text{ns}$ is added to the pulse trigger time, in agreement with the time-resolution results discussed above. All stations are assumed to be identical, which introduces differences between data and Monte-Carlo, in particular for shower core reconstruction.

4.2. Shower reconstruction

The shower reconstruction determines the shower core location (x_C, y_C) and the arrival direction (zenith θ, azimuth ϕ). In a first step, an approximate shower direction is obtained by fitting the shower front arrival times with a plane wave. This direction in turn is used for a 2-steps shower-core determination: starting from a first guess for x_C and y_C by a center of gravity

method (weighted average of the signal amplitudes recorded in each triggered station), the core position is then obtained by fitting the amplitudes with the amplitude distance function (ADF) using the shower direction, as discussed in [4].
A precision shower direction reconstruction is then obtained by fitting the shower front with an analytical model for the arrival times at each station [12]. The fit is performed recursively, excluding stations with time residuals greater than 3σ ($\sigma = 0.5$ ns, see sect.3); the number of excluded hits is ≤ 2 for the $\sim 92\%$ of the events (≤ 1 for $\sim 60\%$). Figure 5(a) shows an example of a fitted event. The distribution of the fit residuals (Fig.5(b)) follows a gaussian with $\sigma_{data} = 0.64$ ns, which gives $\sigma_{WR} \sim 0.57$ ns according to eq.2. We note, that this is close to $\sigma \sim 0.5$ ns obtained for shower reconstruction in DAQ-1 (which includes a pulse shape analysis). We find stability of the rms from individual stations, over the 10-day run period considered here - showing no time dependence of the WR-synchronization. Figure 6 gives the reconstructed zenith and azimuth distributions, for data and simulated events. We find good agreement between data and MC, given the above mentioned approximations. Small deviations from the MC assumed flat zenith-angular response of the stations sensitivity are currently under investigation.

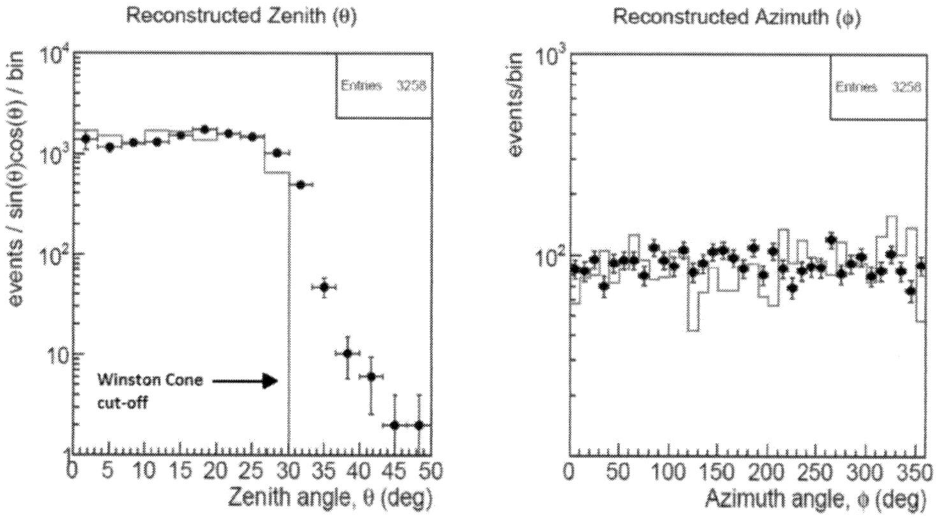

Figure 6: Reconstructed shower direction: Zenith (left), Azimuth (right). Black dots: data events; Red line: simulated events. The cut-off in the reconstructed theta distribution in the Monte-Carlo is due to the used Winston Cone reflection efficiency, $\epsilon_{WC} = 0$ for $\theta \geq 28°$.

5. Conclusion

The HiSCORE detector represents a new concept based on the non-imaging Cherenkov technique for cosmic-ray studies and γ-ray astronomy in the PeV energy regime. The HiSCORE-9 array has been operating successfully since October 2013. A dedicated DAQ system based on the WhiteRabbit technology for sub-nsec timing has been taking data. The analysis of LED calibration runs gives an upper limit on the time resolution of the timing system of $\sigma_t \leq 0.45$ ns, including all experimental effects. This is compatible with standalone laboratory and field results of a 0.2 ns WR-clock jitter.
A first reconstruction of the air showers detected with the HiS-9 array DAQ-2 has been performed. From the shower fit residual distribution we calculate an upper limit on the WR-time resolution of $\sigma_{WR} \leq 0.57$ ns, consistent with the LED results. From Monte-Carlo studies [4] we

expect for high energies this timing resolution to result in a pointing error of the order of 0.1°. Both the time calibration analysis and shower reconstruction have been performed using only the time information from the WhiteRabbit system, i.e. with 1 ns granularity. An improvement is expected after introducing the corrections due to correct pulse amplitude calibration [6] and PMT pulse shape analysis.

Acknowledgements
This work was supported by the Russian Federation Ministry of Education and Science (agreements N 14.B25.31.0010, 2014/51, project 1366, zadanie N 3.889.2014/K), the Russian Foundation for Basic research (grants 13-02-00214, 15-02-10005, 13-02-12095), the Helmholtz Association (Grant HRJRG-303) and the Deutsche Forschungsgemeinschaft (Grant TL51-3).

[1] M. Tluczykont et al. (TAIGA Collabroation), Towards Gamma-Ray Astronomy with timing array, ECRS-2014, Kiel (Germany), these proceedings
[2] M. Tluczykont, D. Hampf, D. Horns, D. Spitschan, L. Kuzmichev, V. Prosin, C. Spiering and R. Wischnewski, Astroparticle Physics 56 (2014) 4253
[3] N. Budnev et al. (TAIGA Collaboration), The Tunka experiment: from cosmic ray to gamma astronomy, ECRS-2014, Kiel (Germany), these proceedings
[4] D. Hampf, M. Tluczykont, D. Horns, Nuclear Instruments and Methods in Physics Research A 712 (2013) 137146
[5] M. Tluczykont, et al., Advances in Space Research 48 (2011) 1935.
[6] S. Epimakhov et al. (Tunka-HiSCORE Collaboration), Amplitude Calibration with the HiSCORE-9 array, ECRS-2014, Kiel (Germany), these proceedings
[7] J. Serrano et al., The White Rabbit Project, ICALEPCS 2009, accelconf.web.cern.ch/accelconf/icalepcs2009/papers/tuc004.pdf
[8] M. Brückner and R. Wischnewski, A White Rabbit setup for sub-nsec synchronization, timestamping and time calibration in large scale astroparticle physics experiments, Proceed. of the 33rd ICRC, 2013, Rio de Janeiro, Brazil, ID-1146, and White Rabbit in Cosmic Ray Physics and Gamma Astronomy, 8th White Rabbit Workshop, CERN, October 2014, http://www.ohwr.org/projects/white-rabbit/wiki/Oct2014Meeting
[9] M. Brückner et al (Tunka-HiSCORE Collaboration), Results from the WhiteRabbit sub-nsec time synchronization setup at HiSCORE-Tunka, Proceed. of the 33rd ICRC, 2013, Rio de Janeiro, Brazil, ID-1158
[10] Paul Scherrer Institute, Villigen, Switzerland, DRS4 Evaluation Board, http://www.psi.ch/drs/evaluation-board.
[11] K. Bernlöhr, Simulation of Imaging Atmospheric Cherenkov Telescopes with CORSIKA and sim telarray, Astroparticle Physics 30 (2008) 149158. [arXiv:0808.2253]
[12] V. Stamatescu, et al., Astroparticle Physics 34 (2011) 886896

Towards gamma-ray astronomy with timing arrays

M Tluczykont[5], I Astapov[8], N Barbashina[8], S Beregnev[11], A Bogdanov[4], D Bogorodskii[1], V Boreyko[9], M Brückner[2], N Budnev[1], A Chiavassa[6], O Chvalaev[1], A Dyachok[1], S Epimakhov[5], T Eremin[11], A Gafarov[1], N Gorbunov[9], V Grebenyuk[9], O Gress[1], T Gress[1], A Grinyuk[9], O Grishin[1], D Horns[5], A Ivanova[1], N Karpov[11], N Kalmykov[11], Y Kazarina[1], V Kindin[4], N Kirichkov[1], S Kiryuhin[1], R Kokoulin[4], K Kompaniets[4], E Konstantinov[1], A Korobchenko[1], E Korosteleva[11], V Kozhin[11], M Kunnas[5], L Kuzmichev[11], V Lenok[1], B Lubsandorzhiev[10], N Lubsandorzhiev[11], R Mirgazov[1], R Mirzoyan[7,1], R Monkhoev[1], R Nachtigall[5], A Pakhorukov[1], M Panasyuk[11], L Pankov[1], A Perevalov[1], A Petrukhin[4], V Platonov[1], V Poleschuk[1], M Popescu[12], E Popova[11], A Porelli[3], S Porokhovoy[9], V Prosin[11], V Ptuskin[8], V Romanov[9], G Rubtsov[10], M Rueger[2,3], E Rybov[1], V Samoliga[1], P Satunin[10], A Saunkin[1], V Savinov[1], Yu Semeney[1], B Shaibonov(junior)[10], A Silaev[11], A Silaev (junior)[11], A Skurikhin[11], M Slunecka[9], C Spiering[3], L Sveshnikova[11], V Tabolenko[1], A Tkachenko[9], L Tkachev[9], A Veslopopov[1], E Veslopopova[1], D Voronov[1], R Wischnewski[3], I Yashin[4], K Yurin[4], A Zagorodnikov[1], V Zirakashvili[8] & V Zurbanov[1]

[1]Institute of Applied Physics, Irkutsk State University, Irkutsk, Russia
[2]Institute for Computer Science, Humboldt-University Berlin, Rudower Chaussee 25, 12489 Berlin, Germany
[3]DESY, Platanenallee 6, 15738 Zeuthen, Germany
[4]National Research Nuclear University MEPhI (Moscow Engineering Physics Institute), Moscow, Russia
[5]Institut für Experimentalphysik, Luruper Chaussee 149, 22761 Hamburg, Germany
[6]Dipartimento di Fisica Generale Universiteta di Torino and INFN, Torino, Italy
[7]Werner Heisenberg Institut, Föhringer Ring 6, 80805, München, Germany
[8]IZMIRAN, Troitsk, Moscow Region, Russia
[9]Joint Institute for Nuclear Research, Joliot-Curie 6, 141980 Dubna, Moscow region, Russia
[10]Institute for Nuclear Research of the Russian Academy of Sciences 60th October Anniversary st., 7a, 117312, Moscow, Russia
[11]Skobeltsyn institute for Nuclear Physics, Lomonosov Moscow State University, 1 Leninskie gory, 119991 Moscow, Russia
[12]Institute of Space Science, Bucharest, Romania

E-mail: `martin.tluczykont@physik.uni-hamburg.de`

Abstract. The gamma-ray energy regime beyond 10 TeV is crucial for the search for the most energetic Galactic accelerators. The energy spectra of most known gamma-ray emitters only reach up to few 10s of TeV, with 80 TeV from the Crab Nebula being the highest energy so far observed significantly. Uncovering their spectral shape up to few 100 TeV could answer the question whether some of these objects are cosmic ray Pevatrons, i.e. Galactic PeV accelerators.

Journal of Physics: Conference Series **632** (2015) 012042 doi:10.1088/1742-6596/632/1/012042

Sensitive observations in this energy range and beyond require very large effective detector areas of several 10s to 100 square-km. While imaging air Cherenkov telescopes have proven to be the instruments of choice in the GeV to TeV energy range, very large area telescope arrays are limited by the number of required readout channels per instrumented square-km (due to the large number of channels per telescope). Alternatively, the shower-front sampling technique allows to instrument large effective areas and also naturally provides large viewing angles of the instrument. Solely measuring the shower front light density and timing (hence timing-arrays), the primary particle properties are reconstructed on the basis of the measured lateral density function and the shower front arrival times. This presentation gives an overview of the technique, its goals, and future perspective.

1. Introduction

At its beginning, the main astrophysical motivation for gamma-ray astronomy was the search for Galactic cosmic ray accelerators. The current generation of major air Cherenkov telescopes reach down to an integral flux detection sensitivity for point sources of about 10^{-12}erg cm^{-2} s^{-1} in the TeV energy regime after 50 h of observaton time. This lead to the discovery of many new sources of different object types in the past decade. However, the question of the origin of cosmic rays is not solved yet. While many supernova remnants (the prime candidates for cosmic ray acceleration) were detected in the TeV regime, in most cases it remains unclear whether the observed gamma-ray emission originates from cosmic ray induced hadronic processes (π^0-decay), and leptonic emission scenarios cannot be excluded (see e.g. [1]). Some of the known sources today were observed up to few 10s of TeV, partly exhibiting hard powerlaw spectra close to a differential index of -2. A measurement of the continuation of these energy spectra towards higher energies not covered by current generation instruments will help understanding the underlying acceleration mechanism of the sources. Furthermore, it remains to be seen if these objects can accelerate cosmic rays up to the (presumably) highest energies of Galactic accelerators (up to 10^{17} eV). A detection of hard gamma-ray spectra up to few 100 TeV and with an exponential cutoff that can be approximated by the form $\exp(-\sqrt{E_\gamma/300\,TeV})$ would represent a smoking-gun signature of the acceleration of cosmic rays up to the knee-feature in the all-particle cosmic ray spectrum ($E_p \approx 3 \cdot 10^{15}$ eV). Such Galactic PeV accelerators, or Pevatrons [2], have not been detected up to now. Beyond the knee-feature, in the energy range from 10^{17}-10^{18} eV, a transition is commonly believed to occur from a Galactic origin to an extragalactic origin of cosmic rays. While a detection of gamma-rays at several 10 TeV to few 100 TeV from extragalactic objects is challenging (absorption by pair-production of gamma-rays in low energy photon fields), such observations also bear potential for an indirect measurement of the extragalactic background light in the 10 μm wavelength range.

Due to the typical powerlaw shape of the energy spectra of cosmic gamma-ray sources, large effective detection areas are needed in order to access higher energies. Accessing the energy range from 10 TeV up to several 100 TeV requires an area of the order of 10 km^2, or more. The Cherenkov Telescope Array (CTA) [3] is planned to extend the energy range up to 100 TeV and to push down the point-source detection sensitivity to few times 10^{-14}erg cm^{-2} s^{-1} in the TeV range. However, CTA is optimized for TeV energies and will suffer from its limited collection area above 100 TeV. Even the strongest sources are expected to exhibit only few gamma-ray photons per km^2 per year above 100 TeV, making spectroscopic studies a difficult proposition.

The non-imaging air Cherenkov technique is a complementary approach which allows larger collection areas of several square kilometers at a comparatively moderate cost in number of read-out channels. Its operating principle is based on the sampling of the density and timing (arrival-time and spread) of the Cherenkov photons from the air-shower front with distributed arrays of detector stations. Basically, this shower front sampling method is not limited to the air Cherenkov technique, but can also be done with different detector types using the secondary shower particles, their fluorescence light, or radio emission. An overview of the most relevant parameters of different detection techniques is given in Table 1.

Table 1. Comparison between different shower-front sampling techniques. For each method, approximate values are given for the following parameters: energy threshold E_{thr}, angular resolution $\Delta\theta/\theta$, energy resolution $\Delta E/E$, quality factor of gamma-hadron separation $Q_{\gamma/h} = \epsilon_\gamma/\sqrt{\epsilon_h}$ (with the gamma-ray and hadron survival probabilities after cuts ϵ_γ, ϵ_h), and the duty cycle. A comparison between the methods always depends on the precise setup, detector multiplicity, and location of the experiment. For the sake of simplicity and in order to give an overview, only approximate typical values are given here for representative experiments such as Tibet air shower array, KASCADE-Grande, HAWC (water Cherenkov), CTA, Pierre-Auger Observatory (Fluorescence).

Method	E_{thr}	$\Delta\theta/\theta$	$\Delta E/E$	$Q_{\gamma/h}$	Duty cycle
Particles/Scintillator	3 TeV	~1°	20-50 %	~1	100 %
Particles/Water-Cherenkov	100 GeV	~0.5°	30-50 %	~6	100 %
Air Cherenkov imaging	5 GeV	~0.1°	10-15 %	~6	10 %
Air Cherenkov timing	10 TeV	~0.1°	10-15 %	~1.5-2	10 %
Fluorescence	10^{17} eV	>1°	10-15 %	?	10 %
Radio	10^{17} eV	<1°	10-15 %	?	100 %

Precise angular reconstruction is a key-ingredient for gamma-ray *astronomy*, and is only possible with precise timing. The particle shower front has a disk width of 30 ns at 100 m distance to the core. With less than 10 ns, the air Cherenkov photon disk width is much smaller, justifying the name *timing arrays* for air Cherenkov detectors that make use of the timing information. Further requirements of γ-ray astronomy are a good energy-resolution, a low energy-threshold, and good γ-hadron separation power, leading us back to the air Cherenkov technique as the method of choice for γ-ray astronomy with timing arrays.

While in principle, imaging air Cherenkov telescopes can also use the timing information of the Cherenkov photon arrival, their reconstruction power stems from the analysis of the air-shower image. A comparison of the imaging and timing air Cherenkov techniques in Table 2 makes clear that both techniques are complementary to each other, measuring partly independent air shower properties. A further notable difference between both techniques is that timing arrays

Table 2. Comparison of observables used for reconstruction of direction, energy and particle type for Imaging air Cherenkov telescopes and for timing arrays. Both methods are largely complementary to each other.

	Imaging	Timing
Direction	Air shower image orientation	Air shower front arrival times
Energy	Total light in image	Total light collected on ground
Particle type	Image shape	Lateral density function
		Time width & arrival times

are detectors with a large field of view (FOV) which are ideal for source surveys, but which come at the cost of a reduced signal-to noise ratio and high energy threshold. Imaging telescopes have a small FOV, and much larger light collection area (i.e. mirror area), resulting in low noise levels per pixel and a low energy threshold.

Early on, experiments such as Jakutsk [4], Themistocle [5], or AIROBICC [6], have made first attempts to use timing arrays for γ-ray astronomy. Due to their small effective areas, these experiments could not reach the sensitivity level required for the detection of γ-ray sources in the multi-TeV energy range. Today, the Tunka-133 [7] array operates as a timing array designed for cosmic ray astrophysics. Each Tunka-133 station consists of a PMT inside a cylindrical box, covering a field of view of 50° half-opening angle. 7 such stations are connected in a cluster with a distance of 80 m between the stations. Based on a similar detection principle but a different detector design, the HiSCORE (Hundred*i Square-km Cosmic ORigin Explorer) concept [8] aims at optimizing the timing-array technique for dedication to γ-ray astronomy beyond 10 TeV. The main design differences are a factor 16 larger light collection area per station (4 PMTs with light concentrators per station, lowering the energy threshold), and fast electronics with digitization in the GHz regime. The HiSCORE detector concept was recently implemented for the first time within the Tunka-HiSCORE experiment in the Tunka-valley in Siberia. This experiment has now evolved into the TAIGA collaboration (Tunka Advanced Instrument for Gamma-ray and cosmic ray Astrophysics, also see [9], this conference), which aims at combining a timing-array with imaging telescopes. In future, also particle detectors for the measurement of the muonic component of the shower are planned.

2. The HiSCORE concept

2.1. The detector

HiSCORE is a concept for a distributed large-area (up to 100 km^2) array of wide-angle (0.6 sr) air Cherenkov detector stations covering the gamma-ray energy range from 10 TeV to several PeV, and cosmic ray physics beyond 100 TeV [8]. Important questions of astroparticle and particle physics can be addressed by air shower observations with such an array [10], the most important question being the abovementioned search for the Galactic cosmic ray Pevatrons. Each detector station consists of 4×8 inch photomultiplier tubes (PMTs), each equipped with a 40 cm diameter light concentrator (Winston cone) with a half opening angle of 30°. The effective solid angle of one station was computed taking into account the Winston cone efficiency by integrating the solid angle weighted with the zenith angle dependent Winston cone transmission probability, resulting in 0.6 sr. With 4 channels (one channel = PMT+cone), the total light collection area of one station adds up to 0.5 m^2. The resulting station threshold allows to detect a 100 TeV γ-ray air shower in the outer part of its lateral light density function (LDF) up to 350 m from the shower core impact position. The station multiplicity (number of stations in one event) is the one parameter with the strongest impact on reconstruction quality. For a given choice of the distance between detector stations, the energy threshold at reconstruction level (station multiplicity >3) can be inferred from the LDF. In [11], the LDF parametrization takes the form of an exponential function in the inner part (typically the inner 120 m) and a powerlaw in the outer part ($>$ 120 m) of the LDF

$$\mathrm{LDF}(r) = \begin{cases} p\,\exp(d\,r) & (r < 120\,\mathrm{m}) \\ q\,r^k & (r > 120\,\mathrm{m}) \end{cases} \tag{1}$$

where p and q are scaling factors. Using the most straight-forward station distribution in a regular square grid with a distance of 150 m between detector stations (referred to as standard-configuration see Figure 8) results in an average distance of ≈ 100 m between the air shower core and a station. When assuming a primary γ-ray energy of 50 TeV, on average 3 stations are expected to trigger within the inner 120 m of the LDF (see [8]). The good quality reconstruction-level threshold of such a configuration can therefore be defined to be of the order of 50 TeV.

Below this energy, events are still detected and successfully reconstructed - at lower performance, however, as illustrated by the dropping effective area (Figure 10, Section 2.4) and deteriorating angular resolution (Figure 11). Adopting a more classical definition of the energy threshold as the energy at which 50% of the effective area is reached, results in a value of about 30 TeV for the standard configuration. A further discussion of the energy threshold is given in Section 2.4. In order to obtain a large dynamic range in energy, one or two PMT-dynodes can be read out in addition to the anode. The station trigger is built using the sum of the 4 clipped analog signals in each station. The signal clipping prior to summing allows to suppress strong upward contributions of individual channels from uncorrelated night-sky background (NSB) induced signals. As opposed to NSB photons, the Cherenkov light from an air shower is - to a good approximation - distributed equally over the 4 detector station channels. The amplitude of the sum of the clipped signals is required to be above a given threshold in mV during 7 ns to issue a trigger. The chosen threshold corresponds to a voltage equivalent to 100 p.e.. This condition results in an effective signal threshold of 180 p.e. for the amplitude of the Cherenkov pulse. The readout system is based on fast analog electronics in the GHz regime, allowing to resolve the Cherenkov light pulse with at least 10 samplings (close to the core) and more than 100 samplings far from the core. Further details on the detector concept and a full detector simulation were presented in [8].

2.2. Reconstruction techniques

Methods for γ-ray and cosmic ray data reconstruction for a HiSCORE detector were presented in [11]. The basic underlying techniques described there follow the same basic principles as methods presented earlier in [7] for the Tunka-133 cosmic-ray array. The observables are the Cherenkov light density and its time-distribution measured at the individual detector stations. Figure 1 shows a Cherenkov photon arrival time distribution for a station at a distance of 200 m from the shower core of a 500 TeV simulated γ-ray air shower. The basic reconstruction parameters

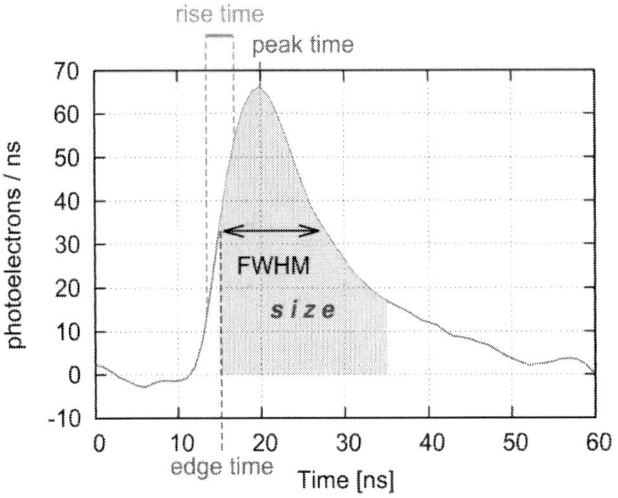

Figure 1. The simulated Cherenkov light pulse of a 500 TeV γ-ray recorded by a HiSCORE detector station in a distance of 200 m from the shower core. Simulated signal assuming a FWHM of 7 ns for the single-photo-electron response, and including noise fluctuations from light of the night-sky background. Source: [11].

are the 50 % pulse edge time, the pulse rise time (from 20 % to 80 %), the pulse amplitude, and the time integrated light signal (sometimes called the size of the pulse).

The direction is reconstructed by a fit of detector station arrival times to a model of the shape of the air shower front. Phenomenological air shower front parametrizations as well as analytical models of the Cherenkov arrival times can be used here. As shown in [11], a relative inter-station time-calibration of better than 1 ns is required in order to limit the impact of the time-resolution on the reconstruction quality.

A first approximation for the shower core impact position is obtained from the center of gravity of the measured light distribution. For a station multiplicity of more than 5, a fit of data to a lateral density function parametrization such as given in Equation 1 yields better results.

The fitted LDF can also be used as an estimate of the total light arriving at the observation level. While the slope of the LDF depends on the shower depth, this dependency is minimal around distances of 200 m from the shower core. The value of the LDF at 220 m is used as energy estimator in [11]. A similar approach is used by [7], where the LDF value at 175 m was chosen. For spectral reconstruction, a normalization method is used based on the steepness of the LDF and a reference spectrum [12]. This method uses the logarithm of the shower-size/energy ratio which was shown to be proportional to the steepness of the LDF, i.e. the ratio between the functional value of the LDF at 100 m (Q_{100}) and its value at 200 m (Q_{200}), as illustraded Figure 2. The observed proportionality does not show a significant dependency

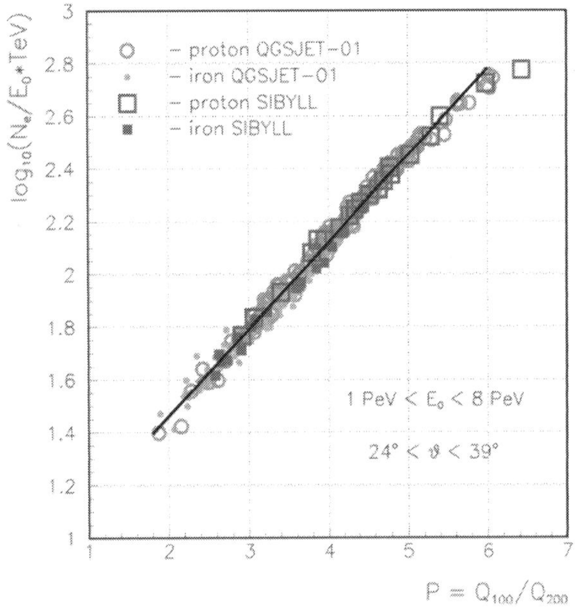

Figure 2. Proportionality between the logarithm of the shower-size/energy ratio and the steepness of the LDF, given by the ratio of its values at 100 m and 200 m from the shower core. Using this proportionality, the primary energy can be reconstructed using the measurement of the shower size and the steepness of the LDF. Source: [7].

on the used interaction model or the simulated hadronic particle type. In order to take into account variations in the light yield of the detector array and atmospheric conditions at the Tunka site, the integral energy spectrum of recorded cosmic rays is normalized to a reference spectrum previously obtained by the QUEST experiment [12].

Air showers initiated by heavier primaries develop their maximum at a smaller shower depth than lighter primaries of identical energy. Therefore, the measured shower depth as a function of the air shower energy can be used for particle separation. Methods for depth reconstruction are based on the ratio of normalization between the inner and outer part of the LDF, or on the FWHM of the time distribution of the Cherenkov light pulse as a function of the shower core impact position. The latter dependency is illustrated in Figure 3.

In a hadron induced air shower, many secondary hadrons are produced which can move further down the axis of shower development, overtaking the Cherenkov front and refueling the

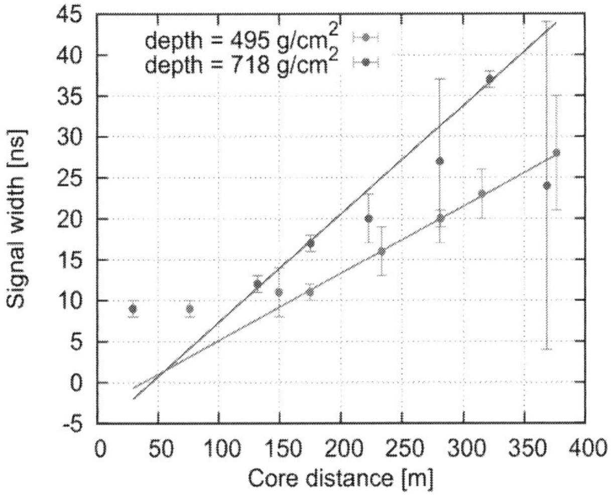

Figure 3. Signal widths (FWHM) versus core distance reconstructed from simulated data with different depths of the shower maximum. The slope of linear fits beyond 150 m core distance is used to reconstruct the depth of the shower maximum. Source: [11].

electromagnetic cascade deeper in the Atmosphere. The resulting Cherenkov light from these subshowers arrives earlier than the bulk Cherenkov light, effectively increasing the rise-time of the pulse. As shown in Figure 4 this effect can be used to a certain degree for γ-hadron separation.

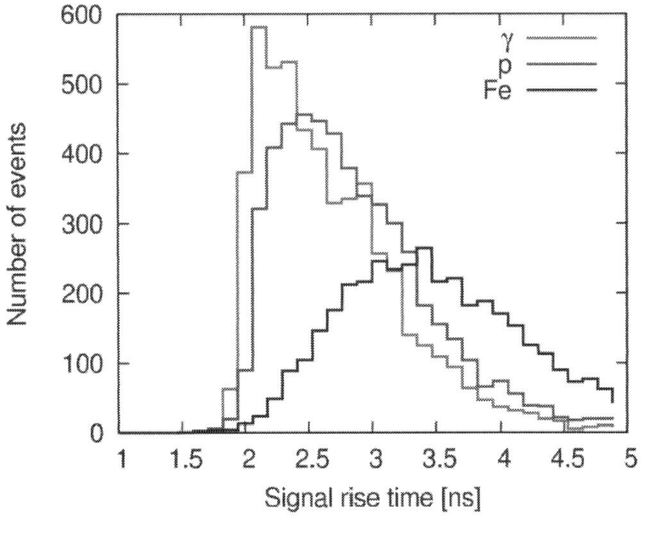

Figure 4. Rise-time distributions for γ-rays, protons and iron nuclei. The difference in rise-time is due to secondary hadrons that overtake the Cherenkov front and refuel the electromagnetic cascade deeper in the atmosphere. Source: [11].

A combination of the γ-hadron separation methods mentioned here results in an overall quality factor for γ-hadron separation of up to 2 at high energies. The reconstruction quality obtained for γ-rays in [7] is summarized in Table 3.

Table 3. Summary of reconstruction quality for the standard configuration HiSCORE array (simple regular grid, 150 m) at different energies.

	100 TeV	300 TeV	1 PeV
angular resolution	$0.35°$	$0.1°$	$<0.1°$
energy resolution	20 %	10 %	<10 %
Xmax resolution	$70\,g/cm^2$	$50\,g/cm^2$	$30\,g/cm^2$
$Q_{\gamma/h}$	1.2	1.5	2.0

2.3. Sensitivity to γ-rays

Effective areas were calculated after acceptance cuts (reconstructed angle less than 25°, core impact contained inside array), reconstruction, and γ-hadron separation cuts. Using these effective areas and a parametrization of the expected cosmic ray background, the point-source sensitivity to γ-rays was calculated. In Figure 5 the sensitivity is shown for a $100\,km^2$ array after 1000 h of observation time. The shown sensitivity curve was obtained using a regular

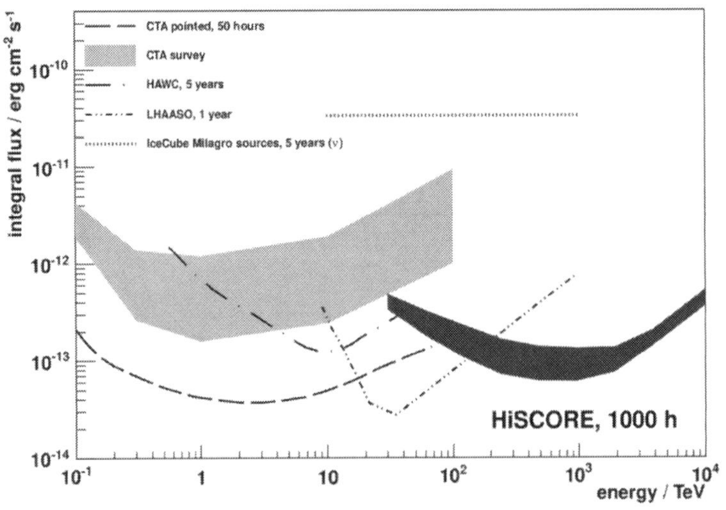

Figure 5. Point-source survey sensitivity of a $100\,km^2$ HiSCORE array after 1000 h of observation time (dark grey area). For comparison, the point-source *survey* sensitivities of CTA [13], of a search for neutrinos by IceCube [14] (Milagro source stacking), HAWC [15], and LHAASO [16, 17] (adapted to a minimum of 50 gamma-rays) are shown. For reference, also the 50-hour *pointed-observation* sensitivity of CTA [18] is given. Source: [8].

grid of detector stations with a spacing of 150 m between stations, shown in Figure 8. For illustration, this Figure also shows the footprint of the detected Cherenkov light of a simulated gamma-ray event. The grey-scale indicates the Cherenkov photon density in arbitrary units. A comparison of the sensitivity curve with other instruments, especially to pointing instruments, is not straight forward. A wide-angle instrument such as HiSCORE always operates in *survey mode*, with exposure accumulating for all objects visible during darkness time within the visible cone of the instrument. The accumulated exposure time for a given position is derived for event directions within 25° of the zenith [19]. The sky coverage is determined by the location of the detector on Earth. For a northern detector site at the latitude of the Tunka-133 cosmic ray array an exposure depth of up to 400 h per source in the FOV is reached over one year, as illustrated in Figure 6. The total sky covered in one year at more than 200 h of observation time is of the order of π sr. Observations in *tilting mode* are foreseen, where all detector stations

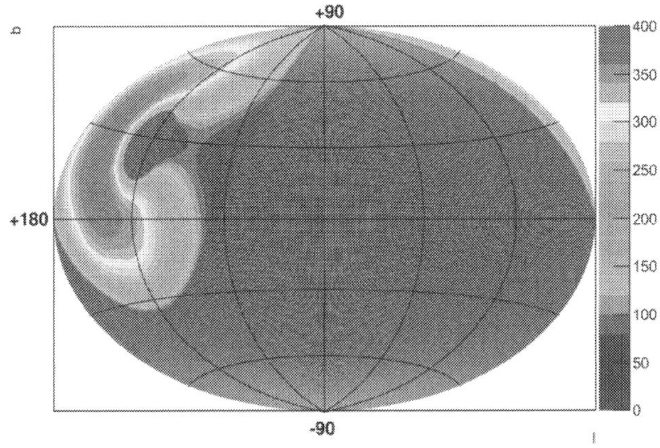

Figure 6. Sky exposure in hours and Galactic coordinates after one year of observation time at a northern site (Tunka-133 experiment site). the deepest exposures in this observation mode reach 400 h per source.

are inclined in declination towards the north or south, therewith acessing different regions of the sky at different exposure times. A schematical representation of the station tilting is shown in Figure 7. When located at a northern site, tilting towards the north results in a deeper average exposure per source while at the same time reducing the total sky coverage. Tilting towards south results in a smaller average exposure but in a large total sky coverage. For illustration, the tilted south exposure times are shown in Figure 7 for a tilting angle of 25° south at the same observation site. In the tilting south mode, the Crab-Nebula is accessed at an exposure depth

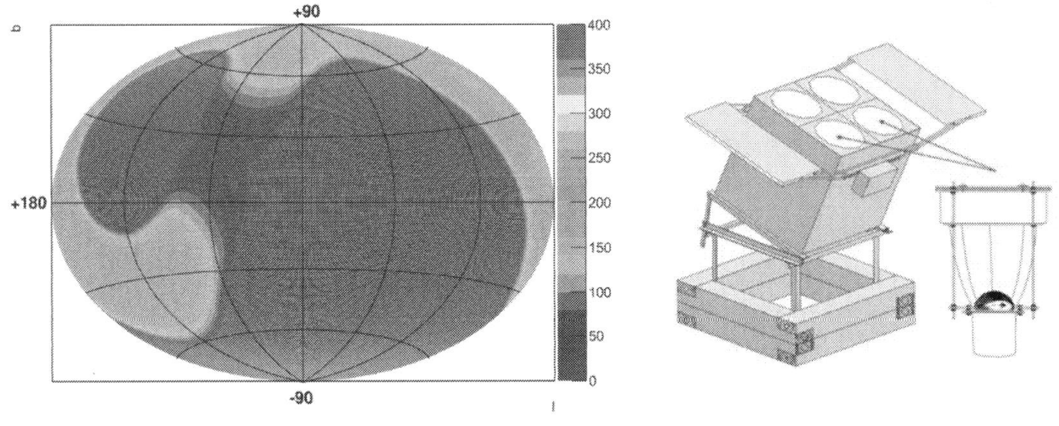

Figure 7. Left: the exposure reached at the Tunka-133 detector site for observations in the tilted south mode. Right: schematical description of the station tilting.

after efficiency correction for weather effects of more than 110 h per year at this observation site.

2.4. Optimization of array design

Alternatively, the stations can be distributed in a graded array (see Figure 9) with small inter-station distances in the center and progressively wider distances towards the edge. Such a

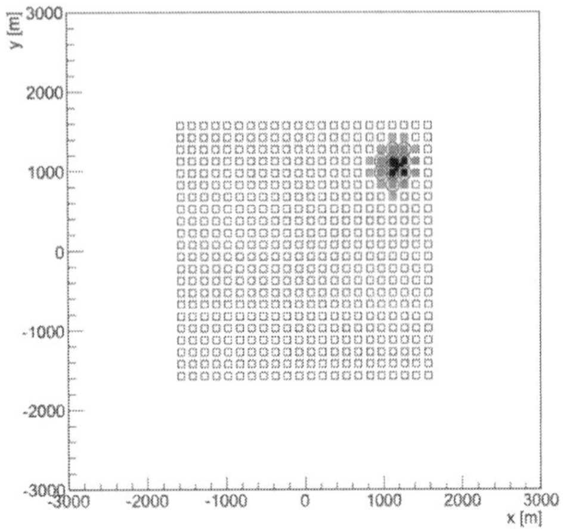

Figure 8. Station layout (484 stations) used in sensitivity calculations, and referred to as *standard layout* in the text. This array covers $10\,\mathrm{km}^2$.

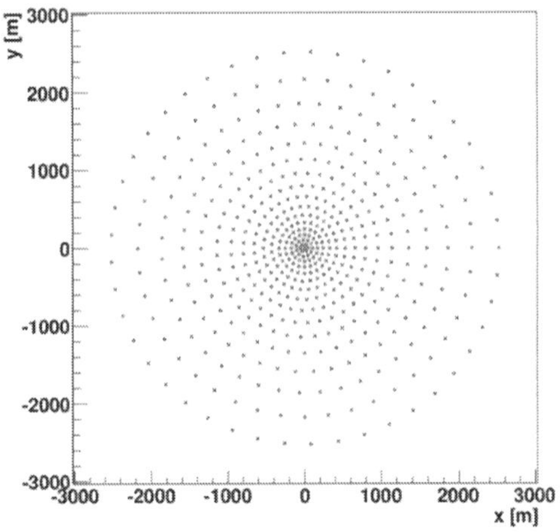

Figure 9. Geometry of a graded array (493 stations) with an instrumented area of $20\,\mathrm{km}^2$ [8]. The gain in area is a factor of 2 with almost identical station numbers.

geometry is of benefit over the whole energy range. At low energies, the low inter-station distances maximize the station multiplicity (improving the angular resolution and reconstruction quality at threshold). At high energies, the larger spacing towards the edge of the array maximizes the instrumented area (compensating the rapidly decreasing γ-ray fluxes).

Using the graded array geometry, the instrumented area is a factor of 2 larger than the standard configuration (regular grid, 150 m spacing) for the same number of stations (Figure 10). At the same time, the close spacing in the inner part of the graded array allows to access lower energies, reducing the energy threshold.

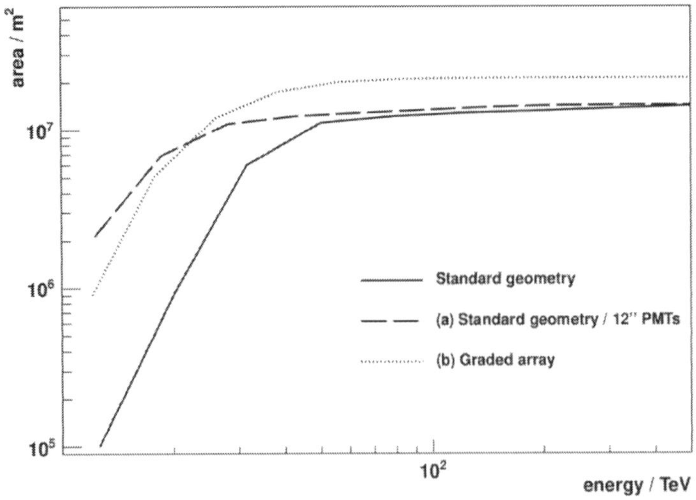

Figure 10. Effective detection area of the standard configuration as compared with two alternative layouts: (a) replacing 8 inch with 12 inch PMTs, (b) graded array. While equipping all stations with 12 inch PMTs is a costly proposition, the graded array design allows both, a gain in area (factor 2) and a lower energy threshold. Source: [8]

The effect of the higher station multiplicities of the graded array for low energies is also visible in Figure 11 showing an improved angular resolution below 300 TeV. Above this energy, the resolution deteriorates slightly, because a large fraction of high energy events are reconstructed closer to the edge of the array, where large spacings result in decreasing station multiplicities as compared to the standard layout. The potential for improvement when using larger PMTs (also equipped with larger Winston cones) is also shown in these Figures. However, while the graded array comes at practically no additional cost, larger PMTs do.

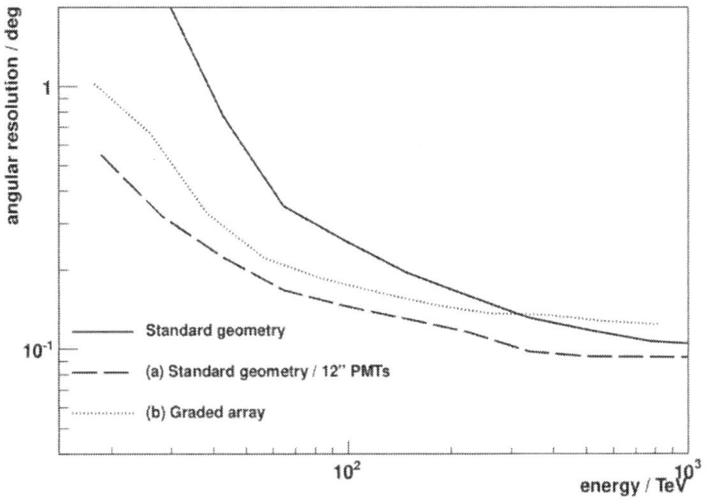

Figure 11. Angular resolution of the standard configuration array layout as compared with two alternative layouts: (a) replacing 8 inch with 12 inch PMTs, (b) graded array. The effect of the improved station multiplicity due to small inter-station distances in the central part of the array is clearly seen at low energies Source: [8]

2.5. Minimization of Energy threshold

In the previous Section, the influence of the array layout and the choice of large area PMTs is illustrated. In this Section, we discuss the minimum achievable threshold and limiting aspects of its minimization. The energy threshold is mainly influenced by the choice of the individual station thresholds (Cherenkov light density), the array geometry (station multiplicity), and the height of the observation level. An observation site located at higher altitudes places the observer closer to the air shower, reducing the absorption of Cherenkov photons in the atmosphere. Due to the conical emission of Cherenkov light, a higher altitude also means a smaller light pool extension. On the one hand this increases the Cherenkov photon density, on the other hand a smaller light pool also reduces the station multiplicity per event. Simulations at higher altitudes show that a benefit at low energies is only achieved when using a smaller inter-station spacing, and thus a smaller overall instrumented area [19].

For a given array geometry, the lowest achievable energy threshold of the individual detector stations will optimize the energy threshold of the array and maximize the reconstruction quality (station multiplicity). The station energy threshold is limited by the imperative that the station signals be distinguishable from the noise fluctuations.

The number of photoelectrons generated by a Cherenkov pulse with photon density ρ is given by

$$N_{\text{Cherenkov}} = \rho \, A \, \epsilon,$$

where A is the detector area ($0.5\,\text{m}^2$), and ϵ is the efficiency of the system, including the transmission through the entrance window, the Winston cone, and the PMT quantum efficiency ($\epsilon \approx 0.14$). The number of photoelectrons generated in one station by uncorrelated NSB photons with a flux of $\phi = 3 \times 10^{12}\,\text{photons}\,\text{m}^{-2}\,\text{s}^{-1}$ is given by

$$N_{\text{NSB}} = \phi \, A \, t \, \Omega \, \epsilon,$$

where $\Omega = 0.6\,\mathrm{sr}$ is the effective solid angle of the detector station. Assuming the NSB in the 4 individual channels is uncorrelated, the noise fluctuations of the analog sum are given by

$$\sigma_{\mathrm{NSB}} = \sqrt{\phi A t \Omega \epsilon}.$$

Here, the time t is the noise integration time, which can be set to the length of the Cherenkov pulse. For the inner 100 m of the LDF t is less than 10 ns. However, this value has an experimental minimum value, due to the pre-amplifier response shape. A fast preamplifier can be preferable since it allows a shorter integration time t which would allow to set a shorter time-over-threshold trigger condition. On the other hand, setting a too short time would limit the sensitivity in the outer part of the LDF, since the signal time width rises with the core impact distance.

The signal-to-noise ratio is defined as $S/N = N_{\mathrm{Cherenkov}}/\sigma_{\mathrm{NSB}} = N_{\mathrm{Cherenkov}}/\sqrt{N_{\mathrm{NSB}}}$. Requiring $S/N = 3$ at threshold, results in a station threshold of 100 p.e., and a minimum detectable Cherenkov photon flux of $\rho_{thr} = 1500\,\mathrm{photons\,m^{-2}}$. This value corresponds to the light density of a 10 TeV air shower at a distance of 100 m from the shower core. For air showers at less than 100 m from the detector station, the resulting station energy threshold is lower than 10 TeV. Note that the effective event reconstruction threshold is defined at a station multiplicity of 3 and therefore strongly depends on the chosen array geometry. Consequently, when choosing a graded array layout with higher station densities in the central part of the array, the reconstruction threshold can be pushed down to the aimed at value of 10 TeV.

A practical limitation is imposed by the dead-time of the system, which rises with rising trigger rate, i.e. with lower energy thresholds. The goal must therefore be to build a read-out system which is capable of handling large data rates at minimal dead-time. After receiving the data in a central processing farm (small PC-farm), an online filtering system can provide a second stage trigger, allowing to reduce the data-flow, i.e. the size of the stored raw data. The most efficient way to reduce data flow without loss of air shower events is the requirement of a next-neighbour trigger within a given time-window. Air showers can only be reconstructed with at least 3 stations in one event. Therefore, requiring 3 neighbouring stations during online-filtering does not affect the detection rate of good air shower events while at the same time reducing the night-sky background induced trigger rate by several orders of magnitude.

3. Tunka-HiSCORE

Since 2013, Tunka-HiSCORE operates a 9-station prototype array in the Tunka-valley in Siberia, on the same site as the Tunka-133 cosmic ray array. The current array of 9 detector stations is arranged on a regular grid of 3x3 stations, with a side length of 300 m (see Figure12). Each individual detector station is built following the description in section 2. Currently, Hamamatsu 10-stage PMTs are used. In order to reduce the current due to night-sky background, the gain is limited by short-circuiting the last 3 dynode stages with the anode, effectively reducing the number of stages to 6. In the next extension, PMTs with only 6 stages will be used (ElectronTubes). Each PMT is equipped with an individual Winston cone built from segments of lightweight reflective sheets (Alanod 4300UP). The analog signals of the 4 PMT channels (only anode signals) of each station are fed into an analog summator/fan-in/fan-out module. This summator board sums the analog signals of the anodes and provides 6 output signals: the 4 analog anode signals, the signal sum, and the sum of the clipped signals. Currently, the signal sum is used for triggering. In future also the clipped signal sum will be used. At this stage of the experiment, two parallel DAQ systems were installed in all 9 stations. The first DAQ system [9] uses an analog sampling board based on the Domino Ring Sampler (DRS 4) chip, using 8 channels for signal sampling. A 9th channel is used for time calibration using a clock signal sent over separate fibers. Data are read out using an Xilinx evaluation board and sent via an optical fibre ethernet link to a central DAQ-PC. The second DAQ system [20] is based on the DRS 4 evaluation board, coupled with a raspberryPI mini-PC, and the WhiteRabbit time-synchronization (time-synchronization and GBit data over a single monomode fiber) [21]. The raspberryPI reads out the DRS 4 evaluation board and sets one of its onboard I/O pins to *high*, signalling readiness for the next trigger to the WhiteRabbit board. The latter is generating the trigger from the analog sum using its FPGA module. After readout, the raspberryPI sends data via optical fibre link to a central DAQ-PC. The previously mentioned crucial requirement of sub-ns relative precision of time-synchronization between stations was achieved with both DAQ and synchronization systems (see [21] and [9]). A recent study presented in [20] uses the residuals of the fit of an air-shower front arrival time model to data to estimate the relative time synchronization between stations, yielding a resolution of 0.6 ns.

The array geometry planned to be completed until 2015 is shown in Figure 13. With the additional stations (shown in red) in total 34 stations will be in operation, covering an area of about 0.25 km^2.

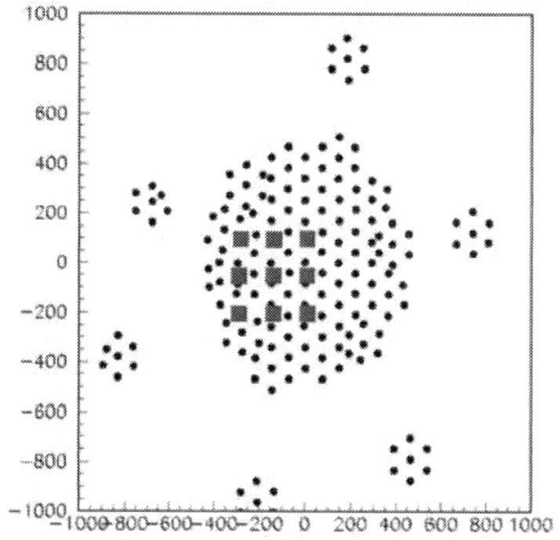

Figure 12. Station layout of the Tunka-HiSCORE 9-station prototype array. The Tunka-HiSCORE detector stations (blue squares) are embedded in the Tunka-133 array (black circles), and covers a total area of roughly 0.1 km^2.

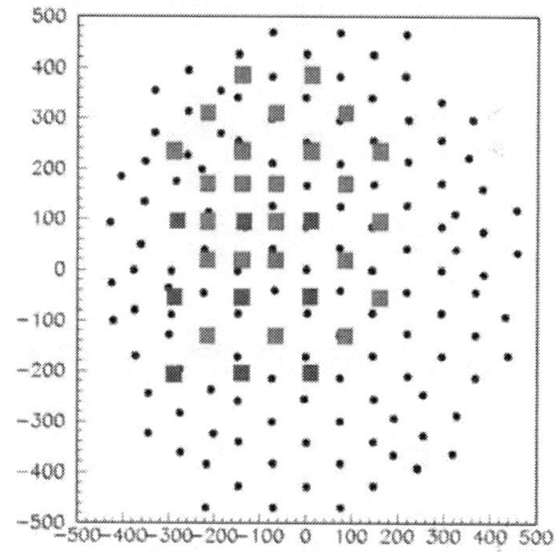

Figure 13. Planned layout for deployment in Autumn 2014. 25 additional stations (red) will reduce the station spacing in the central part of the array (increasing station multiplicity) and increase the total area covered by the array to 0.25 km^2.

4. Combining imaging and timing: TAIGA

The TAIGA experiment [9] is planning to combine the HiSCORE timing array with 10 small sized imaging air Cherenkov telescopes. Each IACT will consist of a 34-segment tesselated mirror dish in Davies-Cotton design, with a diameter of the individual mirror facets of 60 cm, a focal length of 4.75 m, a mirror dish diameter of 4.3 m, and a PMT camera. The PMTs will be arranged with a 30 mm center-to-center distance in 11 hexagonal rings around the central pixel. The energy threshold of such IACTs is predicted to be about 500 GeV. The telescopes can be exploited in different modes: (1) as stand-alone telescopes performing monitoring observations (using the classical imaging technique) of transient sources in the energy range of 500 GeV up to few 10 TeV, and providing a wide energy range for inter-calibration; (2) as integral part of the timing array, providing high-quality amplitude and shower front timing measurements;

(3) in the promising hybrid timing/imaging mode, described below. The timing stations and the individual imaging telescopes are planned to trigger independently from each other. Data merging will be performed using the synchronized (sub-ns) event time-stamps.

The basic idea for a combination of the imaging and timing techniques is to use the timing array to determine the air shower core position and the direction of the primary particle. A combination of these parameters with the image of the IACT is a promising new hybrid approach.

A large instrumented area is the key element for optimizing the array for the energy range from 10 TeV to few 100 TeV. Therefore, the telescopes will be placed at a distance of 600 m from each other. In this configuration all telescopes will operate in monoscopic mode for zenith angles up to 60°. Combining the monoscopic telescopes with the HiSCORE timing array in a hybrid array will result in an improved reconstruction and gamma-hadron separation quality as compared to monoscopic telescopes alone, as outlined below (also see [22], these proceedings).

Image scaling The width of the air shower image as measured by an IACT depends on the impact position of the air shower core, the zenith angle and the energy of the air shower event. Using Monte Carlo simulations, lookup tables can be generated for the expected width as a function of the shower core impact distance (R_c), the zenith angle (θ) and the amplitude (A, also called size) of the image (correlated with the energy), $w_{MC}(R_c, \theta, A_{\rm IACT})$. If these three parameters are known, the measured image width $w_{\rm IACT}$ can be scaled with the expected width:

$$scw = \frac{w_{\rm IACT}}{w_{MC}(R_c, \theta_{\rm IACT}, A_{\rm IACT})},$$

ultimately providing a means for gamma-hadron separation, since hadronic images tend to be wider than their purely electromagnetic counterparts. Typically, this scaled width parameter is used very successfully by *stereoscopic* IACT systems, which provide a good core position reconstruction (R_c), e.g. from the intersection of the image axes of two or more telescopes. Since more than one width measurement is available in stereoscopic systems, the mean of all scaled widths ($mscw$) is used.

In TAIGA however, the telescopes are planned to be placed up to 600 m apart in order to cover a very large area. Therefore, the telecopes will mainly be operating in *monoscopic* mode. Each single telescope taken alone will not be able to reliably measure the core impact for the majority of events. This is where the timing array comes into play: arranged in an array around the IACTs, the HiSCORE timing-array stations will provide their measure of the core impact position $R_{\rm HiSCORE}$ and direction $\theta_{\rm HiSCORE}$ to each telescope, allowing the determination of a hybrid scaled width, $hscw$, as given in Equation 2 without recurring to the stereoscopic technique

$$hscw = \frac{w_{\rm IACT}}{w_{MC}(R_{\rm HiSCORE}, \theta_{\rm HiSCORE}, A_{\rm IACT})}. \tag{2}$$

Preliminary simulations show that the gamma-hadron separation quality factor (defined as the ratio of gamma-efficiency ϵ_γ to the square-root of hadron efficiency ϵ_h) of a cut on the here-described $hscw$ parameter is $Q_{hscw} = \epsilon_\gamma/\sqrt{\epsilon_h} \geq 2$.

Image orientation In stereoscopic systems, the direction of the air shower is reconstructed following the same principle as for the core position. With the large separation of the individual TAIGA telescopes, this reconstruction technique cannot be applied and the precise unambiguous directional information is lost for the bulk of the telescope events. Instead, the timing array will again provide the lost information to the telescopes, resulting in a strong reduction of the background below the gamma-ray signal. Furthermore, the orientation of the elliptical shower image combined with the direction from the timing array can provide an improved directional reconstruction as compared to the pure timing-array reconstruction. Such a combination might also provide a gamma-hadron separation criterion. The major axis from an IACT will intersect the direction of the timing array in case of gamma-rays, while in the case of hadron-induced air showers, the irregular nature of the IACT image will tend to misalign the axis.

Reconstruction quality Finally, integrating the timing and amplitude information of each individual telescope to the measurements of the timing-array will effectively add photon density and timing measurements with very good sensitivity (large mirror area). A telescope would then provide an additional high-quality amplitude measurement for the lateral density function fit, and an additional high-quality timing measurement for the arrival time fit. Overall, this will lead to an improvement of the reconstruction quality of the direction, the energy and the shower depth as compared to the reconstruction quality of the individual detector components alone.

Hybrid array sensitivity By operating the telescopes in monoscopic mode with distances of the order of 600 m between telescopes the total area covered per telescope is larger than the area that could be covered using the same number of telescopes as a stereoscopic system (requiring distances of roughly 300 m in the 10–100 TeV energy regime). This is a key aspect for acessing the energy range from 10 TeV to several 100 TeV (powerlaw spectra). The timing array alone was shown to only allow poor gamma-hadron separation at low energies (10–100 TeV), with a quality factor of the order of unity below 100 TeV and only reaching 2 above few 100s of TeV. As opposed to that, using the hybrid scaled width and image orientation as described above will provide an overall gamma-hadron separation quality of at least 2 at 10 TeV, and possibly as good as 3. The improved quality of the hybrid reconstruction (direction, energy) will also have a beneficial effect on the sensitivity in the overlapping energy range.

Between 10 TeV and 100 TeV, and for a given instrumented detector area, the angular resolution and the gamma-hadron separation are the dominant factors for the point-source sensitivity shown in Figure 5. Above 250 TeV, the gamma-hadron separation alone becomes the dominant factor. Summarizing, the hybrid array reconstruction technique will allow a maximization of the area and an optimization of the sensitivity to gamma-rays above 10 TeV.

5. Summary

Accessing the gamma-ray energy range from 10 TeV to several 100 TeV with spectroscopic measurements is of prime importance for understanding the continuation of known gamma-ray sources and for the search of the Galactic cosmic ray Pevatrons. The timing-array Cherenkov technique is a promising approach for covering this energy regime at the required sensitivity level. With Tunka-HiSCORE, first steps were taken toward a timing-array dedicated to gamma-ray astronomy. Within the TAIGA collaboration, this concept will be realized on a larger scale of the order of km^2 within the coming years. With the hybrid detector concept, combining the air Cherenkov timing-array technique with the air Cherenkov imaging technique, TAIGA will deploy a unique detector, optimizing the sensitivity to gamma-rays above 10 TeV.

5.1. Acknowledgments

We acknowledge the support of the Russian Federation Ministry of Education and Science (agreements N 14.B25.31.0010, zadanie No 3.889.2014/K), the Russian Foundation for Basic research (13-02-00214, 13-02-12095, 15-02-10005, 15-02-05769), the Helmholtz association (grant HRJRG-303), and the Deutsche Forschungsgemeinschaft (grant TL 51-3).

References

[1] Morlino G, Amato E and Blasi P 2009 *MNRAS* **392** 240–250
[2] Gabici S and Aharonian F A 2007 *The Astrophysical Journal Letters* **665** L131–L134

[3] Acharya B S, Actis M, Aghajani T, Agnetta G, Aguilar J, Aharonian F, Ajello M, Akhperjanian A, Alcubierre M, Aleksić J and et al 2013 *Astroparticle Physics* **43** 3–18
[4] Dyakonov M N, Knurenko S P, Kolosov V A *et al.* 1986 *Nuclear Instruments and Methods in Physics Research A* **248** 224
[5] THEMISTOCLE Collaboration, Baillon P, Behr L, Danagoulian S *et al.* 1993 *Astroparticle Physics* **1** 341
[6] Karle A, Merck M, Plaga R *et al.* 1995 *Astroparticle Physics* **3** 321
[7] Berezhnev S F, Besson D, Budnev N M *et al.* 2012 *Nuclear Instruments and Methods in Physics Research A* **692** 98
[8] Tluczykont M, Hampf D, Horns D, Spitschan D, Kuzmichev L, Prosin V, Spiering C and Wischnewski R 2014 *Astroparticle Physics* **56** 42–53
[9] Budnev N *et al.* 2015 The Tunka experiment: from cosmic ray to gamma-ray astronomy; To appear in IOP Conference Series, ECRS 2014, Kiel, Germany (These proceedings)
[10] Tluczykont M, Hampf D, Horns D *et al.* 2011 *Advances in Space Research* **48** 1935
[11] Hampf D, Tluczykont M and Horns D 2013 *Nuclear Instruments and Methods in Physics Research A* **712** 137–146
[12] Korosteleva E E, Prosin V V, Kuzmichev L A and Navarra G 2007 *Nuclear Physics B (Proc. Supp.)* **165** 74–80
[13] Dubus G, Contreras J, Funk S *et al.* 2013 *Astroparticle Physics* **43** 317
[14] Gonzalez-Garcia M C, Halzen F and Mohapatra S 2009 *Astroparticle Physics* **31** 437
[15] Sinnis G 2005 HAWC: A Next Generation VHE All-Sky Telescope *Heidelberg Gamma-Ray Symposium* (*American Institute of Physics* vol 745) ed Aharonian F A, Völk H J and Horns D p 234
[16] Cao Z 2010 *Chinese Physics C* **34** 249
[17] Cui S, Liu Y, Liu Y and Ma X 2014 *Astroparticle Physics* **54** 86 – 92
[18] CTA Consortium, Actis M, Agnetta G, Aharonian F, Akhperjanian A *et al.* 2011 *Experimental Astronomy* **32** 193
[19] Hampf D, Tluczykont M and Horns D 2010 Simulation of the expected performance for the proposed gamma-ray detector HiSCORE proceedings of Science (Texas 2010) 245
[20] Porelli A *et al.* 2015 Timing calibration and directional reconstruction for Tunka-HiSCORE; To appear in IOP Conference Series, ECRS 2014, Kiel, Germany (These proceedings)
[21] Brückner M, Wischnewski R *et al.* 2013 Results from the whiterabbit sub-nsec time synchronization setup at hiscore-tunka; To appear in Proc. of 33RD ICRC 2013, Rio de Janeiro, Brazil
[22] Kunnas M *et al.* 2015 Simulation of the Tunka Advanced International Gamma-ray Advanced experiment (TAIGA); To appear in IOP Conference Series, ECRS 2014, Kiel, Germany (These proceedings)

Diffuse flux results from the ANTARES neutrino telescope

Erwin Visser on behalf of the ANTARES collaboration
NIKHEF, Science Park 105, 1098 XG Amsterdam, The Netherlands
E-mail: erwinv@nikhef.nl

Abstract. The ANTARES neutrino telescope is located in the Mediterranean Sea off the coast of France and offers a high visibility of the Galactic plane. Its main scientific goal is the detection of cosmic neutrinos, which is achieved by measuring the Cherenkov light emitted by the products of neutrino interactions. A limit on the all-sky diffuse neutrino flux has been set using muon-neutrinos, and a new analysis provides an improved sensitivity by also using showers created in electron- and tau-neutrino interactions. In addition, the high visibility of the Galactic plane is exploited in a dedicated analysis focussing on the Galactic plane. In this analysis the diffuse muon-neutrino flux of a region around the Galactic plane is compared with the flux from multiple equivalent off-source regions. Results of these analyses using five years of data will be reported.

1. Introduction

The recent observation of a cosmic neutrino flux by the IceCube neutrino telescope [1] is the start of a new era in neutrino astronomy. Due to the low amount of detected neutrinos and also due to the limited angular resolution of IceCube, the origin of this flux of neutrinos is not yet clear. Measurements of the neutrino flux with other detectors would help in better understanding the emission origin. The ANTARES detector [2] is currently the largest deep-sea neutrino telescope in the Northern Hemisphere and will be succeeded by the next generation Mediterranean Neutrino Telescope, KM3NeT (see also the contribution of R. Coniglione at this conference).

Neutrinos only interact via the weak interaction by exchanging a weak boson with a target nucleus. Neutrino telescopes detect neutrinos indirectly through the interaction products created in these interactions. The exact signature that is produced in the detector depends on the flavour of the neutrino and the type of the interaction, which can be either charged current (CC) or neutral current (NC). In a neutral current interaction, the neutrino breaks up the nucleon (N), resulting in a hadronic shower (X) and a lower energy neutrino:

$$\nu_l(\bar{\nu}_l) + \mathrm{N} \rightarrow \nu_l(\bar{\nu}_l) + \mathrm{X}, \tag{1}$$

with $l = e, \mu, \tau$ the neutrino flavour. In a charged current interaction, a relativistic charged lepton is produced in addition to a hadronic shower:

$$\nu_l(\bar{\nu}_l) + \mathrm{N} \rightarrow l^-(l^+) + \mathrm{X}. \tag{2}$$

The emerging lepton inherits the flavour of the incident neutrino.

In a charged current interaction of an electron-neutrino, an electron is produced, which in turn produces an electromagnetic shower by bremsstrahlung and pair production. In a CC muon-neutrino interaction a muon is produced, which can travel a considerable distance before decaying. Around an energy of 1 TeV, the typical track length in water is a few kilometres, providing a long lever arm for direction reconstruction. In a CC interaction of a tau-neutrino a tau is produced, which predominantly decays to hadrons, producing a second hadronic shower.

In transparent media, like water or ice, the light produced in each of the interactions described above can be detected with an array of sensitive optical sensors. Since neutrinos only interact via the weak interaction (and thus have a low interaction cross section), large detector volumes are required.

In order to reduce the background from atmospheric muons, which are produced in cosmic ray interactions with the Earth's atmosphere, neutrino telescopes are hosted in deep water or in deep Antarctic ice, where several thousands of metres of water (ice) reduce the flux of atmospheric muons by several orders of magnitude. Also, since neutrinos are the only particles that can pass through the whole Earth, neutrino telescopes mainly use upgoing neutrinos originating from the opposite hemisphere to remove the remaining atmospheric muon background.

In addition to atmospheric muons, also neutrinos are produced when cosmic rays interact with our atmosphere. These upgoing atmospheric neutrinos constitute an unavoidable background, since an atmospheric neutrino by itself can not be distinguished from a cosmic one. However, the energy spectrum of atmospheric neutrinos is softer than the expected spectrum of cosmic neutrinos, which can be used to discriminate between them by reconstructing the energy of the neutrinos.

First the ANTARES neutrino telescope will be described in section 2, after which the results of recent diffuse flux analyses will be presented in sections 3, 4 and 5. A conclusion is given in section 6.

2. The ANTARES detector

The ANTARES neutrino telescope is located 40 km offshore from the French city of Toulon, at a depth of 2475 metre below sea level. It is the first operational neutrino telescope in the Mediterranean Sea and was completed in May 2008. The detector consists of 12 vertical detection lines, which each contain 25 storeys with a triplet of so-called Optical Modules (OMs), except for line number 12 which contains only 20 storeys. The top 5 spots of this line contain acoustic detection devices, which are used for the acoustic detection of very high energy neutrinos. The distance between storeys is 14.5 m, with the first storey being located 100 m above the sea floor. The OMs look downward under an angle of 45° in order to optimise the detection of light from upgoing muons. A schematic overview of the detector is shown in figure 1.

An OM consists of a glass sphere, which contains a 10 inch PhotoMultiplier Tube (PMT) that detects photons via the photo-electric effect. The PMT is magnetically shielded against the Earth's magnetic field using a μ-metal cage. The PMT and the μ-metal cage are held in place in the OM with special optical gel. A schematic view of an OM is shown in figure 2.

In addition to three OMs, a storey also contains the Local Control Module (LCM), which houses the offshore electronics. Each line is anchored by the Bottom String Socket (BSS) and held vertical by a buoy. The BSS consists of a dead weight which acts as an anchor, and contains the String Control Module (SCM), which contains the electronics steering the LCMs.

An interlink cable links each line to the Junction Box (JB), which in turn is connected to the shore station by the Main Electro-Optical Cable (MEOC). The MEOC is a standard telecommunications cable and provides the electrical power link and the optical data link between the detector and the shore station.

All data that is recorded by the OMs is sent to shore, where a PC farm runs the trigger algorithms responsible for selecting interesting physics events.

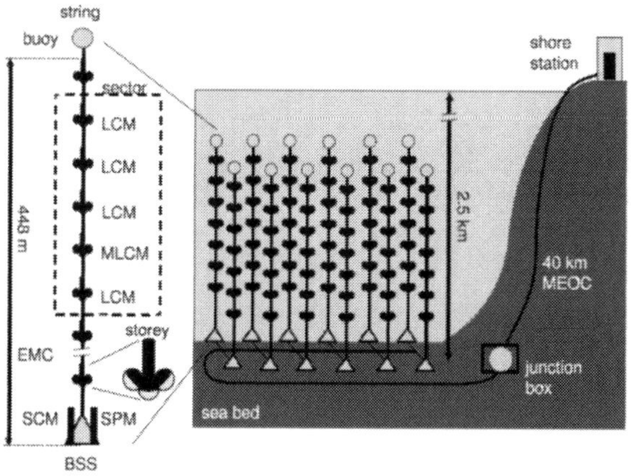

Figure 1. Schematic overview of the ANTARES detector.

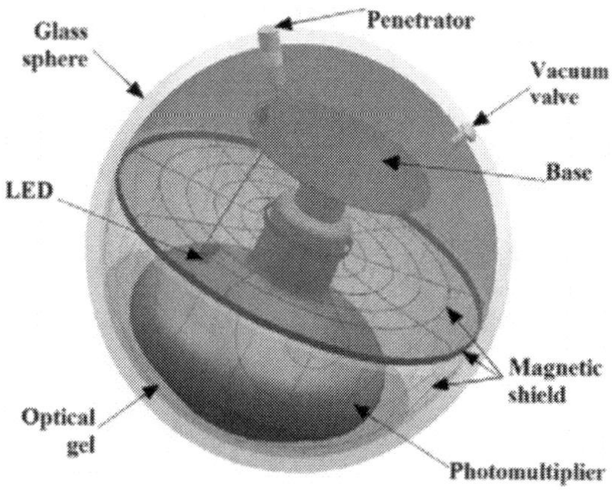

Figure 2. Schematic view of an ANTARES OM.

3. Muon-neutrino diffuse flux analysis

The goal of diffuse flux analyses is to look for an excess of events in the flux integrated over the full field of view. The excess would become apparent above a certain energy threshold, which depends on the spectral index of the cosmic flux and its size compared to the atmospheric neutrino background. The first ANTARES analysis in which a search for a diffuse flux of cosmic neutrinos is performed used data taken during 2008 and 2009 [3]. In the updated analysis two more years of data are used, giving a total livetime of 885 days [4].

In this analysis only muon-neutrino events are considered, which are reconstructed using the track fitting algorithm described in [5]. This algorithm provides the quality parameter Λ, which is derived from the likelihood and the uncertainty β on the reconstructed track direction. These parameters can be used to select well reconstructed events. To reduce the atmospheric muon background and reject badly reconstructed tracks, only upgoing events are considered which have $\beta < 0.5°$. After these cuts are applied, the muon background is reduced to below the 1% level using a combined cut on Λ and the number of hits correlated with the track (N_{hit}). Applying these cuts to the data shows a 28% deficit compared to the simulated atmospheric neutrino flux (using the Bartol flux [6]). The shape of the two distributions is the same (see also [7]) and the deficit is well within the systematic uncertainties on the theoretical expectation. For these reasons the simulation is thus normalised to the data.

The muon energy is reconstructed based on an estimation of the muonic energy loss by using the number of detected hits and the reconstructed track length (dE/dX-estimator [8]). The cut on the reconstructed energy is then optimised to obtain the best flux sensitivity using the Model Rejection Factor (MRF) technique [9]. After normalisation, 8.4 atmospheric events are expected for the optimised cut ($E_{dE/dX} > 1.4$ TeV) and 8 events are observed in data after unblinding. Using the method described in [10] this translates into a 90% CL flux upper limit of:

$$\Phi^{90\%}_{\nu_\mu+\bar{\nu}_\mu} = 5.1 \cdot 10^{-8}\,(E_\nu\,[\text{GeV}])^{-2}\,\text{GeV}^{-1}\,\text{cm}^{-2}\,\text{sr}^{-1}\,\text{s}^{-1}, \tag{3}$$

in the energy range from 45 TeV to 10 PeV (containing the central 90% of the signal). The uncertainties on the signal and background estimation are taken into account in the limit as a systematic error. The upper limit is shown in figure 4 (blue dashed line) and is compared to the atmospheric neutrino flux predictions, upper limits from other experiments and a fit to the neutrino flux measured by IceCube.

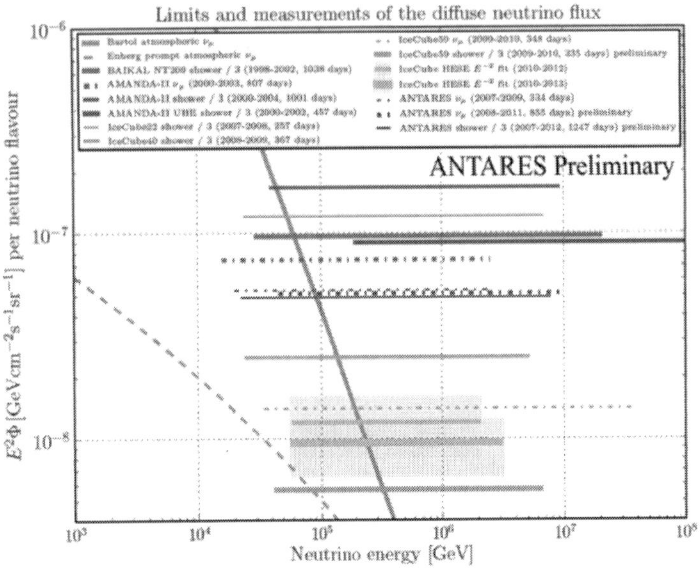

Figure 3. ANTARES 90% CL flux upper limits for an E_ν^{-2} spectrum versus neutrino energy (shown in blue). Also shown are the atmospheric neutrino flux predictions by Bartol [6] and Enberg [11] (in grey), the fits to the neutrino flux measured by IceCube [1,12] (in yellow) and upper limits from other experiments: AMANDA [13–15] (in purple), Baikal [16] (in red) and IceCube [17–20] (in cyan).

4. All flavour diffuse flux analysis using showers

The sensitivity to a diffuse flux can be improved by using electron- and tau-neutrino interactions in addition to muon-neutrinos. As described before, the signature of these types of neutrino interactions is different, since they create hadronic/electromagnetic showers. In the all flavour diffuse flux analysis, the shower reconstruction method described in [21] is used. The reconstruction is a two-step maximum likelihood fit that makes use of simulated likelihood tables. In the first step the shower vertex is reconstructed, after which this is used as input for a second likelihood fit reconstructing the energy and direction of the incoming neutrino. In the (shower) energy range from 1 TeV to 1 PeV the algorithm achieves 0.2-0.3 resolution in $\log_{10}(E_{\mathrm{shower}})$ with a slight underestimation. The median direction resolution for neutrinos is about 6° for showers below 100 TeV and about 25° for 1 PeV showers. The worsening of the direction resolution at the highest energies has to do with the spatial extension of the shower compared to the size of the ANTARES detector. The numbers quoted here are for well reconstructed events that pass an upper cut on the vertex log-likelihood, with an efficiency ranging from 10% at 1 TeV to 60% at 1 PeV.

For the analysis data from 2007 to 2012 were used, with a total livetime of 1247 days. The background of atmospheric muons and track like events (from charged current muon-neutrino interactions) is reduced by using a cut on the vertex log-likelihood and requiring that at least 3 detector lines were used by the reconstruction algorithm. Additionally, it was required that the reconstructed vertex position was not closer than 15 m to any of the OMs to exclude events that are produced by sparks in the OMs.

The cuts on the reconstructed zenith angle θ_{rec} and the shower energy are optimised by again using the MRF technique, resulting in $\theta_{\mathrm{rec}} > 94°$ and $E_{\mathrm{shower}} > 10$ TeV. After unblinding of the data 8 events were observed, where 5 ± 3 were expected from the atmospheric background simulation (atmospheric muons, conventional atmospheric neutrinos using the Bartol flux and prompt atmospheric neutrino using the Enberg flux [11]). The measurement corresponds to an excess over the background of 1.5σ.

The observed number of events can again be translated into a 90% CL flux upper limit of:

$$\Phi^{90\%}_{\nu+\bar{\nu}} = 4.9 \cdot 10^{-8}\,(E_\nu\,[\mathrm{GeV}])^{-2}\,\mathrm{GeV}^{-1}\,\mathrm{cm}^{-2}\,\mathrm{sr}^{-1}\,\mathrm{s}^{-1}, \tag{4}$$

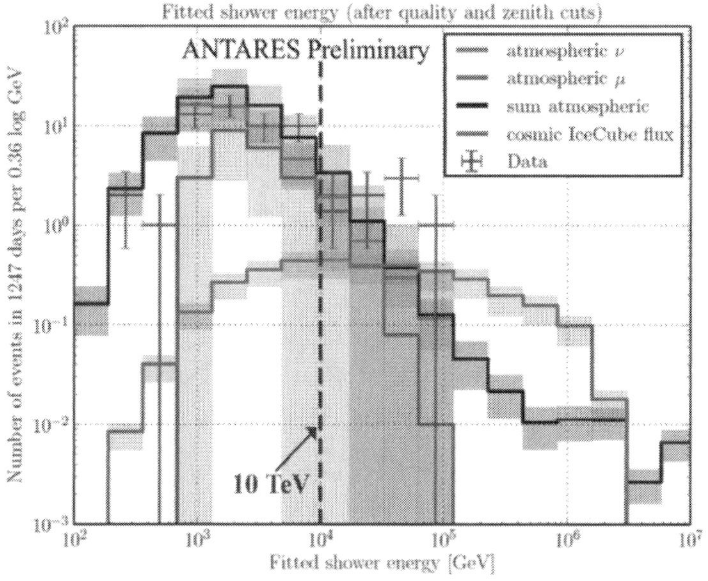

Figure 4. Distributions of the reconstructed shower energy for data and simulation after all cuts are applied, the dashed blue line shows the cut at 10 TeV. The atmospheric neutrino simulation contains both the conventional (using the Bartol flux) and the prompt (using the Enberg flux) contribution. Also shown is the expectation from the fit to the measured IceCube flux with a 2 PeV cutoff [12].

per neutrino flavour, in the energy range from 23 TeV to 7.8 PeV. Again, the systematic uncertainties on signal and background estimation are taken into account. This upper limit is shown in figure 4 as the solid blue line.

5. Galactic plane diffuse flux analysis

In addition to the all sky diffuse flux searches, there are analyses which focus on specific regions on the sky from which a higher diffuse flux is expected. An example of such an analysis is the Fermi Bubble analysis that is described in [22]. Here, the results are reported for another analysis of this kind, which focusses on the Galactic plane region. Since only a part of the sky is used as signal region, multiple equivalent background regions can be defined which can be used to also perform a measurement of the background from the data.

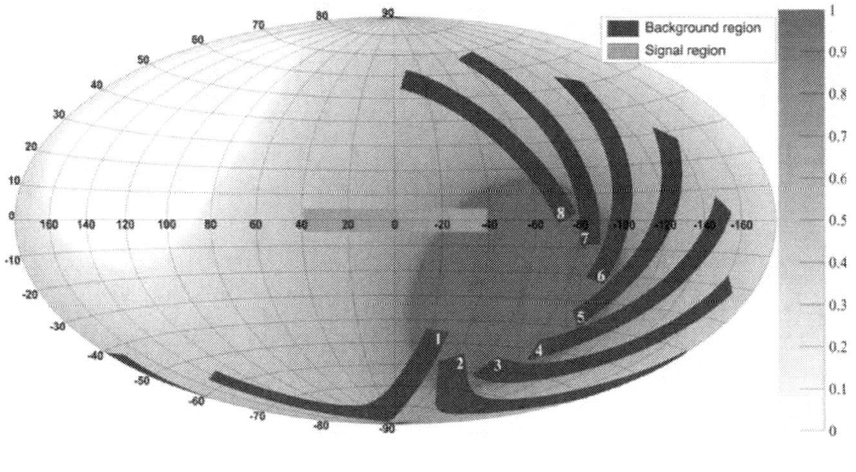

Figure 5. Signal region (in green) and background regions (in dark blue) used in the Galactic plane diffuse flux analysis. Also shown is the ANTARES visibility.

A guaranteed neutrino flux is expected from cosmic ray interactions with the interstellar matter in the Milky Way, which will be maximal in a region around the Galactic Centre (GC) since the matter density peaks there. To model the fluxes, three different theoretical models are considered, which differ in the assumptions that are made. The NoDrift_simple [23] and NoDrift_advanced [24] models assume a constant cosmic ray (CR) flux in the Galaxy and take a uniform matter density of 1 nucleon/cm^3. They differ in the parameterisation used for the CR flux. In the Drift model [25], the assumed magnetic field gives rise to the drift of high energy CRs to the GC, where also the matter density peaks, resulting in an enhanced neutrino flux from the GC. Besides these theoretical models, the γ-ray data as measured by Fermi-LAT [26] are used to estimate the neutrino flux based on the assumption that both originate (partly) from pion decay.

Using these signal models, the size of the signal region is optimised using the MRF technique. This results in an optimal region of 78° in Galactic longitude and 9° in Galactic latitude, centred on the GC. For this region size, 8 background regions can be defined, which trace the same path in local detector coordinates, so that the same number of events are expected in each and most systematic effects cancel out. The signal and background regions are shown in figure 5.

The background of atmospheric muons and atmospheric neutrinos is reduced by only considering upgoing events and optimising the cuts on the track parameters and the reconstructed energy using the MRF technique again. This results in an event selection with $\Lambda > -5.6$, $\beta < 8.0°$ and $E_{\mathrm{ANN}} > 350$ GeV, where the energy is estimated using Artificial Neural Networks [8]. In addition a cut has been placed on a parameter related to the clustering of the hits, which has been found to help in rejecting atmospheric muons [27].

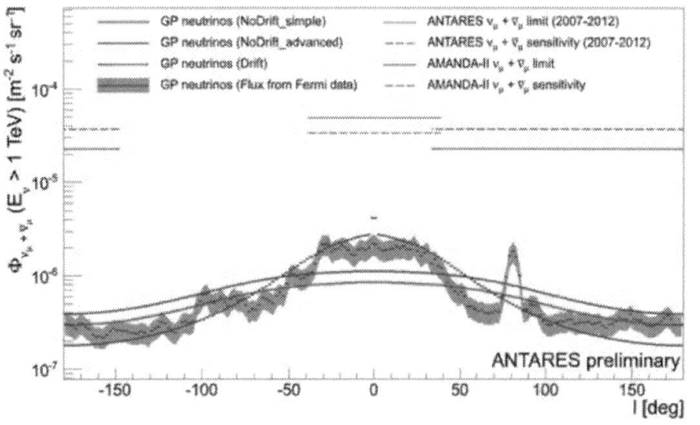

Figure 6. ANTARES 90% CL sensitivity and flux upper limit (for an $E_\nu^{-2.7}$ flux) versus Galactic longitude, integrated above 1 TeV and averaged over Galactic latitude from −4.5° to +4.5°. Also shown are the fluxes predicted by the theoretical models [23–25] and calculated from the Fermi-LAT data [26] and the sensitivity and flux upper limit obtained by AMANDA-II [28].

Data from 2007 to 2012 were used, with a total livetime of 1288 days. After unblinding of the data 177 events were observed in the signal region and on average 166 events from each of the 8 background regions. Using the method described in [29] this corresponds to an excess with a significance of 0.8σ.

Since no significant excess has been observed, upper limits on the flux are placed. Following the method described in [30], the neutrino flux for a spectral index of γ is written as:

$$\frac{E^{\gamma-2}}{\lambda^{\gamma-2}}\Phi_{\nu_\mu+\bar{\nu}_\mu} = \mathcal{F}\cdot(E_\nu\,[\mathrm{GeV}])^{-2}\,\mathrm{GeV}^{-1}\,\mathrm{cm}^{-2}\,\mathrm{sr}^{-1}\,\mathrm{s}^{-1}, \tag{5}$$

with $\lambda = 1\,\mathrm{GeV}$, so that the sensitivities and upper limits can be presented in the same units as used in equations 3 and 4. The 90% CL sensitivities (denoted with a bar) and upper limits on the flux constant $\mathcal{F}$ are presented in table 1 for several values of the spectral index. The sensitivities and limits are shown versus Galactic longitude in figure 6, in which also the model predictions are shown.

Table 1. Sensitivities, obtained flux upper limits and energy validity range (central 90% of the signal) for different spectral indices.

Spectral index γ	$\overline{\mathcal{F}}^{90\%}$ [$\mathrm{GeV^{-1}\,cm^{-2}\,sr^{-1}\,s^{-1}}$]	$\mathcal{F}^{90\%}$	Valid energy range [TeV]
2.5	$1.4 \cdot 10^{-4}$	$2.0 \cdot 10^{-4}$	0.24 - 96
2.6	$3.2 \cdot 10^{-4}$	$4.6 \cdot 10^{-4}$	0.18 - 71
2.7	$7.1 \cdot 10^{-4}$	$1.1 \cdot 10^{-3}$	0.15 - 52

6. Conclusions

The ANTARES neutrino telescope is currently the largest neutrino telescope in the Northern Hemisphere and is in its 7th year of operation. Despite its moderate size, but thanks to its location and excellent angular resolution it is yielding competitive diffuse flux sensitivities and the best limits for the diffuse Galactic neutrino flux from the Galactic plane region. In the near future the results can be improved by adding more data and perform joint track and shower analyses. ANTARES also paves the way for the next generation neutrino telescope, KM3NeT, which will provide more than an order of magnitude improvement in sensitivity and will complement the IceCube detector located on the South Pole.

References

[1] Aartsen M G *et al.* 2014 *Phys. Rev. Lett.* **113** 101101
[2] Ageron M *et al.* 2011 *Nucl. Instrum. Methods A* **656** 11–38
[3] Aguilar J A *et al.* 2011 *Phys. Lett. B* **696** 16–22
[4] Kulikovskiy V for the ANTARES collaboration 2014 *Proc. VHEPU2014, Quy Nhon, Vietnam (to be published)*
[5] Adrián-Martínez S *et al.* 2012 *ApJ* **760** 53
[6] Agrawal V, Gaisser T K, Lipari P and Stanev T 1996 *Phys. Rev. D* **53** 1314
[7] Adrián-Martínez S *et al.* 2014 *Eur. Phys. J. C* **74** 2701
[8] Schnabel J 2013 *Nucl. Instrum. Methods A* **725** 106–109
[9] Hill G C and Rawlins K 2003 *Astropart. Phys.* **19** 393–402
[10] Conrad J, Botner O, Hallgren A and Pérez de los Heros C 2003 *Phys. Rev. D* **67** 012002
[11] Enberg R, Reno M H and Sarcevic I 2008 *Phys. Rev. D* **78** 043005
[12] Aartsen M G *et al.* 2013 *Science* **342** 1242856
[13] Achterberg A *et al.* 2007 *Phys. Rev. D* **76** 042008
[14] Abbasi R *et al.* 2011 *Astropart. Phys.* **34** 420–430
[15] Ackermann M *et al.* 2008 *Astrophys. J.* **675** 1014–1024
[16] Dzhilkibaev Zh-A *et al.* 2009 *Proc. 31st ICRC, Lodz, Poland* (*Preprint* `astro-ph/0909.5562`)
[17] Abbasi R *et al.* 2011 *Phys. Rev. D* **84** 072001
[18] Aartsen M G *et al.* 2014 *Phys. Rev. D* **89** 102001
[19] Aartsen M G *et al.* 2014 *Phys. Rev. D* **89** 062007
[20] Aartsen M G *et al.* 2013 *Proc. 33rd ICRC, Rio de Janeiro, Brazil* (*Preprint* `astro-ph/1309.7003`)
[21] Folger F 2014 Ph.D. thesis Friedrich-Alexander-Universität Erlangen-Nürnberg
[22] Adrián-Martínez S *et al.* 2014 *Eur. Phys. J. C* **74** 2701
[23] Ingelman G and Thunman M 1996
[24] Candia J and Roulet E 2003 *JCAP* **09** 005
[25] Candia J 2005 *JCAP* **11** 002
[26] Ackermann M *et al.* 2012 *ApJ* **750** 3
[27] Visser E Ph.D. thesis Universiteit Leiden *(to be published)*
[28] Kelley J L *et al.* 2005 *Proc. 29th ICRC, Pune, India* (*Preprint* `astro-ph/0509546`)
[29] Li T and Ma Y 1983 *ApJ* **272** 317–324
[30] Spurio M 2014 *Phys. Rev. D* **90** 103004

Phase distribution of the first harmonic of the cosmic ray anisotropy during the initial phase of Forbush effects

M Abunina, A Abunin, A Belov, E Eroshenko, V Oleneva, V Yanke

Pushkov Institute of Terrestrial Magnetism, Ionosphere and Radio Wave Propagation of the Russian Academy of Sciences (IZMIRAN), Kaluzhskoe Hwy 4, Troitsk, Moscow, 142190, Russia

E-mail: abunina@izmiran.ru, erosh@izmiran.ru

Abstract. Phase distribution and amplitude-phase dependence of the first harmonic of the cosmic ray anisotropy in the initial phase of Forbush decreases are studied in the events during the time period 1957-2013. Statistical analysis of all Forbush decreases with a sudden onset for this period showed that the specific features of the phase distribution of the first harmonic of the cosmic ray anisotropy exist throughout the initial phase of the Forbush decreases, starting one hour before the shock, and ending at the time of maximum amplitude of anisotropy. Amplitude of the cosmic ray vector anisotropy is higher than in quiet periods already before the shock arrival, and it gradually increases as Earth enters deeper into the interplanetary disturbance, which creates the Forbush decrease.

1. Introduction

Numerical experimental data and convective-diffusive model [1, 2, 3] tell about advantageous directions of cosmic rays (CR) vector anisotropy. The specific directions in the interplanetary space are determined by the solar wind velocity and field line configuration of the interplanetary magnetic field (IMF). In addition, the observed radial gradient of CR intensity near the Earth is almost continually positive in sign and moderate in size. A consequence of these circumstances is the non-uniform phase distribution of vector anisotropy [4, 5].

Long-term changes of phase distribution and amplitude-phase interrelation of the first harmonic of galactic CR have been studied in [6]. It was shown that these changes mainly follow 11- and 22-year solar cycles, and that non-uniform phase distribution and considerable amplitude-phase dependence are almost continual. The aim of this work is to study phase distribution of the first harmonic of CR anisotropy not only during the quiet but in disturbed periods as well, at the beginning of a Forbush effect – 1 hour prior to the shock arrival and in the first couple (or several) hours after its arrival. Forbush effects are the results of two types of interplanetary disturbances: recurrent and sporadic [10, 11, 12]. The first group of FEs arises from the impact of high speed plasma streams from coronal holes; the second one are caused by coronal mass ejections (CMEs) from the Sun. We use a large statistical sample (the events during 1957-2013) collected in IZMIRAN databases, the description of which is given in the next section.

2. Data and Method

In this work we use IZMIRAN databases, created by the Cosmic Ray Department under the management of A. Belov. The databases provide a selection of CR events with a number of parameters. After processing hourly data from world wide neutron monitor network (www.nmdb.eu; http://cr0.izmiran.ru/mosc) by the Global Survey Method (GSM) [9] we obtain hourly values of the CR density and anisotropy above the Earth atmosphere and magnetosphere, specially calculated for 10GV rigidity. These data are supplemented by corresponding solar, interplanetary and geomagnetic characteristics taken from http://omniweb.gsfc.nasa.gov/ow.html, ftp://ftp.gfz-potsdam.de/pub/home/obs/kp-ap/wdc, and http://www.swpc.noaa.gov/ to form the first database "on the CR variations", containing data in the time period covering more than 50 years of observations (1957-2013) [7, 15]. The second database "on interplanetary disturbances and Forbush effects (FE)" contains characteristics of FEs (see figure 1 and description) on more than 6000 events and corresponding solar wind disturbance parameters, indices of geomagnetic activity and information on solar sources [8, 12]. The information on solar wind is taken from the OMNI database (http://omniweb.gsfc.nasa.gov).

For the analysis the FEs were selected from the FE database using following criteria: (1) the onset of FE is simultaneous with the shock arrival, which was determined using Sudden Storm Commencement (SSC); (2) The FE onset is separated from the following event (onset) at least for 36 hours, and from the beginning of the previous one – for 60 hours and more when the previous event was larger than 1,4%; (3) We also didn't use the hours with observed ground level enhancement (GLE) – all such hours were removed from the analysis. The selection resulted in a total of 817 events.

To understand better which parameters and characteristics of the FEs we use for analysis, below in figure 1 a typical behavior of CR density in the main phase of shock-associated FE is presented. We consider FEs, caused by sporadic interplanetary disturbances accompanied by the interplanetary shock arrival. These FEs are characterized by a sudden decrease followed by the gradual recovery of the CR density that usually lasts for several days [13]. In the text below one can find a description of each parameter which is easy to get from the database.

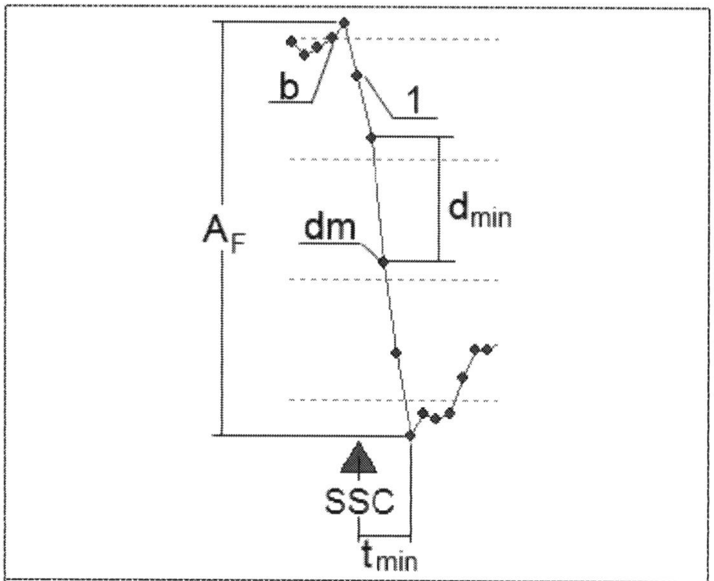

Figure 1. Behavior of the CR density in time at the beginning of FE. For explanation see text below.

In figure 1, obtained by GSM, FE characteristics are shown: A_F – magnitude of FE (the maximum variation of CR density for 10 GV rigidity); d_{min} – maximum one-hour decrease in CR density (maximum decrement, which is always negative); t_{min} – full time of the main phase of FE (time from the onset of event – SSC to the maximum decrease of CR density). The measured timings of the investigated FE time period are: zero hour (shock arrival, *i.e.* SSC), hour preceding the zero hour, denoted "b", and the hour following the zero hour, denoted "1". In addition, the timing of the maximum one-hour decrease in CR density, t_{dm}, and the timing of the maximum amplitude of the equatorial component of CR anisotropy ($A_{xymax}(p)$), t_{max} (not shown in the figure) with respect to zero epoch are considered. Correspondingly, we also consider phase dependence of $A_{xymax}(p)$ where p is the phase of anisotropy, and the phase distribution of the number of hours with maximum amplitude of CR anisotropy, $N_{max}(p)$. All of these parameters are found in the second IZMIRAN database for each FE. The time period between timing "b" and "1" is 2 hours for each of the events in the sample, whereas t_{dm} and t_{max} differ from one event to another. On the average for all of the chosen events, t_{dm} and t_{max} are 10.8 and 14.2 hours, respectively.

3. Results and discussion

Amplitude-phase dependence $A_{xyb}(p)$ and phase distribution $N_b(p)$ before the shock arrival, averaged for each phase by the 817 events, are shown in figure 2 in comparison with respective distributions during the quiet periods $A_{xyq}(p)$ and $N_q(p)$ (averaged for 158140 hours).

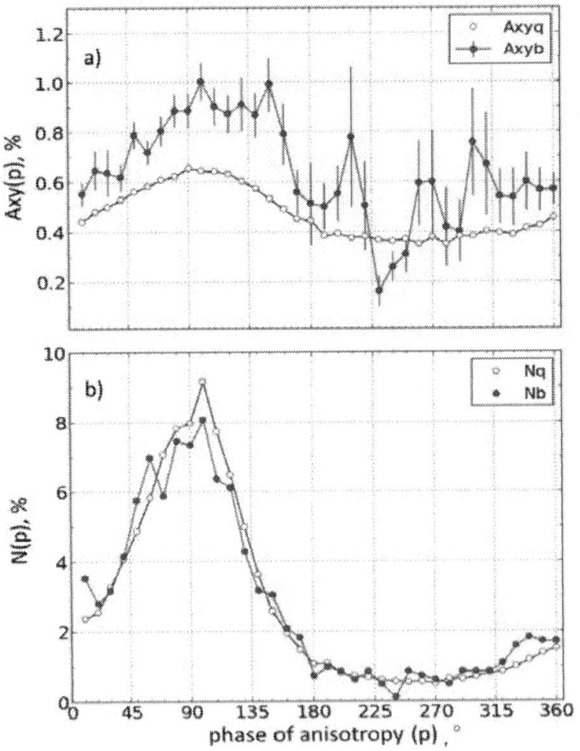

Figure 2. Amplitude-phase dependence (a) and phase distribution (b) one hour before the shock arrival (A_{xyb}, N_b) and during the quiet periods (A_{xyq}, N_q)

The hours of quiet periods were chosen from the first (CR variations) database by the following criteria: IMF intensity B<10 nT, geomagnetic activity index Ap<20 (2nT) and solar wind velocity

V<450 km/s. After excluding GLE effects from our sample 158140 quiet hours remained. One can see that number of hours for quiet parameters is much more than the number of the studied FEs, therefore errors in determination of q parameters are much less. This allows us to use more fine scale for plotting.

One can see that one hour before the shock arrival the amplitude of CR anisotropy ($A_{xyb}(p)$) substantially increased as compared with the usual quiet periods, and this increasing is observed at almost all longitudes (figure 2a). Therefore, it is reasonable to conclude that the interplanetary environment before FE is more disturbed, than during the quiet periods. Given the criteria for selection of quiet time periods, it is reasonable to assume a dependence of amplitude of CR anisotropy on intensity of IMF (B), solar wind velocity (V) and geomagnetic indices (e.g., Ap).

We aim to answer a question: Is the interplanetary medium more disturbed before the FEs than in the quiet periods? Thus, we calculated mean values of space parameters before (b) the FEs (see the table) and compared them to quiet (q) period values. It seems that conditions before the FE are somewhat more disturbed than in quiet periods. Although there seems to be connection between the change in the interplanetary conditions and A_{xy}, this should be further tested.

Table. Mean values of space parameters before the FEs and at quiet period.

	b	q	q_2
Ap_mean	9.3	5.6	
V_mean, km/s	410	370	
B_mean, nT	8.9	5.3	
A_{xy}, %	0.77	0.56	0.63

Will the anisotropy be the same amplitude if we select from quiet hours only the hours with characteristics close to those in the period "before" FEs? To answer this question we selected only quiet hours with parameters close to average values of parameters before the FEs, *i.e.* with a speed of solar wind V_{q2} = 400 - 420km/s, IMF intensity B_{q2} = 7 - 9 nT, and Ap_{q2} = 8 – 10 (2nT). A selection resulted in 629 hours. For this new sample the average value of amplitude of CR anisotropy is A_{xyq2}= 0.63% which is a little higher than A_{xyq} but substantially less than A_{xyb}. Thus, it is obviously, that revealed distinctions in interplanetary conditions are not enough to explain a difference in A_{xy} amplitudes.

Thus, only some part of a difference (A_{xyb}–A_{xyq}) can be explained with the change in the interplanetary conditions. The bigger part of increase in anisotropy just before the FE is most likely the precursor of arrival of interplanetary disturbance. This result confirms and completes the conclusions from [12]. If the disturbances are the same and interplanetary conditions before their arrival were constant, then we would see an increase of CR anisotropy in specific directions only. Rather large anisotropy at all phases indicates a variety of disturbances and situations before the interplanetary shock arrival. It characterizes pre-shock time period and results in a variety of precursors seen in CR measurements [e.g., 14].

In figure 2a we observe an increase in the amplitude of anisotropy corresponding to large phases. This implies the appearance of Sun-directed CR flows towards the disturbed region where the Forbush decrease occurs before the arrival of interplanetary disturbance. Nevertheless, in figure 2b it can be seen that phase distributions for quiet hours and the hour preceding the shock arrival are very similar. The behavior of the phase distribution, N(p) and amplitude-phase dependence, $A_{xy}(p)$ during the passage of the disturbance over the Earth was also investigated.

We compare the two for several different timings before and within the FE main phase: one hour before the shock arrival (A_{xyb} and N_b), one hour after the shock arrival (A_{xy1} and N_1), at the hour of the maximum one-hour decrease in CR density within the FE (A_{xydm} and N_{dm}) and at the hour of the maximum value of the CR anisotropy (A_{xymax} and N_{max}). We remind the reader that the hour of shock arrival is considered as zero hour. The results are presented in figure 3 (a, b).

One can see (figure 3a) that the maximum amplitude of the anisotropy ($A_{xymax}(p)$) is higher than all the others for all of the phases (2-3 times). It is high even in comparison with $A_{xydm}(p)$ value, which is observed already deeply in interplanetary disturbance. $A_{xymax}(p)$ has a large value in all directions, even in unusual anisotropy directions..

To explain such anisotropy within the simplified version of the convective-diffusive model a large negative radial gradient of CR density would be necessary [2; 3]. However in reality such anisotropy results from contribution of all components of the CR gradient in the IMF which can strongly differ from a quasispiral field of a quiet solar wind.

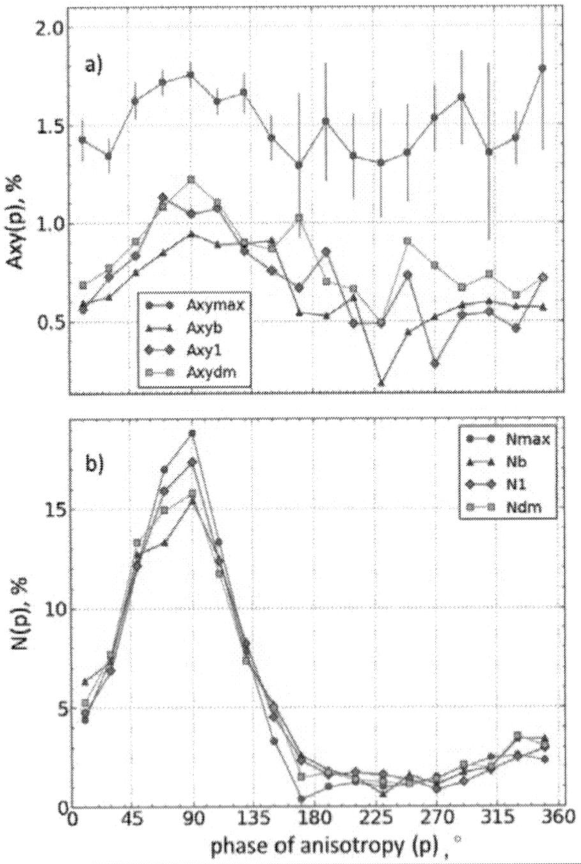

Figure 3 (a, b). Amplitude-phase dependence (a) and phase distribution (b) one hour before the shock arrival (A_{xyb}, N_b), one hour after the shock arrival (A_{xy1} and N_1), at the hour of maximum one-hour change in CR density (A_{xydm} and N_{dm}) and at the hour of maximum value of the anisotropy (A_{xymax} and N_{max}).

For phases up to 130° the anisotropy amplitudes are more or less consecutively increased as the interplanetary disturbance passes over the Earth ($A_{xyb}<A_{xy1}<A_{xydm}<A_{xymax}$). We note that the maximum of A_{xy} is considerably late with respect to the timing of the fastest fall of CR density (on average for 3.4 hours, see Section 2), which in turn has to be close in time with the maximum radial component of the CR gradient. Although it would be interesting to investigate this phenomenon, as well as the behavior of the latitudinal and azimuth components of the CR gradient and the role of the IMF

changes in the increase of the A_{xy}, this would require further analysis which is beyond the scope of this paper.

It can be seen in figure 3b that the phase distributions for all selected FE timings differ very little, however, we see that $N_b(p)$ is somewhat flatter than others, whereas $N_{max}(p)$ is the least flat distribution. We note that decreases in amplitude of anisotropy (figure 3a) and in number of events (figure 3b) around 165° occur not only due to FEs with amplitude $A_F<2\%$, but also due to large FEs.

The majority of distributed number of hours of the phase distributions are found within the phase range 45° - 135° (figure 3b): 69.3% for N_{max}, 61.4% for N_b, 63.2% for N_{dm}, and 66% for N_1. Similar calculations were carried out in [6] for phase distributions over the time period 1957 – 2010, where the majority of distributed number of hours (60.4%) were found within the phase range 40° - 130°. In general, the main features of the distribution remain the same at the entrance into the disturbance, as well as in the continuing hours throughout the main phase, even at the maximum amplitude of anisotropy, i.e. the preferred directions of anisotropy exist almost continually, even during the passage of the disturbance.

4. Conclusions

In the hours preceding an arrival of a shock wave, the amplitude of the CR anisotropy considerably increases compared to quiet periods, and this increase occurs in almost all of the phases of anisotropy.

The maximum amplitude of anisotropy (A_{xymax}) in a Forbush decrease is substantially higher than the anisotropy in an hour before the shock arrival, in an hour after the shock arrival and in an hour of maximum one-hour decrease of CR density.

For phases up to 130° the amplitude of anisotropy increases with time as the interplanetary disturbance passes over the Earth ($Axyb<Axy1<Axydm<Axymax$). In general, the main features of phase distribution are the same just before the FE and throughout its main phase, i.e. the preferred directions of the anisotropy exist even inside of interplanetary disturbances.

5. References

[1] Krymsky G 1964. Diffusion mechanism of daily variation of cosmic rays *Geomagnetism and Aeronomia* vol 4 N **6** pp 977-986

[2] Krymsky G, Kuzmin A, Krivoshapkin P, Samsonov I, Skripin I, Transky I, Chirkov N 1981 Cosmic rays and solar wind *Novosibirsk, Nauka* p 224.

[3] Forman M, Gleeson L 1975 Cosmic-ray streaming and anisotropies *Astrophysics and Space Science* vol 32 pp 77-94

[4] Belov A 1987 The first harmonic of CR anisotropy in the convection-diffusion model *Proc. 20-th ICRC (Moscow, USSR)* vol 4 pp 119-122

[5] Belov A, Dorman L, Eroshenko E, Oleneva V 1987 Cosmic ray gradient variations in 1974 *Proc. 20-th ICRC (Moscow, USSR)* vol 4 pp 137-139

[6] Abunina M, Abunun A, Belov A, Eroshenko E, Oleneva V, Yanke V 2013 Long-Term variations of amplitude-phase interrelation of the first harmonic of cosmic ray anisotropy *Geomagnetism and Aeronomia* vol 53 N 5 pp 601–610

[7] Belov A, Eroshenko E, Yanke V 1999 Global and local indices of cosmic ray activity *Proc. 26th ICRC (Salt Lake City)* vol 6 pp 472-475

[8] Belov A, Eroshenko E, Oleneva V, Struminsky A, Yanke V 2001 What determines the magnitude of Forbush decreases? *Adv. Space Res.* vol 27 N 3 pp 625-630

[9] Belov A, Baisultanova L, Eroshenko E, Mavromichalaki H, Yanke V, Pchelkin V, Plainaki C, Mariatos G 2005 Magnetospheric effects in cosmic rays during the unique magnetic storm on November 2003 *J. Geophys. Res.* **110** doi:10.1029/2005JA011067

[10] Lockwood J 1971 Forbush decreases in the cosmic radiation *Space Sci. Revs.* vol 12 N 5 pp 658-715

[11] Cane H 2000 CMEs and Forbush decreases *Space Sci. Revs.* vol 93 N 1-2 p 55

[12] Belov A 2009 Forbush effects and their connection with solar, interplanetary and geomagnetic phenomena *Universal Heliophysical Processes, Proc. Intern. Astronomical Union, IAU Symp.* vol 257 pp 439-450
[13] Abunin A, Abununa M, Belov A, Eroshenko E, Oleneva V, Yanke V 2012 Forbush-effects with sudden and gradual onset *Geomagnetism and Aeronomia* vol 52 N 3 pp 1-8
[14] Belov A, Dorman L, Eroshenko E, Iucci N, Villoresi G, Yanke V 1995 Search for Predictors of Forbush Decreases *Proc. 24th ICRC (Rome)* vol 4 pp 888-891
[15] Asipenka A.S., Belov A.V., Eroshenko E.F., Klepach E.G., Yanke, V.G. 2009 Interactive database on the cosmic ray anisotropy, *Advances in Space Research*, V. 43, Issue 4, p. 708-716

Acknowledgments
The authors are thankful to collaborators of the world wide network of CR stations providing data of continuous monitoring of the neutron component: http://cr0.izmiran.ru/ThankYou/. We acknowledge the NMDB database (www.nmdb.eu), founded under the European Union's FP7 programme (contract no. 213007) for providing data. We are also thankful to Bernd Heber (as referee) whose thoughtful comments contributed significantly to this paper.

Interrelation between energy and time distributions of high-energy electrons during the observation of the particle bursts in the near-Earth space

S.Yu. Aleksandrin[1], **A.M. Galper, S.V. Koldashov, T.R. Zharaspayev**
National Research Nuclear University MEPhI (Moscow Engineering Physics Institute)
Kashirskoye shosse 31, Moscow, 115409, Russian Federation

E-mail: SYAleksandrin@mephi.ru

Abstract. Many satellite experiments showed interrelation between changes of particle fluxes in the near-Earth space and various magnetospheric and geophysical phenomena. In this report we focus on temporal and energy characteristics of bursts of high-energy electrons in the inner zone of the Earth's magnetosphere ($L<2$). In order to study the variations of electron characteristics during the observation of the bursts, caused by local disturbances of the radiation belt (e.g. lightning or seismic events), numerical modelling the propagation of particle cloud formed by electrons, precipitated from radiation belt, has been carried out. It was shown a relationship between energy distribution and temporal profile of electrons of burst in case of their local precipitation. The results of simulation are analyzed and compared with the data obtained in ARINA and VSPLESK satellite experiments.

1. Introduction

High-energy electron bursts in the near-Earth space have been actively studied for the past 20 years. According to completed studies most of the particle bursts have shown to have geophysical origin and are associated with seismic and thunderstorm activities [1-9]. It is necessary to mention a number of satellite experiments: MARIYA, MARIYA-2 [1,2], DEMETER, PET/SAMPEX, POES, ARINA and VSPLESK [10-16], which are aimed to study electron bursts and geophysical effects causing them.

Physical model of electron burst formation is the following. Low-frequency electromagnetic emission is known to be generated a few hours prior to powerful earthquakes in their epicentres and above them. Propagating through atmosphere and ionosphere, it may cause the local disturbance of electron trajectories in the radiation belt (RB), making electrons to precipitate below RB into altitudes of low Earth orbital (LEO) satellite. The formation (cloud) of precipitated particles drifts along L-shell around the Earth if the altitudes of their local mirror points higher than ~100 km [17]. The disturbance envelopes entire L-shell with time period of longitudinal drift. For high-energy particles this time is ranging from a few dozens of seconds to several minutes. When a LEO satellite crosses the disturbed L-shell, on-board particle instruments detect a sharp short-term increase in particle flux (a particle burst). In this case the L-shell parameter of bursts coincides with the L-coordinate of the local RB disturbance. In fact, as shown in [2, 4, 14], study of particle burst characteristics gives the possibility to determine the location of RB disturbance and thus to search for the place of geophysical event (thunderstorm or earthquake).

[1] To whom any correspondence should be addressed.

Two experiments ARINA and VSPLESK are important to be distinguished from the others, because their acceptance are more than an order of magnitude than acceptance of instruments in other satellite experiments. It allows to identify weak bursts, and thus to increase the number of detected bursts. The ARINA and VSPLESK experiments are carried out on board the low-orbit spacecrafts. The ARINA instrument operates on board of Resurs-DK1 satellite with an altitude of 350-600 km and an orbit inclination of 70° since 2006[18]. The VSPLESK instrument measurements were carried out from 2008 to 2013 on the International Space Station (altitude is 350-400 km, orbit inclination is 51°) [14].

2. Numerical simulation of electron burst formation

As it was described above, the part of precipitated particles drifts along the L-shell around the Earth. Longitudinal drift velocity of particles depends on their energy. The average drift period (i.e., the time required to perform a complete rotation around the Earth) is [19]

$$T_{dr}^{earth} \cong K_1 \frac{(1+\varepsilon)}{\varepsilon\cdot(2+\varepsilon)}\cdot\frac{K_2}{L} \qquad (1)$$

where ε is the ratio of the kinetic energy of the particle to its rest energy, L – drift shell coordinate. In this work we used $K_1 = \frac{4}{3}\pi\frac{Z|e|}{m_0}\frac{M}{c^3 r_0}$ where M – the Earth's magnetic dipole moment, m_o – mass of the particle, $Z|e|$ – electric charge of the particle, r_o – distance from centre of the Earth. K_2 - the ratio of the drift period for particles with their mirror points at φ_m to that for particles moving in the equatorial plane only. The quantity of K_2 depends on latitude of a particle's mirror point φ_m, it varies from 1 to 1.5. K_2 =1.0 as it is on geomagnetic equator.

Particle drift time (T_{dr}) from the precipitation region to the place of registration of particles by a satellite is defined by the formula (2),

$$t_{dr}(\Delta\lambda) = T_{dr}^{earth}\frac{\Delta\lambda}{360}, \qquad (2)$$

where $\Delta\lambda$ is the longitudinal distance passed by the particle, $\Delta\lambda$ – in degrees. A drifting cloud of precipitated electrons contains the particles with various energies. Their energy spectrum is defined by interaction processes resulting to precipitation of electrons. In this work spectrum was set by power function $E^{-\gamma}$. Simulation of processes of electron cloud propagation and its registration by satellite, basing on the Monte-Carlo technique, include the following assumptions. The number of precipitated electrons, space position of particle precipitation zone and satellite orbit are given. The trace of each electron of the cloud from region of precipitation to place of its registration by satellite is calculated. RB disturbance is local. Particle precipitation from RB occurs instantaneous, it means that the time of precipitation (Δt_{pr}) is much less than longitudinal drift time ($\Delta t_{pr} \ll t_{dr}$). The energy range 3–30 MeV was chosen for simulation to compare its results with ARINA and VSPLESK experimental data, which have such energy range. Period of the longitudinal drift for electrons in this energy range is in the frame of 2-20 minutes. The spectrum of precipitated electrons is defined by power function $E^{-\gamma}$, with γ = 3.0, as it is observed in ARINA experiment [20]. Width of disturbed L-shell was taken $|\Delta L| = 0.07$ [5]. Under this condition the satellite crosses disturbed L-shell for 2-5 minutes in dependence on L value. Albedo electrons form background flux, which is taking into account in modeling. By tracing the motion of each electron, various statistical distributions of drifting cloud particles were evaluated. In particular, time profile and energy spectrum of a particle burst were analyzed.

Figure 1 is presented an example of temporal profile of simulated burst of electrons. Such shape of profile can be observed, if the satellite could be fixed in space at given place of disturbed L-shell. Longitudinal distance $\Delta\lambda$ between precipitation zone and place of burst registration by satellite, width (ΔL) disturbed L-shell has influence on total burst duration. In particular, the burst duration of a cloud of electrons passing 180° longitudinal distance to the registration point will be about 12 minutes. But in a reality the satellite moves and crosses fast the region of a drifting electron cloud less than several minutes [21]. In this case only a small part of cloud electrons can be registered by satellite. Figure 2 shows the modeling burst of electrons for crossing of disturbance L-shell by satellite.

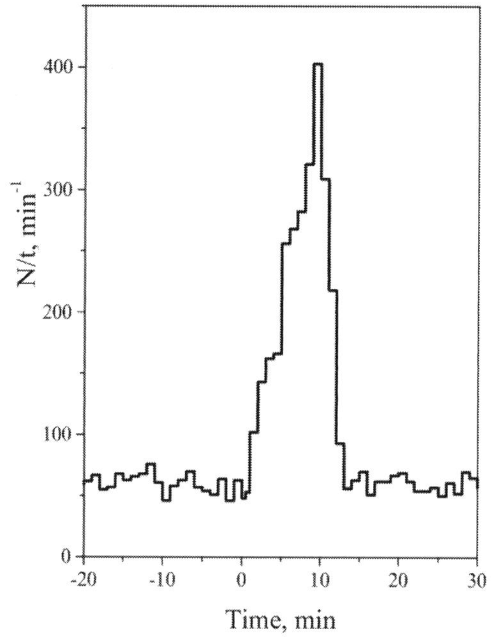

Figure 1. Temporal profile of an electron burst in ideal observation conditions for immovable satellite approximation (simulation result, L=1.2, $\Delta\lambda$=180°, T=0 - electron burst registration start time).

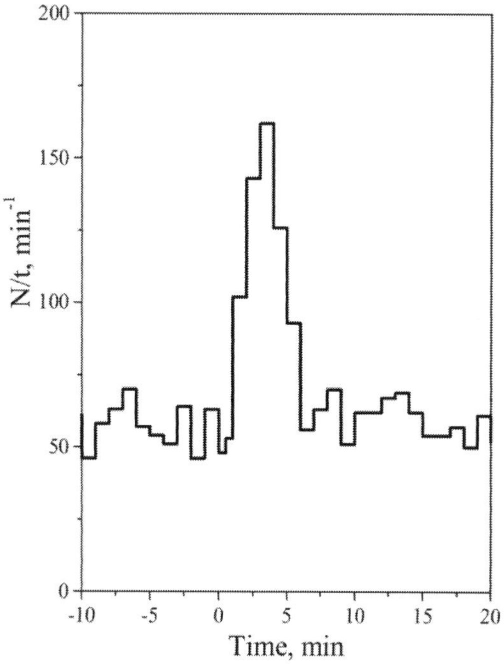

Figure 2. Temporal profile of an electron burst observing by orbiting satellite (simulation result, altitude 600km, inclination 70°, L=1.2, $\Delta\lambda$=180°, T=0 - electron burst registration start time).

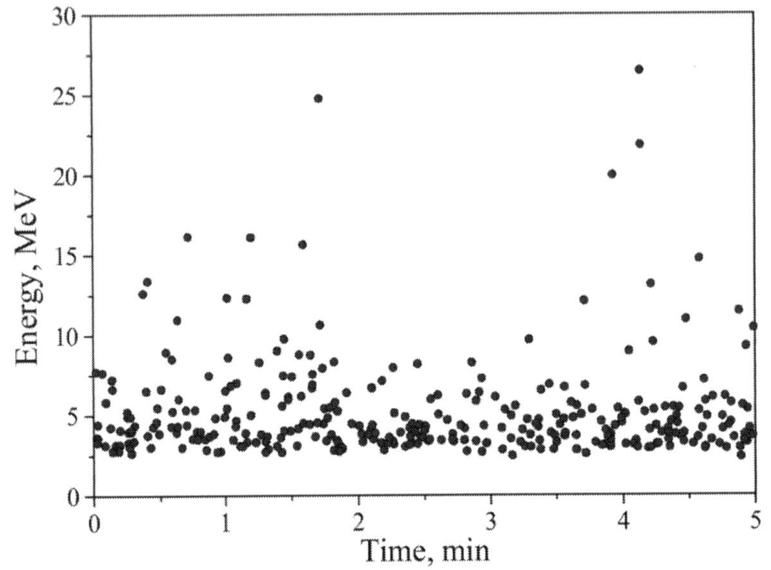

Figure 3. E-t electron distribution of albedo background.

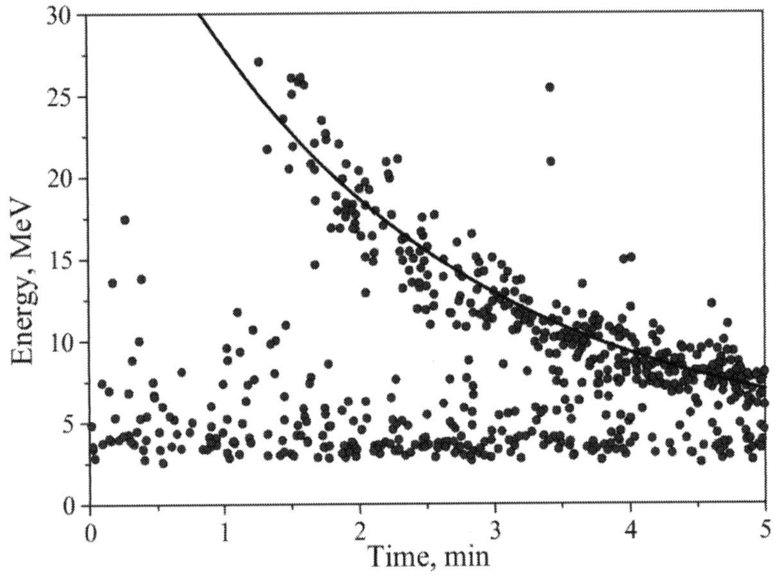

Figure 4. E-t electron distribution of simulated burst (L=1.2, $\Delta\lambda$=180°, T=0 – the moment of electron precipitation on L-shell).

In this work time profiles and electron distributions in E-t space for their various simulation parameters (L, $\Delta\lambda$, γ) were calculated. Analysis of the simulation results and their comparison with the experimental data is convenient to analyze in a two-dimensional energy (E) and time (t) space. The E-t distribution of albedo electrons is shown in Figure 3 [22], as you may see there is no correlation between electron energy and its registration time by instrument on board of the satellite. However, in case of burst particles their drift period is known to be dependent on their energy (1). Difference in particle drift velocity for different energies lead to expansion of particle cloud along the L-shell.

Evaluations show that the particles of burst can be well distinguished from background albedo particles (Figure 4) because high-energy burst electrons are registered on satellite first, and group along the line which is define by formula (2). Also it necessary to take into account that there is no accurate way to separate burst electrons from background albedo particle flux at the end of the burst.

3. Events selection criteria and experimental data analysis

The results of numerical simulation allowed developing basic criteria for burst selection in experimental data in order to plot time and energy distributions of particle bursts, their relationship with location of RB disturbance.

ARINA and VSPLESK experiments have collected database of several hundred electron bursts. A few bursts being of interest for correlation analysis *E-t* distributions were found in experimental data. As an example, in Figure 5 an electron burst observed with the ARINA instrument in August 24, 2009 (29° N, 2^oE, L=1.55) lasting about a minute is shown. Not high statistic in case of real experiment did not allow to plot *E-t* – diagram like in Figure 4. So it is necessary to calculate the average time of registration electrons for each energy channel of ARINA instrument. This *E-t* plot for the burst is presented in Figure 6. The free parameter ($\Delta\lambda$) of curve function (2) is longitudinal distance between location of RB disturbance and position of burst registration on board of satellite. The moment of particle precipitation is unknown in experiment, but the method only depends on characteristics of differential change of registered particles energy in time, so we selected the time $t_0 = 0$ to be the beginning of the burst and introduced additional free parameter t_{off} - time offset between start of burst registration and some unknown time of particles precipitation. An approximation curve on Figure 6 connecting particle registration time with its energy was obtained by regression analysis (ordinary least squares) [23].

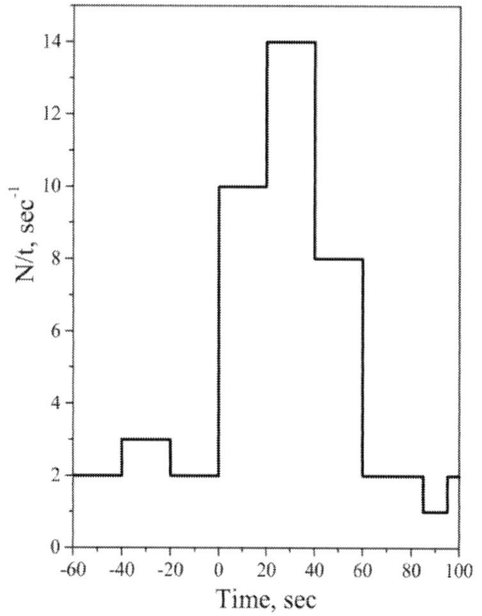

Figure 5. *E-t* electron registered distribution during the burst (the 24th of August, 2009) (29° N, 2^oE, L=1.55).

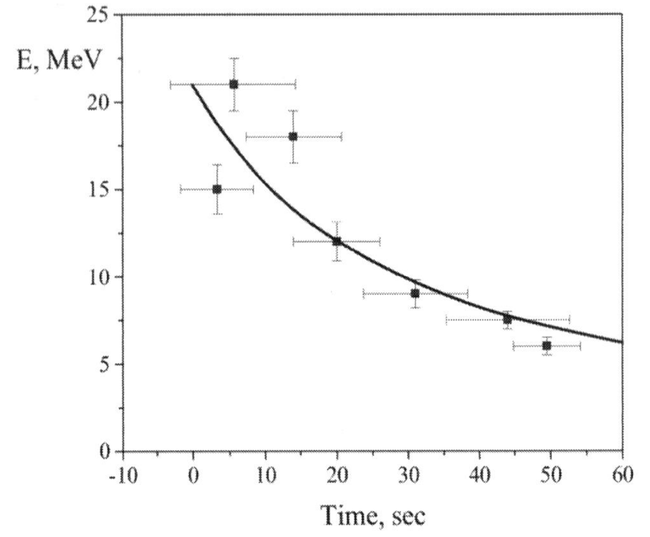

Figure 6. *E-t* electron distribution for the burst. (the 24th of August, 2009) (29° N, 2° E, L=1.55)

Figure 7 shows electron burst (marked with a black square on the Northern Africa), *L*-shell (grey line in the southern and northern hemispheres), which corresponds to the burst registration location and two estimated magnetically conjugated RB disturbance regions (marked in grey). One of them contains a part of the Mid-Atalntic ridge [24] and the other one includes region with high thunderstorm activity in South America [25]. Its relatively large size is defined by instrument characteristics (acceptance defines low particle statistics, and instrument energy resolution defines time of registration errors, etc.) and the width of disturbed *L*-shell where the burst was registered.

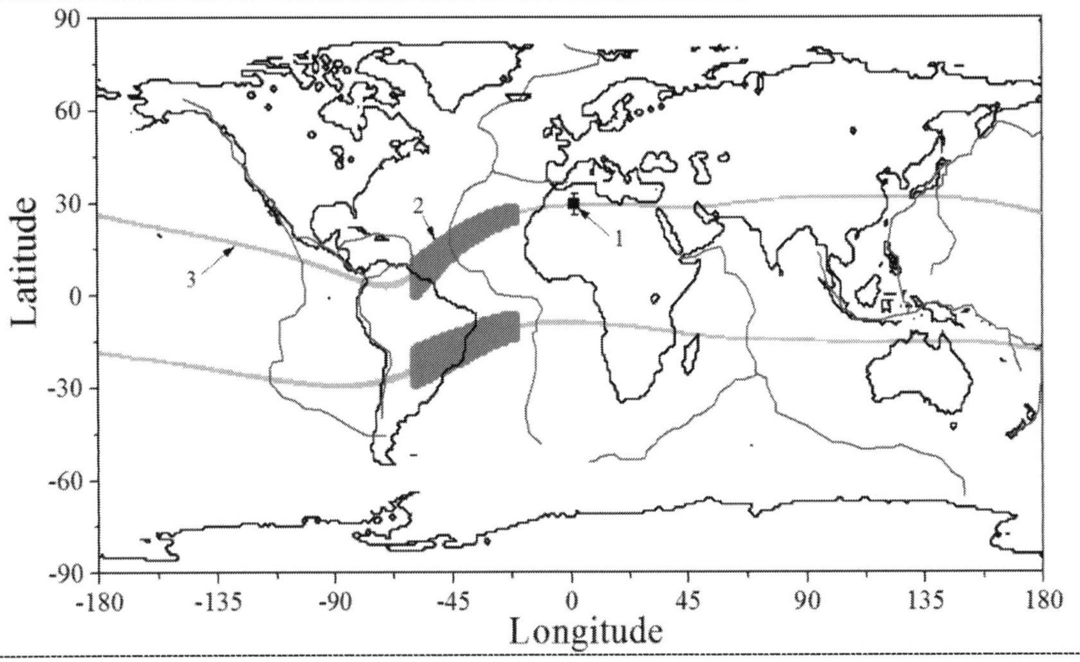

Figure 7. Black square (1)– location of electron burst registration (29° N 2° E) August 24, 2009, grey bands (2) –estimated places of particle precipitation, grey line (3) – *L*-shell (1.55), corresponding to the burst registration location.

4. Conclusion

Time and energy characteristics of high-energy electron bursts generated by the local RB disturbance of geophysical origin have been studied by numerical simulation of electron propagation in the magnetosphere. The possibility to determine RB disturbance region by measuring variation of particle energy distribution in time during the burst observation was examined.

The analysis of the electron burst obtained by ARINA experiment on the 24th of August, 2009 over the Northern Africa region was carried out. Using simulation results for this burst the magnetically conjugate regions of local RB disturbance were determined. Both of them are geophysically active. The north zone contains seismic Mid-Atlantic tectonic ridge, and another one is the region in South America characterized by heavy lightning activity.

References

[1]. Voronov S.A., Galper A.M., Kirillov-Ugryumov V.G., et al. *Registration of Sporadic Increase of High Energy Particle Flux Near Brasilia Magnetic Anomaly Region.* // Proc. 20-th Int. Cos. Ray Conf., 1987. V.4. P.451-454.

[2]. Aleshina M.E., Voronov S.A., Galper A.M., et al. *On the Relationship Between the Locations of Earthquakes Centers and High-energy Particles Precipitation Areas Under the Radiation Belt.* // Cosmic Research, 1992 v.30. pp.79-82.

[3]. Galperin, Yu.I., Gladyshev, V.A., Jorjio, N.V., Larkina, V.I., Mogilevsky, M.M. *Energetic particles precipitation from the magnetosphere above the epicentre of approaching earthquake.* // 1992, Cosmic Research, 30, 89– 106.

[4]. Boskova, J., Smilauer, J., Triska, P., and Kudela, K.: *Anomalous behavior of plasma parameters as observed by Intercosmos 24 satellites prior to the iranian earthquake of 20 June 1990,* // 1994, Studia Geophys. Geod., 38, 213–220.

[5]. Galper A.M., Koldashov S.V., Voronov S.A. *High Energy Particle Flux Variations as Earthquake Predictors.* // Advance Space Research. 1995. V.15. P.131-134.

[6]. Aleksandrin S.Yu, Galper A.M., Voronov S.A. et al. *Space and Time Characteristics of High-energy Charged Particles Bursts in Near-Earth Space.* // RAS Izvestiya, physical series. 2003 v.67. №4. pp.79-83.

[7]. Aleksandrin S.Yu., Galper A.M., Koldashov S.V., et al. *Nigh-energy Charged Particle Bursts in the Near-Earth Space as Earthquake Precursors.* // Annales Geophysicae. 2003. V.21. P.597-602.

[8]. Galper A.M., Koldashov S.V., Ulitin A.A. *Local Perturbations of Earth's Radiation Belt During the Seismic Event Progress in Japan, in March 11, 2011*//Brief Physical Reports, LPI, 2011. № 7. p. 41

[9]. G. C. Anagnostopoulos1, E. Vassiliadis1, S. Pulinets. *Characteristics of Flux-time Profiles, Temporal Evolution, and Spatial Distribution of Radiation-belt Electron Precipitation Bursts in the Upper Ionosphere before Great and Giant Earthquakes.* Annals of Geophysics, 55, 1, 2012

[10]. Sgrigna, V., Carota, L., Conti, L., et al. *Correlations Between Earthquakes and Anomalous Particle Bursts from SAMPEX / PET Satellite Observations.* // J. Atm. Solar-Terrestrial Phys. 2005. V.67. P.1448-1455.

[11]. Cristiano Fidani, Roberto Battiston, William J. Burger. A *Study of the Correlation between Earthquakes and NOAA Satellite Energetic Particle Bursts*, Remote Sensing 2010, 2, 2170-2184

[12]. M. Parrot, J.J. Berthelier, J.P. Lebreton, J.A. Sauvaud, O. Santolı́k, J. Blecki. *Examples of Unusual Ionospheric Observations Made by the DEMETER Satellite over Seismic Regions.* Physics and Chemistry of the Earth 31 (2006) 486–495

[13]. Shailesh Raghuwanshi1, D. K. Sondhiya1, Rahul Shrivastava1, S. K. Vijay and A. K. Gwal. Satellite *Monitoring of Ionospheric Disturbances Associated With Volcanic Eruption.*

International Journal of Engineering Research & Technology, Vol. 2 Issue 9, September - 2013

[14]. Aleksandrin S.Yu., Galper A.M., Koldashov S.V. et al. *Study of Local Radiation Belt Perturbations in the Satellite Experiments ARINA and VSPLESK.* Proceedings of the 31st National Conference on Cosmic Rays, Moscow, Moscow State University, 2010, GEO_9

[15]. Fidani C., Battiston R., Burger W.J. A *Study of the Correlation between Earthquakes and NOAA Satellite Energetic Particle Bursts.* // Remote Sensing, 2010. V.2. P.2170-2184.

[16]. Nithin Sivadas, Akshay Gulati, Deepti Kannapan, Ananth Saran Yalamarthy, Ankit Dhiman, Arjun Bhagoji, Athreya Shankar, Nitin Prasad, Harishankar Ramachandran, R. David Koilpillai, *A Nano-satellite Mission to Study Charged Particle Precipitation from the Van Allen Radiation Belts Caused due to Seismo-electromagnetic Emissions*. arXiv:1411.6034

[17]. Galper A.M., Koldashov S.V., Murashov A.M., *Numerical Simulation of Physical Processes at the Atmospheric Boundary of the Radiation Belt*, Cosmic Research, 2000, vol. 38, no. 1, pp. 102–107.

[18]. Bakaldin A.V., Batischev A.G., Galper A.M., et al. *Satellite experiment "Arina" to study the seismic effects in high-energy particle fluxes in the Earth's magnetosphere*. Cosmic Research, 2007, vol. 45, no. 5, p. 471.

[19]. H. Elliot, *The Van Allen Particles*, Reports on Progress in Physics, 26, pp.145-180, 1963.

[20]. Aleksandrin S.Yu., Bakaldin A.V., Batischev A.G., Bzheumikhova M.A., Galper A.M., Koldashov S.V., Solodskikh P.I., Sharonova N.D., Zharaspayev T.R., *Dynamic of flux of high-energy electrons and protons in the Earth's magnetosphere by satellite experiment data.* Proceeding of the XV Scientific Session of NRNU MEPhI, 2013, V.2, p.25

[21]. Aleksandrin S.Yu. *Space and Time Characteristics of Local Disturbances of High-energy Electron Flux under the Earth Radiation Belt.* Moscow, PhD thesis, MEPhI, 2011.

[22]. O.R. Grigoryan, M.I. Panasyuk, V.L. Petrov, V.N. Sheveleva, A.N. Petrov *Spectral characteristics of electron fluxes at L<2 under the Radiation Belts.* Advances in Space Research, V.42, №9 2008 pp. 1523-1526

[23]. Hastie T., Tibshirani R., Friedman J., *The Elements of Statistical Learning*, Springer New York Inc., 2001

[24]. Smith, D. K., J. Escartin, H. Schouten, and J. R. Cann, *Fault Rotation and Core Complex Formation: Significant Processes in Seafloor Formation at Slow-spreading Mid-ocean Ridges (Mid-Atlantic Ridge, 13-15°N),* Geochem. Geophys. Geosys., v. 9, 2008

[25]. Thomas, J. N., M. J. Taylor, D. Pautet, M. Bailey, N. N. Solorzano, R. H. Holzworth, M. P. McCarthy, M. Kokorowski, F. Sao Sabbas, O.Pinto Jr., S. A. Cummer, N. Jaugey, J. Li, N. J. Schuch. *A Very Active Sprite-production Storm Observed over Argentina*. EOS, Transaction, American Geophysical Union, Vol. 88, No. 10, 6 March 2007

Earth magnetic field as analyzer of the cosmic ray ion charge in MONICA experiment

A V Bakaldin, S A Voronov

National research nuclear university MEPhI, 31 Kashirskoe shosse, RU-115409, Moscow, Russia

E-mail: bakaldin71@mail.ru

Abstract. The future onboard experiment MONICA (monitor of cosmic ray nuclei and ions) is aimed for studying the cosmic ray ion fluxes from H to Ni in the energy range 10-300 MeV/n in vicinity of Earth. It's main scientific objective is the measurement of ion ionization states, as well as elemental, isotope composition and energy spectra of Solar Energetic Particle (SEP) fluxes. The method of the ion ionization state measurement based on the usage of Earth magnetic field as a separator of particle charge will be utilized in this experiment. The present contribution is dedicated to investigate the possibilities and features of this method for the MONICA experiment. The real time dependences of inverse square of cut-off geomagnetic L-shell (L_C^{-2}) on magnetic rigidity (R) in rigidity range 640-2780 MV were studied experimentally during powerful SEP events December 13-15 2006. The possibility to simulate these dependences using the modern geomagnetic field models is analyzed as well.

1. Introduction

New onboard experiment MONICA is aimed for studying the cosmic ray ion fluxes from H to Ni in the energy range 10-300 MeV/n in vicinity of Earth [1]. The experiment's main scientific objective is the measurement of ion charge states, as well as elemental and isotope composition and energy spectra of SEP fluxes for individual solar events and to study the evolution of these characteristics over time. At the same time MONICA will investigate ionization states and isotope composition of anomalous and galactic cosmic rays.

Ion charge states of SEP observed in near-Earth space are considered to be sensitive probes for conditions in acceleration region, such as, particle acceleration dynamics, solar plasma mean electron density and temperature [2, 3]. The ionization state of iron ions is the most informative and sensitive indicator for these processes [4]. It is known that the mean ionization state of iron ions with energies more then 10 MeV/n is about Q=+14 for "gradual" SEP events and more then Q=+20 for "impulsive" events [5, 6]. Thus, Fe mean charge state can be used for identification of SEP event type.

The investigations will be carried out with high-acceptance multilayer silicon telescope-spectrometer MONICA installed onboard satellite which will be launched into low Earth polar orbit with altitude about 600 km [1]. Instrument will be pointed to zenith. The spectrometer will detect nuclei in energy ranges: H and He – from 7 to 70 MeV/n, CNO group – from 10 to 150 MeV/n, Fe group – from 15 to 300 MeV/n. Instrument will measure nuclear charge (Z), mass, energy, and incident angle for each detected particle. The identification of registered nuclei will be implemented by modified ΔE-E method [7]. Spectrometer acceptance is about 100 cm^2sr, the field of view is ±45°.

In MONICA experiment the method of ion charge (Q) measurement based on the usage of Earth magnetic field as a particle charge separator will be realized. This method is unique for investigated energy range of ions. The direct measurement of charge state of cosmic ray ions with energies more than several MeV/n is impossible. The ions with initial charge state Q and energy >10 MeV/n that enter into the spectrometer will be stripped of orbital electrons within several μg/cm^2 of the instrument material, i.e. really inside the thin kapton windows installed above the first silicon detector of multilayer spectrometer. The ion stripping cross sections can be found in [8].

This contribution is dedicated to study the possibility and features of using Earth magnetic field as a separator of charge states of cosmic ray ions with energies from about 10 MeV/n up to several hundreds MeV/n for the MONICA experiment.

2. Method of ion charge measurement

Today the direct SEP ion charge measurements were carried out only for ions with energies less than 5 MeV/n. The electrostatic ion charge analyzers were used for these measurements (for example, ULEZEQ onboard ISEE-1, ISEE-3 [9] and SEPICA onboard ACE [10]). As a rule, these sensors combine the determination of the electrostatic deflection of incoming ions in a collimator-analyzer assembly by the measurement of the impact position on the detector plane and a dE/dx - E telescope. A direct measurement of the charge state is impossible for ions with energies more than 10 MeV/n because of the required electric field strength has to be unrealistic high and the analyzer acceptance has to be unrealistic large.

The unique method for the measurement of ion charge states for energies more than 10 MeV/n is using of Earth magnetic field as a separator.

It is well known that penetration depth of ions into the magnetosphere depends on their gyro radius. On a polar-orbiting satellite with zenith-pointing charge particle detector onboard, moving from a pole to the equator, this can be observed as a flux cut-off for particles with given magnetic rigidity R at a corresponding magnetic L-shell L_C. This effect is depicted in Fig. 1. This figure shows the dependence of registration efficiency $Eff(\Delta E)$ for ions detected by instrument, moving from the geomagnetic equator to pole, in energy range ΔE on L-shell. This efficiency is defined as the ratio of number of ions detected inside the magnetosphere $N_{in}(\Delta E)$ to their number detected outside it in interplanetary space $N_{out}(\Delta E)$.

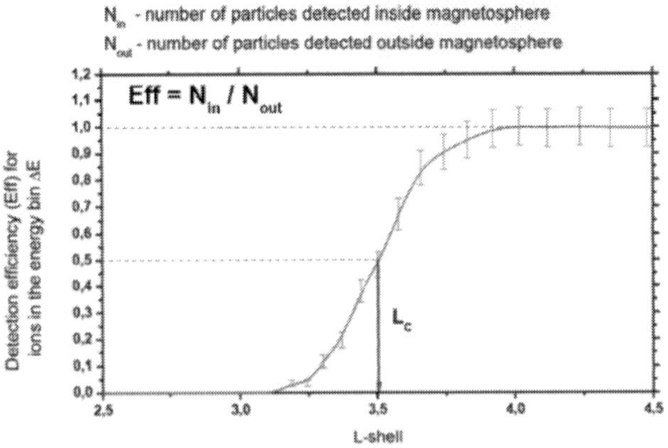

Figure 1. Detection efficiency of ions registered by instrument versus L-shell.

As it was assumed in [4], for the real magnetosphere the inverse square of this cut-off L-shell value L_C linearly depends on the particle rigidity R:

$$1/L_C^2 = aR + b\,.$$

This dependence can be used to determine the mean charge of the particles Q:

$$QR = P = \sqrt{(2M+E)E}\,,$$

$$Q = \frac{a\sqrt{(2M+E)E}}{1/L_C^2 - b}\,,$$

where E, $M=m_0c^2$ and $P=pc$ are the measured ion kinetic energy, the mass at rest and momentum in energy units (eV) correspondingly.

The main features of the method described above are:

1) The high statistics of detected ions is needed for the reconstruction of mean ion charge, and therefore the high-acceptance spectrometer is required.
2) It is necessary to know the geomagnetic cut-off rigidity in real time (i.e. a and b values), as Earth magnetosphere is distorted very often especially during SEP events.

The method of geomagnetic separator was successfully tested in SAMPEX experiment [4] for the solar events of October 30 - November 7, 1992. The mean values of ion charge of helium, carbon, oxygen, neon, magnesium and iron in the energy range 0.3-70 MeV/n were obtained. However, for ion energies more than 70 MeV/n, this method has not been experimentally tested.

In order to check the possibility of separator method using for MONICA energy range 10-300 MeV/n the following investigations have been carried out: 1) experimental study of real-time $1/L_C^2(R)$ dependences in the rigidity range corresponding to the energy interval of iron ions (the most informative solar energetic ions) detected by MONICA; 2) analysis of possibility to simulate real-time $1/L_C^2(R)$ dependences using the modern geomagnetic field models.

The rigidity interval of Fe ions (mean ionization state Q(Fe)=+14) detected by MONICA is 670-3230 MV during "gradual" SEP events and it is 470-2260 MV during "impulsive" events (mean ionization state Q(Fe)=+20). Since cut-off L-shell L_C depends on ion rigidity R only the proton (i.e., hydrogen ions with known charge Q(H)=+1) flux in the energy range 200-2000 MeV (corresponding to the rigidity interval 640-2780 MV) registered by PAMELA spectrometer [11] during SEP events on December 13-15, 2006 have been used for the study of real-time $1/L_C^2(R)$ dependences.

3. Experimental study of $1/L_C^2(R)$ dependence during the solar events with PAMELA data

Experimental study of the $1/L_C^2(R)$ dependence was carried out with the PAMELA experiment data [11]. PAMELA experiment is developed for investigation of cosmic ray antiproton, proton, positron, electron and nuclei fluxes in a wide energy range. Energy range of detected protons is 80 MeV - 700 GeV. The experiment has been carrying out since 15 June 2006 until the present time onboard of the Resurs DK1 satellite. Its orbit is elliptical and semi-polar with an inclination 70.4° and an altitude varying between 350 and 650 km. Magnetic spectrometer PAMELA has a field of view of ±20° and geometrical factor of 21 cm^2sr. The instrument is pointing to zenith. The detailed description of instrument is given in [11].

To study the geomagnetic cut-off rigidity the proton counting rates obtained by PAMELA during the powerful SEP events 13-15 December 2006 were used. This Solar event provided the necessary high statistics of detected protons for reconstruction of $1/L_C^2(R)$ dependencies and allowed to study particle penetration into the Earth's magnetosphere under different levels of disturbance.

In Fig. 2 (top panel) the dependence of the proton counting rate detected by PAMELA instrument is shown, as well as the integral proton flux measured by the GOES-11 satellite versus time during the period of December 13-15 2006. The same figure (bottom panel) illustrates the dependences of D_{st}-index, interplanetary magnetic field components B_y, B_z in GSM coordinates and dynamical pressure of solar wind P_{dyn} over time. As it can be seen from the graphs, these powerful SEP events were accompanied by a severe geomagnetic storm with maximum D_{st} variation of ~-150 nT that occurred at the end of the day December 14 when CME-driven interplanetary shock impacted on the Earth's magnetosphere. Time profile of PAMELA proton counting rate in the polar caps (peaks in counting rate in Fig. 2) qualitatively reproduces the proton flux profile obtained by GOES-11. Due to the PAMELA memory overflow the experimental data for about a day after the beginning of SEP event were lost.

The counting rate of protons in the six energy bins: 200-300, 300-500, 500-700, 700-1000, 1000-1400, 1400-2000 MeV for the reconstruction of $1/L_C^2(R)$ experimental curves were analyzed. The selection of these bins is defined by the SEP proton spectrum. For example, the SEP proton spectrum observed during the crossing of North polar cap at the SEP event beginning on 13.12.2006 is presented in Fig. 3. Galactic proton spectrum obtained for solar quiet period in December 2006 is also entered in the figure. The graph shows that at energies up to 2 GeV SEP protons dominate, while above 2 GeV GCR protons do. PAMELA counting rate of protons with energy more than 2 GeV is too low for our analysis. Hence the upper boundary of 2 GeV for the studied energy range was defined. The lower boundary of 200 MeV was chosen being corresponded to the minimum proton energy when the PAMELA instrument detection efficiency increases up to 1.

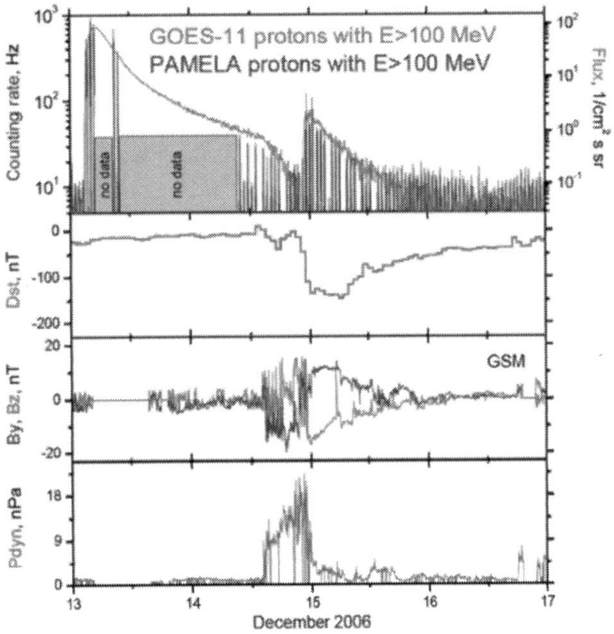

Figure 2. First (from top) graph: counting rate of protons detected by PAMELA instrument and integral proton flux for E>100 MeV measured by GOES-11 vs. time. Second graph: D_{st}-index vs. time. Third graph: interplanetary magnetic field components B_y, B_z in GSM coordinate system vs. time. Fourth graph: dynamical pressure of solar wind P_{dyn} vs. time.

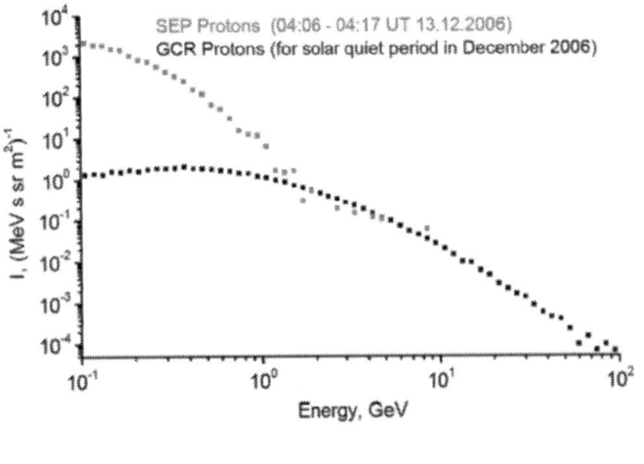

Figure 3. SEP proton spectrum observed by PAMELA during the crossing of polar cap at the SEP event beginning (red) and GCR proton spectrum for solar quiet period in December 2006 (black).

The experimental values of geomagnetic cut-off L-shell L_C were determined from obtained dependences of the proton detection efficiency in the Earth magnetosphere (Eff) on the L-shell (see Fig. 4). These dependencies on the L-shell were approximated by Boltzmann function:

$$Eff = \frac{A_1 - A_2}{1 + e^{(L-L_C)/\Delta L}} + A_2,$$

where A_1 and A_2 are the levels of the lower and upper plateaus (in this case 0 and 1 respectively), L_C - inflection point at half height, ΔL - transition distance from A_1 to A_2. To determine the mean rigidities R for the six selected energy intervals the proton spectra obtained in the polar caps were taken into account.

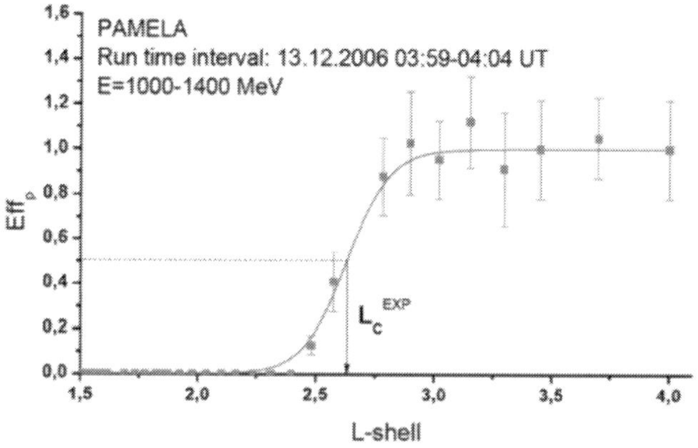

Figure 4. Experimental proton detection efficiency of PAMELA instrument in the Earth magnetosphere vs. the L-shell for run 03:59÷04:04 UT 13.12.2006.

In Fig. 5 the experimental $1/L_C^2(R)$ dependence for run 03:59÷04:04 UT 13.12.2006 is presented. As it can be seen from the plot, the dependence is practically linear in the studied range of rigidities 640-2780 MV. Dependences for all other runs have the similar shapes. As a result all experimental $1/L_C^2(R)$ dependences were approximated by linear functions: $1/L_C^2(R) = a \cdot R + b$. Table 1 shows the obtained a and b values (columns 3 and 4). Runs 1-7 comply with the terms of the undistorted Earth magnetosphere. Runs 8-13 were carried out during a strong magnetosphere disturbance produced by the CME-driven interplanetary shock that reached the Earth at the end of the day 14.12.2006.

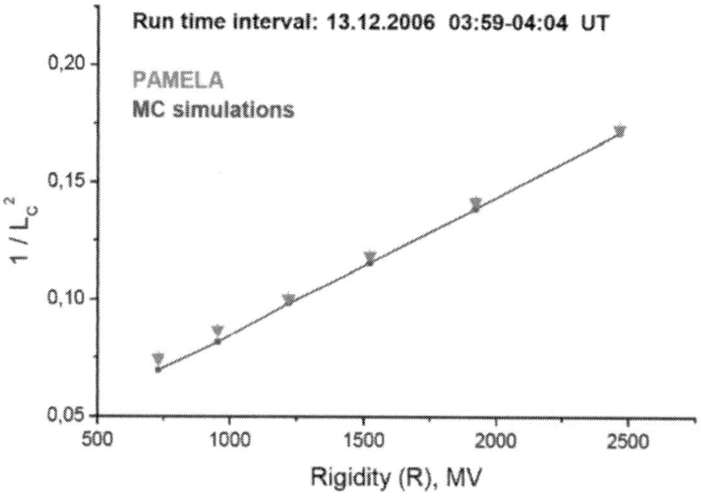

Figure 5. Experimental $1/L_C^2(R)$ dependence for run 03:59÷04:04 UT 13.12.2006.

Table 1. Experimental and simulated values *a* and *b*, obtained for the runs during solar events of 13-15 December, 2006.

Number of run	Run time interval, UT	a_{exp} ($\times10^{-5}$), 1/MV	b_{exp} ($\times10^{-2}$)	a_{sim} ($\times10^{-5}$), 1/MV	b_{sim} ($\times10^{-2}$)
1	13.12.2006 03:08÷03:15	5.3±0.2	3.0±0.2	5.7±0.2	2.7±0.1
2	13.12.2006 03:59÷04:04	5.7±0.2	3.2±0.2	6.0±0.2	2.4±0.1
3	13.12.2006 04:19÷04:24	5.3±0.2	3.2±0.2	6.1±0.14	2.0±0.2
4	13.12.2006 08:36÷08:42	6.0±0.3	2.3±0.3	6.3±0.2	1.7±0.1
5	13.12.2006 09:00÷09:05	6.0±0.3	2.9±0.2	6.6±0.2	1.3±0.1
6	14.12.2006 09:44÷09:50	7.5±0.4	1.4±0.4	6.5±0.2	1.8±0.2
7	14.12.2006 10:05÷10:11	5.9±0.6	3.4±0.5	6.5±0.2	1.6±0.1
8	14.12.2006 22:57÷23:03	6.2±1.6	4.5±1.4	5.0±0.2	8.5±0.2
9	14.12.2006 23:26÷23:31	6.0±1.4	5.6±1.3	5.6±0.2	6.8±0.2
10	14.12.2006 23:49÷23:54	4.8±2.3	8.9±2.3	6.1±0.5	6.8±0.6
11	15.12.2006 00:18÷00:23	4.6±1.8	6.7±1.6	5.6±0.3	6.3±0.5
12	15.12.2006 01:22÷01:28	-	-	6.0±0.4	6.3±0.6
13	15.12.2006 01:54÷01:59	5.0±2.5	7.1±2.1	5.6±0.4	5.5±0.6

4. Investigation of $1/L_C^2(R)$ dependence during solar events with Monte-Carlo simulations

The linear $1/L_C^2(R)$ dependence was also verified by simulations. Efficiency dependences of SEP proton detection in the Earth's magnetosphere on L-shell *Eff(L)* during the solar events on 13.12.2006 – 15.12.2006 were simulated by means of Monte-Carlo method.

Implemented calculations were based on the assumption that SEP proton flux in studied point located inside magnetosphere is equivalent to the omnidirectional flux outside magnetosphere and is isotropic. The influence of the Earth magnetic field was reduced to the appearance of forbidden directions only. This assumption is founded on Liouville's theorem. Therefore, evaluation of *Eff(L)* was reduced to determine the fraction of particles from allowed directions in the telescope field of view.

The evaluation of *Eff(L)* was carried out by the following algorithm. At the first step the isotropic flux of protons in PAMELA instrument field of view in the studied experimental energy interval (see previous paragraph) was modeled using the real proton spectrum measured with PAMELA during crossing the investigated polar cap (for example, see Fig. 3). As the result of this the statistics of occasional proton momentum vectors was collected. Then the penetration and physics interactions of the protons with modeled momentum vectors in PAMELA spectrometer frame were simulated with GEANT 4. The only events corresponding to trigger generating were taken into account. For each of six studied energy bins and each investigated polar cap the number of selected events $N_{out}^{sim}(\Delta E)$=200 was collected.

At the second step the fraction of events with allowed trajectory in geomagnetic field $N_{in}^{sim}(\Delta E)$ was defined among simulated events $N_{out}^{sim}(\Delta E)$. This was implemented for each part of satellite orbit corresponding to the edge of investigated polar caps. To define the allowed events among the simulated ones the tracing of particle trajectories in geomagnetic field was carried out. Since the path of a positively charged particle of a specific magnetic rigidity is identical (except for the sign of the velocity vector) to that of a negatively charged particle reaching the same location in space, the common method of calculating cosmic ray trajectories in the Earth's magnetic field is to calculate the trajectory in the reverse direction. Thus the starting point and initial direction for the reverse trajectory calculation were given by the PAMELA geographic coordinates and by the opposite direction of simulated particle momentum vector, correspondingly. The tracing of particle trajectories was fulfilled by the numerical integration of charged particle motion in geomagnetic field equation with fourth order Runge-Kutta method. The geomagnetic field was modeled as superposition of internal (defined by IGRF model) and external (defined by Tsyganenko96 model) sources. The presence of Earth's solid body and it's atmosphere was took in to account. The real values of solar wind pressure, D_{st}-index and the interplanetary magnetic field components B_y, B_z for the investigated period (see Fig. 2) were used as input parameters for the Tsyganenko96 model. For each investigated point in the satellite orbit and the energy bin ΔE we have simulated $N_{out}^{sim}(\Delta E)$=200 particle trajectories. A calculated trajectory was considered as forbidden if particle was lost in Earth's atmosphere or could not escape magnetosphere for the fixed maximum path length, otherwise the event was considered to be allowed. Finally the simulated efficiency of SEP proton detection in the Earth's magnetosphere by PAMELA instrument was obtained as: $Eff(\Delta E) = N_{in}^{sim}(\Delta E)/N_{out}^{sim}(\Delta E)$.

Fig. 6 shows the simulated dependence *Eff(L)* for a period corresponding to one of the experimental runs. It is seen from Fig. 4 and Fig. 6, that experimental and simulation curves have a similar shape.

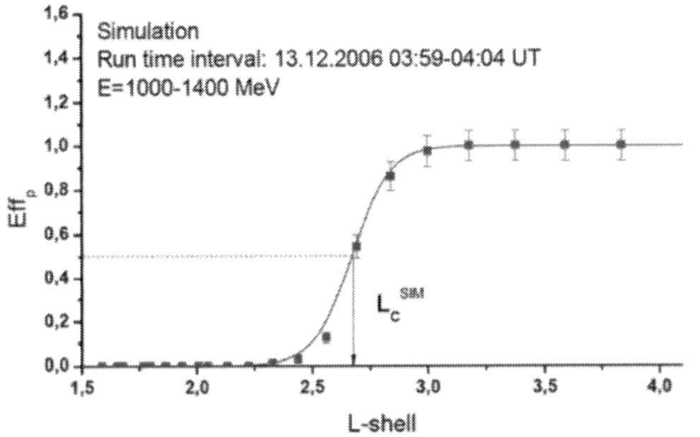

Figure 6. The simulated dependence of *Eff* vs. *L* for 03:59÷04:04 UT, 13.12.2006.

In the Fig. 5 the simulated $1/L_C^2(R)$ dependence is presented together with experimental one, obtained for the same run 03:08÷03:15 UT 13.12.2006. As it can be seen from the graph, the simulated dependence is also linear and practically coincides with the experimental one. Table 1 (columns 5 and 6) summarizes the *a* and *b* coefficients obtained from simulations.

5. Analysis of obtained results

Experimental and simulation $1/L_C^2(R)$ dependences obtained for the undistorted conditions, and under conditions of strongly disturbed Earth magnetosphere, are close to be linear in the studied range of rigidities 640-2780 MV. The linearity of these curves gives the possibility to use the geomagnetic separator method for charge measurement of heavy ions with energies up to several hundreds of MeV/n.

Fig. 7 presents the dependence of experimental and calculated *a* (left) and *b* (right) coefficient values on time during the solar events of December 13-15, 2006. Grey color on the graph marks Earth magnetosphere strong distortion period (D_{st}<-60 nT). These dependences show that magnetosphere distortion has moderate influence on the coefficient *a* value (i.e., slope), and at the same time has a significant impact on the coefficient *b* value (i.e., the initial intercept).

As it can be seen from the Fig. 7 plots, the slopes *a* of experimental and calculated $1/L_C^2(R)$ dependences are in agreement. The coefficients *b* are also in agreement, but are not so good. Thus obtained in the simulation $1/L_C^2(R)$ dependences can be used for practice in realization of geomagnetic separator method. To precise (calibrate) the initial intercept of evaluated dependences *b* in real experiment the proton counting rate for at least the only one low-energy bin can be used. In MONICA experiment for this calibration the counting rate of protons will be measured with simple proton monitor [1] in three energy (rigidity) bins: 20-30 MeV (195-239 MV), 80-90 MeV (396-421 MV), 110-120 MeV (467-489 MV). Particularly this correction is needed during the measurements under conditions of strongly disturbed Earth magnetosphere. Some discrepancy between the experimental and simulated data can be associated with the use of the external Earth's magnetic field model Tsyganenko-96 in simulations, which does not take into account the dynamic magnetosphere distortion parameter variations depending on changes in the external environment.

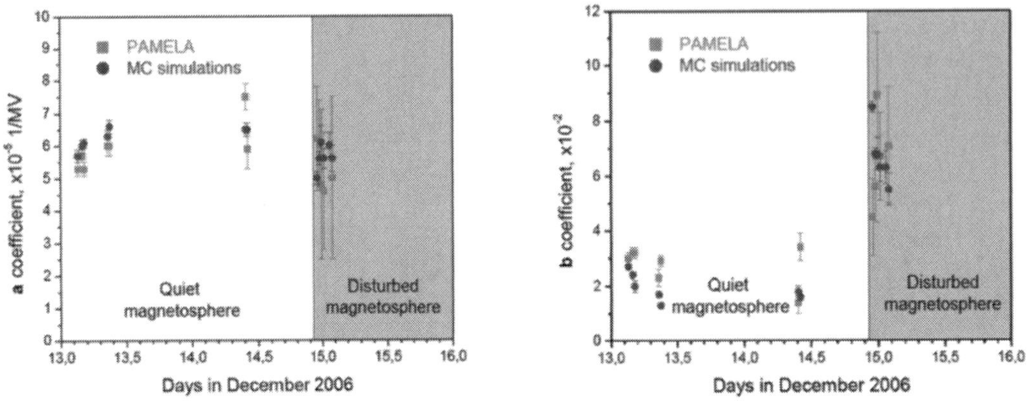

Figure 7. The experimental and simulated a (left) and b (right) coefficient values versus time obtained during SEP events 13-15 December 2006.

6. Conclusion

In this paper the in-depth analysis of ion detection features in the Earth magnetosphere was carried out. The behavior of the dependence of inverse square of geomagnetic cut-off L-shell L_C on the particle rigidity R during the powerful SEP events 13-15 December, 2006 was studied experimentally and by simulation for different magnetosphere disturbance degrees. It was shown that the $1/L_C^2(R)$ dependence is linear in the studied range of rigidities 640-2780 MV. This suggests the possibility of geomagnetic separator method using for charge measurement of heavy ions with energies from 10 MeV/n up to several hundreds of MeV/n.

Good agreement of the experimental and simulated $1/L_C^2(R)$ dependences suggests that dependences obtained by simulation using the latest Earth's magnetic field models, can be used in MONICA experiment as well as in other future experiments for the measuring of the cosmic ray ion charge with the geomagnetic separator method.

This work was supported by the RFBR, grant 13-02-00477.

References

[1] Bakaldin A, et al. 2009 *Proc. 31st ICRC* http://icrc2009.uni.lodz.pl/proc/pdf/icrc0387.pdf
[2] Ostryakov V, Stovpyuk M 1999 *Solar Physics* **189** 2 357
[3] Kartavykh Y, Ostryakov V 1999 *26th ICRC* **6** 272
[4] Oetliker M, et al. 1997 *Ap.J.* **477** 495
[5] Reames D 1999 *SpaceSci.Rev.* **90** 413
[6] Reames D 2013 *SpaceSci.Rev.* **175** 1-4 53
[7] Bakaldin A, et al. 2010 *Instruments and Experimental Techniques* **53** 4 490
[8] Shevelko V, et al. 2009 *J. Phys. B: At. Mol. Opt. Phys.* **42** 065202
[9] Hovestadt D, et al. 1978 *IEEE Trans. Geosci. El.* GE-**16** 166
[10] Mobius E, et al. 1998 *SpaceSci.Rev.* **86** 447
[11] Bongi M, et al. 2004 *IEEE Trans. Nuc. Sci.* **51** 3 854

Upgrade of Apatity Neutron Monitor

Yu Balabin, E Vashenyuk, B Gvozdevsky, and A Germanenko
Polar Geophysical Institute, Apatity, Russia.
E-mail: balabin@pgia.ru

Abstract. The neutron monitor (NM) in Apatity has been deeply upgraded in the end of 2013. We developed and installed new amplifier-discriminators. The detecting tubes of NM were tested and calibrated with additionally using of a pulse-amplitude analyzer. Due to this operation electric noise and interfering pulses are reduced. The NM was equipped with a new rapid data acquisition system. The system registers each NM pulse with time accuracy of 1 microsecond. This gives a possibility to investigate such fast phenomena as, for example, multiplicities in NM. Moreover, using these detailed data, it is possible to produce not only a standard NM count rate (number of pulses per minute) but (if necessary) a count rate with any high time resolution. Based on the detailed data we implemented the software calculation of so called "large dead time" data, which previously was done by hardware.

1. Hardware modernization

The neutron monitor (NM) in Apatity is a conventional 18-NM-64 instrument with SNM-15 tubes filled with BF_3. Here we omit well-known facts about SNM-15 (BP-28 analogue). Using of SNM-15 discharged tube is not deviated from producer recommendations: negative high voltage supply, low input impedance of an amplifier.

In the end of 2013 the NM has been upgraded. As a first step of the modernization, new amplifiers were made in order to replace the outdated ones. New amplifiers are designed using modern electronic components. Because of this, the performance and noise immunity of the amplifier has increased significantly.

The second step was a calibration of all counters/amplifiers. The usual method of NM calibration is to obtain the “voltage – count rate” characteristics, like the one shown in Fig.1. The operating high voltage must be set at the middle of the characteristics plateau (around 2300 V in this case).

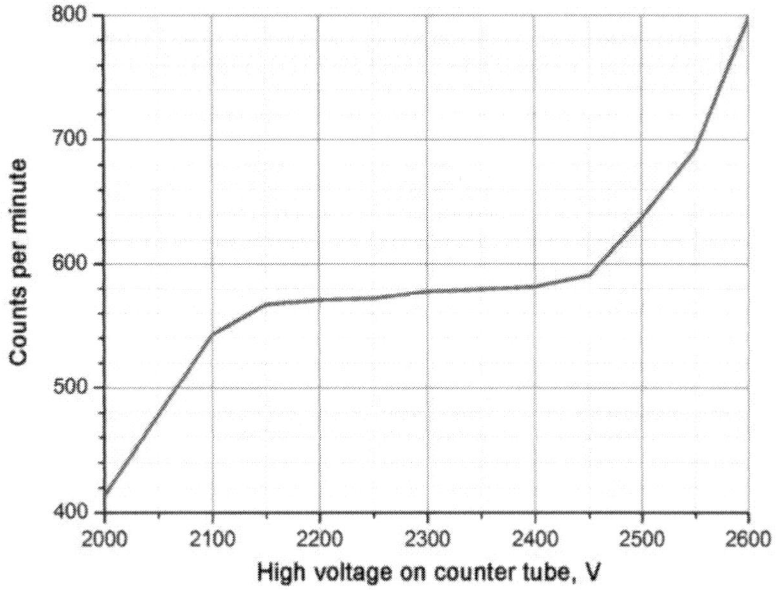

Figure.1. The “high voltage – count rate” characteristics of a NM counter (tube).

We developed a new calibration methodics. Measurement and analysis of an amplitude spectrum of output pulses were included. The spectrum was obtained by a pulse-amplitude analyzer. The sample of the amplitude spectrum is shown in Fig.2. A NM amplifier-discriminator has a cutoff level of pulse amplitude to reduce noise. When the gain factor is optimal, the cutoff level (it is shown red line) falls on the gap between noise and real pulses. Higher gain factor expands the blue graph to the right, and some noise pulses pass through the discriminator because their amplitudes become above 1 V. Lower gain factor presses the blue graph to the left, and some real NM pulses become less then cut off level. Such mismatch between the gain factor and the cutoff level can be a source of latent long-term NM instability. A small variation of any parameter (gain factor, cutoff level, low voltage supply etc.) causes unexplained NM data floating. The cutoff level of a conventional NM is 1 V (red line in Fig.2), while the gain factor is set uniformly ~10^3. Usually these two values are not consistent with individual tube parameters. So for any amplifier an individual adjustment of the gain is necessary. Only after pulse-amplitude spectrum measurement on each tube the gain factor can be set exactly into the gap, as it shown in Fig.2.

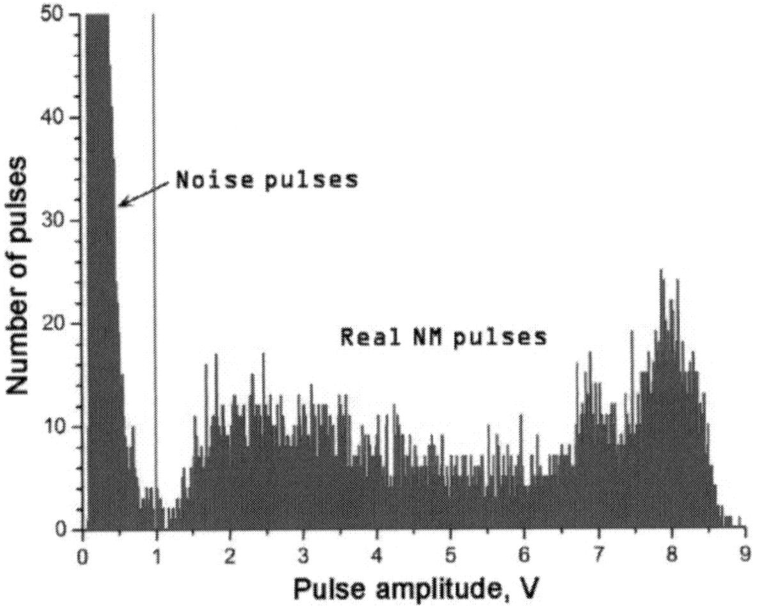

Figure.2. The amplitude-spectral characteristics of a NM counter. The discriminator cutoff threshold is set to 1 V (red vertical line). When the cut off level and gain factor are consistent, the discriminator cutoff threshold is on the gap on the pulse-amplitude spectrum.

The two above mentioned characteristics allow us to set the high voltage and gain factor exactly to stable operating point.

2. Advanced data acquisition system

The traditional way to gather NM data is to register a count rate, i.e. a total number of pulses per time unit (typically one minute). These standard data do not contain information on pulse sequences and distribution within a time unit interval. We call this way 'integral way' or standard.

The second way is to register exact time of each pulse coming. We equipped our registration computer with a multichannel high-speed digital input card Adlink PCI-7348, and our data collecting program registers each NM pulse: in which tube it occurred and how many microseconds elapsed since the previous pulse. The time resolution is 1 μs. These detailed data give much more information about cosmic rays and processes in a neutron monitor. We call this way 'differential way' or detailed. Schematic it is shown in Fig.3.

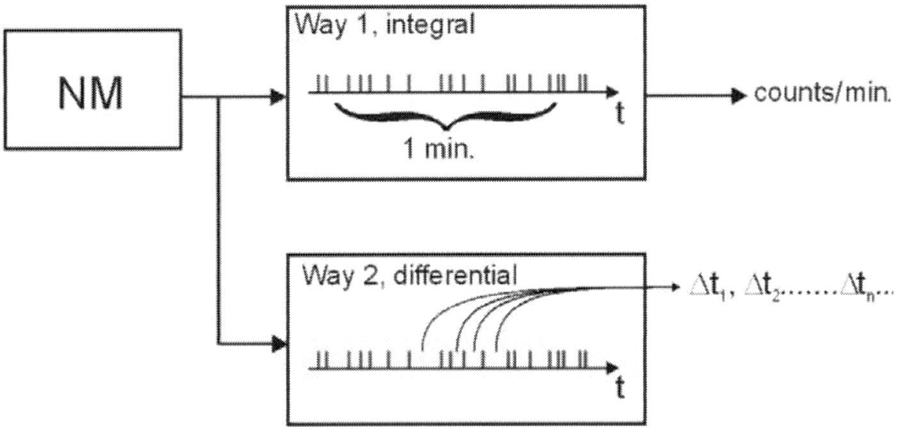

Figure.3. Two ways of NM data registration: integral and differential. The integral way produces minimum information – just a count rate. The differential way gives full information on NM pulses.

The NM in Apatity now registers both standard and detailed data. At the present moment the same registration system is set also in Baksan (Northern Caucasus) and Barentsburg (Spitsbergen) stations. The system has several advantages.

1) The 'integral way' doesn't allow to obtain NM data with time resolution less than standard registration interval (typically 1 min). Original data of the 'differential way' can be processed (anytime later) and count rate with any time resolution down to 1 second or less, if necessary, can be derived. For example, GLE69, 20 Jan 2005, was so huge and sudden (increase was up to thousands percents within some minutes, for example [1]) that time resolution less than 1 minute could be useful.

2) The detailed data are useful for study of multiplicity events (multiple productions of neutrons in NM). A multiplicity event in NM data is a cluster of pulses with short intervals between them. Number of pulses in each multiplicity event, distribution of pulses in time and space (distribution on NM tubes) can be investigated [2, 3]. Now such study is conducted on three equipped stations only. It would be good to setup our system on some stations in other places: low latitude, high mountain etc.

3) Most other NMs register data with a small dead time (SDT) 10 μs. Apatity station from the its very beginning, since 1969, registers also large dead time (LDT) data. Large dead time is 1200 μs. The LDT data more correctly represent the primary cosmic ray flux. Earlier LDT data were produced by hardware. But this hardware is difficult to adjusting and not used on other stations. We decided not to interrupt this long data series. Now it is automatically calculated by software using detailed data. Our system is setup on three stations and LDT processing may be done on their data.

4) Examining detailed data we can reveal some specific false NM pulses. For example, a sequence of pulses with equal time interval between them – periodic interference. Such false pulses give a negligible impact on the total count rate, so they cannot be seen in the standard data. But they are serious obstacle for the multiplicity study when a series of pulses

(multiplicity event) is searched. Due to the detailed data the periodic interference can be easily recognized and rejected.

5) It is easy to integrate any additional measuring unit to our system and join other data with accuracy 1 μs. For example, on the Baksan station a NM and Extended Air Shower (EAS) detector are set in one building. Pulses of these two units are registered by our system and can be compared with high accuracy. Pulses from EAS detector are used as a marker of EAS among NM pulses. In this data flow a hadronic core of EAS falling down to NM can be separated and studied [4].

3. Conclusion

The data acquisition system of the NM in Apatity is significantly upgraded. All electronic equipment from the tube-detector to the registering computer was replaced. The modern high-speed amplifiers, the new circuit design and the high-speed digital I/O cards were used. The software was upgraded too. The main goal of the modernization is to register 'detailed' information about pulses. The registration of the 'standard' NM information is not changed. With a new high-speed system registering the detailed information the capabilities of our neutron monitor as a cosmic ray detector are significantly expanded.

Acknowledgment. The work is supported by the RFBR grant 12-02-01305-a.

References

[1] Perrez-Peraza J, Vashenyuk E, Gallegos-Cruz A, Balabin Yu, Miroshnichenko L 2008 Relativistic proton production at the Sun in the 20 January 2005 solar event *Advances in Space Research* **41** 947–954.

[2] Balabin Yu, Gvozdevsky B, Maurchev E, Vashenyuk E, Dzhappuev D 2011 Fine structure of neutron multiplicity on neutron monitors *Astrophys. Space Sci. Trans.* **7** 283–286

[3] Balabin Yu, Gvozdevsky B, Germanenko A, Vashenyuk E 2013 Observing of atmospheric hadronic shower on the Barentsburg neutron monitor *Journal of Physics: Conference Series* **409** 012053

[4] Balabin Yu, Gvozdevsky B, Vashenyuk E, Dzhappuev D 2011 EAS hadronic component as registered by a neutron monitor *Astrophys. Space Sci. Trans.* **7** 507–510

EPHIN anisotropy measurement capability

S Banjac[1], R Gómez-Herrero[2], B Heber[1], P Kühl[1] and C Terasa[1]

[1] Christian-Albrechts-Universität zu Kiel, Germany, [2]Universidad de Alcalá de Henares, Spain

E-mail: banjac@physik.uni-kiel.de

Abstract. The EPHIN instrument (*Electron Proton Helium INstrument*) forms a part of the COSTEP experiment (*COmprehensive SupraThermal and Energetic Particle Analyzer*) within the CEPAC collaboration on board of the SOHO spacecraft (*SOlar and Heliospheric Observatory*). The EPHIN sensor is a stack of six solid-state detectors surrounded by an anti-coincidence. It measures energy spectra of electrons in the range 250 keV to > 8.7 MeV, and hydrogen and helium isotopes in the range 4 MeV/nuc to > 53 MeV/nuc. In order to improve the isotopic resolution, the first two detectors have been segmented: 5 sectors form a ring enclosing a central segment. This does not only allow to correct the energy-losses for particles with different path-lengths in the detectors, but allows also an estimation of the arrival direction with respect to the sensor axis. For that purpose we developed a method that allows for inferring the angle of incidence and angular distribution for ions. Here we describe the method and apply it to the November, 3, 2011 event. Due to the lack of magnetic field measurements and the restricted view cone of 83°, it is not possible to derive a real pitch angle distribution during this event. However, we can show that the particle distribution is anisotropic for several hours with a symmetry axis that deviates by about 20° from the sensor axis.

1. Introduction

A careful analysis of *solar energetic particle* (SEP) events allows in principle to derive fundamental information about the acceleration, injection into and propagation in the interplanetary medium. Observation, not only of intensity-time profiles, energy spectra and the elemental and ionic charge composition, but also of the temporal evolution of the pitch angle distribution are determined by the combination of the above mentioned steps. Although often neglected, detailed measurements of the pitch angle distribution are essential and have to be considered by the time the energetic particles have reached the Earth. On a spinning spacecraft a full pitch angle coverage is achieved by dividing a full rotation into several sectors. 3-axis stabilized spacecraft require many sensors or sectored detectors to measure the pitch angle distribution.

1.1. Instrumentation

The NASA/ESA *SOlar and Heliospheric Observatory* (SOHO) was launched in 1995 and is in orbit about the L1 Lagrangian point since 1996. The scientific payload of the spacecraft consists of several remote and in-situ instruments including the *Electron Proton Helium Instrument* (EPHIN, Müler-Mellin [1]), a particle telescope with a field of view of about 83° that measures electrons with energies between 0.25 and above 8.4 MeV as well as protons and Helium in the energy range of 4.3 to beyond 53 MeV/nucleon [1].

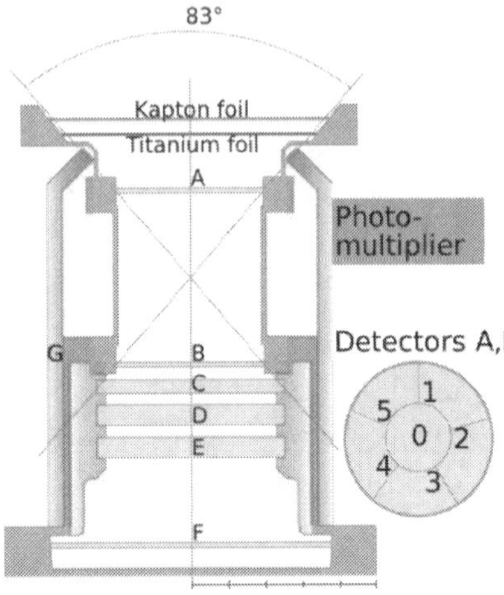

Figure 1. EPHIN schematics, reproduced from [3]

Figure 2. 1MeV electrons are shown in red, 20MeV protons are shown in blue, secondary gammas in green.

A schematic view of the EPHIN instrument is shown in figure 1. The instrument consists of a stack of six solid-state-detectors (labeled A-F) that are surrounded with a scintillation detector (G), which acts as active anti-coincidence detector. Stopping ions are identified by applying the $\frac{dE}{dx} - E$-method [2]. Since the data rate is limited, energy losses in detectors A-E can only be transmitted for a statistical ensemble of all particles counted by EPHIN that needs to be normalized with respective count rates. A crude separation between electrons, protons and helium has been introduced for different penetration depths, i.e. different coincidences by introducing different energy-loss thresholds in detector A, for details see [1].

In order to improve the isotopic resolution and to decrease the deadtime of the instrument during solar energetic particle events with high count rates, the first two detectors (A and B) consist of six individual sectors (see figure 1, right). Knowing which detector sectors were penetrated can be exploited to reduce the path length uncertainty. Here we show that those sectors can be used to distinguish in a qualitative manner if a particle distribution is anisotropic or not. To decrease the deadtime, at times of high count rates, the outer ring segments are disconnected. This reduces the geometric factor from $5.14cm^2sr$ to $0.18cm^2sr$ (first two proton and helium channels).

1.2. Data Structures

The EPHIN data are provided as different levels and file formats. Our analysis uses the Level-2 rate and pulse height files. The rate files contain counting rates for individual particle species plus an additional channel for penetrating particles (the integral channel). In addition, the rates are divided according to the energy channel and - in case of protons and helium isotopes - according to trajectory inclination relative to the sensor axis. The energy channels are defined by the penetration depth in the sensor, e.g. the channel P4 predominantly counts protons that stop in the second detector. The inclinations are given by the extension GM, GR and S.

Here P4 GM designates the protons with energy between 4.3 and 7.8 MeV which, when passing detectors A and B, have passed through the central sectors (sectors A0 and B0 as designated in Fig 1). Notice that only particles with trajectories whose inclination Θ relative to the sensor axis is less then 21° fit into this category. The GR group contains rates of trajectories passing through corresponding sectors A1 and B1 or ... or A5 and B5 having Θ between 0 and

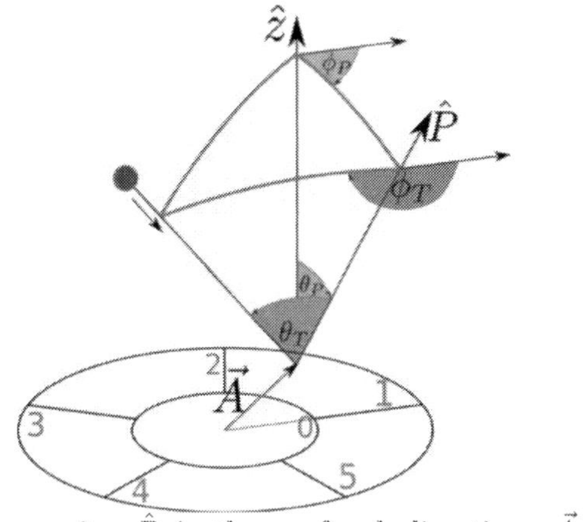

Figure 3. $\hat{P}$ is the prefered direction, $\vec{A}$ represents the penetration point on det. A

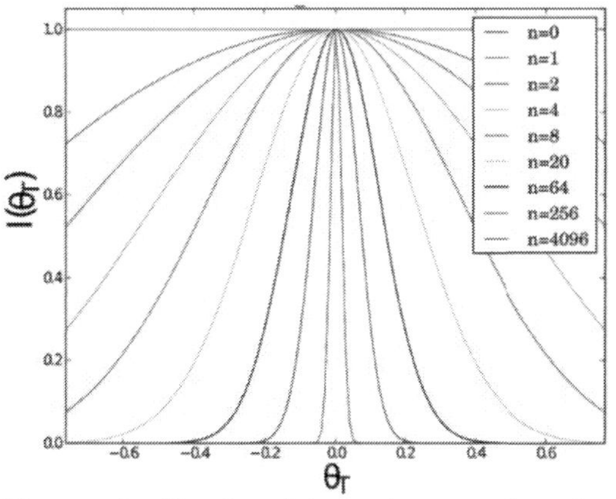

Figure 4. Simulated intensity as a function of θ_T (relative to θ_P, ϕ_P) for various n

28 degree. Finally, the S rates contain trajectories with Ax, By where $x \neq y$ and Θ between 0 and 42 degree. To improve the isotope separation, for Helium, the S rate has been split into S1 and S23 rates. S1 contains trajectories where either A0 or B0 (but not both at once!) with $\Theta \in \{0° \dots 34°\}$. S23 contains trajectories with Ax and By where $x, y \in \{1 \dots 5\}; x \neq y$. In total, there are 16 Helium count rates per observation interval.

The pulse height files contain detailed energy deposition per detector and the detector A and B segment information. Number of pulse-height analyzed particles is only a statistical sample. Helium events are given priority (designated with the priority flag).

The pulse height and count rate files are interelated, as are the developed methods. In most cases we use both to achieve results.

2. Directional response function

To determine the directional response function of the instrument we developed a model based on a GEANT4 [4] Monte Carlo simulation. For this purpose we modeled the instrument and exposed it to an isotropic particle flux. We then modified the sample flux by making it anisotropic by introducing a preferred direction of incidence, defined by angles θ_P and ϕ_P, and a polar angle distribution relative to the preferred incidence direction $\cos^n \theta$. In this model, the particle flux contains trajectories defined with θ_T, ϕ_T relative to the prefered direction of incidence which are selected randomly with the total flux having a polar angle distribution defined by n (see figure 3).

This has been done systematically for $\theta_P \in \{0°, 10°, 20°, 30°\}$, $n \in \{0, 1, 2, 4, 8, 20, 64, 256, 4096\}$ and $\phi_P \in \{0°, 36°\}$. The value of n defines how "beamlike" the flux is, as shown in figure 4. An isotropic flux is given by $n = 0$. A flux in which all particles have the same direction would have $n = \infty$, since $\lim_{n\to\infty} \cos^n(x) = 0$ for $x \notin [0, \pm\pi, \dots]$. This particular set of n's defines several intermidiate fluxes . θ_P modifies the inclination of the preferred incidence direction relative to the sensor axis. $\phi_P = 0°$ represents a flux with preferred azimuthal direction of incidence parallel to the border of sectors 1 and 5 , while $\phi_P = 36°$ describes a flux bisecting the first sector (see figure 1).

We assumed that the particles move along straight lines. That is, when a particle deposits energy in a detector, its direction is not changed. Simulation results show that electrons within the measurement range are strongly scattered when interacting with the detector matter (compare electron trajectories shown in red with the one of protons shown in blue in figure 2). Thus the developed method is not applicable to the electron channels, since it does not consider scattering processes. The simulation shows that the change of direction for ions is negligible

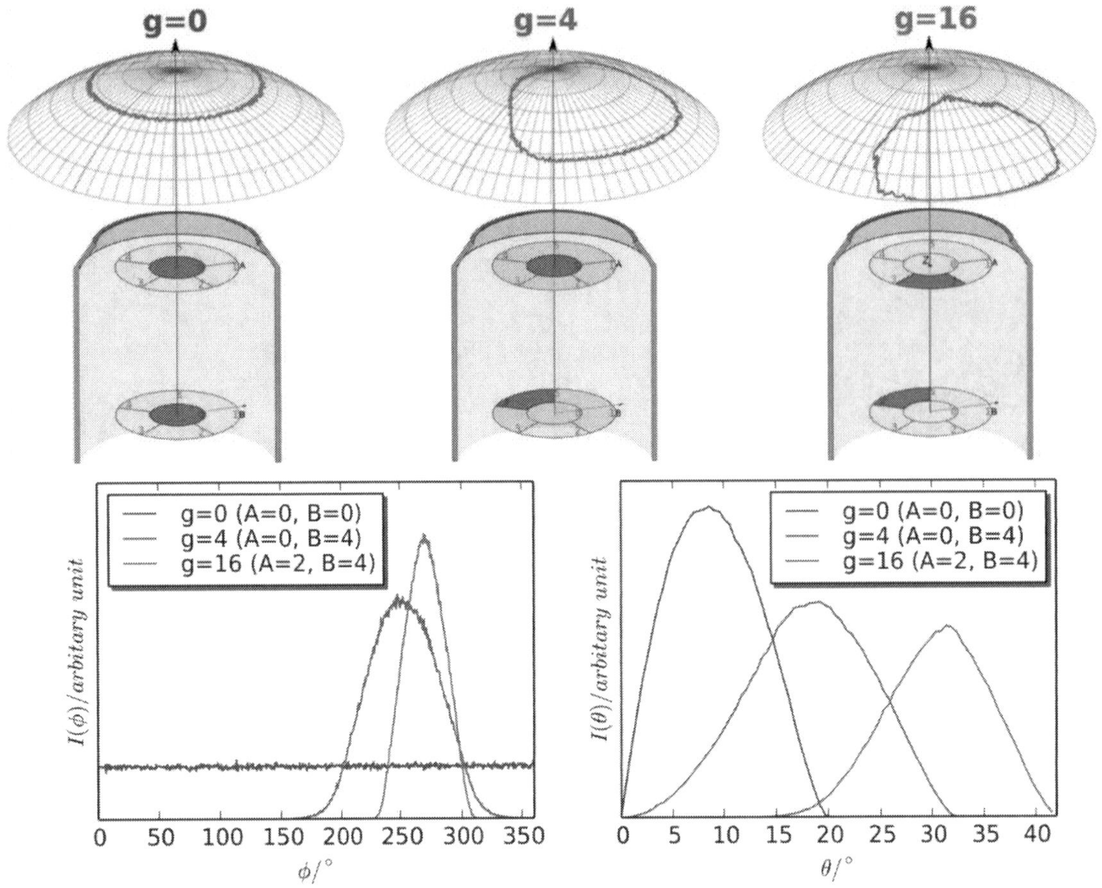

Figure 5. The aperture of 3 virtual detectors is projected onto a spherical surface at infinite distance. The virtual detectors are designated with g (see quation (2)). The lower panels show the ϕ and θ sensitivities of the virtual detectors.

(blue trajectories in figure 2). Simulations have been performed using non-interacting particles (geantinos). The method is an extension of the Monte Carlo determination of geometric factor, as described by Sullivan [5].

Our guiding idea is that separate counters can be viewed as individual particle telescopes.

2.1. The rate data

Figure 5 shows on top the areas covered on the sky by 3 different sector combinations. The lower left panel shows that for A=0,B=0 no sensitivity in ϕ is found. However, the combinations A=0,B=4 and A=2,B=4 cover similar ϕ ranges with a peak at 230° and 260° respectively. From the right panel it is obvious that all 3 combinations cover different θ ranges. Thus when combining all sector combinations with A=0,B $\neq$ 0, the ϕ dependence is lost. The same is true for all sector combination groups found in the rate files. Calculating all different combinations and grouping them accordingly to the EPHIN data gives the result in figure 6. We have also divided the relative intensity by the number of sector combinations within each group.

The next step involved reproducing the EPHIN data products. We used our simulation results to calculate the ratios of the three proton variables and tried to map them to the flux parameters θ_P, ϕ_P, n. In figure 6 we see that the sensitivities of the GM and GR groups are very similar. Therefore we combine them in a single count rate G. The proton flux is now characterized by only two variables: the countrates G and S. We derived a single quantity - the *count rate factor*

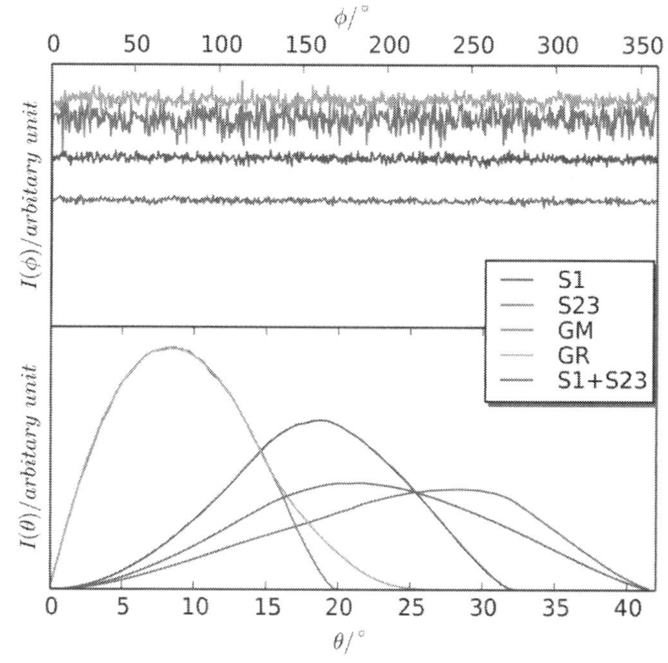

Figure 6. ϕ and θ sensitivity divided by the number of sector combinations per group

F - suitable for visualization.

$$F \equiv \exp\left[-\frac{G}{S}\right] = \exp\left[-\frac{GR + GM}{S1 + S23}\right]. \tag{1}$$

Figure 7 shows how F depends on the simulation parameters n, θ_P. The simulation showed no significant dependence of F on the third simulation parameter ϕ_P.

For some sets of simulation parameters, F is unique enough to allow direct mapping. This is the case for areas designated U and W, which correspond to strong particle beams. Areas V and Y have similar F but correspond to completely different sets of simulation parameters. Finally in the area between U, V and Y we have to use the pulse height data, since no information about θ_P or n can be recovered from the rate data.

2.2. The pulse height analysis (PHA) data

In figure 7, for $\theta_P = 20°$ we have very similar F for $n = 1$ and $n = 8$. Using the detailed information that are available for a statistical sample of all measured events, we can discriminate between such cases. A particle measured in segment x and y of detector A and B, respectively, could be assigned to group:

$$g = 6 \times x + y. \tag{2}$$

When grouped in this way, we have defined 36 virtual telescopes. For each of those, a sensitivity matrix R_g representing the response similar to figure 5 has been generated. Each response matrix holds information about the relative directional acceptability of each virtual telescope. Figure 5 shows for three combinations the sensitive region in the sky and its response function in theta as well as in phi.

This *pulse file method* can help us in cases were the *rate file method* leads to ambiguity. If we multiply the response matrix R_g of each sector combination g with the actual count rate n_g

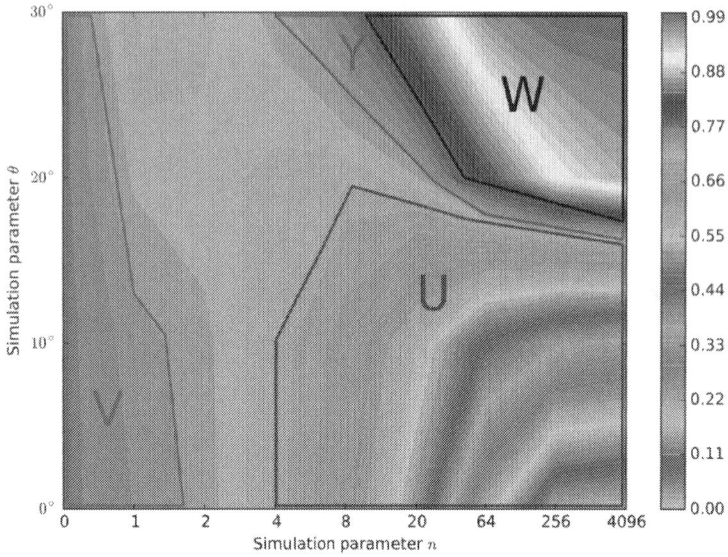

Figure 7. The simulated count rate factor F as a function of simulation parameters n and θ_P. Due to the decreasing sensitivity for $\theta > 32°$ (see lower right panel of figure 5), the analysis has been limited to $\theta \in 0°, 10°, 20°, 30°$.

Table 1. RL2 analysis of the 03.11.2011 SEP event

Period	F	$\theta/°$	n	θ_{ACE}
A	0.73 ± 0.03	$\in [15, 25]$	$\in [14, 42]$	$(33 \pm 5)°$
B	0.74 ± 0.02	$\in [15, 25]$	$\in [14, 42]$	$(35 \pm 15)°$
C	0.76 ± 0.04	$\in [0, 5]$	$\in [0, 1]$	$(127 \pm 13)°$

during some observation interval, we get the flux intensity as a function of the azimuthal and polar angles ϕ, θ. An example of this procedure can be seen in figure 12. A flux with values of n higher than 2 has to show dominant peaks in the intensity plot. Based on this, we can investigate if the *count rate factor* F is assigned to area V or Y.

It is sometimes helpful to observe only the ϕ profile of a given flux, namely $\sum n_g \cdot \Phi_g$ with the associated poissonian error $\sum \sqrt{n_g} \cdot \Phi_g$. The error is indicated in figures 9-11 by the thickness of the graph. Careful analysis shows that this is an error overestimation, since several of the 36 virtual detectors have similar sensitivity matrices (e.g. GR group) and their count rates could be added together to improve statistics. By fitting a $\cos^n(\phi)$ function within the thick graph of such ϕ flux profiles, we can approximately determine n.

The final algorithm exploits both data sets. First the count rate factor F is calculated from the rate data. This usually reduces the number of possible simulation parameters n, θ_P to which the data can be mapped. Then, the *phi* flux profile is reconstructed from the PHA data and finally n is approximated. After crossreferencing the results, we are able to characterize the incident flux. This is successful in cases with good statistics, i.e. high count rates. Sometimes the observation interval has to be increased at the cost of resolution. If pulse height analysis was available for all events, we would have to work only with the PHA data.

3. Application

3.1. SEP event of 3th of November, 2011

As an example we analyzed the proton flux of the SEP November 3th 2011 event (measured with EPHIN on 04.11.2011). This event has been discussed by Gómez-Herrero (see [6]). Figure

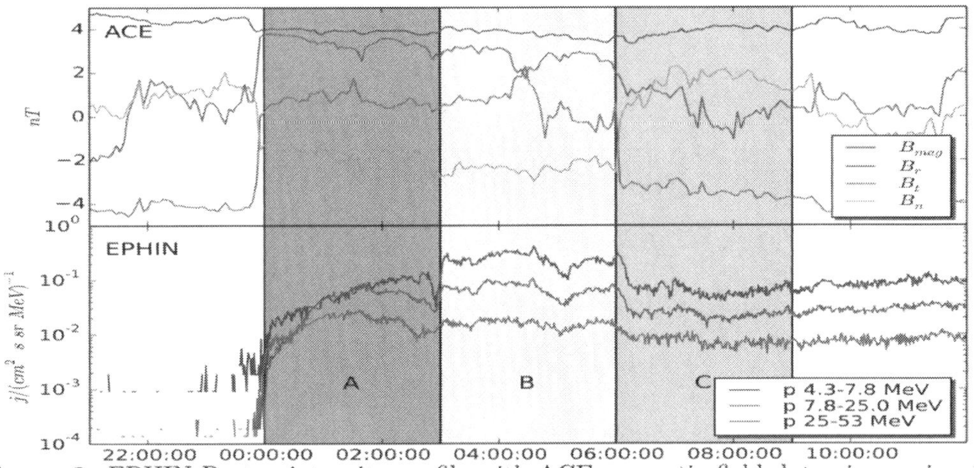

Figure 8. EPHIN Proton intensity profile with ACE magnetic field data given using the RTN (Radial Tangential Normal) coordinate system

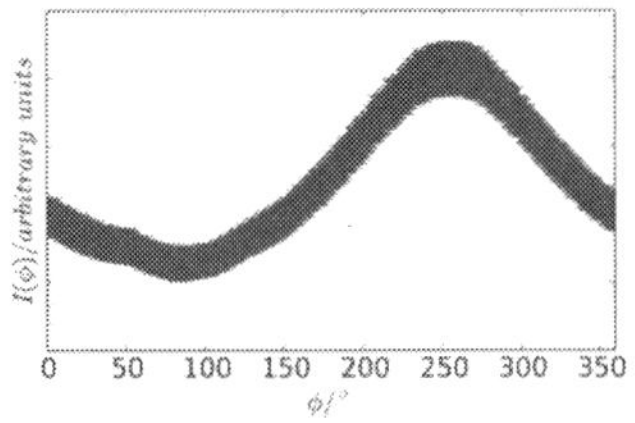

Figure 9. A [00:00-03:00]

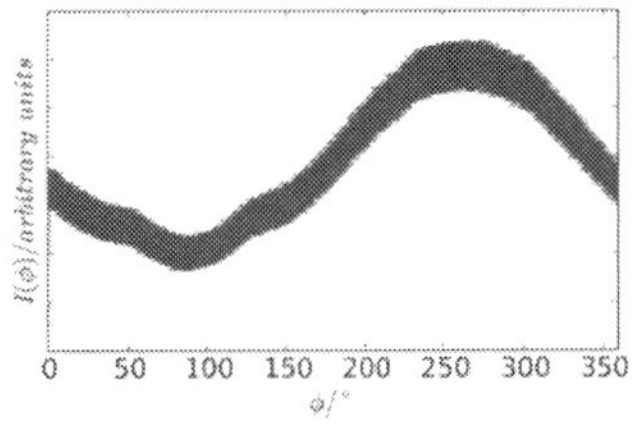

Figure 10. B [03:00-06:00]

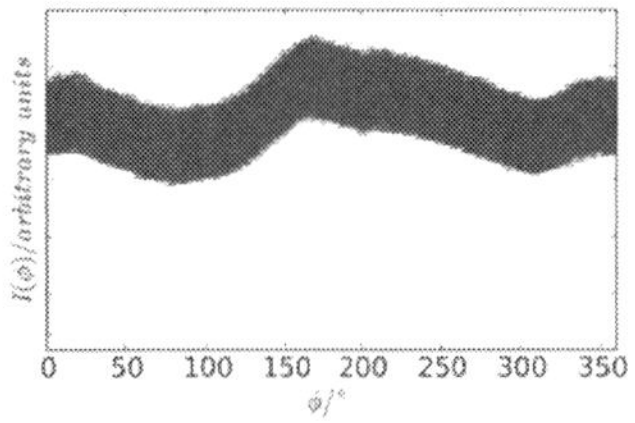

Figure 11. C [06:00-09:00]

8 displays at the bottom the intensity profile of the event. The top shows the local magnetic field at L1 measured by ACE (Advanced Composition Explorer). Measurements with WIND spacecraft (not shown here) indicate very similar magnetic field directions. It is important to note that ACE, SOHO and WIND are all at L1, but not exactly at the same location. Since the field configurations are similar at ACE and WIND, it is plausible to assume, that the magnetic field at SOHO is similar too.

The analysis has been conducted with a three-hour accumulation interval. In that case more than 4000 counts were available for the calculation of the azimuthal angle intensity profile. In table 1 we summarized the measured *count rate factor* as well as the determined preferred θ and n. This event is remarkable, because the anisotropy is maintained and non-varying over a several hour period. The event has been divided according to measured anisotropy into three parts: A,B and C (see figures 9, 10 and 11). These figures show the relative intensity (arbitrary units) as a function of the azimuthal angle ϕ. The determination of the flux parameters during the interval C was only possible because both data sets have been used. Results shown in figure 12 show the reconstructed angular distribution. The bluish circles show the ACE magnetic field direction.

Note that our method depends considerably on the assumption that the preferred angle of incidence does not change during a single accumulation interval. The validity of this assumption is supported if the azimuthal intensity plots (figures 9 and 10) show only one dominant peak. Lack of such peaks indicates intervals with nearly isotropic particle flux (figure 11). Shortening of the accumulation interval sometimes removes additional dominant peaks, on cost of the statistics.

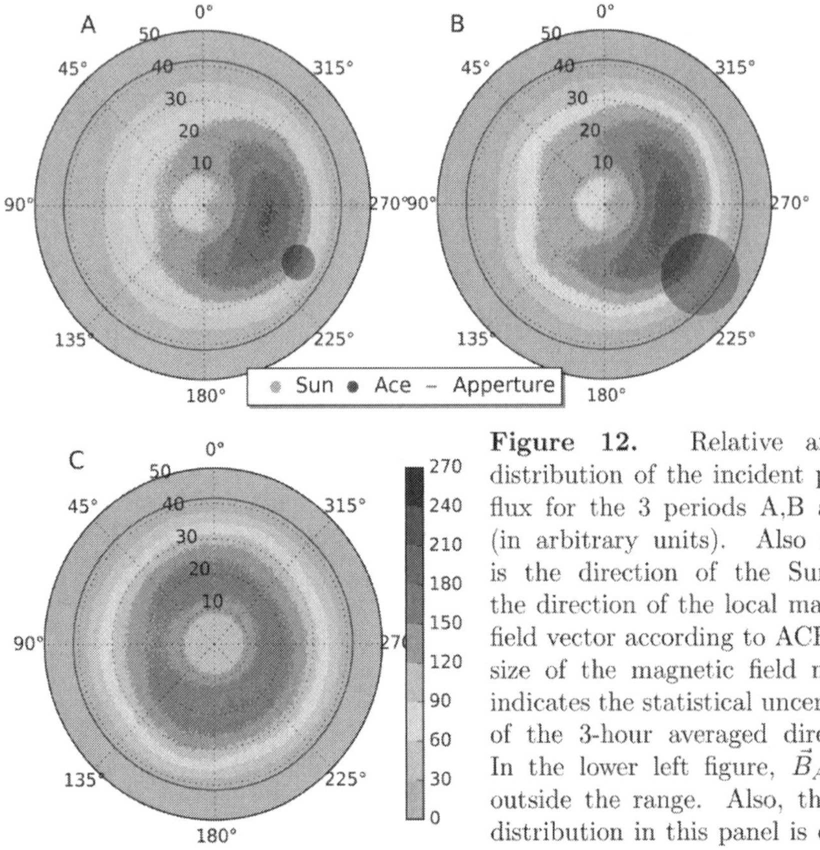

Figure 12. Relative angular distribution of the incident proton flux for the 3 periods A,B and C (in arbitrary units). Also shown is the direction of the Sun and the direction of the local magnetic field vector according to ACE. The size of the magnetic field marker indicates the statistical uncertainty of the 3-hour averaged direction. In the lower left figure, $\vec{B}_{ACE}$ is outside the range. Also, the ring distribution in this panel is due to the geometry of the telescope [5]

4. Conclusion and outlook

We have successfully developed a method which enables anisotropy measurements using EPHIN. This is limited to periods where all detector segments are operational. Soon we plan to swich to a new data set in which the the electron contamination of proton and helium chanels is removed. The developed method serves as a proof of concept: multisector detector stacks can indeed be used to gain information about the angular distribution of the incident particle flux and, if the local magnetic field is known, the determination of the pitch angle distribution. Our method could be used in future for optimizing *in-situ* particle telescopes.

References

[1] Müller-Mellin R, Kunow H, Fleißner V, Pehlke E, Rode E, Röschmann N, Scharmberg C, Sierks H, Rusznyak P, McKenna-Lawlor S, Elendt I, Sequeiros J, Meziat D, Sanchez S, Medina J, Del Peral L, Witte M, Marsden R and Henrion J 1995 *Solar Physics* **162** 483–504

[2] McDonald F B and Ludwig G H 1964 *Physical Review Letters* **13** 783

[3] Gómez-Herrero R 2003 *PhD Thesis*

[4] Geant4-Collaboration 2003 *Nuclear Instruments and Methods in Physics Research Section A: Accelerators, Spectrometers, Detectors and Associated Equipment* **506** 250 – 303 ISSN 0168-9002 URL `http://dx.doi.org/10.1016/S0168-9002(03)01368-8`

[5] Sullivan J D 1971 *Nuclear Instruments and methods* **95** 5–11

[6] Gómez-Herrero R, Dresing N, Klassen A, Heber B, Lario D, Agueda N, Malandraki O E, Blanco J J, Rodriíguez-Pacheco J and Banjac S 2015 *The Astrophysical Journal* **799** 55 URL `http://stacks.iop.org/0004-637X/799/i=1/a=55`

Local anisotropy of muon flux during Forbush decreases from URAGAN data

N Barbashina[1], N Ampilogov[1], I Astapov[1], V Borog[1], A Dmitrieva[1], A Kovylyaeva[1], R Kokoulin[1], K Kompaniets[1], G Mannocchi[2], Yu Mishutina[1], A Petrukhin[1], O Saavedra[3], V Shutenko[1], O Sit'ko[1], G Trinchero[2], E Yakovleva[1], I Yashin[1]

[1]National Research Nuclear University MEPhI (Moscow Engineering Physics Institute), 115409 Moscow, Russia
[2]Istituto di Fisica dello Spazio Interplanetario, INAF, 10133 Torino, Italy
[3]Dipartimento di Fisica dell' Universita di Torino and INFN, 10125 Torino, Italy

E-mail: NSBarbashina@mephi.ru

Abstract. The approach to the analysis of spatial-angular characteristics of the muon flux variations at different phases of Forbush decrease development according to the muon snapshots (muongraphies) obtained using muon hodoscope URAGAN, as well as the analysis results are presented.

Introduction

One of the perfect examples of the influence of solar activity on the cosmic rays is the Forbush effect (FE), which is a sharp decrease of CR intensity due to the propagating large-scale solar wind disturbances [1], and, therefore, is often called a Forbush decrease (FD). Study of the FD using the muon flux has the following features: the first, muons are more sensitive to higher energies of primary cosmic rays (PCR) than neutrons, and so allow study heliospheric disturbances responsible for the high-energy PCR modulation; the second and more important, muons retain the primary particle direction and this enables to obtain the spatial and angular picture of modulations of cosmic rays in the near-Earth space and to study their dynamics in a wide range of zenith and azimuth angles using a single muon hodoscope. In this paper we consider anisotropy of muon flux on the observation level only, since all necessary calculations for transition from muon energy and direction to primary proton energy and direction were made earlier. In paper [2] the mean and median energies of primary particle in dependence to the zenith angle were calculated. These values were used for the calculations of asymptotic directions [3]. The samples of transition of the local muon flux distribution for the Moscow location point (55.7° N, 37.7° E, 173 m a.s.l.) to the primary proton distribution at the boundary of magnetosphere presented in the GSE system were given in our other papers (see f. e. [4]). But our investigations have shown that many features of helispherical processes can be studied in a local system.

Muon hodoscope URAGAN (URAGAN) [5] operates as a part of the unique scientific facility NEVOD [6] since 2006. Hodoscope consists of four supermodules (SMs) with high spatial and angular accuracy of muon registration and registers muon flux variations both in the integral mode and in the hodoscopic mode, i.e. when the variations of the muon flux from different directions of the celestial hemisphere are measured simultaneously. The main feature of the muon hodoscope technique is the real time reconstruction of muon trajectories that can be distributed in the cells of angular matrix of any type. The analysis of these matrices enables to study zenith-angular and azimuthal dependences, as well as the muon flux anisotropy. A similar approach was used in the narrow angle muon telescope at the Mt. Norikura [7], but with lower angular accuracy (about 7°) in comparison with the URAGAN (about 1°). The advantage of the considered approach can be illustrated by the comparison with the multidirectional muon telescopes (MMT) combined in the Global Muon Detector Network (GMDN) [8]. MMT register integral muon flux in rather wide and fixed intervals of zenith and azimuthal angles which, in addition, are overlapped [9]. In the contrast with the GMDN, the muon hodoscope enables to obtain practically differential distributions on zenith and azimuthal angles of arriving muons.

Every minute one SM of the URAGAN registers about 80,000 muons and records their arrival directions in two-dimensional angular matrices. To analyze the muon flux intensity variations during the FD, the matrices averaged over the data of three supermodules obtained during one hour expositions are used. The statistical reliability of every matrix is about 15 million events. Sequence of such matrices-snapshots (muongraphies) provides information about variations in spatial and angular distributions of the muon flux intensity during the FD. The FDs registered by the URAGAN in the period from 2007 to 2012 are analyzed in the paper. Over this period 185 FDs were observed by the integral counting rate. Among them, 44 FDs with amplitude > 0.5%, which are well separated and do not overlap with each other, were selected. To estimate the FD amplitude, the 10-minute data on the muon flux integral intensity corrected for the barometric and temperature effects were used [10, 11]. To study the muon flux relative anisotropy, the muon snapshots and quantitative parameters describing them were used. Each analyzed event was divided into different phases: before the decrease beginning (24 hours), phases of decrease, minimum and recovery (not more than 24 hours).

2. Experimental data and analysis technique

Experimental data obtained with the supermodules of the URAGAN setup are binary files that contain information about one minute snapshots. Supermodule response represents the information on strips triggered in each of two projection planes *XZ* and *YZ*. Track parameters (two projection angles) are reconstructed in real time mode by the software which is based on histogramming of the hits in each projection plane. Projection angles are accumulated in three types of two-dimensional matrices (θ, φ), (θ_X, θ_Y) and (tgθ_X, tgθ_Y) during the minute interval. Different matrices are used to solve various tasks.

To analyze the muon flux intensity variations during the FD, the hourly matrices (tgθ_X, tgθ_Y) summed over three SM are used. Each matrix (figure 1 a) is normalized to the average matrix obtained by averaging minute matrices for the last day (figure 1 b). The result is a matrix (figure 1 c), each bin of which contains deviations from the average one in units of their rms values (σ). To smooth sharp fluctuation peaks in separate bins, all values of the matrix are smoothed using the Gaussian filter (figure 1 d). To optimize the presentation of information considering low statistics in the matrix corners, it is reasonable to present it in the form of a circle limited by zenith angle of 75° (figure 1 e). Such matrix is a "muon snapshot" of the upper hemisphere bounded by the aperture of the detector.

Thin lines in figure 1 e show N-S and E-W directions. Circles highlight zenith angles of 30°, 45°, 60° and 75°. Statistical reliability of such matrix is about 5 million events. The scale in the upper right corner of figure 1 e shows the deviation from the average in sigma, red color corresponds to the excess of muons, blue color – to the deficit of muons. Additionally, for each matrix the following auxiliary information is displayed: the beginning and the end of the snapshot exposure; average atmospheric pressure in the analyzed time interval; matrix type (in this case, the matrix of projection angle tangents) and the value of deviation from the average integral intensity (δ); length of the horizontal projection of the muon flux relative local anisotropy vector (r, relative to the normalization and divided by the statistical error) and its direction (φ_r) indicating the muon flux increase.

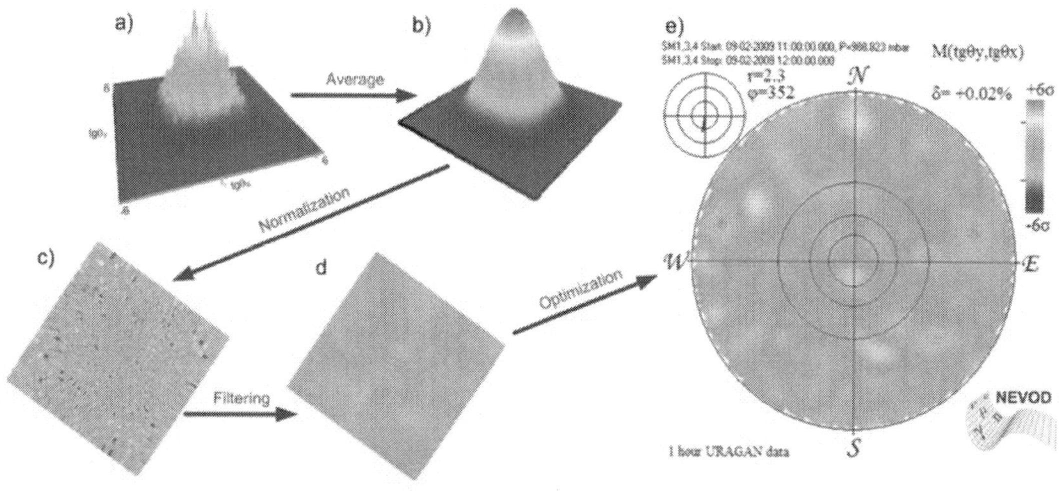

Figure 1. Angular distribution of muons registered during one minute (the matrix of tangents of projection angles): a) initial matrix; b) after averaging; c) two-dimensional image after normalization; d) smoothed angular matrix; e) resulting angular matrix.

Local anisotropy vector $\vec{A}$ characterizes the zenith-azimuthal distribution of the muon flux during some period of time and represents the sum of unit vectors normalized to the number of events [3]. The direction of each unit vector is obtained at a single muon track reconstruction. The summarized vector indicates the average direction of the muon flux. For calculations, the matrix (θ, φ) with the cell dimensions of 1° for zenith (θ) and 4° for azimuthal (φ) angles are used. The vector can be expanded in directions. Particularly, in this work the projections on the geographic South (A_S) and East (A_E) were used. To study the deviations from the mean direction of the anisotropy vector, the relative anisotropy vector $\vec{r}$ and its projections to the South (r_S) and East (r_E) were used:

$$\vec{r} = \vec{A} - \langle \vec{A} \rangle,$$

$$r_S = A_S - \langle A_S \rangle \text{ and } r_E = A_E - \langle A_E \rangle,$$

where $\langle \vec{A} \rangle$ is the average anisotropy vector calculated for a long period of time. Vector $\vec{r}$ and its projections to the South and East indicate the direction of the maximum of muon flux variations. Correspondingly, minimum is on the opposite side. Projections of the vectors r_S and r_E are calculated in geographic coordinates. For the primary protons (considering asymptotic trajectories) the geographical directions N-S and W-E correspond to the asymptotic directions E-W and N-S, respectively. These quantities are convenient for the study of the regularities in the muon flux anisotropy variations on the azimuthal direction during the FD and of the correlation between them for the quantitative description of the muon flux angular distribution variations (for the description of the muon snapshots) [12-14].

3. Examples and results of the analysis of the muon flux relative anisotropy variations

For each phase of 44 selected FDs the muon snapshots and correlations between the projections of the vector $\vec{r}$ to the South and East (r_S and r_E) were plotted.

For example, figure 2 shows the muon snapshots for the FD on October 24, 2011. This is a large enough decrease of the counting rate which was observed in many neutron monitors. According to the data of the URAGAN and the Moscow Neutron Monitor [15], its amplitude was about 1.5% and 4.5%, respectively. Each row of the snapshots in figure 2 corresponds to the different phase of the FD development: the first row reflects the period before the beginning of Forbush decrease (24/10/2011

8:00 – 24/10/2011 18:00, every 3 hours); the second, phase of decrease (24/10/2011 20:00–25/10/2011 6:00, every 3 hours); then, phase of minimum (25/10/2011 7:00– 25/10/2011 17:00, every 3 hours); the last row corresponds to the phase of recovery (25/10/2011 18:00– 26/10/2011 13:00, every 6 hours).

In figure 2, the muon flux variations at different angles during the FD are clearly seen. At the first phase of the FD, the increase of the counting rate in the range of zenith angles from 0 to 45 degrees, which damps before the FD beginning, is observed. At the phase of decrease and minimum, the muon counting rate decrease begins in the same range of zenith angles, but with the N-E azimuth displacement. By 5:00 on 25/10/2011, the counting rate decrease spread on the particles registered under zenith angles more than 60°. At the phase of minimum, the decrease of the flux of registered particles was revealed also in a wide range of zenith angles from 0 to 60 degrees. At the recovery phase the counting rate was changed from S-W decreasing to S-E increasing.

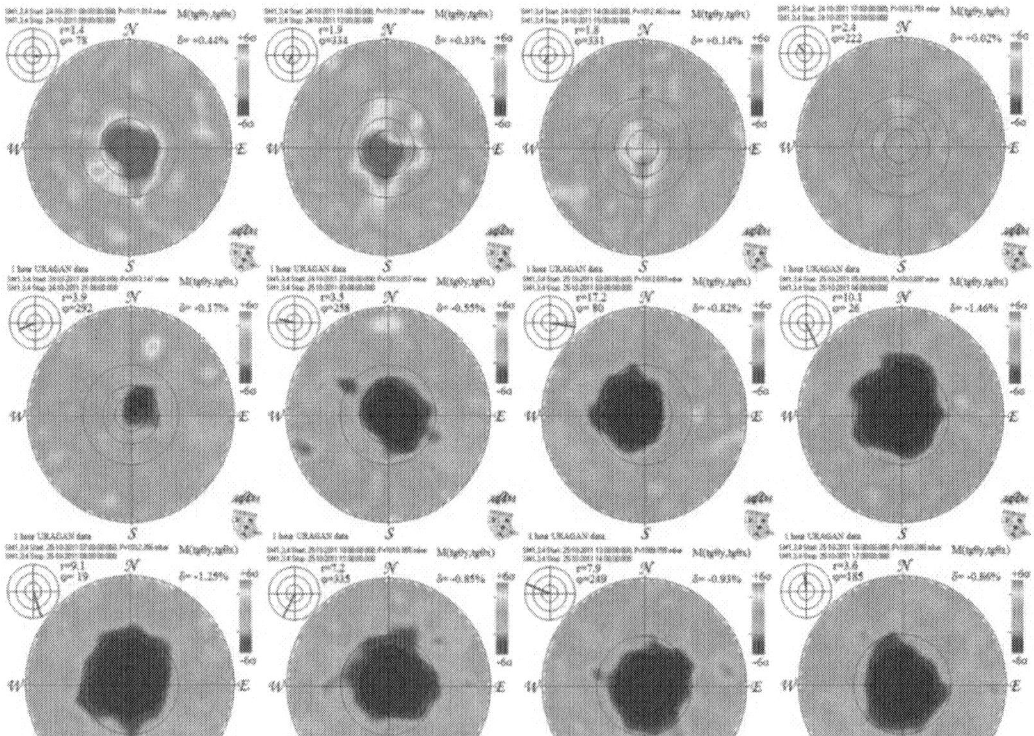

Figure 2. Muon snapshots of FD on October 24, 2011.

Correlations between the projections of the relative anisotropy vector r_S and r_E during the FD on October 24, 2011 are shown in figure 3. Circles in the graphs corresponds to the r_S and r_E variation limits equal to 0.52×10^{-3} during the quiet period from 7 to 10 of February 2009 which is characterized by the absence of disturbances in the heliosphere and the Earth's magnetosphere. The rms values of r_S and r_E during this period are $\sigma(r_S) = 1.30\times10^{-4}$ and $\sigma(r_E) = 1.28\times10^{-4}$, respectively. Thus, the exceedance of the circle boundaries corresponds to the 4σ exceedance of the r_S and r_E variations.

As seen from the figure 3, during the first FD phase the relative anisotropy variations are not observed. At the other phases, the excess of the muon flux relative anisotropy is seen in comparison with the normal value: at the phase of decrease it predominates in the E-W azimuthal direction, at the phase of minimum it moves from the South to the North, at the phase of recovery it shifts to the geographical North.

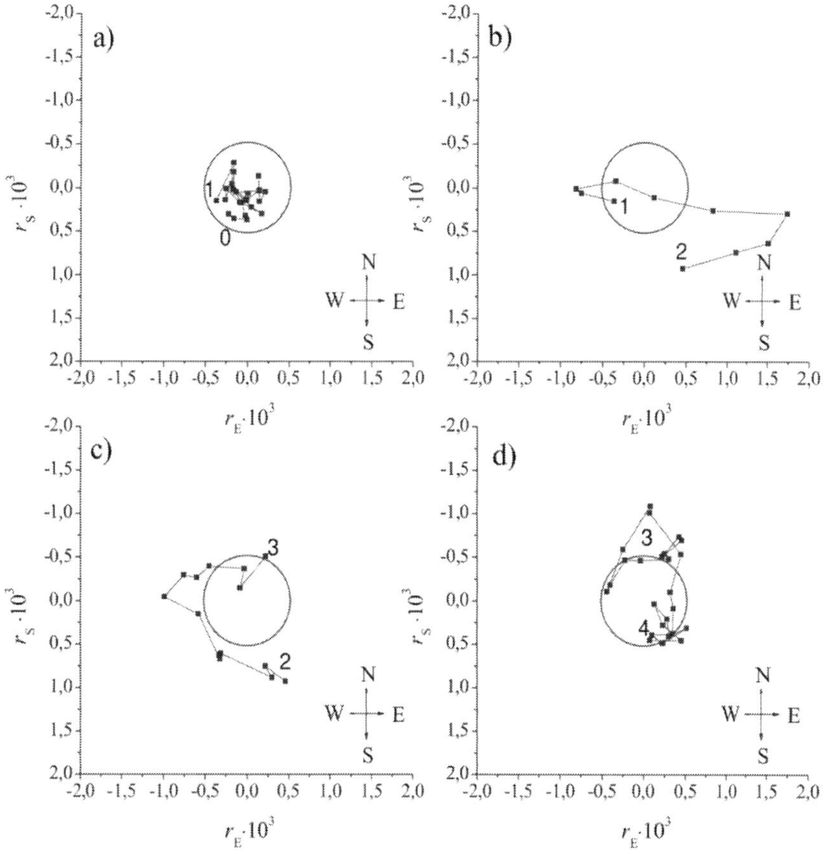

Figure 3. Correlations between r_S and r_E for the FD on October 24, 2011 at four phases of the event development: (a) 0 – 23/10/2011 20:00, 1 – 24/10/2011 20:00; (b) 1 – 24/10/2011 20:00, 2 – 25/10/2011 5:00 UT; (c) 2 – 25/10/2011 5:00, 3 – 25/10/2011 18:00 UT; (d) 3 – 25/10/2011 18:00 UT, 4 – 26/10/2011 18:00 UT.

The analysis of 44 FDs showed that the relative anisotropy exceeding the r_S and r_E variations more than 4σ is observed: at all phases of 12 FDs, at three phases of 11 FDs, at two phases of 7 FDs, at one phase of 6 FDs, and is absent at all phases of 8 FDs. The study of each phase separately showed that the significant relative anisotropy is observed: in the period before FD in 25 FDs, at the phase of decrease in 27 FDs, at the phase of minimum in 25 FDs, at the phase of recovery in 24 FDs. To study the regularities in the behavior of the relative anisotropy along the N-S and E-W geographical directions, the distributions of hourly values of the r_S and r_E for all phases of FDs (figure 4, top and bottom) normalized to the number of points at each phase were plotted.

Distributions of the r_S and r_E values before the FD beginning are symmetrical and are mainly in the range from -1.0×10^{-3} to 1.0×10^{-3}. At the phase of decrease a wider range for both parameters is observed: $r_S = (-1.2 \div 2.0)\times10^{-3}$ and $r_E = (-1.7 \div 1.7)\times10^{-3}$. At that, the values of r_S are obviously shifted towards the South: $\langle r_S \rangle = (0.18 \pm 0.02)\times10^{-3}$. This means that the minimum of the flux is observed more frequently from the North which, in its turn, corresponds to the asymptotic East. At the phase of minimum $r_S = (-1.3 \div 1.0)\times10^{-3}$ and $r_E = (-1.2 \div 1.0)\times10^{-3}$, the mean values practically coincide, r_S is shifted more towards the North. At the recovery phase, the distributions of the relative anisotropy parameters are almost symmetrical.

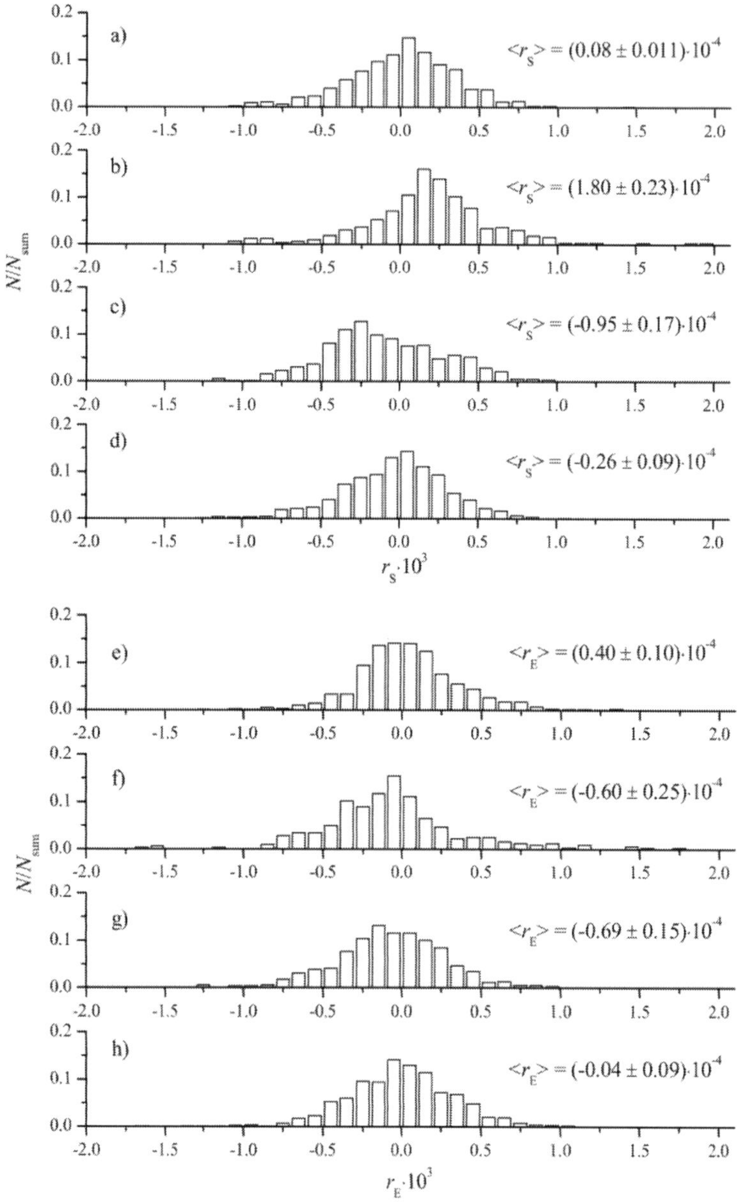

Figure 4. Distribution of values of r_S (top) and r_E (bottom) parameters at four FD phases: (a) and (e) – before the FD beginning; (b) and (f) – phase of decrease; (c) and (g) – phase of minimum; (d) and (h) – phase of recovery.

Thus, on average, at the phase of the counting rate decrease during the FD, the muon flux from the geographic South decreases weaker than from the North, at the phase of minimum vice versa, which corresponds to the variation of the E-W anisotropy of primary cosmic rays. At the other phases, the relative anisotropy is almost symmetrical.

4. Conclusions

Muon hodoscope URAGAN allows reconstruct trajectories of the detected particles with high angular and spatial accuracy and obtain muon snapshots (muongraphies). Such snapshots give qualitative information about the dynamics of the decrease or increase of the intensity of the muon flux from different directions. Usage of the correlations between the projections r_E and r_S of the relative anisotropy vector allows quantitative studying of temporal variations of the muon flux anisotropy at different phases of the FD development. The analysis of 44 Forbush decreases registered by the URAGAN showed that the changes in the relative anisotropy of the muon flux variations are observed mainly at the phases of decrease and minimum and, on the average, correspond to the E-W primary cosmic ray anisotropy variation.

Acknowledgments

This work was performed at the NEVOD Scientific and Educational Center with the financial support from the State provided by the Russian Ministry of Education and Science (government task and project No. RFMEFI59114X0002) and the grant of the Leading Scientific School NSh-4930.2014.2.

References

[1] Belov A V *et al* 2001 *Izv. RAN. Ser. Fiz.* **65** 373
[2] Yakovleva E I *et al* 2009 *Bull. Russ. Acad. Sci. Phys.* **73** *357*
[3] Shutenko V V *et al* 2009 *Bull. Russ. Acad. Sci. Phys.* **73** *347*
[4] Yashin I I et al. 2013 J. Phys.: Conf. Ser. **409** 012192
[5] Barbashina N S *et al* 2008 *Instrum. Experim. Techn.* **51** 180
[6] Unique scientific facility NEVOD: http://ununevod.mephi.ru/en/
[7] Fujimoto K *et al* 2001 *Proc. 27th ICRC (Hamburg, Germany)* **9** 3523
[8] Rockenbach M *et al* 2014 *Space Sci. Rev.* **182**, 1
[9] Yasue S *et al* 2003 *Proc. 28th ICRC (Tsukuba, Japan)* **6** 3461.
[10] Barbashina N S *et al* 2009 *Bull. Russ. Acad. Sci. Phys.* **73** 343
[11] Dmitrieva A N *et al* 2011 *Astroparticle Physics* **34** 401
[12] Shutenko V.V. *et al* 2011 *Proc. 32nd ICRC (Beijing, China)* **11** 276
[13] Barbashina N.S. *et al* 2013 *Bull. Russ. Acad. Sci.: Phys.* **77** 532
[14] Shutenko V V *et al* 2013 *Geomagn. Aeronomy.* **53** 571
[15] Database of Moscow neutron monitor: http://cr0.izmiran.rssi.ru/mosc/main.htm

Correlation of the quasi-biennial oscillations in galactic cosmic rays and in the solar activity indices

G A Bazilevskaya, M S Kalinin, M B Krainev, V S Makhmutov, Y I Stozhkov, A K Svirzhevskaya and N S Svirzhevsky

Lebedev Physical Institute of Russian Academy of Sciences, Leninsky prospect, 53, 119991, Moscow, Russia

E-mail: `bazilevs@sci.lebedev.ru`

Abstract.
Quasi-biennial oscillation (QBO) is a well-known variation in solar activity, interplanetary parameters, geomagnetic disturbances and cosmic rays. Solar QBO is translated to the space via open magnetic flux and modulates intensity of cosmic rays. The highest negative correlation exists in the QBO of cosmic rays with QBO in the heliospheric magnetic field strength B as well as with QBO in the scalar product BV, where V is the solar wind velocity, cosmic ray being delayed by $\approx$ 1 month. During $\approx$ 50 years of cosmic ray monitoring the QBO periods demonstrated some intermittency. It is argued that the Gnevyshev Gap effect and the step-like changes in the cosmic ray intensity appeared to be a part of QBO in cosmic rays.

1. Introduction

Quasi-biennial oscillation (QBO) is the conventional name of temporal variations with characteristic time of $\approx$ 0.5-4 yrs. QBO appears to be the most prevalent quasi periodicity shorter than the 11-yr cycle in solar activity phenomena [1]. Figure 1 demonstrates the QBO in the sunspot area (`http://solarscience.msfc.nasa.gov/greenwch.shtml`). The 11-year cycle is clearly seen in the monthly averaged and 7-months smoothed data. The lower red curve shows the QBO. Time intermittent period can be seen, and a clear 11-year modulation with higher QBO during solar maxima periods is observed.

The solar QBO has been extensively studied. Their main features can be summarized as follows [2]. The QBO is ubiquitous: it is observed at all levels of the solar atmosphere and even below the photosphere. The QBO is intermittent - there is no stable period, although its amplitude is modulated by the 11-year solar cycle being highest at periods close to the solar maximum. There are some specific features of the solar activity QBO: it is independent in the northern and southern solar hemispheres, however synchronous at each level of the solar atmosphere, it is intermittent and has signs of stochasticity [3]. Many solar researchers believe that QBO is intrinsic to the solar dynamo mechanism [4, 5, 6].

The solar QBO is translated to the heliosphere. It is transferred via open magnetic flux which also demonstrates QBO although not synchronous with the other solar activity indices including solar magnetic field. That is why some researchers stated that there is no correlation between QBO on the Sun and QBO in heliosphere, such as cosmic ray (CR) QBO e.g., [7]. A relevant proxy of open magnetic flux is heliospheric magnetic field [8].

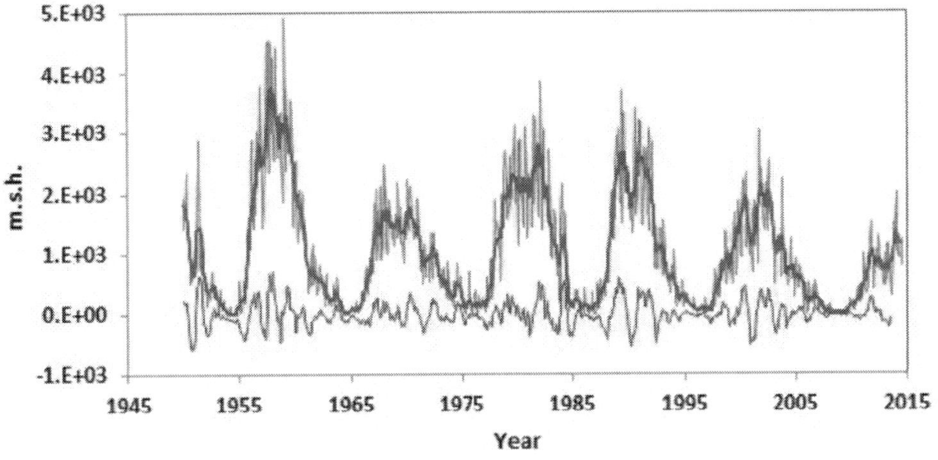

Figure 1. Sunspot area in millionths of solar hemisphere (http://solarscience.msfc.nasa.gov/greenwch.shtml): monthly-averaged (green line) and 7-months smoothed (brown line) data. Lower red curve shows the QBO retrieved as is described in Section 2.

Majority of papers devoted to the QBO in CR are focused on a certain periodicity, such as 1.2 yr, 1.7 yr or 2 yr [9, 10, 11, 12, 13, 14]]. Kato et al. [15] found ≈ 1.7 yr and ≈ 1.3 yr periodicities in the Voyager 1 and 2 data on >70 MeV protons, which were distinct up to 40 AU and similar to the QBOs observed at neutron monitors. In this paper we do not try to isolate a certain CR periodicity but address the QBO as a special type of the CR variability and stress the characteristics similar to those of the solar QBO, namely, the permanent existence although with intermittent period and changing phase. We show a close relation between QBO in CR and in the interplanetary magnetic field. The well-known properties of the CR modulation - Gnevyshev effect and the step-like changes in CR intensity are shown to be appearances of QBO.

2. Retrieving of QBO

The QBO is significantly lower than the 11-year variation and besides it is non-stable, therefore it is isolated from the data sets via various technics including rather sophisticated such as empirical mode decomposition analysis [16] and independent component analysis [5]. Here, we use a simple pass-band filter which consists of 7-months and 25-months smoothing of the monthly data with consequent subtraction the latter from the former. This procedure passes 1.5-1.7 yr signal without distortion (i.e. it has around a 100% frequency response at 1.5-1.7 yr) and above a 50% response between 1 and 3.3 yrs. The output results are, in general, consistent with results of other filtration. The results shown in this paper were found to be stable to different smoothing periods in between 3-37 months: the amplitudes of the QBO were observed to change by an insignificant amount while phases were not violated.

3. Data selection

The QBO in different solar indices (such as sunspot number and area, H-alpha flares and X-burst numbers, coronal lines intensity) is highly correlated [2]. Because CR are affected by magnetic fields we use here as a solar QBO characteristic an energy index of the photospheric magnetic field $B2=B^2/8\pi$. It can be calculated for each solar rotation using the data of observations being performed by the Wilcox Solar Observatory. The sets of the harmonic coefficients completely determining field components are available at the (http://wso.stanford.edu). The details of calculations are described in [17].

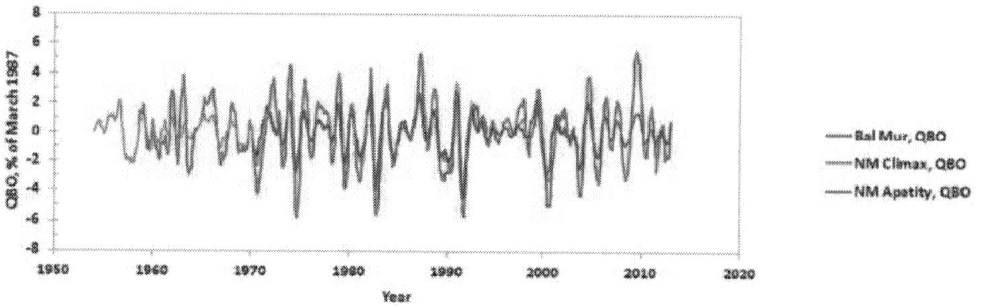

Figure 2. QBO in CR as observed in the stratosphere [18] and with the ground-based neutron monitors (http://www.nmdb.eu/).

As the QBO of interplanetary parameters we examine the heliospheric magnetic field (HMF) strength B, the solar wind velocity V (http://omniweb.gsfc.nasa.gov/ow.html), and the tilt angle of the heliospheric current sheet, classic version (http://wso.stanford.edu/Tilts.html).

We use the data of CR balloon observations in the Earth's atmosphere [18] and the data of the ground level neutron monitors (NM) [http://www.nmdb.eu/]. The phases of the CR QBO as retrieved from different observations are similar to each other as it can be seen in Figure 2. Here a coherency of balloon (Murmansk) and two neutron monitors (Climax and Apatity stations) is shown. The monthly data were normalized to March 1987 before passing through the filter. The QBO at other NMs behaves similarly. As examples, coefficients of correlation between CR QBO of different observations are R=0.72 (Murmansk vs. NM Huancayo/Haleakala), R=0.88 (Murmansk vs. NM Moscow), R=0.90 (Murmansk vs. NM Climax). Therefore, to examine correlation between CR and solar/interplanetary activity we further use the NM data. The QBO amplitudes depend on CR energy and will be discussed elsewhere.

4. Correlation between QBO in cosmic rays and solar/interplanetary indices

Many papers were devoted to search for a link between CR QBO and QBO in solar activity. It was found that relationship was not stable and rather complicated eg., [7]. The contradiction was solved with understanding of the role of the open magnetic flux of the Sun which has its proxy in the HMF strength [8]. The solar activity indices and the open magnetic flux correlate well for the long-term patterns and badly for the QBO. The correlation coefficient between the 25-month smoothed $B2$ index and HMF B in 1977.63-2013.38 is R=0.87±0.01 while for the QBO at the same time R=0.25±0.04 (without a lag). That means a good translation into space of a long-term constituent of the solar magnetic field and a significant distortion of the QBO at transfer from lower levels of the solar atmosphere to space. Upper panel of Figure 3 shows QBO in CR and $B2$. In general correlation is low and it becomes better at the CR delay of 3 months. This may be important for understanding transfer from the solar magnetic field to the open magnetic flux and needs a further study. The best correlation is observed between the QBO in CR and HMF strength B or between CR and the product BV, where V is the solar wind speed, the CR delay being 1 month. The link of CR QBO to the tilt angle QBO (Figure 3, lower panel) is not simple. There are time periods with good coherency, e.g. from mid 1981 to mid 1987, but afterwards a coherency was violated. It is necessary to study this point further taking into account the tilt angle value.

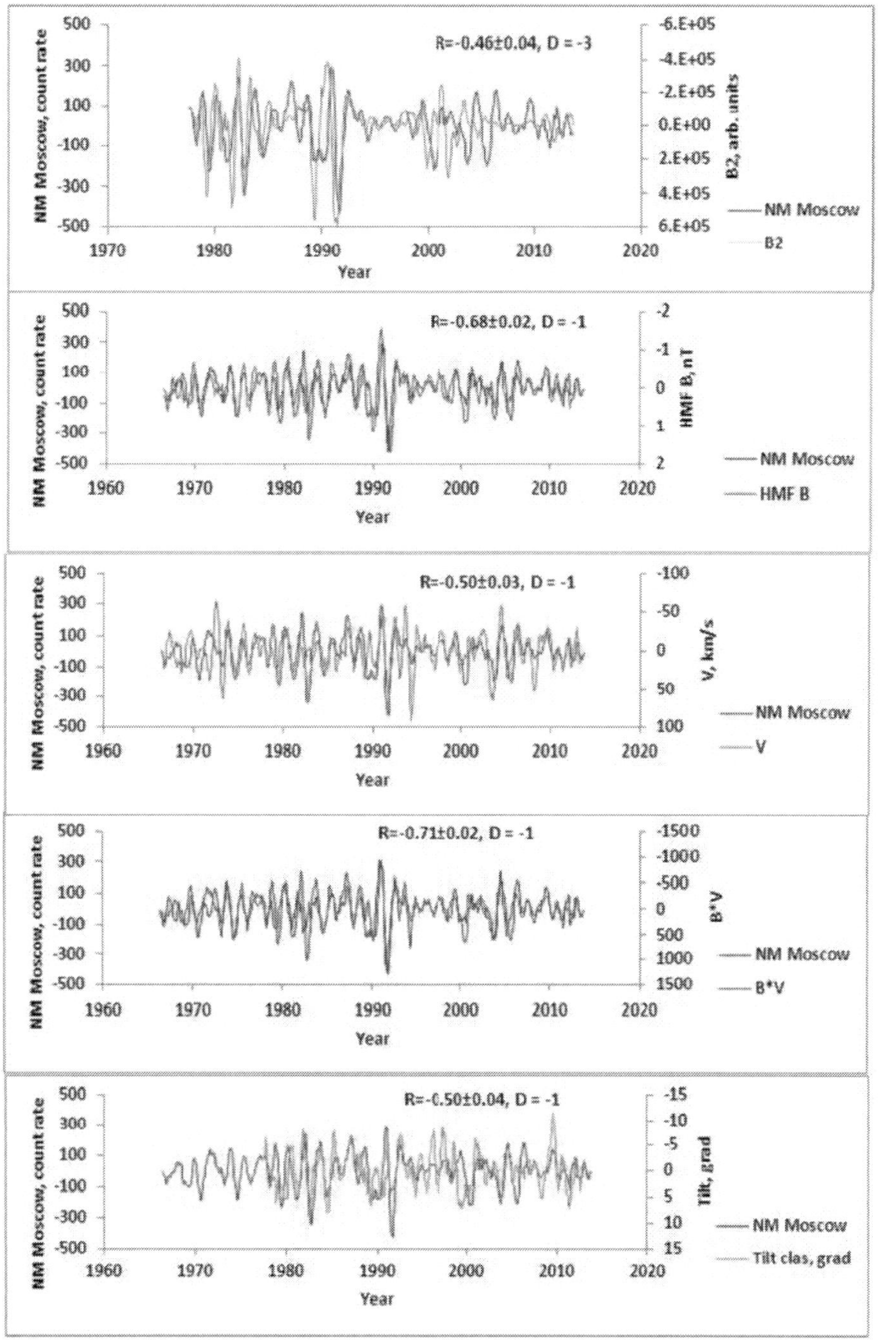

Figure 3. QBO in cosmic rays and in different indices, from top to bottom panels: $B2$ index of photospheric magnetic field, heliospheric magnetic field B, solar wind velocity V, the scalar product of B and V, tilt angle of heliospheric current sheet. R denotes a highest correlation coefficient, D is a delay of CR relative to an index in months.

Correlation between the QBO in CR and HMF B in different solar cycles is presented in table 1. It is seen from table 1 that there is no significant difference from one cycle to another.

Table 1. Correlation between QBO in CR and HMF (different cycles).

Start	*End*	*Solar cycle*	*Delay*	*R*	*error*
1967.54	1976.46	20	1	-0.74	0.04
1976.54	1986.79	21	2	-0.70	0.05
1986.87	1996.37	22	0	-0.74	0.04
1996.46	2008.87	23	1	-0.66	0.05
2008.96	2012.87	rise of 24	1	-0.74	0.07

Table 2. Correlation between QBO in CR and HMF (different polarities of the solar magnetic field).

Start	*End*	*A sign*	*Delay*	*R*	*error*
1970.87	1980.04	>0	1	-0.79	0.04
1980.12	1990.87	<0	1	-0.67	0.03
1990.96	2000.87	>0	0	-0.79	0.03
2000.96	2012.96	<0	1	-0.70	0.04

This result does not confirm a conclusion of [19] about low correlation of the CR QBO with HMF B in the solar cycles 21 and 23. No visible changes in the CR and HMF B QBO during the whole period of observations are seen also in Figure 3 (second panel from the top). We believe a reason of discrepancy with [19] to be in their method of the QBO isolation, which is more complicated than our way.

Table 2 demonstrates the same as table 1 but for periods of positive and negative polarities of the solar magnetic field (`http:/wso.stanford.edu`). Here, as usual, A>0 refers to positive radial component of magnetic field in the northern solar hemisphere. We see that the QBO correlation is higher during the A>0 periods, however, statistics is too small.

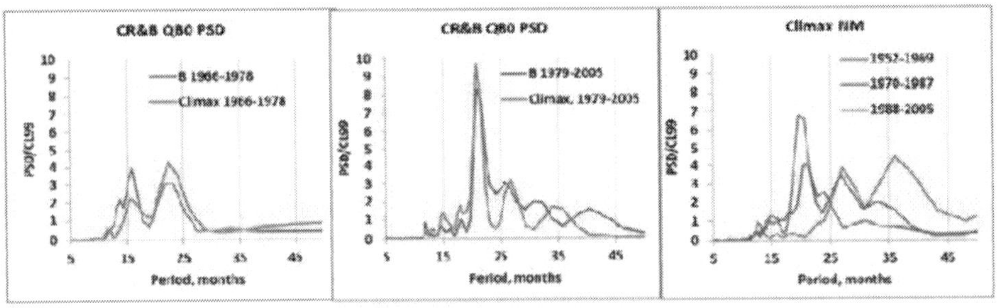

Figure 4. Left panel: power spectra density of QBO vs. period in months for the Climax neutron monitor (green line) and heliospheric B (blue line) in 1966-1978; middle panel: the same but in 1979-2005. Right panel: PSD spectra for the Climax NM in three different periods of time. PSD is given in the units of 99% confidence level.

Further evidence of a close link between QBOs in CR fluxes and HMF B is shown in Figure 4 (left and middle panels). The power spectrum densities (PSD) of CR and HMF B are plotted for 1966-1978 and 1979-2005. For convenience, the PSD is given versus period in months.

5. QBO and 1.7 year variation in cosmic rays

A majority of researchers draw attention to a certain periodicities in CR such as 1.7 yr or 1.3 yr variations [10, 11, 12, 15, 20] and others. We believe that QBO in CR are caused by the QBO in the solar activity. It was shown [2, 3] that solar QBO is intermittent and have signs of stochasticity. On the one hand, the CR QBO may be more stable than solar one, and on the other hand the period of CR monitoring is too short in comparison with solar activity observations which prevents from finding any stochasticity. To examine the problem we took the Climax NM data (`http://ulysses.sr.unh.edu`) - the oldest CR data set (QBO can be extracted from 1952-2005) - divided it in 3 subsets and calculated the PSD spectra. The results are plotted in the right panel of Figure 4. The main periods are 2.25 yr and 3 yr in 1952-1969; 1.7 yr in 1970-1987; 1.8 yr and 2.2 yr in 1989-2005. The same periodicities exist in HMF B as it is depicted on the left and middle panels of Figure 4 for the time of simultaneous observations of Climax and HMF. We conclude that the CR QBO is actually dynamic, although, some selectivity in the CR QBO may be present.

6. QBO, Gnevyshev Gap and the step-like variations in cosmic rays

Gnevyshev Gap is a temporal damping of solar activity just around of maximum of the 11-year solar cycle [21]. These gaps are clearly seen in the sunspot area (upper part of Figure 1). Similarly to QBO, Gnevyshev Gap is observed virtually in all solar and solar-terrestrial indices [22] and references therein. Moreover, like QBO, Gnevyshev Gap appearance is rather intermittent as it is also seen in Figure 1. Bazilevskaya et al. [23] suggested that Gnevyshev Gap in solar activity is an appearance of QBO. In particular, absence of Gnevyshev Gap in some solar cycles is explained by independence of QBO in the northern and southern solar hemispheres and their occasional anti-phase around solar maximum. Later it was supported by [24], although an alternative explanation exists e.g., [25]. Similar to any other variation of solar activity Gnevyshev Gap modulates the CR intensity, where it looks as a temporal weakening of modulation, i.e. as a temporal increasing of CR intensity in the maximum of solar activity. The Gnevyshev effect is clearly seen in Figure 5, where the data of the Moscow NM for three solar cycles are plotted. Gnevyshev Gap is indicated by brown arrows. The red curves is the QBO extracted from the CR intensity. It is seen that Gnevyshev Gap in CR is one of the QBO maximum. The QBO in HMF (values are multiplied by -3) is also shown by the grey curves, thereby confirming that Gnevyshev effect in CR is a part of QBO.

The step-like behavior of CR intensity, which is also clearly seen in Figure 5, was discussed in several papers. In the beginning the steps were considered as a result of CR modulation by the global merged interaction regions (GMIR) of HMF which were formed by coronal mass ejections (CME), shocks and fast-speed flow of solar wind at distance of ≈ 10 a.u. [26]. However, Cane et al. [27] proved the step-like events in CR to be simultaneous with the solar magnetic field variations derived from photospheric observations alone. Wibberenz and Cane [28] formulated main features of the "medium-term modulation events" as related to variations in the open magnetic flux carried by solar wind. Wibberenz and Cane [28] showed also that such events are distributed throughout the 22 year cycle and are not restricted to phases of high solar activity. Actually, these findings manifest the same features as the CR QBO. Green arrows in Figure 5 prolonged by the light grey lines indicate the step-like changes in the CR intensity. Similarly to Gnevyshev Gap they appear to be a part of QBO. Similar conclusion was drawn by Vecchio et al. [29].

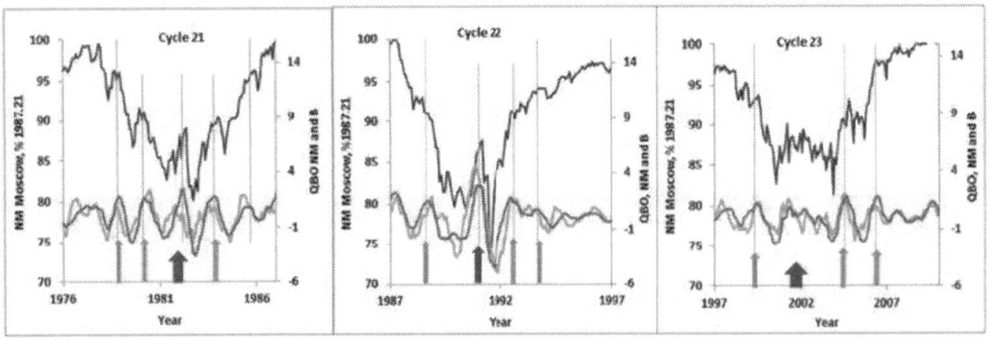

Figure 5. Cosmic ray intensity as observed at the Moscow neutron monitor (blue lines) and QBO in cosmic rays (red curves) and HMF B (multiplied by -3, grey curves). Brown arrows indicate the Gnevyshev effect in CR, green arrows mark the step-like changes in the cosmic ray intensity.

7. Discussion

The results of this work clearly show a generic link between the solar QBO and the CR QBO. Both variations exist permanently; they have irregular intermittent periods of ≈ 0.4-4 yr and stochastic changes of the phase. The amplitude of the QBO in CR is less modulated by an 11-year cycle as can be seen from comparison of figures 1 and 2. However, negative correlation with the QBO in HMF strength, which is a good proxy for the open magnetic flux of the Sun, is seen in figure 3 and is confirmed by the calculated correlation coefficients throughout the whole period of observation. This result does not agree with conclusions of [19], where a low correlation if any was found between QBO in CR and HMF strength in the solar cycles 21 and 23 as well as during the solar maximum phases 1978-1982 and 1998-2002. Indeed, values of correlation coefficient depend on the time period taken for analysis. However, our estimation of correlation coefficients yields R=-0.68±0.07 with CR delay by 3 months in 1978-1982, and R=-0.54±0.09 with no lag in 1998-2002. The cause of discrepancy with [19] in our opinion is a different approach to the QBO retrieval. Our method is rather simple but it is robust and shows an overall correlation between the CR and HMF QBOs. On the other hand, the empirical mode decomposition technique applied by [19], may stress absence of correlation at selected periodicities (see their Table 1). Further analysis is necessary to understand a nature of the found discrepancy.

A mechanism of the QBO modulation may have some difference from mechanisms of the 11-year modulation. The best correlation of QBO in CR and the scalar product BV is similar to results of [31], where the long-term behavior of the CR intensity and diurnal anisotropy was analyzed. BV gives the magnitude of the interplanetary electric field carried by the solar wind and leads to CR convective flow away from the Sun with a speed V. The found correlation underlines a role of convection in the QBO generation.

Difference in the correlation coefficients for the A>0 and A<0 periods may be indicative of the drift influence. It can be noticed in figure 3 that a lack of correlation between the QBO in CR and heliospheric tilt angle was observed before 1981, in 1988- mid 1990, and in 2005-2008, which could be connected with the intermittent drift contribution to the QBO in CR. However, it is interesting that at the same time periods correlation between the QBO in the HMF strength and heliospheric tilt angle was also violated. Study of the QBO dependence on heliospheric latitude would be elucidating.

Small delays of the CR QBO relative to HMF B QBO argues for a diffusion mechanism of the QBO acting within ≈10 a.u. However, Kato et al. [15] found significant QBO in the CR intensity observed by Voyagers 1 and 2 in the outer heliosphere in 1980s and 1990s, i.e. at the distance of ≈10-40 AU from the Sun. Kato et al. [15] suggested that the long-lived GMIR still may play a role in the QBO propagation through the heliosphere.

The QBO in CR opens some other points for further investigation. Transfer from the solar magnetic field to the open magnetic flux on different time scales is an open question [30], and relation between variabilities in the open magnetic flux and CR is not understood also. A question about possible selectivity of periods in the CR QBO remains to be open. Also, a peculiarity may exist in the relation of CR - HMF B QBO during the solar magnetic field reversal [32].

7.1. Acknowledgments
This work is fulfilled with support of the RFBR grants 14-02-00905a, 14-02-10006k, 13-02-00585, 13-02-00931, 15-02-10070k and of the Fundamental Properties of Matter and Astrophysics Program of the Presidium of the RAS. We thank the unknown reviewers for their comments which helped to improve the manuscript.

7.2. Reference lists
[1] Knaack R, Stenflo J O and Berdyugina S V 2005 *Astronomy and Astrophysics* **438** 1067
[2] Bazilevskaya G, Broomhall A -M, Elsworth Y and Nakariakov V M 2014 *Space Sci. Rev.* **186** 359
[3] Wang Y -M and Sheeley Jr. N R 2003 *ApJ.* **590** 1111
[4] Benevolenskaya E E 1998 *ApJL.* **509** L49
[5] Ruzmaikin A, Cadavid A C and Lawrence J 2008 *J. Atmosph. Solar-Terrestr. Phys.* **70** 2112
[6] Popova E P and Yukhina N A 2013 *Astronomy Letters.* **39** 729
[7] Kane R P 2005 *Sol. Phys.* **229** 387
[8] Lockwood M 2001 *J. Geophys. Res.* **106A** 16021
[9] Rybak J, Antalova A and Storini M 2001 *Space Sci. Rev.* **97** 359
[10] Kudela K, Rybak J, Antalova A and Storini M 2002 *Sol. Phys.* **205** 165
[11] Kudela K, Mavromichalaki H, Papaioannou A and Gerontidou M 2010 *Sol. Phys.* **266** 173
[12] Mursula K, Zieger B and Vilppola J H 2003 *Sol. Phys.* **212** 201
[13] Valdes-Galicia J F and Velasco V M 2008 *Adv. Space Res.* **41** 297
[14] Okhlopkov V P 2011 *Moscow University Physics Bulletin.* **66** 99
[15] Kato C, Munakata K, Yasue S, Inoue K and McDonald F 2003 *J. Geophys. Res.* **108A** 1367
[16] Vecchio A, Laurenza M, Carbone V and Storini M 2010 *ApJL.* **709** L1
[17] Krainev M B, Bazilevskaya G A and Makhmutov V S 1999 *Proc. Conf. "Large-scale structure of solar activity: achievements and perspectives", Pulkovo, St.-Petersburg.* 121
[18] Stozhkov Y I, Svirzhevsky N S, Bazilevskaya G A, Kvashnin A N, Makhmutov V S and Svirzhevskaya A K 2009 *Adv. Space Res.* **44** 1124
[19] Laurenza M, Vecchio A, Storini M and Carbone V 2012 *ApJL.* **749** L167
[20] Chowdhury P, Khan M and Ray P C 2010 *Planetary and Space Science.* **58** 1045
[21] Gnevyshev M N 1977 *Sol. Phys.* **51** 175
[22] Storini M, Bazilevskaya G A, Fluckiger E O, Krainev M B, Makhmutov V S and Sladkova A I 2003 *Adv. Space Res.* **31** 895
[23] Bazilevskaya G A, Krainev M B, Makhmutov V S, Fluckiger E O, Sladkova A I and Storini M 2000 *Sol. Phys.* **197** 157
[24] Hathaway D H 2010 *Living Reviews in Solar Physics.* **7** 1
[25] Georgieva K, Kirov B 2011 *J. Atmosph. Solar-Terrestr. Phys.* **73** 207
[26] Burlaga L F, McDonald F B and Ness N 1993 *J. Geophys. Res.* **98A** 1
[27] Cane H V, Wibberenz G I, Richardson G and von Rosenvinge T T 1999 *Geophys. Res. Lett.* **26** 565
[28] Wibberenz G and Cane H V 2000 *J. Geophys. Res.* **105A** 18315
[29] Vecchio A, Laurenza M, Storini M and Carbone V 2012 *Advances in Astronomy.* **2012** 834247
[30] Zurbuchen T H 2007 *Annual Review of Astronomy and Astrophysics.* **45** 297
[31] Ahluwalia H S 2003 *Geophys. Res. Lett.* **30** 1133
[32] Krainev M B and Kalinin M S 2014 *arXiv:1411.7532 [astro-ph.SR].*

Modeling the behavior of the cosmic ray density in magnetic clouds

A Belov[1], E Eroshenko[1], A Papaioannou[2,3], A Abunin[1], M Abunina[1], V Oleneva[1], H Mavromichalaki[2], V Yanke[1]

[1]Pushkov Institute of Terrestrial Magnetism, Ionosphere and Radio Wave Propagation RAS (IZMIRAN), Kaluzhskoe Hwy 4, Troitsk, Moscow, 142190, Russia
[2]Nuclear and Particle Physics Department, Faculty of Physics, National and Kapodistrian University of Athens, Athens, Greece
[3]IAASARS, National Observatory of Athens, Penteli, Greece

E-mail: abelov@izmiran.ru

Abstract. Our study focuses on the behavior of the density of cosmic ray particles at 10 GV rigidity in a magnetic cloud at Earth. It is shown that it can be mostly described by a simple parabolic dependence over distance from the centre of the cloud, when measured in gyroradii. The majority of magnetic clouds modulate cosmic rays, gecreasing their density. However, there is a group of events (about 1/5 part of the total sample) during which the cosmic ray density increases within the magnetic cloud. The factors that contribute to the model description are considered, and estimates of their influence are carried out and discussed.

1. Introduction

In this work we considered only those solar wind (SW) disturbances (Interplanetary Coronal Mass Ejections - ICMEs) in which magnetic clouds (MCs) are observed. It should be noted that such disturbances with well identified magnetic clouds make only a small part of the disturbances capable to cause Forbush effect (FE).

Magnetic clouds encounter the Earth, keeping the elements of the solar filament structure [1]. As it was shown, for example, in [2], all ICMEs that reach the Earth, have a "flux rope" structure, but it cannot be completely observed in all events. There is a geometrical selective effect and the "observed manifestation" of a magnetic cloud strongly depended on the passage of the observing satellite through the ICME structure [3]. Having MC observations, we practically witness filaments which lay on the Sun – with a large regular magnetic field, which significantly differs from the field in usual SW. Most of MCs show a large Bz component at Earth. It is well known the role of this component in strong magnetospheric responses. Therefore the main part of geomagnetic storms are initiated while passing of MCs, and as result, indices of geomagnetic activity (Dst, Kp etc.) during these periods have the high values.

Richardson and Cane [4] made the Coronal Mass Ejections (CMEs)-ICMEs catalog over the time period 1996 to 2009, with an identification of MCs, which is the most complete at present. They showed that CMEs with MCs are more geoeffective, than other transient events. Apparently, considerable part of the MCs observed at Earth, have quasi-cylindrical geometry. This is supported by current ideas that the internal part of the CMEs is the original solar filament arranged as the long cylinder with “flux rope” structure [5]. This also agrees with observations of the SW disturbances at Earth, and with successful attempts to model MCs as cylindrical formations [6]. Small near-Earth part of a MC can be presented in the form of a quasi cylinder. In this work we consider only the isotropic part of the CR variations, i.e. CR density variations, inside the regions where the MC model is applicable.

It is natural to assume that exactly within a MC (owing to its expansion and relative isolation) the mechanism leading to Forbush decreases (FDs), is especially effective. However, it should be noted that FEs are created in the whole disturbance, and in many FDs associated to the MC, the CR minimum may appear inside and also outside of cloud, e.g., between the driven shock and MC nose [e.g., 4, 14]. We consider only the part of FEs which is directly related to the MC. In the simplest case this part of disturbance can be presented in the quasi cylinder form. Accordingly, the minimum of the CR density is expected to be observed near the central axis of this structure, while the CR density is expected to increase close to the surface of the quasi-cylinder. A simple function, capable to display such a distribution is the parabola. This parabolic function will be suitable not only for cylindrical, but for any distribution of CR with one extreme inside or even out of a MC. In the theoretical models the decision is given by more complex functions, but it is possible to show that in the first approximation they can be described with a parabola.

2. Data and Methods

To select the events for analysis we used the catalog of ICMEs [4] for the time period 1996 to 2009 – (the only complete catalog of interplanetary disturbances for these years). It includes key parameters of interplanetary disturbances, their solar sources and the accompanying geomagnetic effects. We selected those ICMEs which were included in the list of magnetic clouds of WIND (http://wind.nasa.gov/mfi/mag_cloud_pub1.html) and/or are listed in [7] and/or [8]. The CR parameters (variations of CR density and anisotropy) on the chosen events were taken from databases created in IZMIRAN. In the first database on “CR variations”, hourly values of the CR density and anisotropy above the Earth atmosphere and magnetosphere, specially calculated for 10GV rigidity by the Global Survey Method (GSM) [9] using data from world network of neutron monitors (www.nmdb.eu; http://cr0.izmiran.ru/mosc), are combined with solar, interplanetary and geomagnetic data [10] (http://omniweb.gsfc.nasa.gov/ow.html; ftp://ftp.gfz-potsdam.de/pub/home/obs/kp-ap/wdc; http://www.swpc.noaa.gov/). The second database “on interplanetary disturbances and Forbush effects (FEs)” contains characteristics of FEs on more than 6000 events, and parameters of associated with FEs, solar wind disturbances, indices of geomagnetic activity and information on solar sources [8, 13]. The information on SW is derived from the OMNI database (http://omniweb.gsfc.nasa.gov). In this work the influence of the variations of the magnetosphere on the CR density is investigated during geomagnetic storms. Furthermore, the contribution to the model of various coefficients and features leading to the density distribution of CRs in a MC is being discussed.

3. Discussion of the results

The selection of ICMEs with MCs observed at Earth within solar cycles 23-24 resulted at 99 events all of which are grouped by the presence of MCs, but all of them being very different.

As it was mentioned above, the simplest function, capable to model the distribution of the CR density variations in the quasi-cylindrical structure of the MC, is a parabola. Discrepancy of real distribution with a parabola is possible, but other circumstances which we consider below, are even more essential when modeling the behavior of the CR density in the MC.

1. We use data of the CR density variations obtained in each hour by the GSM. As the plasma from ICMEs extend almost radially, we assume almost a radial section (to define more precisely: a puncture) when studying the CR density hourly data. This puncture can pass through a cloud differently, but it doesn't limit the applicability of the parabolic representation of the CR density. Nonetheless, when modeling hourly averaged data, it is necessary to

remember that Earth, when crossing the interplanetary disturbance, passes for an hour different distance according to the speed of this part of the disturbance. For charged particles the distance is better expressed neither in kilometers (km) nor in astronomical units (AU), but in Larmor radii (gyroradii) ρ, which depends on the rigidity of particles (R) and on the hourly intensity (B) of the interplanetary magnetic field. It is possible to write down that for an hour the observer passes in the interplanetary disturbance (in particular, in a MC) a distance of $X_\rho = cVB/R$ gyroradii (this is equation of conversion of time into distance), where V is the radial speed of the SW. Upon transition from the linear sizes to gyroradii it is possible to expect more symmetric changes of the CR density in the cloud. This is confirmed in figure 1.

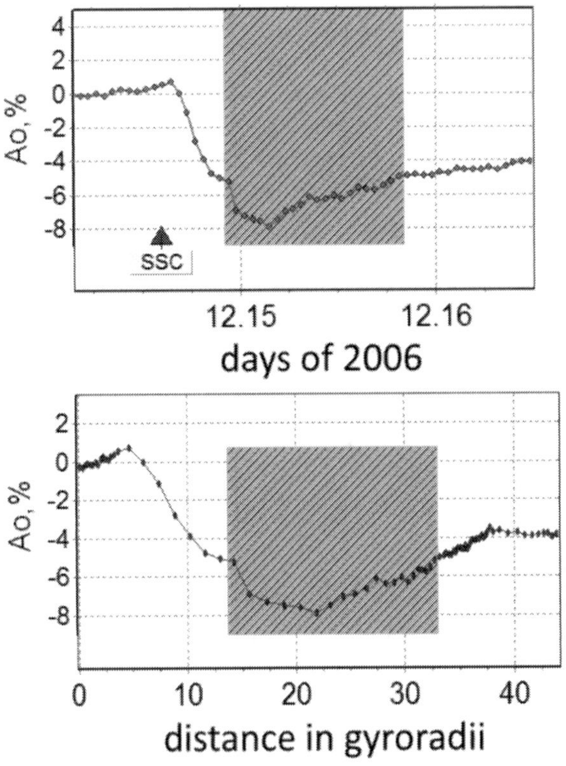

Figure 1. Example of the MC effect in cosmic rays: CR density variations (A0) plotted versus time (top panel) and depending on the gyroradii (low panel) for particles of 10 GV rigidity. Shaded area shows the MC period in this event (2006.12.15). Triangle (SSC) marks the beginning of the geomagnetic storm.

2. The second important fact which should be considered when studying the influence of MCs on the CR, using the ground level neutron monitor data, is the magnetospheric variations during the magnetic storms [11]. The main part of these variations is caused by the change of the geomagnetic cut off rigidity at the points of observation, thus it is variable, depending on the level of geomagnetic activity, resulting in an increase of the NM count rate. This magnetospheric variation partially remains in the changes of CR density obtained by the GSM. It was repeatedly noted [11] that variations of cut off rigidity and the corresponding variations of CR counting rate are closely correlated with changes of the Dst-index of the geomagnetic activity. This correlation seems to be extended also on the variation of CR density obtained by the GSM. Thus, the expected variation of the CR density in a MC can be written as:

$$A0= a+b_1X+ b_2X^2 +b_dDst \qquad (1)$$

where a is a constant, b_1 – trend coefficient, b_2 content the main part of the influence of MC on the CR, b_d defines a contribution of the magnetosphere. X is a distance in giro-radii from the onset of disturbance. Dst is the index of geomagnetic activity.

We applied this simple model to all 99 events of our selection, each time determining parameters a, b_1, b_2, b_d by the RMS method. So, we have a set of these parameters for 99 events. An example of the model simulation of the CR density variations (A0) in a MC is presented in Figure 2, where a good agreement between of the calculated and experimental data is clearly seen. Correlation coefficient for this event is 0.996, b_2 is 0.058± 0.006%/gyroradii.

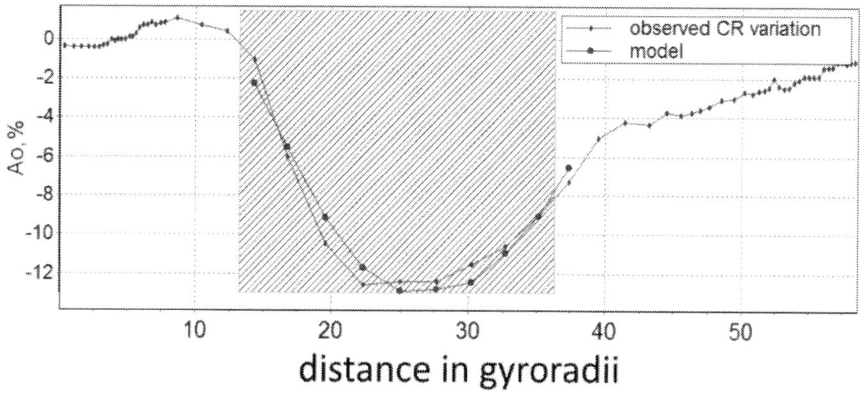

Figure 2. Example of modeling of the CR density in the MC in the event of 2004.07.26. Dependence of the CR density variations (A0) in the magnetic cloud on the gyroradii of particles with rigidity 10 GV: triangles –observable CR density variations obtained by GSM, points – model results of the CR density distribution in this event.

Good or, at least, satisfactory consent was present in the majority of the 99 events to what the high coefficient of correlation testifies. However agreement between the model and actual recorded behavior of the CR density was not always satisfactorily. A number of reasons may be accounted for such discrepancies. One of the main reasons is that the simplified representation of MCs in the form of a quasi-cylinder isn't obviously suitable in some cases. In these cases the dispersion in the rms method appears to be large and the correlation coefficient is small. But it is necessary to point out, that even low dispersion in data not always corresponds to good results. The number of hours the Earth spent within an MC varies in our selection from 6 to 64. Evidently, 6 hours (6 points) aren't enough to determine the four model parameters reliably. However, the mean duration of the MC is 21 hr with the range 6-64 hr, therefore the number of points is sufficient in most the cases examined in this work.

4. Statistical results of the MC modelling

Let's estimate statistical results of the modelling. We did not consider the MCs of length <4ρ and the cases with large dispersion between observable and calculated values in rms method (root mean square standard deviation) - σ^2 (when σ >0.3%). These criteria reduced our selection to a total of 74 events. The coefficient b_d defining a contribution of the magnetosphere variations in the CR density was not always possible to be determined precisely as the magnetosphere variations of CRs inside the MC are often too small. Those eleven events in which it was defined are plotted in figure 3 where b_d coefficient is presented against the minimum of the Dst-index at those hours when Earth was in the

MC. An averaged coefficient b_{dm} = 0.0136 ± 0.0016 %/nT and is plotted in figure 3 by the solid horizontal line. Shaded region is ± σ_{bd} – where σ_{bd} is standard statistical error of b_d coefficient.

It is possible to see that the distribution of points in figure 3 and the errors of separate coefficients are within the limits of the standard deviation σ. There is no dependence of b_d on the value of the Dst-index. The point dropping out at -103 nT, belongs to an event in February, 1998. We had no formal reasons to exclude it from the sample. But, most likely, in that event the structure of the MC was closer to toroidal-shaped, than to cylindrical and the parabolic dependence is an artifact due to casual correlation of CR density changes to the Dst-index. Thus, this coefficient (about 0.27% / nT), apparently is an overestimation and it is better not to consider it, further.

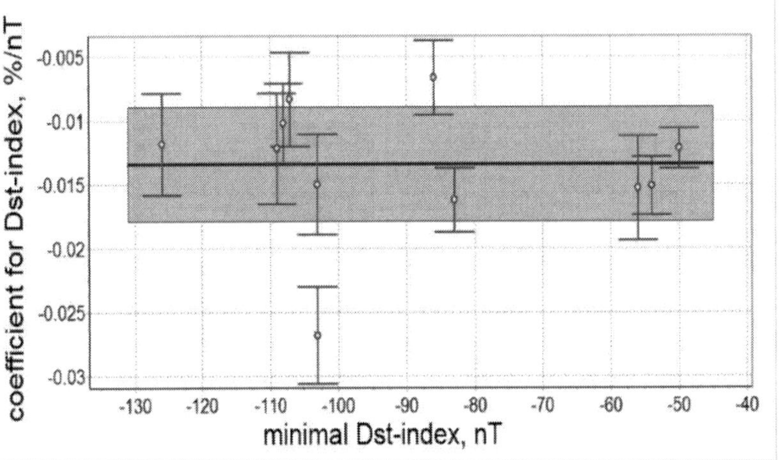

Figure 3. Magnitude of b_d coefficient versus the minimum of Dst index in the magnetic cloud.

Found dependence of CR variations of magnetospheric origin (see section 3, item 2) on Dst index is a magnetospheric effect, its size isn't related directly with type of interplanetary disturbances, whether it be ICME with MC, ICME without MC or a high-speed stream from coronal hole.

Thus, the obtained results allow one to assume that in all events (or, at least, in their majority) the relation of magnetospheric variations with changes of CR density is approximately the same and is sufficiently defined by the average coefficient of b_{dm} specified here above. In spite of the fact that the magnetic storms associated with MC were very different (long and short, large and small) the b_d coefficient was constant for various type of the storms. It gives promise that the obtained dependence is applicable not only to MC, but also to other structures caused CR variations of a magnetospheric origin.

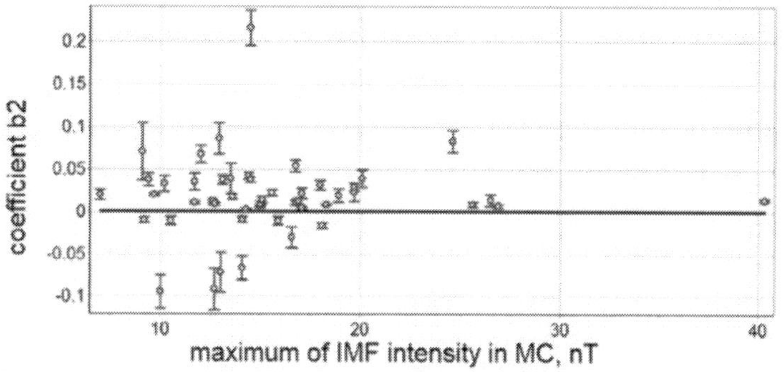

Figure 4. Dependence of b_2 coefficient on the IMF intensity (Bmax) inside the magnetic cloud

The coefficient b_2 (Eq. 1) is the basic for parabolic model. This coefficient displays the main part of a MC influence on the CR density. A positive coefficient corresponds to fall of CR density in model. If the value of coefficient b_2 is small, comparable with a statistical error, it, as a rule, has to mean that this MC poorly influences CRs near Earth. Since FEs are large scale heliospheric phenomena, this influence can be more effective in other locations in the heliosphere [11, 12].

For further analysis we selected events, for which $b_2/\sigma_{b2}>2$, (where σ_{b2} is a standard statistical error of b_2 definition) to provide statistically significant contribution of b_2 in the model. This criterion reduced our sample to 40 events. In figure 4 the distribution of the coefficient b_2 obtained from parabolic model by the 40 events, vs the maximum intensity of IMF (*Bmax*) in the MC, is presented.

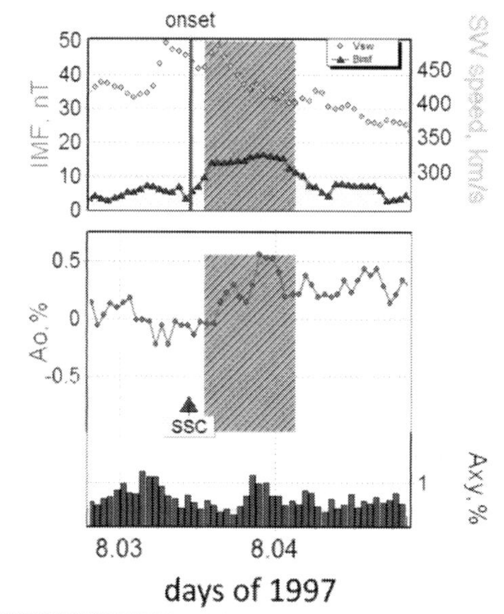

Figure 5. CR density (A0) and anisotropy (Axy) variations with effect of increasing density inside the MC (low panel) for event on August 3-4, 1997. Solar wind speed and IMF intensity are presented on the upper panel.

Prevalence of positive b_2 (i.e. events with the local minimum of CR density in the MC) is obvious. Among of 40 events for which the parabolic model is applicable and the influence of a MC on CRs is noticeable, only 10 events have $b_2<0$. Those are the events, in which within a MC a local maximum instead of minimum of CR density is revealed (see figure 5), and they are the minority [15]. Those can't be explained with bad data or magnetospheric effects. These are real increases of CR density in MCs and, as a rule, they are observed in MCs with relatively weak magnetic field. One of the possible explanations of such anomalous effect might be the following: while in propagation, in some MCs the magnetic field weakens, and they stop being an effective quasi trap for high-energy CR. But regular IMF (fluxrope structure) connects a point of observation to rather remote western areas of a heliosphere where FEs are weaker and correspondingly the CR density higher.

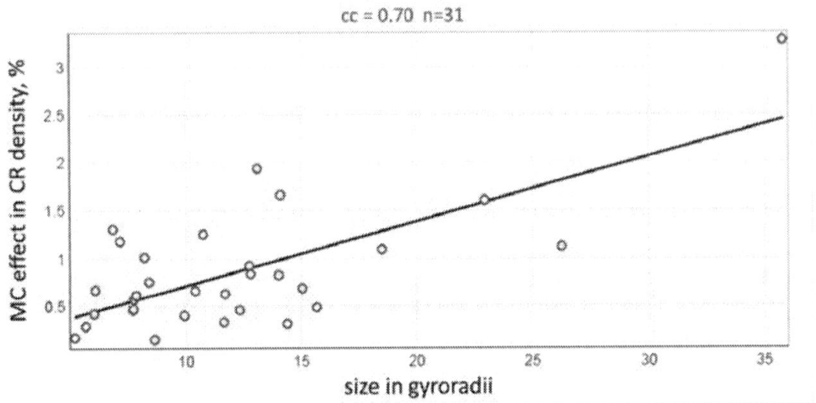

Figure 6. Magnitude of the MC effect on cosmic ray density (amplitude of Forbush decreases) versus the size of magnetic cloud (in gyroradii).

Accordingly to results in figure 4 an obvious dependence of $b_2$on Bmax isn't observed, but it is possible to note that all negative b_2 were observed in relatively weak field, and for Bmax>18 nT all b_2 are positive. Hence, in our selection, in cases of strong magnetic field we always have the normal effect MC on the CR with a local minimum of the CR density within the MC.

In some works [4, 14] is shown that the FE production seems to be relevant with a duration of the MC influence. We think that it is possible to assume that the influence of a MC on the CR density is defined also by its size (in gyro radii).This is confirmed by figure 6 where the magnitude of the influence of the MCs on the CR density is presented for 31 events within which the MCs led to a decrease of the CR density. We see that correlation exists but it is not so large (correlation coefficient cc= 0.70) as the size is not the only parameter which influences of the FE production. It should be noted that the influence of the majority of MCs on the CR density variations is insignificant. In 41 out of the 74 events this influence is <0.3%, and in 50 <0.5%. It is clear that such small effects can't almost be revealed by the data of a single detector. It is no surprise that some researchers came to a conclusion that MCs don't influence CRs. However, it seems to us that all CMEs/ICMEs influence CRs, especially, when MCs are observed at Earth. But effects of ICMEs and MCs often use to be hidden by the influence of other perturbations, so they are often small and difficult for detection.

The extremes (minimum or maximum) of CR density settle down closer to the cloud centre more often than to its edges. This is visible in figure 7. Let's divide all MCs into two parts: central (25-75% of all length) and lateral. Among all 40 events, 34 fall into the central zone and only 6 in lateral. One can see from figure 7 that events with a positive effect (empty circles) have a maximum, generally in the leading part of a MC. In the events with a negative effect (full circles) minima density is distributed more evenly, but tend to be grouped in the tail part of a MC.

5. Conclusions

In large part of events the CR density changes in a magnetic cloud, providing an almost symmetric picture, with the CR density minimum close to the MC center, allowing one to assume its quasi cylindrical structure. Events in which the behavior of the CR density, remain regular but becomes more complicated, with alternations within areas of the MC with an increase and/or lower density, are rather frequent also. Those can be manifestations of the quasi-toroidal structure of some MCs.

In most cases (but not in all) the behavior of the CR density within MCs at Earth can be described by a simple parabolic dependence on distance expressed in gyroradii. The majority (but not all) of MCs modulate CRs, decreasing their density, although there is a group of events (about 1/5) in which the CR density increases near the MC center.

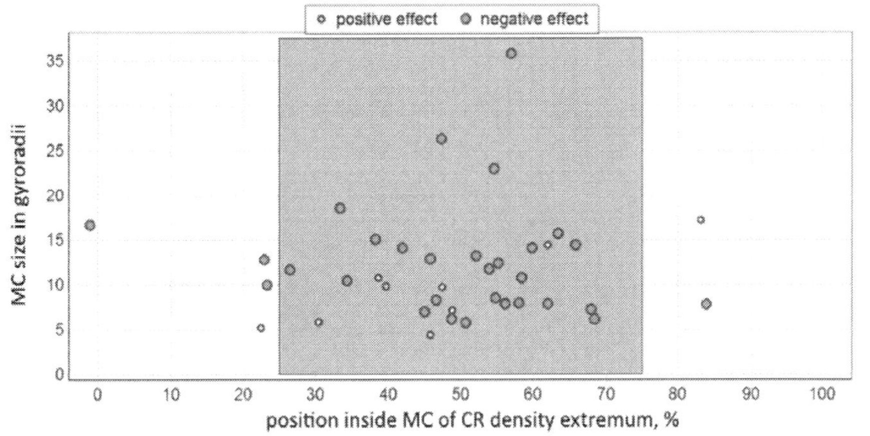

Figure 7. The relative position of extreme CR density within MC. It is defined in a % of all MO: 0% correspond to Earth entrance and 100% -to Earth exit from the cloud

A quantitative relation of Dst index and contribution of a magnetosphere into the CR density variations derived by GSM, was determined via the averaged coefficient b_d.

An obvious dependence of a contribution to CR density variations on Bmax - the maximum intensity of IMF measured in a magnetic cloud isn't observed. But it is possible to note that all positive effects (CR density increases in MCs) were observed in rather weaker field, and for Bmax> 18 nT all variations of CR density are negative, i.e., in cases of a strong magnetic field we always have "normal" FDs with a local minimum of CR density in a MC in our selection.

The extremes (minimum or maximum) of CR density more often settle down closer to the cloud center, than to its edges. Events with a positive effect have a maximum, generally in the leading part of a MC. In the events with a negative effect minima density is distributed more evenly, but tend to be grouped in the tail part of the MC.

Acknowledgements: The authors are thankful to collaborators of the world wide network of CR stations providing data of continue monitoring of neutron component: http://cr0.izmiran.ru/ThankYou/. We acknowledge the NMDB database (www.nmdb.eu), founded under the European Union's FP7 programme (contract no. 213007) for providing data.
Furthermore, we are thankful to the anonymous referees whose thoughtful comments contributed significantly to this paper.

References

[1] Burlaga L, Sittler E, Mariani F and Schwenn R 1981 *J. Geophys. Res.* **86** 6673
[2] Zhang S, Hess P and Poomvises W A 2013 *Solar Phys.* **284** 89
[3] Yashiro S, Gopalswamy N, Makela P and Akiyama S 2013 *Solar Phys.* **284** 5
[4] Richardson I and Cane H 2010 *Solar Phys.* **264** 189
[5] Bothmer V and Schwenn R 1998 *Ann. Geophys.* **16** 1
[6] Kuwabara T, Bieber J, Evenson P, et al. 2009 *J. Geophys. Res.* **114** A5
[7] Gosling J.T. 1990 *Geophys. Monogr. Ser.* **58** 343

[8] Huttunen K, Schwenn R, Bothmer V and Koskinen H 2005 *Ann. Geophys.* **23** 625
[9] Belov A, Baisultanova L, Eroshenko E, Yanke V, et al., 2005 *J. Geophys. Res.* **110** 1
[10] Asipenka A.S., Belov A.V., Eroshenko E.F., Klepach E.G., Yanke, V.G. 2009 *Advances in Space Research*, V. 43, Issue 4, p. 708-716
[11] Dorman L. I. 2010 *Cosmic Rays in Magnetospheres of the Earth and other Planets* (Dordrecht: Springer)
[12] Belov A. 2009 *Proc. IAU257 Symposium. Ed. N. Gopalswamy and D. Webb.* 257 439-450
[13] Eroshenko E., Belov A., Mavromichalaki et al., 2009 *Proc. IAU257 Symposium. Ed. N. Gopalswamy and D. Webb.* **257**. 451-457
[14] Blanco et al, A&A 556, A146 (2013)
[15] Abunin, A., Abunina,M., Belov, A., et al. 2013 *ICRC2013-0198*

CaLMa Neutron Monitor: current status, first observations and future improvements

J. J. Blanco[1,2], J. Medina[1,2], O. García-Población[1,2], R. Gómez-Herrero[1,2], I. García-Tejedor[1,2], S. Ayuso[1,2] and E. J. Catalán[1,2]

[1] Space Research Group, University of Alcalá, Spain
[2] Castilla-La Mancha Neutron Monitor, Parque Científico y Tecnológico de Castilla-La Mancha, Spain

E-mail: juanjo.blanco@uah.es

Abstract. Castilla-La Mancha Neutron Monitor (CaLMa) is a 15 boron trifluoride counter neutron monitor located in Guadalajara, Spain ($40°38'N$, $3°9'W$) at 708 m above sea level (asl) and a vertical rigidity cut-off of 6.95 GV. CaLMa is able to measure neutrons produced by primary cosmic rays with energies above 6 GeV if protons are these primary cosmic rays. CaLMa has been continuously monitoring solar activity since October 2011, having reported more than 25 Forbush decreases. This work presents the preliminary results inferred from CaLMa's count rate variations, shows the latest improvements carried out in CaLMa and outlines our future plans to improve the station

1. Introduction

Neutron monitors are ground level instruments, and the vertical cut-off rigidity at its geographical location together with the atmospheric cut-off energy impose an energy threshold to the primaries that are able to produce secondary cosmic rays recorded that each neutron monitor is able to record. A neutron monitor measures secondary cosmic rays produced by primary cosmic rays with energies up to about 100 GeV with a maximum ranged between 4 and 7 GV at sea-level and shows a strong dependence on the solar cycle [1].

The Castilla-La Mancha neutron monitor (CaLMa) is a 15NM64 station located in Guadalajara, Spain ($40°38'N$, $3°9'W$) at 708 m asl with a vertical cut-off rigidity of 6.95 GV (`http://www.calmanm.es`). It has been operating since 26 October 2011 in a commissioning phase reaching its full operational status (i.e. 15 counters) on 11 July 2012 [2].

A dedicated new registration system based on a field-programmable gate array (FPGA) card has been designed for recording pulses from CaLMa's counters [3]. This system controls 15 counters, the high accuracy BM35 barometer device (`http://www.meteolabor.ch/fileadmin/user_upload/pdf/meteo/WX/bm35_e.pdf`) and three additional channels which are used to test new sensors that could be added to CaLMa in the future. This new registration system has been operating at the Kiel neutron monitor in parallel with the old one, for which it has been shown to be an adequate replacement [3]. On the other hand, the acquisition system is able to interface via internet with the Neutron Monitor Database (NMDB) [4] uploading measurements every minute, both count rate and pressure, so that real time data is available to the end users. The NMDB is a database where many neutron monitor stations upload their measurements, making them available to scientists and other interested parties (`http://www.nmdb.eu`). CaLMa data are reduced according to the median filter algorithm procedure described by Belov et al. [5].The median rates are subsequently corrected from atmospheric pressure effects [6] applying a factor $\exp(-\beta(P - P_0))$, where $P_0 = 935 \pm 6hPa$ is the station standard pressure and $\beta = -0.0067 \pm 0.0002\ hPa^{-1}$ is the barometric correction factor.

2. Following the solar activity with CaLMa

CaLMa measures mainly secondary neutrons produced by cosmic rays with rigidities higher than 6.95 GV. At such energies, cosmic rays are affected by solar activity. This lets us infer the effect of solar activity on Earth, based on temporal variations in CaLMa's count-rate. Since CaLMa has been in operation for almost 4 years, long term and short term variations in solar activity can be studied using its measurements. It is well established that solar activity follows a cycle of eleven years directly related with a higher (maximum) and lower (minimum) sunspot number. Monthly sunspot numbers along the last solar cycle are shown in Fig. 1 (continuous blue line). According to Fig. 1, the last solar activity minimum occurred around 2009 when the lowest sunspot number was recorded. Then, a first maximum happened at the end of 2011, followed by a relative low and then almost flat solar activity from the beginning of 2012 up to the end of 2013. Then, the solar activity grew up reaching a second maximum during early 2014. It can be said that this maximum presents a Gnevyshev gap [7] as did the precedent solar maximum. During this period the solar activity is, on average, lower than during other periods, and this affects the solar magnetic field, the rate of solar flares, coronal mass ejections and solar energetic particle events [8].
CaLMa started to operate during the current solar maximum. CaLMa count-rate shows a behaviour similar to other neutron monitors; for example, CaLMa in Fig.1 is compared to Rome neutron monitor in monthly averaged count rate (blue and green lines respectively). Both neutron monitors have similar vertical cut-off rigidities, 6.27 GV in Rome and 6.95 GV in CalMa, and because of that, similar cosmic ray variations are expected to be measured by both stations. CaLMa and Rome count-rates do not clearly anti-correlate with the sunspot number. Maybe this is caused by the short recording period of only four years, due to which neutron monitor count rates could be more dependent on the number of interplanetary coronal mass ejection arriving at the Earth than the sunspot number. On the other hand, a four year period is not long enough to make a meaningful assessment related to the eleven year solar cycle.

2.1. Short term solar activity

Since the work by Forbush in 1938 [9] it is well known that solar activity can produce decreases in the number of cosmic rays reaching Earth. These decreases are characterized by a fast decreasing phase with respect to the background count-rate level previous to the Forbush decrease (FD), followed by a slow recovery phase which finishes when the cosmic ray count-rate shows again small variations and, in most the cases, recovers the measured background values before the beginning of the FD. The decrease phase usually takes less than one day, while the recovery phase can last from several days up to three weeks [10]. These FDs observed with neutron monitors are caused by the interaction between solar wind transients such as interplanetary coronal mass ejections (ICME), and stream interaction regions (SIR), with the geomagnetosphere [11]. Magnetic clouds (MC) are observed in ICMEs very often. A MC is characterized by large magnetic field intensity and smooth magnetic field rotation. An effective interaction between MCs and Earth magnetosphere can be expected because of its magnetic topology. An example of a FD observed by CaLMa at the end of February 2014 is depicted in Fig.2. This FD was produced by a shock+MC+MC combo. CaLMa count-rate decreased by 4%. The decrease

Journal of Physics: Conference Series 632 (2015) 012052 doi:10.1088/1742-6596/632/1/012052

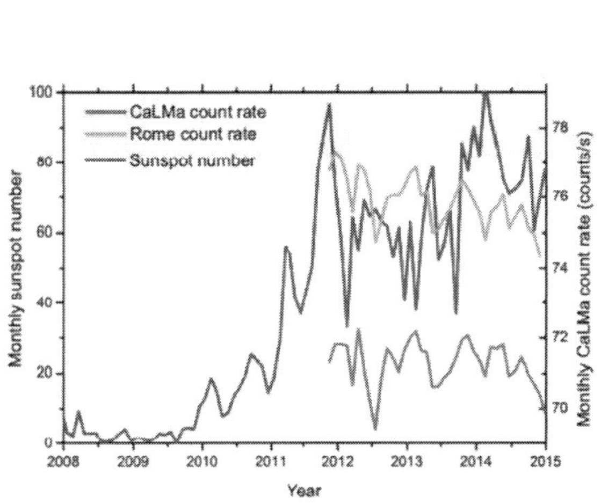

Figure 1. Monthly averaged sunspot number (continuous blue line), monthly averaged CaLMa count-rate (continuous red line) and monthly averaged CaLMa count-rate (continuous green line). Rome's count-rate has been divided by a factor 2 to make its count-rate visually comparable with CaLMa's data.

started with the shock arrival to Earth (first dotted vertical line), followed by a fast reduction in the count-rate during the passage of the first MC and the minimum was reached in coincidence with the second MC. The MCs borders are marked by the second and the third dotted vertical lines, nose and rear of the first MC respectively, and the third and fourth dotted vertical lines show the borders of the second MC. The recovery phase started after the second MC arrival, lasting about five days. The magnetic clouds are confirmed by the relative strong magnetic field intensity and the rotation of the magnetic field components.

The same event also produced a strong geomagnetic storm as can be observed in the Dst and Kp index variations showed in the second and third panels in Fig.2. The geomagnetic storm started later than the shock arrival and in coincidence with the first MC's nose. The minimum in Dst index was detected in coincidence with the first MC. The recovery phase of the geomagnetic storm begun with the second MC cloud arrival.

CaLMa has observed 29 FDs up to now. Twenty seven of them were induced by ICMEs with clear signatures of magnetic cloud, i.e. high magnetic field strength and smooth magnetic field rotation, both with other solar wind transients as shocks or SIR, or simply isolated ICMEs. Only in one case the solar wind transient associated to the FD could not be identified. About 50% of the events were observed with geomagnetic storms (Table 1). This Table 1 lists the solar wind structure observed with each FD, i.e. interplanetary shocks (sh), magnetic clouds (MC) stream interaction regions (SIR) or no clear when there are not clear signatures of the previous solar wind structures in the first column, the date when the FD minimum is observed in second column, the percentage of decrease in CaLMa count-rate in third column and finally, if a geomagnetic storm is observed with the FD (Yes/No) in the fourth column. It is clear from the table that only some of the FDs happened in coincidence with strong geomagnetospheric responses.

The strongest energetic solar events such as flares and very fast coronal mass ejections can accelerate a huge amount of particles up to energies of some GeV. These particles, solar cosmic rays, can produce an increment in the count-rate recorded by neutron monitors. These enhancements in neutron monitor count-rates are referred as ground level enhancements (GLE). Only one GLE (17 May 2012) has been clearly observed during the current solar cycle by the worldwide network of neutron monitors. A recent work about this GLE can be found in [12].

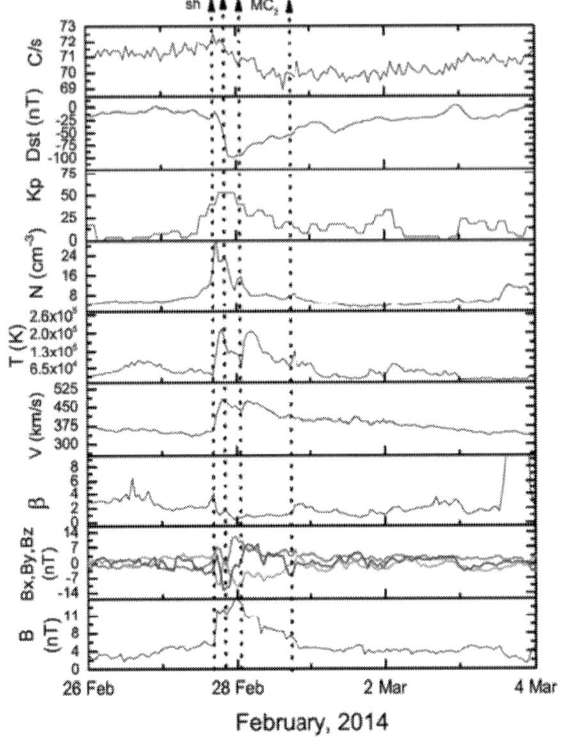

Figure 2. FD observed by CaLMa. From top to bottom CaLMa count-rate, Dst index, Kp index, solar wind proton density, temperature and speed, plasma beta, GSE magnetic field components and field intensity at 1 AU.

However, this GLE was not observed in CaLMa because of the high cut-off rigidity of our station.

3. Close future work

CaLMa is nowadays a fully operational 15NM64 station. Nevertheless, this is not the final stage of CaLMa. The station is going to be complemented with a muon telescope which will give information about the muon component of showers produced by primary cosmic rays. It is expected to obtain valuable information about cosmic ray anisotropy, especially during a FD. In parallel with the muon telescope implementation, a new data acquisition system will be developed to control both instruments, the neutron monitor and the muon telescope. This new system will be able to record individual events with a resolution of a few nanoseconds and will combine counts from both instruments, detect multiplicities in the neutron monitor and in the muon telescope, and store directional muon information from a new coincidence logic system.

For the first operational test, we have built a muon telescope laboratory model made up of a pile of two 30x30x3 cm plastic scintillators which are 12 cm appart, with a lead block with dimensions of 30x30x5 cm in between. The prototype is presented in Fig.3. The upper scintillator and photomultiplier tube (PMT) box can be seen on the upper left corner of the picture, the high power supply is close to it and also on the table. The bottom scintillator

Table 1. FDs in percentage observed by CaLMa from November 2011 to June 2014. gmtorm=geomagnetic storm and sh=shock

Event	Date	FD	gmtorm
sh+MC+SIR	25/01/2012 00:00	2.3	Yes
sh+MC+sh+MC	08/03/2012 23:00	8.9	Yes
sh+MC	13/03/2012 07:00	3.1	No
MC	06/04/2012 02:00	4.7	No
sh+SIR+MC	26/04/2012 14:00	1.9	Yes
sh+MC	17/06/2012 00:00	3.0	Yes
sh+MC	14/07/2012 16:00	3.0	Yes
sh+MC+MC	05/09/2012 03:00	2.9	Yes
sh+MC	13/11/2012 15:00	3.6	Yes
sh+MC + SIR	24/11/2012 23:00	3.3	No
MC	17/01/2013 13:00	1.9	No
sh+SIR+sh+MC	18/03/2013 20:00	5.1	Yes
sh+MC	14/04/2013 18:00	3.7	No
sh	24/06/2013 07:00	2.3	No
sh+MC	28/06/2013 08:00	2.1	Yes
sh+MC	13/07/2013 09:00	2.4	No
MC	03/09/2013 20:00	2	No
SIR+sh+MC	02/10/2013 10:00	2.5	Yes
MC	15/12/2013 19:00	2.9	No
No clear	10/01/2014 02:00	2	No
sh+MC+MC	15/02/2014 16:00	2.7	No
sh+MC+MC	28/02/2014 09:00	4	Yes
sh+MC	05/04/2014 20:00	2	No
SIR+MC+sh+MC?	18/04/2014 21:00	4.5	No
sh+MC	07/06/2014 21 00	3.3	No
sh+MC	13/07/2014 19:00	1.9	No
sh+MC	07/09/2014 08:00	2	No
sh+MC	13/09/2014 01:00	4.1	Yes
sh+sh+MC	23/12/2014 19:00	6	Yes

box is placed just below. On the right side of the picture, the data acquisition chain composed by NIM modules and a computer to visualize the multichannel analyser output is shown. The scintillators are operated in coincidence, while the upper one and the lead block can be moved horizontally allowing different fields of view for testing the directionality of muon flux. The observed angular distribution of muons is plotted in Fig.4.

Photons from the scintillators are collected by a photomultiplier tube connected to a typical Canberra chain, i.e. pre-amplifier, amplifier connected to an analogical digital converter (ADC) and to a single channel analyser (SCA) both connected (ADC and SCA) to a coincidence module. The final output from the ADC is recorded by a dedicated acquisition system [13] with a Beaglebone black (BBB) as core of the system (upper block diagram in Fig.5). We refer this acquisition chain as hardware coincidence. The BeagleBone Black is a low-cost, community-supported development platform for developers that runs linux (`http://beagleboard.org/black`). Using the BBB capabilities, the acquisition chain has been highly simplified taking out the SCA and the coincidence module from the initial acquisition

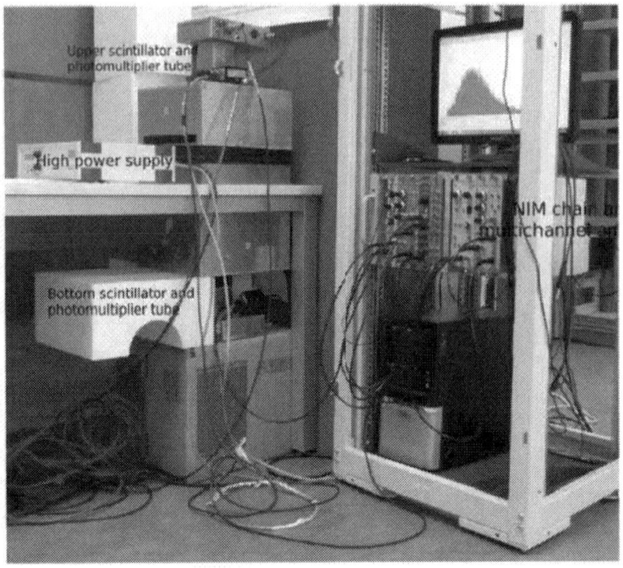

Figure 3. Muon telescope laboratory model.

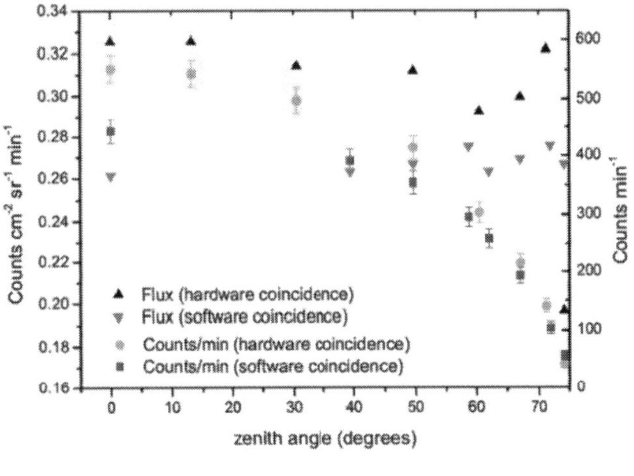

Figure 4. Observed angular distribution of muons arriving CaLMa.

chain (bottom block diagram Fig.5). We refer this acquisition chain as software coincidence. Both these acquisition systems were operated alternatively. The software coincidence chain recorded count-rates (green circles in Fig.4) similar to those recorded by the hardware coincidence chain (black squares in Fig.4). When considering the geometric factor, a clear angular dependence of muon flux arriving to the detector is not observed. This is expected for an isotropic muon flux. Both the coincidence hardware chain (black pointed up triangle) and the coincidence software chain (red pointed down triangle) obtain the same result. We have to point out that no atmospheric corrections have been done yet and therefore differences between muon path length caused by different angles of incidence are not considered.

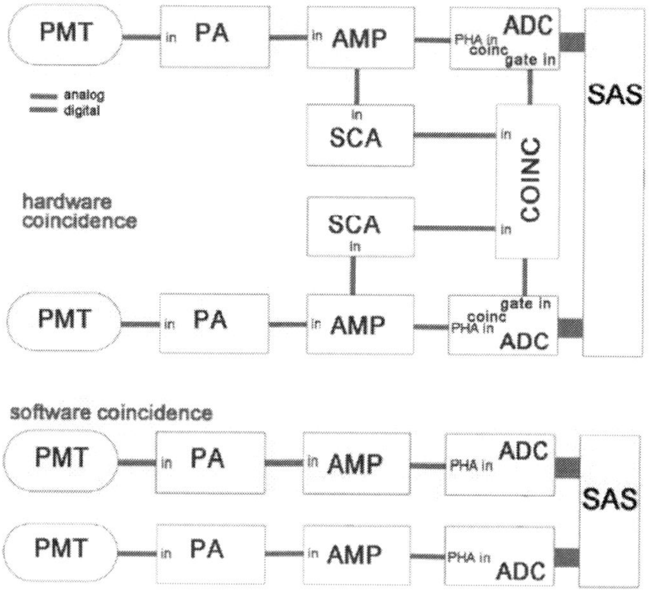

Figure 5. Block diagram of the muon telescope acquisition chain. From the photomultiplier tube (PMT) through the preamplifier (PA), amplifier (AMP), analog to digital converter (ADC), single channel analyser (SCA) and a coincidence module (COINC). The chain ends in a dedicated acquisition system based on a Beaglebone black (BBB) named Software-coincidence Adquisition System (SAS).

4. Conclusions

We have presented the current status and close future plans for the Castilla-La Mancha neutron monitor (CaLMa). CaLMa has been operational since October 2011 producing one minute resolution measurements. These counts are uploaded in real time to the NMDB allowing the use of CaLMa data as source for space weather alarms. CaLMa is observing both solar cycle variations and short term variations of cosmic rays produced by the solar activity. Up to now, CaLMa has detected 29 Forbush decreases, most of them caused by a shock+MC combo. Only 50% of the FDs detected by CaLMa happened during strong geomagnetic storms. CaLMa is growing to be a dual station composed by a neutron monitor and a muon telescope and controlled by a new data acquisition system which is now under development. The muon telescope will add directional information allowing us to study anisotropies in cosmic rays during a Forbush decrease.

Acknowledgments

The authors would like to thank the OMNIweb team for the use of their database and Dr. M. Storini for the use of Rome neutron monitor data. This work has been partially supported by Ministerio de Ciencia y Tecnologia through the project AYA2012-39810-C02-01 The authors want also to acknowledge the Observatoire Midi-Pyrenees, Karl-Ludwig Klein from the L'Observatoire de Paris and Francis Beigbeder from the Observatoire Midi-Pyrenees for putting the BP28 tubes at our disposition,enabling us to start our station. The authors would like to thank for putting the BP28 counters at our disposition, enabling us to start our station. Also thank to Christian T. Steigies, Bernd Heber, Eckart Böhm and Cesar Martín from the Christian Albrechts University in Kiel,Germany, for their help in our research and new improvements.

References

[1] Clem J M and Dorman L I 2000 *Space Sci. Rev* **93** 335–359
[2] Medina J, Blanco J J, García O, Gómez-Herrero R, Catalán E J, García I, Hidalgo M A, Meziat D, Prieto M, Rodríguez-Pacheco J and Sánchez S 2013 *Nucl. Instrum. Meth. A* **727** 97–103
[3] Población O G, Blanco J J, Gómez-Herrero R, Steigies C T, Medina J, Tejedor I G and Sánchez S 2014 *JINST* **9** T08002
[4] Mavromichalaki H, Papaioannou A, Plainaki C, Sarlanis C, Souvatzoglou G, Gerontidou M, Papailiou M, Eroshenko E, Belov A, Yanke V, Flückiger E O, Bütikofer R, Parisi M, Storini M, Klein K L, Fuller N, Steigies C T, Rother O M, Heber B, Wimmer-Schweingruber R F, Kudela K, Strharsky I, Langer R, Usoskin I, Ibragimov A, Chilingaryan A, Hovsepyan G, Reymers A, Yeghikyan A, Kryakunova O, Dryn E, Nikolayevskiy N, Dorman L and Pustil'Nik L 2011 *Adv.Space Res.* **47** 2210–2222
[5] Belov A V, Blokh Y L, Klepach E G and Yanke V G 1988 *Kosm.Luchi* **25** 113
[6] Lockwood J A and Calawa A R 1957 *J.Atmos.Sol.Terr.Phys.* **11** 23–30
[7] Gnevyshev M N 1967 *Sol. Phys.* **1** 107–120
[8] Rodríguez-Pacheco J, Blanco J J, Heber B and Gómez-Herrero R 2012 *Sol. Phys.* **281** 491–499
[9] Forbush S E 1938 *Phys. Rev* **54**(12) 975–988
[10] Usoskin I G, Braun I, Gladysheva O G, HöRandel J R, JäMséN T, Kovaltsov G A and Starodubtsev S A 2008 *J. Geophys. Res. (Space Physics)* **113** A07102
[11] Blanco J J, Catalán E, Hidalgo M A, Medina J, García O and Rodríguez-Pacheco J 2013 *Sol. Phys.* **284** 167–178
[12] Mishev A L, Kocharov L G and Usoskin I G 2014 *J. Geophys. Res. (Space Physics)* **119** 670–679
[13] Ayuso S *Tarjeta de adquisición de datos para cadenas de instrumentación nuclear* Master's thesis Alcala University

What are the causes for the spread of GLE parameters deduced from NM data?

R Bütikofer[1,2] and E Flückiger[1,2]

[1] University of Bern, Physikalisches Institut, Switzerland
[2] International Foundation High Altitude Research Stations Jungfraujoch and Gornergrat, Bern, Switzerland

E-mail: rolf.buetikofer@space.unibe.ch

Abstract. Investigations have shown that the analysis results of ground level enhancements (GLEs) based on neutron monitor (NM) data for a selected event can differ considerably depending the procedure used. This may have significant consequences e.g. for the assessment of radiation doses at flight altitudes. The reasons for the spread of the GLE parameters deduced from NM data can be manifold and are at present unclear. They include differences in specific properties of the various analysis procedures (e.g. NM response functions, different ways in taking into account the dynamics of the Earth's magnetospheric field), different characterisations of the solar particle flux near Earth as well as the specific selection of NM stations used for the analysis. In the present paper we quantitatively investigate this problem for a time interval during the maximum phase of the GLE on 13 December 2006. We present and discuss the changes in the resulting GLE parameters when using different NM response functions, different model representations of the Earth's magnetospheric field as well as different assumptions for the solar particle spectrum and pitch angle distribution near Earth. The results of the study are expected to yield a basis for the reduction in the spread of the GLE parameters deduced from NM data.

1. Introduction

The worldwide network of neutron monitors (NMs) together with the geomagnetic field acts as a giant spectrometer and enables to determine the spectral variations of the galactic cosmic rays near Earth and the characteristics of sporadic solar cosmic ray (SCR) events in the energy range of ~500 MeV to ~15 GeV. Since the introduction of NMs in the 1950s about 70 SCR events were observed by the ground-based cosmic ray detectors. The analysis of these so-called ground level enhancements (GLEs) is essential to understand the particle acceleration mechanisms at or near the Sun and the transport of the SCRs in the interplanetary and near-Earth space. Beside this aspect in fundamental research, the investigation of GLEs is also of particular interest as e.g. the energetic solar cosmic rays may significantly increase the radiation dose rates at flight altitudes.

The results of GLE analysis based on NM data are used as input for e.g. the determination of the radiation doses along flight routes during a GLE. However, investigations in the past [1, 2] have shown that for a selected event the results of different GLE NM analysis procedures may differ considerably. The comparison of published GLE characteristics e.g. during the maximum phase of the GLE on 20 January 2005 shows differences between the highest and lowest published solar proton intensity in the direction of maximum flux at 1 GV of almost two orders of magnitude [2]. The reasons for the spread of the GLE parameters deduced from NM data are at present unclear and may be manifold. They include differences in specific properties of the various analysis procedures, different characterisations of the solar particle flux near Earth as well as the specific selection of NM stations used for the analysis.

In this paper we describe recent work where we investigated the influence of the differences in specific properties of the various analysis procedures on the resulting GLE characteristics during the maximum phase of the GLE on 13 December 2006 (~0300 UT). This solar cosmic-ray event is ranked among the largest in solar cycle 23. The NM stations Oulu with 90% and Apatity with 78% measured the highest relative count rate increases.

2. Investigations

In the following we present the effects on the GLE results when using different NM response functions (section 2.1), different descriptions of the pitch angle distribution (section 2.2), different considerations of the dynamics of the Earth's magnetospheric field (section 2.3), and different selection and number of used NM stations (section 2.4).

The response of a NM to relativistic solar protons can be expressed in the following simplified form:

$$\Delta N_i(t) = \sum_{P=P_c^i(t)}^{\infty} S_i(P) \cdot J_{||}(P,t) \cdot F(\delta_i(P,t),t) \cdot \Delta P \qquad (1)$$

where

$\Delta N_i(t)$	count rate increase at the NM station i due to solar protons as function of time t,
P	particle rigidity,
$P_c^i(t)$	effective vertical geomagnetic cutoff rigidity at the location of the NM station i,
$S_i(P)$	yield function of NM station i,
$J_{\|\|}(P,t)$	solar particle intensity near Earth in the direction of maximum flux (source direction, usually the field vector of the interplanetary magnetic field near Earth),
$F(\delta,t)$	pitch angle distribution of solar particles,
δ	pitch angle, i.e. angular distance between the velocity direction of the particle and the magnetic field vector,
$\delta_i(P,t)$	angular distance between the source direction (usually the field vector of the interplanetary magnetic field near Earth) and the rigidity dependent particle arrival direction outside the geomagnetosphere (asymptotic direction) for particles of vertical incidence at the top of the atmosphere above the location of the NM station i,
ΔP	rigidity interval.

The GLE characteristics are determined by minimizing the sum of squared differences between $\Delta N_{calc.}$ and $\Delta N_{obs.}$ for the selected set of NMs using a trial and error procedure.

For the investigations under sections 2.1, 2.3 and 2.4 an exponential form for the pitch angle distribution, $F(\delta,t)$, with two free parameters $A(t)$ and $B(t)$ was used as proposed by Bombardieri [3]:

$$F(\delta,t) = \exp\left(\frac{-0.5 \cdot (\delta - \sin(\delta) \cdot \cos(\delta))}{A(t) - 0.5 \cdot (A(t) - B(t)) \cdot (1 - \cos(\delta))}\right) \qquad (2)$$

The trajectories of the solar cosmic ray protons (asymptotic directions, effective vertical cutoff rigidities) are computed with MAGNETOCOSMICS [4]. The inner geomagnetic field was described by the IGRF field (epoch 2005.0) [5] and the outer field by the Tsyganenko model 1989 [6] with the K_p-index on 13 December 2006 at 0300 UT ($K_p = 3$) for the sections 2.1, 2.2 and 2.4. We assumed a pure power law in rigidity for the differential rigidity spectrum of the solar cosmic ray protons. The data of the following 36 NM stations of the worldwide network were used for the GLE analysis presented in sections 2.1, 2.2 and 2.3: Alma Ata, Apatity 18NM64, Athens, Cape Schmidt, Fort Smith, Hermanus, Inuvik, Irkutsk, Irkutsk2, Irkutsk3, Jungfraujoch NM64, Jungfraujoch IGY, Kerguelen, Kiel, Kingston, Larc, Lomnický Štít, Magadan, Mawson, McMurdo, Moscow 24NM64, Moscow 6NM64, Nain, Norilsk, Novosibirsk, OLC, Oulu, Peawanuck, Potchefstroom, Rome, Sanae, Terre Adelie, Tixie Bay, Tsumeb, Thule, and Yakutsk.

2.1. Effect of different NM response functions

We considered the following three NM response functions: coupling function by Dorman and Yanke [7] (including modifications for rigidities <2.78 GV according to Belov and Struminsky [8]), the specific yield function by Debrunner at al. [9] and the parameterized yield function by Flückiger et al. [10]. The coupling function derived by Dorman and Yanke [7] is based mainly on NM measurements during latitude surveys, whereas the yield functions by Debrunner at al. [9] and Flückiger et al. [10] are determined by Monte Carlo simulations of the transport of the particles in the atmosphere and of the detection of incident particles in the NM. The pitch angle distribution was described by the formula by Bombardieri [3]. The asymptotic directions as well as the effective vertical cutoff rigidities were computed with the IGRF combined with the Tsyganenko 1989 model with K_p-index equal to 3. All 36 NM stations listed in section 2 were used for the GLE analysis.

Figures 1 and 2 show the deduced differential rigidity solar proton spectra and the resulting pitch angle distributions for the time interval 0305-0310 UT during the GLE on 13 December 2006.

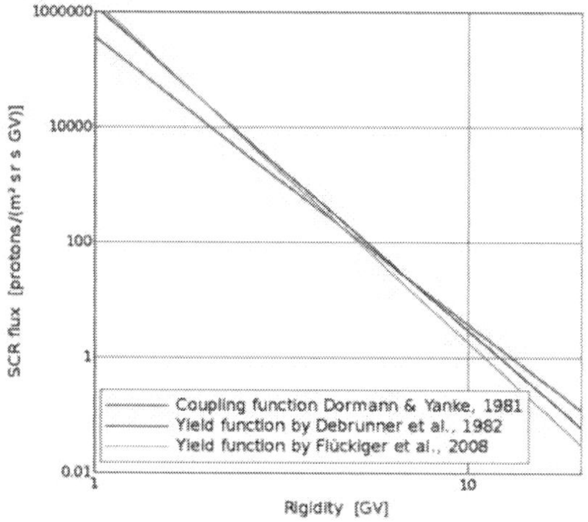

Figure 1. Deduced SCR spectra during the GLE on 13 December 2006 for the time interval 0305-0310 UT by using different NM response functions.

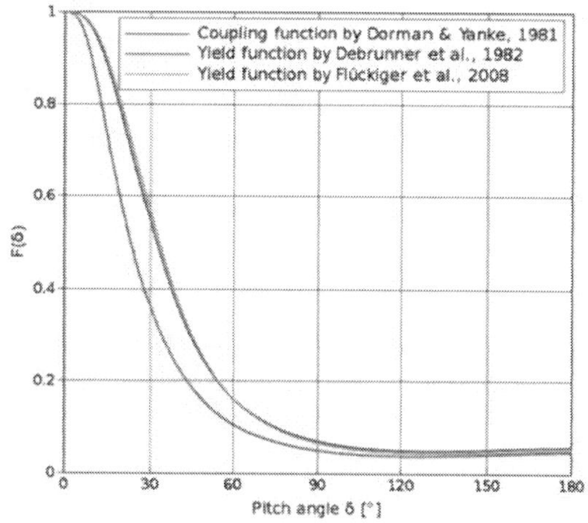

Figure 2. Deduced pitch angle distributions during the GLE on 13 December 2006 for the time interval 0305-0310 UT by using different NM response functions.

The spectrum obtained with the coupling function by Dorman and Yanke [7] is harder than the ones derived with the two yield functions by Debrunner et al. [9] and by Flückiger et al. [10]. The relative difference between the solar proton flux in direction of maximum intensity $J_{||}^{Flückiger\ et\ al.}$ and $J_{||}^{Dorman\ and\ Yanke}$ is ~300% at 1 GV, ~120% at 2 GV and ~-10% at 5 GV, whereas it is only ~20% at 1 GV, ~1% at 2 GV, and ~-10% at 5 GV between $J_{||}^{Flückiger\ et\ al.}$ and $J_{||}^{Debrunner\ et\ al.}$. The additional radiation dose rate[1] produced by the SCR latitudes 55°–90° and at a typical cruise altitude of ~10.5 km (250 g/cm^2) during the GLE maximum is up to ~30 μSv/h according to the results with the coupling function by Dorman and Yanke [7] and maximal ~65 μSv/h with the yield function by Debrunner et al. [9] and ~75 μSv/h with the yield function by Flückiger et al. [10].

2.2. Effect of different descriptions of the pitch angle distribution

For this part of the analysis we used the yield function by Debrunner et al. [9], the solar cosmic ray proton trajectories computed based on the geomagnetic field models IGRF combined with the Tsyganenko 1989 model with K_p-index = 3, and the data of all 36 NM stations listed in section 2.

According to the literature different formulas for the description of the pitch angle distribution $F(\delta, t)$ are adopted. We investigated the effect of the following five functional forms of the pitch angle distribution on the deduced GLE parameters:

- Bombardieri [3] used an exponential form with two free parameters, A and B, see Formula 2.
- Vashenyuk et al. [11] utilized an exponential form with only one free parameter, c:

$$F(\delta, t) = \exp\left(\frac{\delta^2}{c(t)}\right) \tag{3}$$

- Plainaki et al. [12] also used an exponential form with a single free parameter, n_a:

$$F(\delta, t) = \exp\left(-n_a^2(t) \cdot \sin^2(\frac{\delta}{2})\right) \tag{4}$$

- The Bern group used a piece by piece linear function with typically five interpolation values, see the pink line in Figure 3.
- Matthiä [13] used a simple linear dependence:

$$F(\delta, t) = \begin{cases} 1 - b(t) \cdot \delta & : \text{ if } b(t) \cdot \delta < 1 \\ 0 & : \text{ otherwise} \end{cases} \tag{5}$$

The resulting GLE parameters show only small differences. Figure 3 shows the determined solar cosmic ray proton flux in the direction of maximum flux and Figure 4 the deduced pitch angle distributions. The difference in the solar proton intensity in direction of maximum flux is ~30% at 1 GV, ~8% at 2 GV, and ~2% at 5 GV.

2.3. Effect of different consideration of dynamics of the Earth's magnetic field

For this investigation we used the yield function by Flückiger et al. [10], the pitch angle distribution by Bombardieri [3], and the data of all 36 NM stations as listed in section 2.

The effect of different considerations of the dynamics of the Earth's magnetic field on the GLE analysis was investigated in determining the asymptotic directions and the effective vertical cutoff rigidities by using the following three different models of the Earth's magnetic field models:

[1] In this paper the term "radiation dose rate" is used for "ambient dose equivalent rate".

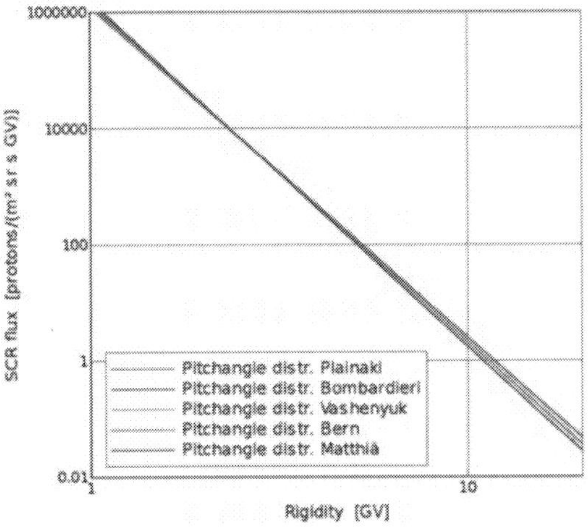

Figure 3. Deduced SCR spectra during the GLE on 13 December 2006 for the time interval 0305-0310 UT by using different forms of pitch angle distributions.

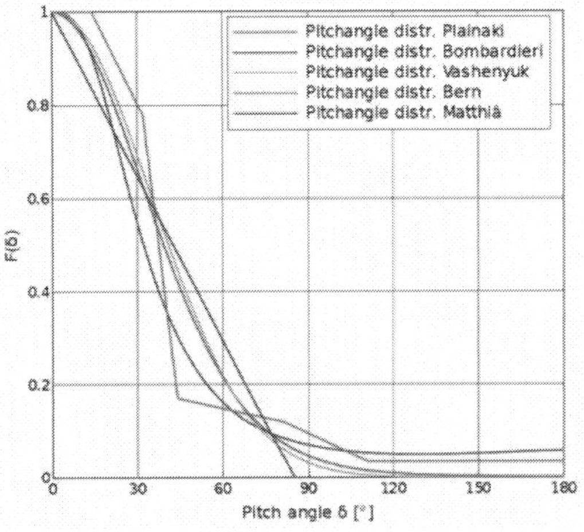

Figure 4. Determined pitch angle distributions during the GLE on 13 December 2006 for the time interval 0305-0310 UT by using different forms of pitch angle distributions.

(i) IGRF [5] + Tsyganenko model 1989 [6], $K_p = 3$
(ii) IGRF [5] + Tsyganenko model 1989 [6], $K_p = 0$
(iii) only IGRF [5]

Figure 5 shows as an example the asymptotic geographic latitude and longitude, as function of the rigidity, for protons of vertical incidence arriving at the top of the atmosphere above the NM station Terre Adelie. The computations were made with MAGNETOCOSMICS [4] for the different assumptions of the geomagnetic field models.

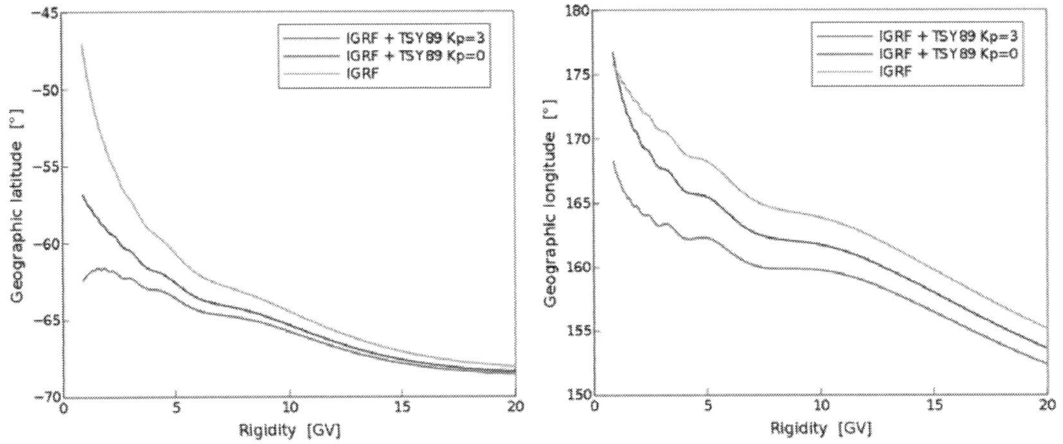

Figure 5. Asymptotic latitude (left) and longitude (right) in geographic coordinates for protons of vertical incidence at the top of the atmosphere above the NM station Terre Adelie on 13 December 2006, 0300 UT and for different assumptions of the geomagnetic field models.

The deduced solar proton spectra and pitch angle distributions are plotted in Figures 6 and 7. The solar proton flux in the rigidity range 1–5 GV is 50-30% higher when only the IGRF geomagnetic field is used for the trajectory computations compared to the deduced values when using the IGRF model [5] combined with the model by Tsyganenko 89 [6] with Kp=3. The differences in the GLE parameters are marginal when using IGRF + Tsyganenko 89 model with K_p=0 and with K_p=3.

The resulting differences in the radiation dose rates at high geographic latitudes and at a typical flight altitude of 10.5 km are on average only small, ∼10-20%. Only on some selected locations the absolute deviations of additional SCR contribution are up ∼20 μSv/h.

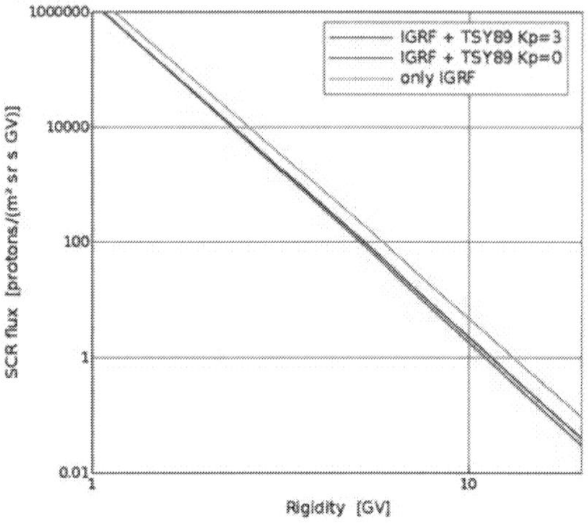

Figure 6. Deduced SCR spectra during the GLE on 13 December 2006 for the time interval 0305-0310 UT by using different geomagnetic field models for the computations of the asymptotic directions and the effective vertical cutoff rigidities. For details see the text.

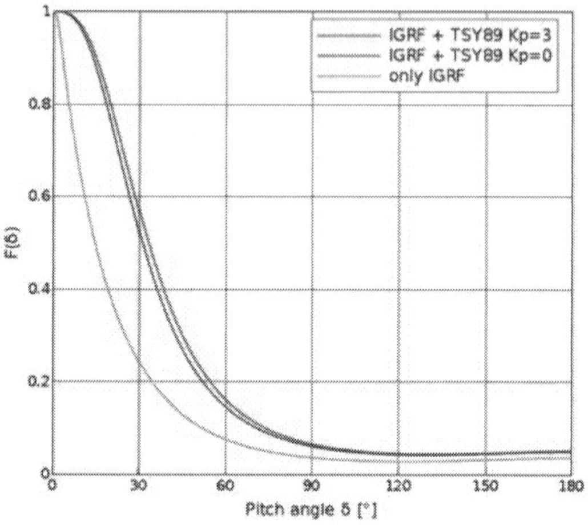

Figure 7. Deduced pitch angle distributions during the GLE on 13 December 2006 for the time interval 0305-0310 UT by using different geomagnetic field models for the computations of the asymptotic directions and the effective vertical cutoff rigidities. For details see the text.

2.4. Effect of the selection and the number of used NM stations

For this investigations the GLE analysis was executed by removing the data of individual or of a group of NM stations such as NMs at mountain altitudes. The following NM stations were considered as mountain stations: Alma Ata, Irkuts2, Irkutsk3, Jungfraujoch NM64, Jungfraujoch IGY, Lomnický Štít. The yield function by Debrunner et al. [9] and the pitch angle distribution by Bombardieri [3] were used for the computations. The deduced results for the solar proton spectra and the pitch angle distributions are shown in Figures 8 and 9.

At an altitude of 10.5 km (atmospheric depth ∼250 g/cm^2) and at restricted locations the radiation dose rate caused by solar cosmic rays in the time interval considered is maximal ∼65 μSv/h, if all NM stations, as listed under section 2, are included in the GLE analysis and at most ∼130 μSv/h when the data of the NM stations Apatity and Oulu, i.e. the NM stations showing the largest count rate increase during this GLE, were not considered for the GLE analysis.

3. Summary and Conclusions

We have attempted to identify the reasons for the differences in the results of GLE analysis by the different GLE analysis procedures. The results of the investigations show that especially the

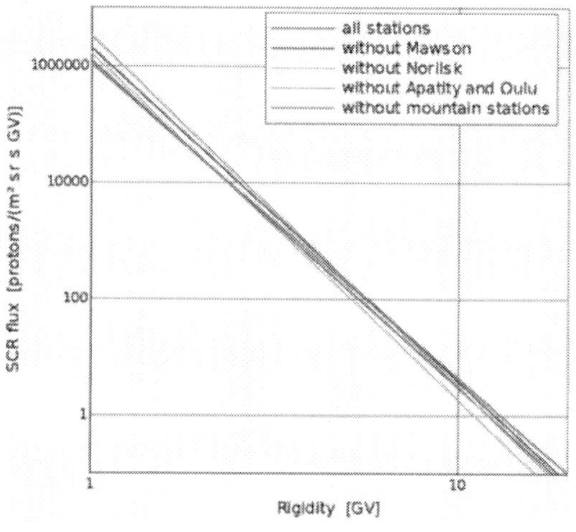

Figure 8. Deduced SCR spectra during the GLE on 13 December 2006 for the time interval 0305-0310 UT by using different grouping of NM stations. For details see the text.

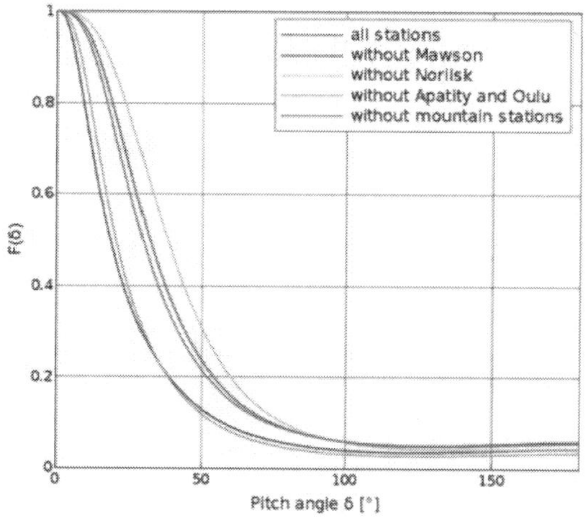

Figure 9. Deduced pitch angle distributions during the GLE on 13 December 2006 for the time interval 0305-0310 UT by using different grouping of NM stations. For details see the text.

used response functions as well as the used geomagnetic field models with different considerations of the dynamics of the geomagnetic field for the trajectory computations have a relevant effect on the deduced GLE parameters. Also the selection of the used NM data may significantly affect the results of a GLE analysis. However, the outcome of this work does not conclusively reveal the reasons for the sometimes diverse results of GLE analysis published in the literature. In addition, other causes not addressed in this study like e.g. different trajectory tracing techniques in the determination of cutoff rigidities and asymptotic directions may also have an effect. Therefore, more detailed exchange of information on the different GLE analysis procedures between the specialists is encouraged and appreciated in view of a harmonization of the GLE analysis procedures. In particular there is a need to reduce the uncertainties in the response function of NMs mainly at the lower rigidity regime, i.e. $\lesssim$2 GV. Reliable and undisputed results of GLE analysis are particularly important, as non cosmic ray specialists use these results in space weather applications, as e.g. the determination of radiation dose assessments at flight altitudes.

Acknowledgments

This research was supported by the University of Bern and by the International Foundation High Altitude Research Stations Jungfraujoch and Gornergrat. We thank the investigators of the following NM stations for the data that we used for this analysis: Alma Ata, Apatity 18NM64, Athens, Cape Schmidt, Fort Smith, Hermanus, Inuvik, Irkutsk, Irkutsk2, Irkutsk3, Kerguelen, Kiel, Kingston, Larc, Lomnický Štít, Magadan, Mawson, McMurdo, Moscow 24NM64, Moscow 6NM64, Nain, Norilsk, Novosibirsk, OLC, Oulu, Peawanuck, Potchefstroom, Rome, Sanae, Terre Adelie, Thule, Tixie Bay, Tsumeb, and Yakutsk.

References

[1] Bütikofer R and Flückiger E O 2013 *Journal of Physics Conference Series* **409** 012166
[2] Bütikofer R and Flückiger E O 2013 *Proc. 33rd Int. Cosmic Ray Conf., Rio de Janeiro*
[3] Bombardieri D J 2008 *PhD thesis, University of Tasmania, Hobart, Australia*

[4] Desorgher L 2004 `http://cosray.unibe.ch/~laurent/magnetocosmics` last access: 06 January 2015

[5] IAGA Division V, Working Group V-MOD, IGRF Model `http://www.ngdc.noaa.gov/IAGA/vmod/igrf.html` last access: 06 January 2015

[6] Tsyganenko N A 1989 *Planetary and Space Science* **37** 5-20

[7] Dorman L and Yanke V 1981 *Proc. 17th Int. Cosmic Ray Conf.* **4** 326

[8] Belov A V and Struminsky A B 1997 *Proc. 25th Int. Cosmic Ray Conf., Durban* **1** 201

[9] Debrunner H, Lockwood J A and Flückiger E 1982 *8th European Cosmic Ray Symposium*

[10] Flückiger E O, Moser M R, Bütikofer R, Desorgher L and Pirard B 2008 *Proc. 30th International Cosmic Ray Conference* **1** 289-292

[11] Vashenyuk E V, Balabin Y V, Gvozdevsky B B and Shchur L I 2008 *Geomagnetism and Aeronomy* **48** 149-153.

[12] Plainaki C, Mavromichalaki H, Belov A, Eroshenko E and Yanke V 2009 *Advances in Space Research* **43** 474-479

[13] Matthiä D 2009 *PhD thesis, Mathematisch-Naturwissenschaftliche Fakultät der Christian-Albrechts-Universität zu Kiel*

Temperature effect corrections for URAGAN based on CAO, GDAS, NOAA data

A Dmitrieva[1], N Ampilogov[1], I Astapov[1], N Barbashina[1], V Borog[1], D Chernov[1], A Kovylyaeva[1], R Kokoulin[1], K Kompaniets[1], G Mannocchi[2], Yu Mishutina[1], A Petrukhin[1], O Saavedra[3], V Shutenko[1], O Sit'ko[1], G Trinchero[3], E Yakovleva[1], I Yashin[1]

[1]National Research Nuclear University MEPhI (Moscow Engineering Physics Institute), Moscow, Russia
[2]Istituto di Fisica dello Spazio Interplanetario – INAF, Torino, Italy
[3]Dipartimento di Fisica dell' Universita di Torino et INFN, Torino, Italy

E-mail: ANDmitriyeva@mephi.ru

Abstract. For the analysis of muon flux variations caused by extra-atmospheric processes it is necessary to introduce corrections for meteorological effects. For temperature effect (TE) correction it is necessary to know the temperature profile of the atmosphere. As a rule, this profile is measured by meteorological balloons two or four times a day. Alternative sources are satellite observations and data obtained from models of atmosphere used for weather forecasting. Vertical temperature profiles obtained from NOAA satellites, GDAS (Global Data Assimilation System) and CAO data (Central Aerological Observatory, Russia) for standard isobaric levels were compared. Mean value of temperature difference for most levels does not exceed 1 K. Comparison of URAGAN data corrected for TE with CAO information, satellites and GDAS shows a good agreement. Counting rate and anisotropy of the muon flux corrected for meteorological effects for 2007-2014 are presented.

1. Muon hodoscope URAGAN

URAGAN [1] (Moscow, Russia, 55.7° N, 37.7° E, 173 m a.s.l.) is a wide-aperture precision muon hodoscope (Figure 1) which is used to study atmospheric and heliospheric processes responsible for variations in the muon flux at the Earth surface [2-4]. The hodoscope consists of separate horizontal assemblies (supermodules) with the area of 11.5 m^2 each. Three supermodules of hodoscope (SM) are now under operation in the exposure mode. The supermodule detects muons with high spatial and angular accuracies (1 cm and 1°, respectively) over a wide range of zenith angles (0–80°). Every minute, angular distribution of muons is recorded in a two-dimensional angular matrix $M(\theta,\varphi)$ (θ and φ are zenith and azimuth angles for matrix cell centers, $\Delta\theta = 1°$, $\Delta\varphi = 4°$), which represents a snapshot of the upper hemisphere with 1-minute exposition. One-minute matrix contains ~ 70-80 thousand events.

For the analysis of muon flux variations caused by extra-atmospheric processes it is necessary to introduce corrections for meteorological effects (barometric and temperature) [5].

Figure 1. Muon hodoscope URAGAN.

2. Corrections for atmospheric effects
Barometric effect is the anticorrelation of cosmic ray intensity with the pressure at the observation level. Temperature effect is caused by changes of the temperature at all altitudes of the atmosphere. Corrected angular matrix $M^{corr}(\theta,\varphi,t,\Delta t)$ can be calculated by the following way:

$$M^{corr}(\theta,\varphi,t,\Delta t) = M(\theta,\varphi,t,\Delta t) + \Delta M_P(\theta,t,\Delta t) + \Delta M_T(\theta,t,\Delta t), \quad (1)$$

where θ and φ are zenith and azimuth angles for matrix cell centers; $M(\theta,\varphi)$ is the number of reconstructed events in a cell (θ,φ) of the matrix M; t is the time of the beginning and Δt is the time interval of the matrix accumulation; ΔM_T and ΔM_P are corrections for temperature and pressure effects.

$$\Delta M_P(\theta,t,\Delta t) = B(\theta)\cdot(P(t,\Delta t) - P_0), \quad (2)$$

where P is the current pressure at registration level, $P_0 = 993$ mbar is the over a long period averaged pressure at the registration level, $B(\theta)$ are barometric coefficients.

$$\Delta M_T(\theta) = M_0(\theta)\cdot\sum_i W_T(h_i,\theta)\Delta T(h_i)\Delta h_i / 100\%, \quad (3)$$

where $W_T(h,\theta)$ are differential in altitude temperature coefficients (DTC) [6], $M_0(\theta)$ is the over a long period averaged number of reconstructed events for zenith angle θ, $\Delta T(h)=T_{SMA}(h)-T(h)$ is the change of the temperature, h is the atmospheric depth, $\Delta h = 0.05$ atm, $T(h)$ is the current temperature profile of the atmosphere, $T_{SMA}(h)$ is the temperature profile for standard model of the atmosphere [7].

Information about temperature profile of the atmosphere can be obtained from the following sources.

1. Information from direct measurements of air temperature with help of meteorological balloon flights. For example, for URAGAN data correction the information from Central Aerological Observatory (Russia, Dolgoprudny) is used [8,9]. Usually temperature profile is measured by meteorological balloons two times a day: 00:00 and 12:00 UT. Temperature of air is measured at random on altitude points. Unfortunately, sometimes launches of meteorological balloons are not carried out, or balloons do not raise high enough.

2. Another source is the information from meteorological satellites. NOAA-18 and NOAA-19 are weather forecasting satellites run by National Oceanic and Atmospheric Administration (NOAA) [10]. They fly over the Moscow region on average 4 times a day with intervals from 10 minutes to 16 hours. Satellite data about atmosphere can be obtained with the help of the receiving station "Alice-SC", which was installed in MEPhI on the roof of the experimental building of the Scientific and Educational Centre NEVOD in November 2013. Available data of NOAA18 and NOAA19 include:

– 42 constant pressure levels at which the retrieved values are calculated (in hPa);
– Date and time of measurement;
– Latitude and longitude of the current measurement point;
– Retrieved temperature profile (in K);
– Temperature guess profile (in K);
– Water vapor retrieval (in g/kg);

– Dew point temperature retrieval (in K);
– Total-column ozone;
– Cloud fraction;
– Cloud top pressure (in hPa);
– Cloud top temperature (in K);
– Clear/cloud index;
– Effective cloud amount;
– Total precipitable water (in mm);
– Additional data used for retrieval.

3. The alternative sources are weather forecasting models of the atmosphere. One of them is the numerical forecasting model of atmosphere GDAS (The Global Data Assimilation System). GDAS output data are available 4 times a day (at 00, 06, 12, and 18 UTC) for the whole globe (1 degree latitude-longitude grid) and for 23 constant pressure levels (from 1000 to 20 mbar). Retrospective archive data (from 2005 year) are in an open access [11].

3. The comparison of GDAS and NOAA data with CAO data

Vertical profiles of atmosphere temperature obtained from meteorological satellites NOAA and from forecasting model GDAS were compared with direct measurements by meteorological balloons. In the Figure 2, results of two measurements are shown as an example: 23.12.2013 00:00 UTC (on the left) and 22.01.2014 12:00 UTC (on the right). NOAA data is in good agreement with CAO data. But, as one can see, NOAA temperature profile is smoothed and does not reproduce sharp changes in temperature dependence on the pressure level. Comparison of GDAS data with CAO data shows very good agreement.

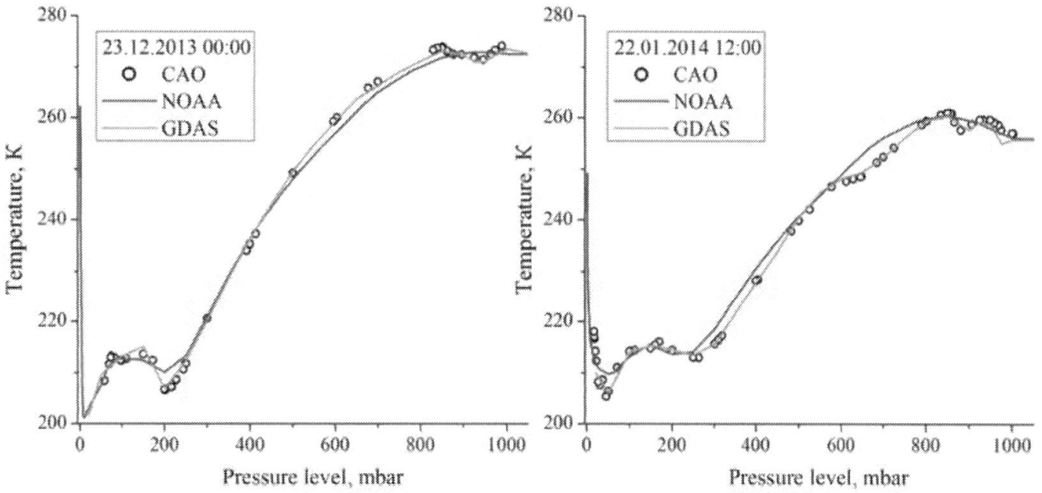

Figure 2. The dependence of air temperature on the pressure level.

Dependence of atmosphere air temperature on time for two pressure levels (50 and 500 mbar) from CAO data and from other sources is shown in the Figure 3. In winter NOAA data is in good agreement with CAO data. But in summer the difference between NOAA and CAO data becomes higher. Comparison of CAO and GDAS data again shows very good agreement.

The dependence of mean value of the temperature difference between data of alternative sources (AS) and CAO ($\Delta T(h,t) = T_{AS}(h,t) - T_{CAO}(h,t)$) on the pressure level is shown in the Figure 4 for NOAA (on the left) and for GDAS (on the right). The shaded region outlines one r.m.s. deviation. For most pressure levels $<\Delta T>$ does not exceed 1 K. So, the temperature profiles of these sources can be used for correction of the muon hodoscope data for the temperature effect.

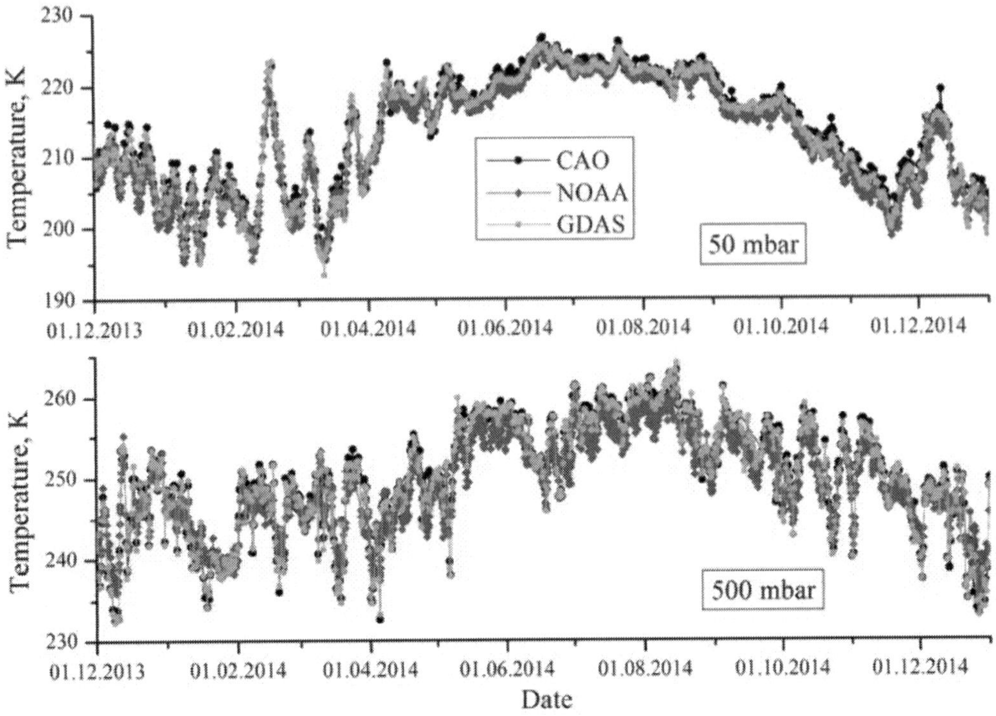

Figure 3. Dependence of atmosphere air temperature on time for two pressure levels.

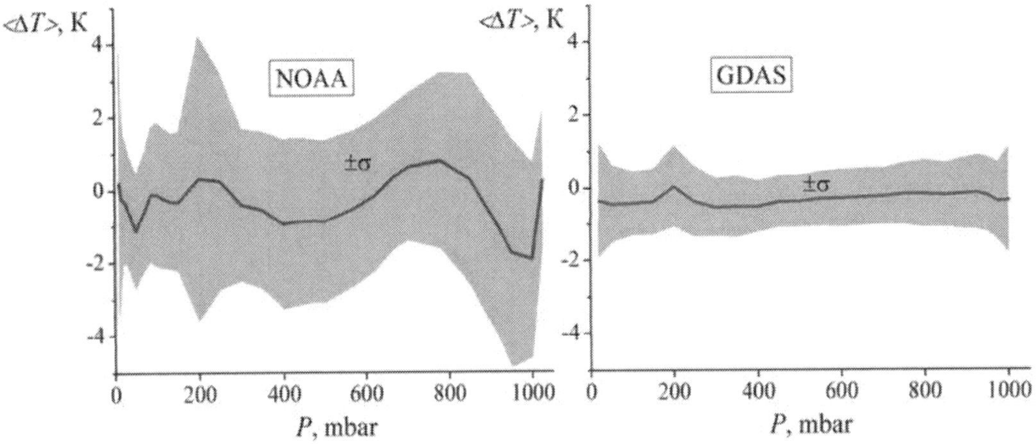

Figure 4. Mean value of temperature difference between NOAA, GDAS and CAO data.

4. Counting rate of the URAGAN hodoscope: 2007-2014

In the Figure 5 (on the top), 1 h average counting rate of the URAGAN hodoscope without and with corrections for meteorological effects is shown. Barometric coefficients for URAGAN slightly depend on zenith angle and are about ~ 0.18%/mbar. After correction for barometric effect, annual variations caused by temperature effect (~ 8 %) become well visible. After correction for temperature effect, variations caused by extra-atmospheric processes appear. For comparison, in the bottom part of Figure 5 the counting rate of Moscow neutron monitor (MNM) corrected for pressure effect [12] is shown. Temperature effect for counting rate of neutron monitors is negligible. URAGAN and MNM are located at the same latitude and have the same threshold rigidity (2.45 GV). But they have different asymptotic directions for primary particles, and besides, muons are sensitive to higher energies (compared to neutrons) of primary cosmic rays. So, muon hodoscopes open new possibilities for studying the heliospheric perturbations responsible for the modulation at high energies.

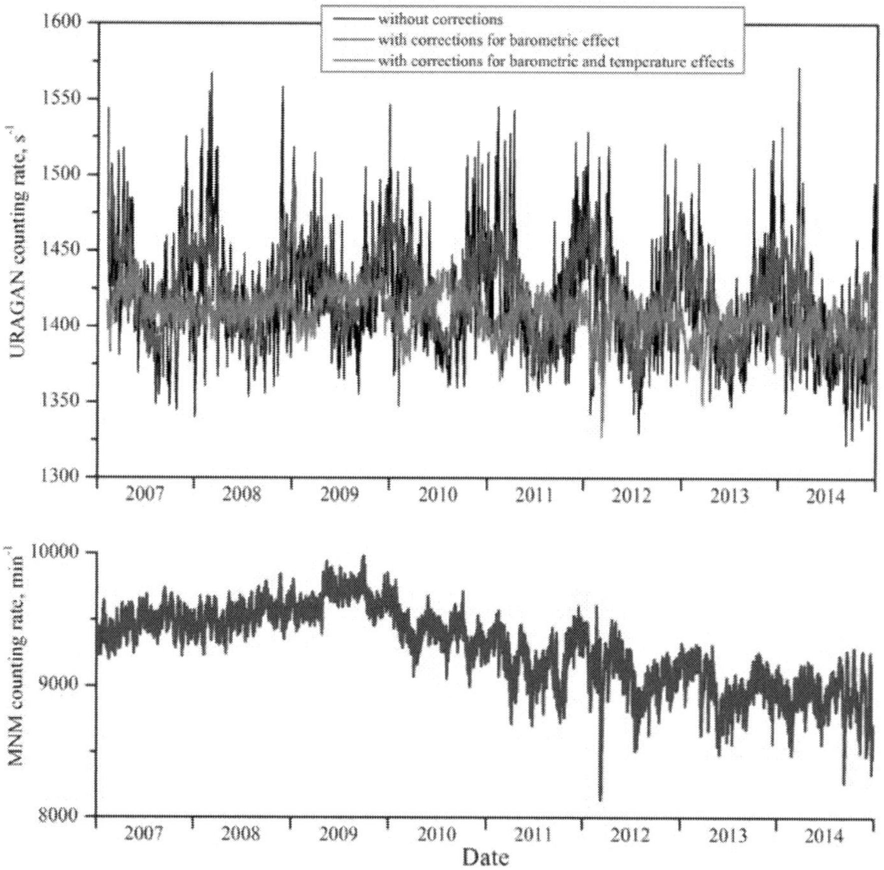

Figure 5. On the top: 1 h average counting rate of the URAGAN hodoscope without and with corrections for meteorological effects. On the bottom: counting rate of Moscow neutron monitor.

URAGAN allows to obtain the vectors of the particle arrival directions and we can use their vector sum. The summary vector normalized to the total number of muons (the vector of local anisotropy) will characterize the angular distribution of the detected particles. The projections of the vector of local anisotropy (A_{South}, A_{East} and A_Z, Figure 6) can be generally defined from the original matrix M data as follows [13]:

$$
\begin{aligned}
A_{\text{South}} &= \frac{1}{N}\sum_{\theta}\sum_{\varphi} M(\theta,\varphi)\cos\varphi\sin\theta\,, \\
A_{\text{East}} &= \frac{1}{N}\sum_{\theta}\sum_{\varphi} M(\theta,\varphi)\sin\varphi\sin\theta\,, \\
A_{Z} &= \frac{1}{N}\sum_{\theta}\sum_{\varphi} M(\theta,\varphi)\cos\theta\,, \\
N &= \sum_{\theta}\sum_{\varphi} M(\theta,\varphi)\,, \qquad A = \sqrt{A_{\text{South}}^2 + A_{\text{East}}^2 + A_Z^2}\,.
\end{aligned}
\tag{4}
$$

Here N is the total number of events in a given range of angles; A_{South} is the projection on the north–south axis, A_{East} is the projection on the west–east axis; A_Z is the vertical projection. Horizontal projections A_{South} and A_{East} are calculated using the original matrices M without corrections for barometric and temperature effects. These corrections are annihilated when the values from the opposite azimuthal angles are added, while the use of corrections in the total value (N) has almost no effect on the result. The A_Z value was calculated using the values of the matrix without and with corrections for barometric and temperature effects (Figure 7). As one can see, annual variations do not disappear after corrections. According to preliminary estimations, the A_Z projection can be used for the analysis of changes of muon energy spectrum.

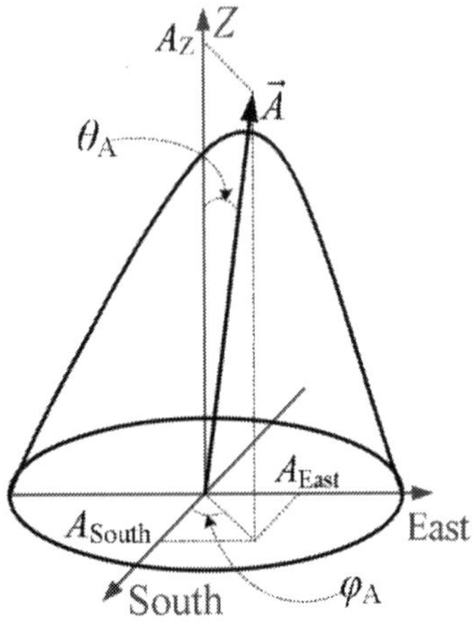

Figure 6. The projections of the vector of local anisotropy.

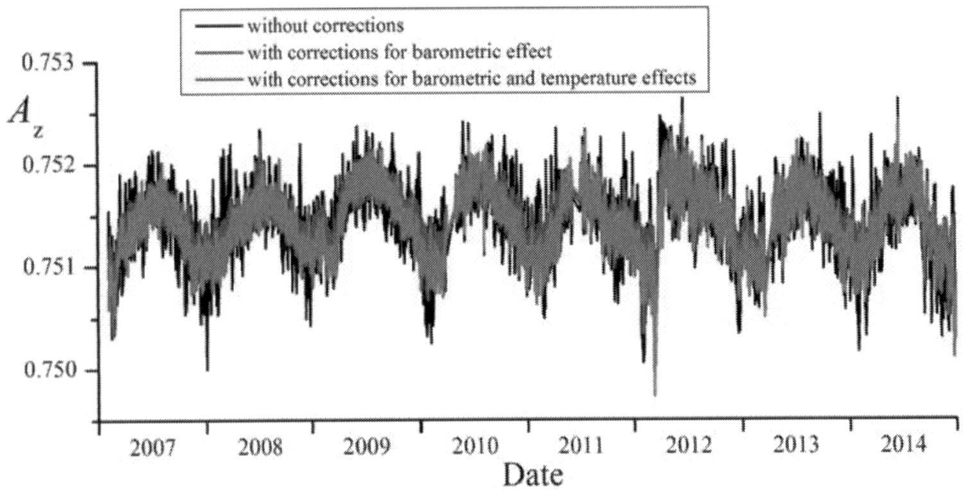

Figure 7. Dependence of vertical projection of the vector of local anisotropy on time.

The comparison of corrections for the temperature effect with the help of satellite data with correction based on data from direct measurements (Figure 8) shows good agreement. So, satellite data can be used for the fast preliminary correction.

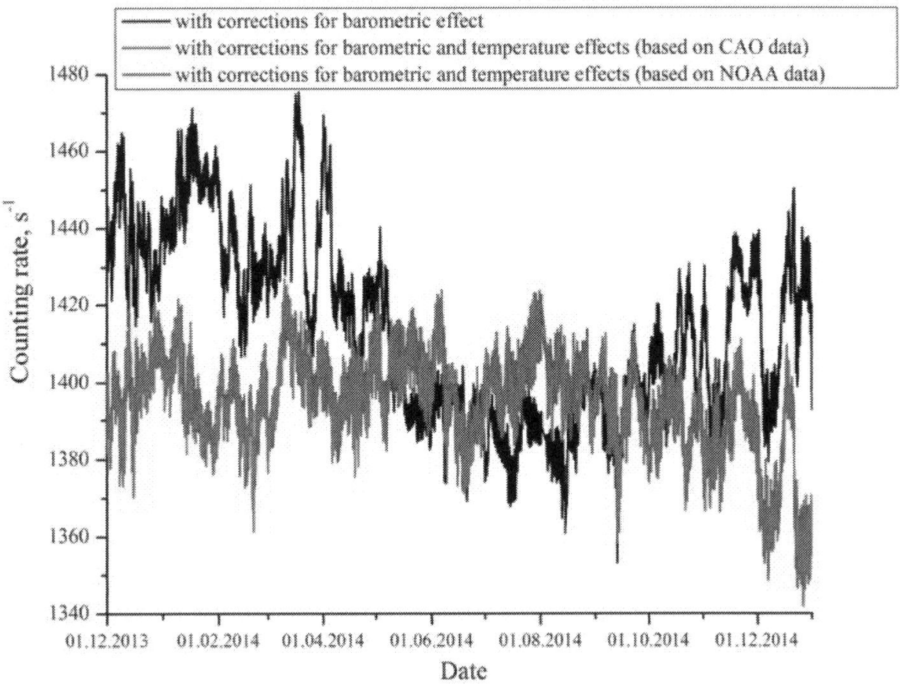

Figure 8. URAGAN counting rate with temperature effect correction based on CAO and NOAA data.

Conclusion

Data of the discussed various sources of the atmosphere temperature profile are sufficiently reliable for the correction of the muon hodoscope counting rate for the temperature effect. Additional retrospective information from CAO and GDAS can be used for temperature effect correction for previous years (since 2005). Satellite data can be used for the preliminary correction in the mode close to real-time (with a delay of only few hours).

Acknowledgments

The research has been performed in the Scientific and Educational Centre NEVOD with the State support from the Ministry of Education and Science of the Russian Federation (government task and project RFMEI59114X0002) and the grant of the Leading Scientific School NSh-4930.2014.2.

References

[1] Barbashina N S *et al* 2008 *Instrum. Experim. Techniques* **51** 180.
[2] Barbashina N S *et al* 2011 *Bulletin of the Russian Academy of Sciences: Physics* **75** 814.
[3] Timashkov D A *et al* 2007 *Proc. of the 30th ICRC in Merida, Yucatan, Mexico* **1** 685.
[4] Ampilogov N V *et al* 2011 *Astrophysics and Space Sciences Transactions* **7** 435.
[5] Dorman L I 1972 *Meteorological Effects in Cosmic Rays* (in Russian, Moscow: Nauka).
[6] Dmitrieva A N *et al* 2011 *Astroparticle Physics* **34** 401.
[7] Glagolev Yu A 1970 *Reference Book on Physical Parameters of the Atmosphere*, Gidrometeoizdat, Leningrad, (in Russian).
[8] Central Aerological Observatory (Russia, Dolgoprudny): http://www.aerology.org/
[9] Department of Atmospheric Science of the University of Wyoming: http://weather.uwyo.edu/
[10] NOAA satellites: http://www.noaa.gov/satellites.html
[11] NOAA Air Resources Laboratory (ARL), 2004, http://ready.arl.noaa.gov/gdas1.php, Tech. rep.
[12] Database of Moscow neutron monitor: http://cr0.izmiran.rssi.ru/mosc/main.htm
[13] Shutenko V V *et al* 2013 *Geomagnetism and Aeronomy* **53** 571.

The Neutron Monitor Control Panel

O García-Población[1,2], H Ivanov[2], I García-Tejedor[1,2], J J Blanco[1,2], J Medina[1,2], R Gómez-Herrero[1,2], E Catalán[1,2] and D Radchenko[2]

[1] Space Research Group, University of Alcalá, Spain

[2] Castilla-La Mancha Neutron Monitor, Parque Tecnológico de Guadalajara, Spain

E-mail: oscar.gpoblacion@uah.es

Abstract.
This work presents the current status and future plans of the Neutron Monitor Control Panel (NMCP), a new software developed to aid the operator in typical station maintenance and configuration operations. This software is integrated with the new so-called NOAS data acquisition system and it can be accessed using a supported web browser. It features a visual inspection tool to help the operator to identify spikes in the data, trace the origin of the spike back to the raw readings of each counter tube and pressure reading, and mark the data as invalid in the Neutron Monitor Database if desired. The software also provides information about station operation status, some descriptive statistics about current data being recorded and, in the future, will provide an interface to configure station parameters.

1. Introduction

Neutron Monitor measurements are widely used by researchers in different research fields, such as space weather and/or geomagnetic studies. That's why the Neutron Monitor Database (NMDB)[1] puts effort into delivering data with the best quality possible. The data is also used in real-time GLE alarm systems and other real-time applications; therefore, a data quality protocol must be applied in real-time, although it will be eventually revised by a human supervisor to ensure its correct behaviour. By quality protocol we refer to all the methods and techniques used, such as:

- Detection of abnormal data, commonly referred to as spikes.
- Detection of inactivity, i.e, if there is no data uploaded in NMDB for the last minutes. This will lead to a notification to the team responsible for the Neutron Monitor station.
- Neutron Monitors are formed by several counter tubes, typically eighteen. A malfunction in a single tube shouldn't make the entire station stop reading cosmic ray measurements. While it is true that missing counters degrades the quality of the measurement by increasing its variability, if the number of working tubes is high enough the result can still be acceptable. This leads us to the need of monitoring and detection of malfunctions separately in each of the tubes.
- Detection of changes with respect to the historical activity. Since changes in the immediate environment or the instruments can cause overall changes in the measured values of a Neutron Monitor, detecting and correcting those changes is desirable.

- Comparing the data from one station to the data of different nearby stations. Detecting isolated events could indicate that there is a malfunction in that station.

Once the corrupted data is detected, the cause of such data corruption can often be traced down to its cause. Most of the time these malfunctions are caused by the instruments forming the data acquisition system, so detecting the corrupt data helps us improve and learn more about its cause. The figure 1 shows schematically the main sources of errors. Most of them are Electro-Magnetic Interferences (EMI) in the preamplifiers (A), in the transmission lines from the amplifiers to the data acquisition system (B) and in the data acquisition hardware itself (C). Also malfunctions in the software (D) can introduce undesirable artifacts in the data.

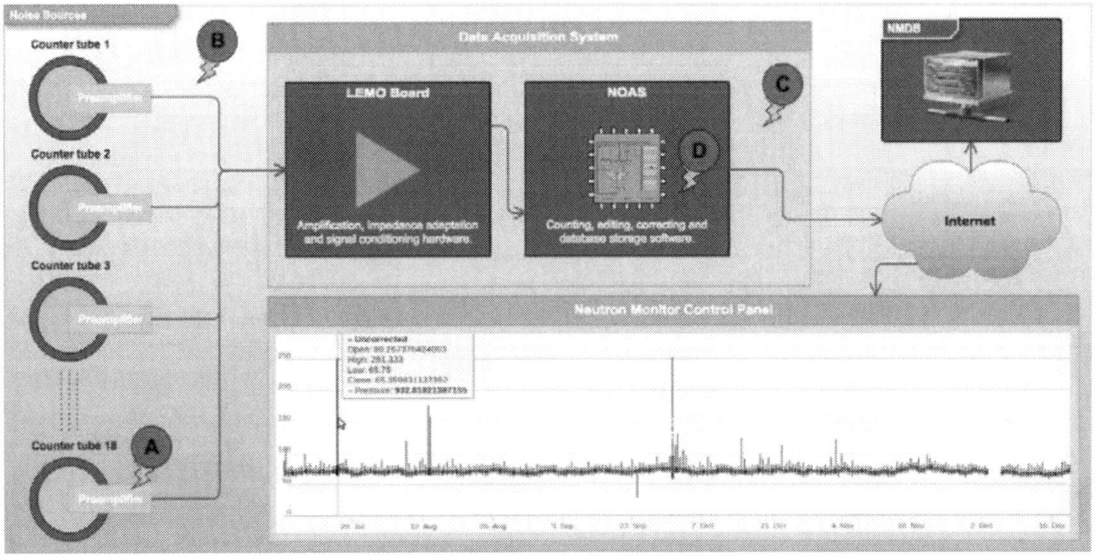

Figure 1. Possible error sources identified in our acquisition process.

In order to facilitate the detection of the corrupt data we are working in the development of a web tool which will help us detect these corruptions. The web application makes use of dynamically generated plots which are designed to highlight possible corrupted data.

This software is currently running for testing in the second release of the data acquisition system[2] designed for the CaLMa[3] and KIEL2 stations.

2. System architecture

There are two main elements in the architecture: a client that requests and consumes data using a web browser and a system that collects and serves the data, which in this system is the data acquisition system. This separation of roles follows the client-server design pattern[4]. The figure 2 shows a block diagram with all the components involved along with its relationships. Each block and relationship will be described below.

The software architecture is based on the Model-View-Controller (MVC) software design pattern[5]. This pattern encourages the separation between the data, the data processing and the data representation.

The view component is in charge of the data representation and the user interface. In our system, this component is mainly executed by the client in a web browser, which is not only a convenient graphical output device but also provides a way to interact with the user. In response to these interactions, the browser makes requests to a server which answers back with new data to be represented. All the software running in this component has been written in JavaScript, and makes intensive use of two libraries: HighStock[6] to draw plots and charts, and Sencha ExtJS[7] which is a very powerful framework that handles among other things the graphical user interface.

The model component is located on the server side, in the data acquisition system. This model comprises all the data captured from each counter tube and the algorithms used to process this data, such as atmospheric pressure corrections, sanity checks, editors to combine all the readings into a single one, etc. There are two databases involved in this component, one in the data acquisition system itself and another in a regular server. The first one is a small and light SQLite database, running in the data acquisition system on a BeagleBone Black embedded system[2]. This database stores per detector count rates, atmospheric pressure and voltage levels along with a time stamp. The data from this database are edited, corrected and sent to a second database server running MySQL[8]. This is an enterprise grade relational database server, used also by NMDB. The software needed to implement the editors, corrections, and transfer between databases has been written in the PHP programming language, using the Zend[9] as support framework.

Several controllers adapt and serve the model data as REST-like web services, using JSON[10] for data representation. REST is a well known software architecture that constraints the design of components to increase compatibility and isolation, standardising the way that services can be consumed. This web services are designed to be a documented Application Programming Interface (API) for accessing the station information and services, allowing third party applications to be further developed. The main consumer of this API right now is the spike tool, which is used to track and mark anomalous surges in the data from the station. The next section describes this tool in detail.

3. The spike tool

Currently the Spike Tool consists of three modules, `Spike`, `SpikeCorrected` and `ChannelStats`. `Spike` and `SpikeCorrected` are very similar, but `SpikeCorrected` works with a different set of data and has some extra functionality.

The `Spike` module offers an interactive chart in which the spikes are easy to trace. As shown in the figure 3, initially the chart displays a large set of data, and some kind of data grouping is needed. Just averaging the data flattens the spikes. The Candlestick chart is used to prevent the aforementioned problem, as this way the max and min values can be easily distinguished. Eventually, when a short interval of data is requested and no grouping is needed to display it, the chart will automatically switch to a line chart.

As mentioned above the chart is interactive, and offers a very intuitive zooming functionality. Clicking and dragging through an interval of the chart will raise a zoom event, and depending on the drag direction the event will be zoomed In or Out. The chart also offers a navigator which allows us to navigate forwards or backwards in time. Datetime input fields are available too, which allows the user to plot a specific time interval.

At first the uncorrected data are used to generate the charts, but we are given the chance to choose between Uncorrected data and Efficiency and Pressure corrected data. The atmospheric pressure is also plotted, which helps us to evaluate how the atmospheric pressure affects the measurements of our Neutron Monitor station.

Once we have a spike located, we can see the raw reading of the counter tubes, which allows us to better track the spike. This option is only activated when representation time scale is small enough to allow plotting every single value for each counter without neither overplotting nor cluttering the chart.

As said before, the `SpikeCorrected` module is very similar to the `Spike` module which was explained above. First of all, this module makes use of a different set of data. The `Spike`

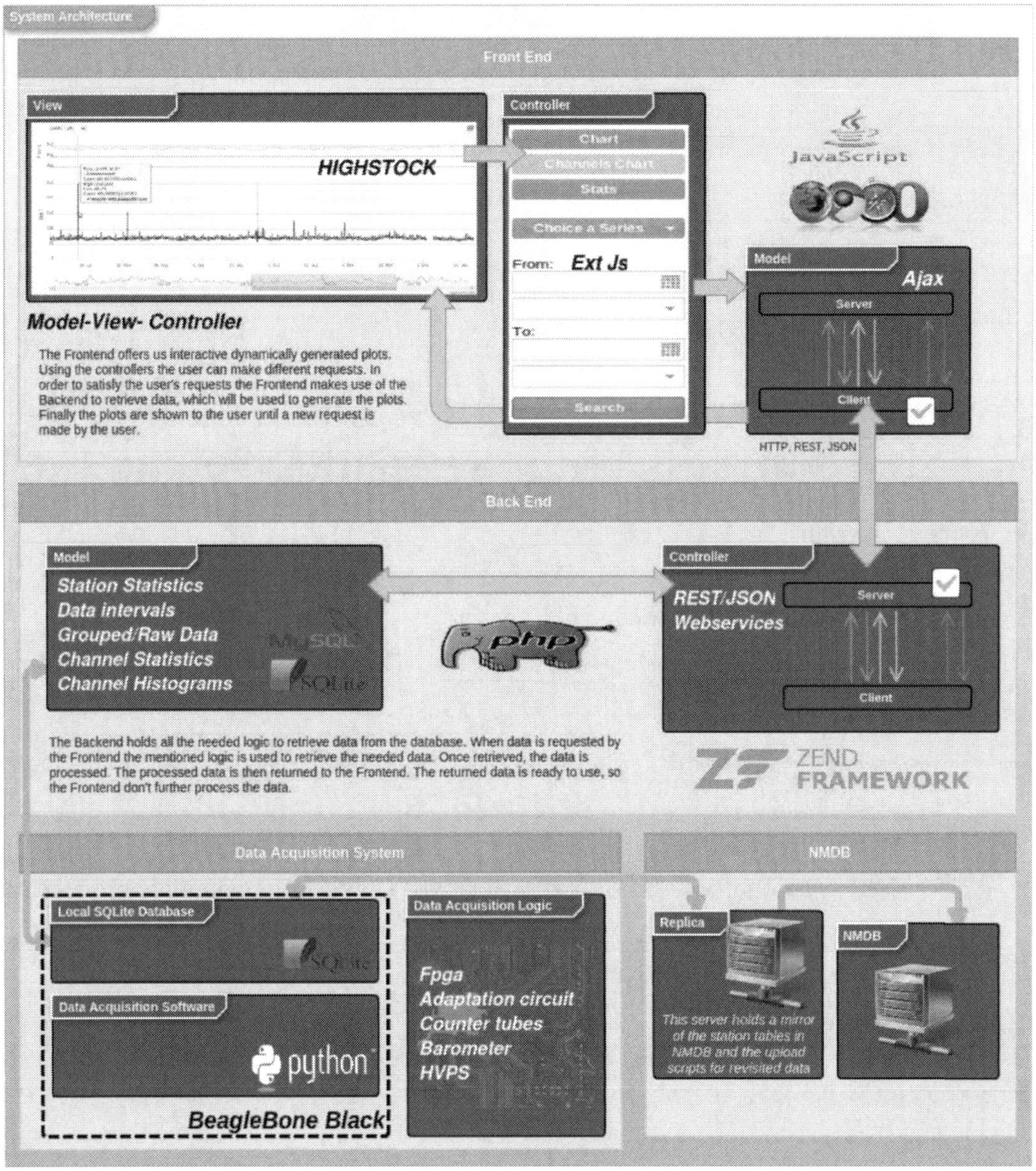

Figure 2. System architecture

module uses the raw data which comes from the data acquisition system while `SpikeCorr` uses the revised data, and also provides the ability to update the revised data.

Clicking on a value in the chart will mark the clicked data, which is stored in a table. The data in the table can be later submitted to the revised set of data, which means that the mentioned data will be considered invalid and treated as void in the future.

The table which contains the marked data can be seen on a separate window, which can be shown and hidden when needed. The window also contains all the controls needed for submitting and unmarking the marked data.

The last module, `ChannelStats`, is intended to give information about the different channels separately, so that we can identify malfunctions on isolated channels. The module offers various statistics for user defined time intervals, among which the simplest are the average value, minimum/maximum value and standard deviation.

The `ChannelStats` module offers distribution charts, too. Histograms with the channels' raw data are plotted in the requested time intervals. The chart is highly interactive, and the series can be shown, hidden and highlighted. This module is currently under development.

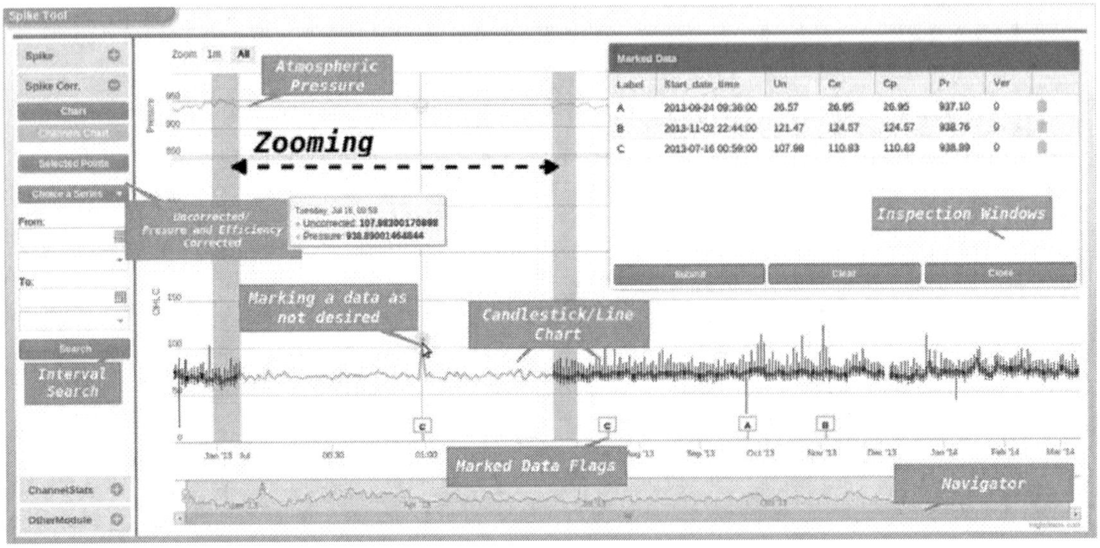

Figure 3. Spike Tool

4. Future work

The Neutron Monitor control panel is a natural add-on to the new data acquisition system designed for Neutron Monitors and is under active development to add new useful features in the near future. Some of these future features are summarized below.

Online statistics The control panel will include a module for descriptive statistics that will be calculated in real-time from the counter tubes raw readings. This will help to identify drifts in the behavior of the detectors and to apply the appropriate actions to maintain data quality.

Station reconfiguration This will allow the operator to disconnect a counter from the station without affecting global station count. This can be useful to perform maintenance operations such as counter tube diagnostics.

Station parameter setup New electronics allow the remote control of some parameters, for example the high voltage power supply set point, the remote database for data upload, backup policies, etc.

Alarms and push notifications When connectivity allows it, it will be possible to configure alarms and notifications to recipients using the new push mobile technologies. For example, if the station is no longer uploading data to NMDB or if its data is out of quality standards.

Detector response histogram A new design from University of New Hampshire of the front-end amplifiers provides a pulse from which the width is proportional to the energy of the incident particle. The new data acquisition system NOAS features a different programmed core running in its Field Programmable Gate Array (FPGA) device. This core enables the system to read these pulses and to generate a histogram with the distribution of the pulse energy over a given time interval. This will enable a mechanism for non-intrusive diagnostics of the counter tubes. This means that diagnostics can be carried out without disconnecting the detector, changing its biasing voltage or, in general, altering its normal operational parameters. Therefore there is no need to interrupt the radiation measurements while the tests are being conducted.

Even though this tool is focused on spikes, there are many other data anomalies that affect data quality, like data drifts, steps, missing data, and so on. Dealing with each one of these requires a very different approach and further study is therefore required. When available, the software designed to address these problems could be easily integrated in the NMCP as a new service.

Acknowledgments
The authors would like to thank the Guadalajara Science and Technology Park for its support for this project.

References

[1] Mavromichalaki H, Papaioannou A, Plainaki C, Sarlanis C, Souvatzoglou G, Gerontidou M, Papailiou M, Eroshenko E, Belov A, Yanke V, Flückiger E O, Bütikofer R, Parisi M, Storini M, Klein K L, Fuller N, Steigies C T, Rother O M, Heber B, Wimmer-Schweingruber R F, Kudela K, Strharsky I, Langer R, Usoskin I, Ibragimov A, Chilingaryan A, Hovsepyan G, Reymers A, Yeghikyan A, Kryakunova O, Dryn E, Nikolayevskiy N, Dorman L and Pustil'Nik L 2011 *Advances in Space Research* **47** 2210–2222
[2] Población Ó G, Blanco J J, Gómez-Herrero R, Steigies C T, Medina J, Tejedor I G and Sánchez S 2014 *Journal of Instrumentation* **9** T08002
[3] Medina J, Blanco J J, García O, Gómez-Herrero R, Catalán E J, García I, Hidalgo M A, Meziat D, Prieto M, Rodríguez-Pacheco J and Sánchez S 2013 *Nuclear Instruments and Methods in Physics Research A* **727** 97–103
[4] Wikipedia Client–server model — wikipedia, the free encyclopedia [Online; accessed 22-January-2015] URL `http://en.wikipedia.org/wiki/Client-server_model`
[5] Wikipedia Model–view–controller — wikipedia, the free encyclopedia [Online; accessed 22-January-2015] URL `http://en.wikipedia.org/wiki/Model-view-controller`
[6] Highcharts js [Online; accessed 23-March-2015] URL `http://www.highcharts.com`
[7] The extjs framework [Online; accessed 23-March-2015] URL `http://www.sencha.com`
[8] The mysql database server [Online; accessed 23-March-2015] URL `http://http://www.mysql.com`
[9] The zend framework [Online; accessed 23-March-2015] URL `http://http://framework.zend.com`
[10] Wikipedia Json — wikipedia, the free encyclopedia [Online; accessed 23-March-2015] URL `http://en.wikipedia.org/wiki/JSON`

Tensor anisotropy of cosmic rays by data of neutron monitors and muon telescopes

P Yu Gololobov, G F Krymsky, P A Krivoshapkin, S K Gerasimova, V G Grigoryev

Yu.G. Shafer Institute of Cosmophysical Research and Aeronomy of SB RAS, 31 Lenin Ave., Yakutsk, Russia, 677980

E-mail: s_k_gerasimova@ikfia.sbras.ru

Abstract. The long-term data of neutron monitors of the world-wide network have been processed using the global survey method. In addition to the symmetrical diurnal variation, the semidiurnal and antisymmetric diurnal variations reflecting the availability of tensor anisotropy have been stood out. Results are analyzed jointly with the data of muon component from the st. Nagoya for the time period 1971 to 2012 and physical mechanisms responsible for anisotropy are discussed.

Introduction

The intensity of galactic cosmic rays registered by ground-based detectors is isotropic in the first approximation. However, there are small deviations from the isotropy which are indicative of the presence of some directional motion of cosmic rays as a whole at each moment of time. This anisotropy is described by the first spherical harmonic by directions of particle motions and can be represented by a three-dimensional vector. Along with the vector anisotropy there exists an anisotropy of another kind which should be described by the second spherical harmonic by directions of particle motions. Five components of this anisotropy in the coordinate system with polar and azimuthal angles θ, ψ look like:

$$\Omega_1 = \sqrt{3}(cos^2\theta - 1/3)$$

$$\Omega_2 = sin2\theta cos\psi$$

$$\Omega_3 = sin2\theta sin\psi$$

$$\Omega_4 = sin^2\theta cos2\psi$$

$$\Omega_5 = sin^2\theta sin2\psi$$

One can represent the corresponding coefficients of decomposition $(a_2^0, a_2^1, b_2^1, a_2^2, b_2^2)$ in the form of five-dimensional vector of anisotropy $\vec{A}$. If we add the isotropic part to the distribution of intensity described by the second spherical harmonic we will gain one more geometrical form of anisotropy i.e. a triaxial ellipsoid. The ellipsoid orientation in relation to the coordinate system is described by three Eulers angles and the ellipsoid itself - by two eccentricities.At last, this anisotropy can be represented by a symmetrical matrix T with the given spur which also requires

Journal of Physics: Conference Series **632** (2015) 012056 doi:10.1088/1742-6596/632/1/012056

five independent components. As the system of angular coordinates changes, the components of matrix T also changes, that is it is a tensor. The mentioned triaxial ellipsoid is a surface which is described by the following equation:

$$\vec{\Omega}T^{-1}\vec{\Omega} = 1,$$

and is also termed as the second tensor ellipsoid.

A method of finding of the anisotropy

The cosmic ray intensity registered with a device, depends on a distribution of intensity by the solid angle $J(\Omega)$ and also on the directional pattern $R(\Omega)$:

$$I = \int J(\Omega)R(\Omega)d\Omega.$$

If we expand the directional pattern of device in spherical functions and take expansion coefficients belonging to the 2nd spherical harmonic $(x_2^0, x_2^1, y_2^1, x_2^2, y_2^2) \equiv \vec{R}$, we will obtain the contribution of tensor anisotropy into the intensity registered with a device:

$$\delta I = \vec{R} \cdot \vec{A}.$$

If there is a set of devices with the R_i vectors then the vector $\vec{A}$ can be found from the linear equation system by their indications δI_i.

The five-dimensional vector $\vec{R}$ termed as a receiving vector, should be determined in the coordinate system related to the rotating Earth where it conserves to be constant. At the same time the vector $\vec{A}$ arises owing to dynamic processes in the solar wind and should be determined in the coordinate system related to the Sun.

A transfer from one system to another is made by means of the matrix M of 5×5 dimensions. Generally the matrix arises as the effect of 3 sequential rotations around axes of coordinates corresponding to Eulerian angles. In order for both vectors $\vec{A}$ and $\vec{R}$ vectors to be transformed by the same matrix, it is necessary for the matrix to be orthogonal. It is achieved by that normalization of components Ω_i which is stated above. As in the expression for δI both vectors should be expressed in the unified coordinate system, at least, one of them will depend on the time of a day and consequently $\delta I(t)$ will contain an antisymmetric daily (a_2^1, b_2^1) and semidiurnal (a_2^2, b_2^2) variations. Corresponding components of a vector $\vec{A}$, can be found by indications of even one device if the anisotropy does not change within a day. If in the solar coordinate system the anisotropy is constant, then parametres of a daily variation will normally change during a year because the orientation of the Earths rotation axis changes.Thus, the slowly changing anisotropy can be found using a method of daily variations. In a case of prompt changes it must be determined by indications of the set of several devices. Such method of determination of anisotropy by the worldwide neutron monitor network has been realized in [1] and has been titled as a "method of global survey".

It should be make some remarks concerning the construction of vectors $\vec{R}_i$. The directional diagramme $R(\Omega)$ taken as a basis is constructed with the account of a geographical location of each device, its aperture, taking into account a direction of cosmic ray trajectories in the geomagnetic field which is different for particles of different energies with the account of a spectrum of anisotropy and coupling coefficients. The detailed results of such calculations are given in [2],[3],[4],[5],[6].

Mechanisms of tensor anisotropy

The most obvious mechanism of tensor anisotropy is a cosmic ray screening [7],[8],[1]. The interplanetary magnetic field formed by the solar wind can contain open magnetic tubes or the closed tubes in the form of magnetic loops. Those and others are diagrammatically depicted in

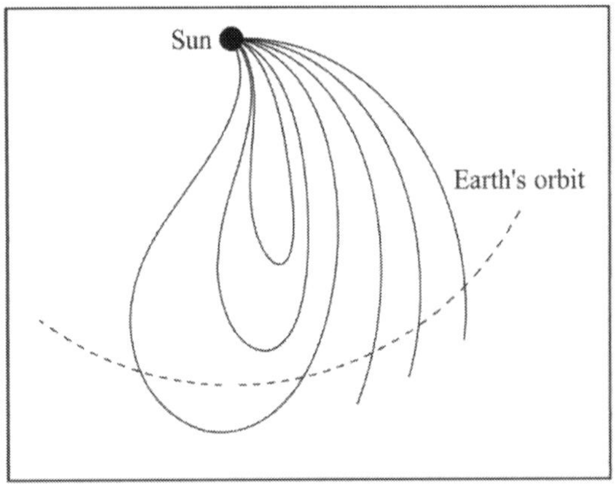

Figure 1. The geometry of magnetic tubes.

Fig.1. Loop-like tubes must form in a relatively narrow helio-latitudinal layer adjoining a plane of the solar equator, from whence the solar wind with lower velocity is emitted. The high-speed wind from higher helio-latitudes carries out open tubes. In each closed tube a deficit of particles moving along a magnetic field appears as the tube expands. A corresponding anisotropy in the system coordinate with the polar axis directed along the field will be depicted by a vector with the following components:

$$(a_2^0, 0, 0, 0, 0), \text{ where } a_2^0 < 0.$$

The physical reason of this effect is the magnetic shielding of particles: particle moving with small pitch-angles cannot penetrate into a tube.

Other mechanism is a shear flow in the inhomogeneous solar wind [9], [10].Cosmic ray scatterings on the moving magnetic inhomogeneities lead to the appearance of directional particle fluxes with the excess intensity

$$\delta I = (\gamma + 2)(\vec{u}\vec{v})/c^2$$

where $\gamma = 2.5$ is an index of cosmic ray spectrum, $\vec{u}, \vec{v}, c$ are velocities of the solar wind, particle and light, respectively. If there exists the shear of velocity, i.e. it is various in the adjacent tubes of flow, then a tensor anisotropy appears. Its parameters have been calculated. Let the velocity is directed along the axis x and depends on z as:

$$u_x(z) = gz, \text{ where } g = const$$

and the cosmic ray particles have the free path before scattering, equals to λ.

Let us introduce a polar coordinate system with the axis along z Then for the particles arriving at the observation point from the direction θ, ψ, we shall have the additional intensity

$$\delta I = \frac{\gamma + 2}{\lambda c^2} \int_0^\infty e^{-r/\lambda} (\vec{v}(\theta, \psi)\vec{u}(r, \theta, \psi))dr$$

As the velocity $\vec{v}$ is directed along a position vector r, it is necessary for us to find a radial velocity component $\vec{u}$. It is equal to $u_r = u_x cos\psi sin\theta = grcos\theta sin\theta cos\psi$. Supposing $v = c$, we can find:

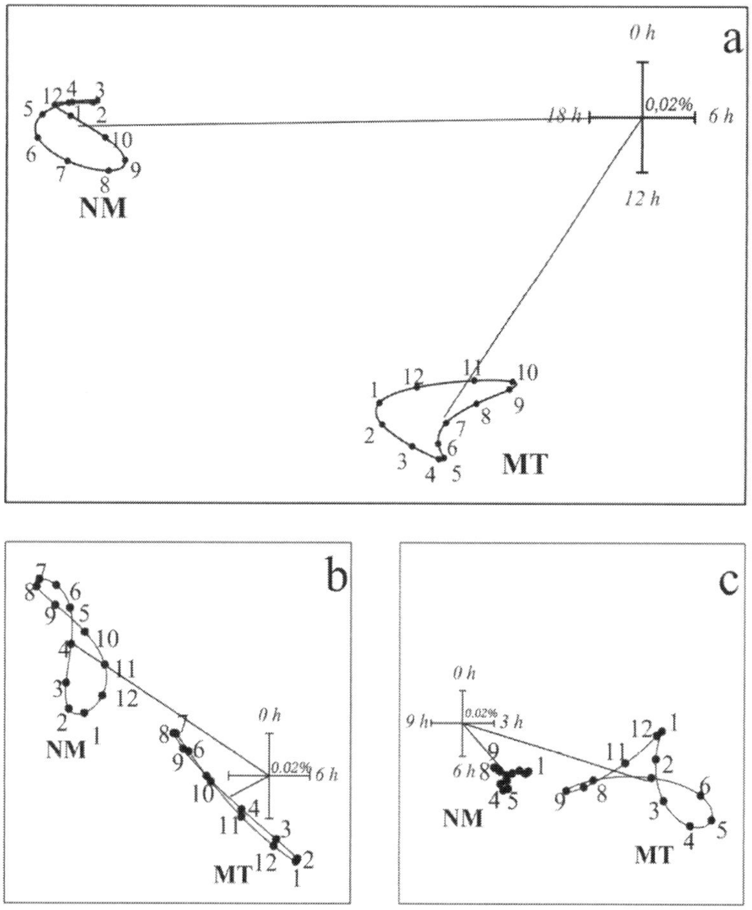

Figure 2. Annual course of the vector anisotropy (symmetric daily variation) (**a**) and tensor anisotropy components: antisymmetric daily variation (**b**), semidiurnal (**c**) by data of the multidirectional muon telescope of Nagoya (MT) and neutron monitor network (NM). Months are represented by numbers near the points on the curve. Error bars are not presented because of their small value.

$$\delta I = \frac{\gamma + 2}{2} \frac{\lambda g}{c} sin2\theta cos\psi$$

Thus, we have the tensor anisotropy with the following components:

$$\vec{A} = (0, a_2^1, 0, 0, 0);\ a_2^1 = \frac{\gamma+2}{2}\frac{\lambda g}{c}.$$

Tensor anisotropy in the neutron component

In [11] we obtained parameters of the antisymmetric daily and semidiurnal variations by data of muon telescopes at Yakutsk and Nagoya. The regular annual change of these parameters is caused by two possible reasons: the changes of orientation of the Earth's axis in the solar coordinate system and shift of the Earth in heliolatitude. These changes differ in phase and, consequently, can be divided in the observations. The main part of semidiurnal variation is

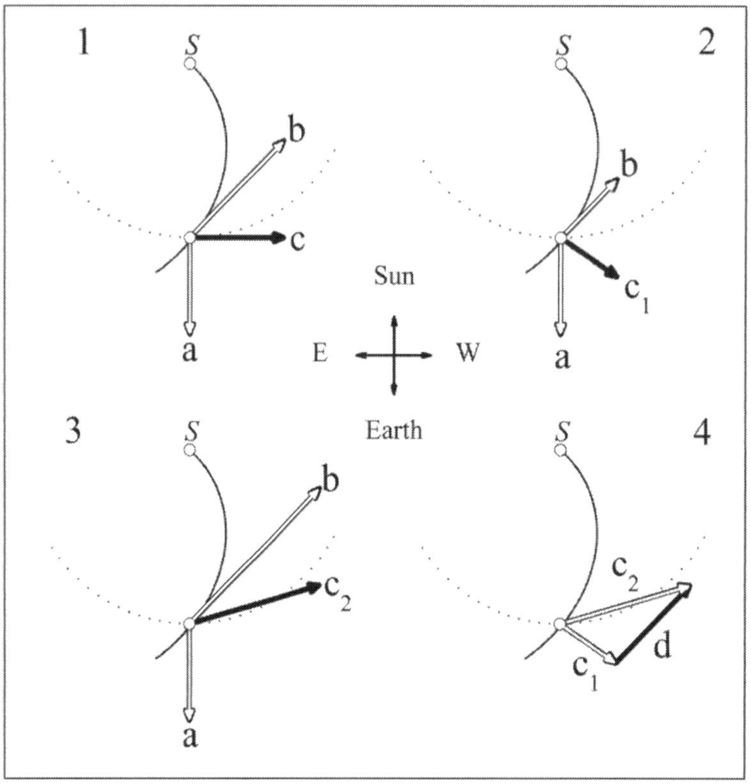

Figure 3. Balance of convective (**a**) and diffusive (**b**) currents and their local disturbance. 1 is the correct balance, 2,3 are the deficit and excess of diffusive current, 4 is the difference of currents (**d**) indicating to the occurrence of current gradient along the magnetic field. **c**, $\mathbf{c_1}$, $\mathbf{c_2}$ are the resultant currents for each case. The dotted curve indicates to the Earth's rotation orbit. The solid curve is the field lines of the interplanetary magnetic field.

created by the screening mechanism, and antisymmetric daily is mainly by the shear flow. It turned out that the extremum of heliolatitudinal changes falls on the $\approx 4.5°$ heliolatitude, but not on the equator. The known fact of southward shift of a low-latitudinal layer in the solar wind [12] confirms it.

Here the results of treatment of observations registered by the worldwide neutron monitor network are given. Figure 2 presents results on the 24 and 12 hour dials. The changes of parameters during a year are shown. The data are smoothed by the 1st and 2nd harmonics (annual and semi-annual variation).

At the same Figure the results of muon measurements from [11] are shown. The general correspondence of those and other data with the account of the fact that the mean particle energies making a contribution into the neutron intensity are considerably lower than in muons is observed. Because the tensor anisotropy spectrum increases with the energy, the neutron component reveals variations which are smaller in value than in the muons. The noticeable difference is revealed in the annual average antisymmetric variation which are considerably higher in the neutrons. Let us discuss the nature of this effect. The above described tensor anisotropy mechanism caused by the solar wind shear flow is the example of more general phenomenon. The tensor anisotropy will occur every time when there is a current gradient of cosmic rays. The reason of this gradient can be not only the shear flow but also any conditions creating the solar wind inhomogeneity, for example, the presence of magnetic mirrors. In [13] the effect of decrease of cosmic rays in a zone of interaction of prompt and slow solar wind streams, caused by the occurrence of magnetic mirror, was considered. Such interaction occurs in the vicinity of neutral sheet of the interplanetary magnetic field. From this it follows that in this field the radial removal of cosmic rays and deficit of compensating diffusive current along the field should prevail. At the boundaries of interaction zone the excess diffusive current which is required for the balance of cosmic rays, on the contrary, should be observed.

The current of cosmic rays creating the ordinary (vector) anisotropy arises as a result of balance of the convective and diffusive currents (**a** and **b** in Figure 3). When the balance is broken, for example, owing to heliolatitudinal differences, it remains unchangeable only on the average. It corresponds to the situation depicted in 2nd and 3rd parts of Figure. The arising current gradient (4th part of Figure) as the north-southern current asymmetry corresponds to the direction of the interplanetary magnetic field. The mentioned effect, apparently, is a reason of annual average of antisymmetric daily variation in the neutron component.

Conclusions

The tensor anisotropy by neutron monitor observations corresponds, as a whole, to the anisotropy observed in the muon component if we take into account the distinction in the energy of particles. However, the annual average vector of antisymmetric daily variation has appeared to be unexpectedly major. In the work it is shown that this property can be understood as the result of interaction of prompt and slow solar wind streams under conditions that its structures are southward shifted. In this case the known mechanism of the tensor anisotropy which is associated with the shear flow of the wind has been generalized.

Acknowledgments

We acknowledge the NMDB database (www.nmdb.eu), founded under the European Union's FP7 programme (contract no. 213007) for providing data. The work has been supported by the Russian Fund for Basic Research (grant nos. 13-02-00989-a, 13-02-00585-a), the Program of Fundamental Research of RAS "High energy physics and neutrino astrophysics" and the grant of the President of Russian Federation on the state support of leading scientific schools (NSh-1741.2012.2).

References

[1] Krymsky G F, Kuzmin A I, Krivoshapkin P A, Samsonov I P, Skripin G V, Transky I A and Chirkov N P 1981 *Cosmic Rays and Solar Wind (in Russian)* (Novosibirsk: Nauka)
[2] Krymsky G F, Kuzmin A I, Chirkov N P, Krivoshapkin P A, Skripin G V and Altukhov A M 1966 *Geomagnetism and Aeronomy* **6** 991–996
[3] Krymsky G F, Kuzmin A I, Chirkov N P, Krivoshapkin P A, Skripin G V and Altukhov A M 1967 *Geomagnetism and Aeronomy* **7** 11–15
[4] Chirkov N P, Altukhov A M, Krymsky G F, Krivoshapkin P A, Kuzmin A I and Skripin G V 1967 *Geomagnetism and Aeronomy* **7** 620–631
[5] Krymsky G F, Altukhov A M, Krivoshapkin P A, Kuzmin A I, Skripin G V and Chirkov N P 1967 *Geomagnetism and Aeronomy* **7** 794–798
[6] Chirkov N P, Krymsky G F and Kuzmin A I 1968 *Canadian Journal of Physics* **46** 614–616
[7] Skeipin G V, Krivoshapkin P A, Krymsky G F and Kuzmin A I 1968 *Canadian Journal of Physics* **46** 973–975
[8] G F Krymsky P A Krivoshapkin A I K 1968 *Canadian Journal of Physics* **46** 959–961
[9] Berezhko E G and Krymsky G F 1981 *Astronomy Letters* **7** 352
[10] Berezhko E G 1981 *Pis'ma v JETP (in Russian)* **33** 416–419
[11] Gerasimova S K, Gololobov P Y, Krivoshapkin P A and Krymsky G F 2013 *Bulletin of the Russian Academy of Sciences. Physics* **77** 526–528
[12] Krymsky G F, Krivoshapkin P A, Mamrukova V P and Gerasimova S K 2007 *JETP* **104** 196–200
[13] Krymsky G F, Krivoshapkin P A, Gerasimova S K and Gololobov P Y 2014 *Astronomy Letters* **40** 230–233

Mini neutron monitor measurements at the Neumayer III station and on the German research vessel Polarstern

B Heber, D Galsdorf, K Herbst, J Gieseler, J Labrenz,
Christian-Albrechts-Universität zu Kiel

C Schwerdt, M Walter,
Deutsches Elektronen-Synchrotron DESY, D-15738 Zeuthen

G Benadé, R Fuchs, H Krüger and H Moraal
Center for Space Research, North-West University, Potchefstroom 2520, South Africa

Abstract. Neutron monitors (NMs) are ground-based devices to measure the variation of cosmic ray intensities, and although being reliable they have two disadvantages: their size as well as their weight. As consequence, [1] suggested the development of a portable, and thus much smaller and lighter, calibration neutron monitor that can be carried to any existing station around the world [see 2; 3]. But this mini neutron monitor, moreover, can also be installed as an autonomous station at any location that provides "office" conditions such as a) temperatures within the range of around 0 to less than 40 degree C as well as b) internet and c) power supply. However, the best location is when the material above the NM is minimized. In 2011 a mini Neutron Monitor was installed at the Neumayer III station in Antarctica as well as the German research vessel Polarstern, providing scientific data since January 2014 and October 2012, respectively. The Polarstern, which is in the possession of the Federal Republic of Germany represented by the Ministry of Education and Research and operated by the Alfred Wegener Institute, Helmholtz Centre for Polar and Marine Research and managed by the shipping company Laeisz, was specially designed for working in the polar seas and is currently one of the most sophisticated polar research vessels worldwide. It spends almost 310 days a year at sea usually being located in the waters of Antarctica between November and March while spending the northern summer months in Arctic waters. Therefore, the vessel scans the rigidity range below the atmospheric threshold and above 10 GV twice a year. In contrast to spacecraft measurements NM data are influenced by variations of the geomagnetic field as well as the atmospheric conditions. Thus, in order to interpret the data a detailed knowledge of the instrument sensitivity with geomagnetic latitude (rigidity) and atmospheric pressure is essential. In order to determine the atmospheric response data from the Neumayer III station here we will perform comparisons to other polar stations, resulting in an atmospheric pressure coefficient of 7.5‰/hPa, while the rigidity dependence will be determined experimentally by utilizing several latitude scans. Thereby, the atmospheric pressure and temperature correction will be discussed in more detail and the results of the latitude scan performed between October 2012 and March 2013 will be presented. Moreover, we will show that this latitude scan can also be described by using the yield function of [4].

1. Introduction

Ground-based measurements of Galactic Cosmic Rays (GCRs) have been performed since their discovery by Viktor Hess in 1912. Since the 1950s neutron monitors (NMs) are utilized for these kind of measurements [5; 6]. Because NMs are integral counters with a threshold energy defined by the minimum of the magnetic cut-off energy and the shielding by the atmosphere, several NMs have been installed world-wide from which more than for forty are accessible through the NM data base(www.nmdb.eu). This should allow to determine the energy spectra from about 400 MeV to above 10 GeV of GCRs as well as solar energetic particles (SEPs) during Ground Level Enhancements, strong SEP events which can be detected on the ground. In order to calibrate the world-wide distributed NMs against each other Moraal et al. [7] proposed the development of a calibration NM, which has been realized at the North-West-University [1; 2; 8–10]. This NM relies on a ^{3}He counter in order to achieve count rates of 1 count/second at sea level allowing a statistical accuracy on the percent level by using hourly averages. This device is well calibrated for temperature as well as atmospheric pressure and has been used at several locations [10]. Based on this idea, Krüger and Moraal [3] suggested to use such a mini NM as a mobile station that can be placed at almost any location in the world that can provide an office, power supply and internet access. In order to make the data easily accessible, they are provided to the NM database (www.nmdb.eu). Due to the price of the ^{3}He counter, recently a new concept on the basis of a $^{10}BF_3$-tube has been developed [3]. Such a device was successfully installed at the German research station Neumayer III located at the Ekström Shelf Ice, Atka Bay, north-eastern Weddell Sea (70°40'S, 008°16'W) in January 2014. Here we will report on the first results.

2. Instrumentation

The design of the mini NM (as installed at the Neumayer III station) has been described in detail in Krüger and Moraal [3]. The mini NM has a length of ~80 cm and a radius of 20 cm. In September 2011 the International Panel on Dangerous Goods increased the maximum atmospheric pressure of $^{10}BF_3$ counters from ~ 300 to ~ 900 hPa, thus by increasing the diameter of the tube from 0.5 to 0.89 cm the count rates of the modified mini NM are comparable to the ones using the ^{3}He-tube. The counter itself has a length of 63 cm (one third of the standard length of the LND25373 and NM64 counters). It is surrounded by a 2 cm thick moderator made of paraffin wax, which is surrounded by a 5 cm lead ring acting as a producer of neutrons. In addition, around the lead ring there is an outer paraffin wax reflector with a thickness of 9.5 cm. In this configuration the mass of the whole unit is estimated to be 220 kg of which 170 kg are contributed by the lead and 25 kg by the paraffin wax. With these dimensions and mass it could be conveniently placed at the Neumayer III station and the Polarstern.

3. Data reduction and validation

The temporal resolution of the data recorded by the mini NM at Neumayer III is in the millisecond range allowing to determine the registration of multiple hits by the corresponding time stamp. From left to right Fig. 1 displays the temperature and atmospheric pressure. While the temperature only varies between ~12 and ~28 degree C atmospheric pressure spans a range of ~950 to ~1020 hPa, with a mean value of about 980 hPa. According to [3] the temperature correction coefficients for $^{10}BF_3$ counters vary between -0.05%/degree C and 0.05%/degree C. Thus, in a first approximation a smooth variation in the temperature can be neglected. Since the total atmospheric pressure correction is more than 20 times more important than the temperature one, a careful analysis for atmospheric pressure changes will be postponed.

3.1. Temperature and atmospheric pressure corrections

The standard method to determine the barometric coefficient is to find a time period where the GCR flux outside the magnetosphere is constant and the magnetosphere shows no strong disturbances [11; 12]. Paschalis et al [12] provide an application within the NM database project to calculate the barometric coefficient based on hourly data. For the Neumayer III station the statistical accuracy of the hourly count rate is less than the one for the one minute averages obtained by standard NMs. In order to make use of the whole data set, we calculate the daily averaged ratio of a atmospheric pressure corrected NM that has a cut-off rigidity below 0.4 GV and is located at altitudes below 100 m, i.e. Terre Adelie (TERA), Inuvik (INVK), McMurdo (MCMU), Nain (NAIN), Peawanuck (PWNK), and Thule (THUL).

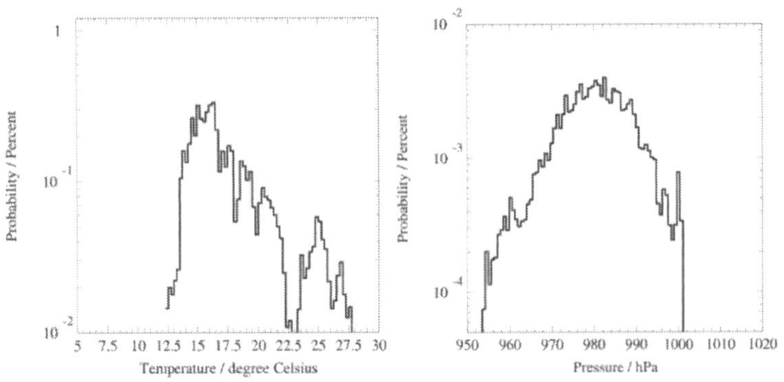

Figure 1. From left to right the temperature and atmospheric pressure from doy 40 to 260 of 2014 is shown. While the temperature only varies between ~12 and ~28 degree C the atmospheric pressure spans the range from ~950 to ~1020 hPa, with a mean value of about 980 hPa.

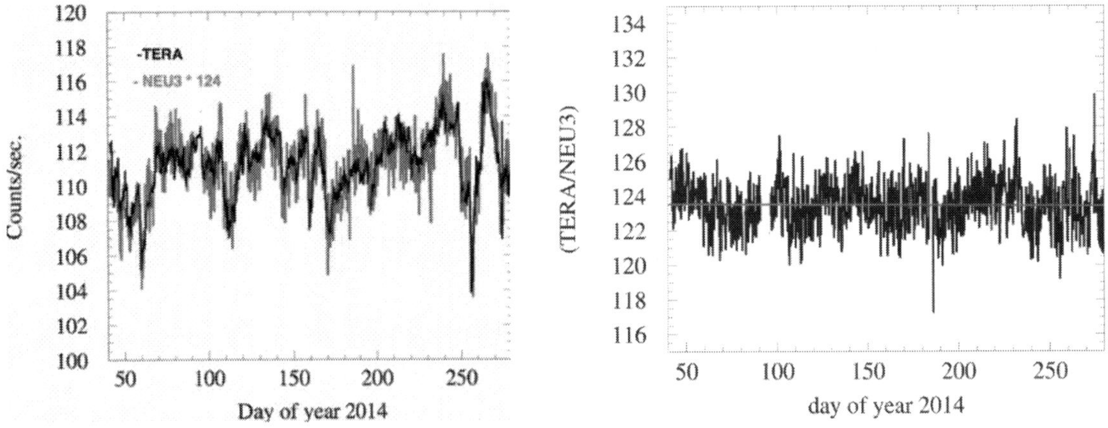

Figure 2. From left to right: Six hour averaged counting rates and ratio of the atmospheric pressure corrected count rates of Terre Adelie and Neumayer III. Here we used $f_p = 7.5‰$/hPa and $f_T = 0.6‰$/degree C for the mini NM. The red line in the right panel displays the fit of $r(t)|_{f_T} = r_0 + p_1 \cdot t$ with $r_0 = 123.5 \pm 0.2$ and $p_1 = -(0.1 \pm 0.1) \cdot 10^{-4}$/day.

In what follows we compare the six-hour averaged count rates obtained by the mini NM with the ones obtained by the different stations as summarized in Table 2. The left panel of Fig. 2 displays the atmospheric pressure corrected counting rates of the NM at Terre Adelie (black curve) and Neumayer III (red curve) while the right panel gives the corresponding ratio. The Neumayer III corrections are based on the minute raw count rates $C_u(t)$ by applying:

$$C_c(t) = C_u(t) \cdot \exp(f_p \cdot (p(t) - p_0)) \cdot \exp(f_T \cdot (T(t) - T_0)) \tag{1}$$

This procedure results in corrected time profiles $C_c(t)$. Herein, f_p and f_T give the barometric and temperature correction factor and $p(t)$, p_0, $T(t)$ and T_0 the measured as well as the reference atmospheric pressure and temperatures. We choose $p_0 = 980$ hPa and $T_0 = 20$ degree C as reference values. The plot gives NM count rates from the Neumayer III station for $f_p = 7.5‰$/hPa and $f_T = 0.45‰$/degree C. The red line in the right panel of Fig. 2 is a fit of the function

$$r(t)|_{f_T} = r_0 + p_1 \cdot t \tag{2}$$

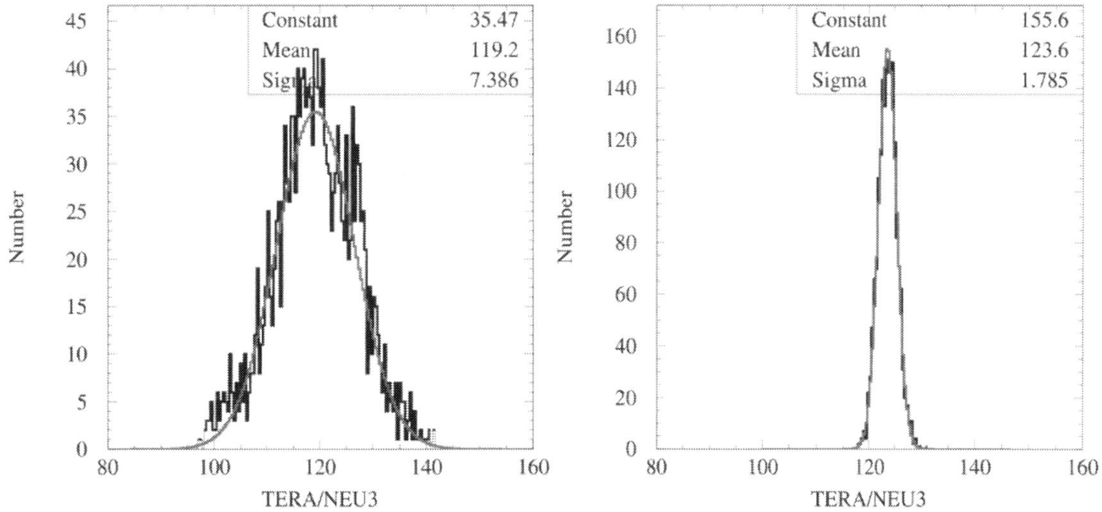

Figure 3. Ratio of six-hour averaged ratio of the counting rates of Terre Adelie and Neumayer III using a atmospheric pressure and temperature coefficient of 0‰/hPa and 0‰/degree C (left panel) and 7.5‰and 0.6‰/degree C (right panel), respectively. For details see text.

Table 1. Regions in the $f_p - f_T$-plane where a minimum of σ can be identified

Barometric coefficient f_p (‰/hPa)	Temperature coefficient f_T (‰/deg)
7.05 to 7.25	-0.65 to -0.5
7.05 to 7.25	0.45 to 0.75
7.25 to 7.55	0.2 to 0.45

to the data. If $p_1 = 0$, there will be no difference on the long term trend between Neumayer III and the reference NM (here Terre Adelie). Because the atmospheric pressure varies on short time scales, the spread of the ratio is mostly determined by the atmospheric pressure correction. Thus, in order to obtain the best agreement between the reference NM and the mini NM at Neumayer III, we vary the atmospheric pressure and temperature correction coefficient and determine the width σ of the ratio distribution. Fig. 3 displays the ratio of the six-hour averaged counting rates of Terre Adelie and Neumayer III using a atmospheric pressure and temperature coefficient of 0‰/hPa and 0‰/degree C (left panel) and 7.5‰/hPa and 0.41‰/degree C (right panel), respectively. The red line displays the result of a fit by a Gaussian to the distribution. While the width of the distribution in the left panel is about 7.4 it decreases to less than 1.8 for the parameter chosen in the right panel. In order to determine the best values for both corrections, we systematically vary the coefficient f_p between 0 and 10‰/hPa and f_T between -1 and 1‰/degree C using forty steps for both parameters. In our further analysis we only use these parameters for which the width σ becomes minimal. The left and right panels of Fig. 4 show the results of this analysis by using Terre Adelie as reference NM. The right panel corresponds to a detailed view of the left panel giving only the values between $f_p = 6.8$ ‰/hPa and $f_p = 7.8$ ‰/hPa. As expected, the width σ is sensitive to changes in f_p but not to the ones in f_T. In the right panel of Fig. 4 regions in the $f_p - f_T$-plane can be identified which result in a minimum of σ. The corresponding locations are summarized in Table 1. In order to distinguish between these results only the one is taken for which the long term trend of Neumayer III agrees best with the one of the reference NM. In our example the best values can be achieved for f_p between 7.0 and 7.6‰/hPa and f_T between -1.3 and 0.45 ‰/degree C, with an average

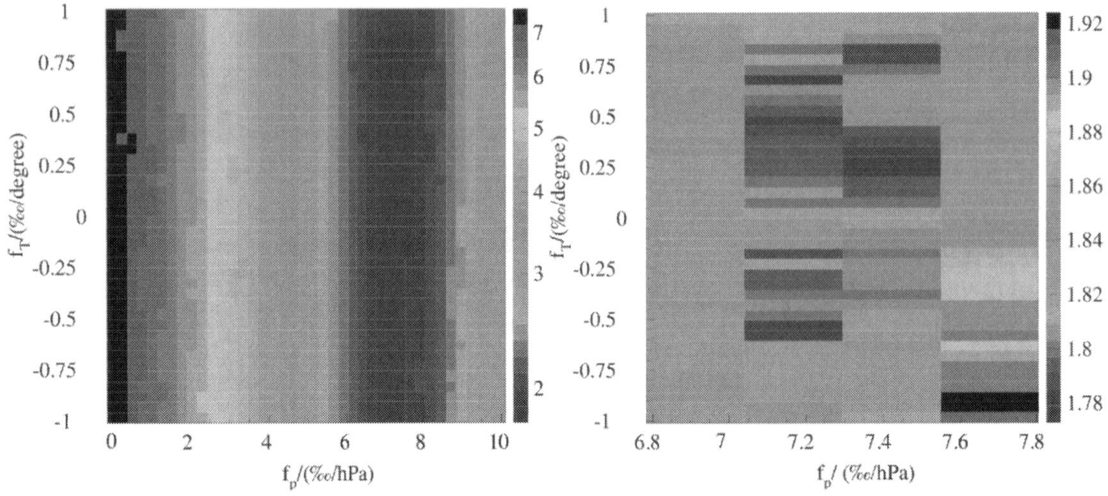

Figure 4. Color coded sigma values (see Fig. 3) as function of the atmospheric pressure p_c and the temperature t_c coefficients. The right panel shows the range between 6.8‰/hPa and 7.8‰/hpa in more detail.

Table 2. Atmospheric pressure and temperature correction coefficients for different NMs located at cut-off-rigidities less than 0.4 GV and altitudes less than 100 m. Temperature corrections are only applied for the mini NM. Note, the in contrast to the Neumayer mini NM the Polarstern mini NM utilizes a ^{3}He-counter tube [13].

Station	Temperature coefficient f_T (‰/deg)	Barometric coefficient f_p (‰/hPa)
TERA	0.6	7.4
INVK	0.8	7.5
MCMU	0.25	7.2
NAIN	-1.3	7.4
PWNK	-1.	7.5
THUL	1.6	7.6
Average		7.5
Polarstern	1.18	7.8

value of f_p = 7.5 ‰/hPa and $f_T \sim -0.4$ ‰/degree C. In addition, the same analysis has been applied to the other NMs, the results are summarized in Table 2. While the barometric coefficients determined by the comparison with the other NMs agree within less than 10%, the temperature coefficients differ by more than a factor of two. The calculation of the later relies on the long term stability of these NMs and thus we expect that an analysis on a longer data interval will improve the accuracy of f_T. The average values of f_p = 7.5‰/hPa and f_T = −0.4‰/degree C are in reasonable agreement with the ones given in the literature [9] and are used to correct the minutely and hourly averages as provided to the NM data base (www.nmdb.eu). For comparison reasons, the last row of Table 2 gives the correction coefficients of the ^{3}He-counter tube as used by the mini NM aboard the Polarstern [9].

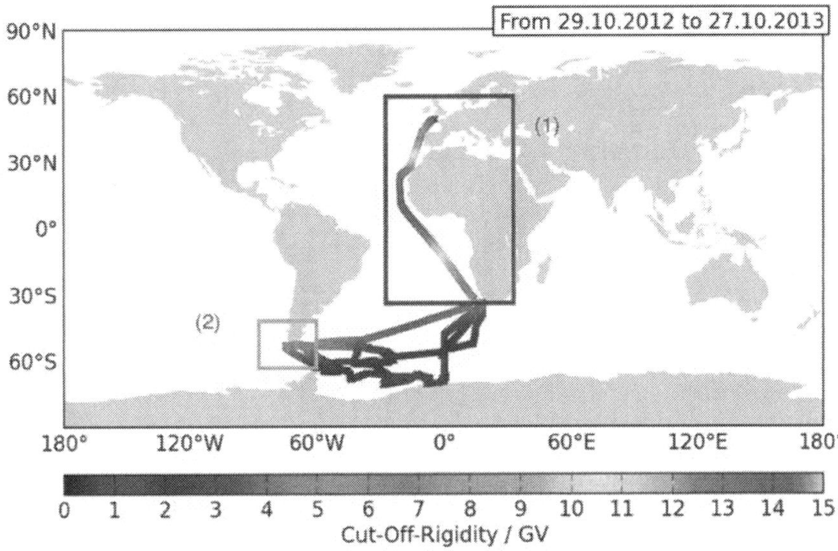

Figure 5. Cut-off-rigidity calculated along the route of the Polarstern from October 29, 2012 to October 27, 2013. The calculation is based on the IGRF dipole model. The cut-off-rigidity ranges from less than 1 GV at Neumayer to more than 14 GV in equatorial regions [14].

3.2. Variation with the effective vertical cut-off-rigidity

Latitude surveys have been used in the past to determine the differential response function of a NM [see e.g. 4]. As stated above, the intensity of GCRs entering the atmosphere strongly depends on the cut-off-rigidity, and with it on the geomagnetic location of the NM. A charged particle, with momentum p and charge q, interacting with the Earths magnetic field $\vec{B}$ has a rigidity R defined as $R = p/q = r_L B$, with r_L representing the Larmor radius. The effective vertical cut-off rigidity is the minimum rigidity that permits a charged particle to arrive from the vertical direction at a given latitude and longitude. In order to compute R_C in an arbitrary magnetic field, numerical computations are mandatory [15; 16]. For our studies we use the simulation code PLANETOCOSMICS [17]. The simulations were carried out for the IGRF model as well as for a magnetic field perturbed by the solar wind according to the Tsyganenko89 model [18]. Figure 5 displays the result of these calculations using an intermediate k_p index along the route of the Polarstern from October 29, 2012 to October 27, 2013. The values range from less than 1 GV at Neumayer to more than 14 GV in equatorial regions [see 14]. Marked by box (1) and (2) are two important regions that are discussed in the following. Figure 6 in its top left panel displays the calculated cut-off-rigidities for the time period between October 29, 2012 and end of March 2013. During the first period the Polarstern was going from Bremerhaven via the equator to Cape Town and continued its journey towards Antarctica (marked by (1) in Fig. 5). In 2013 the Polarstern went to Punta Arenas in Chile which is located at significant higher cut-off-rigidities (marked by (2) in Fig. 5) leading to the two plateau-like structures in the cut-off-time profile. The black and red curve in the left panel of Fig. 6 display the corresponding atmospheric pressure corrected mini NM counting rates and the one measured by the reference NM [red curve, see also 14]. While for most of the time the NMs

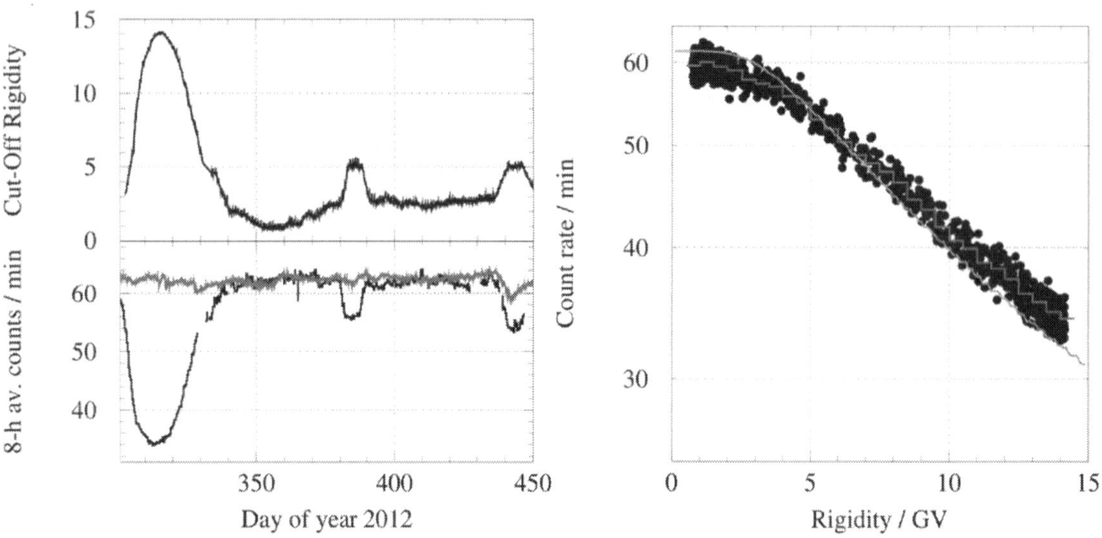

Figure 6. Left panel: Shown from top to bottom are the calculated cut-off-rigidities and the atmospheric pressure corrected count rates (black curve) of the mini NM from end 2012 to March 2013. The red line displays the measurements at Terre Adelie. Right panel: Atmospheric pressure corrected counting rates as function of the cut-off-rigidity (filled dots). The red histogram gives the average in intervals of 0.5 GV width. While above 3 GV the counting rate is well correlated with the cut-off-rigidity, the variation below 2 GV is mostly controlled by the atmospheric shielding.

agree well with each other there are three periods for which we find systematic disagreements. The first one from DOY 300 to 345 in 2012 which is attributed to the systematic change in the cut-off-rigidity. The other two coincide with the excursion towards Chile. In order to investigate this effect in more detail the six-hour averaged atmospheric pressure-corrected counting rates as function of the cut-off-rigidity during the corresponding time is displayed in the right panel of Fig. 6. Here the red and green line gives the mean value in intervals of 0.5 GV and the expectation using the yield function from [4]. While an obvious correlation is seen at rigidities above 3 GV, the atmospheric shielding leads to a flattening below 2 GV.

4. Summary

Galactic cosmic rays are high-energy charged particles, mainly protons, doubly ionized helium, and other fully ionized nuclei originating in the galaxy and bombarding the Earth from all directions. Measurements by various particle detectors have shown that the intensity varies on different timescales, caused by the Suns activity but also geomagnetic and atmospheric (mainly atmospheric pressure) variations. Most of the long term research on GCR intensity variations is based on the analysis of ground-based NMs. In order to enlarge the number of such stations a mini NM was developed [1]. In 2011 two mini NMs were installed, one aboard the research vessel Polarstern and one at the Neumayer III station in Antarctica acting as reference NM. Due to data acquisition problems an upgraded and improved version of the mini NM replaced the one at Neumayer III in 2014, thus the counting statistic now has improved by a factor of three. With a counting rate of ~60 counts/minute a statistical accuracy of <1% is reached for counts integrated over 90 minutes. This is sufficient enought to study most cosmic ray variations. Although the counting rate is more than 100 times smaller compared to the ones measured by e.g. the NM at Terre Adelie or other station at similar locations the time profiles agree well with each other if the reference NMs are used to determine the atmospheric pressure and temperature coefficients, thus the mini NM provides reliable measurements. However, for the latitude survey of the Polarstern between

October 2012 and March 2013 the reference NM at Neumayer III could not be used due to technical problems. For our analysis we, therefore, compared the Polarstern NM with the NM at Terre Adelie. We could show that the count rates of both NMs agree well with each other during times when the vessel is situated in regions with (computed) cut-off-rigidities below 2 GV while differing during times when the vessel is in regions with cut-off-rigidities above 2 GV (e.g. in fall 2012). The latitude survey has been used to determine the count rate variations with cut-off-rigidities in the range from 0 GV to below 15 GV.

Acknowledgements
We acknowledge the NMDB database (www.nmdb.eu), founded under the European Union's FP7 programme (contract no. 213007) for providing data. The data from McMurdo were provided by the University of Delaware with support from the U.S. National Science Foundation under grant ANT-0739620. Kerguelen and/or Terre Adelie NM data were kindly provided by the French Polar Institute (IPEV, Brest) and by Paris Observatory.

References
[1] Moraal H, Benadie A, de Villiers D, Bieber J W, Clem J M, Evenson P E, Pyle K R, Shulman L, Duldig M L and Humble J E 2001 *Proceedings of the 27th International Cosmic Ray Conference. 07-15 August* p 4083 (document), 1, 4

[2] Moraal H, Krueger H, Benadie A and de Villers D 2003 *Proceedings of the 28th International Cosmic Ray Conference. July 31-August 7* p 3453 (document), 1

[3] Krüger H and Moraal H 2013 *Journal of Physics: Conference Series* **409** 2171 (document), 1, 2, 3

[4] Caballero-Lopez R A and Moraal H 2012 *Journal of Geophysical Research (Space Physics)* **117** 12103 (document), 3.2, 3.2

[5] Shea M A and Smart D F 2000 *Space Science Reviews* **93** 229–262 1

[6] Simpson J A 2000 *Space Science Reviews* **93** 11–32 1

[7] Moraal H, Belov A and Clem J M 2000 *Space Science Reviews* **93** 285–303 1

[8] Krüger H, Moraal H, Bieber J W, Clem J M, Evenson P A, Pyle K R, Duldig M L and Humble J E 2003 *Proceedings of the 28th International Cosmic Ray Conference. July 31-August 7* p 3441 1

[9] Krüger H 2006 *A calibration neutron monitor for long-term cosmic ray modulation studies / H. Krüger* Ph.D. thesis North-West-University, Potchefstroom campus, ZA 3.1, 3.1

[10] Krüger H, Moraal H, Bieber J W, Clem J M, Evenson P A, Pyle K R, Duldig M L and Humble J E 2008 *Journal of Geophysical Research (Space Physics)* **113** 8101 1

[11] Carmichael H, Bercovitch M, Shea M A, Magidin M and Peterson R W 1968 *Canadian Journal of Physics. Vol. 46* **46** 1006 3.1

[12] P Paschalis P, Mavromichalaki H, Yanke V, Belov A, Eroshenko E, Gerontidou M and Koutroumpi I 2012 Online application for the barometric coefficient calculation of nmdb stations Tech. rep. 3.1

[13] Clem J M and Dorman L I 2000 *Space Science Reviews* **93** 335–359 2

[14] Galsdorf D 2014 *Cutoff-Steifugkeiten und Asymptotische Richtungen sowie ihre Parameter* Ph.D. thesis Christian-Albrechts-Universität zu Kiel 5, 3.2, 3.2

[15] Herbst K, Kopp A and Heber B 2013 *Annales Geophysicae* **31** 1637–1643 3.2

[16] Pilchowski J, Kopp A, Herbst K and Heber B 2010 *Astrophysics and Space Sciences Transactions* **6** 9–17 3.2

[17] Desorgher L, Flückiger E O and Gurtner M 2006 *36th COSPAR Scientific Assembly. Held 16 - 23 July 2006* **36** 2361 3.2

[18] Tsyganenko N A 1989 *Planetary and Space Science (ISSN 0032-0633)* **37** 5–20 3.2

Solar extreme events

Hugh S. Hudson

Space Sciences Laboratory, University of California, 7 Gauss Way, Berkeley CA 94720-7450, USA

E-mail: hhudson@ssl.berkeley.edu

Abstract. Solar flares and CMEs have a broad range of magnitudes. This review discusses the possibility of "extreme events," defined as those with magnitudes greater than have been seen in the existing historical record. For most quantitative measures, this direct information does not extend more than a century and a half into the recent past. The magnitude distributions (occurrence frequencies) of solar events (flares/CMEs) typically decrease with the parameter measured or inferred (peak flux, mass, energy etc. Flare radiation fluxes tend to follow a power law slightly flatter than S^{-2}, where S represents a peak flux; solar particle events (SPEs) follow a still flatter power law up to a limiting magnitude, and then appear to roll over to a steeper distribution, which may take an exponential form or follow a broken power law. This inference comes from the terrestrial ^{14}C record and from the depth dependence of various radioisotope proxies in the lunar regolith and in meteorites. Recently major new observational results have impacted our use of the relatively limited historical record in new ways: the detection of actual events in the ^{14}C tree-ring records, and the systematic observations of flares and "superflares" by the Kepler spacecraft. I discuss how these new findings may affect our understanding of the distribution function expected for extreme solar events.

1. Introduction

Flares and coronal mass ejections pose well-recognized hazards for human society on Earth and its outposts in space. The damage inflicted by extreme solar events,[1] also known as "superflares" and perhaps "super CMEs" will by definition exceed the levels suffered in the historical past. Mission planning for manned space flight, for example, requires adequate knowledge of the fluence distribution in Solar Particle Events (SPEs) and the likelihood of the occurrence of such events. Our knowledge of the distribution function in magnitude of the events generally points to scale-free power laws (e.g., [1]), as derived from direct measurements of particle or photon fluxes. In this paper we point out progress made in recent years that adds new proxy information that may augment the direct measurements. This new information in principle could change the assumptions needed for forecasting the occurrence of an extreme event, and may eventually make predictions qualitatively more reliable.

The approximately power-law distribution of flare peak fluxes has been known since the Van Allen era of the space age [2] or even before [3]. Note that these papers deal with soft X-ray and microwave bursts, respectively, but observations in many other spectral regions show the same property. Thus we often assume that this power-law characteristic refers basically to the flare energy, as seen through various proxies (such as these), even though any one of them may deviate

[1] This paper adopts the working definition of "extreme event" as one more powerful than any heretofore observed; see Section 2 for discussion.

systematically from this fundamental property. As is well known, such featureless power-law distributions appear in many other situations in nature – Zipf's Law; the Gutenberg-Richter Law; Paneto's Law, etc. (*e. g.*, [4]).

The particle data come directly (at low cosmic-ray energies) from counters deployed in interplanetary space. Van Hollebeke *et al.* [5] presented an early comprehensive study of fluence distributions for energetic protons in the range 20-80 MeV, finding a power-law fit for maximum particle intensity I (particles cm^{-2} s^{-1} sr^{-1} MeV^{-1}) of the form $dN/dI \sim I^{-\alpha}$, with $\alpha = 1.15 \pm 0.1$. The modern data also include ground-based observations from neutron monitors (e.g., [6]). These respond at higher energies and so do not directly measure the most damaging components of the particle spectrum; the neutron monitors also only detect the most powerful events. On the other hand, it is just these extreme events that dominate the particle fluence, and so the information the neutron monitors provide is essential.

The flat power-law distribution of peak particle intensities, carried over to a distribution of fluences, poses a problem for predictions. Such a flat power law leads to uncertainty in estimates of fluences at the top end of the scale; for a flat distribution, the one or two greatest events in a given solar cycle may well provide the bulk of the particle fluence for that entire cycle. The question of occurrence probability then requires an understanding of the likelihood of an event with as great a fluence as any observed, or a still greater one of a magnitude not yet directly recorded, and yet still permissible theoretically.

Modern data have greatly improved our knowledge of the non-extreme events, and one objective of this paper is to put this new knowledge in the context of predictions of occurrence probability. Schrijver *et al.* [7] give the starting point for the present review; this paper surveys flare energetics, and notes the strong evidence for the existence of an upper cutoff in total event energy. Flares (and CMEs) do not directly scale with solar particle events (SPEs); the relationships are strong but not simple. In this review we start by discussing the basically new information from tree rings and from the Kepler stellar observations (Section 3). Section 4 describes the bearing of this information on the relationship between SPE magnitude to total event energy, or flare magnitude. This inspires a "search for the break" (Section 5), assessing our current knowledge of behavior of the occurrence distribution beyond our direct empirical knowledge. Section 6 attempts to reconcile these disparate observations, emphasizing the importance of the rollover to large event magnitudes. The existence of this break energy would basically resolve the prediction issue posed by the flat distribution function apparent in the data of Van Hollebeke et al. [5], for example, in that integration of the distribution now can lead to a reduction in our best guess about the occurrence probability of an extreme solar event.

2. What is an extreme event?

The flat power-law distributions of observed flare magnitudes and particle fluences mean that the major events dominate the total within the distribution, which in statistical jargon has a "heavy tail." These observations therefore provide an antithesis of the "nanoflare" hypothesis [8], in which the accumulation of small events dominates the total. From other considerations an extreme solar event would be one that could do far more damage to our environment than ones previously experienced; such an event would have a vanishing probability of occurrence until it actually happened and thus instantly became an unavoidable reality. Thus the manner in which we extrapolate our observational experience can have a profound influence on assets with vulnerabilities; on an arguably small scale one can point to power-grid failures and satellite outages as costly and unexpected failures of large-scale assets (e.g., [9]) on Earth and in the near-Earth space environment.

The main immediate interest in damge produced by solar extreme events is in the behavior of the solar energetic particles (SEPs or "solar cosmic rays; see e.g. [10, 11] for recent reviews). We must bear in mind, however, that our paradigm derived from weaker events may not apply accurately to much stronger ones; in addition to the SEPs there might also be substantial impact from hard electromagnetic radiations (EUV, X-rays, γ-rays) or from destabilization of the Earth's magnetosphere by the impact of accompanying interplanetary CME.

The Carrington flare (SOL1859-09-01), at the very boundary of the historical record, could have approached the actual occurrence of a solar extreme event [12], in terms of later experience, although this interpretation as the source of an extreme geomagnetic storm may have been based on a misunderstanding of the geomagnetic signatures recorded at low latitudes [13, 14]. The flare and inferred CME did not greatly exceed, by most measures, other more recent and better-studied major events such as SOL2003-11-04 [14], illustrated below, or SOL2012-07-23 [15]. So no clear evidence for a "Black Swan"[2]. seems to exist on a time scale now exceeding one century.

3. Newer developments

3.1. Events discovered via analysis of tree rings

Living things ingest carbon, and this contains a time stamp in terms of the immediate abundance of the ^{14}C radioisotope. Its radioactivity then lets us date a biological sample, a property used in many studies for different purposes. Prior to the publication of the first direct event detection via the ^{14}C proxy in tree rings [18], we had only statistical information about the time dependence of the production rate of the ^{14}C isotope. This first event, which occurred in about 775 AD based on the tree-ring chronology ("dendrochronology"), has generated substantial discussion and certainly could have been solar in origin. The same group soon uncovered a second event [19], dated about 993 AD. The detection of these events required the improvement of time resolution for the radioisotope signature to about one year, as compared with the few-year resolution previously available. The time resolution of such observations, in terms of ^{14}C rate, suffers because of the complicated processes involved in circulation and uptake into the trees; this of course limits the degree to which such an event could be linked to some other phenomenon. The initial discovery of the event was in annual samples of cedar trees from Yaku Island, Japan, and its presence was soon confirmed in annual data obtained from trees in Germany, Austria, the U.S., Russia, and New Zealand, as well as via ^{10}Be records from Antarctica [20, 21] and Chinese corals [22]. This widespread effect provides further evidence for an extraterrestrial origin.

The tree-ring records have provided our first radioisotope evidence for discrete high-energy transients in the terrestrial atmosphere. Could the Sun have been responsible [20]? Similar increases cannot be detected positively in the isotope records for known major flare events, such as SOL1859-09-01 (the Carrington event), and so this would imply that the two new events fall in a new hypothetical category of solar "superflares." The significance of these events would then be that they constrain the distribution of solar events on longer time scales than the extent of the historical record, and thus help define the probability of occurrence of less frequent but more powerful events. Even if a non-solar explanation prevails [23, 24, 25], the absence of other detections still provides upper limits for solar event occurrence on the relevant time scales. The tree-ring events could be consistent with a solar origin, given the roll-off and with an appropriate spectral distribution [26, 23], but would reflect a much larger event than the F_{30} cutoff discussed below in Section 4.2; here F_{30} refers to the fluence of solar particles above 30 MeV, a standard categorization. The cutoff occurs near $F_{30} = 10^{10}$ protons cm^{-2}, but a proton energy of 200 MeV may be more appropriate to allow for reduced model-dependence on the spectral distribution, especially for the ^{10}Be isotope [27]. The rare appearance of such a solar event, over such a long time span, might not conflict with our interpretation in terms of the steepening of the distribution Section 5.

[2] A "Black Swan" is extreme event [16], in popular terminology now adopted by some disaster planning offices. "Such rare events are observed in exceptionally strong windgusts, monster or rogue waves, earthquakes, and financial crashes" [17] – and possibly in stellar magnetic activity as well

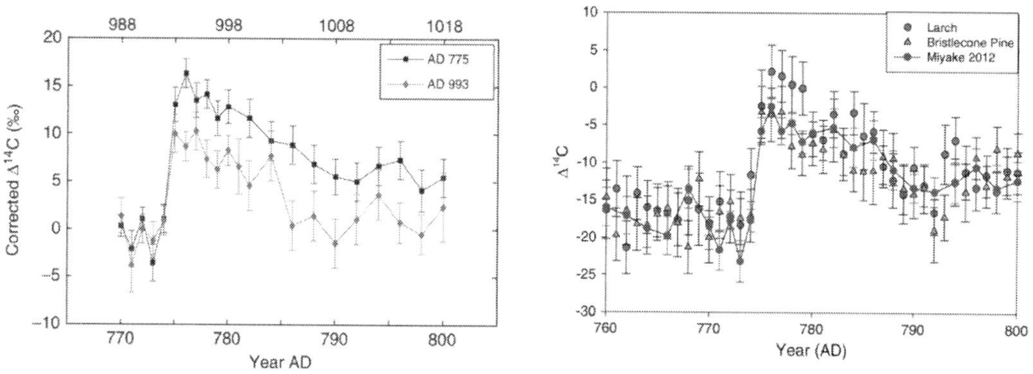

Figure 1. Left, the two tree-ring radioisotope events from Japanese cedars, from annual samples [19]. At this resolution these time series each show an unresolved sudden increase, followed by a slow recovery consistent with terrestrial transport and the uptake of ^{14}C into the tree. Right, earlier event also appeared in tree rings from larches, kauri, bristlecone pines, and oaks, and from both N and S hemispheres [21].

3.2. Stellar flares

Stars of many spectral classes, but most obviously the UV Ceti "flare star" category (dMe), produce flares that more or less match the paradigms of solar flares (e.g., [28]). In general we lack the breadth of diagnostics for these stellar flares that we have for the solar ones, and in particular we do not have systematic observations of hard X-rays and γ-rays that could reveal analogous particle acceleration – this is a matter of sensitivity.

Recently the *Kepler* satellite has opened a database on precise stellar time-series photometry with enormous qualitative and quantitative improvements [29] in the statistics of solar-type flare occurrence on other stars. The new data reveal myriads of flares [30, 31], including "superflares" on solar-type stars [32, 33, 34] that may help us to understand extreme solar events. Figure 2 shows an example of one of the flaring solar-type *Kepler* stars. The large, slow variations (of order 1%) suggest the presence of large-scale magnetic structures on the stellar photosphere (solar-type spots?); superposed on these variations one sees the impulsive increases of the stellar flares.

How similar are the variations seen in the *Kepler* stars and the Sun itself? We note the contrast in the behavior of the two stars compared in Figure 2 (KIC 11764567, via broad-band photometry, and the Sun as viewed in total solar irradiance). The *Kepler* observations directly suggest rotational modulation with a simple structure, whereas the solar timeseries has more complexity. The solar spots appear to be far more concentrated, and appear in the timeseries as downward excursions lasting for about a quarter of the rotational period in a distinctly non-sinusoidal manner (the "dips" [36]). The distinction between these two morphological patterns probably does not depend upon the inclination of the stellar rotation axis, since exact modeling based on known sunspots shows little effect in the solar case for out-of-the-ecliptic viewing [38]. From the Figure, it is difficult to understand how the solar paradigm applies to the stellar case, given the distinctly different morphologies.

How well do the *Kepler* superflares describe what the Sun might do, in the case of an extreme event? The Kepler photometry has only one broad passband (approximately 400-900 nm) and poor time resolution (typically 30 min, though sometimes 1 min), but excellent photometric precision [39] and therefore sensitivity. The broad passband for this photometry incorporates many chromospheric emission lines, in particular Hα, but also could include continuum analogous

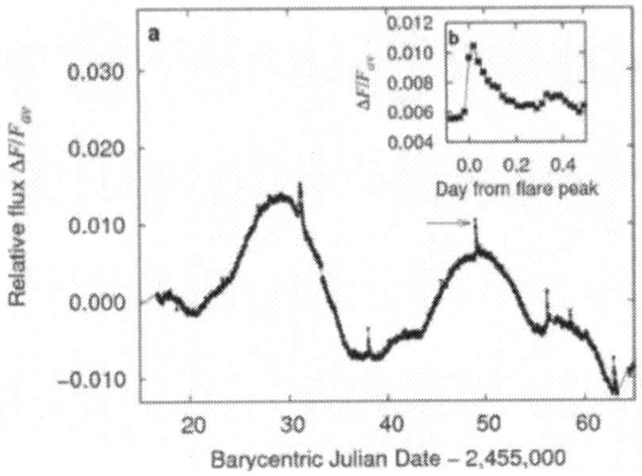

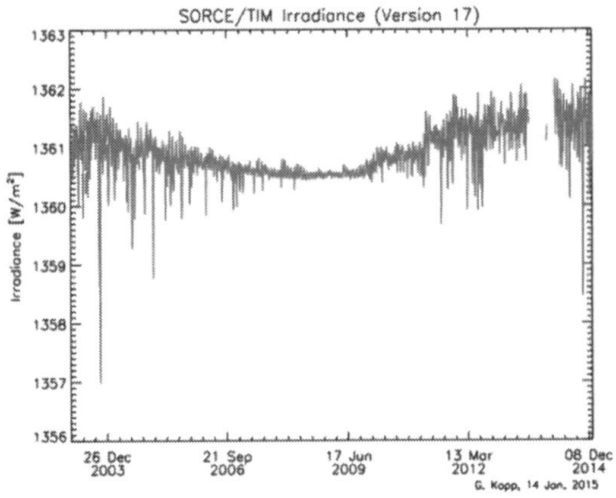

Figure 2. Left, flaring observed by *Kepler* on the slowly-rotating G-type main sequence star KIC 11764567 [32]; (a) shows a 50-day time series with slow modulations attributed to large starspots (e.g., [35]), and the inset (b) for a fraction of a day. Right, a full 11-year cycle of solar bolometric data (courtesy G. Kopp; for an early view of such solar variability via the *Solar Maximum Mission's* ACRIM radiometer, see [36]). Here the background variability has much shorter time scales and much smaller amplitude; there are no flare-related increases [37] and instead, pronounced "dips" from individual large sunspot groups. Note the contrast between these rapid dips (about 1/4 of the rotation period, and not sinusoidal) and the slower and larger modulations of KIC 11764567. The photometric behavior of these two stars could hardly be more different.

to that of a solar white-light flare. The UV extension of this continuum, in the case of a solar flare, may not make an order-of-magnitude difference in estimates of a flare's radiant energy, but this conclusion is based mainly on the observations of total solar irradiance (TSI) in a very few flares [40, 41]. If we make the assumption that the continuum ("white light flare") component dominates both solar and Kepler flare signals, then we meet a morphological discrepancy: solar white-light flares, including those detected via TSI [40, 41, 42], do not have such long durations. On the Sun, true white-light flares, as seen in visible continuum emission, match the impulsive-phase time profile of hard X-rays well, usually not more than ten minutes in duration. In rare cases (SOL2003-11-04), as seen in Figure 3) the emission can continue for tens of minutes, but a solar white-light flare with the duration of the flare seen in the inset of Figure 2 has never been reported. We note the possibility that the Kepler flares shine by scattered light, a rare phenomenon in the solar case sometimes referred to as a "white light prominence" [43, 44, 45, 46]. This radiation presumably results from Thomson scattering in coronal material ejected by the flare process from the lower atmosphere, and trapped in large-scale loop systems that extend visibly above the limb of the star. This continuum, perhaps the source of the extended radiation from SOL2003-11-04, persisted for tens of minutes. It would be easy to imagine a larger-scale version of this in a stellar flare, in which an ejected mass of order 10^{18} g might produce a Thomson-scattered photospheric signature competitive with the direct continuum emission. Via this mechanism, the total energy required for a Kepler superflare might turn out to be less than previously estimated [32, 33].

If we disregard these reservations and simply accept the Kepler superflares as an extension of the solar statistics, we find a distribution function roughly similar to that of solar flares, with a power-law index near 2 [32, 34], and an occurrence rate of one event per 800-5000 years at a total energy of 10^{34-35} erg [33]. Adopting a solar rate of roughly one Carrington event per 80-

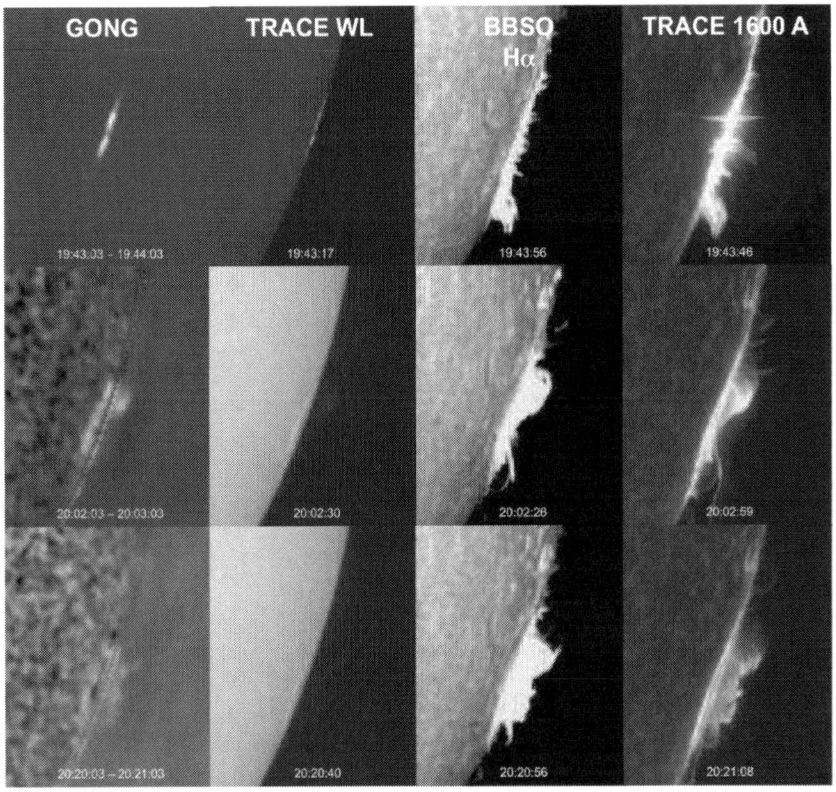

Figure 3. Observations of the limb flare SOL2003-11-04, which saturated GOES at the X17 level. From left to right, the columns show GONG narrow-band continuum; TRACE white light, Hα, and TRACE 1600 Å. The GONG signal has a background profile subtracted, and the above-the-limb portion amplified 10$\times$ (adapted from [44]).

year interval [47], this seems about right for an integral probability scaling roughly as 1/energy (equivalent to $\alpha = 2$ in the differential power law; see Section 4.1). This would conflict with the solar requirement for a roll-off to greater magnitudes (Section 5), but only weakly (Section 6).

Finally, do we have any theoretical guidance about whether the Sun can make superflares? The answer to this question is a clear "no", except in the most general sense. On the Sun, there is at most a weak correlation between spot area and flare magnitude (e.g., [48]), so weak that it difficult to recognize even on solar-cycle time scales (e.g., [49]). There is also virtually no correlation between inferred spottedness and Kepler flare magnitude (see Fig. 9 of [50], where the opposite conclusion is drawn from the observed scatter). The magnetic morphology of the sunspots also plays a major, though ill-understood role [51, 48]. In spite of this the existing theoretical ideas tend to emphasize the role of sunspot size [52, 53, 54, 55], and the relatively smooth lightcurves of the Kepler superflare stars do suggest the presence of large-scale magnetic surface structures. Nevertheless major Hα solar flares can and do occur in spotless or minimally spotted regions [56, 57], as can powerful X-class white-light flares (e.g., SOL1991-12-03 [58]). Accordingly we should not attribute too much specificity to the relationship between the rotational modulation and superflare occurrence.

Nevertheless we might anticipate simple theoretical limits on the magnitudes of extreme events. For example, the supergranulation scale [59] has well-defined properties in the solar photosphere, and sunspot umbrae typically do not exceed this scale in area. The coronal magnetic energy of this scale might be estimated from $W = \varepsilon L^2 H B^2/8\pi$, where ε would be the efficiency of field annihilation (say, 10%), L the network horizontal scale (say, 3×10^9 cm), H the magnetic scale height (say, L/2), and B an estimate of the magnetic intensity of a maximal sunspot, say 5000 G. Other rough estimates could be made, but this generous one gives 1.3×10^{33} erg. This would correspond to an event only an order of magnitude greater than that estimated for major flares or CMEs [60]. Another estimate could come from the total energy accumulated during a solar cycle's worth of dynamo action, assuming that the coronal energy storage resets to zero at each solar minimum [61]. This argument suggests an upper cutoff at the observed magnitude of a once-per-cycle event, and therefore not an extreme event. If this reset of energy accumulation did not cease after each solar cycle, i.e. if the magnetic energy could be stored in the solar interior rather than in the corona, then the maximum energy of an event might increase proportionally with the accumulation time [53] as needed to obtain an extreme event.

4. Solar flares and solar energetic particles

We now turn to more direct information about event distributions, namely the solar record itself. Solar extreme events can influence geospace via enhanced electromagnetic radiation, plasma flows, or high-energy particles. These phenomena all have close relationships to solar flares (low corona) and/or CMEs (middle to high corona), and as a rule flares and CMEs occur in synchrony so close that distinguishing them physically, rather than observationally (flares occur in the solar atmosphere, CMEs in the corona), becomes difficult (e.g. [62]). In near-Earth space (including the Moon), the solar energetic particles (SEPs, also known as "solar cosmic rays" [10]) provide the best proxy records and hence give us our best shot at characterizing the solar event distribution prior to the modern era. Flares and CMEs have several clearly distinguishable kinds of particle acceleration: in the flare itself we have electrons at tens of keV detected via hard X-radiation in the flare impulsive phase, MeV-range electrons in flares associated with γ-ray emission [63], the energetic ions of the flare γ-ray emission itself [64], the long-duration high-energy γ-ray sources (*e. g.*, [65] now observed by *Fermi* [66]; and then the particles underlying many coronal radio emissions [67] and of course the SPEs themselves. The distinct morphological properties of these high-energy signatures point to a variety of mechanisms for high-energy particle acceleration fundamental to both flares and CMEs.

We begin here with the distributions of flare radiation, and then contrast with the distributions of SEP events for which there is a proxy record. One basic question relates to the nature of the relationship between the distributions of events of these very disparate types – electromagnetic radiation from the base of the corona (or the photosphere itself), and particle fluxes closely related to the interplanetary medium at some great distance from the flare site, and with an ill-understood relationship to it.

4.1. Distributions of flares

For any observable of a flare, and they extend from long radio wavelengths well into the γ-ray domain, one can construct a histogram of event numbers as a function of peak flux or time-integrated fluence. For most of the 20th century, the Hα observations defined flaring, but such ground-based observations presented obvious problems in terms of homogeneity. When routine radio observations became available, especially in the microwave range, it became clear that a simple power law could describe the occurrence distribution rather well [3]. Then we obtained access to space and could make X-ray observations, and these have now supplanted Hα as the basic metric for flare occurrence (e.g., [68]). The earliest long sequences of solar soft X-ray observations came from the Geiger-counter experiments of the Van Allen group, from which again power-law distributions provided good fits [2]. For the past few decades we have had a series of standardized X-ray photometers on board the GOES spacecraft (see [69] for a recent description of these data); these systems have provided almost complete coverage with reasonably uniform calibrations for several decades (note that systematic microwave solar monitoring began in 1947 (e.g., [70]), but with lower sensitivity relative to the soft X-ray signatures. The flux levels from the GOES photometers unfortunately do not characterize the actual flare energy so directly; with the advent of TSI observations (as mentioned above) it turned out that the GOES soft X-ray energy amounts to less than 1% of the total radiated energy [42]. The use of such a minor constituent as a measure of flare energy obviously could lead to confusion, and this applies pretty much to any individual observable except of course for the TSI, which unfortunately has extremely limited dynamic range – only the most powerful events have been detected individually in this way. Thus the theoretical desire to use the data to characterize total event energies basically has systematic problems, which however are becoming more tractable with modern data (e.g., [71, 60]).

The differential number N of events at observed peak flux S, per unit peak flux, follows $dN/dS \propto S^{-\alpha}$. The slope α usually falls below 2.0; the original radio result from [3] found a good power-law fit at 1.8, for example, and decades later a large sample of hard X-ray data led to a value $\alpha = 1.732 \pm 0.008$ [72]. Soft X-ray microflares also have flat distributions with $\alpha \approx 1.4 - 1.5$ as observed in *Yohkoh* soft X-ray images [73]. The flatness of these distributions clearly points to a requirement for a break downwards at high peak fluxes, in order to avoid an infinity [1], as had been noted earlier for stellar flares [74]. The flatness also implies that major events contribute as strongly as minor events (microflares or nanoflares [8]) to the total energy budget of the process.

4.2. Distributions of SPEs

The direct measurements tend to show power-law distributions of particle intensities or event fluences, as noted above. Belov et al. [75] find $\mathrm{dN/dI} \propto I^{-1.22\pm0.05}$ for well-connected GOES particle events >10 MeV (protons), and $\mathrm{dN/dI} \propto I^{-1.34\pm0.02}$ for all events. Here I denotes the peak particle intensity, often given in "proton flux units" (PFUs) of protons $(\mathrm{cm}^2\ \mathrm{s})^{-1}$ above 1 MeV. This confirms the original result of a very flat distribution by [5], significantly flatter than even the flare distribution and therefore deeper still into the problem of convergence. The flare radiative fluxes and SEP peak fluxes have very different distribution functions, which implies something about the physical mechanisms involved, since flares and SPEs can have such a close relationship. Given this discrepancy, can we use the proxy records for SEPs to gain any insight into actual flare energies?

The observational situation has recently gotten some clarification: those flares associated with SPEs turn out to have a distribution function consistent with that of the SPEs themselves [76], as shown in Figure 4. The inference from this work is that more powerful flares have a greater propensity for particle acceleration, consistent with the finding that CME associations increase with flare magnitude (e.g. [77]) and with the flatter distributions of the hard X-ray peak fluxes. This distinguishes the SPEs qualitatively even at the flare site, establishing the requirement for eruption and hinting at a threshold effect for particle acceleration [78]. Note that this finding does not reveal the physics underlying this behavior, nor does it solve the problem of the divergence for extreme events. We thus still have the problem of explaining a (dimensional) break point in the event distribution function.

5. The search for a break

No published dataset has yielded convincing evidence for a roll-over in Sun-as-a-star observations, for various reasons, but distributions constructed for flare occurrence in specific active regions via images may show such a phenomenon [79].

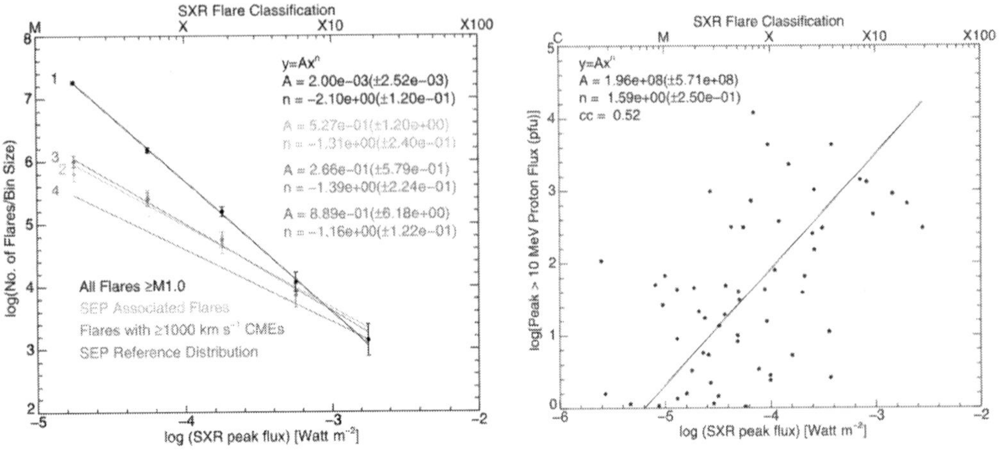

Figure 4. Left, distribution functions for GOES soft X-ray observations in the 1-8 Å band (black) and SEPs (purple), showing the substantial difference between them. Red and blue show flare distributions for the subset of events associated with SEPs and with fast CMEs. Note that the fit function in the figure legend defines our slope α as $-n$. Right, the correlation between peak proton and soft X-ray fluxes for the associated events (figures adapted from [76]).

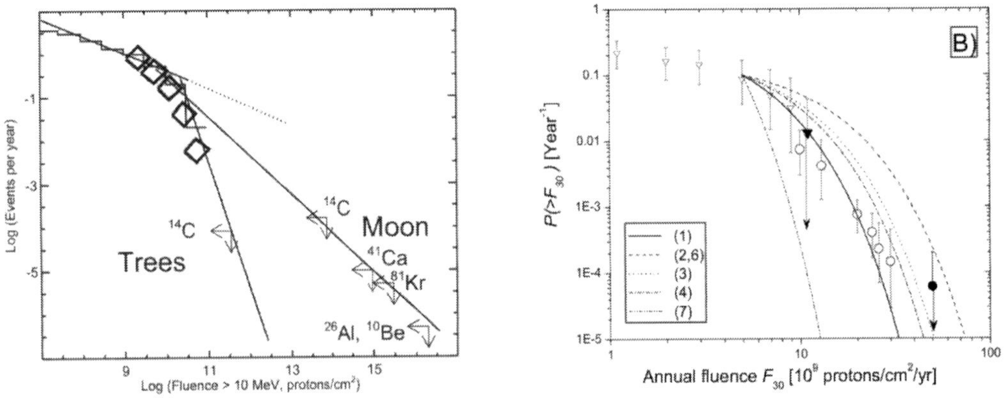

Figure 5. Left, the inference of the SEP fluence distribution from proxy radioisotope records (adapted from [81]). Right, an updated version of this information showing the terrestrial data as triangles and the limits for different isotopes according to their halflives as other symbols; the apparent break is at one event per decade at 5×10^9 protons (cm^2 yr)$^{-1}$ above 30 MeV (adapted from [82]; see this paper for interesting details).

Even prior to the identification of discrete events in the tree-ring record (Section 3.1), the fossil radioisotope record pointed clearly to the existence of a downward break in SEP event fluences [80]. Figure 5 (left panel) shows an improved version of this result, from [81]. The isotope record includes both ^{14}C from tree rings, and several other isotopes from the lunar regolith: ^{41}Ca, ^{81}Kr, ^{36}Cl, ^{26}Al, ^{41}Ca, ^{10}Be, and ^{53}Mn. The half-lives of these isotopes range from 5.73×10^3 to 3.74×10^6 years, and these time scales individually establish effective integration times for each proxy isotope.

The right panel of Figure 5 incorporates information from different isotopes from the lunar regolith, in the form of power-law functions consistent with a given isotope and anchored at a fixed reference point [82]. We identify the isotopes and the analyses as (1) ^{14}C [83]; (2) ^{41}Ca [84] and ^{26}Al [85]; (3) ^{81}Kr [86]; (4) ^{36}Cl [87]; and (7) ^{10}Be and ^{26}Al from [88]. Each of these papers described analyses of the depth dependence in lunar and meteoritic samples of the specific activities, often making use of accelerator experiments and numerical modeling to understand the cross-sections and the activation histories of the samples.

The downward break apparently occurs at a fluence of a few $\times 10^{10}$ protons/cm^2 (>30 MeV) [27]. According to the results discussed above, this fluence could correlate with a specific GOES magnitude and total flare energy, but the physics of SEP fluence and flare radiation are so different that this poses problems theoretically as well as observationally. As discussed above, the morphology rules out any identification of the SEPs and high-energy particles actually at the flare site, although we have not identified the acceleration mechanisms for the flare particles yet. See [11] for a recent systematic overview of the observational material on SEPs and the physical problems of their interpretation. We observe the particles at a given point in space, or sometimes multiple points, but the particles that arrive at that point may have had a complex history of acceleration and propagation within the heliosphere, giving the appearance of diffusive storage in a heliospheric of a "reservoir" on a scale greater than one AU [11]. The existence of the reservoir may be inferred from the observations, but its understanding in terms of heliospheric field structure, even at the orbital distance of Jupiter [89] remain unclear in detail. Thus even the most direct measurements require model assumptions in their interpretation. In terms of fluence, the physics of the diffusive shock acceleration theory (e.g., [11]) also may play a confusing role, since the process may saturate and thus produce a nonlinear relationship (even if one were expected) in terms of flare energy, for example. These uncertainties may contribute to the large scatter seen in the correlation of particle fluxes with flare magnitudes (Figure 4, right panel), but the figure is consistent with a simple correlation. In an interesting wrinkle here, the SEPs themselves (MeV particles) may contain an energy greater than that in the GOES soft X-ray sources, amounting in some cases to more than 10% of the total flare energy [90] rather than a small "test particle" fraction.

That the SEPs produced in a solar eruptive event may constitute a large fraction of the total event energy, as estimated via flare radiation or CME kinetic energy, poses an interesting theoretical problem. It encourages us to think of the event energy release as a proxy for an event's capacity for particle acceleration, but almost all existing theoretical work on flares and CMEs relies upon the single-fluid ideal MHD development. In general there seems no alternative to the identification of the energy source with the coronal magnetic field, but its partition into the observable forms is known only from modeling [91] or, even more crudely, from the observations [60]. The information regarding any individual event is invariably incomplete, and the interpretation of any observable in terms of total energy remains entangled with uncertain and model-dependent assumptions. Accordingly much work on occurrence distributions really deals with simple observables, such as the GOES soft X-ray flux or the peak particle intensity detected by a particular instrument, and an attempted conversion from any such measure to a theoretical construct often introduces new systematic errors.

6. The occurrence of extreme events

Can we do better than a simple extrapolation of the current data and obtain a better prediction of the occurrence of an extreme event? See [47] for a recent example of such a straighforward approach. The proxy evidence on SEP fluences strongly suggests a downturn, and so a simple extrapolation of a flat power-law will definitely overestimate the probability. The problem is that we have no precise information about the location of the break point in the distribution, nor any tangible hint that it might be detectable in the distribution of flare magnitudes (e.g., [92]). This Riley prediction [47] does not invoke the proxy information we have discussed in

Section 5, instead relying simply on the extrapolation of the power law observable in the direct measurements into the realm of the extreme events. This leads to a ten-year probability estimate of one 0.12 Carrington events per decade. Including the rollover would reduce this probability by some factor that we has not been well quantified at present; note that either exponential or Band functions work equally well with the existing data [82]. This paper cannot really improve, therefore, on the 12% solution for probability [47], except qualitatively. The incorporation of a rollover in the distribution could reduce planning concern about a true Black Swan event, i.e. one of substantially greater magnitude than the Carrington event.

The new and interesting discoveries noted in Sections 3.1 and 3.2 are as yet unable to help use substantially in defining the distribution of solar extreme events. The two rapid ^{14}C increases found in tree rings by [18, 19] have no corroborating evidence that would allow us to lock down the solar distribution except in an upper-limit sense; Miyake et al. and other authors had invoked various possible sources for these events, and we don't know for sure that they were solar in origin [26]; on the other hand, from the point of view of astronauts or space hardware at risk, the origin of the events may not matter to first order. These two events could in principle be evidence for a Black Swan phenomenon, describable by a power law without a termination, at some lower amplitude – either solar, following a different paradigm, or non-solar in origin.

Similarly, the Kepler superflares may actually reflect a different paradigm, as seems to have been the case of the celebrated Elatina varves [93, 94] with their possible prehistorical record of the solar cycle. In such a case, and as noted above there are striking dissimilarities between the Sun and the Kepler superflare stars, and we cannot blithely use them to assist in pinning down the actual solar event distribution (e.g., as described in [53]), in spite of the ready availability of a solar analog [95, 30, 32]. The finding of a correlation between starspots and flares in the Kepler data [96] would be consistent with standard ideas about magnetic energy storage and release in the corona [54], but the term "superflare" may connote the wrong paradigm.

A recent comprehensive discussion of the probability of an extreme event, identified with the Carrington flare and geomagnetic storm, Riley [47], concluded that "...our results overall suggest that the likelihood of another Carrington event occurring within the next decade is 12%" (for reference an independent assessment of essentially the same information led to an estimate corresponding to 3% [7], giving a feeling for the uncertainties here). These analyses did not factor in the evidence for a rolloff to greater magnitudes above a certain limiting case, but this information can be incorporated into a Bayesian estimation, for example. For example, following a method consistent with Bayesian or frequentist methods [97] these two events would obtain 0.016 events per decade, with a 2σ uncertainty range of [0.003, 0.09] events per decade.

7. Conclusions

The occurrence of solar extreme events, by definition those flares and/or CMES of a magnitude outside the historical record, must remain a matter of extrapolation from the known events, or a subject for theoretical guidance. The idea that we are at the mercy of a scale-free event distribution that allows for the unpredictable occurrence of a "Black Swan" superflare had originally been suggested long before the Kepler observations began [98]. In the meanwhile, evidence has accumulated establishing the existence of a downward break in the distribution of SEP fluences, suggesting a safer extrapolation based mainly on the fossil record – but we really do not understand the relationship between flare energy and SEPs in this domain.

Theoretical work on flare occurrence remains far from a capability for prediction, even though (not discussed in this review) short-term anticipation of flares does achieve some measure of success on a few-day time scale (e.g., [99]). On longer time scales the occurrence patterns have strong correlations in space and time, such as the solar cycle, but the eruption of magnetic flux otherwise remains a mystery. The organization of surface flux into spots may not play much of a role in flare occurrence, though of course spots and flares both involve the physics of magnetized plasmas.

This paper has discussed two important new developments related to the occurrence of solar extreme events, namely the successful identifications of transient events in the radioisotope data, on a thousand-year timescale, and the rapidly growing database of "superflares" on solar-type stars from *Kepler*. The radioisotope events clearly have an extraterrestrial origin in some high-energy phenomenon, such as a solar flare. If in fact solar in origin, these events immediately extend require an extension of the observed distribution to higher event energies, and to suggest less steepening above the observed roll-over fluence. Similarly the superflare occurrence on solar-type stars also suggests that the roll-over may not be so severe as though. Neither of these new inputs, though, demands a major change in the basic picture of solar event occurrence, namely that the distribution is a flat one that requires a roll-over or downward break of some sort. The nature of the roll-over in the distribution has proven difficult to assess partly because of small-number statistics, but also because of the necessity of linking disparate and ill-understood datasets in a credible manner. We have enough information now to make a firmer (and lower) estimate of the probability of a solar super-event, but no formal literature on such an estimate yet seems to exist in the literature.

Acknowledgments
I thank NASA for support under contract NAS 5-98033 for RHESSI, as well as the International Space Science Institute (Bern) and the University of Glasgow for hospitality during the preparation of this work.

References
[1] Hudson H S 1991 *Solar Phys.* **133** 357–369
[2] Drake J F 1971 *Solar Phys.* **16** 152–185
[3] Akabane K 1956 *Publ. Astron. Soc. Japan* **8** 173
[4] Aschwanden M J 2013 *Self-Organized Criticality Systems* (Open Academic Press)
[5] Van Hollebeke M A I, Ma Sung L S and McDonald F B 1975 *Solar Phys.* **41** 189–223
[6] Simpson J A 2000 *Space Sci. Revs.* **93** 11–32
[7] Schrijver C J, Beer J, Baltensperger U, Cliver E W, Güdel M, Hudson H S, McCracken K G, Osten R A, Peter T, Soderblom D R, Usoskin I G and Wolff E W 2012 *Journal of Geophysical Research (Space Physics)* **117** A08103 (*Preprint* 1206.4889)
[8] Parker E N 1988 *Astrophysical Journal* **330** 474–479
[9] Tsurutani B T and Lakhina G S 2014 *Geophys. Res. Lett.* **41** 287–292
[10] Miroshnichenko L I and Perez-Peraza J A 2008 *International Journal of Modern Physics A* **23** 1–141
[11] Reames D V 2013 *Space Sci. Revs.* **175** 53–92 (*Preprint* 1306.3608)
[12] Tsurutani B T, Gonzalez W D, Lakhina G S and Alex S 2003 *Journal of Geophysical Research (Space Physics)* **108** 1268
[13] Siscoe G, Crooker N U and Clauer C R 2006 *Advances in Space Research* **38** 173–179
[14] Cliver E W and Dietrich W F 2013 *Journal of Space Weather and Space Climate* **3** A31
[15] Baker D N, Li X, Pulkkinen A, Ngwira C M, Mays M L, Galvin A B and Simunac K D C 2013 *Space Weather* **11** 585–591
[16] Taleb N 2007 *The black swan : the impact of the highly improbable* (London: Allen Lane) ISBN 9780713999952
[17] Kleinert H 2014 *Foundations of Physics* **44** 546–556
[18] Miyake F, Nagaya K, Masuda K and Nakamura T 2012 *Nature* **486** 240–242
[19] Miyake F, Masuda K and Nakamura T 2013 *Nature Communications* **4** 1748
[20] Usoskin I G, Kromer B, Ludlow F, Beer J, Friedrich M, Kovaltsov G A, Solanki S K and Wacker L 2013 *Astron. Astrophys.* **552** L3 (*Preprint* 1302.6897)
[21] Jull A J T, Panyushkina I P, Lange T E, Kukarskih V V, Myglan V S, Clark K J, Salzer M W, Burr G S and Leavitt S W 2014 *Geophys. Res. Lett.* **41** 3004–3010
[22] Liu Y, Zhang Z F, Peng Z C, Ling M X, Shen C C, Liu W G, Sun X C, Shen C D, Liu K X and Sun W 2014 *Scientific Reports* **4** 3728
[23] Cliver E W, Tylka A J, Dietrich W F and Ling A G 2014 *Astrophysical Journal* **781** 32
[24] Neuhäuser R and Hambaryan V V 2014 *Astronomische Nachrichten* **335** 949 (*Preprint* 1408.2934)

[25] Pavlov A K, Blinov A V, Vasil'Ev G I, Vdovina M A, Konstantinov A N and Ostryakov V M 2014 *Astronomy Letters* **40** 640–647
[26] Usoskin I G and Kovaltsov G A 2012 *Astrophysical Journal* **757** 92 (*Preprint* 1207.5932)
[27] Kovaltsov G A, Usoskin I G, Cliver E W, Dietrich W F and Tylka A J 2014 *Solar Phys.* **289** 4691–4700
[28] Gershberg R E 2005 *Solar-Type Activity in Main-Sequence Stars*
[29] Koch D G, Borucki W J, Basri G, Batalha N M, Brown T M, Caldwell D, Christensen-Dalsgaard J, Cochran W D, DeVore E, Dunham E W, Gautier III T N, Geary J C, Gilliland R L, Gould A, Jenkins J, Kondo Y, Latham D W, Lissauer J J, Marcy G, Monet D, Sasselov D, Boss A, Brownlee D, Caldwell J, Dupree A K, Howell S B, Kjeldsen H, Meibom S, Morrison D, Owen T, Reitsema H, Tarter J, Bryson S T, Dotson J L, Gazis P, Haas M R, Kolodziejczak J, Rowe J F, Van Cleve J E, Allen C, Chandrasekaran H, Clarke B D, Li J, Quintana E V, Tenenbaum P, Twicken J D and Wu H 2010 *Astrophysical Journal Lett.* **713** L79 (*Preprint* 1001.0268)
[30] Walkowicz L M, Basri G, Batalha N, Gilliland R L, Jenkins J, Borucki W J, Koch D, Caldwell D, Dupree A K, Latham D W, Meibom S, Howell S, Brown T M and Bryson S 2011 *Astronomical Journal* **141** 50 (*Preprint* 1008.0853)
[31] Balona L A 2012 *MNRAS* **423** 3420–3429
[32] Maehara H, Shibayama T, Notsu S, Notsu Y, Nagao T, Kusaba S, Honda S, Nogami D and Shibata K 2012 *Nature* **485** 478–481
[33] Shibayama T, Maehara H, Notsu S, Notsu Y, Nagao T, Honda S, Ishii T T, Nogami D and Shibata K 2013 *Astrophysical Journal Supp.* **209** 5 (*Preprint* 1308.1480)
[34] Wu C J, Ip W H and Huang L C 2015 *Astrophysical Journal* **798** 92
[35] Berdyugina S V 2005 *Living Reviews in Solar Physics* **2** 8
[36] Willson R C, Gulkis S, Janssen M, Hudson H S and Chapman G A 1981 *Science* **211** 700–702
[37] Hudson H S and Willson R C 1983 *Solar Phys.* **86** 123–130
[38] Vieira L E A, Norton A, Dudok de Wit T, Kretzschmar M, Schmidt G A and Cheung M C M 2012 *Geophys. Res. Lett.* **39** L16104
[39] Gilliland R L, Chaplin W J, Dunham E W, Argabright V S, Borucki W J, Basri G, Bryson S T, Buzasi D L, Caldwell D A, Elsworth Y P, Jenkins J M, Koch D G, Kolodziejczak J, Miglio A, van Cleve J, Walkowicz L M and Welsh W F 2011 *Astrophysical Journal Supp.* **197** 6 (*Preprint* 1107.5207)
[40] Woods T N, Eparvier F G, Fontenla J, Harder J, Kopp G, McClintock W E, Rottman G, Smiley B and Snow M 2004 *Geophys. Res. Lett.* **31** 10802–+
[41] Woods T N, Kopp G and Chamberlin P C 2006 *Journal of Geophysical Research (Space Physics)* **111** 10–+
[42] Kretzschmar M 2011 *Astron. Astrophys.* **530** A84 (*Preprint* 1103.3125)
[43] Hiei E, Okamoto T and Tanaka K 1983 *Solar Phys.* **86** 185–191
[44] Leibacher J W, Harvey J W, Kopp G, Hudson H and GONG Team 2004 *American Astronomical Society Meeting Abstracts #204* (*Bulletin of the American Astronomical Society* vol 36) p 669
[45] Martínez Oliveros J C, Krucker S, Hudson H S, Saint-Hilaire P, Bain H, Lindsey C, Bogart R, Couvidat S, Scherrer P and Schou J 2014 *Astrophysical Journal Lett.* **780** L28 (*Preprint* 1311.7412)
[46] Saint-Hilaire P, Schou J, Martínez Oliveros J C, Hudson H S, Krucker S, Bain H and Couvidat S 2014 *Astrophysical Journal Lett.* **786** L19 (*Preprint* 1402.7016)
[47] Riley P 2012 *Space Weather* **10** S02012
[48] Sammis I, Tang F and Zirin H 2000 *Astrophysical Journal* **540** 583–587
[49] Hudson H S 2007 *Astrophysical Journal Lett.* **663** L45–L48 (*Preprint* 0707.1118)
[50] Notsu Y, Shibayama T, Maehara H, Notsu S, Nagao T, Honda S, Ishii T T, Nogami D and Shibata K 2013 *Astrophysical Journal* **771** 127 (*Preprint* 1304.7361)
[51] Künzel H 1960 *Astronomische Nachrichten* **285** 271
[52] Mullan D J 1975 *Astrophysical Journal* **200** 641–645
[53] Shibata K, Isobe H, Hillier A, Choudhuri A R, Maehara H, Ishii T T, Shibayama T, Notsu S, Notsu Y, Nagao T, Honda S and Nogami D 2013 *Publ. Astron. Soc. Japan* **65** 49 (*Preprint* 1212.1361)
[54] Aulanier G, Démoulin P, Schrijver C J, Janvier M, Pariat E and Schmieder B 2013 *Astron. Astrophys.* **549** A66 (*Preprint* 1212.2086)
[55] Candelaresi S, Hillier A, Maehara H, Brandenburg A and Shibata K 2014 *Astrophysical Journal* **792** 67 (*Preprint* 1405.1453)
[56] Dodson H W and Hedeman E R 1970 *Solar Phys.* **13** 401–419
[57] Sersen M and Valnicek B 1993 *Solar Phys.* **145** 339–345
[58] Hudson H S, Acton L W, Hirayama T and Uchida Y 1992 *Publ. Astron. Soc. Japan* **44** L77–L81
[59] Rieutord M and Rincon F 2010 *Living Reviews in Solar Physics* **7** 2 (*Preprint* 1005.5376)
[60] Emslie A G, Dennis B R, Shih A Y, Chamberlin P C, Mewaldt R A, Moore C S, Share G H, Vourlidas A and Welsch B T 2012 *Astrophysical Journal* **759** 71 (*Preprint* 1209.2654)

[61] Mullan D J and Kent D W 1979 *International Cosmic Ray Conference* (*International Cosmic Ray Conference* vol 5) p 323
[62] Temmer M, Veronig A M, Vršnak B, Rybák J, Gömöry P, Stoiser S and Maričić D 2008 *Astrophysical Journal Lett.* **673** L95–L98
[63] Shih A Y, Lin R P and Smith D M 2009 *Astrophysical Journal Lett.* **698** L152–L157
[64] Ramaty R, Mandzhavidze N, Kozlovsky B and Murphy R J 1995 *Astrophysical Journal Lett.* **455** L193
[65] Hudson H and Ryan J 1995 *Ann. Revs. Astron. Astrophys.* **33** 239–282
[66] Ajello M, Albert A, Allafort A, Baldini L, Barbiellini G, Bastieri D, Bellazzini R, Bissaldi E, Bonamente E, Brandt T J, Bregeon J, Brigida M, Bruel P, Buehler R, Buson S, Caliandro G A, Cameron R A, Caraveo P A, Cecchi C, Charles E, Chekhtman A, Chiang J, Chiaro G, Ciprini S, Claus R, Cohen-Tanugi J, Cominsky L R, Conrad J, Cutini S, D'Ammando F, de Palma F, Dermer C D, Desiante R, Digel S W, Silva E d C e, Drell P S, Drlica-Wagner A, Favuzzi C, Focke W B, Franckowiak A, Fukazawa Y, Fusco P, Gargano F, Gasparrini D, Germani S, Giglietto N, Giommi P, Giordano F, Giroletti M, Glanzman T, Godfrey G, Grenier I A, Grove J E, Guiriec S, Hadasch D, Hayashida M, Hays E, Horan D, Hou X, Hughes R E, Inoue Y, Jackson M S, Jogler T, Jóhannesson G, Johnson A S, Johnson W N, Kamae T, Knödlseder J, Kocevski D, Kuss M, Lande J, Larsson S, Latronico L, Longo F, Loparco F, Lott B, Lovellette M N, Lubrano P, Mayer M, Mazziotta M N, McEnery J E, Michelson P F, Mizuno T, Moiseev A A, Monte C, Monzani M E, Morselli A, Moskalenko I V, Murgia S, Murphy R, Nakamori T, Nemmen R, Nuss E, Ohno M, Ohsugi T, Omodei N, Orienti M, Orlando E, Ormes J F, Paneque D, Panetta J H, Perkins J S, Pesce-Rollins M, Petrosian V, Piron F, Pivato G, Porter T A, Rainò S, Rando R, Razzano M, Reimer A, Reimer O, Roth M, Schulz A, Sgrò C, Siskind E J, Spandre G, Spinelli P, Takahashi H, Thayer J G, Thayer J B, Thompson D J, Tibaldo L, Tinivella M, Tosti G, Troja E, Usher T L, Vandenbroucke J, Vasileiou V, Vianello G, Vitale V, Werner M, Winer B L, Wood D L, Wood K S and Yang Z 2014 *Astrophysical Journal* **789** 20
[67] Kundu M R 1965 *Solar radio astronomy*
[68] Fletcher L, Dennis B R, Hudson H S, Krucker S, Phillips K, Veronig A, Battaglia M, Bone L, Caspi A, Chen Q, Gallagher P, Grigis P T, Ji H, Liu W, Milligan R O and Temmer M 2011 *Space Sci. Revs.* **159** 19–106 (*Preprint* 1109.5932)
[69] White S M, Thomas R J and Schwartz R A 2005 *Solar Phys.* **227** 231–248
[70] Covington A E and Harvey G A 1958 *Journ. Roy. Astr. Soc. Canada* **52** 161
[71] Emslie A G, Dennis B R, Holman G D and Hudson H S 2005 *Journal of Geophysical Research (Space Physics)* **110** 11103–+
[72] Crosby N B, Aschwanden M J and Dennis B R 1993 *Solar Phys.* **143** 275–299
[73] Shimizu T 1995 *Publ. Astron. Soc. Japan* **47** 251–263
[74] Collura A, Pasquini L and Schmitt J H M M 1988 *Astron. Astrophys.* **205** 197–206
[75] Belov A, Kurt V, Mavromichalaki H and Gerontidou M 2007 *Solar Phys.* **246** 457–470
[76] Cliver E W, Ling A G, Belov A and Yashiro S 2012 *Astrophysical Journal Lett.* **756** L29
[77] Yashiro S, Akiyama S, Gopalswamy N and Howard R A 2006 *Astrophysical Journal Lett.* **650** L143–L146 (*Preprint* astro-ph/0609197)
[78] Hudson H S 1978 *Solar Phys.* **57** 237–240
[79] Kucera T A, Dennis B R, Schwartz R A and Shaw D 1997 *Astrophysical Journal* **475** 338–347
[80] Lingenfelter R E and Hudson H S 1980 *The Ancient Sun: Fossil Record in the Earth, Moon and Meteorites* ed Pepin R O, Eddy J A and Merrill R B pp 69–79
[81] Reedy R C 1996 *Solar Drivers of the Interplanetary and Terrestrial Disturbances* (*Astronomical Society of the Pacific Conference Series* vol 95) ed Balasubramaniam K S, Keil S L and Smartt R N p 429
[82] Kovaltsov G A and Usoskin I G 2014 *Solar Phys.* **289** 211–220 (*Preprint* 1305.5088)
[83] Jull A J T, Cloudt S, Donahue D J, Sisterson J M, Reedy R C and Masarik J 1998 *Geochimica et Cosmochimica Acta* **62** 3025–3036
[84] Fink D, Klein J, Middleton R, Vogt S, Herzog G F and Reedy R C 1998 *Geochimica et Cosmochimica Acta* **62** 2389–2402
[85] Grismore R, Llewellyn R A, Brown M D, Dowson S T and Cumblidge K 2001 *Earth and Planetary Science Letters* **187** 163–171
[86] Reedy R C 1999 *Lunar and Planetary Science Conference* (*Lunar and Planetary Science Conference* vol 30) p 1643
[87] Nishiizumi K, Arnold J R, Kohl C P, Caffee M W, Masarik J and Reedy R C 2009 *Geochimica et Cosmochimica Acta* **73** 2163–2176
[88] Michel R, Leya I and Borges L 1996 *Nuclear Instruments and Methods in Physics Research B* **113** 434–444
[89] Struminsky A, Heber B, Gomez-Herrero R and et al 2008 *International Cosmic Ray Conference* **1** 131–134
[90] Mewaldt R A, Cohen C M S, Haggerty D K, Mason G M, Looper M L, von Rosenvinge T T and Wiedenbeck

M E 2007 *American Institute of Physics Conference Series* (*American Institute of Physics Conference Series* vol 932) pp 277–282
[91] Birn J, Fletcher L, Hesse M and Neukirch T 2009 *Astrophysical Journal* **695** 1151–1162
[92] Tranquille C, Hurley K and Hudson H S 2009 *Solar Phys.* **258** 141–166
[93] Williams G E and Sonett C P 1985 *Nature* **318** 523–527
[94] Williams G E 1990 *Royal Society of London Philosophical Transactions Series A* **330** 445–457
[95] Schaefer B E, King J R and Deliyannis C P 2000 *Astrophysical Journal* **529** 1026–1030 (*Preprint* arXiv:astro-ph/9909188)
[96] Notsu Y, Shibayama T, Maehara H, Notsu S, Nagao T, Honda S, Ishii T T, Nogami D and Shibata K 2013 *Astrophysical Journal* **771** 127 (*Preprint* 1304.7361)
[97] Love J J 2012 *Geophys. Res. Lett.* **39** L10301
[98] Wdowczyk J and Wolfendale A W 1977 *Nature* **268** 510–512
[99] Bloomfield D S, Higgins P A, McAteer R T J and Gallagher P T 2012 *Astrophysical Journal Lett.* **747** L41 (*Preprint* 1202.5995)

Pressure and temperature effect corrections of atmospheric muon data in the Belgrade cosmic-ray station

M Savić, D Maletić, D Joković, N Veselinović, R Banjanac, V Udovičić, A Dragić
Institute of Physics, University of Belgrade
Pregrevica 118, 11080 Belgrade, Serbia

E-mail: yokovic@ipb.ac.rs

Abstract. We present results of continuous monitoring of the cosmic-ray muon intensity at the ground and shallow underground level at the Belgrade cosmic-ray station. The cosmic-ray muon measurements have been performed since 2002, by means of plastic scintillation detectors. The scintillator counts are corrected for atmospheric pressure for the whole period of measurements and, as well, for vertical temperature profile for the period of the last six years. The results are compared with other correction methods available. One-hour time series of the cosmic-ray muon intensity at the ground level are checked for correlation with European neutron monitors, with emphasis on occasional extreme solar events, e.g. Forbush decreases.

1. Introduction

The Belgrade cosmic-ray station, situated in the Low-level Laboratory for Nuclear Physics at Institute of Physics, Belgrade, have been continuously measuring the cosmic-ray intensity since 2002. The station is at near-sea level at the altitude of 78 m a.s.l.; its geomagnetic latitude is 39° 32' N and geomagnetic vertical cut-off rigidity is 5.3 GV. It consists of two parts: the ground level lab (GLL) and the underground lab (UL); the UL is located at a depth of 12 metres below the surface, i.e. 25 metre water equivalent. At this depth practically only the muonic component is present. The cosmic-ray muon measurements are performed by means of plastic scintillation detectors, a pair of which is, along with instrumentation modules for data acquisition, placed in both the GLL and the UL. The set-up is quiet flexible, as the scintillators could be arranged in different ways, which allows conducting different experiments. The analyses of the measurements yielded some results on variations of the cosmic-ray muon intensity and on precise values of the integral muon flux at the ground level and at the depth of 25 m.w.e. [1,2,3,4].

2. Experimental set-up

The experimental set-up in both the GLL and the UL consists of a large plastic scintillation detector (rectangular shape, 100cm x 100cm x 5cm) and a data acquisition system (DAQ). The scintillator is polystyrene based UPS-89, with four 2-inch photomultiplier tubes attached to its corners, so that each PM tube looks at the rectangle diagonal. Preamplifier signals from two PM tubes looking at the same diagonal are summed in one output signal, thus two output signals are led to the DAQ from each scintillator.

The summed signals from the PM tubes on the same diagonal of the detectors are stored and digitized by the DAQ, which is based on 4-channel flash analog-to-digital converters (FADC), made by CAEN (type N1728B), with 100 MHz sampling frequency. The FADCs are capable of operating in the event-list mode, when every analyzed event is fully recorded by the time of its occurrence and its amplitude. This enables the correlation of events, both prompt and arbitrarily delayed, at all four inputs with the time resolution of 10 ns. Single and coincident data can be organized into time series within any desired integration period. The FADCs can also be synchronized with each other for the additional coinciding of the events in the GLL and the UL.

For both the GLL and the UL detector, two input channels on the corresponding FADC are reserved for events recorded by each of detector's diagonals. The cosmic-ray events recorded by a single diagonal are drown in the background. Coinciding of the prompt events from two diagonals within a narrow time window gives the resulting experimental spectrum of the plastic scintillator, which is the energy deposit (ΔE) spectrum of the cosmic-ray particles (figure 1). Interpretation of the experimental spectra and their features as well as their calibration have been done using Geant4 based Monte Carlo simulation [4,5]. The spectra peak at ~11 MeV and have the instrumental thresholds at ~4 MeV. Comparing the spectra of the GLL detector and the UL detector one can notice the obvious difference in their shape, especially in the low-energy part below ~6 MeV. This difference points to the contribution of the cosmic-ray electrons and gammas (electromagnetic component) to the ΔE spectra at the ground level, which is absent in case of the underground detector.

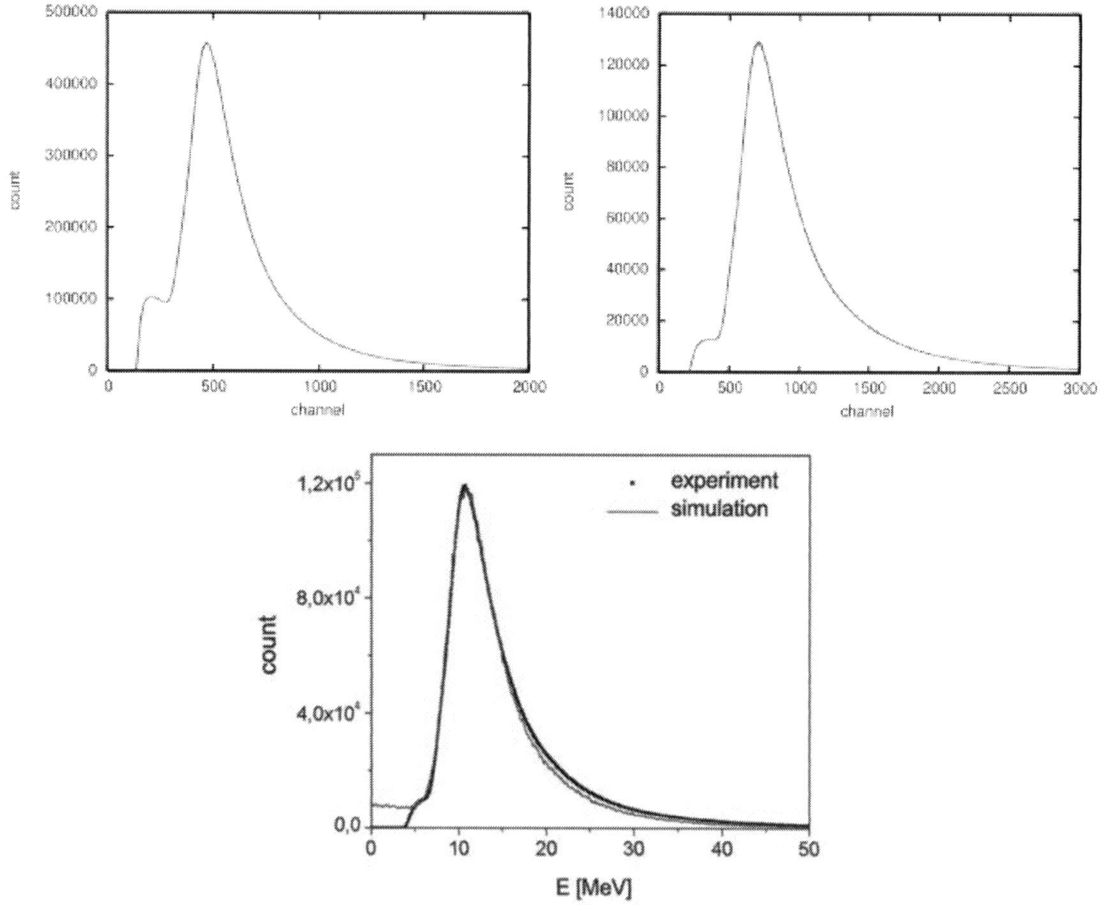

Figure 1. The cosmic-ray ΔE spectra of the GLL detector (top left) and the UL detector (top right). Experimental and simulated ΔE spectra of the UL detector (bottom).

3. Results and discussion

The cosmic-ray intensity data are automatically processed, using a web-based "robot" developed for this purpose, and published online at www.cosmic.ipb.ac.rs/muon_station. The online available data are raw scintillator counts in time series with resolution of 5 min or 1 h. Time series of the raw data are corrected for pressure and temperature effect; pressure corrections have been done for the whole data taking period and temperature effect corrections have been done for the the time period of the last six years.

3.1. Efficiency corrections

The first data corrections are related to detector assembly efficiency. As mentioned, the instrumental thresholds cut the spectra at ~3 MeV. However, the thresholds may vary, thus changing the initial spectrum and resulting in fluctuations of the integral spectrum count. Related to this, the necessary correction has been done by means of constant fraction discriminator (CFD) function (figure 2); with use of the CFD cut the spectrum fluctuations decreased significantly. The CFD is based on cut on chosen height as a percentage of peak height where the spectrum is cut. The simulation tells us that, for the underground detector, ~6% of muon events is also cut (figure 1).

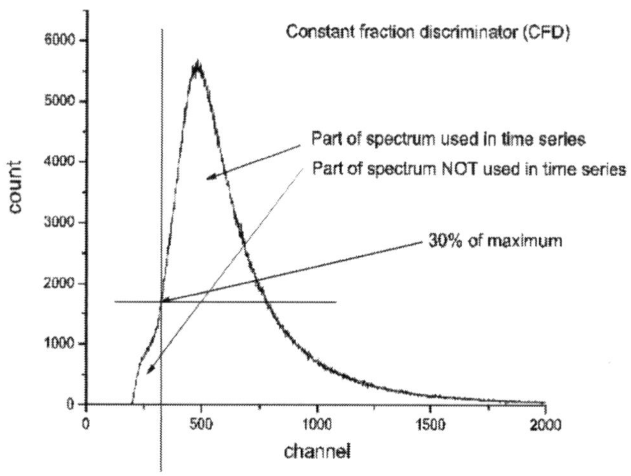

Figure 2. Constant fraction discriminator (CFD) applied in efficiency corrections. The obtained truncated spectrum is used for calculating time series.

The next step in the efficiency corrections is a correction of 5-min count values that are clearly lower than a mean 5-min count in surrounding time intervals. This undershoot comes at the beginning/end of runs, where events are not collected for all 5 min of measurement. The last and smallest correction is a correction of fluctuations of spectrum due to fluctuation in amplification which influence the cut on diagonals and efficiency of coincidence of two diagonals. We found that the CFD cut is proportional to efficiency of coincidence.

3.2. Corrections for atmospheric pressure and for temperature

Significant part of variation of cosmic ray muon component intensity can be attributed to meteorological effects. Here, two main contributors are barometric and temperature effect [6].

Barometric effect is caused by variation of the atmospheric mass above the detector. These pressure corrections are done by finding the linear regression coefficient, using only International Quiet Days, i.e. time series data from periods with more or less constant intensity of galactic cosmic rays, for creation of the distribution of scintillator counts vs. atmospheric pressure. Atmospheric pressure data are available due to on-site continuous measurement.

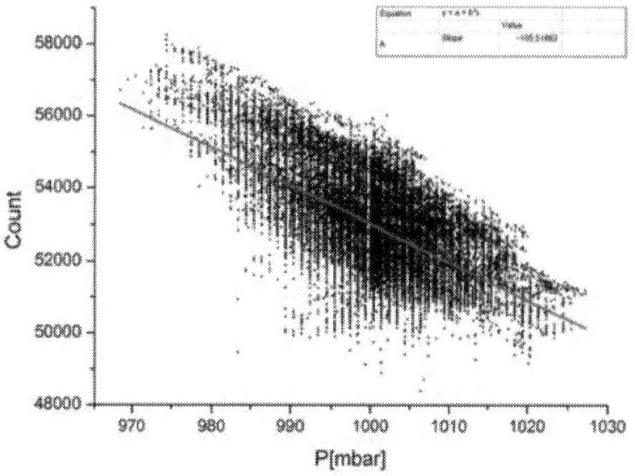

Figure 3. Dependence of 5-min counts on atmospheric pressure.

The temperature effect is related to the variation of the atmospheric temperature profile. The effect is two-fold, as it affects pion decay (positive contribution) as well as muon ionization losses and possible decay (negative contribution). To correct for these effects, integral correction method was applied [6,7]. The variation of the muon intensity due to temperature variations is calculated by using the formula:

$$\delta I_T = \int_0^{h_0} \alpha(h) \cdot \delta T(h) \cdot dh$$

where δI_T is the variation of the muon intensity due to the temperature effect, $\delta T(h)$ is the variation of the atmospheric temperature, which is calculated in reference to the mean temperature value for a given time period (denoted by index M): $\delta T(h) = T_M(h) - T(h)$, where h is atmospheric depth. Temperature coefficient densities $\alpha(h)$ are calculated according to [6].

Available meteorological models make it possible to have hourly atmospheric temperature profiles for 17 standard isobaric levels at the geographic position of the Belgrade muon station, necessary for application of formula shown above. The procedure used here is as described in [7]. Temperature profiles have been obtained from ftp://cr0.izmiran.rssi.ru/COSRAY!/FTP_METEO/blgd_Th/, courtesy of IZMIRAN laboratory.

3.3. Time series of the cosmic-ray intensity

In Figure 4 the count rate time series is shown for all corrections. First, the corrected count rate for efficiency corrected data is shown. Also, the atmospheric pressure and combined atmospheric pressure and temperature corrections time series of count rates are shown.

One-hour time series of the cosmic-ray muon intensity at the ground level are checked for correlation with European neutron monitors (NM), with emphasis on occasional extreme solar events, e.g. Forbush decreases.

In Figure 5 the comparison of time series of pressure corrected and pressure and temperature corrected count rates for the Belgrade muon station and Jungfraujoch, Rome, Baksan and Oulu neutron monitors is presented for Forbush candidate in March 2012. The count rates of neutron monitors are shifted to be close to each-other for visibility. The count rate for the Belgrade station is shown in percentages with additional shift down for visibility. The count rate drop for the neutron monitors is clearly more pronounced than for Belgrade muon monitor.

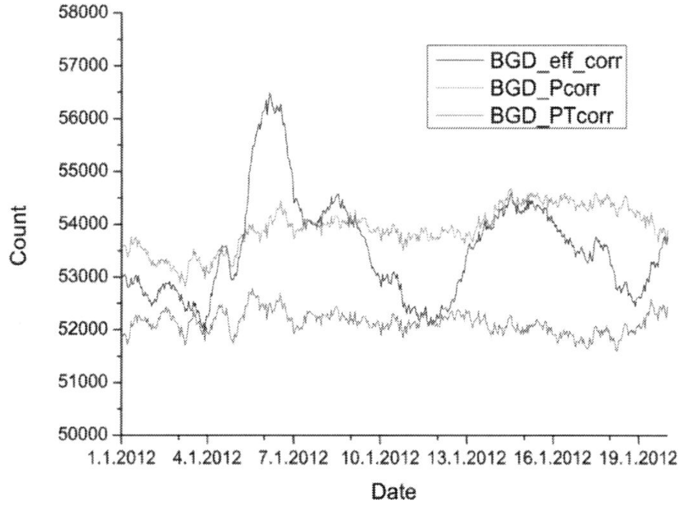

Figure 4. Time series of efficiency corrected, pressure corrected and pressure and temperature corrected counts.

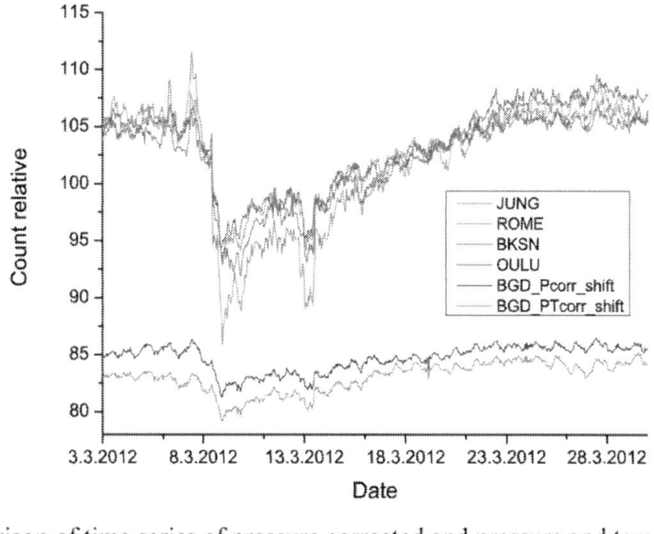

Figure 5. Comparison of time series of pressure corrected and pressure and temperature corrected count rates for the Belgrade muon monitor station and neutron monitors. Count rates are shifted for comparison.

In Figure 6 the comparison of time series of pressure corrected count rates for the Belgrade muon station Jungfraujoch, Rome, Baksan and Oulu neutron monitors is presented. The count rates of neutron monitors are shifted to be close to each-other for visibility. The count rate for Belgrade station is scaled in the way that the drop in count rate is similar to most of the stations (except Jungfraujoch, which is at high altitude). The visual comparison shows the good correlation of the count rates of Belgrade muon monitor and neutron monitors, previously noticed using correlative analyses of count rates. The pressure corrected count rates from Belgrade muon monitor is only dataset used for visual comparison, since neutron monitor data are also only pressure corrected. This was also observed previously using correlative analyses of count rates.

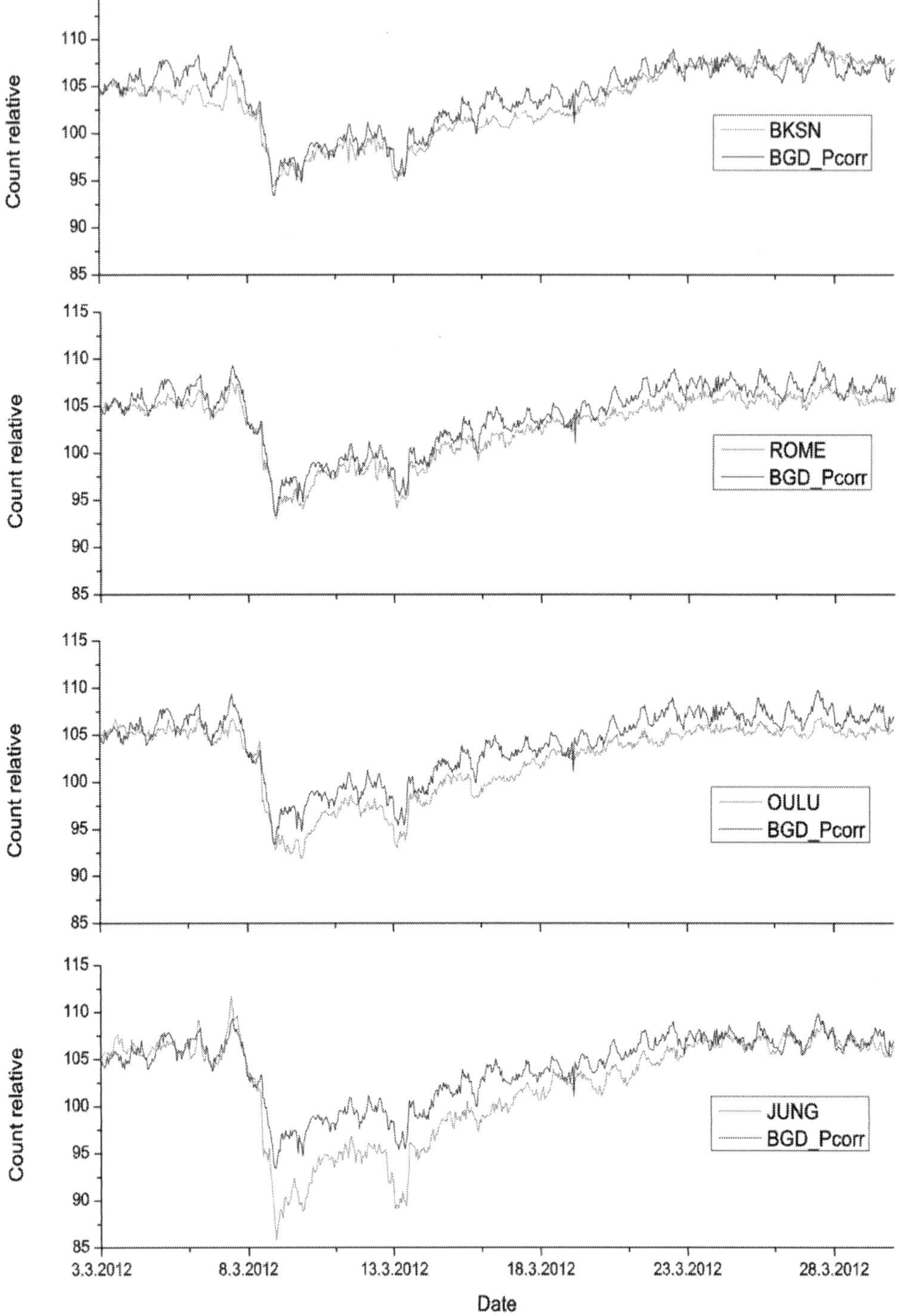

Figure 6. Comparison of time series of pressure corrected count rates for the Belgrade muon monitor station and neutron monitors. Count rates are shifted and scaled for comparison.

4. Conclusions

The results of continuous monitoring of the cosmic-ray muon intensity at the ground and shallow underground level at the Belgrade cosmic-ray station are presented. The scintillator counts are corrected for atmospheric pressure for the whole period of measurements and, as well, for vertical temperature profile for the period of the last six years. The results are compared with other correction methods available and showed excellent agreement. One-hour time series of the cosmic-ray muon intensity at the ground level are checked for correlation with European neutron monitors, with emphasis on occasional extreme solar events, e.g. Forbush decreases. As a result of correlative analysis, the Forbush candidate in March 2012 is the best choice to be used for visual comparison presented in this work. The comparison showed high correlation of the Belgrade muon monitor with neutron monitors, especially geographically closer neutron monitors such as Rome NM. In some specific time periods, like during the Forbush candidate in March 2012, we showed that our muon measurement system has sensitivity comparable to European neutron monitors in this period, but still not as efficient as NM with better geographical position (at high altitude), e.g. Jungfraujoch in the Swiss Alps.

5. Acknowledgements

This work is supported by the Ministry of Education, Science and Technological Development of Republic of Serbia, project no. OI171002. We acknowledge the NMDB database (www.nmdb.eu), founded under the European Union's FP7 programme (contract no. 213007) for providing data. We acknowledge the Department of Physical Sciences, University of Oulu, Finland, as the source of data used in publications. Jungfraujoch neutron monitor data were kindly provided by the Cosmic Ray Group, Physikalisches Institut, University of Bern, Switzerland. We acknowledge the ROME neutron monitor data. SVIRCO NM is supported by IFSI/INAF-UNIRoma3 collaboration. We acknowledge the Baksan Neutrino Observatory of Institute for Nuclear Research of Russian Academy of Science, Moscow, for providing data.

6. References

[1] Dragić A, Banjanac R, Udovičić V, Joković D, Aničin I and Puzović J 2005 *Int. J. Mod. Phys.* A **20** 6953

[2] Dragić A, Banjanac R, Udovičić V, Joković D, Puzović J and Aničin I 2008 *Proc. 21st European Cosmic Ray Symposium (Košice, Slovakia)* p 368

[3] Dragić A, Banjanac R, Udovičić V, Joković D, Aničin I and Puzović J 2009 *Proc. 31st Int. Cosmic Ray Conf. (Lodz, Poland)*

[4] Dragić A, Joković D, Banjanac R, Udovičić V, Panić B, Puzović J and Aničin I 2008 *Nucl. Instrum. Meth.* A **591** 470

[5] Joković D, Dragić A, Udovičić V, Banjanac R, Puzović J and Aničin I 2009 *Appl. Radiat. Isot.* **67** 719

[6] Dorman L.I. 2004 *Cosmic Rays in the Earth's Atmosphere and Underground* (Berlin: Springer)

[7] Berkova M, Belov A, Eroshenko E and Yanke V 2012 *Astrophys. Space Sci. Trans.* **8** 41

Study of deuteron spectra under radiation belt with PAMELA instrument

S A Koldobskiy[1], O Adriani[2,3], G C Barbarino[4,5], G A Bazilevskaya[6], R Bellotti[7,8], M Boezio[9], E A Bogomolov[10], M Bongi[2,3], V Bonvicini[9], S Bottai[3], A Bruno[7,8], F Cafagna[8], D Campana[5], P Carlson[11], M Casolino[12,13], G Castellini[14], C De Donato[12,15], C De Santis[12], N De Simone[12], V Di Felice[12,16], V Formato[9,17,*], A M Galper[1], A V Karelin[1], S V Koldashov[1], S Y Krutkov[10], A A Kvashnin[6], A N Kvashnin[6], A A Leonov[1], V V Malakhov[1], L Marcelli[15], M Martucci[15,18], A G Mayorov[1], W Menn[19], M Merge'[12,15], V V Mikhailov[1], E Mocchiutti[9], A Monaco[7,8], N Mori[3,20], R Munini[9,17], G Osteria[5], F Palma[12,15], B Panico[5], P Papini[3], M Pearce[11], P Picozza[12,15], M Ricci[18], S B Ricciarini[3,14], R Sarkar[8,], V Scotti[4,5], M Simon[19], R Sparvoli[12,15], P Spillantini[2,3], Y I Stozhkov[6], A Vacchi[9], E Vannuccini[3], G I Vasilyev[10], S A Voronov[1], Y T Yurkin[1], G Zampa[9], N Zampa[9]**

[1] National Research Nuclear University MEPhI, RU-115409 Moscow, Russia
[2] Department of Physics and Astronomy, University of Florence, I-50019 Sesto Fiorentino, Florence, Italy
[3] INFN, Sezione di Florence, I-50019 Sesto Fiorentino, Florence, Italy
[4] Department of Physics, University of Naples "Federico II", I-80126 Naples, Italy
[5] INFN, Sezione di Naples, I-80126 Naples, Italy
[6] Lebedev Physical Institute, RU-119991 Moscow, Russia
[7] Department of Physics, University of Bari, I-70126 Bari, Italy
[8] INFN, Sezione di Bari, I-70126 Bari, Italy
[9] INFN, Sezione di Trieste, I-34149 Trieste, Italy
[10] Ioffe Physical Technical Institute, RU-194021 St. Petersburg, Russia
[11] KTH, Department of Physics, and the Oskar Klein Centre for Cosmoparticle Physics, AlbaNova University Centre, SE-10691 Stockholm, Sweden
[12] INFN, Sezione di Rome Tor Vergata, I-00133 Rome, Italy
[13] RIKEN, Advanced Science Institute, Wako-shi, Saitama, Japan
[14] IFAC, I-50019 Sesto Fiorentino, Florence, Italy
[15] Department of Physics, University of Rome Tor Vergata, I-00133 Rome, Italy
[16] Agenzia Spaziale Italiana (ASI) Science Data Center, Via del Politecnico snc, I-00133 Rome, Italy
[17] Department of Physics, University of Trieste, I-34147 Trieste, Italy
[18] INFN, Laboratori Nazionali di Frascati, Via Enrico Fermi 40, I-00044 Frascati, Italy
[19] Department of Physics, Universitt Siegen, D-57068 Siegen, Germany
[20] Centro Siciliano di Fisica Nucleare e Struttura della Materia (CSFNSM), Viale A. Doria 6, I-95125 Catania, Italia
[*] now at INFN, Sezione di Perugia, I-06123 Perugia, Italy
[**] now at Indian Centre for Space Physics, 43 Chalantika, Garia Station Road, Kolkata 700084, West Bengal, India

E-mail: `koldobskiy@gmail.com`

Abstract. This paper presents the results of measurements of proton and deuteron fluxes of albedo radiation in the Earth vicinity, obtained in the PAMELA experiment. PAMELA is an international experiment meant to study cosmic rays. PAMELA is carried out on board the satellite Resurs-DK1. High-precision equipment of the experiment allows registration and identification of cosmic ray particles of different varieties in a wide energy range. The albedo deuteron spectrum and albedo deuteron-to-proton fluxes ratio in the energy range 70 – 600 MeV/nucleon at altitude of 350 – 600 km for different geomagnetic latitudes is presented.

1. Introduction

Albedo radiation is a product of the interaction of primary cosmic rays (PCR) with the nuclei of the residual atmosphere by different processes. After their generation secondary deuterons move in the magnetic field of the Earth forming albedo fluxes under the radiation belt. The composition of atmosphere and cosmic rays are well known. Therefore the different mechanisms of deuteron generation can be studied by measuring the deuteron spectrum. In addition, knowing of the albedo deuteron flux is necessary for evaluating the radiation conditions at low altitudes in space.

The Earth's radiation belt and albedo radiation were discovered in 1957 by American scientist James Van Allen [1] and Soviet scientist S. N. Vernov [2]. It has been found that particles of albedo are predominantly protons [3, 4, 5]. Events, corresponding to heavier nuclei and electrons were detected as well in subsequent experiments. In [6] it was mentioned the registration of deuteron albedo radiation for the first time, in [7] the results of the unique attempt to measure the spectrum of deuterons under radiation belt in the AMS-01 experiment was presented. These results were compared with the calculated models described in [8, 9].

Results of the experiment AMS-01 [7] do not coincide with the computational model, which is based on the predominance of deuteron generation by means of nuclear reactions of the incident cosmic ray particles with the nucleons of the atmospheric nuclei. In view of the paucity of data on the measurement of deuteron fluxes under radiation belt new measurements, as well as verification and refinements, are undoubtedly required. In addition, these measurements are needed to precisely estimate the radiation environment in near Earth orbits.

In this paper we present the first results of data analysis for measuring the albedo deuteron spectra and d/p ratio for different geomagnetic latitudes. Data of the PAMELA experiment for period from 2006 till 2008 were used.

2. PAMELA experiment

PAMELA instrument [10] (Fig. 1) is a magnetic spectrometer (MS), composed by a permanent magnet hosting a tracking system and equipped with a time-of-flight system (ToF) and the calorimeter. The tracking system consists of six double-sided micro-strip silicon sensors. MS allows to measure the particle charge and rigidity (momentum over charge) with high precision. The time of flight system consists of three planes of double-layer plastic scintillators (S_1, S_2 and S_3) and serves for precise measurements of particle velocity and energy losses. It provides the main trigger for the experiment. PAMELA apparatus also includes a tail shower scintillation detector S_4 and neutron detector. The magnet is surrounded by scintillation counters operating in the anticoincidence mode (AC) used to select events with particles entering the spectrometer in its field of view.

Detailed description of the scientific equipment PAMELA detectors can be found in [10, 11, 12, 13].

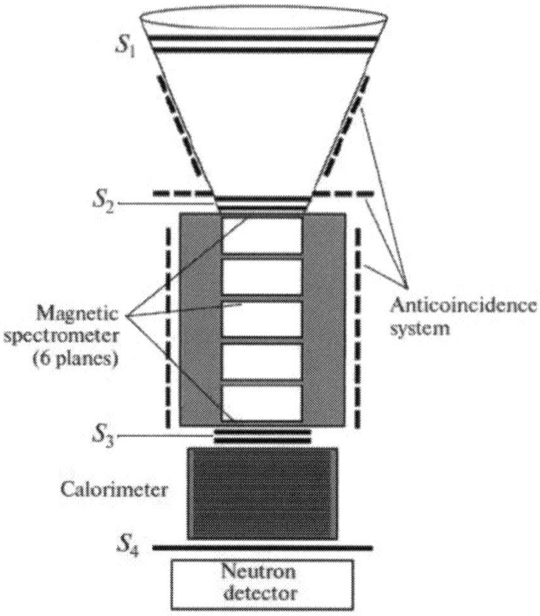

Figure 1. The schematic view of PAMELA experiment.

3. Particle identification

At the first stage of data analysis the selection of "good" events was carried out because only events with correct detector signals were used for subsequent event identification procedure. For this purpose, the analysis of the event characteristics was done to identify hydrogen isotopes in the energy range from 60 MeV/nucleon to 2 GeV/nucleon. The special set of criteria for implementing the so-called "basic" selection has been developed. Events that passed "basic" selection had no signals from anticoincidence detectors, corresponded to particle arriving inside the aperture of apparatus, with no interactions in the detector material, and characterized by well approximated trajectory in MS.

The most important and complex task was authentic identification of deuterons on the huge background of protons with similar detector signals. For the identification of detected particle nature, as well as the recovery of its energy the multivariate correlation analysis of the signals from the PAMELA detectors was used. The procedure of event identification was applied using values of particle rigidity R, measured by MS, its velocity β, measured by ToF and multiple measurements of the energy losses of ToF and tracker detectors. Multiple measurements of energy losses were used to suppress the proton background in $1/\beta$ distributions where deuteron amount obtaining was implemented by direct counting for energies below 300 MeV/nucleon or by distribution approximation by two Gaussians for higher energies (examples of steps in particle identification procedure are in Fig. 2). Employment of this identification procedure allowed to distinguish deuterons in energy range from 100 to 600 MeV/nucleon.

Verification of particle identification procedure and selection efficiencies were fulfilled by comparing of flight data with Monte Carlo simulation results. The simulation was performed by means of the software package GEANT [14].

More detailed description of particle identification procedure can be found in [13, 15]

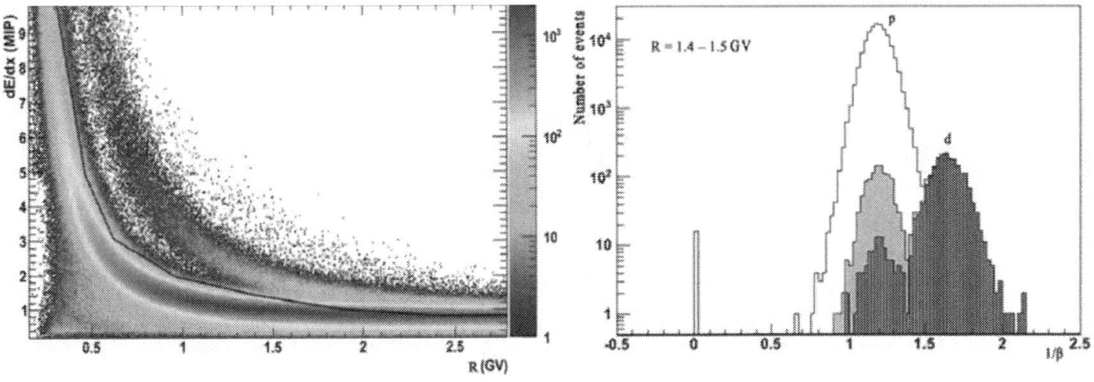

Figure 2. Left panel: Mean energy losses in MS versus rigidity distribution for protons and deuterons mix; black line is a cut used for proton suppression. Right panel: $1/\beta$ distributions for narrow rigidity bin $1.4 < R < 1.5$ GV without proton suppression (white distribution), using proton suppression with energy losses in MS (grey distribution) and in MS and ToF (blue distribution).

4. Albedo deuteron spectra

The deuteron spectrum in each narrow energy interval ΔE for different geomagnetic coordinate intervals were calculated as follows:

$$\frac{dJ}{dE} = \frac{N(\Delta E, E)}{\Delta E \times \Gamma(E) \times \varepsilon(E) \times t_{live}},$$

where $N(\Delta E, E)$ – the number of particles with energy E registered in the energy interval ΔE, $\Gamma(E)$ – the geometric factor of the instrument, $\varepsilon(E)$ is the total efficiency of the event selection and t_{live} – "live" time of the spectra measurements. The cut on magnetic field value $B < 0.23$ G was implemented to exclude radiation belt particles from analysis. Figure 3 shows the measured spectra of albedo deuterons for different geomagnetic latitudes ϕ_M. Following geomagnetic latitude bins are considered: 0.2 – 0.6; 0.6 – 0.7; 0.7 – 0.8; 0.8 – 0.9 (radians), which correspond to the L-coordinate intervals $1.04 < L < 1.46$; $1.46 < L < 1.71$; $1.71 < L < 2.06$; $2.06 < L < 2.56$. In the same figure the results of calculations made in [9] as well as the results of the experiment AMS-01 [7] are shown. Figure 4 presents the measured deuteron-to-proton albedo fluxes ratio for the same intervals of geomagnetic latitude.

As mentioned above, in the area under radiation belt deuterons are born as a result of interaction of cosmic ray particles with the nuclei of the residual atmosphere. In Fig. 3 solid line corresponds to a model where the coalescence is the dominant reaction of albedo deuterons generation [9]. PAMELA experiment data are well agreed with the calculation in the entire energy range. At low energies, where there is no calculation result deuteron spectrum has an indication to be flatter. For energy above 200 MeV/nucleon PAMELA data are in good agreement with AMS-01 results [7]. In the range of L-coordinate 2.06 – 2.58 the contribution of galactic deuterons in the obtained albedo spectrum becomes noticeable. The different galactic deuteron contribution in AMS-01 and PAMELA spectra can be explained by the difference of the mean geomagnetic cutoff in the measurement areas and by different apertures of instruments.

The radiation belt zone was investigated as well and the deuterons were registered in the wide energy range from 100 MeV/nucleon to 600 MeV/nucleon. However it was not possible to reconstruct the spectrum due to sharp flux anisotropy and low statistics for different angles. Nevertheless, if one can suppose the equal selection efficiencies for protons and deuterons it is

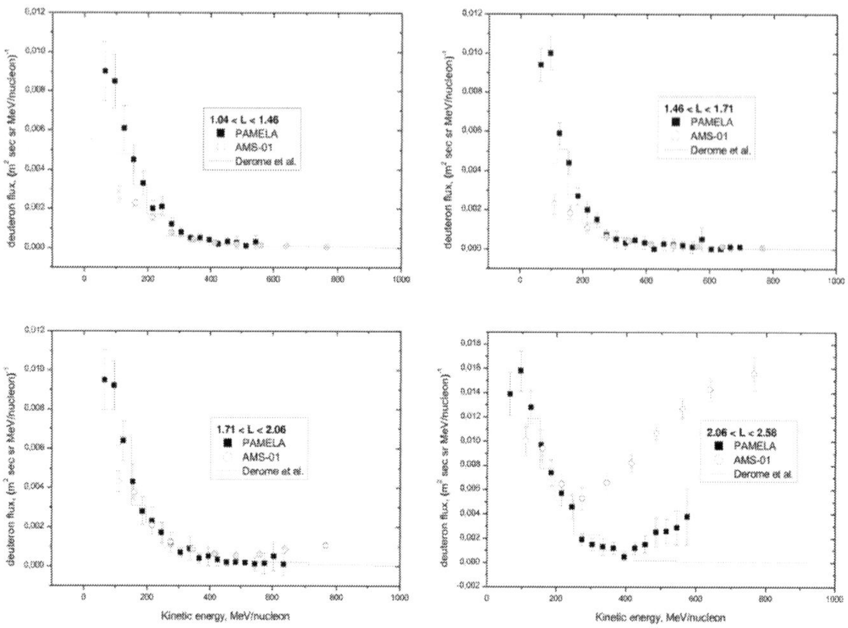

Figure 3. The deuteron albedo spectrum obtained in the PAMELA experiment for different L-geomagnetic coordinate intervals and geomagnetic field less than 0.23 G (black squares). Red circles represent AMS-01 data [7]. Blue solid lines shows the theoretical calculation results [8, 9].

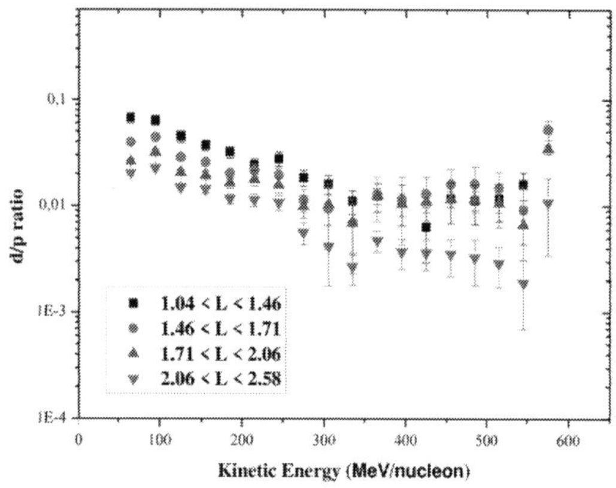

Figure 4. The albedo deuteron-to-proton fluxes ratio obtained in the PAMELA experiment for magnetic field value $B < 0.23$ G and different L-values.

possible to get rude evaluation of deuteron-to-proton ratio of order 0.02 for energy range 120 – 350 Mev/nucleon.

5. Conclusion

As a result of this work the spectrum of deuterons and deuteron-to-proton ratio under radiation belt in the energy range 70 – 600 MeV/nucleon is measured with PAMELA experiment. The measured spectra are agreed with the theoretical model results, where it is assumed that deuterons under radiation belt are generated mainly by nuclear fusion reactions of the incident cosmic ray particles and nucleons of the nuclei of the atmosphere and consequent movement in the Earth's magnetic field.

Acknowledgments

This work was supported by the Russian Science Foundation (grant 14-12-00373) and the grant of the President of the Russian Federation MK-4599.2014.2. The authors would like to thank the reviewers for their comments that help improve the manuscript.

References

[1] Van Allen J A 1959 *Sci. Am.* **200** 39
[2] Vernov S, Ginzburg V, Kurnosova L V *et al.* 1958 *Proc. 8th Intern. Astron. Congr.* p 464
[3] Intriligator D S 1968 *Phys. Rev. Lett.* **20** 1048
[4] Barwick S W *et al.* 1998 *J. Geophys. Res.* **103** 4817
[5] Bidoli V *et al.* 2003 *J. Geophys. Res.* **108** 1211
[6] Looper M, Blake J, Cummings J and Mewaldt R 1996 *Radiation Measurements* **26** 967
[7] Lamanna G *et al.* 2001 *Proc. of the 27th Intern. Cosmic Ray Conf.* p 1614
[8] Derome L *et al.* 2000 *Phys. Lett. B* **489** 1
[9] Derome L and Buenerd M 2001 *Phys. Lett. B* **521** 139
[10] Picozza P, Galper A M, Castellini G *et al.* 2007 *Astropart. Phys.* **27** 296
[11] Ricciarini S *et al.* 2008 *Nucl. Instrum. Methods Phys. Res. A* **582** 892
[12] Papini P, Adriani O, Ambriola M *et al.* 2008 *Nucl. Instrum. Methods Phys. Res. A* **588** 259
[13] Voronov S A, Danilchenko I A and Koldobskiy S A 2011 *Instr. And Exp. Tech.* **54** 752
[14] Agostinelli S *et al.* 2003 *Nucl. Instrum. Methods A* **506** 250
[15] Adriani O e a 2013 *Astrophysical Journal* **770** 2

On the method of the GCR partial intensities related to the main physical processes of solar modulation

Mikhail Krainev

Lebedev Physical Institute, RAS, Leninsky prosp., 53, 119991, Moscow, Russia

E-mail: mkrainev46@mail.ru

Abstract. We discuss in more details the earlier suggested method to decompose the calculated GCR intensity into the partial intensities connected with the main physical processes (the diffusion, convection, adiabatic cooling and magnetic drift) of the solar modulation. The expressions for the partial intensities connected with the convection and adiabatic cooling are improved and some results are illustrated.

1. Introduction

This work deals with some additional calculations that can be made in the course of the usual numerical solution of the boundary problem for the GCR intensity. The aim of these calculations is to improve our understanding of behavior of this intensity in the heliosphere.

But first, why are we not fully satisfied with the conventional approach and look for some additional means? Of course, all physics of modeling the GCR modulation lies in the Transport Partial differential Equation (TPE) with the boundary and "initial" conditions for the intensity and in the models of the TPE coefficients used. So one solves the boundary problem, compares the results with the observations and makes the conclusions about the degree of understanding of the modulation process. However, observations of the GCR intensity and heliospheric characteristics are very incomplete; besides, even in the simplest case the calculated GCR differential intensity $J(\vec{r}_0, T_0, t_0)$ for the position $\vec{r}_0$, energy T_0 and time t_0 is the result of complex interplay between several main physical processes (the diffusion, convection, drift, adiabatic cooling) for all higher energies ($T > T_0$) and in previous moments ($t < t_0$) in the whole heliosphere. And the details of this interplay are still unrevealed in the calculations, only its result. But to understand this interplay, in particular, the contribution of different processes to the calculated intensity, is very interesting and important.

In [1, 2] we suggested the method to decompose the calculated GCR intensity into the partial intensities connected with the main physical processes. In [3] the method was applied to understand the mechanisms forming the time profiles of the GCR intensity near the Earth during the last three periods of low solar activity. In this paper, first, we shall briefly discuss the usual process of modeling the GCR intensity and specify which additional calculations could be made and how they can be used for our purpose. Second, the expressions used earlier for the partial intensities connected with the convection and adiabatic cooling will be improved. Then some results will be illustrated and discussed.

2. Modeling the GCR intensity

2.1. The boundary–value problem

The TPE for the cosmic ray intensity J in its conventional form was derived by Parker [4, 5], Krymsky [6], Jokipii et al. [7] (if we list few names) in the first twenty years of the theoretical study of the GCR modulation. Usually it deals with the distribution function $f(\vec{r}, p, t) = J(\vec{r}, T, t)/p^2$, where p is the momentum of particles. For the stationary case the TPE balances the spatial divergences of the diffusion, drift and convection fluxes and of that due to adiabatic cooling in the momentum space:

$$-\frac{\partial f}{\partial t} = \underbrace{-\nabla(K\nabla f)}_{\text{diffusion}} + \underbrace{\vec{V}^{sw}\nabla f - \frac{\nabla\vec{V}^{sw}}{3}p\frac{\partial f}{\partial p}}_{\text{convection + adiabatic loss}} + \underbrace{\vec{V}^{dr}\nabla f}_{\text{drift}} = 0, \tag{1}$$

where $\vec{V}^{sw}$, $\vec{V}^{dr}$ and K are the solar wind and magnetic drift velocities, and diffusion tensor, respectively. Note that we combined the second and third terms as the sum for the convection and adiabatic cooling processes because each of these terms by itself is not equal to the divergence of the convection and adiabatic fluxes, respectively (see subsection 2.4). For the case of axial symmetry in a spherical system of coordinates the boundary–value problem, i. e., the transport equation with boundary and "initial" conditions, appears as follows

$$\underbrace{-\frac{1}{r^2}\frac{\partial}{\partial r}\left(K_{rr}r^2\frac{\partial f}{\partial r}\right) - \frac{1}{r^2\sin\vartheta}\frac{\partial}{\partial\vartheta}\left(K_{\vartheta\vartheta}\sin\vartheta\frac{\partial f}{\partial\vartheta}\right)}_{\text{diffusion}} +$$

$$\underbrace{V_r^{sw}\frac{\partial f}{\partial r} - \frac{\nabla\vec{V}^{sw}}{3}p\frac{\partial f}{\partial p}}_{\text{convection+adiabatic loss}} + \underbrace{V_r^{dr}\frac{\partial f}{\partial r} + \frac{1}{r}V_\vartheta^{dr}\frac{\partial f}{\partial\vartheta}}_{\text{drift}} = 0 \tag{2}$$

$$\left.\frac{\partial f}{\partial r}\right|_{r=r_{min}} = 0, \qquad f|_{r=r_{max}} = f_{nm}(p), \qquad \left.\frac{\partial f}{\partial\vartheta}\right|_{\vartheta=0,\pi} = 0 \tag{3}$$

$$f|_{p=p_{max}} = f_{nm}(p_{max}), \tag{4}$$

where $f_{nm}(p)$ is the nonmodulated distribution function corresponding to nonmodulated GCR intensity $J_{nm}(T)$ and r_{min}, r_{max}, p_{max} are the radii of the inner and outer boundaries of the modulation region and the momentum above which there is no modulation. Here for us it is important that only adiabatic loss of the energy is considered, without any acceleration (e. g., at the termination shock of the solar wind or in the heliosheath). Then with respect to the momentum or energy it is the Cauchy problem with the "initial condition" (4).

In the following we discuss the numerical solution of the boundary–value problem (2-4), some additional calculations and their results. Note that nowadays there are research groups dealing with much more sophisticated 3D and nonstationary models of the heliosphere and electromagnetic field – charged particle interaction (e. g., [8]). However we believe that the main features of the long-term (11– and 22–year) GCR intensity variations in the inner and intermediate heliosphere ($r < 70$ AU) can be described and understood using much simpler 2D and quasi–stationary tools like our boundary–value problem (2-4) with rather simple TPE coefficients. Besides, we are interested in two types of tasks. The first task is the study of the space–energy distribution of the GCR intensity with some specific set of the main parameters characteristic for the given phase of solar cycle (e. g., the sunspot minimum) and different heliospheric magnetic field (HMF) polarity (the sign of the HMF radial component B_r in the N–hemisphere, $A = \pm 1, 0$). The second task is the study of the time profile of the GCR intensity for some spatial locations and energies. In this paper we deal mainly with the first task.

2.2. Solving the boundary–value problem

To specify which additional calculations can be made for our purpose we should briefly discuss the well-known method of solution of the boundary–value problem and the widely used approximations of the differential operators by the finite differences on the grid (see, e. g., [9]).

If we change the variables $X = X(r)$, $Y = Y(\vartheta)$, $\tau = \tau(p)$ and the unknown function $U = f/f_{nm}(p) = J/J_{nm}(T)$ so that for new variables normalized to [0,1] the grids are homogeneous ($X_i = i\Delta X$, $Y_j = j\Delta Y$, $\tau_k = k\Delta\tau$), our TPE for U, which we call the normalized intensity, can be rewritten in the operator form:

$$\underbrace{\frac{\nabla\vec{V}^{sw}}{3}p\tau'\frac{\partial U}{\partial \tau}}_{A_\tau L_\tau U} = \underbrace{-\frac{X'}{r^2}\frac{\partial}{\partial X}\left(K_{rr}r^2X'\frac{\partial U}{\partial X}\right) + V_r^{sw}X'\frac{\partial U}{\partial X} + V_r^{dr}X'\frac{\partial U}{\partial X}}_{L_X U}$$

$$\underbrace{-\frac{Y'}{r^2\sin\vartheta}\frac{\partial}{\partial Y}\left(K_{\vartheta\vartheta}\sin\vartheta Y'\frac{\partial U}{\partial Y}\right) + \frac{1}{r}V_\vartheta^{dr}Y'\frac{\partial U}{\partial Y}}_{L_Y U}\underbrace{-\frac{\nabla\vec{V}^{sw}}{3}p\tau'(\ln f_{nm})'U}_{A_0 U} \tag{5}$$

and the differential operators can be approximated ($\approx$) by their difference analogues to the second order for ΔX, ΔY and to the first order for $\Delta\tau$:

$$L_X U = AX_0\left\{-\frac{\partial}{\partial X}\left[AXX\frac{\partial U}{\partial X}\right] + AXV\frac{\partial U}{\partial X}\right\} \approx \mathcal{L_X}^k\mathcal{U}_{i,j}^k \tag{6}$$

$$\mathcal{L_X}\mathcal{U}_i = \mathcal{AX}_0\left\{-\frac{\mathcal{AXX}_{i+\frac{1}{2}}(\mathcal{U}_{i+1}-\mathcal{U}_i) - \mathcal{AXX}_{i-\frac{1}{2}}(\mathcal{U}_i-\mathcal{U}_{i-1})}{\Delta^2 X} + \mathcal{AXV}\frac{(\mathcal{U}_{i+1}-\mathcal{U}_{i-1})}{2\Delta X}\right\} \tag{7}$$

$$A_\tau L_\tau U \approx \mathcal{A}_\tau\frac{(\mathcal{U}^{k+1}-\mathcal{U}^k)}{\Delta\tau}, \tag{8}$$

where $\mathcal{U}_{i,j}^k \approx U(X_i, Y_j, \tau_k)$, the coefficients $\mathcal{AX}_0$, $\mathcal{AXV}$, $\mathcal{A}_\tau$ in (7, 8) are equal to their counterparts in (5, 6) in the nodes of the grid, while $\mathcal{AXX}_{i\pm\frac{1}{2}}$ in (7) is equal to AXX in (6) at halfway between nodes X_i and $X_{i\pm1}$. The indices j, k are not repeated in (7), as well as i, j in (8). The approximation for L_Y is similar to that for L_X. Then the initial TPE can be also approximated by the finite differences on the grid

$$\mathcal{A}_\tau\frac{\mathcal{U}^{k+1}-\mathcal{U}^k}{\Delta\tau} = \sigma_X\mathcal{L_X}^{k+1}\mathcal{U}^{k+1} + \sigma_Y\mathcal{L_Y}^{k+1}\mathcal{U}^{k+1} + (1-\sigma_X)\mathcal{L_X}^k\mathcal{U}^k + (1-\sigma_Y)\mathcal{L_Y}^k\mathcal{U}^k$$
$$+ \sigma_X\mathcal{A}_0^k\mathcal{U}^k + (1-\sigma_X)\mathcal{A}_0^{k+1}\mathcal{U}^{k+1}, \tag{9}$$

with the weights $0 \le \sigma_{X,Y} \le 1$. For each step in τ (or momentum, or energy) using the Alternating Direction Implicit (ADI) method the system of the algebraic equations (9) with the appropriate boundary conditions can be solved. So $\mathcal{U}_{i,j}^{k+1}$ and hence the distribution function and intensity for the current energy in the whole heliosphere is calculated from the already known $\mathcal{U}_{i,j}^k$ (starting from $k = 0$, $\mathcal{U}_{i,j}^0 = 1$) and hence intensity for a little bit higher energy.

Usually after this point the process is repeated for even smaller energy and so on. However we here stop for a while to calculate and memorize (for some r_0, ϑ_0) the final difference approximation of any TPE term (diffusion, convection, drift) and of all necessary coefficients (e. g., $\mathcal{A}_\tau$, $\nabla\vec{V}^{sw}$ etc.) for the current energy.

2.3. Decomposition of the calculated J into the partial intensities

Note that TPE (5) in (r_0, ϑ_0) can be considered as the Transport Ordinary differential Equation (TOE) (10) with respect to $f(\tau)$ and in ∇–notations with the natural initial condition (11):

$$\frac{df(\vec{r}_0,\tau)}{d\tau} = \underbrace{\frac{3}{\nabla\vec{V}^{sw}}\nabla(K\nabla f)}_{T^{diff}(\vec{r}_0,\tau)} - \underbrace{\frac{3}{\nabla\vec{V}^{sw}}\vec{V}^{sw}\nabla f}_{\breve{T}^{conv}(\vec{r}_0,\tau)} - \underbrace{\frac{3}{\nabla\vec{V}^{sw}}\vec{V}^{dr}\nabla f}_{T^{drift}(\vec{r}_0,\tau)} \qquad (10)$$

$$f|_{\tau=0} = f_{nm}(p_{max}), \qquad \tau = \ln\frac{p_{max}}{p} \qquad (11)$$

This Cauchy problem could be solved if one knew all terms $T^{diff}, \breve{T}^{conv}, T^{drift}$ in (r_0, ϑ_0) for $\tau > 0$ $(T < T_{max})$. That is why we have already calculated and memorized the approximations of these terms while solving the algebraic equations (9) for U. Then integrating the initial value problem (10,11) we can calculate the partial "intensities" related to diffusion, convection and drift and their sum $\breve{J}_p^{dcd}$ (12).

$$\breve{J}_p^{dcd}(\vec{r}_0,T) = \underbrace{p^2\int_0^{\tau} T^{diff}d\tau}_{J_p^{diff}} + \underbrace{p^2\int_0^{\tau}\breve{T}^{conv}d\tau}_{\breve{J}_p^{conv}} + \underbrace{p^2\int_0^{\tau}T^{drift}d\tau}_{J_p^{drift}} \qquad (12)$$

$$\breve{J}_p^{adiab} = J - \breve{J}_p^{dcd} \qquad (13)$$

$$J = J_p^{diff} + \breve{J}_p^{conv} + J_p^{drift} + \breve{J}_p^{adiab} \qquad (14)$$

But we can't find the partial "intensity" related to adiabatic loss, as we used part of the related term as the left–hand side of (10). However, knowing the total J, we can calculate the partial "intensity" related to adiabatic energy loss as the difference between J and $\breve{J}_p^{dcd}$. So the decomposition (14) is what we tried to get. In this paragraph we marked the partial intensities with the inverted commas to emphasizes their strangeness, as they can be negative if the process leads to a reduction in intensity. In the following we omit the inverted commas. Besides, we marked by the breve (˘) the terms and partial intensities, related to the convection and adiabatic energy loss, which we consider incorrect and try to improve (see our note in 2.1 and the next subsection).

It is easily seen that both diffusion and drift partial intensities can be further decomposed into their components. As $K_{rr} = K_\parallel \cos^2\chi + K_{\perp,r}\sin^2\chi$ (with χ being the regular HMF Parker spiral angle) and $K_{\vartheta\vartheta} = K_{\perp,\vartheta}$, the diffusion partial intensity can be decomposed into those related to the diffusion along the regular HMF in the radial direction and perpendicular to the regular HMF in the radial and latitudinal directions, $J_p^{diff} = J_{\parallel,r}^{diff} + J_{\perp,r}^{diff} + J_{\perp,\vartheta}^{diff}$.

$$J_p^{diff} \propto \int_0^{\tau}\frac{1}{r^2}\frac{\partial}{\partial r}\left(K_\parallel\cos^2\chi r^2\frac{\partial J}{\partial r}\right)d\tau + \int_0^{\tau}\frac{1}{r^2}\frac{\partial}{\partial r}\left(K_{\perp,r}\sin^2\chi r^2\frac{\partial J}{\partial r}\right)d\tau$$
$$+ \int_0^{\tau}\frac{1}{r^2\sin\vartheta}\frac{\partial}{\partial\vartheta}\left(K_{\perp,\vartheta}\sin\vartheta\frac{\partial J}{\partial\vartheta}\right)d\tau \qquad (15)$$

Similarly as $\vec{V}^{dr} = \vec{V}^{dr}_{reg} + \vec{V}^{dr}_{cs}$, the drift partial intensity can be decomposed into those related to the drift in the regular HMF in the radial and latitudinal directions and along the heliospheric current sheet in the radial direction, $J_p^{drift} = J_{reg,r}^{drift} + J_{cs,r}^{drift} + J_{reg,\vartheta}^{drift}$. The terms corresponding to these parts of the diffusion and drift partial intensities are written in (15-16).

$$J_p^{drift} \propto \int_0^{\tau} -V_r^{dr,reg}\frac{\partial J}{\partial r}d\tau + \int_0^{\tau} -V_r^{dr,cs}\frac{\partial J}{\partial r}d\tau + \int_0^{\tau} -\frac{1}{r}V_\vartheta^{dr,reg}\frac{\partial J}{\partial \vartheta}d\tau \qquad (16)$$

2.4. On the convection and adiabatic energy loss partial intensities

The transport equation for the cosmic ray distribution function in the form (1), first suggested probably in [7], is simple to solve. However, as we mentioned in subsection 2.1, each of its second and third terms by itself is not equal to the divergence of the convection $\vec{S}^{conv}$ and adiabatic energy loss S^{adiab} fluxes, respectively. As was shown in [10] (see also the references therein and [11]) the right expression for these fluxes in terms of f are

$$\vec{S}^{conv} = 4\pi p^2 C_{CG} f \vec{V}^{sw} \qquad (17)$$

$$S^{adiab} = \langle \dot{p} \rangle = \frac{1}{3} p \vec{V}^{sw} \vec{G}, \qquad (18)$$

where $C_{CG} = -1/3\partial(\ln f)/\partial(\ln p)$ and $\vec{G} = \nabla f/f$ are the Compton-Getting factor and the relative gradient of the distribution function, respectively. So the TPE for f with each its term equal to the divergence of the corresponding flux is

$$-\frac{\partial f}{\partial t} = \underbrace{-\nabla(K\nabla f)}_{\text{diffusion}} + \underbrace{\nabla\left(C_{CG} f \vec{V}^{sw}\right)}_{\text{convection}} + \underbrace{\frac{1}{3p^2}\frac{\partial\left(p^3\vec{V}^{sw}\nabla f\right)}{\partial p}}_{\text{adiabatic loss}} + \underbrace{\vec{V}^{dr}\nabla f}_{\text{drift}} = 0 \qquad (19)$$

The TPE in form (19) looks quite different from (1). However as was shown in [10] and [7] these two forms are equivalent. So the right expression for the convective partial intensity would be

$$J_p^{conv} = p^2 \frac{3}{\nabla \vec{V}^{sw}} \int_0^{\tau} -\nabla\left(C_{CG} f \vec{V}^{sw}\right) d\tau \qquad (20)$$

instead of that implied in (10, 12). As to the adiabatic loss partial intensity it still can be calculated using (13), where however J_p^{dcd} includes the proper J_p^{conv} (20).

3. The illustrations

To solve the boundary–value problem (2-4) we need the models of the coefficients of the transport equation and the GCR nonmodulated spectrum. Here we shall not consider them as we used the same coefficients and $J_{nm}(T)$ as in [1, 2, 3, 12, 13, 14], where they (and even their questionable features) were described and discussed. As to the physical task we are most interested in understanding the role of the magnetic drift and other processes associated with the dominating HMF polarity A in the long–term variations of the GCR intensity. So we choose for illustration the same task as in [12, 14] that is the calculation of the space-energy distribution of the GCR intensity during the consecutive solar minima 21/22 (between solar cycles 21 and 22) and 22/23 characterized by similar behavior of the solar activity and the opposite HMF polarity.

In [12, 14] we first chose the parameters of the models in such a way that the calculations basically described the latitude, radial and energy dependencies of the observations for $A < 0$ (1987) and $A > 0$ (1997) solar minima. Then we repeated the calculations with the same parameters but switched off the drift, that is omitted the drift terms in (9) or used $A = 0$ as

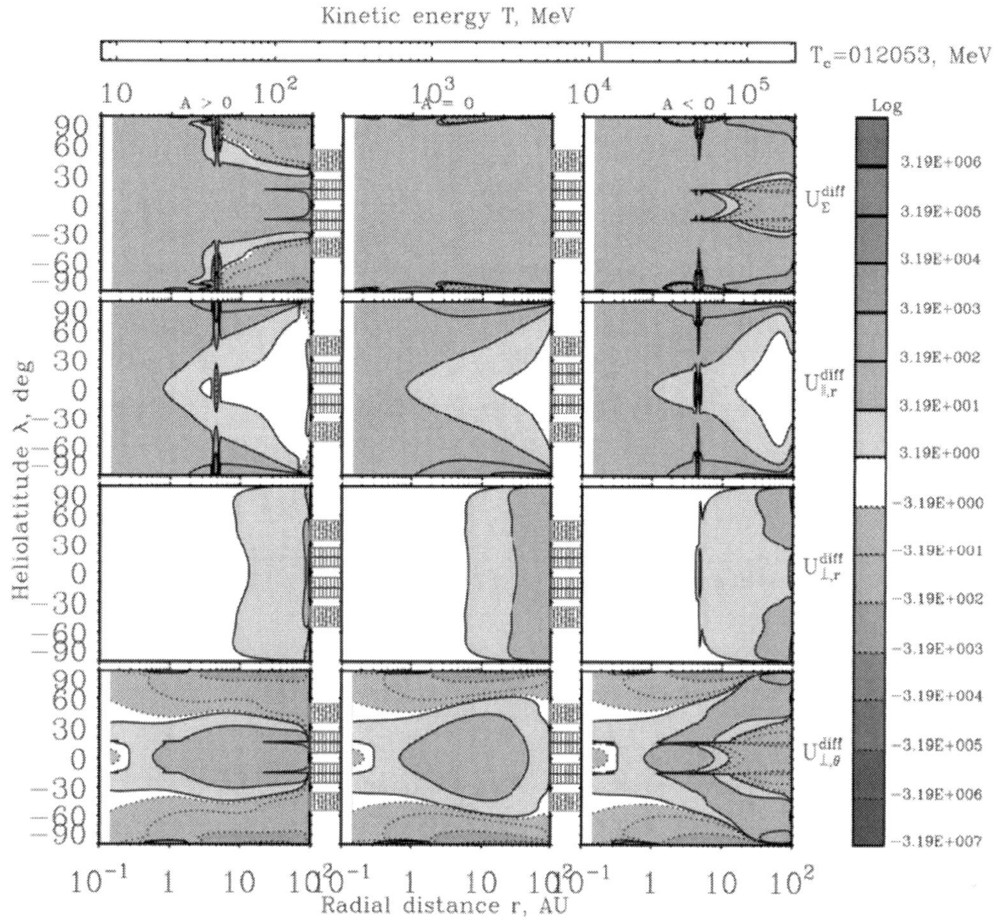

Figure 1. The spatial distribution of the normalized diffusion partial intensity and its three components for $T \approx 12$ GeV. Three columns of panels are for cases $A > 0$, $A = 0$ and $A < 0$. The upper row is for the total normalized diffusion partial intensity U_p^{diff}, while the next three rows are for $U_{\parallel,r}^{diff}$, $U_{\perp,r}^{diff}$, $U_{\perp,\vartheta}^{diff}$. The energy scale is the upper horizontal bar while the normalized intensity scale is in the right vertical panel. The horizontal bands between the columns mark the latitudes where the TPE coefficients (the solar wind and drift velocities and diffusion tensor) greatly change (see [3, 12, 13, 14]) for details.

these terms are proportional to A. And we noted that the calculated intensity almost everywhere decreased by a factor of 3-5 when compared with the intensities for both ($A > 0$)– and ($A < 0$)–periods. In [3] we showed that the same is true for all other phases of solar cycle except the years near solar maxima. So we made the conclusion that in the models used the drift contribution is large for both polarities, that is it strongly influences both 11- and 22-year cycles in the GCR intensity. It should be mentioned that the active role of drift in forming the 11-y variation of the GCR intensity followed from several model calculations [15, 16]. Note that in both of these works as well as in our calculations the first generation drift models were used. We used the simple model for the regular and current sheet drift velocities as in [17].

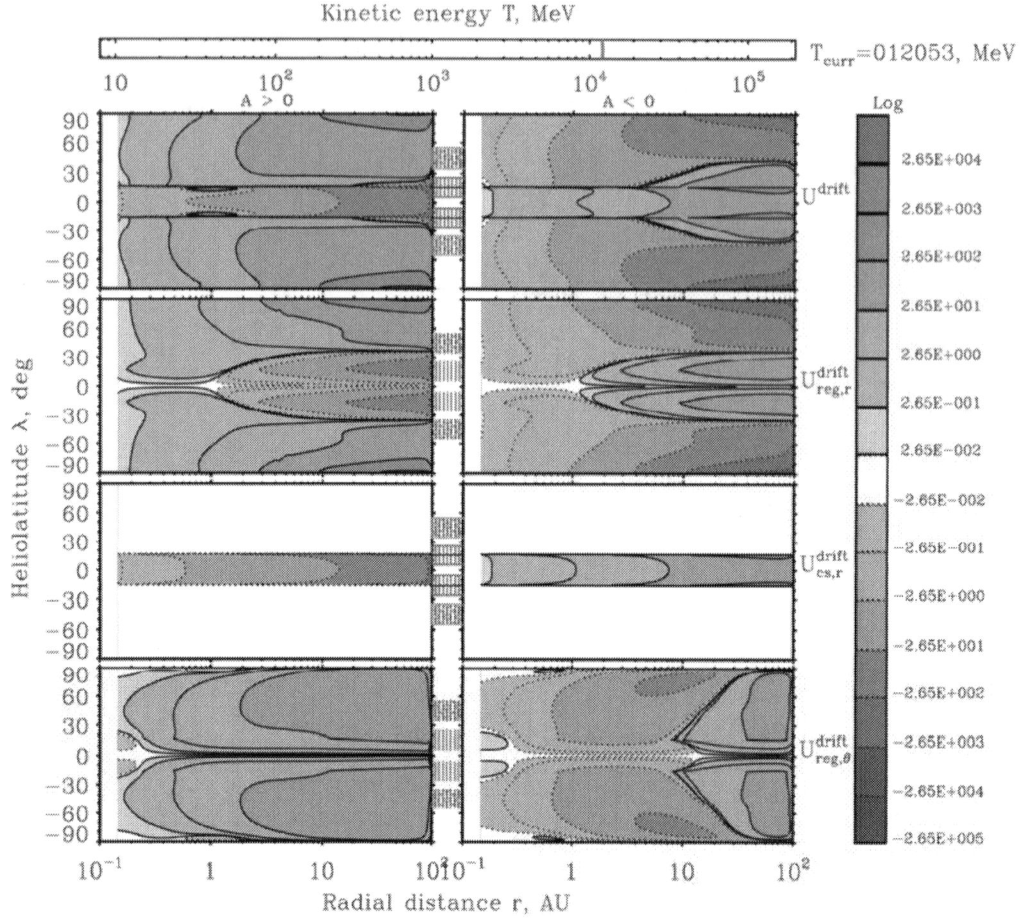

Figure 2. The same as in Fig.1 but for the drift partial intensity and without $A = 0$ column.

In this paper we try to prove that the partial intensities can be useful for revealing the mechanisms of the GCR intensity modulation and the heliospheric structure. The detailed picture of the space, energy and time distributions of different partial intensities will be discussed elsewhere. Here only two topics will be considered, namely, the spatial distributions for neutron monitor energy ($T \approx 12$ GeV) and all three cases of the HMF polarity: 1) of the diffusion and drift partial intensities and their parts and 2) of the drift partial intensity compared with the sum of all other partial intensities.

In Fig.1 the normalized diffusion partial intensity and its parts (as in (15)) are shown as functions of the radial distance and latitude for all $A > 0$, $A = 0$ and $A < 0$ cases. It should be noted that the vertical feature seen at a glance at $r = 5$ AU in the upper three rows of Fig.1 is due to somewhat poor choice of $X(r)$, when $X(r)$ is continuous but its derivative, $X'(r)$, breaks at $r = 5$ AU. So the terms including its second derivative (such as those connected with the diffusion in the radial direction) demonstrate such unusual feature. It turns out that for such energies the total diffusion partial intensity (that is the contribution of the diffusion in the calculated intensity), which in case of the absence of drift ($A = 0$) is positive almost everywhere (except small regions near the poles) is negative in the far from the Sun regions for the case $A > 0$ in the middle and high latitudes and for the case $A < 0$ in the equatorial latitudes.

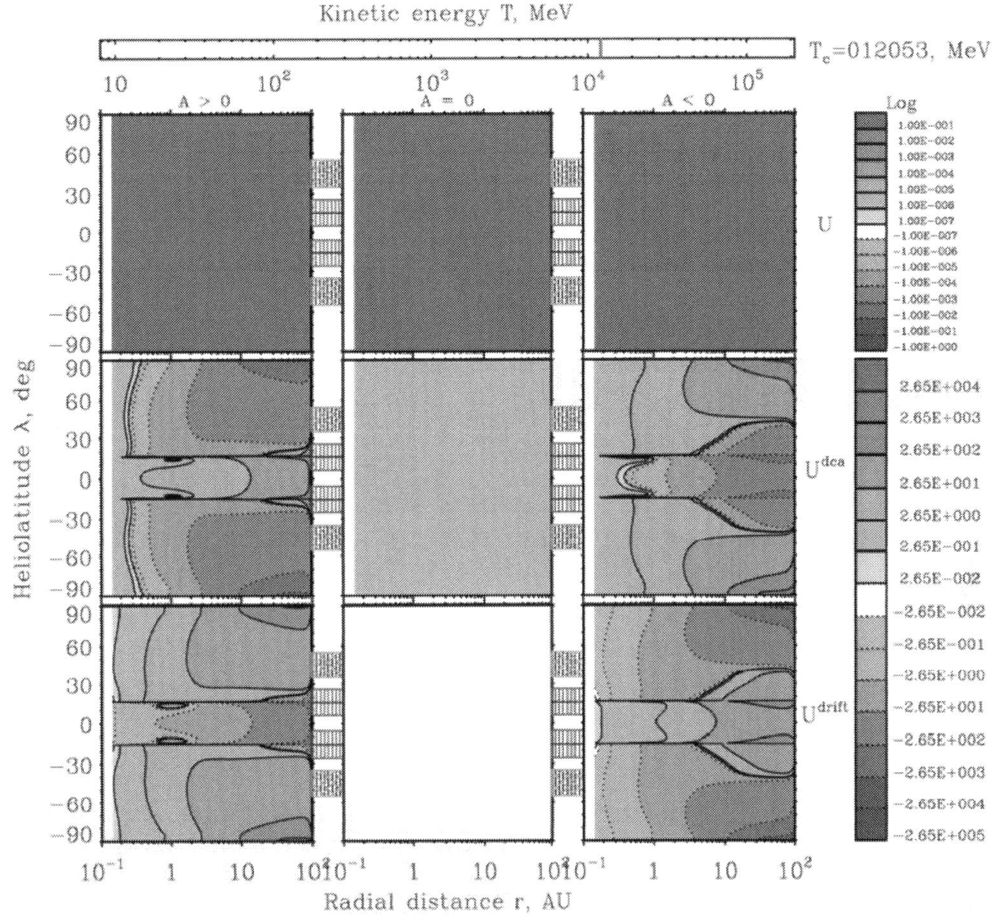

Figure 3. The balance between drift and sum of other partial intensities. Three columns of panels are for cases $A > 0$, $A = 0$ and $A < 0$. The upper row is for the total calculated intensity U, while the next two rows are for U_p^{dca} and U_p^{drift}. The energy scale is the upper horizontal bar while the intensity scales (different for U and U_p^{dca}, U_p^{drift}) are in the right vertical panels.

Besides, it turns out that the negative contribution of the diffusion for such energies for all cases (positive, zero and negative HMF polarities) is due entirely to the diffusion perpendicular to the regular HMF in the latitudinal direction. Of course, it is also easily seen that this picture reveals very detailed and unexpected heliospheric structure with respect to the diffusion of the GCRs.

Similar analysis can be also fulfilled for the drift partial intensity (see Fig.2). It also reveals the details of the drift contribution in the total calculated GCR intensity and the heliospheric structure with respect to the drift of the GCRs.

As to the main task — understanding the role of drift — in Fig.3 we separate the calculated normalized intensity U (positive (red) everywhere) into two parts much greater in absolute values and of different signs in different regions depending on the HMF polarity: the sum of the partial intensities due to the overall contribution of diffusion, convection and adiabatic cooling (U_p^{dca}) and the drift partial intensity (U_p^{drift}). It can be easily seen that there are heliospheric regions where the contribution of drift dominates over that of the combined action of the diffusion, convection and adiabatic cooling — (positive (red) drift partial intensity) at the middle and high latitudes for $A > 0$ and at equatorial latitudes for $A < 0$ — while outside these regions the combined action of the diffusion, convection and adiabatic cooling dominates.

4. Conclusions

The decomposition of the calculated GCR intensity into the partial intensities related to the main physical processes can be useful in understanding the physics of the GCR modulation (e. g., the relative significance of different processes in the long-term variations of the GCR intensity) and in the theoretical study of the heliospheric structure.

There are some unresolved problems and difficulties:

- What is the detailed physical meaning of the suggested mathematical procedure (e. g., why do the diffusion, convection and drift partial intensities depend on $\nabla\vec{V}^{sw}$)?
- How to find the partial intensities in case of any GCR acceleration (e. g., at the termination shock)?
- Is it possible to find the partial intensities in case of any other method of solving the boundary–value problem (e. g., by the stochastic modeling the GCR trajectories,[18])?

It is important that to calculate the partial intensities the errors of the grid approximation and of the solution of the boundary–value problem should be rather small (as we are interested in spatial derivatives of the intensity). These problems are even more important if we want to consider the gradients of these partial intensities.

Acknowledgments

We thank for help the Russian Foundation for Basic Research (grants 12-02-00215, 13-02-00585) and the Program of the Presidium of the Russian Academy of Sciences "Fundamental Properties of Matter and Astrophysics".

References

[1] Krainev M B and Kalinin M S 2013 On the description of the 11- and 22-year cycles in the GCR intensity *J. Phys.: Conf. Series* **409** 012155
[2] Krainev M B and Kalinin M S 2013 On formation of the spot and magnetic cycles in the galactic cosmic ray intensity in the solar activity minima *Bulletin of the Lebedev Physics Institute* **77** 510-512
[3] Krainev M B etal 2013 *Proc. 33-rd ICRC* Rio-de-Janeiro Brasil icrc2013-0305.pdf arXiv:1411.7526
[4] Parker E N 1958 *Phys. Rev.* **110** 1445
[5] Parker E N 1965 *Planet. Space Sci.* **13** 9–49
[6] Krymsky G F 1964 *Geomagnetism and Aeronomy* **4** 977–985 (in Russian)
[7] Jokipii J R, Levy E H and Hubbard W B 1977 *Astrophysical Journal* **213** 861–861
[8] Strauss R D, Potgieter M S and Ferreira S E S 2012 *Adv. Space Res.* **49** 392–407
[9] Samarsky A A 1977 *Theory of finite-difference schemes* Nauka Moscow 656p (in Russian)
[10] Gleeson L J and Webb G M 1978 *Astrophysics and Space Science* **58** 21-39
[11] Potgieter M S 1984 *PhD Thesis* Potchefstroom University for Christian Higher Education
[12] Kalinin M S and Krainev M B 2013 The formation of the sunspot and magnetic cycles in the GCR intensity in the heliosphere *J. Phys.: Conf. Series* **409** 012156
[13] Krainev M B etal 2013 On the status of the sunspot and magnetic cycles in the galactic cosmic ray intensity *J. Phys.: Conf. Series* **409** 012016
[14] Kalinin M S etal 2013 *Proc. 33-rd ICRC* Rio-de-Janeiro Brasil icrc2013-0297.pdf arXiv:1411.7519
[15] Kóta J and Jokipii J R 1983 *Astrophysical Journal* **265** 573
[16] Le Roux J A Potgieter M S 1995 *Astrophysical Journal* **442** 847
[17] Krainev M B and Kalinin M S 2010 *AIP Conf. Proc.* **1216** 371–374
[18] Zhang M 1999 *Astrophysical Journal* **513** 409–420

Recurrent and sporadic Forbush-effects in deep solar minimum

O Kryakunova[1], A Belov[2], A Abunin[2], M Abunina[2], E Eroshenko[2], A Malimbayev[1], I Tsepakina[1], V Yanke[2]

[1] Institute of Ionosphere, Kamenskoe Plato, Almaty, 050020, Kazakhstan
[2] Pushkov Institute of Terrestrial Magnetism, Ionosphere and Radio Wave Propagation (IZMIRAN), Moscow, Troitsk, 142190, Russia

E-mail: krolganik@yandex.ru

Abstract. The effects of high-speed solar wind streams from low-latitude coronal holes and coronal mass ejections (CMEs) on cosmic ray intensity in 2007 are studied. The database on Forbush effects created at IZMIRAN, with cosmic ray density and anisotropy calculated by the Global Survey Method (GSM) on the basis of Neutron Monitor network data has been used. The behaviour of the mean characteristics by all the Forbush-effects in 2007 caused by coronal holes (interplanetary magnetic field intensity and solar wind velocity, 10 GV cosmic ray density and equatorial component of the cosmic ray anisotropy) is calculated by the epoch method. Features of the Forbush-effects caused by high-speed solar wind streams from low-latitude coronal holes and coronal mass ejections are described.

1. Introduction

The Forbush effect (FE) result from the influence of coronal mass ejections (CMEs and ICMEs) and/or high speed streams of the solar wind from the coronal holes on the background cosmic rays [1-5]. Thus, interplanetary disturbances create the Forbush-effects (FEs) are both of sporadic and recurrent nature. It would be desirable to leave only one class of sources and not consider recurrent phenomena as FEs. But this is practically impossible because the solar wind disturbance is often a result of the interaction of different factors [5,6], both of sporadic (coronal ejections, flares, filament disappearances) and recurrent (coronal holes, streamer structure) origin. It is reasonable to divide the FEs into groups by their solar sources: sporadic and recurrent [1], where the first group is being connected with disturbances of the interplanetary medium caused by coronal mass ejections (ICMEs), while the events in the latter group are usually caused by streams of high-speed solar wind from the low-latitude coronal holes (CHs). Solar wind recurrent high speed streams may be the main reason of many FEs, but these effects are never too large. It seems the limiting value of FE caused by coronal holes does not exceed 5%. Herewith in the large recurrent FEs it is also possible to find the contribution of CMEs. We may certainly state that all FEs of large magnitude and the majority of FEs of middle and small magnitude are caused by CMEs. During the minimum solar activity the large CMEs are rare, so the observable FEs are mainly caused by CHs. The goal of the present investigation is to study the influence of high-speed solar wind streams (HSSs) from low-latitude CHs and coronal mass ejections (CMEs and ICMEs) on cosmic rays (CRs).

2. Data and methods

The parameters used in this analysis include cosmic ray density (A0) and anisotropy (Axy equatorial component of the first harmonic of the anisotropy), solar wind data (velocity and density), interplanetary magnetic field intensity (B_{IMF} and Bz) as well as solar data and geomagnetic activity indices (Kp and Dst). Using the global survey method (GSM, [7,8]) for the calculation of the CR density variations and anisotropy vectors we operated the whole bulk of experimental material through the year 2007. The CR parameters were supplemented by the information on the interplanetary medium taken from the OMNI database (http://omniweb.gsfc.nasa.gov/ow.html), and by the data of geomagnetic activity (ftp://ftp.gfz-potsdam.de/pub/home/obs/kp-ap/wdc). The CR parameters were calculated for particles with a rigidity 10 GV which is close to the effective rigidity of the particles being registered by the world wide neutron monitor network.

One of the advantages of the year 2007 for our study is a possibility to observe the high-speed streams from CHs without significant CME influence. The background velocity of the solar wind during this year was usually so low that the streams of medium-speed (and sometimes also below than that) were noticeable resulting in an interaction with the low-speed solar wind and in additional modulation of the CR. Also, this year falls on a long period of a large number of CHs which has started at the end of 2002. The numerous (nearly 50) and diverse CHs were the main source of interplanetary perturbations in 2007, while the CME-caused disturbances were a minority, and the CMEs connected with a big solar flares were nearly absent. The events caused by solar filament disappearance were a little more frequently in 2007, but not so often as from CH.

3. Analysis of the events

Let us consider FE caused by HSS from recurrent coronal hole and CME. The north coronal hole CH299 (figure 1) has appeared on the disk of the visible Sun and came up to a geoeffective position on November 14-15 2007. The amplitude of the effect in cosmic rays was AF = 2.0, the maximum equatorial anisotropy was Axy = 1.83. As in many such events the increase of the solar wind density and the interplanetary magnetic field were observed (figure 2). The short and deep Forbush effect was due to the CME partial halo. It is interesting, that this Forbush effect is not accompanied by a disturbance of the Earth's magnetic field (Kp_{max} = 3). Then the CR intensity decreased gradually, which was the result of the impact of the overlay effects from CME and CH299, because the coronal hole actually consists of two parts, which led to a slightly increased solar wind HSS. The sharp turn of the vector of equatorial anisotropy occurred before the start of the FE on 19th November (figure 3). It may be noted that the anisotropy behaves quite smoothly in FEs caused by the HSSs from CHs, than in FEs caused by CMEs. As a rule the cosmic ray anisotropy sharply changes its direction under the influence of CME.

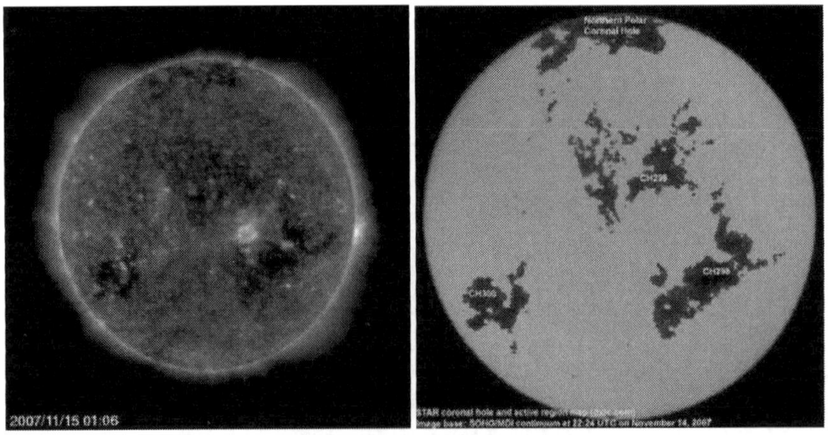

Figure 1. Coronal hole CH299 (adopted from (http://www.solen.info/solar)

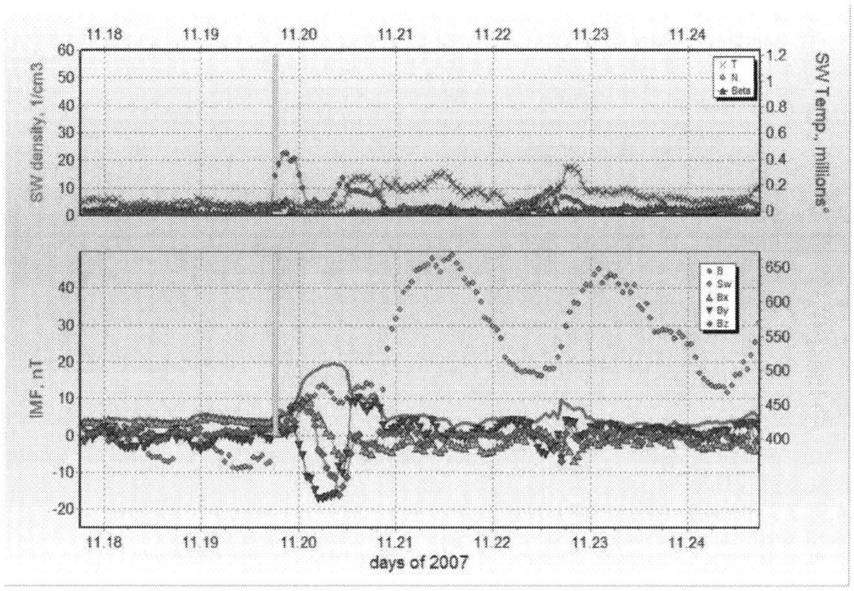

Figure 2. Behaviour of the solar wind parameters: IMF intensity, solar wind velocity, density and temperature during November 17-24, 2007.

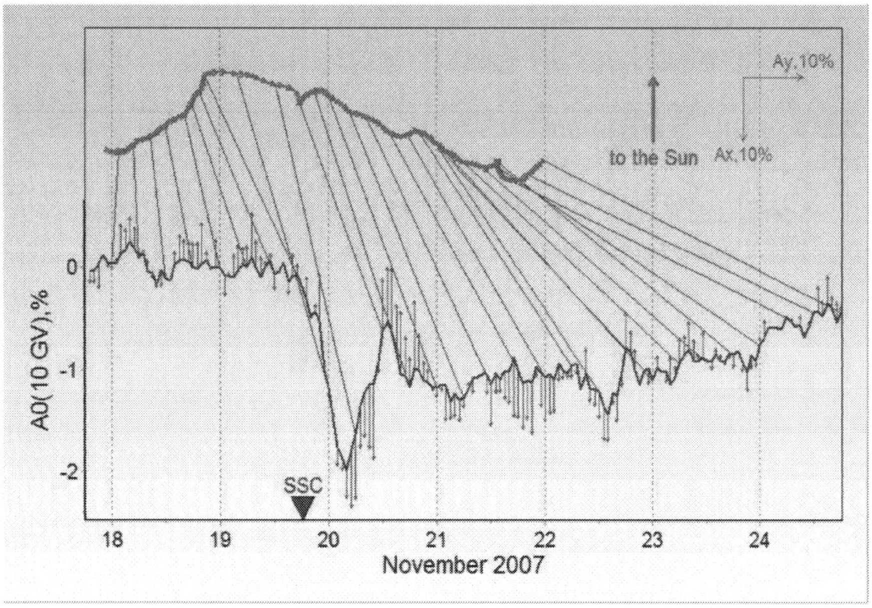

Figure 3. Behaviour of the 10 GV CR density (A0, brown line) and equatorial component of the first harmonic of CR anisotropy (Axy – dark green vector diagram) during November 17-24, 2007. Vertical green lines mean the North-South anisotropy for each hour. Thin solid lines connect parts of the vector diagram and the curve of density every 6 hours. The triangle SSC (Sudden Storm Commencement) marks the moment of shock arrival and the onset of the Forbush effect.

As we defined earlier [9], the magnitude of the IMF significantly affected the magnitude of the FE. In the event of 19-20 November, 2007 it is clearly visible: the large magnitude of effect for this class of FEs the (2%) is associated with a large magnitude of the IMF (20 nT). The configuration of the interplanetary magnetic field on November 20 is one of the signs of a magnetic cloud. This CME was the result of the disappearance of the solar filament (DSF) on 15 November.

The mean characteristics averaged by all the events in 2007 caused by coronal holes, are shown in figure 4. All parameters are averaged by the epoch method where the 0-day is the day of the Forbush effect onset.

From the analysis of 48 FEs that occurred in 2007 and are associated with low latitudes coronal holes, it can be noted that the HSS from low-latitude coronal holes leads to disturbances of the magnetosphere to the level of minor (and rarely moderate) magnetic storm, on average, these events are accompanied by a disturbed level of the magnetic field (with Kp=4). The maximum speed HSS from coronal holes was 567 km/s, the maximum value of the solar wind velocity was V_{SW}=698 km/s. The magnitude of FEs in these events varies from 0.3% to 2.3%, the average value of FE is 1.03 ± 0.06%. The maximum values of the equatorial part of the first harmonic of the anisotropy of CR reaches 1.02 ± 0.03, and it varies from a minimum of 0.64 to a maximum of 1.51.

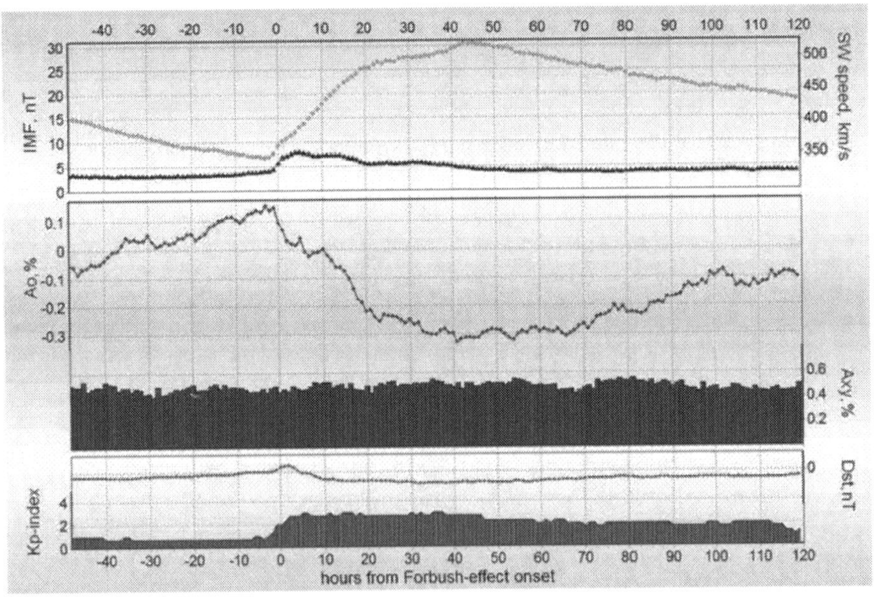

Figure 4. Behaviour of the characteristics averaged by all the FDs in 2007 that are caused by coronal holes: IMF intensity (umber line) and solar wind velocity (yellow lie, upper panel); 10 GV CR density (A0, red line) and equatorial component of the CR anisotropy Axy (blue stakes, middle panel); Kp (green stakes) and Dst-indexes (violet line, lower panel) of geomagnetic activity.

For a comparison with CH group events, the average characteristics of all events from our database including a CME, separated each from other by not less than 48 hours, are plotted in figure 5 [10]. It means that in calculation we use only those events that separated each from other by not less than 48 hours, otherwise FEs may overlap. 349 CME-caused events we picked up from the whole database for the years 1996-2010. We selected only those CME-caused events in which the values of the solar wind velocity and the interplanetary magnetic field intensity were closed to the values in the group of CH-caused events.

The comparison shows that high-speed streams of a solar wind from coronal holes created a small (on average, 1.03±0.06%), slowly developing FE. In the CME caused events the maximum of the IMF intensity (t_{Hmax}), the maximum of a solar wind speed (t_{Vmax}) and the minimum of CR density are located in a compact group in the range of several hours, whereas in the CH associated events the maximum of the solar wind speed V_{max} and the CR density minimum are delayed essentially relatively to the maximum of IMF intensity (t_{Hmax}).

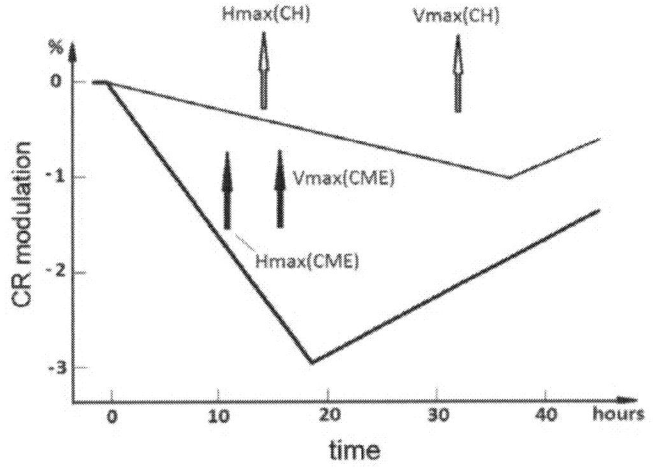

Figure 5. Average characteristics of FEs in 2007 caused by coronal holes (48 events, upper line), and events, created by CMEs (349 events, lower line). X axis - time from FE onset, Y axis - magnitude of CR modulation.

4. Conclusions

Prolonged CR decreases characterized by a late minimum in 2007. The maximum speed of HSS from coronal holes, averaged for 48 events, was 567 km/s, and the maximum IMF intensity was 11,6 ± 0,5 nT, the Kp index of geomagnetic activity was 4,08 ± 0,13, the equatorial component Axy of the vector CR anisotropy varies very slightly.

The example of 2007 demonstrates that even at very quiet conditions, i.e., reduced activity of the Sun, disturbances resulting to FEs are present; these disturbances are the CHs. The data of the global Neutron Monitor network, processed by GSM, allows us to study small CR effects. A distinctive feature of the FEs caused by CHs, is their long duration, the minimum intensity of cosmic rays greatly lags the IMF intensity. FEs caused by coronal holes are quite different from the events caused by CMEs: they have a shorter duration of the FE, the time distribution - late minimum in CR intensity, which is close to the time of the solar wind velocity maximum and delayed with respect to the IMF maximum. There are large differences between recurrent and sporadic FEs in the derived CR anisotropy: the value of the anisotropy is significantly smaller for recurrence than for sporadic FEs and furthermore, it presents a more smooth, behaviour.

Acknowledgments
We acknowledge the NMDB database (www.nmdb.eu), founded under the European Union's FP7 programme (contract no. 213007) for providing data and personally Dr. Christian T. Steigies for the

maintenance of the NMDB database and coordination of the NMDB consortium work. The authors wish to acknowledge all teams providing continued ground level CR monitoring (http://cr0.izmiran.rssi.ru/ThankYou/). This work was supported by Kazakhstan Aerospace Committee under program 076 and under grant N0014/GFZ.

References

[1] Belov A 2009 *Universal Heliophysical Processes Proc. Int. Astronomical Union Symposium* **257** 439
[2] Lockwood J A *et al* 1991 *J. Geophys. Res.* **96** pp 11587–11604
[3] Cane H V 2000 *Space Sci. Rev.* **93** pp 55–77
[4] Cane H V *et al* 1994 *J. Geophys. Res.* **99** A11 pp 21429–21441
[5] Richardson I G and Cane H 2011 *Solar Phys.* **270** pp 609–627
[6] Crooker N U and Cliver E W 1994 *J. Geophys. Res.* **99** 23383
[7] Belov A *et al* 2005 *J. Geophys. Res.* **110** A09S20 doi:10.1029/2005JA011067
[8] Asipenka A S *et al* 2007 *Proc. 30th Int. Cosmic Ray Conf. (Mexico)* Id 1006
[9] Belov A V and Ivanov K G 1997 *Proc. 25th Int. Cosmic Ray Conf. (Durban)* **1** 421
[10] Kryakunova O *et al* 2013 *J. Phys.: Conf. Series* **409** 012181

Possible ground level enhancements at the beginning of the maximum of Solar Cycle 24

A Belov[1], E Eroshenko[1], O Kryakunova[2], N Nikolayevskiy[2], A Malimbayev[2], I Tsepakina[2], V Yanke[1]
[1] Pushkov Institute of Terrestrial Magnetism, Ionosphere and Radio Wave Propagation (IZMIRAN), Moscow, Troitsk, 142190, Russia
[2] Institute of Ionosphere, Kamenskoe Plato, Almaty, 050020, Kazakhstan

E-mail: krolganik@yandex.ru

Abstract. Three solar energetic proton events in the beginning of 2012 (January and March) revealed significant increases of the integral proton fluxes with energies >100 MeV were analyzed by the data from subpolar neutron monitors (>500 MeV). As it was found the event on January 27, 2012 was followed by cosmic ray enhancement of about 2% at several subpolar and high latitude neutron monitors which coincided with the onset of the intensity increase in the GOES channels. This was also confirmed with the GOES/HEPAD data that allow us to consider the event on January 27, 2012 as a Ground Level Enhancement. The events in March 2012 (7.03 and 13.03) occurred under very complicated situation in the interplanetary space and a more detailed analysis is needed. Nevertheless, these events may contain some contribution of solar cosmic rays in the ground level observations.

1. Introduction

Proton events (enhancement of the flux of accelerated charged particles – protons and nuclei – in the Earth's atmosphere and near Sun space environment) are one of the most important and most dangerous phenomena of space weather, which needs careful studying (www.swpc.noaa.gov). Time distribution of proton events is very inhomogeneous (figure 1) with the majority of them occurring during the maximum of the solar cycle as well as within its descending phase. In order to study proton increases it is useful to unite data of the Geostationary Operational Environmental Satellite (GOES) in which uniform and full information on X-ray flares and protons fluxes at various energy ranges are accumulated, with data of the world neutron monitor (NM) network.

Currently, it is considered that in the 24th cycle of solar activity only 2 ground level enhancements (GLEs) of solar cosmic rays (SCR) are registered: on May 17, 2012 [1, 2, 3, 4] and on January 6, 2014 [5]. Both in quantity and in power of GLEs and in solar activity as well the current cycle concedes to the previous [6, 7, 8]. The number of solar proton events (SPEs) registered on satellites are much higher than the number of GLEs recorded on the ground; however, as a rule, GLEs tend to occur more frequently in higher level storms and thus have a considerable radiation impact. In 2012 around 30 SPEs were marked. We decided to analyze the behaviour of the cosmic ray (CR) intensity recorded by the world wide NM network during the events of the 24th solar activity cycle when a significant increase of the integral proton flux with energy > 100 MeV was recorded by satellite experiments. This turned out to be directly in the events on January 27, March 7, and March 13 of 2012. As GLE should be considered an event in which at least on one NM at sea level statistically significant increase of count rate coinciding on time with increase in the satellite data would be observed. Perhaps, the decisive word here is in modeling of increase on a network of stations. This criterion is satisfied by the event in January 2012. And other events considered in this work in March 2012 need to be studied and analyzed more detailed with using some modeling approach. It was not noticed earlier because nobody paid attention to so small events.

The aim of our work is to show that all above mentioned events may be considered as candidate GLEs. Apparently, in 24th cycle more GLEs than it is widely recognized were observed.

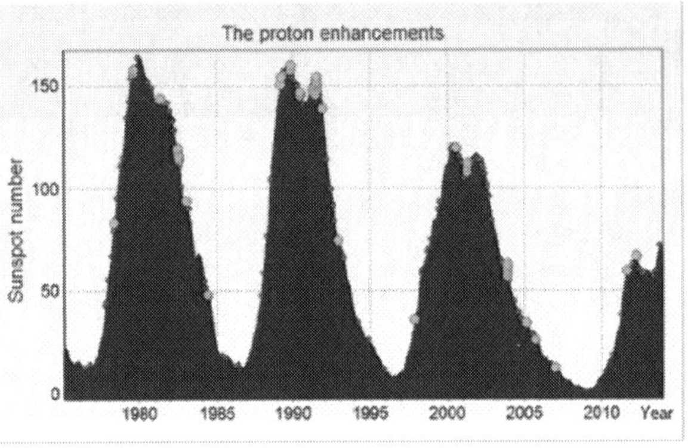

Figure 1. Proton increases in 1976-2013 on the background smoothed numbers of solar spots. Small red circles mark the time of increase of the protons of >10 MeV energy higher 10 pfu, large yellow circles mark the time of GLEs of SCR.

2. Data and method

For this analysis, data of the neutron component of CRs recorded at subpolar and high latitude NMs with time resolution of 1-, 5-, and 30 minutes have been used. Those were taken from the NMDB database (www.nmdb.eu) for January and March 2012. These data were compared with measurements of soft X-Ray (GOES15) and integral fluxes of protons of different energies (>10, >50 and >100 MeV) obtained onboard GOES13. Data from High Energy Proton and Alpha Detector (HEPAD) on the high energy channels was also involved in the analysis. Parameters of the solar activity and the interplanetary space were accessible by http://www.swpc.noaa.gov/ and OMNI data base – (http://nssdc.gsfc.nasa.gov/omniweb/ow.html).

3. Event on January 27, 2012

The first large proton event in 2012 started on January 23, after the flare of M8.7 class from the region AR11402 (N27W71), located at western side of solar disk. The particle flux with energies of >100 MeV was 2.3 pfu. On January 24 the flux of particles of >10 MeV reached 3400 pfu (radiation storm of 3 level), on the 25 January – 6300 pfu (radiation storm of 2 level). On January 26 proton enhancement continued in the level 1 of radiation storm. This proton increase did not come to the end as on January 27, after the first 2012 flare of the highest class (X1.7), the next enhancement had begun. The source of this flare was in the same active region AR11402 (N27W71) which by this time came to the western limb of the visible solar disk. At 17:27 UT on January 27 UT in this area the flare of the X1.7 class occurred and reached maximum at about 18:37 UT (peak time). Time of the CME lift-off was 18:28 UT, type II radio burst started at 18:30 UT. At this time a large proton increase for particles >10 MeV (800 pfu), >50 MeV and >100 MeV (11.9 pfu) occurred. For particles of >100 MeV, the event was close to background on the next day 28 January 2012. For particles with energies >10 MeV the event lasted much longer and finished on the 1st of February, 2012.

In figure 2 the data from the NMs South Pole B (SOPB), South Pole (SOPO) and sea level station Mirny (MRNY) with a time resolution of 30 minutes (baseline time interval is 17 – 18 UT) are plotted together with 5-minute GOES 13 data for protons with energies >10 MeV, >50 MeV, >100 MeV.

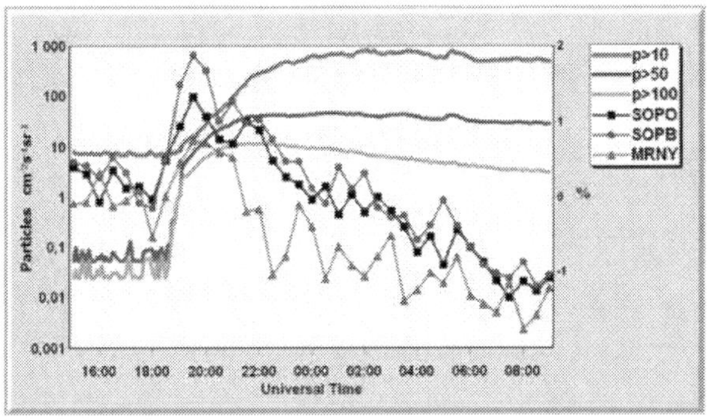

Figure 2. CR variations (right scale) recorded at NMs South Pole B (SOPB), South Pole (SOPO) and Mirny (MRNY) with 30-minute resolution and 5-minute data of proton fluxes recorded at GOES 13 on 27 January 2012, 1500 UT – 28 January 2012, 0900 UT.

A small increase of count rate (about 1%) with the onset around 18:25 UT was also observed at other high latitude ground level NMs: Thule (THUL), Inuvik (INVK), McMurdo (MCMU), Terre Adelie (TERA), Fort Smith (FSMT) (see figure 3 for FSMT and INVK). At all these stations an increase of NM count rates in percentages relatively base line reaches about 1-2% that corresponds to 8-10 standard sigma, so, this enhancement is statistically provided. Duration of GLE may be ranged from one hour to about 48 hours. In our case we see a significant difference in duration for southern NMs (figure 2) and northern ones (figure 3) that may tell about large north-south anisotropy in this event. It is not clear what is the reason of this north-south anisotropy and so long enhancement at the northern stations. This needs further investigations. But we can suppose that the signal that is seen by the NMs Fort Smith and Inuvik, may be caused by solar cosmic rays only during the initial phase.

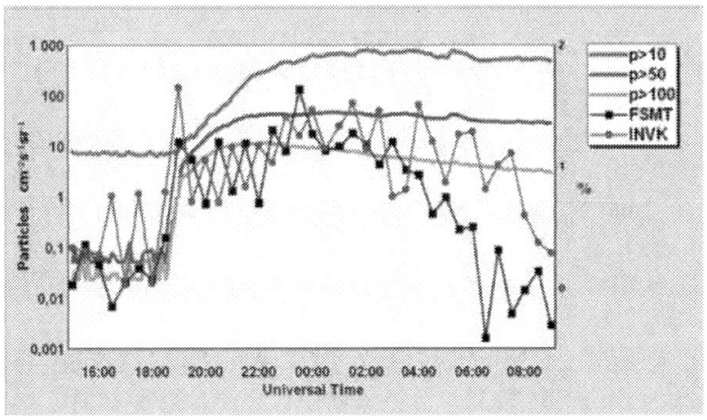

Figure 3. CR variations (right scale) recorded at NMs Fort Smith (FSMT) and Inuvik (INVK) with 30-minute resolution and 5-minute data of protons by the data of GOES 13 on 27 January 2012, 1500 UT – 28 January 2012, 0900 UT.

In figure 4 data from HEPAD for proton fluxes averaged by the energies within each HEPAD channel: Proton P8 (350 to 420 MeV), Proton P9 (420 to 510 MeV), Proton P10 (510 to 700 MeV) are presented with 10 minute resolution. For simplicity sake these channels are marked as 375, 465 and 605 MeV in figures 4, 6, 9. A coincidence in the onset of enhancements in both groups of data is evident. This completely satisfies criteria for GLE and allows us to consider the event on 27.01.2012 as a GLE one. However, it is worthy further studying and modeling [9, 10]: definition of asymptotic directions, integral and differential spectra and so on. Modeling by different groups of scientists carried, combine data of all ground level detectors and allows one to define spectra and angular distribution of SCR. In result, more definite conclusions may be obtained than those derived from the analysis of data from separate detectors. Unfortunately, nobody carries modeling of such small GLEs, since usually more powerful events are preferred.

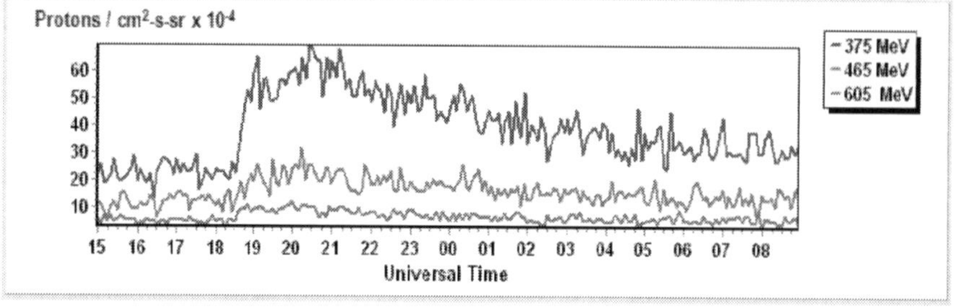

Figure 4. Average fluxes of 375, 465 and 605 MeV protons recorded on HEPAD on 27 January 2012, 1500 UT – 28 January 2012, 0900 UT.

4. Event on March 7, 2012

Next the powerful burst of the solar activity occurred in March 2012 when AR 11429 appeared on the visible side of solar disk, having complicated magnetic configuration and extending over a large area (www.swpc.noaa.gov). During the period from March 2 to March 8, 2012 this group produced 11 flares of class M and 2 flares of the highest class X. The peak of the flare activity fell on March 7, 2012. On this day there were two flares of class X: the first one, of X5.4 importance, began at 00:02 UT in AR11429 (N17E27) with peak time 0:24 UT. The second flare began at 01:05 UT in AR 11430 (N22E12) and ended at 01:30 UT. This X1.3 flare was associated to a fast CME of 1825 km/s velocity at 01:30 UT. The first solar flare (X5.4) was associated to an even faster CME at 00:24 UT (with a speed of 2684 km/s). As it was shown by Richardson et al. [11] the SEP (Solar Energetic Particle) event of March 7, 2012 is the result of the first flare-CME pair (e.g. X5.4 and 2684 km/s velocity of CME) and that the possible interaction among the two CMEs is excluded. A large and fast CME was observed starting from 0:24 UT and radio bursts of II and IV type started at 0:17 UT. As a result, on March 7 the new, more powerful proton event was associated with solar flare of X5.4 from AR 11429. Proton flux with energy >10 MeV started to grow at about 4:24 UT and reached 6000 pfu, and for energy >100 MeV - 70 pfu. For protons with energy >100 MeV, the event ended on March 11, 2012, for protons with energy >10 MeV, it proceeded much more long, gradually weakening. In figures 5 and 6 the behaviour of proton fluxes by the GOES data shows at the beginning the profile characteristics for an eastern source. The fluxes of low energy protons firstly increased very gradually (from 4:30 to 10:00 UT) and from 10:00 UT rapid growth of the proton fluxes is clearly visible. It coincides with the time interval between the arrival of two disturbances: the first one was a shock wave (defined by SSC) registered at 4:20 UT and associated with CME from March 4, 2012, and second one – a substantial increase of solar wind velocity (from 400 to 600 km/s in 10:00 – 16:00 UT interval) and associated with CME on March 5, 2012.

In figure 5 the NM data are presented from the stations: Tixie Bay (TXBY), Fort Smith (FSMT), South Pole B (SOPB), South Pole (SOPO) and Thule (THUL) at 30-minute resolution (baseline time interval is 23 – 24 UT). Also 5-minute proton data with the energies >10 MeV, >50 MeV, >100 MeV recorded at GOES 13, are shown. Small increase of CR intensity (about 2%), registered by ground level stations SOPO and SOPB at 8:00 -10:00 UT, is coincided with the increase of the protons at GOES13 and in the average fluxes of 375, 465 and 605 MeV protons from HEPAD. One can see that the behavior of count rate at subpolar NMs repeats the behavior of the high energy particle fluxes on HEPAD that confirms a contribution of solar CR in the ground level observations.

But still we cannot definitely refer this event to the GLE because of very complicated interplanetary situation at that time, as it seen in figure 7.

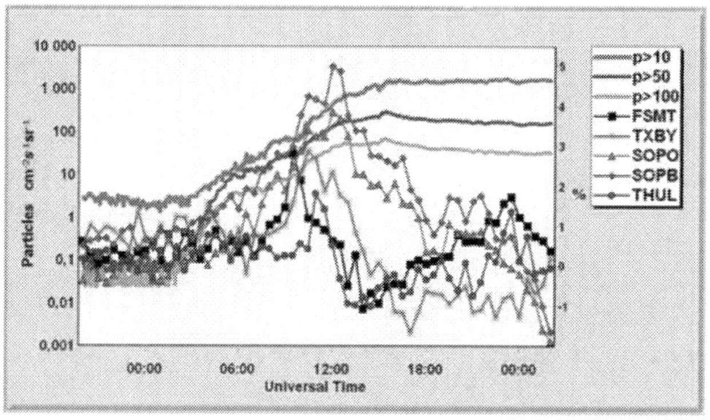

Figure 5. CR intensity recorded at NMs Tixie Bay (TXBY), Fort Smith (FSMT), South Pole B (SOPB), South Pole (SOPO) and Thule (THUL) with 30-minute resolution and 5-minute data of protons by the data of GOES 13 on 6 March 2012, 2000 UT – 8 March 2012, 0200 UT.

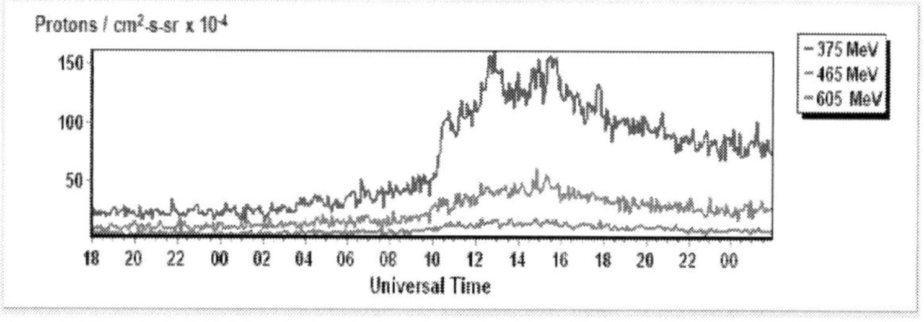

Figure 6. Average fluxes of 375, 465 and 605 MeV protons recorded on HEPAD on 6 March 2012, 1800 UT – 8 March 2012, 0200 UT.

These complications arose due to the solar ejections (halo CMEs) occurred on 4 – 5 of March in those AR that generated later the flares of X importance.

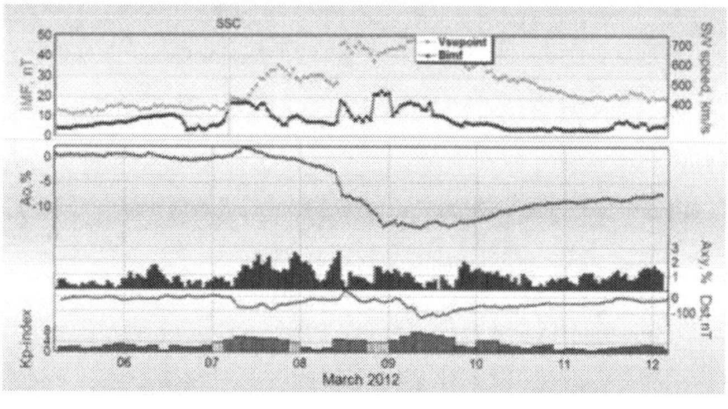

Figure 7. Behaviour of 10 GV CR density (A0) and equatorial component Axy of the vector CR anisotropy (middle panel), relevant parameters of the interplanetary magnetic field (IMF): IMF intensity and solar wind velocity (upper panel), Kp-index and Dst of geomagnetic field (lower panel) during the period March 5-12, 2012. Vertical line means the shock arrival (SSC).

At the beginning of March 7 (4:20 UT) the SSC assumingly associated with halo CME observed on March 4, arrived at Earth. After this the IMF intensity increased significantly (up to 20 nT) but solar wind velocity remained at not high level some time. It started to grow after 10:00 UT and reached value of 600 km/s in some hours. Perhaps this increase of velocity was caused by arrival of the other interplanetary disturbance associated with the halo CME on March 5 from the same AR11429. Speed of this CME was higher, than the speed of the previous one, and near Earth the second front caught up with the first that could favor to additional acceleration of energetic charged particles which were already observed near the Earth at this time. The possible enhancements of the solar particles of ~1 GeV is very difficult to select on the background of the strong CR modulation. At the same time geomagnetic storm also occurred. It should not be expected that this magnetospheric effect causes large variations of cutoff rigidities at high latitude stations, but asymptotic directions may be strongly changed, that creates additional problem for modeling of this event. Specific situation arose with two approaching fronts of interplanetary disturbances. It was difficult for particles to go out from this region between two fronts, but they could be accelerated and accumulated there.

We suggest that a contribution of SCR to the increase of count rate at some sea level NMs on March 7 exists, however model calculations are necessary for more certain conclusions.

5. Event on March 13, 2012

The situation on March 13 is much simpler than on March 7. The flare M7.9 was registered on March 13 at 17:12 UT in the AR11429 (N19W59) and reached peak time at 17:41 UT. The start of CME observations was at 17:36 UT, radio burst type II started at 17:15 UT and type IV – at 17:17 UT. Initial CME velocity was 1884 km/s.

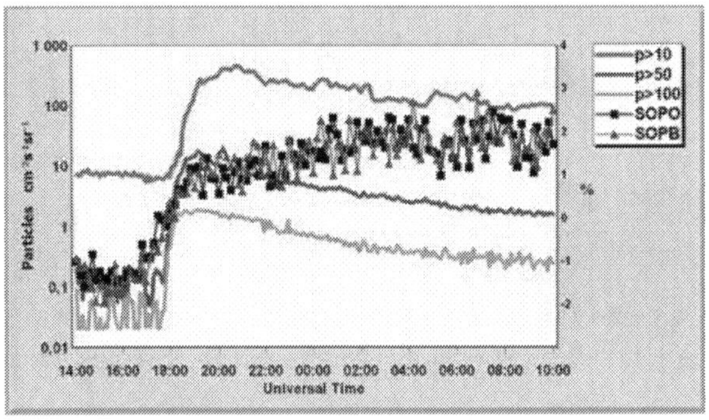

Figure 8. CR intensity recorded at NMs South Pole B (SOPB) and South Pole (SOPO) with 10-minute resolution and 5-minute data on the protons with the energy >10 MeV, >50 MeV, >100 MeV by the data on satellite GOES 13, for the time interval 13 March 2012, 1400 UT – 14 March 2012, 1000 UT.

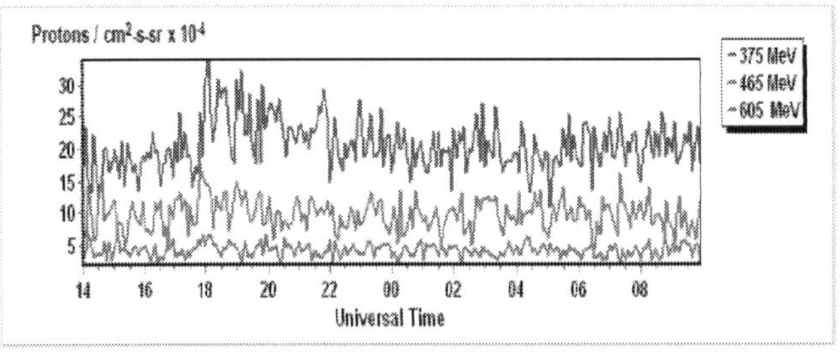

Figure 9. Average fluxes of 375, 465 and 605 MeV protons recorded on HEPAD on 13 March 2012, 1400UT – 14 March 2012, 1000 UT.

In figure 8 data of NMs at the stations (SOPB) and (SOPO) with 10-minute resolution (baseline time interval is 16 – 17 UT) and 5-minute data of protons with the energies >10 MeV, >50 MeV, >100 MeV from GOES 13 are plotted. One can see that growth of CR intensity at ground level stations began close to increase of particle fluxes on the GOES 13 that makes possible a contribution of SCR to the observed effect. A small increase was recorded also at sea level stations Terre Adelie (TERA) and Thule (THUL) with the onset time identical to the enhancement recorded in GOES 13. In figure 9 data of average fluxes of 375, 465 and 605 MeV protons on HEPAD are presented for this event. It is seen that the onset of SCR enhancement 17 UT is not in contradiction by observations of all detectors (GOES 13, HEPAD, NMs). But after the time of maximum proton fluxes we do not see a decrease in the count rate of the NMs.

Some difficulties of analysis for this event are caused by its occurrence in the recovery phase of Forbush effect (FE) which started on March 12. In figure 10 parameters of interplanetary space (IMF intensity and solar wind velocity) are plotted together with CR characteristics derived by GSM [12] and index of geomagnetic activity Kp. It is seen that after the disturbance arrival the solar wind velocity increased up to 800 km/s and planetary Kp index reached 7 (the level of strong magnetic storm). After the shock arrival a long and large (~5%) Forbush effect began which lasted about 2 days. At the same SSC moment a strong increasing of the equatorial anisotropy (up to ~3%) was observed, that occurs seldom and is a characteristic feature of this event. Apparently, because of the recovery phase of FE the CR intensity at polar stations continue to grow (figure 8) after the clear enhancement simultaneously with protons measured at GOES 13 and HEPAD.

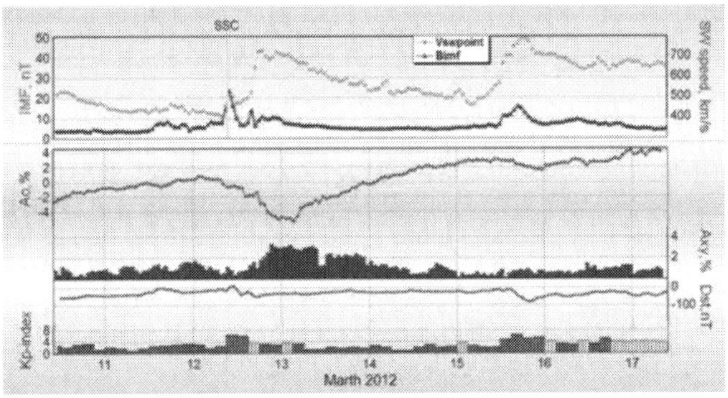

Figure 10. Behaviour of 10 GV CR density (A0) and equatorial component Axy of the vector CR anisotropy (middle panel), IMF intensity and solar wind velocity (upper panel), and Kp-index and Dst of geomagnetic activity (lower panel) during the March 10-17, 2012.

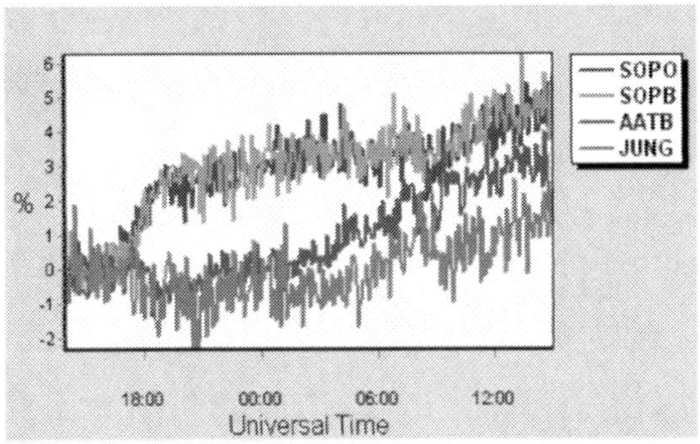

Figure 11. CR intensity recorded at NMs Alma-Ata B (AATB), Jungfraujoch (JUNG), South Pole B (SOPB) and South Pole (SOPO) on 13 March 2012, 1200 UT – 14 March 2012, 1500 UT.

If compare the CR intensity at South Pole (SOPO), South Pole B (SOPB) and mid latitude stations Alma-Ata B (AATB) and Jungfraujoch (JUNG), then, directly at the moment of the onset of proton enhancement the CR intensity at polar stations increases as compared with mid latitude stations (figure 9). Such discordance between data of two groups of NMs might be explained by the presence of low energy (solar) CR which are registered at polar stations.

6. Conclusions

All three events at the beginning of 2012 may be considered as candidate GLEs of SCR.

On January 27, 2012 an increase of the counting rate of ~2%, coincides in time with a proton increase at the integral intensity recorded onboard the GOES satellite (E>100 MeV) and also with increases of the average fluxes of 375, 465 and 605 MeV protons recorded on HEPAD. Such an increase is also registered on several sea level NMs that allow one to speak about GLE for this event.

On March 13, 2012 an increase at the counting rate similar in size to the event on January 27 is observed on several high-latitude NMs at the same time that an increase in GOES high energy protons, is presented. A GLE, possibly, was observed here. If there was no Forbush decrease at this time the situation, most probably, would been clearer.

The most difficult and most interesting situation that deserves a detailed analysis, modeling and further investigation, took place on March 7. In this event, the arrival of SCR to Earth coincided in time with large modulation effects in galactic CRs and with significant magnetospheric disturbance. Our analysis shows that in this event there is a possible influence of SCR on the counting rate of subpolar NMs.

Acknowledgments

Authors wish to acknowledge to all teams providing continued ground level CR monitoring (http://cr0.izmiran.rssi.ru/ThankYou/) and acknowledge the NMDB database (www.nmdb.eu), founded under the European Union's FP7 programme (contract no. 213007) for providing data. (www.nmdb.eu). This work was supported also by Kazakhstan Aerospace Committee under program 076 and under grant N0014/GFZ.

References

[1] Papaioannou A, Souvatzoglou G, Paschalis P, Gerontidou M, Mavromichalaki H 2014 *Solar Phys.* **289** 423 DOI 10.1007/s11207-013-0336-2
[2] Takao K, Bieber J, Clem1 J, Evenson P, Gaisser T, Pyle R, Tilav S 2013 *Proc. 33nd Int. Cosmic Ray Conf. (Rio de Janeiro)* SH21A-2183
[3] Augusto C R A, Kopenkin V, Navia C E , Tsui K H, Feliciano A C, Pinto A C , de Oliveira M N et al. 2013 *Proc.33nd Int. Cosmic Ray Conf. (Rio de Janeiro)*
[4] Li C, Firoz Kazi A, Sun Ling P, Miroshnichenko L 2013 *Astrophys. J.* **770** 34
[5] Thakur N, Gopalswamy N, Xie H, Mäkelä P, Yashiro S, Akiyama S, and Davila J M 2014 *Astrophys. J. Lett.* **790** 1
[6] Gopalswamy N, Xie H, Akiyama S, Pertti A, Mäkelä P and Yashiro S 2014 *Earth, Planets and Space* **66** 104
[7] Sharma A, Verma S R 2013 *International J. of Astronomy and Astrophys.* **3** 212
[8] Chapman G A, de Toma G, Cookson A M 2013 *Solar Phys.* **289** 3961
[9] Plainaki C, Belov A, Eroshenko E, Mavromichalaki H, Yanke V 2007 *J. Geophys. Res.* **112** No A4 CiteID A04102
[10] Plainaki C, Belov A, Eroshenko E, Mavromichalaki H, Yanke V 2008 *Adv. Space Res.* **41** doi:10.1016/j.asr.2008.07.011
[11] Richardson I G et al. 2014 *Solar Phys.* **289** 3059
[12] Belov A et al. 2005 *J. Geophys. Res.* **110** A09S20 doi:10.1029/2005JA011067

Magnetospheric transmissivity for cosmic rays during selected recent events with interplanetary/geomagnetic disturbances

I Parnahaj[1], P Bobík[2], K Kudela[2]

[1]Faculty of Science, Pavol Jozef Šafárik University, Košice, Slovakia
[2]Institute of Experimental Physics, Slovak Academy of Science, Košice, Slovakia

E-mail: kkudela@kosice.upjs.sk

Abstract. The variability of cosmic rays (CRs) observed at selected European neutron monitors (NMs) around moderate geomagnetic disturbances, namely during the intervals (a) DOY 49-51 in 2014, (b) DOY 58-59 in 2014, (c) DOY 238-240 in 2014 and (d) DOY 6-8 in 2015 is discussed. Assuming the primary spectra of the CREME96 model, the yield function and geomagnetic transmissivity changes provided by the Tsyganenko96 model, the expected increases at the mid-latitude station Lomnický štít are compared with the observed ones. The examples stress the importance of including anisotropy of the CR flux in interplanetary space, the use of other geomagnetic field models and other yield functions to the computations in future analysis.

1. Introduction

During geomagnetic field disturbances the transmissivity of the magnetosphere for the cosmic rays is changing (e.g. [1,2] and references therein). The structure of forbidden and allowed trajectories as well as the asymptotic directions for the allowed ones obtained by numerical tracing of CR particle trajectories differ for different geomagnetic field models. There exist various geomagnetic field models with external current systems (e.g. [3]). Checking the transmissivity improvement during geomagnetic storms is not easy, since in many cases (not in all, e.g. [4]) such events are accompanied by Forbush decreases (FD, studied long time since paper [5]) which mask the geomagnetic effect.

The recent period is characterized by not many irregular changes in CR intensity observed by NMs. Four events with increases in the CR flux observed by selected European NMs (data from PIs and/or from http://nmdb.eu) during the Dst depressions not accompanied by FDs are used to compare the expected effects of the magnetospheric transmissivity change with that observed by NM measurements. The Dst index represents the axially symmetric disturbed magnetic field at the dipole equator on the Earth's surface [12]. During strong magnetic storms the Dst is reaching several hundreds of nT (negative values). For that we use simplified assumptions: neglected anisotropy; vertical geomagnetic cut-off rigidities only; primary CR spectra approximated by the CREME96 model [6]; Tsyganenko96 and 05 geomagnetic field models [7,8] and the yield function given by [9]. CREME96 is a code providing (among other components of near-Earth ionizing radiation) the primary CR energy spectra for various elements and various phases of the solar cycle activity. There are several studies of changes of cut-offs during geomagnetic and interplanetary disturbances (e.g. [1,2]). Review of geomagnetic models for CR trajectory computations is in paper [10] or in monograph [11].

2. Description of events

Figures 1a - 1d show four intervals with variations of CR records at selected european NMs with different vertical geomagnetic cut-offs: Rome, R_{eff} = 6.27 GV; Jungfraujoch 1, R_{eff} = 4.5 GV; Lomnický štít, R_{eff} = 3.84 GV and Oulu, R_{eff} = 0.8 GV. Here the lowest Dst values were around < - 60 nT whereas the IMF was directed southward. The hourly values of the count rates of each NM station are normalized to the mean of the first twelve hours of the event. Along with the Dst index (available at http://wdc.kugi.kyoto-u.ac.jp/) the hourly values of the solar wind pressure (p), the solar wind density and velocity (NSW, VSW) as well as the magnitude of the IMF (B) and its latitude in GMS coordinates are plotted (data downloaded from http://omniweb.gsfc.nasa.gov NASA site).

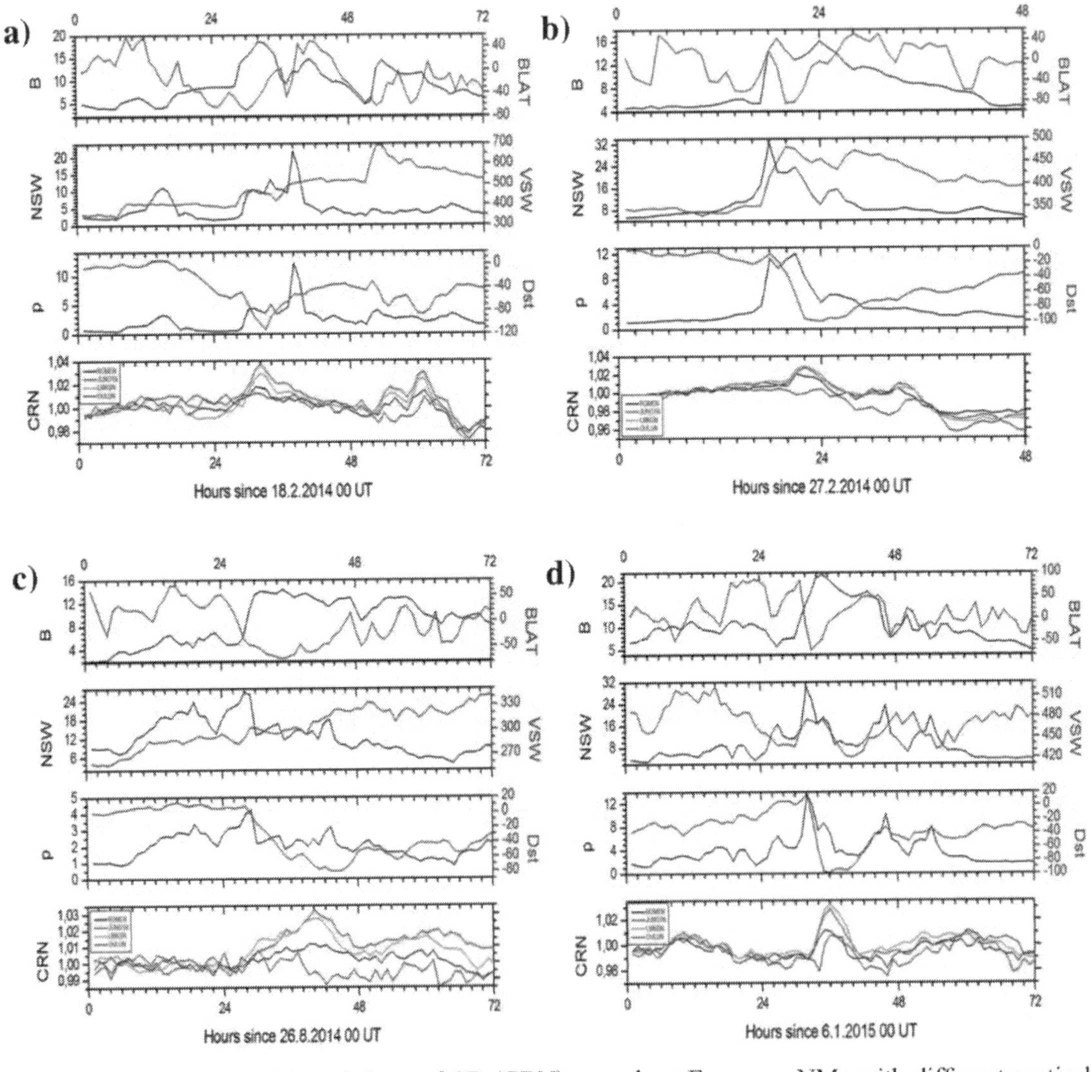

Figure 1. Four events with variations of CR (CRN) records at European NMs with different vertical geomagnetic cut-offs (Rome, Jungfraujoch 1, Lomnický štít, Oulu). Event 1 (upper left panel, a): DOY 49-51 in 2014; event 2 (upper right panel, b): DOY 58-59 in 2014; event 3 (lower left panel, c): DOY 238-240 in 2014; event 4 (lower right panel, d): DOY 6-8 in 2015. B is in nT, BLAT in degrees (GSM system), NSW in cm^{-3}, p in nPa, Dst in nT, VSW in $km.s^{-1}$. The values CRN are normalized to the mean of the first twelve hours of the event. Local peaks in time profile of high mountain stations count rate (red and green line in lowest parts of the panels) are supposed to correspond to Dst decreases.

3. Discussion

In all four events the CR increases during the Dst depression are most pronounced at the two mid-latitude high mountain stations, Jungfraujoch 1 and Lomnický štít. The smallest increases (in event 3 even a decrease) are observed at the low cut-off station Oulu. At Oulu the decrease of geomagnetic cut-off is not producing any strong increase in the CR flux due to an improvement of the geomagnetic transmissivity, since it is close to the atmospheric cut-off.

The expected/computed CRN profile was based on simplified assumptions: the energy spectra of primary CR is not changing during the event and it is estimated using the CREME96 model [6]; the yield function is taken from the recent paper [9]; the geomagnetic field model Tsyganenko96 [7] and for one event also Tsyganenko05 [8] is used. A basic description of the CREME96 code can be found at https://creme.isde.vanderbilt.edu/CREME-MC/help/what-is-creme96. We downloaded the differential energy spectra for protons for the period near solar activity maximum. The vertical geomagnetic cut-off rigidities have been computed by the tool available at http://www.geomagsphere.org/geomag/ [9]. The relative changes in the count rate profiles due to the variability in the geomagnetic cut-off values were estimated by integrating the product of primary CR flux and the yield function over the rigidity range > R_{eff} when R_{eff} is changing.

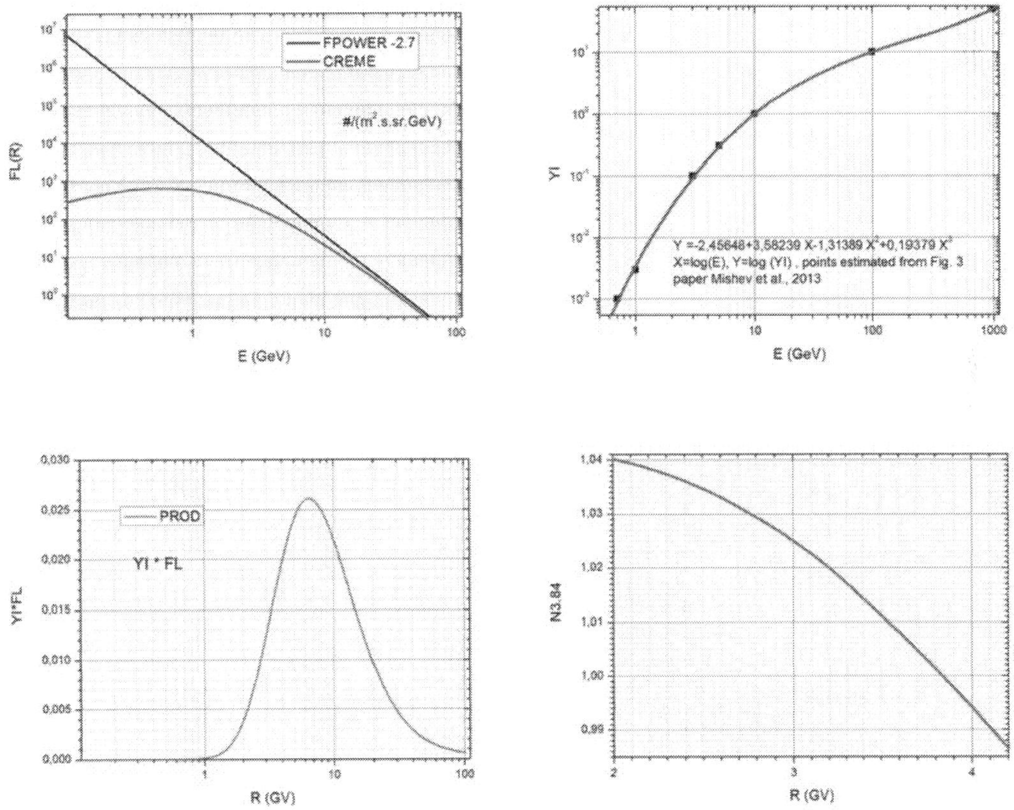

Figure 2. Upper left panel: FL (in #/(m^2.s.sr.GeV)) is differential energy spectra of protons from CREME 96 (for solar maximum) using [6]. Upper right panel: approximated yield function from figure 3 in paper [9] (in arbitrary units). Lower left panel: product of differential spectra and the yield function (both recalculated to rigidity scale). Lower right panel: integral of the function plotted in lower left panel (integration above the given rigidity R, and normalized to unity for Lomnický štít position with geomagnetic vertical nominal cut-off rigidity 3.84 GV).

Using the procedure illustrated in Figure 2, a comparison of the estimated and measured CR profiles at NM Lomnický štít is shown in Figure 3 for all four events.

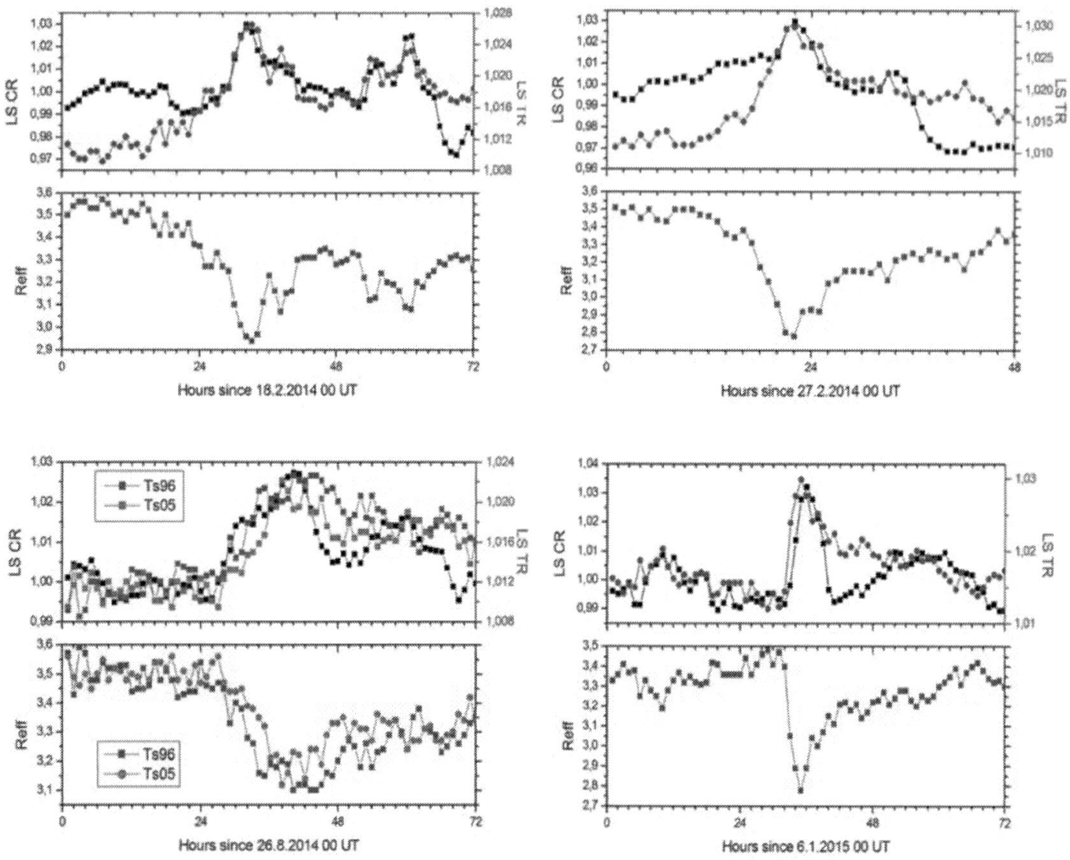

Figure 3. Effective cut-off rigidity Reff computed on hourly basis and comparison of measured (LSCR) and expected (LSTR, different scale, on the right hand side) normalized count rate at Lomnický štít NM for events 1 to 4 of Figure 1. In all cases the Tsyganenko96 model is used. For one event (3) in addition a comparison with model Tsyganenko05 is given.

In all four events the time profile of the NM count rate at Lomnický štít is similar to that estimated due to the geomagnetic transmissivity change characterized by the effective cut-off rigidity. The main CR increases observed in the events 1, 2 and 4 correspond to times of improved geomagnetic transmissivity (depression of Reff). Event 3 is more complicated and corresponds to a slower decrease of the Dst index in comparison with the other three events. For that event the use of the Tsyganenko05 model is matching slightly better the measured count rate during the recovery phase of the geomagnetic disturbance than model Tsyganenko96.

Although the measured profiles (LSCR) correspond qualitatively to the expected (LSTR) ones based on Tsyganenko96 model and the yield function used, the amplitudes of the observed increases is usually higher than those expected due to the simplifications used here.

4. Summary

Four events during the past two years when geomagnetic disturbances were not accompanied by Forbush decreases, observed at four stations with different geomagnetic cut-off, have been checked for the effect of geomagnetic transmissivity changes. The time profiles of NM count rates around the time of the minimum Dst for the positions where the geomagnetic cut-off is clearly above the atmospheric one, shows *qualitative agreement* with the profiles obtained from computations of the changing effective cut-off rigidity. Around the time of the minimum Dst, rather strong increases are observed at two mid-latitude high mountain stations. The values of the observed increases at one of them (Lomnický štít) are usually higher than those expected by computations based on simplified assumptions of primary spectra, yield function and one geomagnetic field model.

The examples stress the importance of including anisotropy of CR flux in interplanetary space, other geomagnetic field models and various yield functions to the computations in future analysis.

Acknowledgement
We acknowledge CREME96 project; Ch. Steigies and colleagues for keeping and updating http://nmdb.eu data base; PIs of NM Oulu, Jungfraujoch and Rome for the preliminary data. VEGA grant agency project 2/0040/13 and VVGS-PF-2014-445 are acknowledged for support. The authors acknowledge two anonymous reviewers for their comments and corrections.

References
[1] Tyasto M I, Danilova O A, Ptitsyna N G et al. 2004 Cosmic Ray Cutoff Rigidities during Geomagnetic Storms: A Comparison of Magnetospheric Models *Geomagn. Aeronomy* **44** pp270-276
[2] Kudela K, Bučík R and Bobík P 2008 On transmissivity of low energy cosmic rays in disturbed magnetosphere *Adv. Space Res.* **42** 1300-1306
[3] Desorgher L et al. 2009 Comparison of Earth's magnetospheric magnetic field models in the context of cosmic ray physics *Acta Geophysica* **57** Issue 1 pp 75-87
[4] Kudela K, Brenkus R 2004 Cosmic ray decreases and geomagnetic activity: list of events 1982–2002 *J. Atm. Solar Terr. Phys.* **66** 1121–1126
[5] Forbush S 1937 On the effects in cosmic-ray intensity observed during the recent magnetic storm *Phys. Rev.* **51**:1108-09
[6] Tylka A J, Adams J H Jr., Boberg P R et al. 1997 CREME96: A Revision of the Cosmic Ray Effects on Micro-Electronics Code, *IEEE Trans. Nucl. Sci.* Vol **44** No 6
[7] Tsyganenko N A 1995 Modeling the Earth's Magnetospheric Magnetic Field Confined Within a Realistic Magnetopause *J. Geophys. Res.* **100** 5599-5612
[8] Tsyganenko N A and Sitnov M I 2005 Modeling the dynamics of the inner magnetosphere during strong geomagnetic storms *J. Geophys. Res.* **110** A3 10.1029/2004JA010798
[9] Mishev A L, Usoskin I G and Kovaltsov G A, 2013 Neutron monitor yield function: New improved computations *Journal of Geophysical Research: Space Physics* Volume **118** Issue 6 pages 2783–2788
[10] Smart D F, Shea M A and Flückiger E 2000 Magnetospheric models and trajectory computations *Space Sci. Reviews* **93** 1-2 305-333
[11] Dorman L I 2009 *Cosmic Rays in Magnetospheres of the Earth and other Planets* (Springer)
[12] http://wdc.kugi.kyoto-u.ac.jp/dstdir/dst2/onDstindex.html

Cosmic ray intensity for about five solar cycles

M Storini[1], F Signoretti[1], F Re[1], P Diego[1], M F Marcucci[1], M Laurenza[1], S. Massetti[1] and M Parisi[2]

[1] IAPS/INAF, Via del Fosso del Cavaliere, 100 – 00133 Roma, Italy
[2] Dipartimento di Matematica e Fisica/UNIRomaTre, Via della Vasca Navale, 84 – 00146 Roma, Italy

E-mail: storini@iaps.inaf.it or monica.laurenza@iaps.inaf.it

Abstract. Continuous records of the cosmic ray nucleonic component have been achieved at Rome (SVIRCO Group) by using data from different types and locations of neutron monitors (first at La Sapienza University: 41.90°N, 12.52°E, altitude about 60 m a.s.l., and then at Roma Tre University: 41.86°N, 12.47°E, about sea level). The normalized data, covering the whole period from July 1957 to June 2014, are available to the scientific community by simple request. Here we illustrate some useful results derived from them.

1. Introduction

The Cosmic Ray (CR) Physics has been developed until about 1955 by using terrestrial measurements in specific research projects (mainly mini-networks of ionization chambers, balloons, ship and aircraft surveys). Only after the start of the Geophysical International Year (1957-1958) and the Space Age the field was world-wide faced. The last five decades gave considerable progress towards the understanding of the solar wind/CR relationship through the study of solar wind macrostructures and wind turbulence in the heliosphere. On the other hand, with the improvement of our knowledge on the terrestrial atmosphere and magnetospheric physics (thanks to satellite measurements), it was possible to quantitatively describe atmospheric and geomagnetic effects on CRs arriving at the different locations of the Earth.

A remarkable synergy between CR studies and space measurements led to investigate in detail CR variations such as the 11-yr (22-yr) modulation, 27-day recurrences, Forbush decreases, CR anisotropies, solar energetic particle (the so-called SEP/GLE) events. The possibility to extend the research at low CR energies in the interplanetary space not only helped the investigation of SEP events but also brought to the discovery of relevant heliospheric phenomena related to CR physics (e.g. charged particle/shock interactions, creation/propagation of pick-up ions and anomalous CRs, corotating ions, 3D-transport of charged particles). Nevertheless, along with the great amount of information obtained from space missions, ground-based instrumentations (ionization chambers, Geiger-Muller counters, neutron monitors, muon telescopes, air shower arrays,...) continue to be required for many research works, particularly for the new scientific branches of the dynamical heliospheric environment. There exist many links between such branches and CRs; for example, it is possible to use CR signatures to forecast/nowcast Space Weather events, to rebuilt Space Climate data series or to help International Alert Services for the so-called dangerous events (including the ones directly connected with the CR interaction with matter such as the biological tissues).

In other words, research during this century still needs continuous CR records but with high-temporal resolution and low statistical errors (as underlined by [1]). In this context, efforts have been made at Rome (Italy) to maintain a *long data series of homogenous CR intensities* derived from measurements performed with different types and locations of neutron monitors (see section 2). The neutron monitor (NM) data has been employed over the years to investigate several research topics by using different mathematical tools, such as the Descriptive Statistics (DS [2]; subsection 2.1), the Wavelet Technique (WT [3]; subsection 2.2), or the Empirical Mode Decomposition (EMD [4]; subsection 2.3). We discuss some results (section 3) to emphasize the relevance of a homogeneous CR dataset for Space Research.

2. SVIRCO Observatory and Terrestrial Physics Laboratory of Rome

A research team, aiming to investigate cosmic ray intensity variations (Italian acronym: SVIRCO), was formed at the start of the second half of the last century in Rome (see [5] for details). The SVIRCO station (now observatory) for continuous measurements of the nucleonic component was opened on 1 July 1957 with the so-called IGY neutron monitor (figure 1, left upper panel) at the Physics Department of La Sapienza University (RM1: 41.90°N, 12.52°E, altitude 60 m a.s.l.). It was replaced with a NM-64 type during 1966 (figure 1, middle upper panel). In May 1997 the NM-64 was moved into the Physics Department of Roma Tre University (RM3: 41.86°N, 12.47°E, altitude about s.l.; see right upper panel of figure 1). Since then it has been continuously running at the new location (http://webusers.fis.uniroma3.it/svirco/).

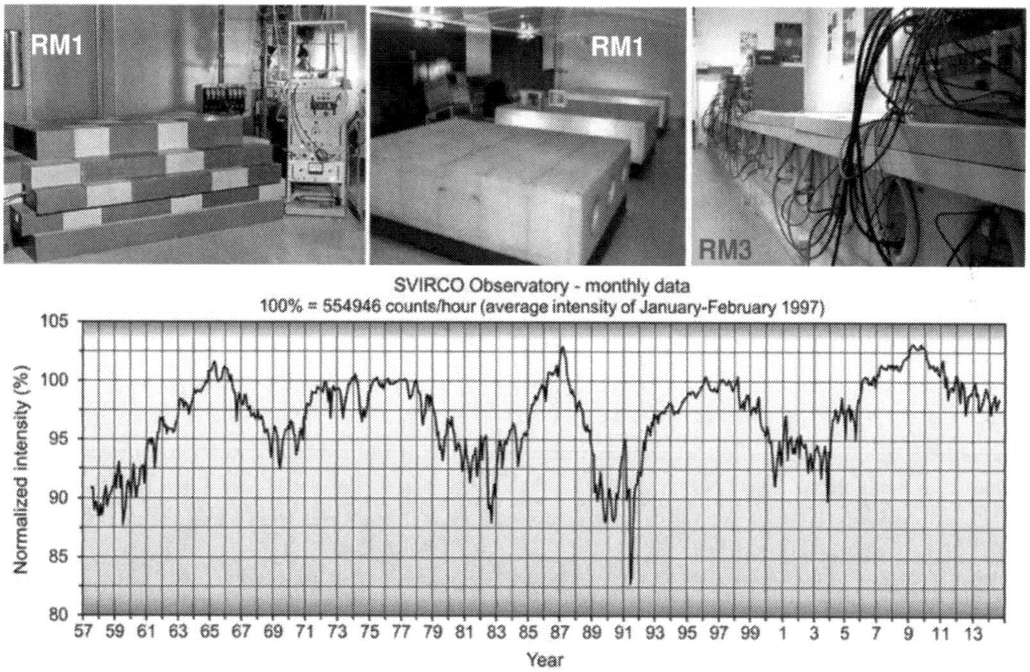

Figure 1. *Upper panel:* La Sapienza University (RM1: IGY detector, left; NM-64, center) and Roma Tre University (RM3: NM-64, right). *Lower panel:* Monthly averages of the nucleonic component as derived from Rome experimental data for the July 1957 to June 2014 time interval.

Different detector configurations were used over the years (see table 1 reported by [5]).

Annual data reports are made available to the scientific community by using revised hourly, daily and monthly records (e.g. [6] and references therein). Moreover, monthly prompt reports with 5-min data are published since 2003. Also efforts to evaluate the diurnal wave at Rome are performed and results are published as soon as they are ready (see [7] as an example).

Particular attention was paid to create a homogeneous dataset on hourly, daily and monthly basis (we will call it RM set) for long term studies. As an example, the lower panel of figure 1 shows the monthly averages obtained for July 1957 - June 2014.

2.1. DS of Rome CR data

Descriptive Statistics (DS, [2]) was used during the nineties to investigate general aspects of CR modulation. In fact, DS allows to characterize a data set in general terms such as mean, variance, standard deviation, extreme values, skewness, kurtosis, and so on. The monthly means and the associated parameters, derived from the homogenous RM set, allowed correctly to follow the CR modulation from one solar activity cycle to the next and to identify difference between even and odd cycles. It was also possible to confirm the existence of the Gnevyshev Gap (identified in the polar neutron monitor records by [8]) in the time history of a middle latitude measurement site [9].

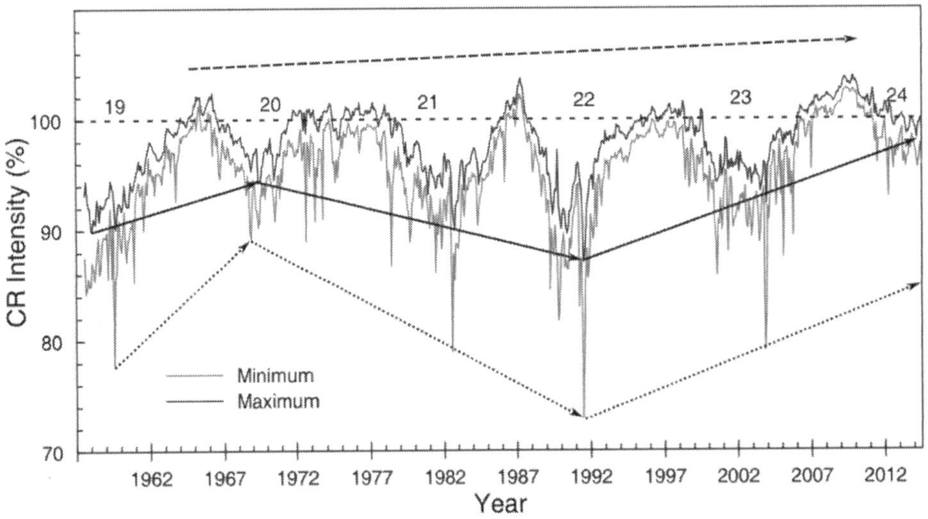

Figure 2. Monthly extreme values of the nucleonic component of galactic cosmic rays derived from the RM set on daily basis. See the text for arrows meaning.

The lower panel of figure 1 (were the 100% of intensity was arbitrary fixed as the average value derived from January-February 1997 pressure corrected data) shows that the CR modulation level is practically the same during the transition from an even to an odd solar activity cycle (i.e. 20/21 and 22/23). On the contrary, transitions from odd to even cycles demonstrate a tendency for decreasing levels of modulation (19/20 → 21/22 → 23/24). Such features are better seen in figure 2 where extreme daily values inside each monthly data subset were selected. Briefly, DS analysis suggests:

- equivalent intensity during the plateau phases of CR modulation.
- an effective residual difference between maximum and minimum daily intensity levels.
- a tendency for a slow increase of RM intensity at the end of odd modulation cycles, from cycle 19 to 23 (see dashed line).
- periodicities higher than the one of the Schwabe cycle (as continuous and dotted arrows underline).

The study of solar periodicity transmission in the heliospheric environment deserves special attention. From such kind of study it is possible not only to separate fundamental periods from the harmonics but also to investigate phenomena associated with CR modulation behaviour. Certainly, more cycles covered by continuous CR records will help to decide if Sun exhibits a quasi-periodic (stochastic) or intermittent (deterministic chaos) behaviour, but presently some insight into the matter can at least be reached.

2.2. WT on Rome CR data

The Wavelet Transform (WT, [3]) is generally used to investigate the time evolution of periodic or quasi-periodic trends of experimental data sets (e.g. [10] and [11] for details). It allows to analyze non-stationary processes containing multi-scale features. In fact, the technique is able to determine both the dominant variability modes and how those modes vary in time, by decomposing a time series into the time-frequency space.

The possibility to evaluate a local spectrum by means of the WT distinguishes the wavelet analysis from the different Fourier approaches or mathematical filtering techniques. Main periodicities such as quasi-annual, quasi-biennial, about a decade and about two decades were identified in the RM set of monthly averages covering 1957-2004 and discussed within the heliospheric context (see, among others, [12] and [13]).

Presently, it is very interesting to study the 5-minute averages of the RM dataset. The analysis of the 10 September 2005 Forbush decrease (FD) has suggested that such kind of fine data resolution contains a signal for interplanetary storms approaching the Earth up to 9 hours before the onset of the main phase of the event [14]. This is a feature related to the CR 3D-propagation in the inner heliosphere (as underlined in the last section of this paper), that can be easily observed by analysing the time behaviour of the 8-h CR periodicity derived from data recorded by a single detector with near equatorial particle asymptotic directions and the systematic use of the WT code (available at http://paos.colorado.edu/research/wavelet). The exploitation of this method, described in detail by [14], should be useful for Space Weather purposes.

Also Space Climate can take advantage from long time series of CR measurements. For example, to create lists of homogenous interplanetary plasma streams it is possible to apply the WT code on the 5-minute averages of the RM set and search for periodicities in the range 24 h to 120 h. Figure 3 exemplifies in its upper panel (including four graphs) results for two consecutive transient solar wind streams (**TS-1** and **TS-2**) observed during May 2005. They produce two intense Forbush decreases (**FD-1** and **FD-2,** of about 4 % each one). Three CR periodicities (24 h, 55 h and 74 h) were identified evaluating significant peaks in the global WT power spectra against a white noise background at the 95% statistical confidence level.

The fourth graph of the upper panel of Figure 3 illustrates for each periodicity the time history of the power/noise signal (R). Assuming a ratio $R \geq 5$ it is possible to observe that such signal starts to be significant before the beginning of the CR event and cover at least the complete Forbush decrease phenomenon. This is particularly true if the high periodicities (55 h and 74 h) are considered. It is interesting to notice that for the 74 h periodicity the R trend never decreases to no significant levels.

Figure 3 also reports in its lower panel results obtained by analyzing the 21 January – 8 February 2000 period, characterized by corotanting plasma streams (**CS-1** and **CS-2**) in the near Earth space. In this case the solar wind perturbations are causing small or moderate CR decreases (**D-1** and **D-2**); only the 24 h and 36 h periodicities were singled out, imposing the above significance level for peaks in the global WT power spectra.

The range of variability for the power/noise signals (fourth graph of the lower panel of figure 3) resulted to be small for corotanting streams if compared with those obtained for transient streams (fourth graph of the upper panel). Setting again the threshold of $R \geq 5$ for the power/noise of periodicities only the 24 h signal is significant, even before the stream arrival in the near Earth space and its time history is consistent with past results for the diurnal wave [15]. In fact, during the transit of the trailing edge of the corotanting stream the periodicity signal went to zero significance (see the arrow shown in the fourth graph of the lower panel).

Comparing the R time history of May 2005 with the one of January-February 2000 for the identified periodicities, we conclude that transient and corotating stream effects on CRs are very different and can be inferred from CR records.

The described features should help to discriminate the passage in the near-Earth space of transient from corotating solar-wind streams, particularly when no interplanetary data are available.

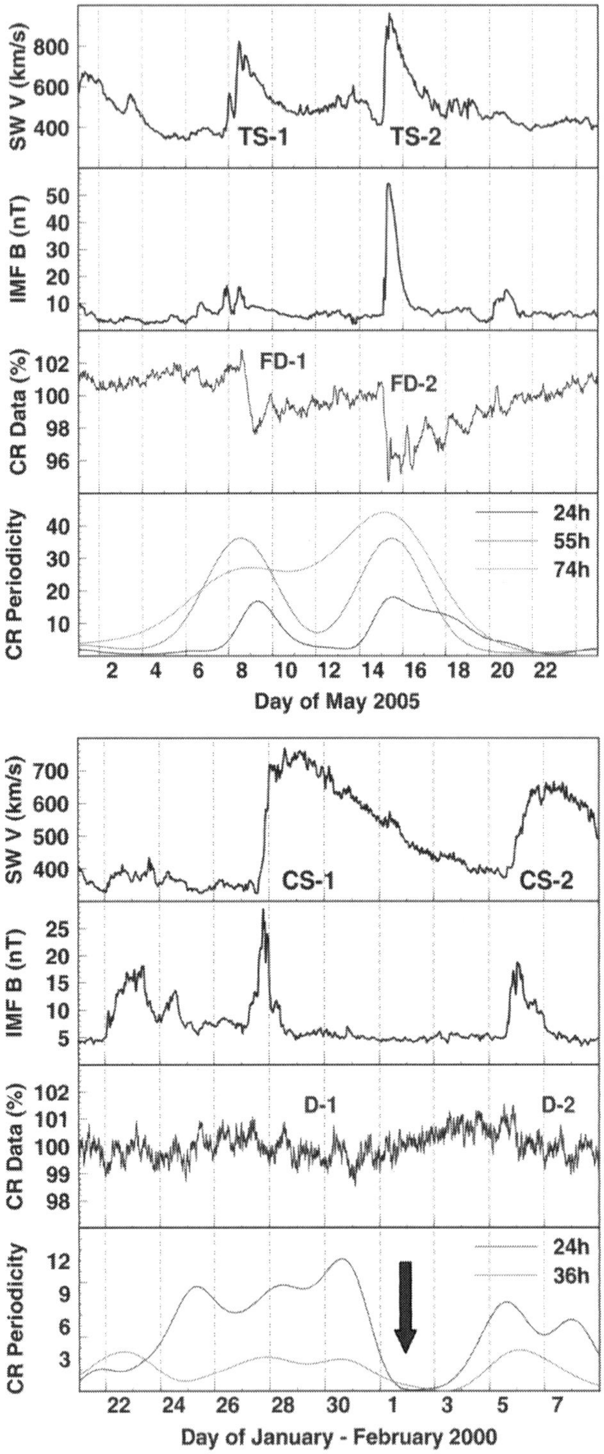

Figure 3. Power to noise of CR periodicity, percentile of the CR variability, interplanetary magnetic field (IMF) intensity B and solar wind (SW) velocity V (as measured in the near-Earth space: http//omniweb.gsfc.nasa.gov/) for transient (*upper panel*) and corotating (*lower panel*) interplanetary perturbations.

2.3. *EMD for Rome CR data*

The variability of the monthly RM set, covering 1957-2004, was also analyzed through the Empirical Mode Decomposition (EMD, [4]) analysis, which is a proper method for the description of non-linear and non-stationary processes containing multiscale features. It is a mathematical tool able to split a dataset into a finite number of components (called intrinsic mode functions), representing C_j oscillatory modes in the time domain, with the signal described by a time dependent amplitude and phase function.

We identified the significant periodicities (through the C_j modes) in the RM set (and their relative amplitude) by deriving the eigenmodes from the signal without a priori conditions. Several interesting results, able to connect CR modulation phenomena with the heliospheric environment, were obtained (in particular we highlight results discussed in papers [16] to [20]; see figure 2 reported by [18] for C_j significance test). Briefly:

(i) the C_3 (typical period: 1.02 ± 0.02 yr) mode is related to the annual variation.

(ii) the C_4 (1.6 ± 0.1 yr) mode characterizes Quasi Biennial Oscillations (QBOs) during 1965-1976 and 1987-1995 (i.e. during even solar cycles) and the C_5 (typical period: 2.7 ± 0.1 yr) mode during 1978-1986 and 1995-2004 (odd solar cycles).

(iii) the QBO contribution (obtained by summing up modes C_4 and C_5) to the global signal is not the same during successive solar cycles, although the amplitude ratio QBO/11 yr is at least 30 % (~ 11 yr mode identified as C_7: 10.4 ± 0.4 yr).

(iv) QBOs of CR data are delayed with respect to sunspot activity but the time lag (τ) is shorter for the positive polarity states of the heliosphere (the so-called $A > 0$ periods of even sunspot cycles; $\tau < 5$ months) than for the negative ones ($A < 0$ periods of odd cycles; $\tau > 5$ months).

(v) the superposition of the ~ 11 yr and QBO modes well reproduces the experimental data set during the maximum phase of the Schwabe cycles as can be seen in figure 4 (compare the magenta with the green curve). It also seems to account for the Gnevyshev Gap phenomenon.

(vi) the observed step-like (medium term) CR modulation can be partly described by the superposition of QBOs to the long-term trend of ~ 11 yr.

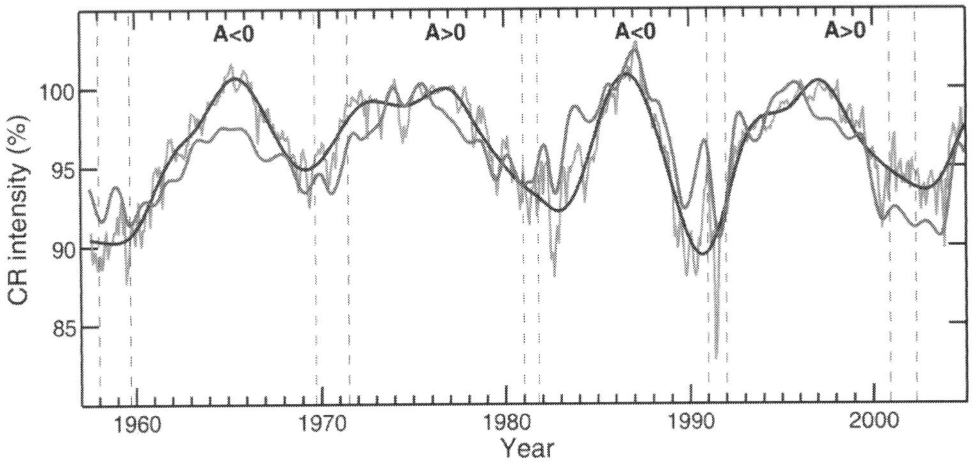

Figure 4. CR intensity as derived from experimental data recorded at Rome (green line), superposition of the ~ 11 yr and QBOs (magenta line) and superposition of the C_6, C_7, and ~ 22 yr modes (blue line) computed by using EMD. Vertical dashed lines delimit the short time period when there was no well-defined heliomagnetic field polarity.

(vii) the C_6 (5.6 ± 0.5 yr) mode is correlated with the heliospheric current sheet (HCS) latitudinal excursion (from the equatorial to polar solar latitudes and vice versa) and its role in charged particle drift phenomena. Nevertheless, a threshold for the HCS tilt angle there exists: over the threshold value the CR flux recorded at a measurement point is unaffected, below the threshold the CR flux is strongly modulated by the tilt angle variations.

(viii) the C_6 mode also accounts for the flat/peaked CR maximum during even/odd cycles (compare the blue with the green curves reported in figure 4, where a good agreement between experimental data and data reconstruction including C_6 is obtained).

The detailed study of C_1 and C_2 modes will be performed in the future, as the C_8 (or higher) mode which requires a CR set covering more sunspot cycles of data.

3. Discussion and conclusion

Space Weather/Climate investigations require complete data series to extract correct features able to characterize phenomena under study. Often proxy data are used for filling procedures aiming to avoid data gaps in the investigated dataset. Among the proxy data used for solar-terrestrial relation studies, CR intensity records were the ones often tested in the past. They resulted to be a relevant tool in such kind of procedures, provided that homogeneous CR datasets are used.

Some results derived from the use of 5-minute, daily and monthly CR intensities of Rome NM are illustrated in section 2. Here we briefly discuss aspects of the heliospheric environment at the origin of the reported CR modulation phenomena.

Inhomogeneous regions of the interplanetary space, such as those present during traveling interplanetary perturbations generated by solar flares and CME ejections are able to change CR paths in the heliosphere; signatures of the moving 3-D structure of transient plasma streams are transferred to the near-Earth space before the stream arrival by the relativistic CRs. Hence, promising results can be obtained by using CR records in forecasting codes, not only by using contemporary hourly data from several CR detectors but also from a single detector if fine time resolution records are used.

Moreover, the CR paths inside/outside the complex plasma structures generate intense periodicities in the terrestrial CR records, which are not present in a quasi-stationary inner heliosphere. This is the case for corotating plasma streams traveling from the Sun to the Earth with stable (or slowly varying) macro structures, able to induce practically only azimuthal CR variability.

The signatures identified by using data with fine time resolution help to discriminate transient from corotating solar-wind streams and to build up lists of homogenous type of interplanetary perturbations even when no in situ data are available.

Medium- and long-term characteristics of CR modulation can be derived from a detailed study of CR periodicities with an appropriate mathematical tool. The use of the EMD on CR records clarified the relevance of QBOs, which are together with the one of the Schwabe cycle the outstanding variabilities present in CR datasets. QBOs seem also to be at the origin of the Gnevyshev Gap and CR step-like phenomena. Moreover, a relationship between the HCS position and the CR flux reaching the terrestrial environment was found.

Due to space limitations we would like only to remark that the obtained results were strengthened by including in most of the investigations other NM data series such as those recorded at:

- Climax and Huancayo/Haleakala: [12-13] and [18-20];
- Calgary, Climax, Lomnický Štít and Huancayo/Haleakala: [21];
- Lomnický Štít and LARC: [22].

These locations have different rigidity thresholds and they offer the possibility to look for the energy dependence of the investigated phenomena.

The analysis of the energy dependence for several CR modulation items was also performed by other researchers in the past, although the question of the data reliability has been often raised. For example, by using Climax, Huancayo, Moscow, Kiel and Calgary, the authors of [23] were forced to use 12-point moving averages of monthly data in wavelet analysis; as a consequence, the time precision got lost.

The necessity to use long data series from several NMs with different cut-off rigidities was also underlined by [24], after a work performed to estimate CR periodicities during solar cycle 23, where only Oulu NM data were employed.

Also [25] investigated in detail the time behaviour of CR intensity registered at Oulu for the period 1965-2011. Interesting results were obtained but the rigidity dependence of them could not be established, because only a single measurement point was used.

Finally, the valuable work of the "modulation potential reconstruction", extended back till July 1936 [26], demonstrated a better reliability for the obtained results when data from several NMs are used (Goose Bay, Oulu, Kerguelen, Kiel, Hermanus, Rome, Climax, Mt. Washington).

This reinforces the relevance of good long-term series of CR data from different terrestrial locations.

In conclusion we remind that CR detectors are not only an outstanding heritage (as it is often believed), but also the potential sources of data for novel research.

Acknowledgements
Authors are indebted with the different Institutions supporting SVIRCO since its establishment. Presently SVIRCO Observatory is supported by the INAF/UNI Roma Tre collaboration and is sending real-time data to the High-resolution Neutron Monitor Database (NMDB project founded under the European Union's FP7 Program). MS thanks IAPS/INAF for her associated research position and the Italian Space Agency (ASI contract n° I/022/10/0) for SIXS/BepiColombo activities. Also Finnish colleagues are acknowledged for their kind collaboration.

References

[1] Storini M 2000 *Proc. 26th Int. Cosmic Ray Conf. (Salt Lake City, USA, 17-25 August 1999) (AIP Conf. Proc.* vol 516*)* ed B L Dingus et al (Melville: American Institute of Physics) p 120
[2] Press W H, Teukolsky S A, Vetterling W T and Flannery B P 1992 *Numerical Recipes in FORTRAN: the art of scientific computing - 2nd edition* (New York: Cambridge University Press)
[3] Torrence C and Compo G P 1998 *Bull. Am. Meteor. Soc.* **79** 61
[4] Huang N E, Shen Z, Long S R, Wu M C, Shih H H, Zheng Q, Yen N –C, Tung C C and Liu H H 1998 *Proc. R. Soc.* A **454** 903
[5] Storini M and Signoretti F 2009 *Adv. Space Res.* **44**(10) 1221
[6] Signoretti F, Re F, Massetti S, Storini M, Laurenza M and Parisi M 2014 *SVIRCO DATA: January to December 2013 – Report IAPS-2014-14* (Rome: Institute for Space Astrophysics and Planetology)
[7] Re F, Signoretti F and Parisi M 2014 *Cosmic ray diurnal wave – SVIRCO data: 2012-2013 – Report IAPS-2014-13* (Rome: Institute for Space Astrophysics and Planetology)
[8] Storini M and Pase S 1995 *STEP GBRS News* **5** (Special Issue) 305
[9] Storini M, Pase S, Sýkora J and Parisi M 1997 *Solar Phys.* **172** 317
[10] Daubechies I 1990 *IEEE Transaction on* **36**(5) 961
[11] Kumar P and Foufoula-Georgiou E 1997 *Rev. Geophys.* **35**(4) 385
[12] Storini M, Laurenza M and Cliver E W 2005 *Proc. 29th Int. Cosmic Ray Conf. (Pune, India,3-10 August 2005)* vol 2 ed B Sripathi Acharya et al (Mumbai: Tata Institute of Fundamental Research) p 259
[13] Laurenza M, Giangravè S, Storini M and Moreno G 2008 *Proc. 30th Int. Cosmic Ray Conf. (Mérida, Mexico, 3-11 July 2007)* vol 1 ed R Caballero et al (Mexico City: Universidad Nacional Autónoma de México) p 513

[14] Diego P and Storini M 2009 *Proc. 31st Int. Cosmic Ray Conf. (Łódź, Poland , 7-15 July 2009)* Conf CD (ISBN 978-83-929057-1-4) or http://icrc2009.unilodz.pl/proc/pdf/icrc0044.pdf
[15] Iucci N, Parisi M, Storini M and Villoresi G 1983 *Nuovo Cim. C* **6** 145
[16] Vecchio A, Laurenza M, Carbone V and Storini M 2010 *Astrophys. J. Lett.* **709** L1
[17] Laurenza M, Vecchio A, Signoretti F, Storini M and Carbone V 2011 *Proc. 32nd Int. Cosmic Ray Conf. (Beijing, China, 11-18 August 2011)* DOI: 10.7529/ICRC2011/V11/0227 available on line vol 11 (http://www.ihep.ac.cn/english/conference/icrc2011/paper) p 162
[18] Laurenza M, Vecchio A, Storini M and Carbone V 2012 *Astrophys. J.* **749:167** (11pp)
[19] Laurenza M, Vecchio A, Storini M and Carbone V 2012 *Astrophys. J.* **754:155** (1p)
[20] Laurenza M, Vecchio A, Storini M and Carbone V 2014 *Astrophys. J.* **781:71** (12pp)
[21] Kudela K, Storini M, Antalová A and Rybák J 2001 *Proc. 27th Int. Cosmic Ray Conf. (Hamburg, Germany,7-15 August 2001)* vol 9 ed W Dröge et al (Katlenburg-Lindau: Copernicuss Gesellschaft 2001) p 3773
[22] Storini M, Kudela K and Diego P 2005 *Proc. 29th Int. Cosmic Ray Conf. (Pune, India,3-10 August 2005)* vol 2 ed B Sripathi Acharya et al (Mumbai: Tata Institute of Fundamental Research) p 413
[23] Ma L H, Han Y B and Yin Z Q 2009 *Solar Phys.* **255** 187
[24] Chowdhury P, Khan M and Ray P C 2010 *Planet. and Space Sci.* **58** 1045
[25] Aslam O P M and Badruddin 2014 *Solar Phys.* **289** 2247
[26] Usoskin I G, Bazilevskaya G A and Kovaltsov 2011 *J. Geophys. Res.* **116** A02104 (doi:10.1029/2010JA016105) (9pp)

The Weibull functional form for SEP event spectra

M Laurenza[1], G Consolini[1], M Storini[1] and A Damiani[2]

[1]INAF-IAPS, Via del Fosso del Cavaliere 100, 00133, Roma, Italy
[2]Japan Agency for Marine-Earth Science and Technology 3173-25 Showa-machi, Kanazawa-ku, Yokohama, 236-0001, Japan

E-mail: monica.laurenza@iaps.inaf.it

Abstract. The evolution of the kinetic energy spectra of two Solar Energetic Particle (SEP) events has been investigated through the Shannon's differential entropy during the different phases of the selected events, as proposed by [1]. Data from LET and HET instruments onboard the STEREO spacecraft were used to cover a wide energy range from ~ 4 MeV to 100 MeV, as well as EPAM and ERNE data, on board the ACE and SOHO spacecraft, respectively, in the range 1.6 – 112 MeV. The spectral features were found to be consistent with the Weibull like shape, both during the main phase of the SEP events and over their whole duration. Comparison of results obtained for energetic particles accelerated at corotating interaction regions (CIRs) and transient-related interplanetary shocks are presented in the framework of shock acceleration.

1. Introduction

The processes of solar energetic particle (SEP) acceleration are thought to be associated with solar flares and/or with shocks driven by coronal mass ejections (CMEs) in the corona and interplanetary space. The relative contribution of flares and CME-driven shocks and the details of the involved acceleration processes are still under debate (e.g., [2-3]). Particle acceleration to high energies occurs in a part of the solar corona that is not yet accessible to in situ measurements and until now, remote sensing has not enabled to disentangle the different mechanisms at play during the acceleration process. Information on them can be obtained through the time evolution of the SEP energy spectra derived from the particle flux registered in the interplanetary space. These energy spectra can be representative of the source spectra provided that the processes of particle release and particle transport in the interplanetary medium do not distort their original shape. During impulsive SEP events, particles should propagate fast enough that energy changes can be neglected; hence, their spectrum should be representative of the source spectrum (e.g. [4]). On the other hand, during gradual SEP events, the particle spectrum should be related to the characteristics of the CME-driven shock, and it can be highly variable due also to propagation effects. In case of the hybrid events, the phenomena leading to SEP events might involve both flare and shock acceleration [3] or a variable shock geometry acting on different seed particle populations [2].

The common theory to explain gradual SEP events is the diffusive shock acceleration (e.g., [5]), which predicts a power law spectrum, the spectral index and shock compression ratio (regardless of species) being linearly related. It was proposed that the power law spectra should roll over at high energies due to increasing diffusion coefficient with energy [6]. Nevertheless, the double power law is often used to explain the observed SEP spectra [2], although there have been few attempts to explain why the spectral breaks result in such function. Recently, [7] found that the Weibull distribution is the best fit, with respect to the previous models, for the energy spectrum both at the Earth's shock passage and for the main phase of the 4 April 2000 SEP event. These authors suggested that the Weibull distribution might be associated with shock acceleration, explaining it in terms of a stochastic multiplicative process.

High energy particles can also be produced by shock waves in the heliospheric environment, such as the energetic storm particle (ESP) events, related to the passage of a CME driven shock wave past the Earth, and those associated with the shocks bounding corotating interaction regions (CIRs), produced by the interaction between fast and slow solar wind streams (e.g., [8]). At low heliographic latitudes, such CIRs are typically bounded by forward and reverse waves on their leading and trailing edges, respectively, that steepen into shocks generally at heliocentric distances greater than 2 AU [9]. In situ observations have indicated that such corotating shocks can accelerate ions up to 20 MeV/nucleon in energy [10-12]. We stress that transients and corotating shocks are systems where shock acceleration hypothesis can be tested against in situ observations.

In order to get insight in the SEP acceleration phenomena, we analyzed the evolution of particle spectra during properly selected SEP events, including their ESP phase, through the Shannon entropy method as described in section 2. Moreover, we tested the Weibull type shape both to SEP events and CIR events in section 3 and 4, respectively. Results are compared and discussed in section 5, where conclusions are also drawn.

2. Selection of SEP events

2.1. Data used and method of analysis

In order to minimize the effects related to the particle transport in the interplanetary space, we selected the 21 March 2011 and 26 December 2001 SEP events, because they are isolated, preceded by almost undisturbed solar wind (no magnetic clouds/CMEs) and magnetically well connected with the solar source. Data used to study the former event are 1 minute averaged proton fluxes measured by the Low Energy Telescope (LET) and the High Energy Telescope (HET) on STEREO A in 23 energy channels in the range 4 – 100 MeV [13-14]. For the latter event we retrieved 5 minute proton measurements from both ERNE [15] aboard SOHO and EPAM [16] aboard ACE at L1 location to cover the energy range from 1.6 to 112 MeV for a total of 26 energy channels. For both events, data are averaged on 1 hour basis. A proper calibration procedure was applied by combining the ERNE channels from 1.8 MeV to 5.1 MeV to obtain the ACE equivalent energy channel 1.9 - 4.8 MeV. Then, the ACE fluxes were rescaled by about a factor 2, obtained by performing a linear regression between the two ERNE and ACE equivalent channels.

For each event we define the relative weight functions $p_i(E, t)$, for particles having energy E between E_i and E_{i+1} at time t and differential flux $f_i(E, t)$ in each energy channel i (from 1 to the number of channels N):

$$p_i(E,t) = \frac{f_i(E,t)\Delta E_i}{\sum_i f_i(E,t)\Delta E_i}$$

where $\Delta E_i = E_{i+1} - E_i$ is the channel energy bin.

The weight functions are greater than zero and normalized to unity. Then, we compute the Shannon entropy SE [17] on the basis of the normalized quantities $p_i(E, t)$ as follows:

$$SE = -\sum_i p_i(E,t)\ln\big(p_i(E,t)\big)$$

which is not affected by particle density variations. Indeed, it is sensitive only to changes in the particle spectral shape [1]. In particular, an increase of the Shannon entropy corresponds to a flattening of the SEP spectrum and viceversa. The normalized Shannon entropy (Δ) can be also computed by diving SE to its maximum value $ln(N)$.

In order to understand the different processes leading to the SEP event, we studied how the spectrum evolves in its different phases, based on the concept of the SE, as it can be used as a proxy for the spectrum changes as discussed by [1]. We remark that, as the SE is able to detect a change in the particle spectrum, it can be useful to identify changes in the particle acceleration mechanisms (such as the shock passage or a vanishing well connection to the shock nose). Nevertheless, the SE is not linked to the nature of such a mechanism, which has to be investigated by other quantities.

2.2. The 21 March 2011 SEP event

A solar eruption on the far side of the Sun (W132°) as viewed from Earth was observed on 21 March 2011. It produced an X-ray flare and a fast halo CME (speed >1341 km/s) at 02:24 UT [18], accompanied by Type II and Type IV radio sweeps, indicating the presence of a propagating interplanetary shock. The LET and HET instruments on board STEREO A (located at W88° with respect to Earth) recorded a sudden increase in the proton differential fluxes at energies between 4 and 100 MeV (see upper panel of Figure 1).

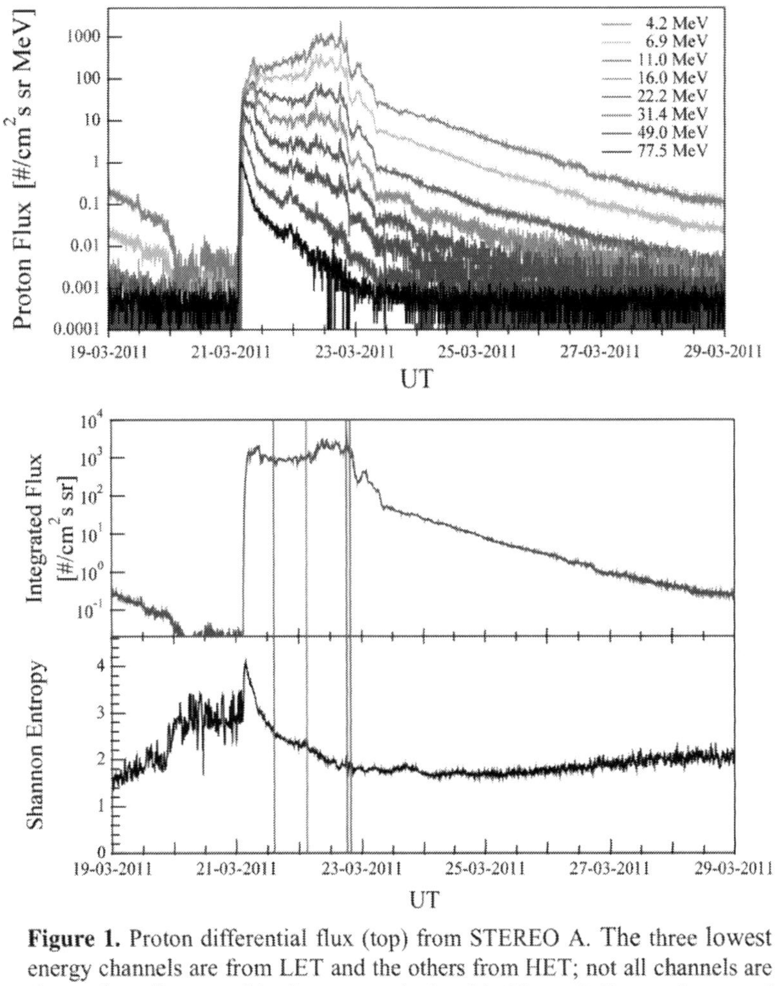

Figure 1. Proton differential flux (top) from STEREO A. The three lowest energy channels are from LET and the others from HET; not all channels are shown, but all are used in the spectra depicted in Figure 3. Energy integrated flux in the range 4-100 MeV (middle) and Shannon's entropy (bottom) for the 21 March 2011 SEP event. Vertical lines indicate the time intervals during which the spectra shown in Figure 3 are derived. The shock passage is inside the period marked by the two violet lines.

The flux of high-energy protons (60–100 MeV) at STEREO A rises and falls over a 12 hr period but the flux of lower-energy protons rises steadily until ~18:00 UT on March 22 when an interplanetary shock passes by STEREO A, with a discontinuous change in the solar wind speed from 420 to 770 km s^{-1} [18]. The SEP peak flux is seen at the shock to form the ESP phase of the event. The SEP event can be classified as long-lasting gradual and was also observed at the Earth and L1 point, although much weaker than at STEREO A, which was more favourably located with respect to the associated solar flare (being the nominal azimuthal distance from the flare location less than 50° [19]). The > 10 MeV proton flux at geosynchronous orbit rose above 10 pfu on 21 March at 19:50 UT, reached a maximum of 14 pfu on 22 March at 01:30 UT and ended that day at 03:35 UT.

The lower panel of Figure 1 shows the time evolution of SE computed from STEREO A data. The SE time variation indicates that the SEP spectral shape continuously changes throughout the considered SEP event. In particular, a fast rise of the Shannon entropy can be observed during the prompt phase, because of the lack of low energy particles due to the velocity dispersion effect. Soon after, a relatively constant SE is observed, in a time interval (indicated by the vertical green lines in Figure 1) where also the energy integrated flux (middle panel of Figure 1) is almost flat. In this "plateau phase", the spectrum can be considered to be almost stable.

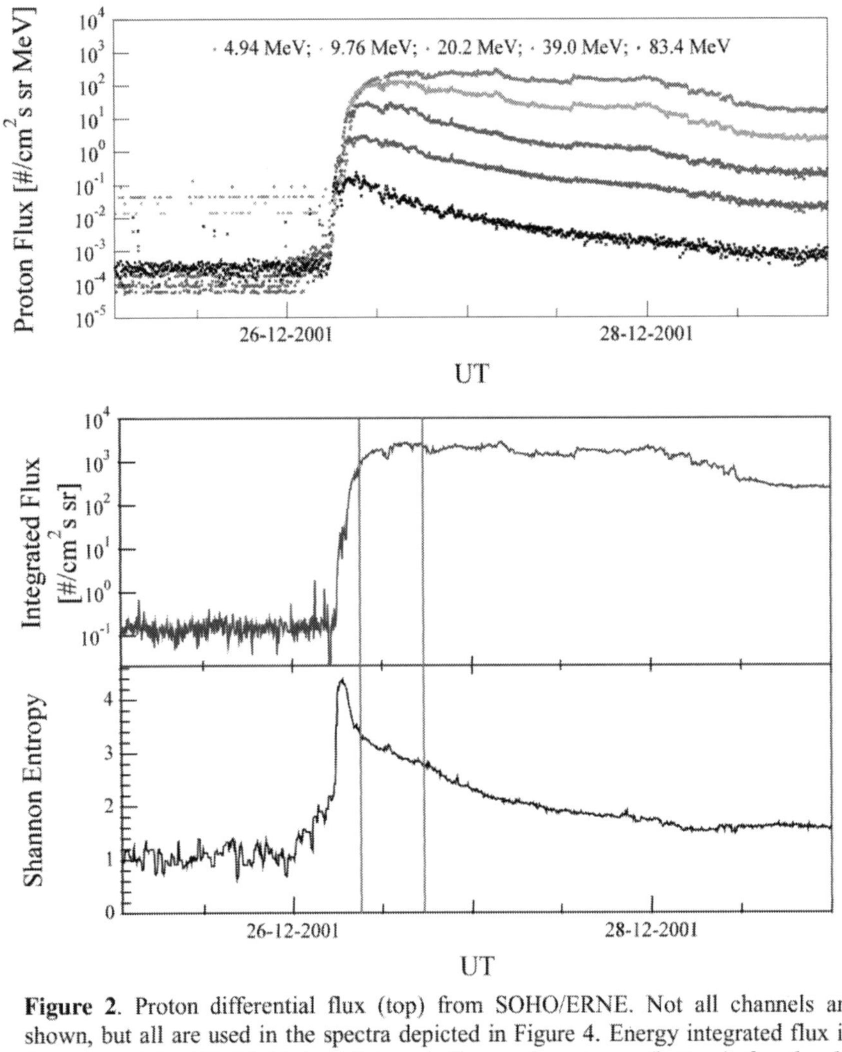

Figure 2. Proton differential flux (top) from SOHO/ERNE. Not all channels are shown, but all are used in the spectra depicted in Figure 4. Energy integrated flux in the range 1.6 -112 MeV (middle) and Shannon's entropy (bottom) for the 26 December 2001 SEP event. Vertical green lines indicate the time during which the spectrum in Figure 4 is derived.

2.3. The 26 December 2001 SEP event

On 26 December 2001, the Sun released an M 7.6 X-ray flare, peaking at 05:36 UT, and a partial Halo CME (speed = 1446 km/s) at 05:30 UT (from the active region 9742, located at W54°), that were associated with a proton event, which extended to very high energies. It was recorded at geosynchronous orbit at energies greater than 100 MeV and 10 MeV. The greater than 100 MeV event began at 05:55 UT, reached a peak of 50 pfu at 07:20 UT, and ended at 19:40 UT. The greater than 10 MeV event began at 06:05 UT, reached a peak of 779 pfu at 11:15 UT, then ended at 10:40 UT on 28 December.

The 26 December 2001 SEP event can be classified as gradual and produced a ground level enhanchement (numbered as GLE63), recorded by the worldwide network of neutron monitors. A sudden increase was measured also by the ERNE instrument in all the proton channels, without any electron contamination or saturation. The top panel of Figure 2 shows the flux profiles for some energy channels: the proton flux rise is fast and followed by a slower decay, which is still ongoing on 28 December. The energy integrated flux (middle panel of Figure 2) shows an almost flat profile after the prompt phase, while the SE (bottom panel of Figure 2), derived from ERNE data, gradually decreases after the first peak due to the velocity dispersion effect. The plateau phase is identified where SE changes less than 0.5 and it is delimited by the vertical green lines in Figure 2.

3. Model Fit

In order to study the energy content of the considered SEP events, we computed the particle spectrum both averaged over the whole duration of the event and in the plateau phase, where it is supposed to be relatively constant, such as the SE. This can be assumed as the main phase of the SEP event, that should also contain the main source acceleration processes imprint, although the transport processes may play a significant role that can modify the source acceleration spectrum. The main transport effects are supposed to be the following, as suggested by [20]: i) strong suppression of upstream ion intensities occurs at the plateau for energies near ~1 MeV amu^{-1}, although GLEs with low intensities of 10–100 MeV protons on the plateau show spectra of ions that rise monotonically toward low energies with no peaking and no suppression of low-energy ions; ii) the temporal plateau formation does not necessarily imply a streaming limited spectrum (for instance the 1998 May 2 event studied by those authors has a reasonably well-defined temporal plateau even though strong streaming effects are absent). We tried to minimize these effects by using: 1) proton energies > 1 MeV, in order to avoid the suppression at low energies; 2) medium size SEP events for which the wave amplification by the streaming protons is reduced; 3) events for which the spectra obtained for the temporal plateau phase do not show any plateau at lower energies; and 4) well-connected events. Moreover, considering that the spectra are not constrained by other factors such as scattering by ambient pre-event waves [20] and that the interplanetary conditions before the events were almost unperturbed, we can assume that the transport effects are reduced and that in the considered temporal plateau phase, the spectra still contains the imprint of the acceleration processes at the solar source.

For the 21 March 2011 SEP event, the spectrum was derived also for the ESP phase around the shock arrival, where particles are assumed to be locally accelerated by the shock, although the ESP phase may also contain particles trapped around the shock that were accelerated before the shock arrival.

Figure 3 depicts the spectra for the 21 March 2011 SEP event. As expected, the ESP phase of the event has a softer energy spectrum than the SEP plateau phase (left panel in Figure 3), as low energy particles are more easily accelerated at the shock passing the spacecraft location. Nevertheless, they can both be fitted with the same functional form, i.e., the Weibull like shape (also known as the two-parameter stretched exponential - see [21]) *W(E)* defined as:

$$W(E) = kE^{b-1}e^{-\left(\frac{E}{E_0}\right)^b}$$

The Weibull functional form gives a good fit also to the spectrum derived for the whole duration of the SEP event (right panel of Figure 3). Similar results (see Figure 4) were obtained for the 26 December 2001 SEP event, in case of the whole event. In the plateau phase, consistency between the spectrum and the Weibull distribution is also found, because the data uncertainties for ERNE data are high. Nevertheless, at high energies the Weibull function departs from the actual spectrum, but we cannot rule out possible instrumental problems affecting the ERNE data.

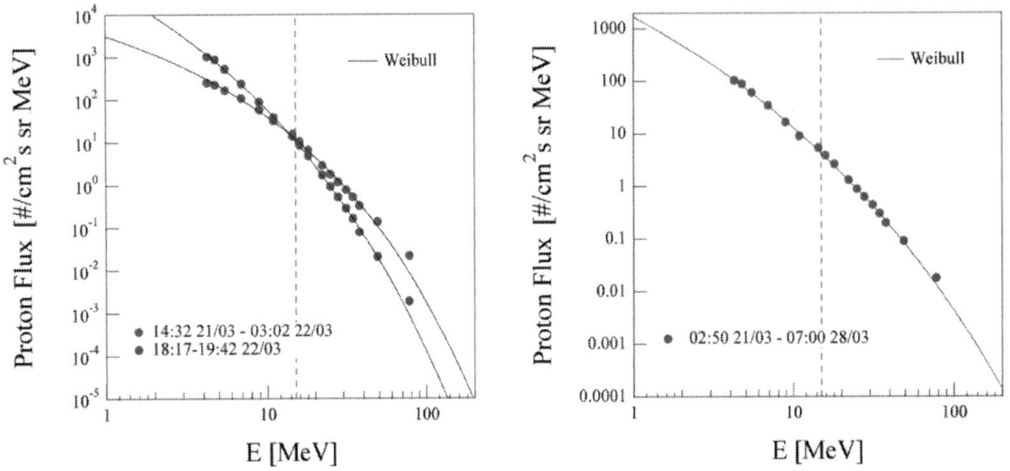

Figure 3. Time averaged spectrum over the plateau phase (green dots, left panel), at the shock arrival (violet dots, left panel) and over the whole duration of the event (right panel) obtained for the 21 March 2011 SEP event. Black curves are the Weibull function used to fit the spectra. The dashed line separates data from LET and HET instruments. Data errors are inside the marker size.

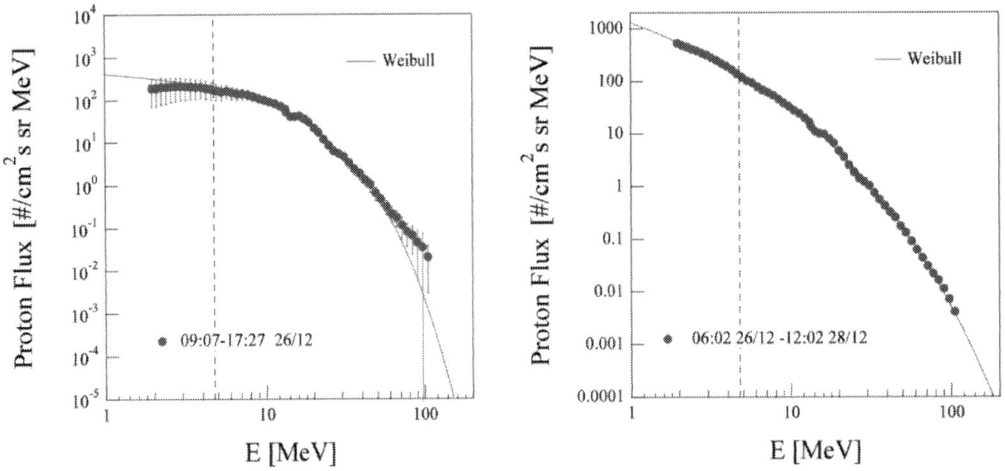

Figure 4. As in Figure 3 for the 26 December 2001 SEP event. Blue curves are the Weibull function used to fit the spectra. The dashed line separates data from ACE/EPAM and SOHO/ERNE instruments.

4. CIR events

We examined the temporal profiles and particle spectra for the 9 August 2008 and 8 January 2008 CIR events. A proton enhancement was recorded by the EPAM instrument onboard ACE in the energy range 0.047 – 4.75 MeV (upper left panel of Figure 5). The proton peak is found close to the compression region trailing edge (10 August, 03:46 UT), which was identified as a reverse shock [22]. The SE (lower left panel of Figure 5) suddenly decreases before the trailing edge, indicating a spectral change due to the low energy population increase. The spectrum (right panel of Figure 5) was derived in the period (delimited by the vertical green lines in Figure 5) where SE varied slowly. Left upper panel of Figure 6 shows a CIR associated energetic proton enhancement at STEREO A, registered by the SEPT [23] and LET instruments in 34 energy channels from 0.10 – 12 MeV. A calibration procedure was applied by comparing the SEPT channel from 2.2 MeV to 6.5 MeV with the LET energy channel 4.0 - 4.5 MeV, because they have a comparable geometric mean energy (3.8 MeV and 4.2 MeV, respectively). Then, the LET fluxes were rescaled up by about a factor 5, obtained by performing a linear regression between data from the two channels. It can be seen that low energy (< 1 MeV) particle fluxes present two peaks, while the higher energy ones only one peak near the trailing edge (at 01:48 UT on 8 January) identified as a reverse shock [22]. Again, the SE (lower left panel) is slowly changing in the period around the shock passage (between the vertical green lines), where the spectrum is derived (right panel).

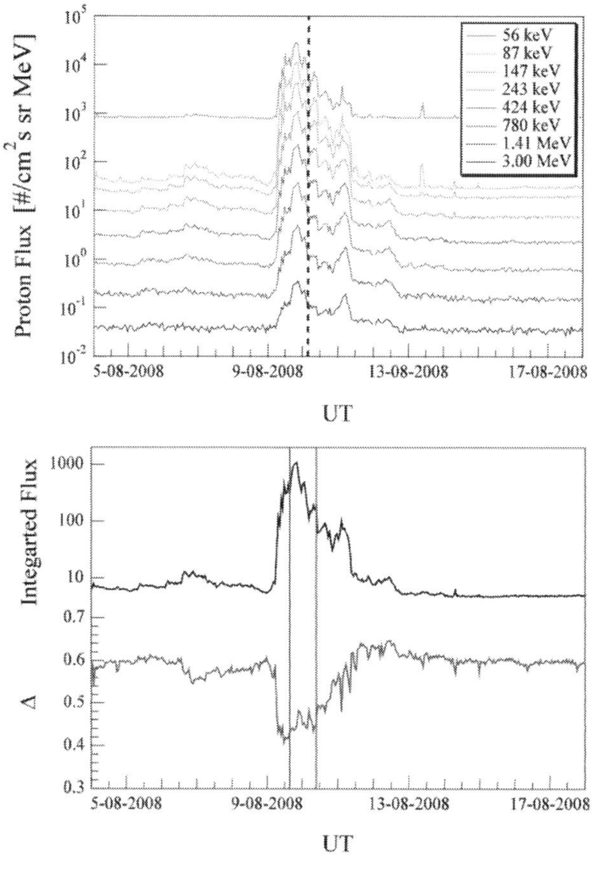

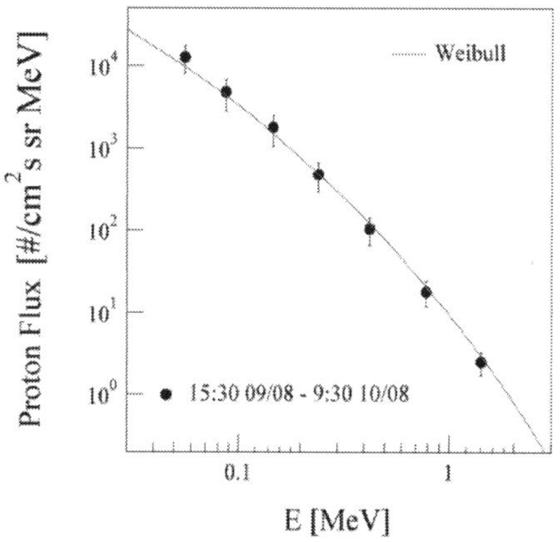

Figure 5. Proton flux recorded from ACE/ EPAM (top left panel); the dashed line indicates the time of the shock passage. Time history of the energy integrated flux in the range 0.047 – 4.75 MeV (black) and normalized Shannon's entropy Δ (with respect to its maximum value *ln(N)*, blue) computed for the 9 August 2008 CIR event (bottom left panel). Time averaged spectrum (right panel) for the 9 August 2008 during the time interval indicated by the green vertical lines.

5. Discussion and Conclusion

The SE has been used as a quantity able to describe changes in the SEP event spectral shape and identify the different phases of the event dynamics. For two SEP events we singled out a main phase, that we refer to as the plateau phase, which is not far from the prompt phase and for which we assume it may still contain the imprint of the acceleration at the solar source. Moreover, in the plateau phase the spectrum can be considered to be almost stable, given the constancy of the SE. In case of the 21 March 2011 SEP event, it was recognized an ESP phase, which is due to the particle acceleration in concomitance of the shock passage at STEREO A location. We applied the SE method also to two CIR events, for which a reverse shock was identified to exist at the trailing edge [22].

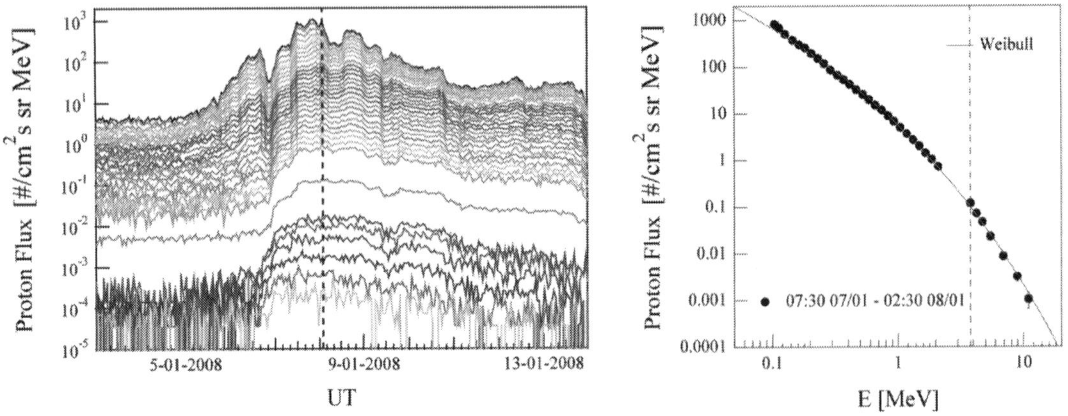

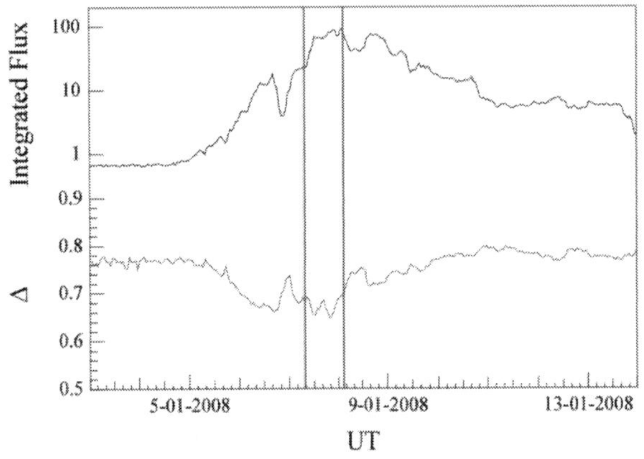

Figure 6. Proton flux (top left) recorded on board STEREO A from the SEPT (energy range 0.10 – 6.5 MeV, first 28 lines) and LET (energy range 4 – 12 MeV, last 6 lines) instrument. The tick dashed line indicates the time of the shock passage. Time history energy integrated flux in the range 0.10 – 12 MeV (black) and normalized Shannon's entropy Δ (with respect to its maximum value *ln(N)*, blue) computed for the 8 January 2008 CIR event (bottom left). Time averaged spectrum (right) during the time interval indicated by the green vertical lines. Dashed line separates The dashed line separates data from SEPT and LET.

Table 1. Fit parameters, by using the Weibull function, for the considered events.

Event	**Event Phase**	**b**	**E_0 (MeV)**
26 Dec. 2001 - SEP	Plateau phase	0.89±0.04	6.1±0.5
26 Dec. 2001 - SEP	Whole	0.46±0.04	0.51±0.02
21 Mar. 2011 - SEP	Plateau phase	0.450±0.005	0.30±0.02
21 Mar. 2011 - SEP	ESP	0.35±0.02	0.02±0.01
21 Mar. 2011 - SEP	Whole	0.30±0.03	0.02±0.01
9 Aug. 2008 - CIR	Reverse shock	0.30±0.01	0.001±0.0004
8 Jan. 2008 - CIR	Reverse shock	0.30±0.01	0.002±0.0004

A phase of slowly varying SE was identified behind the reverse shock, indicating that the spectrum is not greatly changing. Comparing the SE computed for CIR and SEP events an opposite trend is observed: the SE shows an abrupt increase in the SEP event prompt phase (due to the arrival first of the highest energy particles), whereas at the start of CIR events, the SE suddenly decreases due to the increase of the lowest energy population. We derived the spectra both in the identified phases and for the whole duration of the considered events. The same functional form (Weibull) was used to perform a fit, whose results are listed in Table 1, where b is the exponent and E_0 the rollover energy. In all cases the Weibull function is found to reproduce the observed energy spectra. This suggests that the acceleration mechanism at work could be the same for SEP, ESP and CIR events under study here.

The good consistency with the Weibull form in case of the ESP phase of the 21 March 2011 SEP event and of the reverse shocks in CIR events, confirms that shock acceleration observed in situ can be associated with the Weibull form, as proposed by [7]. This can be understood by considering that the first order Fermi acceleration, which is supposed to be effective at shock waves, is a stochastic multiplicative process. In the theory of large/extreme deviations, the Weibull shape is associated with the probability density function resulting from a multiplicative process of random variables [21], when assuming a threshold to mimic the system size limitation. Hence, the Weibull shape, which has not to be confused with a limited power law with an exponential roll-over, could provide a good description for particle acceleration occurring in a finite region and also resulting from a finite number of generations in the multiplicative process. In fact, the *b* parameter of the Weibull shape is inversely related to the number of generations involved in the multiplicative process [24]. Hence, the *b* parameter could be related to the permanence time in the acceleration region (or to the particle escape probability) as well as the shock strength. Thus, the proposed function might take into account important corrections to the classical diffusive shock acceleration scenario, from which a pure power-law is derived (assuming an infinite plane shock implying a huge number of generations in a multiplicative process).

The Weibull distribution shape is also able to reproduce the SEP spectrum in the plateau phase of SEP events, which supports a shock acceleration mechanism in the high corona. Nevertheless, in the GLE event an additional acceleration source cannot be excluded. In fact, during its main phase (where we assume the source acceleration imprint is still present) the Weibull function departs from the actual spectrum at high energies (although the fit is still good because of high error bars). This could be due to instrumental problems affecting the ERNE data or to a contribution from another acceleration source. On the other hand, in the case of 21 March 2011 SEP event, the Weibull provides a very good fit in all phases of the event (also even during the prompt phase). It seems not probable that flare and shock components are both present. In case of a flare contribution to the acceleration, i.e., if two accelerators are at work, spectra are expected to harden with increasing energy, as the flare should be more prolific than the shock at producing the highest energy particles. Hence, a break in the spectrum is expected at some energy as proposed by [4], instead of a function with sub-exponential decay such as the Weibull. According to [4]: "Of course, it is possible that the relative strengths and spectral shapes of the two components could be such that they minimize the appearance of a spectral break. But if the two components truly are independent, there is no reason to believe that such an arrangement should generally be the case. This spectral hardening should be particularly acute in event-integrated spectra, where the highest energy particles are produced by the flare for a brief time, while the shock continues to pump up the softer, lower end of the spectrum over an extended period." This is not the case in the considered SEP events (something similar happens only in the plateau phase of the 26 December 2001 SEP event). Nevertheless, the good agreement between the spectrum derived over the whole SEP events and the Weibull function, suggests that both SEP events are dominated by the shock acceleration process.

Finally, we suggest that both b and E_0 parameters of the Weibull function could be related to the shock strength, and hence to the acceleration efficiency, as they are lower for the reverse shocks and ESP phase than for the plateau phase, where the coronal shock is supposed to be more intense. This can be supported by comparing the values of the Mach number obtained for the considered ESP and the 8 January CIR event (1.9 and 1.28, respectively; available at http://www-ssc.igpp.ucla.edu/~jlan/STEREO/Level3/STEREO_Level3_Shock.pdf) with their b values (0.35±0.02 and 0.30±0.01, respectively). However, more work is needed to provide a theoretical evidence for the link between this spectral shape and particle acceleration mechanisms.

Acknowledgements

We thank the Referees for their valuable comments. This work was partly supported by the Italian Space Agency (ASI contract n° I/022/10/0). AD was supported by the SOUSEI program, MEXT, JAPAN. Proton data were retrieved at http://www.srl.utu.fi/erne_data/main_english.html;

http://www.srl.caltech.edu/ACE/ASC/level2; http://www.srl.caltech.edu/STEREO. Information about the solar sources of SEP events were obtained from ftp://ftp.swpc.noaa.gov/pub/warehouse/.

References

[1] Laurenza M, Consolini G, Storini M and Damiani A 2012 1 *Astrophys. Space Sci. Trans.* **8** 19
[2] Tylka A J, Cohen C M S, Dietrich W F, Lee M A, Maclennan C G, Mewaldt R A, Ng C K and Reames D V 2005 *Astrophys. J.* **625** 474
[3] Cane H V, Mewaldt R A, Cohen C M S and von Rosenvinge T T 2006 *J. Geophys. Res.* **111** A06S90
[4] Dröge W 2000 *High Energy Solar Physics Workshop - Anticipating HESSI (ASP Conf. Series* vol 206) eds R Ramaty and N Mandzhavidze (San Francisco: Astronomical Society of the Pacific) p 191
[5] Lee M A 1983 *J. Geophys. Res.* **88** 6109
[6] Ellison D C and Ramaty R 1985 *Astrophys. J.* **298** 400
[7] Laurenza M, Consolini G, Storini M and Damiani A 2013 *SOLAR WIND 13: Proc. of the Thirteenth Int. Solar Wind Conf. (Big Island, Hawaii, 17–22 June 2012) (AIP Conf. Proc.* vol 1539) eds G Zank et al (Melville: American Institute of Physics) p 219
[8] Gosling J T 1996 *Annu. Rev. Astron. Astrophys.* **34** 35
[9] Hundhausen A J 1976 *J. Geophys. Res.***81** 1436
[10] McDonald F B, Teegarden B J, Trainor J H and von Rosevinge T T 1975 *Astrophys. J.* **203** L149
[11] Barnes C W and Simpson J A 1976 *Astrophys. J.* **210** L91
[12] Tsurutani B T, Smith E J, Pyle K R and Simpson J A 1982 *J. Geophys. Res.* **87** 7389
[13] Mewaldt R A et al 2008 *Space Sci. Rev.* **136** 285
[14] Von Rosevinge T T et al 2008 *Space Sci. Rev.* **136** 391
[15] Torsti J 1995 *Sol. Phys.* **162** 505
[16] Gold R E et al 1998 *Space Sci. Rev.* **86** 541
[17] Shannon C E 1948 *Bell Syst. Tech.* J. **27** 379
[18] Rouillard et al. 2012 *Astrophys. J.* **752:44** (20pp)
[19] Gardini A, Laurenza M and Storini M 2011 *Adv. Space Res.* **47** 2127
[20] Reames D V and Ng C K 2010 *Astrophys. J.* **723** 1286
[21] Frisch U and Sornette D 1997 *J. Phys. I* **7** 1155
[22] Ebert R W, Dayeh M A, Desai M I and Mason G M 2012 *Astrophys. J.* **749:73** (13pp)
[23] Müller-Mellin R et al 2008 *Space Sci. Rev.* **136** 363
[24] Sornette D 2000 *Critical Phenomena in natural sciences* Springer-Verlag Berlin Heidelberg

Present state of cosmic ray stations of the Institute of solar-terrestrial physics of Siberian branch of the Russian Academy of sciences (ISTP SB RAS)

A Lukovnikova[1]
Institute of Solar-Terrestrial Physics, Russian Academy of Sciences, Siberian Branch
Russia, 664033, Irkutsk p/o box 291; Lermontov st., 126a

E-mail: luk@iszf.irk.ru

Abstract. The paper describes means for improving the operation of cosmic ray (CR) stations of the Institute of Solar-Terrestrial Physics of Siberian Branch of Russian Academy of Sciences (ISTP SB RAS) in 2012–2014.
We consider changes taken place in functioning CR stations of ISTP SB RAS in 2012–2014. Instrumental measures for improving real time data presentation of the CR stations (IRKUTSK, IRKUTSK2, IRKUTSK3 and NORILSK) are presented. There are some problems for each CR station. We show ways for solving these problems to provide data without failures and omissions. Introduction

There are four CR stations at our Institute. They are IRKUTSK, IRKUTSK2, IRKUTSK3 and NORILSK (Table 1).
All stations are equipped with a recording device including one 8-ch Counter/Timer PCI Card (PCI-1780) and Russian barometers BRS-1M. PCI-1780 records data of the CR neutron component. PCI-1780 is a general-purpose multi-channel counter/timer card for the PCI bus. It uses the AM 9513 Counter Chip and provides eight 16-bit counter channels, 8 digital outputs and 8 digital inputs; the time base is up to 20 MHz. The data is recorded onto the hard disc minutely.
To record atmospheric pressure, the recording computers are equipped with barometers BRS-1M. The measuring range of BRS-1M is 600–1100 hPa; the absolute accuracy is ±33 (20 for BRS-1M-2) Pa; the data is displayed on the screen and recorded onto the hard disc using RS-232 interface unit.
The data are transmitted from the NM stations to the server in Irkutsk by a satellite link every 3–5 min (1-min values) and every hour (hourly values).
There the data are processed and transmitted in real-time to the World Data Center (WDC) and to the Neutron Monitor Data Base (NMDB).
Primary data checking is carried out by the Super Editor in registering computers at each station.

[1] To whom any correspondence should be addressed.

Table 1. Characteristics of the cosmic ray stations IRKUTSK, IRKUTSK 2, IRKUTSK 3 and NORILSK

Name	Altitude, m	Geographic latitude/ longitude	Number counters, CNM-15 [1]	Barometric coefficient, %/mb	Standard atmoshperic pressure P_0, mb	Effective vertical cutoff rigidity R_c, GV
IRKT	475	52.47N, 104.03E	18	–0.713	960	3.56
IRK2	2000	51.37N, 100.55E	12	–0.713	800	3.56
IRK3	3000	51.37N, 100.55E	6	–0.72	715	3.56
NRLK	0	69.26N, 88.05E	18	–0.7	1000	0.56

For operative measures and correction of eventual errors which cannot be corrected automatically, Irkutsk server should have not only preprocessing files but also the information from each recording channel separately.

The local databases at each CR station as well as the server at the institute in Irkutsk are continuously synchronized.

All databases are functioning on the basis of the Oracle Data Manager [2].

Creation of a database of CR stations of ISTP SB RAS has been determined by the necessity of collecting data from CR stations IRKUTSK, IRKUTSK2, IRKUTSK3 and NORILSK at the shared center in order to perform on-line control of data quality, fault diagnosis and program failures [3].

The free version of Oracle Express Edition is used at CR stations of ISTP SB RAS. This edition has the following constraints: its hardware requirements are 1 GB RAM and a single operating CPU, and the maximum database size of Oracle Database XE is 12 GB. This includes the space of 0.5–0.9 GB for the data dictionary, internal schemas, and temporary space, which leaves just 11.0 GB for user's data.

Afterward the data can be corrected manually.

1. Irkutsk

The detector housing is not well isolated and the roof of the building is flat. The temperature inside varies from +5 to +25 °C in winter and from +20 to +27 °C in summer.

In winter, there is often snow on the roof, and it must to be cleaned.

At IRKUTSK CR station, there is a special handwritten registry with the indication of the date and time (accuracy is ± 10 min) when the snow is removed from the roof. We do not correct CR intensity data of IRKUTSK station for the snow effect because the roof is cleaned every morning when snowing.

2. Irkutsk2

The problem at IRKUTSK2 is a dirty electrical power supply, i.e. the alternating voltage is not a pure sinus wave, and there are frequent power-offs.

There is a diesel generator at the observatory but the diesel motor must be started manually that is not always possible as the station is not continuously operated by technicians.
The WLAN devices DWL-2700 and DWL-2000 need a reboot after sticking. DWL-2000AP+ is a multi-purpose wireless access point. It provides a stable connection with 802.11g wireless devices at a rate of up to 54 Mbit/s. The transmitter power is 100 mW.
Outdoor wireless access point D-Link AirPremier DWL-2700AP that conforms to IEEE 802.11b/g standards covers a large operating distance providing wireless Internet access at transfer speeds of up to 54 Mb/s. DWL-2700AP is an ideal instrument for adverse climatic conditions due to its solid, watertight housing and built-in heater with a temperature sensor, so this AP is environment-proof. Power supply over Ethernet cable enables an access point to be installed at places with no AC power sockets available.

3. Irkutsk3

Under strong wind, the measured atmospheric pressure values can vary abruptly up to 1 hPa within one minute (e.g., 31.03.2014, 17:00 LT, IRKUTSK3, IRKUTSK2).
The reason is that at this altitude, the building construction is not solid and wind can easily penetrate. The building walls at IRKUTSK3 station are made of 100×18×18 cm wooden pieces connected with metal staples. Under strong wind, the walls shake hard. This is not typical for IRKUTSK2 station (the altitude is 2000 m) built in a similar way.
There are many problems with IRKUTSK3 CR station. In winter the temperature inside the detector housing is cold. We use the central power supply for electric heating. There are frequent failures. We can do nothing against this fact.
The low temperature in the detector housing affects the counting rate of the neutron monitor. At present, we examine the dependence of IRKUTSK, IRKUTSK2, IRKUTSK3 and NORILSK data on the temperature inside the buildings.
Conductor sag, lightning protection failures and insulator breaking at 10 kV power lines feeding CR IRKUTSK3 station can occur during thunderstorms, strong winds and snow. So, the replacement of the step-down transformer is required. The ascent to the IRKUTSK3 CR station every week is very difficult because we use horses at any weather.

4. Norilsk

The problems at the station NORILSK are similar to those at the other stations. Power-offs are frequent and long because of old line accidents. In the autumn of 2014, it was planned to replace the burned-out substation. In winter, Norilsk CR station did not operate because of –45 °C outside temperature and strong wind. Besides, the inside temperature was 0 °C. There was also a failure of the barometer that is installed inside the building of the CR station. For the barometer BRS-1M, the operating temperature is + (15–35) °C. In the summer of 2014, windows at the CR station were sealed. After a power-off, there can be sticking of the transmitting equipment (router or access point). There is the ground connection between the CR station building and the administrative building. The distance between buildings is 800 meters. Besides, induced electromagnetic fields can occur.
Yearly, the Norilsk station is supported by the engineer. In winter, the buildings are not easy of access because of snowdrifts up to 2.5 m. Skis and a snowmobile are needed to reach the cosmic ray station and the doors of the station must be cleaned from the snow.

Acknowledgments

A technician mounts to the stations IRKUTSK2 (2000 m) and IRKUTSK3 (3000 m) stations once a week if necessary to check the equipment, uninterrupted power supply and data recording and after

power-offs. Remote CR stations need to enlarge the Internet satellite channel to increase the data transfer rate [3].
Power lines and the electric power supply systems at IRKUTSK2, IRKUTSK3 and NORILSK CR stations also need to be reconstructed. The main reason is that because of power-offs great information content can be lost.

References

[1] http://www.nmdb.eu/?q=node/142
[2] http://www.oracle.com/technetwork/database/database-technologies/express-edition/overview/index.html
[3] Lukovnikova A et al. J. Phys.: Conf. Ser. 409 (2013)
[4] Aleshkov V, Dvornikov V, Lukovnikova A, Sdobnov V International Symposium: fundamental Science and Applied Aspect (Athens, Greece, 2007) 396 (2007).

The Ground Temperature Effect on Cosmic-Ray Muons at Mid latitude City

A Maghrabi[*1]; **R Alotaibi**[1,2], **M Almutayri**[1], **M Garawi**[2], **and M Baig**[2]

(1) National Centre For Applied Physics, King Abdulaziz City For Science and Technology, Riyadh 11442, Saudi Arabia.
(2) Physics and Astronomy department, King Saud University, Riydah, Saudi Arabia

E-mail: amaghrabi@kacst.edu.sa

Abstract. The investigation of meteorological effects is of a great importance to the analysis of the cosmic ray variations. In this paper, we study the effect of the ground temperature on the cosmic ray recorded by KACST detector. This detector has monitored secondary cosmic ray muon since 2002 at Riydah, Saudi Arabia; (lat 24 43; long. 46 40; alt. 613 m; *Rc* ~14 GV).

1. Introduction

The rate of the detected cosmic ray muons depends mainly on the atmospheric pressure and temperature [1-4]. Corrections for the changes in these factors have to be done to know the properties of the primary cosmic rays.

Variations in the atmospheric temperature cause changes in its density; this varying density affect the rate of the detected muons. In summer, the atmosphere is warmer, higher, and less dense. In this case, the first cosmic ray interactions occur at a higher altitude; thus muons will travel longer distance before they reach the detector and larger muons will decay. During cold periods, the atmosphere is shallower and denser. The first interactions happen near the surface and fewer number muons will be detected. [5].

Several studies have been carried out in different parts of the world to study the effects of ground temperature on cosmic-ray muons [6]. Substantial variations of the temperature coefficients have been reported by different investigators [5-7].

The objective of this paper is to investigate the influence of the ground temperature on the detected muons using eleven years of observations recorded by KACST detector.

2. Instrumentation and Methods

Cosmic ray measurements were obtained from a single channel muon detector installed at King Abdulaziz City for Science and Technology (KACST) Riyadh (lat. 24 43 ; long. 46 40; alt. 613 m Rc =14 GV) Saudi Arabia for the period between 2002 and 2012. The detailed descriptions of the detector and its calibration procedures are given in [8-10].

Basically it consists of a 4×250 mm^2 of plastic scintillator, PMT, and its associated electronic system. Data were acquired every 15 minutes with a total of 96 data records each day. Hourly meteorological data for the considered period were obtained from Saudi Meteorological Agency.
The variations of the secondary cosmic ray intensities with the ground temperature can be expressed by: [4, 11]

$$\frac{dI}{I} = \alpha dT; \qquad \frac{I_i - I_0}{I_0} = \alpha(T_i - T_0) \qquad (1)$$

Where I_0 and T_0 are, in respective, the muons mean intensity and temperature for the considered period of time (11-years); *i*th is hour reading of intensity I_i and temperature T_i. Regression analysis were carried out between the hourly values of the pressure corrected muon intensities and the ground temperature to obtain the daily values of α. These daily means of α were averaged to get the monthly values.

3. Results and Discussion

Figure (1) shows the frequency distribution of the calculated daily values of α for 1882 days. The Skewness and the Kurtosis for this distribution were 1.03 and 13.89 respectively. α ranges between -0.19 %/°C and 0.33 %/°C with a mean value of 0.025 ± 0.00061 % / °C.

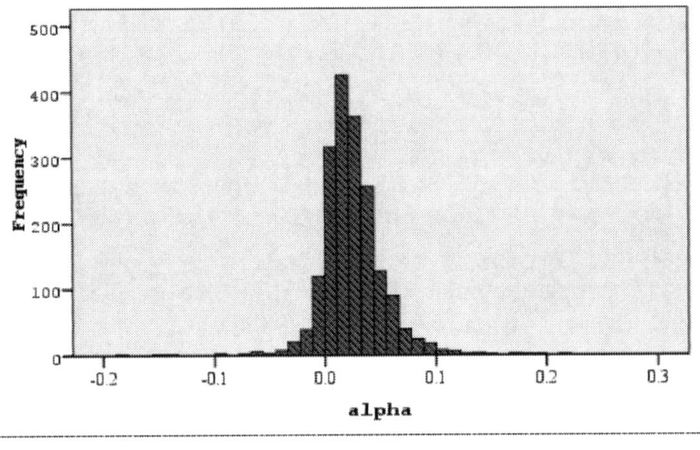

Figure 1. The frequency distribution of the calculated daily values of the temperature coefficients, α, for 1882 days.

Figure (2) shows the relationship between the (a) daily and (b) monthly; values of the cosmic muons and the ground temperatures during the study period. Although the spread in the daily data is more apparent than the monthly data, the general tendency between the two parameters (inverse relation) is evident, which is in agreement with the previously established results [5,7]. The temperature coefficient from both graphs was -0.023 ± 0.00053 % /°C.

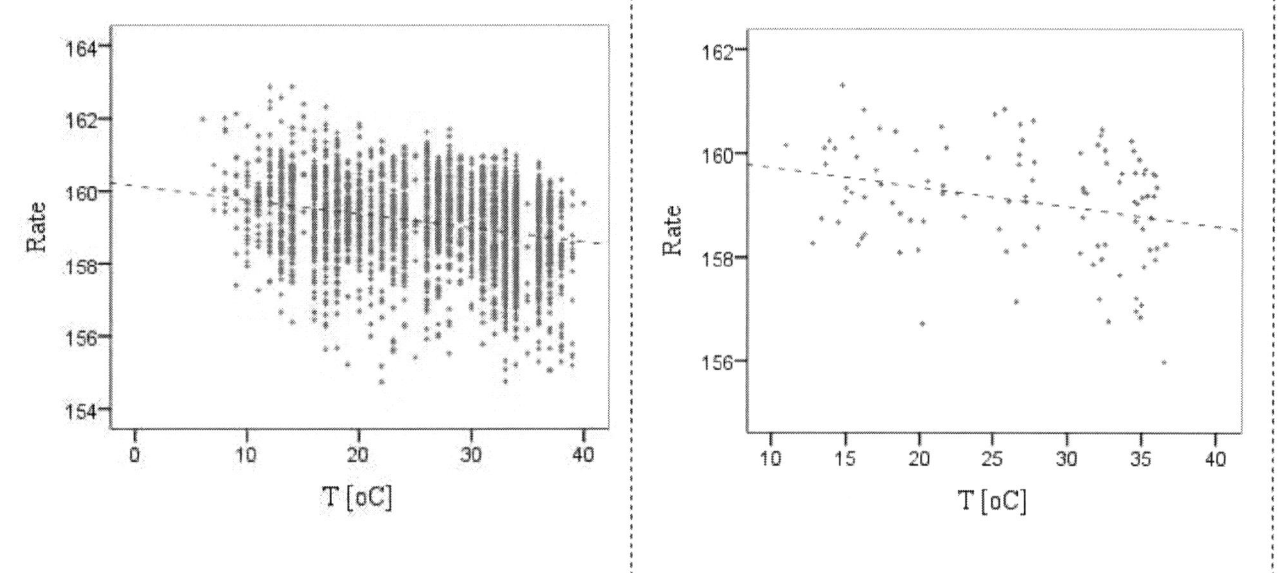

Figure 2. Scatter plot between the ground temperature and the muon count rate for (left) daily and (right) monthly based on the daily values ; the straight dotted lines is the best fit line in which the slope is the temperature coefficient found in equation 2.

However, when the obtained results are closely checked, several complications appear to arise.
Firstly, the temperature coefficient demonstrated wide range of values. This variability may be attributed to the local changes in the ground level or upper air atmospheric parameters on the observed cosmic ray muons. *Secondly*, we have found that the obtained α values for Riyadh are much smaller than those previously calculated by several researchers [for instance 4, 5, and7]. However, theoretical calculations conducted by [11] using CORSIKA code revealed that, for the threshold energy of KACST detector, altitude, and zenith angle, the obtained α values are comparable with those calculated in the present study. *Thirdly*, large numbers of the daily αvalues are positive; muons increases as the temperature increases. This positive relationship is contradicting the formerly established experimental results. This discrepancy may be due to the occurrence of unusual meteorological, environmental, or atmospheric changes, such as temperature inversions which are very common in Riyadh.
We expect that these difficulties would disappear if the mean atmospheric temperature was used instead of the ground temperature. This is the topic of on-going research by our group.
Figure (3) shows the monthly variations of the cosmic ray muons and the ground temperature. The inverse relationship between the two parameters is clearly evident. Figure (4) is a scatter plot between the monthly observed muons and the ground temperature. In this case, less scatter in the data (compared with figure 2) is noticeable and the relation between the two parameters is clearly evident. The correlation coefficient of this data is 0.75, and α is -0.024 % / ^{o}C.

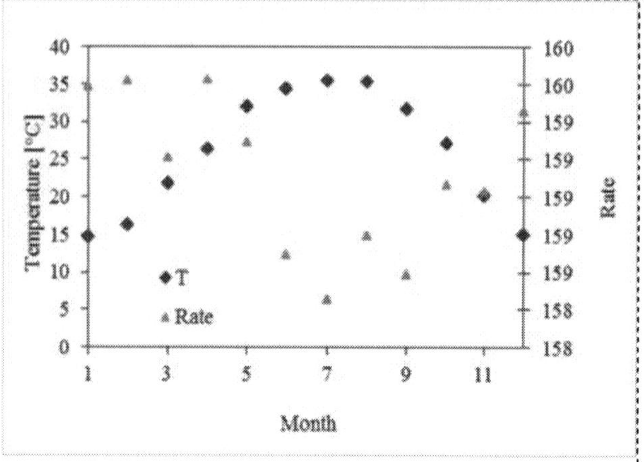

Figure 3. Shows the monthly variations of ground temperature (T) and cosmic Ray intensity over Riyadh for the whole study period.

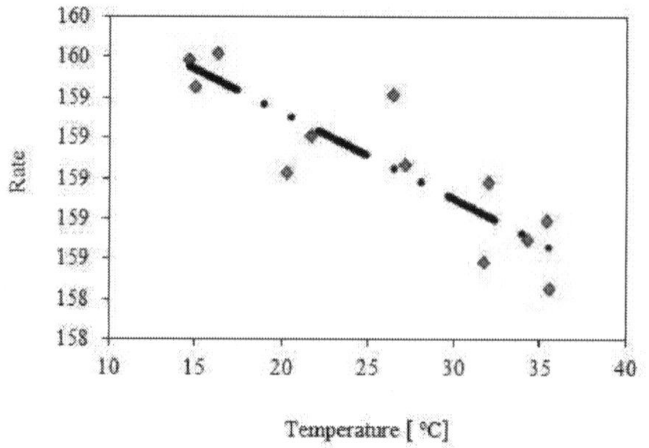

Figure 4. Scatter plot between the Twelve-months binned ground temperature and the muon ; the straight dotted lines is the best fit line.

4. Conclusion

In this paper, we studied the relationship between the ground temperature and cosmic rays using eleven years of data collected from KACST muon detector. Inverse relationship between the two parameters was established with a slope, (temperature coefficient)α, ~ - 0.023 % / °C. Surprisingly, in some occasions, positive correlations between the temperature and the muon intensities were found. Yet, no clear explanations were reached for this behaviour and further investigations are required. We also, found that the calculated temperature coefficients, presented in this study, were smaller in comparisons with those obtained by different investigators. We expect that some of the found difficulties would disappear if the mean atmospheric temperature used for the correlation with cosmic-ray intensity.

Acknowledgments

The authors thank the Saudi Meteorological Agency for providing the surface weather data. The authors would also like to thank King Abdulaziz City for Science and Technology (KACST) for supporting this work.

5. References

[1] Bercovitch M 1967 *Proc. Int. Conf. on Cosmic Rays, Calgary*, **Part A**, 269-344
[2] Kudela K, Firoz K, Langer R. and Koll'ar V 2009a, *In Proc. of 21st ECRS, Kosice*, **2008**, 374
[3] WangC. P., and A. H. Lee 1967 *J. Geophys. Res.*.**72**(23), 6107
[4] DeMendonça R, Raulin S, Echer E, Makhmutov V,and Fernandez A 2013 *J. Geophys. Res* **118**, 1403.
[5] Berkova M, Belov A, Eroshenko A, and Yanke V.2011 *Bull. Russ. Acad. Sci. Phys.*, **75**(6), 820.
[6] Yanke V, Asipenka A, Berkova R, de Mendonca R, Raulin J, Bertoni F, Echer G, Fernández, and Makhmutov V 2011 *Proc. Int. Conf. on Cosmic Rays* 32, **11**, 377.
[7] Duperier A 1949 , *Proc. Phys. Soc. A*, **62**(11), 684
[8] Maghrabi A, Alharbi H, Alghamdi A, Alwaheeb A, Almuteri M 2013, *Proc. Int. Conf. on Cosmic Ray 33th*
[9] Maghrabi A, Al Harbi H, Al-Mostafa Z, Kordi M, S. Al-Shehri M 2012, *Advances in Space Research* **50** ,700.
[10] Maghrabi A, Alghamdi A, Alotaibi R, Almutari M and Garawi M 2014 *Proc. SPIE* 9154, **91542K**
[11] Kovylyaeva A , Dmitrieva A, Tolkacheva N, Yakovleva E 2013 , *In Proc. of 23rd ECRS (and 32nd Russian Cosmic ray Conf.* **409**, 012128

Time variations of proton flux in Earth inner radiation belt during 23/24 solar cycles based on the PAMELA and the ARINA data.

V V Malakhov[1], S V Koldashov[1], A G Mayorov[1], M A Mayorova[1], V V Mikhailov[1], S Yu Aleksandrin[1], O Adriani[2,3], G C Barbarino[4,5], G A Bazilevskaya[6], M Boezio[9], E A Bogomolov[10], M Bongi[2,3], V Bonvicini[9], S Bottai[3], A Bruno[7,8], F Cafagna[8], D Campana[5], P Carlson[11], M Casolino[12,13], G Castellini[14], C De Donato[12,15], C De Santis[12], N De Simone[12], V Di Felice[12,16], V Formato[9,17,*], A M Galper[1], A V Karelin[1], S Yu Krutkov[10], A A Kvashnin[6], A N Kvashnin[6], A A Leonov[1], L Marcelli[15], M Martucci[15,18], W Menn[19], M Merge[12,15], E Mocchuuitti[9], A Monaco[7,8], N Mori[3,20], R Munini[9,17], G Osteria[5], F Palma[12,15], B Panico[5], P Papini[3], M Pearce[11], P Picozza[12,15], M Ricci[18], S B Ricciarini[3,14], R Sarkar[8,], V Scotti[4,5], M Simon[19], R Sparvoli[12,15], P Spillantini[2,3], Yu I Stozhkov[6], A Vacci[9], E Vannuccini[3], G I Vasilyev[10], S A Voronov[1], Yu T Yurkin[1], G Zampa[9], N Zampa[9]**

[1] National Research Nuclear University MEPhI, RU-115409 Moscow, Russia
[2] Department of Physics and Astronomy, University of Florence, I-50019 Sesto Fiorentino, Florence, Italy
[3] INFN, Sezione di Florence, I-50019 Sesto Fiorentino, Florence, Italy
[4] Department of Physics, University of Naples "Federico II", I-80126 Naples, Italy
[5] INFN, Sezione di Naples, I-80126 Naples, Italy
[6] Lebedev Physical Institute, RU-119991 Moscow, Russia
[7] Department of Physics, University of Bari, I-70126 Bari, Italy
[8] INFN, Sezione di Bari, I-70126 Bari, Italy
[9] INFN, Sezione di Trieste, I-34149 Trieste, Italy
[10] Ioffe Physical Technical Institute, RU-194021 St. Petersburg, Russia
[11] KTH, Department of Physics, and the Oskar Klein Centre for Cosmoparticle Physics, AlbaNova University Centre, SE-10691 Stockholm, Sweden
[12] INFN, Sezione di Rome Tor Vergata, I-00133 Rome, Italy
[13] RIKEN, Advanced Science Institute, Wako-shi, Saitama, Japan
[14] IFAC, I-50019 Sesto Fiorentino, Florence, Italy
[15] Department of Physics, University of Rome Tor Vergata, I-00133 Rome, Italy
[16] Agenzia Spaziale Italiana (ASI) Science Data Center, Via del Politecnico snc, I-00133 Rome, Italy
[17] Department of Physics, University of Trieste, I-34147 Trieste, Italy
[18] INFN, Laboratori Nazionali di Frascati, Via Enrico Fermi 40, I-00044 Frascati, Italy
[19] Department of Physics, Universitt Siegen, D-57068 Siegen, Germany
[20] Centro Siciliano di Fisica Nucleare e Struttura della Materia (CSFNSM), Viale A. Doria 6, I-95125 Catania, Italia
[*] now at INFN, Sezione di Perugia, I-06123 Perugia, Italy
[**] now at Indian Centre for Space Physics, 43 Chalantika, Garia Station Road, Kolkata 700084, West Bengal, India
E-mail: vvmalakhov@mephi.ru

Abstract. The PAMELA and the ARINA experiments are carried out on the board of satellite RESURS-DK1 since 2006 up to now. Main goal of the PAMELA instrument is measurements of high energy antiparticles in cosmic rays while the ARINA instrument is intended studying high energy charged particle bursts in the magnetosphere. Both of these experiments have a possibility to study trapped particles in the inner radiation belt. Complex of these two instruments covers proton energy range from 30 MeV up to trapping limit (E= ~2 GeV). Continuous measurements with the PAMELA and the ARINA spectrometers include falling and rising phases of 23/24 solar cycles and maximum of 24th one. In this report we present temporal profiles of proton flux in the inner zone of the radiation belt ($1.11 < L < 1.18$, $0.18 < B < 0.22$G). Dependence of proton fluxes on a magnitude of the solar activity was studied for various phases of 23/24 solar cycles. At that it was shown that proton fluxes at the solar minimum are several times greater than at the solar maximum.

1. Introduction

Measurements of trapped and quasi-trapped radiation of high energy (E>30 MeV) at the low altitudes (h<1000km) have a great importance as well for theoretical models as for engineering purposes. These altitudes are heavily populated with spacecrafts including ISS and manned spacecrafts. It is known that high energy particles (greater than 30 MeV) in radiation belt have negative impact on electronics [1-4]. For example in [3] and [4] a correlation between passing through the inner radiation belt area and a degradation of solar cells was shown. Moreover interactions of particles with residual atmosphere cause difficulties for theoretical prediction of trapped fluxes due to variations of its density. New data measured on a boundary of particle trapping could help in better understanding of particle loss mechanisms. At low L-shells fluxes of trapped particles vary with time, as it is shown for example in [5]. In this paper also an anticorrelation between fluxes and radio flux at wave length 10.7 cm characterizing the solar activity was shown. Usually this anti correlation is explained by variations in residual atmosphere density in the near Earth space [6]. Today we have a lack of long period accurate measurements of high energy (E>100 MeV) particle fluxes at low altitudes (h<1000 km) continuing at least one full solar cycle. Widely known POES-program lasts several solar cycles but it has a problem with systematic uncertainties [7]. And though methods for taking into account factor of degradation of detectors were developed [8] new measurements are necessary for independent checking of those results. Now de-facto standard models for calculation in radiation belt AP8 max and AP8 min [9] have known restrictions for lower edge of inner belt. They have been discussed many times, see [10] – [18]. The most important ones are that it is based on measurements of 60-70th years and not considering changing Earth's magnetic field. Magnetic dipole moment is decreasing and magnetic shells sink into atmosphere causing more extensive particle lost. There are also other empirical models such as PBS97 [19], Low Altitude Trapped Radiation (LATR) [20], Trapped Proton Model (TPM-1) [21], Combined Release and Radiation Effect Satellite Proton Model (CRESSPRO) [22], NOAAPRO model (more detailed survey see in [23]) But all of them are static and do not cover full solar cycle. There are also theoretical models [24],[25], which include the solar activity variations, but their applicability for nowadays should be confirmed by modern measurements. Detailed analysis of trapped proton flux time variation based on POES (NOAA-10, NOAA-15) data for 21st-23rd solar cycles has been made in work [26]. A strong anti correlation between atmosphere density, which in turn depends on solar activity, and proton fluxes at near Earth L-shells were obtained. At present the PAMELA and the ARINA instruments have been operating on the orbit for more than 8 years. They collected great amount of data at different periods of solar activity from the minimum of 23rd cycle to maximum of 24rd cycle. Main goal of this work is to present variations of trapped proton fluxes at the lower edge of the inner radiation belt with the PAMELA and the ARINA spectrometers along with the solar activity changes.

2. Experiments and instruments

2.1. The RESURS-DK1 satellite

The Resurs-DK1 satellite which carries the PAMELA and the ARINA instruments was launched on 15th June 2006 into an elliptical orbit with altitudes of 350-610 km and an inclination of 70.4 degrees. The orbit was circularised on 10 Sep 2010 to constant 573 km altitude. At these altitudes the instrument crosses the radiation belt in the South Atlantic Anomaly zone 6-8 times a day, so it allows collecting enough data in this region to study trapped particle fluxes with high statistical accuracy.

2.2. The PAMELA spectrometer

The PAMELA instrument was designed for study of antiparticles in cosmic rays, but it also has possibilities to measure trapped particle flux of energies greater than ~80 MeV. The PAMELA apparatus, as shown in figure 1, compounds of following subdetectors: a time of flight system (ToF), a magnetic spectrometer, an electromagnetic calorimeter, an anticoincidence system (CAS, CAT, CARD), a shower bottom scintillator and a neutron detector. In this analysis ToF and anticoincidence system was used. ToF produces a trigger signal if all of its planes are hit (i.e. it mainly registers protons with energy more than 80 MeV) and determines particle velocity and direction. For more detailed information see [27].

2.3. The ARINA spectrometer

The ARINA spectrometer originally was designed to study high-energy charged particle bursts in the magnetosphere. It consists of 10 scintillator plates and has capability to identify electrons and protons and to measure energy of particles by its range in a multilayer detector. Scheme of the ARINA instrument is shown in figure 2. Energy range of ARINA is 30 – 110 MeV. For more detailed information see in [28].

Characteristics of both instruments are shown in table 1. It is seen that they complement each other by energy range. Different orientation of their axes also allows enhancing angular cover. That fact that both apparatuses are installed onto the same satellite and have the same external conditions and overlapping energy ranges gives an opportunity to crosscheck results on its boundaries of sensitivity.

Table 1. Characteristics of Instruments.

	PAMELA	ARINA
Orientation on the board	Pointing to zenith	Perpendicular to orbit plane
Field of view	16° *19°	20°*20°
Geometrical factor	25 cm^2Sr	10 cm^2Sr
Angular resolution	±2°	±7°
Available trapped protons energies	80 MeV – trapping limit	30 – 110 MeV

3. Experimental data

For analysis data for period from June 2006 till November 2014 were used. For both apparatuses trapped particles were considered. For the PAMELA spectrometer downgoing particles producing signals in S1, S2 and S3 detectors of ToF (it corresponds to protons with E>80 MeV) were taken into analysis. In the ARINA spectrometer protons stopped in 8th or 9th layers and producing no signal in C10 plane have been considered. That corresponds to proton energy of 80 – 110 MeV. For both spectrometers the same geomagnetic areas and pitch-angles of their main axis equal to 85 – 95° were chosen. The South Atlantic Anomaly region was selected by L<1.20 and B<0.23G criterion, where L is McIlwain parameter describing set of the Earth magnetic field lines and B is intensity of the Earth magnetic field, for their calculation the IGRF11 model was used [28]. Because fluxes strongly depend

Journal of Physics: Conference Series **632** (2015) 012069 doi:10.1088/1742-6596/632/1/012069

on L and B parameters, the L,B space was divided into small bins and for each of them proton fluxes depending on time were obtained. L interval was divided into bins dL=0.005 and B into dB=0.001 for PAMELA and dL=0.02 and dB=0.005 for ARINA. To separate equatorial fluxes a parameter B/B0 also was calculated with the IGRF11 model. Contamination of galactic cosmic ray component is minimal for PAMELA as the geomagnetic area selected is below cutoff rigidity 10 GeV and populated mostly with secondary low energy particles. That part of galactic cosmic rays, which is still included in the flux, is not affected by solar modulation and thereby contributes in the radiation belt rate evenly over whole observation period and does not influence the time profile. For ARINA there is no contamination by galactic cosmic rays at all since it has upper energy limit about 110 MeV.

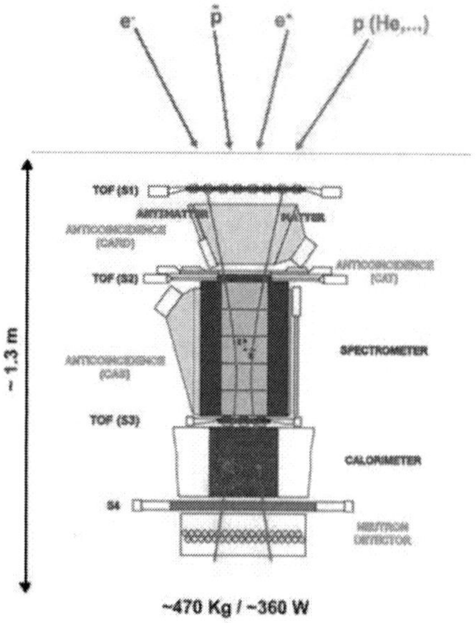

Figure 1. Layout of the PAMELA spectrometer

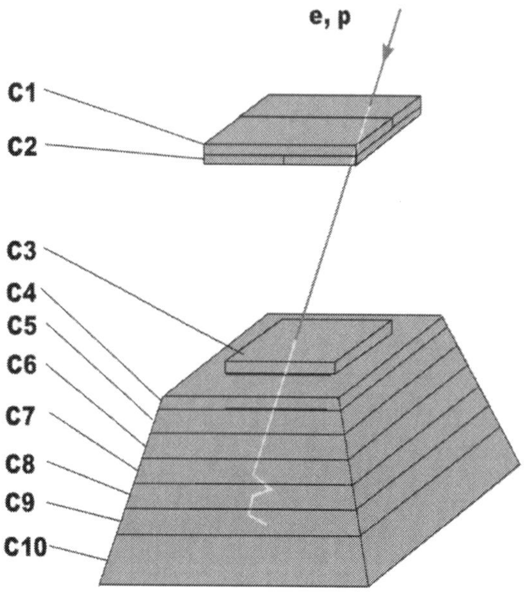

Figure 2. Layout of the ARINA spectrometer

4. Results and discussion

In this article proton fluxes of the PAMELA and the ARINA instruments are shown for period from June 2006 to November 2014 including solar minimum of 23/24 cycle and current solar maximum of 24th cycle. This time span is very interesting because solar activity of current maximum of the 24th solar cycle is relatively low and characterised by 2 years time delay from expected one. In figure 3 as example time profiles of proton flux on L=1.14-1.16 and B/B0=1.00-1.07 i.e. equatorial flux are shown, the same results were obtained for each cell in LB space described above. Green line presents smoothed f10.7 cm wave length flux intensity characterising solar activity level. It is seen that the PAMELA and the ARINA data are in agreement and both of them anti correlate with f10.7 flux intensity. Different spread in fluxes values between solar minimum and maximum measured by the PAMELA and the ARINA spectrometers can be explained by different mean energy of particles taken into analysis. In figures 4 and 5 proton fluxes of ARINA, PAMELA, count rates of NOAA-15(MEPED) and AP8-MIN and AP8-MAX model calculation are shown depending on the Earth magnetic field B (corresponding B/B0 interval is 1.0 – 1.07) for the solar minimum and the solar maximum respectively. The NOAA-15 data were normalized on the ARINA fluxes. In these figures a good agreement also is seen. Moreover the PAMELA and the ARINA data go to the lower B, where AP8 does not give any results.

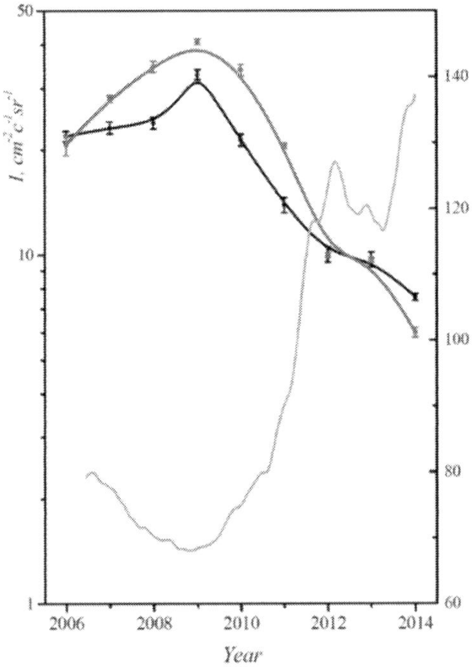

Figure 3. Proton flux time variations measured by PAMELA (black curve) and ARINA (red curve) for L-shells 1.14 – 1.16 (B/B0=1.0-1.07, i.e. equatorial flux) and f10.7 cm wavelength solar radio flux (green line)

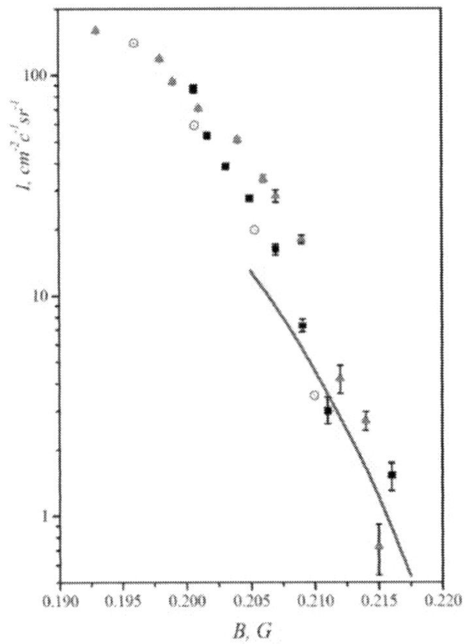

Figure 4. Proton flux vs B dependence measured by the ARINA (red triangles), the PAMELA (black squares), MERED on NOAA-15 (blue circles)[26], and AP8MIN calculations (blue curve), for solar minimum and L=1.14 – 1.16

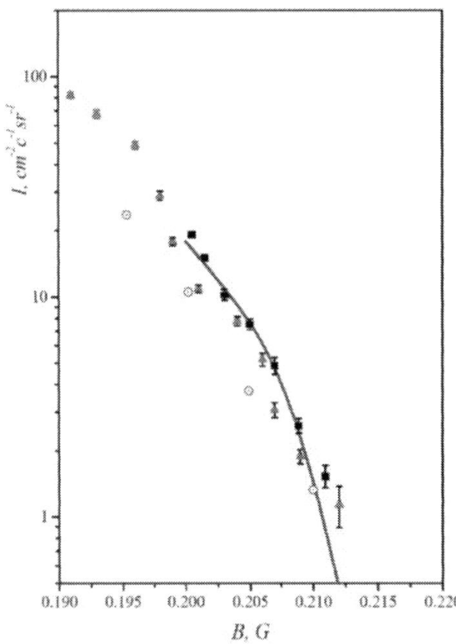

Figure 5. Proton flux vs B dependence measured by the ARINA (red triangles), the PAMELA (black squares), MERED on NOAA-15 (blue circles)[26], and AP8MIN calculations (blue curve), for solar maximum and L=1.14 – 1.16

5. Conclusion

In this work time variations of proton fluxes with energies E>80 MeV at the low edge of the inner radiation belt with PAMELA and ARINA instrument were obtained. An anti correlation between flux and level of the solar activity was shown. Flux dependence on B for different L-shells in interval 1.12 - 1.20 has been obtained. Good agreement between PAMELA and ARINA data is seen.

Acknowledgments

This work was supported by the Russian Science Foundation (grant 14-12-00373) and by RF Presidential Grant MK-6271.2015.2.

References

[1] Knoll G F 1989 Radiation Detection and Measurement *JohnWilley &Sonic Inc.*
[2] Kazama Y, Goka G, 2008 A new modelling method of solar energetic proton events for ISO specification *Adv. Space Res.* **42** 1293–99
[3] Miyake W, Miyoshi Y, Matsuoka A, 2014 On the spatial of proton radiation belt from solar cell output variation of the Akebono satellite *Adv. Space Res* **53** 1603-9
[4] Ishikawa H, Miyake W, Matsuoka A, 2013 Variation of proton radiation belt deduced from solar cell degradation of Akebono satellite *Earth planets space 65* 121-5
[5] Huston S L, Kuck G A, Pfitzer K A 1997 Low Altitude Trapped Radiation Model Using TIROS/NOAA data *Radiation Belts:Models and Standards*
[6] Miyoshi Y, Morioka A, Misawa H 2000 Long Term Modulation of Low Altitude Proton Radiation Belt by the Earth's Atmosphere *Geophysical Research Letters* **27 14** 2169-72
[7] Galand M, Evans D 2000 Radiation Damage of The Proton MERED Detector on POES (TIROS/NOAA) *Technical report OAR* **456-Sec 42**
[8] Asikainen T 2012 Correction of detector noise and recalibration of MERED/NOAA energetic proton fluxes *Journal of Geophysical Research* **117** A0924
[9] Sawyer D M, Vette J I 1976 AP8 Trapped Proton Environment for Solar Maximum and Solar Minimum *NSSDC/WDC-A-R&S* 76-06, NASA Goddard Space Flight Center, Greenbelt, Maryland
[10] Vette J I, Chan K W, Teague M J 1978 Problems in modelling the Earth's trapped radiation environment *NASA TM-80447*
[11] Abel B, Thorne M, Vampola A L 1994 Solar cyclic behavior of trapped energetic electrons in Earth's inner radiation belt *J.Geophys. Res.* **99** 19,427-19,431
[12] Heynderickx D Comparison between methods to compensate for the secular motion of the south Atlantic anomaly *Rad. Meas.* **26**
[13] Dyer C S 1998 Space radiation environment dosimetry *IEEE **NSREC Short Course Notes,*** Ch. **2,** Newport Beach, CA, USA
[14] Daly E J, Evans H D R 1996 Problems in radiation environment models at low altitudes *Rad. Meas.* **26** 363-8
[15] Blake J B, Kolasinski W A, Fillius R W, Mullen E G 1996, Injection of electrons and protons with energies of tens of MeV into L<3 on Mar. 24, 1991 *Geophys. Res. Letr.* **18** 821
[16] Mazur J 2002 The radiation environment outside and inside a spacecraft, *IEEE NSREC Short Course Notes,* Ch. 2, Phoenix, AZ, USA
[17] Armstrong T W, Colburn B L 2000 Evaluation of trapped radiation model uncertainties for spacecraft design *NASNCR-2000-2* 10072
[18] Gussenhoven M S, Mullen E G, Brautigam D H 1994 Near-earth radiation model deficiencies as seen on CRRES, *Adv. Space Res.* **14** 927-41

[19] Heynderickx D, Kruglanski M, Pierrard V, Lemaire J, Looper M D and Blake J B 1999 A low altitude trapped proton model for solar minimum conditions based on SAMPEX/PET data *EEL Tram. Nucl. Sci.* **46**
[20] Huston S L 2002 Space Environments and Effects: Trapped Proton Model, *NASNCR-2002-2 I* 1784
[21] Boscher D M, Bourdarie S A, Friedel R H W, Belian R D 2003 Model for the geostationary electron environment: POLE, *IEEE Trans. Nucl. Sci.* ,**50** 2278-83
[22] Huston S L, Pfitzer K A, 1998 Space Environment Effects: Low-Altitude Trapped Radiation Model, NASNCR-1998-208593
[23] Lauenstein J-M, Barth J L 2005 Radiation Belt Modeling for Spacecraft Design Model Comparisons for Common Orbits" NSREC 2005 Data Workshop Paper # W-16
[24] Selesnick R S, Looper M D, Mewaldt R A, 2007, A theoretical model of the inner proton radiation belt, *Space weather* 5 S04003
[25] Lodhi M A K, Diaz A B, Wilson T L, 2005 Simplified solar model of inner trapped belt proton flux as a function of atmospheric density, *Radiation Measurements*, **39** 391-99
[26] Kuznetsov N V, Nikolaeva N I, Panasyuk, 2010, Variation of the trapped proton flux in the inner radiation belt of the Earth as a function of solar activity, *Cosmic Research,***48** 80-85
[27] Picozza P, Galper A, Castellini G et al, 2007 PAMELA: Payload for Antimatter Matter Exploration and Light nuclei Astrophysics, *Astropart. Phys.* **27** 296-315
[28] Bakaldin A et al, 2007, Satellite experiment ARINA for studying seismic effects in the high-energy particle fluxes in the Earth's magnetosphere, *Cosmic Research*, **45** 445-8
[29] Finlay C C, Maus S, Beggan C D et al 2010, International Geomagnetic Reference Field: the eleventh generation, *Geophysical Journal International* **183** 1216-30

Heliospheric Energetic Particles and Galactic Cosmic Ray Modulation

Olga Malandraki

IAASARS, National Observatory of Athens, GR-15236, Penteli, Greece

E-mail: omaland@astro.noa.gr

Abstract. The paper presents an overview of the SH 'Solar and Heliospheric cosmic rays' session of the 24th European Cosmic Ray Symposium (ECRS), Kiel, Germany, 2014. It covers the topics of Solar Energetic Particle (SEP) origin, acceleration and transport at the Sun and in the interplanetary medium, also from the aspect of multi-spacecraft observations, as well as the Galactic Cosmic Ray (GCR) short- and long-term variations and the Jovian electron variations in the heliosphere. Relevant instruments and methods presented are also covered by this review. The paper is written from a personal perspective, emphasizing those results that the author found most interesting.

1. Introduction

Solar Energetic Particles (SEPs) and Galactic Cosmic Rays (GCRs) are two high-energy particle populations in the heliosphere [1]. Both populations are important components of the space radiation environment [2], also posing risks to flight crews and instruments and constituting a major space weather concern in planning for long-duration manned missions [3]. Ground level enhancements (GLEs), the high-energy SEP component, provide the opportunity to detect energized coronal material from the Sun that reaches Earth's atmosphere within a matter of ~10 minutes [4]. In this paper the results of the SH session 'Solar and Heliospheric cosmic rays' of the 24th European Cosmic Ray Symposium (ECRS) are reviewed. During this session 30 oral contributions and 34 poster contributions were presented and considered in this review. Throughout the paper a particular contribution is cited by reference to the name of the first author and the corresponding Abstract Number as in the ECRS Programme (SH***). It should be noted that the results considered as most interesting by the author are highlighted in the paper.

The paper is organized, based on my division of the presented papers, as follows: in section 2 the results on Solar Energetic Particles are presented, discussing the origin, acceleration and transport of SEPs, also based on multi-spacecraft observations as well as space weather and space climate results from the SEP perspective. In section 3, Forbush decreases and GCR short term variations are discussed, in section 4 the results presented on the long term variations of GCRs are highlighted, in section 5 Jovian electron variations in the heliosphere are discussed and in section 6 the instrumentation and methods are summarized. In section 7 the results are briefly summarized and future prospects are discussed.

2. Solar Energetic Particles

2.1. Origin of Solar Energetic Particles

The origin of SEPs has been an issue under debate for decades. Potential accelerators in different regions of the corona exist, namely magnetic reconnection in the flaring active region, as well as shock waves driven out from the Sun by fast Coronal Mass Ejections (CMEs) [5]. Many attempts were made in the past to identify a unique accelerator by establishing statistical relationships between SEP parameters, particularly their peak intensity measured near 1 AU and one of the quantities describing the importance of the associated eruptive activity – flare radiation on the one hand, and the speed of the CME on the other [6,7,8,9,10].

Klein et al. (SH423) carried out a comparative analysis of the correlation between SEP intensities – electrons between a few tens and a few hundreds of keV, protons at some tens of MeV- with quantities characterizing the eruptive solar activity: CME speed, and quantities describing the flare-related energy release: peak flux and fluence of the soft X-ray (SXR) emission and fluence of the microwave emission. Their sample comprises 38 SEP events associated with flares of SXR M and X-class at western longitudes during the period 1997-2006, as listed in [7], in which the flare-related particle acceleration is accompanied by radio bursts indicating electron escape into the interplanetary space [11]. The novel approach of this study compared to earlier work is that besides the classical Pearson correlation coefficient the partial correlation coefficients are calculated and the effects of correlations between the solar parameters themselves are disentangled. The partial correlation analysis shows that the only parameters that significantly affect the SEP intensity are the CME speed and SXR fluence, with no additional contribution by the SXR peak flux and the microwave fluence [12]. The authors concluded that their findings bring statistical evidence that both flare acceleration and CME shock acceleration contribute to the deka-MeV proton and near-relativistic electron population in large SEP events.

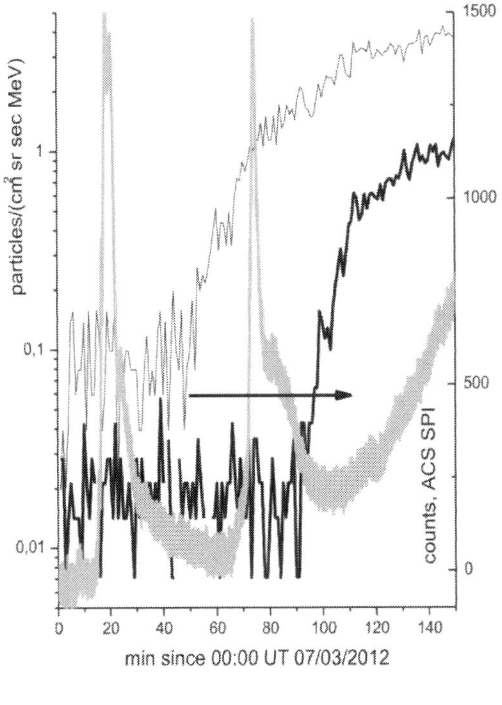

Figure 1. Onset of the SEP event observed on 7 March 2012 by HET/STEREO B (thin black line – 1.4-2.8 MeV electrons, thick black line – 29.5-33.4 MeV protons) in comparison with ACS SPI count rate (grey line). The arrow marks the first Fermi time window (Struminsky et al. SH316).

Investigating the origin of SEPs with new earlier unavailable instruments, Struminsky et al. (SH316) considered 18 solar flares in which high energy > 100 MeV γ-emission was registered by the Large Area Telescope (LAT) instrument onboard the Fermi Gamma-Ray Observatory (FermiGRO) [13, 14]. They checked for these events intensity time profiles observed in SXR (GOES) (http://www.ngdc.noaa.gov/stp/satellite/goes/) and hard (INTEGRAL/ACS SPI) X-rays (HXR) [15, 16, 17] and fluxes of SEPs registered in the heliosphere. The STEREO mission consists of two nearly identical space-based observatories (STEREO-A and STEREO-B) launched on October 25, 2006 [18]. Figure 1 presents the onset of the SEP event observed by the HET/STEREO-B on March 7, 2012 in comparison with ACS SPI count rate. The arrow marks the first Fermi time window i.e. the time interval of first high energy gamma ray observations in the event of March 7, 2012 (see Table 1 of [14]). The ACS SPI data can show processes of energy releases and particle acceleration, which are not visible by instruments aboard the RHESSI [19] or FermiGRO spacecraft due to peculiarities of their low circular orbits. These authors found that bursts of solar HXR emission were observed in the impulsive phase of the 12 events detected by ACS SPI under study, whereas HXR-bursts were not detected during the prolonged decay phase of high energy γ-emission in these 12 solar flares (or in some cases were masked by SEP events). Four solar flares with >100 MeV γ-rays were found not to be associated with SEP events even considering data from three different point in the heliosphere, i.e. with STEREO A/B and SOHO [20]. Thus, these authors concluded that if particles interacting in the solar atmosphere and propagating in the heliosphere are the same populations, then processes of particle release are very important. Also, continuous high energy X-ray and gamma solar observations with better spatial, temporal and energy resolution for progress of the SEP origin issue are needed.

2.2. *Solar Energetic Particle Acceleration and Transport*

2.2.1. Acceleration and Transport Modelling. Petukhova and Petukhov (SH192) studied the efficiency of surfing acceleration [21] and concluded that it depends on the value of a potential jump, stream velocity, and particle distribution function injected into the acceleration process. The transport of Cosmic Rays (CRs) in the heliosphere is dominated by the interaction with magnetic irregularities (Spanier et al. (SH149)). Kocharov et al. (SH170) presented a semi-transparent shock model they developed for gradual SEP events. Based on plasma observations, the turbulence energy levels in neighbouring magnetic tubes of solar wind may differ from each other by more than one order of magnitude. These authors considered the effect of such intermittency on the acceleration and transport of protons in travelling shocks. [22] have modelled the diffusive shock acceleration of the sub-MeV seed particles available upstream of the shock and the transmission through the Interplanetary (IP) shock wave of high-energy ions from coronal sources situated behind the shock.

They performed new modelling, shown in figure 2, of the acceleration and transport of both the high-energy ions escaping into the IP medium and the ions precipitating into the solar chromosphere, where they produce via nuclear interactions the secondary emissions, γ-rays and neutrons [14, 23]. They modelled particle acceleration in the spherical IP shock wave expanding through the radial structure that comprises the highly turbulent central core (the acceleration core θ_A in figure 2) embedded into a wider cone of quiet plasma via which the accelerated particles can escape (the particle escape sector) and propagate towards or away from the Sun. The model presented also accounts for stochastic re-acceleration by shock-amplified turbulence behind the shock (shaded region in figure 2). Figure 3 shows the time-integrated energy spectra of accelerated protons for the two modelled cases: without (A) and with (B) stochastic re-acceleration of the shock accelerated particles. Interacting and interplanetary distributions are the time-averaged distributions of particles escaping at the bottom and the top of the simulation box, respectively. They found that including the shock downstream re-acceleration can increase the interacting-to-interplanetary proton ratio by an order of magnitude more than would be expected for original shock acceleration (closed red points in case B versus points of case A). They also concluded that the characteristics of the cross-field transport and seed particle distribution are crucial for such modelling. Dalla et al. (SH387) showed that drift motions of particles associated with gradient and curvature of the Parker spiral magnetic field lead to the particles experiencing a deceleration, which they quantified through full orbit test particle simulations of protons.

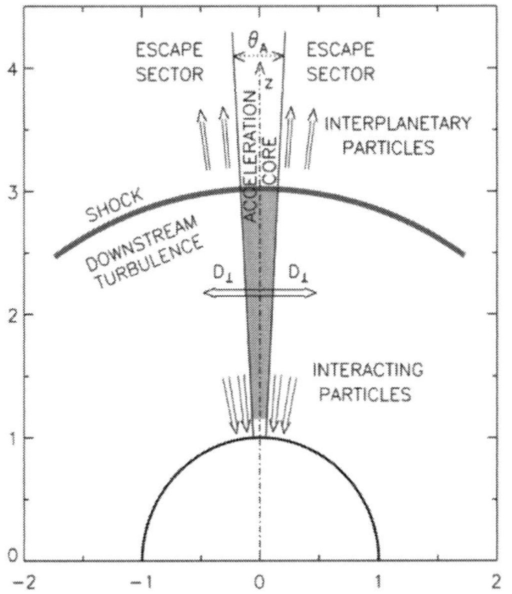

Figure 2. Particle acceleration and transport model. The model also accounts for stochastic re-acceleration by shock-amplified turbulence behind the shock (shaded region) (see text) (Kocharov et al. SH170).

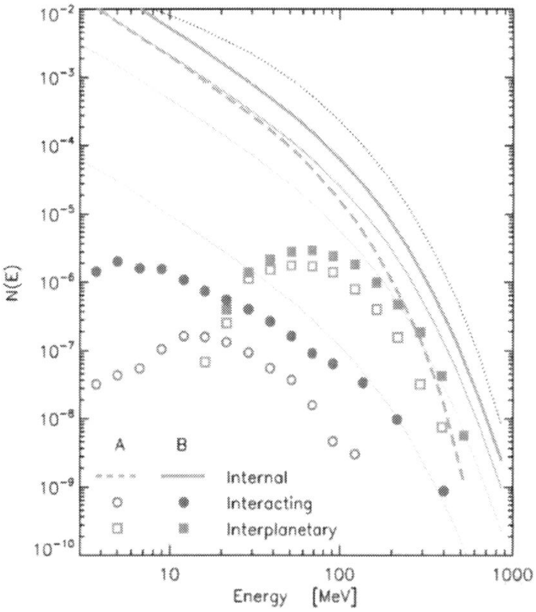

Figure 3. Energy spectra of accelerated protons for the two modelled cases: (A) without stochastic re-acceleration of the shock-accelerated particles and (B) with stochastic re-acceleration (Kocharov et al. SH170).

2.2.2. Multi-spacecraft Observations. The STEREO twin spacecraft have provided multi-spacecraft observations available to the SEP research community [19]. In February 2011 the two STEREO spacecraft reached a separation of 180 degrees in longitude, offering a complete view of the Sun for the first time ever. SOHO [20], ACE [24] and Wind [25] near-Earth observations can be used to offer a third vantage point (Zelina and Dalla, SH394, Klassen et al. SH447). Of particular importance for the extraction of conclusions on the particle acceleration and transport is the analysis of the particle anisotropy characteristics by Dresing et al. (SH287), who carried out a statistical survey of wide spread events in the heliosphere. Their main selection criterion for these events is that the longitudinal separation between active region and spacecraft magnetic footpoint is at least 80 degrees for the widest separated spacecraft. They investigated the events in a statistical manner in terms of maximum intensities, onset delays, and rise times and determined their longitudinal extent. They also investigated energetic electron anisotropies (see [26] for anisotropy definition) to disentangle source and transport mechanisms leading to the observed wide particle events. These authors divided the events in three classes, presented in figure 4. In Class 1 significant anisotropy is observed at a well-connected spacecraft ($\varphi < 60°$) but almost no anisotropy at a far separated one ($A < 0.6$ at $\varphi > 60°$, with φ being the longitudinal separation). In the events of Class 2 the highest anisotropy is observed by the best connected spacecraft but significant anisotropy ($A > 0.6$) is still observed at far separated positions ($\varphi > 60°$). In the Class 3 events the highest anisotropy is not observed by the best-connected spacecraft but by a further-separated one. Figure 4 (top part) shows the three sketches presented by these authors, in agreement with the variable anisotropy distributions observed. While in the first sketch, corresponding in the Class 1 events, just a small source region at the Sun (flare) is present (yellow star), the scenarios for the other two Classes 2 & 3 comprise an extended source region at the Sun (red arc). In all three diagrams red areas mark regions of strong anisotropy, because here a good connection to the source is present. In the Class 1 sketch, just a well-connected observer detects an anisotropic increase but far-separated spacecraft observe an isotropic event. Thus, SEPs reach far-separated positions just due to strong perpendicular diffusion, with vanishing anisotropies (represented by the gray areas). The presence of perpendicular transport in the IP medium is represented by the wavy magnetic field lines in the sketches. In the Class 2 scenario the SEPs are injected over a much wider angular range at the Sun and under normal conditions particles arrive at the spacecraft with a noticeable anisotropy. This results to a significant anisotropy detected over a larger longitudinal extent correlated with the extent of this source region. In the Class 2 events, the best-connected observer

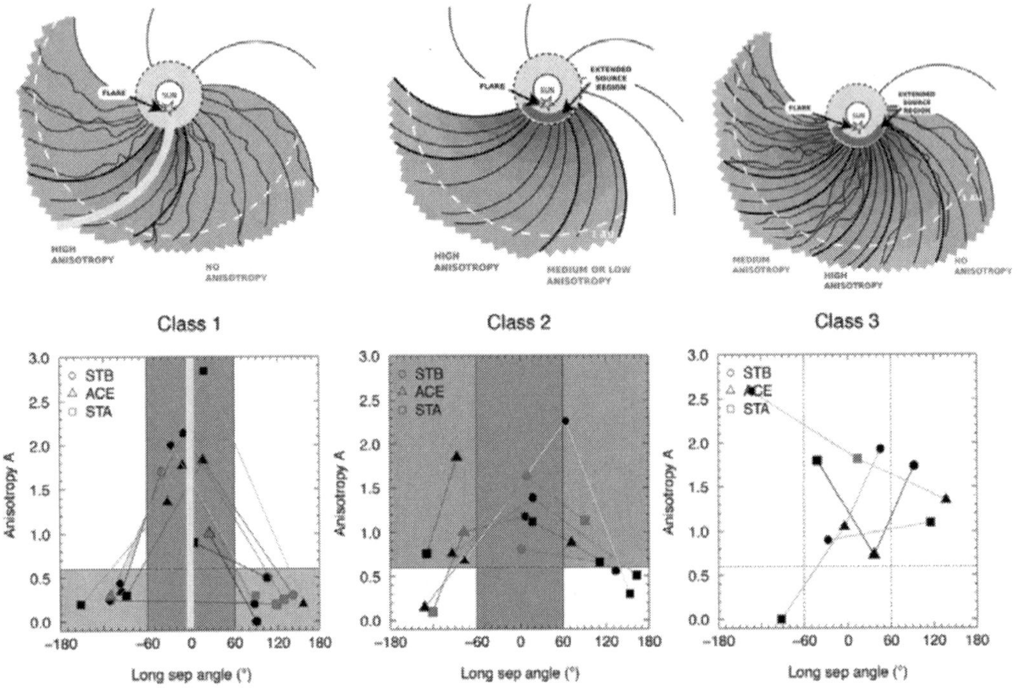

Figure 4. (Top) Sketches for each Class in agreement with the variable anisotropy distributions observed (Bottom) Maxima of absolute anisotropies observed during the early phases of the events for the three event classes. Black symbol fillings denote that the pitch angle coverage was sufficient whereas red symbol fillings denote that the anisotropy may be underestimated due to poor pitch angle coverage. The color fillings mark the borders defining the different anisotropy classes (see text) (Dresing et al. SH287).

detects the largest anisotropy but further-separated spacecraft still observe significant anisotropies during the rising phase of the event. The events in Class 3 are the more challenging to explain because the best-connected spacecraft does not observe the highest anisotropy but a further-separated one does. In the Class 3 sketch an extended source region is also present but the propagation conditions in the IP medium vary so that a nominally best-connected spacecraft may not observe the highest anisotropy because of a smaller parallel mean free path in this region. A neighbouring spacecraft, however, may detect the highest anisotropy because of a rather scatter-free transport to its position. However, also pre-event CMEs or large IP flux rope structures could produce Class 3 observations, when a better connection to a further-separated spacecraft is provided than the nominally best-connected one [27, 28]. Dresing et al. (SH287) concluded that they found evidence for both efficient perpendicular transport in the IP medium and a broad SEP spread close to the Sun. A mixture of processes is thus very likely.

The kinetic energy spectra of several SEP events were investigated by Laurenza et al. (SH344) through the Shannon's differential entropy [29, 30] during different phases of the selected events and the implications for the shock acceleration mechanisms were discussed. As an example, figure 5 presents the energy integrated flux (upper panel) and Shannon's entropy S(t) computed for the 21 March 2011 SEP event observed by STEREO A (lower panel). The change of the Shannon entropy indicates that the spectral shape continuously changes throughout all the considered SEP event. During the period indicated by the two vertical lines in figure 5 a relatively constant S(t) is observed, where also the energy integrated flux is almost flat (plateau phase). In this plateau phase, the spectrum can be considered to be almost stable and should also contain the main source acceleration processes. At the shock arrival particles are accelerated locally (violet vertical lines). Figure 6 presents time averaged spectra over the plateau phase (green dots, left panel), at the shock passage (violet dots, left panel) and over the whole duration of the event (right panel in figure 6). As [31] previously also found for the 4 April 2000 SEP event the Weibull distribution is the best fit for the energy spectrum both at the shock passage at the Earth and for the plateau phase. These authors elaborated that this can be explained in terms of a stochastic multiplicative process, such as the first order Fermi acceleration, which is supposed to be effective at shock waves. In the theory of extreme deviations, the Weibull shape is associated with the Probability Density Function (PDF) of the multiplication of random variables, when assuming a threshold to mimic the system size limitation [32, 33]. The same functional form was also found to fit the spectra at reverse shocks of Corotating Interaction Region (CIR) events by these authors. As they stated more work is needed to provide a theoretical evidence for the link between this spectral shape and particle acceleration.

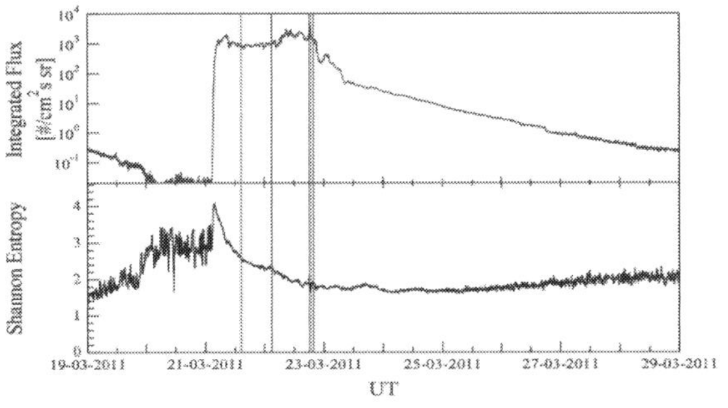

Figure 5. Energy integrated flux (upper panel) and Shannon's entropy computed for the 21 March 2011 SEP event (lower panel). Vertical lines indicate the time intervals during which the spectra in Figure 6 have been derived (Laurenza et al. SH344).

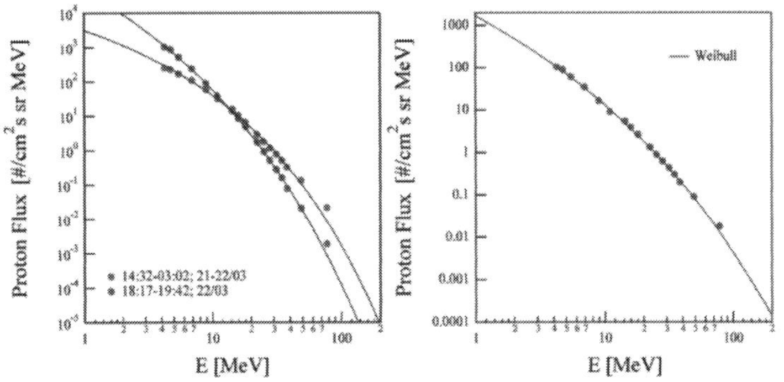

Figure 6. Time-averaged spectra over the plateau phase (green dots, left panel) at the shock passage (violet dots, left panel) and over the whole duration of the event (right panel) (Laurenza et al. SH344).

2.3 GLE Events: Observations and Modelling

GLE events are related to the most energetic class of SEP events, associated with both solar flares and CMEs, and requiring acceleration processes that produce particles with energies $\geq$ 500 MeV upon entry in the Earth's atmosphere. GLE data recorded by the worldwide Neutron Monitor (NM) network are a useful resource for space weather modelling during solar extreme events. Plainaki et al. (SH568) reviewed how their Neutron Monitor Based Anisotropic GLE Pure Power Law (NMBANGLE PPOLA) model [34] can lead to the derivation of the properties of the relativistic solar protons during a GLE. They presented results for two test cases, GLE60 and GLE71. 5-min GLE data for GLE60, on 15 April 2001, and 29 NM stations for GLE71, on 17 May 2012, widely distributed around the Earth were incorporated to fit the equations of the model giving the time variations $\Delta N/N_0$ of the total neutron counting rate N_o, observed at cut-off rigidity R_c, at level h in the atmosphere at some moment t, and the equation of A, the dimensionless normalized function that describes the spatially anisotropic arrival of the Solar Cosmic Rays (SCRs) at 1 AU, [34, 35]. The Levenberg-Marquadt non-linear optimization algorithm was applied for these events. The SEP spectrum obtained by these authors is presented in figures 7 and 8. Figure 7 shows the spectral index evolution (upper panel) and SEP fluxes (lower panel) on 15 April 2001 at an altitude of ~20 km, with a hard spectrum of accelerated protons during the initial phase of GLE60 and a rather soft spectrum in later phases i.e. after 14:00 UT (γ ~-5.5). During the initial phase of GLE71 a spectral index γ ~-2.1 is obtained (figure 8) whereas in later phases i.e. after 02:20 UT, the result is a rather soft spectrum of accelerated protons (γ ~-3.8). These authors deduced that their results for both events are in good agreement with estimations based on other works, e.g. [36] and the results for GLE71 are consistent with the typical range found by [37] for shock wave acceleration in case of relativistic SEP events, although a direct flare contribution cannot

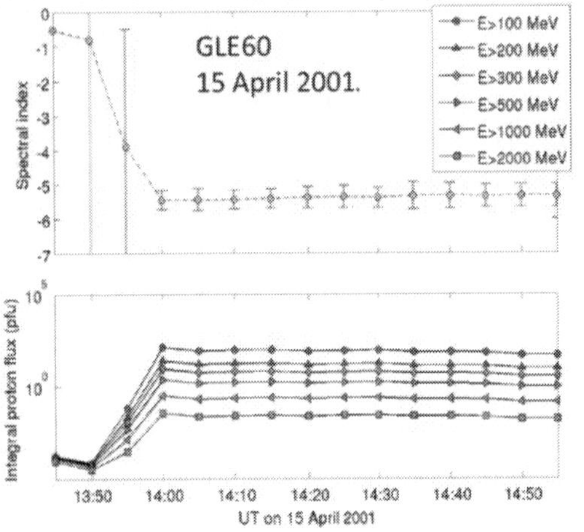

Figure 7. Spectral index evolution (upper panel) and SEP fluxes (lower panel) on 15 April 2001, at an altitude of ~20 km (Plainaki et al. SH568).

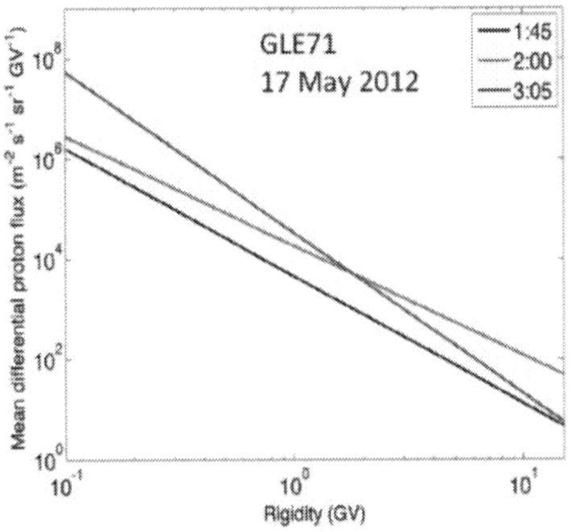

Figure 8. SEP spectrum on 17 May 2012, at an altitude of ~20 km (see text) (Plainaki et al. SH568).

be excluded. Comparison by these authors of the integral SEP fluxes calculated by the NMBANGLE PPOLA model with GOES observations showed that in the higher energy range the model results can provide realistic estimation of the SEP fluxes [35].

Computation of radiation dose rate at flight altitudes is of particular importance [38]. The dose rate can be strongly affected by the spread of GLE parameters deduced from NM data by different groups [35, 39, 40], the causes of which were investigated by Bütikofer and Flückiger (SH296). They performed detailed calculations of the GLE parameters for the GLE70 event, on 13 December 2006 using different NM response functions (figure 9) and found, that the ambient radiation dose rate caused by SCRs at high geomagnetic latitude and at typical cruise altitude (~11 km) is ~2 μSv/h, when they used the coupling function of [41] and ~5 μSv/h when they used the other two yield functions [42, 43], thus showing significant differences. Other possible reasons in the derived GLE parameters having significant consequences in the computed radiation dose rates were found to be the asymptotic directions and cut off rigidities computed with different geomagnetic field models, as well as the selection of NM stations. Based on their calculations, the assumed form of the particle pitch angle distribution wasn't found to play a significant role. These authors highlighted the need for more detailed exchange of information on the different GLE analysis procedures between specialists and suggested a forum for this to materialize (EURADOS WG11).

Kryakunova et al. (SH176) analysed the behaviour of the Cosmic Ray (CR) intensity at the worldwide NM network in the events of 2012 when there was a significant increase of the integral

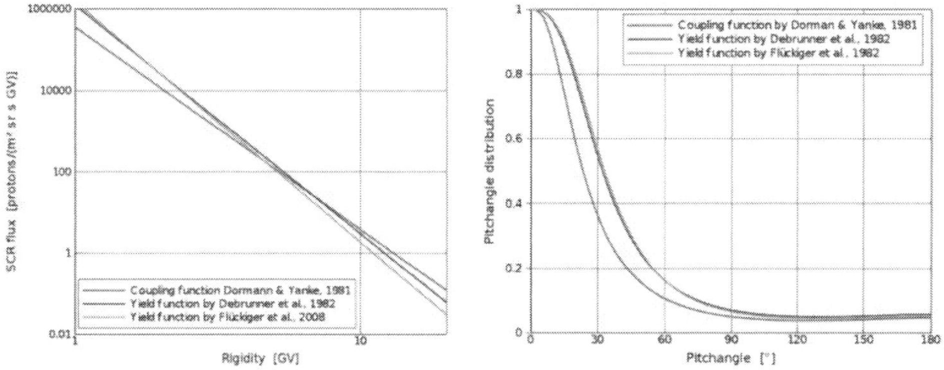

Figure 9. Deduced SCR spectra (left) and pitch angle distributions (right) by using different NM response functions (Bütikofer and Flückiger SH296).

proton flux with energies > 100 MeV, namely in the events of January 27, March 7, and March 13, 2012. According to them all these events can be considered as candidates for the GLEs of SCRs, increasing the number of GLEs observed in solar cycle 24 than widely recognised. Kühl et al. (SH271) presented the extended measurement capabilities of the EPHIN instrument onboard SOHO providing particle spectra up to 1 GeV. The solar source of GLEs, their propagation conditions as well as their impact on the Earth environment during the last two solar cycles was investigated by Catalán et al. (SH214) using NM and SOHO data.

2.4 Space Weather and Space Climate: The SEP Perspective

As Usoskin and Kovaltsov (SH510) pointed out a typical quantity for a Solar Particle Event (SPE) is the fluence of SEPs with energy above 30 MeV, F_{30} [44]. There is sufficient knowledge of SPEs over the space era since the mid-1950s [45] with only several events with F_{30} = (1-10) x 10^9 cm^{-2} and hundreds of weaker SPEs observed. Figure 10, shows the probability of occurrence of the annual fluence (> 30 MeV) exceeding the given value F_{30}, with open triangles showing the measured annual fluences for the space era (1956-2008) and the black triangle reflecting the fact that no SPE with fluence F_{30} > 10^{10} cm^{-2} was observed [44]. However, it is important to know, both for purely theoretical aspects of solar/stellar physics and for technical applications, the statistics of extreme SPEs with F_{30} > 10^{10} cm^{-2}, a study possible only by using indirect proxy data, as carried out by these authors. They used data on the cosmogenic isotopes ^{14}C and ^{10}Be in terrestrial archives spanning the timescale from centuries to 11 millenia. Figure 10 summarizes their obtained results. In the annually resolved ^{10}Be data, four events with F_{30} = (1-1.5) x 10^{10} cm^{-2} (open red star in figure 10 corresponds to these 4 candidate SPE events) and no events with F_{30} > 2 x 10^{10} cm^{-2} (filled red star corresponding to this) were identified since 1400 AD. From more roughly resolved data, they identified 20 SPE candidates with F_{30} = (1-3) x 10^{10} cm^{-2}, (represented in figure 10 by the open blue circles) and clearly no event with F_{30} > 5 x 10^{10} cm^{-2} (represented by the filled blue circle) over the last 11,400 years. They found that practical limits can be set as F_{30} ~ 1, 2-3 and 5 x 10^{10} cm^{-2} for occurrence probabilities ~10^{-2}, 10^{-3}, and 10^{-4} yr^{-1}, respectively [44]. The authors also revisited these assessments on a timescale of up to 1 Myr, based on measurements of cosmogenic radionuclides in lunar rocks and concluded that the three timescales yield a consistent distribution, with the data suggesting a strong roll-over of the occurrence probability, so that SPE events with F_{30} > 10^{11} cm^{-2}, are not expected on a Myr timescale [46]. Herbst et al. (SH323), besides the variation of cosmogenic radionuclides due to the modulation of GCRs, investigated the influence of 58 out of the 71 GLEs which occurred within the past 5 solar cycles and discussed the possibility to detect such events in present ice-core and tree-ring records. The investigations showed that none of the modern GLE events has had a detectable impact on the global ^{14}C, ^{10}Be and ^{36}Cl production. However, two major points were given: 1) surprisingly, most sensitive to the influence of GLEs is the radionuclide ^{36}Cl and 2) in order to increase the production about 15%,

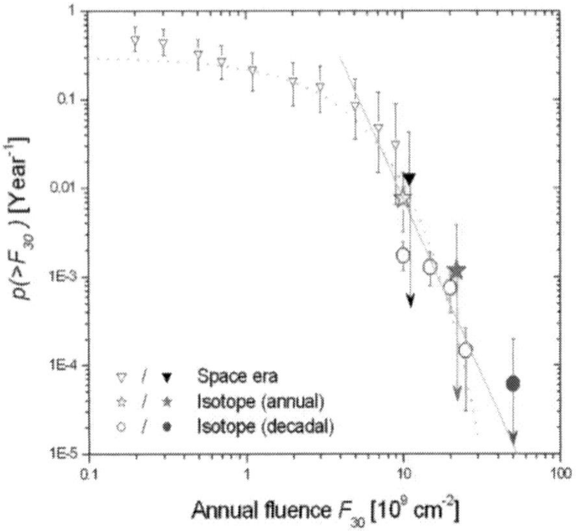

Figure 10. Summary of the obtained results based on cosmogenic isotope ^{14}C and ^{10}Be data (see text) (Usoskin and Kovaltsov SH510).

like e.g. during the 775AD event, a GLE flux up to 10 times higher than the strongest GLE ever measured (23rd February 1956) would have been necessary.

Following a different approach, Miroshnichenko (SH376) analysed all available data on the largest SPEs for the period from 1561 up to now, identifying a number of physical and methodological limitations that are important for the estimation and prediction of hazardous SCR radiation fluxes [47]. Very promising in this direction is the combination with the 'Upper Limit Spectrum' (ULS) [48, 49] they carried out to develop a new approach to the 'worst-case' concept with the Carrington event, providing a crucial normalization point for this goal.

As Bakaldin et al. (SH127) pointed out the direct ion charge measurement of CRs with energies more than several MeV/nucleon is impossible. They presented the main results of their experimental analysis and Monte-Carlo simulations of the geomagnetic separator technique possibility for the study of CR ion charge state composition in the energy range from several MeV/nucleon to hundreds of MeV/nucleon. Protopopov et al. (SH440) presented the exploitation results of the Roscosmos space radiation exposure on electronic components of engineering Monitoring System elements. Calculation of different particle contribution in dose rate increases were carried out and presented. These authors showed that solar particle exposure can be a cause of considerable dose rate increase and consequently electronic equipment failure. But the main cause of dose rate increasing events at the circular orbit ~ 20000 km is electron exposure.

3. Forbush Decreases and GCR Short Term Variations

During the passage of Interplanetary Coronal Mass Ejections (ICMEs) decreases in energetic particle fluxes are measured in a very wide energy range. These decreases were first observed by ground-based detectors and were termed Forbush decreases (FDs) [50]. They are characterised by a sharp decrease followed by a slow recovery phase. These decreases measured by spacecraft are often called FDs [51] in a similar way to the decreases observed by ground level detectors [50]. Blanco and Hidalgo (SH272) using unique measurements from the Helios mission (https://www2.mps.mpg.de/de/projekte/helios/) from 0.3 to 1 AU studied the effect of Magnetic Clouds (MCs) on the GCR background and the MC properties in the inner heliosphere. The E6 experiment onboard Helios is the energetic particle detector able to measure electrons, protons, and alphas in the range of 300 keV/n to > 50 MeV/n [52]. It has been shown previously that, in absence of strong SEP events the single detector rates of the E6 anti-coincidence (A) and sapphire Cherenkov detectors are sensitive to CRs with rigidities about GV.

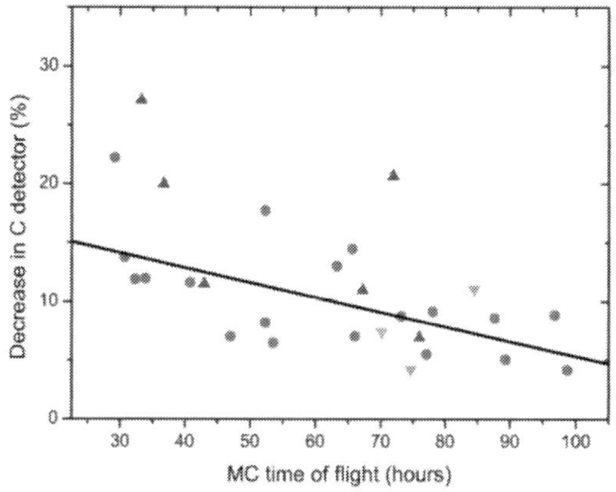

Figure 11. Decrease in the C detector versus MC time of flight. The solid line represents a linear fit to the type 1, 19 events (red circles). The other symbols represent events of other types as classified in [54] (Blanco and Hidalgo SH272).

Because their statistical precision is in the order of hundreds of counts per second, both detectors are very well-suited for studying the short-term decreases observed in their count rates during MC passages. These authors identified a total of 35 MCs at the Helios locations. They used the model of [53] as fit function to the measured magnetic field with the aim to confirm a flux rope structure, to establish the flux rope limits and thus the MC limits, as well as determine the MC orientation. 19 MCs were found to be free of SEP contamination and were used for the investigation of the effect of MCs on the GCR flux [54]. These authors found that the FD depth is linearly related to the magnetic field strength of the MC by a correlation factor of 0.66 and to the MC associated rigidity by a factor of 0.66 [54]. However, they found a poor linear relationship with the MC speed. Figure 11 presents the decrease in the detector C versus the MC time of flight. The solid line represents a linear fit to the 19 events (red circles) which the authors termed as type 1, in which the MCs were associated to count-rate decreases measured by the A and C detectors simultaneously. A negative linear relationship (Pearson coefficient=-0.64) was found with the time of flight of the MC, which implies a constant incoming rate of GCRs into MCs, and supports the idea of magnetically closed MCs i.e. with the two legs rooted back to the Sun. Their analysis of the MC properties showed that the MC diameter grows with the distance to the Sun, confirming the MC expansion during its travel in the IP medium. The CME speed did not show a clear dependence in the distance to the Sun. As these authors state, if these observations are confirmed, the deceleration process has to take place at close distances to the Sun $<$ 0.3 AU [54]. These authors also presented results of an analysis of a pool of FDs detected at 1 AU by NMs, and produced by ICMEs with MCs and ICMEs with flux ropes and found that MCs are more effective than flux ropes in producing deeper FDs. Because the flux ropes were mostly aligned along the ICME legs, they concluded that the spacecraft path through ICME can affect the FD strength because the spacecraft is observing different regions of the ICME [55]. Hidalgo et al. (SH504) studied the effect of MC disturbances on the SEP spectra with the aim to evaluate the role of the arrival direction of the SEP into the MC.

Belov et al. (SH248) modelled the GCR density variations in MCs and estimated the contribution of different parameters into the model. They utilised the database of FDs in IZMIRAN (http://www.izmiran.ru/), based on the density and anisotropy variations of CR particles with 10 GV rigidity derived by the global survey method (GSM) [56], using data from the world network of NMs. These authors found that in most cases the behaviour of CR density within MCs at Earth can be described by a simple parabolic dependence on distance expressed in gyroradii. An important finding is that the majority of MCs modulate CRs, decreasing their density, although there is a group of events (~1/5) in which the CR density increases within a MC. The extremum (minimum or maximum) of CR density more often settles down closer to the cloud center than its edges. They also derived a successful quantitative relation of the variation of CR density of a magnetospheric origin with the Dst-index. The obtained coefficient can be applied to variations of CR density during any periods.

Utilising the IZMIRAN database on FDs, Kryakunova et al. (SH409) studied the recurrent and sporadic FDs in deep solar minimum. They concluded that FDs caused by fast wind streams associated with Coronal Holes (CHs) are of a shorter duration compared to those caused by ICMEs. Regarding the distribution of times, a late minimum in CR intensity close to the solar wind speed maximum is observed, delayed with respect to the Interplanetary Magnetic Field (IMF) maximum. Furthermore, there are large differences between recurrent and sporadic FD in the CR anisotropy: in the recurrent case it is significantly smaller than for the sporadic case and exhibits a more smooth behaviour. Abunina et al. (SH253) carried out a statistical analysis of all FDs with a sudden onset during the period 1957-2012 and showed that the main features of the phase distribution of the first harmonic of the CR anisotropy persist throughout the main phase of the FDs, starting from the hour before the shock wave until the hour with the maximal amplitude of anisotropy.

Extending previous work [57], Wawrzynczak and Alania (SH329) analysed the temporal changes of the rigidity spectrum of FDs of the GCR intensity in different energy ranges, based on the hourly data from the worldwide network of NMs and muon telescopes. They confirmed that the rigidity spectrum of FDs of the GCR intensity is observed to depend on energy also for the FDs in November 2004 and October-November 2003. They analysed the time-evolution of the state of the turbulence of the IMF in various frequency ranges during FDs. The analysis showed that the decrease of the exponent ν of the Power Spectral Density ($PSD \sim Pf^{-\nu}$, where P is the power and f is frequency) with decreasing frequency resulted in the softer rigidity spectrum of FD for GCR particles with higher energies.

Stealth CMEs, that are CMEs with no apparent solar surface association and their IP counterpart have become a subject in recent studies of solar activity [58]. Whether all of such stealth CMEs can drive a FD is difficult to investigate on the basis of NM measurements, because these measurements not only reflect the GCR intensity variation in IP space but also the variation of the geomagnetic field as well as the conditions in the Earth's atmosphere. Heber et al. (SH393) studied FDs associated to 11 stealth CMEs using data by the Electron Proton Helium Instrument (EPHIN) onboard the spacecraft SOHO and Chandra, which is capable to measure CR variations of less than 0.1% when using 30 minute average single detector count rates of the anticoincidence A and the last semiconductor F [59]. None of the 11 events were accompanied by low energy particles, making single detector rates ideally suited to investigate the time profiles of the corresponding FDs. All stealth ICMEs were found to cause a FD with amplitudes between 0.6 and 3.5%. None of them can be found in the NM data unambiguously. The best case is displayed in figures 12. Figure 12 shows from top to bottom the measured count rates of the EPHIN detector F aboard SOHO and Chandra (blue, black color traces on top of this figure), together with the ones from the NM stations Tera Adelaide (black), Apatity (red), Norlisk (green), and Oulu (blue). While the FD is visible in the spacecraft data, there is no evidence at the NM stations used, apart from a small decrease observed in Tera Adelaide. This indicates either a strong rigidity dependence or effects caused by the Earth's magnetic field or by the atmosphere.

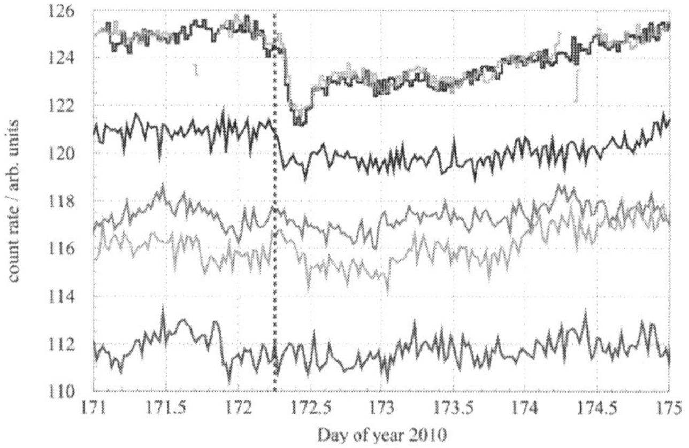

Figure 12. 30-min averaged count rates of the anticoincidence detector (EPHIN) in comparison with the neutron monitor count rates from different neutron monitors. The dotted line represents the start time of the count rate decrease (see text) (Heber et al. SH393).

The Belgrade CR station at the Institute of Physics in Belgrade has been continuously measuring the CR muon intensity since 2002. Srećković et al. (SH438) presented comparative analyses of extreme solar events - GLEs and FDs using CR muon measurements and very-low frequency (VLF) radio wave measurements by the CR and VLF stations in Belgrade. They demonstrated how such measurements can be used as a new method for investigations of high-energy transient phenomena of astrophysical importance. Barbashina et al. (SH383) analysed FDs registered by the muon hodoscope URAGAN [60] during the period 2007-2012. During this period, 185 well-separated FDs, including 44 FDs with amplitude > 0.5%, were found in the integral counting rate. One supermodule of the hodoscope registers and records in 2-dimensional angular matrix the arrival directions of about 80 thousand muons every minute. The analysis of such matrices allowed to study zenith-angular and azimuthal dependences and anisotropy of the muon flux using the single setup. These authors concluded that in the phase of the counting rate decrease, the relative local anisotropy of muon flux increases in the geographical directional from North to South, while in the phase of minimum as well as in the phases before the beginning of decrease and recovery it remains symmetric. Furthermore, the analysis of correlations between the projections of the relative local anisotropy vectors provide additional opportunities for the study and identification of various heliospheric disturbances.

The quasi-biennial oscillations (QBOs) appear to be the most prevalent quasi-periodicity shorter than the 11 yr cycle in solar activity phenomena. They are observed at all levels of the solar atmosphere and are transmitted to the heliosphere via open solar magnetic flux. However, the nature of the QBOs at the Sun and their influence on the heliosphere remain not well-understood and need further investigation. Towards this goal, Bazilevskaya et al. (SH200) studied the correlation of the QBOs in GCRs and the solar activity indices, in the context of contemporary ideas on the GCR modulation. Their main results are that the QBOs in CRs are due to the QBOs in solar activity, however, they are transmitted to the heliosphere via open magnetic flux. That is why the QBOs in CRs do not well correlate with the QBOs in solar indices, and are rather coherent with the QBOs in the Heliospheric Magnetic Field (HMF) strength B. Also, contrary to the solar QBOs, the QBOs in CRs are hardly modulated by the 11-year cycle. The CR QBO amplitudes and duration do not correlate with those of the 11-year cycle. Moreover, there is no apparent difference in the CR QBOs between the odd and even 11-year solar cycles, as well as between the A>0 and A<0 magnetic polarity periods. As these authors stated, they aim to perform future work on the search for the relation of QBO with Forbush decreases, and will attempt to understand the difference found in the 11-year and QBO CR modulation.

Analyzing data of the worldwide NM network, Balabin et al. (SH419) found a considerable annual variation of CR flux in 2011-2013. The variation was observed in all stations, circumpolar, mid-latitude and subequatorial. They found that it is present in the CR density changes obtained by the global survey. The phase variation was found to be the same for all NM with a maximum in December-January and a minimum in June-July. The amplitudes of the variation was more than 1%. These authors showed that, in the same period 2011-2013, quasi-annual variations of the IMF, other parameters of the IP medium (solar wind speed and dynamic pressure) and the parameters of the solar magnetic field are observed. They concluded that these variations are in good agreement with the identified CR variations. They suggested as possible agents of these variations the features of solar activity or the asymmetry of the heliosphere.

Paouris et al. (SH448) studied the GCR intensity modulation based on solar and heliospheric indices for solar cycle 23 (1996-2008). In previous works a number of different indices such as the sunspot number, the CME-index (as defined in [61]) the interplanetary magnetic field strength and the heliospheric current sheet tilt were selected to be the most appropriate ones in order to describe the CR intensity of 10 GV observed by the NM network. Their new approach was the extension to the influence of the solar magnetic field parameters, the mean magnetic field strength and the polar magnetic field strength. With use of the wavelet analysis method they confirmed a major periodicity of about 20-21 years, which indicated the existence of the 22-year cycle in CR variations. They derived the best empirical relation, as they state, of the CR modulation taking into account the sunspot number

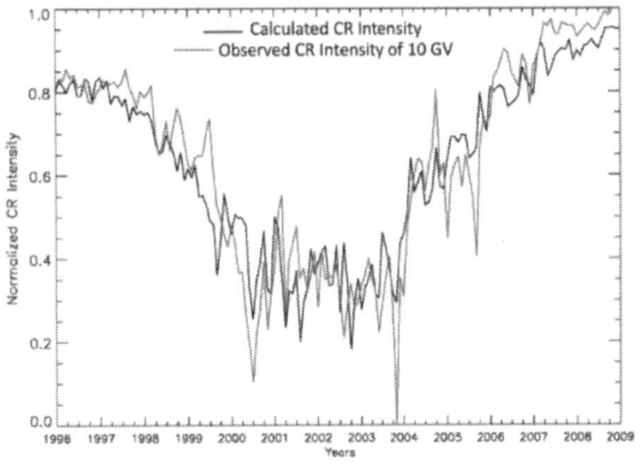

Figure 13. The observed values of CR intensity of 10 GV (red line) and those calculated by the derived new empirical relation (black line) (Paouris et al. SH448).

the CME-index, the mean magnetic field of the sun and the solar polar field, which was improved significantly to a relative Root Mean-Square Deviation (RMSD) of 8.7%, instead to the previous one of about 10% between the observed and the calculated CR intensity values (figure 14).

Various works studied the 27-day variation of the GCR intensity. Alania et al. (SH354) studied the period 2005-2008 of solar cycle 23 using NM data corrected and uncorrected for geomagnetic disturbances and the Dst index. They showed that the amplitudes of the 27-day variations of GCR intensity calculated for corrected and uncorrected data do not differ for NMs with cut-off rigidities less than 4-5 GV, while the difference is as notable as cut-off rigidity increases. Figure 15 presents their results on the rigidity spectrum of the 27-day variations of GCR intensity, calculated as described in [57], for corrected and uncorrected NM data, which they found to be generally soft, while it is hard for the corrected data case. For both cases they observed a clear tendency of a softening of the temporal changes of the rigidity spectrum from 2005-2008 (approaching solar activity minimum). These authors concluded that a study of peculiarities of the 27-day variations of the GCR intensity should be carried out by NMs with cut-off rigidities with less than 4-5 GV.

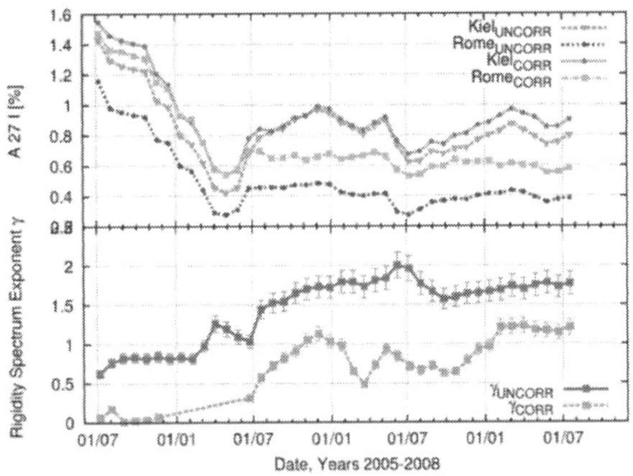

Figure 14. (Top part) Temporal changes of the amplitudes of the 27-day variation of the corrected and uncorrected GCR intensity for Kiel and Rome NMs. (Bottom part) The exponent γ of the rigidity spectrum of the 27-day variation of the corrected and uncorrected data GCR intensity in 2005-2008 (Alania et al. SH354).

[62, 63] had found a new class of quasi-periodicities. Alania and Gil (SH397) further investigated this effect and concluded that the 3-4 Carrington Rotation Periods (CRP) recurrence found is shaped by combined processes of the turbulent solar magnetic dynamo and differential rotation of the Sun leading to the conversion of the poloidal magnetic field into toroidal (α-ω effect). As the most appropriate candidate of the topological structure considered to be a cause of the 3-4 CRP effect they considered the diffusion of the current sheet in the Sun's atmosphere. Gil et al. (SH372) continuing the study of this phenomenon found quasi-periodicities with periods less than, and larger than 3-4

Journal of Physics: Conference Series **632** (2015) 012070 doi:10.1088/1742-6596/632/1/012070

CRP. The 27-day variation of the 3-dimensional solar anisotropy of GCRs during the period 1965-2012 was studied by Modzelewska and Alania (SH399). They showed that the GG index, calculated by Nagoya telescope data (being a measure of the North-South asymmetry), is highly correlated with the Bx and By components of the IMF. Also they found that the 27-day variation of the GG index varies in accordance to the solar cycle with a period of 11-years, being in good agreement with the 27-day variation of the Az component of the GCR anisotropy calculated by the IZMIRAN group. They also demonstrated that the anisotropy component of the 3D GCR anisotropy is larger for positive polarity periods than for negative polarity periods.

Alania et al. (SH400) studied the annual distribution of the visually observed cloudless days and cloudless nights at Abastumani Astrophysical Observatory in Georgia in 1957-1993, in relation with the GCR intensity variations. They showed that the inter-annual distribution of the geomagnetic Ap index for cloudless days and nights are different, a result which they used in the construction of a 2D non-stationary model of GCR propagation in the heliosphere. For the modelling they also took into account new data of primary GCR proton spectrum in the local interstellar medium (LISM). Uchaikin and Sibatov (SH557) following a new fractional approach presented a more realistic, as they stated, compound model of CR diffusion, and compared their results with those of other authors, discussing the reasons of discrepancy.

From a theoretical viewpoint, Wawrzynczak et al. (SH443) applied stochastic methodology to model the GCR transport in the heliosphere, based on the solution of the Parker's Transport Equation (PTE). They presented their model of the FD and the 27-day variation of the GCR intensity based on this stochastic approach. The modelling results are in good agreement with the NM data. They concluded that the models obtained based on the solution of the stochastic differential equations allow to reflect the stochastic character of the GCR particle distribution in the heliosphere. Gerasimova et al. (SH208) studied the tensor anisotropy of CRs, using long-term observations from the muon intensity of CRs at Nagoya and Yakutsk and also the observations of the NM worldwide network, which revealed amplitude-phase oscillations of the semidiurnal variation during a year and also oscillations of antisymmetric diurnal variation. These effects reflect the properties of a tensor anisotropy which these authors determined with simple geometric models. They concluded that the main mechanism of the formation of the CR anisotropy tensor is associated with loop structures in the IMF.

4. GCR Long Term Variations

Cosmic Rays entering in the heliospheric region propagate into the solar wind, i,e. an expanding medium out-flowing from the Sun. This causes particles to diffuse into the heliosphere losing energy causing the so-called solar modulation, exhibited in the decrease of CR differential intensity below 30 GeV. This variation is strongly related to solar activity, the greater the solar activity the less is the CR intensity. Boschini et al. (SH368) using the HelMod Monte-Carlo code (http://www.helmod.org), which is based on the Parker's transport equation, which contains diffusion, convection, particle drift and energy loss [64], explored the effect of high solar activity on CR intensity as function of time and at different energies during the present solar maximum. They implemented a short-term description of the diffusion coefficient based on NM measurements. The comparison they carried out between their model and the AMS-02 proton data from July 2011 up to May 2013, during the maximum of the present solar cycle 24, showed that their solutions well agree with the experimental data.

Gieseler et al. (SH308) investigated the radial and latitudinal gradients of GCR protons and alphas, using measurements by the Kiel Electron Telescope (KET) onboard the Ulysses spacecraft [65], the first ever to explore the 3D structure of the heliosphere, as well as 'baseline' measurements delivering the temporal variation for a stationary observer by the PAMELA space borne experiment launched in June 2006 [66]. For helium they found G_r(He)~2.2%/AU and G_θ (He)~0 or slightly negative and the gradients for protons: G_r(p)~3%/AU and G_θ(He)~-(0.05-0.12)%/deg. As Munini (SH285) showed the PAMELA experiment has measured the electron spectrum at Earth in great detail, extending up to ~100 GeV and now, with a special effort, down to ~70 MeV. The yearly GCR electron spectrum measured during the A<0 solar minimum of solar cycle 23 (2006-2009) was presented and the author discussed future plans to extend the measure to positrons for the investigation of the drift effect and to data collected after 2009. Krainev et al. (SH364), using the Ulysses solar wind and HMF data for all three Ulysses orbits around the Sun, compared these characteristics in and outside of the HMF sector structure zone. Using a simple GCR model they considered how the differences found in the HMF distribution may influence the calculated GCR intensity in the heliosphere and compared the results of the calculations with the observations.

Stozhkov et al. (SH436) presented long-term observations of CRs in the Earth's atmosphere. The available data cover more than five 11-year cycles of solar activity from July 1957, i.e. the middle of the 19th solar cycle, till August 2014, i.e. the middle of the 24th solar cycle. These authors showed CRs play the main role in the Earth's atmospheric electricity processes. Sorokin et al. (SH445) performed an analysis of the relationship between CR fluxes observed in the atmosphere and solar activity. They carried out a comparison of the CR fluxes measured in positive and negative phases of the 22-year solar magnetic cycles for the period 1957 – 2014 and found that the CR drift fluxes do not exceed 10% from the total GCR flux. The long-term CR behaviour at a rigidity threshold of ~6 GV was studied by Storini et al. (SH345) using data for about 5 solar cycles. They derived results from a detailed analysis performed by using the descriptive statistics technique and discussed them in the context of variations in the heliospheric conditions. Krainev et al. (SH199) further discussed their method to decompose the calculated intensity into the partial 'intensities' related to the main physical processes: diffusion, convection, adiabatic cooling and magnetic drift as well as its limitations in more detail.

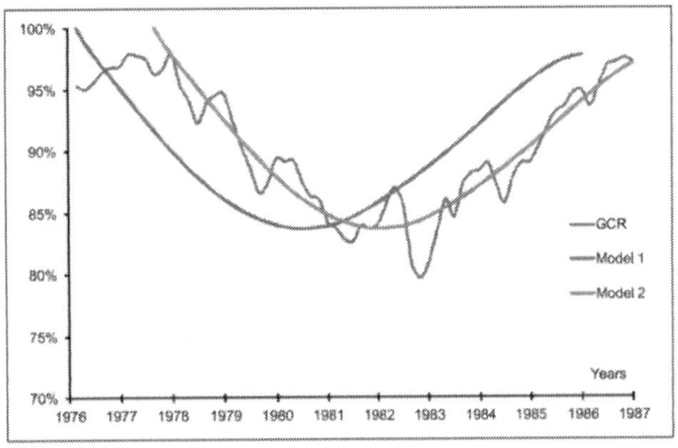

Figure 15. Changes of the amplitude of the 11-year variation of the GCR intensity in [%] registered by the Moscow NM (red line). Model 1: the solution of the Parker's non-stationary transport equation (blue line) Model 2: this solution shifted by 18 months (green line) (Siluszyk et al. SH458).

A new 2D time dependent model describing the long term variations of the GCR intensity has been developed by Siluszyk et al. (SH458). Their model implements in the Parker's transport equation the smoothed parameters characterizing temporal changes of the exponent γ of the Power Spectral Density (PSD) of the IMF turbulence, B magnitude of the IMF, tilt angle of the heliospheric current sheet for solar cycle 21 and changes of the drift effect of the GCR particles versus solar activity. Figure 16 shows how their model results compare with the observations for the period under study. As these authors concluded a delay time ~18 months can be accepted as an effective delay time caused by the combined influence of all parameters implemented in their 2D model. Iskra et al. (SH463) studied the relation of an exponent γ of the power law rigidity spectrum of the long period variations of the GCR intensity and the exponent ν_y and ν_z of the PSD of B_y and B_z components of the IMF turbulence. They showed that an inverse correlation between γ and ν_y and between γ and ν_z are universal, as expected from the Quasi-linear Transport theory for the rigidities of the GCR particles to which NMs respond.

5. Heliospheric Jovian Electron Variations

Early investigations of Jovian electrons assumed that they reach the Earth along IMF lines mainly during the period of direct connection between Earth and Jupiter (occurring every 13 months) [67, 68]. In the existing models, in the absence of direct Earth-Jupiter connection, the appearance of Jovian electrons near Earth and their 27-day variation during the following 5-6 months were explained by perpendicular electron diffusion near Jupiter, and by CIRs limiting their angular spread [69, 70]. The interest to Jovian electrons resumed after the extremely quiet minimum of solar activity in 2007-2008, when electrons were observed near the Earth for 14 successive solar rotations. Kecskeméty et al. (SH306) suggested another propagation model to account for the recent observations. Their explanation of the long-term occurrence of electrons near the Earth is the possible formation of magnetic traps in the IP medium filled by electrons when they pass by Jupiter and then detected when passing near Earth.

According to the proposed model, Jovian electrons can be observed in any period irrespective of the phase of the Earth-Jupiter connection, whereas the 27-day variation of these electron fluxes is explained by the rotation of traps along with the Sun. These authors presented observational evidence during the quiet time of 2007-2008 which showed that, under the absence of stability of the solar wind structure the trap is unstable as well, with the periodicity of electron fluxes observed to be then broken. Furthermore, they showed an anti-correlation between the fluxes of trapped electrons and galactic protons. The Jovian electron flux maxima were found to coincide with galactic proton minima, indicating their entrance to the central part of the trap is inhibited by the enhanced magnetic

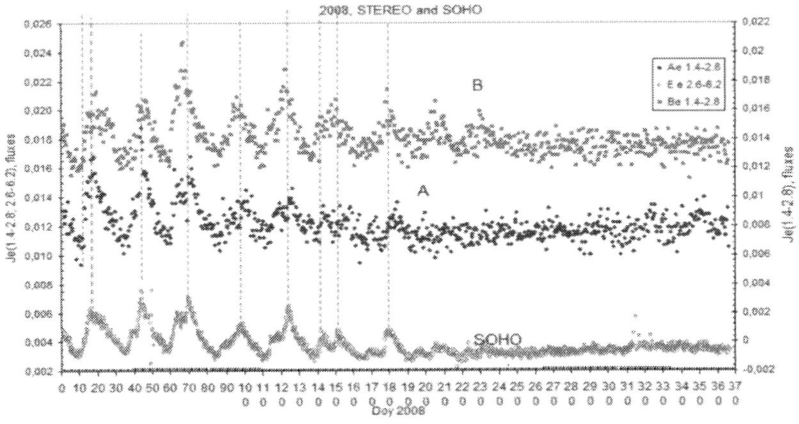

Figure 16. Due to rotation of the trap together with the Sun Jovian electrons are registered first at STEREO B, then at SOHO and then at STEREO A. The red dotted vertical lines indicate the times of maxima at the various spacecraft. The high background of HET makes the definition of exact times of maxima difficult, but the sequence is confirmed (Kecskeméty et al. SH306).

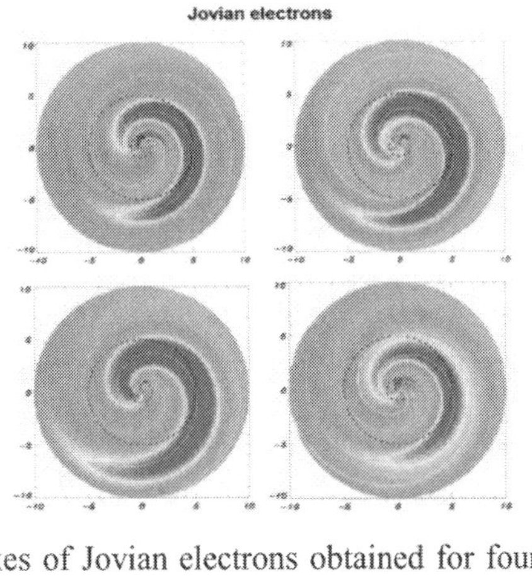

Figure 17. Contour plots of fluxes of Jovian electrons obtained for four different phases of the CIR model [69]. The position of Jupiter is at 3 hr (Kecskeméty et al. SH306).

field inside the trap. The temporal shifts of the profiles between STEREO A and B were found to be compatible with the rotation of the trap (figure 17). These authors carried out numerical simulations in which Jupiter is a moving source in the co-rotating frame and perfect co-rotation is assumed. The results of the simulations showed (figure 18) the appearance of traps under different time variations of solar wind speed in 2008, and confirm the hypothesis of stronger magnetic field behind Jupiter might trap electrons from escaping the region resulting to trap formation [71].

Using two different approaches, Vogt et al. (SH281) modelled the modulation of Jovian MeV-electrons by CIRs using the VLUGR3-Code [72, 73] to solve the Parker's transport equation. They modelled the diffusion coefficients as well as the solar wind speed from 0.1 to 50 AU and also compared their simulation results with the IMP-8 electron count rates.

6. Instrumentation and Methods

For many years, there have been astrophysical muon detector systems operated at the University of Adelaide, originally for teaching purposes. As Clay et al. (SH355) showed, the detectors have recently been rebuilt and upgraded with a new data acquisition system, to become H.E.A.M.S., the Adelaide High Energy Astrophysics Muon System (http://www.physics.adelaide.edu.au/astrophysics/muon/). These authors discussed the properties of the new system, including its various pressure coefficients, with illustrations of short term variations and solar wind sidereal anisotropies. Yashin (SH379), on behalf of the URAGAN Team, presented the real-time data recorded by the muon hodoscope URAGAN for each hour and also the results of their analyses which are available at the URAGAN website. They also discussed some peculiarities of the application of these data for the analysis of various events in the heliosphere, magnetosphere and atmosphere. Makhmutov et al. (SH337) presented the CARPET cosmic ray device at the astronomical complex CASLEO in Argentina [74, 75] and discussed the CR variation and solar flare activity observed in October-November 2003.

As Blanco et al. (SH124) presented, CaLMa, the Castilla-La Mancha NM [76], is a NM integrated by 15 boron trifluoride counters located in Guadalajara, Spain. With a vertical rigidity of 6.95 GV it is able to measure neutrons produced by primary CRs with energies > 6 GeV/nucleon. CaLMa has been monitoring the solar activity continuously since October 2011, and has reported more than 15 FDs. These authors presented preliminary results inferred from the CaLMa count rate variations and showed last improvements as well as future plans. García-Población et al. (SH125) presented mmPanel, a new software tool developed in CaLMa to control the station operation (http://www.calmanm.es/start). The application provides a tool to identify anomalies in data and allows the operator to trace its origin, and since the application is NMDB (Neutron Monitor Database)-aware, it also features automatic uploading of revisited data to the network. García-Tejedor et al. (SH126) presented the work in progress for the construction of a muon telescope in the same facilities as CaLMa, in order to use both instruments simultaneously to measure neutrons and muons produced by primary CRs at the same physical location and also analyse the possible correlation between both measurements. These authors presented the proposed muon telescope physical configuration, data acquisition system, local architecture (hardware and software), as well as the long term data storage and publication and the projected muon detection capabilities that this system should be able to perform.

Changes in the CR stations ISTP SB RAS of the Siberian Branch of the Russian Academy of Science for 2012-2014 were described by Lukovnikova (SH404). Proposed ways of solving problems that exist in each CR station were presented in order for data to be provided without failures and omissions every minute. Recent research applications at the Athens NM station were presented by Mavromichalaki et al. (SH457). These include the optimized automated GLE Alert Plus, currently a service under testing and evaluation by the European Space Agency, a simulation tool of the CR showers in the atmosphere, based on Geant4 and named DYnamic Atmospheric Shower Tracking Interactive Model Application (DYASTIMA) (http://cosray.phys.uoa.gr/apps/DYASTIMA/DYASTIMAmanual.pdf) as well as a Space Weather Forecasting Center which provides a 3-day geomagnetic activity report on a daily basis.

e-CALLISTO (Compound Astronomical Low-cost Low-frequency Instrument for Spectroscopy and Transportable Observatory) is a worldwide network with 24 hours a day solar radio burst monitoring (http://www.e-callisto.org/). As Russu et al. (SH437) presented, in 2013 the University of Alcalá joined the e-CALLISTO network with the installation of two Solar Radio Telescopes (SRTs), the Melibea and the EA4RKU SRTs. The Spanish e-CALLISTO SRTs routinely provide data to the network. These authors presented a list of type II radio bursts as observed by Melibea during its first year of operation and results of their study of their relation with SXR flares observed by GOES, and CMEs and SEP events observed by the SOHO spacecraft. In space, the Energetic Particle Detector (EPD) suite for the European Space Agency's Solar Orbiter mission will provide key measurements to address particle acceleration at and near the Sun. As Kulkarni et al. (SH416) highlighted, the EPD suite consists of four sensors (STEP, SIS, EPT and HET). The University of Kiel in Germany is responsible for the design, development and construction of EPT, HET, as well as STEP. The Electron-Proton Telescope (EPT) is designed to cleanly separate and measure electrons in the energy range from 20-400 keV and protons from 0.02 - ~15 MeV. EPT relies on the magnet/foil-technique. The High Energy Telescope (HET) will measure electrons from 300 keV up to 15 MeV, protons from 10-100 MeV, and heavy ions from ~20 to 200 MeV/nucleon. These measurement capabilities are accomplished by a combination of solid-state detectors and a scintillator calorimeter which allows use of the dE/dx vs. total E technique for particle identification and energy measurement. These authors also presented the current status of the development of the EPT and HET units focusing on test calibration results obtained with engineering model as well as future activities.

7. Summary and Future Prospects

An overview of the SH 'Solar and Heliospheric cosmic rays' session of the 24th European Cosmic Ray Symposium (ECRS), Kiel, Germany, 2014 was presented in this paper, with results highlighted in the various topics covered, based on a personal perspective of the author. The debate of the SEP origin in flares or CME-driven shocks is still undecided, with both mechanisms found to provide contributions in the observed SEPs. Methods of imaging HXRs and gamma spectroscopy aboard the RHESSI and Fermi observatories can lead to improvements of the investigation of the SEP sources.

Detailed analysis of multi-spacecraft observations by the twin STEREO spacecraft as well as near-Earth observatories have shed new light in the propagation processes of SEPs that lead to the observed longitudinal distributions in the heliosphere. This important research is expected to continue intensely in the next year and provide new insights on the acceleration and transport of SEPs. Evidently, these new observations comprise a challenge to the current models in order to account for them. More work is also needed to provide a theoretical evidence for the link between the SEP spectral shapes and particle acceleration.

The computation of radiation dose rate at flight altitudes is of particular importance but the dose rate can be strongly affected by the spread of GLE parameters as deduced currently from NM data by different groups. As highlighted, there is a need for more detailed exchange of information on the different GLE analysis procedures between specialists and this is expected to occur in the near future.

The study of CRs will continue to be carried out by networks of ground based NMs as well as muon detectors. The investigations of the CR spectrum from hundreds of MeV up to tens of GeV will continue to be implemented using direct measurements in space by PAMELA, Fermi, AMS and SOHO detectors.

Acknowledgments

The author, OM, thanks the organisers of the 24th European Cosmic Ray Symposium for the invitation and for the travel support provided as well as for an interesting and well-organised conference.

References

[1] Reames DV 1999 *Space Sci. Rev.* **90** 413
[2] Vainio R et al. 2009 *Space Sci. Rev.* **147** 187

[3] PourArsalan M, Townsend LW, Schwadron NA, Kozarev, Dayeh MA and Desai M 2010, *Space Weather* **8** S00E05 doi:10./1029/2009SW000533
[4] Gopalswamy N, Xie H, Yashiro S, Akiyama S, Mäkelä P, Usoskin IG 2012 *Space Sci. Rev.* **171** (1-4) 23
[5] Reames DV 2013 *Space Sci. Rev.* **175** (1-4) 53
[6] Gopalswamy N, Yashiro, S, Lara A, Kaiser ML, Thompson BJ, Gallagher PT and Howard RA, 2003 *Geophys. Res. Lett.* **30** (12) 8015 doi:10.1029/2002GL016435
[7] Cane HV, Richardson IG and von Rosenvinge TT 2010 *J. Geophys. Res.* **115** 8101 doi:10.1029/2009JA014848.
[8] Miteva R, Klein K-L, Malandraki O, Dorrian G 2013 *Solar Phys.* **282**, 579 doi:10.1007/s11207-012-0195-2.
[9] Richardson IG, von Rosenvinge TT, Cane HV, Christian ER, Cohen CMS, Labrador AW, Leske RA, Mewaldt RA, Wiedenbeck ME and Stone EC 2014 *Solar Phys.* doi:10.1007/s11207-014-0524-8
[10] Dierckxsens M, Tziotziou K, Dalla S, Patsou I, Marsh MS, Crosby NB, Malandraki OE, Tsiropoula G 2014 *Solar Phys.* in press
[11] Klein K-L, Trottet G, Samwel S and Malandraki O 2011, *Solar Phys.* **269** 309 doi:10.1007/s11207-011-9710-0
[12] Trottet G, Samwel S, Klein K-L, Dudok de Wit and Miteva R 2014 *Solar Phys.* in press
[13] Atwood WB et al 2009 *Astrophys. J.* **697** 1071
[14] Ackermann M et al 2014 *Astrophys. J.* **787** 1 15
[15] Vedrenne G et al 2003 *Astron. Astrophys.* **411** L63-L70
[16] Winkler C et al 2003 *Astron. Astrophys.* **411** L1-L6
[17] von Kienlin A et al 2003 *Astron. Astrophys.* **411** L299-L305
[18] Kaiser ML, Kucera TA, Davila JM, St. Cyr OC, Guhathakurta M and Christian E 2008 *Space Sci. Rev.* **136** 5
[19] Lin RP et al 2002 *Sol. Phys.* **210** 3
[20] Domingo V, Fleck B and Poland AI 1995 *Sol. Phys.* **162** 1
[21] Shapiro V and Üçer D 2003 *Planetary and Space Science* **51** 665
[22] Kocharov L, Laitinen T, Usoskin I and Vainio R 2014 *Astrophys. J. Lett.*, **787** L21
[23] Ajello M et al 2014 *Astrophys. J.* **789** 1 20
[24] Stone EC, Frandsen AM, Mewaldt RA, Christian ER, Margolies D, Ormes JF and Snow F 1998 *Space Sci. Rev.* **86** 1
[25] Lin RP et al 1995 *Space Sci. Rev.* **71**, 125
[26] Dresing N, Gómez-Herrero R, Heber B, Klassen A, Malandraki O, Dröge W and Kartavykh Y 2014, *Astron. Astrophys.* **567** A27
[27] Gómez-Herrero R, Klassen A, Müller-Mellin R, Heber B and Wimmer-Schweingruber RF 2006 SOHO-17. 10 Years of SOHO and Beyond *in ESA Special Publication,* Vol. 61
[28] Leske RA, Cohen CMS, Mewaldt RA, Christian ER, Cummings AC, Labrador AW, Stone EC, Wiedenbeck ME and von Rosenvinge TT 2012 Solar Phys. **281** 1 301
[29] Shannon CE 1948 *A Mathematical Theory of Communication*, *Bell Syst. Tech. J.* **27** 379
[30] Laurenza M, Consolini G, Storini M, Damiani 2012 *Astrophys. Space Sci. Trans* **8** 19
[31] Laurenza M, Consolini G, Storini M, Damiani 2013 *Proc. of the 13th Intern. Solar Wind Conf.* AIP Conference Proiceedings, Vol 1539 p 219
[32] Frisch U. and D. Sornette 1997 *J. Phys. I France* **7** 1155
[33] Sornette D. 1998 *Phys. Rew.* **E 57** 4811
[34] Plainaki C, Mavromichalaki E, Belov A, Eroshenko E, Andriopoulou M and Yanke V 2010 Solar Phys. **264** 239
[35] Plainaki C, Mavromichalaki E, Laurenza M, Gerontidou M, Kanellakopoulos A and Storini M 2014 *Astrophys. J.* **785** 160
[36] Gopalswamy N, Xie H, Yashiro S, Akiyama S, Yashiro S, Usoskin I, and Davila JM 2013

Astrophys. J. Lett., **765** L30
[37] Ellison DC and Ramaty R 1985 *Astrophys. J.* **298** 400
[38] Mishev AL, Adipbour F, Usoskin I and Felsberger E 2014 *Adv. Space Res.* in press
[39] Bombardieri D J, Michael K J, Duldig M L and Humble J E 2007 *Astrophys. J.* **665** 813
[40] Matthiä D 2009 PhD thesis, Mathematisch-Naturwissenschaftliche Fakultät der Christian-Albrechts-Universität zu Kiel
[41] Dorman L and Yanke V 1981 *Proc. 17th Int. Cosmic Ray Conf.* Vol. 4 p 326
[42] Debrunner H, Lockwood J A and Flückiger E 1982 *8th European Cosmic Ray Symposium*
[43] Flückiger E O, Moser M R, Bütikofer R, Desorgher L and Pirard B 2008 *Proc. 30th International Cosmic Ray Conference* Vol. 1 p 289
[44] Usoskin I and Kovaltson GA, 2012 *Astrophys J.* **757** 92
[45] Smart DF, Shea MA, Spence HE and Kepko L 2006 *Adv. Space Res.* **37** 1734
[46] Kovaltson GA and Usoskin I 2014 *Solar Phys.* **289** 211
[47] Miroshnichenko LI and Nymmik RA 2014 *Radiat. Meas.* **61** 6
[48] Miroshnichenko LI, 1994 *Geomagnet. Aeron.* **34** 4 29
[49] Miroshnichenko LI, 1996 *Radiat. Meas.* **26** 3 421
[50] Forbush SE 1938 *Phys. Rev.* **54** 975
[51] Richardson IG and Cane HV 2011 *Solar Phys.* **270** 609
[52] Kunow H et al. 1977 *J. Geophys. Z. Geophys.* **42** 615
[53] Hidalgo MA and Nieves-Chinchilla T 2012 *Astrophys. J.* 748 109
[54] Blanco JJ, Hidalgo MA, Gómez-Herrero R, Rodríguez-Pacheco J, Heber B, Wimmer-Schweingruber R and Martín C 2013a *A&A*, 556, A146
[55] Blanco JJ, Catalán E, Hidalgo MA, Medina J, García O and Rodríguez-Pacheco J 2013b *Solar Phys.* **284** 167
[56] Belov A, Baisultanova L, Eroshenko E, Yanke V, Mavromichalaki E, Plainaki C, Mariatos G and Pchelkin V 2005 *J. Geophys. Res.* 110 doi:10.1029/2005JA011067
[57] Alania MV and Wawrzynczak A 2012 *Adv. Space Res.* **50** 725
[58] Howard TA and Harrison RA 2013 *Solar Phys.* **285** 1 269
[59] Kühl P et al 2014 *A&A* submitted
[60] Barbashina NS et al. 2008 *Instruments and Experimental Techniques,* **51** 180
[61] Mavromichalaki H, Paouris E and Karalidi, T 2007, *Solar Phys.* **245** 369
[62] Gil A and Alania MV 2011 *J. Atmos. Sol-Terr. Phys.* **73** 2-3 294 doi:10.1016/j.jastp.2010.01.013
[63] Gil A and Alania MV 2012 *Solar Phys.* **278** 2 447, doi:10.1007/s11207-012-9944-5
[64] Bobik P et al 2012 *Astrophys. J.* **745** 132
[65] Wenzel KP, Marsden RG, Page DE and Smith EJ, *Astron. Astrophys. Supplem.* **92** 207
[66] Adriani O et al. 2014 *Physics Reports* **544** 323
[67] Teegarten BJ et al 1974 *J. Geophys. Res.* **79** 3615
[68] Trainor JH et al 1974 *J. Geophys. Res.* **79** 129
[69] Chennet DJ, Conlon TF, and Simpson JA 1974 *J. Geophys. Res.* **79** 3551
[70] Conlon TF 1978 *J. Geophys. Res. A* **83** 541
[71] Kecskeméty K, Daibog EI, Kota J, Logachev YUI and Kudela K 2013 *Proc. 33rd International Cosmic Ray Conference* Vol. 1 p 289
[72] Blom JG and Verwer JG 1994 *Appl. Numer. Math.* **16** 129
[73] Blom JG and Verwer JG 1996 *ACM Trans. Math. Softw.* **22** 329
[74] De Mendonca R, Raulin J-P, Makhmutov V, Stozhkov YI, Kvashnin AN, Maksumov OS, Mizin SV and Fernandez G 2009 *Bull.Russian Academy Sci.* **73** (3) 404
[75] De Mendonca R, Raulin J-P, Bertoni F, Echer E, Makhmutov V and Fernandez G 2011 *JASTP* **73** 1410
[76] Medina J et al 2013, *Nuclear Instruments and Methods in Physics Research A*, **727**, 97

Recent Research applications at the Athens Neutron Monitor Station

H Mavromichalaki[1,*], M Gerontidou[1], P Paschalis[1], A Papaioannou[1,3], E Paouris[2], M Papailiou[1] and G Souvatzoglou[1,4]

[1] Nuclear and Particle Physics Department, Faculty of Physics, National & Kapodistrian University of Athens, Athens, Greece
[2] Astrophysics, Astronomy and Mechanics Department, Faculty of Physics, National & Kapodistrian University of Athens, Athens, Greece
[3]IAASARS, National Observatory of Athens, Greece
[4]ISNeT Company, Athens, Greece

* E-mail: emavromi@phys.uoa.gr

Abstract. The ground based neutron monitor measurements play a key role in the field of space physics, solar-terrestrial relations, and space weather applications. The Athens cosmic ray group has developed several research applications such as an optimized automated Ground Level Enhancement Alert (GLE Alert Plus) and a web interface, providing data from multiple Neutron Monitor stations (Multi-Station tool). These services are actually available via the Space Weather Portal operated by the European Space Agency (http://swe.ssa.esa.int). In addition, two simulation tools, based on Geant4, have also been implemented. The first one is for the simulation of the cosmic ray showers in the atmosphere (DYASTIMA) and the second one is for the simulation of the 6NM-64 neutron monitor. The contribution of the simulation tools to the calculations of the radiation dose received by air crews and passengers within the Earth's atmosphere and to the neutron monitor study is presented as well. Furthermore, the accurate calculation of the barometric coefficient and the primary data processing by filtering algorithms, such as the well known Median Editor and the developed by the Athens group ANN Algorithm and Edge Editor which contribute to the provision of high quality neutron monitor data are also discussed. Finally, a Space Weather Forecasting Center which provides a three day geomagnetic activity report on a daily basis has been set up and has been operating for the last two years at the Athens Neutron Monitor Station.

1. Introduction

The cosmic rays are high energetic particles, mainly protons, alpha particles and heavier nuclei that originate from galactic and stellar sources. The cosmic rays modulated and in some cases enhanced by the solar activity, penetrate the Earth's magnetic field and reach the atmosphere. The nuclear interactions of these particles with the molecules of the atmosphere produce cascades of secondary particles consisting of muons, neutrinos, electrons, positrons, gamma and the hadronic component which means neutrons, protons, $\pi+$ and $K\pm$. These particles can reach the Earth's surface and be detected by ground based detectors.

The neutron monitors are the ground based detectors that continuously measure the flow of the cosmic rays that reach the Earth's surface. The measurements of the neutron monitors are of great importance since they contribute to the study of several scientific fields, such as the solar activity and the prediction of the space weather. Since the beginning of neutron monitor operation about 100 neutron monitors have been installed at different places in the world creating the Worldwide Neutron Monitor Network, in order for their measurements to be easily accessible. A wide research field uses the neutron monitor measurements, which can be divided into three main categories [1, 2]

- Studies of anisotropy, which requires neutron monitors with well-defined and narrow cones of acceptance for charged particles, covering all directions as evenly as possible;
- Solar neutron measurements, which require neutron monitors at high altitude in equatorial regions;
- Spectral measurements, which require an optimal distribution along cut-off rigidities.

The advantage of the neutron monitors in the network is their high statistical accuracy due to their high counting rate, which allows them to observe small events that are not possible to be detected with satellites.

On the other hand, the neutron monitors record galactic and solar relativistic cosmic rays and can play a key-role in space weather forecasting, as a result of their interaction with interplanetary disturbances. Note that space weather is an environmental concept that refers to the dynamic conditions in the space contiguous to Earth, but also at an interplanetary and interstellar space scale. A wide variety of physical phenomena influence space weather including solar events as coronal mass ejections (CMEs) and solar flares (SFs), populations of galactic (GCRs) and solar (SCRs) cosmic rays, geomagnetic storms, ionospheric disturbances and geomagnetically induced currents at the Earth's surface. Cosmic rays also directly affect the terrestrial environment and serve as indicators of solar variability and non-anthropogenic climate changes on Earth at present and in the distant past. Despite decades of tradition, neutron monitors remain the state-of-the-art instrumentation for measuring GeV cosmic rays that cannot be measured in the same simple, inexpensive, and statistically accurate way by space experiments [3]. The continuous monitoring of the cosmic ray intensity near Earth by neutron monitors since the International Geophysical Year 1957/58 represents the longest continuous, high-time resolution series of particle radiation measurements in space science.

In this work the modern Athens Neutron Monitor Station (ANeMoS) operating at the Faculty of Physics of the National and Kapodistrian University of Athens at Greece, as well as its main research applications that have been developed and currently provided to the scientific community is being presented.

2. Athens Neutron Monitor Station

ANeMoS was established in November 2000 with a standard Super 6NM-64 at an altitude of 260 m above sea level located at 37.58° N, 23.47° E and with a vertical cut-off rigidity of 8.53 GV (figure 1). The Athens station operates continuously 24h per day and provides in real time high resolution data (1-hour, 1-min) in graphical and digital format to the internet. It was the fourth station worldwide which operated in real time. This station is important for the estimation of the maximum particle energy during great proton events due to the fact that frequently the maximum energy of these particles ranges within 5–10 GeV, i.e. very close to the minimal energy of particles recorded in Athens [4]. The measurements of the station are being processed automatically in order to be compatible with data from other stations. This is significant because in order to study in detail CR variations and space weather conditions, it is necessary to compare a number of high rigidity stations with good quality data. The resolution of the measurements reaches values as low as one second –with a mean counting rate ~55 counts/s.

Figure 1. Infrastructure of registration system of ANeMoS consisting of 6NM-64 counters.

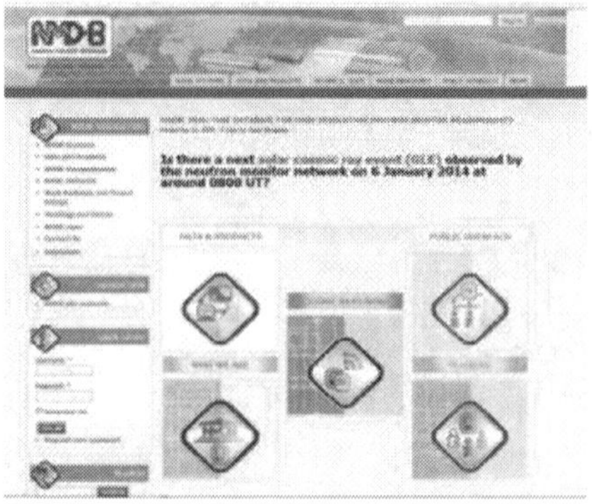

Figure 2. The webpage of the High Resolution Neutron Monitor Database Project -NMDB.

In 2003 the Athens cosmic ray station has developed a data collection system capable of gathering data from remote stations and able to present their data to the Internet via FTP or HTTP servers, named Athens Neutron Monitor data Processing Center (ANMODAP). It is created with the aim to make feasible the use of the neutron monitor network data in real time for the space weather tasks. The modern methods applied to these data and the obtained results gave a way for a possible forecasting of the arrival at Earth of the powerful disturbances from the Sun. This center was collecting and processing ground based data from twenty-five real time NM stations as well as satellite data [5]. At the same time with some other Centers (IZMIRAN, BARTOL University) was gathering data to detect possible abrupt changes in the cosmic rays associated with the real-time solar wind and geomagnetic disturbances. This center put the basis of a world wide data collection system in real time from the neutron monitor network, adjusted software capable of the real time data processing and worked on the elaboration of forecasting models.

Once in 2008 a wide European collaboration for the implementation of the first real time database of Neutron Monitors with the participation of 12 countries and more than 50 Neutron Monitor Stations has begun. Twelve different countries have cooperated within the Seventh Framework Programme of the European Commission, in order to create a real-time database with high resolution neutron monitor data [6, 7]. The central database comprises all neutron monitor data acquired in about the last 50 years and new continuously updated observations from 17 NMs with 1-min and 1-h resolution, operated by the institutes that constitute the Neutron Monitor Database (NMDB) consortium (figure 2). The key point of this project is the development of a flexible database with important easy to use applications (http://www.nmdb.eu). ANeMoS is one of the fundamental partners of the network of the NMDB database sending data in real time to this database and it furthermore maintains and manages one mirror server of this project.It is important to note that the NMDB had to overcome four important challenges:

(a) To design a database capable of hosting data of 1-min and 1-h resolution, in real-time, together with historical ones and meta-data.

(b) To provide an access point for all NM data providers.

(c) To develop user tools with which NM data will be uploaded to the database and by which users will get access to the NMDB for downloading data.

(d) To compute comprehensive parameters from the NM data in near real-time as CR distribution function in space; ionization rates in the atmosphere *vs.* altitude and cut-off rigidity; radiation dose rates in space, magnetosphere and in atmosphere; possible changes of cut-off rigidities.

Therefore, the need for high quality real time data is imperative. The quality of the data is handled by different correction algorithms that filter the real time measurements for undesired instrumental variations. In order to achieve this essential aim, a number of different correction algorithms has been developed and used in ANeMoS data, the output of which is provided in real time by the corresponding web form (figure 3): a) the established Median Editor and the Median Editor Plus that constitutes an improvement version of Median Editor, b) the ANN algorithm based on neural networks and the improved ANN Plus, c) the Edge Editor that is currently used for the announcement of the data to the NMDB [8]. Moreover, a new online tool is developed for the barometric coefficient calculation by using a reference station, the output of which reflects the data quality of the stations [9] (figure 4).

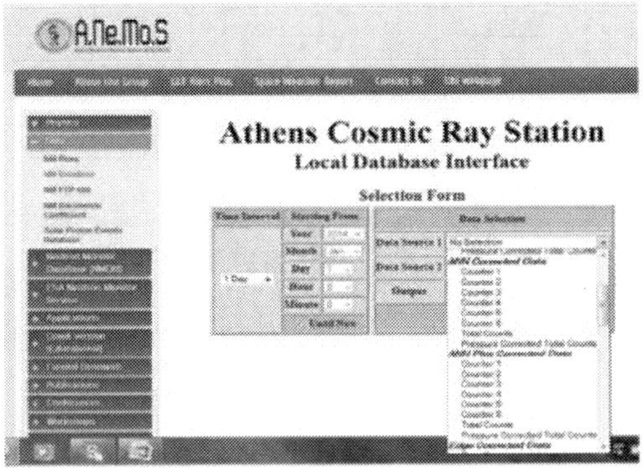

Figure 3. Athens NM station provides cosmic ray corrected data with five different algorithms in graphical and digital form.

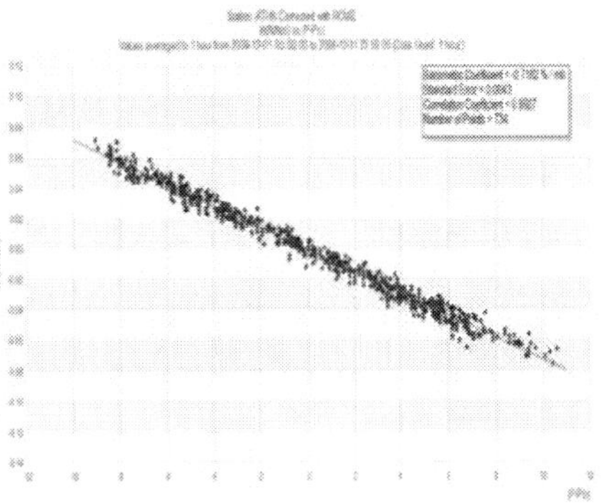

Figure 4. The barometric coefficient, necessary for the pressure correction of data, can be calculated for every NMDB station via the new online tool. The accuracy of the calculation, expressed via the correlation coefficient of the linear regression, reflects the data quality of the stations.

3. Research Applications

During the past years ANeMoS has developed and implemented several research applications that can play a key role in space and solar terrestrial studies as well as for space weather tasks. The most important applications provided at this time are presented below:

3.1 ESA Services

The Cosmic Ray group of ANeMoS has developed and implemented the establishment of the European Neutron Monitor Service available via the ESA SSA SWE portal (http://swe.ssa.esa.int) in the frame of the Space Situational Awareness (SSA) Program of the European Space Agency (ESA). The Athens Cosmic ray group is considered as an expert group of the Space Radiation Center and provides to ESA two products: [a] Access to the Neutron Monitor Data of Multiple Stations and [b] Reliable and timely GLE Alerts (http://swe.ssa.esa.int/web/guest/space-radiation).

3.1.1 GLE Alert Plus

Ground level enhancements (GLEs) are observed as significant cosmic ray intensity increases at neutron monitor measurements, caused by an intense solar flare and/or a very energetic coronal mass ejection. Due to their space weather impact it is crucial to establish a real-time operational system that would be in place to issue reliable and timely GLE Alerts. Such Neutron Monitor Service named GLE Alert Plus has been developed. The core of this Neutron Monitor Service is the GLE Alert software that is an upgrade of the existing GLE Alert software [10, 11] and aims to minimize the probability of false alarms.

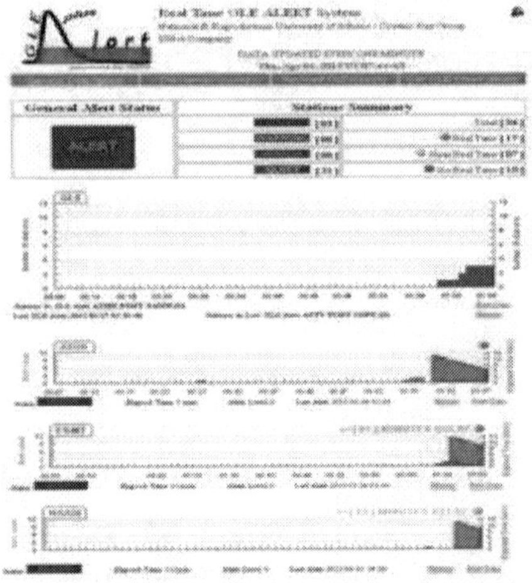

Figure 5. An artificial example of the evolution of the GLE Alert Plus issuing an Alert signal.

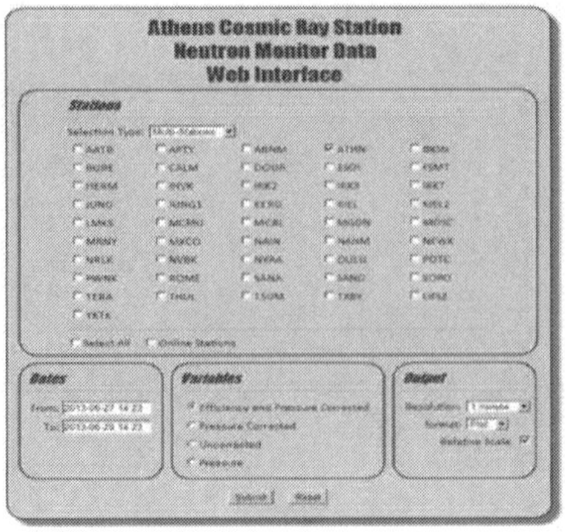

Figure 6. Multi-station NM Data Service.

The recordings of each NM station providing data to the NMDB are the input data for the GLE Alert Plus [12]. For every 1 minute it calculates the moving average of the previous hour (i.e. 60 1-min measurements) and the threshold that represents the upper limit for which each NM station is considered to be at 'Quiet' mode. If three consecutive 1-min measurements exceed this threshold, the particular NM station is considered to be at a ***'Station Alert'*** mode and an elapsed time window of 15 min is being triggered. In case three NM stations, independently of each other, enter the ***'Station Alert'*** mode within the aforementioned time window a General ***'GLE Alert'*** is being marked and an Alert is issued (figure 5). The operational real-time Alert Code of the ANeMoS via NMDB issued an Alert signal at 17.05.2012 at 02:13UT. The relevant Alert from GOES 100MeV protons issued by NOAA at 02:52UT (http:// www.swpc.noaa.gov/alerts/warnings_timeline.html) [13].

3.1.2 Multi Station Data

The multi-station interface provides an easy way to access the data that are stored in the NMDB database. The interface connects to the NMDB mirror server located at the Athens station. The user has to select the stations, the variables, the time interval and the resolution of the exported data. The output can be obtained in both plot and ascii format (figure 6). Moreover, a feature that allows the retrieval of data in csv file has been implemented, allowing further processing of the data by the user.

3.2 Simulations

A simulation model, named Dynamic Atmospheric Shower Tracking Interactive Model Application (DYASTIMA), based on the well known Geant4 toolkit has been developed in ANeMoS [14]. The model allows the study of the cosmic ray showers that are developed when primary cosmic ray particles enter the atmosphere. DYASTIMA supports the definition of a beam consisting of particles that vary in type, energy and direction. The output data are exported in a detailed or in a synoptic form, ready to be used for plots, histograms or contour figures. The spatial distribution of neutrons in several heights when a vertical proton beam of 10 GeV enters the atmosphere is shown in figure 7. The neutron monitor response to the several cosmic ray particles that reach the Earth's surface can be determined by using the simulation application of the 6NM-64 providing information concerning the detection efficiency [15].

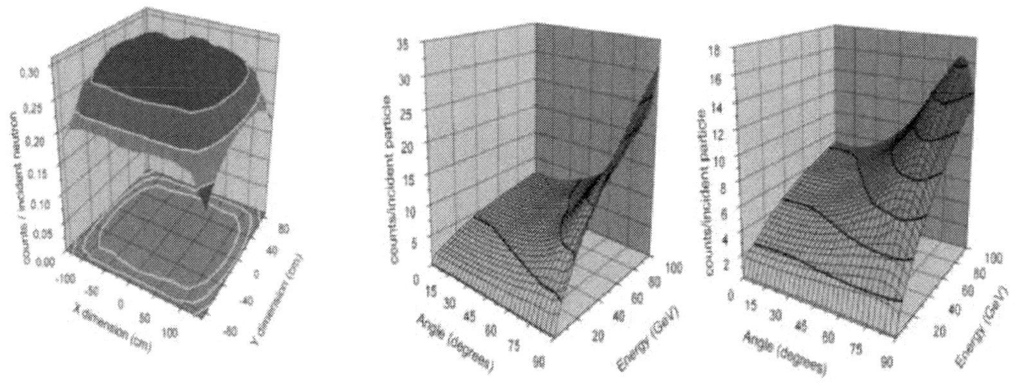

Figure 7. These diagrams illustrate the detection efficiency in case of neutron beam for the different regions of the neutron monitor surface (left) and for the different energies and incoming angles when the azimuth of the beam is perpendicular (middle) or parallel (right) with the counters [15].

3.2 Space Weather Forecasting Centre

Space Weather is the term applied to disturbances that occur in the Earth's magnetosphere and ionosphere as a consequence of interaction of solar wind with the Earth's main field.

From the beginning of the year 2012, a new service named "*Athens Space Weather Forecasting Centre*" is operating at the ANeMoS [16]. This service provides a 3-day geomagnetic activity forecast report on a daily basis at the website of the station *(http://cosray.phys.uoa.gr).* Due to the fact that the space environment and the conditions that are prevalent in it, result into disturbances at the Earth's magnetosphere and ionosphere, which are the direct consequences of the interaction of solar wind and transient magnetic fields of coronal mass ejections (CMEs) with the Earth's magnetic field, an estimation of the disturbances of the Earth's magnetic field is usually being quantified with indices. The Ap index is a measure of the magnetic activity and is the only global planetary magnetic index [17]. The forecasting process is based on a set of rules that include a number of known parameters/properties of the Ap index, as well as current observations of the Sun and near-Earth space [16]. The first item to note in the forecasting agenda is the yesterday's Ap index. If a trend is being marked in Ap's behavior, i.e. if the Ap has been decreasing or in contrast if it has been increasing for several days, this trend may continue, as it may be due to a recurrent behavior of the Sun or the passage of a transient magnetic field.

The next item is the recurrent behavior of the Sun, which is dominant within the heliosphere and acts as the ruler of the near-Earth environment conditions. The magnetic activity 27-days before, when due to the solar rotation the same active region was facing the Earth is also considered. Moreover, the phase of the solar cycle is taken into account.

3.3 Galactic Cosmic Ray Spectrum

According to the coupling coefficient method [18], secondary cosmic ray measurements can be linked to the primary incident cosmic ray particles via specific mathematical functions that take into account the acceptance vectors for each detector (neutron monitor), based on its local characteristics [19]. In order to study analytically the temporal changes of the rigidity spectrum of the GCR intensity during Forbush decreases, daily averaged data of neutron monitor stations obtained from the High-resolution Neutron Monitor Database–NMDB are used. In order to pinpoint the spectral index γ, the difference between the mean amplitude and the amplitude of the specific detector, is considered [20]. Moreover, the Galactic cosmic ray spectrum during the ground level enhancements (GLEs) is modeled using the PPOLA NMBANGLE model [21]

3.4 Forbush Decreases Precursors

It is suggested in many studies that the pre-increases or pre-decreases of the cosmic ray intensity (known as precursors) which usually precede a Forbush decrease could serve as a useful tool for studying space weather effects. A number of Forbush decreases, which were associated with western solar flares and followed by geomagnetic activity, were studied. It was concluded that before geomagnetic storms and Forbush effects, caused by CMEs with a source in the western portion of the solar disk, precursors can be seen in galactic cosmic rays as pre-increases and/or pre-decreases in certain longitudes [22]. Growing amplitude of the first harmonic of cosmic ray anisotropy by neutron monitor network data is connected, as a rule, with the appearance of the precursor. Sometimes the size of the precursor is so big and the cosmic ray intensity distribution is so unusual and specific, that this precursor may serve as a basis for prognosis of a significant interplanetary disturbance arriving to Earth, even in the absence of other data.

3.5 Galactic Cosmic Ray Modulation

Cosmic ray modulation is a complex phenomenon which occurs all over the heliosphere and depends on many factors. No single solar or heliospheric index, however sophisticated, can account for cosmic ray variations. Different scientists have proposed empirical relations describing the long-term cosmic ray variations based on the joint use of solar and/or heliospheric indices (see [23] and its references). Athens Cosmic Ray group developed a model taking into account monthly values of the sunspot number, the CME index based on the characteristics of CMEs [24], the interplanetary magnetic field (IMF) and the heliospheric current sheet tilt (HCS). This model is applied in the 10 GV data of cosmic ray intensity obtained from all neutron monitors worldwide with very good results giving a standard deviation $<10\%$ between the observed and the calculated by the model values [25].

Conclusions

As the recordings of the ground based neutron monitors constitute a useful tool for space weather studies, the ANeMoS has developed a lot of research applications based on these data. This kind of applications is important at first for fundamental studies of Cosmic Ray and Solar-terrestrial Physics as well as for applied research. For example, the applications of the Galactic Cosmic Ray Spectrum and the simulation model DYASTIMA can be used for the calculation of the ionization rate and the radiation dose in the atmosphere that affects the crew and passengers during flights. Moreover, the services of GLE Alert Plus and Multi-Station provided at the ESA Space Radiation Centre portal (http://swe.ssa.esa.int), as well as the Athens Forecasting Centre and Forbush decreases Precursors provided by the ANeMoS (http://cosray.phys.uoa.gr) serve for a trustworthy and timely alert of a great solar particle event and as a basis for prognosis of a significant interplanetary disturbance or/and arriving to the Earth.

The Galactic Cosmic Ray Spectrum as well as the Forbush Decreases Precursors are described in detail in a number of published works [20, 21, 22] and are useful for space weather effects monitoring. The influence of the magnetic fields of the CMEs in the galactic cosmic ray modulation can be an important factor in order to improve the forecasting models of the geomagnetic conditions near Earth. This will be the upcoming new service of the Athens Forecasting Centre. All of the above applications are currently provided to the scientific community.

Acknowledgements: Athens Neutron Monitor station is supported by the Special Research Account of Athens University (70/4/5803). The research of the European Neutron Monitor Services has received funding under ESA Tender: RFQ/3-13556/12/D/MRP. We acknowledge the NMDB database (www.nmdb.eu), founded under the European Union's FP7 programme (contract no. 213007) for providing cosmic ray data. Thanks are due to the anonymous referee for the comments improving significantly this article.

References

[1] Mavromichalaki H, Papaioannou A, Plainaki C et al. 2011 *Adv. Space Res.* **47** 2210
[2] Dorman L 1974 Cosmic ray variations and Space Explorations *North-Holland Publishing Company*-Amsterdam
[3] Simpson J 2000 *Space Sci. Rev.* **93** 11
[4] Mavromichalaki H, Sarlanis C Souvatzoglou G, Tatsis S, Belov A et al. 2001 Proc. 27th ICRC 4099
[5] Mavromichalaki H., Souvatzoglou, G., Sarlanis, C., Mariatos G, Gerontidou M *et al.* 2005 *Ann Geophys.* **23** 3103
[6] Steigies C 2008 *Geophys. Res. Abstr.* 10 EGU2008-A
[7] Mavromichalaki H 2010, *EOS AGU* 305
[8] Paschalis P, Mavromichalaki H 2013 *NIM A* **714** 38
[9] Paschalis P, Mavromichalaki H, Yanke V, Belov A, Eroshenko E et al. 2013 *New Astron.* **19** 10
[10] Mavromichalaki H, Souvatzoglou G, Sarlanis C et al for the NMDB team 2010 *New Astron.* **15** 744
[11] Mavromichalaki H, Papaioannou A, Plainaki C et al for the NMDB team 2011 *Adv Space Res.* **47** 2210
[12] Souvatzoglou G, Papaioannou A, Mavromichalaki H, Dimitroulakos J, Sarlanis C 2014 *Space Weather* **12** 633
[13] Papaioannou A, Souvatzoglou G, Paschalis P, Gerontidou M, Mavromichalaki H 2014 *Solar Phys.* **289** 423
[14] Paschalis P, Mavromichalaki H, Dorman L, Plainaki C, Tsirigkas D 2014 *New Astron.* **33** 26.
[15] Paschalis P, Mavromichalaki H, Dorman L 2013 *NIM A* **729** 877
[16] Abunina M, Papaioannou A, Gerontidou M, Paschalis P, Abunin A *et al.* 2013 *Jour. Phys.: Conf. Ser.* **409** 012197
[17] McPherron R 1999 *Physics and Chemistry of the Earth* Part C **24** 45
[18] Dorman LI 2004 *Astrophys. Space Sci. Libr.* **303**
[19] Wawrzynczak A and Alania M V 2008 Adv. Space Res. **41** 325
[20] Livada M, Papaiannou A, Mavromichalaki H 2013 Proc. Helas Conf. S_1
[21] Plainaki C, Mavromichalaki H, Laurenza M, Gerontidou M, Kanellakopoulos A and Storini M 2014 *Astrophys. J.* **785** 160
[22] Papailiou M, Mavromichalaki H, Belov A, Eroshenko E, Abunina M, *et al.* 2013 *Solar Phys.* **283** 557
[23] Mavromichalaki H, Paouris E, Karalidi T 2007 *Solar Phys.* **245** 369
[24] Paouris E 2013 *Solar Phys.* **284** 589
[25] Paouris E, Mavromichalaki H, Belov A, Gushchina R, Yanke V 2012 *Solar Phys.* **280** 255

Three dimensional solar anisotropy of galactic cosmic rays near the recent solar minimum 23/24

R. Modzelewska[1], M.V. Alania[1,2]

[1]Institute of Mathematics and Physics, Siedlce University, Poland.
[2]Institute of Geophysics, Tbilisi State University, Tbilisi, Georgia.

E-mail: renatam@uph.edu.pl, alania@uph.edu.pl,

Abstract.
Three dimensional (3D) galactic cosmic ray (GCR) anisotropy has been studied for 2006-2012. The GCR anisotropy, both in the ecliptic plane and in polar direction, were obtained based on the neutron monitors (NMs) and Nagoya muon telescopes (MT) data. We analyze two dimensional (2D) GCR anisotropy in the ecliptic plane and north-south anisotropy normal to the ecliptic plane. We reveal quasi-periodicities - the annual and 27-days waves in the GCR anisotropy in 2006-2012. We investigate the relationship of the 27-day variation of the GCR anisotropy in the ecliptic plane and in the polar direction with the parameters of solar activity and solar wind.

1. Introduction

Diurnal variation, being the measure of the galactic cosmic ray (GCR) anisotropy, arises from the Earth's rotation as a ground-based detector's viewing direction through the sky during each 24-hours period. Three-dimensional (3D) anisotropy vector is determined by the distribution of the stream of cosmic rays in the 3D heliosphere. An average anisotropy vector has been explained based on the diffusion-convection theory of GCR modulation in the heliosphere [1]-[3]. That is a consequence of the equilibrium established between the radial convection by solar wind and the inward diffusion of GCR particles along the interplanetary magnetic field (IMF) owing to the radial gradient of GCR.

Scientific staff of IZMIRAN's cosmic ray laboratory (http://helios.izmiran.troitsk.ru/cosray/main.htm), have calculated the components Ar, Af, and A_θ of the 3D anisotropy by Global Spectrographic Method (GSM) [4]-[6] based on hourly data from all operating neutron monitors (NMs). Unfortunately, the derivation of the A_θ component is possible with an accuracy up to constant, and so, a value of the north-south anisotropy A_θ obtained by GSM method is not accurate (http://cr20.izmiran.rssi.ru/AnisotropyCR/index.php). That is, sorrowfully, a deficiency of the GSM related to the nature of NMs data.

In general determination of the two dimensional (2D) anisotropy in the ecliptic plane is feasible based on establishing the radial Ar and tangential Af components by the harmonic analysis method for an individual detector (e.g., NM or Muon Telescope (MT)). Alania et al., [7] demonstrated that the 2D GCR anisotropy calculated by the radial and tangential components

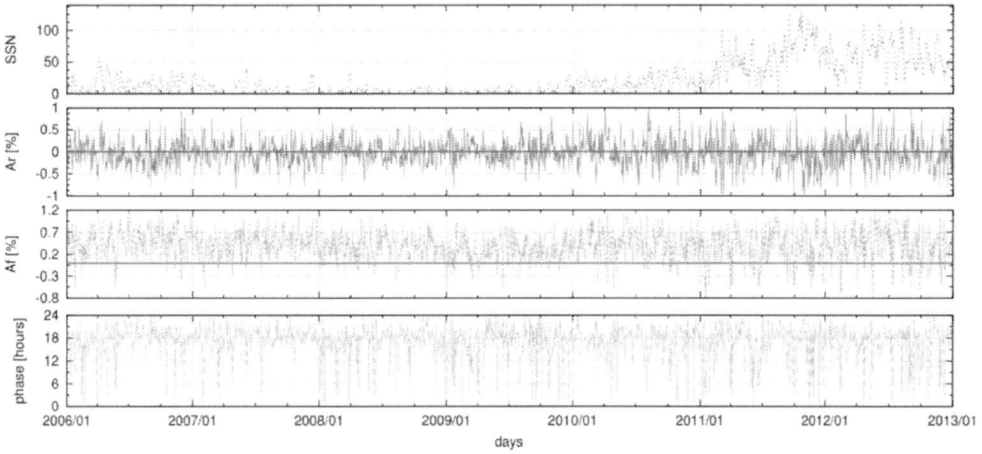

Figure 1. Daily data of the sunspot number SSN, radial Ar and tangential Af components and phase of the 2D GCR anisotropy for Oulu NM in 2006-2012.

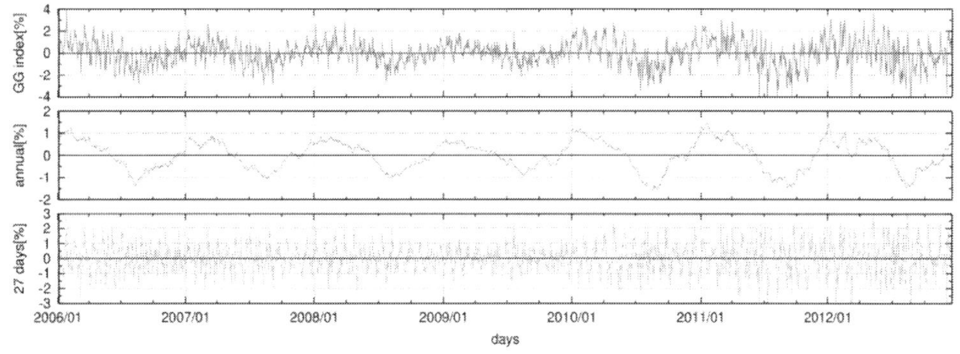

Figure 2. Temporal changes of the (top) daily GG index for 2006-2012, (midle) annual trend in GG index and (bottom) daily GG index with excluded annual trend.

determined using GSM basically do not differ from the values of anisotropy found by harmonic analysis method for individual NM with cut-off rigidities < 5 GV near the solar minimum epoch.

Swinson [8] first suggested that the north-south anisotropy could be related to the cosmic ray flow caused by a positive heliocentric radial density gradient **Gr** of cosmic rays and the **By** component of the interplanetary magnetic field (IMF) **B**, as expressed by the vector product, **By** x **Gr**. Since the average **By** and the **Gr** lie in the ecliptic plane, the direction of the vector product **By** x **Gr** (or the north-south anisotropy) is expected to be perpendicular to the ecliptic plane and to reverse direction with changes of the magnetic field direction. It is directed upward when **By** is positive (IMF away from the Sun-$A > 0$) and vice versa when **By** is negative (IMF toward the Sun-$A < 0$).

Papers [9]- [12] studied the cosmic ray anisotropy vector in three dimensions. They have determined north-south anisotropy using data from polar located NMs. They concluded that a magnitude of the north-south anisotropy varies with $\sim$ 11-years period and there is not any dependence on solar magnetic polarity.

An alternative method to study north-south anisotropy is proposed by Mori and Nagashima

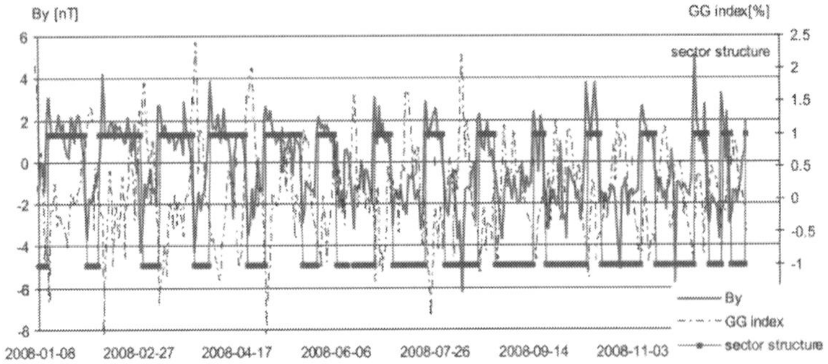

Figure 3. Daily GG index, By component and the sector structure of the IMF in 2008

[13]. They have introduced an index GG calculated from Nagoya MT data as:

$$GG = \frac{1}{2}[(49N - 49S) + (49N - 49E)] \quad (1)$$

The terms represent the counting rate differences for telescopes pointing in North, South, and East directions at 49^0 zenith angle. The GG index is free of noise in isotropic intensity caused by Forbush decreases, periodic variations, atmospheric temperature effects, and geomagnetic cut-offs. GG index mainly reflects change of difference between the intensity from north polar direction and that from north and parallel to the equatorial plane directions. Counting rate differences of the north-south and north-equatorial could not be precisely contained the same type of information. In spite, a GG is accepted by cosmic ray community as a good alternative index to study the north-south (or north-equatorial) asymmetry of the cosmic ray flux e.g., [14].

The present paper investigates the annual and 27-day waves of the GCR anisotropy based on ground observations of secondary cosmic rays by NMs and MTs near the recent solar minimum 2006-2012. We investigate the relationship of the 27-day variation of the GCR anisotropy in the ecliptic plane and in the polar direction with parameters of solar activity and solar wind.

2. Data

Figure 1 presents the temporal changes of the daily sunspot numbers (SSN), daily radial Ar and tangential Af components, and phase of the 2D GCR anisotropy obtained by Oulu NM for 2006-2012. Figure 1 shows that components of the daily ecliptic plane GCR anisotropy obtained from hourly data of NM experience large dispersions during the Sun's rotation period. However, one can recognize that the average daily radial component Ar oscillate near zero, which is caused by a sign dependent drift effect in well established sector structure of the IMF; the tangential component Af ($\sim 0.4\%$) represents diffusive corotational anisotropy and phase (time of maximum intensity) indicates that GCR stream is directed averagely from $\sim 18h$.

Figure 2 (top panel) presents daily data of the GG index for 2006-2012. Figure 2 shows that a clear annual trend is seen in changes of the GG index.

Our aim is to study both annual wave and 27-day variation in GG index. We reveal annual trend by smoothing daily data over 27 days (Fig. 2 middle panel). Finally for studying the 27-day variation of the GG index we exclude the annual trend and present detrended daily GG in the bottom panel of Fig. 2.

As far the formulation of the north-south anisotropy is based on the drift model, it has been established that there is a good correlation between GG index and the polarity of the IMF

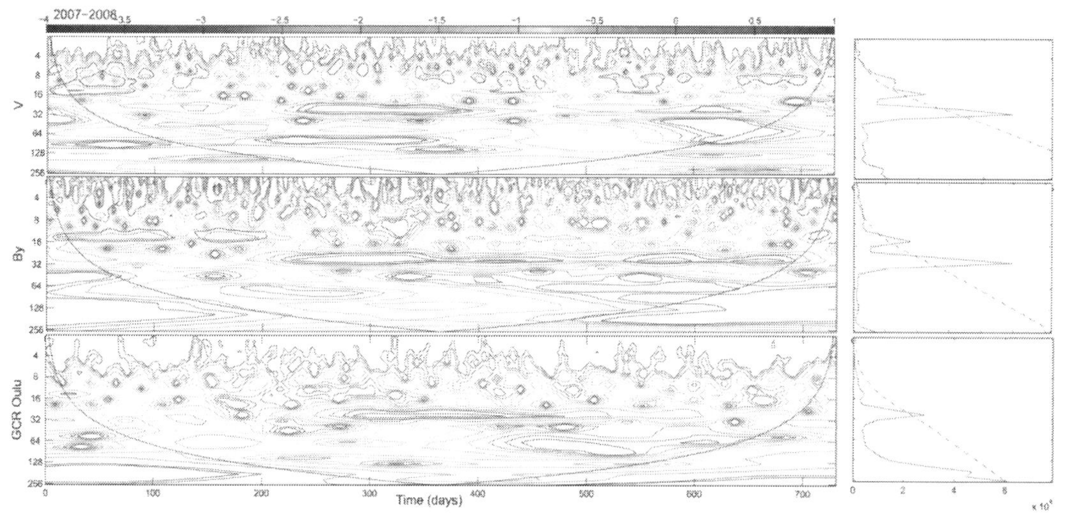

Figure 4. Wavelet analysis of the daily solar wind velocity V (a), By component of the IMF (b), GCR intensity for Oulu NM (c) for 2007-2008.

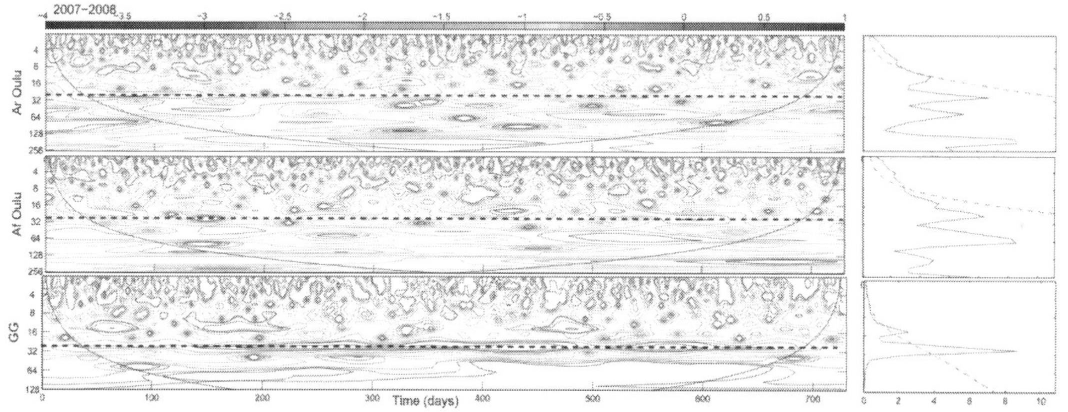

Figure 5. Wavelet analysis of the Ar (a) and Af (b) components of the 2D GCR anisotropy and daily GG index (c) for 2007-2008. Dashed line designates the period of 27 days.

during each solar rotation [13]. Consequently the GG index being the measure of the north-south asymmetry is inversely related with By component of the IMF. It is also well pronounced in the recent solar minimum 23/24. As an example Fig. 3 presents daily GG index, By component and the sector structure of the IMF in 2008.

3. 27-day wave in GCR anisotropy near the solar minimum 23/24: 2007-2012

The 27-day variation of the GCR anisotropy has been studied less intensively up to present, in general. It partially is connected with the complexity of the reliable revealing of the 27-day variation of the GCR anisotropy by means of small amplitudes of the diurnal variations of GCR ($< 0.3\%$, measured by NMs), and with a large dispersion of the amplitudes of diurnal variation comparable with the accuracy of hourly data of NMs. In papers Alania et al., [7], [18] studied

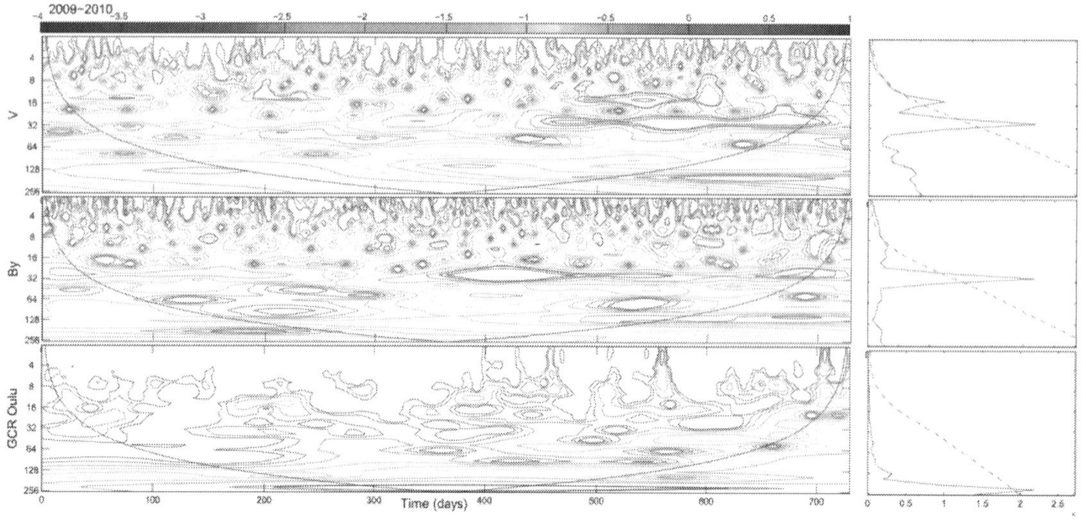

Figure 6. Wavelet analysis of the daily solar wind velocity V (a), By component of the IMF (b), GCR intensity for Oulu NM (c) for 2009-2010.

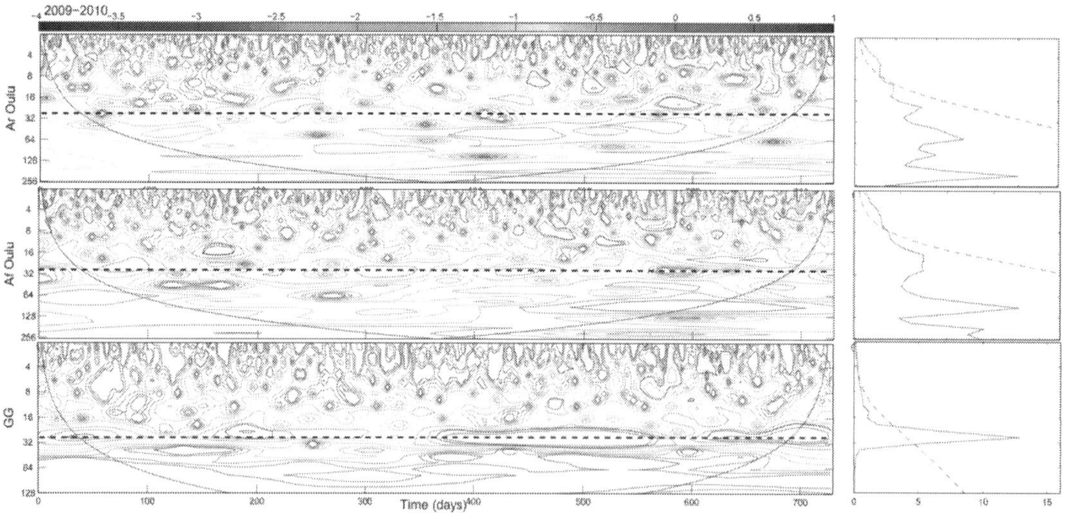

Figure 7. Wavelet analysis of the Ar (a) and Af (b) components of the 2D GCR anisotropy and daily GG index (c) for 2009-2010. Dashed line designates the period of 27 days.

the 27-day variation of the 2D GCR anisotropy in the ecliptic plane. They demonstrated that the average amplitude in the minimum epoch of solar activity is polarity dependent, as it is expected from the drift theory.

The 27-day variation of the north-south anisotropy was studied in the series of papers by Swinson and coauthors [20]-[22]. They showed that 27-day variation of the north-south anisotropy is correlated with solar activity and this correlation is not clearly dependent upon solar magnetic polarity.

The recent solar minimum provided an unique opportunity to study recurrent variations under

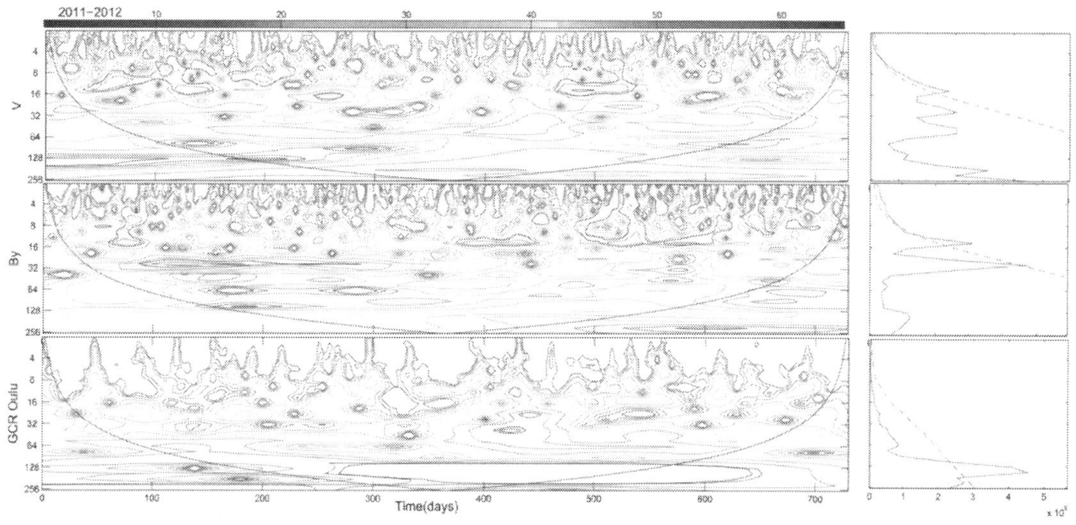

Figure 8. Wavelet analysis of the daily solar wind velocity V (a), By component of the IMF (b), GCR intensity for Oulu NM (c) for 2011-2012.

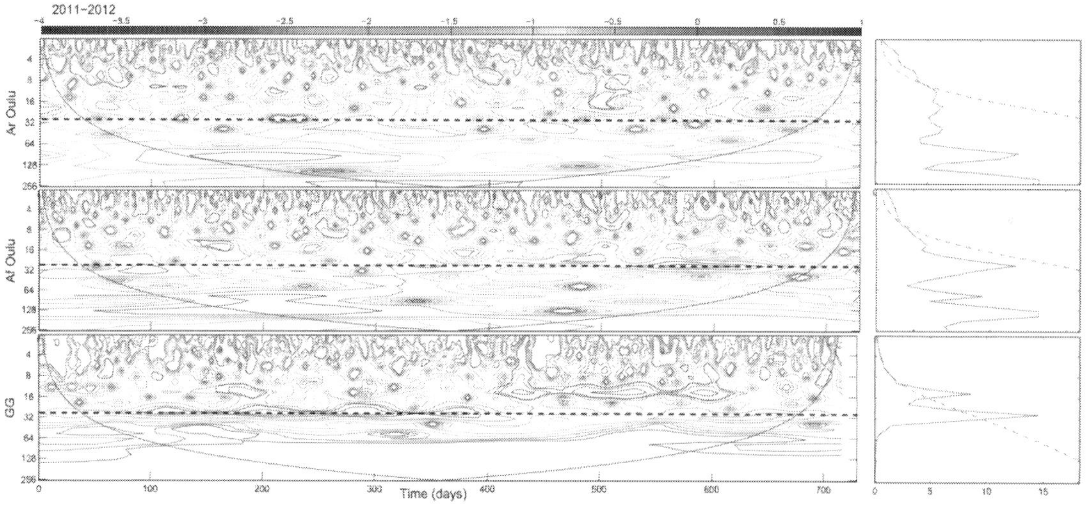

Figure 9. Wavelet analysis of the Ar (a) and Af (b) components of the 2D GCR anisotropy and daily GG index (c) for 2011-2012. Dashed line designates the period of 27 days.

relatively stable low solar activity conditions. Recurrent variations connected with corotating structures ($\sim$ 27 days) are clearly established in all solar wind and interplanetary parameters. Consequently the 27-day recurrent variations of cosmic ray intensity were clearly seen in a variety of cosmic ray counts detected by neutron monitors (e.g., [15], [16]) and space probes (e.g., [17]). Recently, Yeeram et al., [23] studied recurrent 'trains' (trend) of enhanced GCR anisotropy under influence of corotating solar wind structures near the recent solar minimum.

To reveal the quasi-periodic variation of the 2D GCR anisotropy in ecliptic plane and north-south asymmetry of GCR stream by means of GG index near the recent solar minimum 2007-2012, we employ the wavelet time-frequency spectrum technique developed by Torrence and Compo [24]. The wavelet software is available at the website [http://paos.colorado.edu/research/wavelets/software.html]. In our calculation we used the Morlet wavelet mother function.

We performed wavelet analysis for the daily solar wind velocity **V**, By component of the IMF, GCR intensity for Oulu NM, GG index, Ar and Af components of the 2D GCR anisotropy during 2007-2012 considering as a sampling time interval equaling two years (2007-2012). In our case a time interval of two years gives a good enough statistics as far our aim is to reveal recurrences $\leq$ 27-day period, which is $\sim 4\%$ from whole sampling period of 26 solar rotations (two years).

Results of calculation for **V**, By and GCR intensity are presented in Fig. 4 for 2007-2008, Fig. 6 for 2009-2010, Fig. 8 for 2011-2012, respectively. Wavelet analysis of Ar and Af components of the 2D GCR anisotropy for Oulu NM and GG index are presented in Figs. 5 for 2007-2008, Fig. 7 for 2009-2010, Fig. 9 for 2011-2012, respectively. Figures 5, 7, 9 panel (c) present very clear quasi-periodic changes in GG index related to the Sun's rotation ($\sim$ 27 days) for almost whole time interval 2007-2012. Similar quasi periodic character is clearly visible in **V** and By component of the IMF. 27-day variation of GCR intensity is well established at the end of 2007 and in 2008. Although recurrent variations connected with corotating structures ($\sim$ 27 days) are clearly established in almost all solar wind parameters, 2D GCR anisotropy shows a weak 27-day variation, only in some periods. This is connected with large dispersion of daily Ar and Af components of the 2D GCR anisotropy.

4. Annual waves in GG index near the solar minimum 23/24: 2006-2012

We have also studied a feature of the north-south asymmetry based on the GG index depending on the position of the Earth ($\pm 7.5^0$ in September and March) during its annual motion around the Sun. For this purpose we calculated 3 month average of the GG index for each year during 2006-2012. In each year 1st point is the average of November-December-January, 2nd point is the average of February-March-April, 3rd point is the average of May-June-July and 4th point is the average of August-September-October. Results of calculations for superimposed GG index are presented in Fig. 10. Figure 10 shows that GG index exhibits clear annual wave with evident maximum phase near the February-March-April period. We suppose that it is an indication of the asymmetry of a heliolatitudinal distribution of solar activity having minimum in south hemisphere near 7.5^0; being some source of north-south asymmetry of the GCR flux in 2007-2012. Unfortunately, we could not make a precise quantitative estimation due to some uncertainty character of GG data used in this paper. At the same time, one can suppose that this finding is an indication of an existence of north-south asymmetry of the GCR flux.

5. Conclusions

(i) We study recurrent variations of the GCR anisotropy in the ecliptic plane and in the polar direction connected with corotating structures observed in the heliosphere near the recent solar minimum in the period 2007-2012.

(ii) 2D GCR anisotropy generally does not show a clear evidence of the 27-day variation, but in some periods weak recurrent character is visible. This is partially connected with large dispersion of daily Ar and Af components of the 2D GCR anisotropy.

(iii) Using wavelet time-frequency method we reveal clear 27-day waves in the GG index, being the measure of the north-south anisotropy, for almost whole analyzed period 2007-2012. Similar recurrent properties are observed in solar wind velocity and interplanetary magnetic field.

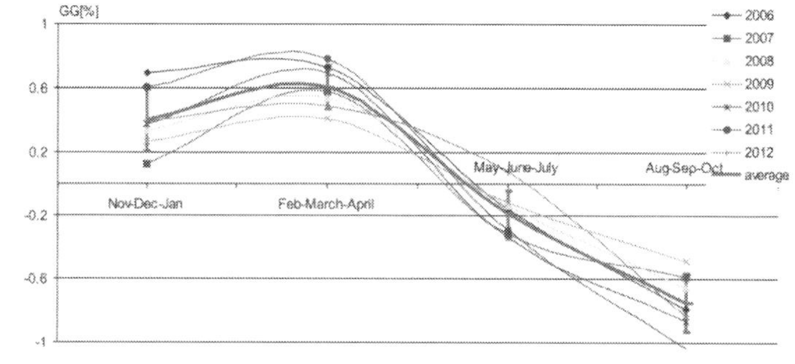

Figure 10. Superposition of the 3 month average *GG* index for each year during 2006-2012

(iv) We show that *GG* index exhibits a clear annual wave with an evident phase of maximum near the February-March-April period. One can suppose that possibly it is an indication of the asymmetry of a heliolatitudinal distribution of solar activity having minimum in south hemisphere near $\sim 7.5^0$; this heliolatitudinal asymmetry can be considered as a some source of north-south asymmetry of the GCR flux in 2007-2012.

Acknowledgments
We thank the principal investigators of Oulu neutron monitor and Nagoya muon telescope for the ability to use their data. We acknowledge authors of the wavelet software which is available at URL: http://paos.colorado.edu/research/wavelets/. We would like to thank the reviewers for helpful suggestions.

References
[1] Ahluwalia H S and Dessler A J 1962 *Planetary and Space Science* **9** 195-210
[2] Krymsky G F 1964 *Geomagnetism and Aeronomy* **4** 763-769
[3] Parker E N 1964 *Planetary and Space Science* **12** 735-749
[4] Belov A V Baisultanova L Eroshenko E A et al., 2005 *Journal of Geophysical Research* **110** A09S20
[5] Krymski G F Kuzmin A I Chirkov N P et al., 1966 *Geomagnetism and Aeronomy* **6** 991-996
[6] Krymski G F Kuzmin A I Chirkov N P et al., 1967 *Geomagnetism and Aeronomy* **7** 11-16
[7] Alania M V Gil A and Modzelewska R 2008 *Adv. Space Res.* **41** 280-286
[8] Swinson D B 1969 *Journal of Geophysical Research* **74** 5591
[9] Duggal S P and Pomerantz M A 1977 *J. Geophys. Res.* **82** 2170-2174
[10] Pomerantz M A Owens A J Tolba M F Tsao C H and Duggal S P 1982 *J. Geophys. Res.* **87** 10325-10330
[11] Bieber J W and Pomerantz M A *Astrophys. J.* **303** 843-848
[12] Chen J and Bieber J W *Astrophys. J.* **405** 375-389
[13] Mori S and Nagashima K 1979 *Planetary and Space Science* **27** 39-46
[14] Munakata K Kozai M Kato C Kota J *Astrophys. J.* **791** 22, doi:10.1088/0004-637X/791/1/22
[15] Alania M V Modzelewska R and Wawrzynczak A 2010 *Adv. Sp. Res.* **45** 421-428
[16] Modzelewska R and Alania M V 2013 *Solar Physics* **286** 593-607
[17] Leske R A Cummings A C Mewaldt R A and Stone E C 2013 *Space Science Reviews* **176** 253263
[18] Alania M V Gil A Iskra K and Modzelewska R 2005 *Proceedings of the 29th International Cosmic Ray Conference* **SH3.4** 215-218
[19] Swinson D B and Yasue S-I 1991 *Proceedings of the 22nd International Cosmic Ray Conference* **SH** 481-484
[20] Swinson D B and Yasue S-I 1992 *Journal of Geophysical Research* **92** A12 19 149-19 155
[21] Swinson D B Yasue S and Fujii Z 1993 *Proceedings of the 23rd International Cosmic Ray Conference* **SH** 671-674
[22] Swinson D B and Fujii Z 1995 *Proceedings of the 24th International Cosmic Ray Conference* **SH** 576-579
[23] Yeeram T Ruffolo D Saiz A Kamyan N and Nutaro T 2014 *Astrophys. J.* **784** 136-147
[24] Torrence C and Compo G P 1998 *Bull. Am. Meteorol. Soc.* **79** 6178

Journal of Physics: Conference Series **632** (2015) 012073 doi:10.1088/1742-6596/632/1/012073

Solar modulation of GCR electrons over the 23rd solar minimum with PAMELA

R. Munini[1,2], V. Di Felice[3], O. Adriani[4,5], G. C. Barbarino[6,7], G. A. Bazilevskaya[8],R. Bellotti[9,10], M. Boezio[2], E. A. Bogomolov[11], M. Bongi[4,5], V. Bonvicini[2], S. Bottai[5], A. Bruno[9,10], F. Cafagna[10], D. Campana[7], P. Carlson[12], M. Casolino[3,14], G. Castellini[15], C. De Donato[3], C. De Santis[13], N. De Simone[3], V. Formato[1,2], A. M. Galper[16], A. V. Karelin[16], S. V. Koldashov[16], S. Koldobskiy[16], S. Y. Krutkov[11], A. N. Kvashnin[8], A. Leonov[16],V. Malakhov[16], L. Marcelli[13], M. Martucci[13,17], A. G. Mayorov[16], W. Menn[18], M Merge'[3,13], V. V. Mikhailov[16],E. Mocchiutti[2], A. Monaco[9,10], N. Mori[5], G. Osteria[7], F. Palma[3,13], B. Panico[7] P. Papini[5], M. Pearce[12], P. Picozza[3,13], M. Ricci[17], S. B. Ricciarini[15], R. Sarkar[2], V. Scotti[6,7], M. Simon[18], R. Sparvoli[3,13], P. Spillantini[4,5], Y. I. Stozhkov[8], A. Vacchi[2], E. Vannuccini[5], G. Vasilyev[11], S. A. Voronov[16], Y. T. Yurkin[16], G. Zampa[2], N. Zampa[2], M. S. Potgieter [21], E.E. Vos [21].

[1]University of Trieste, Department of Physics, I-34147 Trieste, Italy
[2]INFN, Sezione di Trieste, I-34149 Trieste, Italy
[3]INFN, Sezione di Rome "Tor Vergata", I-00133 Rome, Italy
[4]University of Florence, Department of Physics, I-50019 Sesto Fiorentino, Florence, Italy
[5]INFN, Sezione di Florence, I-50019 Sesto Fiorentino, Florence, Italy
[6]University of Naples "Federico II", Department of Physics, I-80126 Naples, Italy
[7]INFN, Sezione di Naples, I-80126 Naples, Italy
[8]Lebedev Physical Institute, RU-119991, Moscow, Russia
[9]University of Bari, Department of Physics, I-70126 Bari, Italy
[10]INFN, Sezione di Bari, I-70126 Bari, Italy
[11]Ioffe Physical Technical Institute, RU-194021 St. Petersburg, Russia
[12]KTH, Department of Physics, and the Oskar Klein Centre for Cosmoparticle Physics, AlbaNova University Centre, SE-10691 Stockholm, Sweden
[13]University of Rome "Tor Vergata", Department of Physics, I-00133 Rome, Italy
[14]RIKEN, Advanced Science Institute, Wako-shi, Saitama, Japan
[15]IFAC, I-50019 Sesto Fiorentino, Florence, Italy
[16]National Research Nuclear University MEPhI, RU-115409 Moscow
[17]INFN, Laboratori Nazionali di Frascati, Via Enrico Fermi 40, I-00044 Frascati, Italy
[18]Universität Siegen, Department of Physics, D-57068 Siegen, Germany
[19]INFN, Sezione di Perugia, I-06123 Perugia, Italy
[20]Agenzia Spaziale Italiana (ASI) Science Data Center, I-00044 Frascati, Italy
[21]North-West University, Centre for Space Research,2520 Potchefstroom, South Africa

E-mail: `Riccardo.Munini@ts.infn.it`

Abstract. The satellite-borne PAMELA experiment has been continuously collecting data since 15th June 2006, when it was launched from the Baikonur cosmodrome to detect the charged component of cosmic rays over a wide energy range and with unprecedented statistics. The apparatus design is particularly suited for particle and antiparticle identification. The PAMELA experiment has measured the electron spectrum at Earth in great detail, extending up to about 100 GeV and down to about 200 MeV. The galactic cosmic ray electron spectra for 2007 and 2009, i.e. measured during the A<0 solar minimum of solar cycle 23, are presented. These fluxes provide important information for the study of charge dependent solar modulation effects.

1. Introduction

The study of the precise measurements of the energy spectrum of cosmic ray electrons provide fundamental information regarding their origin and propagation. However the majority of the measurements where obtained with experiments well inside the heliosphere. Therefore the effects of the propagation and modulation of galactic cosmic rays in the heliosphere cannot be neglected. As cosmic rays traverse the turbulent magnetic field embedded into the solar wind particles are scattered by its irregularities and undergo convection and adiabatic deceleration in the expanding solar wind. Gradient and curvature drifts have also an effect, that is dominant during periods of minimum solar activity. Cosmic rays with rigidities up to tens of GV are affected but the largest effect is seen at low rigidities (less than a few GV). Furthermore, drift models predict a clear charge-sign dependence for the modulation of cosmic rays [1], whose effects are expected to be particularly evident at energies below a few GeV. During so-called A [1] < 0 polarity cycles like solar cycle 23, when the heliospheric magnetic field is directed toward the Sun in the northern hemisphere, negatively-charged particles will drift inward primarily through the polar regions of the heliosphere. Conversely, positively-charged particles will drift inward primarily through the equatorial regions of the heliosphere, encountering the wavy heliospheric current sheet in the process. The situation reverses when the solar magnetic field changes its polarity at each solar maximum. This was shown by the electron ($e^- + e^+$) and proton measurements from few up to several hundreds MeV by the KET instrument on board the Ulysses spacecraft [2] that explored the high latitude regions of the inner heliosphere from 1990 to 2009. Similar conclusion can be drawn by the electron ($e^- + e^+$) data collected by the balloon-born experiment LEE over several decades [3].

The most recent period of solar minimum activity and the consequent minimum modulation conditions for cosmic rays were unusual. It was expected that the new activity cycle would begin early in 2008. Instead solar minimum modulation conditions continued until the end of 2009 when the largest fluxes of galactic cosmic rays since the beginning of the space age were recorded [4]. This period of prolonged solar minimum activity is well suited to study the modulation processes that affect the propagation of galactic cosmic rays inside the heliosphere.

PAMELA (Payload for Antimatter Matter Exploration and Light-nuclei Astrophysics) is a satellite-borne experiment designed to make long duration measurements of the cosmic radiation [5]. Results on the effects of the solar modulation on the energy spectra of galactic cosmic-ray protons in the period July 2006-December 2009 have already been published [6] by the PAMELA collaboration. Here we present new results on the long-term variation in the energy spectrum of galactic cosmic-ray electrons (e^-) measured down to 200 MeV and with no atmospheric overburden. This set of measurements allows detailed analysis of the propagation of cosmic rays in the heliosphere and the study of particle drift effects to be performed.

[1] In the complex sun magnetic field the dipole term nearly always dominates the magnetic field of the solar wind. A is defined as the projection of this dipole on the solar rotation axis.

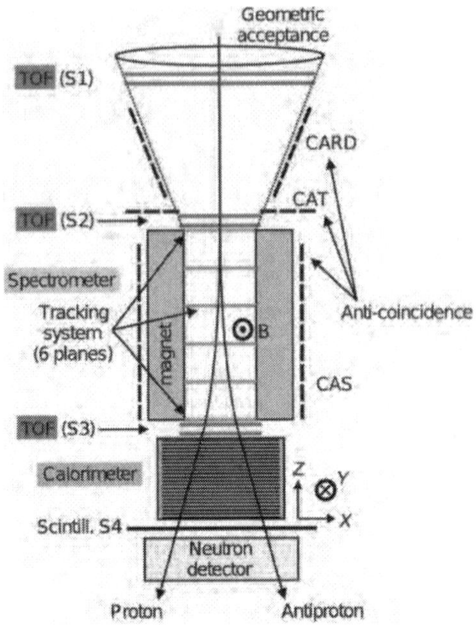

Figure 1. PAMELA and its sub-detectors.

2. The PAMELA instrument

The PAMELA experiment was launched on June 15th 2006 from the Bajkonur cosmodrome on-board of the Resurs DK1 satellite and, since then, it has been almost continuously taking data. The instrument is following a high inclination (70°) orbit, ideal condition to observe the solar modulation of galactic cosmic rays down to very low energies (< 100 MeV).

The PAMELA spectrometer [5] was designed and built to study the antimatter component of cosmic rays from tens of MeV up to hundreds of GeV and with a significant increase in statistics with respect to previous experiments. To reach this goal the apparatus was optimized for the study of Z = 1 particles and to reach a high level of electron-proton discrimination.

The apparatus comprises the following subdetectors, arranged as shown schemativally in Figure 1 (from top to bottom): a time-of-flight system (TOF S1, S2, S3); a magnetic spectrometer; an anticoincidence system (CARD, CAT, CAS); an electromagnetic imaging calorimeter; a shower tail catcher scintillator (S4) and a neutron detector.

The central components of PAMELA are a permanent magnet and a tracking system composed of six planes of double-sided silicon sensors, which form the magnetic spectrometer. The main task of the magnetic spectrometer is to measure the particle rigidity ρ = pc/Ze (p and Ze being respectively the particle momentum and charge, and c the speed of light) and the ionization energy losses (dE/dx). The rigidity measurement is done through the reconstruction of the trajectory based on the impact points on the tracking planes and the resulting determination of the curvature due to the Lorentz force. The Time-of-Flight (ToF) system comprises three double layers of plastic scintillator paddles with the first two placed above and the third immediately below the magnetic spectrometer. The ToF system provides the measurements of the particle velocity combining the time of passage information with the track length derived from the magnetic spectrometer. By measuring the particle velocity, direction and curvature the spectrometer can distinguish between downgoing particles and upgoing splash [2] albedo particles and separate negatively from positively-charged particles.

[2] Upgoing secondary particles resulting from the interaction of high energy CRs with the atmosphere.

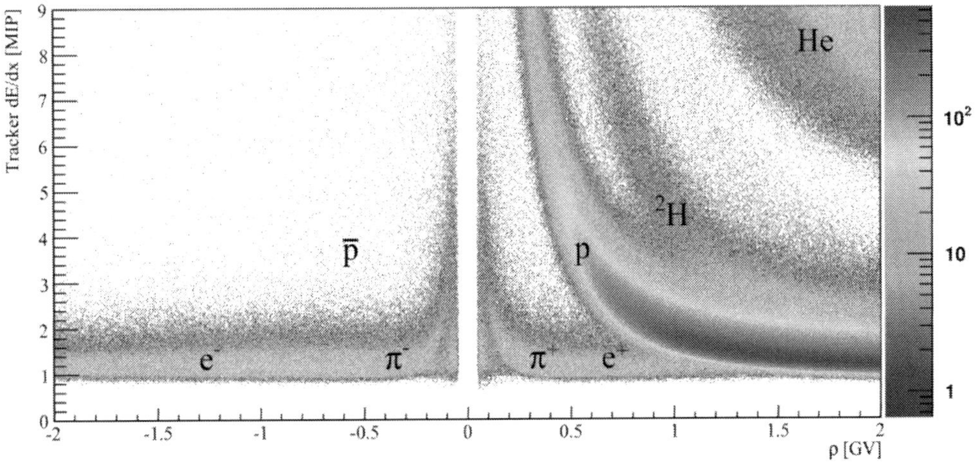

Figure 2. Rigidity distribution for the mean energy release on the tracker planes expressed in minimum ionizing particle units, MIP. On the positive side protons dominate. Below 300 MV is clearly visible the secondary pion contamination.

The sampling imaging calorimeter (16.3 radiation lengths, 0.6 interaction lengths) is used for hadron-lepton separation, using topological and energetic informations about the shower development in the calorimeter. The shower tail catcher and the neutron detector beneath provide additional information for the discrimination. An anticoincidence system is used to reject spurious events in the off-line phase.

The total weight of PAMELA is 470 kg while the power consumption is 355 W. A more detailed description of the instruments and the data handling can be found in [5].

3. Data analysis

A first selection on the goodness of the reconstructed track, expressed in terms of the χ^2 of the fit, was made in order to obtain a sample of events with a reliable reconstructed rigidity. Only single track events were selected and a minimum of three impact points on the non bending view and four impact points on the bending view in the tracking system were required. Furthermore to increase the spectrometer performance the track was required to be reconstructed inside a fiducial volume bounded 0.15 cm from the magnet cavity walls. Downward going particles were selected requiring that their velocities, $\beta = v/c$, measured by the ToF system were greater than zero, hence rejecting splash albedo particles.

Figure 2 illustrates the rigidity distribution for the mean ionization losses in the tracker planes for events selected as described above. On the positive side the protons signal dominates. For the identification of the electrons three sources of contamination were studied:

- Anti-protons: the galactic anti-proton component represents a contamination of a few percent over the entire rigidity range.
- Pions: this secondary component is locally produced by cosmic rays interacting with the PAMELA structure or pressure vessel. This component had already been studied for the anti-proton analysis [7]. Pion contamination reaches a maximum value around 200-300 MV amounting to a few percent of the electrons signal and decrease rapidly becoming negligible, according to simulation, above 1 GV. This component is clearly visible below 300 MV in figure 2 both for positive and negative rigidities.

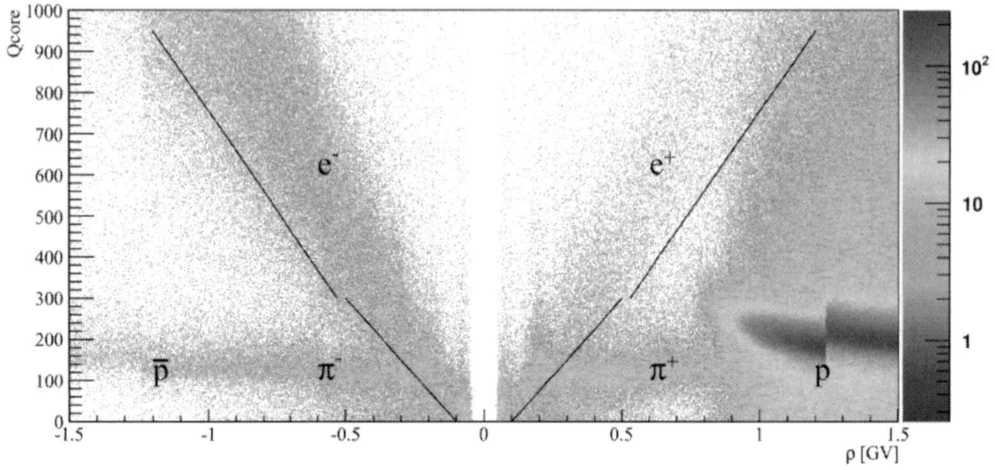

Figure 3. Qcore rigidity distribution. On the positive side above 1 GV the proton component dominates. Pions are present on both sides and overlap with electrons and positrons below 300 MV. This residual contamination was removed using selections on other calorimeter variables. The black lines represent the electron selection on Qcore.

- Spillover protons: this contamination consists of high energy (> 10 GV) protons reconstructed as low energy (< 1 GV) negatively-charged particles. Because of the presence of spurious hits in the tracker planes, a wrong sign of the curvature is assigned to the reconstructed tracks of these events. These events are significant at energies below 1 GeV and amount to a few percent of the electrons signal.

Requiring $\beta > 0.9$, tracker dE/dx < 1.8 MIP and ToF dE/dx< 3 MIP, anti-protons and pions were rejected up to 1 GeV and 250 MeV respectively. Below 1 GeV the negatively charged events remaining after the velocity and dE/dx selections were predominately electrons with a small residual pion contamination. The majority of the pion events had hits in the top anticoincidence scintillators, therefore they were removed requiring no activity on CARD and CAT. Being relativistic, spillover protons were unaffected by these selections.

The residual pion and proton contaminations were removed using the calorimeter informations. The calorimeter selections were developed using a Monte Carlo simulation of the PAMELA apparatus based on the GEANT4 code [8]. The simulation reproduces the entire PAMELA apparatus, including the pressure vessel, and was validated using particle beam data. The longitudinal and transverse segmentation of the calorimeter allow leptonic showers to be selected with high efficiency and small contamination above 300 MeV. The calorimeter electron selection was based on variables that emphasizes the differences between the leptonic and hadronic shower like the multiplication with increasing calorimeter depth and the collimation of the electromagnetic cascade along the track.

Figure 3 shows the distribution of one of these variables, Qcore, for the events passing the ToF, tracking and anticoincidence selections. This quantity, related to the multiplication of the leptonic shower, turned out to have large values for leptons, lower for non interacting and late interacting hadrons because of the limited longitudinal and transverse dispersion of the hadronic shower. The black line on figure 3 indicates the lower limit for electron selection based on this quantity. Combining several of these variables anti-proton and pion contamination were reduced to a negligible amount up to tens of GV [9].

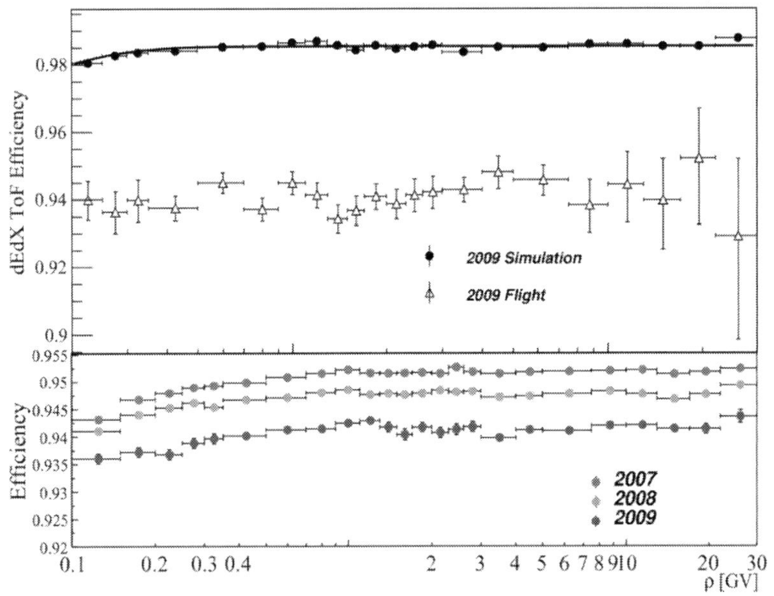

Figure 4. Top panel: comparison between the simulated and flight data ToF ionization losses efficiency. Black line represent a fit to the simulated data. Bottom panel: temporal evolution of the ToF ionization losses efficiency from 2007 to 2009.

The spillover protons were rejected using calorimeter variables that exploited the energy deposit in the last planes of the calorimeter. Electromagnetic showers below 1 GeV usually develop in the first half of the calorimeter instead of high energy protons that traverse the entire volume. A cut on the number of strips hit in the last 6 calorimeter planes rejected spillover protons without affecting significantly the electron signal.

After all these selections, the residual contamination of pions, anti-protons and spillover protons was estimated to be negligible over the entire energy range.

4. Results

The fluxes $\phi(\rho)$ are evaluated as follows:

$$\phi(\rho) = \frac{1}{\epsilon(\rho) \times G \times T \times \Delta\rho} \times N \tag{1}$$

where N is the number of selected events, $\epsilon(\rho)$ the overall selections efficiency, G the geometrical factor, T the live-time and $\Delta\rho$ the width of the rigidity bin.

The various efficiencies were estimated using both simulated and flight data. Tracker efficiency was obtained entirely from simulation. The anticoincidence and ToF ionization losses efficiencies were evaluated with simulation using the real data both as cross-check and to evaluate possible normalization factors. Figure 4 shows the efficiencies for the ToF ionization losses as a function of rigidity. The upper panel displays the comparison between the 2009 simulated and flight data. As can be seen from Figure 4 a constant normalization factor ($\sim 4\%$) is needed in order to match the two sets of data. It should be noted that the shape of the flight data is well reproduced by simulation. The black line is a fit to the simulated data. The efficiencies of the β and calorimeter selections were entirely evaluated using flight data.

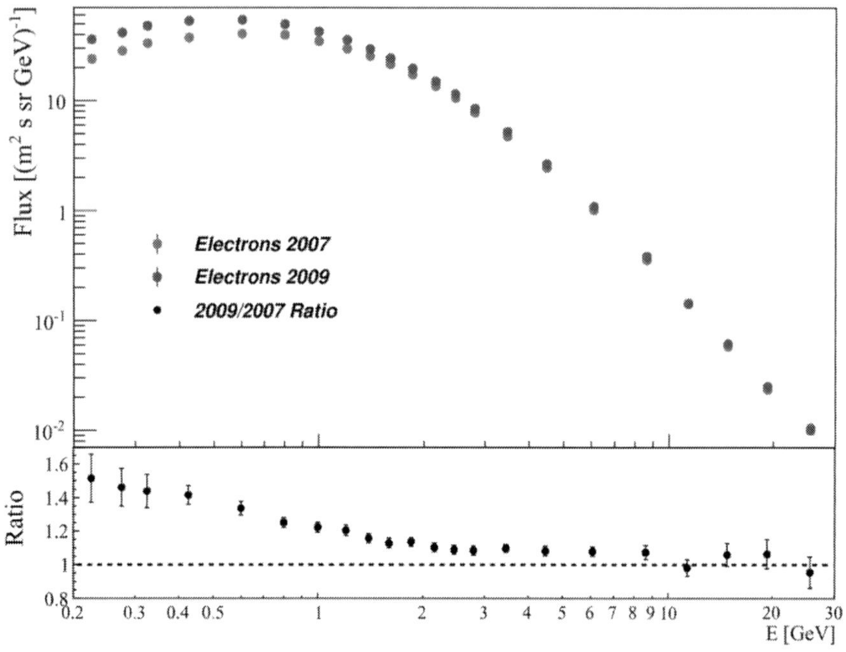

Figure 5. Top panel: 2007 and 2009 electron energy spectra as measured by PAMELA. Bottom panel: ratio between the 2009 and 2007 electron fluxes. The solar modulation is clearly visible and became negligible above $\sim$ 20 GeV. The error bars indicate only statistical uncertainties.

The temporal dependence of the efficiencies was carefully studied. As an example the lower panel of figure 4 show the dEdX ToF efficiencies for three different time period. A small dependence on time is present due to vary calibrations of the read-out signals over the years. The tracker efficiency was estimated using the simulation and it was found to decrease with time due to a progressive deterioration of the read-out chips [6].

The live time was provided by an on-board clock that timed the periods during which the apparatus is waiting for a trigger. The geometrical factor was estimated with simulation to be constant at 19.9 cm^2 sr in the energy range of interest.

Electron rigidity spectra were obtained for different intervals of vertical geomagnetic cutoff rigidities, estimated in the Störmer approximation [10] using the satellite orbital information. Then spectra were unfolded using a Bayesian unfolding procedure [11] in order to take into account energy losses in the instrument and the tracker resolution on the rigidity reconstruction. In this way the energy spectra at the top of the payload were obtained. After the unfolding procedure the spectra were combined accounting for the proper live times and acceptance using only the fluxes at energies that exceeded 1.3 times the maximum vertical geomagnetic cutoff at each cutoff interval, thus excluding contamination from re-entrant albedo particles.

Figure 5 shows the electron energy spectra for 2007 and 2009 (top panel) and the ratio between the two sets of fluxes (lower panel) clearly showing the time dependence of the fluxes. It follows from this ratio that the low-energy electron flux increased by a factor of about 1.5 from 2007 to 2009. Protons at these energies, on the other hand, increased by a factor of about 2.5 over this period [6], indicating the effect of particle drifts.

5. Conclusions

We have presented new results on the electron energy spectrum obtained by the PAMELA experiment during the past extra-ordinary solar minimum period that ended in late 2009 - beginning of 2010. An analysis for low energies electrons (down to 70 MeV) as well as a more detailed study of the time dependence of the electron spectrum and its systematic uncertainties in progress and will be the topic for a future publication.

Acknowledgments

We acknowledge support from The Italian Space Agency (ASI), Deutsches Zentrum für Luft- und Raumfahrt (DLR), The Swedish National Space Board, The Swedish Research Council, The Russian Space Agency (Roscosmos) and The Russian Science Foundation. M Potgieter and E. Vos acknowledge the partial financial support from the South African Research Foundation (NRF) under the SA-Italy Bilateral Programme.

References

[1] M. S. Potgieter 2014 *Adv. Space Res.* **53** 1415-1425
[2] D. Heberm, M. S. Potgieter, 2006 *Space Science Reviews.* **127** 117
[3] P. Evenson, J. Clem, 2011 *Proc. of the 32rd Int. Cosmic Ray Conf.* (Beijing. China) **11** 52
[4] M. S. Potgieter et al. 2013 *Proc. of the 33rd Int. Cosmic Ray Conf.* (Rio de Janeiro. Brazil) arXiv:1308.1617 [astro-ph.SR]
[5] P. Picozza, et al. 2007 *Astropart. Phys.* **27** 296
[6] O. Adriani, et al. 2013 *Astrophys. J.* **765** 91
[7] O. Adriani, et al. 2009 *Phys. Rev. Lett.* **102** 051101
[8] S. Agostinelli et al. 2003 *Nucl. Instrum. Meth. A* **506** 250
[9] O. Adriani, et al. 2009 *Nature* **458** 607-609
[10] M. A. Shea, D. F. Smart, and L. Gentile 1987 *Physics of the Earth and Planetary Interiors* **48** 200
[11] G. D'Agostini 1995 *NIM* **A362** 487-498

The Solar Polar Field in the Cosmic-Ray Intensity Modulation

E Paouris[1], H Mavromichalaki[1], A Belov[2], E Eroshenko[2] and R Gushchina[2]

[1] Faculty of Physics, National and Kapodistrian University of Athens, Athens, Greece
[2] Pushkov Institute of Terrestrial Magnetism, Ionosphere and Radio Wave Propagation Russian Academy of Sciences (IZMIRAN), Moscow, Russia

E-mail: evpaouris@phys.uoa.gr, emavromi@phys.uoa.gr, abelov@izmiran.ru

Abstract. In this work the modulation of galactic cosmic-ray intensity based on solar and heliospheric indices for the solar cycle 23 (1996-2008), is studied. In previous works a number of different indices such as the sunspot number, the CME-index, the interplanetary magnetic field and the heliospheric current sheet tilt were selected to be the most appropriate ones in order to describe the cosmic ray intensity of 10 GV observed by the network of neutron monitors. The new approach is an extension of this study to the influence of the solar magnetic field parameters, such as the mean magnetic field and the polar magnetic field, on the cosmic ray (CR) modulations. Using the wavelet analysis method a major periodicity of about 20-21 years was confirmed in the solar magnetic field data, indicating the existence of the solar magnetic cycle in CR variations. The best empirical relation of the cosmic ray modulation taking into account the sunspot number, the CME-index, the mean solar magnetic field and the solar polar field was improved significantly: a relative root mean square deviation (RMSD) between the observed and the calculated cosmic ray intensity values is found to be 8.7% instead to the previous one of about 10%.

1. Introduction

The modulation of galactic cosmic rays has been studied for several decades with both theoretical and empirical work focused on assessing the influence of solar and heliospheric influences on cosmic rays (see [1] and [2] and the references there in). In these empirical models a number of solar and heliospheric parameters, such as the sunspot number, the number of solar flares, the number of CMEs, the geomagnetic index Ap, the interplanetary magnetic field, the heliospheric current sheet tilt etc., have been used. Using solar and heliospheric parameters such as the sunspot number, the geomagnetic index Ap, the number of CMEs per month and the heliospheric current sheet tilt, the cosmic ray intensity through the empirical modulation with the lowest standard deviation of 10.8% between the observed and the calculated values of the cosmic ray intensity for the years 1996-2006 was produced [1].

This study was improved using two contributed factors. The first one was the data of 10 GV cosmic ray intensity instead of the data of separate detectors and the second one was the introduction of a solar activity parameter (P_i-index) based on the CMEs which are important for the cosmic ray modulation. The cosmic ray variations are most accurately measured for the rigidity of 10 GV because data from many neutron monitors can be used. As a result, the study of the cosmic ray long-term modulation at the rigidity of 10 GV is used in this work, as it is independent from the cut-off rigidity phenomena, and we can say that these calculated cosmic ray time series present the cosmic ray variations at 1 AU outside the magnetosphere and atmosphere. The combination of these improvements led to a new empirical relation with the following parameters of sunspot number, the CME-index, the heliospheric current sheet tilt and the interplanetary magnetic field.

In a recent work [2] two suggestions were underlined as possible improvements. The first one was the examination of the P_i-index data as a function of the angular width of CMEs and the second one was the possible relation of the magnetic fields, such as the solar polar field strength with the galactic cosmic ray modulation. In this work the study of the P_i-index as a function of the angular width presented in section 2.1, was concluded that the best results occurred for data from CMEs with angular width greater than 15 degrees showing that narrow CMEs are less effective for cosmic ray modulation [3]. The wavelet toolkit was applied in section 3 in order to examine further the periodicities of the solar polar field and the cosmic-ray intensity of 10 GV. The magnetic fields taking into account in the section 4 were examined in relation to the cosmic ray modulation giving the best results up to now.

2. Data Collection

In order to study the long-term cosmic-ray modulation for the solar cycle 23 through the years 1996-2008, monthly values of cosmic-rays of 10 GV beyond the magnetosphere were used. These data were kindly provided by IZMIRAN group using the global survey method (GSM). This method uses data from as many ground based detectors (*e.g.* neutron monitors) as possible and it is actually a global method providing useful and reliable information on the conditions of the space environment. It is conceptually a version of spherical analysis [4], [5], [6] and different versions of this method have been evolved and improved at different stages of data processing [7], [8]. The variations of 10 GV cosmic rays with respect to the level of the year 1976 were calculated.

In this study we have also used data of the mean monthly sunspot number (Rz) which have been taken from the National Geophysical Data Center (ftp://ftp.ngdc.noaa.gov/STP/space-weather/solar-data/solar-indices/). The intensity of the interplanetary magnetic field (IMF) was obtained from the OMNI database (http://omniweb.gsfc.nasa.gov/). The data on the tilt of the heliospheric current sheet (HCS), on the mean magnetic field of the Sun (MF) and on the solar polar field strength (PF) were obtained from the Wilcox Solar Observatory database (http://wso.stanford.edu/). The data of CMEs for the formation of the coronal mass ejection index (Pi) were taken from the SOHO/LASCO CME catalog (http://cdaw.gsfc.nasa.gov/CME_list/). Unfortunately the SOHO database has no data for CMEs for the months of July, August, and September of 1998 and January of 1999. In order to fill these data gaps, a smoothing method has been used. This method based on the polynomial fit uses the monthly number of CMEs and the mean linear plasma speed from January 1998 up to December 1999 as input information.

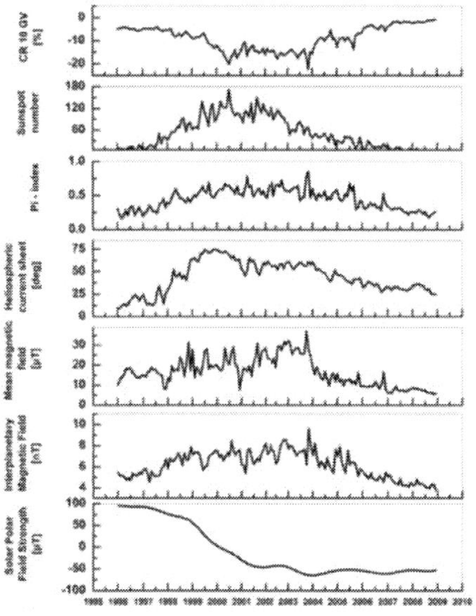

Figure 1. Time profiles of different solar and interplanetary variables used in this study.

The time profiles of solar, interplanetary, and geomagnetic parameters used in this work for all these phases of the interval 1996 to 2008 are presented in Figure 1. It is noted that all the above examined parameters present a clear 11-year variation, except of the solar polar field strength. The sunspot number R_z and the heliospheric current sheet HCS have their maximum values around 2000-2001, while the P_i-index, the mean field MF and the interplanetary magnetic field IMF have their maximum values at the end of 2003. The solar polar field strength has its maximum at the beginning of the solar cycle 23 and the minimum at the end of 2003. It means that this parameter has a different behavior in comparison with the others.

2.1. The CME-index

In a previous work [3] a dimensionless index concerning the properties of CMEs such as the angular width, the monthly number of CMEs and the mean linear speed, named P_i-index, was introduced and a proof of the ineffectiveness of narrow CMEs with angular width less than 30 degrees for cosmic ray modulation was presented. For the examined period of the present work the correlation between this index and the cosmic ray intensity was examined and a step of 5° was applied to the data of the P_i-index starting from the total available data of CMEs (width > 0°) up to data of CMEs with angular width greater than 60°. The results from this analysis are presented in Figure 2. From this figure it is obvious that the best cross correlation coefficient (r = -0.86) between the P_i-index and cosmic-ray intensity occurred for the set of data with angular width of CMEs greater than 15°. At the same figure the right Y-axis is for the cross correlation coefficient between this index and the sunspot number where it is noticeable the fact that when data based on greater CMEs are used, the coefficient is higher (r > 0.80 for CMEs with width > 50°). This suggests a possible relation between the angular width of the CMEs and the sunspot numbers which needs a further study.

The values of P_i based on the data of CMEs with angular width greater than 15° are produced by the monthly number of CMEs (Nc) and the mean plasma velocity (Vp) with the following relation:

$$P_i = \alpha \cdot \frac{Nc}{Nc_{\max}} + \beta \cdot \frac{Vp}{Vp_{\max}} \qquad (1)$$

where $Nc_{\max} = 173$ and $Vp_{\max} = 856.9 \text{ km} \cdot \text{s}^{-1}$. The factors α and β are obtained by seeking for the best cross correlation in a linear fit between the cosmic ray intensity and the CME index P_i, where α+β=1, α,β>0. This P_i index was applied to the examined time period 1996-2008 and the factors α and β have been found as 0.23 and 0.77, respectively. These values are the best ones which maximize the correlation coefficient (r) between the P_i index and the cosmic-rays of 10 GV (r = -0.86) and with the sunspot number as well (r = 0.78).

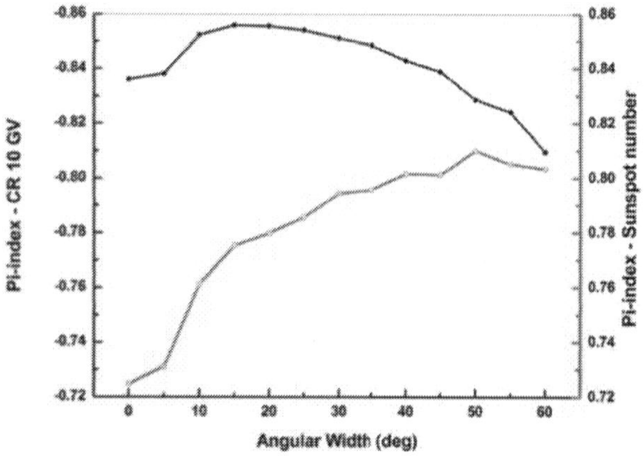

Figure 2. Correlation coefficient of P_i-index with CR intensity (left-y axis) and sunspot number (right-y axis) as a function of the angular width of the CMEs.

3. Time lag and Wavelet Analysis

The 11-year modulation of the cosmic-ray intensity shows some time lag behind the solar activity which is a kind of hysteresis effect [9], [10], [11] and also in [1].

In our previous work [2] the time lag of the cosmic-ray intensity behind various solar and heliospheric activity parameters concerning the period of 1996-2010, was investigated. In this work an analysis of the correlation between the monthly values of the cosmic-ray variations at 10 GV and the parameters R_z, P_i, IMF, HCS, and the MF, PF as well studied for the first time, has been carried out for the solar cycle 23 (1996-2008).

3.1. Correlation coefficients and Time Lags

In order to calculate the time lag of each parameter in reference to the cosmic-ray intensity [1], [2], the cross-correlation coefficients between these parameters with varying time lags from 0 to ±120 months for the time interval 1996 – 2008 have been calculated. The maximum cross-correlation coefficients and the corresponding time lags are given in table 1.

Table 1. Cross-correlation coefficients and the corresponding time lags of the examined parameters

Indices	Correlation coefficients (r) (95% significance level)	Time lags (months)
Sunspot number R_z	-0.87 ± 0.01	+13
Coronal mass ejections index P_i	-0.86 ± 0.01	0
Heliospheric current sheet HCS	-0.77 ± 0.01	+4
Mean Magnetic Field MF	-0.78 ± 0.01	+1
Interplanetary magnetic field IMF	-0.85 ± 0.01	0
Solar Polar Field Strength PF	-0.93 ± 0.01	+67

High correlation values are found between cosmic rays and R_z (-0.87), P_i index (-0.86), IMF (-0.85), and HCS (-0.77) which are in accordance with our previous work [2]. The correlation value of the mean magnetic field MF is r = -0.78 with a time lag of one month. The most interesting results are the very high correlation value of the strength of the solar polar field PF with r = -0.93 and the extremely large time lag of 67 months (~5.58 years). As it is obvious in Figure 1 the maximum of PF coincides with the minimum of solar activity and/or the maximum of cosmic ray intensity at the beginning of the cycle 23 at the year 1996. The minimum of PF is at the end of the cycle at the year 2008 implying that the PF has a cycle of around 22 years that means two solar cycles. This is a result which is in accordance with the periodicity of 22 years of the solar magnetic field (Hale cycle) [12], [13].

The obtained results for sunspot number, CME-index and HCS are in accordance with our previous works. The MF is close to our previous results for IMF, so the time lag of one month is acceptable. The PF shows the most interesting result as it has a time lag of 67 months with the best value of r = -0.93 among the examined variables.

3.2. Wavelet Analysis

In order to study further the periodicities firstly of the solar polar field and secondly of the other parameters, the wavelet toolkit was applied. Non-stationary time series were expanded in terms of time-localized waves, or wavelets. We obtained a compact two dimensional representation (see [14] and [15]):

$$y(t,t',f) = \exp(2i\pi ft)\exp\left(-f^2\frac{(t-t')^2}{2}\right), \tag{2}$$

where f is the frequency, t' the time delay and

$$\exp\left(-f^2\frac{(t-t')^2}{2}\right), \tag{3}$$

is the Gaussian support.

The Morlet wavelet is the most common function used in astrophysical signals expansions; this makes easier the comparison with previously published work. Furthermore, due to its Gaussian support, the Morlet wavelet expansion inherits optimality as regards the uncertainty principle [14]. The global wavelet spectrum [15], which is the time-averaged wavelet spectrum over all the local wavelet spectra, is given by

$$\overline{W}^2(s) = \frac{1}{N}\sum_{n=0}^{N-1}|W_n(s)|^2 \tag{4}$$

where $W_n(s)$ is the wavelet power and N the number of local wavelet spectra. From this, we obtained an unbiased and consistent estimation of the true power spectrum of a time series.

The new perspective in this work is the wavelet analysis of the cosmic-ray intensity of 10 GV, the heliospheric current sheet and the polar field strength. The wavelet analysis of the sunspot number has been presented also by previous researchers [16], [17]. For this wavelet study data for four solar cycles were obtained for cosmic-ray intensity of 10 GV, of sunspot numbers, of polar field strength and heliospheric current sheet tilt angle.

- For the data of cosmic-ray intensity of 10 GV the most important periodicity is that one of 128 m (10.67y). Minor periodicities are spotted around 32 m (2.67 y), 16 m (1.33y), 12 m (1y) and 8 m (0.67 y).
- For the sunspot number it is obvious that the most important periodicity is the one of 128 m (10.67 y). Minor periodicities are noticed around 32 m (2.67 y), 16 m (1.33y) and 8 m (0.67 y).
- For the polar field strength the most important periodicity is spotted around the 245 m (20.42 y). Minor periodicities are visible around 16 m (1.33 y).
- For the heliospheric current sheet tilt angle, it has almost the same periodicities with the R_Z. The most important seems to be the one of 128 m (10.67 y). Then other periodicities are present also around 50 months = 4.17 y and 16 m (1.33 y).

It is interesting that all parameters present the 11-year variation as well as the 1.33 year one. These periodicities for sunspot number and for the cosmic ray intensity have also been spotted in previous works [18], [19], [20]. The new point is that only the solar polar field presents the periodicity of 20.47 y that corresponds to the 22-y variation which is the magnetic cycle periodicity. The comparison of the wavelet figures with the periodicities are presented in Figure 3. The contour levels are chosen so that 75%, 50%, 25%, and 5% of the wavelet power is above each level, respectively.

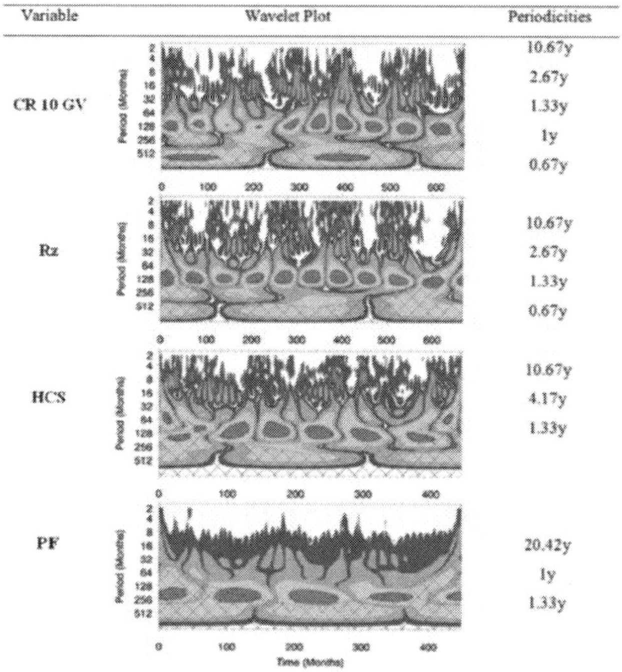

Figure 3. Wavelet analysis diagrams and list of the determined periodicities

4. Galactic Cosmic Ray Modulation

In a previous work [2] an empirical relation of the cosmic ray modulation for the time period from January 1996 up to December 2009, covering the solar cycle 23 and the extended minimum between solar cycles 23 and 24 was applied with very good results.

According to the previous model, the modulated cosmic-ray intensity I is expressed by a constant C and the sum of a few source functions appropriately selected from the solar and interplanetary indices that affect cosmic-ray modulation. This relation is given by the following expression:

$$I = C - 10^{-3}\left(a_1 X + a_2 Y + a_3 Z + a_4 W\right) \quad (5)$$

where C is a constant, X, Y, Z and W are the selected time-lagged solar-heliospheric parameters, and a_i (i=1 to 4) are coefficients calculated by the RMS-minimization method. Constant C is linearly correlated to the cut-off rigidity of each station according to the relation:

$$C = 0.95 + 0.005P \ \ [\mathrm{GV}] \quad (6)$$

where P is the cut-off rigidity for each neutron monitor station [20]. In this work, using data of the cosmic ray variations of 10 GV obtained from the network of all neutron monitors around the world, constant C is found to be equal to 1 that means rigidity independent.

This model was derived by a combination of solar and heliospheric parameters such as the sunspot number (Rz), the CME-index (Pi), the heliospheric current sheet tilt angle (HCS), and the interplanetary magnetic field (IMF). This work suggested two possible improvements for future studies. The first one is the data which are used for the formation of the CME-index taking into account only the wider CMEs except the narrow ones, which are studied in the present work in section

2.1. The second one is the use of another solar parameter such as the Sun' s polar magnetic field which is implemented in the present work.

The previous empirical modulation with the minimum root mean square deviation (RMSD) of 9.96% was expressed by the following relation:

$$I = C - 10^{-3}\left(a_1 R_z + a_2 P_i + a_3 \mathrm{IMF} + a_4 \mathrm{HCS}\right) \tag{7}$$

where constant C was found equal to 1, R_z, P_i, IMF (in μT) and HCS (in deg) are the solar-interplanetary parameters incorporating the time lag. Coefficients a_i, were found equal to 3.5, 395.5, 2.5, and 0.5, respectively.

The next model which examined in the present work took in to account the CME-index (P_i), the heliospheric current sheet (HCS) tilt angle, the mean magnetic field (MF) and the polar field (PF) strength by the equation:

$$I = C - 10^{-3}\left(a_1 P_i + a_2 \mathrm{HCS} + a_3 \mathrm{MF} + a_4 \mathrm{PF}\right) \tag{8}$$

where constant C was found equal to 1, P_i, HCS (in deg), MF (in μT) and PF (in μT) are the solar-interplanetary parameters incorporating the time lag. Coefficients a_i equal to 227.75, 2.75, 5.50 and 1.50 respectively, were found. This version gave a RMSD of 11.81%, the most important result was that the minimum of the modulated cosmic-ray intensity coincides with the observed one and this contribution is due to the CME-index which is a more reliable solar activity index especially for the periods of solar extreme events [3] than sunspot number, where the sunspot number is reliable for the overall solar cycle. Set of similar parameters as in equation (8) demonstrated the best results on the long period 1976-2007 [21].

However in solar cycle 23 the best approximation of modulation produced by the equation:

$$I = C - 10^{-3}\left(a_1 R_z + a_2 P_i + a_3 \mathrm{MF} + a_4 \mathrm{PF}\right) \tag{9}$$

where constant C was found equal to 1, R_z, P_i, MF (in μT) and PF (in μT) are the solar-interplanetary parameters incorporating the time lag. Coefficients a_i equal to 3.00, 268.60, 3.40, and 0.80, respectively, were found. This modulation gave a RMSD of 8.72%, which is a better result than the results from (7) or (8). In future works this simulation in order to determine the possible extension of this modulation in other periods such as in the solar cycle 24, will be applied. The observed (red line) and the modulated cosmic-ray intensity values (black line) are presented in Figure 4.

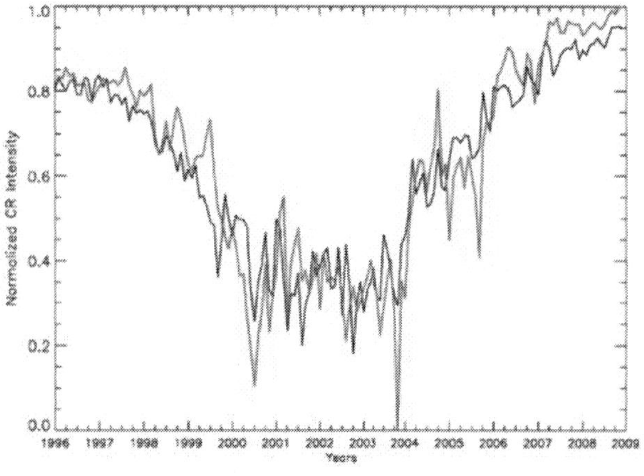

Figure 4. The observed values of cosmic ray intensity of 10 GV (red line) and those calculated by (9) (black line).

5. Conclusions

The cosmic ray modulation is a complex phenomenon, which occurs all over the heliosphere and depends on many factors, so it is important to study as many as possible solar and heliospheric variables. In this work new variables, such as the mean magnetic field and the polar field strength of the Sun, have been studied in relation to the long-term modulation. The solar polar field as a solar parameter has also been presented in previous works, such as in [21].
The CME-index as one of the most important indices for the solar activity formatted using only the wider CMEs except the narrow ones with angular width w<15°.
Applying a wavelet analysis for first time to the cosmic ray data of 10 GV, the well known periodicities of the 11-year as well as the one of 1.33-year, were presented in all examined here variables. It is interesting to note that the solar polar field presents a variation of 20.47 y that seems to be corresponded to the 22-years solar magnetic cycle.
Moreover, the contribution of the solar magnetic fields to the cosmic ray modulation was also outlined from the fact that the best modulation was obtained using the including these variables equation (9) giving a RMSD of 8.72% improving all the previous suggested models.

Acknowledgements: We are grateful to the providers of the solar, interplanetary, cosmic ray and geomagnetic data used in this work. Data of coronal mass ejections for the formation of CME index (P_i) are taken from the SOHO/LASCO CME catalogue (http://cdaw.gsfc.nasa.gov/CME_list/). This CME catalogue is generated and maintained at the CDAW Data Center by NASA and The Catholic University of America in cooperation with the Naval Research Laboratory. SOHO is a project of international cooperation between ESA and NASA. Thanks are due to the anonymous referee for useful comments.

References

[1] Mavromichalaki, H., Paouris, E., Karalidi, T.: 2007, *Solar Phys.* **245**, 369.
[2] Paouris, E., Mavromichalaki, H., Belov, A., Gushchina, R., Yanke, V.: 2012, *Solar Phys.* **280**, 255.
[3] Paouris, E.: 2013, *Solar Phys.* **284**, 589.
[4] Krymsky, G. F., Kuzmin, A. I., Chirkov, N. P., Krivoshapkin, P. A., Skripin, G. V., Altukhov, A. M.: 1966, *Geomagn. Aeron.* **6**, 991.
[5] Dvornikov, V. M., Sdobnov, V. E.: 1997, *J. Geophys. Res.*, **102**, 24209.
[6] Belov, A.V., Gushchina, R.T., Yanke, V.G.: 1999, *Proc. 26th ICRC*, **7**, 175.
[7] Baisultanova, L. M., Belov, A. V. and Yanke, V. G.: 1995, *Proc. 24th ICRC*, **4**, 1090.
[8] Belov, A., Baisultanova, L., Eroshenko, E., Mavromichalaki, H., Yanke, V., Pchelkin, V. C. Plainaki, C., Mariatos, G.: 2005, *J. Geophys. Res.*, **110**, A09S20.
[9] Moraal, H.: 1976, *Space Sci. Rev.* **19**, 845.
[10] Usoskin, I. G., Alanko, K., Mursula, K., Kovaltsov, G. A.: 2002, *Solar Phys.* **207**, 389.
[11] Kane, R.P.: 2011, *Solar Phys.* **269**, 451.
[12] Hale, G.E.: 1927, *Nature* **119**, 708.
[13] Babcock, H.W.: 1961, *Astrophys. J.* **133**, 572.
[14] Morlet, J., Arens, G., Forgeau, I., Giard, D.: 1982, *Geophysics* **47**, 203.
[15] Torrence, C., Compo, G.P.: 1998, *Bull. Am. Meteorol. Soc.* **79**, 61.
[16] Prabhakaran Nayar, S.R., Radhika, V.N., Revathy, K., Ramadas, V.: 2002, *Solar Phys.* **208**, 359.
[17] Katsavrias, C., Preka-Papadema, P., Moussas, X.: 2012, *Solar Phys.* **280**, 623.
[18] Kudela, K., Rybak, J., Antalova, A., Storini, M.: 2002, *Solar Phys.* **205**, 165.
[19] Kudela, K., Mavromichalaki, H., Papaioannou, A., Gerontidou, M.: 2010, *Solar Phys.* **266**, 173.
[20] Mavromichalaki, H., Marmatsouri, E., Vassilaki, A.: 1990, *Solar Phys.* **125**, 409.
[21] Gushchina, R.T., Belov, A.V., Obridko, V.N., Shelting, B.D.: 2009, *Bull. of the Russian Acad. of Scien.: Physics* **73**, 3, 334.

24th European Cosmic Ray Symposium (ECRS2014) IOP Publishing
Journal of Physics: Conference Series **632** (2015) 012075 doi:10.1088/1742-6596/632/1/012075

A Novel Forecasting System for Solar Particle Events and Flares (FORSPEF)

A Papaioannou[1], A Anastasiadis[1], I Sandberg[1], M K Georgoulis[2], G Tsiropoula[1], K Tziotziou[1], P Jiggens[3], A Hilgers[3]

[1]National Observatory of Athens, Institute for Astronomy, Astrophysics, Space Applications and Remote Sensing, Greece
[2]Academy of Athens, Research Center for Astronomy and Applied Mathematics, Greece
[3]European Space Agency, European Research and Technology Center, The Netherlands

E-mail: atpapaio@astro.noa.gr

Abstract. Solar Energetic Particles (SEPs) result from intense solar eruptive events such as solar flares and coronal mass ejections (CMEs) and pose a significant threat for both personnel and infrastructure in space conditions. In this work, we present FORSPEF (Forecasting Solar Particle Events and Flares), a novel dual system, designed to perform forecasting of SEPs based on forecasting of solar flares, as well as independent SEP nowcasting. An overview of flare and SEP forecasting methods of choice is presented. Concerning SEP events, we make use for the first time of the newly re-calibrated GOES proton data within the energy range 6.0–243 MeV and we build our statistics on an extensive time interval that includes roughly 3 solar cycles (1984–2013). A new comprehensive catalogue of SEP events based on these data has been compiled including solar associations in terms of flare (magnitude, location) and CME (width, velocity) characteristics.

1. Introduction

Solar Energetic Particle (SEP) radiation storms occur when solar eruptive events (solar flares and/or CMEs) accelerate particles, turning them into a significant risk for satellites, communication and power systems and human life [1, 2]. Due to the sporadic nature of SEP events and the far from complete understanding of the mechanisms (i.e. acceleration and propagation) involved in their generation, their forecasting and/or nowcasting is no trivial task. Adding to the problem are the complex nature of the underlying physical processes and the lack of detailed, spatially distributed in–situ observations. Moreover, successful solar flare prediction consists of identifying the host active region (AR) and projecting of the onset time and size of the flare(s) to be triggered in that region. Credible forecasting of solar flares is a major step forward, as high-energy flare photons entail risk, or even danger, for sensitive space-borne electronics and humans engaging in extravehicular activities with only nominal space-suit protection. As there is no early-warning window for flare photons, flares need to be ideally predicted in advance of their actual occurrence. Over the course of time, the scientific community managed to overcome certain past obstacles and developed forecasting tools to issue notifications on both solar flare and SEP events and thus provide forewarning. It is important to note, however, that from the modeling point of view solar flares, fast CMEs, and SEPs are closely related to each other, with fast CMEs causing SEP events more often than not, associating to a flare and a host AR.

1.1. *Solar Flare Forecasting*

The nature of the flare prediction problem has resulted in a relatively limited number of researchers applying to this purpose a diverse array of techniques. The interested user should refer to [3]. Several studies focus on the multiscale behavior (fractality and multifractality), and the turbulence in photospheric active–region magnetic fields, indicating that flaring active regions exhibit distinct, distinguishable complexity. Therefore, such information holds significant flare–predictive capability or, at least, it can be utilized to identify flaring active regions before they actually flare [4]. Moreover, morphological flare–prediction methods tend to exploit features of eruptive solar active regions. These methods rely on parameters stemming from these regions' morphological and topological characteristics such as the photospheric magnetic–polarity inversion lines (PILs) or photospheric properties in general [5, 6, 7, 8, 9]. Another morphological parameter is the effective connected magnetic field strength (B_{eff}), originally proposed by [7]. B_{eff} is inferred by means of a magnetic connectivity map in ARs that, by construction, favours and emphasizes connections alongside PILs. Another approach is the use of statistical methods relying on historical data, based on the argument that flare occurrence globally in the Sun is a time-dependent Poisson process [10]. Furthermore, machine–learning, computer–based exploitation of active–region properties shows a predictive ability that is based on either a single or a combination of active-region properties. A notable effort in this respect is the Automated Solar Activity Prediction (ASAP) tool, operating at the University of Bradford [11]. Other machine learning and supervised techniques include methods of logistic regression [12], and multi-resolution predictor terms [13] that may hold more promise in flare prediction than individual predictor terms [14]. At another approach, using a master equation for flares with some minimum energy E that also includes an energy–input factor $\alpha(E, E')$, [15], investigated flare–like analytical solutions and derived a size distribution of flare energy. Furthermore, there are methods that aim to address whether it is possible to know if an AR will flare prior to its emergence? According to [16, 17] it is, and a way to achieve this is by local helioseismology, providing the subsurface kinetic helicity and results of ring-current analysis, respectively. From a different angle, an interesting advance warning of major solar flares was suggested by [18] focusing on the nuclear decay rates of the radioactive nuclide ^{54}Mn. The study reported significant (several σ) anomalous decreases of these rates at Earth that appeared well correlated with flaring activity in the Sun.

Finally, despite developments in fully automated solar flare forecasting, NOAA/Space Weather Prediction Center (SWPC) insists on a human–interactive, expert–based flare prediction system. Forecasters on duty inspect a lookup table and issue a prediction probability that depends on climatology, persistence, and the forecasters experience to extract flare probabilities [19].

1.2. *Solar Energetic Particle Forecasting*

There are several prediction schemes of SEPs and energetic storm particles (ESPs) currently available. One large category includes physics-based numerical models of particle injection, shock evolution, and SEP acceleration at shocks, such as the SOLar Particle Engineering Code (SOLPENCO) [20]. SOLPENCO can also accurately capture the arrival at 1 AU of low–energy ESP ions travelling along with the CME shock and is a tool able to predict the flux and the cumulative fluence profiles of gradual SEP events associated with interplanetary shocks, originating anywhere between the solar western limb and far eastern locations. However, such efforts have not yet matured to a level that suffices for operational purposes.

On the other hand, empirical models that rely on observations of associated precursor phenomena, have been developed over the years. In particular, examples of such operational systems include: the *UMASEP* (University of Malaga Solar Energetic Particle) system [21], that predicts in real time the time interval on which the integral proton flux is expected to meet or surpass the SWPC-set SEP threshold $J(E > 10MeV) = 10pfu$ ($1pfu = pr\ cm^{-2}sr^{-1}s^{-1}$) and the intensity of the first hours of SEP events. Furthermore, using solar flare observations, *PPS* – an empirical SEP prediction system [22] – predicts, for > 5, 10 and 50 MeV SEP events at 1 AU, characteristics such as the occurrence, timing, intensities, spectra and elemental abundances. To fulfil its purpose, *PPS* is based on empirical correlations between the properties of a large number of SEP events observed by the IMP satellites and their correlated flare signatures (GOES soft X-rays, $H\alpha$ flare location, and GOES proton fluxes) [23]. Moreover, based on the assumption that there is a relationship between solar flare emissions and SEP events the *PROTONS* system assumes that the particles are produced at or near the flare site and subsequently diffuse through interplanetary space. This is an operational proton prediction tool in use at the NOAA/SWPC [24]. At a higher energy limit of SEP events (E > 433 MeV), *GLE Alert Plus* – another operational forecasting tool – issues alerts pertaining to the start time of a ground level enhancement (GLE) event utilizing the real-time data from the Neutron Monitor Database (NMDB) (www.nmdb.eu) [25]. Furthermore, *RELeASE* (Relativistic Electron Alert System for Exploration) system is an operational empirical system that issues short-term forecasting of the appearance and intensity of solar ion events by means of relativistic, near light-speed electrons [26]. It utilizes the recordings of the Solar and Heliospheric Observatory (SoHO)/Comprehensive Suprathermal and Energetic Particle Analyzer (COSTEP) electrons within the energy range 0.3–1.2 MeV and issues a forecast for the expected proton intensity of 30–50 MeV. Finally, the technique proposed by [27] yields an estimate of the probability of occurrence of a SEP event whose integrated intensity surpasses the SWPC SEP threshold of 10 pfu. It is based on the "big flare syndrome" [28], i.e. the fact that the larger the flare, the more probable it is to produce a significant SEP event at 1 AU. Hence, it predicts the probability of a SEP occurrence only for flares higher than M2. It also aims at maximizing warning time and for this reason it uses real-time data available at SWPC such as flare location ($H\alpha$ data), flare size (GOES soft X-rays) and SEP acceleration and escape (type III radio burst data). To summarize, most of the current operational prediction schemes assume a relation between SEPs and their parent solar events and are based on empirical and/or statistical relationships between the SEP parameters and the characteristics of associated precursor parent solar events [29, 24, 27, 30].

However, it is necessary to improve the forecasting capabilities for both solar flares and SEP events and at the same time to combine independent tools into one integrated solution. Such an integrated solution has been developed, aiming at forecasting X- and M-class flares, CMEs, and SEP events. That is MAG4, which forecasts the AR's rate of producing each kind of event from a proxy of the AR's free magnetic energy. This proxy is measured from a magnetogram of the AR, for ARs within 30 heliocentric degrees from the disk center [31, 32].

The subject of this study is the FORSPEF activity, whose objectives are to advance our knowledge on the underlying physical mechanisms that drive solar radiation storms and to develop reliable and interconnected tools for the prognosis of SEPs, solar flares and CMEs, providing forecasts evolving with time as more information becomes available.

2. The FORSPEF System

The FORSPEF (Forecasting Solar Particle Events and Flares) system will provide nowcasting and forecasting of solar eruptive events, such as solar flares with a projection to CMEs (occurrence and velocity), likelihood of a SEP event as well as the complete SEP profile - including its peak flux and duration [33]. To this end, forecasts will be made available up to 24 hours in advance and near real-time assessments (typically ~15-20 minutes) of the near Earth environment will be made available, including a qualitative description of significant changes in the space environment (such as solar flares, CMEs) which in turn will be used as input for SEP warnings and alarms. Moreover, the outputs of FORSPEF will put interested parties on alert, and consequently help lower the risk for spacecraft and payloads. They may further be used for planning critical orbital manoeuvres by launch operators. FORSPEF offers a 24–hour forecast of SEPs, up to 70° EW covering practically the entire course of the AR toward the limb (up to $\sim 85^o$), under the assumption that the active region does not change significantly over this course. Finally, FORSPEF, providing the expected SEP flux will also meet the requirement of spacecraft operators by enabling them to take preemptive and recovery measures in case of predicted significant disruptions.

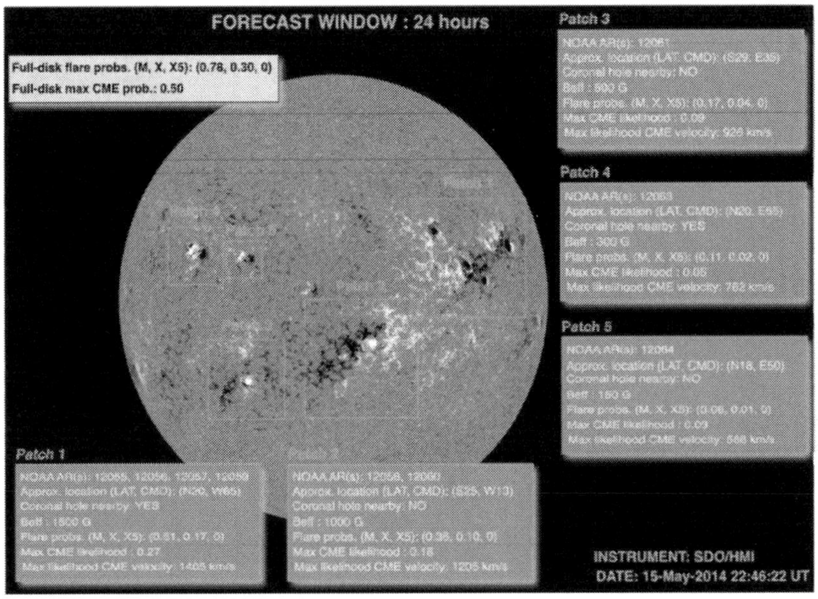

Figure 1: The output of the FORSPEF flare prediction module with a 24-hour forecast window on a recent full-disk SDO/HMI magnetogram.

2.1. Prediction of Solar Flares to Automatically Detected Solar Active Regions

The Solar Flare Prediction will rely primarily on a *morphological method* (see subsection 1.1), namely on the *effective connected magnetic field strength* (B_{eff}) prediction metric [7] and secondarily, from 50° to 70° EW, to a proxy of B_{eff} provided by the existing statistical association between the unsigned flux of the AR and the respective B_{eff} value. The Solar Flare Prediction algorithm will be part of a module designed for automatically detected solar active regions. Detection will be performed on periodically downloaded latest full-disk line-of-sight (LOS) solar magnetogram data from Solar Dynamic Observatory (SDO) / Helioseimisc and Magnetic Imager (HMI) [34]. A typical forecast window will be of the order 24 hours, with forecasts renewed every three (3) hours or more.

Figure 1 depicts a mock-up image of the flare-prediction process with a 24–hour forecast window on a full-disk SDO/HMI magnetogram. This is the output that the user of FORSPEF will obtain from the online service. The background includes the envisioned Active Region Identification Algorithm (ARIA) result on the magnetogram and each of the orange inset boxes contains the result for each identified active–region patch on the solar disk. The B_{eff} value for each patch (calculated separately) was assumed for the purposes of the probability calculation while the probabilities and CME velocities rely on existing statistics and scaling laws. Cumulative, full-disk probabilities for flares and CMEs within the given forecast window are shown in the cyan inset box.

Figure 2 depicts the derived conditional–probability histogram for flares of all sizes (C1.0 – X10) for a given AR target, under the condition of a B_{eff} equal to 1500 G (red histogram). The flare probabilities are to be understood as cumulative, in the sense that they refer to a given flare size "and above" (for example, the M1.0 conditional flare probability corresponds to flares of class $\geq$ M1.0). Figure 2 also shows the respective CME–likelihood for each flare class (blue histogram) that corresponds to the superposition of two sigmoidal-fitting curves, namely, that of the conditional flare probability and that of the flare–CME association. Statistically, it expresses the likelihood that a given flare (of a given class and above) is eruptive, i.e., associated with a CME. Notice from the figure that the superposition of the two fitting curves gives rise to a peak CME likelihood of ~0.27 for a flare-class of M2.0 and above - this can be viewed as the maximum CME likelihood. Besides the CME likelihood, a projected near–Sun CME velocity relying on the given B_{eff} value has been inferred independently by [35] and is shown in the inset. A plot, such as the one of Figure 2, will be provided for each automatically identified AR in the earthward solar hemisphere at the time of the processing.

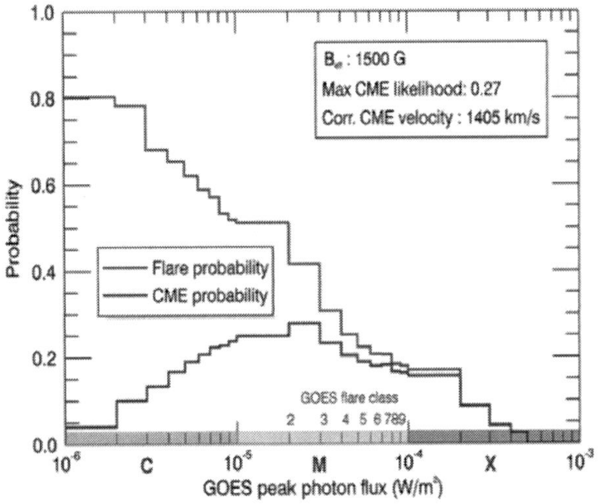

Figure 2: Predictions for an assumed B_{eff}= 1500 G: The derived range of probabilities for different flare classes (red histogram). Also shown is the respective CME likelihood curve (blue histogram).

2.2. *Prediction of Solar Energetic Particle Events*

The core of the FORSPEF SEP events prediction is based on a purely statistical approach that has been proven to be the key ingredient for SEP prognosis in a number of cases (see subsection 1.2). For the first time, our statistics are built from recently re-calibrated GOES proton data and cover an interval of nearly 30 years, from 1984 to 2013 [36]. A new database of solar flares, CMEs and SEPs was implemented by exploiting all available information within this time period.

The database was constructed on the basis of soft X-ray measurements from GOES and CME information from SoHO/ Large Angle and Spectrometric Coronagraph Experiment (LASCO). Our initial sample consisted of 35306 C, M and X solar flares and 22143 CME events. By applying temporal criteria [37], a set of 3680 CME-flare pairs, with a relative timing indicative of a causal relationship, was established. Moreover, the application of spatial criteria [38] led to a sub–sample of 888 CME–flare pairs that fulfilled both criteria (out of which 68 were X-class, 235 M-class and 585 C-class flares). The cross–calibrated (with IMP-8) GOES data [36] cover an energy range of 6–243 MeV. GOES proton data of $E_{eff} \sim 10$ MeV have been scanned and 314 well defined SEP events have been identified. For each SEP event, velocity dispersion analysis (VDA) has been performed in order to identify the estimated solar release time of the particles. Details on the mathematical formulation of VDA can be found at [39]. This methodology was used to associate SEP events to their parent solar sources. The improved quality of the data used in the analysis, the carefully inspected solar associations and the large number of the resulting SEP events have significantly improved the statistics and thus the

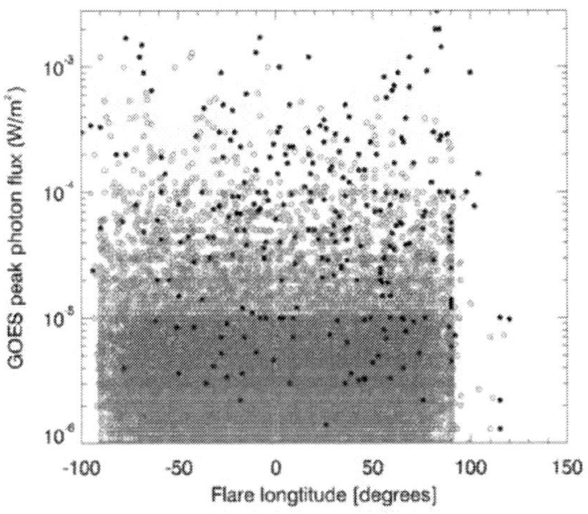

Figure 3: Peak flare photon flux vs. flare longitude. Flares associated with SEP events are presented by filled black circles while non-SEP flares by open red circles.

resulted forecasting/nowcasting of SEPs. Hence, the FORSPEF database is one of the most complete databases currently available.

In Figure 3 we present the peak photon flux of solar flares as a function of longitude, using the total sample of well defined events in our newly constructed database (i.e. 20429; this number was deduced from the initial flare sample, excluding flares with missing information). One may notice the relative lack of SEPs in the case of eastern and relatively weak flares whereas the majority of strong western flares are SEP–related. This result is anticipated and highlights the fact that the FORSPEF database performs as expected.

FORSPEF provides the probability of SEP occurrence based on information of solar flares with known characteristics. For a given location of an AR, the information stored in the FORSPEF database is used to derive the Distribution Functions (DFs) of the flare magnitudes associated and/or not associated to SEPs. The latter DFs are folded with the solar forecast (see subsection 2.1) leading to a weighted forecast. Using this scheme, statistical outputs (e.g. percentiles) for the SEP occurrence are derived. Moreover, the prediction can be further refined by including forecasts (and/or nowcasts) of CME occurrence. Preliminary outputs indicate that the occurrence of a SEP and the corresponding generated probability are controlled by the occurrence (or not) of a CME, based on our FORSPEF event database.

2.3. FORSPEF's Forecasting Work Scheme

The actual FORSPEF's forecasting work scheme is presented in Figure 4. For a given AR (in this case AR12157), the probabilities of flare and CME occurrence are first derived, (right top panel of Figure 4). Results stemming from the FORSPEF database are presented in the bottom left panel of Figure 4. The abscissa corresponds to the flare peak photon flux and the ordinate corresponds to the derived probabilities of SEP occurrence. These probabilities are also plotted with respect to the longitude of the flare for six (6) color-coded bins, from -89 to 89 degrees. This clearly illustrates the longitudinal dependence of SEP occurrence, with the probability increasing when moving from towards more western longitudes. The probabilities derived by the Solar Flare Prediction are folded with the probabilities derived by the FORSPEF database. This is illustrated at the bottom right panel of Figure 4 which has similar axis and colour–coded bins as with the previous panel. FORSPEF provides the maximum probability of SEP occurrence from that bottom right panel of Figure 4.

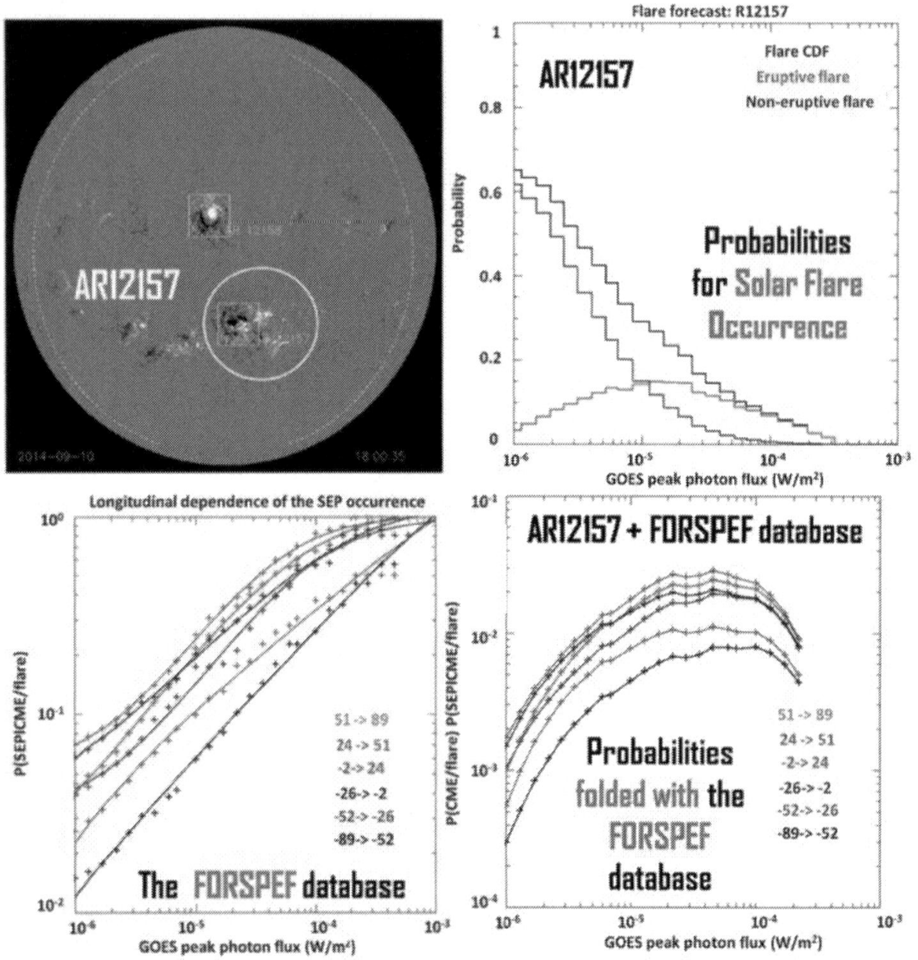

Figure 4: The FORSPEF forecasting work scheme.

3. Conclusions

We hereby introduce FORSPEF, a new forecasting system for eruptive solar flares and SEP events that is targeted towards launch operators and will soon be released. For the Solar Flare Prediction module, FORSPEF incorporates the *effective connected magnetic field strength* (B_{eff}) prediction metric that was never before set into an operational mode[1]. Furthermore, for the SEP prediction a new database of SEP events, solar flares and CMEs has been built. It makes use of a homogeneous SEP dataset (GOES), covering a large time span (1984–2013). This homogeneous data set uses the cross–calibrated GOES data (with IMP-8) and is augmented with the identification of the SEP characteristics (fluence, peak flux) in the time series at well defined effective energies. Both are used in the analysis for the first time. Due to its dual approach, FORSPEF will provide: a) forecasting of SEPs, including the probability of occurrence, peak flux and duration, using solar event data (i.e. flare and CME characteristics), as input and b) nowcasting of SEPs using solar event observations as input. This approach enables the refinement of predictions upon the availability of new data that characterize changes on the Sun and

[1] The Athens Effective Solar Flare Forecasting (A-EFFort), an ESA/SSA service under implementation that is targeted exclusively on solar flare prediction, is also using B_{eff}. The service will be finalized within 2015.

the interplanetary space. Moreover the combined use of solar flare and SEP forecasting methods upgrades FORSPEF to an integrated forecasting solution.

Acknowledgments

Work was performed under the ESA Contract No 4000109641/13/NL/AK.

References

[1] Feynman J and Gabriel S 2000 *Journal of Geophysical Research: Space Physics (1978–2012)* **105** 10543–10564
[2] Shea M and Smart D 2012 *Space Science Reviews* **171** 161–188
[3] Georgoulis M K 2012 *The Sun: New Challenges* (Springer) pp 93–104
[4] Georgoulis M K 2012 *Solar Physics* **276** 161–181
[5] Falconer D, Moore R and Gary G 2006 *The Astrophysical Journal* **644** 1258
[6] Schrijver C J 2007 *The Astrophysical Journal Letters* **655** L117
[7] Georgoulis M K and Rust D M 2007 *The Astrophysical Journal Letters* **661** L109
[8] Leka K and Barnes G 2007 *The Astrophysical Journal* **656** 1173
[9] Mason J and Hoeksema J 2010 *The Astrophysical Journal* **723** 634
[10] Moon Y J, Choe G, Yun H and Park Y 2001 *Journal of Geophysical Research: Space Physics (1978–2012)* **106** 29951–29961
[11] Qahwaji R and Colak T 2007 *Solar Physics* **241** 195–211
[12] Song H, Tan C, Jing J, Wang H, Yurchyshyn V and Abramenko V 2009 *Solar Physics* **254** 101–125
[13] Yu D, Huang X, Wang H and Cui Y 2009 *Solar Physics* **255** 91–105
[14] Huang X, Yu D, Hu Q, Wang H and Cui Y 2010 *Solar Physics* **263** 175–184
[15] Wheatland M 2008 *The Astrophysical Journal* **679** 1621
[16] Reinard A, Henthorn J, Komm R and Hill F 2010 *The Astrophysical Journal Letters* **710** L121
[17] Haber D A, Hindman B W, Toomre J, Bogart R S, Larsen R M and Hill F 2002 *The Astrophysical Journal* **570** 855
[18] Jenkins J H and Fischbach E 2009 *Astroparticle Physics* **31** 407–411
[19] Crown M D 2012 *Space Weather* **10**
[20] Aran A, Sanahuja B and Lario D 2005 *Advances in Space Research* **36** 2333–2338
[21] Núñez M 2011 *Space Weather* **9**
[22] Smart D and Shea M 1989 *Advances in Space Research* **9** 281–284
[23] Kahler S, Cliver E and Ling A 2007 *Journal of Atmospheric and Solar-Terrestrial Physics* **69** 43–49
[24] Balch C C 2008 *Space Weather* **6**
[25] Souvatzoglou G, Papaioannou A, Mavromichalaki H, Dimitroulakos J and Sarlanis C 2014 *Space Weather* **12**
[26] Posner A 2007 *Space Weather* **5**
[27] Laurenza M, Cliver E, Hewitt J, Storini M, Ling A, Balch C and Kaiser M 2009 *Space Weather* **7**
[28] Kahler S 1982 *Journal of Geophysical Research: Space Physics (1978–2012)* **87** 3439–3448
[29] Garcia H 2004 *Space Weather* **2**
[30] Dierckxsens M, Tziotziou K, Dalla S, Patsou I, Marsh M, Crosby N, Malandraki O and Tsiropoula G 2014 *arXiv preprint arXiv:1410.6070*
[31] Falconer D, Barghouty A F, Khazanov I and Moore R 2011 *Space Weather* **9**
[32] Falconer D A, Moore R L, Barghouty A F and Khazanov I 2014 *Space Weather* **12** 306–317
[33] Anastasiadis A, Sandberg I, Papaioannou A, Georgoulis M, Tsiropoula G, Tziotziou K, Katsiyannis A C, Jiggens P and Hilgers A 2014 *European Space Weather Week 11*
[34] Scherrer P, Schou J, Bush R, Kosovichev A, Bogart R, Hoeksema J, Liu Y, Duvall Jr T, Zhao J, Schrijver C *et al.* 2012 *The Solar Dynamics Observatory* (Springer) pp 207–227
[35] Georgoulis M K 2008 *Geophysical Research Letters* **35**
[36] Sandberg I, Jiggens P, Heynderickx D and Daglis I 2014 *Geophysical Research Letters* **41** 4435–4441
[37] Vršnak B, Sudar D and Ruždjak D 2005 *Astron. Astrophys* **435** 1149–1157
[38] Youssef M 2012 *NRIAG Journal of Astronomy and Geophysics* **1** 172–178
[39] Vainio R, Valtonen E, Heber B, Malandraki O E, Papaioannou A, Klein K L, Afanasiev A, Agueda N, Aurass H, Battarbee M *et al.* 2013 *Journal of space weather and space climate* **3** A12

Derivation of relativistic SEP properties through neutron monitor data modeling

C Plainaki[1,2*], M Laurenza[1], H Mavromichalaki[2], M Storini[1], M Gerontidou[2], A Kanellakopoulos[2], M Andriopoulou[3], A Belov[4], E Eroshenko[4], V Yanke[4]

[1]INAF-IAPS, Via del Fosso del Cavaliere, 00133, Rome, Italy
[2]Nuclear and Particle Physics Section, Physics Dpt., National and Kapodistrian University of Athens, Greece
[3]Space Research Institute, Austrian Academy of Sciences, Graz, Austria
[4]Institute of Terrestrial Magnetism, Ionosphere and Radio Wave Propagation by Pushkov (IZMIRAN), Moscow, Russia

E-mail: christina.plainaki@iaps.inaf.it

Abstract. The Ground Level Enhancement (GLE) data recorded by the worldwide Neutron Monitor (NM) network are useful resources for space weather modeling during solar extreme events. The derivation of Solar Energetic Particles (SEPs) properties through NM-data modeling is essential for the study of solar-terrestrial physics, providing information that cannot be obtained through the exclusive use of space techniques; an example is the derivation of the higher-energy part of the SEP spectrum. We briefly review how the application of the Neutron Monitor Based Anisotropic GLE Pure Power Law (NMBANGLE PPOLA) model (Plainaki et al. 2010), can provide the characteristics of the relativistic SEP flux, at a selected altitude in the Earth's atmosphere, during a GLE. Technically, the model treats the NM network as an integrated omnidirectional spectrometer and solves the inverse problem of the SEP-GLE coupling. As test cases, we present the results obtained for two different GLEs, namely GLE 60 and GLE 71, occurring at a temporal distance of ~ 11 years.

1. Introduction

Ground-level enhancement (GLE) events are related to the most energetic class of solar energetic particle (SEP) events, associated with solar flares and/or coronal mass ejections (CMEs), and requiring acceleration processes that produce particles with energies ≥ 500 MeV upon entry in the Earth's atmosphere. This class of SEPs can produce showers of secondary particles with sufficient energy to be detected by ground-level Neutron Monitors (NMs) and with intensities that exceed the Galactic Cosmic Ray (GCR) background (see, among others, [1]).

Several techniques for modeling the dynamical behavior of GLEs are presently available (e.g. [2 - 11]). Typically, the responses of ground level NMs to SEPs are modeled to determine both the best-fit spectrum and the flux spatial distribution of the solar particles during a GLE event. Usually, a least squares procedure is applied in order to define the values of the parameters that best fit the GLE model in use. The functions considered to perform the fit represent the physical processes involved in the

* Corresponding author

particle rigidity distribution and propagation as well as the atmospheric response to the incoming particle fluxes. One of the most intriguing open questions, to which GLE modeling can give important insights, is the determination of the acceleration mechanism that is responsible for the production of relativistic SEPs. Whereas numerous techniques, based on the analysis of satellite data, have been successfully applied in order to get insights into which acceleration mechanism is the responsible one for a specific event (i.e. acceleration at the flare reconnection sites or acceleration at the shocks driven by CMEs propagating through the solar corona and in the interplanetary space), the incorporation of ground-based methods can integrate these analyses by providing information also on the high-energy component of the SEP population. In order to determine the direction of arrival of the solar particles at the Earth during a GLE, models can assume either a pitch-angle distribution function (see [12 - 13]) or, simply, a latitude and longitude-dependent function (that represents the spatial diffusion of the solar particles around the apparent source direction), without inserting directly the interplanetary magnetic field (IMF) direction information in the model (see [4-5], [7], [10-11], [14]). Whereas the alignment of the solar particle velocity with the IMF could be an indication of propagation along the magnetic field lines almost without scattering ([12]), [15] argued that there is no reason why the magnetic field measured by a satellite at some point would be the same as the average field sampled by the particles over their orbit. Such was the case of GLE 59 (on 2000 July, 14), when the derived asymptotic latitude of the anisotropy in the Geocentric Solar Ecliptic (GSE) coordinate system deviated from the IMF latitude up to almost 100^0 (opposite hemisphere) during the event maximum [15]. In this view, it can be generally stated that the model-assumption of particle propagation around the local IMF vector is not always a quantity to be inserted a priori as a pre-defined input while modeling GLEs.

Current modeling techniques incorporate realistic geomagnetic field models (e.g.: Tsyganenko, 1989 – T89; 1996 – T96) which take into account possible geomagnetic disturbances; the choice of the magnetic field model to be incorporated inside a GLE model has a major weight in the determination of the actual SEP parameters. Indeed, [16] showed that the use of different magnetic field models (T89 and T96) inside the NMBANGLE PPOLA model, influence the derived results to an extent depending on the SEP energy. For the case of GLE 71, [16] found that the modeled solar proton spatial distribution (for rigidities of ~1 GV) is less diffused around the apparent source direction when the T96 (rather than the T89) model is considered. In particular, the angular distribution around the apparent source has a width equal to 19^0 and 32^0 when T96 and T89 are considered, respectively. For the 2 GV particles, both models gave similar results. The accurate modeling of a GLE event depends also on two other important factors: the number of the NMs used in the analysis and their spatial distribution around the world; for example, in order to avoid a biasing of the modeling-results, it is important to use data originating from NMs that are almost equally distributed between the two hemispheres (see also [17]).

In this work we perform a brief comparative examination of the results obtained from the NMBANGLE PPOLA model application to GLE 60 (on 2001 April, 15) and GLE 71 (on 2012 May, 17) that took place at a temporal distance of ~11 years, both occurring in the maximum solar activity phase, although the sunspot number in 2012 was significantly lower than its value in 2001. The modeling results presented here are based exclusively on the use of ground-based NM data hence our technique is completely independent from the satellite measurements analysis. Consequently, an a posteriori comparison of model-derived quantities with those obtained from the satellite measurements (e.g. GOES energetic particle data) can be used to validate the model itself. The scope of this comparative study is to show that GLE modeling based exclusively on the use of NM-data, can provide: a) the SEP spectrum reconstruction, b) an explanation of the form of the GLE time-intensity profiles on the basis of the solar particle arrival directions and the NM asymptotic directions of viewing, and c) quantitative information on the SEP flux evolution and direction upon arrival in the Earth's atmosphere. In Section 2, we provide a brief description of the NMBANGLE PPOLA modeling technique and a short summary of two GLEs under study. In Section 3 we perform a comparative analysis of the obtained results and finally, in Section 4, we give the main conclusions of the current study.

2. Modeling technique and events description

2.1. *The NMBANGLE PPOLA model*

The NMBANGLE PPOLA model is a modified version of the original NMBANGLE model [7], which is based on the coupling coefficient method [18] applied numerous times in the past (e.g., [4-5], [7], [14]). The NMBANGLE PPOLA model couples SEPs (primary particles) at a selected altitude in the Earth's atmosphere with the secondary ones detected by ground-level NMs during GLEs. This model dynamically calculates the SEP spectrum and the SEP flux spatial distribution at some altitude in the atmosphere. A power-law SEP spectrum with two free parameters (spectral index and amplitude) is assumed. SEPs are assumed to be protons in the model's current version. The details of the physics considered in the NMBANGLE and the NMBANGLE PPOLA models have been provided in the past (see [7], [10-11]). In this Section we only give a very brief description of the model.

The NMBANGLE PPOLA model uses as inputs the response of the worldwide NM network to the high-energy solar protons (i.e., ≥ 500 MeV) and the disturbance level of the geomagnetic field (through the use of the k_p index). The NM network is treated as an integrated omnidirectional spectrometer able to measure the characteristics of the relativistic primary flux at a selected altitude (h_0=20 km) in the Earth's atmosphere. GLE data from NM stations widely distributed around the world are incorporated (for a detailed description of the format of the GLE data used in the model, see [10]). In this context, the modeling of the NM response to an anisotropic SEP flux and the solving of the inverse problem can provide the actual characteristics of the relativistic SEPs that are responsible for the event. The results of the application of both NMBANGLE and NMBANGLE PPOLA models to past GLEs have been, in general, in good agreement with space observations, when available ([7], [10-11]). We note that due to the particle motion inside the geomagnetic field, each ground level detector is capable of recording secondary particles produced by primaries arriving from a set of directions in space, depending on their energy, known as the station's asymptotic directions of viewing. For the evaluation of the NM directions of viewing, the method described by [19] is applied; the assumption that the Earth's magnetospheric field can be adequately described by the T89 model ([20]) is considered. In the particle trajectory calculation, the relativistic SEPs have been considered to have a vertical incidence at each neutron monitor location.

According to the NMBANGLE PPOLA model, the possible time variations, $\Delta N/N_0$, of the total neutron counting rate, N_0, observed at cut-off rigidity R_c, geographic latitude ϕ and longitude λ, at level h in the atmosphere and at some moment t, are determined by the following expression ([7], [10]):

$$\frac{\Delta N(R_c,h,t,t_0)}{N_0(R_c,h,t_0)}=\int_{R_c}^{R_u}\frac{W(R,h,t_0)\cdot\Delta I(R,\Omega(R,t),t)}{I_0(R,t_0)}dR=\int_{R_c}^{R_u}\frac{W(R,h,t_0)\cdot A(R,\Omega(R,t),t)b(t)R^{\gamma(t)}}{I_0(R,t_0)}dR \tag{1}$$

where $W(R, h, t_0)$ is the rigidity-dependent coupling function (properly normalized to have values in the 0-1 range) between secondary and primary particles arriving at altitude h_o (with $h_o > h$) in the Earth's atmosphere (see [7] for analytical expression); $\Delta I(R,\Omega(R,t),t)=I(R,\Omega(R,t),t)-I_0(R,t_0)$ is the increase of the primary flux due to the arrival of SEPs at altitude h_o, assumed to follow a power law in rigidity ([7]), i.e. $\Delta I(R,\Omega(R,t),t)=A(R,\Omega(R,t),t)\cdot b(t)R^{\gamma(t)}$; γ (t) is the exponent of the power-law SEP spectrum; $A(R, \Omega, t)$ is a dimensionless normalized function that describes the spatially anisotropic SEP arrival at altitude h_o, with $\Omega(R, t)$ being the angle defined by the direction of viewing for vertical incidence at a NM for a rigidity R and the vertical direction of the apparent SEP source at altitude h_0, as defined in [7]; R_u is the theoretical upper limit for the rigidity of the SEP particles, a parameter that in the model is specified a priori (here 8 GV) without a big loss in the estimation accuracy of the other parameters ([4]); $b(t)$ is the amplitude of the SEP differential flux (in protons m^{-2} s^{-1} sr^{-1} GV^{-1}); and I_0 (R, t_0) is the GCR differential flux (in protons m^{-2} s^{-1} sr^{-1} GV^{-1}). In our model, we define $A(R, \Omega(R, t), t)$ as in [7]:

$$A(R,\Omega,t) = \exp\left(-n_a(t)^2 \sin^2 \frac{\Omega(R,t)}{2}\right) \tag{2}$$

where $n_a(t)$ is a dimensionless parameter that characterizes the width of a primary solar particle beam, arriving at altitude h_o, around a specific location, whose latitude and longitude are free parameters for the model. Through the selection of the mathematical form described in Equation (2), the NMBANGLE PPOLA model parameterizes the level of the primary particle flux anisotropy by the use of variable $n_a(t)$, which is dynamically determined after each model run. Big values for $n_a(t)$ mean that the arriving SEP flux is narrowly distributed around a specific location at altitude h_0, whereas smaller values for $n_a(t)$ mean that the SEP flux is more widely distributed in longitudes and latitudes. This parameter is determined independently from the magnitude of the primary SEP intensity. However, it is the product of the anisotropy function with the primary spectrum that gives the total behaviour of the primary flux responsible for the event registered at ground level. A least-square fitting technique based on the Levenberg Marquardt algorithm allows an efficient derivation of the optimal solution for each time interval and the determination of the GLE parameters evolution. Five-minute GLE data from NM stations of the worldwide network are incorporated to fit Equation (1). Each time represents the start of a five minute integrated time interval.

2.2. Description of GLE 60, GLE 71 and related SEP events

On 2001 April 15, a strong flare (X14.4/2B) was observed at the west limb of the solar surface at the position S20W85, associated with a fast CME (>1200 km/s) (see [21], [22] for details). Following the detection of gamma and X-rays, the High Energy Proton and Alpha Detector on board GOES 10 satellite recorded sudden increases in relativistic protons (510-700 MeV) between 13:50 UT and 13:55 UT [6]. High energy protons and possibly neutrons, associated with the above mentioned solar events, were detected by the ground-level NMs of the worldwide network, starting at about 13:50 UT, in 5-min NM data (see Fig. 1). The event was seen by polar and mid-latitude NMs, as well as by some low-latitude NMs (see [10] for a full description). The maximum NM % count increase was 225.4%, at South Pole NM.

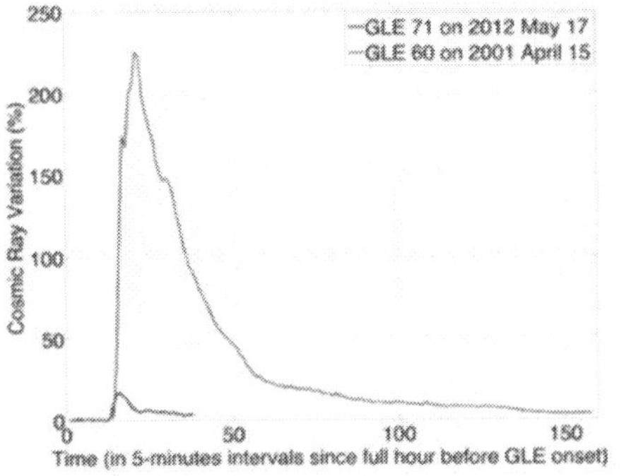

Figure 1: Relative counting rate variation of the South Pole neutron monitor during GLE 60 (in blue) and GLE 71 (in black). The x-axis unit is 5-minutes intervals since the full hour before GLE onset.

Early on 2012 May 17, the Energetic Particle Sensor (EPS) onboard GOES spacecraft recorded an increase in the proton and electron channels, extended also to high energies. The > 100 MeV proton event began at 02:00 UT, reached a maximum of ~20 pfu at 02:30 UT and ended at 17:25 UT. The SEP event is associated with the M5/1f flare (peak time on May 17 at 01:47 UT) occurring in the Active Region (AR) 11476, located at N11W76. Based on the flare longitude the SEP event was relatively "well-connected" to the source region, also considering the observed solar wind speed of ~400 Km/s preceding the event. Secondary particles, associated with this SEP event were detected by the ground-level NMs of the worldwide network (e.g. [9],[11], [23]), with a maximum increase of ~17% at South Pole NM. Among the numerous mid-latitude NMs, GLE71 was registered at Kiel, at Kerguelen and at Yakutsk. The low-latitude NMs (high magnetic cut-off rigidity, R_c) did not register GLE 71 (e.g., [24]). Figure 1 illustrates the percentage variation of the counting rate registered at the NM of South Pole during these two events as a function of time (measured in 5-minutes intervals since full hour before GLE onset). The pre-increase baseline period used to derive the percentage is 301.622 counts/s for GLE71 (at 0:00 – 01:00 UT) and 243.361 counts/s for GLE60 (at 12:00 - 13:00 UT).

3. Brief comparative analysis - Discussion

3.1 SEP spectrum

The modeled SEP rigidity spectra are computed for both GLEs, every 5 minutes during their evolution. In Fig. 2, we present the SEP rigidity spectra derived by the application of the NMBANGLE PPOLA model to GLE 60 and GLE 71, at the main phase of each event (i.e. the time intervals 2001 April 15 14:25-14:30 UT and 2012 May 17 02:15-02:20 UT, respectively). We underline that we avoided the spectra comparison at the initial event phases because our model results have bigger uncertainties at that time ([10], [11]). As far as the GLE 71 (GLE 60) is concerned, no NM with geomagnetic cutoff rigidity above ~3 GV (~5 GV) has observed count rate increases. Therefore, the GLE analysis provides less reliable information on the SEP spectrum at rigidities > ~3 GV (~5 GV) during GLE 71 (GLE 60). We first note that the spectrum at 1 GV for GLE 71 is in good agreement with the SEP flux in the direction of anisotropy obtained by [12]. Nevertheless, the SEP intensity in the direction of anisotropy obtained by [9] at 1 GV is about one order of magnitude higher. Moreover, for GLE71, the predicted proton spectrum continues to exceed the GCR background (dotted line in Figure 3) even at rigidities > 3 GV (red line). At 3 GV our modeled intensities are almost one order of magnitude higher than those shown in [12] and [9]. Perhaps the anisotropic direction of arrival of solar particles might play a role in the high intensities estimated by our model. In a detailed study of GLE 71 using the NMBANGLE PPOLA tool, [11] estimated the expected anisotropy in the arrival direction, through the calculation of the parameter n_α (see *Section 2.1*), which was found to take a non zero value in the time interval 2012 May 17 02:15-02:20 UT. It is underlined that this parameter expresses exactly the spread of the primary flux over latitudes and longitudes (for a complete explanation of the n_α meaning see [7]). Therefore the high rigidity particles (if present) could be sensed through their secondaries only by those NMs that had direction of viewing (corresponding to these high rigidities, i.e. in the range 3-10 GV) located close to the maximum flux of the apparent particle source. The small magnitude of this event in any case can be an additional reason for which the increases in the NMs were not significant and the latitude effect was absent. The latter has been pointed out also by other researchers as well (see Section 4.3 in [25]). A further analysis regarding these issues is, in any case, necessary and intended in the near future.

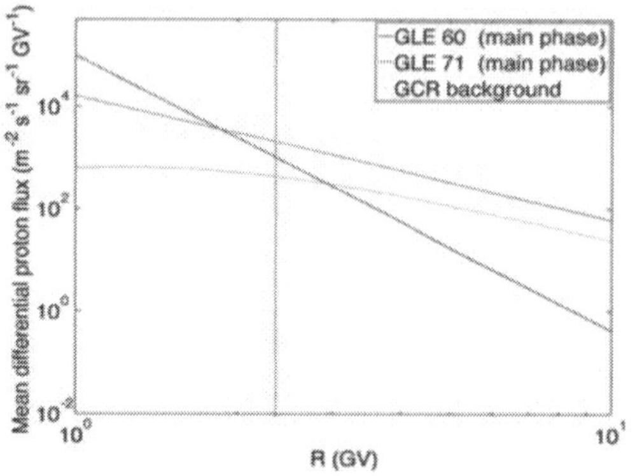

Figure 2: GLE 60 and GLE 71 spectra at the main phases of each event, obtained from the application of the NMBANGLE PPOLA model. The red line marks the 3 GV rigidity value over which the GLE71 spectrum is less reliable. Dot line represents the GCR background computed through the CRÈME model (https://creme.isde.vanderbilt.edu/) for May 17, 2012.

According to Fig. 2, the NMBANGLE PPOLA model application gives a harder spatially averaged spectrum for GLE 71 than for GLE 60: in the rigidity range from 2 to ~ 4 GV the mean differential flux of the primary SEPs during GLE 71 is higher than its values during GLE 60, by up to one order of magnitude. This result seems to contradict the fact that during GLE 60, SEPs with higher maximum energy than those during GLE 71 have been observed [11]. However, this result can be explained on the basis of the following arguments. The above comparison of the two SEP spectra refers to the main phase of each event, i.e. after the registration of the maximum NM intensity variation, and not to the main SEP acceleration phase that had taken place during the first 5-min intervals after the flare in each case. The high rigidity SEPs during GLE 60 were registered mainly during this initial phase (see for example [10]). Therefore a relatively soft SEP spectrum in the main phase of GLE 60 is reasonable. Moreover, although the GLE 60 shows intensity variations in the ground observations significantly higher than during GLE 71, the spatial distribution of the primary flux is more anisotropic, consistently with NM observations [10]. As a consequence, the peak secondary flux (at the ground) can be measured mainly at specific locations that are magnetically favorable with respect to the SEP source. Therefore, the spatially averaged modeled-SEP spectrum in Fig. 2 is less representative of the actual GLE 60 event evolution, since the averaged SEP fluxes smoothen too much the actual flux variations in rigidity. At this point we would like to emphasize that an averaging of the differential primary flux over all longitudes and latitudes is necessary in order to obtain the mean SEP spectrum per solid angle unit; such a quantity can be easily compared with the one obtained from the omnidirectional measurements of different kind or from other models. Finally, GLE 60 took place inside a a Forbush Decrease (that was indeed modeled in NMBANGLE PPOLA [10]) where the GCR intensity (I_0 in Eq. 1) was lower than the one during GLE 71, that took place in a relatively quiet period. This might have affected the computation of the modeled spectra.

In the low rigidity range ($\leq$ 2 GV), the mean differential flux of the SEPs during GLE 60 is bigger than its values during GLE 71, by up to one order of magnitude. This result is in good agreement with the NM observations that are higher during GLE 60 (by more than one order of magnitude at the South Pole NM) than during GLE 71. We note that in the case of GLE 60, in the lower rigidity range (i.e. < 2 GV), the NMBANGLE PPOLA spectrum is in good agreement with the spectrum calculated by other models (e.g. [6]). On the other hand, in the higher rigidity range, the estimated NMBANGLE PPOLA intensities seem to be overestimated: for GLE 60 they are in general higher than those estimated by [6], where a modified power law spectrum was assumed. Nevertheless, spacecraft measurements at rigidities ≥1GV are not always available (e.g. for GLE 60) and modeling the ground based-data remains an important tool for the derivation of the SEP characteristics at 1 AU, integrating also the information obtained by satellite instruments that cover the lower rigidity range (e.g. proton detectors onboard GOES). Since NMs are distributed in different geographical latitudes and longitudes, they provide information related to a different part of the SEP spectrum. Therefore, NM-data from a network of stations distributed geographically in the most possible balanced way is necessary for analyses of this kind.

3.2 SEP fluxes

Comparison of the extrapolated model-derived flux of lower-energy (> 100 MeV) SEPs with 5-min GOES observations, for both GLE 60 and GLE 71 events, has shown that in general the NMBANGLE PPOLA model simulates adequately the real SEP flux in the main event-phase. For example, in the case of GLE 71, for the time period after 02:00 UT, the > 100 MeV modeled and GOES fluxes differ by a factor lower than ~3 (see Fig. 6 in [11]). Similarly, in the case of GLE 60, for the time period after 14:00 UT, the > 100 MeV modeled and GOES fluxes differ by a factor lower than ~4 (see Fig. 9 in [10]). On the other hand, in the initial time period of each event the model based on neutron monitor data gives in general higher SEP fluxes than the ones observed by satellite detectors, mainly at low rigidities. To explain this difference, one should consider the streaming limit effect, which currently has not been integrated in the NMBANGLE PPOLA model. During a SEP event, the intensity of the MeV-energy particle fluxes can exhibit a plateau of about 100 $cm^{-2}\,s^{-1}\,MeV^{-1}$, varying by a factor of two among different events [27]. This phenomenon, known as the 'streaming limit' [27], can result in suppressing the low-energy particle arrival below the levels expected from higher-energy spectral responses. It would be very useful, therefore, if during future GLE modeling efforts the streaming effect were taken into account while extrapolating the model-derived data to the lower energy ranges.

3.3 SEP anisotropy direction

The direction of the apparent source of solar particles at a selected altitude in the Earth's atmosphere, a quantity that is generally difficult to determine, is a dynamical output of the NMBANGLE PPOLA model. The time-dependent variation of the position of this apparent source can explain the differences in the profiles of NMs with similar cut-off rigidities, located in different places though. In case of GLE 71, in the initial event phase, the apparent SEP source direction was located in the Northern hemisphere, whereas with time, it moved to lower latitudes. This result is consistent with the ground-based observations [11]. In case of GLE 60, during the initial phases of the event, the latitude of the apparent SEP source, as obtained by the NMBANGLE PPOLA model, varies significantly around the ecliptic plane. However, in that period, the model results contain large uncertainties due to the extremely anisotropic direction of propagation of the solar particles and due to the big differences in the counting rates recorded between different NMs. In later phases of the GLE 60 event, the apparent source moves northwards. An important reason that most models fail during the rising phase of a GLE is that the SEP flux changes are too dynamic and the models would need data of sufficient statistical accuracy, with sampling times as short as a minute. These data are not always available, since many NM stations have registration systems with larger sampling times. Moreover, it often happens that there is very poor representation of southern viewing directions (see for example [10]). This can significantly bias the modeling results, especially those considering the position of the anisotropy source, which seems to be the less reliably modeled parameter in the NMBANGLE PPOLA model.

4. Conclusions

The SEP spectra during GLE 60 and GLE 71 as well as their spatial distribution have been estimated through modeling of the ground-based NM data. The main results can be summarized as follows.

- A hard rigidity spectrum of accelerated protons was found during the initial phase of GLE 60 and a rather soft spectrum in later phases, i.e. after 14:00 UT ($\gamma \sim -5.5$).
- During the main phase of GLE 71, the rigidity spectrum index γ was estimated to be ~ -2.1; in later phases, i.e. after 02:20 UT, a softer spectrum of accelerated protons ($\gamma \sim -3.8$) has been derived. The corresponding values for the energy spectrum are -1.55 and -2.4, respectively (see [11]).
- The results for GLE 71 are consistent with the typical range found by [28] for shock wave acceleration in case of relativistic SEP events (see also [29]), although a direct flare contribution cannot be excluded (see also [11]).
- The model-results can provide realistic estimation of the SEP fluxes in the energy range where NM increases are registered.
- For both GLEs, the model seems to overestimate the spatially averaged SEP spectrum in the high rigidity range, where no NM increases are registered.
- Comparison of the results obtained for the two events is valid mainly in the rigidity range 1-3 GV. The spectrum computed in the event main phase results to be harder for GLE 71. This could be due to the different SEP flux spatial distribution with respect to the NM direction of viewing and the different GCR background level (that is reduced during GLE60);
- The integral SEP fluxes calculated by the NMBANGLE PPOLA model are in good agreement with GOES observations if extrapolated to the lower energy range.

The NMBANGLE PPOLA results will be more reliable during future model-applications to GLEs, if the NM data-quality is high hence the NM data used in the model should be provided with the same quality standards. The NMBANGLE PPOLA results can be also used for radiation dose calculations (as in [17]) in the context of space weather perspectives. In our model, the position of the GLE source outside the atmosphere has been considered to be the most significant factor determining the propagation and final registration of the secondaries at ground level. Up to now, this assumption has been considered also in other modeling efforts for different GLEs (see [6], [8], [12]). However, other phenomena, like the diffusion of the secondary particles inside the atmosphere could also have a role in the interpretation of the ground-level intensity time profiles during a GLE and should be taken in detail into account during future analysis.

Acknowledgements
The authors acknowledge all colleagues at the NM stations who kindly provided us with the data used in this study and the Referees for useful comments. Thanks are also due to the High-resolution Neutron Monitor NMDB database, founded under the European Union's FP7 Program (contract No. 213007) for providing cosmic ray data.

References

[1] Lopate C 2006 In: Gopalswamy, N, Mewaldt, R, Torsti, J (eds) *Solar Eruptions and Energetic Particles,* Geophysical Monograph Series **165,** AGU Press, Washington 283
[2] Shea M A and Smart D F 1982 *Space Sci. Rev.* **32** 251
[3] Humble J E, Duldig M L, Smart D F and Shea M A 1991 *Geophys. Res. Lett.* **18** 737.
[4] Belov A, Eroshenko E, Mavromichalaki H, Plainaki, C and Yanke V 2005, *Ann. Geophys.* **23** 2281
[5] Belov A V, Eroshenko E A, Mavromichalaki H, Plainaki C and Yanke VG 2005 *Proc. 29th Inter. Cosmic Ray Conf.* vol 1, p 189.
[6] Bombardieri D J, Michael K J, Duldig M L and Humble J E 2007 *Astrophys. J.* **665** 813
[7] Plainaki C, Belov A, Eroshenko E, Mavromichalaki H and Yanke V 2007 *J. Geophys. Res.* **A 112** 4102
[8] Vashenyuk E V, Balabin Y V and Gvozdevsky B B 2011 *ASTRA* **7** 459
[9] Mishev A L, Kocharov L G and Usoskin I G 2014 *J. Geophys. Res.* **119** 670
[10] Plainaki C, Mavromichalaki H, Belov A, Eroshenko E, Andriopoulou M and Yanke V 2010 *Sol. Physics* **264** 239
[11] Plainaki C, Mavromichalaki H, Laurenza M, Gerontidou M, Kanellakopoulos A and Storini M 2014 *Astrophys. J.* **785** 160
[12] Balabin, Y V, Germanenko A V, Gvozdevsky B B and Vashenyuk E V 2013 *J. of Physics: Conference Series* **409** 012053
[13] Bütikofer R, Flückiger E O, Desorgher L, Moser M R and Pirard B 2009 *Adv. Space Res.* **43** 499
[14] Plainaki C, Mavromichalaki H, Belov A, Eroshenko E and Yanke V 2009 *Adv. Space Res.* **43** 474
[15] Bieber J W, Droge W, Evenson P A, Pyle R, Ruffolo D, Pinsook U, Tooprakai P, Rujiwarodom M, Khumlumlert T and Krucker S 2002 *Astrophys. J.* **567** 622
[16] Kanellakopoulos A, Plainaki C, Mavromichalaki H, Laurenza M, Gerontidou M, Storini M and Andriopoulou M 2014 *EGU General Assembly 2014* **16**, EGU2014-461
[17] Bütikofer R and Flückiger E O 2013 *J. of Physics: Conference Series* **409** 012166
[18] Dorman LI 2004 *Astrophys. Space Sci. Libr.* **303**
[19] Plainaki C, Mavromichalaki H, Belov A, Eroshenko E and Yanke V 2009 *Adv. Space Res.* **43** 518
[20] Tsyganenko N A 1989 *Planet. Space Sci.* **37** 5
[21] Muraki Y, Matsubara Y, Masuda S, Sakakibara S, Sako T, Watanabe K, Bütikofer R, Flückiger E O, Chilingarian A, Hovsepyan G, Kakimoto F, Terasawa T, Tsunesada Y, Tokuno H, Velarde A, Evenson P, Poirier J and Sakai T 2008 *Astropart. Phys.* **29** 229
[22] Gopalswamy N, Lara A, Yashiro S and Howard R A 2003 *Astrophys. J. Lett.* **598** 63
[23] The IceCube Collaboration 2013 *Science* **342** 6161
[24] Signoretti F & Storini M 2013 *INAF Report IAPS-2013–05* (Rome, Italy: Institute for Space Astrophysics and Planetology)
[25] Li, C, Firoz, K A, Sun, L P, and Miroshnichenko, L I. 2013 *Astrophis. Journal* **770** 34
[26] Lovell J L, Duldig M L, Humble J E 1998 *J. Geophys. Res.* **103** 23733
[27] Reames D V and Ng C K 1998 *Astrophys. J.* **504** 1002
[28] Ellison D C and Ramaty R 1985 *Astrophys. Journal* **298** 400
[29] Mewaldt R A et al 2012 Space Sci. Rev. **171** 97

Solar particle events contribution in the space radiation exposure on electronic equipment

Vasily S Anashin[1], Grigory A Protopopov[1, 3], Olga S Kozyukova[1], Ninel N Sitnikova[2]
[1] Branch of JSC "URSC"-"ISDE", Moscow, Russia
[2] JSC " Academician M.F.Reshetnev" Information Satellite Systems", Zheleznogorsk, Russia

E-Mail: conf@niikp.org

Abstract. The article presented the on-board measurement results of dose sensors. We carried out the estimation of the contribution of solar protons in a dose rate value during solar events. We determined the solar proton events conditions when these events can markedly affect the reliability of on-board equipment operation.

1. Introduction

Space radiation is the most critical natural influencing factor which limits spacecraft's electronic equipment life time. The main components of space radiation are particles from Van Allen belts, solar and galactic cosmic rays [1]. Space radiation accounts from 30 to 50 per cent of qualified failure (because of total ionizing dose, displacement damage and single event effects), even though this percentage is higher due to the stimulation of other types of failures, first of all, electrostatic [2].
At the present time there is a need for carrying out permanent monitoring of space radiation characteristics for the determination of the degree of space radiation exposure on electronic equipment and for the forecast of spacecraft operability. For these purposes, the monitoring system of space radiation exposure on spacecraft electronic equipment was made on demand of Roscosmos [3].

2. Space radiation exposure on the radio-electronic equipment engineering monitoring system

2.1. The monitoring system aim and tasks

The main aim of the monitoring system is the increase of spacecraft active shelf life. The monitoring system is not the alternative and is an addition to the existing scientific monitoring system and is used, generally, for:

- Measurement of space radiation impact on the spacecraft electronic equipment.
- Calculation and control of the spacecraft remaining lifetime.
- Management of the structural and algorithmic approach for active lifetime increase.
- Forecasting of space radiation impact changes (including dangerous) on the spacecraft radio electronic equipment.
- Receiving the flight data on spacecraft electronic equipment hardness.

[3] To whom any correspondence should be addressed.

- The understanding of space radiation influence mechanisms on the spacecraft electronic equipment.
- Improvement of charge particles flux models.
- Forecasting of "space weather" characteristics.

2.2. The monitoring system structure and elements

The monitoring system structure is presented in figure 1. The monitoring system includes space borne and ground segments. The main elements of the space borne segment are sensors which are based on the MNOS-dosimetry principle (metal-nitride-oxide-semiconductor transistor is used as a sensitive element). The sensors were calibrated at the certified stand "GU-200" with ^{60}Co isotopes [3]. The ground segment includes several stations for the input, data processing and information output among other the on-board measurements At the moment there are 42 sensors on the board of 21 spacecrafts developed by JSC "Information Satellite Systems" which are functioning on the circular orbit ~ 20000 km (outer radiation belt). The engineering sample of the monitoring system ground segment was developed (website www.kosrad.ru).

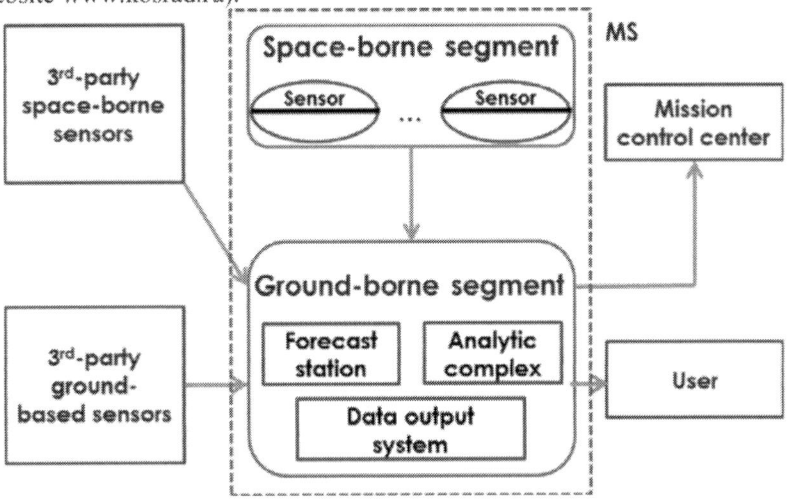

Figure 1. The monitoring system structure

3. Experimental and calculation results

3.1. Abrupt increase of dose rate event

The analysis of the received data from October 2008 to September 2013 allows us to find out the periods with sharp increase of dose rate and definite value of such an increase. We also analysed space radiation characteristics data from other monitoring systems (such as GOES [4]) in dates of anomalous increase of dose rate. As you can see from the table 1, the increase of high-energy electron fluxes is observed on all dates and also geomagnetic storms are observed (Kp index is from 5) for all dates, while there are only two solar proton events which can contribute to the dose rate increase.

Table 1. The characteristics of space weather on the date of sharp increase of dose rate

Date	Dose rate before event (a.u. s^{-1})[a]	Dose rate after event (a.u. s^{-1})[a]	Dose rate increase (times)	GOES proton fluxes >10 MeV (cm^{-2} day^{-1} sr^{-1}) [4]	GOES electron fluxes > 2 MeV[b] (cm^{-2} day^{-1} sr^{-1}) [4]
05.04.2010	$1.22\ 10^{-5}$	$1.34\ 10^{-2}$	1098	$1.6\ 10^{4}$	$5.4\ 10^{9}$
03.08.2010	$2.67\ 10^{-4}$	$3.08\ 10^{-3}$	11	$1.1\ 10^{5}$	$2.5\ 10^{8}$
01.03.2011	$1.29\ 10^{-4}$	$1.02\ 10^{-3}$	79	$1.4\ 10^{4}$	$7.8\ 10^{8}$
02.05.2011	$1.06\ 10^{-4}$	$1.89\ 10^{-3}$	17	$1.2\ 10^{4}$	$6.8\ 10^{8}$
09.09.2011	$1.26\ 10^{-4}$	$2.18\ 10^{-3}$	17	$4.2\ 10^{4}$	$3.2\ 10^{8}$
23.01.2012	$2.01\ 10^{-5}$	$1.59\ 10^{-4}$	8	$1.4\ 10^{8}$	$2.5\ 10^{8}$
08.03.2012	$3.61\ 10^{-4}$	$2.35\ 10^{-3}$	7	$2.0\ 10^{8}$	$5.9\ 10^{8}$
08.10.2012	$6.08\ 10^{-4}$	$9.90\ 10^{-3}$	16	$1.4\ 10^{4}$	$4.8\ 10^{8}$

[a] a.u. are arbitrary units
[b] Maximum value during the week

3.2. Computation algorithm of dose rate value from solar protons

In this work we examined the largest solar particle event during the period of absorbed dose measurements (08.03.2012). The following algorithm was applied to assess the contribution of this flare in dose rate on the considered orbit of 20000 km.

- The integral spectrum of solar protons on the Geostationary Earth Orbit (GEO) was determined for the period from 5.03.2012 to 17.03.2012. For this purpose, the daily data of the solar proton fluxes for different energy ranges were used [4].
- The differential spectrum of solar protons was calculated for this event. We considered the energy range from 1 to 100 MeV (protons with this energy give the main contribution in the absorbed dose). The differential spectrum, (in the MeV s^{-1} cm^{-2}) was estimated for the range from 1 MeV to 10 MeV on the formula (1), for the range from 10 MeV to 100 MeV – (2) and from 100 MeV – (3):

$$f_1 = \frac{F_1 - F_{10} - F_{100}}{13.2\pi} \quad (1)$$

$$f_2 = \frac{F_{10} - F_{100}}{36\pi} \quad (2)$$

$$f_3 = \frac{F_{100}}{360\pi} \quad (3)$$

  Where F_1, F_{10}, F_{100} are the corresponding integral proton fluxes with energy more than 1MeV, more than 10 MeV and more than 100 MeV, respectively.
- We recount the differential spectrum of solar protons for the 20000 km orbit with the help of correction indices for the consideration of geomagnetic cuttoff rigidity (Rc). These indices are the ratio of solar protons fluxes for the 20000 km orbit and for GEO at same energies. Calculations were carried out using the COSRAD software [5]. The software is the main certified tool in Russia for space radiation environment calculation and is used for the establishing radiation hardness requirements for electronic equipment developed by the order of Roscosmos. Table values of Rc are used in the software for different altitudes and inclinations, which were precalculated for average geomagnetic activity (K index ~2-3). So changes of the Earth's geomagnetic field are not taken into account, and that is the limitation of our method. Figure 2 shows the correction indices.

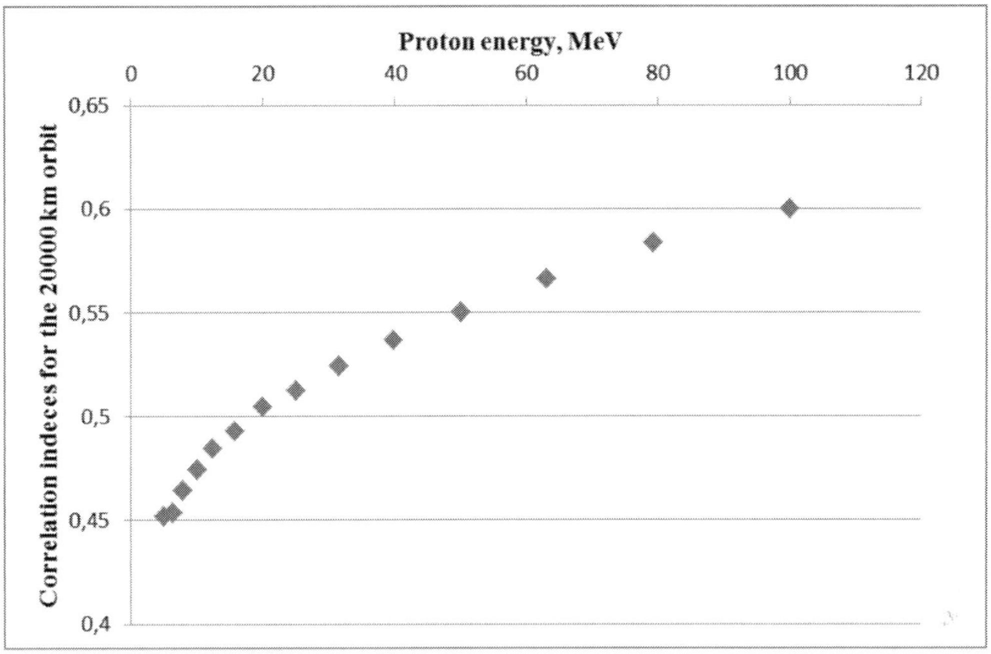

Figure 2. Correction indices for proton fluxes with certain energy from the 36000 km orbit to 20000 km orbit

- We calculated dose rate values behind the shielding of 1 g cm^{-2} (the value of the effective thickness shielding was determined taking the spacecraft's frame and MNOS transistor's case into account) during the period of solar proton flare with received spectrum by the «OMERE» software [6]. The calculation results are presented in table 2.

Table 2. The calculation results of dose rate values

Data	GOES proton flux > 10 MeV, proton$(cm^2\ s\ sr)^{-1}$	GOES proton flux > 100 MeV, proton$(cm^2\ s\ sr)^{-1}$	Dose rate from solar protons, a.u. s^{-1}	Contribution to the total dose %
25.09.2001	12900	30	$4.3\ 10^{-4}$	~5[a]
06.11.2001	31700	200	$9.52\ 10^{-4}$	~10[a]
24.11.2001	18900	4	$1.59\ 10^{-4}$	~1[a]
02.10.2001	2360	0,2	$1.08\ 10^{-5}$	~0,1[a]
28.10.2003	29500	200	$1.15\ 10^{-3}$	~11[a]
24.01.2012	6310	2	$4.28\ 10^{-5}$	~0,5[a] (27[b])
08.03.2012	6530	70	$4.82\ 10^{-4}$	~6[a] (21[b])

[a] comparison with the maximum dose rate value for whole period of measurements (~10^{-2} a.u. s^{-1} according to Table 1)
[b] comparison with the dose rate value measured for certain day

The calculated value of dose rate from solar protons on the circular orbit 20000 km was 7.12 10^{-5} a.u. s^{-1} during the period from 5.03.2012 to 17.03.2012 (the maximum increase of solar proton flux during the whole period of dose measurement). If we compare the data of table 2 (right column) with the data of table 1, it may be concluded that during this flare the solar protons didn't give the considerable contribution to the dose rate increase.

4. Contribution to dose rate values of solar protons according to different models

We also made the analysis of contribution to dose rate values of model events which are used in the software «OMERE» and in software «OSOT» [7] (Nymmik model [8]). Figure 3 shows the dose rate from solar protons reached the value of ~ 10^{-2} to 10^{-1} a.u. s^{-1} which is comparable to and even more than trapped electron contribution to the dose rate value.

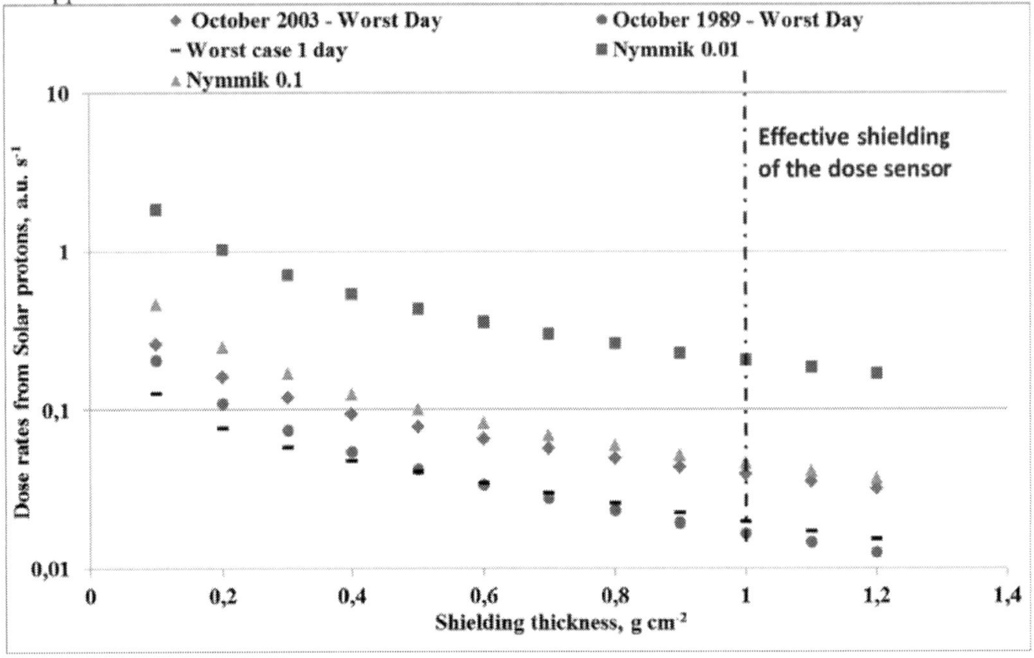

Figure 3. The dependence of dose rate value from solar protons vs. shielding thickness for different modeled events

5. Determination of critical solar proton flare class

In terms of obtained data it is possible to determine the classes of solar proton events which can deposit important contribution in dose rate increase on the 20000 km orbit (more than 10 % of the maximum dose rate value 10^{-2} a.u. s^{-1}; note that the average dose rate for the 20000 km orbit and shielding 1 g cm^{-2} is approximately equal to 10^{-3} a.u. s^{-1}). These classes are S4 and S5 [10], but with the condition that the proton flux at energies above 100 MeV exceeds the value of 200 proton (cm^2 s $sr)^{-1}$. Characteristics and a probability of these events are presented in the table 3.

Table 3. The scale of solar proton events and their occurrence frequency

Solar proton event	Integral flux >=10 MeV[a]	The annual event occurrence probability	
		Observed between 1994-2014 [9][b]	All events
S5 Extreme	100000	–	< 0.1 times during the year
S4 Severe	10000	0.2 times during the year	0.3 times during the year

[a] Average value of integral flux within 5 minutes, proton(cm^2 s $sr)^{-1}$
[b] Upon condition proton flux >100 MeV more than 200 proton(cm^2 s $sr)^{-1}$

The dose rate abrupt increase can cause the abnormal operation of on-board equipment because of the total ionizing dose effect in case an electronic component has accumulated dose which is close to its failure level. Also the stimulation of other effects (internal charging and etc.) is possible. It should be noted that there are orbits (low-altitude, polar) where the protons influence exceed the electrons one. There are also orbits which are more shielded by the Earth's geomagnetic filed.

6. Conclusion

The analysis of space-borne measurements of dose rate on the middle-Earth orbit and proton and electrons fluxes from the GEO for 2008 – 2013 showed that the main contribution to the dose rate increase is given by trapped electrons. However, two proton events were seen (with class S3) which gave the contribution more than 20%. The direct in-flight measurements and calculations showed that it is necessary to consider the proton flux not only with the energy >10 MeV (which is used for event type classification), but with greater energy (first of all, 100 MeV) upon condition of flux more than 200 proton $(cm^2\ s\ sr)^{-1}$. In this case, solar protons give the contribution more than 10% in comparison with the maximum measured dose rate and future probable events will be able to contribute more than 50% in comparison with the same value. Consequently, the assurance of reliable operation of on-board equipment demands the control of the S4-class events, when the >100 MeV proton flux exceeds the value of 200 proton $(cm^2\ s\ sr)^{-1}$. Further, it is necessary to estimate the solar proton contribution to the dose rate on the other orbits (low-altitude orbit, polar orbit) on which dominates the influence of protons, not electrons.

7. References

[1] J. Mazur "The radiation environment outside and inside a spacecraft". IEEE NSREC 2002 Short Course, pp. II-1 – II 69, 2002.
[2] Koons H.C. et al. The Impact of the Space Environment on Space Systems // AEROSPACE REPORT NO.TR-99 (1670)-1, 1999.
[3] V. S. Anashin, G. A. Protopopov, and Y. A. Milovanov, "Monitoring of space radiation in Russian federal space agency," *12th European Conference on Radiation and Its Effects on Components and Systems,* Sep. 2011.
[4] http://www.swpc.noaa.gov
[5] http://cosrad.sinp.msu.ru/manual.html
[6] http://www.trad.fr/OMERE-Software.html
[7] Anashin Vasily, Protopopov Gregory, and Milovanov Yury, "SEE testing for exposure of space ionizing radiation on radio-electronic equipment," *Proceedings of 5th International Conference on Recent Advances in Space Technologies - RAST2011*, Jun. 2011.
[8] R. A. Nymmik, "Probabilistic model for fluences and peak fluxes of solar energetic particles," *Radiation Measurements*, vol. **30**, no. 3, pp. 287–296, Jun. 1999.
[9] http://umbra.nascom.nasa.gov/SEP/
[10] http://www.swpc.noaa.gov/noaa-scales-explanation

A year of operation of Melibea e-Callisto Solar Radio Telescope

A Russu[1,2], R Gómez-Herrero[1,2], M Prieto[1,2], C Monstein[3], H Ivanov[2], J Rodríguez-Pacheco[1,2], J J Blanco[1,2]

[1] Space Research Group (SRG), University of Alcalá, Alcalá de Henares, Spain
[2] Parque Científico y Tecnológico de Guadalajara (PCyT), Guadalajara, Spain
[3] Institute for Astronomy, Swiss Federal Institute of Technology Zurich (ETHZ), Switzerland

E-mail: Andres.Russu@edu.uah.es

Abstract. The e-CALLISTO (Compound Astronomical Low-cost Low-frequency Instrument for Spectroscopy and Transportable Observatory) is a worldwide radio-spectrograph network with 24 hours a day solar radio burst monitoring. The e-CALLISTO network is led by the Swiss Federal Institute of Technology Zurich (ETHZ Zurich), which work up collaborations with local host institutions. In 2013 the University of Alcalá joined the e-CALLISTO network with the installation of two Solar Radio Telescopes (SRT): the EA4RKU-SRT that was located at the University of Alcalá from January 2013 till June 2013 and the Melibea-SRT that is located at Peralejos de las Truchas (Guadalajara) in operation from June 2013. The Spanish e-Callisto SRTs provide routine data to the network. We present examples of type III and type II radio-bursts observed by Melibea during its first year of operation and study their relation with soft X-ray flares observed by GOES and Coronal Mass Ejections (CMEs) and Solar Energetic Particle (SEP) events observed by space-borne instrumentation.

1. Introduction

Solar Radio Telescopes (SRT) can be developed with a low-cost budget in comparison with other solar instrumentation. Solar radio burst associated with solar cosmic rays are observed from Earth with radio equipment based on commercial components (spectrograph + antenna).

CALLISTO (Compound Astronomical Low-cost Low-frequency Instrument for Spectroscopy and Transportable Observatory) is rather a very good patrol device allowing to monitor solar radio emission around the day (24 hours). These spectrometers allow to recognise the main types of solar radio bursts, like type III bursts as signatures of accelerated electrons and type II bursts as signatures of shock waves produced by flares or/and CMEs. Nevertheless, most of the CALLISTO-based telescopes have limitations in frequency and/or time resolution, and/or in sensitivity to study some kinds of faint radio emissions. These limitations are mainly due to low effective area of antennas and noisy locations of the stations.

The Space Research Group (SRG) of the University of Alcalá joined the e-CALLISTO network thanks to the set up of the EA4RKU-SRT in January 2013. The EA4RKU-SRT was initially installed at the Technical School building of the University of Alcalá de Henares [1]. The EA4RKU-SRT was in operation for 6 months and served as test bench of the final e-CALLISTO SRT. After this period, the instrument performances as remote operated SRT were satisfactory, and then, it was moved to the Peralejos de las Truchas Astronomical Observatory (OAPT from the Spanish acronym).

2. e-CALLISTO Worldwide Network

e-CALLISTO is an European-led worldwide network which aims 24 hours monitoring of solar radio emission [2]. The e-CALLISTO network is led by the Swiss Federal Institute of Technology Zurich (ETHZ Zurich), which works up collaborations with local host institutions, which operates an e-CALLISTO SRT. Thanks to the Instrument Deployment Program managed by the International Space Weather Initiative (ISWI), e-CALLISTO is able to continuously observe the solar radio spectrum. Figure 1 shows the geographical distribution of the e-CALLISTO stations on December 2014, currently more than 69 instruments in more than 38 locations. However, there are two issues which should be mentioned. On one hand, the coverage on America, Africa and Asia, sometimes, depends on a unique station. On this regard, the network is placing efforts on increasing the number of stations in these three continents. Particularly, the SRG is working on two parallel projects which are focussed on the installation of e-CALLISTO stations: one on Argentina (Catamarca desert) and, the other one, on Chile (Atacama desert). On the other hand, some of the stations are located on an area with very high background radio emission from terrestrial emitters, which reduce the quality of the observed data.

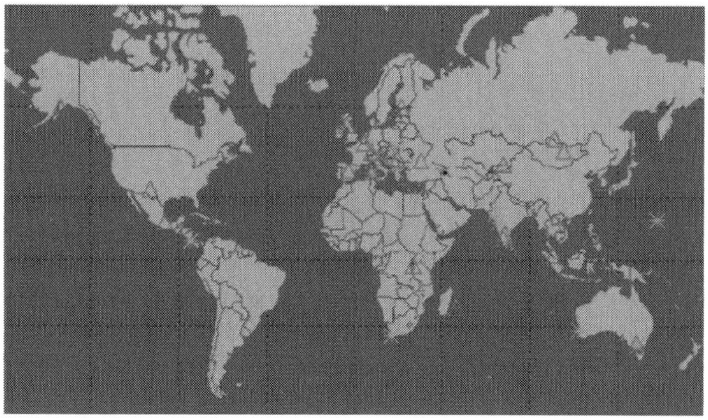

Figure 1. Map of current distribution of CALLISTO instruments in December 2014. Red triangles: stations providing data, orange star: stations do not providing data yet/anymore (http://www.e-callisto.org).

The CALLISTO spectrometer [3] was built in the framework of IHY2007 (International Heliophysic Year) and the ISWI by former Radio and Plasma Physics Group (Principal Investigator Christian Monstein) at Institute for Astronomy, ETH Zurich. This spectrometer is considered the standard for the e-CALLISTO network. The instrument operates between 45 and 870 MHz using a modern, commercially available broadband cable-TV tuner. The CALLISTO operational spectral range can be tuned within the full range. This tuning capability is useful when the antenna frequency response window does not cover the full CALLISTO range or when certain frequency windows are strongly affected by RFI. Table 1 details the main characteristics of the CALLISTO spectrometer. CALLISTO provides good quality data for scientific purposes with a remarkable low cost. As an example, data from the CALLISTO SRT at Birr Castle in Ireland [4] has been recently used in a possible explanation for a quasi-periodic acceleration of electrons by a plasmoid-driven shock in the solar atmosphere [5].

Table 1. CALLISTO spectrometer main characteristics.

Option	Description
`Frequency range`	45.0 - 870.0 MHz (34 cm - 6.7 m) (tunable)
`Frequency step size`	62.5 kHz
`Integration Time`	1 msec
`Dynamic range`	higher than 50 dB
`Noise figure`	lower than 10 dB
`Measuring rate`	800 pixels/sec maximum
`Sweep length`	1-400 Channels
`Power consumption`	12 V 225 mA (2.7 W)
`Cost`	Hardware around 300 Euro (only spectrometer)

The e-CALLISTO network presents areas which are covered by several SRTs (e.g. Europe and India). This network feature is known as geographic redundancy. The geographic redundancy replicates systems between two geographically distant stations so that the observed data are acquired from several stations. That assures the observation of a possible event in case of failure of one of the stations. This geographic redundancy is very reliable from the point of view of both, solar science and space weather. As an example, Europe has large geographic redundancy providing high reliability of the network (see Figure 1).

3. Spanish contribution to e-CALLISTO Network

In 2012, several Radio Frequency Interference (RFI) (background radio emission from terrestrial emitters) measurements were performed on several places of the province of Guadalajara (central Spain) in order to find the most suitable location of an e-CALLISTO SRT. The lowest RFI measurement was found at Peralejos de las Truchas (200 km East from Alcalá de Henares). During the last quarter of 2014, new RFI measurements were done on the area. The Figure 2 shows the 2014 RFI measurement once the instrumental background has been subtracted. The left plot of the figure shows the measurement at the University of Alcalá with strong interference by terrestrial radio emitters. The University of Alcalá is nearby Madrid city (33 km) and Barajas International Airport (10 km). Therefore, this location is not suitable for radio solar observations. The right plot shows the RFI at the OAPT. This location is more appropriate for solar observations due to the lower RFI, except for the frequency range for the commercial Frequency Modulation (FM). The FM frequency range can be rejected using a notch filter in order to avoid cross modulation (harmonics of FM frequency range). The background measurements were done with a system composed by a 25-1300 MHz Discone type antenna connected to a CALLISTO spectrometer.

At the beginning of 2013, the SRG developed a prototype of an e-CALLISTO based SRT which was in operation for 6 months at Alcalá de Henares. The prototype was designed as EA4RKU-SRT and their data are available on the e-CALLISTO repository. After the testing period, the prototype EA4RKU-SRT was moved to its final location at the OAPT. The name of this new station is Melibea-SRT.

The Melibea-SRT is operating at OAPT since May 2013. The location of the Melibea-SRT is N40°35' W01°15' world geodetic coordinates and 1240 metres above sea level near the village of Peralejos de las Truchas (Guadalajara).

The Melibea-SRT is based on a horizontally polarized log-periodic antenna mounted on an automatic Sun-tracking system (see figure 3 left). The antenna has a forward gain of 10-12 dBi.

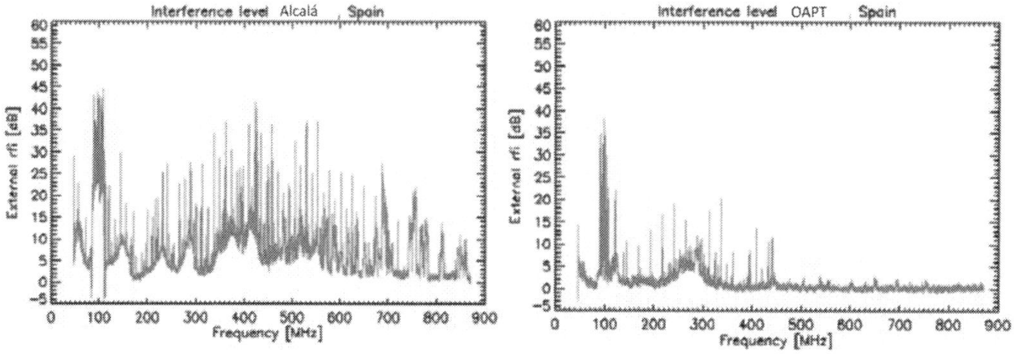

Figure 2. RFI measurements at central Spain with discone antenna and CALLISTO spectrometer. Left: RFI measurements in the University of Alcalá Campus (Alcalá de Henares). Right: RFI measurements in OAPT (Peralejos de las Truchas)

The antenna is pointed by a Sun tracking system based on two (azimuth and elevation) rotors which are driven by an AVRSCOM control system. The software used for the sun tracking is the SUM provided with the SatPC32 software developed for radio amateur operators. The antenna is connected to a CALLISTO spectrometer by 25 meters of coaxial cable (attenuation: 93 dB/km@200 MHz; 138 dB/km@400 MHz). The CALLISTO spectrometer operates in a frequency range between 45 MHz and 870 MHz with 200 channels on 1.25 ms time resolution per channel. The system is configured by the CALLISTO V116 software. All the software is running in a unique Windows 7 workstation. The Melibea SRT antenna is installed on the roof of the observatory building (see figure 3 right). In order to reduce RFI, the antenna will be moved to ground in a near future.

Figure 3. Melibea Solar Radio telescope at Peralejos de las Truchas Astronomy Observatory: left antenna and right its location on the Observatory building

As an example of the scientific measurements provided by CALLISTO, we show a type IV radio burst observed by Melibea-SRT. This burst was associated with an X-ray solar flare with M1.3 intensity at 08:14 UT on 5th June 2013. The X-ray solar flare was identified by the NOAA Space Weather Prediction Center (SWPC, http://www.swpc.noaa.gov/) from the 1762 Active Region (AR), located at heliographic coordinates S31W52. This flare was accompanied by a type III radio-burst and a partial halo CME. Accelerated electrons are bounced between side-to-side in large magnetic arches producing the type IV emission.

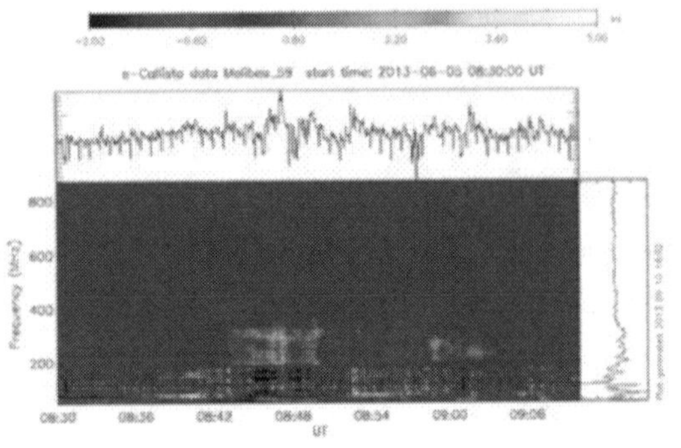

Figure 4. Type IV radio burst associated with the M1.3 X-ray flare from active region 1762. This event was observed the 5th June 2013 by Melibea-SRT on the OAPT, Guadalajara.

4. Applications in the field of Solar Energetic Particles

Observations of solar radio bursts provide relevant information for the understanding of particle acceleration during solar eruptive phenomena. In order to illustrate the relevance of the e-CALLISTO network for the understanding of Solar Energetic Particle (SEP) events, we will briefly discuss observations during 18th April 2014 SEP event.

The event on 18th April 2014 corresponds to an X-Ray solar flare with M7.3 intensity observed at 12:31 UT in NOAA AR 2036, located at S20W34 heliographic coordinates. The event was accompanied by several radio bursts (Type II, III, IV and V) which were clearly observed by several stations of the e-CALLISTO network, including Melibea-SRT. Type III and type II radio bursts were observed by many of the radio spectrometer telescopes of the e-CALLISTO network in Europe.

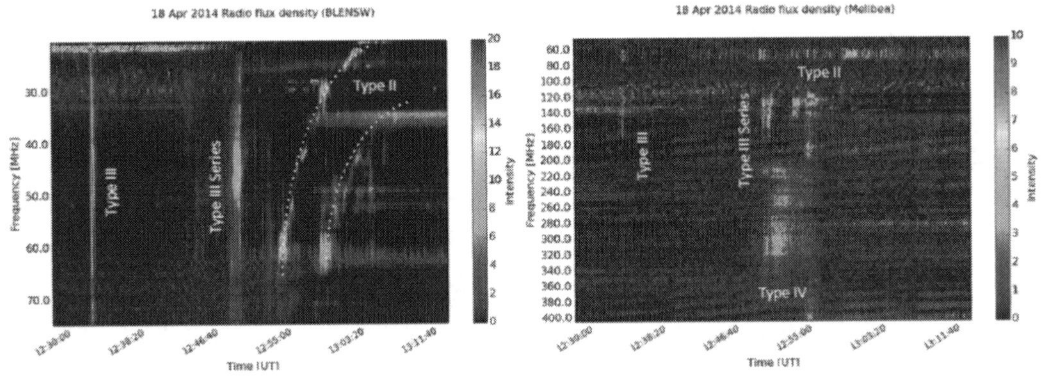

Figure 5. Sequence of Radio Bursts of 18th April 2014 event observed by e-CALLISTO stations.

The figure 5 shows the associated radio signatures of the 18th April 2014 solar event. An early Type III, seen at 12:30 UT, was not accompanied by a significant X-Ray flux increase nor a CME (clearly shown in left part of each plot). The M7.3 X-ray flare was accompanied by a series of Type III radio bursts, followed by a Type IV radio burst (clearly seen on the right figure).

The different configuration (antenna, frequency range and integration time) of each of the radio telescopes on the e-CALLISTO network provides large information about prompt phases of the electrons acceleration. Moreover, geographic redundancy is available for this event with several SRTs; see figure 6. The e-CALLISTO stations are synchronized and provide data between 12:30-13:15 UT over 20 MHz to 1430 MHz spectral range. The KRIM e-CALLISTO SRT only have data available until 13:00 UT.

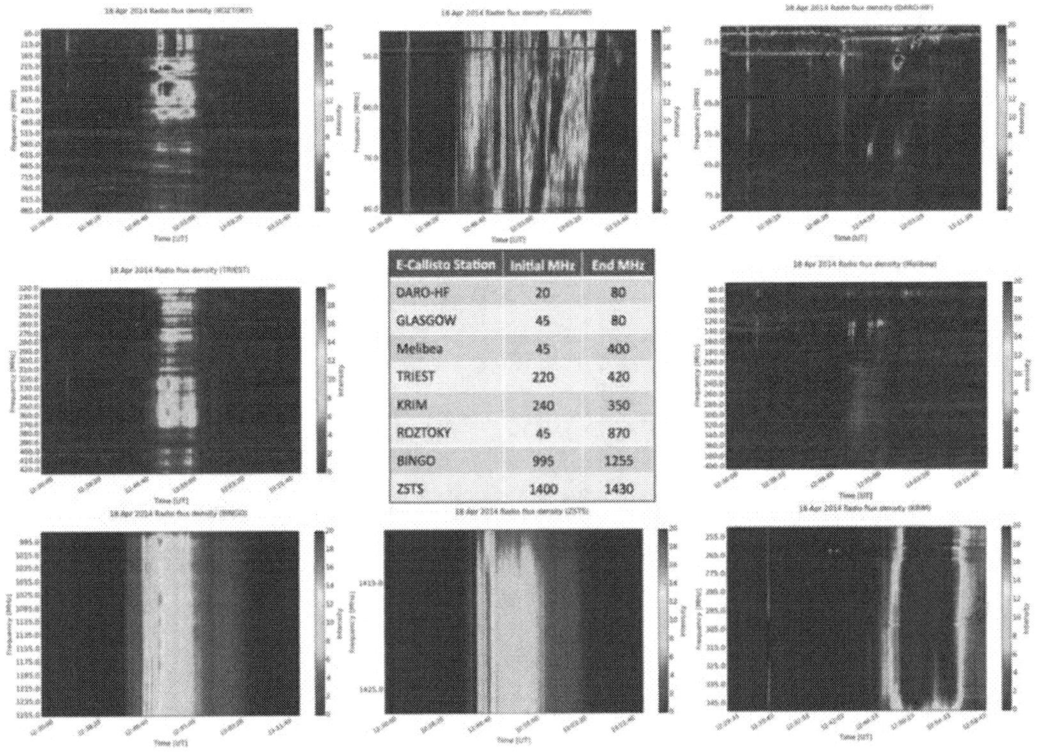

E-Callisto Station	Initial MHz	End MHz
DARO-HF	20	80
GLASGOW	45	80
Melibea	45	400
TRIEST	220	420
KRIM	240	350
ROZTOKY	45	870
BINGO	995	1255
ZSTS	1400	1430

Figure 6. Solar radio event observed by e-CALLISTO instruments over 20 MHz to 1430 MHz on 18th April 2014. The KRIM e-CALLISTO SRT only have data available until 13:00 UT

These e-CALLISTO radio spectra are produced by a web-based tool currently under development by the SRG/UAH for quick look plotting. A preliminary version is available at http://www.sorbete.srg.uah.es/sorbitools/ecallisto.

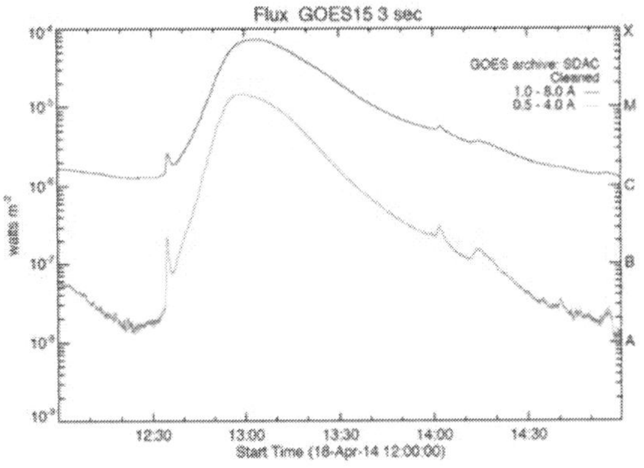

Figure 7. GOES15 3 seconds X-Ray flux of the observed event on 18th April 2014: pink line 0.5 - 4 A and black line 1.0 - 8.0 A.

The soft X-ray flux was measured by GOES satellites during this period, the observation is correlated with the type III radio burst measured by the e-CALLISTO stations. Figure 7 shows

Journal of Physics: Conference Series **632** (2015) 012078 doi:10.1088/1742-6596/632/1/012078

soft X-ray measurements during this period by the GOES 15 satellite in the 1-8 Angstroms (0.1-0.8 nm) and 0.5-4.0 Angstroms (0.05-0.4 nm) passband.

This event was also accompanied by a fast CME propagating southward, which was observed by the COR2 coronograph onboard STEREO-B Spacecraft [6] and by LASCO [7] onboard SOHO [8]. Figure 9 shows a sequence starting by LASCO-C2 observation at 13:48 UT and COR2 observations for the time interval 13:54-15:54 UT. According to the CACTUS catalog [9], the CME had a position angle of 194° (114° angular width) and a median velocity of 812 km/s. The estimated CME onset time at 13:25 UT.

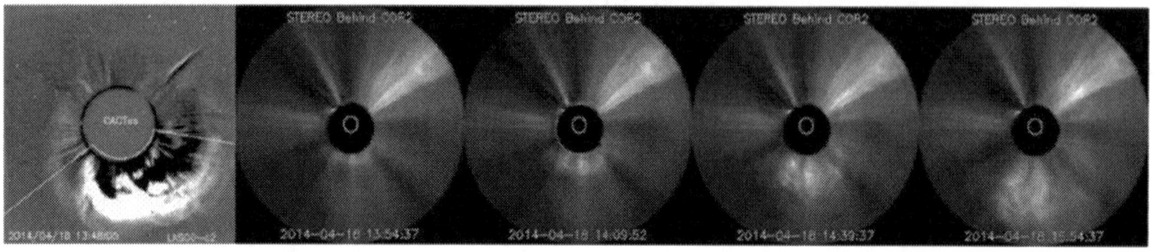

Figure 8. Left: LASCO-C2 observation of the observed event on 18th April 2014 . Right: STB/COR2 sequence of COR2 observations of the same event

Figure 9 left shows a summary of 10-minute averaged solar energetic particle observations by the solar Electron Proton Instrument (EPHIN) [10] onboard SOHO. Temporal profiles of 250-700 keV electrons, 4.3-7.8 MeV protons and 4.3-7.8 MeV/n Helium intensities are shown as red, green and blue lines, respectively. SOHO was relatively well connected to the flare and it observed a clear SEP event, reaching proton energies above 50 MeV (not shown in the figure). The electron flux started to increase at SOHO location at 13:00 UT +/- 2 min.

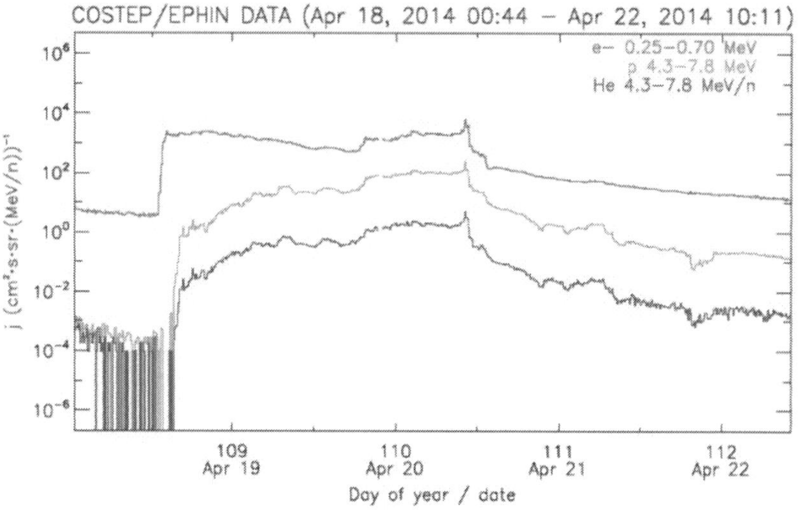

Figure 9. 18th April 2014 10-minute averaged solar energetic particle observations by the solar Electron Proton Instrument (EPHIN) onboard SOHO

5. Summary and future work

The Melibea-SRT is already in operation and integrated as a part of the e-CALLISTO network by the University of Alcalá. The e-CALLISTO stations have been proved to be sensitive to the most intense solar radio bursts with large geographic redundancy, with significant spectral and temporal resolution and with low cost. We have also shown how the system can provide valuable data for the study of SEP events. Such data are particularly useful in order to determine the physical phenomena responsible for the particle acceleration in the Sun (e.g. coronal shocks and solar flares). As future activity, the Celestina-SRT is already under development for covering the frequency range between 20-90 MHz for better data link with the data provided by the space-based instruments. Celestina-SRT will be installed in the OAPT during 2015 (near the location of Meliea-SRT).

Acknowledgements

This work has been supported by Solar Orbiter Energetic Particle Detector (EPD) project funded by MINECO AYA2012-39810-C02-01 and ESP2013-48346-C2-1-R. The Melibea-SRT is funded by a collaboration between PCyT of Guadalajara and the University of Alcalá. And developed with the collaboration of Peralejos de las Truchas Town Hall. Thanks to the University of Alcalá Radio Club (EA4RKU) for the continuous support with knowledge, materials and facilities. We thank the e-CALLISTO network for providing data and their collaboration. Thanks to STEREO and GOES for the data and plots (SolarMonitor.org). Thanks to NOAA and SWPC for all the information provided in the realization of this research.

References

[1] Russu A et al 2013 EA4RKU-SRT, the first e-CALLISTO Solar Radio Telescope operational in Spain *ICRC 2013 proceedings* Pending of publication.
[2] Benz A O et al 2009 A world-wide net of Solar Radio Spectrometers - e-CALLISTO *Earth Moon and Plantes* **104** 277-285
[3] Benz A O and Monstein C 2005 CALLISTO - A new concept for Solar Radio Spectrometers *Sol. Phys.* **226** 143
[4] Zucca P et al 2012 Observations of Low Frequency Solar Radio bursts from the Rosse Solar-Terrestrial Observatory *Sol. Phys.* **280 Issue 2** 591-602
[5] Carley E P et al 2013 Quasi-periodic acceleration of electrons by a plasmoid-driven shock in the solar atmosphere *Nature Phys.* **9** 811.
[6] Kaiser M L et al 2008 The STEREO Mission: An introduction *Space Sci. Rev.* **136** 1-4
[7] Brueckner G E et al 1995 The Large Angle Spectroscopyc Coronagraph (LASCO) *Sol. Phys.* **162** 357-402
[8] Domingo V, Fleck B and Poland A I 1995 The SOHO mission: an Overview, *Sol. Phys.* **162** 1-37
[9] Gopalswamy N et al 2009 The soho/lasco cme catalog *Earth, Moon, and Planets* **104 1-4** 295-313
[10] Müller-Mellin R et al 1995 COSTEP - Comprehensive Suprathermal And Energetic Particle Analyser *Sol. Phys.* **162** 483-504

Rigidity spectrum of the long-period variations of the galactic cosmic ray intensity in different epochs of solar activity

K Iskra[1] , M Siluszyk[1] and M Alania[1,2]

[1] Institute of Mathematics and Physics, Siedlce University, Poland
[2] Ivan Javakhishvili Tbilisi State University, Nodia Institute of Geophysics, Tbilisi, Georgia

E-mail: marek.siluszyk@uph.edu.pl

Abstract. We investigate character of correlations between the rigidity spectrum exponent γ of the 11-year variations of the Galactic Cosmic Ray (GCR) intensity and the exponent ν of the Power Spectral Density (PSD) of the Interplanetary Magnetic Field (IMF) turbulence in period 1968-2012 using data of the IMF and carefully selected neutron monitors. We show that a clear inversely correlation between γ and ν is observed when a resonance frequency range f_1–f_2 of the IMF turbulence (using for calculation of ν), are determined taking into account changes of the solar wind velocity V_{sw} and strength B of the IMF versus solar activity. We assume that an essential contribution to the scattering of GCR particles in the heliosphere is setting up generally by the turbulence of B_y and B_z components of the IMF. Therefore, in this paper we use the exponents ν_y and ν_z of PSD of the B_y and B_z components of the IMF turbulence. Generally, both γ and ν can be accepted as the essential proxies to study GCR propagation near earth orbit in heliosphere for the rigidity range of GCR particles to which neutron monitors and ground muon telescopes respond

1. Introduction and Motivation

In [1], [2] and [3] have demonstrated that about 75–80% changes of amplitudes of the 11-year variations of the Galactic Cosmic Ray (GCR) intensity measured by neutron monitors can be explained in the first approximation by the anisotropic diffusion–convection model of GCR transport. Consequently, if that is the case, changes of the solar wind velocity and diffusion coefficient should play a vital roles in formation of the 11–year variation of the GCR intensity near earth orbit in heliosphere. On the other, data of the solar wind velocity does not show any regular valuable changes versus the 11–year cycle of solar activity. So, contributions of the convection and adiabatic cooling of the GCR particles (directly related with the solar wind velocity) in long period variation of the GCR intensity can be basically ignored. Thus, one can declare that a character of diffusion of the GCR particles versus solar activity should play an important role in the formation of the 11–year variation of the GCR intensity. However, if that is so, there exists a natural question, what of the parameters characterized solar wind and solar activity contributed in complex changes of diffusion coefficient can be generally responsible for the alternation of diffusion coefficient versus solar activity. In order to answer to this essential question we have to take into account one of the vital arguments, a peculiarity of the temporal changes of the power law rigidity R spectrum exponent γ ($\delta D(R)/D(R) \propto R^{-\gamma}$, $D(R)$ is GCR primary spectrum) of the amplitudes of the 11–year variations of the GCR intensity measured by neutron monitors.

We have found in [2], [3], [4], [5] and [6] that the exponent γ is larger (γ=1.2±0.1) in the maxima epochs (rigidity spectrum is soft) than in minima epochs (γ=0.6 ±0.1) of solar activity (rigidity spectrum is hard).

We also have established that an exponent ν of the Power Spectral Density (PSD) of the Interplanetary Magnetic Field (IMF) turbulence (In resonance frequency range f_1–f_2, to which neutron monitors respond) varies versus solar activity. Also, temporal changes of exponents γ and ν versus solar activity are in clear inversely correlation. Thus, we have two very essential physically reasonable parameters, (1) the rigidity spectrum exponent γ of GCR variations, and (2) the exponent ν of the PSD of the IMF turbulence. Thus, one can declare that both γ and ν can be accepted as the very essential proxies to study GCR propagation near earth orbit in heliosphere. Other than, despite we find an apparent correlation between changes γ and ν based on the experimental data of neutron monitors and IMF, there is a question of interest what is the theoretically expected relationship between changes of γ and ν based on the modelling of GCR transport in heliosphere.

We have numerically solved in [3], [5] and [6] transport equation next in turn for physically acceptable changes of parameters included in parallel diffusion coefficient in [7]. We found that the similar relationships between γ and ν obtained by experimental data of neutron monitors and IMF turbulence is gained from theoretical modelling when parallel diffusion coefficient K_{II} depends on the parameter R^{α} ($K_{II} \propto R^{\alpha}$, where $\alpha=2-\nu$) [11]. So, one can conclude that a changeable IMF turbulence versus solar activity plays a decisive role in creation of the 11–year variation of GCR intensity.

An essential contribution to the scattering of GCR particles in the heliosphere is generally setting up by the turbulence of B_y and B_z components of the IMF in [7], [8] and [3], although quantitative contributions each of them in creation of the electromagnetic conditions at the near earth orbit in heliosphere are different. In spite, between the temporal changes of the exponents ν_y and ν_z there is a precious correlation, the average power P of the PSD of the B_y component is about twice larger than the average power of the B_z component in [3]. In paper [9] we have studied the time interval 1968–2002 dividing it into three successive periods: 1968–1976 (I), 1977–1989(II), and 1990–2002(III). We used data of neutron monitors, B_y and B_z components of the IMF, and sunspot numbers. We have compared changes of an annual values of exponent γ, and ν_y and ν_z *for* each three periods. Results are presented in figure 1 of paper [9]. We have found a clear inversely correlation between γ and ν_y and between γ and ν_z in period 1977–1989; however for periods 1968–1976 and 1990–2002 relations between γ and ν_y and between γ and ν_z are not so substantial.

It should be noted that the annual average values of ν_y and ν_z of the PSD of the B_y and B_z components of the IMF were calculated in [9] for the same constant frequency interval $\Delta f=f_2-f_1=3^x 10^{-6} Hz$ (from here called as the first method). Calculations of the constant frequencies $f_1=1x10^{-6}Hz$ and $f_2=4^x10^{-6}Hz$ were carried out using the constant solar wind velocity $V_{sw}=432.4$ *km/s* and the constant strength $B=4.6nT$ of the IMF for the whole period 1968–2002. So, the frequencies f_1 and f_2 which determine the frequency intervals $\Delta f=f_2-f_1\approx(3\text{-}4)x10^{-6}Hz$ of the IMF turbulence (responsible for the scattering of GCR particles to which neutron monitors respond) are considered as the constants for the whole period 1968–2002 including a few 11–year cycles of solar activity. On the other, the solar wind velocity V_{sw} and strength B of the IMF vary from year to year for the whole period 1968–2002, and correspondingly the resonant frequencies f_1 and f_2 using for calculation of the interval Δf are changed. Therefore, it is of interest how changes correlations between (γ and ν_y) and between (γ and ν_z) are calculated for the frequency intervals $\Delta f=f_2-f_1\approx(3\text{-}4)x10^{-6}Hz$ obtained based on the alternated magnitudes of the solar wind velocity V_{sw} and strength B of the IMF from year to year (from here called as the second method). Here we consider last solar cycle #23 with unusual prolonged minimum. We calculated when ν_y and ν_z using the second method. A purpose of this paper is to carry on a study of relations between (γ and ν_y) and between (γ and ν_z) in the period 2002–2012 consisting of the ending epoch of the unusual minimum period of the 11–year solar cycle #23 and the beginning of solar cycle #24. The period 2002–2012 will be further called as the fourth (IV) period of the whole time interval 1968–2012.

Thus, our analyse will be dealt to investigate relations between (γ and ν_y) and between (γ and ν_z) in the time interval 1968–2012 consisting of four successive periods: 1968–1976(I), 1977–1989(II), 1990–2002(III), and 2003–2012(IV).

2. Experimental Data and discussion

We present in figures 1ab changes of the solar wind velocity V_{sw} (figure 1a) and the strength B of the IMF (figure 1b) for period 1968–2012. Figures 1ab show that the strength B has a noticeable quasi 11–year periodicity, while this periodicity is absent in the changes of the solar wind velocity V_{sw}.

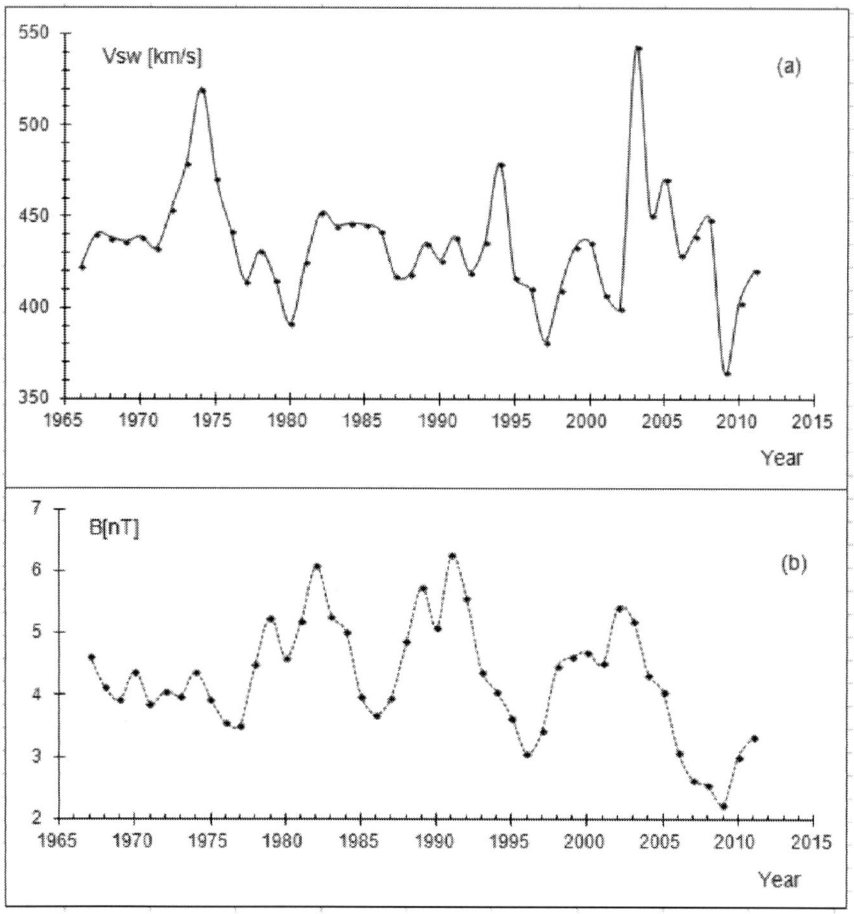

Figure 1ab. Temporal changes of the solar wind velocity V_{sw} (a) and strength B of the IMF (b) in 1968 – 2012.

To calculate a resonant frequencies f_{res} and corresponding frequency intervals Δf for changeable V_{sw} and B upon solar activity we employ the formula $f_{res} = \dfrac{300}{2\pi}\dfrac{V_{SW} \cdot B}{R}$ from [7]. We used for calculations yearly averages of the solar wind velocity V_{sw}, the strength B of the IMF and rigidity of GCR particles in GV. Temporal changes of the f_{res} for rigidity 10 GV of GCR particles are presented in figure 2.

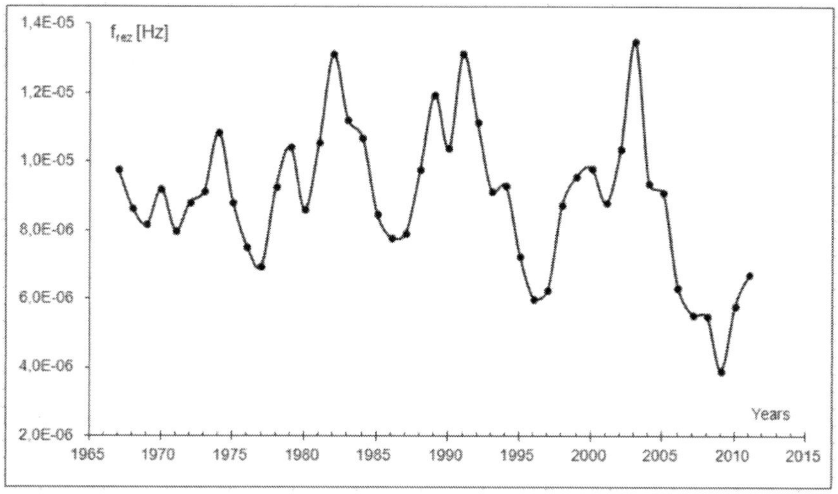

Figure 2 Annual changes of the resonant frequency f_{res} for rigidities 10 GV of GCR particles in different period of solar activity from 1968 to 2012.

Comparing figures 1ab and figure 2 one can recognize that changes of B from year to year which has clearly expressed quasi 11–year cycling, is responsible for time profile of f_{res}, while the role of solar wind velocity is not significant. Correlation coefficient r between f_{res} and B is equal to 0.95 ± 0.01, but between f_{res} and U_{sw} that is equal to 0.31 ± 0.03.Resonant frequencies and corresponding intervals $\Delta f = f_1-f_2 = \sim(3-4)^x 10^{-6} Hz$ for changeable V_{sw} and B in period 1968–2012 are presented in table 1.

Table 1. Frequency ranges Δf determined for period 1968 to 2012.

No.	Periods	Frequency range [Hz]	Δf [μHz]
I	1968	7.9*10^-7---3.8*10^-6	3,0
	1969-1970	2.1*10^-6---5.8*10^-6	3,7
	1971-1975	7.9*10^-7---3.8*10^-6	3,0
	1976-1977	7.6*10^-7---3.8*10^-6	3,0
II	1978-1980	1.7*10^-6---5.8*10^-6	4,1
	1981-1982	1.0*10^-6---4.0*10^-6	3,0
	1983	1.7*10^-6---5.8*10^-6	4,1
	1984-1987	7.6*10^-7---3.8*10^-6	3,0
	1988	1.0*10^-6---4.0*10^-6	3,0
III	1989-1992	1.7*10^-6---5.8*10^-6	4,1
	1993-1998	7.0*10^-7---3.7*10^-6	3,0
	1999-2000	1.7*10^-6---5.2*10^-6	3,5
	2001	1.0*10^-6---4.0*10^-6	3,0
	2002	1.7*10^-6---5.2*10^-6	3,5
IV	2003-2005	7.0*10^-7---3.7*10^-6	3,0
	2006	4.1*10^-7---3.0*10^-6	2,6
	2007	4.4*10^-7---3.0*10^-6	2,5
	2008	4.1*10^-7---2.8*10^-6	2,6
	2009	4.1*10^-7---3.5*10^-6	3,1
	2010-2012	1.0*10^-6---4.0*10^-6	3,0

We assume that an essential contribution to the scattering of GCR particles in the heliosphere is generally setting up by the turbulence of B_y and B_z components of the IMF, so for investigation we use in this paper exponents ν_y and ν_z of PSD of the B_y and B_z components of the IMF. Disposing resonant frequencies f_{res} and frequency intervals Δf (see table 1) of the turbulence of the IMF (which are responsible for scattering of GCR particles to which NM respond) we have calculated the corresponding annual values of the exponents ν_y and ν_z. Changes of the rigidity spectrum exponent γ (solid blue line), and exponents ν_y of the PSD of the IMF turbulence calculated by the first method (dot black line) and by the second method (dashed red line) in periods I, II, III and IV are presented in figure 3; the same but for ν_z are presented in figure 4.

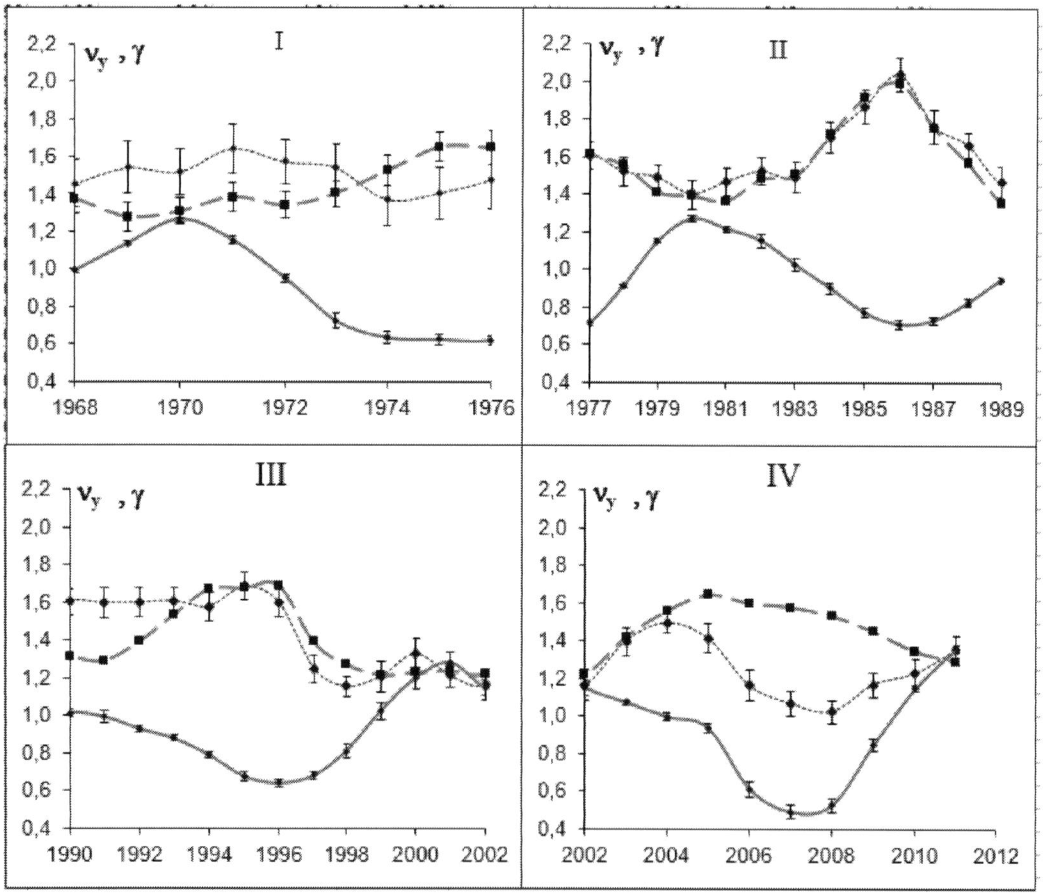

Figure 3. Temporal changes of the rigidity spectrum exponent γ (solid blue line) and exponents ν_y calculated by the first method (dot black line) and by the second method (dashed red line) in periods I, II, III and IV.

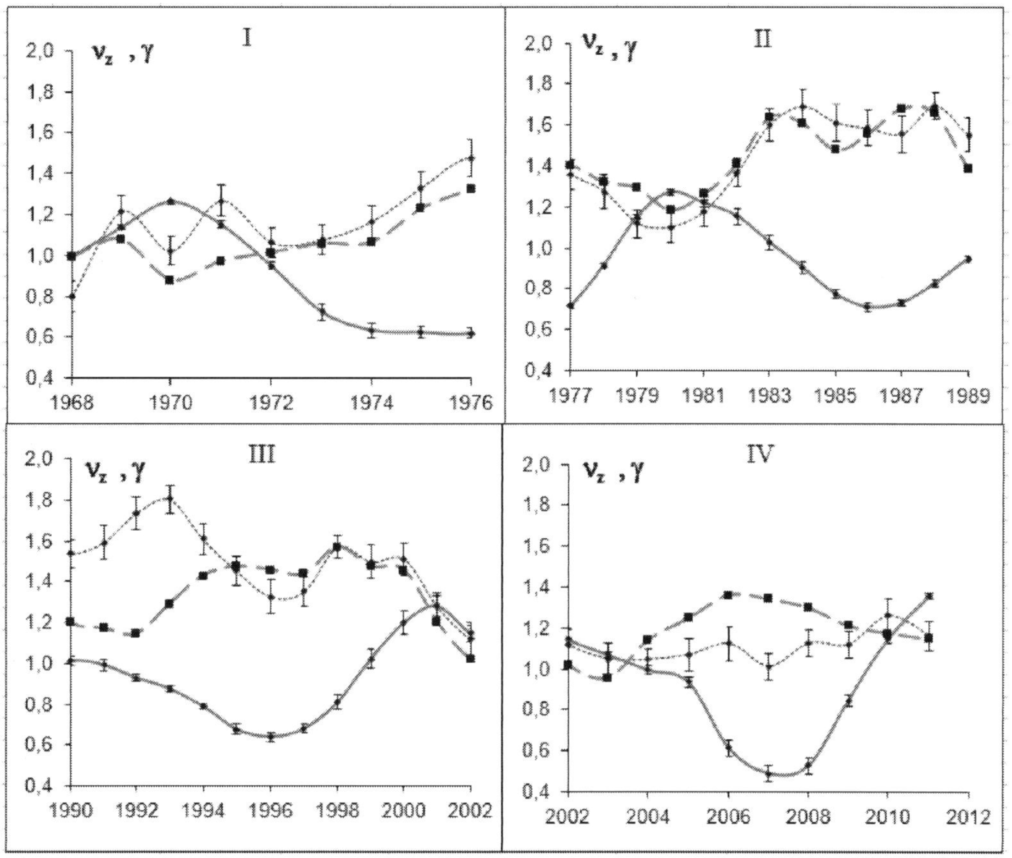

Figure 4. Temporal changes of the rigidity spectrum exponent γ (solid blue line), and exponents ν_z calculated by the first method (dot black line) and by the second method (dashed red line) in periods I, II, III and IV.

To more detail study of relations between (γ and ν_y) and between (γ and ν_z), we calculate correlation coefficients r_1 $(\gamma ; \nu_y)$ and $r_1(\gamma ; \nu_z)$ when the values ν_y and ν_z were obtain by the first method and $r_2(\gamma ; \nu_y)$ and r_2 $(\gamma ; \nu_z)$ calculated by the second method for periods: I, II, III and IV for whole analysed period 1968–2012. Results are presented in table 2.

Table 2. Correlation coefficients r_1 and r_2 between (γ and ν_y) and between (γ and ν_z)

	Period	r_1 (γ ; ν_y)	r_2 (γ ; ν_y)	r_1 (γ ; ν_z)	r_2 (γ ; ν_z)
I	1968-1976	0.64 ± 0.17	-0.86 ± 0.11	-0.43 ± 0.20	-0.78 ± 0.14
II	1977-1989	-0.78 ± 0.11	-0.78 ± 0.11	-0.69 ± 0.13	-0.66 ± 0.13
III	1990-2002	-0.43 ± 0.16	-0.78 ± 0.12	-0.19 ± 0.18	-0.54 ± 0.15
IV	2003-2012	0.72 ± 0.16	-0.73 ± 0.13	0.41 ± 0.17	-0.74 ± 0.14

A high inverse correlation is observed in 1977–1989 (the II period, solar cycle 21) between temporal changes of (γ and ν_y) and between (γ and ν_z). A level of high correlation does not depend whether the ν_y and ν_z are calculated by the first method or by the second method. At the same time a clear inverse correlation takes place between temporal changes of (γ and ν_y) and (γ and ν_z) for shifted frequency range versus solar activity corresponding to the alternation of resonant frequencies in periods I, III and IV (solar cycles: # 20, # 22 and #23/24). Likely, 11–year solar cycles from point of view of the IMF turbulence structure are really individual. An existence of an inverse correlation between γ and ν during solar cycles #20, #22 and #23/24 for shifted resonant frequencies versus level of solar activity should be considered as a normal phenomenon matching to the theory of scattering of GCR particles in heliosphere. However, the solar cycle #21, when an inverse correlations between (γ and ν_y) and between (γ and ν_z), are high and do not depend on the IMF strength alternation versus solar activity should be considered as an exclusion one.

We demonstrate that changes of ν_y and ν_z versus solar activity play a central role in the formation of the features of the long period variations of the GCR intensity, especially its rigidity dependence. Thus, one can conclude that inverse relations between (γ and ν_y) and between (γ and ν_z) are regular. We consider that the found relations are expected from the QLT for GCR particles to which neutron monitors respond [7] and [10].

Summary

1. We believe that temporal changes of the exponents ν_y and ν_z of the PSD of the IMF obtained by in situ measurement of B_y and B_z components in local space satisfactorily reflects the average structure of the IMF turbulence in the vicinity of the heliosphere where modulation of the long period variations of the GCR intensity takes place; both ν_y and ν_z have clearly expressed 11–yr cycling and can be considered as the central parameters shedding light on the modulation of GCR in the range of rigidity of cosmic ray particles to which neutron monitors respond.
2. We find that for the GCR particles of rigidity 10 GV the resonant frequencies regularly are lower in minimum than in maximum epochs of solar activity for 1968–2012; in these changes a decisive role belongs to quasi 11–yr cycle of the strength B of the IMF.
3. The significant high inverse correlation is observed between temporal changes of (γ and ν_y) and between (γ and ν_z) in 1977–1989 (solar cycle # 21). A range of correlation does not depend whether the ν_y and ν_z are calculated for constant B and U_{sw} corresponding to average resonant frequencies for individual ascending and descending periods or for alternated of resonant frequencies of the IMF turbulence versus solar activity. So, ν_y and ν_z are unchanging when a range of frequency $\Delta f = \tilde{}(3\text{–}4)^x 10^{-6}\ Hz$ is sliding versus time during #21 solar cycle. Possible, a turbulence of the IMF during the solar cycle #21 can be characterized as the completely homogeneous and highly isotropic.
4. At the same time for the solar cycles #20, #22, and #23/24 a clear inverse correlation between (γ and ν_y) and between (γ and ν_z) take place for shifted frequency range versus solar activity corresponding to the alternation of resonant frequencies of turbulence.
5. We assume that an existence of an inverse correlation between γ and ν during the solar cycles #20, #22, and #23/24 for shifted resonant frequencies versus level of solar activity should be considered as a normal case matching to the theory of scattering of GCR particles in interplanetary space; however, the solar cycle #21, when an inverse correlation between γ and ν is high and does not depend on the IMF strength alternation versus solar activity should be considered as an exclusion event.

Acknowledgements

We are very grateful to the providers of the solar, interplanetary, neutron monitor and geomagnetic data used in this study. Authors thank investigators of the websites:
http://spidr.ngdc.noaa.gov/spidr/,
http://omniweb.gsfc.nasa.gov/index.html,
http://cr0.izmiran.rssi.ru/moscow/main.htm
http://www.nmdb.eu

References

[1] Dorman L I 2001 *Adv. Space Res.* **27** 601
[2] Alania M V 2002 *Acta Phys. Polonica B* 33 **4** 1149
[3] Alania M V, Iskra K and Siluszyk M 2010 *Adv. Space Res.* **45** 1203
[4] Alania M V, Iskra, K 1995, *Adv. Space Res.* 16 **9** 241
[5] Alania M V, Iskra K and Siluszyk M 2008a *Adv. Space Res.* **41** 26
[6] Alania M V, Iskra K and Siluszyk M 2008b *Acta Phys. Polonica B* 39 **11** 2961
[7] Jokipii J R 1971 *Rev. of Geoph. and Space Physics.* **9** 27
[8] Bieber J W, Mathaeus W H and Smith C W 1994 *Astrophys. J* **420** 294
[9] Siluszyk M, Iskra K and Alania, M V 2014 *Solar Physics* 289 **11** 4297
[10] Shalchi A 2009 *Nonlinear Cosmic Ray Diffusion Theories*, Springer, Berlin 180
[11] Siluszyk M, Wawrzynczak A and Alania M V 2011 *J. of atm. and S-T Physics* 73 **13** 1923

2–D Modelling of Long Period Variations of Galactic Cosmic Ray Intensity

M Siluszyk[1] · K Iskra[1] and M Alania[1,2]

[1] Institute of Mathematics and Physics, Siedlce University, Poland
[2] Ivan Javakhishvili Tbilisi State University, Nodia Institute of Geophysics, Tbilisi, Georgia

E–mail: marek.siluszyk@uph.edu.pl

Abstract.
A new two–dimensional (2–D) time dependent model describing long-period variations of the Galactic Cosmic Ray (GCR) intensity has been developed. New approximations for the changes of the magnitude B of the Interplanetary Magnetic Field (IMF), the tilt angle δ of the Heliospheric Neutral Sheet (HNS) and drift effects of the GCR particles have been included into the model. Moreover, temporal changes of the exponent γ expressing the power law – rigidity dependence of the amplitudes of the 11–year variation of the GCR intensity have been added. We show that changes of the expected GCR particle density precedes changes of the GCR intensity measured by the Moscow Neutron (MN) monitor by about 18 months. So ~18 months can be taken as an effective delay time between the expected intensity caused by the combined influence of the changes of the parameters implemented in the time-dependent 2–D model and the GCR intensity measured by neutron monitors during the 21 cycle of solar activity.

1. Introduction and Motivation

Generally, to model a propagation of the GCR in the heliosphere is rather complicated problem. Difficulties are related to an accurate implementation of the temporal changes of parameters (among them obtained from direct measurements) into the transport equation that determines fundamental processes in the heliosphere and causes a modulation of GCR. An additional difficulty is related with the selection of the length of the modelling time interval, due to the existence of a varying delay time τ between the changes of the GCR intensity, on the one hand, and various parameters characterizing electro-magnetic conditions in the heliosphere, on the other. We present in figure 1 the temporal changes of the monthly smoothed sunspot number SSN (top panel), the GCR intensity *I(CR)* observed by the Moscow neutron monitor (middle panel), and the exponent γ of the rigidity spectrum of the GCR intensity variations [1] for the period of 1968–2012 (bottom panel). In table 1 we present the maximal invers correlation coefficients corresponding to the delay times τ for the pair of *I(CR)* and SSN and for the pair of *I(CR)* and the exponent γ. The *SSN* is a relative index describing the level of solar activity, but it is not in a straight line usable as a quantitative parameter to implement it in the transport equation of the GCR propagation. However, bearing in mind that the magnitude B of the IMF is proportional to the SSN and an inverse relationship between I(CR) and B ("CR-B") is valid everywhere inside the termination shock [2], one can make use of correlation between the parameter SSN and I(CR) as an indication of the GCR modulation.
For the positive (A>0) polarity period 1968-1976, delay time τ is 2 months, and for the period 1990-2002 that is zero. Hence, there is a clear polarity dependence of delay times τ for various 11–year cycles of solar activity. Delay time τ between SSN and I(CR) for periods 1977–1987 (A>0 (4 years) and A<0 (7 years)) is 10 months and 14 months for the negative (A<0) polarity periods 2003-2012. However, we think that this problem needs a more careful study. Delay times τ between I(CR) and exponent γ for all 11–year cycles of solar activity is near zero; that was expected, as far γ expresses the rigidity dependence of the amplitudes of the long-period variations found directly from neutron monitors data for given period of consideration. Thus, it must be noted that a caution about an existence of difficulties related to the selection of a length of the modelling time interval because of dynamical changes of delay time τ, e.g., between SSN and I(CR) is not

groundless; to model the GCR propagation for any time interval needs a careful preliminary study of the dynamical changes of delay time τ for the chosen time interval, and what is an extremely important, in model should be implemented different parameters with corresponding different delay time τ.

Table 1. Correlation coefficients r and delay time τ (in months) between the parameter SSN and I(CR) and between the changes of I(CR) and parameter γ for period 1968 to 2012.

No.	Periods	r(SSN ; I(CR))	$\tau(I(CR), SSN)$	r(γ ; I(CR))	$\tau(I(CR), \gamma)$
I	1968-1976	-0.87 ± 0.01	2 ± 1	-0.94 ± 0.01	1 ± 1
II	1976-1987	-0.87 ± 0.01	10 ± 1	-0.71 ± 0.01	0 ± 1
III	1990-2002	-0.93 ± 0.01	0 ± 1	-0.76 ± 0.01	0 ± 1
IV	2003-2012	-0.88 ± 0.01	14 ± 1	-0.58 ± 0.02	1 ± 1

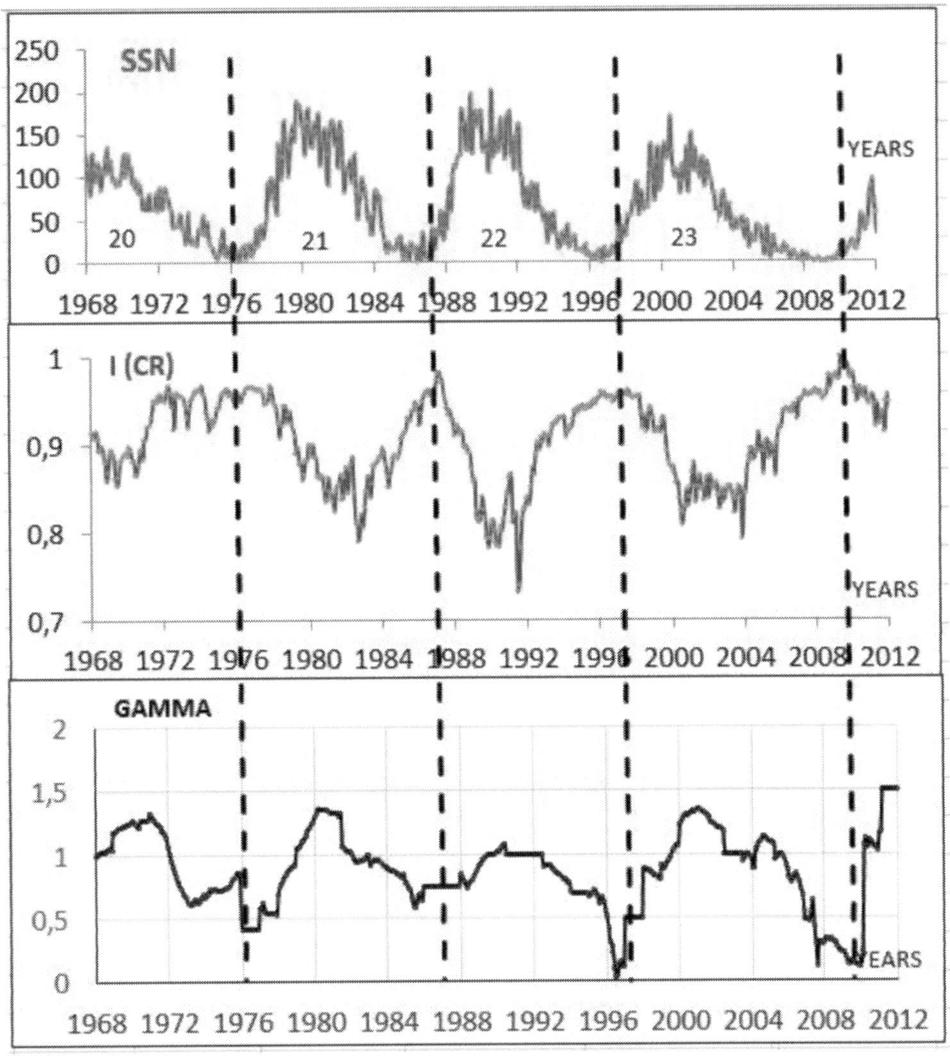

Figure 1. Temporal changes of the monthly smoothed *SSN* – sunspot number (top panel), I(CR) intensity observed by the Moscow NM (middle panel) and rigidity exponent $\gamma(t)$ of the GCR intensity variations (bottom panel) for the period of 1968–2012.

Our aim in this paper is: (1) to construct time dependent 2–D model of the 11–year variations of the GCR intensity and algorithm solved by difference scheme method in C# programming language, (2) to consider 11–year variation of the GCR intensity during the period 1976–1987 by the monthly smoothed data of the Moscow NM, in such way there are excluded fluctuations of the GCR intensity shorter than one month, (3) to implement in the Parker's transport equation the parameters characterizing temporal changes (for the case of constant solar wind velocity *U=400 km/s*) of the magnitude B of the IMF, tilt angle δ of the HNS and changes of drift effect of the GCR particles. We assume that drift effects have a maximum value normalized to 100% in the minimum epoch of solar activity (drift dominated epoch), and 20% in the maximum epoch (almost diffusion dominated period), and (4) we also implement in the model changes of the exponent γ (for the first time) of the rigidity R spectrum of the long-period variations of the GCR intensity. The importance of the latter is demonstrated in [3], [4] and [5], where it is shown that in the case of almost constant solar wind velocity, a central role in the formation of the 11–year variation of the GCR intensity can be ascribed to the changes of the diffusion coefficient versus the solar activity. So, changes of the character of the rigidity dependent diffusion remain an essential source of the 11–year variation of the GCR intensity, playing a vital role in the formation of the rigidity dependence of the amplitudes of the GCR intensity variations.

2. Model of the 11–year variation of GCR: 1976 – 1987.

For a modelling we take the period of 1976–1987 (11–year cycle #21). Cycle #21 was chosen because of, (1) almost a symmetric changes of the monthly smooting of the GCR intensity were observed for ascending (A>0) and descending (A<0) epochs of the 11–year cycle of solar activity, and (2) during the cycle #21 the largest number (~7 events) of clearly expressed step like changes of the GCR intensity were observed comparing to any other 11–year cycle during last 50 years. These step like changes of the GCR intensity can be related to the Forbush decreases (Fds) [5]. However the roles of Merged Interaction Regions (MIRs) and Global Merged Interaction Regions (GMIRs) [6] as barriers for the propagation of the GCR particles in the heliosphere cannot be excluded [6] and [7]. Moreover in cycle #21 a pure inverse correlation between the temporal changes of the rigidity R spectrum exponent γ and of the exponent ν_y of the power law spectral density of the B_y component of the IMF turbulence is observed. Our calculations [3] show that the turbulence of the IMF have a Gaussian distribution and the GCR particles propagation can be considered as normal diffusion for whole 11-year cycle #21; of course we accept that delay times τ between changes of the GCR intensity, on the one hand, and various parameters characterizing electro-magnetic conditions in the heliosphere remain constant during the solar cycle #21, on the other.

To model the 11–year variations of GCRs we use Parker's non–stationary transport equation [8] and [9]

$$\frac{\partial N}{\partial \tau} = -\left(U + \langle v_d \rangle\right) \cdot \nabla N + \nabla \cdot \left(K_{ij}^S \cdot \nabla N\right) + \frac{1}{3}(\nabla \cdot U)\frac{\partial N}{\partial \ln R} \tag{1}$$

where N and R are the omnidirectional distribution function and rigidity of the GCR particles, respectively; τ –the time, U –the solar wind velocity and v_d is the drift velocity. We set up the dimensionless density $f = \frac{N}{N_0}$, the time $t = \frac{\tau - \tau_0}{\tau_S - \tau_0}$ and the distance $r = \frac{\rho}{\rho_0}$; N_0 is the density in the Local Interstellar Medium (LISM) accepted as $N_0 = 4\pi I_0$, where the intensity I_0 in the LISM has the form: $I_0 = 21.1T^{-2.8}/\left(1 + 5.85T^{-1.22} + 1.18T^{-2.54}\right)$ in [10], [11]; T is kinetic energy in GeV ($T = \sqrt{R^2 \cdot e^2 + 0.938[GeV]^2} - 0.938[GeV]$), e is an elementary electric charges, ρ and ρ_0 are the radial distance and size of the modulation region; $\tau_0 -$ is the characteristic time corresponding to the changes in the heliosphere for the certain class of the GCR variation; we considered in model $\tau_0 = 1976$ and $\tau_S = 1987$.

The size of the modulation region ρ_0 equals 100AU, and the solar wind velocity $U = 400 km \cdot s^{-1}$ is used throughout the heliosphere. The equation (1) for the dimensionless variables *f*, *R* and *t* in the 2–D spherical coordinate system (r, θ) can be written, as:

$$A_0 \frac{\partial f}{\partial t} = A_1 \frac{\partial^2 f}{\partial r^2} + A_2 \frac{\partial^2 f}{\partial \theta^2} + A_3 \frac{\partial^2 f}{\partial r \partial \theta} + A_4 \frac{\partial f}{\partial r} + A_5 \frac{\partial f}{\partial \theta} + A_6 f + A_7 \frac{\partial f}{\partial R} \quad (2)$$

where $A_0 = \frac{\rho_0^2}{\tau_0 K_{II}}$ and the coefficients A_1, A_2, ..., A_7 are functions of the spherical coordinates (r, θ), the rigidity *R* of the GCR particles and the time *t*. The anisotropic diffusion tensor of GCR $K_{ij} = K_{ij}^{(S)} + K_{ij}^{(A)}$ consists of the symmetric $K_{ij}^{(S)}$ and $K_{ij}^{(A)}$ – antisymmetric parts. We implement a drift velocity of the GCR particles in the model as, $< v_{D,i} >= \frac{\partial K_{ij}^{(A)}}{\partial x_j}$ [12]. This expression is equivalent to the standard formula for $< v_D >$ [13]. The heliospheric magnetic field vector $\vec{B}$ is given as in [14] and [15]:

$$\vec{B} = (1 - 2H(\theta - \theta'))\left(B_r \vec{e}_r\right) \quad (3)$$

where *H* is the Heaviside step function changing the sign of the global magnetic field in each hemisphere θ' corresponds to the heliolatitudinal position of the HNS and e_r is the unite vector directed along the component B_r of the 2D Parker's field [16]. Parker's spiral heliospheric magnetic field is implemented through the angle $\psi = \arctan(\Omega r \sin\theta \cdot U^{-1})$ in the anisotropic diffusion tensor for the GCR particles where ψ is the angle between magnetic field lines and radial direction in the equatorial plane.

As an ad hoc assumption, a quasi linear theory (QLT) [21] formally arising from hard-sphere scattering (or the billiard ball diffusion) is considered as a more reasonable and simple tool for describing a propagation of GCR in heliosphere. The QLT is not generally the best approximation [17], [18], [19] and [20] for describing a propagation of large energy range of GCR particles, but it works well in the energy range to which neutron monitors and muon telescopes respond (for rigidities R>(10–15)GV) [22]. So, as an ad hoc assumption we employ the ratios of the perpendicular $K_\perp$ and drift K_d diffusion coefficients to the parallel $K_\parallel$ diffusion coefficient $\beta = \frac{K_\perp}{K_{II}}$ and $\beta_1 = \frac{K_d}{K_{II}}$ for the cosmic ray particles of rigidities R≥ 10GV, as follows,

$$\beta = \frac{1}{1 + \omega^2 \tau_1^2}, \qquad \beta_1 = \frac{\omega \tau_1}{1 + \omega^2 \tau_1^2}, \quad (4)$$

where for rigidities R= 10GV is accepted that $\omega\tau_1$=3, i.e. $\beta = \frac{K_\perp}{K_{II}} = 0.1$ at the earth orbit. Then, changes of $\omega\tau_1$ is determined by the Parker's spiral magnetic field in the whole heliosphere.

A parallel diffusion coefficient used in the model is expressed, as:

$$K_{II} = K_0 K(r) K(t) K(R, \gamma(t)), \quad (5)$$

where $K_0 = 1.9\times10^{19} cm^2/s$, $K(r) = 1+50r$; a function *K(t)* is introduced to make a consistent change of the diffusion coefficient K_{II} throughout the 11–year cycle of solar activity. The expression $K(t) = 2.3\cdot\exp(4t)$ implemented in equation (1), is presented in figure 2. $K(R,\gamma(t))$ contributes to the changes of the parallel diffusion coefficient K_{II} due to dependence on the GCR particles rigidity *R*. In the QLT this dependence is expressed as $K(R,\gamma(t)) = R^{\gamma(t)}$ which is valid for rigidities R>10 GV [21], [22], [23] and [24]. The analytical expression of the $\gamma(t)$ (figure 3) implemented in the model is shown in the figure caption.

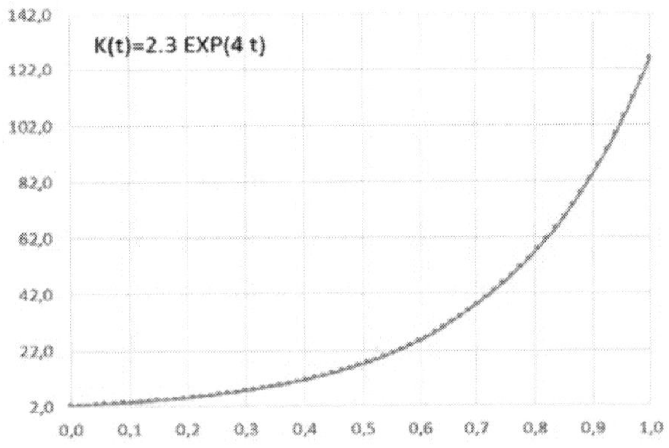

Figure 2. Temporal changes of the normalized parallel coefficient *K(t)* used in modeling.

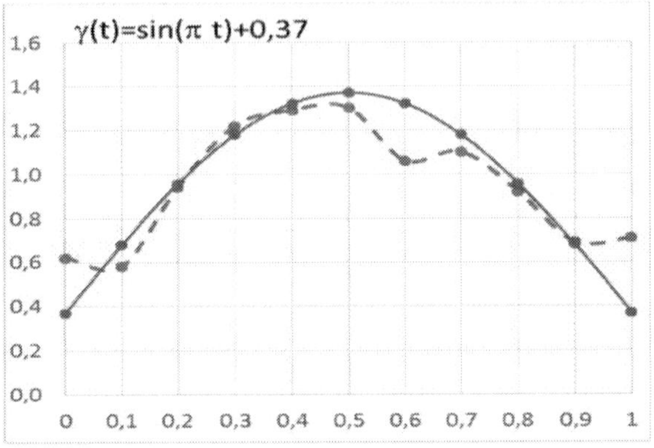

Figure 3. Changes in the exponent *γ(t)* in the period: 1976–1987 normalized to <0;1>. The 1–year averages (dashed red line) and the trigonometrical approximation (solid blue line) are included in the model.

The neutral sheet drift was taken into account according to the boundary condition method [26]. The delta function at the HNS is a consequence of the abrupt change in sign of the IMF. Changes of the magnitude B of the IMF (dashed line) in the considered period and its approximation (solid line) are shown in figure 4.
We assume that the drift effect *D(t)* has maximum value (100%) in the minima epochs, 1976–1977 and 1986–1987, and it is scaled down to 20% in maximum epoch of solar activity (figure 5). The temporal changes of the observed tilt angle δ of the HNS in the range of ~4^0 up to ~70^0 (dashed red line) and its approximation $\delta(t) = 408.6\cdot t^3 - 759\cdot t^2 + 360.7\cdot t - 1.4$ (solid blue line) are presented In figure 6. In addition, in the model the waviness of the HNS is implemented, using the formula of [25] $\theta' = \frac{\pi}{2} + \delta(t)\sin\left(\frac{r\Omega}{U}\right)$.

The equation (2) was transformed to the algebraic system of equations using the implicit finite difference scheme, and then solved by the Gauss–Seidel iteration method (e.g. [26]) using the following boundary conditions:

$$f\Big|_{r=100AU} = 1, \quad \frac{\partial f}{\partial r}\Big|_{r=0} = 0, \quad \frac{\partial f}{\partial \theta}\Big|_{\theta=0} = \frac{\partial f}{\partial \theta}\Big|_{\theta=\pi} = 0, \tag{6}$$

as well as the initial conditions $f\big|_{R=100GV} = 1$ and $f(r,\theta,R_k,t)\big|_{t=0} = f(r,\theta,R_k)$ [26]. We start the calculations from the inner radius *(r=0)* of the spherical system.

The solutions for each k-layer of rigidity R_k where (R_k=*100, 90, 80,....,10 GV*) for the stationary case are considered as an initial conditions for the non-stationary case for the given rigidity *R* and at time *t*=0.

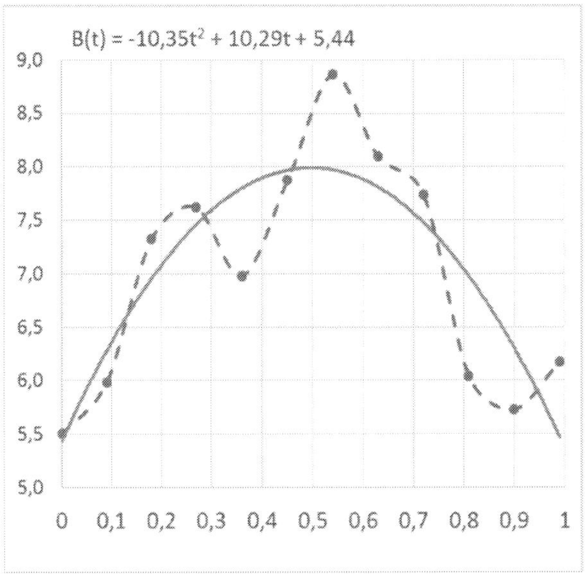

Figure 4. Magnitude B of the IMF in the period between 1976 and 1987 normalized to <0;1>. The yearly averages of data from SPIDR (dashed red line)

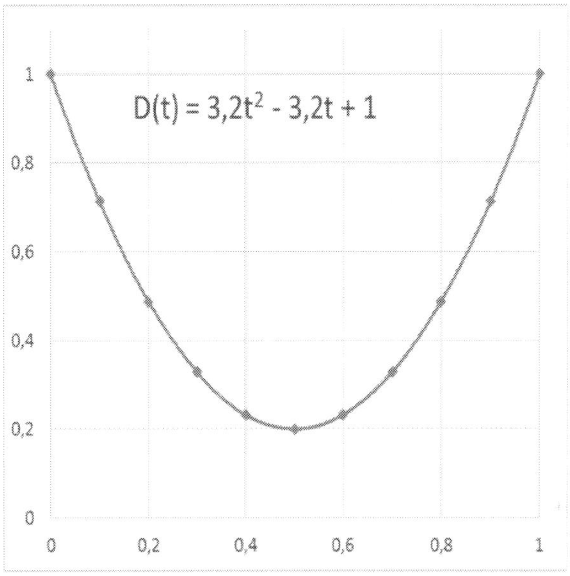

Figure 5. Changes of the drift ratio D(t) normalized to the minimum epochs implemented in the model is shown by polynomial approximation.

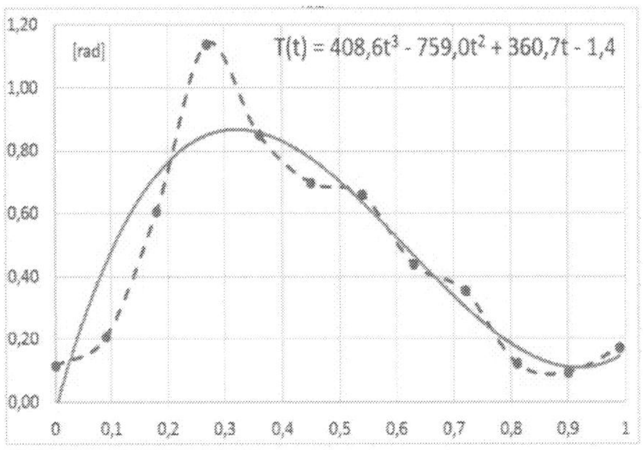

Figure 6. Yearly averages of the HNS tilt angle in the period 1976 and 1987 (dashed red line) and the approximation included in the model (solid blue line) normalized to < 0;1 >.

The results of the numerical solution of equation (2) for rigidity R=10 GV (dashed blue line), monthly averaged changes of the GCR intensity (solid red line) observed by the Moscow neutron monitor and its second order polynomial approximation (dotted red line) are demonstrated in figure 7. Figure 7 shows that the changes of the expected (dashed line) GCR particles density precede the smoothed by the second order polynomial approximation (dotted line) of the GCR intensity. To make clearer the existence of delay time between them we present in figure 8 the results of the numerical solution of equation (2) for rigidity R =10 GV (dashed line) shifted for 18 months with respect to the second order polynomial approximation (dotted line) of monthly averaged changes of the GCR intensity observed by the Moscow neutron monitor.

Our analysis shows that the delay time is ~18 months [27]. So, ~18 months can be accepted as an effective delay time caused by the combined influence of all parameters implemented in the 2–D model for the period 1977–1987.

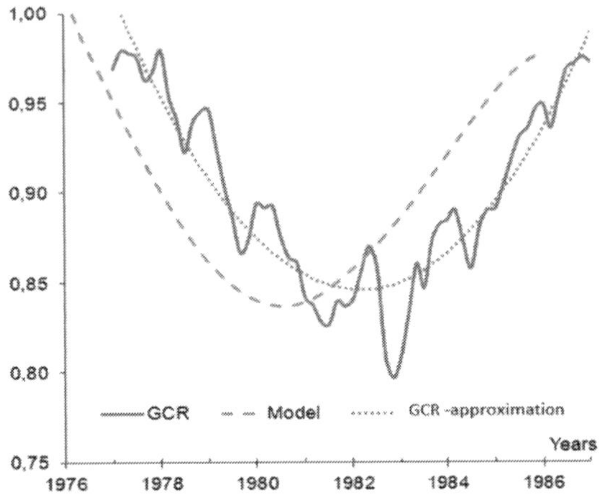

Figure 7. Changes of the amplitude of the 11–year variation of the GCR normalized intensity (red solid line) and approximated observed changes of the GCR intensity (red dotted line). Model – the solution of the equation (1) for the rigidity of 10 GV (blue dashed line)

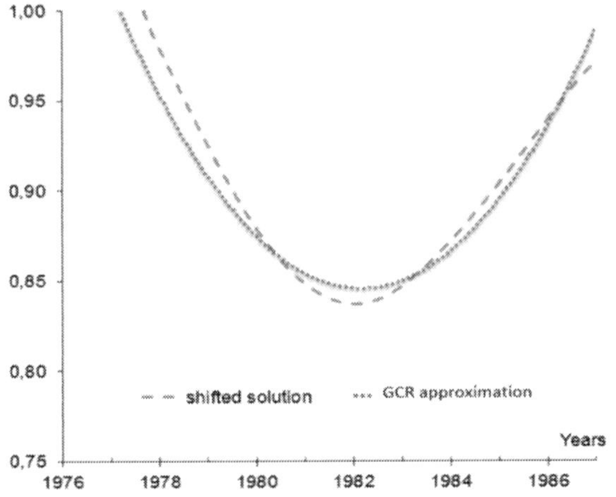

Figure 8 Shifted solution of the equation (1) by 18 months (dashed green line) with respect to the approximated observed changes of the GCR intensity (red dotted line).

Recently Bobik et al.,[28] developed a 2-D stochastic MC code for GCR particles propagation across the heliosphere, and have received an acceptable agreement between modeling results of antiproton to proton ratios (aP/P) and observations by BESS and PAMELA. Also, they estimate two specific times - (1) tsw ~ 14 months, time needed by solar wind to expend from the outer corona up to 100AU with a speed 400 km/s, and (2) a) time τev ~ 1month and b) time τev ~ few days needed for stochastic evolution of particles in heliosphere with energies of 10 GV and 200 MeV, respectively. Besides, in paper is touched upon an important problem dealing with GCR modulation, namely, a necessity to consider a whole heliosphere as a consisting from small regions with different electromagnetic conditions, e.g., determined by monthly averaged parameters. Similar study has been performed in various papers, among them in [29], where was taken into account that the free path of GCR scattering is controlled by the real distribution of the Sun's coronal green line intensity (CGLI).

Conclusions

1. A new 2–D time dependent model of the 11–year variation was developed. This model implements the parameters characterizing the temporal changes of the magnitude B of the IMF, tilt angle δ of the HNS for the period of 1976–1987 of cycle #21 and the changes of drift coefficient of the GCR particles versus the solar activity. The drift coefficient of the GCR particles has a maximum value ~ 100% in the minimum epoch (drift dominated period), and 20% in the maximum epoch (almost diffusion dominated period).
2. In the model temporal changes of the rigidity spectrum exponent γ, characterizing a rigidity dependence of amplitudes of the 11–year variations of the GCR intensity, were implemented for the first time.

3. The temporal changes of the parameters implemented in the 2–D model have different delay times with respect to the temporal changes of the smoothed experimental data of the GCR intensity observed by Moscow neutron monitor. We show that an acceptable compatibility is kept for the period of 1976–1987 (solar cycle #21), when the minimum of the expected temporal changes of the GCR particles density is shifted by 18 months with respect to the minimum of the temporal changes of the smoothed experimental data of the GCR intensity.
4. We conclude that a delay time ~18 months can be accepted as an effective delay time caused by the combined influence of all parameters implemented in the 2–D model. Generally, a direct implementation of the delay time τ in modeling is a important problem needing additional study.

References

[1] Siluszyk M, Iskra K and Alania, M V 2014 *Solar Physics* 289 **11** 4297
[2] Richardson J D and Burlaga L F 2013 Space Sci Rev **176** 217
[3] Alania M V, Iskra K and Siluszyk M 2010 *Proc. of 38th COSPAR* D11 **0005** 10
[4] Siluszyk M, Iskra K, Modzelewska R and Alania M V 2005 *Adv. Space Res.* **35**, 4, 677
[5] Alania M V, Modzelewska R and Wawrzynczak A 2014 *JGR* **119** 6 4164
[6] McDonald F B and Burlaga L F 1997 *Global Merged Interaction Regions in the Outer Heliosphere*, University of Arizona, 389
[7] le Roux J A and Potgieter M S 1995 *Astrophys J* **442** 847
[8] Parker E N 1965 The passage of energetic charges particles through interplanetary space *Planet. Space Sci.* **13** 9
[9] Manuel R, Ferreira S E S and M S Potgieter 2014 Solar Physics **289** 2207
[10] Webber W R, Lockwood J A 2001 *JGR* **106** 9 323
[11] Caballero–Lopez R A and Moraal H 2004 *JGR* **109** A01101
[12] Jokipii J R, Levy E H and Hubbard W B 1977 Astrophysical Journal 1 **213** 861
[13] Rossi B and Olbert S 1970 *Introduction to the physics of space* New York McGraw–Hill
[14] Jokipii J R and Kopriva D A 1979 *Astrophys J* **234** 384
[15] Kota J and Jokipii J R 1983 **1** 265 573
[16] Parker E N 1958 *Astrophys. J* **128** 664
[17] Parhi S, Burger R A, Bieber J W and Matthaeus W H 2001 *Proc. of the 27th ICRC* **3670**
[18] Parhi S, Bieber J W, Matthaeus W H and Burger R A 2003 *Astrophys. J* **585** 1 502
[19] Parhi S, Bieber J W, Matthaeus W H and Burger R A 2004 *JGR* **109** A1
[20] Shalchi A, Li G and Zank G P 2010 *Astrophysics and Space Science* **325** 1 99
[21] Jokipii J R 1971 *Rev. of Geoph. and Space Physics.* **9** 27
[22] Shalchi A and Schlickeiser R 2004 *Astrophys. J* **604** 861
[23] Bieber J W 2003 *Adv. Space Res.* **32** 4 549
[24] Shalchi A 2009 *Nonlinear Cosmic Ray Diffusion Theories* Springer Berlin 180
[25] Jokipii J R and Thomas B 1981 *Astrophysical Journal* **243** 1115
[26] Kincaid D and Cheney W 2002 *Numerical Analysis* American Mathematical Society
[27] Siluszyk M, Wawrzynczak A and Alania M V 2011 *J. of atm. and S-T Physics* **73** 13 1923
[28] Bobik P, Boschini M J, Consolandi C, Della S, Torre, Gervasi M, Grandi D, Kudela K, Pensotti S and. Rancoita P. G 2011 Astrophys. Space Sci. Trans. **7** 245
[29] Alania M V, Gil A, Gushchina R T, Iskra K, Siluszyk M, 2003 Izviestia RAN ser. fiz. **67** 4 506

24th European Cosmic Ray Symposium (ECRS2014) IOP Publishing
Journal of Physics: Conference Series **632** (2015) 012081 doi:10.1088/1742-6596/632/1/012081

Observation of solar high energy gamma and X-ray emission and solar energetic particles

A Struminsky[1]and W Gan[2]

[1] Space Research Institute, RAS, Profsoyuznaya st., 84/32, Moscow 117927, Russia
[2] Purple Mountain Obervatory, CAS, 2 West Beijing Road, Nanjing 210008, China

E-mail: astrum@iki.rssi.ru

Abstract. We considered 18 solar flares observed between June 2010 and July 2012, in which high energy >100 MeV γ-emission was registered by the Large Area Telescope (LAT) aboard FermiGRO. We examined for these γ-events soft X-ray observations by GOES, hard X-ray observations by the Anti-Coincidence Shield of the SPectrometer aboard INTEGRAL (ACS SPI) and the Gamma-Ray burst Monitor (GBM) aboard FermiGRO. Hard X-ray and π^0-decay γ-ray emissions are used as tracers of electron and proton acceleration, respectively. Bursts of hard X-ray were observed by ACS SPI during impulsive phase of 13 events. Bursts of hard X-ray >100 keV were not found during time intervals, when prolonged hard γ-emission was registered by LAT/FermiGRO. Those events showing prolonged high-energy gamma-ray emission not accompanied by >100 keV hard X-ray emission are interpreted as an indication of either different acceleration processes for protons and electrons or as the presence of a proton population accelerated during the impulsive phase of the flare and subsequently trapped by some magnetic structure. In-situ energetic particle measurements by GOES and STEREO (High Energy Telescope, HET) shows that five of these γ-events were not accompanied by SEP events at 1 AU, even when multi-point measurements including STEREO are taken into account. Therefore accelerated protons are not always released into the heliosphere. A longer delay between the maximum temperature and the maximum emission measure characterises flares with prolonged high energy γ-emission and solar proton events.

1. Introduction

At present two competitive paradigms of solar energetic particle (SEP) origin exist – acceleration directly in the flare site or by shock wave driven by a coronal mass ejection (CME) [1]. Active discussions on a relative role of flares and CME's for SEP acceleration and propagation are continuous till now. Additional arguments supporting one or another paradigm can be obtained investigating new events and using new instruments earlier unavailable.

More than 20 years ago the prolonged high energy γ-emission was discovered in solar flares of June 1991 by using sensitive detectors aboard the ComptonGRO and GAMMA-1 spacecraft designed for astrophysical studies [2-6]. The registration of π^0 decay γ-emission during several hours after the flare proves a possibility of prolonged proton interaction at the Sun and is an additional argument in favor of the flare origin of SEP. The launch of the Fermi Gamma-Ray Observatory (FermiGRO) into the orbit, which is the new generation of gamma-observatories, in 2008 allowed hoping that new observations of prolonged high energy γ-emission would be performed in the 24th solar cycle.

The Large Area Telescope (LAT) aboard FermiGRO observed high energy >100 MeV solar γ-emission in 18 flares [7]. For the first time such emission was attributed to solar flares of X-ray classes of C and M. The record duration, more than 20 hours, of the solar >100 MeV γ-emission was registered on March 7, 2012[8]. The π^0 decay is the most possible mechanism of its origin [7, 8]. Such a prolonged solar γ-emission raises three questions: "Are protons producing π^0 accelerated in flare impulsive phase and then trapped? Are they accelerated in several episodes during decay phase? Are they a possible source of solar protons propagating in the heliosphere?" In order to answer first two questions we need to check solar observations in SXR and HXR wave bands. Light curves of SXR emission may show episodes of additional energy release (plasma heating). Bursts of HXR emission will show episodes of electron acceleration and roughly maximal energy of electron acceleration. Note, that in one loop evaporation model of solar flares maxima of time derivatives of SXR intensity correspond to maxima of HXR intensity (the Neupert effect [9]). An answer to the third question would be positive, if solar high energy γ-emission is observed during solar proton events.

In this work we consider the 18 FermiGRO solar γ-events [7] in soft X-rays (SXR) by data of GOES spacecraft and hard X-rays (HXR) by ACS SPI and GBM data (table 1) as well as an association of these events with solar proton events observed in the heliosphere (table 2) to answer three questions stated above. The GBM registers solar HXR and gamma emission from few keV to about 1 MeV by NaI scintillators and from ~150 keV to ~30 MeV by BGO crystals. The ACS SPI is a BGO crystal detector and registers primary (solar) and secondary (caused by Solar Energetic Particles (SEP)) hard X-rays >100 keV. Methodological problems of ACS SPI data application for studies of solar flares were considered recently in [10]. The INTEGRAL spacecraft is in an orbit with perigee of 10000 km and apogee of 153000 km and has an orbit period of 72 hours, it is out of the radiation belts most of the time.

Data of ACS SPI may show processes of energy release and interaction of accelerated particles, which are not observed by the FermiGRO and RHESSI satellites due to their night time or passing the radiation belts. In this regard we underlined three results [11-13] obtained by using ACS SPI data. In the solar event on October 26, 2003 the HXR burst after > 90 min since the impulsive phase was discovered by the ACS SPI data [11], which was not accompanied by any SXR response (burst on tail – BOT). Decay phases of the giant events of the 23rd solar cycle remind those of the June 1991 events with prolonged γ-emission [12]. It was shown in [13] that time profiles of plasma temperature and HXR intensity logarithm were similar in the beginning of the December 6, 2006 event. This correlation disappeared during the decay phase, i.e. significant electron fluxes may not be able to heat effectively the expanding plasma [13].

2. Observations

An automated data analysis pipeline was created by the authors of [7] to monitor the high-energy γ-ray flux from the Sun throughout the Fermi mission. The time intervals in which the Sun is less than 60^0off axis for the LAT were selected for the analysis [7] and they are called the Fermi Time Windows (FTW). Each interval corresponds to the maximum time with continuous Sun exposure, and the durations of these intervals vary as the Sun advances along the ecliptic and as the orbit of Fermi precesses. The high energy >100 MeV γ-emission was registered by the LAT/FermiGRO instrument in 18 solar flares occurred between June 2010 and July 2012. Since statistics of solar high energy γ-photons is low within FTW's data are averaged for each FTW.

Table 1 summarizes observations by LAT, GBM and ACS SPI for dates in the first column. The second column shows start (UT) and duration (min) of FTW's for LAT solar observations [7]. The third column is a maximal energy of solar emission registered by GBM/FermiGRO. Next columns present characteristics of ACS SPI count rate enhancements: start (UT), duration (min) and maximal value (counts) due to primary HXR; start (UT) due to secondary HXR (SEP arrival). The ACS SPI detector was switched off in 4 from the 18 LAT/FermiGRO events. Bursts of solar HXR >100 keV emission were observed by ACS SPI in the impulsive phase of 13 events. Note that only in six cases FTW's coincide in time with HXR bursts registered by ACS SPI. In these six cases GBM measured HXR emission >300 keV. Time intervals, when prolonged high energy γ-emission was registered by LAT, were not accompanied by HXR >100 keV emission according to the GBM and ACS SPI data

Journal of Physics: Conference Series **632** (2015) 012081 doi:10.1088/1742-6596/632/1/012081

Table 1. The Fermi LAT high energy γ-events [7] and solar HXR bursts (Fermi GMB http://sprg.ssl.berkeley.edu/~tohban/browser/) and ACS SPI (http://isdc.unige.ch/~savchenk/spiacs-online/). Six cases, when Fermi Time Windows (FTW) and ACS SPI bursts coincide in time, are in bold.

	FTW	GMB	ACS SPI				
	Start, duration	keV	Start, (min)	duration	Max(UT, counts)		SEP (start, UT)
2010Jun12	**00:55, 0.8**	**1000**	**00:55:40.8**	**2**	**00:55:50.8**	**4196**	**no**
2011Mar7	20:15, 25 23:26, 36 02:38,35 05:49, 35	<300 no <50 <50	19:58:00	15	20:01:54	223	Mar07/ 2143
2011Jun2	09:43, 45	<50	No enhancements				
2011Jun7	07:34, 53	<50	06:24:00	18	06:25:49.9	107	Jun 07/ 0646
2011Aug4	04:59, 34	<25	03:50:00	9	03:51:18.6	204	Aug 04/ 0515
2011Aug9	**08:01, 3.3**	**<300**	**08:00:30**	**6.5**	**08:02:10.8**	**732**	**Aug 09/ 0812**
2011Sep6	**22:17, 0.2** **22:13, 35**	 **1000**	**22:18:09**	**6**	**22:19:02.4**	**2075**	**no**
2011Sep7	23:36, 63	<50	22:35:51	3	22:36::18	1441	no
2011Sep24	**09:34, 0.8**	**1000**	**09:35:00**	**3**	**09:36:32.4**	**5385**	**no**
2012Jan23	04:07, 51 05:25, 69 07:26, 16 08:47, 35	<300 <25? no <50?	No ACS SPI data				
2012Jan27	19:45, 11 21:13, 24	<25? <25?	17:38:00	9	17:38:43	185	Jan 27/ 1755
2012Mar5	04:12, 49 05:26, 71 07:23, 28	<300 <50 <50	No ACS SPI data				
2012Mar7	**00:46, 60** 03:56, 32 07:07, 32 10:18, 32 13:29, 32 19:51, 25	<300 **1000** <25 <25 <50? <50? no	00:15:00 **01:07:00**	>25 **30**	00:17:54 **01:07:00**	1174 **1394**	**Mar 07 / 0150**
2012Mar9	05:17, 34 06:52, 35 08:28, 34	<25 <100 no	No ACS SPI data				
2012Mar10	21:05, 30	<25?	No ACS SPI data				
2012May17	02:18, 22	<100?	01:29:00	10	01:35:16	85	May 17/ 0141
2012Jun3	**17:40, 23**	**<300**	**17:53:00**	**1.4**	**17:53:24**	**296**	**No**
2012Jul6	23:19, 52	<50	23:04:30	6.5	23:06:34	23424	Jul 06/ 2332

Table 2. Solar proton events (http://umbra.nascom.nasa.gov/SEP/) and their parent X-ray event.

	Solar proton event		X-ray solar event					
	Day/UT	PFU_{max} /UT	X-class, Coord UT	T(MK) UT	TTM_T min	EM $10^{49}cm^{-3}$, UT	TTM_{EM} min	Δt min
2010Jun12	No		M2.0 00:30-01:02	20.5 00:56	26	1.0 00:58	28	2
2011Mar7	**08/01:05**	**50 08/08:00**	**M3.7 S22W67 19:43-20:58**	**13.4 20:04**	**21**	**2.2 20:13**	**30**	**9**
2011Jun2	no		C2.7 09:42-09:50	9.7 09:47	5	0.27 09:48	6	1
2011Jun7	07/08:20	72 07/18:20	M2.5 S21W64 06:16-06:59	13.6 06:28	12	1.53 06:33	17	5
2011Aug4	04/06:35	96 05/21:50	M9.3 N19W36 03:41-04:04	18.0 03:54	13	4.9 03:57	16	3
2011Aug9	09/08:45	26 09/12:10	X6.9 N17W69 07:48-08:08	32.5 08:03	15	29 08:05	17	2
2011Sep6	no		X2.1 N14W18 22:12-22:24	28.5 22:19	7	9.9 22:21	9	2
2011Sep7	no		X1.8 N14W28 22:32-22:44	21.8 22:37	5	7.8 22:39	7	2
2011Sep24	23/22:55	35 26/11:55	X1.9 N12E60 09:21-09:48	29.5 09:37	16	7.7 09:42	21	5
2012Jan23	**23/05:30**	**6310 24/15:30**	**M8.7 N28W21 03:38-04:34**	**17.8 03:50**	**12**	**4.7 04:01**	**22**	**10**
2012Jan27	**27/19:05**	**796 28/02:05**	**X1.7 N27W71 17:37-18:56**	**17.6 18:27**	**50**	**9.7 18:40**	**63**	**13**
2012Mar5	**STEREO B**		**X1.1 N16E54 02:30-04:43 N17E52**	**18.4 03:37 19.0 03:52**	**67 82**	**5.7 04:11**	**101**	**34 19**
2012Mar7	**07/05:10**	**6530 08/11:15**	**X5.4 N17E27 00:02-00:40** X1.3 N22E12 01:05–1:23	**26.0 00:18** 16 01:15	**16** 10	**23.5 00:25** 7.0 01:16	**23** 10	**7** 1
2012Mar9	Previous SEP event		M6.3 N15W01 03:22-04:18	18.6 03:26 18.0 03:44	4 22	0.9 03:27 3.3 03:53	5 31	1 9
2012Mar10	Previous SEP event		M8.4 17:15-18:30	19.8 17:24 18.2 17:38	9 23	2.8 17:27 4.3 17:46	12 31	3 8
2012May17	**17/02:10**	**255 17/04:30**	**M5.1 N11W76 01:25-02:14**	**15.5 01:38**	**13**	**2.9 01:48**	**23**	**10**
2012Jun3	no		M3.3 17:48-17:57	14.9 17:54	6	1.9 17:56	7	1
2012Jul6	07/04:00	25 07/07:45	X1.1 S18W50 23:01-23:49	25.5 23:07	6	5.5 23:09	8	2

(i.e. electrons were accelerated only up to 100 keV). Therefore, most likely, protons producing π^0 were accelerated during impulsive phase of the flares and then trapped.

Table 2 presents information on related proton events (start (UT), maximum (UT) and maximal proton flux). Proton fluxes are integral 5-minute averages for energies >10 MeV, given in Proton Flux Units (PFU), measured by GOES spacecraft 1 PFU = 1 p cm^{-2} sr^{-1} s^{-1}. The start of proton events is defined as the first of 3 consecutive data points with fluxes greater than or equal to 10 PFU. Next columns present plasma characteristics of parent solar flares deduced from GOES SXR observations (maximal temperature (MK) and its time moment (UT); maximal emission measure and its time moment; time to maximum (TTM) for temperature and emission measure since the beginning of SXR event, a time delay (min) between their values. SEP events were not registered in 5 from the 18 events even considering additionally observations aboard the STEREO A/B spacecraft [14]. Therefore accelerated protons are not always released into the heliosphere. We note that a difference between time to maximum (TTM) of plasma temperature and emission measure is greater for flares with prolonged high energy γ-emission, which are parent flares of powerful proton events (table 2). These cases are in bold in table 2. At present we do not know a physical reason for such delay. Below three SEP events on March 5, March 7 and July 6, 2012 are considered in more details.

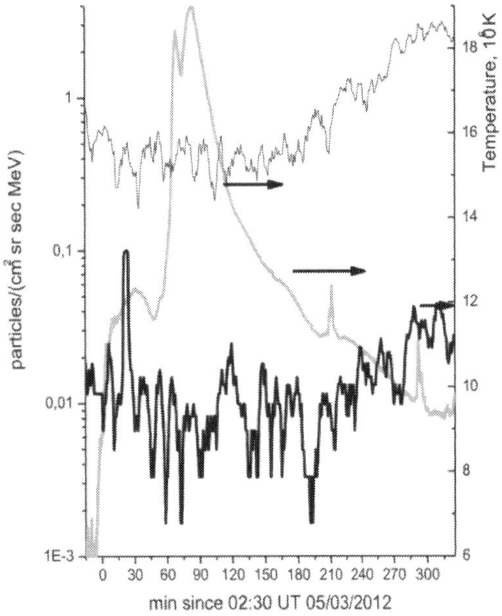

Figure 1. Onset of SEP event (thin black line – electrons of 1.4-2.8 MeV and - thick black line protons of 40-60 MeV) observed by HET/STEREO B on March 5, 2012 in comparison with plasma temperature (grey line). Arrows mark FTW's.

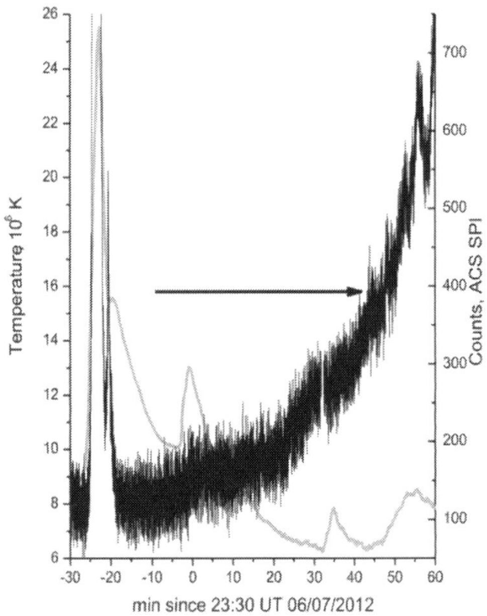

Figure 2. Time profiles of plasma temperature (grey line) and ACS SPI count rate (black) Maximal ACS SPI counts of ~23434 are out of scale for the X1.1 event on July 6, 2012. An arrow marks the FTW (52 min from 23:19 UT).

High energy γ-emission was registered on March 5, 2012 within three FTW's (fig. 1). The first FTW corresponds to the flare decay phase, when HXR emission <300 keV was observed by GMB. Maxima of the plasma temperature deduced from GOES SXR observations correspond to the second and third time FTW's, during which only HXR emission <50 keV was registered by GBM. The high energy γ-photons were emitted during periods of energy release associated with electron acceleration to <50 keV. Data from ACS SPI are not available for this event. The SEP event on March 5, 2012 was observed aboard STEREO B, which was in better position for the parent flare with coordinates N16E54 in comparison with SOHO and STEREO-B, at a background of the event on March 4 [14]. The solar electron and proton intensity started to rise during the second FTW (fig. 1). Solar electrons and protons might be released into the heliosphere during time intervals of the Fermi windows.

The event on July 6, 2012 is outstanding, because the maximal count rate of ACS SPI in its impulsive phase was comparable with maximal values observed in the giant events of October-November 2003 [10]. The enhancement of ACS SPI count rate caused by SEP particles possibly masks delayed bursts of primary solar HXR during periods of high energy γ-ray registration as shown in Fig. 2 for the event on July 6, 2012. The GBM didn't show HXR bursts >50 keV during the decay phase. Note that SEP (secondary γ-rays) were not registered by ACS SPI in 5 events, however even in these cases HXR bursts were not observed during the late decay phase.

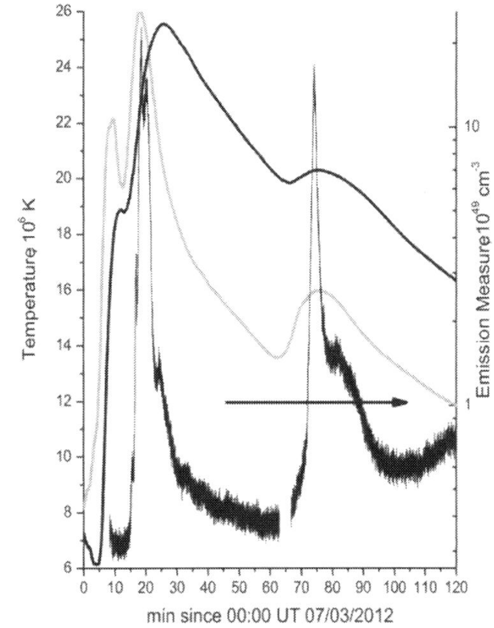

Figure 3. Time profiles of plasma temperature (grey line) and emission measure (black line) in comparison with ACS SPI count rate for the X5.4 and X1.3 events on March 7, 2012. An arrow marks the first FTW (60 min from 00:46 UT).

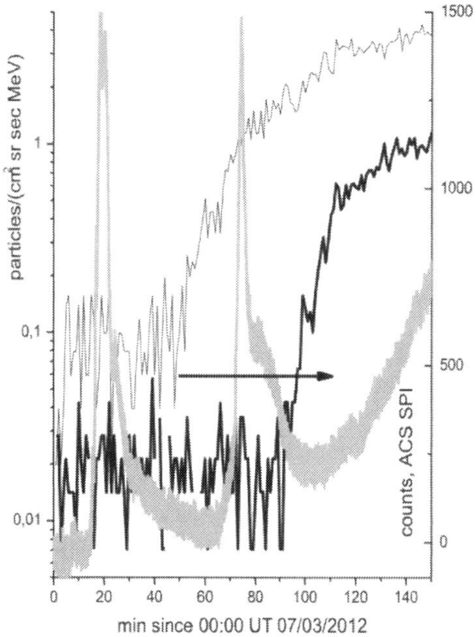

Figure 4. Onset of SEP event (thin black line – electrons of 1.4-2.8 MeV and thick black line - protons of 29.5-33.4 MeV) observed by HET/STEREO B on March 7, 2012 in comparison with ACS SPI count rate.

Two different solar flares (X5.4 onset at 00:02 UT and X1.3 onset at 01:05 UT) accompanied by comparable bursts of HXR with time lag of ~60 min composed the event on March 7, 2012 (Fig. 3). By a ratio of effects observed in SXR and HXR in the first and second flares this event reminds the event on October 26, 2003 [9]. The first flare corresponds to impulsive phase, but the second one to BOT. The FermiGRO spacecraft was at the night side of the orbit during the first X5.4 flare. The high energy γ-emission was registered during the second X1.3 flare and later during next five FTW's in the prolonged decay phase [7, 8]. The GBM didn't observe HXR bursts >50 keV within these five FTW's (i.e. electrons were not accelerated to energy >50 keV). Most likely, protons producing π^0 were accelerated during impulsive phases of these two flares and then trapped.

The authors [14] believe that the well-connected STEREO B observations provide unambiguous evidence of the correct solar association, according to their opinion onset of particle acceleration was associated with the first flare and not with the second one. Their arguments are: 1) there was no additional feature in the intensity–time profile that might be related to the second event; 2) only the first event was accompanied by bright type III radio emissions; 3) interplanetary type II emission commenced at 01:00 UT ahead of the second event. These arguments correspond to our tentative conclusion that flares associated with proton acceleration are characterized by large time delay between maximal temperature and maximal emission measure (table 2). Undoubtedly first electrons registered by STEREO B are associated with the first solar flare as seen in fig. 4. However we may not exclude that protons accelerated in both flares continuously interacted in the solar atmosphere producing π^0 and released into the heliosphere for more than 20 hours.

3. Conclusions

- We do not find any difference between solar flares accompanied by high energy γ-emission and flares without it in other wave bands.
- High energy solar γ-emission observed by LAT/FermiGRO during impulsive phase of flares was accompanied by HXR emission >100 keV observed by GBM/FermiGRO and/or ACS SPI, but not so during prolonged decay phase, therefore processes of proton and electron acceleration should be different in impulsive and prolonged decay phases and/or proton trapping is more effective than electrons.
- Not all flares with high energy γ-emission are associated with SEP events even considering SEP observations at three far-spaced spacecraft in the heliosphere (the STEREO A/B and SOHO spacecraft). If particles interacting in the solar atmosphere and escaping into the heliosphere are of the same population, then processes of particle release into the heliosphere play a crucial role.
- Prolonged observations of high energy gamma and HXR emission with better spatial, temporal and energetic resolution are necessary for progress in the problem of SEP origin.

Acknowledgment
This work was partly supported by the Presidium of the Russian Academy of Sciences (Program P-22) and the joint Russian-Chinese grant (RFBR-13-02-91165 and NNSFC-11233008). The authors thank two reviewers for their valuable comments and suggestions.

References

[1] Reames D V 2013 *Space Science Rev* **175** 1-4 53
[2] Kanbach G, Bertsch D L, Fichtel C E et al. 1993 *A&A. Suppl.* **97** 1 349
[3] Akimov V V, Belov A V, Chertok I M et al. 1993 *23rd ICRC Invited, Rapporteur, and Highlight Papers*. Ed. D A Leahy, R B Hickws, and D Venkatesan. Singapore: World Scientific **3** 111
[4] Kocharov L G, Kovaltsov G A, Kocharov G E et al. 1994 *Solar Phys.* **150** 1-2 267
[5] Murphy R J, Share G H 1993 *23rd ICRC Invited, Rapporteur, and Highlight Papers*. Ed. D A Leahy, R B Hickws, and D Venkatesan. Singapore: World Scientific **3** 99
[6] Struminsky A, Matsuoka M, Takahashi K 1994 *Astrophys. J.* **429** 1 400
[7] Ackermann M, Ajello M, Albert A et al. 2014 *Astrophys. J.* **787** 1 15
[8] Ajello M, Albert A, Allafort A et al. 2014 *Astrophys. J.* **789** 1 20
[9] Neupert W M 1968 *Astrophys. J.* **153** L59
[10] Rodríguez-Gasén R, Kiener J, Tatischeff V et al. 2014 *Solar Phys.* **289** 5 1625
[11] Zimovets I, Struminsky A 2012 *Solar Phys.* **DOI** 10.1007/s11207-0112-8
[12] Struminsky A 2013 *Geomagnetism and Aeronomy* **53** 7 843
[13] Struminsky A B and Zimovetst I V 2010 *Astronomy Letters* **36** 6 430
[14] Richardson I G, von Rosenvinge T T, Cane H V et al. 2014 *Solar Phys.* **289** 8 3059

Modelling the Influence of Corotating Interaction Regions on Jovian MeV-electrons

A. Vogt[1], F. Effenberger[2], H. Fichtner[2], B. Heber[1], J. Kleimann[2], A. Kopp[1], M.S. Potgieter[3], O. Sternal[4], T. Wiengarten[2]

[1] Christian-Albrechts Universität zu Kiel, Germany,
[2] Ruhr-Universität Bochum, Germany,
[3] North-West University, Campus Potchefstroom, South Africa,
[4] Mint-Kolleg Baden-Württemberg, Universitt Stuttgart, Germany

E-mail: vogt@physik.uni-kiel.de

Abstract. Corotating Interaction Regions (CIRs) are recurrent structures in the solar wind characterized by velocity jumps and a magnetic field compression. Since the 1970s it is known that Jupiter is a quasi-continuous source of MeV-electrons dominating the flux in the inner heliosphere. In connection with CIRs, this flux is modulated mainly by changing propagation conditions in the inner heliosphere. In order to model these recurrent variations in Jovian electron intensity the VLUGR3-Code was used to solve Parker's Transport Equation. The diffusion as well as the solar wind speed are modelled from 0.1 to 50 AU. Two different approaches are used, one derived by Kissmann (2002) and another by Giacalone et al.(2002) which was further developed. The simulation results are compared to IMP-8 electron count rates to investigate the differences of the two solar wind models in the propagation code.

1. Introduction

Statistically significant 27-day variations are detected in spacecraft data of the galactic cosmic ray intensities, the electron count rates, the magnetic field and the solar wind speed data. Because these variations are all in phase, it is reasonable to assume a common cause. The most probable of the discussed causes are Corotating Interactions Regions (CIRs) which occur if a coronal hole extents into the ecliptic plane. As the high-speed solar wind originating from the coronal hole runs into a slower preceding solar wind as it is shown in Figure 1, a region of high magnetic flux and particle density is formed. According to [1] and [2] these CIRs are often recurrent structures if the extension of the coronal hole is stable over several solar rotations.

The compression of solar wind influences the heliospheric magnetic field and particle propagation conditions as described in section 2 according to [3, 4]. The two presented model setups by [5, 6] and [7] both vary the particle propagation as well as the magnetic connectivity via the solar wind speed and the variation of the diffusion inside a CIR inverse to the magnetic flux density.

The particle distribution is computed by using the VLUGR3-Code [8, 9] to solve Parker's transport equation [10] on a grid with locally varying parameters.

Apart from that the model setups differ in calculating parameter changes caused by a CIR. The setup by [5, 6] decreases the diffusion coefficient inside a CIR independently from the solar wind speed. Based on [7] the second setup derives the variation of the diffusion tensor from the derivative of the solar wind speed.

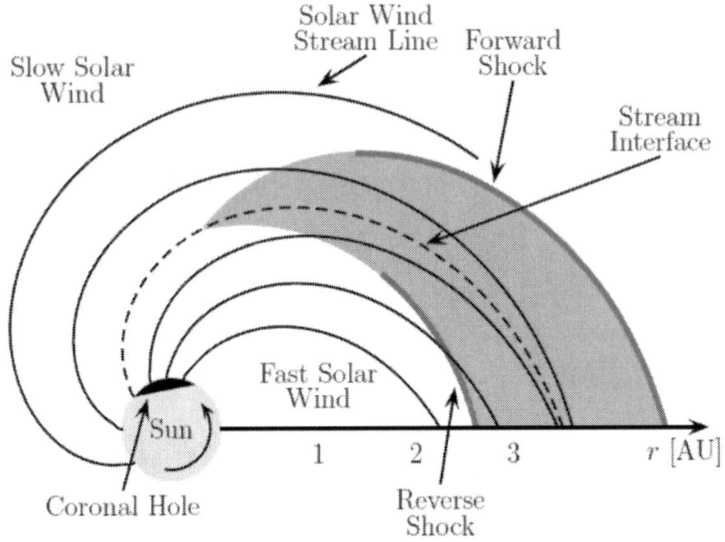

Figure 1. Sketch of the magnetic field line configuration that leads to a CIR, taken from [11].

2. CIRs And Low-Energetic Electrons

2.1. The Structure of CIRs

CIRs are a direct consequence of Parker's heliospheric magnetic field and were predicted to exist already by himself [4]. In Parker's model the magnetic field lines are "frozen" into the solar wind and connect particles of the same source regions. Due to the Solar rotation with an angular velocity of Ω_S the longitudinal position changes with respect to the source region and the connecting magnetic field line is bent to an Archemedian Spiral (called Parker Spiral in this context) with a curvature depending on the radial solar wind speed u:

$$\phi(r) = \phi_0 + \frac{\Omega_S \cdot \sin\Theta_0}{u}(r - R_S) \quad (1)$$

if the source region is defined by the longitudinal angle ϕ_0 and the co-latitudinal angle Θ_0 whereas the radial location of the source region is at the solar radius R_S and the radial distance from the source surface is r.

Considering Eq. (1) higher solar wind speeds lead to lesser bending spiral magnetic field lines. As shown in Figure 1 the fast solar wind cannot overtake the slower one, because their corresponding Parker spirals cannot cross each other. Therefore they align and form a so-called stream interface ([12, 13]), a region of high magnetic and particle density. The area around the stream interface is also called compression region due to the increase of magnetic pressure and particle density and is associated with a rarefaction region located at the angle where the fast solar wind descends to smaller velocities again and the particle and magnetic densities are decreased.

The pressure gradient between the compression region and its surrounding area causes two pressure waves which expand into the fast and the slow solar wind regions and widen the CIR thereby decreasing the velocity gradient.

If the difference between the fast and the slow solar wind speed is higher than the propagation speed of the pressure waves in the plasma, they become shock waves, most of them at a radial distance of $2 - 3$ AU [14].

At even further heliocentric distances of $r = 5 - 8$ AU these structures can merge to so-called Merged

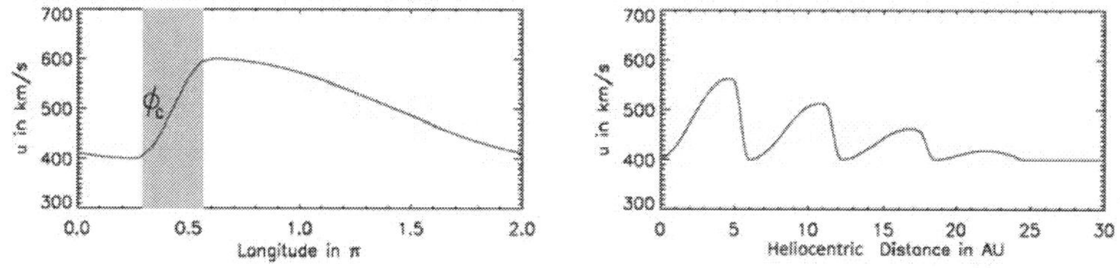

Figure 2. The Kissmann model the solar wind speed. In the left plot over a full solar rotation at Earth's orbit the shaded area indicates the width of the Compression Region, which is referred to as $\Delta\Phi_C$ in Eqs. (3) and (4). The beginning of the CIR is given by the angle ϕ_C. The right plot shows the solar wind speed variation for radial distances between the Solar surface and 30 AU. The radial damping is modelled the function $d = \max((25\ \text{AU}-r)/r, 0)$ to smooth out the variation at distances larger than 25 AU.

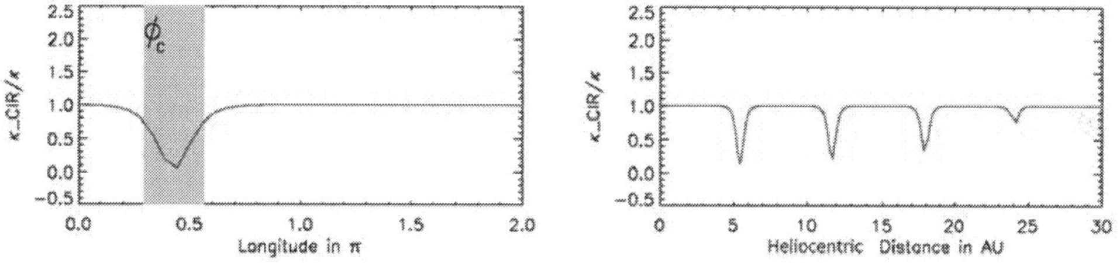

Figure 3. The Kissmann Model for the parallel diffusion coefficient. The left shows the ratio of the varied parallel diffusion coefficient κ_{CIR} as defined in Eq. (5) and the unvaried κ according to [15] over a full solar rotation at Earth's orbit. The width of the compression region $\Delta\Phi_C$ as well as the beginning of the structure ϕ_C are shown similar to Figure 2. The right plot shows κ_{CIR}/κ at a constant longitude. The damping function $d = \max((25\ \text{AU}-r)/r, 0)$ is not directly multiplied with the diffusion tensor but scales the parameter $A = 1 - B_0/B_{CIR}$ via $B_{CIR} \propto 1 + d \cdot f_{CIR}$ where f_{CIR} defines the variation inside the CIR, similar to f_{var} for solar wind speed variations.

Interaction Regions (MIRs). With increasing distance the shock waves weaken and CIRs/MIRs evolve to Corotating Pressure Enhancements (CPEs) as described by [14].

Due to the limitations of the used code shock waves and MIRs could not be considered in our CIR model and therefor don't contribute to the modeling results.

2.2. *Jovian Electron Transport*

The VLUGR3 code, described in [8, 9], solves Parker's Transport Equation numerically in three spatial dimensions and time. Due to this limitation in the number of dimensions it is not possible to calculate energy spectra. Therefore only 7 MeV Jovian electrons were used to investigate the modulation and propagation effects caused by the model CIRs.

Jupiter is known as a source of low MeV electrons since the Pioneer 10 flyby in 1977 (c.f. [16],[17] and [18]) and can even be considered as point-like on larger scales according to [19], [20] and [21] . Due to Jupiter's synodic period of 13 months which determines the magnetic connectivity between the Earth and Jupiter a long-term variation of this period is detected in the electron count rates. Despite this 13 months period also the shape of the Jovian spectrum is well investigated and found to dominate the global spectrum in its low MeV range up to $\approx$ 20 MeV within the inner heliosphere for distances $r \lesssim 20 - 30$ AU by [22].

The described approach has already been used successfully by [6], [11] and [23] for numerical parameter studies as well as for comparison with electron spacecraft data.

The VLUGR3 code computes the differential electron intensities $j = P^2 \cdot f(\vec{r}, P, t)$, via solving the distribution function $f(\vec{r}, P, t)$ given by Parker's Transport Equation [10]

$$\frac{\partial f}{\partial t} = \nabla \cdot (\hat{\kappa} \cdot \nabla f) - \vec{u} \cdot \nabla f + \frac{1}{3}(\nabla \cdot \vec{u}) \frac{\partial f}{\partial lnP} + S \tag{2}$$

which takes the effects into account caused by diffusion (first term) via the diffusion tensor $\hat{\kappa}$ [24] as well as the convection of the particles via the solar wind speed $\vec{u}$ (second term on the right-hand side). The third term represents adiabatic energy changes, while with S additional sources as Jupiter are included. Some of these terms depend on the rigidity $P = pc/|q|$ with the momentum p, the speed of light c and the particle charge q. Further dependencies involved are the position $\vec{r}$ and the time t.

In order to avoid any influence of boundary conditions on our simulation results in the inner Heliosphere's ecliptic plane, the differential electron intensities were simulated form $r_{min} = 0.01$ AU up to $r_{max} = 50$ AU and over the whole latitudinal as well as longitudinal range.

3. Two CIR Models

3.1. The Kissmann Model

The first model being implemented was the approach by [5, 6]. In this setup the longitudinal solar wind speed variation is defined by a sharp increase at the angle $\phi_C(r)$. The width of that region and therefore the slope of the increase is determined by $\Delta\phi_C$ as shown in Figure 2. Characteristic for the setup is a much softer decrease which stretches over the rest of the longitudial range. The variation of $\phi_C(r)$ with radius via a Parker spiral, leads to the radial profile of the solar wind speed that is also shown in Figure 2. To adjust the different velocities in the outer heliosphere a linear radial damping function $d = \max((25\text{ AU}-r)/r, 0)$ is applied, to let the CIR decline.

Technically this solar wind speed profile over ϕ is realised by two equations: the sharp increase inside the CIR is decribed by

$$u(r, \phi) = u_{slow}(1 + d \cdot f_{var} \cdot (1 + \cos(\pi \frac{\phi - \phi_C}{\Delta\phi_C}))) \tag{3}$$

while the long decrease outside the structure is determined by

$$u(r, \phi) = u_{slow}(1 + d \cdot f_{var} \cdot (1 - \cos(\pi \frac{\phi - \phi_C}{2\pi - \Delta\phi_C}))), \tag{4}$$

where f_{var} defines the amplitude of the variation inside the CIR, while u_{slow} is the velocity of slow solar wind at the source latitude.

The magnetic properties of the CIR are not directly taken into account but by their effect to Parker's Transport Equation, precisely by the variation of the diffusion tensor $\hat{\kappa}$. The approach that $\hat{\kappa} \propto 1/B$ by [22] leads in the most simple way to a box-like decrease of the diffusion, and the high magnetic flux density region can be assumed as a diffusion barrier. Inspired by data, [6] uses a smoother decrease symmetric around ϕ_C as shown in Figure 3.

This variation is implemented as a modulation factor to the standard value of $\kappa_{\parallel}$ with the parameter $A = 1 - B_0/B_{CIR}$ scaling the variation to the magnetic flux density variation outside (B_0) and inside (B_{CIR}) the CIR

$$\kappa_{\parallel CIR} = \kappa_{\parallel} \left(1 - A \frac{1}{\cosh(\frac{\phi_r - \phi_C}{\Delta\phi_C})} \right) \tag{5}$$

As the perpendicular component of the diffusion tensor is assumed to be proportional to the parallel one according to [22] and [25] it is changed in the same manner. Because the transport equation is

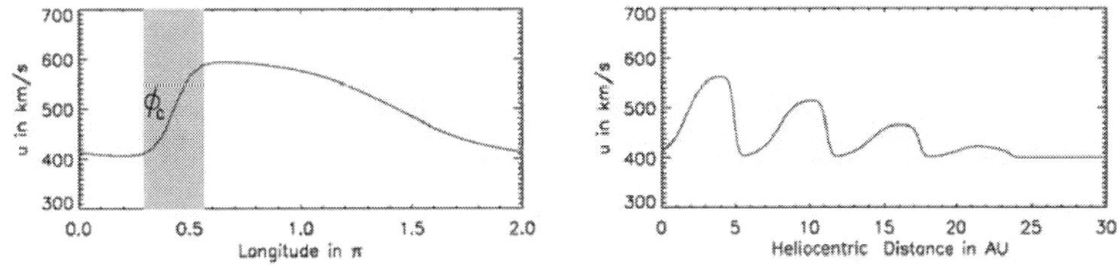

Figure 4. The Giacalone model for the solar wind speed. Similar to Figure 2 the left plot shows the solar wind speed variation over a full solar rotation at the earth orbit with ϕ_C as the beginning of the CIR and a shaded compression region. The right plot shows the radial development described by a similar damping function as for the Kissmann setup added to Eq.(6).

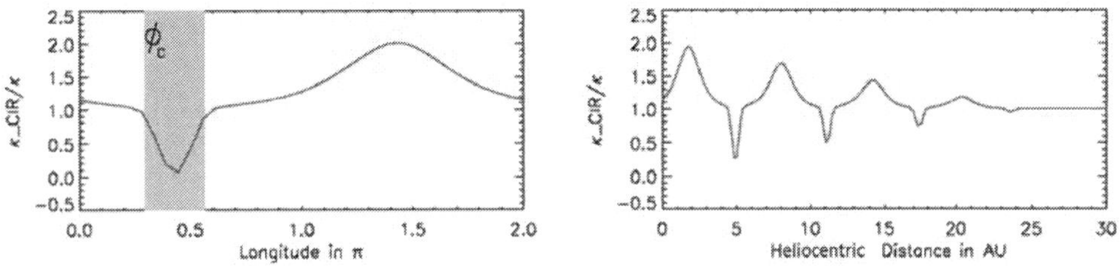

Figure 5. The Giacalone model for the parallel diffusion coefficient. The left plot shows the variation over a full rotation as in Figure 3 for the Kissmann setup but with the significant increase at the rarefaction region. The right plot showing the diffusion coefficient vs. the heliopheric distance differs from the corresponding plot due to this feature. The amplitude damping is linear for the Giacalone setup as it results from the derivative of the linear damped solar wind speed, cf. Eq. (7)

solved time-dependently with a rotating CIR, the diffusion barrier also shows an effect similar to the loss of magnetic connection between the Jovian source and an arbitrary point in space on the opposite side of the barrier.

As the following Giacalone Model in section 3.2, the Kissmann Model is limited to the laditudinal range of 3.5° beneath and above the ecliptic plane where [22] showed the Jovian source to be most dominant. Likewise the limitation of one stream per rotation this range was choosen in order to keep the computational time short and could be extented for further investigations.

3.2. The Giacalone Model

To expand the possibilities of investigation a second solar wind speed and magnetic field model was implemented, based on the approach of [7]. The main difference to the first is, that it only parameterises the solar wind speed changes in the inner heliosphere and derives the magnetic field as a result of the velocity field.

For the solar wind speed variation the modified approach of [7] is used:

$$u(r,\phi) = u_{fast} + \frac{1}{2}\cdot(u_{fast} - u_{slow})\cdot\left(\tanh\left(\frac{\phi_C-\phi-r\Omega_S/W}{\Delta\phi_C}\right) - \tanh\left(\frac{\phi_{RF}-\phi-r\Omega_S/W}{\Delta\phi_{RF}}\right)\right), \quad (6)$$

where W is a constant speed which can be assumed as the speed with which the structure moves radially outward in the inertial reference frame. In Figure 6 the model is shown compared to spacecraft data of the simulated time period.

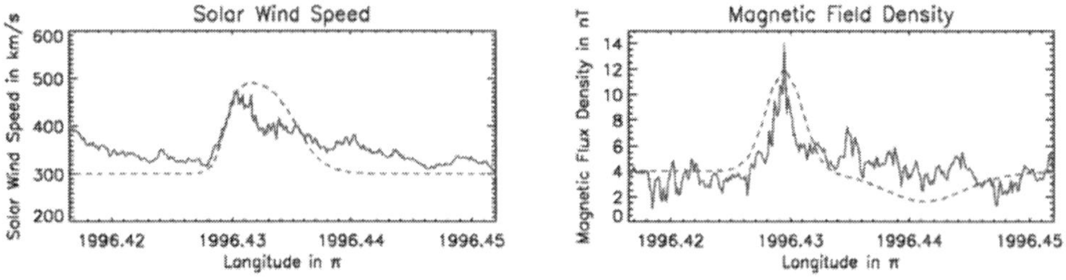

Figure 6. The Giacalone model's solar wind speed and magnetic flux density in comparison to measurements. The data is taken from *ftp://spdf.gsfc.nasa.gov/pub/data/omni/low_res_omni/*

The Giacalone model additionally describes the rarefaction region and therefore includes also the fast solar wind speed u_{fast} as measured inside the CIR, the angle at which the rarefaction region occurs ϕ_{RF} and also the angular width of the rarefaction region $\Delta\phi_{RF}$. Another difference to the model setup by [6] is that the CIR does not cover the whole longitudinal range which allows to vary its width.

In contrast to the original approach by [7], the corresponding magnetic flux density B and therefore the diffusion coefficients are assumed to be antiproportional to the derivative of the solar wind speed

$$B(r,\phi) \propto \left|\frac{\partial}{\partial\phi}u(r,\phi)\right| \tag{7}$$

based on the magnetohydrodynamic continuity equation.

The magnetic flux considered in the model seems to be in agreement with the corresponding data in Figure 6 and with the theoretical concept of CIR evolution because it reproduces the compression as well as the rarefaction region.

This allows to survey the influence of the rarefaction region on the Jovian electron modulation, if the simulated data for the Giacalone model are compared to the Kissmann model results.

4. Results and concluding remarks

The comparison of the IMP-8 daily electron data (gray) with the simulation results in Figure 7 shows, that the computed effects of the Giacalone model setup (red) seem to be in qualitatively good agreement with the observations. While the 13 month Jovian synodic period is already discussed by [6] we focused here on a period of good connection, as this is the period where we expect the greatest effects. The time period used in Figure 7 is one of absolute quiet time conditions, the interplanetary stream and sector structure has been discussed by [26]. Quantitatively the variations had to be intensified by the factor of 10 to be comparable to the results of the Kissmann setup (black).

For both models Figure 7 suggest that they reproduce the magnetic cutoff and its effect on the particle distribution in the surrounding quite reliably. The results of the Kissmann model show this as the wedge shaped intensity decreases every 27th day when the CIR is located between the spacecraft at Earth orbit and the Jovian source.

The different shape of the electron intensity variations in the Giacalone model shows the influence of the rarefaction region properties. The smoother influence of the 13 months period which is dominant in the Kissmann model can be understood via the increased diffusion coefficient in the rarefaction region which allows more efficient particle scattering and leads to a more uniform distribution.

Apart from that the influence of the increased diffusion is visible in the 27-day variation of the Giacalone model. Close to the compression region it appears to enable more particles to penetrate or cross the diffusion barrier what creates the smooth transition between the undisturbed electron flux and the decrease caused by the compression region and the magnetic cutoff.

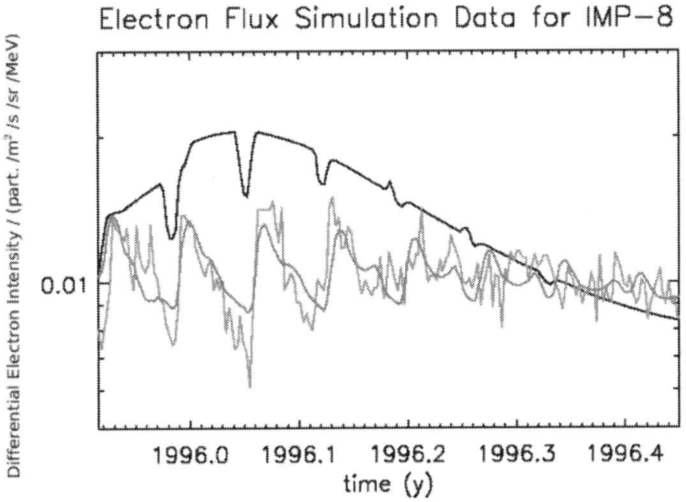

Figure 7. Daily measured cunting rates of the 2 – 12 MeV electron channel of IMP-8 (grey) plotted against the numerical solutions for the Kissmann model (black) and the Giacalone model (red). An offset is applied to the Giacalone results to allow for a better comparison of the three curves. Also the amplitude of the Giacalone results is intensified.

Both models allow for a quantitative treatment of major features in the velocity and magnetic field data as well as in the energetic electron counting rates. Minor features or finer details are reproduced qualitatively. Despite of their analytical simplicity the comparison of both models results allows to conclude that the influence of the rarefaction region's properties on the Jovian electrons is not negligible and needs further investigations.

References

[1] Smith E J and Wolfe J H 1976 *Geophysical Research Letters* **3** 137–140 1
[2] Richardson I G 2004 *Space Science Reviews* **111** 267–376 1
[3] Parker E N 1958 *ApJ* **128** 664 1
[4] Parker E N 1963 *Interplanetary dynamical processes.* 1, 2.1
[5] Kissmann R 2002 *Diploma Thesis. Ruhr-Universität Bochum* 1, 3.1
[6] Kissmann R, Fichtner H and Ferreira S E S 2004 *Astronomy and Astrophysics* **419** 357–363 1, 2.2, 3.1, 3.1, 3.2, 4
[7] Giacalone J, Jokipii J R and Kóta J 2002 *ApJ* **573** 845–850 1, 3.2, 3.2
[8] Blom J G and Verwer J G 1994 *Appl. Numer. Math.* **16** 129–156 1, 2.2
[9] Blom J G and Verwer J G 1996 *ACM Trans. Math. Softw.* **22** 329–347 1, 2.2
[10] Parker E N 1965 *Planet. Space Sci.* **13** 9 1, 2.2
[11] Sternal O 2010 *Ph.D. Thesis. Universität Kiel* 1, 2.2
[12] Forsyth R J and Marsch E 1999 *Space Science Reviews* **89** 7–20 2.1
[13] Wiengarten T, Kleimann J, Fichtner H, Kühl P, Kopp A, Heber B and Kissmann R 2014 *The Astrophysical Journal* **788** 80 (*Preprint* 1406.0293) 2.1
[14] Gazis P R 2000 *JGR* **105** 19–34 2.1
[15] Ferreira S E S 2002 *Ph.D. thesis, North-West University , South Africa* 3
[16] Teegarden B J, McDonald F B, Trainor J H, Webber W R and Roelof E C 1974 *JGR* **79** 3615 2.2

[17] Chenette D L 1980 *JGR* **85** 2243–2256 2.2

[18] Conlon T F and Simpson J A 1977 *Astrophysical Journal,* **211** L45–L49 2.2

[19] McDonald F B, Cline T L and Simnett G M 1972 *JGR* **77** 2213 2.2

[20] Simpson J A, Hamilton D, Lentz G, McKibben R B, Mogro-Campero A, Perkins M, Pyle K R, Tuzzolino A J and O'Gallagher J J 1974 *Science* **183** 306–309 2.2

[21] Ferrando P 1999 *International Cosmic Ray Conference* (*International Cosmic Ray Conference* vol 7) p 135 2.2

[22] Ferreira S E S, Potgieter M S, Burger R A, Heber B and Fichtner H 2001 *JGR* **106** 24979–24988 2.2, 3.1, 3.1, 3.1

[23] Fichtner H 2001 *Space Sc. Rev.* **95** 639–754 2.2

[24] Burger R A, Krüger T P J, Hitge M and Engelbrecht N E 2008 *ApJ* **674** 511–519 2.2

[25] Ferreira S E S, Potgieter M S, Burger R A, Heber B, Fichtner H and Lopate C 2001 *JGR* **106** 29313–29322 3.1

[26] Heber B, Bothmer V, Dröge W, Kunow H, Müller-Mellin R, Posner A, Ferrando P, Raviart A, Raviart A, Paizis C, McComas D, Forsyth R J, Szabo A and Lazarus A J 1997 *Correlated Phenomena at the Sun, in the Heliosphere and in Geospace* (*ESA Special Publication* vol 415) ed Wilson A p 331 4

The connection of the interplanetary magnetic field turbulence and rigidity spectrum of Forbush decrease of the galactic cosmic ray intensity

A. Wawrzynczak[1], M.V. Alania[2]

[1]Institute of Computer Sciences, Siedlce University, Poland,
[2]Institute of Mathematics and Physics, Siedlce University, Poland .

E-mail: awawrzynczak@uph.edu.pl, alania@uph.edu.pl

Abstract. We analyze the temporal changes in the rigidity spectrum of Forbush decrease (Fd) of the galactic cosmic ray (GCR) intensity observed in November 2004. We compute the rigidity spectrum in two energy ranges based on the daily data from the worldwide network of neutron monitors and Nagoya ground muon telescope. We demonstrate that the changes in the rigidity spectrum of Fd are linked to the evolution/decay of the interplanetary magnetic field (IMF) turbulence during various phases of the Fd. We analyze the time-evolution of the state of the turbulence of the IMF in various frequency ranges during the Fd. Performed analysis show that the decrease of the exponent ν of the Power Spectral Density ($PSD \propto f^{-\nu}$, where f is frequency) of the IMF turbulence with decreasing frequency lead to the soft rigidity spectrum of Fd for GCR particles with relatively higher energies.

1. Introduction

A fast decrease of the galactic cosmic ray (GCR) intensity during one-two days followed by its gradual recovery is called a Forbush decrease (Fd) [1]. Forbush decreases (Fds) are formed after outstanding flares on the Sun and intensive solar coronal mass ejecta (CME) [2, 3]. These features appear randomly without any regularity, increasing its frequency in declining phase of a solar sunspot cycle. Cane [3] divided Fds into three basic types: (1) caused by a shock and ejecta, (2) caused by a shock only, and (3) caused by the ejecta only. The amplitude of the Fd is defined as the difference between the GCR intensity at the onset and the minimum point of the Fd. A dependence of the Fd amplitude on the rigidity of GCR particles is one of its essential characteristics, called the rigidity spectrum of the Fd. Lots of papers are devoted to this problem. Cane [3] notes that, if the power law $R^{-\gamma}$ gives the rigidity dependence of the amplitude of the Fds, then γ ranges from about 0.4 up to 1.2. In [4] authors described a new methodology for the study of the rigidity dependence of the transient modulations with use of median rigidity of response of a detector. Their results confirm the power law rigidity dependence of the Fds amplitudes with the negative exponents [5]. In [6] was studied a dependence of the recovery time of Fd on the energy of cosmic rays. Authors found that Fds with magnitudes exceeding 10% show an energy dependence of the recovery time while Fds with lesser amplitudes show no energy dependence. We are confident that an examination of the rigidity dependence of Fd only through the amplitudes for one point of the GCR intensity minimum is not sufficient. So, we consider to study dynamics of the Fd as a vital to establish a

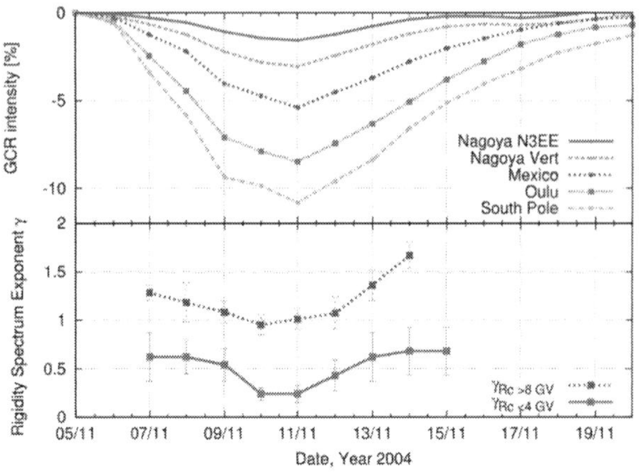

Figure 1. *Top panel*: Three day averaged temporal changes of the GCR intensity for the Mexico, Oulu and South Pole neutron monitors and Vertical and N3EE channels of Nagoya muon telescope during the Fd. *Bottom panel*: Temporal changes in the rigidity spectrum exponent γ based on the data of the stations divided into two groups according to their cut-off rigidities.

rigidity dependence of the GCR intensity in different phases of the Fd. This type of study was presented in [7, 8, 9, 10, 11] and references therein. In these papers it has been shown that a rigidity spectrum $\delta D(R)/D(R) \propto R^{-\gamma}$ of the vast majority of the Fds gradually hardens during the decreasing and minimum phases of the Fd and gradually softens in the recovery phase of the Fd. So, the exponent γ of the rigidity spectrum of the Fd is in the range $\gamma \in [1, 1.6]$ at the beginning phase of the Fd, then it gradually decreases up to the minimum (or near minimum) of the GCR intensity, when it takes the lowest value $\gamma \in [0.2, 0.6]$ and increase again during the recovery phase of the Fd. Moreover, the time profiles of the Fds rigidity spectrum are associated with the changes of the power spectral density (PSD) of the interplanetary magnetic field (IMF) turbulence ($PSD = P(\frac{f}{f_0})^{-\nu}$, where P is power, f is frequency and f_0 is normalization frequency) [7, 8, 9, 10, 11]. Specifically, changes in the exponent γ of the power law rigidity R spectrum explicitly depends on the changes of the exponent ν of the PSD in the range of frequency $f \in [10^{-6}, 10^{-5}]$ Hz of the IMF turbulence, to which neutron monitors and ground muon telescopes respond. This relation can be deduced based on the dependence of the diffusion coefficient K_{II} of GCR particles on the rigidity R, as $K \propto R^{2-\nu}$, where ν is exponent of the PSD of the IMF turbulence e.g.[12, 13, 14].

The relation between the exponents γ and ν entails expectation of the rigidity dependence of the exponent γ arising from the changes in the exponent ν versus frequency. However, to reveal the rigidity dependence of γ it is a difficult task for Fd due to its complexity. In [8] it was presented that during the Fd in September 9-23, 2005 is observed clearly recognized dependence of the exponent γ of the rigidity spectrum of the Fd on the rigidity of GCR particles, when the exponent γ is larger, the larger are the cut-off rigidities of stations used in its calculations; i.e. the rigidity spectrum of the GCR intensity variations during the Fd is hard for lower energy range and is soft for the higher energy range. Correspondingly, the consistent frequency dependence of the exponent ν of the PSD of the IMF components during the Fd was observed.

In this paper, we reinforce the existence of the rigidity dependence of the exponent γ during the Fd by the detailed examination of the Fd in November 2004. We also evaluate the temporal evolution of the state of the turbulence during the Fd by the calculation of the PSD of the running series of IMF components. The gradual increase of the exponent ν during the Fd is clearly manifested. Moreover, we show that the decrease of the exponent ν with decreasing frequency lead to the soft rigidity spectrum of the Fd for GCR particles with higher rigidities.

2. Experimental data and methods

The Sun's activity in November 2004 revealed a series of consecutive CMEs appearing from 4 to November, 10. They were the result of solar flares in the active region of the Sun marked as 10696. The first class C6 flare occurred on November 6, the outcome was a halo CME, which reached the Earth on November 7. According to NOAA, in the following days the halo or part-halo CMEs appeared almost daily [15]. The result was 2-3 shock waves spreading at the same time. The strongest X2.8 solar flare occurred on November, 10 and reached the Earth on November 12 in the form of CME. The highest solar wind speed was 700-800 km/s and the magnetic field strength 45 nT. The result of this activity was a Fd of the GCR intensity observed at the Earth by neutron monitors and ground muon telescopes in the period of November 5-20, 2004.

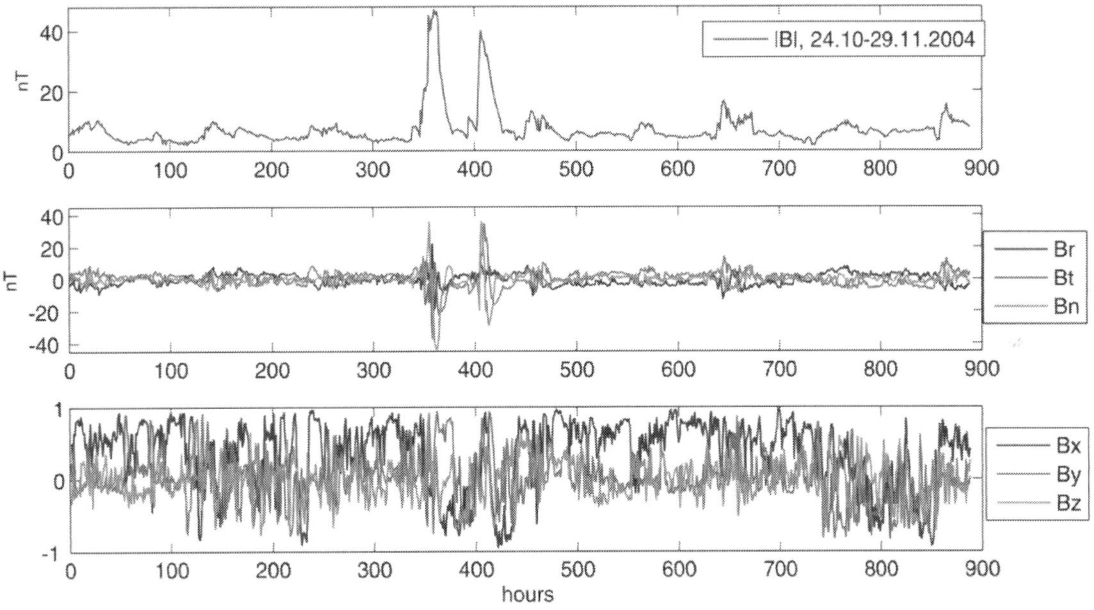

Figure 2. The IMF strength (top panel) and its components in the RTN (middle panel) [18] and in the mean field (bottom panel) coordinate system for the period of October 24 - November 29, 2004.

2.1. Rigidity spectrum of the Fd

The rigidity spectrum of the Fd in November 5-20, 2004 was earlier examined in [7]. We have revealed that the rigidity spectrum of this Fd is soft at the beginning of the Fd ($\gamma \approx 1.15 \pm 0.3$), then it slowly becomes hard during the declining and minimum phases of the Fd ($\gamma \approx 0.7 \pm 0.15$) and gradually becomes soft ($\gamma \approx 1.35 \pm 0.32$) in the recovery phase. Currently, we investigate whether we observe the rigidity dependence of the rigidity spectrum exponent γ for this Fd. For this purpose, we calculate the rigidity spectrum of the Fd based on the GCR intensity from

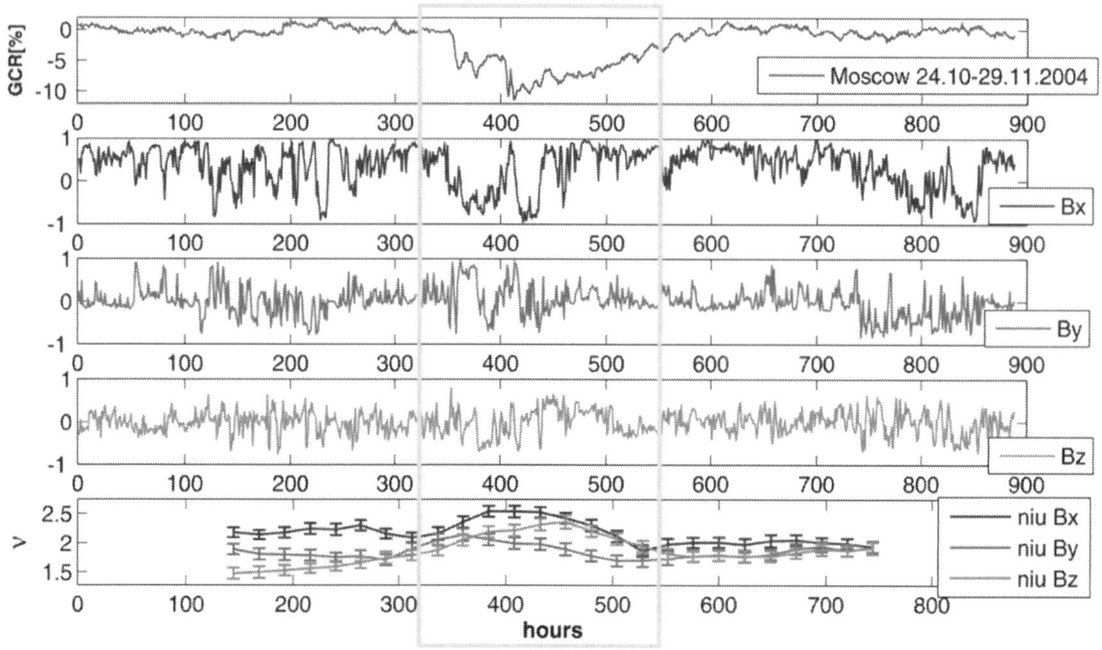

Figure 3. The hourly changes of the GCR intensity measured by the Moscow neutron monitor, IMF components in the mean field coordinate system and the values of the exponent ν of the PSD for IMF components based on the 12 days running series of hourly IMF data in October 24 - November 29, 2004.

neutron monitors and the Nagoya ground muon telescope data divided into two groups according to their cut-off rigidities: for low and high cut-off rigidities. In the group of stations with low ($R_c < 4GV$) cut-off rigidities were stations: Calgary, Cape Smidth, Kergelungen, McMurdo, Oulu, South Pole, whereas in the group of station with high cut-off ($R_c > 8GV$) rigidities were: Haleakala, Mexico and following channels of Nagoya muon telescope: N0VV, N1EE, N3EE, N4NE, N4SE.

The top panel in Fig. 1 presents changes in the GCR intensity smoothed over three days measured by Mexico, Oulu and South Pole neutron monitors, and N3EE channel of Nagoya ground muon telescope. We consider smoothed data in order to reveal reliable average temporal changes in the rigidity spectrum exponent γ and its energy dependence. The rigidity spectrum exponent γ of the power law rigidity spectrum was found using the expression e.g. [16, 17]:

$$\frac{\delta D(R)}{D(R)} = \begin{cases} AR^{-\gamma} & R \leq R_{max} \\ 0 & R > R_{max}, \end{cases} \tag{1}$$

R_{max} designates the upper limiting rigidity beyond which the Fd of the GCR intensity disappears (200 GV). The daily average amplitudes J_i^k of the Fd for the 'i-th' detector (neutron monitor or muon telescope) were calculated, as: $J_i^k = (N_k - N_0)/N_0$; N_k is the running daily average count rate (k =1,2,3,...,days) and N_0 is the 3 days average count rate before the Fd (3-5 November). Based on the method in detail described in [8] we calculated the rigidity spectrum exponent γ of Fd for each day of November 5-20, 2004. The bottom panel in Fig. 1 presents temporal changes in rigidity spectrum exponent γ (k=1,2,3,...,days). The exponent γ was calculated for two groups of stations with $R_c < 4GV$ and $R_c > 8GV$. Fig. 1 shows that time profile of γ is approximately the same for both cut-off rigidity groups. However, the values of γ are larger for the group with higher

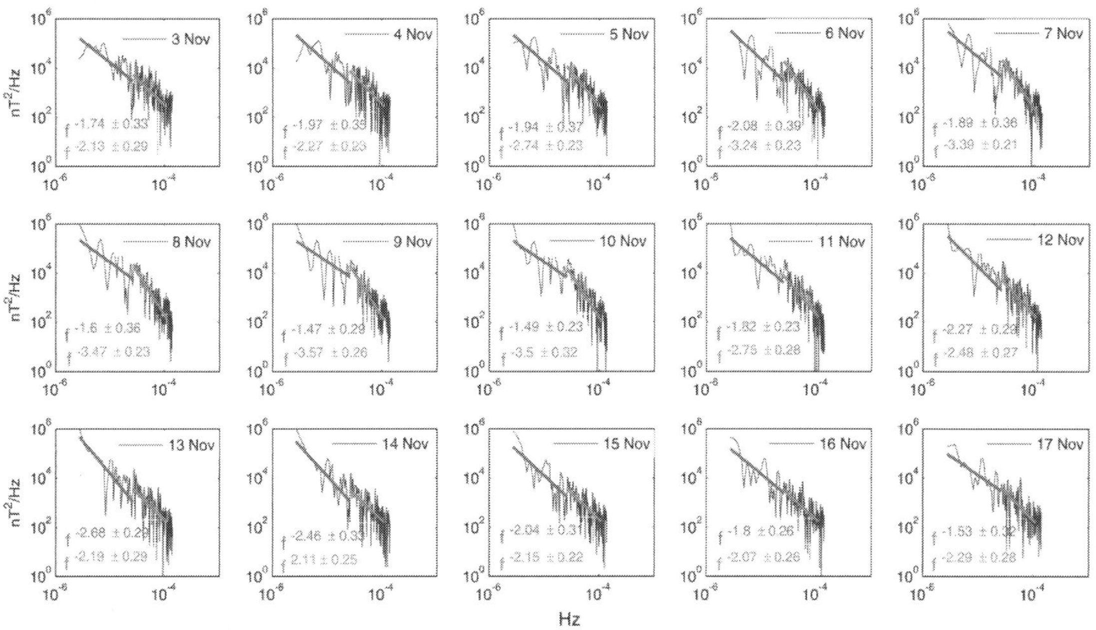

Figure 4. The PSD for the IMF component Bz based on the 12 days running series of hourly data in November 2004.

cut-off rigidities. The value of $\gamma_{R_c>8GV}$ changes from $\gamma_{R_c>8GV} = 1.28 \pm 0.08$ at the beginning of the Fd down to $\gamma_{R_c>8GV} = 0.95 \pm 0.1$ in the minimum of the Fd, and increases again in the recovery phase up to $\gamma_{R_c>8GV} = 1.67 \pm 0.13$. In turn, the exponent $\gamma_{R_c<4GV}$ estimated based on the low cut-off stations data is smaller by factor of 0.6, changing from $\gamma_{R_c<4GV} = 0.62 \pm 0.25$, through $\gamma_{R_c<4GV} = 0.24 \pm 0.19$ up to $\gamma_{R_c<4GV} = 0.68 \pm 0.24$. Consequently, we can conclude that rigidity spectrum of Fd is hard for the lower energy range and is soft for the upper energy range. This conclusion is in agreement with results obtained in [8] for Fd in September 2005.

2.2. Power spectral density of the IMF during Fd

Hitherto in [7, 8, 9, 10] was presented that during the Fd we observe an increase of the exponent ν of the PSD of the IMF components. However, it was done based on the comparison of the single data series with the length predetermined by the duration of the Fd; i.e. the PSDs were calculated for three periods, before, during and after the Fd. This results from the requirement of employing relatively long data series to attain the values of the exponent ν in low frequency range ($f \in [10^{-6}, 10^{-5}]$) responsible for the scattering of the GCR particles to which neutron monitor and ground meson telescopes respond. In this paper, we analyze the time-evolution of the state of the turbulence of the IMF when approaching the Fd. In the calculations, we used the hourly data of the IMF components recorded by the ACE spacecraft [18] for the period of October 20, 2004 - November 29, 2004 (Fig. 2). Before the calculation of the PSD of the IMF components, we convert the IMF from RTN to the mean field reference system [19]. The mean field reference system reduces the problem of cross-talking between the components, because the interplanetary magnetic field is not oriented along the axes of the reference system in which spacecraft perform the measurement. In consequence, any component will experience a contribution from the other ones. Fig. 2 presents the hourly changes in the IMF components in the RTN and the mean field reference system in the considered period. The mean field reference system is oriented as follows: the direction Bx is parallel to the mean field direction, Bz is perpendicular to Bx and radial direction and the By completes the system.

To estimate the PSD of the IMF turbulence in the frequency range responsible for the modulation of the GCR particles registered by neutron monitors and muon telescopes we use the 12-day long series of the 1-hour data (288 hours). As a result, we obtain one value of ν for each IMF component. To note the time changes of the exponent ν we estimated the PSD for 12 days running series of data shifted by 24-hours. The results of these calculations are presented in Fig. 3. The top panel in the Fig. 3 gives the corresponding changes of the GCR intensity registered by the Moscow neutron monitor, the three following panels provide the IMF components used in these calculations. The bottom panel in Fig. 3 presents the obtained values of the exponent ν of the PSD ($PSD \propto f^{-\nu}$). We assigned value of ν in the middle of the data series used in its calculation. One can see the gradual increase in the exponent ν for all IMF components when we approach the period of the Fd (marked in Fig. 3 by yellow box), and then its gradual decrease when Fd is over. We can also observe some discrepancy between the time profile of ν for Bx, By and Bz components. This divergence suggests that the structure of the IMF turbulence was inhomogeneous and was developing differently in parallel and perpendicular directions.

Detailed analysis of bottom panel of Fig. 3 shows that the real structure of the turbulence of the IMF during the Fd can be considered as a 3-D, being a strongly anisotropic and inhomogeneous. However, if it is the case describing a scattering of the GCR particles by the slab/2-D approximation of the IMF turbulence scarcely can be valid [21]. We assume that during the Fd the 3-D turbulence of the IMF is stipulated by fluctuations of the Alfvenic waves in all three spatial directions. The Bx and By components are represented by both- regular and fluctuating parts of the IMF, while Bz component involves only fluctuations. Owing to this, in the $\overrightarrow{z}$ direction the turbulence of the IMF is completely realized by Bz component. During the Fd the enhanced Bz component is a source of a high-level turbulence with preferentially existing the large inhomogeneities (i.e. when the exponent ν is high).

We have also estimated the slope of the PSD of the Bz component in two frequency intervals, the first - I-$[2.7 \times 10^{-6} Hz, 5 \times 10^{-5} Hz]$, and the second one II-$[5 \times 10^{-5} Hz, 10^{-4} Hz]$. Fig. 4 presents that during the main phase of the Fd (6-13 November) we observe the significant increase of the slope in the higher frequency range.

2.3. Relationship between the exponents γ and ν during the Fd

The comparison of the time profiles of the rigidity spectrum of the Fd exponent γ and the exponent ν of the PSD of the IMF components during the Fd in November 2004 is presented in Fig. 5. Fig. 5 confirms that during the Fd in November 2004 we observe a gradual decrease of the exponent γ simultaneously with the progressive increase of the exponent ν, and then its return to the initial level. The correlation coefficient between all variables are presented in Table 1. The various values of correlation for the IMFs components suggests that during this Fd the IMF turbulence was propagating inhomogeneously in space. However, in this case the strongest anticorrelation is recognized for the exponent ν for Bz component of the IMF, the weakest for the By component.

Table 1. Correlation coefficient between the changes of the exponent γ and ν during Fd in November 4-15, 2004.

Correlation Coefficient	ν_{Bx}	ν_{By}	ν_{Bz}
$\gamma_{R_c>8GV}$	-0.88 ± 0.02	-0.46 ± 0.23	-0.94 ± 0.01
$\gamma_{R_c<4GV}$	-0.57 ± 0.16	-0.21 ± 0.32	-0.85 ± 0.03

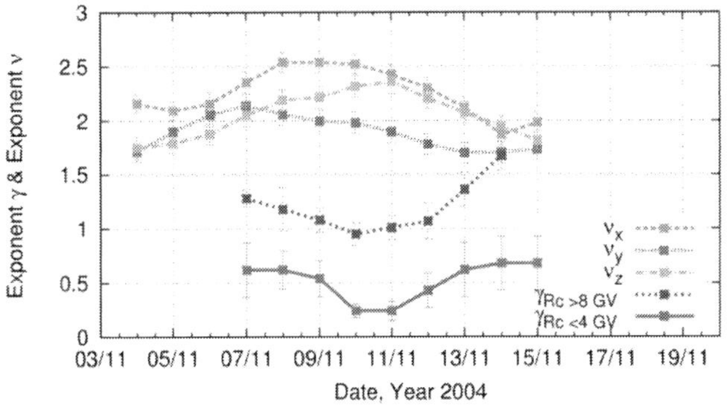

Figure 5. Temporal changes in the exponent γ and ν during Fd in November 4-15, 2004.

The robust relation between the exponents γ and ν is also confirmed by changes of the exponent ν versus frequency presented in Fig. 4. Taking into account the average, for the considered period, solar wind velocity $V = 600$ km/s and the IMF magnitude $B = 5 \times 10^{-5}$ Gs the lower frequency range I is responsible for modulation of the GCR particles with rigidities in the range $[10GV, 60GV]$ and higher frequency range II for modulation of the GCR particles with rigidities in $[5GV, 15GV]$ [20]. Thus, in the lower frequency range in which GCR particles with higher energies are modulated we have smaller exponent ν, than in the upper frequency range in which are modulated lower energy particles. This is in coincidence with the energy dependence of the exponent γ of the rigidity spectrum (Fig. 1); i.e. the growth of the exponent γ for higher energy GCR particles. This evident anticorrelation reflect the role of the exponent ν of the PSD of the IMF in creating the dependence of the amplitudes of Fd on rigidity of GCR particles, i.e. a manifestation of the role of ν in the formation of the rigidity spectrum (changes in γ).

The proved relationship between exponents γ and ν was found from the expectation that the diffusion coefficient K_{II} of GCR particles depends on the rigidity R, as $K \propto R^{2-\nu}$, according to QLT e.g.[12, 13, 14] . Furthermore, the dependence of the parallel diffusion coefficient K_{II} on ν entails a dependence of the perpendicular diffusion coefficient $K_\perp$ on R, as far $K_\perp$ is proportional to K_{II} : $\frac{K_\perp}{K_{II}} = \frac{1}{1+(\omega\tau)^2}$, where ω is particle angular velocity and τ is the time between two sequence GCR particles collisions. For the GCR particles to which neutron monitor respond the $\omega\tau$ can change in the range $3 \leq \omega\tau \leq 5$. In case of a more complicated expression for $\omega\tau$ one can infer that the character of K_{II} is hidden in the $K_\perp$.

3. Conclusion

(i) We presented an apparent dependence of the exponent γ of the rigidity spectrum of the Fd in November 2004 versus the GCR particles energy. The rigidity spectrum exponent γ is the larger, the higher are cut-off rigidities of stations used in calculations.

(ii) We estimated the daily time evolution of the exponent ν of the PSD of the IMF components throughout the Fd. Results show the gradual increase of the exponent ν up to the main phase of the Fd and then its gradual return to the initial level.

(iii) We have confirmed the clear anticorrelation between the exponents γ and ν during the Fd.

(iv) The presented dependence of the exponent ν of the PSD of the IMF turbulence upon frequency during the Fd strengthen the relationship between the rigidity spectrum exponent γ and the exponent ν.

Acknowledgments
We are grateful to the Principal Investigators of the worldwide network of neutron monitors, Nagoya muon telescope and Advanced Composition Explorer spacecraft for access to the data.

References

[1] Forbush S.E. 1937 *Phys. Rev.*, **51** , 11081109
[2] Burlaga L.F. 1995 *Interplanetary Magnetohydrodynamics*, (New York: Oxford Univ. Press)
[3] Cane H.V. 2000 *Space Sci. Rev.*, **93**, 5577
[4] Ahluwalia H.S. and M. M. Fikani 2007 *J. Geophys. Res.*, **112** , A08105
[5] Ahluwalia H.S., R.C. Ygbuhay and M.L. Duldig 2009 *Adv. Space Res.* ,**44**, 58-63
[6] Usoskin I.G., Braun I. and Gladysheva O.G. et al. 2008 *J. Geophys. Res.*, **113** , A7
[7] Alania M.V. and Wawrzynczak A. 2008*Astrophys. And Space Sci. Transactions*, **4** , 59-63
[8] Alania M.V. and A. Wawrzynczak 2012 *Adv. in Space Res.* **50** 725-730
[9] Wawrzynczak A. and Alania, M.V.2008 *Adv. in Space Res.* **41** 2, 325-334
[10] Wawrzynczak A.and Alania M.V. 2010 Adv. *Adv. Space Res.*, **45**, 5, 622-631
[11] Alania M.V., A. Wawrzynczak, V. E. Sdobnov and M. V. Kravtsova 2013 *Solar Phys.* **286** 561-576
[12] Jokipii J.R. 1966 *Astrophys. J.*, **146**, 480-487
[13] Toptygin I.N. 1985 *Cosmic rays in interplanetary magnetic fields*, (Reidel Publishing Company)
[14] Shalchi A. 2009 *Nonlinear Cosmic Ray Diffusion Theories*, (Springer)
[15] $http://www.solen.info/solar/old_reports/2004/november/20041107.html$
[16] Ahluwalia H.S. and J.H. Ericksen 1971 *J. Geophys. Res.*, **76** 6613-6623
[17] Dorman L.I. 2004 *Cosmic Rays in the Earth's Atmosphere and Underground*, (Dordrecht: Kluwer Academic Publishers)
[18] *http://www.srl.caltech.edu/ACE/ASC/level2/index.html*
[19] Bruno R and V. Carbone 2013 *Living Rev. Solar Phys* **10**
[20] Jokipii J. R. and P.J. Coleman 1968 *J. Geophys. Res.*, **73**, 17, 54955503
[21] Bieber, J.W., W. H., Mathaeus, C. W., Smith, et al. 1994 *Astrophys. J.*, 420, 294-306

A stochastic method of solution of the Parker transport equation

A. Wawrzynczak[1], **R. Modzelewska**[2], **A. Gil**[2]
[1]Institute of Computer Sciences, Siedlce University, Poland,
[2]Institute of Mathematics and Physics, Siedlce University, Poland.
E-mail: awawrzynczak@uph.edu.pl, renatam@uph.edu.pl, gila@uph.edu.pl

Abstract. We present the stochastic model of the transport of galactic cosmic ray (GCR) particles in the heliosphere. Based on a solution of the Parker transport equation we developed models of the short-term variation of the GCR intensity, i.e. the Forbush decrease (Fd) and the 27-day variations of the GCR intensity. Parker's transport equation, being the Fokker-Planck type equation, delineates non-stationary transport of charged particles in a turbulent medium. The presented approach of the numerical solution is grounded on solving the set equivalent stochastic differential equations (SDEs). We demonstrate the method of deriving from Parker's transport equation the corresponding SDEs in the heliocentric spherical coordinate system for the backward approach. Features indicative of the preeminence of the backward approach over the forward are stressed. We compare the outcomes of the stochastic models of the Fd and 27-day variation of the GCR intensity with our former models established by the finite difference method. Both models are in agreement with the experimental data.

Introduction

The first stochastic equation (Langevin's) was linked to Newton's principle [1]. From the beginning of the XX century, the stochastic approach became more useful for describing random physical processes. Stochastic differential equations (SDEs), in conjunction with various Monte Carlo technics, are broadly used in many fields, such as physics, finance, biology, chemistry, engineering, or management science. The main statistical characteristic is the representation of the solution of the Fokker-Planck type equation as a probability density distribution. Employment of probabilistic description with Monte Carlo simulations allows to reduce the solution of the partial differential equation (PDE) describing the analyzed phenomena to the integration of SDEs.

We apply stochastic methodology to model the galactic cosmic rays (GCR) transport in the heliosphere. Algorithms used in the particle transport simulations are mainly based on finite difference methods (e.g. [2]). Galerkin methods are used for solving time-dependent high order PDEs (e.g. [3]). The homotopy perturbation method [4] was recently developed for the numerical solution of various linear and nonlinear PDEs. Also, there exists an approach to the PDE solution grounded on particle methods, characterized by low numerical dispersion [5].

During the propagation through the heliosphere, GCR particles are modulated by the solar wind and heliospheric magnetic field (HMF). Modulation of the GCR is a result of an action of four primary processes: convection by the solar wind, diffusion on irregularities of HMF, particles drifts in the non-uniform magnetic field and adiabatic cooling. Transport of the GCR particles in the heliosphere can be described by the Parker transport equation [6]:

$$\frac{\partial f}{\partial t} = \vec{\nabla} \cdot (K_{ij}^S \cdot \vec{\nabla} f) - (\vec{v}_d + \vec{U}) \cdot \vec{\nabla} f + \frac{R}{3}(\vec{\nabla} \cdot \vec{U})\frac{\partial f}{\partial R}, \tag{1}$$

where $f = f(\vec{r}, R, t)$ is an omnidirectional distribution function of three spatial coordinates $\vec{r} = (r, \theta, \varphi)$, particles rigidity R and time t; $\vec{U}$ is the solar wind velocity, $\vec{v}_d$ the drift velocity, and K_{ij}^S is the symmetric part of the diffusion tensor of the GCR particles.

Employing the stochastic approach to solving the Parker transport equation is not the latest idea [7, 8]. However, the majority of models presented in the literature are used to determine the simulated spectra and compare it with experimental observations carried out by space probes such as Voyagers, AMS, BESS and PAMELA (e.g. [9]-[13]). In this paper, we present models in which we do not only reproduce the proton spectra but also simulate the short- time GCR variations i.e. the Forbush decrease (Fd) and 27-day variation. Additionally, we compare the results of the stochastic modeling with our previous well grounded models, developed by using the finite difference method (FDM) to solve the Parker transport equation (e.g. [14]-[16]). To perform the reliable comparison between the two models, we consider the same coefficients and parameters as in the baseline Parker transport equation. Furthermore, the parameters included are obtained by approximation of the experimental data.

The aim of our paper is twofold: at first, to compose a consistent mathematical model of the GCR transport in the heliosphere by means of the SDEs. And then, to employ the created model to simulate the short-term variations being in an agreement with the experimental data.

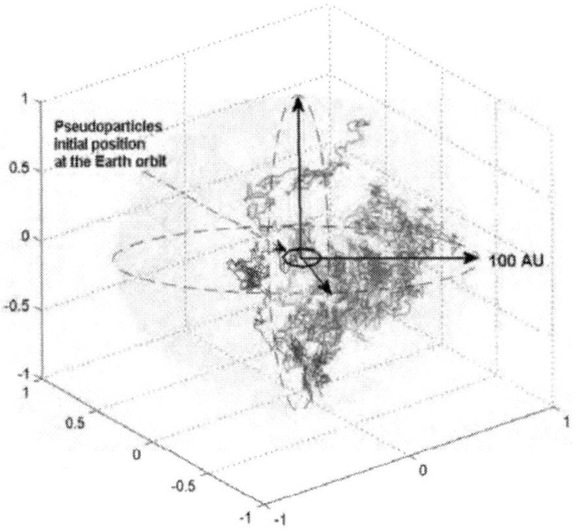

Figure 1. The sample pseudoparticles trajectories within the heliosphere.

2. The stochastic approach

In order to model the GCR transport in the heliosphere using stochastic methods equivalent SDEs must be obtained. In the first step, the Parker transport equation (Eq. 1) must be converted to a general form of the Fokker-Planck equation (FPE). Depending on the direction of integration, FPE must be expressed in two forms [17],

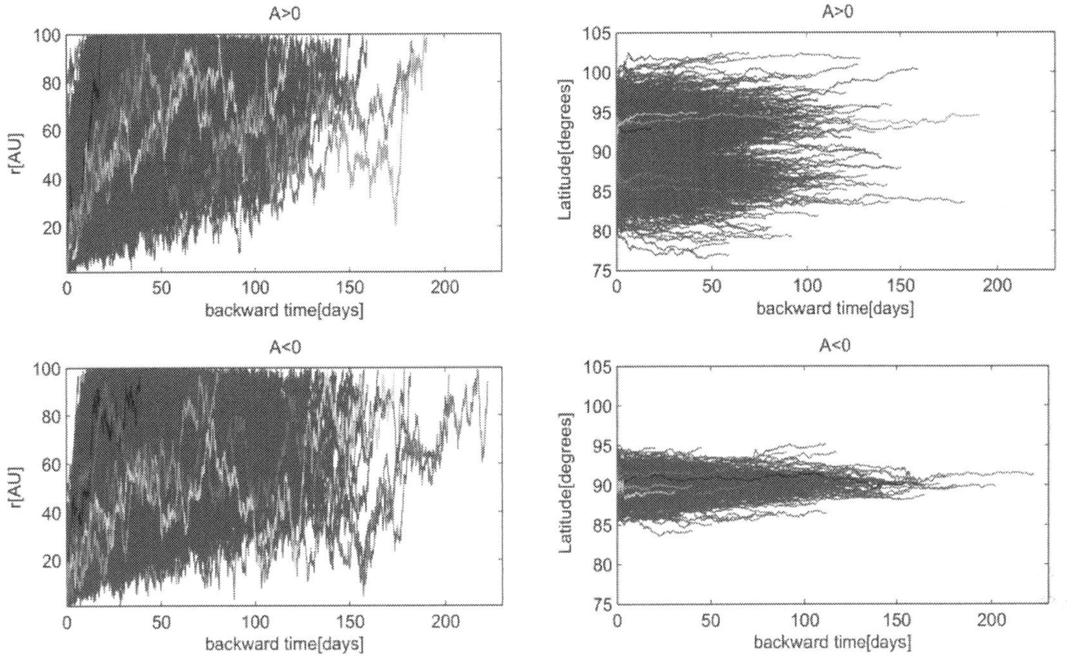

Figure 2. Trajectories of the pseudoparticles with rigidity 10 GV for the A>0 and A<0 solar magnetic cycle. The specific colors highlight the trajectories of the sample pseudoparticles traced backward in time from the heliosphere boundary until they reach the position $r = 1AU$, $\theta = 90°$, $\varphi = 180°$.

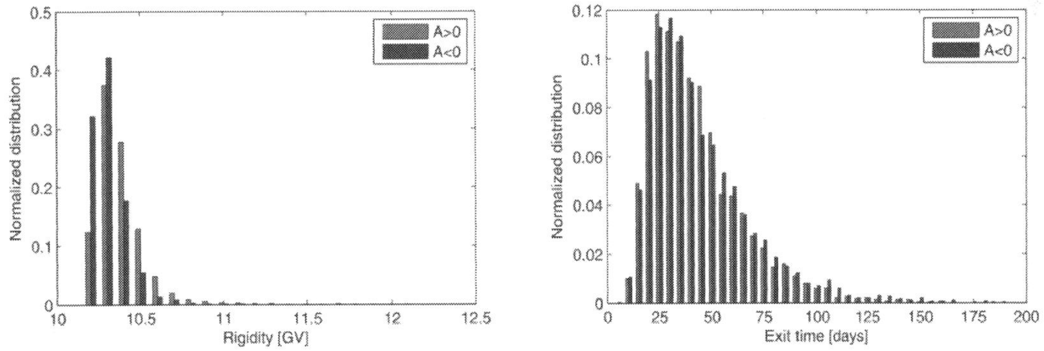

Figure 3. The histograms of the particles rigidity and exit time for the pseudoparticles initialized with rigidity 10 GV from position $r = 1AU$, $\theta = 90°$, $\varphi = 180°$ for the A>0 and A<0 solar magnetic cycle.

the time-backward:

$$\frac{\partial F}{\partial t} = \sum_i A_i \frac{\partial F}{\partial x_i} + \frac{1}{2} \sum_{i,j} B_{ij} B^T_{ij} \frac{\partial^2 F}{\partial x_i \partial x_j}, \tag{2}$$

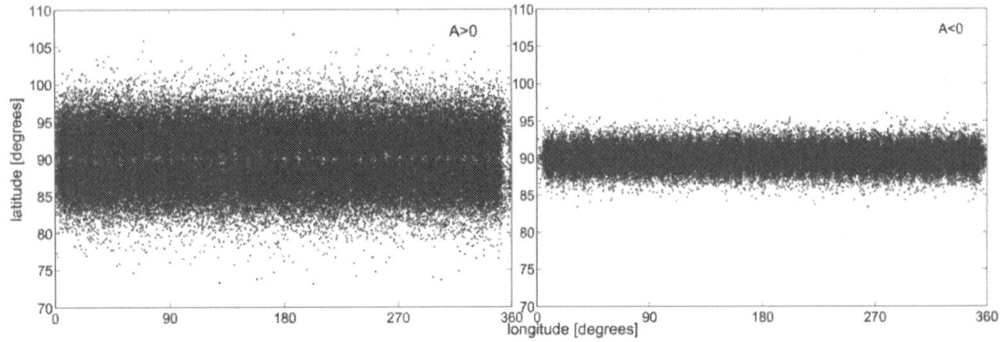

Figure 4. Latitude vs. longitude distribution of simulated pseudoparticles (protons) for the $A > 0$ and $A < 0$ solar magnetic cycle.

and the time-forward:

$$\frac{\partial F}{\partial t} = \sum_i \frac{\partial}{\partial x_i}(A_i \cdot F) + \frac{1}{2}\sum_{i,j} \frac{\partial^2}{\partial x_i \partial x_j}(B_{ij}B_{ij}^T \cdot F). \tag{3}$$

SDEs equivalent to Eq. 2 and 3 has the form (e.g. [17]):

$$d\vec{r} = \vec{A}_i \cdot dt + B_{ij} \cdot d\vec{W}, \tag{4}$$

where $\vec{r}$ is the individual pseudoparticle trajectory in the phase space and dW_i is the Wiener process, commonly written as $dW_i = \sqrt{dt} \cdot dw_i$, with dw_i being the randomly fluctuating term with Gaussian distribution.

To solve Eq. 4 in both cases (backward and forward), at the onset, we initiate pseudoparticles at some starting point in space and time and integrate alongside the pseudoparticles trajectories until they reach the boundary. Choosing between the forward or backward in time approach we need to bear in mind the problem that we want to solve. In the time-forward integration, pseudoparticles start from diverse boundary points, being for the GCR particles the entrance to the heliosphere. After that, their trajectories are traced up to the target position, e.g. 1 AU. Thus, a high number of pseudoparticles has to be initialized in order to obtain a robust statistic because plenty of them do not reach the target position. The time-backward integration is much more effective in the case of the GCR propagation in the heliosphere. In the time-backward approach, the number of 'useless' particles is reduced. Pseudoparticles start from point of interest (e.g. 1 AU) and are traced backward in time until crossing the heliospheric boundary (in this paper this boundary is assumed at 100 AU, Fig. 1). During their travel throughout the heliosphere the pseudoparticles gain/lose their energy/rigidity proportionally to their travel time. In result the value of the particle distribution function, $f(\vec{r}, R)$, for the starting point can be found as an average of $f_{LIS}(R)$ values for pseudoparticles characteristics at the entry positions, $f(\vec{r}, R) = \frac{1}{N}\sum_{n=1}^{N} f_{LIS}(R)$, where $f_{LIS}(R)$ is the cosmic ray local interstellar spectrum taken as in [18] for rigidity R of the nth particle at the exit/entrance point.

The Parker equation (Eq. 1), in the 3-D spherical coordinate system (r, θ, φ), as the time-backward FPE diffusion equation, has the form:

$$\frac{\partial f}{\partial t} = a_1\frac{\partial^2 f}{\partial r^2} + a_2\frac{\partial^2 f}{\partial \theta^2} + a_3\frac{\partial^2 f}{\partial \varphi^2} + a_4\frac{\partial^2 f}{\partial r \partial \theta} + a_5\frac{\partial^2 f}{\partial r \partial \varphi} + a_6\frac{\partial^2 f}{\partial \theta \partial \varphi} + a_7\frac{\partial f}{\partial r} + a_8\frac{\partial f}{\partial \theta} + a_9\frac{\partial f}{\partial \varphi} + a_{10}\frac{\partial f}{\partial R} \tag{5}$$

where coefficients have the form:

$a_1 = K_{rr}^S, a_2 = \frac{K_{\theta\theta}^S}{r^2}, a_3 = \frac{K_{\varphi\varphi}^S}{r^2 sin^2\theta}, a_4 = \frac{2K_{r\theta}^S}{r}, a_5 = \frac{2K_{r\varphi}^S}{r sin\theta}, a_6 = \frac{2K_{\theta\varphi}^S}{r^2 sin\theta},$

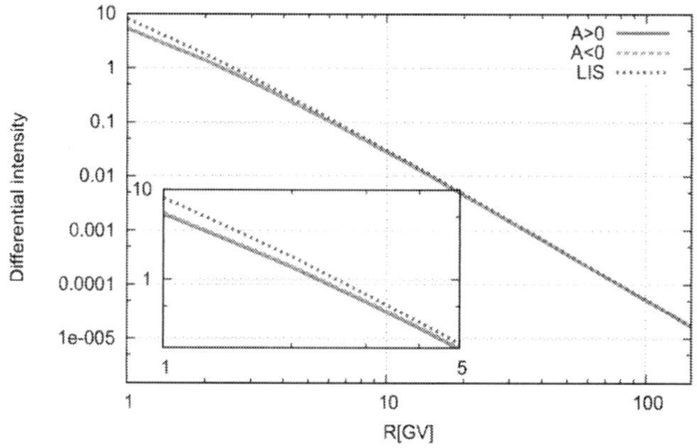

Figure 5. Simulated galactic protons rigidity spectra at Earth with respect to the LIS (blue dotted line) at 100 AU for two cases: the $A > 0$ (red solid line) and the $A < 0$ (green dashed line) polarity epochs.

$$a_7 = \frac{2}{r}K^S_{rr} + \frac{\partial K^S_{rr}}{\partial r} + \frac{ctg\theta}{r}K^S_{\theta r} + \frac{1}{r}\frac{\partial K^S_{\theta r}}{\partial \theta} + \frac{1}{r sin\theta}\frac{\partial K^S_{\varphi r}}{\partial \varphi} - U - v_{d,r},$$
$$a_8 = \frac{K^S_{r\theta}}{r^2} + \frac{1}{r}\frac{\partial K^S_{r\theta}}{\partial r} + \frac{1}{r^2}\frac{\partial K^S_{\theta\theta}}{\partial \theta} + \frac{ctg\theta}{r^2}K^S_{\theta\theta} + \frac{1}{r^2 sin\theta}\frac{\partial K^S_{\varphi\theta}}{\partial \varphi} - \frac{1}{r}v_{d,\theta},$$
$$a_9 = \frac{K^S_{r\varphi}}{r^2 sin\theta} + \frac{1}{r sin\theta}\frac{\partial K^S_{r\varphi}}{\partial r} + \frac{1}{r^2 sin\theta}\frac{\partial K^S_{\theta\varphi}}{\partial \theta} + \frac{1}{r^2 sin^2\theta}\frac{\partial K^S_{\varphi\varphi}}{\partial \varphi} - \frac{1}{r sin\theta}v_{d,\varphi},\ a_{10} = \frac{R}{3}\nabla\cdot U.$$

The drift velocity is calculated as: $v_{d,i} = \frac{\partial K^{(A)}_{ij}}{\partial x_j}$ [20], where $K^{(A)}_{ij}$ is the anisotropic part of the full 3D anisotropic diffusion tensor of the GCR particles $K_{ij} = K^{(S)}_{ij} + K^{(A)}_{ij}$ containing the symmetric $K^{(S)}_{ij}$ and antisymmetric $K^{(A)}_{ij}$ parts presented in [21].
The equivalent to Eq. 5 set of SDEs, with matrix B_{ij}, $(i, j = r, \theta, \varphi)$, has the following form (the same can be found in [22]):

$$\begin{aligned} dr &= a_7\cdot dt + [B\cdot dW]_r \\ d\theta &= a_8\cdot dt + [B\cdot dW]_\theta \quad (6) \\ d\varphi &= a_9\cdot dt + [B\cdot dW]_\varphi \\ dR &= a_{10}\cdot dt, \end{aligned} \qquad B_{ij} = \begin{bmatrix} \sqrt{2a_1} & 0 & 0 \\ \frac{a_4}{\sqrt{2a_1}} & \sqrt{2a_2 - \frac{a_4^2}{2a_1}} & 0 \\ \frac{a_5}{\sqrt{2a_1}} & \frac{a_6 - \frac{a_4 a_5}{2a_1}}{B_{\theta\theta}} & \sqrt{2a_3 - B^2_{\varphi r} - B^2_{\varphi\theta}} \end{bmatrix}.$$

The Euler-Maruyama scheme is used in order to integrate Eq. 6 backward in time. To solve Eq. 6 in spherical coordinates system we are using boundary conditions of the form: $\varphi_i < 0 \rightarrow \varphi_i = \varphi_i + 2\pi$; $\varphi_i > 2\pi \rightarrow \varphi_i = \varphi_i - 2\pi$; $\theta_i < 0 \rightarrow \theta_i = -\theta_i$ & $\varphi_i = \varphi_i \pm \pi$ and $\theta_i > \pi \rightarrow \theta_i = 2\pi - \theta_i$ & $\varphi_i = \varphi_i \pm \pi$ [22]. The reflecting boundary is considered to be the inner radial boundary, $\frac{\partial f}{\partial r} = 0$ at $r = 0.001$ AU. An empty heliosphere constitutes an initial condition, $f_i(0.01AU < r < 100AU, \theta, \varphi, R, 0) = 0$, as was shown in [11]. the performed tests proved that the 3000 pseudoparticles were enough to simulate the short-time variations of GCR particles with rigidity 10 GV.

Fig. 2 shows the trajectories of simulated pseudoparticles with rigidity 10 GV (galactic protons) for the positive ($A > 0$) (upper panels) and for the negative ($A < 0$) polarity epoch (bottom panels), traced backward in time from the heliospheric boundary until they reach the position $r = 1AU$, $\theta = 90°$, $\varphi = 180°$. The left panels in Fig. 2 display the radial position of pseudoparticles versus backward time, while the right panels show the position in latitude versus time. In order to validate our newly developed numerical model, we demonstrate the binned exit rigidity and the corresponding binned propagation times for simulated pseudoparticles in Fig. 3. Moreover, Fig. 3 presents that the most frequently occurring exit time for simulated pseudoparticles varies around 25 days and the energy loss of pseudoparticles, with an initial rigidity equals 10 GV, is rather low. The exit time is a bit shorter for the $A > 0$ than for the

Journal of Physics: Conference Series **632** (2015) 012084 doi:10.1088/1742-6596/632/1/012084

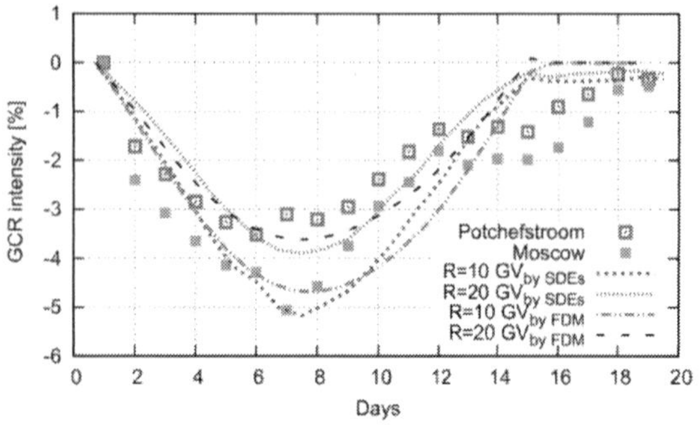

Figure 6. Changes of the expected amplitudes of the Fd of the GCR intensity at the Earth orbit, for the rigidity of 10 and 20 GV based on the solutions of the Parker transport equation by SDEs and FDM in comparison with the GCR intensity registered by Potchefstroom and Moscow neutron monitors during the Fd from 18 March to 4 April 2002.

$A < 0$ because of the large drift speed in the polar regions for $A > 0$. This is in agreement with [7]. Fig. 4 displays latitude vs. longitude distribution of simulated pseudoparticles initiated at Earth's orbit with rigidity 10 GV for $A > 0$ and $A < 0$ cycles. Fig. 4 shows the different character of heliospheric transport depending on various drift patterns in the $A > 0$ and $A < 0$ cycles. In the $A > 0$ cycle, pseudoparticles are transported mainly toward higher latitudes, while in the $A < 0$ epoch pseudoparticles are drifting outward mainly along the neutral sheet at low latitudes (Fig. 2, right panel). Fig. 5 demonstrates simulated rigidity spectra at the Earth's orbit for the $A > 0$ (red solid line), and for the $A < 0$ (green dashed line) polarity epoch. The blue dotted line represents the unmodulated spectrum (LIS) [18] at 100 AU. However, for the rigidity greater than 1 GV the difference is subtle, but still visible, being in agreement with Zhang results (compare with [7], Fig. 3).

In all presented models the heliospheric magnetic field vector $\overrightarrow{B}$ is taken as $\overrightarrow{B} = (1 - 2H(\theta - \theta'))(B_r\overrightarrow{e_r} + B_\varphi\overrightarrow{e_\varphi})$ [19, 2] where H is the Heaviside step function changing the sign of the global magnetic field in each hemisphere, θ' corresponds to the heliolatitudinal position of the heliospheric neutral sheet, $\overrightarrow{e_r}$ and $\overrightarrow{e_\varphi}$ are the unite vectors directed along the components B_r and B_φ of the HMF for the two-dimensional Parker field [26]. Furthermore, we assume that perpendicular diffusion coefficient $K_\perp$ is proportional to parallel one K_{II} : $\frac{K_\perp}{K_{II}} = \frac{1}{1+(\omega\tau)^2}$, where ω is a particle's angular velocity and τ is the time between two sequences of GCR particle collisions. For the GCR particles to which the neutron monitor responds the $\omega\tau$ can change in the range $3 \leq \omega\tau \leq 5$.

3. The model of the Forbush decrease of the GCR intensity

We present the model of the recurrent Fd taking place due to established corotating heliolongitudinal disturbances in the interplanetary space. Corotating interaction regions (CIR) passing the Earth gradually diminish the diffusion at the Earth's orbit, causing larger scattering of the GCR particles, and in effect fewer GCR particles reach the Earth. We simulate this process by the gradual decrease and then the increase of the diffusion coefficient at the Earth's orbit with respect to the heliolongitudes. The diffusion coefficient $K_\|$ of cosmic ray particles has the form: $K_\| = K_0 \cdot K(r) \cdot K(R,\nu)$, where $K_0 = 10^{21} cm^2/s$, $K(r) = 1 + 0.5 \cdot (r/1AU)$. The dependence of the diffusion coefficient on the particles rigidity R, based on the quasilinear theory, e.g.[23, 24], takes the form: $K(R,\nu) = (R/R_0)^{2-\nu}$, where $R_0 = 1GV$ and ν is the exponent of the power spectral density (PSD) of the HMF turbulence. The exponent ν pronounces the increase of the HMF turbulence in the vicinity of space where the Fd is created (for details see e.g. [14, 25]), and is taken as: $\nu = 1 + 0.3\sin(\varphi - 90°)$ for $r < 30AU$ and $90° \leq \varphi \leq 270°$. We assume the existence of two dimensional Parker's spiral heliospheric magnetic field B [26] implemented through the angle $\psi = \arctan(-B_\varphi/B_r)$ in the 3D anisotropic diffusion tensor K_{ij}

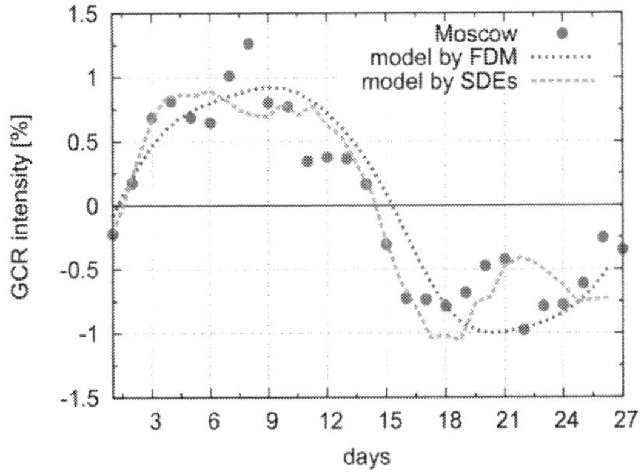

Figure 7. Changes of the expected amplitudes of the 27-day variation of the GCR intensity at the Earth's orbit for the rigidity of 10 GV based on the solutions of the Parker transport equation by SDEs and FDM in comparison with the GCR intensity registered by Moscow neutron monitor for the 7 September - 3 October 2007.

of GCR particles [21]. We compare the results obtained by the solution of the Parker transport equation by SDEs with our previous method of solution by FDM [25] assuming the same changes of all included parameters.

The expected changes of the GCR intensity for the rigidity of 10 and 20 GV during the simulated Fd in comparison with the profiles of the daily GCR intensities, recorded by the two neutron monitors with different cut off rigidities from 18 March to 4 April 2002 presents Fig. 6. One can see that the proposed models are in an acceptable coincidence with the experimental data and, as is expected, the amplitude of the Fd decreases for higher rigidities. Moreover, the model of the Fd obtained based on the solution of the SDEs allows reflecting the stochastic character of the GCR particles distribution in the heliosphere and analyze the pseudoparticle trajectory through the 3D heliosphere (Fig. 1), which is not possible based on the solution of the Parker transport equation by the FDM (e.g. [14]).

4. The model of the 27-day variation of the GCR intensity

The recurrence of the GCR intensity connected with the solar rotation is commonly called the 27-day variation, although the durations can slightly differ. The 27-day variation of the GCR intensity is connected with the heliolongitudinal asymmetry of the electromagnetic conditions in the heliosphere. The recent minimum of solar activity between solar cycles No. 23 and 24 was quite exceptional. Recurrent variations connected with corotating structures ($\sim$ 27 days), at the end of 2007 and almost for the whole year 2008 were clearly established in all solar wind and interplanetary parameters. Consequently, the 27-day variation of cosmic ray intensity was clearly visible in a variety of cosmic ray counts of neutron monitors (e.g., [28, 15]) and space probes (e.g., [27]). We present the model of the 27-day variation of the GCR intensity considering an individual period of solar rotation starting on 2007.09.07. As it was stated by [28, 29], the 27-day variation of the GCR intensity in the minimum epochs is preferentially related to the heliolongitudinal asymmetry of the solar wind velocity.

In the model we apply approximation of the in situ measurements of the solar wind speed as source of the 27-day variation of the GCR intensity described by the formula: $U = U_0(1 - 0.31\sin(\varphi + 6.10) + 0.06\sin(2\varphi + 0.82) - 0.10\sin(3\varphi - 1.04))$, $U_0 = 400km/s$. The diffusion coefficient $K_{\|}$ of cosmic ray particles has the form: $K_{\|} = K_0 \cdot K(r) \cdot K(R)$, where $K_0 = 10^{22}cm^2/s$, $K(r) = 1+0.5\cdot(r/1AU)$ and $K(R) = R^{0.5}$. Fig. 7 compares the results of the solutions of the Parker transport equation by SDEs and by FDM, considering the same changes of all parameters taken into account, for the GCR particles with rigidity R=10 GV. The model of the 27-day wave of the GCR intensity obtained by the SDEs and FDM at the Earth's orbit (1 AU, $\theta = 90^0$) is in reliable agreement with the data of Moscow NM (Fig. 7).

The reason for some discrepancies between the SDE and FDM models lies in the various numerical approaches to solve the Parker transport equation. In the case of the stochastic models we must remember about the role of the randomly fluctuating component. From the other side our previously published models based on the FDM contain the initial condition with respect to the rigidity (e.g., [28]), which cannot be passed one to one into the stochastic solution.

5. Conclusion

- We presented the model of the Fd and the 27-day variation of the GCR intensity obtained based on the stochastic approach to the solution of the Parker transport equation. The modeling results are in a good agreement with the data of neutron monitors. We showed a quite reliable agreement between the stochastic approach results and grounded in our previous papers the finite difference method results.
- The SDEs were integrated backward in time in the spherical coordinates applying the full 3D anisotropic diffusion tensor. The models obtained based on the solution of the SDEs reflect the stochastic character of the pseudoparticles distribution in the heliosphere. Additionally, this approach allows to observe the statistically possible changes of the pseudoparticles trajectories, which is not possible based on the solution of the Parker transport equation by finite difference methods.

Acknowledgments

This work is supported by The Polish National Science Centre grant awarded by decision number DEC-2012/07/D/ST6/02488. We thank the principal investigators of Potchefstroom and Moscow neutron monitors and OMNIweb for the ability to use their data. We thank the referees for the thorough review.

References

[1] Langevin P 1908 *C. R. Acad. Sci.* **146** 530
[2] Kota J and Jokipii R 1983 *Astrophysical Journal* **265** 573-581
[3] Yan J and Shu C-W 2002 *Journal of Scientific Computing* **17** 27-47
[4] He J-H 1999 *Comput. Methods Appl. Eng.* **178** 257-262
[5] Cuia S, Kurganova A and Medovikov A 2014 *Applied Numerical Mathematics* doi:10.1016/j.apnum.2014.04.005 (in press)
[6] Parker E N 1965 *Planetary and Space Science* **13** 9-49
[7] Zhang M 1999 *Astrophys. J.* **513** 409-420
[8] Gervasi M, Rancoita P G, Usoskin I G and Kovaltsov G A 1999 *Nuclear Physics B (Proc. Suppl.)* **78** 26-31
[9] Bobik P, Boella G, Boschini M J, et al. 2012 *Astrophys. J.* **745** article id. 132
[10] Alanko-Huotari K, Usoskin I G, Mursula K and Kovaltsov G A 2007 *J. Geophys. Res.* **112** A08101
[11] Pei C, Bieber J W, Burger R A and Clem J 2010 *J. Geophys. Res.* **115** A12107
[12] Strauss R D, Potgieter M S, Büsching I and Kopp A 2011 *Astrophys. J.* **735** article id. 83
[13] Guo X and Florinski V 2014 *J. Geophys. Res.* **119** 2411-2429
[14] Wawrzynczak A and Alania M V 2008 *Adv. Space Res.* **41** 325-334
[15] Modzelewska R and Alania M V 2013 *Solar Physics* **286** 593-607
[16] Gil A and Alania M V 2014 *Applied Mathematical Modelling* **38** 3260-3268
[17] Gardiner C W 2009 *Handbook of stochastic methods. For physics, chemistry and the natural sciences* Springer Series in Synergetics
[18] Webber W R and Lockwood J A 2001 *Journal of Geophysical Research* **106** 29323-29332
[19] Jokipii J R and Kopriva D A 1979 *Astrophys. J.* **234** 384-392
[20] Jokipii J R, Levy E H and Hubbard W B 1977 *Astrophysical Journal* **1** **213** 861-868
[21] Alania M V 2002 *Acta Physica Pol. B* **33** 4 1149-1166
[22] Kopp A, Büsching I, Strauss R D and Potgieter M S 2012 *Computer Physics Communications* **183** 530-542
[23] Jokipii J.R. 1966 *Astrophys. J.*, **146**, 480-487
[24] Shalchi A. 2009 *Nonlinear Cosmic Ray Diffusion Theories*, (Springer)
[25] Wawrzynczak A and Alania M V 2010 *Adv. Space Res.* **45** 622-631
[26] Parker E N 1958 *Astrophys. J.* **128** 664
[27] Leske, R.A., Cummings A.C.,Mewaldt R.A., Stone E.C. 2013 *Space Sci Rev.*, **176**, 1-4, 253-263
[28] Alania M V, Modzelewska R and Wawrzynczak A 2010 *Adv. Sp. Res.* **45** 421-428
[29] Gil A and Alania M V 2013 *Solar Physics* **283** 565-578

27-day variation of the GCR intensity based on corrected and uncorrected for geomagnetic disturbances data of neutron monitors

M.V. Alania[1], R. Modzelewska[1], A. Wawrzynczak[2], V.E. Sdobnov[3], M.V Kravtsova[3]

[1]Institute of Mathematics and Physics, Siedlce University, Poland
[2]Institute of Computer Sciences, Siedlce University, Poland,
[3] The Institute of Solar- Terrestrial Physics of Siberian Branch of RAS, P.O.Box 291, Irkutsk, Russia.

E-mail: alania@uph.edu.pl, renatam@uph.edu.pl, awawrzynczak@uph.edu.pl, sdobnov@iszf.irk.ru

Abstract. We study 27-day variations of the galactic cosmic ray (GCR) intensity for 2005-2008 period of the solar cycle #23. We use neutron monitors (NMs) data corrected and uncorrected for geomagnetic disturbances. Besides the limited time intervals when the 27-day variations are clearly established, always exist some feeble 27-day variations in the GCR intensity related to the constantly present weak heliolongitudinal asymmetry in the heliosphere. We calculate the amplitudes of the 27-day variation of the GCR intensity based on the NMs data corrected and uncorrected for geomagnetic disturbances. We show that these amplitudes do not differ for NMs with cut-off rigidities smaller than 4-5 GV comparing with NMs of higher cut-off rigidities. Rigidity spectrum of the 27-day variation of the GCR intensity found in the uncorrected data is soft while it is hard in the case of the corrected data. For both cases exists definite tendency of softening the temporal changes of the 27-day variation's rigidity spectrum in period of 2005 to 2008 approaching the minimum of solar activity. We believe that a study of the 27-day variation of the GCR intensity based on the data uncorrected for geomagnetic disturbances should be carried out by NMs with cut-off rigidities smaller than 4-5 GV.

1. Introduction

Magnetic cut-off rigidities for galactic cosmic ray (GCR) significantly changes depending on the level of disturbances of the Earth's magnetosphere. According to the Dungey mechanism [1] relatively high energy particles rush towards the Earth but are diverted around the Earth in circular orbits in the equatorial plane, forming a ring current at several earth radii, which causes significant geomagnetic field reduction. These reductions in the terrestrial magnetic field strength are measured by the Dst index (disturbance storm time index) [2].

An influence of short period changes of the Earth's magnetic field connected with interaction of the Earth's magnetosphere and disturbed vicinity of the interplanetary space (coronal mass ejecta, shocks) on GCR intensity and anisotropy was studied based on corrected and uncorrected neutron monitors (NMs) data in [3, 4, 5] and references therein. At the same time, assessment of the 27-day variation of the GCR intensity by NMs data corrected for disturbances of Earth's magnetic field was not studied at all. It is worth to underline that a survey of the 27-day

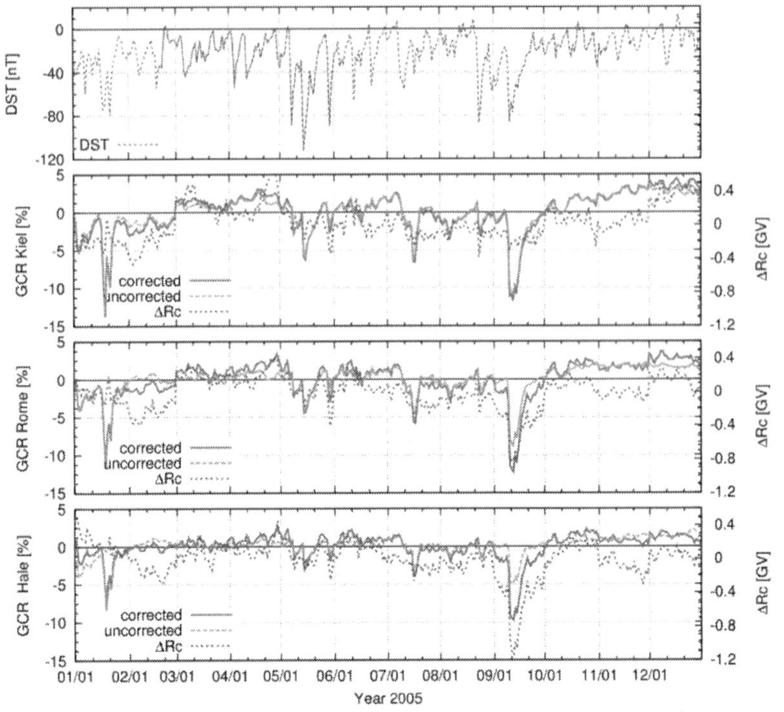

Figure 1. Daily changes of DST index, cut-off variations ΔR_c and the corresponding GCR intensity for Kiel ($R_c = 2.29$ GV), Rome ($R_c = 6.32$ GV), and Halekala ($R_c = 13.3$ GV) NMs corrected and uncorrected for geomagnetic disturbances for year 2005.

variation of the GCR intensity is fundamental to understand the influence of the heliolongitudinal asymmetry of solar wind and solar activity on behavior of the Earth's magnetosphere and atmosphere, and its consequences for the global climate processes. In order to address this issue an aim of this paper is: (1) to estimate how the changes in the cut-off rigidities of various NMs influence the amplitudes of the 27-day variation of GCR intensity and (2) to study changes in rigidity spectrum of the 27-day variation of the GCR intensity in 2005-2008, the time interval of the unusually prolonged deep minimum in the solar activity #23.

The recent solar minimum 23/24 provided a unique opportunity to study recurrent variations of the GCR intensity under relatively stable conditions. Recurrent variations connected with corotating structures ($\sim$ 27 days), at the end of 2007 and almost for the whole year 2008 were clearly established in all solar wind and interplanetary parameters. Consequently, the 27-day recurrent variations of cosmic ray intensity were clearly visible in a variety of cosmic ray counts of neutron monitors (e.g.,[6, 7]) and space probes (e.g.,[8]).

2. Analysis of NMs data corrected and uncorrected for geomagnetic disturbances

To find changes of cut-off rigidities ΔR_c for different neutron monitors during the analysed period of 2005-2008 we have employed the spectrographic global survey (SGS) method developed by [9]. Using ground-based measurements of GCR from the worldwide network of stations the SGS method allow to collect information on energy and pitch-angle distribution of primary GCRs in the interplanetary magnetic field (IMF), as well as on variations in the planetary system of geomagnetic cut-off rigidities per each observation hour. This method provides also the

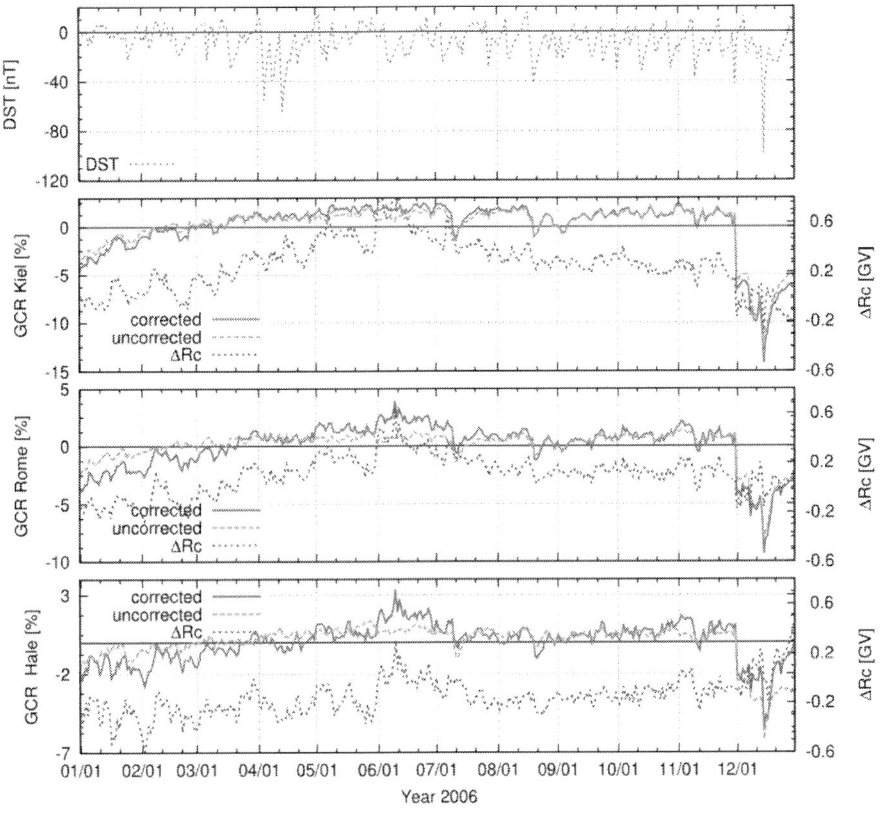

Figure 2. The same as in Fig. 1 but for year 2006.

possibility to use the whole set of ground-based recording equipment (the worldwide network of neutron monitors at any level of the Earths atmosphere, ground-based and underground meson telescopes, etc.) for analysis. Along with phases of the first and second harmonics of the pitch-angle anisotropy, it can determine the rigidity spectrum of the isotropic component and anisotropy, obtain information about the IMF orientation from data on the phase of pitch-angle anisotropy. Moreover, the SGS method can determine variations in the GCR planetary system per each observation hour or per shorter time intervals when the CR intensity near Earth goes up during solar proton events.

The essence of the SGS method is as follows: if amplitudes of variations in primary GCR intensity $\Delta J/J$ are small, then variations in intensity of secondary particles $\Delta I/I$ are connected with $\Delta J/J$ by the following relation:

$$\begin{aligned} \frac{\Delta I_c^i}{I_c^i}(h_l) &= -\int_0^{2\pi}\int_0^{\frac{\pi}{2}} \Delta R_c(\alpha,\beta) W_c^i[(R_c(\alpha,\beta),\alpha,\beta,h_l] sin\beta d\beta d\alpha + \\ &+ \int_0^{2\pi}\int_0^{\frac{\pi}{2}}\int_{R_c(\alpha,\beta)}^{\infty} \frac{\Delta J}{J}(R,\alpha,\beta) W_c^i(R,\alpha,\beta,h_l) sin\beta d\beta d\alpha dR \end{aligned} \quad (1)$$

where, $\frac{\Delta I_c^i}{I_c^i}(h_l)$ is the amplitude of variations in flux of secondary particles of type i observed at geographical point c at height h_l in the Earth's atmosphere at time t; $\frac{\Delta J}{J}(R,\alpha,\beta)$

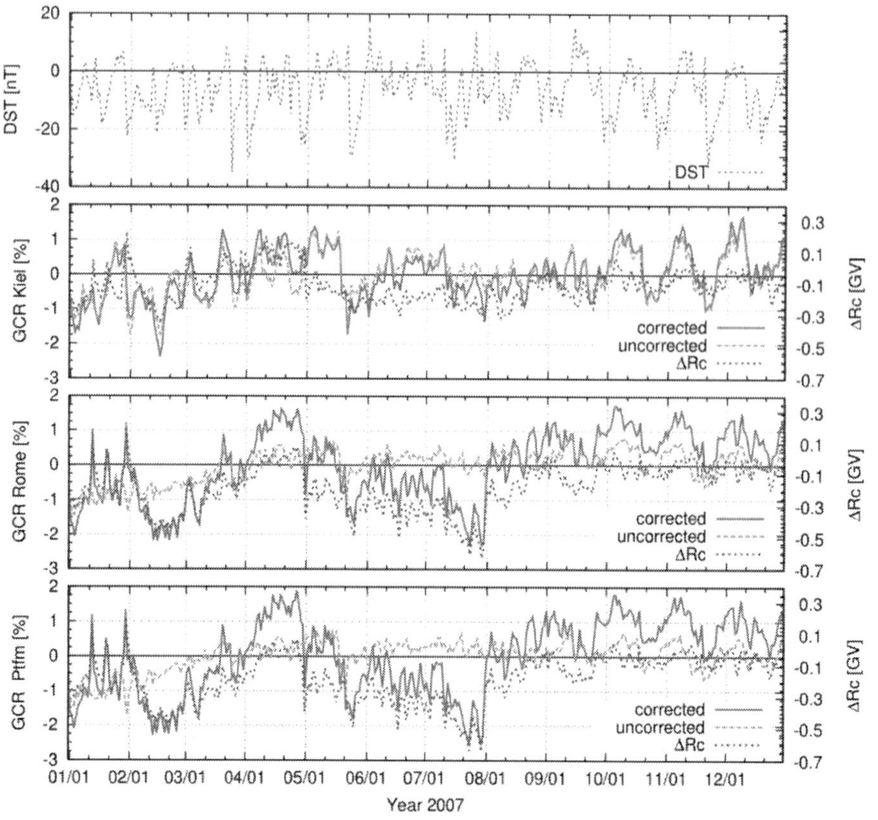

Figure 3. The same as in Fig. 1 but for year 2007. Due to shutdown of the Haleakala station the analogous changes for the Potchefstroom NM ($R_c = 7.10$ GV) are presented.

are the variations in particle intensity at the boundary of the atmosphere at the given point; α, β are the azimuth and zenith angles of primary particles' arrival to the boundary of the atmosphere; R is the magnetic rigidity of particles; $R_c(\alpha, \beta)$ is cut-off rigidity in directions (α, β), and $\Delta R_c(\alpha, \beta)$ are its possible variations. The $W_c^i(R, \alpha, \beta, h_l)$ is the function determining a connection between $\frac{\Delta I_c^i}{I_c^i}(h_l)$ and $\frac{\Delta J}{J}(R, \alpha, \beta)$ satisfying the normalisation condition: $\int_0^{2\pi} d\alpha \int_0^{\frac{\pi}{2}} sin\beta d\beta \int_{R_c(\alpha,\beta)}^{\infty} W_c^i(R, \alpha, \beta, h_l)dR = 1$. Assuming that direction of the particle motion in the given coordinate system is determined by the angles ψ and λ, and the IMF direction by the angles ψ_0 and λ_0, then the particle pitch angle θ can be expressed by following characteristics $\mu = cos\theta = sin\lambda sin\lambda_0 + cos\lambda cos\lambda_0 cos(\psi - \psi_0)$. Simultaneously, we can represent the function $\frac{\Delta J}{J}(R, \psi, \lambda)$ as a series $\frac{\Delta J}{J}(R, \psi, \lambda) = \sum_{n=0}^{m} a_n(R) P_n(\mu)$ where $P_n(\mu)$ are the Legendre polynomials. Taking into account the simplifications and restrictions presented in detail in [10] we obtain

$$\frac{\Delta I_c^i}{I_c^i}(h_l) = -\Delta R_c W_c^i(R_c, h_l) + \sum_{j=0}^{10} \sum_{k=1}^{m_n} A_{jk} B_{jkc}^{il} \qquad (2)$$

where $A_{0k} = a_{0k}$; $A_{1k} = a_{1k} cos\psi_0 cos\lambda_0$, $A_{2k} = a_{1k} sin\psi_0 cos\lambda_0$; $A_{3k} = a_{1k} sin\lambda_0$, $A_{4k} = a_{2k} cos^2\psi_0 cos^2\lambda_0$, $A_{5k} = a_{2k} cos\psi_0 sin\psi_0 cos^2\lambda_0$, $A_{6k} = a_{2k} sin^2\psi_0 cos^2\lambda_0$, $A_{7k} = a_{2k} cos\psi_0 cos\lambda_0 sin\lambda_0$, $A_{8k} = a_{2k} sin\psi_0 cos\lambda_0 sin\lambda_0$, $A_{9k} = a_{2k} sin^2\lambda_0$, $A_{10k} = a_{2k}$. The coefficients

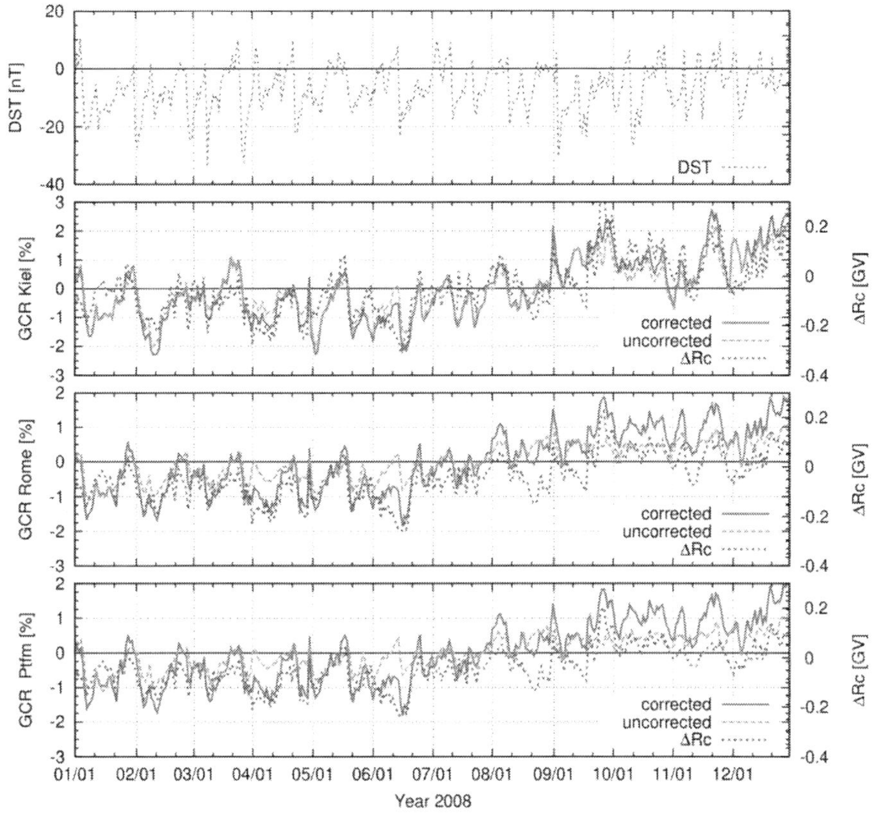

Figure 4. The same as in Fig. 3 but for year 2008.

B^{il}_{jkc} are determined based on the asymptotic angles derived from calculations of the particles trajectory [11]. Summarizing, the problem reduces to estimation of the following unknown parameters: ΔR_c-showing the variation of cut-off rigidity; $a_{01}, a_{02}, a_{03}, a_{11}, a_{12}, a_{21}, a_{22}$ -showing amplitudes and rigidity dependencies of the zero, first and second harmonics of the pitch-angle GCR distribution; ψ_0 and λ_0 - showing the IMF direction averaged over the Larmor circle (accurate to π). To solve the problem in hand, it is essential that the network of stations and detector arrays at stations used in the calculations provide abundance of the system of above mentioned equations and linear independence of coefficients B^{il}_{jkc}. There should be at least two detectors with different $W^i_c(R_c, h_l)$ at the stations where GCR variations are determinable. The details of the SGS method are presented in [10].

Applying the SGS method we have calculated the daily changes of the ΔR_c by means of 27 neutron monitor stations hourly data during years 2005-2006 versus the cut-off rigidity R_c on 10 Jun 2004, and for period of 2007-2008 versus R_c on 5 November 2007. Figs. 1- 4 present daily changes of Dst index, cut-off variations ΔR_c and the corresponding GCR intensity for Kiel, Rome, Potchefstroom and Haleakala NMs corrected and uncorrected for geomagnetic disturbances for 2005 (Fig. 1), 2006 (Fig. 2), 2007 (Fig. 3) and 2008 (Fig. 4). Fig. 1-Fig. 4 show that the corrected and uncorrected GCR intensity do not differ much for NMs with cut-off rigidities smaller than 4-5 GV, while for NMs with higher cut-off rigidities the difference is as notable as cut-off rigidity increases i.e. a level of changes of cut-off rigidities is larger. One can see that the correction for geomagnetic disturbances is the largest when the rapid changes of Dst

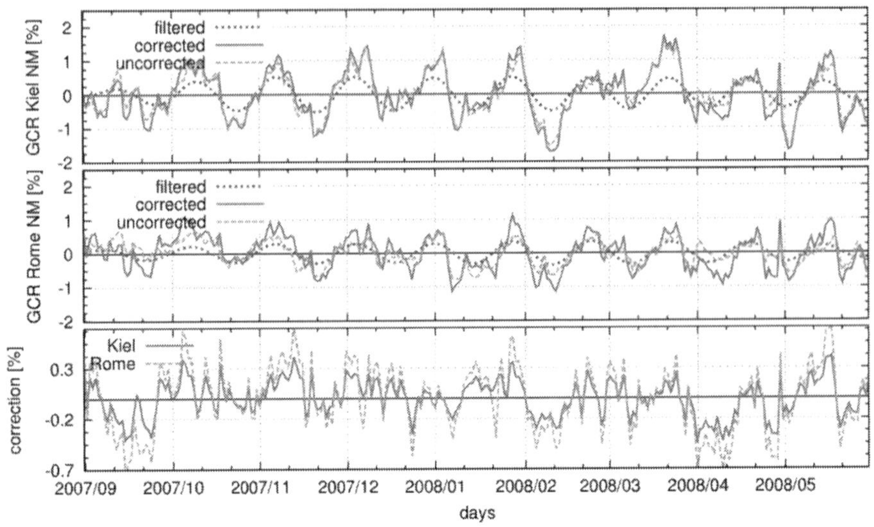

Figure 5. Daily GCR intensity corrected and uncorrected for geomagnetic disturbances with filtered 27-day wave and the corresponding correction for Kiel and Rome NMs in 2007-2008.

index are observed (related with shocks and Forbush decreases), but becomes weaker in quite stable conditions.

In this paper we consider in detail 27-day wave in corrected and uncorrected GCR intensity and the effect of geomagnetic correction on periodic variations. Daily GCR intensity corrected and uncorrected for geomagnetic disturbances with filtered [12] 27-day wave and the corresponding correction for Kiel and Rome NMs in 2007-2008 are presented in Fig. 5. As an example Fig. 5 presents only few Bartel's rotations in 2007-2008 with excellent quasi-periodic changes ($\sim$ 27 days) in the GCR intensity.

3. Rigidity spectrum of the 27-day variation based on the corrected and uncorrected NMs data

Usually, a power law rigidity R spectrum of any classes of the GCR intensity variations is characterized by an exponent γ, as $\sim R^{-\gamma}$. The exponent γ is an important instrument to study a dependence of amplitudes of the GCR intensity changes on energy of primary GCR particles. The value of γ is determined by the structure of the IMF turbulence responsible for the scattering of GCR particles to which neutron monitor respond (e.g.[13, 15, 16]). A character of this dependencies is different for various classes of the GCR intensity variations. In [18] was shown that temporal changes of the exponent γ for the 11-year variations [13] inversely correlates with the changes of the exponent γ for the 27-day variations of the GCR intensity.

In this paper the values of the rigidity spectrum ($\delta D(R)/D(R)$) exponent γ of the 27-day variation of the GCR intensity were calculated using both, corrected and uncorrected for geomagnetic disturbances data of NMs with different cut-off rigidities.

We have calculated the amplitudes of the 27-day variation of the GCR intensity by means of daily data using the harmonic analyses method e.g.[14] during each Bartel's rotation period:

$$x(k\Delta t) = \frac{a_0}{2} + \sum_{n=1}^{N/2}[a_n cos(\frac{2\pi}{T}nk\Delta t) + b_n sin(\frac{2\pi}{T}nk\Delta t)], \quad (3)$$

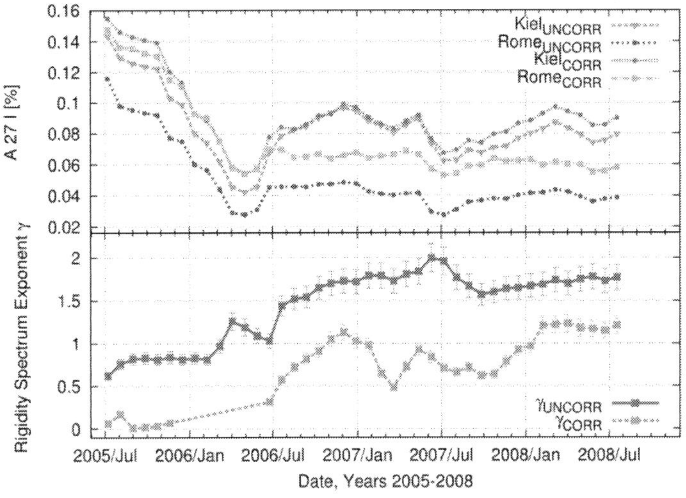

Figure 6. (Top panel)The temporal changes in the amplitudes of the 27-day variation of the corrected and uncorrected GCR intensity for Kiel and Rome NMs; (Bottom panel) The exponent γ of the rigidity spectrum of the 27-day variation of the corrected and uncorrected GCR intensity in 2005-2008.

where $x(k\Delta t)$ designates the daily data of the GCR intensity, $N = 27$ and the coefficients have a form: $a_0 = \frac{2}{N}\sum_{k=1}^{N} x(k\Delta t)$, $a_n = \frac{2}{N}\sum_{k=1}^{N}[x(k\Delta t)cos\frac{2\pi kn}{N}]$, $b_n = \frac{2}{N}\sum_{k=1}^{N}[x(k\Delta t)sin\frac{2\pi kn}{N}]$. Applying the above presented formulas we determine the amplitude of the 27-day variation (A27I) of the GCR intensity as $A27I = \sqrt{a_1^2 + b_1^2}$.

The rigidity spectrum of the quasi-periodic variation was calculated by means of the amplitudes of the 27-day variation of the GCR intensity based on the method presented, e.g. in [17],[18]:

$$\frac{\delta D(R)}{D(R)} = \begin{cases} AR^{-\gamma} & R \leq R_{max} \\ 0 & R > R_{max} \end{cases} \tag{4}$$

R_{max} designates the upper limiting rigidity beyond which the quasi-periodic variation of the galactic cosmic ray intensity disappears (100 GV). Detailed description of the γ calculation is described in [19].

An investigation of the long-period changes of the power law rigidity spectrum of the 27-day variation of the GCR intensity (e.g. in the period 2005-2008) requires a data from neutron monitors with a long term stability. Unfortunately, only a few neutron monitors from worldwide network of stations satisfy this requirement. For this reason, we chose Kiel and Rome neutron monitors data to calculate the rigidity spectrum of the amplitudes of the 27-day variation of the GCR in the period 2005-2008. To use two neutron monitors for the calculation of the reasonably reliable rigidity spectrum exponent γ there must be considerable difference between the cut-off magnetic rigidities (R_c). This demand is satisfactorily fulfilled for Kiel and Rome neutron monitors; for Kiel neutron monitor $R_c = 2.29$ GV and for Rome neutron monitor $R_c = 6.32$ GV (or the corresponding median rigidity of response for Kiel neutron monitor equals 17 GV and for Rome - 23 GV, respectively. A validity of this approach was shown in papers [16, 18].

Fig. 6 (top panel) demonstrates the temporal changes in the amplitudes of the 27-day variation of the corrected and uncorrected GCR intensity for Kiel and Rome NMs; and (bottom panel) the matching exponent γ of the rigidity spectrum of the 27-day variation of the corrected and uncorrected GCR intensity in 2005-2008. Unfortunately, in the first half of the year 2006 calculation of the exponent γ based on the corrected data was not possible due to lack of the discrepancy between the amplitudes of the 27-day variation for the used NMs. Fig. 6 shows that rigidity spectrum of the 27-day variation of the GCR intensity found by uncorrected data of NMs is soft while it is hard for corrected data. For both cases there is visible a clear tendency of the softening of the temporal changes of the rigidity spectrum from 2005 to 2008 (approaching minimum of solar activity) coinciding with results obtained in [16].

4. Summary

(i) Comparison of the amplitudes of the 27-day variation of the GCR intensity calculated based on the NMs data corrected and uncorrected for geomagnetic disturbances showed that these amplitudes do not differ for NMs with cut-off rigidities smaller than 4-5 GV in comparison with NMs of higher cut-off rigidities. This indicates that to study the 27-day variations of the GCR intensity by uncorrected data of NMs it is necessary to use data of NMs with cut of rigidities smaller than 4-5 GV.

(ii) Rigidity spectrum of the 27-day variation of the GCR intensity found by uncorrected data of NMs is soft, while it is hard for data corrected on geomagnetic disturbances. For both cases, there is an apparent tendency of softening temporal changes of the rigidity spectrum from 2005 to 2008 period approaching minimum of solar activity.

Acknowledgments

We are grateful to the Principal Investigators from worldwide network of neutron monitors and OmniWeb database for access to the data.

References

[1] Dungey, J.W. 1961 *Phys. Rev. Lett.*, **6**, 47
[2] Sugiura, M. 1964 *Ann. Int. Geophys. Year*,**35**, 945
[3] Tyasto, M.I., Danilova, O. A., Ptitsyna, N. G., Sdobnov, V. E. 2013 *Adv. Space Res.* **51** 1230-1237
[4] Dvornikov, V. M., Sdobnov, V. E. 2008 *Geomagnetism and Aeronomy* **48** 314-318
[5] Alania, M.V., Wawrzynczak, A., Sdobnov, V.E., Kravtsova, M.V. 2013 *Solar Phys.* **286** 561-576
[6] Alania, M.V., Modzelewska, R., Wawrzynczak, A. 2010 *Adv. Sp. Res.* **45** 421-428
[7] Modzelewska, R., Alania, M.V. 2013, *Solar Phys.*, **286**, 2, 593-607
[8] Leske, R.A., Cummings A.C.,Mewaldt R.A., Stone E.C. 2013 *Space Sci Rev.* **176**, 1-4, 253-263
[9] Dvornikov, V.M., Sdobnov, V.E., Sergeev, A.V. 1983 *Proc. of the 18th International Cosmic Ray Conference*, 3
[10] Dvornikov, V.M.,Sdobnov, V.E. 1998 *Solar Phys.*, **172**, 2, 405-422
[11] Shea, M.A., Smart, D.F., McCracken, K.G. 1965 *J.of Geophys. Res. Atmospheres* **70(17)**,4117-4130
[12] Otnes, R.K., Enochson, L. 1972 *Digital Time Series Analysis*, John Wiley and Sons, New York
[13] Alania, M.V., Iskra, K., Siluszyk, M. 2010 *Adv. Space Res.*, **45** , 10, 1203-1210
[14] Gubbins, D. 2004 *Time Series Analysis and Inverse Theory for Geophysicists*, Chapter 2, Cambridge University Press, Cambridge, 21-26
[15] Wawrzynczak, A., Alania, M.V. 2010 Adv. *Adv. Space Res.*, **45**, 5, 622-631
[16] Gil A., Alania, M.V., 2013 *Solar Phys.*, **283**, 565-578
[17] Dorman, L.I. 1963 *Cosmic rays variations and space exploration* Nauka, Moscow
[18] Gil A., Alania,M.V. 2011 *Journal of Atmospheric and Solar-Terrestrial Physics* **2-3** 294-299
[19] Alania, M.V., Wawrzynczak,A. 2012 *Adv. in Space Res.* **50** 725-730

Journal of Physics: Conference Series **632** (2015) 012086 doi:10.1088/1742-6596/632/1/012086

Real-time experimental data of the muon hodoscope URAGAN accessible in www

I I Yashin, I I Astapov, N S Barbashina, V V Borog, A N Dmitrieva, D V Chernov, R P Kokoulin, K G Kompaniets, Yu N Mishutina, A A Petrukhin, V V Shutenko, O A Sit'ko, E I Yakovleva

National Research Nuclear University MEPhI
(Moscow Engineering Physics Institute), 115409, Moscow, Russia

E-mail: iiyashin@mephi.ru

Abstract. Muon hodoscope URAGAN consists of four independent supermodules with total area 46 m^2 and is under operation in the Experimental complex NEVOD (MEPhI, Moscow) from 2007. At present, the sequence of real-time data of angular variations of muon flux measured in the hodoscopic mode and also the results of their analysis: local angular anisotropy, wavelet frequencies, etc. are available at the URAGAN web-site. Some peculiarities of application of these data for the analysis of various events in heliosphere, magnetosphere and atmosphere are discussed.

1. Introduction

Muon hodoscope URAGAN [1] is located inside the campus of National Research Nuclear University MEPhI (Moscow Engineering Physics Institute), Russia (55.7° N, 37.7° E, 173 m a.s.l.) and represents a wide-aperture large area tracking detector capable of real-time reconstruction of track of each muon arriving from any directions of the upper celestial hemisphere. URAGAN detection system consists of four independent supermodules (SM) with area 11.5 m^2 each. Supermodules detect the angular distribution of the muon flux in a wide range of zenith angles (0-80°) with a high angular resolution (~1°) and with threshold energy of 0.2 to 0.4 GeV. Each supermodule represents an eight-layer assembly of streamer tube chambers with external two-coordinate *X-Y*-strips read-out system. Trigger condition of event detection is the coincidence of signals from the strips of 4 or more detection *X*-planes within the time gate of 250 ns. The main objective of the URAGAN is the study of the variations of the angular distribution of the muon flux caused by different atmospheric and extra-atmospheric processes [2-5]. Three SM (SM1, SM3 and SM4) are continuously under operation, but one (SM2) is used mainly for other methodical and calibration purposes. The exposition of every experimental run is divided into equal one-minute intervals within which monitoring of detection channels (~2 s) and the actual detection (~ 58 s) are carried out. Total operational "live" time is the sum of actual detection time fractions of all one-minute intervals of the exposition except of the sum of dead-times of triggered event data read-out (in total, for one-minute interval it is about 4-5 s and is varied with trigger rate). The fraction of reconstructed events with muon tracks is ~ 91% of the total intensity of 1700-1900 triggers per second. Synchronization of all SMs operation is provided by GLONASS /GPS.

2. Real-time data processing

The experimental data represent sequences of matrices (two-dimensional histograms of 90×90 cells) of the muon angular distributions accumulated during one-minute intervals. There are three matrix types with different angular cells: zenith (θ) and azimuth (φ) angles M (θ, φ), projection angles $M(\theta_X; \theta_Y)$, and also their tangents - $M(\mathrm{tg}\theta_Y, \mathrm{tg}\theta_X)$. Each matrix represents the "muon snapshot" (muonography) of the upper hemisphere limited by the detector aperture with one-minute exposure. The use of three types of matrices is due to the fact that for the subsequent analysis the information about the angular distribution in various forms may be required. Further real-time data processing includes formation of time series of averaged over one minute and 60 minutes atmospheric pressure, counting rates of reconstructed tracks (without the barometric and temperature correction) and "live" time. These series are stored in 1-minute and 60-minute daily files in a text format. Evaluation of the degree of temporary changes in the angular distribution of the muon flux is performed by comparing the value of the current flux with some average value obtained for the preceding 24 hours. Therefore on the basis of data obtained for preceding 24 hours the following values are obtained:

- matrix $M_N(\mathrm{tg}\theta_Y, \mathrm{tg}\theta_X)$: sum of averaged over 24 hours one-minute matrices, cells contain the sum over SM of the average number of events;
- live time $t_{\mathrm{live\_N}}$: sum over SM of the averaged over 24 hours of "live" times of one-minute frames;
- averaged over 24 hours atmospheric pressure P_N.

These values are used for estimation of changes of counting rates in every cell of the current matrix. Every 5 minutes, and at the beginning of each hour an additional processing is carried out. These results are presented on four pages of the web-site of the NEVOD complex:

- 5- and 60-minute variations of the counting rate – http://nevod.mephi.ru/uragan_data.htm;
- angular muon flux anisotropy parameters, 2D Fourier analysis, muonographies (5-minute time series) – http://nevod.mephi.ru/English/atmosphere.php;
- angular muon flux anisotropy parameters, GSE- muonographies (60-minute time series) – http://nevod.mephi.ru/English/heliosphere.php;
- wavelet-analysis – http://nevod.mephi.ru/English/wavurg.htm.

Also on the NEVOD web-site are presented:

- image files of 5- and 60-minute matrices $M(\mathrm{tg}\theta_Y, \mathrm{tg}\theta_X)$ of variations in the angular distributions in the East-North (local geographic) and GSE (Geocentric Solar Ecliptic) coordinate systems;
- image files of obtained time-series graphs.

Formation of 5- and 60-minute time series is determined by the convenience of further physical analysis of atmospheric or heliospheric (magnetospheric) processes.

3. Barometric and temperature corrections

Cosmic ray variations of the extra-atmospheric origin are of the same order as atmospheric ones. Therefore, to study the extra-terrestrial effects it is necessary to introduce corrections related with changes of the atmosphere [6]:

$$M^{cor}(\theta,\varphi,t,\Delta t) = M(\theta,\varphi,t,\Delta t) + \Delta M_P(\theta,t,\Delta t) + \Delta M_T(\theta,t,\Delta t), \quad (1)$$

here: ΔM_T and ΔM_P are corrections on temperature and barometric effects for matrix M cells. Barometric correction (barometric coefficients - B) is determined for long periods of time operation of the hodoscope, during which the atmospheric pressure varied widely. Barometric coefficients B are determined for each cell of the angular distribution matrix, for example, $B(\text{tg}\theta_Y, \text{tg}\theta_X)$. The matrix of cell correction ΔM_P is calculated by the formula:

$$\Delta M_P(tg\theta_Y, tg\theta_X) = B(tg\theta_Y, tg\theta_X) \times (P_0 - P) \times t_{live}, \quad (2)$$

where t_{live} is a "live" time of the data frame.

Unlike the barometric coefficient, for temperature correction it is necessary to know the temperature profile of the atmosphere above the setup location [6]:

$$\Delta M_T(\theta) = M_0(\theta) \cdot \sum_i W_T(h_i, \theta) \Delta T(h_i) \Delta h_i / 100\% . \quad (3)$$

Here M_0 a is the averaged during long period matrix M; $W_T(h,\theta)$ are differential in altitude temperature coefficients (DTC), calculated in [7], h is the depth of the atmosphere, $\Delta h = 0.05$ atm, $\Delta T(h) = T_{SMA}(h) - T(h)$ are changes in altitude profile of the temperature, $T(h)$ is the temperature at a depth h, $T_{SMA}(h)$ is the temperature of a standard model of the atmosphere [8] at the altitude h.

Barometric correction of values of matrices is done separately for each SM using barometric coefficients for every cell of the matrix:

$$M_{ci} = M_i + B_i \cdot (P_0 - P) \cdot t^i_{live} . \quad (4)$$

Here: i is the number of the SM, M_{ci} is the corrected value in the cells of the matrix [s^{-1}]; B_i is the matrix of barometric coefficients for the cells of the current matrix [$\text{mbar}^{-1}\ s^{-1}$]; P is the atmospheric pressure [mbar]; t^i_{live} is "live" time.

In the cells of the matrix, values of changes of counting rates in terms of statistical errors δ_σ are calculated as follows:

$$\delta_\sigma = \frac{M - M_N \cdot t_{live} / t^N_{live}}{\sqrt{M_N \cdot t_{live} / t^N_{live}}} . \quad (5)$$

Here matrix M_N is the sum of averaged over 24 hours one-minute matrices, cells contain the sum over SM of the average number of events; t^N_{live} is the "live" time: sum over SM of the averaged over 24 hours of "live" times of one-minute frames;

Temperature correction is introduced into the one-minute, 5-minute and hourly data. These procedures is carried out in off-line mode and are presented on the web-site with a delay, which is required to obtain data on the temperature profile of the atmosphere and to calculate the differential in altitude temperature coefficients. In particular, the information from Central Aerological Observatory (Russia, Dolgoprudny) is used (http://www.aerology.org/). It based on the temperature profile measured by meteorological balloons two times a day: 00:00 and 12:00 UT. The data is then interpolated and the temperature correction is calculated individually for each time interval of time series data sequence. Additional data about the temperature profile of the atmosphere above the Moscow region is obtained from satellite data (http://nevod.mephi.ru/english/alice.htm).

4. Counting rate variations

On the webpage http://nevod.mephi.ru/uragan_data.htm in the section "Integral intensity mode" the graphs of 5-minute (during the last 24 hours) and 60-minute (over the last 30 days) variations averaged for three supermodules counting rates (figure 1) in terms of deviations (in percent) from the average counting rate are presented.

Counting rate variations with pressure correction are given relative to the following average values of counting rates: 1450 s^{-1}, 1412 s^{-1} and 1414 s^{-1} for SM1, SM3 and SM4 respectively.

The average counting rate of three detectors is formed in a following way:

$$I = \frac{I_1 \cdot t^1_{live} + I_3 \cdot t^3_{live} + I_4 \cdot t^4_{live}}{t^1_{live} + t^3_{live} + t^4_{live}} , \quad (6)$$

here, I_1, I_3, I_4 are counting rates of SM1, SM3 and SM4 with barometric corrections, t^1_{live}, t^3_{live} and t^4_{live} are "live" times of SM1, SM3 and SM4. Barometric coefficients of counting rates for supermodules SM1, SM3 and SM4 are -2.79 $mbar^{-1}s^{-1}$, -2.55 $mbar^{-1}s^{-1}$ and -2.58 $mbar^{-1}s^{-1}$, respectively. Statistical error of 5-minute values of deviations of the average counting rate for all supermodules is ~ 0.1%, and for 60-minute values ~0.03% for average counting rate of three SM, and ~0.05% for supermodules separately.

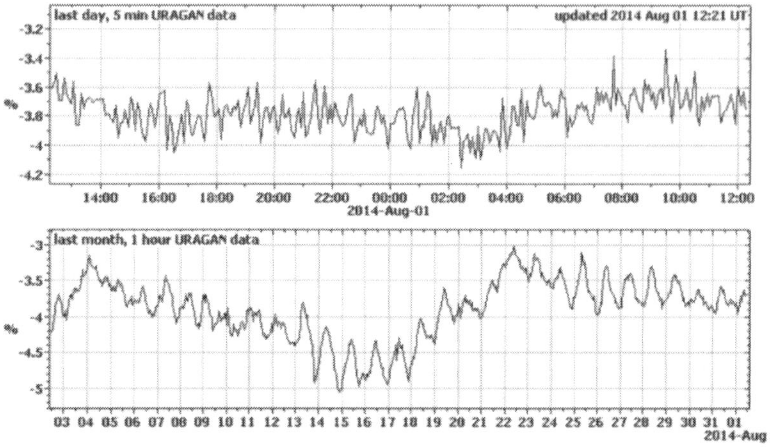

Figure 5. Averaged counting rate variations of URAGAN supermodules, normalized and corrected for atmospheric pressure.

5. Angular muon flux anisotropy

On the webpage http://nevod.mephi.ru/English/atmosphere.php the time series of characteristics of the angular distributions of 5-minute matrices $M(\theta, \varphi)$ and the results of Fourier analysis of the matrices of 5-minute changes in the counting rate δ_σ and 48 their smoothed images (within the last 2 hours) are presented.

For the analysis of the changes in the angular distribution of the muon flux according to the data of three SM the vector of local anisotropy $\vec{A}$ is used [9]. Vector $\vec{A}$ is the sum of the unit vectors of particle tracks, normalized to the number of tracks. The range of zenith angles $\theta = 0 \div 75^\circ$ and of azimuth angles $\varphi = 0 \div 360^\circ$ is used.

Horizontal projections of the local anisotropy vector on the North-South axis A_S and East-West A_E do not depend on the barometric and temperature corrections. At calculating the vertical projection A_Z the barometric correction depending on the zenith angle is used. The additional information provides a difference between the current anisotropy vector $\vec{A}$ and its average value calculated over data collected during some preceding period (in the real-time regime, $\vec{A}_N$ ($A_{N\_S} \approx 0.0031$, $A_{N\_E} \approx 0.0020$, $A_{N\_Z} \approx 0.83$) resulting from the time interval from February to November 2007 is used). The difference between these vectors $\vec{r}$ (relative anisotropy vector) and its horizontal projection r_h:

$$\vec{r} = \vec{A} - \vec{A}_N \,, \qquad r_h = \sqrt{r_S^2 + r_E^2} \tag{7}$$

indicate the direction and level of deformation of the shape of angular distribution of the particle flux.

6. Muonography

Muonography is the graphical representation of the matrices of angular distribution of the counting rate changes obtained for 5-minutes or 1-hour "live" time periods with respect to the normalized matrix during preceding 24 hours, in terms of statistical errors [10]. According to the data of three SM, the following muonographies are formed: unsmoothed, smoothed by Gaussian low-frequency filter, with the correction of the bell-like shape of the angular distribution, in the coordinate systems East-North and GSE (Geocentric Solar Ecliptic System). If to carry out an approximation of the current and the normalized angular distributions using the dependence in the form $C{\cdot}cos^{\alpha}\theta{\cdot}d\Omega$ and obtain for them coefficients (C, α) and (C_N, α_N), respectively, it is possible to harmonize the shape of the normalized distribution in accordance with the shape of the current angular muon flux distribution reflected in the matrix. Conversion coefficient of the form of angular distribution $k(\theta)$ is:

$$k(\theta) = \frac{C \cdot \cos^{\alpha}\theta}{C_N \cdot \cos^{\alpha_N}\theta}. \tag{8}$$

Correction to the bell-shape is done to improve the image detail at possible relatively large changes of the current and normalized matrices.

On the web-page http://nevod.mephi.ru/English/atmosphere.php, a sequence of 24 pairs of images (within the last 2 hours) of 5-minute matrices of angular muon flux changes δ_σ, smoothed with the Gaussian filter (uncorrected and corrected for the shape of the angular distribution) in the East-North coordinate system are shown. Figure 2 shows one of the pairs of images of 5-minute matrices without correction for the shape of the angular distribution and with correction. The projection of matrix M to the map of the Moscow region is done in the following way. Each cell of the matrix M is projected onto the plane located at an altitude of 15 km (muon generation level) above the Earth's surface. Then the coordinates of found cells are projected on a map. Color indicates the deviation in the counting rate in terms of statistical errors.

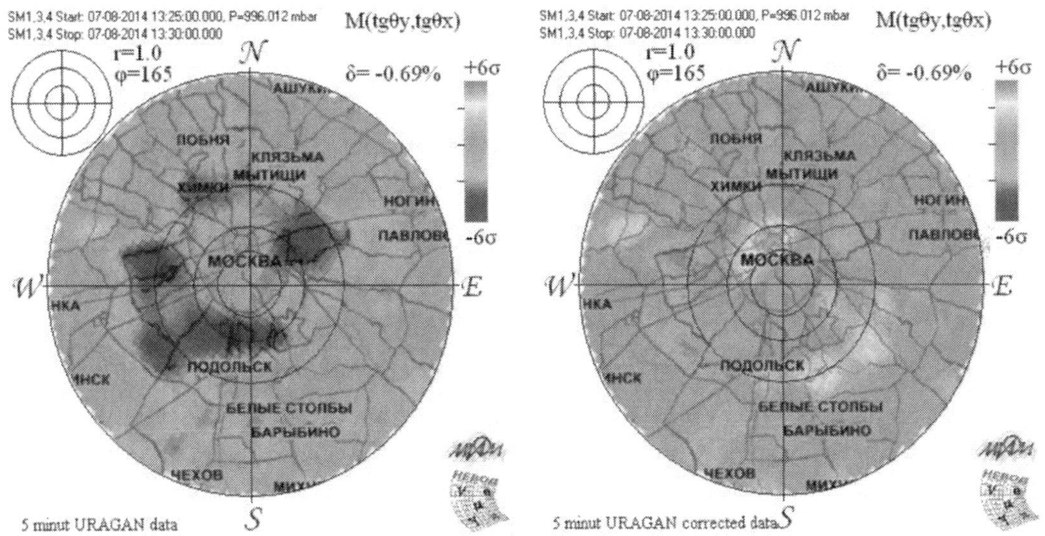

Figure 2. Example of a smoothed 5-minute matrix changes of the angular distribution of the detected particle flux in terms of statistical errors. Left - without correction for the shape of the angular distribution, right - with correction.

The figure also displays: δ which is a change in the counting rate in %, r is value of r_h and φ is the azimuth direction (in degrees) of the vector $\vec{r}$ calculated from the data of 5-minute matrix (0° is direction to the South, 90° is direction to the East). Top left, inside of small circles with radii 1, 2 and 3 (in units of $\sigma_{stat}(r_h) \approx 3.6{\cdot}10^{-4}$), the length and direction of the horizontal projection of the vector $\vec{r}$ are displayed.

On the webpage http://nevod.mephi.ru/English/heliosphere.php, the time series of the angular distribution characteristics for 60-minute matrices M are presented. Also 24 pairs of images (for the last 24 hours) of smoothed 60-minute matrices of counting rate changes δ_σ transformed from the local coordinate system to the GSE coordinate system are presented. For the qualitative representation of the anisotropy of angular distribution of the muon counting rate changes δ_σ caused by the heliosphere disturbances, we made the transformation of the images of matrices from the laboratory system M observed at the detector level to the magnetopause in GSE coordinate system. GSE-mapping of detected muons is formed with the use of the pre-calculated for the location of URAGAN setup asymptotic directions of primary protons [11]. For each angular cell of the original matrix it is possible to introduce four coordinates of the cell corners $(\theta, \varphi)_{lab}$. Taking in to account the table of asymptotic direction for the corners $(\theta, \varphi)_{lab}$ we determined coordinates of the corners $(\theta, \varphi)_{GSE}$ in GSE coordinate system of transformed cell at the level of the magnetopause boundary. Thus, the variations of the muon flux of each cell of the original matrix are reflected from the laboratory system of the detector in the GSE system in the range limited by coordinates $(\theta, \varphi)_{GSE}$.

One of the pairs of GSE images of 60-minute matrices without correcting for the shape of angular distribution and with correction is shown in figure 3. The figure also represents the change in the counting rate in % (δ), the direction of the line of force of the interplanetary magnetic field (indicated by a circle with a diagonal cross), the direction to the Sun (the image of the solar corona) and asymptotic direction for vertical muons (cross).

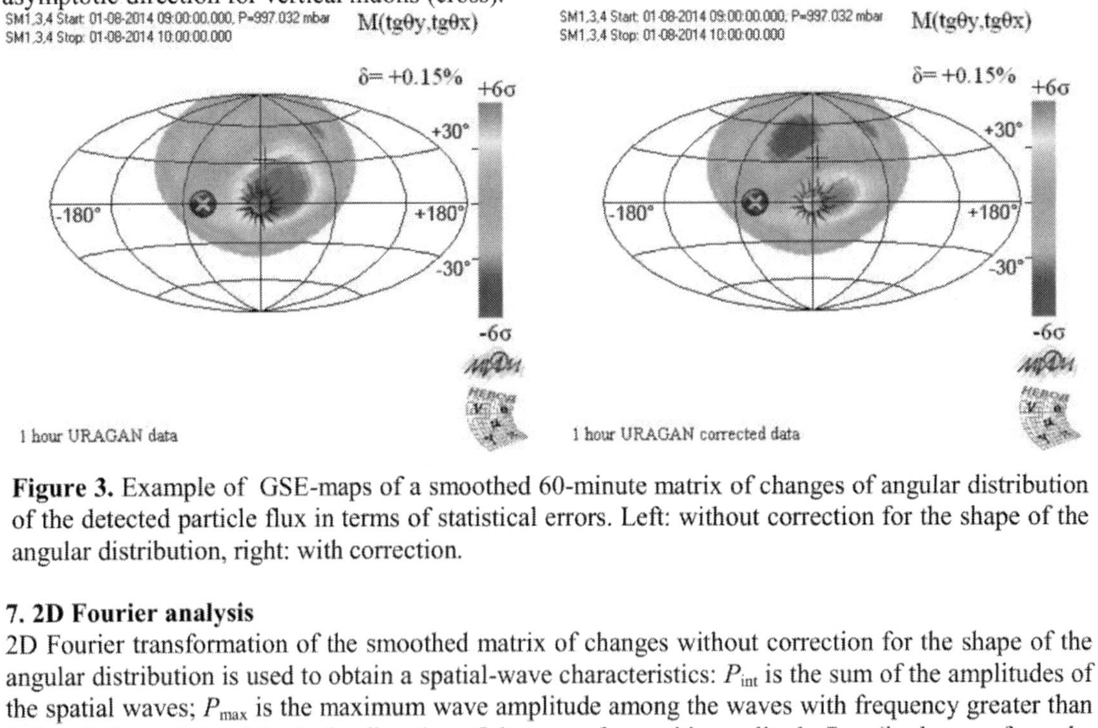

Figure 3. Example of GSE-maps of a smoothed 60-minute matrix of changes of angular distribution of the detected particle flux in terms of statistical errors. Left: without correction for the shape of the angular distribution, right: with correction.

7. 2D Fourier analysis

2D Fourier transformation of the smoothed matrix of changes without correction for the shape of the angular distribution is used to obtain a spatial-wave characteristics: P_{int} is the sum of the amplitudes of the spatial waves; P_{max} is the maximum wave amplitude among the waves with frequency greater than 1 (in relative units); *alpha* is the direction of the wave front with amplitude P_{max} (in degrees from the direction to South toward the direction to the East); *Freq* is the relative frequency of the wave with an amplitude of P_{max}.

8. Wavelet-analysis

For analysis of the different waves presenting in the data of time series of counting rate and characteristics of zenith-angular distributions, the Morlet wavelet processing is performed. Wavelet analysis is carried out immediately after the receipt of the next 5- and 60-minute time series values. At the wavelet coefficients calculating for the current time only the current and preceding time series data are used. Results of wavelet processing are presented as time dependences of amplitudes and power wavelet coefficients for periods ranging from 20 minutes to 32 hours. We use different ranges of periods for more detailed revealing of waves related with different modulation processes in the atmosphere and extra-terrestrial space.

Examples of graphic representation of time series of wavelet coefficients for total hodoscope counting rate I_{sum} for period range from 20 minutes up to 120 minutes presented on the webpage http://nevod.mephi.ru/English/wavurg.htm are shown in figures 4-5.

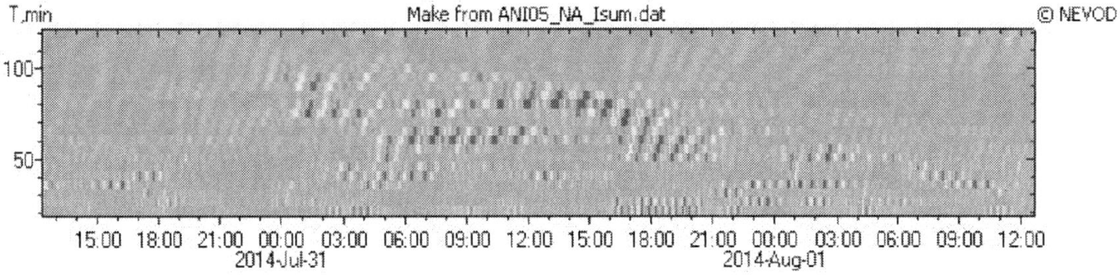

Figure 4. The real part of the wavelet coefficients for frequencies with periods from 20 minutes to 2 hours for 5-minute time series of the total counting rate I_{sum}.

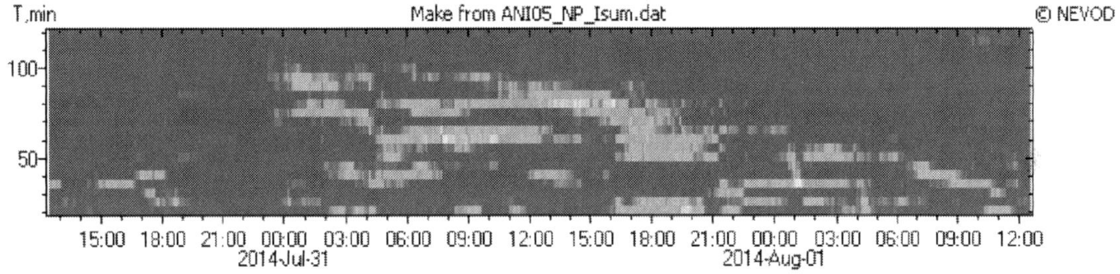

Figure 5. Power of wavelet coefficients for the frequencies with periods from 20 minutes to 2 hours in 5-minute time series of the total counting rate I_{sum}.

Color gradation (from blue to red) reflects the change in the coefficient values (from minimum to maximum) in the displayed time interval for frequencies with periods T.

9. Conclusions

Real-time data of the muon hodoscope URAGAN presented in the NEVOD web-site give possibility to analyze temporal changes not only of the counting rate of the particles but also of the characteristics of their angular distributions.

Variations of characteristics of the muon flux represented by 5-minute data are mainly due to the changes of atmospheric conditions, but they can also be caused by solar proton events, for example, GLE (Ground Level Enhancement). In contrast, variations that present in 60-minute data allow to observe mainly solar events that affect the distribution of galactic cosmic rays in the heliosphere and the Earth's magnetosphere.

The examples of analysis different kinds of atmospheric and extra-terrestrial phenomena by means of URAGAN data can be found in [12].

Acknowledgments

This work was performed at the NEVOD Scientific and Educational Center with the financial support from the State provided by Russian Ministry of Education and Science (government task and project No. RFMEFI59114X0002) and the grant of the Leading Scientific School NSh-4930.2014.2.

10. References

[1] Barbashina N S et al 2008 *Instrum. Experim. Techn.* **51** 701.

[2] Timashkov D A et al 2007 *Proc. 30th Int. Cosmic Ray Conf.* (*Merida, Mexico*) vol 1 (Mexico: Universidad Nacional Autonoma de Mexico) p 209.

[3] Timashkov D A et al 2008 *Astropart. Phys.* **30** 117.

[4] Barbashina N S et al 2009 *Bull. Russ. Acad. Sci. Phys.* **73** 343.
[5] Mikhaylenko A S et al 2011 *Bull. Russ. Acad. Sci. Phys.* **75** 827.
[6] Dorman L I 1972 *Meteorological Effects in Cosmic Rays* (in Russian, Moscow: Nauka) .
[7] Dmitrieva A N, Kokoulin R P, Petrukhin A A, Timashkov D A 2011 *Astropart. Phys.* **34** 401–411.
[8] Glagolev Yu A 1970 *Reference Book on Physical Parameters of the Atmosphere* (in Russian, Leningrad: Gidrometizdat).
[9] Shutenko V V et al 2013 *Geomagn. Aeronomy.* **53** 571.
[10] Shutenko V V et al 2013 *J. Phys.: Conf. Ser.* **409** 012193.
[11] Shutenko V V et al 2009 *Bull. Russ. Acad. Sci. Phys.* **73** 347.
[12] I.I. Yashin et al 2013 *J. Phys.: Conf. Ser.* **409** 012192

The average longitudinal air shower profile: exploring the shape information

R. Conceição[1], S. Andringa[1], F. Diogo[1], M. Pimenta[1,2]

[1] LIP, Av. Elias Garcia, 14-1, 1000-149 Lisboa, Portugal

[2] Departamento de Física, IST, Av. Rovisco Pais, 1049-001 Lisboa, Portugal

E-mail: ruben@lip.pt

Abstract. The shape of the extensive air shower (EAS) longitudinal profile contains information about the nature of the primary cosmic ray. However, with the current detection capabilities, the assessment of this quantity in an event-by-event basis is still very challenging. In this work we show that the average longitudinal profile can be used to characterise the average behaviour of high energy cosmic rays. Using the concept of universal shower profile it is possible to describe the shape of the average profile in terms of two variables, which can be already measured by the current experiments. These variables present sensitivity to both average primary mass composition and to hadronic interaction properties in shower development. We demonstrate that the shape of the average muon production depth profile can be explored in the same way as the electromagnetic profile having a higher power of discrimination for the state of the art hadronic interaction models. The combination of the shape variables of both profiles provides a new powerful test to the existing hadronic interaction models, and may also provide important hints about multi-particle production at the highest energies.

1. Introduction

The origin and nature of ultra high energy cosmic rays (UHECRs) is still unknown. Although it is expected to be of hadronic nature, something between proton and iron, the determination of the UHECRs mass composition is still one of the present biggest challenges on the experiments dedicated to the study of these particles. Their mass composition can only be inferred indirectly through the analysis of the shower that was initiated by the interaction of UHECRs with the atmosphere atoms.

However, the shower development depends also on the hadronic interactions for which there is no first principle description, only different phenomenological models that try to extrapolate the accelerator data obtained at lower energies and different phase space regions.

The analysis of the electromagnetic (e.m.) shower longitudinal profile can be used to determine the primary energy and composition. The primary energy is related with the maximum number of particle, N_{max}, and the primary mass composition can be inferred statistically from the depth at which the maximum occurs, X_{max}. Again, this mass composition observable has also a strong dependence on the primary cross-section. There is, however, more information hidden in the details of the shower longitudinal profile shape [1, 2, 3].

Electromagnetic longitudinal shower profiles are well described by the Gaisser-Hillas function, which can be parametrised in $N' \equiv N/N_{max}$ and $X' \equiv X - X_{max}$ as:

$$N' = \left(1 + \frac{RX'}{L}\right)^{R^{-2}} \exp\left(-\frac{X'}{LR}\right) \tag{1}$$

With the above parametrisation, introduced in [1], the profile can be recognised as a Gaussian, of width **L**, with an asymmetry introduced by non-zero values of **R**. It is worth noting that after translating the maximum to zero all the variations due to the point of the first interaction disappear. In other words, the shape should be insensitive to the primary cross-section.

The shape of the electromagnetic longitudinal profile has been shown to be sensitive to primary mass composition, in an event-by-event basis, using simulated showers [1]. However, from the experimental point of view this is not an easy measurement. The Gaisser-Hillas fit to extract the shape parameters is affected by the statistical uncertainties on the light collection by the fluorescence telescopes and the uncertainty on the determination of X_{max}. Moreover, this kind of optical experiments have several systematic uncertainties such as Cherenkov direct and scattered light estimation, telescope alignment, atmospheric conditions, among others [4, 5]. This makes it extremely difficult to extract the information on the shape, in particular to access the two shape parameters at the same time (see fig. 1 (left)).

A possible strategy would be to look at the average longitudinal profile taking advantage of its universal behaviour, as shown in figure 1 (right). By doing so, the statistical uncertainties should be reduced, so that the two shape parameters, **L** and **R**, can be fit simultaneously. On the other hand, the measurement of the average profile could help to control the experimental systematic uncertainties enabling future event-by-event analysis (for instance, the results between different telescopes should be statistically consistent).

However, such strategies rely on the assumption that the average *universal* electromagnetic longitudinal shower profile still has information about the primary mass composition.

To investigate this, we have generated, using CONEX [6, 7], 10000 showers of proton, helium, nitrogen and iron, for different hadronic interaction models (SIBYLL2.1 [8], QGSJET-II.04 [9] and EPOS-LHC [10, 11]) and different energies being the reference energy through this manuscript $E = 10^{19}$ eV. The zenith angle was fixed at 40° as it was shown previously [1, 12] that the shape do not depends significantly on this quantity.

The first test that one can conceive is to average several electromagnetic shower profiles and then extract the two shape variable through a Gaisser-Hillas fit using equation (1). The fit range was chosen as $X' \in [-300; 200]\text{gcm}^{-2}$. Although the choice of the fit range is somewhat arbitrary, it had as motivation the experimental conditions at which the measurement should be done: on one hand include a range as big as possible to increase the significance of the fit procedure; on the other hand exclude regions such as the tails where the uncertainties are large, for instance due to the Cherenkov scattering light component. The proposed test is shown in figure 2 for different primaries and as a function of the number of events that was used to perform the average.

It can be seen from this figure that in fact, the average e.m. profile is sensitive to the difference between primaries. Moreover, the values of **L** and **R** converge after averaging a few hundred events allowing to distinguish between the different primaries. Another important point to make is that, as expected, both shape parameters contain relevant information about the primary particle.

In the following sections we shall explore the sensitivity of the shape of the longitudinal profile to both primary mass composition and hadronic interaction models. In section 3 we will discuss the same for the average muon production depth profile. Finally, we end with some conclusions.

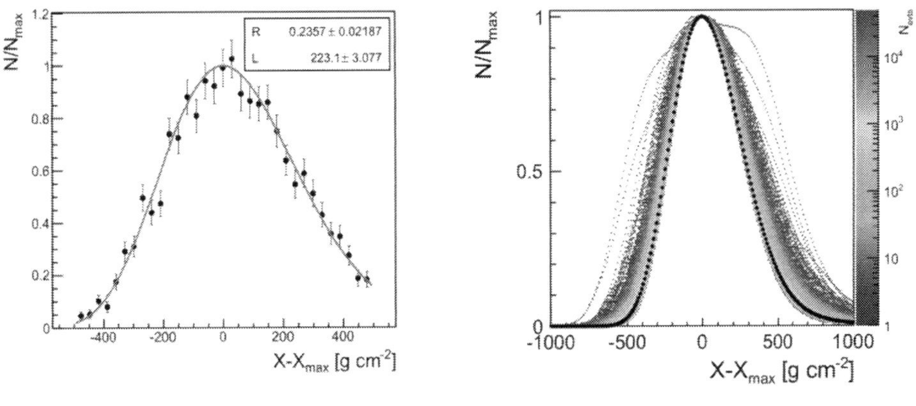

Figure 1. (left) Single electromagnetic longitudinal profile written as a function of $X' \equiv X - X_{max}$ and normalized to its maximum, N_{max}. The error bars represent the statistical uncertainty for a typical UHECR shower event measured by a fluorescence telescope. (right) The black points represent the average electromagnetic longitudinal profile averaged over the 10000 shower profiles displayed in the plot. The color code represents the density of showers in (X', N').

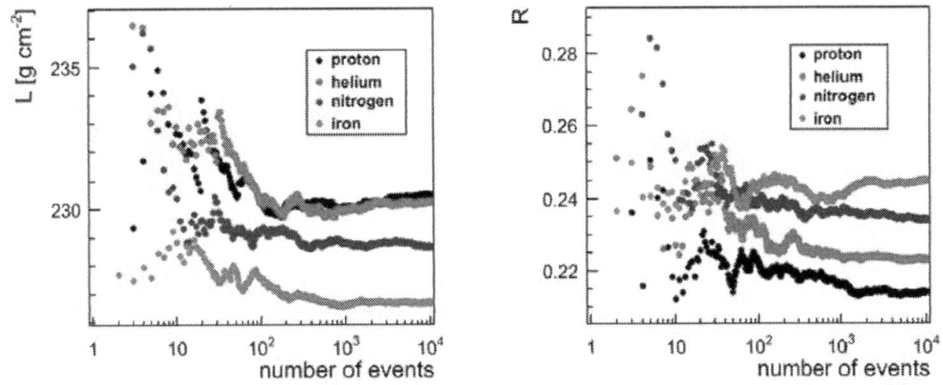

Figure 2. Shape parameters **L** (left) and **R** (right) as a function of the number of events used to get the average shower. The showers were generated using QGSJET-II.04 as the high energy hadronic interaction model. Color codes show different primaries at $\log(E/\mathrm{eV}) = 19.0$.

2. Average electromagnetic profile shape

In this section we shall discuss the properties that can be extracted from the shape of the average electromagnetic profile.

2.1. Sensitivity to primary mass composition

It has been seen in the previous section that the average longitudinal electromagnetic shape variables **L** and **R** are both sensitive to the nature of the cosmic ray that induces the shower. It is then only natural to plot both variables at the same time to assess its sensitivity to the primary mass composition. This is what is shown in figure 3. The crosses represent the values (**L**,**R**) for pure mass composition samples. In this figure it is also shown all the possible mass composition possibilities using four primaries: proton, helium, nitrogen and iron. Each coloured dot represents a particular mass composition scenario and the colour is the average mass logarithm given by,

$$\langle ln(A)\rangle = f_p \ln(1) + f_{He} \ln(4) + f_N \ln(14) + f_{Fe} \ln(56) \tag{2}$$

where f_X is the fraction of the element X.

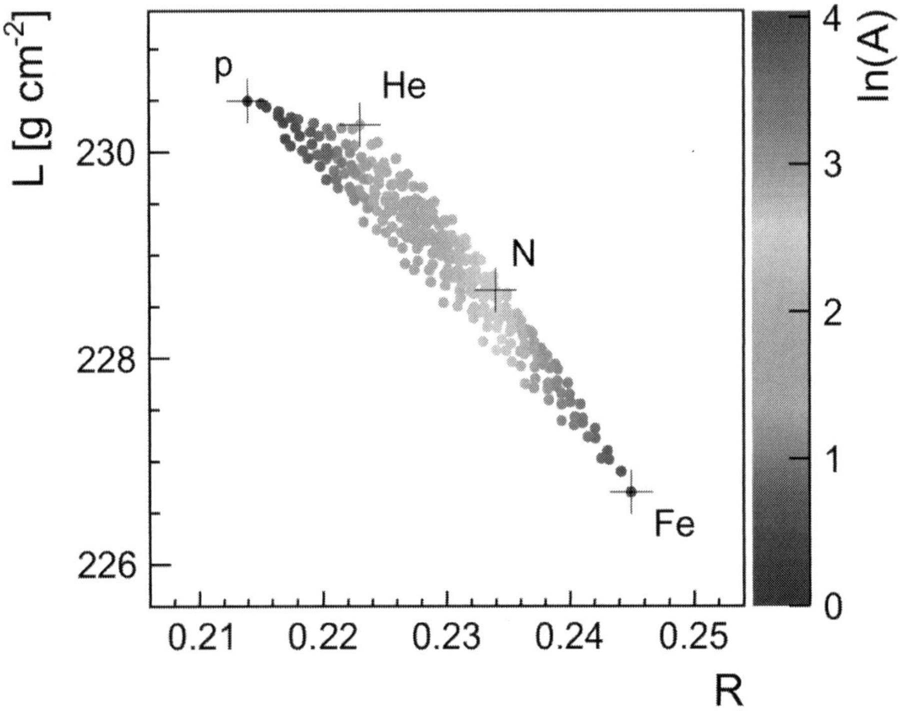

Figure 3. Relation of the shape parameters of the average shower profile with the $\langle ln(A)\rangle$. The showers were generated using QGSJET-II.04 at $\log(E/\text{eV}) = 19$. Color codes show the $\langle ln(A)\rangle$ for different combinations of primaries and crosses mark the results for pure composition samples. Each point was generated averaging 10000 shower longitudinal profiles.

It is possible to conclude from this figure that, within a particular hadronic interaction model, the assessment of the shape quantities **R** and **L** can be used to derive the $\langle ln(A)\rangle$, similar to what is done for X_{max} [13], independently of the first interaction point.

2.2. Sensitivity to Hadronic Interaction Models

The dependence of the shape variables with the hadronic interaction models was also investigated. It can be seen, in figure 4 that there is some discrimination power for the standard hadronic interaction models. Although EPOS-LHC and QGSJET-II.04 have some overlap there are some mass composition choices for which the models occupy a different phase space, for instance for pure proton.

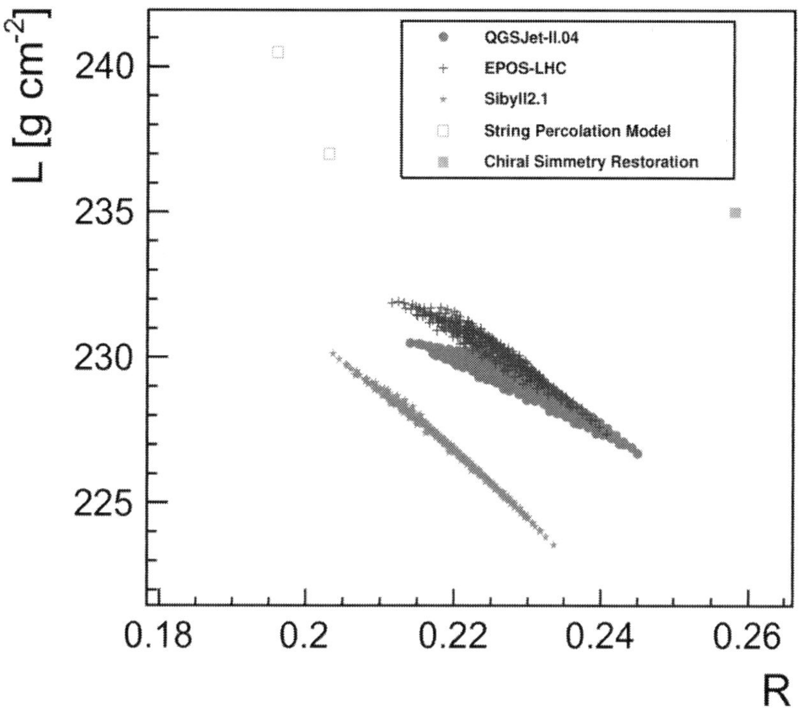

Figure 4. Shape parameters **(L, R)** of the average electromagnetic shower profile , generated at $\log(E/\mathrm{eV}) = 19$, for the different mass composition combinations with different hadronic interaction models (see legend for details).

The predictions for the shape variables of the e.m. average profile from exotic models (models that try to explain the UHECR data through changes on the hadronic interaction physics while the mass composition remains constant) are also displayed on fig. 4. It can be seen from this plot that the shape variables have a strong discrimination power for this kind of models. Moreover, note that these models were tuned to reproduce the X_{max} and the number of muons at ground. Therefore, the measurement of the shape of the average electromagnetic profile is an important test to this exotic scenarios.

The energy evolution of the shape variables phase space for different models was also investigated and is presented in figure 5 (for all the available high energy hadronic interaction models, including older versions). All the models display a similar behaviour: as the energy increases **L** increases and **R** decreases. The increase of **L** is explained by its relation with the shower energy (remember that **L** is the width of the gaussian, and therefore related with the integral which by its turn gives the energy). On the other hand, the decrease of **R** is an indication that the shower is getting less asymmetrical (more gaussian). Moreover, the distance between the models phase space is being reduced as the energy increases. This may be an indication that the shower is becoming more universal and that the difference between models have a lower impact on the shower development. Finally, it is worth to mention that the newer versions of the models seems to be favouring higher **L** and **R** parameters. These models have been re-tuned not only to describe better the new Large Hadron Collider (LHC) data but also to reproduce

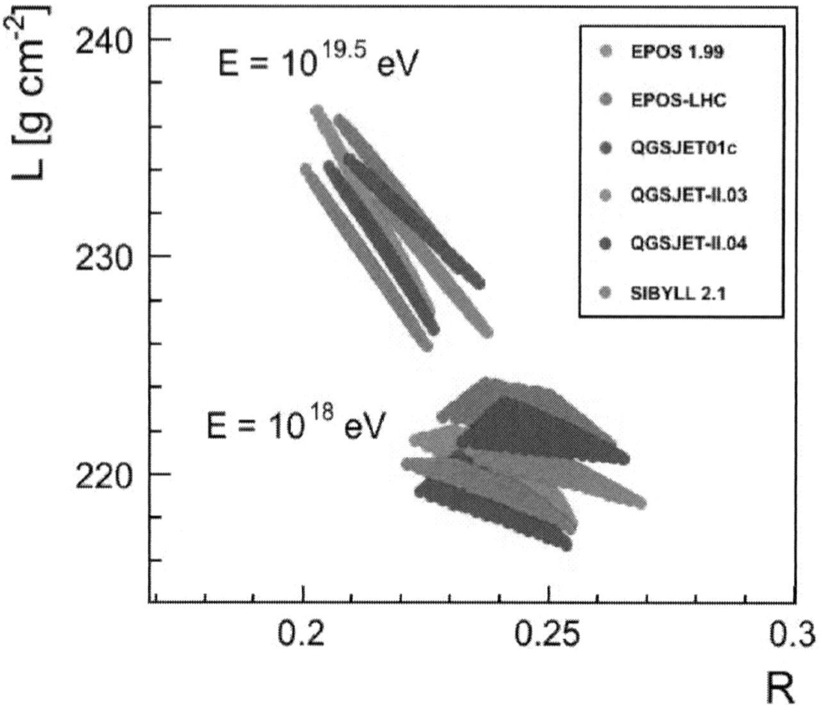

Figure 5. Shape parameters (**L**, **R**) of the average electromagnetic shower profile for the different mass composition combinations with different hadronic interaction models (see legend for details). The results are shown for two different energies: $\log(E/\text{eV}) = 18$ (bottom) and $\log(E/\text{eV}) = 19.5$ (top).

the number of muons at ground measured by UHECR experiments. The increase of the muon content on simulated showers affects the balance between the electromagnetic and hadronic shower component and this should be reflected in the shower average longitudinal shape. This means that the assessment of the **L** and **R** of the average profile may be used to evaluate some of the tunings of the hadronic interaction models.

3. Average muon production depth profile shape

It has been shown in [12] that the muon production depth profile also presents an universal shape when presented in terms of $X' = X - X^{\mu}_{max}$ and $N' = N/N_{max}$. Moreover it has been shown in this same manuscript that this profile can be describe by a Gaisser-Hillas function and its main characteristics (X_{max} and shape) have the same sensitivity to mass composition and hadronic physics as the ones from the electromagnetic profile.

Hence, it is only natural that the previous investigations on the electromagnetic shower average profile are now applied to the muon production depth *average* profile. The results of this investigation can be summarised in figure 6. In this plot we show the shape variables of the average muon production depth profile for three different hadronic interaction models. Similarly, to what was shown for the e.m. profile in fig. 4, each point represents a specific mass composition combination of p, He, N and Fe, and the full area displays the possible phase space within a model.

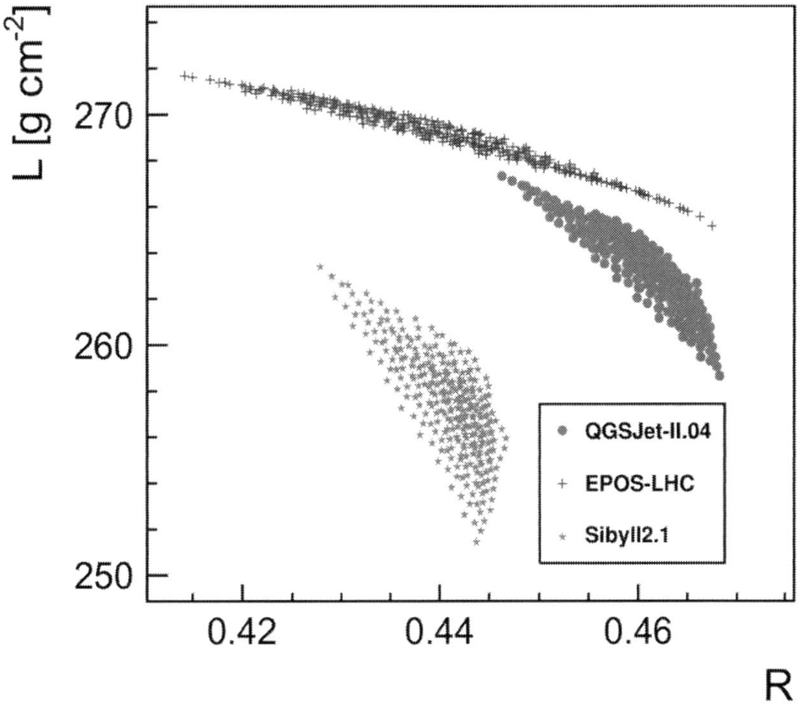

Figure 6. Shape parameters **(L, R)** of the average muon production depth profile , generated at $\log(E/\mathrm{eV}) = 19$, for the different mass composition combinations with different hadronic interaction models (see legend for details).

It is clear from this plot that for all the models, both muonic shape variables have higher values than the ones from the electromagnetic profile, which means that the muonic production profile is broader and more asymmetric. A closer inspection of the plot reveals that the shape variables of this profile have a higher power of discrimination between models than the ones from the e.m. profile. In fact, the models can be distinguished independently of the primary mass composition. However, presently, this profile is more difficult to measure experimentally [14].

Finally, it is important to emphasise that the assessment of the electromagnetic and muonic profile gives insight to distinct components of the shower. The former is related to π^0 production and to the development of the electromagnetic cascade (photons, electrons...), which decouples very early from hadronic shower development, while the later arises essentially from the decay of charged pions and kaons being thus sensitive to all the stages of the hadronic shower.

Although these two profiles may give insight to different processes and stages of the shower development, both are connected by the same primary particle. This means that the interpretation of the shower observables, in particular the shape variables of the average profiles, need to present consistent results in terms of primary mass composition. The failure to achieve this would be a proof that there are inconsistencies with the shower description. Hence, a combined analysis of figures 4 and 6 can enhance our capability to test our knowledge over hadronic interaction models (or other exotic models) independently of the primary mass composition.

4. Conclusions

The interpretation of UHECRs in terms of mass composition is bounded to our understanding of hadronic interactions at the highest energies. It has been shown that the shape of the shower profile provides a new insight over shower mass composition and hadronic interaction physics without depending on the first interaction. Although very promising, these quantities are not easy to access experimentally on an event-by-event basis. Hence, in this work we propose to measure instead the shape of the *average* electromagnetic shower profile. In this way one could measure simultaneously the two shape parameters. We have demonstrated that the shape of the average electromagnetic longitudinal profile is both sensitive to mass composition and to hadronic interaction models. Moreover, we have investigated the shape of the muon production depth profile, using the same procedure, and obtained similar findings. In fact, the shape the average muon production profile allow us to discriminate between hadronic interaction models independently of the considered mass composition. Finally, it is important to note that not only the shape of these two distinct average profiles are offering different insights of the shower development, but they are also bounded to the same primary. Therefore, a combined analysis can provide strong tests to our current knowledge over the shower description by requiring that the shapes of the two profiles are interpreted in terms of the same primary mass composition.

Acknowledgments

This work is partially funded by Fundação para a Ciência e Tecnologia (SFRH/BPD/73270/2010) and (SFRH/BD/89109/2012).

References

[1] Andringa S, Conceição R and Pimenta M 2011 *Astropart.Phys.* **34** 360–367
[2] Matthews J, Mesler R, Becker B, Gold M and Hague J 2010 *J.Phys.G* **G37** 025202 (*Preprint* 0909.4014)
[3] Giller M, Kacperczyk A, Malinowski J, Tkaczyk W and Wieczorek G 2005 *J.Phys.G* **G31** 947–958
[4] Keilhauer B, Bohacova M, Fraga M, Matthews J, Sakaki N *et al.* 2013 *EPJ Web Conf.* **53** 01010 (*Preprint* 1210.1319)
[5] Pekala J, Homola P, Wilczynska B and Wilczynski H 2009 *Nucl.Instrum.Meth.* **A605** 388–398 (*Preprint* 0904.3230)
[6] Pierog T *et al.* 2006 *Nucl. Phys. Proc. Suppl.* **151** 159–162 (*Preprint* astro-ph/0411260)
[7] Bergmann T *et al.* 2007 *Astropart. Phys.* **26** 420–432 (*Preprint* astro-ph/0606564)
[8] Ahn E J, Engel R, Gaisser T K, Lipari P and Stanev T 2009 *Phys.Rev.* **D80** 094003 (*Preprint* 0906.4113)
[9] Ostapchenko S 2011 *Phys.Rev.* **D83** 014018 (*Preprint* 1010.1869)
[10] Werner K, Liu F M and Pierog T 2006 *Phys.Rev.* **C74** 044902 (*Preprint* hep-ph/0506232)
[11] Pierog T, Karpenko I, Katzy J, Yatsenko E and Werner K 2013 (*Preprint* 1306.0121)
[12] Andringa S, Cazon L, Conceição R and Pimenta M 2012 *Astropart.Phys.* **35** 821–827 (*Preprint* 1111.1424)
[13] Kampert K H and Unger M 2012 *Astropart.Phys.* **35** 660–678 (*Preprint* 1201.0018)
[14] Cazon L, Conceição R, Pimenta M and Santos E 2012 *Astropart.Phys.* **36** 211–223 (*Preprint* 1201.5294)

The AMIGA enhancement of the Pierre Auger Observatory

B. Daniel for the Pierre Auger Collaboration[1]
UNICAMP - State University of Campinas - Brazil
E-mail: bdaniel@ifi.unicamp.br

Abstract. AMIGA (Auger Muon and Infill for the Ground Array) is a low energy enhancement for the Pierre Auger Observatory, designed to extend its measurements down to 10^{17} eV to get new insights in the region of the spectrum where the transition between galactic and extragalactic cosmic rays occurs. The current status and preliminary results from AMIGA are presented and discussed in this work.

1. Introduction

While the original proposal for the Pierre Auger Observatory [1] was to study the ultra-high energy cosmic rays from 3×10^{18} eV onwards, AMIGA [2] is an enhancement to detect primary particles above 3×10^{17} eV. It consists of an Infill array of water-Cherenkov detectors, separated by 750 m (the normal spacing for the surface detector is 1500 m) covering an area of 23.5 km^2. Each station is accompanied by a 30 m^2 scintillator counter buried at a depth of 2.3 m to shield from the electromagnetic particles and directly count muons (important information for composition studies). The counter is composed of four independent modules grouped in two with 5 m^2 and two with 10 m^2, as sketched in figure 1. Each module is segmented in 64 plastic scintillator strips (4 cm wide and 1 cm thick to prevent pile-up effects) with optical fibers to collection and transmission of light. The photons are guided to a 64-pixel photomultiplier tube positioned at the center of each module. The scintillators work as counters, i.e., they count signals above a tunable threshold.

2. AMIGA status

The Infill array of water Cherenkov detectors is complete since September 2012, with 61 stations overlooked by three high elevation fluorescence telescopes [3] in the western part of the surface detector (SD) array, shown in the map of figure 2. Data taking started in 2008 and results have already been obtained, like the energy spectrum presented in this work.

A unitary cell (UC) of seven stations with counters forming a hexagon (figure 3) was completed in 2014, but some of the locations still have only one module. The scintillators marked with a yellow circle in the figure are already taking data. Two stations are instrumented with a pair of independent counters called twins (with 30+30 m^2), to allow a direct measurement of the muon counting accuracy. With data from the UC, it will be possible to test the muon

[1] Full author list at http://www.auger.org/archive/authors_2014_10.html

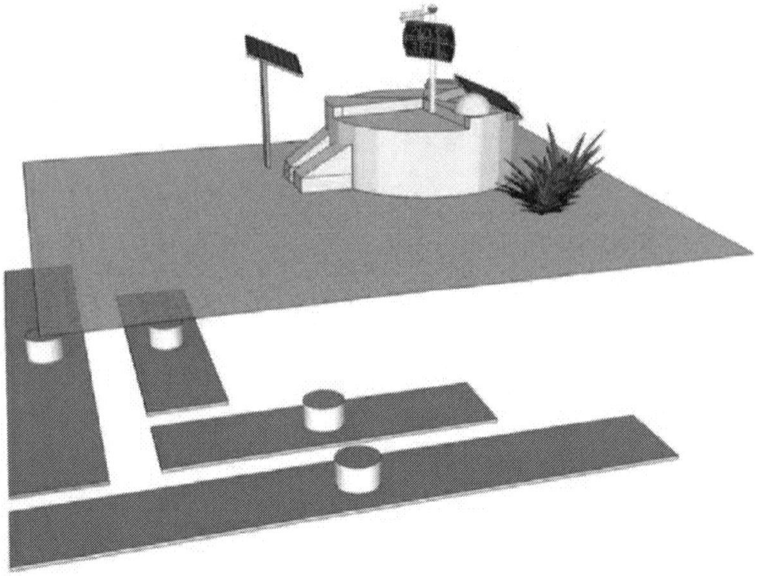

Figure 1. Configuration of the AMIGA stations, with the water Cherenkov detector and buried scintillator modules.

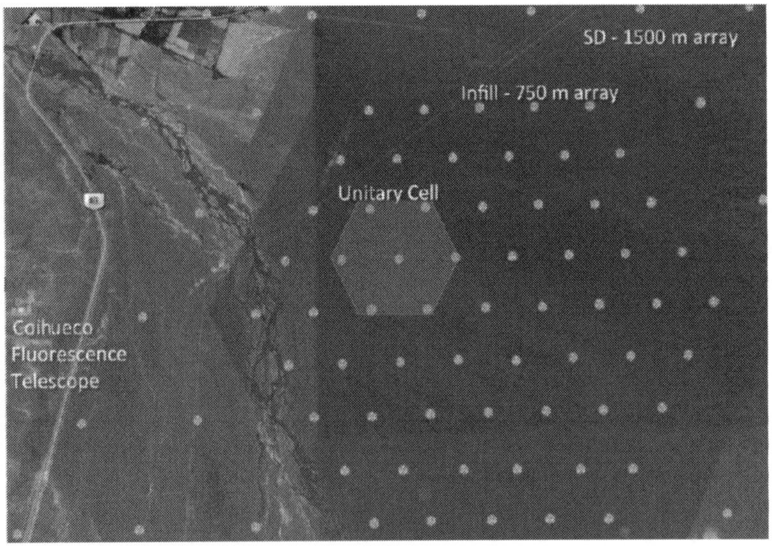

Figure 2. Map of the AMIGA Infill and unitary cell of muon counters.

counting technique, develop the calibration method and validate the detector design, before proceeding to the full deployment.

3. Energy spectrum from the Infill

The energy spectrum from the Infill array [4] was obtained using well reconstructed, high quality events from 08/2008 to 12/2012 (total exposure of (79 ± 4) $\text{km}^2\,\text{sr}\,\text{yr}$), providing the result from figure 4. The energy is evaluated based on the calibration obtained from a subset of hybrid events (observed both by the Infill stations and the fluorescence telescopes). Up to the energy

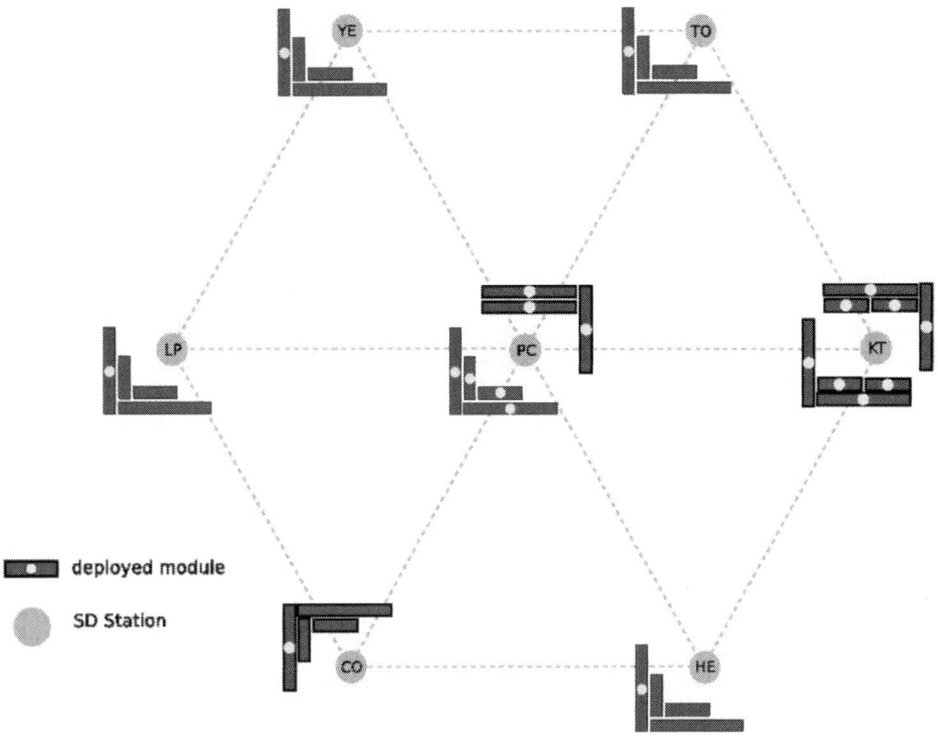

Figure 3. Muon counters of the AMIGA unitary cell. The modules without the yellow dot were not deployed yet.

of the ankle (4×10^{18} eV), a power law fit (red line in the plot) gives a spectral index of ($3.27 \pm 0.02(stat.) \pm 0.07(syst.)$), in agreement with previous results [5]. In figure 5 the spectrum is compared with other results [6] from the surface detector: for the 1500 m array, the inclined events and the hybrid reconstruction. A good agreement is found between them.

4. First results from the buried scintillators

Operation started in March 2013 at the UC. Data from the muon counters are transmitted to the central of acquisition, where they are merged with the surface detector events. The geometry and energy of the shower are taken from the SD, using the usual procedures of reconstruction, and the muon numbers at each station are given by the scintillators. A counting strategy based on laboratory measurements for the duration of the signal of muons from air showers is applied in order to reduce the miscounting due to systematic uncertainties. A pile-up correction is also applied to the total number of muons.

For a given event from the UC, the muonic lateral distribution function (MLDF) is fitted through a KASCADE-Grande like function [7], with particle density given by

$$\rho_\mu(r) = \rho_\mu(450)\left(\frac{r}{r_0}\right)^{-\alpha}\left(1+\frac{r}{r_0}\right)^{-\beta}\left[1+(\frac{r}{10r_0})^2\right]^{\gamma},$$

where $\rho_\mu(450)$ and β are the fitting parameters and the others are fixed at $\alpha = 1$, $\gamma = 1.85$ and $r_0 = 150$ m. In figure 6 an example of MLDF is presented, for an event of energy 2.7×10^{18} eV and zenith angle of 39.9°. In this case, all SD stations from the UC were triggered, but some counters are silent (detected less than three muons) or saturated (received more than 21 muons in a time window of 25 ns).

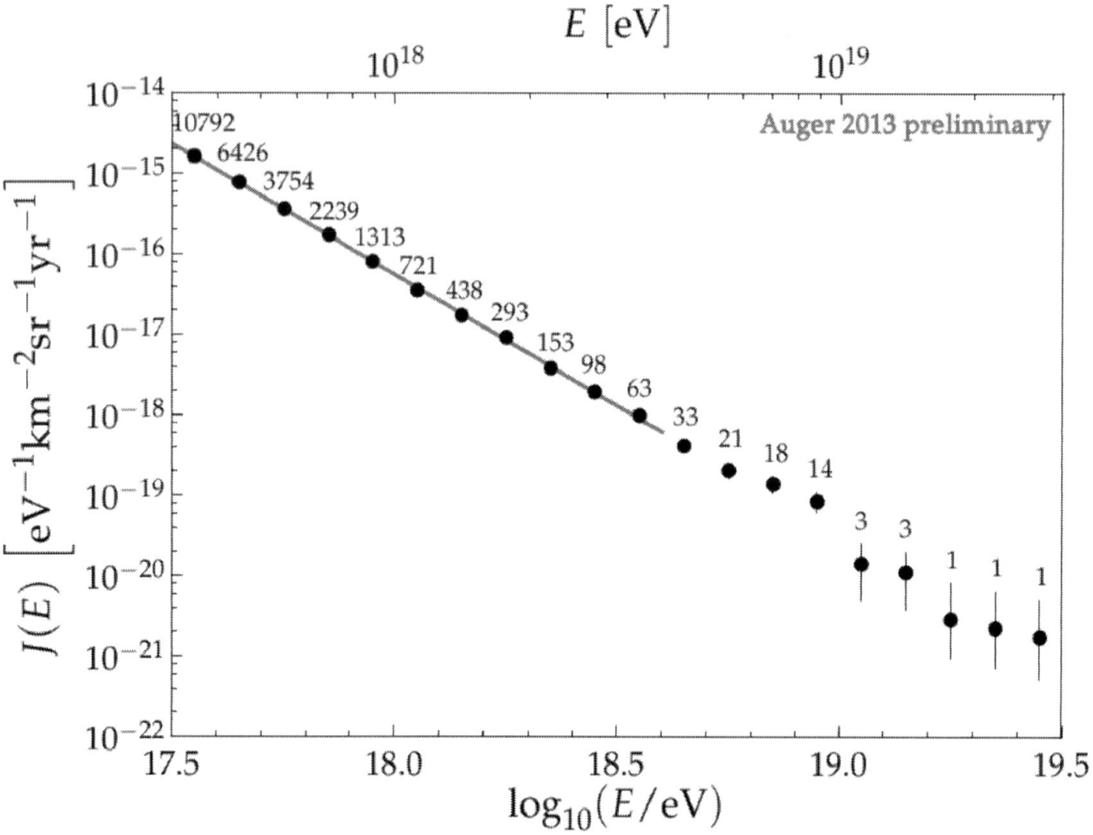

Figure 4. Energy spectrum of events from the 750 m surface detector array.

Despite the still short time of operation, selecting all events with at least two counters, the average measured MLDF (normalized to the fitted muon density at 450 m from the core) was obtained [8] for 230 events with energy above the full efficiency of detection, as shown in figure 7. The points are fitted only in a very limited range of core distance, due to the low number of events. Although preliminary, the radial dependence of the muon densities with the decreasing zenith angle can be seen for the three angular bins evenly separated in $\cos(\theta)\sin(\theta)$.

The accuracy of the muon counters was studied with data from the instrumented twin [9] position. A preliminary comparison of data, with 3000 events above 10^{17} eV, from both counters at the same surface station (with their counts plotted in figure 8) indicates that the detectors are working properly. The red line in the plot represents a perfect behavior, where both detectors count the same number of muons, so the points are expected to be distributed around it. The accuracy seems to increase for higher numbers of muons, as expected, since the poissonian fluctuations decrease.

5. Summary

AMIGA is taking data with the full Infill array of surface stations and a Unitary Cell of buried scintillator counters, which are not all complete yet. The reconstruction of first events from this 750 m array allowed the study of the energy spectra above $10^{17.5}$ eV, giving a spectral index of $3.27 \pm 0.02(stat.) \pm 0.07(syst.)$ up to the ankle, compatible with previous results from the surface detector. The first results from the scintillators show that the detectors are working as expected and indicate that the muon counting accuracy can be evaluated when enough data will

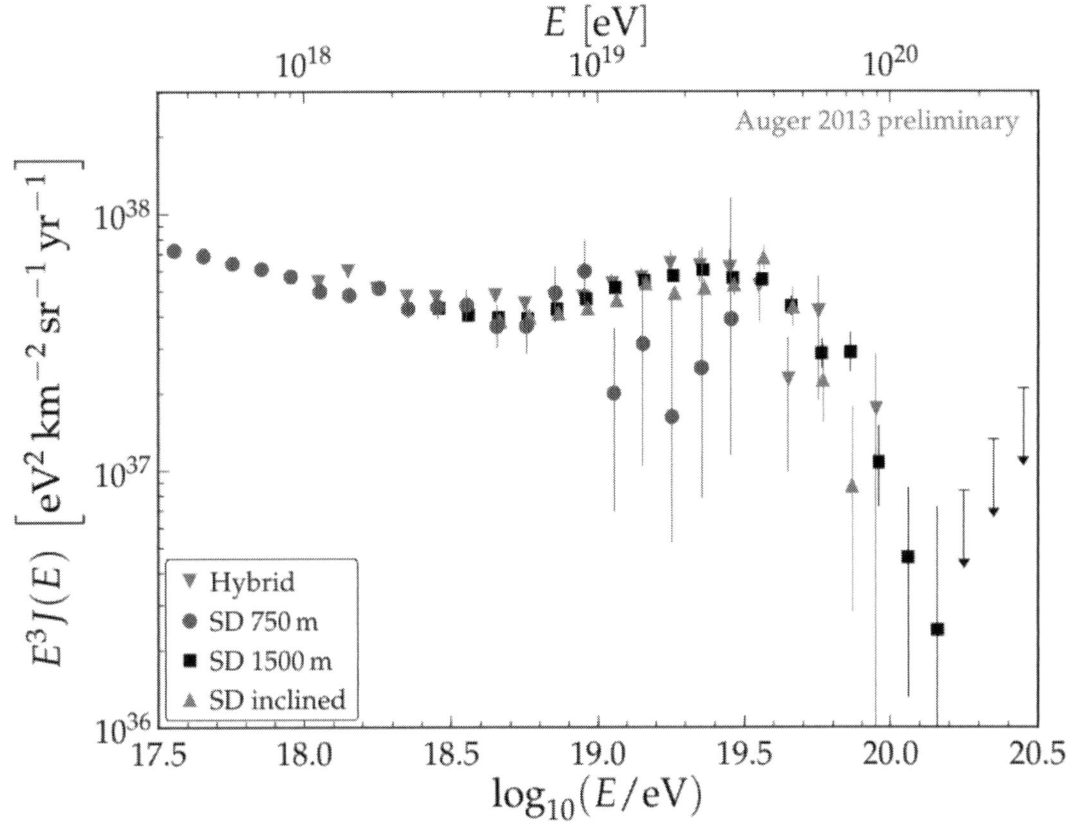

Figure 5. Energy spectra derived from the surface detector and hybrid data.

be collected with the twins.

6. Acknowledgments

This work was supported by the grant 2013/17546-6, from the São Paulo Research Foundation (FAPESP). The opinions, hypothesis and conclusions or recommendations expressed in this text are responsibility of the authors and don't necessarily reflect FAPESP's thoughts.

The authors acknowledge Antonella Castellina and Federico Sanchez for the reading and comments on the manuscript, and also the members of the Pierre Auger Collaboration who contributed to the results presented. The successful installation, commissioning, and operation of the Pierre Auger Observatory would not have been possible without the strong commitment and effort of the technical and administrative staff in Malargüe.

References

[1] The Pierre Auger Collaboration 2011 *Astropart. Phys.* **35** 266
[2] Etchegoyen A for the Pierre Auger Collaboration 2007 *Proc. 30th ICRC* (Merida - Mexico) **5** 1191
[3] Mathes H J for the Pierre Auger Collaboration 2007 *Proc. 30th ICRC* (Merida - Mexico) **3** 153
[4] Ravignani D for the Pierre Auger Collaboration 2013 *Proc. 33rd ICRC* (Rio de Janeiro - Brazil) (*Preprint* arXiv:1307.5059)
[5] Salamida F for the Pierre Auger Collaboration 2011 *Proc. 32nd ICRC* (Beijing - China) **2** 145
[6] Schulz A for the Pierre Auger Collaboration 2013 *Proc. 33rd ICRC* (Rio de Janeiro - Brazil) (*Preprint* arXiv:1307.5059)
[7] KASCADE-Grande Collaboration 2005 *Proc. 29th ICRC* (Pune - India) **6** 301

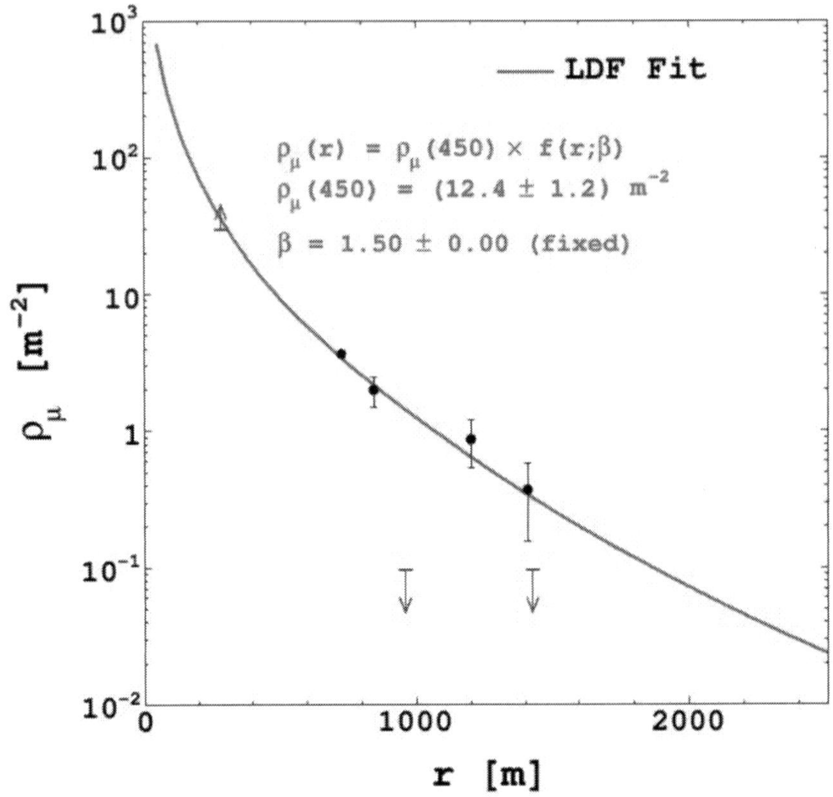

Figure 6. MLDF for an AMIGA event. Blue upward and downward arrows are saturated and silent counters respectively.

[8] Suarez F for the Pierre Auger Collaboration 2013 *Proc. 33rd ICRC* (Rio de Janeiro - Brazil) (*Preprint* arXiv:1307.5059)
[9] Maldera S for the Pierre Auger Collaboration 2013 *Proc. 33rd ICRC* (Rio de Janeiro - Brazil) (*Preprint* arXiv:1307.5059)

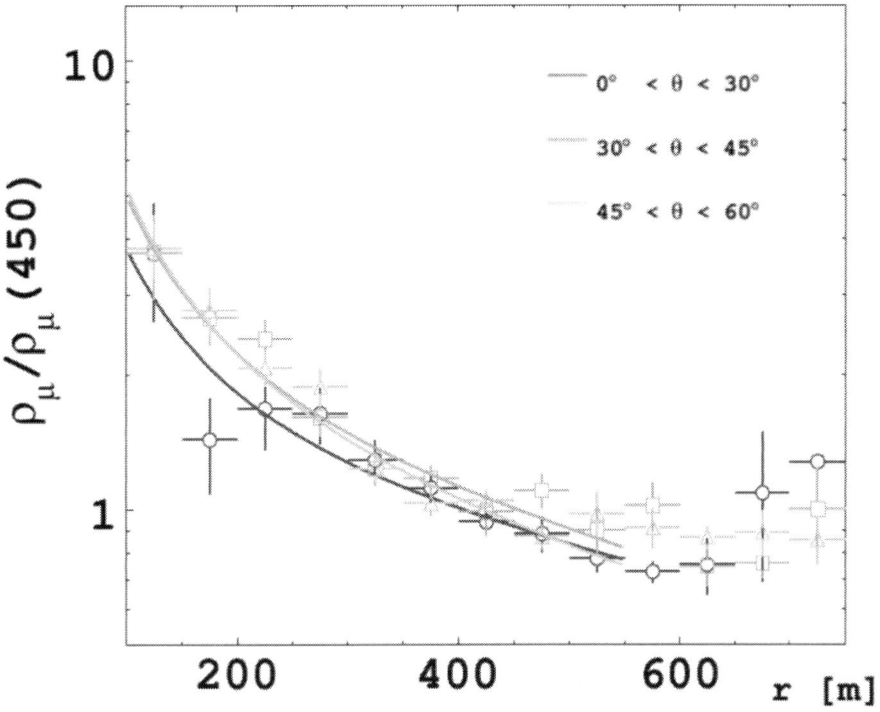

Figure 7. Mean muon lateral distribution function obtained with the AMIGA muon counters.

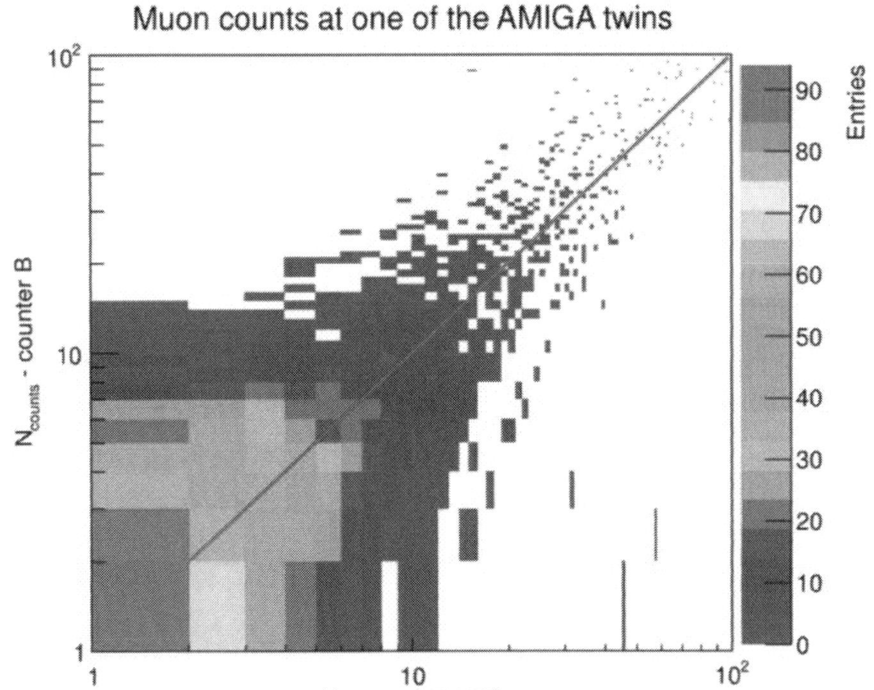

Figure 8. Muon counts in events of a twin muon counter of AMIGA.

Latest results from the ARGO-YBJ experiment

Di Sciascio Giuseppe on behalf of the ARGO-YBJ Collaboration
INFN - Sezione Roma Tor Vergata, Roma, Italy
E-mail: giuseppe.disciascio@roma2.infn.it

Abstract. The ARGO-YBJ experiment has been in stable data taking for 5 years at the YangBaJing Cosmic Ray Observatory (Tibet, P.R. China, 4300 m a.s.l., 606 g/cm^2). With a duty-cycle greater than 86% the detector collected about 5×10^{11} events in a wide energy range, from few hundreds GeV up to about 10 PeV. A number of open problems in cosmic ray physics has been faced exploiting different analyses. In this paper we summarize the latest results in cosmic ray physics and in gamma-ray astronomy.

1. Introduction

Aiming to face the open problems in Galactic cosmic ray (hereafter CR) physics through a combined study of photon- and charged particle-induced extensive air showers (hereafter EAS) with the same detector, ARGO-YBJ has been in stable data taking for more than 5 years at the YangBaJing Cosmic Ray Observatory (Tibet, P.R. China, 4300m a.s.l., 606 g/cm^2, 90.5° East, 30.1° North). The detector acted simultaneously as a wide aperture (~sr), continuosly-operated γ-ray telescope at sub-TeV – TeV photon energies and as a high resolution CR detector in the broad energy range between few TeV and 10 PeV.

The combination of the high elevation of the site, the full coverage of the central carpet and the high granularity of the readout provides very important advantages in various aspects of CR physics. In fact, in addition to the decrease of the threshold energy down to few hundreds GeV, the location of the esperiment above 4000 m a.s.l. ensures that shower fluctuations are small (we are working in the shower maximum region) and that all nuclei produce showers with nearly the same electromagnetic size. The low energy threshold is crucial for the overposition with direct measurements carried out by balloon/satellite-born detectors allowing one to cross-calibrate the different energy scales. The nearly independence of the size on the mass of the primary determinates the size – energy relation to be better defined, allowing the study of the elemental composition around the knee in a very efficient and reliable way.

In this paper the latest results obtained by ARGO-YBJ in CR physics and in gamma-ray astronomy will be briefly summarized.

2. The ARGO-YBJ experiment

ARGO-YBJ is a full coverage air shower detector constituted by a central carpet ~74×78 m^2, made of a single layer of resistive plate chambers (RPCs) with ~93% of active area, enclosed by a guard ring partially instrumented (~20%) up to ~100×110 m^2. The apparatus has a modular structure, the basic data acquisition element being a cluster (5.7×7.6 m^2), made of 12 RPCs (2.85×1.23 m^2 each). Each chamber is read by 80 external strips of 6.75×61.80 cm^2 (the spatial pixels), logically organized in 10 independent pads of 55.6×61.8 cm^2 which represent the time pixels of the detector [1]. The readout of 18,360 pads and 146,880 strips is the experimental output of the detector. In addition, in order to extend the dynamical range up to PeV energies, each chamber is equipped with two large size pads (139×123 cm^2) to collect the total charge developed by the particles hitting the detector [2]. The RPCs are operated in streamer mode by using a gas mixture (Ar 15%, Isobutane 10%, TetraFluoroEthane 75%) for high altitude operation [3]. The high voltage settled at 7.2 kV ensures an overall efficiency of about 96% [4]. The central carpet contains 130 clusters and the full detector is composed of 153 clusters for a total active surface of ~6,700 m^2. The total instrumented area is ~11,000 m^2. For each event the location and timing of every detected particle is recorded, allowing the reconstruction of the lateral distribution and the arrival direction. The trigger is based on a time correlation among the pad signals depending on their relative distance. In this way, all the shower events giving a number of fired pads $N_{pad} \geq N_{trig}$ in the central carpet in a time window of 420 ns generate the trigger. The whole system has been in stable data taking from November 2007 to January 2013, with the trigger condition $N_{trig} = 20$ and a duty cycle ≥86%. The trigger rate is ~3.5 kHz with a dead time of 4%.

Because of the small pixel size, the detector is able to record events with a particle density exceeding 0.003 particles m^{-2}, keeping good linearity up to a core density of about 15 particles m^{-2}. This high granularity allows a complete and detailed three-dimensional reconstruction of the front of air showers at an energy threshold of a few hundreds GeV. Showers induced by high energy primaries (> 100 TeV) are also imaged by the charge readout of the large size pads which allows to study the structure of the shower core region up to particle densities of $\sim 10^4/m^2$ [2].

Details on the analysis procedure (e.g., reconstruction algorithms, data selection, background evaluation, systematic errors) are discussed in [5, 6, 7]. The performance of the detector (angular resolution, pointing accuracy, energy scale calibration) and the operation stability are continuously monitored by observing the Moon shadow, i.e., the deficit of CRs detected in its direction [6, 8]. The measured angular resolution is better than 0.5° for CR-induced showers with energy E > 5 TeV and the overall absolute pointing accuracy is ~0.1°. According to MonteCarlo (MC) simulations the angular resolution for γ-induced events results smaller by 30% - 40%. The absolute rigidity scale uncertainty of ARGO-YBJ is estimated at 10% level in the range 1 - 30 TeV/Z [6, 8].

3. Cosmic Ray Anisotropy

The CR arrival direction distribution and its anisotropy has been a long-standing problem ever since the 1930s. The study of the anisotropy is complementary to the study of their energy spectrum and elemental composition to understand CR origin and propagation. It is also a powerful tool to probe the structure of the magnetic fields through which CRs travel.

The anisotropy in the CR arrival direction distribution hs been observed by different experiments with increasing sensitivity and details at different angular scales. Current experimental results show that the main features of the anisotropy are uniform in the energy range (10^{11} - 10^{14} eV), both with respect to amplitude (10^{-4} - 10^{-3}) and phase ((0 - 4) hr). The existence of two distinct broad regions, one showing an excess of CRs (called "tail-in"), distributed around 40° to 90° in R.A., the other a deficit (the "loss cone"), distributed around 150° to 240° in R.A., has been clearly observed (for a review see, for example, [9]). In the last years different experiments reported evidence of the existence of a medium angular scale anisotropy in the both hemispheres [9].

In Fig. 1 the ARGO-YBJ sky map of medium angular scale (order of 10°) anisotropy in galactic coordinates as obtained with 4.5 years data is shown (for details see [10, 11]). The map center points towards the galactic Anti-Center. The zenith angle cut ($\theta \leq 50°$) selects the declination region $\delta \sim$ -20°÷ 80°. In this zenith angle bin the angular resolution is nearly

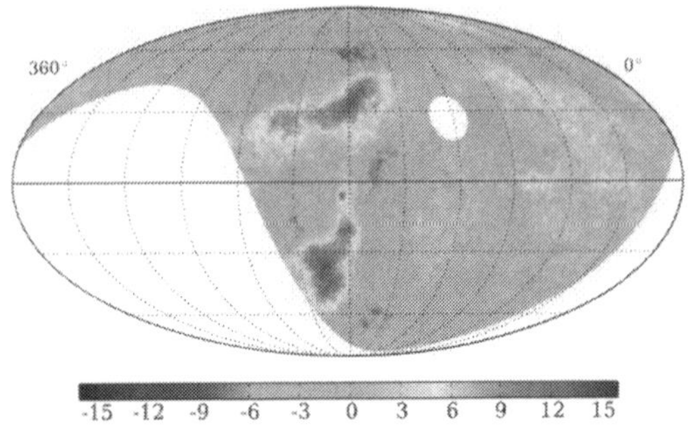

Figure 1. ARGO-YBJ sky-map in galactic coordinates. The color scale gives the statistical significance of the observation in standard deviations. The map center points towards the galactic Anti-Center.

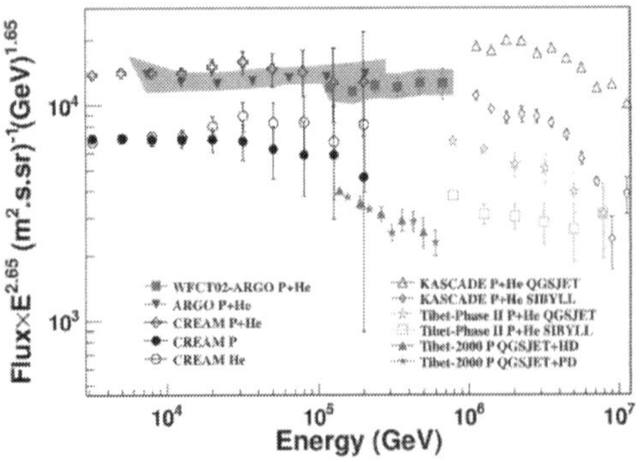

Figure 2. Light component (p+He) energy spectrum of primary CRs measured by ARGO-YBJ compared with other experimental results. The results obtained by the ARGO-YBJ/WFCTA system are shown by the filled red squares [23].

constant and the energy of selected events is in the TeV range. According to the simulation, the median energy of the isotropic CR proton flux is $E_p^{50} \approx 1.8$ TeV (mode energy ≈ 0.7 TeV). No gamma/hadron discrimination algorithms have been applied to the data. Therefore, the sky map is filled with all CRs possibly including photons, without any discrimination.

In spite of the fact that the bulk of Supernova Remnants (SNRs), pulsars and other potential CR sources are in the Inner Galaxy surrounding the Galactic Centre, the excess of CR is observed in the opposite, Anti-Centre direction. As stressed in [12], the fact that the observed excesses are in the Northern and in the Southern Galactic hemisphere, favors the conclusion that the CR at TeV energies originate in sources whose directions span a large range of Galactic latitudes. The right side of the sky map shows different few-degree excesses not compatible with random fluctuations (the statistical significance is up to 7 s.d.). The observation of these structures is reported by ARGO-YBJ for the first time, as discussed in [11].

From the theoretical viewpoint, the anisotropy of CRs is important as a direct trace of potential sources. Nonetheless, models failed to explain the whole set of observations at different angular scales [9]. Unlike predictions from diffusion models, that foresee only a dipole, the CR arrival distribution in sidereal time was never found to be purely dipolar, what translates in needing at least two harmonics to describe the data (the observation of anisotropies down to $\sim 10°$ implies multipoles of order 18). So far, no theory of CRs in the Galaxy exists which is able to explain the origin of these different anisotropies leaving the standard model of CRs and that of the local Galactic Magnetic Field unchanged at the same time. The anisotropy problem is the most serious challenge to the standard model of the origin of galactic CRs from diffusive shock acceleration [13].

4. Measurement of the Cosmic Ray Energy spectrum

There is a general consensus that Galactic CRs up to the all-particle knee originate in SNRs accelerated by the first order Fermi mechanism in shock waves. The theoretical modelling of this mechanism can reproduce, in principle, the measured spectra and composition of CRs. Recently AGILE and Fermi observed GeV photons from two young SNRs (W44 and IC443) showing the typical spectrum feature around 1 GeV (the so called "π^0 bump", due to the decay of π^0) related to hadronic interactions [14, 15]. This important measurement however does not demonstrate the capability of SNRs to accelerate CRs up to the knee and above.

In the standard picture, mainly based on the results of the KASCADE esperiment, the knee is attributed to the steepening of the p and He spectra [16]. However, a number of experiments reported evidence that the bending of the light component (p+He) is well below the PeV and the knee of the all-particle spectrum is due to heavier nuclei (see, for example, [17, 18, 19].

A large number of theoretical papers discussed the highest energies achievable in SNRs and the possibility that protons can be accelerated up to PeVs (for a recent review see [20] and references therein). Therefore, the determination of the proton knee, as well as the measurement of the evolution of the heavier component across the knee, are the key components for understanding origin and acceleration mechanisms of Galactic CRs.

A measurement of the CR primary energy spectrum (all-particle and light component) in the energy range few TeV – 10 PeV is under way with the ARGO-YBJ experiment. To cover this wide energy range different 'eyes' have been used:

- *'digital readout'*, based on the strip multiplicity, i.e. the picture of the EAS provided by the strip/pad system, in the few TeV – 200 TeV energy range [21];
- *'analog readout'*, based on the particle density in the shower core region, in the 100 TeV – 10 PeV range [22];
- *'hybrid measurement'*, carried out by ARGO-YBJ and a wide field of view Cherenkov telescope, in the 100 TeV - PeV region [23].

In the following we summarize results obtained below 100 TeV with the digital readout and describe the hybrid measurement up to the PeV range. Preliminary results obtained with the ARGO-YBJ analog data are described in [22].

4.1. Light component (p+He) spectrum of Cosmic Rays

As described in [21], requiring quasi-vertical showers ($\theta < 30°$) and applying a selection criterion based on the particle density, a sample of events mainly induced by p and He nuclei, with shower core inside a fiducial area (with radius ~28 m), has been selected. The contamination by heavier nuclei is found negligible. An unfolding technique based on the Bayesian approach has been applied to the strip multiplicity distribution in order to obtain the differential energy spectrum of the light component.

The spectrum measured by ARGO-YBJ is compared with other experimental results in Fig. 2 (blue inverted triangles). The ARGO-YBJ data agree remarkably well with the values obtained by adding up the p and He fluxes measured by CREAM both concerning the total intensities and the spectral index [24]. The value of the spectral index of the power-law fit to the ARGO-YBJ data is -2.61±0.04. With this analysis for the first time a ground-based measurement of the CR spectrum overlaps data obtained with direct methods for more than one energy decade.

This measurement has been extended to higher energies exploiting an hybrid measurement with a prototype of the future Wide Field of view Cherenkov Telescope Array (WFCTA) of the LHAASO project [25]. The telescope, located at the south-east corner of the ARGO-YBJ detector, about 78.9 m away from the center of the RPC array, is equipped with 16×16 photomultipliers (PMTs), has a FOV of 14°×16° with a pixel size of approximately 1° [26].

From December 2010 to February 2012, in a total exposure time of 728,000 seconds, the ARGO-YBJ/WFCTA system collected and reconstructed 8218 events above 100 TeV according to the following selection criteria: (1) reconstructed shower core position located well inside the ARGO-YBJ central carpet, excluding an outer region 2 m large; (2) more than 1000 fired pads on the central carpet; (3) more than 6 fired pixels in the PMT matrix; (4) a space angle between the incident direction of the shower and the telescope main axis less than 6°. This selection guarantees that the Cherenkov images are fully contained in the FOV, an angular resolution better than 0.2° and a shower core position resolution less than 2 m.

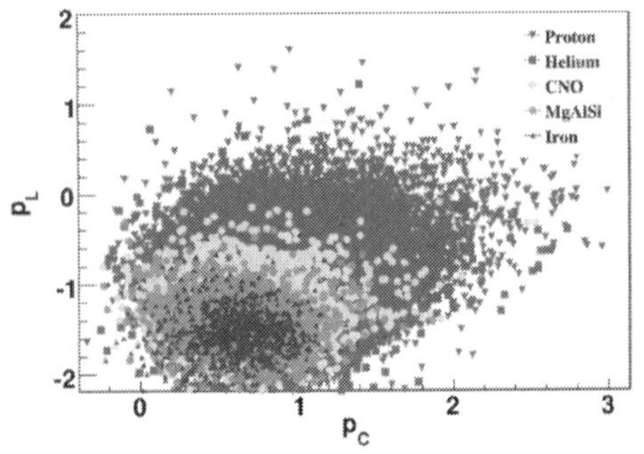

Figure 3. Scatter plot of the parameters p_C and p_L for showers induced by different nuclei. The primary masses have been simulated in the same fraction, assuming a -2.7 spectral index in the energy range 10 TeV – 10 PeV.

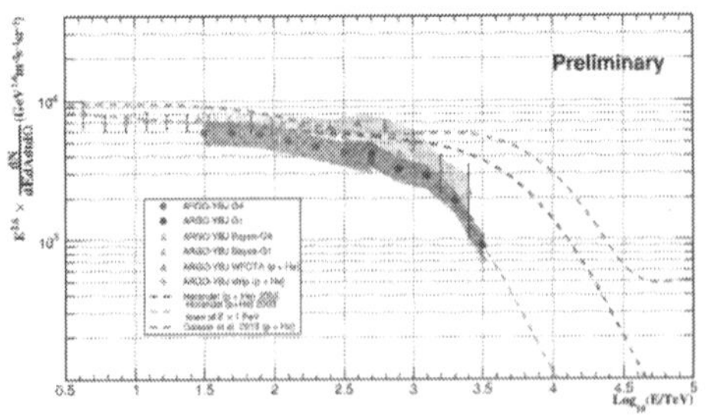

Figure 4. Light (p+He) component energy spectrum of primary CRs measured by ARGO-YBJ with different analyses. The systematic uncertainty is shown by the shaded area and the statistical one by the error bars. The parametrizations given in [28, 29] are shown for comparison. A Horandel-like spectrum with a modified knee at Z×1 PeV is also shown (for details see [22]).

According to the MC simulations, the largest number of particles N_{max} recorded by a RPC in a given shower is a useful parameter to measure the particle density in the shower core region, i.e. within 3 m from the core position. For a given energy, in showers induced by heavy nuclei N_{max} is smaller than in showers induced by light particles. Therefore, N_{max} is a parameter useful to select different primary masses. In addition, N_{max} is proportional to $E_{rec}^{1.44}$, where E_{rec} is the shower primary energy reconstructed using the Cherenkov telescope. We can define a new parameter $p_L = log_{10}(N_{max}) - 1.44 \cdot log_{10}(E_{rec}/TeV)$ by removing the energy dependence [23].

The Cherenkov footprint of a shower can be described by the well-known Hillas parameters [27], i.e. by the width and the length of the image. Older showers which develop higher in the atmosphere, such as iron-induced events, have Cherenkov images more stretched, i.e. narrower and longer, with respect to younger events due to light particles which develop deeper. Therefore, the ratio between the length and the width (L/W) of the Cherenkov image is expected to be another good estimator of the primary elemental composition.

Elongated images can be produced, not only by different nuclei, but also by showers with the core position far away from the telescope, or by energetic showers, due to the elongation of the cascade processes in the atmosphere. Simulations show that the ratio of L/W is nearly proportional to the shower impact parameters R_p, the distance between the telescope and the core position, which must be accurately measured. An accurate determination of the shower geometry is crucial for the energy measurement. In fact, the number of photoelectrons collected in the image recorded by the Cherenkov telescope N_{pe} varies dramatically with the impact parameter R_p, because of the rapid falling off of the lateral distribution of the Cherenkov light. Only an accurate measurement of the shower impact parameters R_p, and a good reconstruction of the primary energy allow to disentangle different effects. A shower core position resolution better than 2 m and an angular resolution better than 0.2°, due to the high-granularity of the ARGO-YBJ full coverage carpet, allow to reconstruct the shower primary energy with a resolution of 25%, by using the total number of photoelectrons N_{pe}. The uncertainty in absolute energy scale is estimated about 10%.

Therefore, in order to select the different masses we can define another new parameter $p_C = L/W - 0.0091 \cdot (R_p/1m) - 0.14 \cdot log_{10}(E_{rec}/TeV)$ by removing both the effects due to the shower distance and to the energy. The values of these parameters for showers induced by different nuclei are shown in the Fig. 3. As can be seen from the figure, a suitable selection in the $p_L - p_C$ space allows to pick out a light composition sample with high purity. In fact, by cutting off the concentrated heavy cluster in the lower-left region in the scatter plot, i.e. $p_L \leqslant -0.91$ and $p_C \leqslant 1.3$, the contamination of nuclei heavier than He is less than 5%. About 30% of H and He survives the selection criteria. In the sample of 8218 events recorded above 100 TeV by the hybrid system, 1392 showers are selected in the (p+He) sub-sample. The aperture of the ARGO-YBJ/WFCTA system is $\sim$170 m^{2}sr above 100 TeV and shrinks to $\sim$50 m^{2}sr after the selection of the (p+He) component.

The light component energy spectrum measured by the ARGO-YBJ/WFCTA hybrid system is shown in the Fig. 2 by the filled red squares. A systematic uncertainty in the absolute flux of 15% is shown by the shaded area. The error bars show the statistical errors only. The spectrum can be described by a power law with a spectral index of -2.63 $\pm$ 0.06. The absolute flux at 400 TeV is $(1.79\pm0.16)\times10^{-11}$ GeV^{-1} m^{-2} sr^{-1} s^{-1}.

This result is consistent for what concern spectral index and absolute flux with the measurements carried out by ARGO-YBJ below 200 TeV and by CREAM. The flux difference is about 10% and can be explained with a difference in the experiments energy scale at level of 4%. The measurement of the west-ward displacement of the Moon shadow under the effect of the geomagnetic field, as a function of the event multiplicity, allowed to calibrate the relation shower size - primary energy, thus calibrating the absolute energy scale of the detector at 10% level below 20 TeV [6]. Above this energy the overposition with CREAM provides a solid anchorage to the absolute energy scale at few percent level.

In order to extend the measurement of the ARGO-YBJ/WFCTA hybrid experiment to the PeV, the selection cuts in the $p_L - p_C$ space have been modified as follows: events for which $p_L \leqslant -1.25$ and $p_C \leqslant 1.1$ are rejected. The aperture is a factor 2.4 larger. The contamination increases and the purity of the p+He sample below 700 TeV reduces to 93%. At 1 PeV the contamination is less than 13% increasing to 44% around 6 PeV. About 72% of p+He events survive the selection criteria.

The preliminary measurement of the light component obtained by ARGO-YBJ with different analyses is summarized in the Fig. 4. As can be seen, all different approaches show evidence of a knee-like structure in the (p+He) spectrum starting from about 650 TeV, well below the all-particle spectrum knee confirmed by ARGO-YBJ at $\sim$3 PeV [22]. With respect to a single power-law with a spectral index -2.63 the deviation is observed at a level of about 6 s.d. .

More sensitive measurements of the energy spectrum and elemental composition in the range from 10^{12} to 10^{18} eV are required to observe the knees of different nuclei and to investigate in detail the end of the spectrum of Galactic CRs and the contradictory results among different experiments.

5. The Fermi Cocoon in the Cygnus region

The Cygnus X region is one of the most luminous region of the Northern γ-ray sky and it is rich in potential CR accelerator sites, e.g. Wolf Rayet stars, OB associations and SNRs. The Fermi collaboration detected a complex extended source, in a position consistent with the source ARGO J2031+4157, attributed to the emission of freshly accelerated CRs interacting with gas and radiation, filling a bubble (a "cocoon") carved by stellar winds and multiple supernovae shock waves [30].

The significance map around ARGO J2031+4157 as observed by ARGO-YBJ is shown in the left plot of Fig. 5. To improve the sensitivity to γ-induced events, an optimizated selection based on the shower core position is applied [31]. A smoothing is applied with an energy-dependent point-spread function evaluated for point γ-ray sources. For comparison, the known TeV sources and the Cygnus Cocoon are marked in the figure. The sizes of markers indicate the

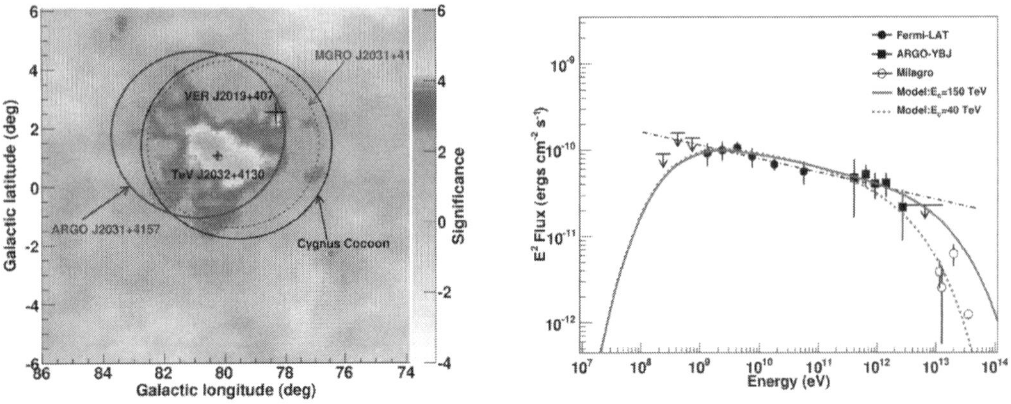

Figure 5. Left plot: The significance map of the ARGO J2031+4157 region observed by the ARGO-YBJ experiment. Right plot: Spectral energy distribution of the Cygnus Cocoon. Different markers stand for the spectra measured by different detectors. The dot-dashed line shows the best fit to the Fermi-LAT and ARGO-YBJ data using a simple power-law function. The red solid (dashed) line is obtained by a hadronic model with a proton cutoff energy at 150 (40) TeV. For references and details see [31].

68% containment size of the extension (for details see [31]).

The energy spectrum of the Cygnus Cocoon measured by Fermi-LAT, ARGO-YBJ and Milagro is shown in the right plot of Fig. 5. The flux determined by ARGO-YBJ appears consistent with the extrapolation of the Fermi-LAT spectrum suggesting that the emission of ARGO J2031+4157 can be identified as the counterpart of Cygnus Cocoon at TeV energies.

The combined spectrum of Fermi-LAT and ARGO-YBJ can be described by a differential power law (dot-dashed line) $dN/dE = (3.46\pm0.33)\times10^{-9}\cdot(E/0.1TeV)^{(-2.16\pm0.04)}$ photons cm^{-2} s^{-1} TeV^{-1}, suggesting the same origin for both GeV and TeV extended gamma-ray emission. Only statistical errors are shown, the systematic errors on the flux are estimated to be less than 30%. The upper limits of Fermi-LAT and ARGO-YBJ indicate the presence of a slope change or cutoff below $\sim$1 GeV and above $\sim$10 TeV, respectively. At a distance of 1.4 kpc, the observed angular extension of about 2° corresponds to more than 50 pc, making the Cygnus Cocoon the largest identified Galactic TeV source. The observation of this large TeV source, not removed in the Milagro analysis of the γ-ray diffuse flux from the Cygnus region, can explain the excess observed by them in the region $65° < l < 85°$, $|b| < 5°$.

Such a large region can be related to different scenarios. As discussed in [30], the favored scenario to explain the emission in the Cygnus Cocoon is the injection of CRs via acceleration from the collective action of multiple shocks from supernovae and the winds of massive stars, which form the Cygnus superbubble. Such superbubbles have been long advocated as CR factories, therefore the Cygnus Cocoon could be the first evidence supporting such hypothesis.

In order to test a possible hadronic origin of the γ-ray emission through the π^0 decay, we considered inelastic collisions between accelerated protons and target gas. We assumed that the primary protons follow a power law with an index similar to the gamma spectrum and with an exponential cutoff at 150 TeV, as suggested in [30] to describe CR acceleration by random stellar winds in the Cygnus superbubble. This energy is the maximum proton cutoff allowed by the ARGO-YBJ upper limit. The resulting spectrum is shown in the right plot of Fig. 5 by the solid red line [31]. It is worth noting that the Milagro data are not described by this model and can be reconciled only with a cutoff of about 40 TeV (dashed red line) non consistent with the

ARGO-YBJ results.

6. Conclusions

The ARGO-YBJ experiment has been in stable data taking for more than 5 years at the YangBaJing Cosmic Ray Observatory. With a duty-cycle greater than 86% the detector collected about 5×10^{11} events in a wide energy range, from few hundreds GeV up to about 10 PeV, exploiting the performance of the RPCs to image the front of atmospheric showers with unprecedented resolution and detail.

Since November 2007 to January 2013 ARGO-YBJ monitored the Northern sky at TeV photon energies with a cumulative sensitivity ranging from 0.24 to ~1 Crab units, depending on the declination. This sensitivity, the lowest obtained so far, is allowing a detailed study of different galactic and extragalactic sources, focusing in particular on flaring and extended emissions.

The high granularity of the readout and the high altitude location of the experiment offer a unique opportunity for a detailed study of several characteristics of the hadronic component of the CR flux up to about 10 PeV, starting from an energy window marked by the transition from direct to indirect measurements. In addition, the analog readout of the detector provides a powerful tool to study, for the first time, the particle distribution in the shower core region. Final analysis with the full statistics of the analog data will give new inputs to the hadronic interaction models currently used to describe particle physics and CRs up to the highest energies.

References

[1] Aielli G. et al. 2006 *NIM* **A562** 92.
[2] Bartoli B. et al. 2015 *Astropart. Phys.* **67** 47.
[3] Bacci C. et al. 2000 *NIM* **A443** 342.
[4] Aielli G. et al. 2009 *NIM* **A608** 246.
[5] Aielli G. et al. 2010 *Astrophys. J. Lett.* **714** L208.
[6] Bartoli B. et al. 2011 *Physics Rev. D* **84** 022003.
[7] Bartoli B. et al. 2011 *Astrophys. J.* **734** 110.
[8] Bartoli B. et al. 2012 *Physics Rev. D* **85** 022002.
[9] G. Di Sciascio and R. Iuppa, *"On the observation of the Cosmic Ray Anisotropy below 10^{15} eV"*, in "Homage to the Discovery of Cosmic Rays, the Meson-Muons and Solar Cosmic Rays", Chapter 9, pagg. 221-257 (Nova Science Publishers, Inc., New York, 2013). Preprint: arXiv:1407.2144.
[10] Iuppa R. and Di Sciascio G., 2013 *Astrophys. J.* **766** 96.
[11] Bartoli B. et al. 2013 *Physics Rev. D* **88** 082001.
[12] Erlykin A.D. and Wolfendale A.W 2013 *JCAP* **04** 006.
[13] Hillas A.M. 2005 *J. Phys. G* **31** R95.
[14] Giuliani A. et al. 2011 *Astrophys. J. Lett.* **742** L30
[15] Ackermann M. et al. 2013 *Science* **339** 807.
[16] Apel W.D. et al. 2009 *Astropart. Phys.* **31** 86.
[17] Amenomori M. et al. 2006 *Phys. lett. B* **632** 58.
[18] Glasmacher M.A.K. et al. *Astropart. Phys.* **12** 1.
[19] Tokuno H. 2008 et al. 2008 *Astropart. Phys.* **29** 453.
[20] Blasi P. 2013 *Astron. Astrophys. Rev.* **21** 70.
[21] Bartoli B. et al. 2012 *Physics Rev. D* **85** 092005.
[22] Di Sciascio G. *et al* 2014 *Frascati Physics Series* **58** 215. Preprint: arXiv:1408.6739.
[23] Bartoli B. et al. 2014 *Chinese Phys. C* **38** 045001.
[24] Yoon Y.S. et al. 2011 *Astrophys. J.* **728** 122.
[25] Cao Z. et al. 2010 *Chinese Phys. C* **34** 249.
[26] Zhang S.S. et al. 2011 *NIM* **A629** 57.
[27] Hillas A.M. 1985 *Proc. 19th ICRC (La Jolla)* **3** 445.
[28] Hörandel J.H. 2003 *Astropart. Phys.* **19** 193.
[29] Gaisser T.K. at al. 2013 *Frontiers of Physics* **8** 748.
[30] M. Ackermann M. et al. 2011 *Science* **334**1103.
[31] Bartoli B. et al. 2014 *Astrophys. J.* **790** 152.

Event-by-event study of CR composition with the SPHERE experiment using the 2013 data

R.A. Antonov[1], T.V. Aulova[1], E.A. Bonvech[1], D.V. Chernov[1], T.A. Dzhatdoev[1,*], Mich. Finger[2,3], Mir. Finger[2,3], V.I. Galkin[4,1], D.A. Podgrudkov[4,1], T.M. Roganova[1]

[1] Skobeltsyn Institute of Nuclear Physics, Lomonosov Moscow State University, Leninskie gory 1-2, 119991 Moscow, Russia

[2] Faculty of Mathematics and Physics, Charles University, Ovocn trh 3-5, 116 36 Praha 1, Prague, Czech Republic

[3] Joint Institute for Nuclear Research, Joliot-Curie 6, 141980 Dubna, Moscow region, Russia

[4] Faculty of Physics, Lomonosov Moscow State University, Leninskie gory 1-2, 119991 Moscow, Russia

E-mail: *timur1606@gmail.com

Abstract. We present an event-by-event study of cosmic ray (CR) composition with the reflected Cherenkov light method. The fraction of CR light component above 5 PeV was reconstructed using the 2013 run data of the SPHERE experiment which observed optical Vavilov-Cherenkov radiation of extensive air showers, reflected from snow surface of Lake Baikal. Additionally, we discuss a possibility to improve the elemental groups separability by means of multidimensional criteria.

1. Introduction

The study of the superhigh energy ($E > 10^{15} eV = 1\ PeV$) cosmic ray (CR) composition is the most difficult experimental problem of CR physics. Indeed, even for the most basic primary nuclei mass-dependent quantity, the mean logarithmic mass number $< lnA >$, the spread of results is very high (almost from proton to Iron at some energies) [1]. A somewhat newer review, at the cost of ignoring a part of older works, shows lesser, but still considerable, scatter of $< lnA >$ values [2]. Available information on the elemental groups spectra [3]–[7] at $E > 3$ PeV is almost exclusively due to particle detectors relying on the electron and muon numbers measurement [2]; this method is highly model dependent [8]. Moreover, results obtained with event-by-event techniques are very scarce [9], and most of the above-mentioned works [3]–[7] relied on the deconvolution method, that may suffer from the ill-posedness of the inverse problem [10].

In the present work we describe an event-by-event study of CR composition using the 2013 run data of the SPHERE experiment. The results of this Cherenkov experiment are much less dependent on high-energy hadronic model than for the case of particle detectors and, additionally, are expected to be more stable than ones obtained with the deconvolution method. The datasample used in our analysis is considered in section 2. The method employed for the composition study is described in section 3. In section 4 the result of the analysis — the fraction of CR light component above 5 PeV — is presented. Finally, in section 5 we show that it is possible to enhance the sensitivity of the method to the primary nuclei mass number with multidimensional criteria.

2. The datasample and low-level data analysis

The SPHERE experiment, stage 2 (2008-2013), employed the SPHERE-2 balloon-borne detector [11] to observe optical Vavilov-Cherenkov radiation ("Cherenkov light") of extensive air showers (EAS), reflected from snow surface of Lake Baikal. The foundations of the method were discussed in [12]. Many details on the detector hardware, observation conditions, simulations, and experimental data analysis could be found in [11]; for brief review see [13]. During the 2013 run, the typical observation altitude (measured by GPS) was in range H= 400–700 m above the snow surface. In total, 3813 events were recorded; 459 of them were recognized as EAS events (for the low-level experimental data analysis methods, see [11], section 5.4). For the next step of the analysis, the lateral distribution function (LDF) reconstruction, we employed a new code, that is generally similar to the code used in our previous works [14], [11], but was written in a completely independent manner. This circumstance allows us to test the robustness of our results against those obtained with the "old" LDF reconstruction method [14].

In total, 421 LDF were obtained with the new method out of the 459 events. The zenith and azimuthal angles of these showers (θ, φ) were measured under an assumption that the shower's front is a plane. Out of the 421 LDF 354 events have estimated zenith angles $\theta < 40°$ and reconstructed axis distance to the center of the detector's field-of-view (FOV) projection to the observation surface $R < R_{Max}$; $R_{Max} = R_{FOV}$+30 m, where $R_{FOV} = 0.4903 \cdot H$ is the FOV projection radius. The present analysis relies on a subsample of the late 354 LDF, 328 events. The simulation, however, was perfomed for the full sample of showers with $\theta < 40°$ and $R < R_{Max}$, that corresponds to 354 events. Given that the experimental sample is nearly complete, in what follows we neglect all possible associated biases.

3. The method to separate elemental groups

3.1. The LDF steepness parameter

For composition study we follow the general approach developed in [15]–[16], [14]. Using a sample of model LDF, we define the criteria of the elemental groups separation for different primary energies, zenith angles and observation altitudes. As in [15]–[16], [14], the parameter sensitive to the primary nuclei mass is the LDF "steepness", i.e. the ratio of the number of Cherenkov photons in the circle with the radius of 70 m with the center in the LDF's axis to the same number in the concentric ring with the radii of 70 m and 140 m.

3.2. Simulations

Simulations of EAS development were carried out for primary energies E_{MC}= 10, 30, 100 PeV by means of full direct Monte Carlo (MC) method using the CORSIKA code [17] with the QGSJET-I high energy hadronic model [18] and the GHEISHA low energy hadronic model [19]. The detector response database, that consists of a large number ($\sim 3 \cdot 10^5$) of model showers, was calculated using the Geant4 code [20]. Each "CORSIKA shower" was used multiple times with different axis coordinates. The axis coordinates were uniformly (randomly) distributed over big square with dimensions $1.5 \cdot H \times 1.5 \cdot H$.

As well, the instrumental trigger response was simulated for a range of energies E_{Trig} =5-200 PeV, and for a wide range of other conditions. For E_{Trig} =5–17.3 PeV, the 10 PeV CORSIKA showers were used, and for E_{Trig} >17.3 PeV — the 30 PeV showers. The model showers with $E_{Trig}/E_{MC} = K$ were obtained by multipying the corresponding responses to the factor K. We

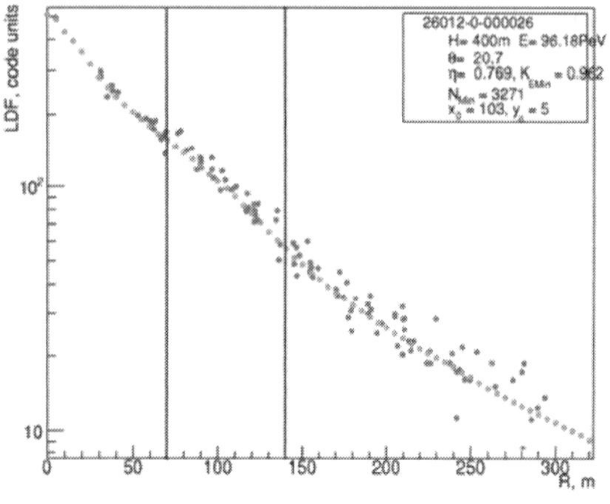

Figure 1. An example of the individual model LDF (red circles) together with a "composite model LDF" (green circles) that fits the individual LDF. The energy of primary nucleus was set to 100 PeV. The legend shows the values of some "true MC" parameters; (x_0, y_0) $[m]$ is the axis position coordinates with respect to the center of the detector's FOV projection to the observation surface. The zenith angle value θ is measured in °. As well, the reconstructed energy value (96.18 PeV) is shown.

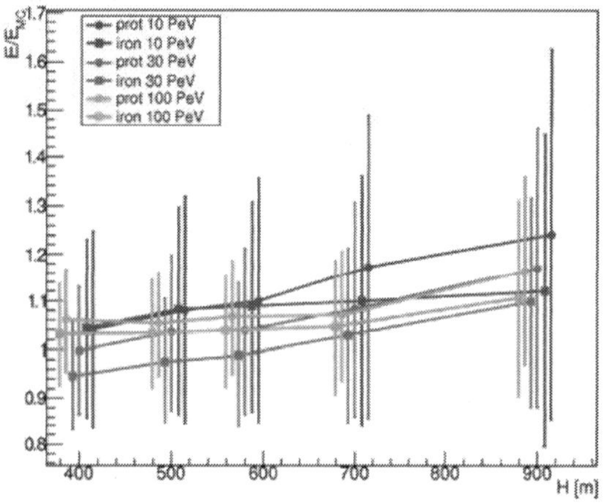

Figure 2. Reconstructed energy for model showers with various primary nuclei, primary energies and observation altitudes.

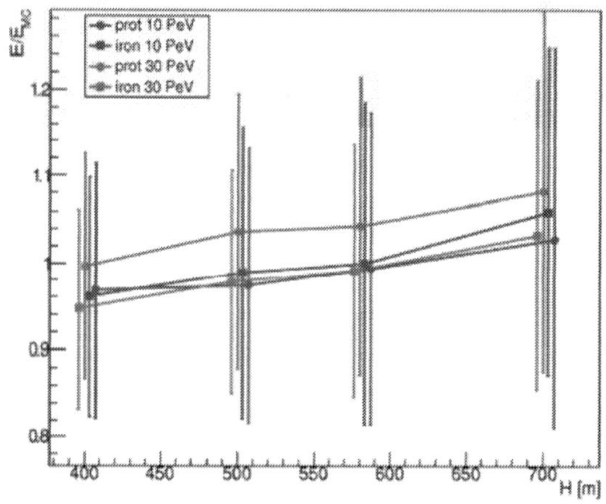

Figure 3. The same, as in figure 2, but only for showers that were registered by the model of the instrumental trigger.

have verified that the E_{MC} values discreteness has a negligible impact on results of the trigger response simulation.

3.3. Estimation of shower parameters

As a first step of the elemental groups separation procedure, we have estimated the LDF steepness parameter η for all model showers, integrated on time, by fitting these "discrete LDF" by a set of specially calculated smooth continious curves. Each curve of the set (we call it a "composite model LDF", CLDF) was composed from many model "discrete LDF" corresponding to the same CORSIKA shower. An example of a CLDF is shown in figure 1 by green circles, and the corresponding individual model LDF is plotted by red circles. The normalisation factor of a CLDF, that is nearly proportional to the primary energy [21], was utilized as the primary energy estimator, and the CLDF steepness parameter — as an estimate of the η value. The regions (0, 70) m and (70, 140) m used for the η parameter estimation are constrained by vertical black lines in figure 1.

The energy reconstruction performances of the described procedure are presented in figures 2-3. Figure 2 deals with all showers in the model sample, irrespectively of the axis position. The mean reconstructed energy values are shown by symbols for proton and Iron primaries for the

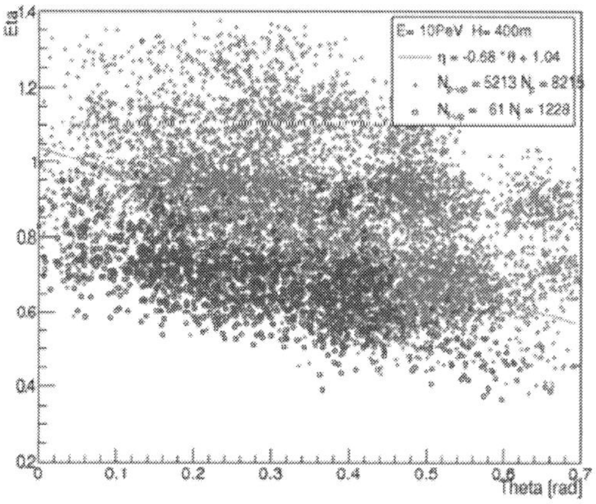

Figure 4. The result of the model showers classification for E_{Trig}= 11.2 PeV.

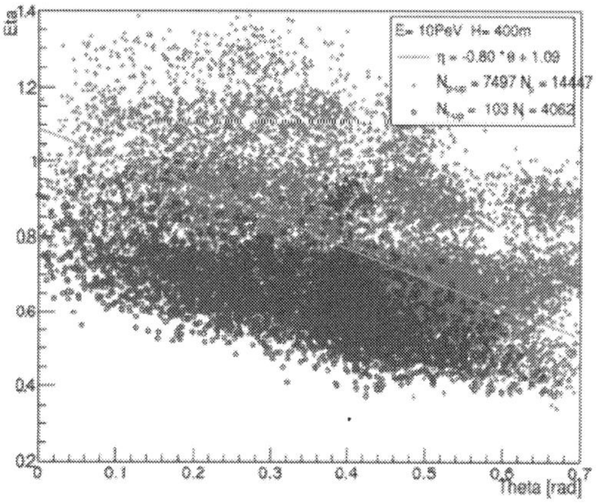

Figure 5. The same, as in figure 4, but for E_{Trig}= 15.9 PeV.

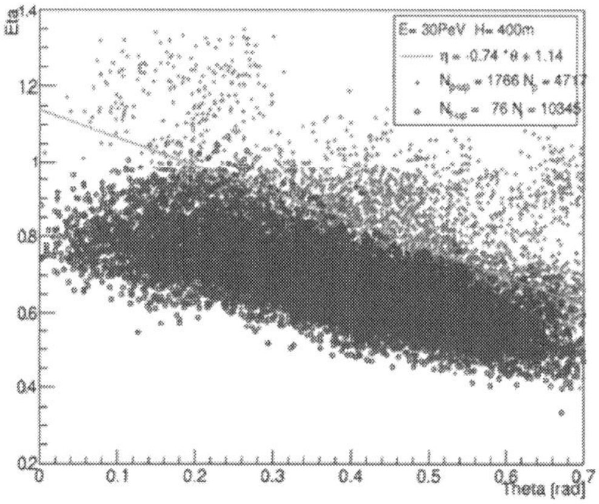

Figure 6. The same, as in figures 4-5, but for E_{Trig}= 20.9 PeV.

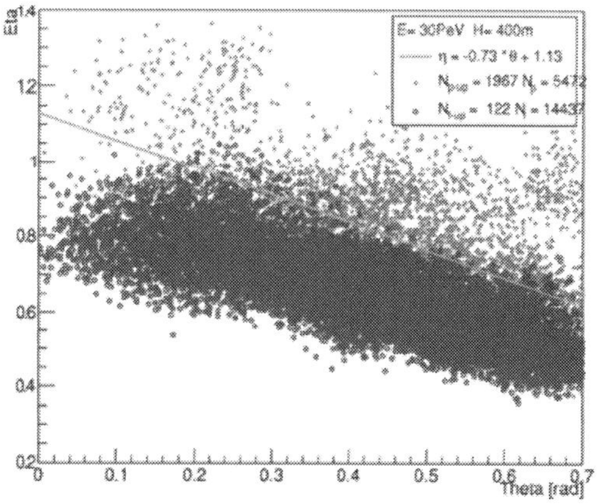

Figure 7. The same, as in figures 4-6, but for E_{Trig}= 29.6 PeV.

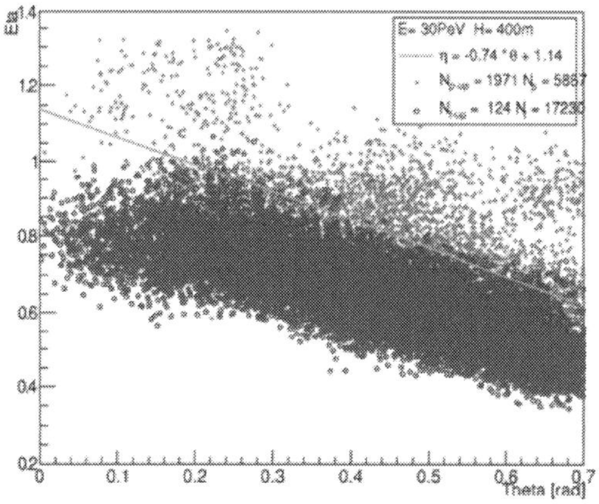

Figure 8. The same, as in figures 4-7, but for E_{Trig}= 41.8 PeV.

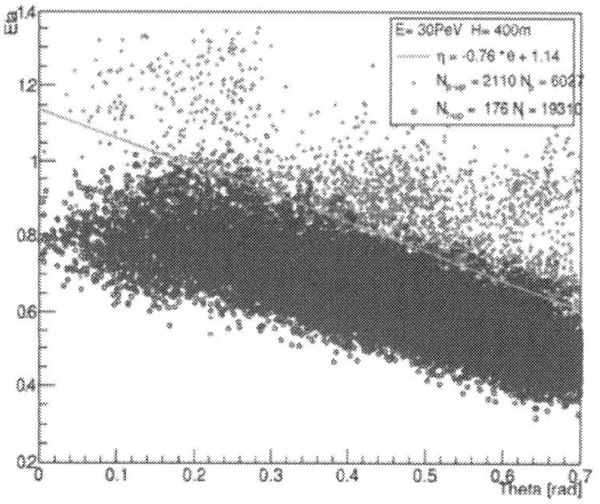

Figure 9. The same, as in figures 4-8, but for E_{Trig}= 102.6 PeV.

primary ("true MC") energy values E_{MC}= (10, 30, 100) PeV, and several altitudes (400, 500, 580, 700, 900) m. The corresponding standard deviations σ_E (σ_E is a measure of the primary energy statistical reconstruction uncertainty) are shown by bars. A small artificial shift between the results for different conditions is introduced on the H axis to make them visible. Similar results for E_{MC}= 10 PeV and 30 PeV are shown in figure 3 for showers that satisfy the trigger condition. We do not show results for 900 m, because the maximal altitude for the 2013 run was ≈700 m. It is evident that the statistical uncertainty of the primary energy value is qiute low (from ∼10 % to ∼20 % depending from the primary energy and the altitude values), and the systematic error is moderate (<5 % for most of the cases).

3.4. Separation of the elemental groups

To study the CR composition, we employed a set of Bayesian classifiers [22] using the above-described model LDF sample as a training set. In the present work, for simplicity, we constrain ourselves to the case of the linear border between the classes of primary nuclei, and two event classes — proton and Iron primary nuclei. Following the famous "Bayes rule", we have set the prior probabilities 0.5 for both classes. The results of the classifier training are shown for the altitude value H= 400 m and different energies E_{Trig} in figures 4-9. The values of the LDF steepness parameter for proton (red crosses) and Iron (blue circles) vs. the primary zenith angle value θ are plotted; the boundary between the classes is shown by green straight line. The E_{MC} parameter value is shown (see subsection 3.1 for explanation), as well as the parameters of the border line equation. The numbers of proton (N_{p-up} and N_p) and Iron (N_{f-up} and N_f) showers above the border, and total numbers of these model showers, respectively, are presented. To suppress statistical fluctuations arising from a small number of CORSIKA showers, we have introduced a small Gaussian smearing on both parameters (θ, η).

The fraction of model proton showers classified as light nuclei is $F_p = N_{p-up}/N_p$= (0.635, 0.519, 0.374, 0.359, 0.335, 0.350) for the E_{Trig} values (11.2, 15.9, 20.9, 29.6, 41.8, 102.6) PeV, respectively, and the corresponding fraction of Iron contamination $F_f = N_{f-up}/N_f$= (0.050, 0.025, 0.007, 0.008, 0.007, 0.009). It is evident from figures 4-5 that Iron nuclei are effectively suppressed by the instrumental trigger effects; at sufficiently low energy the SPHERE-2 detector could observe practically only light nuclei. Overall, the Iron contamination appears to be quite low. Results such as shown in figures 4-9 were obtained for 5 altitudes (400, 500, 580, 700, 900) m and 54 logarithmically spaced energy bins from 5 to 200 PeV.

All experimental showers from the sample defined in section 2 have the parameters (θ, E, η) estimated, and the altitude value H measured. These events were classified as proton-like or Iron-like according to their position above or below the border. The total number of the light CR component showers in 6 energy bins in the energy range 5-200 PeV was reconstructed as $N_{i-light} = N_{i-p-like}/F_p$, where $N_{i-p-like}$ is the number of experimental showers above the border in the ith bin. The light and heavy component intensities were corrected for the instrumental acceptance effects to account for the difference between the detection efficiencies of light and heavy nuclei.

In comparison to [14], in the present work we were able to substantially lower the threshold of the method using calculations of the instrumental acceptance [23],[11]. The first-order correction in the present work uses the 2013-integrated acceptance curves for the case of $R < R_{FOV} + 100$ m, instead of $R < R_{FOV} + 30$ m. The number of available model showers at E= 100 PeV is currently low, therefore the analysis was performed with the 10 PeV and 30 PeV model showers. Corrections for these inaccuracies are made in section 4.

4. Results

The result of our analysis — the estimated fraction of light nuclei in the energy range 5-200 PeV — is shown in figure 10 by red stars with statistical uncertainties. For the case of the

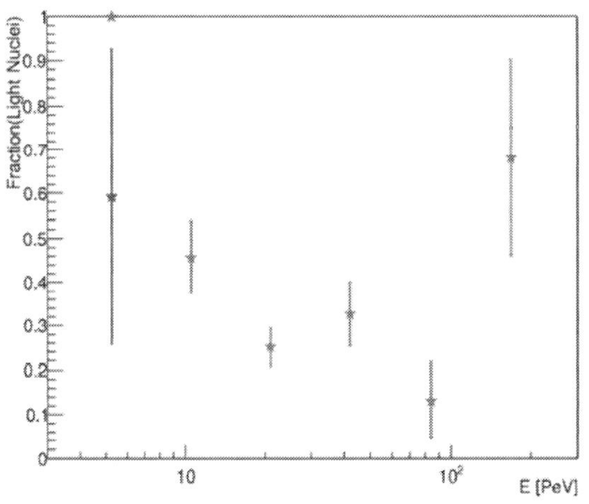

Figure 10. The fraction of light ("proton-like") nuclei reconstructed using the 2013 run data of the SPHERE experiment.

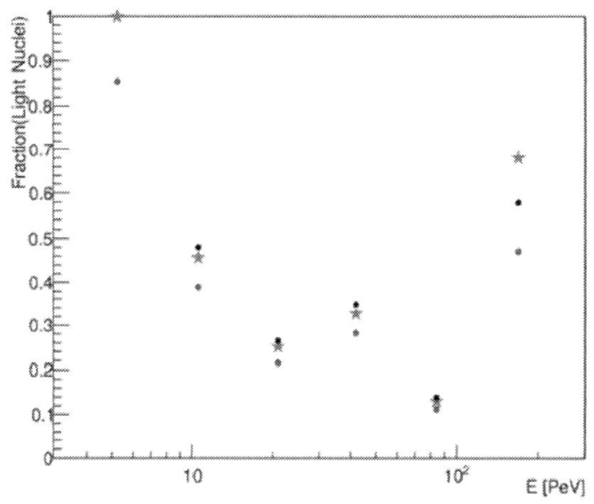

Figure 11. The fraction of light nuclei before (red stars) and after corrections (black and blue circles; see text for more details).

120–240 PeV energy interval the statistical uncertainty is quite high and our results are not constraining. For the case of 3.5–7 PeV energy interval we didn't find any indication of the Iron nuclei presence. To estimate the statistical uncertainty for this bin, we have added one "bonus" Iron shower; the corresponding result is shown in figure 10 by the blue star. The light nuclei fraction behaviour presented in figure 10 is qualitatively similar to the KASCADE-Grande result [7], and doesn't contradict our earlier estimate [14] obtained with the 2012 run data. We have already applied the correction for model energy discreteness here (see section 3).

Several systematic effects do exist that could change our results. A shift of the light nuclei fraction could occur as a result of the energy estimation systematic uncertainty (see figure 3). For some energy bins, this effect tends to enhance the reconstructed light nuclei intensity up to 10–15 % with respect to its true value. Therefore, the corrected fraction of light nuclei in these bins would be lower than plotted in figure 10 (for instance, 25-27 % instead of 30 %). We have estimated and applied corrections for this effect, as well as the acceptance correction discussed at the end of section 3. The new result after these corrections is shown in figure 11 by black circles. As well, we have estimated the LDF reconstruction procedure systematic uncertainty, and the corresponding result including this last correction, as well as all previous ones, is shown in figure 11 by blue circles. Dedicated work to study these and other systematic effects is in progress, and will be reported elsewhere.

5. Multidimensional method for superhigh energy CR composition study

In sections 3–4 were have employed the set of one-dimensional criteria, relying on the $\eta = \eta(A, E_{Trig}, \theta, H)$ parameter. Here we show that it is possible to improve the nuclei classes separability by using multidimensional pattern recognition methods. We demonstrate the case on our model sample (without imposing the trigger effects) by studiyng classification performances for 3 nuclei groups separation — proton, Nitrogen, and Iron. The feature vector was chosen to be 4-dimensional, and was composed of Cherenkov light intensities in concentric rings with the center in the shower's axis and with radii (0, 40) m, (40, 80) m, (80, 120) m, and (120, 160) m. The zenith angle value θ was added as the fifth classification parameter. The multivariate Bayesian pattern recognition technique under assumption of multidimensional Gaussian distribution of features [22] was used for this study. This technique was recently shown to be very effective for the case of CR composition study with lateral-angular distribution of

Cherenkov light [24]–[25].

5.1. Proton–Nitrogen separation

All graphs in this subsection (figures 12-15) show contamination of Nitrogen nuclei (i.e. the fraction of Nitrogen nuclei classified as protons) vs. proton selection efficiency. Black curves are drawn for $H = 400$ m, red curves — for $H = 580$ m, and blue curves — for $H = 900$ m. The primary energy values and zenith angle range are shown in the captions to the graphs.

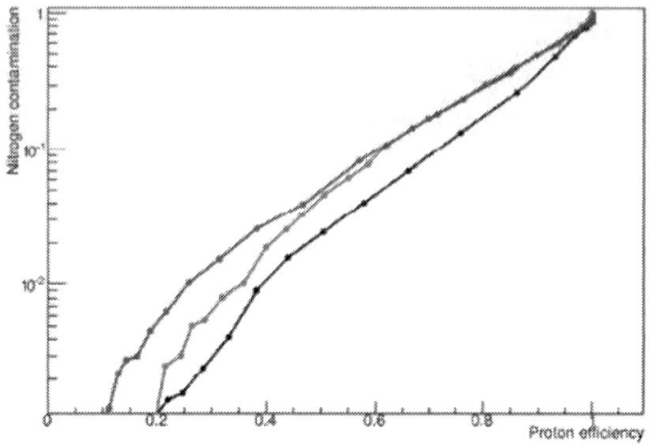

Figure 12. Proton-Nitrogen separation for $E = 10$ PeV and $\theta = 0$-$20°$.

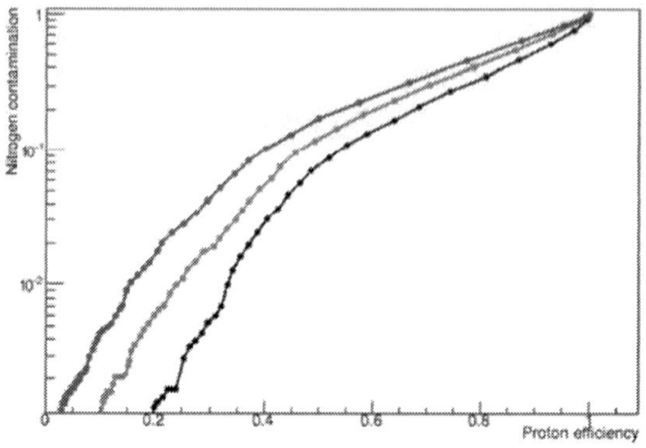

Figure 13. The same, as in figure 12, but for $\theta = 20$-$40°$.

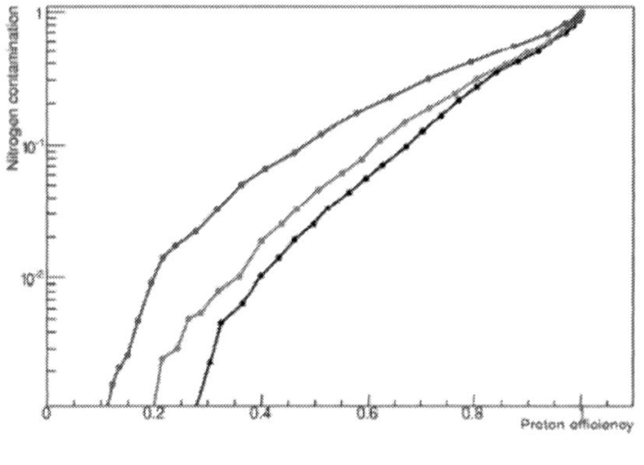

Figure 14. The same, as in figures 12-13, but for $E = 30$ PeV and $\theta = 0$-$20°$.

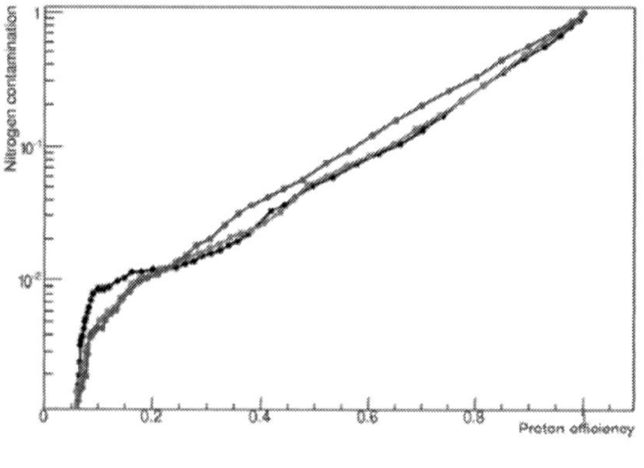

Figure 15. The same, as in figure 14, but for $\theta = 20$-$40°$.

5.2. Proton–Iron, proton–Helium and Nitrogen–Iron separation

The same results as in figures 12-15, but for separation of other elemental groups, were also obtained. For the case of proton-Iron separation (except the 10 PeV and $\theta = 20$-$40°$ case), the typical proton selection efficiency corresponding to ~ 1 % Iron contamination is ≈ 60-90 %, depending on conditions. For the 10 PeV and $\theta = 20$-$40°$ case, the proton selection efficiency is lower, but in this case the Iron showers would be suppressed by the trigger effects, as was explained in section 3. The same applies to other cases where proton selection efficiency is <75 %; therefore, the proton selection efficiency is, in fact, >75 % for ~ 1 % Iron contamination.

As well, we have studied the Nitrogen–Iron classification performance, and obtained the typical Nitrogen selection efficiency value ≈ 40 % for ~ 1 % Iron contamination. To conclude, let us note that proton–Helium separation is, in principle, also possible with the presented multidimensional technique. However, with the criteria considered in the present work, the proton selection efficiency was observed to be always <25 % for ~1 % Helium contamination.

6. Conclusions

We have presented and discussed the method for event-by-event study of superhigh CR composition applicable to the SPHERE experiment working conditions. It was found that a part of the light component showers can be readily selected against the heavy component background. The primary energy can be reconstructed with sufficiently low uncertainty, both statistical and systematic. The result of the analysis — the light nuclei fraction vs. energy — is qualitatively similar to the KASCADE-Grande result, and doesn't contradict to our earlier estimates. As well, we have shown that multidimensional criteria, in principle, allow to enhance the separability of the elemental group classes.

Acknowledgments

The authors are grateful to the technical collaborators of the SPHERE-2 experiment. The work was supported by the Russian Foundation for Basic Research (grants 11-02-01475-a, 12-02-10015-k, 13-02-00470-); the Russian President grants LSS-871.2012.2; LSS-3110.2014.2; the Program of basic researches of Presidium of Russian Academy of Sciences "Fundamental properties of a matter and astrophysics". Calculations were performed using the SINP MSU space monitoring data center computer cluster; we are grateful to Dr. V.V. Kalegaev for permission to use the hardware and to V.O. Barinova, M.D. Nguen, D.A. Parunakyan for technical support. The reported study was supported by the Supercomputing Center of Lomonosov Moscow State University.

References

[1] Tsunesada Y et al. (BASJE) 2008 *Proc. Int. Conf. on Cosmic Rays* vol 4 127
[2] Kampert K-H and Unger M 2012 *APh* **35** 660
[3] Antoni T et al. (KASCADE) 2005 *APh* **24** 1
[4] Antoni T et al. (KASCADE) 2009 *APh* **31** 86
[5] Garyaka A et al. (GAMMA) 2007 *APh* **28** 169
[6] Apel W D et al. (KASCADE-Grande) 2013 *APh* **47** 54
[7] Apel W D et al. (KASCADE-Grande) 2013 *Phys. Rev.* **D87** 081101(R)
[8] Apel W D et al. (KASCADE-Grande) 2014 *Advances in Space Research* **53** 1456
[9] Antoni T et al. (KASCADE-Grande) 2002 *APh* **16** 245
[10] Ter-Antonyan S V 2007 *APh* **28** 321
[11] Antonov R A and et al. 2015 *Physics of Particles and Nuclei* **46** 60
[12] Chudakov A E 1974 *Proc. All-USSR Symp. on Exp. Meth. of UHECR, Yakutsk* (Yakutsk: Russian Academy of Sciences Publisher) (In Russian) 69
[13] Antonov R A and et al. 2015 *DESY Conf. Proc. (PANIC-2014)* (In print)
[14] Antonov R A and et al. 2013 *J. Phys.: Conf. Series* **409** 012088
[15] Anokhina A M and et al. 2009 *Bulletin of the Lebedev Physics Institute* **36** 146
[16] Antonov R A and et al. 2009 *Proc. Int. Conf. on Cosmic Rays (Lodz)* vol HE.1.3 (Lodz:) id. 434
[17] Heck D and et al. 1998 *FZKA Report* **6019**
[18] Kalmykov N et al. 1997 *Nucl. Phys. Proc. Suppl.* B **52** 17
[19] Fesefeldt H C 1985 *Technical Report* **PITHA 85-02 RWTH**
[20] Agostinelli S and et al. 2003 *NIM* A **506** 250
[21] Dedenko L G and et al. 2004 *Nucl. Phys.: Proc. Suppl.* B **136** 1217
[22] Theodoridis S and Koutroumbas K 2003 *Pattern Recognition* 2nd edition (Amsterdam: Elsevier)
[23] Antonov R A and et al. 2013 *Proc. Int. Conf. on Cosmic Rays (Rio de Janeiro)* (Rio de Janeiro:) id. 1185
[24] Galkin V I and Dzhatdoev T A 2011 *Izvestiya Rossiiskoi Akademii Nauk* **75** 338 (In Russian)
[25] Borisov A S and Galkin V I 2013 *J. Phys.: Conf. Series* **409** 012089

Atmospheric influence on space-based observation of high-energy cosmic rays

Stefanie Falk for the JEM-EUSO collaboration
Karlsruhe Institute of Technology, P.O. Box 3640, 76021 Karlsruhe - Germany
E-mail: stefanie.falk@kit.edu

Abstract. High-energy extensive air showers developing in the Earth's atmosphere emit faint UV light that can be detected from space. The impact of varying atmospheric conditions on light emission and transmission has been studied in detail for the space-borne ultra high-energy cosmic ray observatory JEM-EUSO. By these studies, the importance of atmospheric scattering and reflection from ground on the fraction of Cherenkov light as well as fluorescence light received by JEM-EUSO is pointed out. For any telescope measuring UV light from an altitude higher than 40 km, the attenuating influence of the ozone layer cannot be disregarded. Based upon air shower simulation, quantitative numbers of ozone attenuation will be presented.

1. Introduction

The Extreme Universe Space Observatory on board the Japanese Experiment Module (JEM-EUSO) has the capability to explore cosmic rays of energies beyond 50 EeV. On board the International Space Station (ISS), this fluorescence telescope will orbit the Earth in a 90 minutes cycle at an altitude of approximately 400 km. Given a field of view of about 60°, JEM-EUSO will monitor a surface area with a radius of roughly 231 km [1]. JEM-EUSO will observe the night side of the Earth to detect the faint UV light emitted by high-energy air showers. In Figure 1, the observation principle of JEM-EUSO is illustrated. The main signals reaching the telescope are fluorescence photons emitted isotropically along the shower track and scattered Cherenkov light. A significant contribution originates from the reflection of mainly Cherenkov light on the Earth's surface which is visible as characteristic peak in the measured signal (see Figure 1b). For the detection of air showers, knowledge about the state of the Earth's atmosphere is important, since it significantly affects the development of the shower, as well as the UV light emission and transmission. The fluorescence yield as well as the Cherenkov yield are sensitive to changes in the state of the atmosphere [2,3].

Apart from the Atmospheric Monitoring System (AMS) which will be installed together with JEM-EUSO, data from a Global Atmospheric Model (GAM) will be used as important information at the air shower event reconstruction. The AMS of JEM-EUSO will consist of an IR camera and a LIDAR system to monitor especially the cloud and aerosols in the field of view of the detector [1]. In this work, the Global Data Assimilation System (GDAS) has been chosen as GAM [5]. GDAS provides near real-time atmospheric state functions (pressure, temperature, humidity, etc.) as altitude-dependent profiles on a $1° \times 1°$ global grid. GDAS data have been successfully used for several years in data reconstruction by the Pierre Auger Observatory [6]. For the new studies regarding the effects of the ozone, altitude-dependent ozone data have

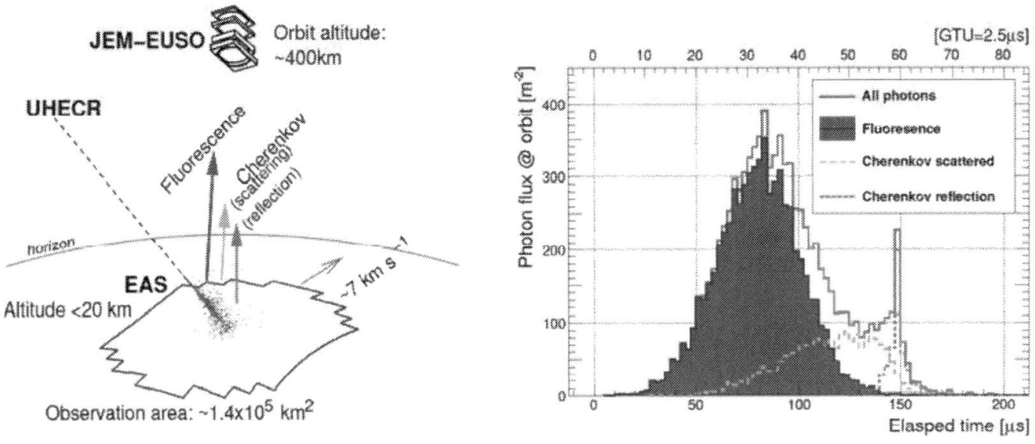

Figure 1: Illustration of the observation principle of JEM-EUSO for a nominal orbit of 400 km [4]. (a) The main signals are fluorescence photons emitted isotropically along the track and (delayed) scattered Cherenkov light; (b) Light components at the aperture of JEM-EUSO as function of time for a simulated EAS of zenith angle 60° and energy 10^{20} eV. The peak from ground or cloud reflected Cherenkov light is visible well.

been selected from the ozone sounding program of the US National Oceanic and Atmospheric Administration (NOAA) [7].

2. Atmospheric transmission

The atmospheric transmission of UV light plays an important role for air shower measurements with the fluorescence as well as the Cherenkov technique. The classical observation from ground level is strongly affected by attenuation due to Rayleigh and Mie scattering of light from its source to the detector. While Rayleigh scattering is caused by the air molecules themselves, Mie scattering is due to aerosols (e.g., dust, fumes, volcanic debris) and hydro meteors (e.g., water droplets). The Mie attenuation length of 53 km within a mixing layer below 3 km has been taken from measurements of the Pierre Auger Observatory [8]. This value corresponds to a *clear* atmospheric condition. The state of the atmosphere is described by the U.S. Standard Atmosphere 1976 (US-StdA 76). Light emission and transmission of an air shower at zenith angle $\theta = 60°$ and primary energy $E = 10^{20}$ eV have been simulated using a version of $\overline{\text{Off}}\underline{\text{line}}$, the Pierre Auger Observatory analysis and simulation framework, adopted for JEM-EUSO [9]. Two scenarios have been studied. For the *classical observation*, the observer has been placed in 40 km distance to the shower core at ground level (see Figure 2a). A strong attenuation within the detection range (290 nm $\leq \lambda \leq$ 450 nm) is visible. At the main fluorescence emission line of 337 nm, the total attenuation is about 98%. The attenuation is dominated by Rayleigh scattering. For the *space-borne observation* the same parameters have been used, though the observer has been placed in a distance of $\sim$ 400 km above the shower core (see Figure 2b). The total attenuation is much less compared to the classical observation, only 50% at 337 nm. For space-borne observation in a clear atmosphere, attenuation by Mie scattering becomes insignificant compared to Rayleigh scattering.

3. Ozone attenuation

Previous studies have shown that ozone absorption can usually be neglected [10]. To be detected from space, the light emitted by an extensive air shower has to pass the total column of

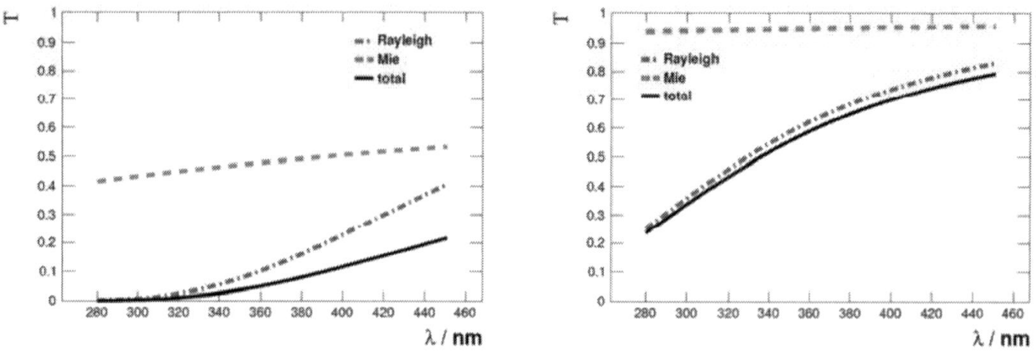

(a)

(b)

Figure 2: Comparison of the transmission factor T for Mie and Rayleigh scattering. A Mie attenuation length is 53 km in a homogeneous mixing layer below 3 km, taken from measurements of the Pierre Auger Observatory. The Rayleigh attenuation length is computed with respect to the US-StdA 76. The detector was placed (a) in 40 km distance at ground level, (b) at the ISS altitude 400 km above ground.

stratospheric ozone which peak at about 40 km altitude is known as *ozone layer*. To analyze the impact of UV absorption by ozone on the space-borne measurement of air showers, several simulation studies have been conducted. Since GDAS does not provide altitude-dependent ozone profiles, ozone sounding data of the Ozone and Water Vapor Group (OZWV) of the global monitoring division of NOAA have been used [7]. The transmittance T_{O_3} is computed based on an ozone cross-section parametrization [11]. For wavelengths above 320 nm, the transmission is above 90%, below 320 nm the transmission drops rapidly. This applies to about 26% of the generated fluorescence light. The impact on the transmission has been studied for example air showers ($\theta = 60°$, $E = 10^{20}$ eV) using US-StdA 76 and different ozone profiles. Average profiles for each month have been computed for three selected sounding locations. Boulder is located in the state of Colorado (USA), Hilo is located on the state of Hawaii (USA) and Pago Pago on American Samoa which is on the southern hemisphere. They represent continental ozone and different maritime concentrations. In Figure 3, the result is shown as relative difference of the number of photons at the JEM-EUSO aperture $\Delta N_\gamma = N_\gamma - N_\gamma^{\mathrm{WOO}}$ and number of photons without considering ozone absorption N_γ^{WOO}. The total amount of light at the aperture decreases on average by $8 - 10\%$. The typical annual change in ozone concentration is reflected by the transmission ratio. The amount of ozone in the stratosphere decreases in the course of summer and recovers in winter [12]. Therefore, the transmission in summer is higher compared to winter. The strongest decrease is seen for Boulder, for the ozone concentration is higher at higher geographical latitudes and lower close to the equator. The monthly variation as well as the difference between the average annual transmission is of the order of $\pm 1\%$. It is shown by these studies that ozone has to be taken into consideration for air shower reconstruction in case of space-borne observation.

4. Impact of varying atmospheric conditions

The influence of varying atmospheric conditions and albedo on the amount of UV light at the aperture of JEM-EUSO has been studied for nine example locations. Atmospheric data from GDAS have been used to compute average profiles for temperature, pressure, and humidity. For each, 100 showers have been processed using $\overline{\text{Off}}\underline{\text{line}}$. An average UV albedo of 2% has been assumed and a Lambertian phase function has been used. In Figure 4a, the resulting

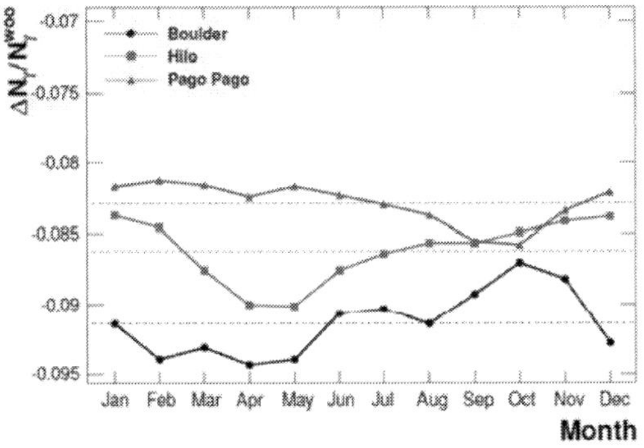

Figure 3: Monthly change of light transmission due to ozone absorption as relative difference $\Delta N_\gamma = N_\gamma - N_\gamma^{\mathrm{WOO}}$ and number of photons without ozone absorption N_γ^{WOO}. The dashed lines indicate the annual sample for the three locations Boulder, Hilo, and Pago Pago.

mean relative deviation from the average of all locations is depicted. The maximum deviation in summer are +2%, −3%, and in winter +4%, −6%. The further away from the equator (Alëutian islands, Timmins (Canada), Boulder (USA), Malargüe (Argentina)), the less light is emitted/transmitted. The closer to the equator (remaining locations), the more light is emitted/transmitted. The most extreme difference is seen for the Alëutian islands which are located at 52° geographical latitude. If the local atmospheric conditions are not addressed properly the energy reconstruction of JEM-EUSO will be significantly biased.

The wide field of view of JEM-EUSO covers at least 5° in geographical latitude/longitude. This does not only include different states of the atmosphere, but also various ground and cloud scenarios. Exemplarily, the position of the detector has been assumed above Malargüe (Argentina). To compare the significance of the different states, data of four surrounding GDAS grid points have been chosen one and two degree to the east and west ($\Delta\Phi$). Cloud has been neglected. The corresponding albedo has been estimated by the local vegetation based on UV albedo values on natural surfaces [13]. For Malargüe and the two locations towards the east this refers to an albedo of 2%. Towards the west one may first encounter a high altitude glacier or snow on top the Andes ($\Delta\Phi = -1$°E; 68% albedo) and forest on the western slope of the mountains ($\Delta\Phi = -2$°E; 7% albedo). The simulation results are depicted relative to the total amount of light at the aperture in Figure 4b. For the locations to the east, the total amount of light varies only by about 1% which is of the same order of magnitude as the statistical uncertainty of the simulation. The largest divergence is found for the Andes, where the surface altitude is above 3 km. While the total number of photons increases by 120% due to the high snow albedo, the number of fluorescence photons decreases by about 6%. A significant fraction of light originates from reflection. The assumed albedo is important for reconstruction of air shower events observed from space and will have to be addressed in the future in more detail.

5. Conclusion

The UV attenuation for a space-borne observatory is much less compared to ground-based observatories. In the space-borne scenario, the attenuation ranges from about 75% at 280 nm to 20% at 450 nm (50% at the main fluorescence line of 337 nm). The maximum attenuation that may occur in case of a ground-based observatory on the horizontal line of sight ranges from almost 100% at 280 nm to 78% at 450 nm. The UV absorption by ozone over the total wavelength range (280-450 nm) becomes significant. For the total amount of light at the JEM-EUSO detector aperture, an absorption of 8 − 10% has been computed. Finally, the influence of varying atmospheric conditions on emission/transmission amounts to ∼ 6%.

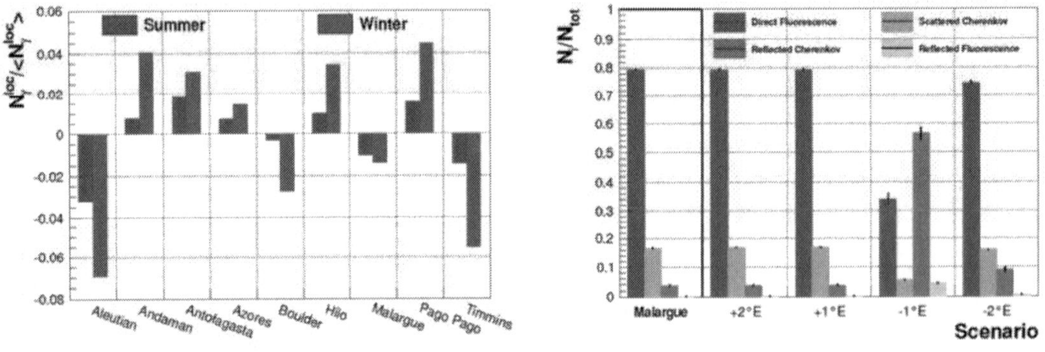

(a) (b)

Figure 4: The ratio of light at aperture for different atmospheric conditions. (a) Mean relative deviation from average of light at aperture. The albedo had been set to Savannah in all cases. Ozone profiles of Hilo, Boulder, and Pago Pago have been used according to month and location. Winter refers to January on northern hemisphere, July on southern, summer vice versa. (b) Malargüe, November 2009, the May ozone profile of Boulder data has been used.

Acknowledgments

This work is supported by the 'Helmholtz Alliance for Astroparticle Physics HAP' funded by the Initiative and Networking Fund of the Helmholtz Association.

References

[1] The JEM-EUSO Collaboration 2012 *ArXiv e-prints* (*Preprint* `arXiv:1204.5065[astro-ph]`)
[2] Ave M *et al.* (AIRFLY Collaboration) 2013 *Astropart. Phys.* **42** 90–102 (*Preprint* `arXiv:1210.6734[astro-ph]`)
[3] Frank I and Tamm I 1937 *C.R. Acad. Sci. URSS* **14** 109–114
[4] Adams J *et al.* (JEM-EUSO Collaboration) 2013 *Astropart. Phys.* **44** 76–90 (*Preprint* `arXiv:1305.2478[astro-ph]`)
[5] National Oceanic and Atmospheric Administration (NOAA) 2004 Global Data Assimilation System (GDAS1) Archive Information Tech. rep. Air Ressources Laboratory (ARL) `http://ready.arl.noaa.gov/gdas1.php`
[6] Abreu P *et al.* (Pierre Auger Collaboration) 2012 *Astropart. Phys.* **35** 591–607 (*Preprint* `arXiv:1201.2276[astro-ph]`)
[7] National Oceanic and Atmospheric Administration (NOAA) 2014 Global Monitoring Devision – Ozone and Water Vapor Group (OZWV) Tech. rep. Earth System Research Laboratory (ESRL) `http://www.esrl.noaa.gov/gmd/dv/iadv/index.php?map_program=ozwv`
[8] BenZvi, SY and Connolly, BM and Matthews, JAJ and Prouza, M and Visbal, EF and others 2007 *Astropart. Phys.* **28** 312–320 (*Preprint* `arXiv:0704.0303[astro-ph]`)
[9] Argiro S, Barroso S, Gonzalez J, Nellen L, Paul T C *et al.* 2007 *Nucl. Instrum. Meth.* **A580** 1485–1496 (*Preprint* `arXiv:0707.1652[astro-ph]`)
[10] Keilhauer B 2004 *Wissenschaftl. Berichte Forschungszentrum Karlsruhe, Technik und Umwelt* **FZKA-6958**
[11] Molina L T and Molina M J 1986 *J. Geophys. Res.* **91** 14,501–14,508
[12] Salby M L 2012 *Physics of the Atmosphere and Climate* (Cambridge University Press Textbooks)
[13] Chadyšiene R and Girgždys A 2008 *J. Environ. Eng. Landsc.* **16** 83–88

Physics Goals and Status of JEM-EUSO and its Test Experiments

Andreas Haungs[1] **for the JEM-EUSO Collaboration**[2]

[1]Institut für Kernphysik, Karlsruher Institut für Technologie (KIT), Germany
[2]see http://jemeuso.riken.jp

E-mail: haungs@kit.edu

Abstract. The JEM-EUSO mission aims to explore the origin of the extreme energy cosmic rays (EECRs) through the observation of air-shower fluorescence light from space. The superwide- field telescope looks down from the International Space Station onto the night sky to detect UV photons (fluorescence and Cherenkov photons) emitted from air showers. Such a space detector offers the remarkable opportunity to observe a huge volume of atmosphere at once and will achieve an unprecedented statistics within a few years of operation. Several test experiments are currently in operation: e.g., one to observe the fluorescence background from the edge of the Atmosphere (EUSO-Balloon), or another to demonstrate on ground the capability of detecting air showers with a EUSO-type telescope (EUSO-TA). In this contribution a short review on the scientific objectives of the mission and an update of the instrument definition, performances and status, as well as status of the test experiments will be given.

1. Introduction

The region of cosmic rays of the highest energies, i.e. the Extreme Energy Cosmic Rays - EECR, of the wide spanning cosmic ray energy spectrum is the one least explored. This is natural as the flux of such particles reaching Earth is extremely low. In addition to the low statistical accuracy, the existing measurements disagree in their flux by a surprisingly large amount (see Fig. 1).

The origin of the differences is still not fully understood and several causes may play a role. One reason could lie in the fact that the energy reconstruction is limited by our current theoretical understanding of extensive air showers (EAS) [1]. These showers are generated by interaction of the impinging cosmic particle with the Atmosphere. The products of this are measured with detectors of large exposure and the data is interpreted in terms of the arrival direction, energy and mass of the primary particle. The features of the first interaction (cross-sections and multiplicity distributions of secondaries) as well as partly the kinematic features (extreme forward direction) of interactions at all energies relevant for the shower development are not well measured by accelerators. Hence, extrapolations are used in simulating the EAS leading to different interpretations if different hadronic interaction models or different observables are used. Another cause could be that the two main experiments at the highest energies (Pierre Auger Observatory [2] and Telescope Array [3]) are located at different hemispheres and therefore, look at different regions of the sky. Several theories suggest that the EECR stem from astronomically nearby ($< 100\,\mathrm{Mpc}$) Active Galactic Nuclei, where it is well known that the distributions of those AGNs are not uniform over the sky [4].

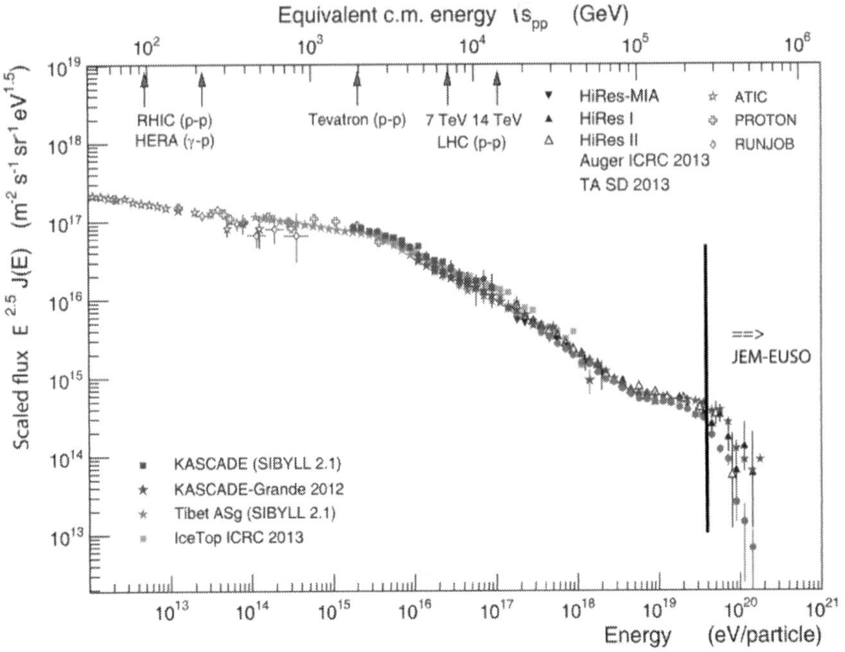

Figure 1. The all-particle spectrum of cosmic rays as measured by various experiments. Largest uncertainties are in the region of the highest energies: the target of the JEM-EUSO mission.

For future experiments, large efforts have to be made to reach significantly higher statistics in measurements as well as a coverage of the whole sky. Both require a much higher exposure than that of existing experiments.

The Extreme Universe Space Observatory (EUSO) (Fig. 2) at the Japanese Module (JEM) of the International Space Station (ISS) is the first space mission devoted to the scientific research of EECR [5–8].

2. JEM-EUSO

2.1. Scientific Goals

Main goal is the exploration of the Universe through the detection of the extreme energy cosmic rays and neutrinos by looking downward from the ISS to detect the fluorescence light of extensive air-showers that they generate in the Earth's atmosphere (Fig. 2).

The main scientific objective is astronomy and astrophysics through the particle channel. This requires to identify the sources of cosmic rays by the reconstruction of the arrival direction and energy spectra with a high collecting power, beyond any other previous or planned experiment so far. In addition, exploratory objectives were defined [9]. This leads to the following physics program of JEM-EUSO:

(i) astronomy and astrophysics through the particle channel at $E > 5 \times 10^{19}$ eV;

(ii) the detection of extreme energy gamma rays;

(iii) the detection of extreme energy neutrinos;

(iv) exploratory studies of the galactic magnetic fields;

(v) fundamental physics studies (e.g. Lorentz invariance tests) at extreme energies;

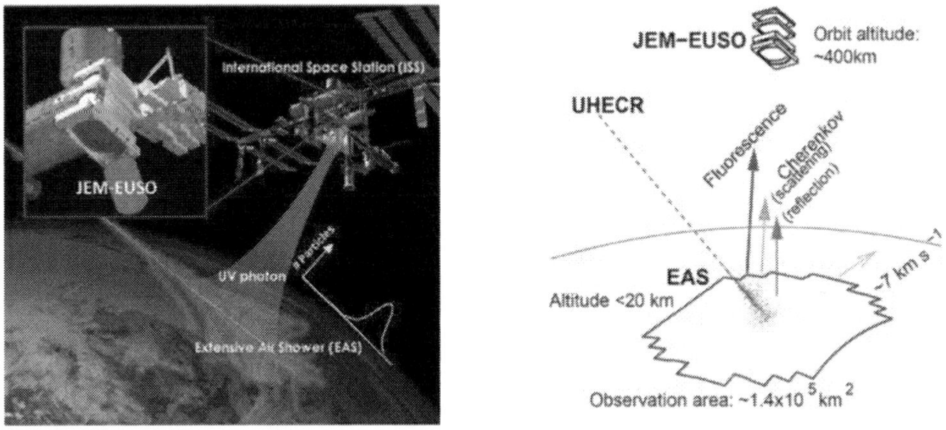

Figure 2. Left: Principle of the JEM-EUSO mission to detect extreme energy cosmic rays via the fluorescence emission of extensive air showers in the Atmosphere. Right: Schematic view of the JEM-EUSO mission: EASs generate fluorescence light that is emitted isotropically as well as Cherenkov light emitted in forward direction. The latter is also scattered and reflected and can reach the aperture of the telescope.

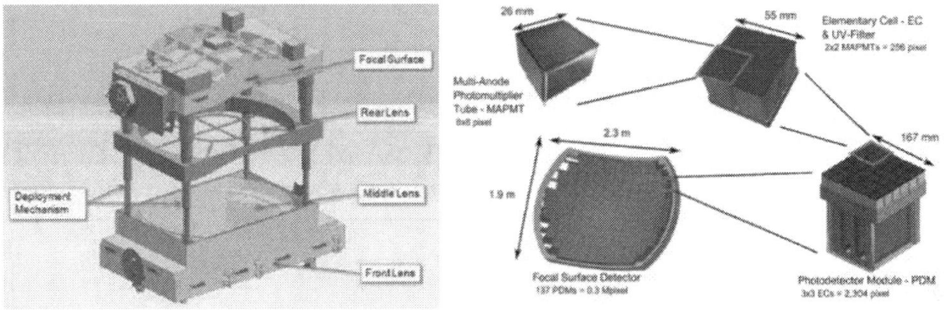

Figure 3. Conceptual view of the JEM-EUSO telescope with its three lenses (left panel). Schematic view of the focal surface and it's components of the JEM-EUSO mission (right panel).

(vi) global survey of nightglows, plasma discharges, lightning, meteors and other terrestrial transient sources in UV light.

The last one is possible due to the fast UV-light monitoring of the entire Atmosphere, which was never performed before.

Detailed simulations have shown that with three years of operation, JEM-EUSO will achieve a statistics of 100 events above 10^{20} eV and that with these events, if exists, nearby sources (even with a heavy dominant elemental composition at those energies) will be identified [10].

2.2. *Instrument*

The JEM-EUSO instrument [8, 11] consists of the telescope, the focal surface, a monitoring system for the atmospheric conditions, and a calibration system. In addition, there will be support and calibration systems on ground as well as at the ISS. The telescope (Fig. 3) consists of four main parts: collecting optics with 3 Fresnel lenses, a focal surface detector, electronics and a structure. JEM-EUSO constitutes a μs-fast and highly pixelized digital camera with a diameter of about 2.5 m and a $\pm 30°$ wide field of view. It works in the near-UV wavelength

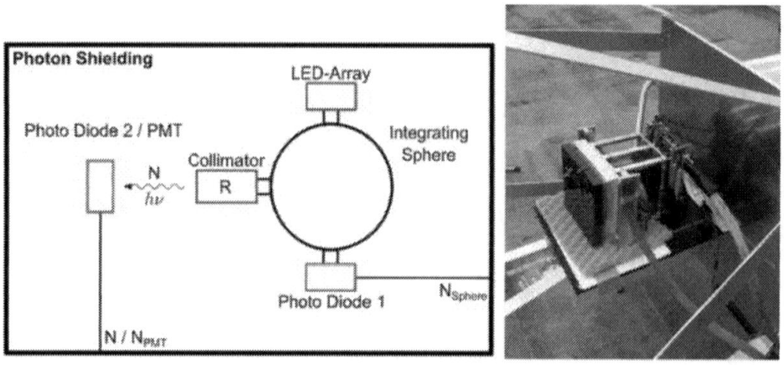

Figure 4. Left: Sketch of the calibration stand for photo sensors. Right: first fully assembled Photo Detector Module (PDM) for JEM-EUSO prototype experiments.

(290 – 430 nm) and enables a single-photon counting.

The focal surface will consist of 4932 64-channel multianode photomultipliers from Hamamatsu. They will be organized in elementary cells of 4 PMs each and in 137 photo detection modules (PDM) of 9 elementary cells each. The focal surface (Fig. 3) comprises therefore more than 300,000 pixels in total. The electronics processes triggers for air-shower or other transient events in the Atmosphere and send necessary data to the ground for further analysis.

The optics with three Fresnel lenses focuses the incident UV photons onto the focal surface with an angular resolution of 0.07°.

The Atmospheric Monitoring System (AMS) [12] consists of an infrared camera [13] and a LIDAR (LIght Detection And Ranging) system, both monitoring the Earth's atmosphere continuously inside the FoV of the JEM-EUSO telescope. A measurement of clouds top heights with an accuracy better than 500 m will be possible.

2.3. Calibration

For fluorescence detection of cosmic rays it is essential to calibrate the detector pre-flight with utmost precision and to monitor the performance of the detector throughout the whole mission time. For that purpose a calibration stand on-ground was built to measure precisely the performance of Hamamatsu 64 pixel MAPMTs that are planned to be used for JEM-EUSO (Fig. 4). It consists of a photon shielding, a uniform light source and readout electronics for the MAPMTs. The uniform light is available thanks to an integrating sphere and the light intensity is monitored in real-time by a NIST-calibrated photo diode attached to the sphere. To investigate the suitability of alternative detector devices, further research is done with state-of-the-art silicon photomultipliers (SiPMs). These will also be tested in the calibration stand and their performance can be compared to conventional photomultiplier tubes. A designed inflight calibration system measures continuously the efficiencies of the optics, the focal surface detector, and the data acquisition electronics [14, 15].

2.4. Performance of JEM-EUSO

Figure 5 shows a typical longitudinal profile of an EAS at the pupil of the telescope. Simulations including the transmission losses of the optical system, the efficiency of the focal surface, as well as the rate of background photons show that JEM-EUSO reaches almost full efficiency already at energies around $3 \cdot 10^{19}$ eV for a restricted subset of events, and full aperture at energies $E > 5 - 6 \cdot 10^{19}$ eV. The annual exposure of JEM-EUSO at 10^{20} eV will exceed by one order

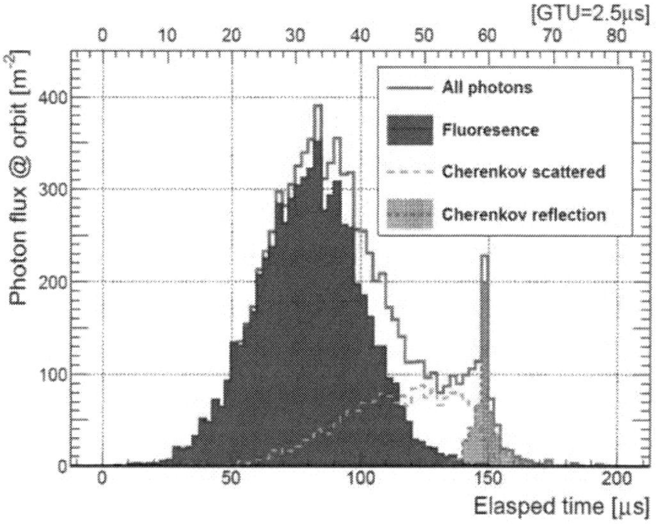

Figure 5. Arrival time distribution of photons at the pupil of the telescope per m^2 from a simulated high-energy EAS in case of a clear atmosphere.

of magnitude the level presently reached by ground-based observatories. For clear atmospheric conditions a reconstruction of the shower geometry and primary energy with good accuracy, as well as a rough estimate of the shower maximum will be possible [16]. In presence of optically thin clouds at high altitudes, often also categorized as cirrus, most of the EAS photons penetrate the layer of clouds and are attenuated only partly. This may lead to a reconstruction with a lower energy, however, the situation can be taken into account as the clouds are monitored by the AMS and the information is known for each triggered EAS. The geometry of the shower axis can still be properly determined by the analysis of the angular velocity of the EAS signal.

One big advantage of JEM-EUSO will be the uniform full sky coverage due to the ISS orbit (Fig. 6, left). JEM-EUSO will be the first experiment to be able to take data of EECRs for both hemispheres to build a full sky map. This will lead to anisotropy studies and identification of sources and source regions of high-energy cosmic rays which is not possible with ground-based observatories.

The second advantage of the JEM-EUSO mission - compared to existing cosmic ray experiments using the fluorescence technique - is the big observation area on ground and therefore the big detector volume. This will enhance the statistics for UHECR events that are expected with a rate of one event per century per steradian per kilometer squared. JEM-EUSO will have a large observation area in nadir mode (studied here, Fig. 6, right panel) and even a roughly ten times bigger one in tilt mode (not yet fully explored, but the design enables the option to tilt the entire instrument at the ISS to have a larger coverage on ground).

JEM-EUSO triggers on the UV light from EAS, measures the intrinsic luminosity of each EAS near its maximum and accurately reconstructs the EAS path so that the arrival direction of the initiating EECR can be reconstructed. Detailed studies [9] indicate that for 68% of the events the arrival direction can be reconstructed to better than 2.5°. Regarding the energy reconstruction, at the current status of development of the reconstruction algorithms, proton showers are reconstructed in clear-sky conditions with a typical energy resolution $\Delta E/E$ of $\approx$25%(20%) at energies around $4 \cdot 10^{19}$ eV (10^{20} eV). In addition, our still preliminary results indicate that the X_{max} resolution can be better than 70 g/cm^2 for $E > 10^{20}$ eV in the central

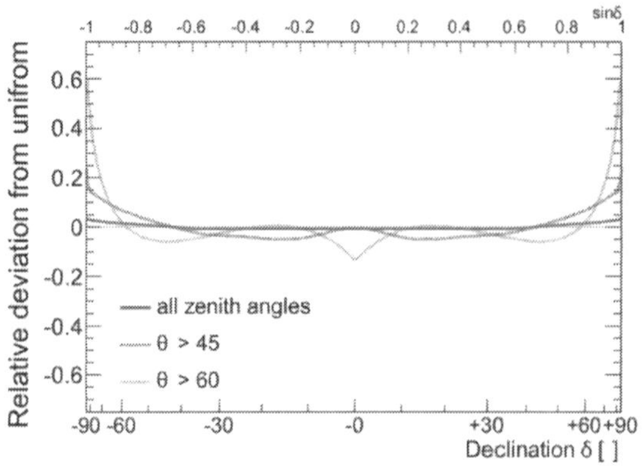

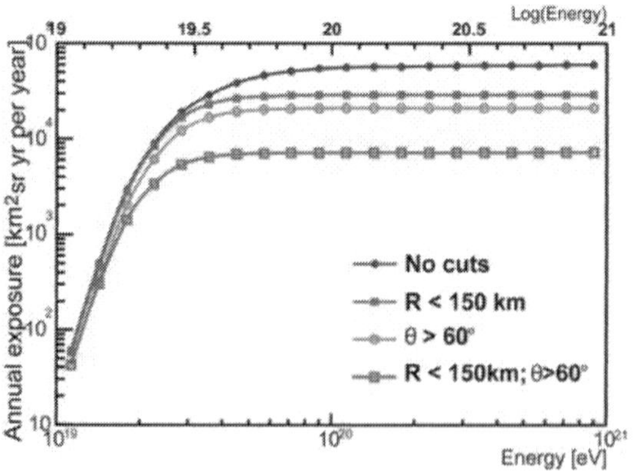

Figure 6. Left: Relative deviation from uniformity of the aperture as a function of sine of declination. The curves show the cases for a selection of events with different zenith angle. Right: Annual exposure of JEM-EUSO (ISS at 400 km) for the full FoV and for the case of extremely high quality cuts applied to the data. For the most stringent cuts, the threshold is low enough for an overlap with ground based experiments with good statistics. Plots adapted from [9].

part of the FoV. With 0.1 - 1 million km^2 sr exposure and the uniform coverage of the full sky, JEM-EUSO will observe all possible source directions at least within several hundred Mpc and will make possible the (i) identification of sources with unprecedentedly high statistics by arrival direction analysis and (ii) the measurement of the energy spectra from individual sources to constrain the acceleration or the emission mechanisms.

3. Test Experiments

On the way to the full instrument ready for launch to the ISS, several test or pathfinder experiments are needed and currently under development or in operation.

A PDM (photon detection module with 2304 pixels, see Figs. 3 and 4) can be seen as an independent unit, where one of which are or will be used for most of the test experiments. However, new optical systems are adapted for the respective purpose.

- **EUSO-Balloon:** EUSO-Balloon [17] will serve as a demonstrator for technologies and methods featured in the space instrument (Fig. 7, right panel). This balloon-borne instrument points towards the nadir from a float altitude of about 40 km. With its Fresnel optics and PDM, the instrument monitors a 12 by 12 degree wide field of view. Key objectives are among others the full-scale end-to-end test of the JEM-EUSO technique, the experimental determination of the effective UV background below 40 km, ground based calibration systems, and, the observation of the first UV-image of an EAS looking down on the Earth's atmosphere. The flights are handled by the balloon division of the French Space Agency CNES. A first flight was performed on August, 25, 2014 at Timmins, Canada. 5 hours of valuable data from the PDM, from the installed infrared camera, as well as Laser pulses sent by a helicopter accompanying the full flight are presently being analysed.
- **EUSO-TA:** Also important information on the capabilities of the full instrument will be provided by EUSO-TA [18], which is a ground-based telescope formed by one PDM (identical to the PDM of EUSO-Balloon) and two Fresnel lenses, which are prototypes of those foreseen in JEM-EUSO. EUSO-TA will be located at the Telescope Array (TA) [3] site in Utah, USA (Fig. 7, left panel). The instrument will measure the UV light in the

Figure 7. Left: photo of the EUSO-TA installation in the front of the Telescope Array fluorescence detector in Utah, US. Right: photo of the EUSO-Balloon payload (with floating devices) just before the launch in August 2014 in Timmins, Canada.

Atmosphere in its ±4 degrees FoV, will be triggered by the fluorescence detectors of TA, and will make use of the LIDAR and the Electron Light Source of TA. The aims are to obtain an end-to-end calibration of the prototype telescope, and an inter-calibration with the fluorescence detector of TA. Everything is prepared to start the measurements in early 2015.

- **K-EUSO:** Currently, the collaboration considers a cooperation with the Russian project KLYPVE. KLYPVE [19] is a mirror based fluorescence telescope foreseen to be installed at the Russian segment of the ISS. Despite its smaller size, KLYPVE could be the first experiment detecting EAS from space. Combined with a Fresnel lens and a JEM-EUSO type focal surface the exposure and quality of the experiment could be considerably improved. However, due to limitations in space and weight, the exposure of the baseline JEM-EUSO design probably can not be reached.
- **Mini-EUSO:** Another project is Mini-EUSO, a small prototype experiment (one PDM and a two-25cm-diameter Fresnel-lens system) foreseen to operate inside the ISS for an observation of the UV emission from the night-Earth through an UV-window. This instrument would map for the first time the Earth in UV, and could study atmospheric phenomena and bio-luminescence at Earth as well as meteors.
- **SiPM-EUSO:** The collaboration has also started to evaluate if SiPM can replace the heavy and expensive Multianode PMT's presently in use. Designing and manufacturing an R&D early prototype of a SiPM based PDM is under study for the use in the test experiments.

4. Outlook

JEM-EUSO is planned as a three to five year mission, where initially the launch was foreseen in 2017. Due to financial reasons this will not be possible. However, the JEM-EUSO collaboration continues the efforts to improve the baseline design and the capabilities and sensitivity of such an instrument with the aim to launch it to space at a later opportunity. Still the idea lives on of a comprehensive, virtual wide-range Cosmic Ray Observatory in space, like outlined by Vernon Jones from NASA, e.g. (Figure 8).

The various test and pathfinder experiments presently under construction or in operation will provide more information and details on the capabilities of a space instrument like JEM-EUSO.

Figure 8. Illustration of four cosmic ray instruments on the ISS, which would comprise complementary studies of a wide field in cosmic ray physics. Picture taken from [20].

Acknowledgments

Supported by the 'Helmholtz Alliance for Astroparticle Physics HAP' funded by the Initiative and Networking Fund of the Helmholtz Association.

References

[1] d'Enterria D *et al.* 2011 *Astropart. Phys.* **35** 98
[2] Abraham J *et al.* (Pierre Auger Collaboration) 2008 *Phys. Rev. Lett.* **101** 061101
[3] Abu-Zayyad T *et al.* (Telescope Array Collaboration) 2013 *Astropart. Phys.* **39** 109
[4] Véron-Cetty M-P and Véron P 2006 *Astron. Astrophys.* **455** 773
[5] Ebisuzaki T *et al.* 2008 *Nucl. Phys. B (Proc. Suppl.)* **175** 237
[6] Takahashi Y *et al.* 2009 *New J. Phys.* **11** 065009
[7] Casolino M *et al.* (JEM-EUSO Collaboration) 2011 *Astrophys. Space Sci. Trans* **7** 477
[8] Casolino M *et al.* (JEM-EUSO Collaboration) 2014 *Exp. Atron.* DOI 10.1007/s10686-014-9418-x
[9] Adams J *et al.* (JEM-EUSO Collaboration) 2013 *Astropart. Phys.* **44** 76
[10] Rouillé d'Orfeuil B *et al.* 2014 *Astron. Astrophys.* **567** A81
[11] Kajino F *et al.* 2010 *Nucl. Inst. Meth. A* **623** 422
[12] Toscano S *et al.* (JEM-EUSO Collaboration) 2014 *Exp. Atron.* DOI 10.1007/s10686-014-9378-1
[13] Morales de los Ríos J A *et al.* (JEM-EUSO Collaboration) 2014 *Exp. Atron.* DOI 10.1007/s10686-014-9402-5
[14] Karus M *et al.* (JEM-EUSO Coll.) 2013 *Proc. 33rd ICRC, Rio de Janeiro* **ID0545** arXiv:1307.7071
[15] Sakaki N *et al.* (JEM-EUSO Coll.) 2013 *Proc. 33rd ICRC, Rio de Janeiro* **ID0546** arXiv:1307.7071
[16] Fenu F *et al.* (JEM-EUSO Coll.) 2011 *Proc. 33rd ICRC, Beijing* **3** 116 arXiv:1204.5065
[17] von Ballmoos P *et al.* (JEM-EUSO Coll.) 2013 *Proc. 33rd ICRC, Rio de Janeiro* **ID1171** arXiv:1307.7071
[18] Casolino M *et al.* (JEM-EUSO Coll.) 2013 *Proc. 33rd ICRC, Rio de Janeiro* **ID1213** arXiv:1307.7071
[19] Panasyuk M *et al.* 2014 these proceedings
[20] Jones W V 2014 *Braz. J. Phys.* **44** 534

Ultra-High Energy Cosmic Rays: A Recap of the Discussions at the European Cosmic Ray Symposium 2014

Andreas Haungs
Institut für Kernphysik, Karlsruher Institut für Technologie (KIT), 76021 Karlsruhe, Germany
E-mail: haungs@kit.edu

Abstract. This contribution summarizes the talks, posters, and discussions of the ECRS 2014, held in Kiel, Germany - related to the research field of ultra-high energy cosmic rays (UHECR). Here, the definition of UHECR is cosmic rays with an energy above approximately 0.1 EeV, i.e. 10^{17} eV, and the corresponding sessions were named HECR-II. Recent experimental results, like the identified heavy knee in the cosmic ray energy spectrum or the hotspot in the arrival direction of cosmic rays with highest energy, will be shown and discussed in relation to the need and requirements of future experimental efforts.

1. Introduction

The spectrum of primary cosmic rays covers a large energy range from a few GeV to above 100 EeV, where the flux decreases very fast with a factor ≈ 1000 per decade of energy. In first order the spectrum can be described by power laws, where three evident features define slight changes of the index of these power laws (see Fig. 1): the knee, the ankle, and the final flux suppression. Whereas at lower energies the measurements are performed in a direct way with compact detectors in space or at high-flying balloons, primary cosmic rays above energies of about 100 TeV have to be investigated by observations of extensive air showers (EAS) [1]. EAS are generated when high-energy cosmic particles enter the atmosphere. Forward-boosted secondary particles as well as emitted light during the development of the EAS in various frequency ranges form the detectable products. Using large-area ground-based detector installations for registering various components of the EAS, the direction, the energy, and the mass of the primaries are estimated. The variety of possible EAS observation techniques is displayed in Figure 2.

By such indirect studies of the primary cosmic rays a steepening of the power-law spectrum at around 35 PeV, known as the knee, has been identified. At higher energies around 3-5 EeV appears to be a further change of the spectral index, called the ankle. Finally, in the energy region above ca. 50 EeV, a cut-off like feature of the cosmic ray spectrum is observed.

Many theoretical models (often also contradictory ideas) exist to describe the origin of the observed features, but for none of them a convincing explanation exists, in particular concerning the details. The gross picture explains cosmic rays as follows: Galactic sources, in particular Supernova Remnants, accelerate fully ionized nuclei to a maximum energy corresponding to the knee-energy. However, the acceleration is more effective for nuclei of higher charge, therefore,

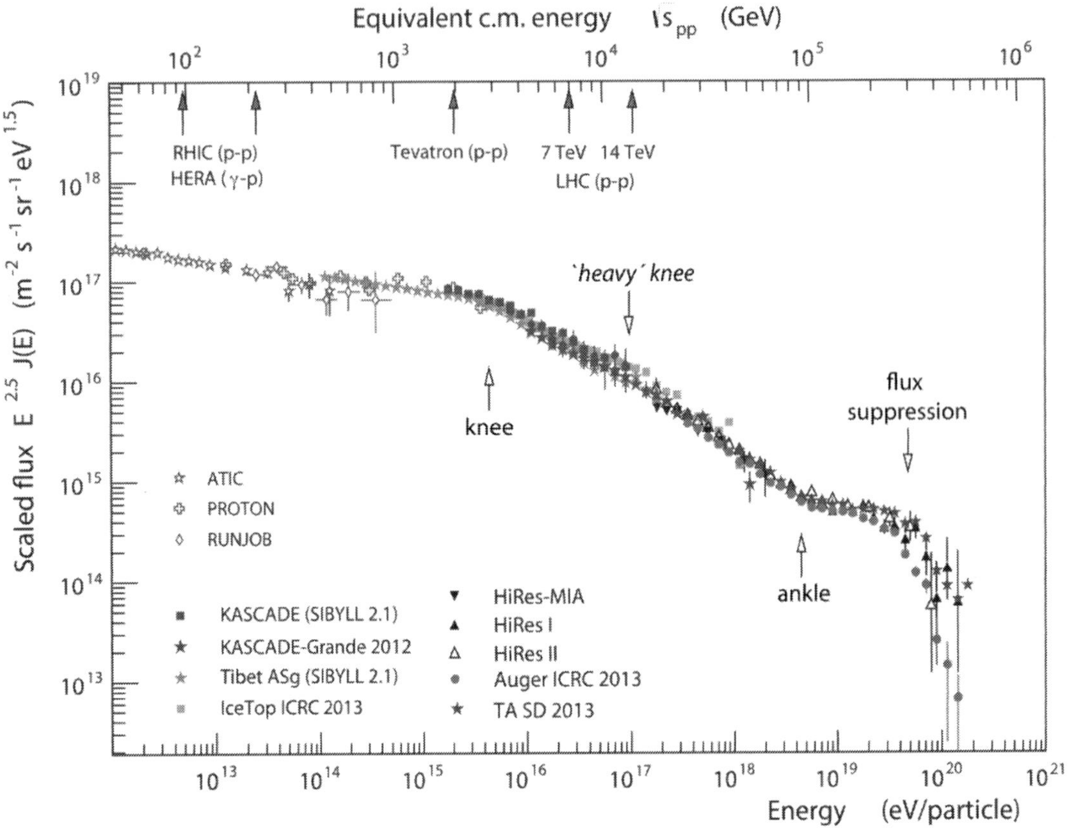

Figure 1. The all-particle spectrum of cosmic rays as measured by various experiments (adapted from [2]).

for iron nuclei this maximum energy is expected at around 100 PeV. The transition to cosmic rays of extra-galactic origin occurs at energies where the galactic component vanishes and the ankle could be the signature of this transition.

Cosmic rays of highest energy will interact with the cosmic microwave background leading to a maximum energy at Earth for extra-galactic sources (Greisen- Zatsepin-Kuz'min, or short GZK cut-off) which is predicted for energies were the third structure is observed, but that could also mark the maximum energy of acceleration of extra-galactic sources.

Any explanation of the features of the cosmic-ray spectrum needs sufficiently detailed knowledge of its shape and of the variation of the mass composition of cosmic rays. In addition, anisotropy studies at various scales, energy ranges, and primary mass groups are needed to solve the puzzle of the origin of UHECR. But, it might be that even that will not help in understanding the astrophysics of the high-energy cosmos. Here, the most promising ansatz for the future is the multi-messenger particle astronomy, i.e. combining information from measurements of many observables from the same sources to understand the acceleration and propagation processes in detail. The information from high-energy cosmic rays, gamma rays, and neutrinos, at best paired with multi-wavelength electromagnetic observations, and even, in far future, of gravitational waves have to be combined in experiment and theory.

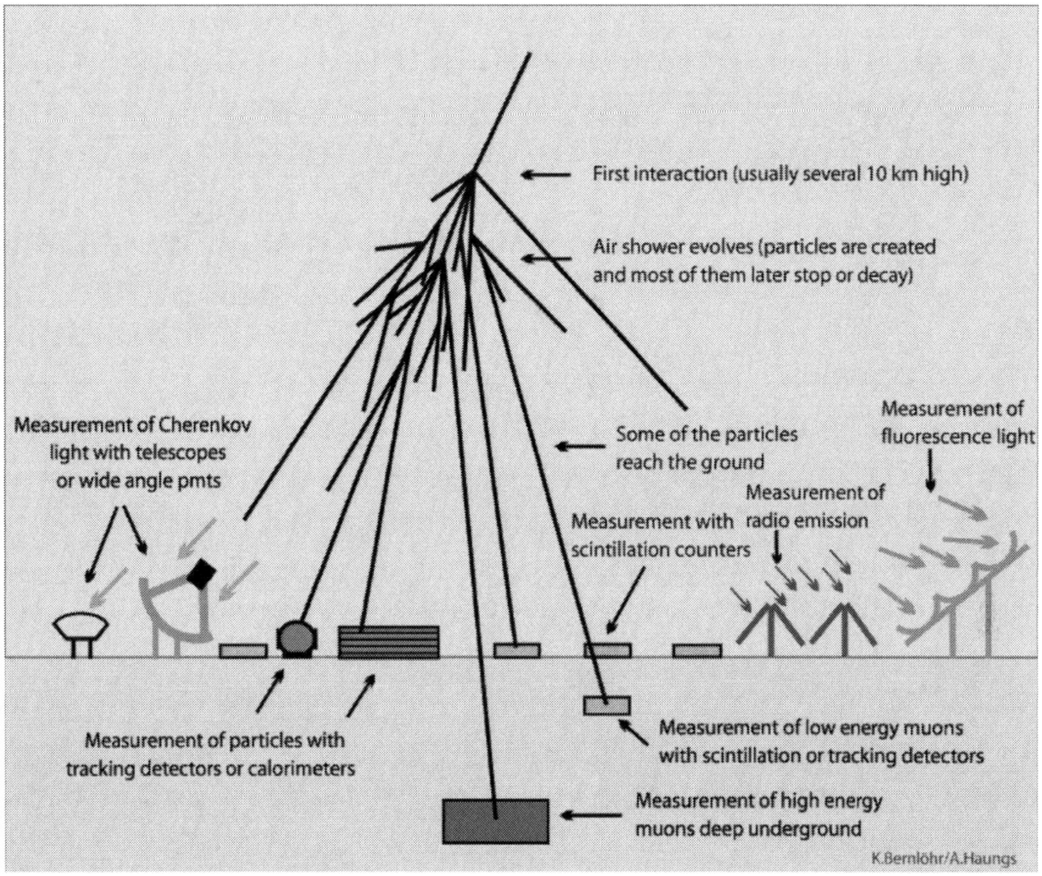

Figure 2. Schematic view of an extensive air shower with its different detection possibilities. To interpret the measurements in terms of the parameters searched for, i.e. energy, mass and arrival direction of the primary particle, as well as to unfold the uncertainties of the hadronic interaction models used for these interpretations, it is envisaged to measure a large number of observables per individual EAS (use of a hybrid detector system). Figure adapted from [1].

In this contribution discussions are focused on the part of the energy spectrum above the energy region of the knee, experimentally investigated by the observation of extensive air showers[1].

2. Experimental Status: Approaching the Ankle

A decade ago, the KASCADE experiment [3] has shown that the knee feature is due to a distinct break in the intensity in the light component of cosmic rays ($Z< 6$), only, where the difference in the energies of the knee features of protons and Helium nuclei is in agreement with the assumption of a rigidity (charge) dependent knee [4]. Compared to the knee energy, the range 10^{16} - 10^{18} eV stayed somehow unexplored until recently, though the study of primary energy spectrum and mass composition in this energy range is of crucial importance for understanding origin and propagation of cosmic rays, as here the transition from galactic to extra-galactic origin of cosmic

[1] Naturally, not all current topics of the field were presented at the conference, and even not all contributions given are covered by this review, which is based on a few very personal impressions. I apologize for the incompleteness and not paying attention to detail, but hope to touch some aspects of interest for a more general readership.

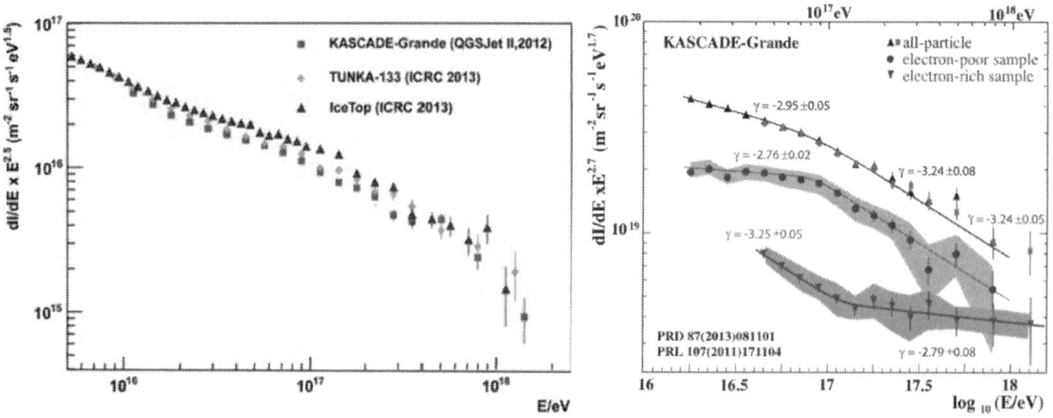

Figure 3. Left panel: The all-particle, cosmic-ray energy spectrum by three different experiments (KASCADE-Grande [10]; Tunka-133 [13]; IceTop [14]). Right panel: All-particle, light-mass enriched, and heavy-mass enriched energy spectra from KASCADE-Grande. One all-particle and the heavy enriched spectra is from one analysis, the other all-particle and the light primary spectrum result from a larger data set with higher energy threshold (from [6]).

rays is expected.

Now, at the ECRS 2014, three experiments presented results for the all-particle energy spectrum, at least. I discuss in the following the spectra of cosmic rays in the range from 10^{16} to 10^{18} eV obtained by KASCADE-Grande [5, 6], Tunka-133 [7, 8], and IceTop [9]. Despite the overall power-law behavior of the all-particle spectrum (see Fig. 3), there are some structures observed, which do not allow to describe the spectrum with a single slope index. Of course, these structures are smaller than the well-known knee or ankle features, but statistically significant, found by KASCADE-Grande and confirmed by Tunka-133 and IceTop. There is a clear evidence that just above 10^{16} eV the spectrum shows a concavity, i.e. a hardening of the spectrum appears. Another feature, also seen by all three experiments, is a small break at around 10^{17} eV, which appears at KASCADE-Grande at a little lower energy than for the other two experiments. Applying a second power law a statistically significant change of the index is similarly obtained at all three experiments.

KASCADE-Grande has shown that the spectral form does not depend on the hadronic interaction model in use [15]. This is not true for the absolute flux, where a difference of 15-20% can appear. As KASCADE-Grande measures the total number of electrons and muons, separately, these differences are related to the absolute normalization of the energy scale by the various models. IceTop has reported results on the basis of SIBYLL and QGSJet, but found the difference very small, which is probably owed to the observation level close to the shower maximum. Tunka-133's measurements are based on a calorimetric method[2]; i.e. the energy calibration is seen as less dependent from the hadronic interaction model (for a more detailed discussion see [16]).

The main difference of the three spectra is the absolute normalization to the energy scale, which moves the found structures slightly in energy and by that also in the absolute flux. But the spectra agree nicely within the given systematic uncertainties. The normalizations depend

[2] Whereas imaging fluorescence or Cherenkov telescopes can resolve the shower development in the atmosphere directly, Cherenkov arrays like Tunka-133 need to analyse the lateral or time distribution of the light to reconstruct the composition sensitive shower maximum.

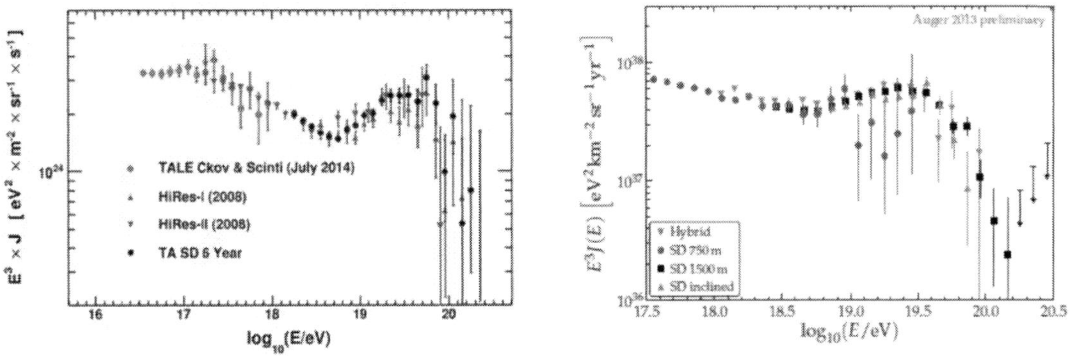

Figure 4. All-particle energy spectra of the low energy extensions of the Telescope Array (left, [17]), and the Pierre Auger Observatory (right, [18]).

on the calibration (hadronic interaction model) as well as on the treatment of including the a-priori unknown elemental composition. Despite the fact that the three experiments use different observation techniques, are located at different observation levels, and use different hadronic interaction models to interpret their data, the agreement within 20% of the total flux is surprisingly good. On one hand this confirms the structures found so far, but on the other hand it also confirms the high quality of the data taken by these modern experiments and also the validity of the hadronic interaction models (at least for this energy range and for the observables measured). At the present conference, Sven Schoo also discussed the recent results of KASCADE-Grande on the investigations of the elemental composition, where the goal is the reconstruction of individual mass group spectra [6]. Structures observed in these individual spectra provide stronger constraints to astrophysical models of origin and propagation of high-energy cosmic rays than the all-particle spectrum or a mean logarithmic mass. The evolution of the mass parameter as a function of energy keeps track of the evolution of the composition, and allows an event-by-event separation between light, medium and heavy primaries, at least. The application of this methodical approach to shower selection and separation in various mass groups is performed and cross-checked in different ways, where the right panel of Figure 3 shows the main results: The reconstructed spectrum of the electron-poor events, i.e. the spectrum of heavy primaries, shows a distinct knee-like feature at about $8 \cdot 10^{16}$ eV [11]. Hence, the selection of heavy primaries enhances the knee-like feature that is already present in the all-particle spectrum. In addition, an ankle-like feature was found in the spectrum of the electron-rich events, e.g. light elements of the primary cosmic rays, at an energy of $10^{17.08\pm0.08}$ eV [12]. This feature can be interpreted as a first experimental hint for an early appearance of an extra-galactic, light mass cosmic-ray component.

Both large observatories, the Telescope Array and the Pierre Auger Observatory have been setup with a denser array on a small part of its total area, called TALE and Auger Infill, respectively. First all-particle spectra have been presented by these installations at the conference [17, 18] displayed in Figure 4. In case of TALE, a reconstruction of the fluorescence telescope data is applied, which takes into account also the measured Cherenkov light which leads to a significantly lower energy threshold. The shown preliminary spectrum confirms the beforehand found spectral features below the ankle.

Denis Allard examined the energy range of the transition and the experimental results from a theoretical point of view [19]. He concluded that the observed light composition ankle at 100 PeV indeed could be the signature of an early appearance of cosmic rays of extra-galactic origin.

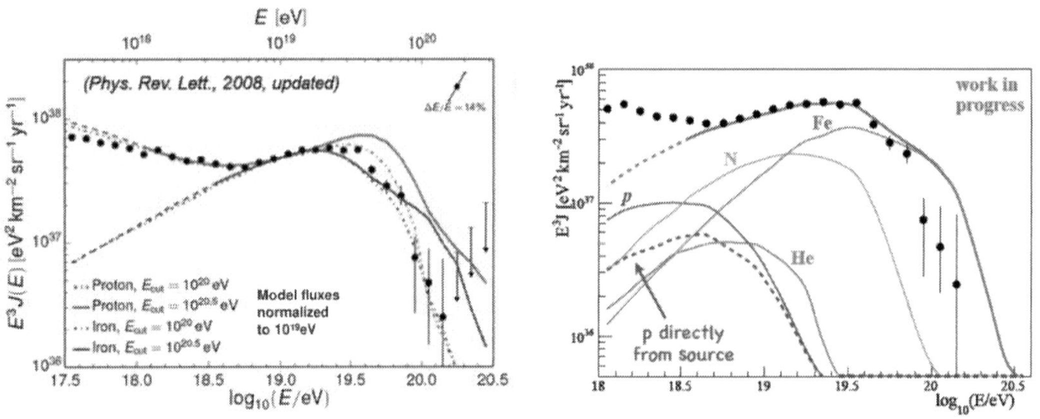

Figure 5. Scenarios to explain the flux and shape of the cosmic-ray energy spectrum at the highest energies measured by the Auger observatory (from [19]). Left: the (GZK-)propagation model, where the cosmic rays are accelerated to the indicated very high energy of and the spectral shape arise from propagation effects; Right: the maximum acceleration model, where the cut-off starts with primary protons at an energy of $E_{max} = 10^{18.4}$ eV and the flux of the other primaries cut according to their charge at $E = Z \cdot E_{max}$.

This is also in agreement with a scenario where the spectrum at the highest energies and the flux suppression (see next chapter) is explained by reaching the maximum acceleration power at the source regions. For a conclusive explanation, however, more experimental data is required at all energies of UHECR. The planned LHAASO experiment [20] at lower energies as well as the TALE and Auger Infill installations will contribute to set more constraints on the theoretical models.

3. Experimental Status: The Highest Energies

Mainly two experiments contribute presently to the investigation of cosmic rays of the highest energies, i.e. above the ankle. These are the Pierre Auger Observatory ([21], and references therein) located in Argentina viewing the Southern Hemisphere of the sky, and the Telescope Array ([22], and references therein) in Utah, US sensitive to the Northern Sky. Both experiments detect air showers by the hybrid technique, i.e. by an array of particle detectors operating 24 hours per day plus fluorescence telescopes, where a calorimetric measurement of the EAS (therefore reaching a good composition sensitivity) is possible, paid by the price of a significantly reduced duty cycle as this technique can only be performed in clear, moonless nights.

The reported [17, 23] all-particle energy spectra of both experiments are displayed in Figures 1 and 4. Both experiments have observed a clear flux suppression at an energy of $\approx 5 \cdot 10^{19}$ eV. Where the absolute flux agrees at both experiments at energies below and around the ankle, the Telescope Array reports a systematically higher flux at the highest energies compared to the Pierre Auger Observatory. Taking into account the statistical and systematic uncertainties of both experiments, however, the difference is still possible by instrumental effects. But, the interesting question is now why such a difference is not visible at lower energies when it is based on systematic uncertainties. As both experiments use similar techniques of detection, the question is allowed if the reason can be due to the fact that the observation happens at different hemispheres of the Earth?

The most important question, however, is related to the origin of the flux suppression. It

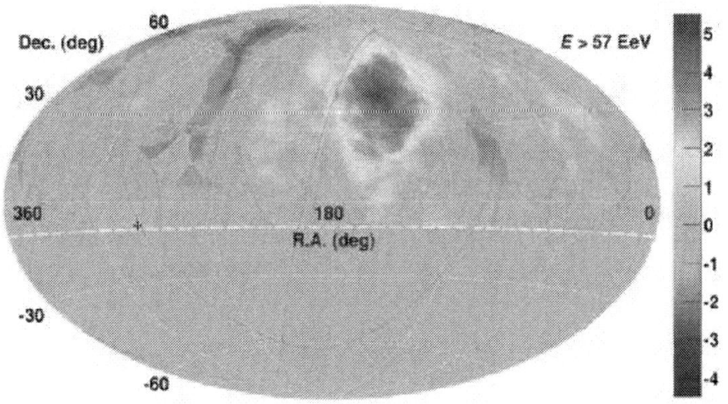

Figure 6. Skymap of the detected cosmic rays of the Telescope Array with an energy > 57 EeV. The distribution is created by using an oversampling of 20 degree. Included are 72 events from 5 years of data taking. The observed hotspot is located south of the the supergalactic plane with a significance of more than five sigma (by Li-Ma calculation). From [17].

could be either due to the predicted and expected GZK-effect [24, 25][3] or by exceeding the maximum energy of acceleration processes at the sources of the cosmic rays. For the latter case a dependence on the charge Z of the maximum energy for the individual primary masses is expected. Also a mixture of both scenarios is possible.

The measured all-particle spectrum is explainable by both scenarios (Fig. 5) [19, 23, 26], or to be more precise, both scenarios are not perfectly in agreement with the data, despite the large uncertainties still present in the measurements. To answer the question of the true scenario a detailed determination of the elemental composition of the cosmic rays is needed. Unfortunately, this is hampered at the highest energies by the low duty cycle of the mass sensitive fluorescence measurements. No new results on composition could be presented at this conference. But, there are methodical improvements to extract composition information also from the surface detectors to overcome the low duty cycle of the fluorescence measurements, presented by Ruben Conceicao [27], e.g.. Further constraints on the composition could be obtained by determining limits of (or detect) primary photons and neutrinos at energies around 10^{18} eV. The concept is based on the fact that protons interacting with the cosmic microwave background generate those gammas and neutrinos (so-called GZK particles). Hence, such limits will constrain the content of primary protons on the total flux at the energy range of the flux suppression. The limits could be improved over the last years, but the necessary sensitivity is not yet reached to additionally constrain the models.

An independent access to the data is the study of the arrival directions of the measured cosmic rays. At the conference an interesting hotspot was discussed observed by the Telescope Array for events above the flux suppression energy (Fig. 6). After six years of data taking a large-area hotspot is visible with a significance of meanwhile more than five sigma [17]. This is of high interest, as at the highest energies it is expected that the deflection of the cosmic rays (at least of the singly-charged protons) is small so that the source can be observed directly.

[3] The Greisen-Zatsepin-Kuz'min limit (GZK limit) is a theoretical upper limit on the energy of cosmic rays coming from distant sources. The limit is set by slowing-interactions of cosmic ray protons (photo-pion production) with the microwave background radiation over long distances. For primary iron nuclei a similar effect (photo-dissociation at giant dipole resonances) is expected to start at roughly the same energy.

However, as there is no clear counterpart of a possible source behind this hotspot the origin is presently neither confirmed nor understood.

In summary, despite the large progress in the detection of UHECR and its connected achievements in the analysis of the data the mystery of the origin of the UHECR is not yet solved. The most urgent problems and challenges for the community can be listed as

- Explaining the results (theory): None of the existing theories is able to explain all the measurements, where this can also be reasoned in the statistical accuracy and quality of the data.
- Statistics at highest energies: Still the statistical accuracy is far too small for detailed analyses at the highest energies.
- Better composition measurements: to decide on the astrophysical origin of the flux suppression the elemental composition has to be determined as accurately as possible.
- Understanding of the air showers: any reconstruction and interpretation of the air-shower measurements relies on the understanding of the shower development in the atmosphere.
- Particle physics with UHECR: Studying the hadronic interaction processes via air-shower observables helps to improve the simulations as well as gives complementary information to accelerator experiments (at a much higher interaction energy).
- What is the best detection technique to reach these goals?
- Multi-messenger particle astronomy: The answer to the big question 'What is the origin of cosmic rays?' can probably only be solved by combining the information of many messengers sent by the sources, i.e. high energy gammas, neutrinos, and cosmic rays.
- And finally: understanding the source, acceleration and propagation of cosmic rays (theory).

In the following I will touch some progress in these challenges as far as reported at the conference.

4. Presented Theoretical Achievements

Only a few theoretical papers related to UHECR were discussed at this conference. There are still too many free parameters to develop a consistent theory of source, acceleration, and propagation of UHECRs describing all (partly also contradicting) observations. For example, it is still not fully excluded (even though implausible[4]) that the entire cosmic ray spectrum is of galactic origin (see the contribution of Volkov et al. [28]). However, achievements were presented how to work with the available data for getting deeper insights. Two examples shall be given here:

Ptuskin [30]) discussed a method to obtain the source spectrum of cosmic rays, i.e. the spectrum after acceleration but before propagation, by means of the measured spectrum on Earth. Taking into account different possible composition scenarios and by solving the inverse transport problem the source spectra are obtained. Interestingly, the relatively small differences in the measured spectra between Auger and TA lead to significantly different source spectra (see Fig. 7).

Alves Batista [31] studied the effect of magnetic fields to the propagation of UHECR and came to the conclusion that a much better understanding of both, the galactic and the extra-galactic magnetic fields as well as the general matter distribution in the Universe is needed for pursuing UHECR astronomy. The studies show that relatively small changes of the magnetic field and/or matter distribution alter the arrival pattern of UHECR on Earth significantly.

[4] The model described in [28, 29] focuses on the possibility to accelerate particles within our Galaxy to 10^{20} eV, but the main argument for an extra-galactic origin of the highest energies is the non-observation of significant anisotropies at 10^{18} eV, which is ignored in this model.

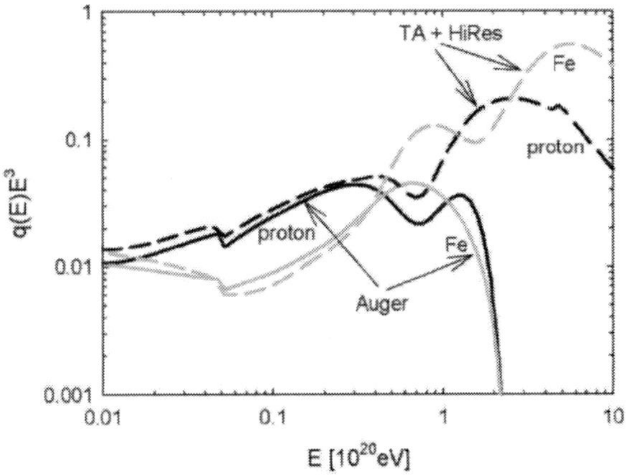

Figure 7. Calculated source spectra in arbitrary units based on an analytical approximation using the Auger (solid lines) and TA+HiRes (dash lines) data. Black lines for proton sources; grey lines for Iron sources. From [30].

5. Do we Understand the Measurements?

Both techniques applied to measure the EAS generated by UHECR, the fluorescence and the particle detection technique do still have large systematic uncertainties in calibrating the absolute energy and mass scale of the primary cosmic ray. Besides the intrinsic shower-to-shower fluctuations these are the main sources hampering a sufficient energy resolution of better than $\approx 15-20\%$.

Concerning the fluorescence detection technique, in the last decade, in particular by laboratory measurements, many efforts were done to understand the transmission of the light through the atmosphere, to control the temperature and pressure dependence of the fluorescence light yield and scattering, to understand effects of aerosols and ozone, and to study the absolute fluorescence yield. A nice and valuable summary of these efforts was given by Bianca Keilhauer [32]. In the conclusions she stated that obviously the understanding of fluorescence light emission and transmission is essential for the air-shower interpretation, but that meanwhile a consistent and comprehensive description of the fluorescence emission is available. However, the absolute scaling (fluorescence yield) is still the main unknown parameter and consequently the largest uncertainty at the interpretation and calibration of the air showers.

An interesting study was presented by Stefanie Falk [33], where the transmission of the light in the atmosphere was compared for telescopes located on Earth and in Space[5]. Figure 8 shows that there are indeed large differences. For example, for a wavelength of 337 nm (a peak in the fluorescence emission spectrum) only 2% of the light is transmitted to the ground telescope, whereas, due to the less dense atmosphere, 50% of the light reaches the telescope in space. However, in space the absorption of the light by the ozone layer plays a non-negligible role and reduces the light by another 6-10% (not shown in Figure 8, but discussed in [33]).

For air-shower measurements with an array of particle detectors at ground, (a non-calorimetric detection), the interpretation depends heavily on the understanding of the air-shower development, i.e. on the reliability of the hadronic interaction models in use [34]. There

[5] Plans exist to install a sophisticated fluorescence telescope at the International Space Station observing the air-shower developments in the atmosphere from above. This project is called JEM-EUSO, see chapter 7 of this paper.

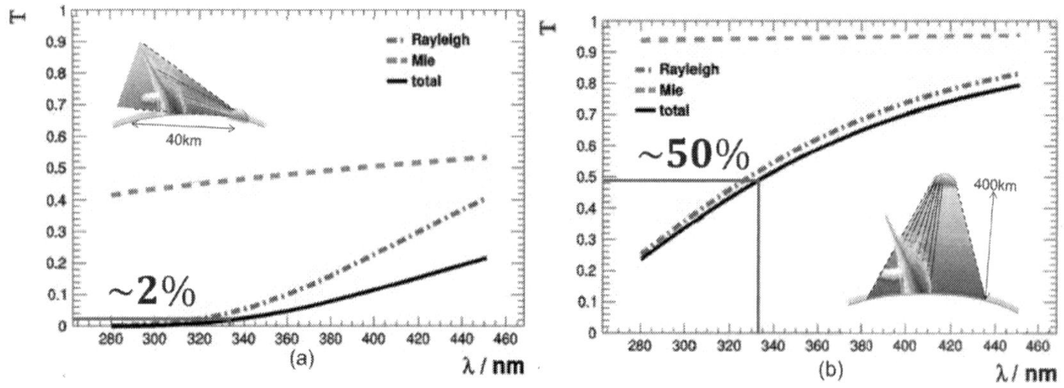

Figure 8. Comparison of the transmission factor T for Mie and Rayleigh scattering. The Mie attenuation length is assumed to 52.645 km in a homogeneous mixing layer below 3 km, taken from measurements of the Pierre Auger Observatory. The Rayleigh attenuation length is computed with respect to the US-StdA 76. The detector was placed (a) in 40 km distance at ground level, (b) at the ISS altitude 400 km above ground (from [33]), see text.

were a lot of improvements of the model descriptions in the last years based on both, multi-observable measurements of EAS (mainly by KASCADE) and by LHC results, and the latest generation of models agree well to each other as well as to a large extend also to the EAS measurements - at least at energies lower than 10^{17} eV. For the highest energies, however, still the interaction physics have to be extrapolated by three orders of magnitude, and also the important phase space for the EAS development (the extreme forward direction) is not really covered by accelerator experiments.

Concerning the latter, there exists the LHCf experiment at CERN, dedicated to measure the impact points and energy of neutral particles in the extreme forward direction. The measurements are already used to check the validity of the present hadronic interaction models [35]. Bongi also discussed in a more general way possible contributions of accelerator experiments to improve the simulation and therefore the reconstruction quality and the interpretation of extensive air-showers. Four topics were identified:

- Inelastic cross-section: Important for the first interaction, as it defines if the EAS starts early in the atmosphere or if it penetrates deep.
- Forward energy spectrum of secondaries: When the spectrum is soft, the shower has an early and rapid development.
- Inelasticity of the interactions: A large inelasticity would also hint to a rapid development of the shower.
- Nuclear interactions: Multiplicity and rapidity distributions of the secondary particles at all energies are an important ingredient to model the shower development.

The higher energy accessible by the next stages of the LHC will further improve the situation. In addition, an LHCf-like detector at RHIC or a Carbon or Nitrogen beam at LHC would significantly contribute to a better understanding of the EAS and by that finally help to solve the puzzle of the origin of high-energy cosmic rays.

Back to the present air-shower measurements, tests of the hadronic interaction models are performed by analyzing multi-parameter observations. There are many hints that the muon

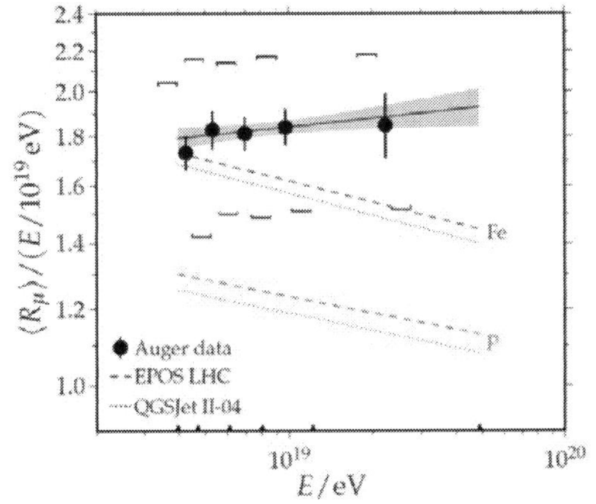

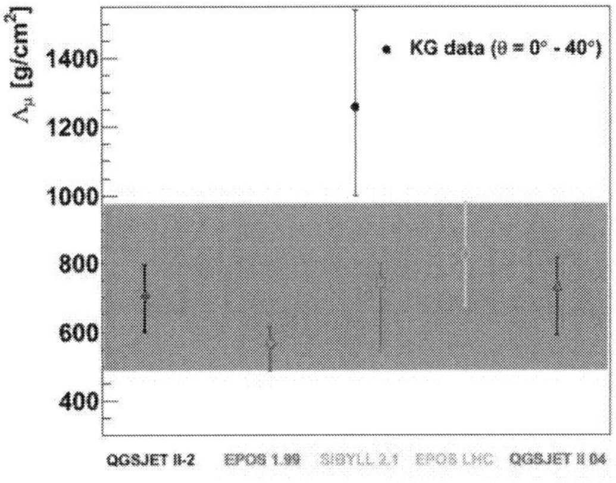

Figure 9. Left panel: Comparison of predicted and measured muon numbers, normalized to the muon number of 10 EeV primary energy (estimated and calibrated by means of the fluorescence telescope measurements) of horizontal air shower events of the Pierre Auger Observatory [36]. Right panel: Muon attenuation length of air showers measured by KASCADE-Grande ($\approx$ 0.1 EeV) compared to the attenuation length predicted by different hadronic interaction models [37]. Both analyses show a disagreement between predictions and measurements of the muon component of EAS.

component in the air-showers are not well described, at least at higher energies. Two examples of such investigations were presented at the conference:

At the Pierre Auger Observatory [36] inclined showers (zenith angle larger than 60 degree) were studied as for them at observation level the muon component is dominant compared to the electromagnetic component. The primary energy is estimated by the total number of muons calibrated with a subsample of showers detected in hybrid mode, i.e. by the fluorescence measurements. In Figure 9, the determined muon numbers are compared to predictions of Monte Carlo simulations of different hadronic interaction models. Obviously, a significantly smaller muon number is predicted than measured. The effect increases with energy and is finally so large that even a pure iron composition (which is not expected at 10^{19} eV!) cannot explain the data. The conclusion is that all the hadronic interaction models currently in use show a muon deficit in their predictions, at least at the highest energies.

At KASCADE-Grande [37], i.e. at an energy two orders of magnitude lower, the muon component predicted by the models is tested by another method. The same method cannot be applied as KASCADE-Grande (i) with its thin scintillators is not as sensitive to inclined showers as large water Cherenkov detectors are, and (ii) an independent energy measurement is also not possible as only particle detectors are available. However, KASCADE-Grande provides 622 m^2 dedicated muon detectors and the method chosen here is to investigate the attenuation of the muon component in the atmosphere by studying the dependence of the muon number at fixed energy with the zenith angle. The resulting attenuation parameter Λ is for the data significantly larger than for the predictions, independent of the choice of the hadronic interaction model (Fig. 9). This means less predicted than measured muons and therefore, is tendentially in agreement with the findings of the Auger Observatory. However, in an absolute scale it leads to a smaller effect in number of muons (at a factor 100 lower energy).

Summary of this chapter: Despite the problems in fixing the absolute scale in energy and mass, the understanding of the process of the evolution of extensive air showers is improving by both, investigations of the multi-parameter air-shower measurements and reaching higher energies and better forward detectors at accelerator experiments. This allows us to be more confident in the observed spectral features of the cosmic ray flux as well as in tendencies (getting heavier or lighter) in the elemental composition in the entire energy range from PeV to at least EeV.

6. Is Radio Detection a Technique for the Future?

The above mentioned problems in detecting and interpreting EAS in terms of energy and mass of the primary, together with the high costs of instrumenting large areas, make the study of new detection techniques very plausible. Over the past decade, in particular, tremendous progress has been made in the field of the digital radio detection of cosmic ray air showers. Impressive results are achieved with experiments like LOPES [38, 39], AERA [40, 41], LOFAR [42, 43], and Tunka-Rex [44], which have been presented at the conference and which exploit the potential of the detection technique.

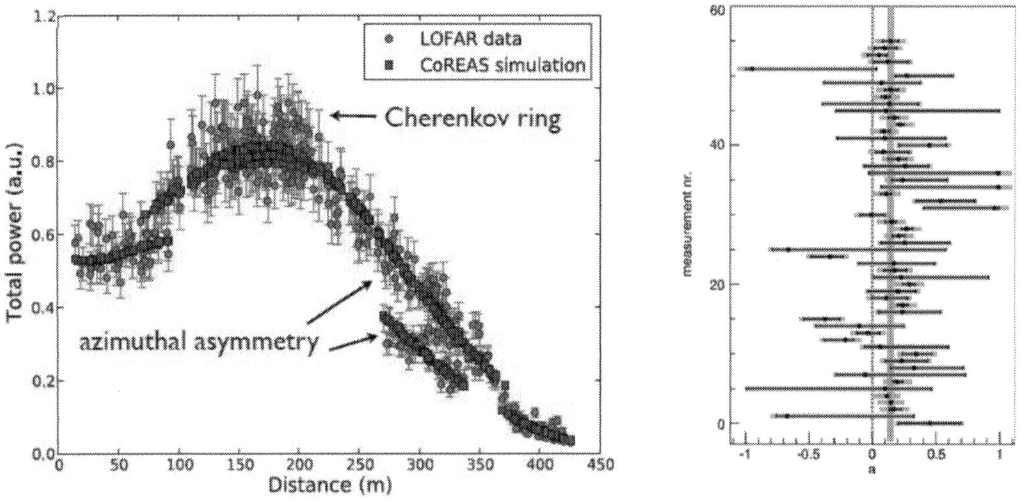

Figure 10. Left panel: Comparison of measurements and simulations for one air shower. Shown is the integrated radio pulse power for simulation (blue squares) and LOFAR measurements (red circles) as a function of distance to shower axis (from [43]). Several hundreds of antennas participated in the measurement of this individual EAS. Right panel: Distribution of the measured value of the a parameter, where the solid line indicates the 68% confidence interval around the mean value of the parameter; the dashed line shows the value of $a = 0$. The parameter a is defining the part of the radial polarized signal in the EAS, i.e. gives the fraction of the signal stemming from the Askaryan effect [41].

In parallel to the experimental activities, simulation codes and analysis strategies have been developed to get knowledge about the sensitivity of the radio technique to primary direction, mass, and energy of UHECR. In addition, a clue is now available how the accuracies compare to the standard techniques, like fluorescence or particle detection. At all experiments and based on this meanwhile solid understanding of the underlying radio emission physics, analyses have been devised to reliably extract the geometry, energy, and even depth of shower maximum of air showers measured with radio antennas.

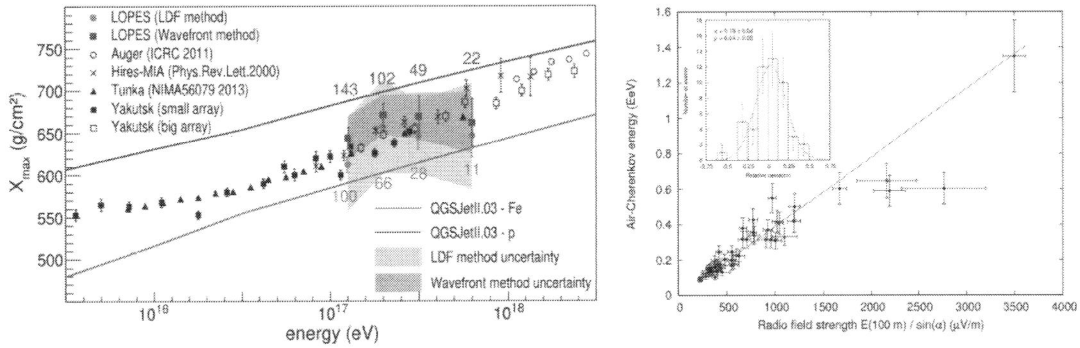

Figure 11. Left panel: Measurements of different experiments of the mean atmospheric depth of the shower maximum vs. energy. The results for LOPES are based on two independent methods [46]. The indicated number of events per bin deviate between both methods due to different quality cuts. Right panel: Correlation between the radio field strength at 100 m normalized by the sine of the geomagnetic angle and the energy reconstructed with the air-Cherenkov measurements of Tunka-133. The uncertainties of the electrical field are from the LDF fit, the uncertainties of the Tunka-133 energy reconstruction is estimated to 15%. The few outliers are probably events measured during thunderstorm conditions. As the bulk of the events are at lower energies, the inlay shows for all events the scatter between the energy reconstructed by Cherenkov and radio [44].

The LOFAR core instrument, built as low frequency astronomical radio antenna array in the Netherlands, is equipped with a small scintillator array enabling to trigger and record air showers with a huge number of antennas per single event. By this (Fig. 10) the irregular shape and the asymmetries of the footprint of the radio signal on ground as expected by the detailed emission simulations is proofed experimentally [43]. The irregular footprints stem from various effects of the radio emission, like the interfering of the two dominant emission mechanisms, geomagnetic effect and Askaryan emission with different polarizations, the geometry of the impinging shower (mainly relative to the Earth's magnetic field), or the used frequency band for detection.

The Auger Engineering Radio Array AERA, covering the largest area and highest energies in radio hybrid experiments could proof the emission mechanisms via detailed polarization measurements. As mentioned before, two radio emission mechanisms, the geomagnetic and the Askaryan effects, are going to contribute mainly to the radio signal from air showers. The electric field induced by the geomagnetic effect points to the direction of $\vec{v} x \vec{B}$, where $\vec{v}$ is a scalar vector describing the direction of air showers and $\vec{B}$ is the geomagnetic field. Whereas the electric field induced by the Askaryan effect has a radial direction pointing towards the core of the showers [45]. For individual showers (Fig. 10) AERA could determine the contribution of the Askaryan mechanisms to the total signal in terms of the a parameter (see figure caption). In average this is approximately 14% [41].

By the improved understanding of the radio signal first analyses of the primary UHECR parameters, like energy and mass are possible now. Figure 11 shows two examples, where the LOPES experiment applied two different methods (using the lateral distribution of the radio signal, and the shape of the wavefront, respectively) to determine the shower maximum [46]. This is the first measurement of the shower maximum by the radio detection technique. This is very promising for the future, despite the fact that at LOPES the systematic uncertainties are very large due to the human-made noise at the industrial environment of KASCADE-Grande. Tunka-Rex, a radio array measuring in coincidence with the Tunka Cherenkov array, has shown a first correlation of the radio estimated energy with the energy determined by the Cherenkov light emission. It shows that an energy resolution of better than 15% can be reached with the radio technique.

The very promising achievements for an application of the radio detection technique in future experiments, however, are accompanied by the cognition of some intrinsic limitations of the radio detection technique. In my view, for the present status of the radio detection technique following statements can be made:

- Radio measurements provide a calorimetric energy measurement of the electromagnetic component of the air shower with nearly 100% duty cycle (except thunderstorm periods).
- Having small experimental uncertainties (i.e. a radio quiet environment) the energy resolution achievable with radio detectors should be better than 10%.
- Radio emission gives information complementary to particle detectors as the pure electromagnetic component is selected. Self-trigger is possible, but difficult due to human made transient signals faking the radio EAS signal. All reconstructions are easier if the geometry of the EAS is known.
- The mass of the primary cosmic rays can be measured with the radio technique via the geometrical distance to the main emission region in the shower, which is closely correlated with the depth of the shower maximum. The distance to the shower maximum leaves an imprint in several observables, like lateral signal pattern, shape of the shower wavefront, or the frequency spectrum of the signal. The resolution achievable, however still needs experimental verification.
- The angular resolution is expected to be better than 0.5 degree for sophisticated and large arrays.
- The radio signal pattern and the detection threshold depends heavily on the geometry of the EAS, but is due to the exact knowledge of the emission mechanisms under control, and the dependencies on the atmospheric parameters are negligible (compared to the fluorescence light measurements).
- Radio detectors can be built cost effective, but this depends on two main aspects, the cost of an individual radio detector station and the needed density of these sensors. It is clear that antennas and digital electronics can certainly be built at much lower prices than those of particle detectors. The difficulty for a cheap radio detector arises, however, when the needed spacing of the detectors is considered.
- A big disadvantage of the radio technique lies in the relatively small footprint of the detectable signal (in terms of the ratio of signal-to-(galactic)noise) for vertically arriving air showers. This footprint, however, increases drastically when more inclined events are measured.
- Still studies have to be continued to make the technique a cost-effective, high sensitive instrument for a large-scale stand-alone or hybrid application of the radio detection technique. This concerns the R&D on the optimization of the radio stations themselves as well as on the reconstruction techniques. An encouraging ansatz is the study of a local hybrid measurement of the radio signal with a water-Cherenkov particle detector together with the corresponding single signal analysis.

7. Go for Highest Energies - with Larger Statistics and Better Mass Sensitivity

Still, cosmic rays of the highest energies of the wide spanning cosmic ray energy spectrum is the one least explored, which is natural as the flux of such particles reaching our Earth is extremely low. Though, it is the most interesting region as the origin is unknown and there is even no convincing theory explaining the acceleration to such high energies. Correspondingly, many and partly exotic explanations exist in literature. As an experimental physicist, I will not go to all these speculations, but will discuss the present and planned efforts to solve the long-standing mystery of the 10^{20} eV cosmic particles.

As earlier mentioned, at these high energies, in addition to the low statistical accuracy, the existing measurements disagree in their flux by a surprisingly large amount (see Fig. 1). The origin of the differences is not understood and several causes may play a role, e.g.:
(i) Statistical fluctuations and experimental systematic uncertainties.
(ii) It could lie in the fact that the energy reconstruction is limited by our current theoretical understanding of extensive air showers (EAS) [34]. The relevant energies as well as partly the kinematic features of the hadronic interactions at all energies are not well measured by accelerators, so that extrapolations are used in simulating the EAS leading to different EAS interpretations in terms of the primary energy and mass.
(iii) It could be that the two main experiments at the highest energies (Pierre Auger Observatory [21] and Telescope Array [22]) are located at different hemispheres and therefore, look at different regions of the sky. The standard picture is that the UHECR stem from astronomically nearby (< 100 Mpc) Active Galactic Nuclei, where we know that the distributions of those AGNs are not uniform over the sky [47].

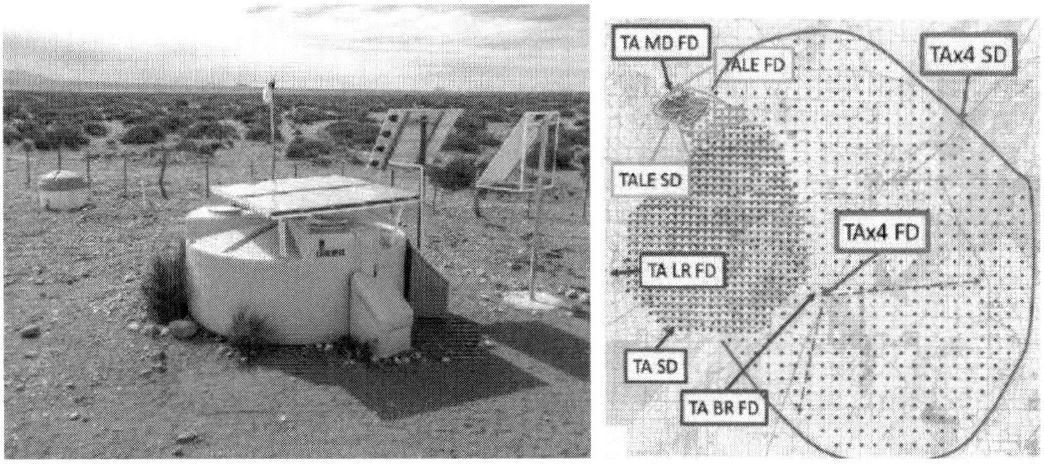

Figure 12. Left: Pierre Auger upgrade prototype of a scintillator detector on top of the water-Cherenkov detector. In the back an access tower to an AMIGA underground muon detector is seen [36]. Right: layout of the planned Telescope Array extension TAx4 [26].

It is obvious, that in future one needs (for the highest energies)

- higher statistics
- better mass composition sensitivity
- observation of the full sky
- multi-messenger information (UHECR, UHE-γ, UHE-ν)
- all of these with small experimental and systematic uncertainties!

to finally solve the puzzle of the origin of the highest energy cosmic rays.

The Pierre Auger Collaboration is presently preparing an upgrade of the Observatory to address the optimization of the primary particle identification [23, 36]. To improve the composition measurement at the highest energies, the Auger Collaboration proposes to enhance the electron-muon separation capabilities of the existing surface detector stations by adding 4 m^2 scintillation detectors on top of each water Cherenkov detector (Fig. 12). In addition, the data readout of the enhanced surface detector stations will be facilitated by replacing the current readout electronics by modern state-of-the-art electronics providing three times faster sampling, a significantly enhanced dynamic range, and enabling enhanced trigger and monitoring capabilities. Further on, a sophisticated underground muon detector array (AMIGA) at a smaller area of the observatory shall be built. With this upgraded observatory following physics program is envisaged [36, 48]:

- measure a factor of 10 in statistics for composition determination (as then possible by the data of the surface detectors measured with 100% duty cycle)
- decision if the flux suppression is due to the GZK effect or due to maximum acceleration, or a combination of both
- by selecting protons perform particle astronomy, e.g. in terms of composition enhanced anisotropy studies.
- study hadronic interactions at an interaction energy of $\sqrt{s} > 70\,\mathrm{TeV}$
- discover or limit ultra-high energy cosmogenic neutrino fluxes
- discover or limit ultra-high energy GZK induced gamma-ray fluxes
- clarify the science questions and the measurement techniques for the next generation UHECR experiment

The last item includes the plans that the Observatory is further used for developing new technologies in detecting high-energy cosmic rays, e.g. the radio detection techniques or applying improved photon sensors using semiconductor based detectors (SiPM). SiPMs are less sensitive for damages by bright light and could therefore enhance the duty cycle of the fluorescence measurements.
Preparations for the upgrade have been started and the plan is to finalize it in 2017.

The Telescope Array is planning to extend the sensitive area by a factor of 4 (Fig. 12). The larger area with a sparse distribution of particle detectors aims for considerably increasing the statistics of the highest energies, mainly to verify and establish the above mentioned hotspot [26].

A totally different concept to reach higher statistics is to observe the fluorescence light of extensive air showers from space [49, 50]. The Extreme Universe Space Observatory (EUSO) (Fig. 13) at the Japanese Module (JEM) of the International Space Station (ISS) is the most prominent space mission devoted to the scientific research of EECR [51][6]. Main goal is the exploration of the Universe through the detection of the extreme energy cosmic rays and neutrinos by looking downward from the ISS to detect the fluorescence light of extensive air-showers that they generate in the Earth's atmosphere. The main scientific objective is astronomy and astrophysics through the particle channel. This requires to identify the sources of cosmic rays by the reconstruction of the arrival direction and energy spectra with a high collecting power, beyond any other previous or planned experiment so far. The JEM-EUSO instrument consists of the telescope, the focal surface, a monitoring system for the atmospheric conditions, and a calibration system. In addition, there will be support and calibration systems on ground as well as at the ISS.

One big advantage of JEM-EUSO will be the uniform full sky coverage due to the ISS orbit. JEM-EUSO will be the first experiment to be able to take data of EECRs for both hemispheres

[6] The JEM-EUSO collaboration invented the term EECR, Extreme Energy Cosmic Rays, to emphasize the focus of the mission to the primary energies above the flux suppression.

Journal of Physics: Conference Series **632** (2015) 012093 doi:10.1088/1742-6596/632/1/012093

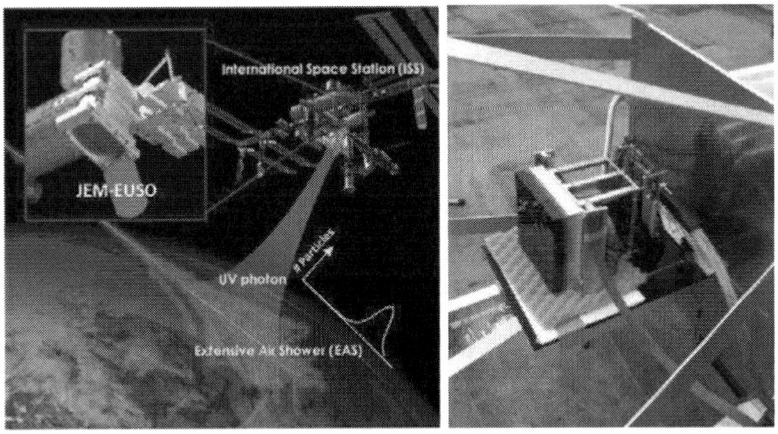

Figure 13. Left: Principle of the JEM-EUSO mission to detect extreme energy cosmic rays via the fluorescence emission of extensive air showers in the atmosphere. Right: first fully assembled Photo Detector Module (PDM) for JEM-EUSO prototype experiments (From [49]). The full instrument will consist of 137 PDMs.

to build a full sky map. This together with the huge exposure (up to a factor 10 compared to the Pierre Auger Observatory) to be reached will lead to anisotropy studies and identification of sources and source regions of high-energy cosmic rays which is not possible with ground-based observatories. JEM-EUSO will have a large observation area in nadir mode and even a roughly ten times bigger one in tilt mode, which is not yet fully explored, but the design enables the option to tilt the entire instrument at the ISS to have a larger coverage on ground.

On the way to the full instrument ready for lunch to the ISS, several test or pathfinder experiments are needed and currently under development or in operation. A PDM (photon detection module with 2304 pixels, see Figure 13) can be seen as an independent unit, where one of those is used or will be used for most of the test experiments. These various test and pathfinder experiments presently under construction or in operation will provide more information and details on the capabilities of such a space instrument like JEM-EUSO:

- One PDM is used for EUSO-Balloon, serving as a demonstrator for technologies and methods featured in the space instrument. A first flight could be performed on August, 25, 2014 at Timmins, Canada. 5 hours of valuable data from the PDM, from the installed infrared camera, as well as Laser pulses send by a helicopter accompanying the full flight are presently being analysed. Further shorter flights are foreseen and a long duration flight at the Southpole is applied for.
- Also important information on the capabilities of the full instrument will be provided by EUSO-TA, which is a ground-based telescope formed by one PDM (identical to the PDM of EUSO-Balloon) and two Fresnel lenses. EUSO-TA is located at the Telescope Array site. The aims are to obtain an end-to-end calibration of the prototype telescope, and an inter-calibration with the fluorescence detector of TA.
- Currently, the collaboration considers a cooperation with the Russian project KLYPVE. KLYPVE [50] is a mirror based fluorescence telescope foreseen to be installed at the Russian segment of the ISS. The pathfinder of KLYPVE is TUS on board of the Russian Lomonosov (launch foreseen in 2015) satellite. Despite its smaller size, KLYPVE (or even TUS) could be the first experiment detecting EAS from space.
- Another project is Mini-EUSO, a small prototype experiment (one PDM and a two-25cm-diameter Fresnel-lens system) foreseen to operate inside the ISS for an observation of the UV

emission from the night-Earth through an UV-window. This instrument would map for the first time the Earth in UV, and could study atmospheric phenomena and bio-luminescence at Earth as well as meteors.

- The JEM-EUSO collaboration has also started to evaluate if SiPM can replace the heavy and expensive Multianode PMT's presently in use. There is a fast commercial development of improved SiPM, where a comparable quantum efficiency to standard PMT's is expected in near future.

JEM-EUSO is planned as a three to five year mission, where initially the launch was foreseen in 2017. Due to financial reasons this will not be possible. However, the JEM-EUSO collaboration continues the efforts to improve the baseline design and the capabilities and sensitivity of such an instrument with the aim to launch it to space at a later opportunity.

8. Final Goal: Multi-Messenger Particle Astronomy

There is the justified hope that in a decade from now a comprehensive study of the high-energy Universe will be possible by combining the data and the information of sophisticated experiments for all three particle channels: cosmic rays, neutrinos, and gamma rays (Fig. 14).

The present status (early 2015) of the most relevant experiments for such a multi-messenger particle astronomy is (see also [52][7]):

- Gamma rays: MAGIC, VERITAS, and H.E.S.S. are already taking valuable data from many of the potential cosmic ray sources. The Cherenkov Telescope Array CTA [56] will be the next generation of this kind of experiments (TeV Imaging Cherenkov Telescope Arrays). Prototype telescopes in different sizes (sensitive to different energy ranges) are already existing, site selection and start of site preparation will happen in 2015.
- Neutrinos: IceCube [57] takes data and has detected first extra-terrestrial PeV neutrinos. An extension of IceCube (IceCube-Gen2) at the South Pole is under discussion and the pendants at the Northern Hemisphere (KM3NeT at the Mediterranean Sea [59, 60] and at a smaller scale GVD [58] at Lake Baikal in Siberia) have started the deployment of its first phase.
- Cosmic Rays: As discussed above, the Auger upgrade will be installed in the coming three years, Telescope Array plans to extend its sensitive area by a factor of four, and there is with the test experiments for JEM-EUSO a large progress for the first fluorescence telescope in Space.

However, there is still a long way to go. For example in UHECR studies, first the requirements of having more precise measurements with higher statistics have to be fulfilled before all information can be combined to perform particle astronomy!

Discussing the future in investigating the high-energy, non-thermal Universe I will not miss to affirm the need on combining all the data available in the various particle and wavelength channels. This will only be possible, if the data is publicly available with a short latency, which is meanwhile also a requirement of many funding agencies. A forerunner in this context is KCDC, the 'KASCADE Cosmic-ray Data Centre', a web portal, where data of astroparticle physics experiments is made available for the interested public [61]. In a first release, with KCDC [62] more than 160 million air showers are provided to the public. In addition, KCDC provides the conceptional design, how the data can be treated and processed so that they are also usable

[7] For all three particle channels there are also other running or planned experiments, partly with different detection techniques, like HAWC [53], LHAASO, TAIGA/HiSCORE [54] for gamma rays; the virtual combination of all neutrino detectors (Global Neutrino Observatory, GNO) for high energy neutrinos [55], or LHAASO and IceTop for cosmic rays. Details about these experiments can be found in other contributions to this conference.

Figure 14. Most relevant (according to the opinion of the author) existing and future experiments for exploring the high-energy, non-thermal Universe. By a combination of all the information a multi-messenger particle astronomy is envisaged.

outside the community of experts in the research field. Detailed educational examples make a use also possible for high- school students and early stage researchers.

In summary, there is a lot of progress in the research field of ultra-high energy cosmic rays, and it makes sense to stay tuned waiting for achievements presented at international conferences in the coming years.

Acknowledgments

The author thanks the organizers of the conference for the invitation on summarizing the interesting research field of high-energy cosmic rays, in particular Christian Spiering and Bernd Heber. Additionally, I thank all the colleagues contributing to the conference and discussions as well as all my colleagues in the research field. Thanks are also given to Sven Schoo for discussions and proof reading of the manuscript. Supported by the 'Helmholtz Alliance for Astroparticle Physics HAP' funded by the Initiative and Networking Fund of the Helmholtz Association.

References

[1] Haungs A, Rebel H, Roth M 2003 *Rep. Prog. Phys.* **66** 1145
[2] Blümer J, Engel R, Hörandel J R 2009 *Prog. Part. Nucl. Phys.* **63** 293
[3] Antoni T et al. - KASCADE 2003 *NIM A* **513** 429
[4] Antoni T et al. - KASCADE 2005 *Astropart. Phys.* **24** 1
[5] Apel W D et al. - KASCADE-Grande 2010 *NIM A* **620** 202
[6] Schoo S et al. - KASCADE-Grande 2015 *Proc. ECRS 2014, Kiel*; these proceedings
[7] Berezhnev S F et al. - Tunka-133 2012 *NIM A* **692** 98
[8] Budnev N et al. - Tunka 2015 *Proc. ECRS 2014, Kiel*; these proceedings
[9] Abbasi R et al. - IceCube 2013 *NIM A* **700** 188
[10] Apel W D et al. - KASCADE-Grande 2012 *Astropart. Phys.* **36** 183
[11] Apel W D et al. - KASCADE-Grande, 2011 *PRL* **107** 171104
[12] Apel W D et al. - KASCADE-Grande, 2013 *PRD* **87** 081101(R)
[13] Prosin V V et al. - Tunka-133 2014 *NIM A* **756** 94
[14] Aartsen M G et al. - IceCube 2013 *Phys. Rev. D* **88** 042004
[15] Apel W D et al. - KASCADE-Grande 2014 *Advances in Space Research* **53** 1456
[16] Haungs A 2015 *Physics Procedia* **61** 425
[17] Thomson G et al. - Telescope Array 2015 *Proc. ECRS 2014, Kiel*; these proceedings
[18] Daniel B et al. - Pierre Auger Observatory 2015 *Proc. ECRS 2014, Kiel*; these proceedings
[19] Allard D 2015 *Proc. ECRS 2014, Kiel*; these proceedings
[20] Cao Z et al. - LHAASO 2014 *NIM A* **742** 95

[21] Abraham J et al. - Pierre Auger 2008 *Phys. Rev. Lett.* **101** 061101
[22] Abu-Zayyad T et al. - Telescope Array 2013 *Astropart. Phys.* **39** 109
[23] Zas E et al. - Pierre Auger 2015 *Proc. ECRS 2014, Kiel*; these proceedings
[24] Greisen K 1966 *Physical Review Letters* **16** 748
[25] Zatsepin G T, Kuz'min V A 1966 *Theoretical Physics Letters* **4** 78
[26] Fukushima M 2015 *Proc. ECRS 2014, Kiel*; these proceedings
[27] Conceicao R et al. - Pierre Auger 2015 *Proc. ECRS 2014, Kiel*; these proceedings
[28] Volkov N 2015 *Proc. ECRS 2014, Kiel*; these proceedings
[29] Lagutin A A et al. 2015 *Bull. Russ. Academy of Sciences: Physics* **79** 322
[30] Ptuskin V 2015 *Proc. ECRS 2014, Kiel*; these proceedings
[31] Alves Batista R 2015 *Proc. ECRS 2014, Kiel*; these proceedings
[32] Keilhauer B 2015 *Proc. ECRS 2014, Kiel*; these proceedings
[33] Falk S et al. - JEM-EUSO 2015 *Proc. ECRS 2014, Kiel*; these proceedings
[34] d'Enterria D *et al.* 2011 *Astropart. Phys.* **35** 98
[35] Bongi M et al. - LHCf 2015 *Proc. ECRS 2014, Kiel*; these proceedings
[36] Valino I et al. - Pierre Auger 2015 *Proc. ECRS 2014, Kiel*; these proceedings
[37] Arteaga J C et al. - KASCADE-Grande 2015 *Proc. ECRS 2014, Kiel*; these proceedings
[38] Falcke H et al. - LOPES 2005 *Nature* **435** 313
[39] Apel W D et al. - LOPES 2014 *Phys. Rev. D.* **90** 062001
[40] Fuchs B et al. - Pierre Auger 2012 *NIM A* **692** 93
[41] Hasankiadeh Q et al. - Pierre Auger 2015 *Proc. ECRS 2014, Kiel*; these proceedings
[42] Schellart P et al. - LOFAR 2013 *A&A* **560** A98
[43] Nelles A et al. - LOFAR 2015 *Proc. ECRS 2014, Kiel*; these proceedings
[44] Kostunin D et al. - Tunka-Rex 2015 *Proc. ECRS 2014, Kiel*; these proceedings
[45] Aab A et al. - Pierre Auger 2014 *Phys. Rev. D* **89** 052002
[46] Schröder F G et al. - LOPES 2015 *Proc. ECRS 2014, Kiel*; these proceedings
[47] Véron-Cetty M-P and Véron P 2006 *Astron. Astrophys.* **455** 773
[48] Kampert K-H et al. - Pierre Auger 2015 scientific community discussions
[49] Haungs A et al. - JEM-EUSO 2015 *Proc. ECRS 2014, Kiel*; these proceedings
[50] Panasyuk M et al. 2015 *Proc. ECRS 2014, Kiel*; these proceedings
[51] Adams J et al. - JEM-EUSO 2013 *Astropart. Phys.* **44** 76
[52] APPEC - Roadmaps, see http://www.appec.org/strategy/roadmap.html, January 2015
[53] Arteaga J C et al. - HAWC 2015 *Proc. ECRS 2014, Kiel*; these proceedings
[54] Tluczykont M et al. - HiSCORE 2015 *Proc. ECRS 2014, Kiel*; these proceedings
[55] Spiering C 2015 *Proc. ECRS 2014, Kiel*; these proceedings
[56] Actis M et al. - CTA 2011 *Experimental Astronomy* **121** 1
[57] Aartsen M G et al. - IceCube 2013 *Science* **342** 1241856
[58] Spiering C et al. - GVD 2015 *Proc. ECRS 2014, Kiel*; these proceedings
[59] Margiotta A et al. - KM3NeT 2014 *NIM A* **766** 83
[60] Coniglione R et al. - KM3NeT 2015 *Proc. ECRS 2014, Kiel*; these proceedings
[61] Haungs A et al. - KASCADE-Grande 2015 *Proc. ECRS 2014, Kiel*; these proceedings
[62] see https://kcdc.ikp.kit.edu, January 2015

The escape model for Galactic cosmic rays

G. Giacinti[1], M. Kachelrieß[2], D. V. Semikoz[3]
[1]University of Oxford, Clarendon Laboratory, Oxford, United Kingdom
[2]Institutt for fysikk, NTNU, Trondheim, Norway
[3]AstroParticle and Cosmology (APC), Paris, France

Abstract. The escape model explains the cosmic ray (CR) knee by energy-dependent CR leakage from the Milky Way, with an excellent fit to all existing data. We test this model calculating the trajectories of individual CRs in the Galactic magnetic field. We find that the CR escape time $\tau_{\rm esc}(E)$ exhibits a knee-like structure around E/Z = few $\times\, 10^{15}$ eV for small coherence lengths and strengths of the turbulent magnetic field. The resulting intensities for different groups of nuclei are consistent with the ones determined by KASCADE and KASCADE-Grande, using simple power-laws as injection spectra. The transition from Galactic to extragalactic CRs happens in this model at low energies and is terminated below $\approx 3\times10^{18}$ eV. The intermediate energy region up to the ankle is populated by CRs accelerated in starburst galaxies. This model provides a good fit to $\ln(A)$ data, while the estimated CR dipole anisotropy is close to, or below, upper limits in the energy range $10^{17} - 10^{18}$ eV. The phase of the dipole is expected to change between 1×10^{17} and 3×10^{18} eV.

1. Introduction

The all-particle cosmic ray (CR) energy spectrum is a nearly featureless power-law between $\sim 10^{10}$ eV and $\sim 10^{20}$ eV, with only a few breaks in its spectral index. The two most prominent ones are the knee at $E_{\rm k} \approx 4$ PeV, and the ankle at $E_{\rm a} \approx 4$ EeV. Another related open question is where the transition from Galactic to extragalactic CRs occurs. For the knee, two main explanations currently remain possible. First, it may be the signature of the maximum energy to which Galactic CR sources can accelerate protons [1]. Second, the knee could be caused by a change in the energy dependence of the CR diffusion coefficient and thence confinement time in the Galaxy [2, 3, 4], if the CR Larmor radius is the order of the coherence length $l_{\rm c}$ of the turbulent Galactic magnetic field (GMF) at $E_{\rm k}$. In Ref. [4], we have studied this possibility—which we denote as the "escape model"—by propagating individual CRs in recent GMF models. We showed that the escape model is viable and can explain the individual fluxes of CR groups as measured by KASCADE and KASCADE-Grande. Moreover, our estimate for the dipole anisotropy in this model was consistent within uncertainties with observations.

Here we extend our previous study [4] and present preliminary results for the entire energy region between 300 GeV/Z and the ankle [5]. In particular, a more detailed study of the Galactic CR primary composition between $E_{\rm k}$ and $E_{\rm a}$ is presented here and compared to observations. We show that any remaining heavy nuclei flux in the sub-ankle region would be dominated by only one or few local sources. We use limits on the iron fraction at $\gtrsim 7 \times 10^{17}$ eV determined by the Auger collaboration together with $\ln(A)$ measurements to constrain the transition energy between Galactic and extragalactic CRs, deducing $\mathcal{R}_{\rm max} = E_{\rm max,Fe}/(26e) \sim 10^{17}$ V as the maximal rigidity $\mathcal{R}_{\rm max}$ to which Galactic CR sources are able to accelerate CRs. The recovery of the proton and helium spectra above $E/Z \sim 10^{16}$ eV in the KASCADE-Grande data is maninly explained by the specific shape of the escape rate $\tau_{\rm esc}(E)$ discovered in [4]. We show also that observational constraints from anisotropy limits are compatible with the escape model. A natural extension of the escape model to other normal galaxies suggests that the intermediate energy region up to ankle is composed of CRs accelerated in starburst galaxies. In both cases, observational constraints from anisotropy and composition measurements are compatible with the escape model.

2. Galactic magnetic field models and CR confinement in the Galaxy

An important constraint on CR propagation models comes from ratios of stable primaries and secondaries produced by CR interactions on gas in the Galactic disk. In particular, the B/C ratio has been recently measured by the AMS-02 experiment up to 670 GeV/nucleon [6]. In order to take advantage of the high-quality and the large energy range of the B/C data from AMS-02, we use now these data to derive the grammage traversed by CRs as function of their energies in two simple models. In Fig. 1, the two sets of grammages deduced are shown with magenta and blue error-bars. Note that the error-bars take into account only the statistical and systematic errors of the AMS-02 measurement, while uncertainties in the cross sections or deficiencies of our approximations are not accounted for. The latter can be estimated by the differences between the results from the two approximations.

In order to compare these measured values to those predicted in the escape model, we inject N cosmic rays at $z = 0$ in the Galaxy and follow their trajectories $\boldsymbol{x}_i(t)$ until they reach the edge of the Galaxy. Then we calculate the average grammage $\langle X \rangle = N^{-1} c \sum_{i=1}^{N} \int dt\, \rho(\boldsymbol{x}_i(t))$ summing up the density along the trajectories of individual CRs. Since the grammage $X(E) \propto E^{-\delta}$ scales as the confinement time $\tau(E) \propto E^{-\delta}$, this quantity serves also as an indicator for changes in the CR intensity induced by a variation of the CR leakage rate.

In Fig. 1, we compare the grammage calculated from simulated CR trajectories with the grammage deduced from B/C measurements. We use the JF model for the Galactic magnetic field [7], choosing as the maximal length of the fluctuations $L_{\rm max} = 5l_{\rm c} = 10$ pc, and consider two values of its root mean square (rms) strength, the original one suggested in [7] ($\beta = 1$) and a second one rescaling it to one tenth of its original value ($\beta = 1/10$). (Similiar results are obtained in the GMF model of Ref. [8].) Because of the large energy reach of the AMS-02 data, the extrapolation required from the lowest energy of our numerical calculations, $E = 10^{14}$ eV, to the measurements has shrunken to two orders of magnitude. Using the JF model with $\beta = 1$ as proposed in [7] would require a constant power spectrum of magnetic field fluctuations, $\mathcal{P}(k) \propto k^{-\alpha}$ with $\alpha = 0$, in the intermediate energy range. Such a power-spectrum is difficult to reconcile with the theoretical understanding of turbulence. Moreover, the CR spectrum is very close to a power-law above $\simeq 200$ GV. This implies that if $D(E)$ would become significantly flatter beyond TeV energies (e.g. changing from $D(E) \propto E^{1/3}$ to $\propto E^0$), then the injection spectrum of sources has to have the exact opposite change of slope (e.g. respectively from $\propto E^{2.4}$ to $\propto E^{2.7}$). Alternatively, a change in the source density should compensate the change in $D(E)$ such that the observed CR intensity remains a nearly featureless power-law [9]. Although such a conspiracy cannot be excluded, it appears to us as a not very attractive option.

We choose a Kolmogorov power-spectrum $\mathcal{P}(k) \propto k^{-5/3}$ as the theoretical model with the smallest slope α, and have therefore to reduce $B_{\rm rms}$ by a scaling factor $\beta < 1$. Next we examine how the shape of the grammage X as function of energy E/Z depends on the two parameters l_c and β. In [4], we discovered a specific shape of $X(E)$ that leads not only to a knee-like feature but reproduced also the recovery of the proton and helium spectra above $E/Z \sim 10^{16}$ eV, visible in the KASCADE-Grande data. It is clear that a too strong turbulent field, $\beta \sim 1$, results in knee-like feature at too high energy. Compensating a relatively strong turbulent field by decreasing the coherence length tapers off both the knee-like feature and the recovery. As a consequence, the allowed range of turbulent field strengths and coherence lengths is correlated and very restricted, $l_c \simeq (1-10)\,\text{pc}$ and $\beta \simeq 1/10 - 1/8$.

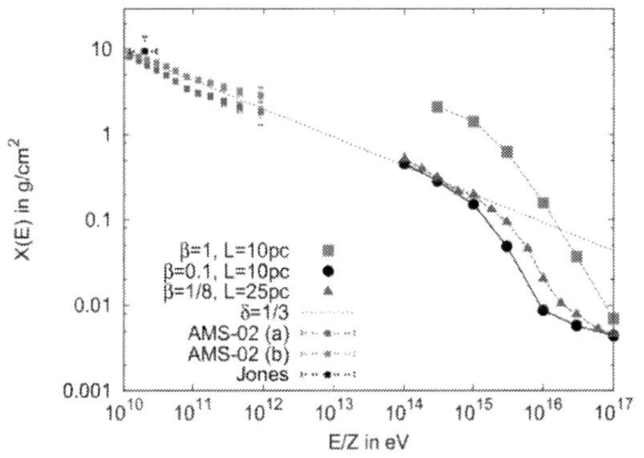

Figure 1. Grammage for different coherence lengths l_c and turbulent fields: red squares $L_{\max} = 10\,\text{pc}$ and $\beta = 1$, black dots $L_{\max} = 10\,\text{pc}$ and $\beta = 0.1$, and blue triangles $L_{\max} = 25\,\text{pc}$ and $\beta = 0.125$; all cases for the JF GMF model [7]. Additionally we show the grammage deduced from B/C data.

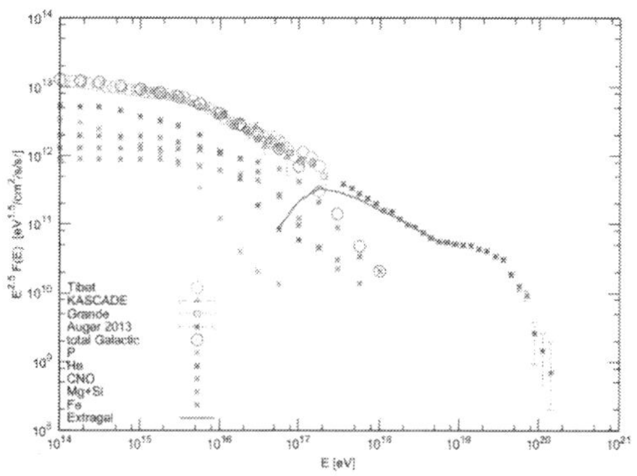

Figure 2. CR flux and the extragalactic component we predict for $\mathcal{R}_{\max} = 1 \times 10^{17}\,\text{eV}$, shown with the data from CREAM [11], KASCADE [16], KASCADE-Grande [17], TIBET [18], and Auger [19] experiments.

3. Fluxes of Galactic CR groups

For the calculation of the CR flux at Earth, we adapt a procedure based on the use of templates as decribed in detail in Ref. [5]. Results for nuclei with charge Z are deduced from the calculations for protons by shifting the energy by a factor Z. We then interpolate the resulting CR nuclei fluxes to the same energies as for protons. At energies below $Z \times 100\,\text{TeV}$, we assume that diffusion in the Kolmogorov turbulence shifts the injection power-law $\propto E^{-\alpha}$ by 1/3 to the spectrum $\propto E^{-\alpha-1/3}$ observed. This is indeed what we observe in our simulations in the energy range $\simeq Z \times (100-300)\,\text{TeV}$. Therefore we assume that the CR spectrum of protons released by sources follows a power-law spectra $\propto E^{-2.4}$; the maximal energy E_p of protons will be fixed later by considering constraints form the resulting dipole anisotropy and the observed nuclear composition of CRs. For all other nuclei, we use power-law spectra with either $\propto E^{-2.17}$ or $\propto E^{-2.22}$ and maximal energy ZE_p. These power-law indices are chosen so as to fit the direct observations from CREAM at low energy. We fix the density of sources by normalizing the flux found in our simulations to the observed one at 100 TeV. On average, we require 130 sources per 100 kyr for a total energy per source of $E_{\text{tot}} = 10^{50}\,\text{erg}$, so as to fit the observed CR spectra.

In Fig. 3, we plot the CR nuclei fluxes, multiplied by $E^{2.5}$, as a function of energy. In the upper left and upper right panels of Fig. 3, we show the proton and helium fluxes, both for turbulent fields with $L_{\max} = 10\,\text{pc}$ and with $L_{\max} = 25\,\text{pc}$. We plot orange lines $\propto E^{-2.4-1/3}$ (upper left panel) and $\propto E^{-2.22-1/3}$ (upper right panel), which represent the slopes expected theoretically at Earth, for our injection spectra with $\alpha = -2.4$ and -2.22 as power indices and Kolmogorov turbulence. Note that the slopes of the injection spectra required for nuclei, $\alpha \simeq 2.2$, coincide with the naive expectations from diffusive shock acceleration. Only the proton injection spectra requires a somewhat harder slope, $\alpha = 2.4$, than expected.

In the two upper panels, we show the experimental data from PAMELA [10] (orange points), CREAM [11] (magenta), KASCADE [12] (green) and KASCADE-Grande [12] (blue). The proton flux reported by KASCADE-Grande is 40% larger than the flux from KASCADE in the

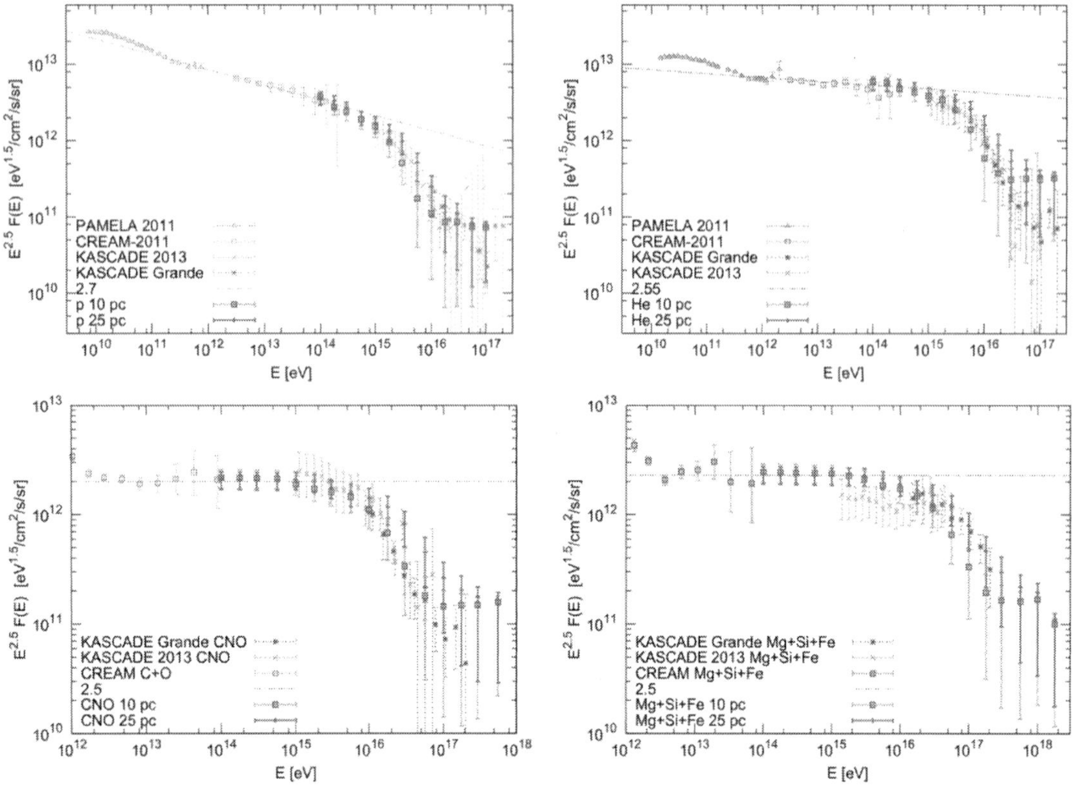

Figure 3. Fluxes of CR protons (upper left), helium (upper right), CNO (lower left) and heavy elements including Mg, Si and Fe (lower right) are shown in red. Errorbars show the variations in time of the fluxes. Experimental data from PAMELA for protons and helium, CREAM [11], KASCADE [12] and KASCADE-Grande [12].

$(10-30)$ PeV region, where errorbars of both experiments are relatively small. In contrast, the helium flux from KASCADE-Grande is below the one measured by KASCADE . This behavior is explained by the insufficient discrimination power between protons and helium in the KASCADE-Grande experiment [13]. For this study, we then choose to reduce the proton flux of KASCADE-Grande by 40%, and add this difference to the helium flux, in same energy bins. By doing so, the CR fluxes of KASCADE-Grande and KASCADE experiments become consistent with each other. In the lower left panel of Fig. 3, we plot the CNO flux, which predominantly consists of carbon and oxygen. We calculate the carbon and oxygen fluxes by normalizing them to the CREAM fluxes interpolated to higher energies with power-laws, and then sum them up. The CREAM flux in this figure is the sum of its carbon and oxygen fluxes, where we use carbon energy bins for the binning, and interpolate the oxygen flux to these bins before summing up. KASCADE and KASCADE-Grande measurements of the CNO flux are directly compared to our fluxes. In the lower right panel of Fig. 3, we show the flux of heavy nuclei, which is dominated by Mg, Si and Fe nuclei. The combined 'heavy nuclei flux' measured by the KASCADE experiment is smooth and agrees well with a simple power-law extrapolation of the CREAM flux to higher energies. It also agrees with the KASCADE-Grande flux. As for the other components, the model presented in this work fits well the heavy nuclei flux too.

As can be seen in Fig. 3 (upper right), the CR proton flux follows a power-law from 300 GeV up to about 1 PeV. It then changes to a steeper slope at the knee, and recovers at $\simeq 10$ PeV to a

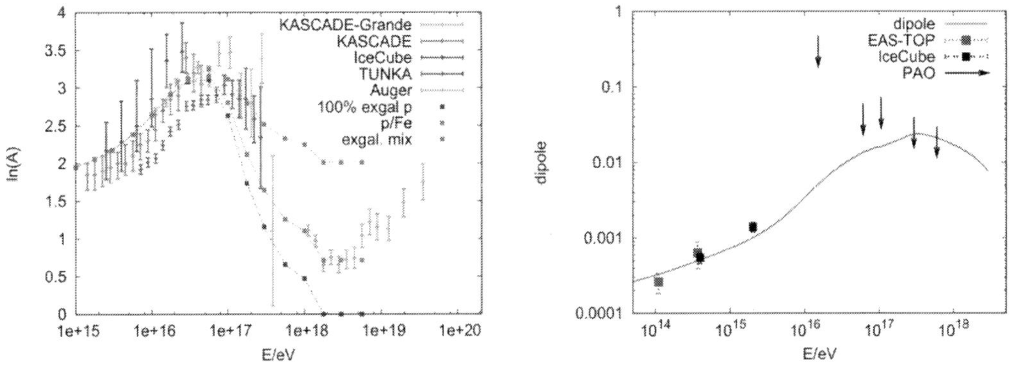

Figure 4. Left: Average of the logarithmic mass ln A predicted by our model for three different assumptions on the composition of extragalactic CRs, versus the experimental data. Right: Dipole amplitude $d(E)$ as a function of energy E in the GMF model of Ref. [7], using a reduced turbulent magnetic field with $\beta = 1/8$ and $L_{\max} = 25\,\mathrm{pc}$.

flatter power-law with index $\alpha \simeq 2.5$. Similar 'knee-like' cutoffs, shifted by factors Z in energy, are visible in the fluxes of all groups of CR nuclei—see the other panels of Fig. 3. These plots demonstrate that the escape model fits very well all these observations. As discussed previously in [4], the knee is due, in this model, to a change in behaviour with energy of the CR diffusion coefficient. The energy of the knee corresponds to the energy at which the Larmor radius of CR protons is of the order of the coherence length of the turbulent magnetic field ($l_c = L_{\max}/5$ for a Kolmogorov spectrum). For the field strengths we consider in this paper, $B_{\rm rms} \simeq 0.3\,\mu$G (or $\beta = 0.1/8$) close to the solar position, we find in our calculations a change in the slope of the CR flux at about 1 PeV, as observed in the *proton* data.

Let us note that, in this model, the flatter part of the CR proton flux above $\simeq 10$ PeV is dominated by recent nearby sources. This is due to the fact that the confinement time of CR protons in the Galaxy quickly drops with energy beyond the energy of the knee. While the fluxes from the few nearby sources are obviously subdominent below the knee, they tend to explain most of the measured flux above 10 PeV. As should be expected in such a scenario, fluctuations with time (on $\simeq (10-100)$ kyr time scales) of the CR proton flux at Earth are large above 10 PeV—what explains the large error bars of the theoretical prediction at high energies.

4. Transition from Galactic to extragalactic CRs

Determining at which energy E_* the CR flux starts to be dominated by extragalactic sources is one of the most important unsolved problems in CR physics. While in the 'dip model' of Refs. [14] the transition energy is as low as $E_* \approx$ a few $\times 10^{17}$ eV, in other models (e.g. [15]) it is expected to occur at $E_* \approx$ a few $\times 10^{18}$ eV. In our model, the maximum rigidity $\mathcal{R}_{\max}$ to which Galactic sources are able to accelerate CRs is a free parameter. The energy of the transition E_* in our model can be determined by using additional observational constraints. One possibility is to constrain the maximum contribution of Galactic sources to the total CR flux by using the observational limits on the anisotropy of the CR flux. Another way to determine E_*, is to study the elemental composition of primary CRs and use the fact that the composition of Galactic and extragalactic CRs should in principle differ from one another.

We start with the latter method and restrict our discussion to the case that the Auger iron limit is obeyed. As a first step, we derive the all-particle CR flux summing up all CR groups and compare it to the experimental data of KASCADE [16], KASCADE-Grande [17], TIBET [18], and Auger [19]. Then we deduce the extragalactic flux for a given $\mathcal{R}_{\max}$ by subtracting the predicted total Galactic flux from the measured total CR flux. The resulting extragalactic flux is shown in the right panel of Fig. 1 with a red dashed line for $\mathcal{R}_{\rm max} = 1.0 \times 10^{17}$ V. Next, we have to fix the nuclear composition of the extragalactic CR flux. As a first approximation, we can assume that its composition is constant in a sufficient small energy interval around E_*. In contrast, the Galactic CR composition is strongly energy dependent between the knee and the cutoff of the Galactic flux. Thus, we expect that an observable like the average of the logarithmic mass number, $\ln(A)$, will be quickly changing for $E \lesssim E_*$, while being approximately constant for energies slightly above E_*.

In the left panel of Fig. 4, we plot measurements of $\ln(A)$ from several experiments, together with the values of $\ln(A)$ calculated within the 'escape model' studied here. The points for KASCADE have been computed by converting the flux measurements given in [12] into $\ln(A)$ values. The most striking feature, namely the peak in $\ln(A)$ around 5×10^{16} eV, is clearly visible in all data sets, although its exact position and strength depend on the experiment. Our model reproduces the trend in the data very well. At higher energies, the composition becomes lighter because of the 'flattening' of the escape time at such energies, see Section 2. For the value of $\mathcal{R}_{\rm max}$ we consider here, extragalactic CRs start to contribute to the observed flux at $\approx 10^{17}$ eV. Consequently, above this energy, the exact value and shape of $\ln(A)$ depends on the assumed composition of the extragalactic flux. In blue, we show $\ln(A)$ for an extragalactic flux made of protons only, in magenta for a mix of 50% p and 50% Fe, and in red for a mix of 60% p, 25% He, and 15% N. Independently of the composition chosen for the extragalactic component, we can identify the energy where $\ln(A)$ stops to decrease with the maximum energy $E_{\rm max,Fe}$ to which Galactic sources can accelerate iron. It corresponds to the rigidity $\mathcal{R}_{\rm max} = E_{\rm max,Fe}/26e$. This transition is clearly visible in the PAO data, around the ankle, and allows us to determine the maximum rigidity as $\mathcal{R}_{\rm max} \simeq 1 \times 10^{17}$ V. For the case of a mixture of 60% p, 25% He, and 15% N (red curve), we obtain a good agreement with the $\ln(A)$ data from PAO up to 2×10^{18} eV. While this choice of composition is not unique, it is consistent with the results from the recent composition study published in [22]. In particular, Ref. [22] found the fraction of iron to be below 20% above 6×10^{17} eV. This agrees well with the results of our model, where the Galactic flux at 6×10^{17} eV consists purely of iron but contributes to only 15% of the total CR flux.

In addition to fitting the above observables, we still have to verify that the model presented here is also consistent with the existing upper limits on the CR anisotropy. In the diffusion approximation, the CR dipole anisotropy d is given by $\boldsymbol{d} = 3D\,\boldsymbol{\nabla}\ln(n)/c$. Following the same procedure as in [4], we compute the average anisotropy and derive the energy dependence of $D(E)$ from the escape rate as calculated previously, setting $D(E/Z) \propto 1/\tau_{\rm esc}(E/Z)$. We fix the proportionality constant by requiring that the dipole amplitude $d = \sum_k f_k d_k$ equals the dipole component $\tilde{d}$ observed by the EAS-TOP collaboration at $E = 1.1 \times 10^{14}$ eV [23]. Here, k labels the groups of nuclei we consider in the Galactic flux plus an extragalactic component. The latter has a dipole amplitude which is independent of its composition and which we set equal to 0.6%, as expected for the extragalactic Compton-Getting effect [27]. The factor f_k corresponds to the fraction the component k contributes to the total CR flux, and $d_k \propto 1/\tau_{\rm esc}(E/Z)$ to their individual dipole.

In the right panel of Fig. 4, we show the resulting dipole amplitude d as a function of energy E. As expected, the amplitude raises below the knee as $E^{1/3}$, while it increases approximately as $E^{0.7}$ until 1×10^{17} eV. At higher energies, the dipole amplitude decreases, which is due to the facts that the Galactic composition becomes heavier and that the extragalactic contribution grows. We also plot the values of $\tilde{d}$ observed by IceCube [24], as well as the 99% C.L. upper limits on $d_\perp$ from the Pierre Auger Observatory [25]. Comparing our estimate for the dipole amplitude with the upper limits in the energy range $10^{17} - 10^{18}$ eV, we should take into account that the approximation $d \propto 1/\tau_{\rm esc}(E/Z)$ starts to break down above $E/Z \gtrsim 10^{17}$ eV, which leads to a sizeable error. We conclude therefore that our prediction is marginally consistent with these limits. The Pierre Auger Observatory should however be able to reach a detection of the dipole anisotropy. Let us also note that the escape model predicts that the phase of the dipole amplitude varies strongly in the energy range between 1×10^{17} and 3×10^{18} eV: This corresponds to the range where the transition from Galactic to extragalactic CRs lies. Such a picture is supported by current observations of the phase of the dipole, see References [23, 24, 25].

In summary, there are two reasons for having an early transition, from predominantly Galactic to predominantly extragalactic CRs, at $E \approx$ a few $\times 10^{17}$ eV. First, the limits on the observed dipole anisotropy requires either a very heavy Galactic composition or a predominantly extragalactic contribution at $E \gtrsim 10^{18}$ eV [28, 25]. The former possibility is however strongly disfavored by the recent composition measurements from the Auger collaboration [21, 22]. Second, identifying the energy where $\ln(A)$ stops to decrease with the maximum energy to which Galactic sources can accelerate iron, $E_{\rm max,Fe} \approx 3 \times 10^{18}$ eV, suggests that the maximal rigidity reached in Galactic sources satisfies $\mathcal{R}_{\rm max} = E_{\rm max,Fe}/(26e) \sim 10^{17}$ V. Note that the cutoff in the Galactic CR spectrum would be at lower energies, if no nearby active CR source exists [5].

5. Conclusions

We have shown in [4, 5] that the knee can be entirely explained by energy-dependent CR leakage from the Milky Way, with an excellent fit to *all* existing data from $E/Z \sim 300$ GeV to 100 PeV. In particular, all deviations from a single power-law behavior that are observed in the CR intensity of individual CR groups in this energy range are consistently explained by rigidity-dependent CR escape. This model requires small coherence lengths of the turbulent field and relatively small turbulent magnetic fields. If these two conditions are fullfilled, then the CR escape time $\tau_{\rm esc}(E)$ exhibits a knee-like structure around $E/Z =$ few $\times 10^{15}$ eV together with a recovery around $E/Z \simeq 10^{16}$ eV.

In this presentation, we have considered only the case that a nearby CR source is present (for the opposite case see [5]). Then the maximal rigidity $\mathcal{R}_{\rm max}$ to which Galactic CR sources are able to accelerate CRs can be determined by identifying $Ze\mathcal{R}_{\rm max}$ with the energy where $\ln(A)$ derived from PAO measurements stops to decrease. This allowed us to deduce $\mathcal{R}_{\rm max} = E_{\rm max,Fe}/(26e) \sim 10^{17}$ V. As a result, the flux ratio of Galactic and extragalactic sources is in this case 1:1 at $E_* \approx 2 \times 10^{17}$ eV, dropping to 0:1 at 2×10^{18} eV. Since the transition from Galactic to extragalactic CRs happens in this model at rather low energies, the estimated CR dipole anisotropy is consistent within uncertainties with upper limits in the energy range $10^{17} - 10^{18}$ eV, while it reproduces the measurements at lower energies from EAS-TOP and IceCube. The dipole phase is expected to change between 1×10^{17} and 3×10^{18} eV, i.e. the energy range of the transition from Galactic to extragalactoc CRs. Such a behavior corresponds to the one observed, providing thus additional evidence for a transition from Galactic to extragalactic CRs in this energy region.

The ankle is interpreted in the escape model as the transition to a new, second extragalactic source class, as e.g. active galactic nuclei or gamma-ray bursts. Moreover, this model suggests that the CR knee in starburst galaxies is shifted by up to two orders of magnitude to higher energies [29]. Therefore, the CR flux in the intermediate energy region up to ankle should be composed mainly of CRs accelerated in starbust galaxies.

Let us note also that the weakness of the turbulent GMF in the "escape model" would have important consequence for the search of UHECR sources: UHECRs from a single source would be mainly deflected by the regular component of the GMF, while the spread of their arrival directions due to the turbulent GMF should be small. As a result, the search for UHECR sources, at least in the case of protons or light nuclei, should be more easily than thought before. Even for heavier nuclei, the deflections in the regular field of the Galaxy can be traced back in those patches of the sky with small turbulent fields [30]. Weaker magnetic fields will also simplify the search of nuclei sources using the methods discussed in Refs. [31, 32]. Thus the results of this work are an additional motivation for future searches of UHECR sources, performed by future all-sky missions as e.g. JEM-EUSO [33] and KLYPVE [34].

Acknowledgments
We would like to thank Andreas Haungs for discussions about the KASCADE–(Grande) data and Michael Unger for sending us data files of the ln(A) values derived in [35]. MK thanks the Theory Group at APC for hospitality. GG acknowledges funding from the European Research Council, DS was supported in part by grant RFBR # 13-02-12175-ofi-m.

References

[1] A. M. Hillas, J. Phys. G **31**, R95 (2005) and references therein.
[2] V. L. Ginzburg and S. I. Syrovatskii, *The Origin of Cosmic Rays* (Pergamon Press, 1964).
[3] V. S. Ptuskin *et al.*, Astron. Astrophys. **268**, 726 (1993); J. Candia, E. Roulet and L. N. Epele, JHEP **0212**, 033 (2002); J. Candia, S. Mollerach and E. Roulet, JCAP **0305**, 003 (2003).
[4] G. Giacinti, M. Kachelrieß and D. V. Semikoz, Phys. Rev. D **90**, R041302 (2014).
[5] G. Giacinti, M. Kachelrieß and D. V. Semikoz, arXiv:1502.01608 [astro-ph.HE].
[6] A. Oliva *et al.* [AMS-02 Collaboration], to appear in proc. of the 33rd ICRC (2013) [ID 1266].
[7] R. Jansson and G. R. Farrar, Astrophys. J. **757**, 14 (2012); Astrophys. J. **761**, L11 (2012).
[8] M. S. Pshirkov, P. G. Tinyakov, P. P. Kronberg and K. J. Newton-McGee, Astrophys. J. **738**, 192 (2011);
[9] M. Kachelrieß and D. V. Semikoz, Phys. Lett. B **634**, 143 (2006) [astro-ph/0510188].
[10] O. Adriani *et al.* [PAMELA Collaboration], Science **332**, 69 (2011).
[11] Y. S. Yoon *et al.*, Astrophys. J. **728**, 122 (2011).
[12] W. D. Apel *et al.* [KASCADE-Grande Collaboration], Astropart. Phys. **47**, 54 (2013).
[13] A. Haungs, private communication.
[14] V. Berezinsky, A. Z. Gazizov and S. I. Grigorieva, Phys. Rev. D **74**, 043005 (2006) [hep-ph/0204357].
[15] C. T. Hill, D. N. Schramm and T. P. Walker, Phys. Rev. D **34**, 1622 (1986); D. Allard, E. Parizot and A. V. Olinto, Astropart. Phys. **27**, 61 (2007).
[16] T. Antoni *et al.* [KASCADE Collaboration], Astropart. Phys. **24**, 1 (2005).
[17] W. D. Apel *et al.* [KASCADE-Grande Collaboration], Astropart. Phys. **36**, 183 (2012).
[18] M. Amenomori *et al.* [TIBET III Collaboration], Astrophys. J. **678**, 1165 (2008).
[19] A. Aab *et al.* [Pierre Auger Collaboration], to appear in proc. of the 33rd ICRC (2013) [arXiv:1310.4620 [astro-ph.HE]].
[20] R. Abbasi *et al.* [IceCube Collaboration], Astropart. Phys. **42**, 15 (2013).
[21] A. Aab *et al.* [Pierre Auger Collaboration], [arXiv:1409.4809 [astro-ph.HE]].
[22] A. Aab *et al.* [Pierre Auger Collaboration], [arXiv:1409.5083 [astro-ph.HE]].
[23] M. Aglietta *et al.* [EAS-TOP Collaboration], Astrophys. J. **692**, L130 (2009).
[24] R. Abbasi *et al.* [IceCube Collaboration], Astrophys. J. **746**, 33 (2012).
[25] P. Abreu *et al.* [Pierre Auger Collaboration], Astrophys. J. **762**, L13 (2012); Astrophys. J. Suppl. **203**, 34 (2012); I. Sidelnik *et al.* [Pierre Auger Collaboration], to appear in proc. of the 33rd ICRC (2013).
[26] M. G. Aartsen *et al.* [IceCube Collaboration], Astrophys. J. **765**, 55 (2013).
[27] M. Kachelrieß and P. D. Serpico, Phys. Lett. B **640**, 225 (2006).
[28] G. Giacinti, M. Kachelrieß, D. V. Semikoz and G. Sigl, JCAP **1207**, 031 (2012).
[29] M. Kachelrieß and S. Ostapchenko, arXiv:1405.3797 [astro-ph.HE].
[30] G. Giacinti, M. Kachelrieß, D. V. Semikoz and G. Sigl, Astropart. Phys. **35**, 192 (2011).
[31] G. Giacinti, X. Derkx and D. V. Semikoz, JCAP **1003**, 022 (2010).
[32] G. Giacinti, M. Kachelrieß, D. V. Semikoz and G. Sigl, JCAP **1008**, 036 (2010).
[33] T. J. J. H. Adams, Jr *et al.* [EUSO Collaboration], Contributions to the 32nd International Cosmic Ray Conference, Beijing (2011) arXiv:1204.5065 [astro-ph.IM].
[34] M. I. Panasyuk *et al.*, "The current status of orbital experiments for UHECR studies," arXiv:1501.06368 [astro-ph.IM].
[35] K. H. Kampert and M. Unger, Astropart. Phys. **35**, 660 (2012).

Measurements of the energy deposit of inclined muon bundles in the CWD NEVOD

R P Kokoulin[1], A G Bogdanov[1], L I Dushkin[1], S S Khokhlov[1], V A Khomyakov[1], V V Kindin[1], E A Kovylyaeva[1], G Mannocchi[2], A A Petrukhin[1], O Saavedra[3], V V Shutenko[1], G Trinchero[2], and I I Yashin[1]

[1] National Research Nuclear University MEPhI (Moscow Engineering Physics Institute), 115409 Moscow, Russia
[2] Istituto di Fisica dello Spazio Interplanetario – INAF, 10133 Torino, Italy
[3] Dipartimento di Fisica dell' Universita di Torino e INFN, 10125 Torino, Italy

E-mail: RPKokoulin@mephi.ru

Abstract. First results of investigations of the energy deposits of inclined muon bundles in the ground-based Cherenkov water detector NEVOD are presented. As a measure of the muon bundle energy deposit, the total number of photoelectrons detected by PMTs of the Cherenkov calorimeter is used. For each event, the local muon density at the observation point and the muon bundle arrival direction are estimated from the data of the coordinate-tracking detector DECOR. Registration of the bundles in a wide range of zenith angles allows to explore the interval of primary particle energies from $\sim 10^{16}$ to $\sim 10^{18}$ eV. Measurement results are compared with CORSIKA based simulations of EAS muon component. It is found that the mean energy of muons detected in the bundles rapidly increases with the zenith angle and reaches about 500 GeV near the horizon.

1. Introduction

In a number of cosmic ray experiments conducted at ultra-high energies of primary particles, an excess of multi-muon events (intensity of muon bundles, muon abundance in EAS) compared to simulations performed with widely used hadron interaction models is observed [1 – 4]. In principle, this excess may be caused by cosmo-physical (changes in the mass composition of primary cosmic rays with the increase of energy) or by nuclear-physical reasons (changes in characteristics of interactions of primary cosmic ray nuclei with the nuclei of air atoms). However, at energies of the order of 10^{18} eV and above, the excess of multi-muon events does not find its explanation in the frame of the existing interaction models even under the assumption of extremely heavy (iron nuclei) primary composition [3 – 5]. Therefore it seems that the contemporary models of the development of the nuclear cascade in the atmosphere have to be revised. The key to the solution of this problem, often referred to as 'the muon puzzle', may come from investigations of the energy characteristics of the EAS muon component and of their variations with primary energy [6].

One of the possible approaches to such investigations is the measurement of the energy deposit in the detector material at the passage of muon bundles. Average specific energy loss of muons almost linearly depends on their energy, and an appearance of an excessive flux of high-energy muons in the bundles has to be reflected in the dependence of the average energy deposit on the energy of primary particles. An experiment on the study of the energy deposit of muon bundles has been started at the NEVOD-DECOR complex in 2012. The complex includes a large-volume Cherenkov water calorimeter [7, 8] and coordinate-tracking detector [9] which ensures reliable identification of multi-muon events. The detection of muon bundles of different multiplicities at various zenith angles gives possibility to explore a very wide range of primary particle energies, from the knee region to few EeV, within a single experiment. The results of the measurements of the average energy deposit of inclined muon bundles obtained during first experimental series and their comparison with CORSIKA based simulations of muon energy characteristics are presented below.

2. Experimental data

The measuring system of the Cherenkov water detector (CWD) NEVOD with the inner volume $9\times9\times26$ m^3 represents a spatial lattice of quasi-spherical optical modules (QSMs) [7]. Each QSM includes six FEU-200 PMTs with flat 15 cm diameter photocathodes directed along rectangular coordinate axes. In total, the measuring system contains 91 QSMs (546 PMTs) arranged in vertical strings with spacing of 2.5 m along the water tank, and 2.0 m across it and over the depth. After a recent modernization [8], the electronics system ensures measurements of PMT signals in a wide dynamic range (from 1 to 10^5 photoelectrons for each PMT) that provides the possibility of calorimetric studies of high-energy cascade showers and muon bundle energy deposits. The coordinate-tracking detector DECOR is specially designed for investigating multi-particle events at large zenith angles and includes 8 supermodules (SMs) deployed in the galleries of the NEVOD building from three sides of the water tank. The sensitive area of each SM is 8.4 m^2. SMs consist of 8 vertical planes of plastic streamer tube chambers and provide spatial and angular accuracies of the muon track localization better than 1 cm and 1° respectively. An example of the muon bundle event detected in the NEVOD-DECOR complex is shown in figure 1.

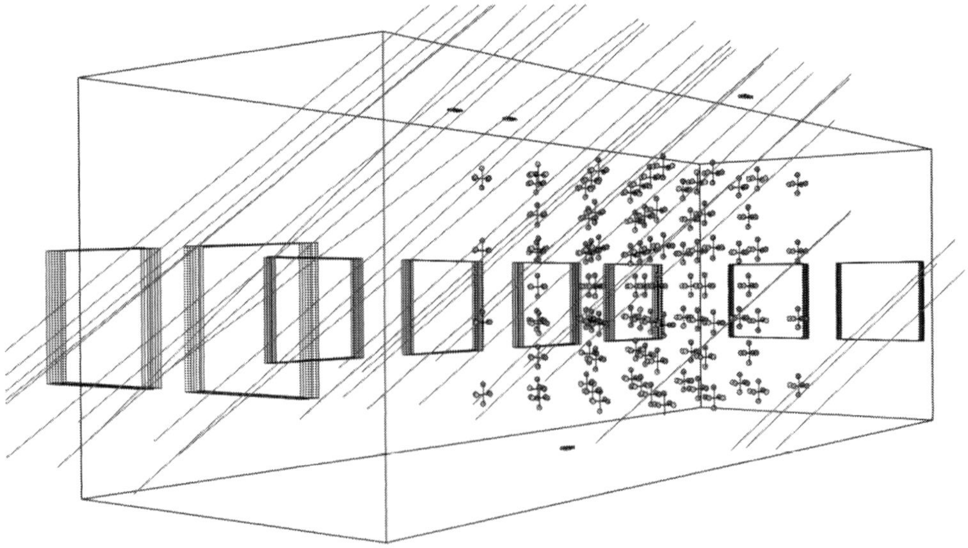

Figure 1. An example of the muon bundle event in the NEVOD-DECOR complex. Thin lines represent reconstructed muon tracks. Circles show hit PMTs of the CWD; big rectangles around CWD are DECOR supermodules; small rectangles on the top and on the bottom are triggered scintillation counters of the calibration telescope system.

In the present analysis, data of two first series of measurements – from May 2012 to March 2013 and from July 2013 to February 2014 – are used. Total 'live' observation time amounted to 9673 h. From these data, we selected 16416 muon bundle events with multiplicities $m \geq 5$, zenith angles $\theta \geq 55°$, in two 60°-wide sectors of azimuth angle where most of DECOR SMs (six of eight) were screened with the NEVOD water volume; data of these six shielded SMs being used for muon track counting. The average threshold muon energy for such selection criteria is about 2 GeV. In addition, from the initial part of the experimental material (for 3253 h observations) the bundles with lower zenith angles ($40^\circ \leq \theta < 55^\circ$, 15084 events) were selected. As a measure of the muon bundle energy deposit we use the sum Σ of all hit CWD PMT signals (in units of photoelectrons, ph.e.). The local muon density D in the event is estimated from the measured bundle multiplicity m with the allowance for the effective detector area $S_{\rm det}$ for the given arrival direction of the bundle. However, a straightforward calculation of the density as $D = m / S_{\rm det}$ would lead to a bias, especially serious at moderate multiplicities, because of statistical fluctuations of the number of detected particles in combination with a rapidly decreasing spectrum of the events in the muon density: $dF / dD = A \times D^{-(\beta+1)}$, where $\beta \approx 2.1$ is the integral density spectrum slope in a considered range of densities and zenith angles [3]. Assuming that *at a fixed density* the number of particles that hit the detector obeys the Poisson distribution with the mean value $< m > = D \times S_{\rm det}$, the mean value of density in the events contributing to the formation of muon bundles with *a fixed multiplicity* may be easily found analytically (see section 4 of [3]):

$$< D > = (m - \beta) / S_{\rm det} . \qquad (1)$$

Further, we use the estimate of the muon density defined by this equation.

3. Results and discussion

The correlation of the measured energy deposit with the muon density estimate for a limited sample of the events with zenith angles $\theta \geq 60^\circ$ is shown in figure 2. As it might be expected, in a first approximation the total energy deposit in the CWD is nearly proportional to the muon density (the solid line in the figure), therefore further we analyze the specific energy deposit Σ / D, i.e. the CWD response normalized to the muon density in the event.

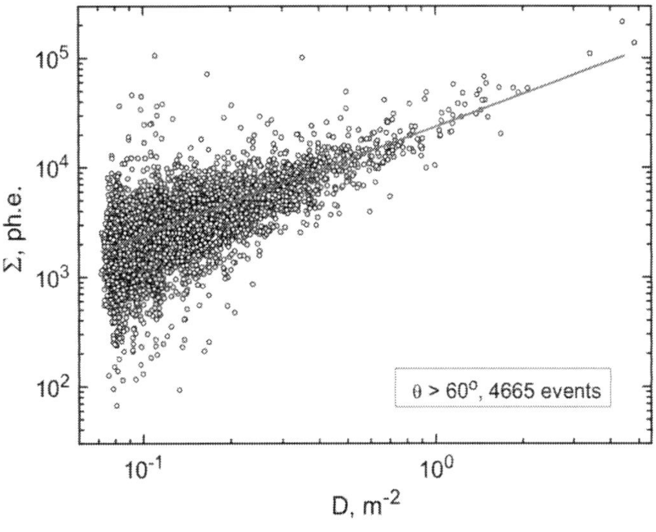

Figure 2. The correlation of the muon density estimate and the measured energy deposit.

Results of the measurements of the dependence of the average specific energy deposit on the zenith angle are presented by circles in figure 3. Arrows in the lower part of the figure indicate typical energies of primary particles (calculated mean-logarithmic values) that contribute to the formation of muon bundles at various angles. At moderate zenith angles ($\theta < 55^\circ$, open circles in the figure), a rapid decrease of the measured energy deposit with the increase of zenith angle is observed. It can be interpreted as a decreasing residual contribution of EAS electromagnetic and hadron components to the response of an unscreened detector located on the Earth's surface. In this angular range, the measured dependence is well approximated by the negative exponent of the atmosphere thickness (the dashed curve in the figure):

$$<\Sigma / D> \sim \exp(-X_0 \sec\theta / \Lambda) + \text{const}\,, \qquad (2)$$

where $X_0 = 1014$ g/cm^2 is the average vertical thickness of the atmosphere layer above the setup at the observation point, $\Lambda = (134 \pm 15)$ g/cm^2 is the attenuation length which is close to the known value of cosmic ray nucleon component absorption path in the atmosphere.

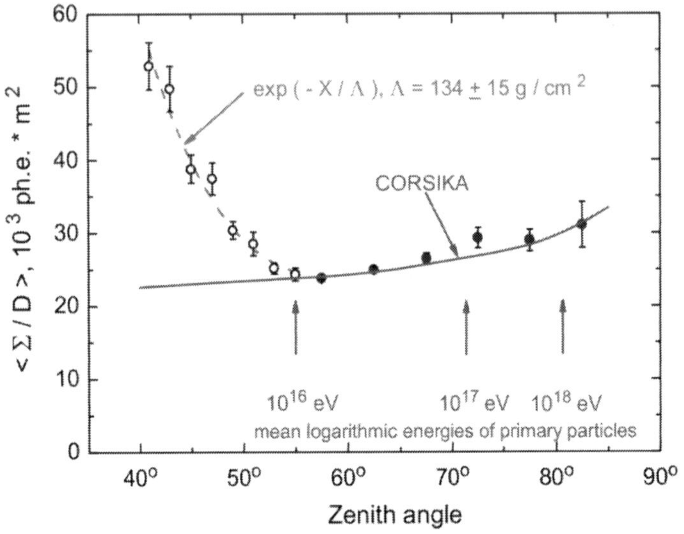

Figure 3. Dependence of the average specific energy deposit of muon bundles in the CWD on the zenith angle. Points are measurement results. Dashed curve is the exponential fit (2) of the data in the range of zenith angles less than 55°; solid curve represents the expected dependence obtained from CORSIKA-based simulations.

At detection of muon bundles with zenith angles more than 55°, practically pure muon component remains. As seen from the figure, the data (solid circles) show a growth of the average specific energy deposit in this angular range, thus indicating an increase of the average muon energy in the bundles. The solid curve in figure 3 represents the expected dependence of the energy deposit of muon bundles obtained on the basis of EAS muon component simulations for primary protons with the CORSIKA code [10]. We used the combination of hadron interaction models SIBYLL + FLUKA for hadron energies more than 80 GeV and below this value, respectively. Influence of the geomagnetic field at the setup location was taken into account. Simulations were performed for a set of fixed zenith angles from 40° to 85° and fixed primary energies E_0 from 10^{15} to 10^{19} eV. The number of simulated EAS varied from 1000 at moderate primary energies to 100 showers at maximal energies. The electron-photon component was not considered, that allowed simulating muon component using full Monte-Carlo (without thinning) with reasonable computer resources. Using the simulated showers, the average lateral distribution functions $\rho(E_0, \mathbf{r})$ for muons with energies more than 2 GeV were constructed. In calculating the mean muon energy in the events selected by muon density (that corresponds to the selection of muon bundles in an experiment with a small area detector), individual muons were taken with a weight proportional to $[\rho(E_0, \mathbf{r})]^{\beta-1}$. Physically, this weight allows taking into consideration that the higher particle density at a given point of the EAS cross section, the higher the frequency of the appearance of these muons in the selected events (for more details, see ref. [3]). The calculated value of the mean muon energy in the bundles rapidly increases at large zenith angles and reaches about 500 GeV near the horizon. The expected zenith-angular dependence of the average specific energy deposit of muon bundles in the CWD (the curve in the figure) was calculated in a following way. First, for each zenith angle the mean muon energy loss <*dE/dX*> was calculated using the energy loss tables [11]. Further, we made an assumption that the total Cherenkov light yield in the calorimeter is proportional to the muon energy loss. Finally, the absolute value of the proportionality coefficient was obtained from the normalization of calculated dependence to the experimental values near 60° zenith angle. The measured dependence of the average specific energy deposit on the zenith angle confirms the increase of the mean muon energy and is in a general agreement with CORSIKA-based expectation, with a possible exception of the point between 70 and 75°, which corresponds to effective energies of primary particles ~ 10^{17} eV.

In order to check whether the observed zenith angular dependence of the energy deposit might be imitated by the properties of the CWD detection system, we analyzed the measured average specific energy deposit as a function of the azimuth angle between the muon bundle arrival direction and the main axis of the NEVOD water tank (figure 4). As it is seen from the figure, the data exhibit a very good uniformity of the detector response for various azimuth angles; thus, a non-ideally isotropic structure of the CWD QSM lattice does not distort the results of measurements of the angular dependence of the energy deposit.

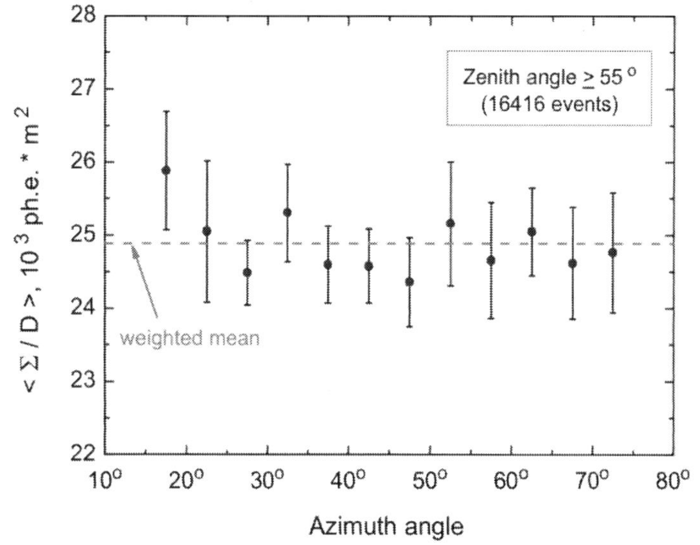

Figure 4. The dependence of the measured average specific energy deposit on the azimuth angle.

In figure 5, the experimental values of the average specific CWD response for muon bundles are presented as a function of the local muon density. All selected events with zenith angles $\theta \geq 55°$, where the residual contribution of electron-photon and hadron components is small, are included in these data. In fact, such an analysis for a fixed interval of zenith angles allows us to follow possible changes of the detector response (and correspondingly, of the average muon energy in the bundles) on the primary energy. Arrows in the bottom part of the figure indicate typical (mean-logarithmic) energies of primary particles contributing to the events with corresponding values of the muon density. The curve in the figure represents the expected dependence of the energy deposit on the local muon density obtained from our CORSIKA-based calculations for the zenith angle of 60°; the same coefficient for normalizing data and calculations as in figure 3 was applied. For the moment, taking into account the statistical uncertainties, it seems too early to speak about any clear dependence of the muon bundle energy deposit on the energy of primary particles, though some indication for its increase beyond the expectation at energies above 10^{17} eV is seen (two last points in figure 5). Higher statistics is necessary for a more detailed consideration, in particular, for the analysis of the behaviour of the average energy deposit in different zenith-angular ranges.

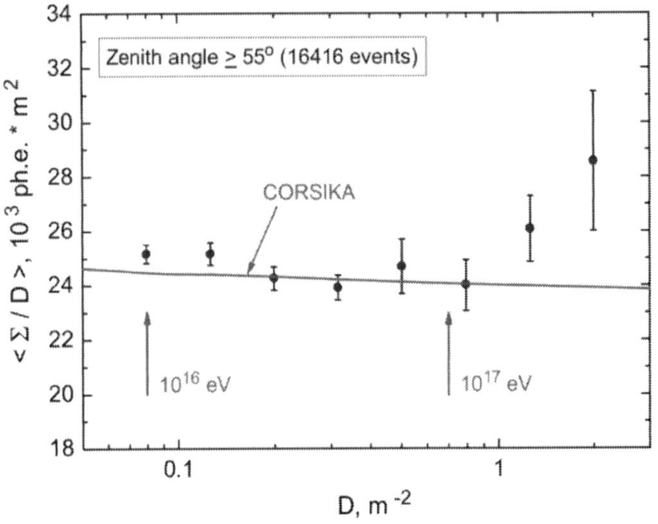

Figure 5. The dependence of the average specific energy deposit on the local muon density.

4. Conclusion

The experiment on the investigation of the energy characteristics of inclined muon bundles formed as a result of interactions of primary cosmic ray particles with energies $10^{16} - 10^{18}$ eV is being conducted at the NEVOD-DECOR complex. The aim of the experiment is the search of possible reasons of the appearance of the excessive flux of multi-muon events in ultra-high energy cosmic rays. First results of the measurements of the zenith-angular dependence of the average specific energy deposit of muon bundles in the Cherenkov water detector are in a reasonable agreement with CORSIKA-based simulations of EAS muon component and confirm the increase of the average energy of muons in the bundles from ~ 100 GeV at moderate zenith angles up to ~ 500 GeV near the horizon. A hint for an excess of the measured energy deposit compared to expectation at primary energies about 10^{17} eV is found; however, for the moment it is not statistically significant. If confirmed with a further increase of statistics, this deviation would be a good evidence for inclusion of a new mechanism of generation of high-energy muons at ultra-high energies of primary particles.

Acknowledgments

The work has been performed in the Scientific and Educational Centre NEVOD with a State support provided by the Ministry of Education and Science of the Russian Federation (government task and project RFMEFI59114X0002) and the grant of the Leading Scientific School NSh-4930.2014.2.

References

[1] Avati V *et al* 2003 *Astropart. Phys.* **19** 513
[2] Abdallah J *et al* 2007 *Astropart. Phys.* **28** 273
[3] Bogdanov A G *et al* 2010 *Phys. Atom. Nucl.* **73** 1852
[4] Nellen L (for the Pierre Auger Collaboration) 2013 *J. Phys.: Conf. Ser.* **409** 012107
[5] Saavedra O *et al* 2013 *J. Phys.: Conf. Ser.* **409** 012009
[6] Petrukhin A A 2014 *Nucl. Instrum. Meth. Phys. Res.* A **742** 228
[7] Aynutdinov V M *et al* 1998 *Astrophys. Space Sci.* **258** 105
[8] Khokhlov S S *et al* 2011 *Astrophys. Space Sci. Trans.* **7** 271
[9] Barbashina N S *et al* 2000 *Instrum. Experim. Techniques* **43** 743
[10] Heck D *et al* 1998 *Forschungszentrum Karlsruhe Report FZKA* 6019
[11] Groom D E, Mokhov N V, Striganov S I 2001 *Atomic Data and Nuclear Data Tables* **78** 183

The Tunka Radio Extension: Latest Analysis Results

D Kostunin[1], P A Bezyazeekov[2], N M Budnev[2], O A Gress[2], A Haungs[1], R Hiller[1], T Huege[1], Y Kazarina[2], M Kleifges[3], E N Konstantinov[2], E E Korosteleva[4], O Krömer[3], L A Kuzmichev[4], R R Mirgazov[2], L Pankov[2], V V Prosin[4], G I Rubtsov[5], F G Schröder[1], R Wischnewski[6], A Zagorodnikov[2]

[1] Institut für Kernphysik, Karlsruhe Institute of Technology (KIT), Germany
[2] Institute of Applied Physics ISU, Irkutsk, Russia
[3] Institut für Prozessdatenverarbeitung und Elektronik, Karlsruhe Institute of Technology (KIT), Germany
[4] Skobeltsyn Institute of Nuclear Physics MSU, Moscow, Russia
[5] Institute for Nuclear Research of the Russian Academy of Sciences, Moscow, Russia
[6] DESY, Zeuthen, Germany

E-mail: dmitriy.kostunin@kit.edu

Abstract. Tunka-133 is an air-Cherenkov array placed in Siberia, near the southern tip of Lake Baikal, which registers air showers induced by cosmic rays from initial particles with energies of 10^{16}–10^{18} eV. After several years of successful data collection, this array was extended by other detector arrays. One of them is the Tunka Radio Extension (Tunka-Rex): the radio array consists of presently 25 antenna stations connected to the data acquisition of Tunka-133. This combination provides hybrid measurements and the possibility for cross-calibration between the air-Cherenkov and radio measurement techniques. The main goal of Tunka-Rex is to determine the precision of the reconstruction of air-shower parameters using the radio detection technique. We present the latest results on the event reconstruction.

1. Introduction

Radio emission from extensive air showers was theoretically predicted [1, 2, 3, 4] and first detected [5, 6, 7] about 50 years ago. The radio detection techniques became popular in the last decade again, because standard detection methods are going to reach technological and economical limits, contrariwise, the digital processing of radio data became possible. Thus, a number of modern experiments [8, 9, 10, 11] aims at obtaining the main properties of extensive air showers using the radio detection technique.

Tunka-Rex, the radio experiment in the Tunka valley, is motivated by the well-developed infrastructure and unique possibility of hybrid air-Cherenkov and radio measurements. These conditions give a chance to show the prospects of a low-cost, scalable radio detector for cosmic rays. The main goal of Tunka-Rex is to determine the precision for the reconstruction of the primary energy, E_{pr}, and the atmospheric depth of the shower maximum, X_{max}, based on the cross-calibration with an air-Cherenkov detector.

The Tunka-133 air-Cherenkov detector [13] is taking data since 2009. In 2012 two new extensions were introduced: HiSCORE [14] and Tunka-Rex [15]. Moreover, the development of a new gamma-astronomy facility named TAIGA started [16]. This new facility will consist of

Journal of Physics: Conference Series **632** (2015) 012096 doi:10.1088/1742-6596/632/1/012096

the existing Tunka-133 and HiSCORE detectors, additional imaging air-Cherenkov telescopes, and new particle detectors. TAIGA partly has started data acquisition in 2014.

Tunka-Rex currently consists of 25 antenna stations attached by cables to the cluster centers of the Tunka-133 photomultiplier array. Each antenna station consists of two orthogonally oriented SALLA antennas [17], aligned in magnetic North-East and North-West directions.

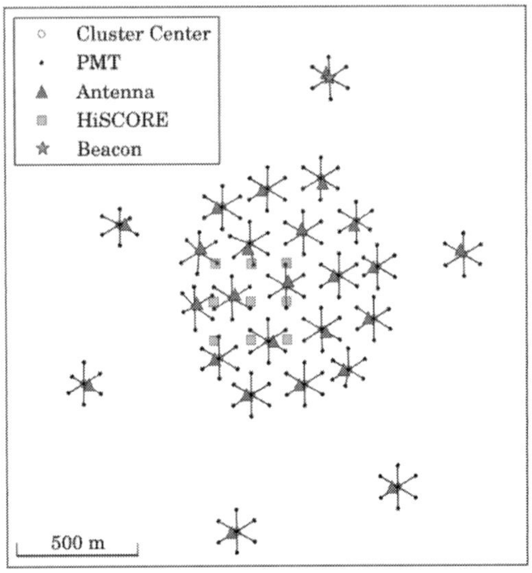

Figure 1. The Tunka facility.
Left: Map of Tunka. One Tunka-Rex antenna station, consisting of two SALLAs [17], is attached to each cluster.
Right: A Tunka-Rex antenna station in front of a Tunka-133 cluster center box, the central PMT of the cluster and a house with scintillators.

2. Event selection and data analysis

In the season 2012/2013 (from October to April), Tunka-133 had 316 hours of effective measurement time. The Tunka-Rex data acquisition is triggered by the air-Cherenkov array, thus this is limiting the its measurement time. For the first season we performed a non-blind analysis, i.e., we used the full reconstruction information from the Cherenkov detector as input for the radio analysis.

For analysis of the radio measurements, we use the radio extension of the Offline software framework developed by the Pierre Auger Collaboration [19, 20]. An example of a reconstructed event is presented in Fig. 2.

Since a full event reconstruction (i.e. reconstruction of E_{pr} and X_{max}) of Tunka-133 is available only for events with zenith angle less than 50° (will be explained below), we restrict the cross-calibration analysis to these events.

In the first step, we search for the signal in a defined time window. The time window is the same for every antenna because of the identical hardware configuration. The main uncertainty comes from hardware delays and shower geometry. A rough estimation can be found in Ref. [21]. Detailed study of traces has shown that the arrival times have relatively small uncertainties and that a window size of 300 ns is sufficient. Noise is defined as a RMS in special window, where the probability of false pulses is smallest. Thus, SNR (signal-to-noise ratio) is defined as a squared fraction between signal and noise. We apply two different SNR cuts on different reconstruction

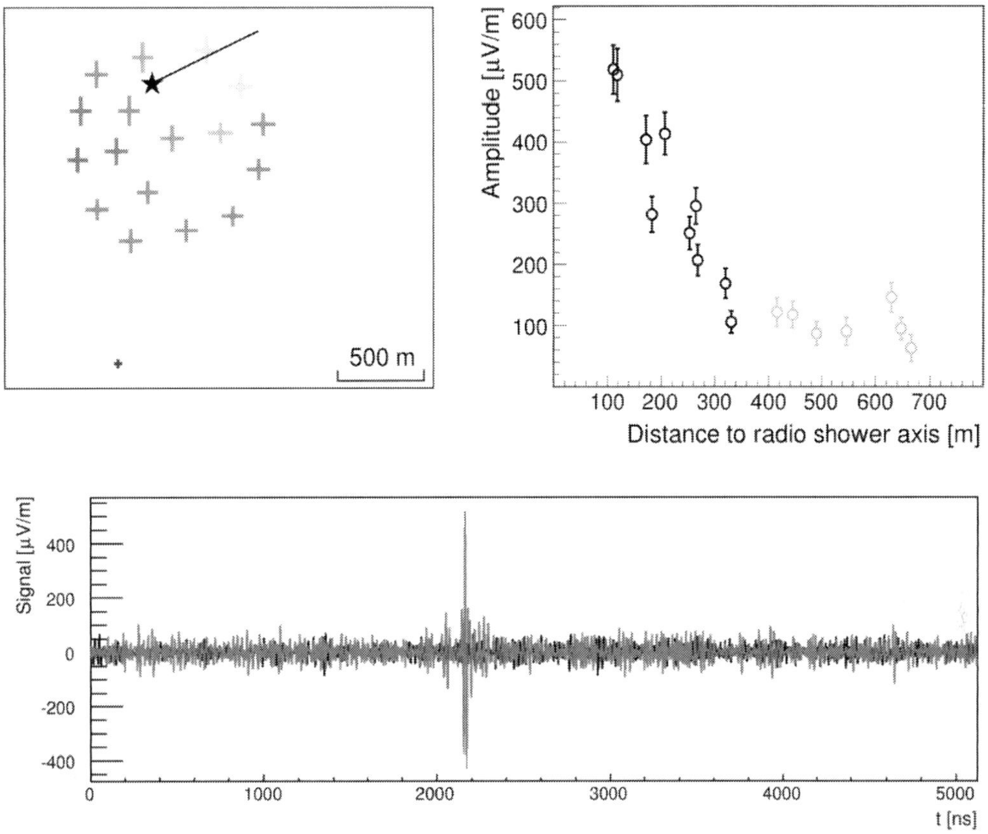

Figure 2. Example for a Tunka-Rex event.
Top left: footprint of the event, where the size of the crosses indicates the signal strength, the color code the arrival time, and the line and the star the direction and shower core, respectively.
Top right: lateral distribution of reconstructed signals. As in the footprint, grey points indicate signals which did not pass the SNR (signal-to-noise ratio) cut.
Bottom: example trace of the reconstructed electric-field strength, we expect the radio pulse around 2000 ns.

levels: the SNR must be larger than 9 in both channels and the SNR must be larger than 6 on the station level (reconstructed electrical field vector). After these cuts we found 117 events, which contain 3 or more antennas with signal.

For the next quality cut we compare the arrival direction obtained from the Cherenkov and radio detectors. We set an upper limit for the reconstruction difference to 5° and compare the values obtained from a plane fit of the radio shower front with the values from Cherenkov reconstruction. After this cut 65 events remain, where the other events likely are contaminated by disturbances.

In the third step we apply a final quality cut: after removing clear outliers in the signal strength from the lateral distribution, we set a limit for the LDF (lateral distribution function, described later) fitting quality of $\chi^2/\mathrm{NDF} < 8.0$. This number was chosen arbitrarily in order to select events, for which an exponential LDF provides a sufficient approximation. In a detailed theoretical study [22], it is shown, that the exponential function can not give a good description for every case, and a detailed analysis taking this into account is in preparation. Assuming that the theoretical limit for the energy precision is about 10% [22], thus, to obtain a resolution of order 20%, it is sufficient to remove events with fitting uncertainty larger than 20%. Under the assumption, that the geomagnetic effect dominates, one event with small geomagnetic angle ($\alpha_g < 0.3$ rad) is also removed[1]. Finally, 54 events remain (quality cuts from third step removed 11 events).

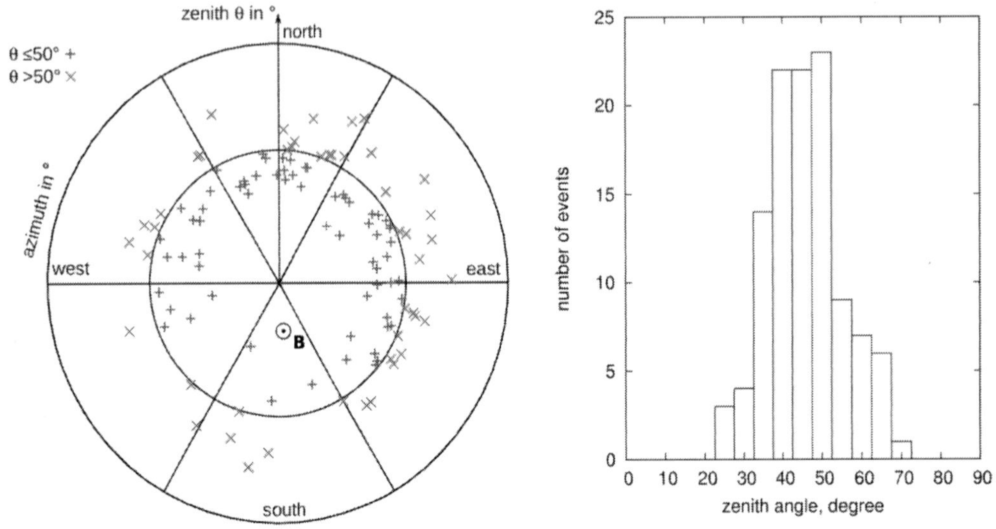

Figure 3. Distribution of the arrival directions of the Tunka-Rex events passing the quality cuts.
Left: sky map of the arrival directions. We present events passing the direction cut and do not apply additional quality cuts for this distribution. Due to the geomagnetic effect, the radio signal is expected to be on average stronger for events coming from North, which explains the asymmetry in the detection efficiency: 46 of the 65 events are in the northern half, and 19 in the southern half.
Right: Zenith angle distribution of detected radio events. The maximum efficiency is reached at the Tunka-133 reconstruction threshold of $\theta \approx 50^\circ$. The statistics at smaller angles is mainly suppressed by the steeply falling lateral distribution of the radio signal, the statistics at larger angles is suppressed by trigger detection capabilities of the air-Cherenkov detector.

The distribution of the arrival directions is shown in Fig. 3. The upper limit for direct Cherenkov light detection is a zenith angle less than 50°, but the zenith angle for triggering can be extended up to 70° due to indirect detection of reflected light. The total number of all inclined events is only about 5% of all Tunka-133 high energy events, since the sensitivity of trigger to big zenith angle decreases rapidly. This means that the detection of inclined events is suppressed by trigger design, and the detection of events with small zenith angles is suppressed

[1] The asymmetry given by the charge excess phenomena is proportional to

$$\frac{\varepsilon \sin \alpha_g}{\sin^2 \alpha_g + \varepsilon^2} \cos \phi \,, \tag{1}$$

where ε is a contrubution of charge excess effect to total signal strength, α_g is a geomagnetic angle, ϕ is an azimuth in geomagnetic coordinate system [22]. E.g., if the contribution of charge excess is 10% and geomagnetic angle is about 0.3 rad, the asymmetry can be up to 30% depending on the azimuth of the antenna station with respect to the shower axis. This can be crucial when the number of antennas is small.

by the geometry of the detector (the spacing between antennas is about 200 m, thus shower need a high energy to illuminate at least three antennas at small zenith angles).

In Fig. 3 we divided events into two groups: vertical ("+" symbol) and horizontal ("×" symbol). For horizontal events the air-Cherenkov detector can reconstruct only the direction, but neither the energy nor the shower maximum. Therefore, we use only the vertical events for cross-calibration.

It is worth mentioning that we also see a North-South asymmetry in the arrival direction as is predicted by the geomagnetic effect. The number of vertical events coming from the Northern and Southern hemispheres of the sky is 46 and 19, respectively.

To explain the features of the lateral distribution one can use the following exponential description of the LDF

$$\mathcal{E}(d) = \mathcal{E}_{d_0} \sin\alpha_{\mathrm{g}} \exp[f_\eta(d - d_0)]\,, \quad f_\eta(x) = \sum_{k=1}^{n} a_k x^k\,, \tag{2}$$

where $\mathcal{E}(d)$ is the strength of the electrical field at the distance d from the shower axis, a_k are parameters describing the shape of the lateral distribution; d_0 is usually selected in a way to degenerate the dependence on the shower maximum of $\mathcal{E}_{d_0}$, i.e. $\mathcal{E}_{d_0}(E_{\mathrm{pr}}, X_{\mathrm{max}}) \approx \mathcal{E}_{d_0}(E_{\mathrm{pr}})$. The geomagnetic angle α_{g} is defined as the angle between the shower axis $\mathbf{A}$ and the geomagnetic field $\mathbf{B}$: $|\mathbf{A}\times\mathbf{B}| = |\mathbf{A}|\cdot|\mathbf{B}|\sin\alpha_{\mathrm{g}}$. The shower parameters such as energy and shower maximum mostly depend on the properties of LDF $\mathcal{E}(d)$: $E_{\mathrm{pr}} = E_{\mathrm{pr}}(\mathcal{E}_{d_0})$ and $X_{\mathrm{max}} = X_{\mathrm{max}}(a_k)$. A rough description of the lateral distribution can already be obtained by setting $n = 1$

$$\mathcal{E}(d) = \mathcal{E}_{d_0} \sin\alpha_{\mathrm{g}} \exp\left[-\eta(d - d_0)\right] \tag{3}$$

where $\eta \equiv a_1$ is an exponential slope parameter [23]. Traditionally we chose $d_0 = 100$ m [24]. After normalizing by the sine of the geomagnetic angle, we check for the correlation between the reconstructed electrical field $\mathcal{E}_{d_0}$ obtained fitting the LDF and the energy of primary particle reconstructed by Tunka-133 (see Fig. 4). The relative deviation between the energy reconstructed by both detectors is about 20%. Thus, we conclude that the two parameter exponential LDF is sufficient the energy estimation with that precision. The uncertainty of the energy reconstruction is mainly defined by the parameter fitting uncertainty. A more sophisticated analysis taking into account the Askaryan effect, and aiming at a better energy precision is in preparation.

3. Conclusion

Tunka-Rex has shown the feasibility of the present implementation of hybrid air-Cherenkov and radio measurements of cosmic rays. Our measurements are compatible with the picture predicted by theoretical description and latest results given by other experiments. The next step in the analysis will be the reconstruction of the shower maximum and unblinding the air-Cherenkov data of the season 2013/2014 for cross-check and analysis with high statistics.

The local infrastructure allows to extend the radio detector and perform super-hybrid measurements of the different components of air-showers: air-Cherenkov, particles on ground and radio. In 2014, 19 new antenna stations were attached to future stations of the scintillator extension and three antenna stations were attached to HiSCORE stations for technical tests. The new scintillator triggered radio extension can increase the duty cycle of Tunka-Rex up to 100%, and, consequently, increase the statistics by one order of magnitude. Consistent with the amplitude calibration it helps to check radio emission models and the shower universality hypothesis.

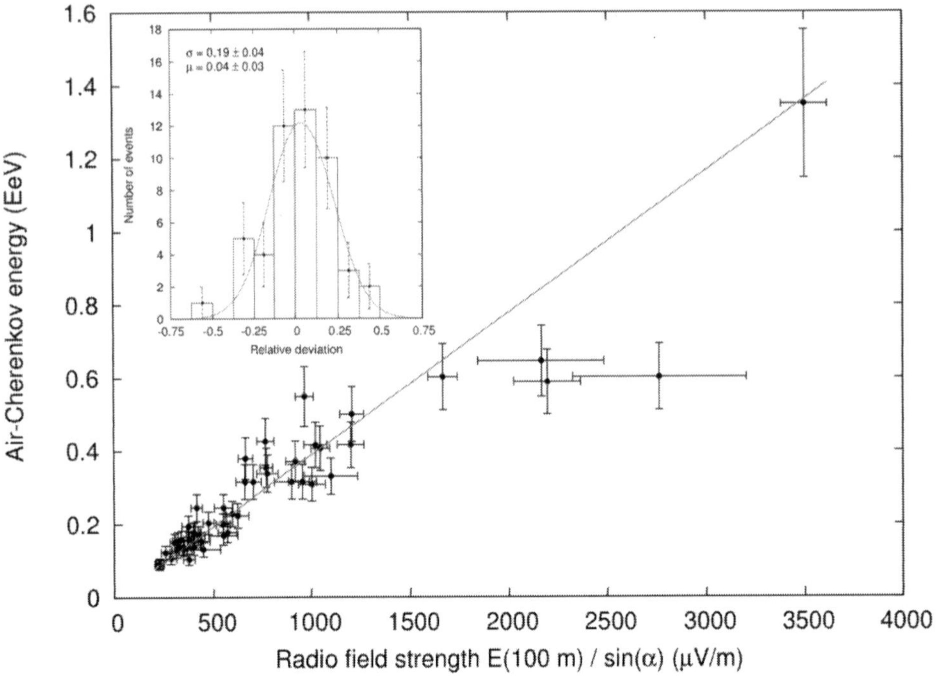

Figure 4. Correlation between the radio field strength at 100 m $\mathcal{E}_{d_0}$ normalized by the sine of the geomagnetic angle and the energy reconstructed with the air-Cherenkov measurements. The uncertainties of the electrical field are from the LDF fit, the uncertainties of the Tunka-133 energy reconstruction are 15%. The histogram shows the scatter between the energy reconstructed by Cherenkov and radio.

Acknowledgments

Tunka-Rex is funded by the German Helmholtz association (grant HRJRG-303) and supported by the Helmholtz Alliance for Astroparticle Physics (HAP). This work was supported by the Russian Federation Ministry of Education and Science (G/C 14.B25.31.0010), the Russian Foundation for Basic Research (Grants 12-02-91323, 13-02-00214, 13-02-12095, 14002-10002) and the President of the Russian Federation (grant MK-1170.2013.2). We thank the Pierre Auger Collaboration for their permission to use the **Offline** analysis software for the Tunka-Rex analysis.

References

[1] Askaryan G A 1961 *Sov. Phys. JETP* **14** 441
[2] Kahn F D and Lerche I 1966 *Proc. Phys. Soc., Sect. A* **289** 206
[3] Castagnoli C, Silvestro G, Picchi P and Verri G 1969 *Nuovo Cimento B* **63** 373
[4] Hough J H 1973 *J Phys. A* **6** 892
[5] Jelley J V, Fruin J H, Porter N A *et al.* 1965 *Nature* **205** 327
[6] Vernov S N, Khristiansen G B, Abrosimov A T *et al.* 1969 *Proceedings on the 11th ICRC*
[7] Allan H R 1971 *Prog. in Elem. Part. and Cosmic Ray Phys.* **10** 171
[8] Falcke H (LOPES Collaboration) 2005 *Nature* **435** 313
[9] Ravel O (CODALEMA Collaboration) 2012 *NIMA* **662** S89-S94
[10] Abreu P *et al.* (Pierre Auger Collaboration) 2012 *JINST* **7** P11023
[11] Schellart P (LOFAR Collaboration) *et al.* 2013 *Astron. Astrophys.* **560** A98
[12] Schröder F G (LOPES Collaboration) 2012 *AIP Conf. Proc.* **1535** 78
[13] Prosin V V (Tunka-133 Collaboration) 2014 *NIMA* **756** 94.

[14] Tluczykont M *et al.* 2014 *Astropart. Phys* **56** 42
[15] Schröder F G (Tunka-Rex Collaboration) 2012 *AIP Conf. Proc* **1535** 111
[16] Budnev N M (TAIGA Collaboration) 2014 *JINST* **9** C09021.
[17] The Pierre Auger Collaboration 2012 *JINST* **7** P10011
[18] Hiller R (Tunka-Rex Collaboration) 2014 *these proceedings*
[19] Argiro S *et al.* 2007 *NIMA* **580** 1485
[20] The Pierre Auger Collaboration 2011 *NIMA* **635** 92
[21] Kostunin D 2014 (Tunka-Rex Collaboration) *NIMA* **742** 89
[22] Kostunin D, Bezyazeekov P A, Hiller R, Schröder F G, Lenok V and Levinson E 2015 arXiv:1504.05083 [astro-ph.HE], submitted to *Astropart. Phys.*
[23] Apel W D (LOPES Collaboration) 2010 *Astropart. Phys.* **32** 294
[24] Apel W D (LOPES Collaboration) 2012 *Phys. Rev. D* **85** 071101

The current status of orbital experiments for UHECR studies

M I Panasyuk[1,2], M Casolino[3,4], G K Garipov[1], T Ebisuzaki[3], P Gorodetzky[5], B A Khrenov[1], P A Klimov[1], V S Morozenko[1], N Sakaki[6], O A Saprykin[7], S A Sharakin[1], Y Takizawa[3], L G Tkachev[8], I V Yashin[1] and M Yu Zotov[1]

[1] D.V. Skobeltsyn Institute of Nuclear Physics, M.V. Lomonosov Moscow State University (SINP MSU), Moscow 119991, Russia
[2] Faculty of Physics, M.V. Lomonosov Moscow State University, Moscow 119991, Russia
[3] RIKEN, 2-1 Hirosawa, Wako, 351-0198 Japan
[4] Istituto Nazionale di Fisica Nucleare – Sezione di Roma Tor Vergata, Italy
[5] APC, Univ Paris Diderot, CNRS/IN2P3, CEA/Irfu, Obs de Paris, Sorbonne Paris Cité, France
[6] Karlsruhe Institute of Technology (KIT), 76021 Karlsruhe, Germany. Now at: Osaka City University, Japan
[7] Central Research Institute of Machine Building (TsNIIMash), Korolev 141070, Russia
[8] Joint Institute for Nuclear Research (JINR), Dubna 141980, Russia

E-mail: panasyuk@sinp.msu.ru

Abstract.
Two types of orbital detectors of extreme energy cosmic rays are being developed nowadays: (i) TUS and KLYPVE with reflecting optical systems (mirrors) and (ii) JEM-EUSO with high-transmittance Fresnel lenses. They will cover much larger areas than existing ground-based arrays and almost uniformly monitor the celestial sphere. The TUS detector is the pioneering mission developed in SINP MSU in cooperation with several Russian and foreign institutions. It has relatively small field of view ($\pm 4.5^\circ$), which corresponds to a ground area of $6.4 \cdot 10^3$ km^2. The telescope consists of a Fresnel-type mirror-concentrator (~ 2 m^2) and a photo receiver (a matrix of 16×16 photomultiplier tubes). It is to be deployed on the Lomonosov satellite, and is currently at the final stage of preflight tests. Recently, SINP MSU began the KLYPVE project to be installed on board of the Russian segment of the ISS. The optical system of this detector contains a larger primary mirror (10 m^2), which allows decreasing the energy threshold. The total effective field of view will be at least $\pm 14^\circ$ to exceed the annual exposure of the existing ground-based experiments. Several configurations of the detector are being currently considered. Finally, JEM-EUSO is a wide field of view ($\pm 30^\circ$) detector. The optics is composed of two curved double-sided Fresnel lenses with 2.65 m external diameter, a precision diffractive middle lens and a pupil. The ultraviolet photons are focused onto the focal surface, which consists of nearly 5000 multi-anode photomultipliers. It is developed by a large international collaboration. All three orbital detectors have multi-purpose character due to continuous monitoring of various atmospheric phenomena. The present status of development of the TUS and KLYPVE missions is reported, and a brief comparison of the projects with JEM-EUSO is given.

Journal of Physics: Conference Series **632** (2015) 012097 doi:10.1088/1742-6596/632/1/012097

1. Introduction

The nature and origin of extreme energy cosmic rays (EECRs, those with energies $\gtrsim$ 50 EeV[1]) remains one of the greatest puzzles of modern astrophysics after more than 50 years since their first registration [1]. In this introduction, we will briefly review a number of important experimental results on the subject that were obtained in recent years. These include results on the energy spectrum, mass composition and anisotropy of ultra-high energy cosmic rays (UHECRs) obtained with the Pierre Auger Observatory (Auger for brevity), located in Argentina (South hemisphere); High Resolution Fly's Eye (HiRes) and the Telescope Array Project (TA), both located in Utah, USA (North hemisphere). Then we recall the main idea behind observation of EECRs from space and review three orbital projects aimed to become the next generation of EECR experiments that are currently under active development mostly focusing on TUS and KLYPVE, but also briefly discussing the JEM-EUSO mission. We aim to explain how these projects are able to open new possibilities to solve the EECR puzzle.

The most important result relates to the all-particle energy spectrum of UHECRs. It was found by HiRes [2, 3], Auger [4] and confirmed by TA [5] that the flux of cosmic rays is strongly suppressed at energies $\gtrsim$ 50 EeV. A joint working group of the Pierre Auger, Telescope Array and Yakutsk Collaborations concluded that the energy spectra determined by Auger and TA are consistent in normalization and shape within systematic uncertainties after energy scaling factors are applied [6]: both spectra demonstrate an "ankle" at around 5 EeV and a cut-off at $\gtrsim$ 50 EeV. It was found though that the cut-off takes place at lower energies in the spectrum by Auger and is steeper than the TA one.

The cut-off was found at the energy scale predicted by Greisen [7] and independently by Zatsepin and Kuz'min [8], who demonstrated that the flux of protons will be strongly suppressed at energies $\gtrsim$ 50 EeV due to the photopion production on the cosmic microwave background radiation. Still, the discrepancy in the energy spectra obtained by Auger and TA does not allow concluding if the observed cut-off is due to the GZK effect, due to the maximum energy attained in accelerators of EECRs, or a combination of both, see [9] for an in-depth discussion. It does not either allow one to make a definite conclusion concerning the mass composition of the primary particles: are they protons or heavier nuclei, e.g. iron.

Another discrepancy between results of the Auger and TA experiments further complicates the situation with the composition of UHECRs. Namely, the Auger data witness in favour of the mass composition becoming heavier at energies above 3 EeV [10] while the TA data are consistent with a light, dominantly protonic composition [11, 12] but incompatible with a pure iron composition. The TA results are also in agreement with previous HiRes stereo measurements. The elongation rate and mean values of the depth of the shower maximum measured at TA are in good agreement with Auger data [12]. Though one should keep in mind that, on the one hand, the measurement of masses of primary particles is one of the most difficult tasks in the physics of UHECRs since it is based on hadronic interaction models at energies far beyond the reach of man-made accelerators like the LHC. On the other hand, Auger and TA employ different analysis techniques and different cuts selecting data for the analysis, so that a direct comparison of the results of the two experiments can be misleading. A work on a detailed comparison of mass composition obtained with Auger and TA is in progress [13].

One more aspect of cosmic ray physics that attracts great attention is anisotropy of arrival directions of EECRs. It is one of the key elements in a search for possible sources but also useful for examining mass composition of EECRs independently of measurements of the shower depth. The most intriguing new result in the field is the discovery of a "hotspot" of cosmic rays with energies above 57 EeV made by the TA experiment in the Northern hemisphere [14]. The hotspot is a cluster of 19 (of 72 registered) events within a 20° radius circle centered at

[1] 1 EeV = 10^{18} eV.

(equatorial coordinates) $(\alpha, \delta) = (146.^\circ 7, 43.^\circ 2)$ with an expected number of events equal to 4.49, which gives a (pre-trial) statistical significance of 5.1σ. The (post-trial) statistical significance of such a hotspot appearing by chance was estimated to be 3.4σ. The hotspot is located near the Supergalactic plane but there are no known specific sources behind it. It was suggested by the TA Collaboration that the hotspot may be associated with the closest galaxy groups and/or the galaxy filament connecting the Milky Way with the Virgo cluster; or, if EECRs are heavy nuclei in accordance with the Auger results, the hotspot events may originate in the Supergalactic plane and be deflected by the magnetic fields. Recently, it has been argued that the nearby starburst galaxy M82 and the bright nearby blazar Mrk 180 (or one of them) are likely sources of the TA hotspot assuming its events have a pure composition [15]. Still, the available statistics, the ambiguity in the mass measurements of EECRs and the insufficient knowledge of extragalactic magnetic fields do not allow making a definite conclusion.

The Pierre Auger Collaboration is also performing intensive studies of anisotropy. In particular, a correlation is reported between the arrival directions of cosmic rays with energies above 55 EeV and the distribution of active galactic nuclei (AGN) within 75 Mpc, among them the Centaurus A (Cen A) radiogalaxy, located at less than 4 Mpc distance [16, 17, 18]. More recently, Auger presented results of a whole number of tests aimed to search for signals of anisotropies of cosmic rays with energies above 40 EeV (with 231 of them having energies ≥ 52 EeV and zenith angles $< 80^\circ$) [19]. None of the tests performed revealed a statistically significant deviation from isotropy but it was found that there is a certain excess of events with energies ≥ 58 EeV around the direction to Cen A and around AGN of the Swift catalogue within 130 Mpc and brighter than 10^{44} erg/s. In both cases, the probability of arising the excess by chance was estimated to be of the order of 1.3–1.4%. No hotspot similar to the one observed by TA was found. On the other hand, a similar study performed by TA using EECRs with energies ≥ 40 EeV registered by the surface detector during the first 40 months of its operation did not reveal any statistically significant correlations with AGN from a number of survey catalogues [20].

Another direction of anisotropy studies is a search for large-scale anisotropy of EECRs. Recently, Auger presented results of two Rayleigh analyses, one in the right ascension and one in the azimuth angle distributions, of the arrival directions of events with energy above 4 EeV with zenith angle up to 80° [21]. The largest departure from isotropy was found in the $E > 8$ EeV energy bin, with an amplitude for the first harmonic in right ascension $(4.4 \pm 1.0) \times 10^{-2}$, which has a chance probability of 6.4×10^{-5}. It was found that assuming the only significant contribution to large-scale anisotropy is from the dipolar component, the observations above 8 EeV correspond to a dipole of amplitude 0.073 ± 0.015 pointing to $(\alpha, \delta) = (95^\circ \pm 13^\circ, -39^\circ \pm 13^\circ)$, thus supporting an earlier result obtained for events with zenith angles $< 60^\circ$ [22].

Telescope Array has also performed an analysis of large-scale anisotropy of cosmic rays at the highest end of the spectrum, though a totally different approach was employed [23]. Assuming pure proton composition, it was found that the distribution of events with energies > 10 EeV and > 40 EeV is compatible with the hypothesis of isotropy and incompatible with the distribution of local large-scale structure (LSS) at smearing angles smaller than $\sim 20^\circ$ and $\sim 10^\circ$ respectively. To the contrary, the data set composed of events with energy above 57 EeV was found to be compatible with the LSS model and incompatible with isotropy at (pre-trial) statistical significance $\sim 3\sigma$.

Anisotropy studies present a complicated task because none of the experiments observes the whole celestial sphere but a proper determination of the full set of multipole coefficients requires full-sky coverage. That is why a joint analysis of large-scale anisotropy of UHECRs with energies above 10 EeV performed by the TA and Pierre Auger collaborations is of special interest [24]. A declination band commonly covered by both experiments $(-15^\circ < \delta < 25^\circ)$ was used for cross-calibrating the fluxes. The multipolar expansion of the UHECR flux obtained in the study allowed performing a series of anisotropy searches but no significant deviation from isotropy was revealed at any angular scale. It was later found that the obtained angular power spectrum contradicts the hypothesis that the distribution of UHECRs at energies above 10 EeV follows the LSS model if a purely protonic primary composition is assumed, in that amplitudes of low multipoles in the data are significantly lower than those calculated from the LSS model [25]. One of the possibilities to explain the result is that the primaries are heavy, as suggested by the Auger collaboration results.

Nowadays, obtaining definite conclusions concerning anisotropy of EECRs becomes one of the most important tasks. As is clear from the above, a considerable part of the difficulties in finding the sources of EECRs is due to (i) the limited statistics of events and (ii) the incomplete and non-uniform coverage of the celestial sphere by any of the modern experiments. Thus one needs to increase the statistics of EECRs providing a uniform exposition of the sky. This is exactly the field where orbital experiments open new horizons of research (not to mention a whole number of other scientific tasks related to astrophysics, fundamental physics and atmosphere sciences, which we do not discuss here) being complementary to ground-based detectors.

2. Observation of EECRs from space

The atmosphere of the Earth acts as a huge calorimeter for detecting UHECRs due to extensive air showers (EAS) produced by primary particles moving nearly at the speed of light. The secondaries of an EAS cascade ionize and excite molecules of nitrogen in the air, which then radiate photons in the range $\approx$ 330–400 nm. This near-ultraviolet (UV) air fluorescence light is produced isotropically and can be used to record the development of an EAS. The number of photons emitted at any point along the EAS track is proportional to the number of charged particles in the cascade, mostly electrons and positrons. Additional information can be obtained from the Cherenkov light diffusively reflected at the surface of the Earth. Both can be used to reconstruct the energy and arrival direction of the primary particle and to obtain information about its nature.

The idea of observing EECRs from space was put forward by Benson and Linsley more than thirty years ago [26].[2] They suggested to look earth-ward from a satellite flying on a circular equatorial orbit at height 500–600 km and equipped with a mirror 36 m in diameter with $10'$ resolution and $\sim$ 5000 photomultiplier tubes 5 cm in diameter located at the focal surface of the mirror. They estimated that the instrument will have a circular field of view about 100 km in diameter (thus covering an area three times larger than that of Auger), a duty cycle of the order of 20–30%, and the energy threshold below 10 EeV.

While the basic idea might look simple, it was realized soon that such an experiment poses a whole number of challenging problems both in technology and science since one has to register a faint flux of photons emitted in the continuously varying conditions of the Earth atmosphere and background spoiled by man-made lights, star light and other phenomena with an instrument working in a harsh space environment and built with strong restrictions on mass, energy consumption, etc. Since the pioneering work by Benson and Linsley, different aspects of registering EECRs from space have been studied in great detail, see, e.g., [27, 28] and references therein. Still, not a single project of an orbital detector of EECRs has been implemented yet. In what follows, we will review the current status of development of the TUS and KLYPVE experiments and compare them with the most advanced of the existing projects, JEM-EUSO.

[2] One can find earlier dates in the literature but we failed to find a published work prior to this one.

3. TUS

The TUS (Tracking Ultraviolet Setup) experiment was started in 2001 [29] as a pathfinder for a more advanced KLYPVE project. Plans of accommodating the TUS detector on a satellite were changed several times and the detector itself was improved in comparison with the initial version [30, 31, 32]. Now it is accommodated on the Lomonosov satellite as a part of instrumentation for studying the Extreme Universe Phenomena [33] and is scheduled for launching in late 2015.

The TUS detector is based on the same optical scheme as the one suggested by Benson and Linsley but with a much more modest size of the elements. It consists of two main parts: a mirror-concentrator with an area of 2 m^2 and a photodetector composed of 256 pixels, located at the mirror focus, see figure 1. The technology of TUS was described elsewhere [34, 35, 36, 37] and only the main instrumental parameters are discussed below.

Figure 1. Detector TUS on board the Lomonosov satellite.

The mirror-concentrator is designed as an ensemble of 11 parabolic rings that focus a parallel beam to one focal point. In this design, the thickness of the mirror construction is small (3 cm) which is important for its deployment on the satellite frame. The focal distance of the mirror is 1.5 m. The mirror is cut into hexagonal segments with diagonal 63 cm. The segments are made of carbon plastic supported by a honey comb aluminum plate so that the whole construction is stable in a wide range of temperatures. The surface of the segments is obtained as plastic replicas of aluminum press forms (one for the central mirror part and one for 6 lateral parts). In a vacuum evaporation process, the plastic mirror surface is covered by aluminum film and protected by a MgF_2 coat. Reflectivity of the mirror surface at wavelength 350 nm (average for the atmosphere fluorescence) is 85%. The mirror passed various space qualification as well as optical tests. Those tests demonstrated sufficient stability of its optical quality in space conditions. Expected life time of the mirror is more than 3 years. The main parameters of the detector are listed in table 1. The mirror placed on the scientific payload frame of the Lomonosov satellite is shown in figure 2.

Optical tests were provided by two methods: 1) independent measurements of each mirror segment by scanning with parallel laser beams (described in detail in [38], 2) measurements with a distant (∼30 m) pinpoint light source. The first method allows one to evaluate the quality of production of the mirror segments surface. By the second method, final results on the point spread function (PSF) of the full size mirror at various beam angles were obtained. They are shown in figure 3.

Figure 2. Mirror-concentrator of the TUS detector attached to the Lomonosov frame.

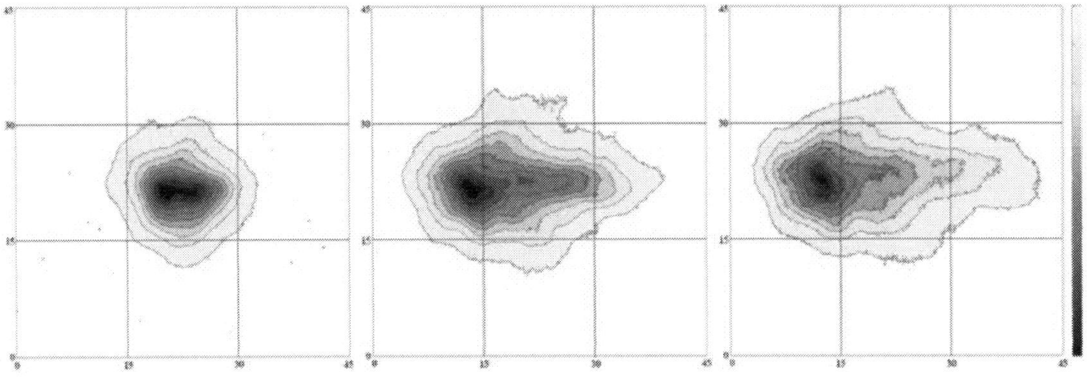

Figure 3. Results of the PSF measurement. Zenith angles from left to right: 0°, 3°, 4.5°. Dimensions are in millimeters. A square of 15×15 mm^2 corresponds to one pixel of the TUS photodetector.

The photodetector pixels are made by photomultiplier tubes (PMTs) R1463 of Hamamatsu with multi-alcali cathode of 13 mm diameter. The quantum efficiency of the PMT cathode is 20% for wavelength 350 nm. The PMTs multi-alcali cathode (instead of usually used bi-alcali one in ground-based fluorescence detectors) was chosen because the cathode operates in a wider

Journal of Physics: Conference Series **632** (2015) 012097 doi:10.1088/1742-6596/632/1/012097

Table 1. Parameters of the TUS detector

Parameter, units	Value
Area of the mirror-concentrator, m^2	2
Mirror focal distance, cm	150
Pixel size, mm	15
Pixel number	256
Time step, μs	0.8
Observable area of the atmosphere at orbit height 500 km, km^2	6400
Pixel size in the atmosphere, km	5

range of temperatures in the linear regime. Light guides with a square entrance (15×15 mm) and circle output adjusted to PMT cathodes were used for making the detector field of view uniformly filled with pixels.

After preliminary testing, PMTs with a similar gain were grouped in 16 clusters. Data from each tube in a cluster are digitized by an analog-digital converter and then analyzed and memorized by a field-programmable gate array (FPGA). The final detector triggering and memorizing of all data is done by the central FPGA. Information volume of one EAS data is ~100 Kbytes. Expected volume of one day EAS data transmitted to the mission center is 250 Mbytes. It is worth noticing that electronics of TUS is developed to measure not only EAS but also other transient phenomena in the atmosphere (lightnings, transient luminous events, meteors). This is achieved by recording waveforms of events with various time of digital integration (from 0.8 μs up to 6.6 ms). The duration of measured events varies from 205 μs up to 1.7 s. The TUS photodetector and one PMT cluster can be seen in figure 4.

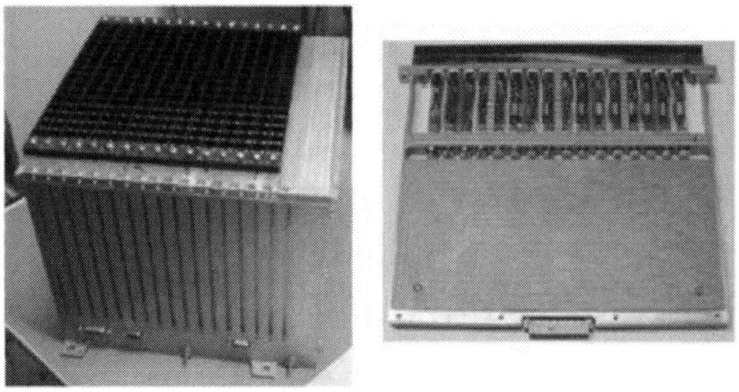

Figure 4. The TUS photodetector (left) and one PMT cluster (right).

Performance of the TUS detector was simulated taking into account parameters of the real TUS mirror-concentrator and TUS electronics [39]. The EECR trigger operates in two levels: at the first level, pixels with signals Z times larger than noise RMS are selected. At the second level, signals triggered on the first level and lined up in space in T consecutive time intervals are selected. Data from all pixels are recorded by a command of the second level trigger. Numbers Z and T will be set from the mission center for the optimal relation between the lowest TUS detector energy threshold and the trigger rate, limited by volume of information to be transmitted to the mission center.

Experimental results on the atmosphere glow were obtained during the Universitetsky-Tatiana-2 mission (Tatiana-2) [40]. The glow varies with the Moon phase and its height above the horizon but the atmosphere emits a glow even at moonless nights. Data of the Tatiana-2 mission, which include approximately $1.9 \cdot 10^4$ minutes of operation in night time, are presented in figure 5 as the operation time versus measured UV background intensity J. The background intensity varies from lower values at moonless nights (at the darkest atmosphere regions: above the Pacific ocean, above deserts and a part of Siberia) with $J = 3 \cdot 10^7$–10^8 ph cm^{-2} s^{-1} sr^{-1} to full-moon nights with $J \sim 3 \cdot 10^9$ ph cm^{-2} s^{-1} sr^{-1}.

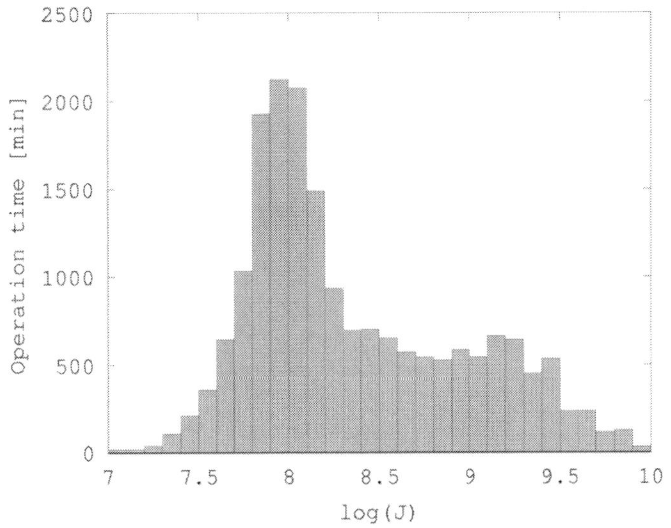

Figure 5. Operation time of Tatiana-2 as a function of the measured atmospheric background intensity J.

The Lomonosov satellite will be launched to a polar orbit close to that of Tatiana-2, so the available data can be used for estimating the TUS operation time with various background intensities. Results of such an estimate are presented in table 2. The number of registered EECRs will be much lower than those given in the table because the efficiency of the triggering system was not taken into account. With the limited exposure, the TUS detector will not make a breakthrough to the problem of EECR origin. Its main aim is to check the EAS fluorescence detector performance in a harsh space environment.

Table 2. Expected TUS performance parameters.

Intensity J, ph cm^{-2} s^{-1} sr^{-1}	$3 \cdot 10^7$–10^8	10^8–$5 \cdot 10^8$	$5 \cdot 10^8$–$5 \cdot 10^9$
One year night time (2047 hrs) vs J, %	33	39	28
TUS one year exposure, km^2 year sr	$3 \cdot 10^3$	$3.7 \cdot 10^3$	$2.7 \cdot 10^3$
Energy threshold (preliminary), EeV	60	150	400
EECR rate, events/km^2 year sr	0.024	0.002	0.0001
EECR event number per year	~ 70	~ 7	< 1

An important source of background events in orbital EECR measurements are UV flashes (duration of 1–100 ms), whose origins are related to electrical discharges in the atmosphere.

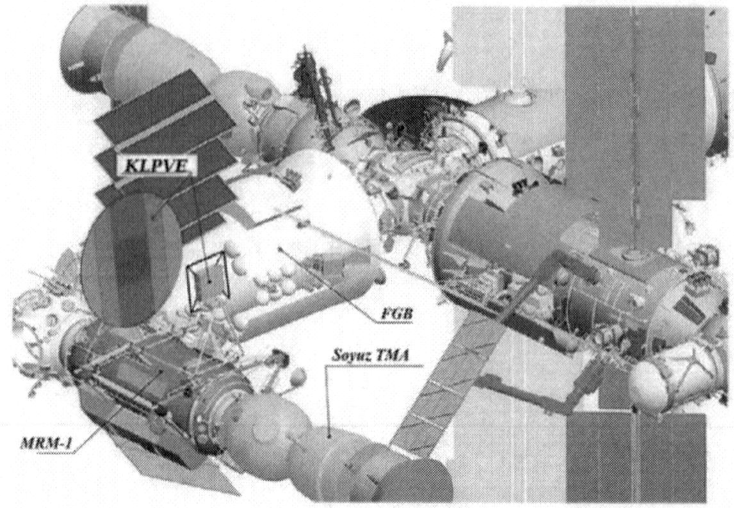

Figure 6. The KLYPVE detector on the MRM-1 module of the ISS.

In this respect, the data on UV flashes from the Tatiana-2 satellite were analyzed [41]. Measurements were done in a wide range of photon number Q in the atmospheric UV flashes: from $Q = 10^{21}$ up to $Q \sim 10^{25}$ where tens of events were registered. The main features of flashes with $Q > 10^{23}$, i.e., their duration of 10–100 ms and their global distribution concentrated in equatorial region above continents, allowed suggesting that those flashes are either lightnings or transient luminous events generically related to lightnings. Those "bright" flashes will be easily separated from EAS fluorescent signals due to their long duration and an enormous number of photons (to compare with EAS parameters: their duration is not more than 0.1 ms and the number of UV photons $Q \sim 10^{16}$ for $E = 100$ EeV).

More dangerous for imitation of EECR events is the background of dim and short flashes ($Q \sim 10^{21} - 10^{23}$, duration ~1 ms) also observed by the Tatiana-2 detector. In contrast to bright flashes, dim ones are distributed more uniformly over the Earth. Rates of both kinds of flashes (about 10^{-4}–10^{-3} km^{-2} hr^{-1}) are much higher than the expected rate of EECR events $\sim 10^{-6}$ km^{-2} hr^{-1}, so that the problem of distinguishing EAS flashes from atmospheric flashes is a complicated one. In this respect, the performance of the TUS detector will be an important test of future space measurements of EECRs.

4. KLYPVE

In 2012, Skobeltsyn Institute of Nuclear Physics of Moscow State University finished the preliminary design stage of the KLYPVE[3] reflector type telescope for EECR measurements from the ISS. It will be located on the outer side of the Russian segment, see figure 6. The main component of the detector is a segmented optical system (OS) with a large entrance pupil. All parts of the telescope are to be delivered to the ISS by the Progress-TM vehicle.

The main goal of KLYPVE is to considerably increase the field of view (FOV) of the detector in comparison with TUS and to decrease the energy threshold. The mirror-concentrator has a reflective surface of about 10 m^2 (entrance pupil diameter 3.6 m, see figure 7). The OS has a focal distance of 3 m and FOV diameter of 15°. The angular resolution of the OS is the same as in the TUS detector, i.e., 5 mrad (pixel size 15 mm × 15 mm).

[3] KLYPVE is a Russian acronym for extreme energy cosmic rays.

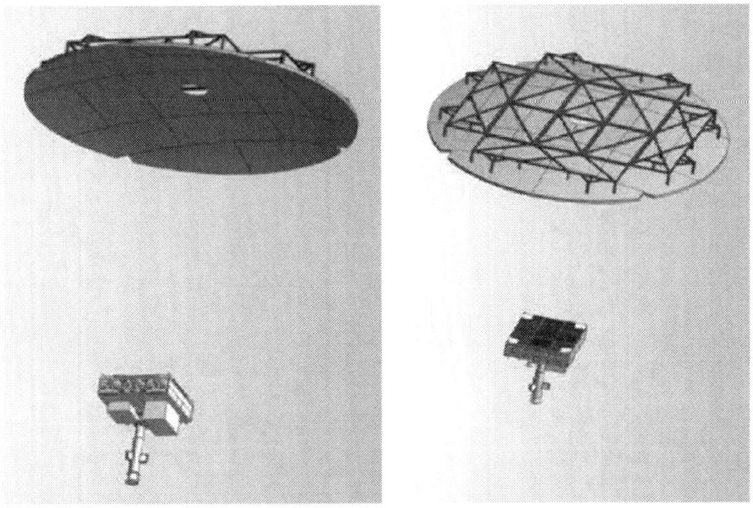

Figure 7. KLYPVE segmented mirror-concentrator and photodetector. Left: view from the front side, right: view from the rear side. A supporting frame is shown at the rear side of the mirror.

However, it became clear during the preliminary design phase that the characteristics of the instrument (observation area and image quality) do not allow solving the problems related to anisotropy of EECRs. The main reason for that is that due to the off-axis aberration of a fast optical system, the spot size increases very rapidly with increasing field angle. For an accurate reconstruction of EECR parameters (primary energy and direction), it is desirable to make a pixel of the photodetector smaller. Thus one has to look for optics with a much smaller spot size. These considerations initiated the development of a new OS for the KLYPVE detector in order to increase the FOV and to improve the spatial and angular resolution and the overall performance of the instrument. This work began in a close collaboration with members of the JEM-EUSO Collaboration in late 2013.

To eliminate the off-axis aberration, an additional corrective element was introduced into the telescope system in form of a Fresnel lens. The thickness of the lens should be sufficiently small, 1 cm in this case. The material of the lens is PMMA-000 (by Mitsubishi Rayon Co., Ltd.), which is a UV-transparent version of polymethyl metacrylate. It has a good UV transmittance (more than 90% for 15 mm thickness layer at wavelength more than 320 nm), with a refractive index of about 1.51 and has been used in space on many occasions. The lens has a radial Fresnel structure with groove depth 1 mm. Lenses with similar characteristics were manufactured at Ohmori Materials Fabrication Laboratory, RIKEN (Japan) in 2009–2011 [42].

Depending on the size and complexity of the forms of individual optical elements of the system, two approaches have been proposed.

4.1. Baseline System

In the so called Baseline system, the diameter of the reflector and the lens-corrector equals 3.4 m and 1.7 m respectively. The total length of the system (more precisely, the axial distance from the pole to the center of the focal surface) is equal to 4 m, the distance from the lens to the focal surface equals 70 cm, see figure 8. In this case, it is possible to expand the FOV up to $\pm 14^\circ$, and the diameter of the image is not larger than 6 mm in the entire FOV. The angular resolution of the system is $\approx 0.057^\circ$, which is equivalent to ~ 0.4 km at ground.

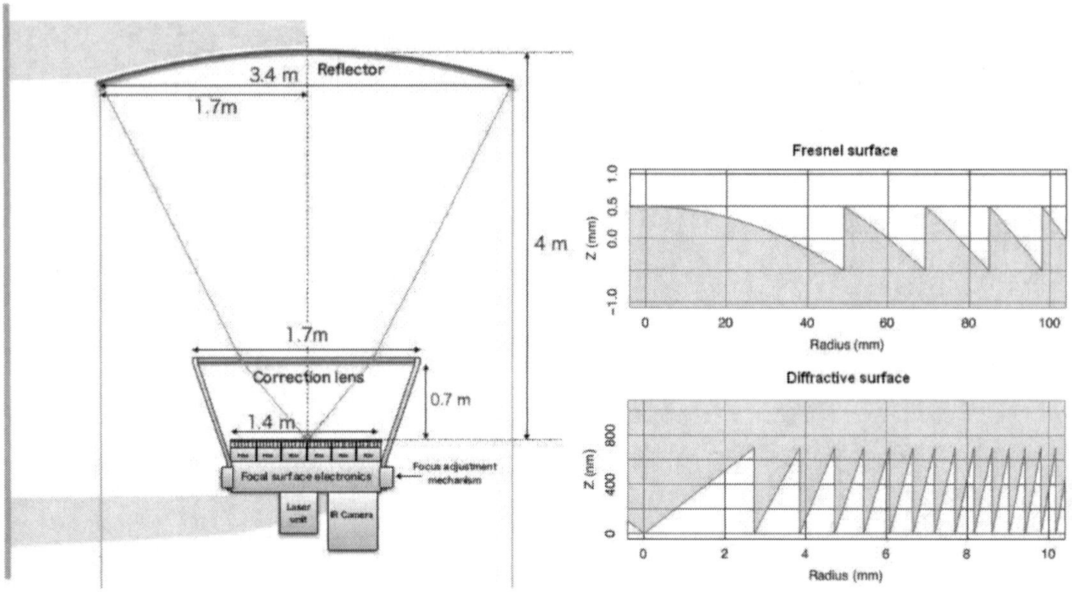

Figure 8. The optics of the Baseline system. Right: grooves of the Fresnel and diffractive structures of the corrective lens.

Table 3. Baseline OS performance characteristics for different field angles.

Field Angle	0°	4°	6°	8°	10°	12°	13°	14°
Spot Size, mm	2.8	2.6	3.2	4.2	4.5	4.3	4.0	3.9
Optical Efficiency, %	62.8	61.7	58.8	57.4	51.4	43.8	37.1	29.6

One disadvantage of this approach is the difficulty of manufacturing of optical elements, especially lenses. To correct aberrations, both lens surfaces should have a small curvature, wherein the front side is a Fresnel spherical surface (with a radial structure of grooves of 1 mm depth), and the back side is a spherical diffraction surface (with grooves depth 700 nm, see the right side of figure 8).

Two main optical performance characteristics are the spot size and optical efficiency, which can be determined by non-sequential ray tracing. Table 3 represents these parameters for different field angles as RMS diameter of the focal surface image and as the ratio of the number of rays in the spot to the number of rays at the entrance pupil.

Delivering such a huge system to the ISS is a complicated task by itself. In case Progress-TM is used, cargo must be first placed inside the ISS and thus pieces of the instrument have to pass through a cylindrical lock of 70 cm diameter and 120 cm length. The currently offered solutions require segmentation of all the major components of the system, including the lens, the mirror and the photodetector, and subsequent deployment in space.

4.2. Multi-Eye Telescope System

The optical system of the KLYPVE is a wide-field large aperture and fast optics system. For such a complex instrument, a more promising option might be a Multi-Eye Telescope System

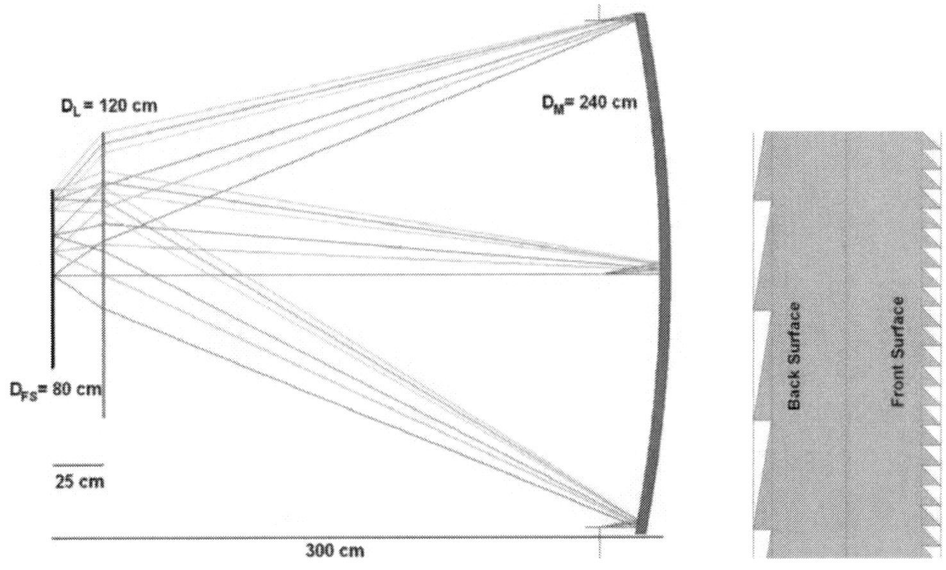

Figure 9. Left: the METS optical scheme. Right: grooves of the Fresnel structures at the edge of the corrective lens.

(METS). The main idea of METS is to divide a wide FOV into several FOVs of identical smaller telescopes. In this case, we have a simplification of an individual telescope: the design is simplified by cutting the aperture and the FOV of an individual telescope compared to the Baseline option thus reducing the requirements for their manufacturing.

This system has several additional advantages:

- smaller dimensions of individual telescopes and their narrower FOVs allow correcting aberrations without the use of a complex surfaces corrector (diffraction, curved Fresnel);
- not only manufacturing of the mirror segments and lenses but testing and adjustment of the overall system might be simpler;
- dimensions of the elements can be chosen basing on the possibility of transporting the individual telescopes assembled.

One of the possible drawbacks of METS is the necessity of a trigger working for all three telescopes simultaneously.

In the optimization calculations, the following dimensions of an individual telescope were obtained: 2.4 m diameter mirror, 1.2 m lens, 0.9 m photodetector, the total axial length 3 m, see figure 9. Correction of aberrations in the FOV $\pm 10°$ (which corresponds to the diameter of overall FOV of three telescopes $\sim 35°$) can be achieved using plane Fresnel surfaces. The angular resolution of METS is $\approx 0.075°$, which is equivalent to ~ 0.5 km at ground.

Spot size and optical efficiency of the METS telescope is presented in table 4. The sharp increase of the spot size on the edge of the FOV is due to the fact that a flat focal surface was used in the calculations.

To control spots within 5 mm, one should slightly bend the focal surface at the edges. Another advantage of the multi-eye system is the ability to use an active (dynamic) configuration. A special support structure of three telescopes can afford one to implement operation of the detector in various modes. In the basic mode, shown in the top row of figure 10, the FOVs of individual telescopes are adjacent (or slightly overlapping for cross-calibration). In the coincidence mode (overlapping FOVs), one can significantly reduce the energy threshold. Finally,

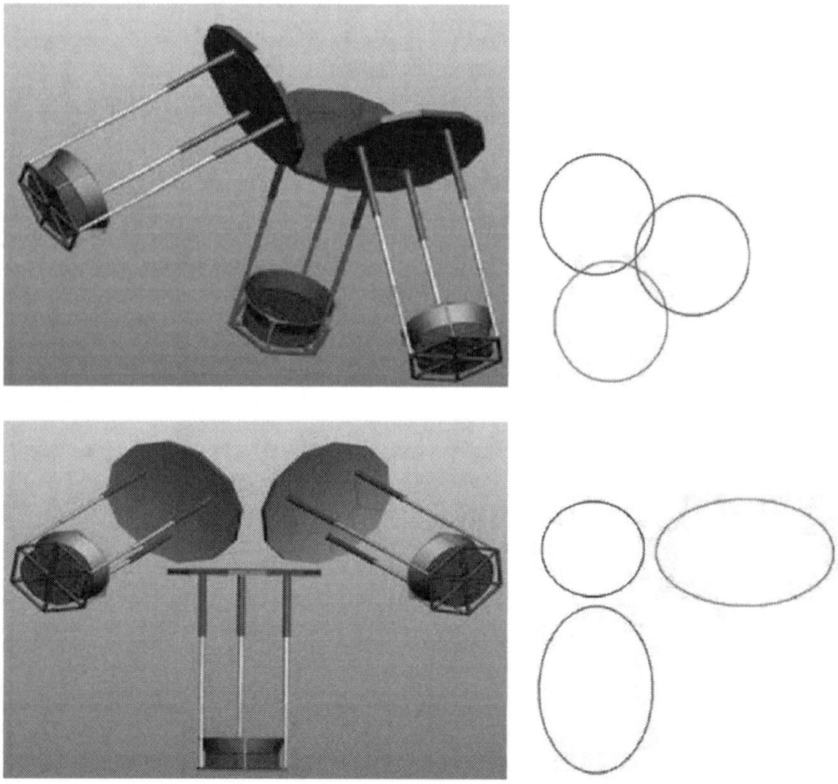

Figure 10. Top: the basic mode of METS operation. The angles of inclination of individual telescopes are exaggerated for clarity. Bottom: The tilted mode of METS operation. The right panels show mutual position of FOVs of separate telescopes.

Table 4. The METS OS performance characteristics for different field angles.

Field Angle	0°	3°	5°	6°	7°	8°	9°	10°
Spot Size, mm	2.0	2.7	3.7	3.7	4.0	4.2	4.5	5.4
Optical Efficiency, %	68.8	67.2	62.0	57.7	55.0	53.3	49.1	41.9

in the tilted mode (FOVs of two or all telescopes are inclined relative to the direction at the nadir, see the bottom row of figure 10), it is possible to increase the exposure, and hence to collect larger statistics of the EECR events.

5. JEM-EUSO

JEM-EUSO (the Extreme Universe Space Observatory on board the Japanese Experiment Module), which aims to be installed at the International Space Station (ISS), is the most ambitious and sophisticated of the three projects, see figure 11.

The instrument consists of a UV telescope and an atmospheric monitoring system. The telescope consists of three Fresnel lenses with an optical aperture of 4.5 m^2 and a focal surface detector formed by 137 photodetector modules composed of $\sim$ 5000 multi-anode photo-multiplier tubes in total. The focal surface detector thus includes $\sim 3 \cdot 10^5$ channels providing a spatial

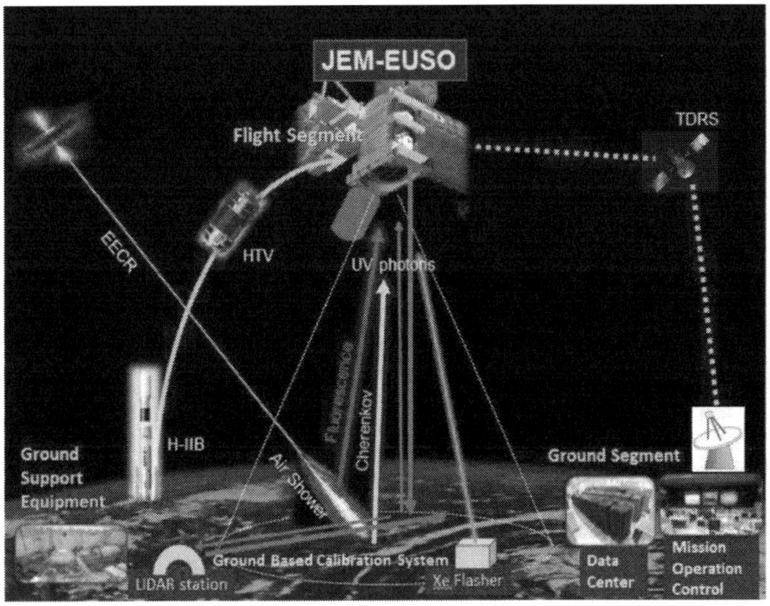

Figure 11. Principle of detecting EECRs with the JEM-EUSO telescope [43].

resolution $\approx 0.074°$, equivalent to ~ 0.5 km at ground. The opening angle of the telescope equals 60° providing an observational area of $\sim 1.4 \cdot 10^5$ km^2 in nadir mode. The annual exposure of JEM-EUSO for EECRs above 100 EeV is estimated to be an order of magnitude larger than that of Auger [44]. Thanks to the ISS orbit, JEM-EUSO is able to survey the entire celestial sphere almost uniformly, with the level of non-uniformity of the exposure on declination and right ascension $\sim 10\%$ [45]. Detailed simulations show that due to these factors JEM-EUSO will be able to detect significant anisotropies of EECRs practically in all feasible astrophysical scenarios [46] thus providing a huge step towards finding sources of EECRs. In December 2013, the Japanese Space Agency (JAXA) refused to deploy the instrument on board the JEM. A detailed discussion of the current status of the project and its pathfinders can be found in [47] and in dedicated contributions by A. Haungs and M. Fukushima in this volume.

6. Discussion

Figure 12 shows the dependence of the effective area S_{eff} of the Baseline and METS optical systems of the modified KLYPVE detector and the JEM-EUSO telescope on the field angle γ. Here, the effective area means the ratio of the luminous flux to irradiance, i.e.,

$$S_{\text{eff}} = \varepsilon_{\text{opt}}(\gamma) S_{\text{geom}}(0°) \cos\gamma,$$

where γ is the field angle, ε_{opt} is the optical efficiency, and the axial geometrical area (the second factor) takes into account the reduction of the entrance pupil due to the central screening by the lens. Thus, this parameter is one of the most important in the design of optical systems of an orbital detector.

The effective area is reduced at the edge of FOV for both systems, primarily due to an increase in the scattering on Fresnel and diffraction (in case of the Baseline system) grooves, which leads to a significant spot smearing at large field angles. As can be seen from figure 12, the effective area of the METS telescope approximately coincides with the JEM-EUSO one. In the case of the Baseline system, it is possible to increase the value of this parameter in average

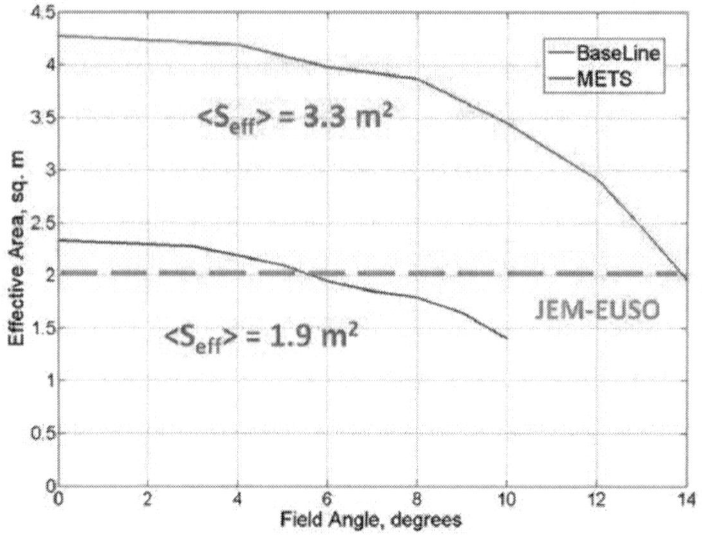

Figure 12. Effective area of the Baseline system and METS compared to JEM-EUSO for the entire FOV. $\langle S_{\text{eff}} \rangle$ denotes average values of the effective area for the isotropic distribution of light.

Table 5. Performance characteristics (spot size d, effective area S_{eff}, FOV Ω and throughput ΩS_{eff}) for 5 different orbital detectors of EECRs.

	d, mm	S_{eff}, m^2	Ω, sr	ΩS_{eff}, m^2 sr
JEM-EUSO	3	2	0.8	1.6
TUS	20	1.2	0.025	0.03
KLYPVE (orig.)	20	5	0.05	0.25
Baseline	3.5	3.5	0.2	0.7
METS	4	2.0	0.3	0.6

over the entire FOV by 70%, and more than twice for events in the cone $\gamma < 10°$. However, it should be noted that a refraction efficiency of the (first) diffraction peak was assumed to be 95% in these calculations. Reaching such values over the entire surface during manufacturing of the lens can pose a problem since the diffractive surface area is about 2.3 m^2. Thus, having undoubted advantages in decreasing the energy threshold, the Baseline system might have the effective FOV much smaller than the total FOV of the METS telescope.

Table 5 summarizes the main characteristics of the five optical systems: the lens system of JEM-EUSO, TUS Fresnel mirrors, the original version of the KLYPVE detector (without any corrector) and the two advanced options, the Baseline system and METS (with three identical telescopes in the basic mode of operation). Scattering of light by Fresnel surfaces was taken into account when calculating the effective area. It is important to underline that the effective area of the Baseline system provides an annual exposure two times larger than that of Auger, see figure 13. Together with an almost uniform exposure of both hemispheres, this gives an additional advantage for anisotropy studies.

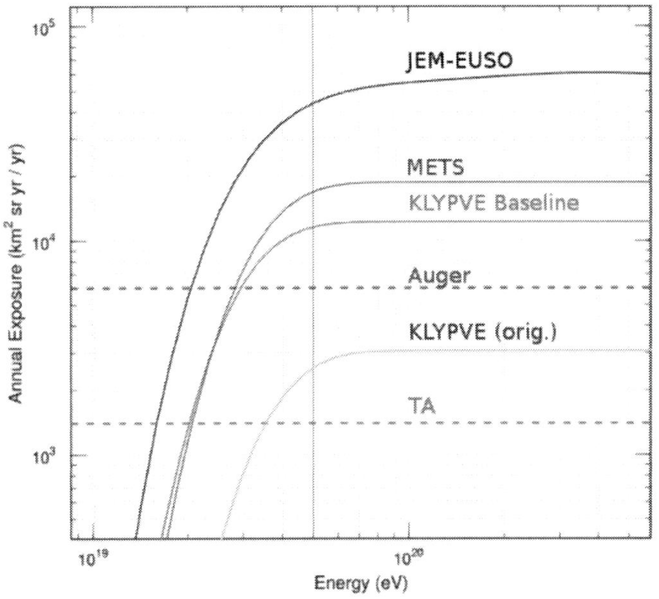

Figure 13. Annual exposure of different ground-based and space experiments.

7. Conclusions

Three orbital projects aimed at detecting extreme energy cosmic rays are being developed these days.[4] TUS is scheduled for launching in late 2015. If successful, TUS will become the pioneer of exploration of EECRs from space and a pathfinder for more advanced missions. It will also provide precious information about the UV background of the Earth atmosphere. The KLYPVE project finished the stage of the preliminary design and is included in the Russian Federal Space Program. It has entered the stage of the design project in its modified version aimed to improve the technical parameters of the instrument. There is a realistic possibility for KLYPVE to be installed on board the Russian Segment of the ISS in the next 6–7 years. The development of the full instrument of the JEM-EUSO project is currently facing certain problems, but the successful flight of EUSO-Balloon[5] on 24 August, 2014, which was used to test a number of components of the telescope and the atmospheric monitoring system, gives a strong ground for the hope that difficulties will be overcome and the most sophisticated of the three projects will be implemented in the future. Both KLYPVE and JEM-EUSO will provide outstanding possibilities for a breakthrough in solving the long-standing puzzle of the nature and origin of the most energetic particles ever detected on Earth.

Acknowledgments

We thank the members of the JEM-EUSO collaboration who are taking an active part in the development of the modified KLYPVE project. Our special thanks are addressed to Andreas Haungs. The work was done with a partial financial support by the Russian Foundation for Basic Research grant 13-02-12175-ofi-m. N.S. was supported by the "Helmholtz Alliance for Astroparticle Physics HAP" funded by the Initiative and Networking Fund of the Helmholtz Association, Germany.

[4] A discussion of the Super-$\mathcal{EUSO}$ [48] and OWL [49] missions goes beyond the scope of the article.

[5] See `http://euso-balloon.lal.in2p3.fr`

References

[1] Linsley J 1963 *Phys. Rev. Lett.* **10**(4) 146–48
[2] Bergman D R and High Resolution Fly's Eye Collaboration 2007 *Nucl. Phys. B Proc. Suppl.* **165** 19–26 (*Preprint* astro-ph/0609453)
[3] Abbasi R U *et al.* 2008 *Phys. Rev. Lett.* **100** 101101 (*Preprint* arXiv:astro-ph/0703099)
[4] Abraham J *et al.* 2008 *Phys. Rev. Lett.* **101** 061101 (*Preprint* arXiv:0806.4302)
[5] Abu-Zayyad T *et al.* 2013 *Astrophys. J. Lett.* **768** L1 (*Preprint* arXiv:1205.5067)
[6] Dawson B R, Mariş I C, Roth M, Salamida F, Abu-Zayyad T, Ikeda D, Ivanov D, Tsunesada Y, Pravdin M I and Sabourov A V 2013 The energy spectrum of cosmic rays at the highest energies *European Physical J. Web of Conf.* vol 53 p 1005 (*Preprint* arXiv:1306.6138)
[7] Greisen K 1966 *Phys. Rev. Lett.* **16**(17) 748–50
[8] Zatsepin G T and Kuz'min V A 1966 *Soviet J. of Experimental and Theoretical Physics Letters* **4** 78
[9] Harari D 2014 *Comptes Rendus Physique* **15** 376–83 (*Preprint* arXiv:1406.1117)
[10] Letessier-Selvon A 2014 *Brazilian J. of Physics* **44** 560–70 (*Preprint* arXiv:1310.4620)
[11] Tsunesada Y 2013 Study on mass composition of ultra-high energy cosmic rays by Telescope Array *Proc. 33rd Int. Cosmic Ray Conf. (Rio de Janeiro)*
[12] Abbasi R U *et al.* 2015 *Astropart. Phys.* **64** 49–62 (*Preprint* arXiv:1408.1726)
[13] Abu-Zayyad T *et al.* 2013 Pierre Auger Observatory and Telescope Array: Joint contributions to the 33rd Int. Cosmic Ray Conf. (ICRC 2013) (*Preprint* arXiv:1310.0647)
[14] Abbasi R U *et al.* 2014 *Astrophys. J. Lett.* **790** L21
[15] He H N, Kusenko A, Nagataki S, Yang R Z and Fan Y Z 2014 The possible extragalactic source of ultra-high-energy cosmic rays at the Telescope Array hotspot (*Preprint* arXiv:1411.5273)
[16] Pierre Auger Collaboration 2007 *Science* **318** 938–43 (*Preprint* arXiv:0711.2256)
[17] Pierre Auger Collaboration 2008 *Astropart. Phys.* **29** 188–204 (*Preprint* arXiv:0712.2843)
[18] Abreu P *et al.* 2010 *Astropart. Phys.* **34** 314–26 (*Preprint* arXiv:1009.1855)
[19] Pierre Auger Collaboration 2014 Searches for anisotropies in the arrival directions of the highest energy cosmic rays detected by the Pierre Auger Observatory (*Preprint* arXiv:1411.6111)
[20] Abu-Zayyad T *et al.* 2013 *Astrophys. J.* **777** 88 (*Preprint* arXiv:1306.5808)
[21] Pierre Auger Collaboration 2014 Large scale distribution of ultra high energy cosmic rays detected at the Pierre Auger Observatory with zenith angles up to 80° (*Preprint* arXiv:1411.6953)
[22] de Almeida R M 2013 Constraints on the origin of cosmic rays from large scale anisotropy searches in data of the Pierre Auger Observatory *Proc. 33rd Int. Cosmic Ray Conf. (Rio de Janeiro)* 0768 (*Preprint* arXiv:1307.5059)
[23] Fukushima M, Ivanov D, Kido E, Pshirkov M, Rubtsov G, Sagawa H, Thomson G, Tinyakov P, Tkachev I and Urban F 2013 Search for large-scale anisotropy of ultra-high energy cosmic rays with the Telescope Array *Proc. 33rd Int. Cosmic Ray Conf. (Rio de Janeiro)* 0935
[24] Aab A *et al.* 2014 *Astrophys. J.* **794** 172 (*Preprint* arXiv:1409.3128)
[25] Tinyakov P G and Urban F R 2014 Full sky harmonic analysis hints at large UHECR deflections (*Preprint* arXiv:1411.2486)
[26] Benson R and Linsley J 1981 Satellite observation of cosmic ray air showers *Proc. 17th Int. Cosmic Ray Conf. (Paris)* vol 8 pp 145–48
[27] Bugrov D *et al.* 2001 Modeling of space telescopes for observation of tracks produced in the Earth's atmosphere by extremely high energy cosmic rays *Mathematical modeling of complex information processing systems* ed Sadovnichii V and Doger Guerero E (Lomonosov Moscow State University Press) pp 98–118
[28] Pallavicini M, Pesce R, Petrolini A and Thea A 2012 *Astropart. Phys.* **35** 402–20 (*Preprint* arXiv:1105.2129)
[29] Khrenov B A *et al.* 2001 Space program KOSMOTEPETL (project KLYPVE and TUS) for the study of extremely high energy cosmic rays *Observing Ultrahigh Energy Cosmic Rays from Space and Earth* (*American Inst. of Physics Conf. Series* vol 566) ed Salazar H, Villasenor L and Zepeda A pp 57–75
[30] Khrenov B A *et al.* 2004 *Physics of Atomic Nuclei* **67** 2058–61
[31] Abrashkin V *et al.* 2006 *Adv. in Space Research* **37** 1876–83
[32] Abrashkin V *et al.* 2008 *Adv. in Space Research* **41** 2079–88
[33] Panasyuk M *et al.* 2012 *J. of Cosmology* **18** 7964–77
[34] Garipov G K *et al.* 2001 Electronics for the KLYPVE detector *Observing Ultrahigh Energy Cosmic Rays from Space and Earth* (*American Inst. of Physics Conf. Series* vol 566) ed Salazar H, Villasenor L and Zepeda A pp 76–90
[35] Tkachev L *et al.* 2009 *Nucl. Phys. B Proc. Suppl.* **196** 243–46
[36] Tkachenko A *et al.* 2011 Photo receiver of the orbital ultra high energy cosmic rays detector TUS *Proc. 32nd Int. Cosmic Ray Conf. (Beijing)* vol 11 p 272
[37] Tkachev L *et al.* 2011 The TUS Fresnel mirror production and optical parameters measurement. *Proc. 32nd*

Int. Cosmic Ray Conf. (Beijing) vol 3 p 124

[38] Grinyuk A, Slunecka M, Tkachenko A, Tkachev L, Klimov P and Sharakin S 2014 *Nucl. Inst. Methods in Phys. Res.* A **763** 604–609

[39] Khrenov B A *et al.* 2013 Pioneering space based detector for study of cosmic rays beyond GZK limit *European Physical J. Web of Conf.* vol 53 p 9006

[40] Vedenkin N N *et al.* 2011 *Soviet J. of Experimental and Theoretical Physics* **113** 781–90

[41] Garipov G K *et al.* 2013 *J. of Geophysical Research (Atmospheres)* **118** 370–79

[42] Hachisu Y *et al.* 2011 JEM-EUSO lens manufacturing *32nd International Cosmic Ray Conference (Beijing)* vol 3 p 184

[43] Ebisuzaki T, Medina-Tanco G and Santangelo A 2014 *Advances in Space Research* **53** 1499–1505

[44] Adams J H *et al.* 2013 *Astropart. Phys.* **44** 76–90 (*Preprint* arXiv:1305.2478)

[45] Bertaina M, Bobik P, Fenu F and Shinozaki K 2014 *Experimental Astronomy* URL http://link.springer.com/article/10.1007%2Fs10686-014-9376-3

[46] Rouillé d'Orfeuil B, Allard D, Lachaud C, Parizot E, Blaksley C and Nagataki S 2014 *Astron. & Astrophys.* **567** A81 (*Preprint* arXiv:1401.1119)

[47] Bertaina M and Parizot E for the JEM-EUSO Collaboration 2014 *Nucl. Phys. B Proc. Suppl.* B **256** 275–86

[48] Petrolini A 2011 *Nucl. Inst. Methods in Phys. Res.* A **630** 131–35 (*Preprint* arXiv:0909.5220)

[49] Krizmanic J F, Mitchell J W and Streitmatter R E 2013 Optimization of the Orbiting Wide-angle Light Collectors (OWL) mission for charged-particle and neutrino astronomy *Proc. 32nd Int. Cosmic Ray Conf. (Beijing)* ID 1085 (*Preprint* arXiv:1307.3907)

Mass Composition of Cosmic Rays at Ultra High Energies by Yakutsk Data

Stanislav Knurenko and Igor Petrov

Yu. G. Shafer Institute of Cosmophysical Research and Aeronomy SB RAS, Lenin ave. 31, Yakutsk, Russia

E-mail: knurenko@ikfia.sbras.ru, igor.petrov@ikfia.sbras.ru

Abstract. The paper describes methods for the analysis of cosmic rays mass composition and results for over a large time span. The data were obtained at Small Cherenkov array over 20 year period of continuous observation and 40 - years of observations at the main Yakutsk array. The experimental data indicates changes in the MC in the energy range 10^{16} - 10^{18} eV which confirmed by results obtained by other EAS arrays.

1. Introduction

In this review, we used Yakutsk group work in which MC at ultra-high energies estimated [1 - 5]. In these studies characteristics of longitudinal and lateral development of EAS were analyzed and reconstructed according ot observations at the Yakutsk array. This primarily refers to measurements of Cherenkov light of EAS and muons with a threshold energy ≥ 1 GeV. These components considered to be most sensitive air shower characteristics to atomic weight on the basis of obtained results. The results were obtained using different models of hadronic interactions [6 - 8].

One can assume that all presented results on the mass composition are rather indicative, because its are not direct but indirect measurements of the mass composition and depend on many factors: 1) the accuracy of the measurement of the basic characteristics (they are all different for different arrays), 2) the conditions of registration, the techniques used, mathematical processing and selection of events, 3) and, finally, from a large uncertainty in the choice of a single model of hadron interactions to describe the development of extensive air showers at ultrahigh and, especially in the region of huge energies.

However, it is necessary to make the estimation of CR mass composition in the region of ultra-high energies and compare these results with direct measurements, so called normal composition, which obtained at high energies from satellite and balloon measurements. This leads to a refinement of our knowledge of the nature of cosmic radiation and a better understanding of the physics of EAS development in the field of the highest energies.

2. Mathematical methods of analysis and results.

2.1. The method of joint analysis of the average characteristics of the longitudinal development of EAS and their fluctuations: X_{max}, $\sigma(X_{max})$, dE/dX_{max}

In paper [1] was suggested that composition of primary particles consist of a mixture of protons and iron nuclei. The analysis also used the superposition hypothesis, it was assumed that the collapse of the primary nucleus did not occur on the top of the atmosphere, but at a depth corresponding to a run for the collision of nuclei in the air. Therefore, the average depth of the shower maximum for this superposition is modified by the value of the path is lower than X_{max} calculated from the diffusion equations of nuclear cascade process. In the method were used the hydrodynamic model with $n_{ch} \sim E_{1/3}$ [6].

Let the distribution of the maximum depth of showers from primary nuclei has an exponential form with the first moment equal to their run for the nuclear interaction. Then, the average depth of the shower maximum for the sum of two exponential functions will be equal.

$$\bar{X}_{max} = \eta \cdot X_p + (1 - \eta) \cdot X_{Fe} \quad (1)$$

Where η - the fraction of protons in the primary cosmic radiation, X_p and X_{Fe} - the depth of maximum development of the primary proton and iron nuclei by the chosen model of the EAS. For the dispersion we have X_{max} expression

$$D(X_{max}) = \beta\{\eta \cdot \lambda_p^2 + \eta(1-\eta) \cdot (X_p - X_{Fe})^2 + (1-\eta) \cdot \lambda_{Fe}^2\} \quad (2)$$

Here λ_p and λ_{Fe} respectively the path for the nuclear interaction of the proton and iron nuclei. β multiplier takes into account the increase of the dispersion X_{max} due to fluctuations of the inelasticity coefficient and is taken to be $(1 \text{ -1 } <k>)^{-1}$, where $<k>$ - the mean value of the inelasticity coefficient of the leading particle. If we assume that the ratio $\lambda_{Fe}(E)$ / λ_p (E) is constant and known, then equations (1) and (2) in the framework of a two-component composition can determine the proportion of protons in the primary radiation and the cross section of the proton - nucleus of an atom from the experimental values X_{max} and $D(X_{max})$. Technically, the above simplification can be extended to multi-component composition of the primary particles, but the accuracy of formulas of the type (1) and (2) for X_{max} and D (X_{max}) will be slightly worse.

We also found an indication of a gradual increase in the energy of the protons in the region of energy $3 \cdot 10^{17}$ $3 \cdot 10^{18}$ eV.

2.2. Method of comparing the asymmetry of distribution at different Xmax at different fixed energies

This method does not depend from model of air shower development. Since the distribution of X_{max} at fixed energy is formed by nuclei of different masses, therefore, its shape will reflect their contribution to a statistical contribution of X_{max}, g / cm^2. This is understandable, since showers produced by particles of different masses have either rapid development, for example, the iron nucleus or slow development as it happens, if we consider the proton. In paper [2] the ideology of asymmetry of X_{max} distributions at different energies were used. Its essence is as follows. The value of the effective cross section for inelastic collisions of protons with air nuclei on the distribution of heights of the maxima of the EAS in the energy range of the primary particles 10^{17} - 10^{18} eV and 10^{18} - 10^{19} eV was studied. Right-hand side of such distributions is determined mainly by the effective cross section for inelastic collisions of protons with air nuclei. Height of the maximum development of the shower was determined by the spatial distribution of the Cherenkov light at the distance range of 100 - 600 m from the shower axis.

Next, assuming that at high energies only protons are presented and rationing distribution at lower energies by the proton (deeper than 700 g / cm^2), simply by subtracting the estimated fraction of nuclei in the primary radiation with energy $\sim 10^{17}$ - 10^{18} eV.

Thus the indication that in the energy range 10^{16} - 10^{19} eV observed systematic increase in the fraction of protons: $\sim 1.2 \cdot 10^{16}$ eV - (43±5), $\sim 9 \cdot 10^{16}$ eV - (50±6), $\sim 5 \cdot 10^{17}$ eV - (60±10)% and $\sim 5 \cdot 10^{18}$ eV - (90±10)% was obtained.

2.3. Distribution of X_{max} shape analysis jointly with calculated distribution using QGSJET model by maximum likelihood method.
In this method we used experimental data of X_{max} at energies 10^{15} - 10^{19} eV and simulated showers according to QGSJET 01 model [8]. Joint analysis of showers allowed to obtain quantitative estimates of the mass composition of primary CR, using the distribution of X_{max} at fixed energy [3]. To do this, we compared the experimental data and theoretical predictions of predictions according QGSJET for different primary nuclei with applied criterion χ^2. The value was determined by the equation χ^2.

$$\chi^2(X_m) = \sum_n (N_e(X_{max}) - N_T(X_{max}))^2 / N_T(X_{max}) \tag{3}$$

where N_e (X_{max}) - experimental number of showers in the range ΔX_{max}. N_T (X_{max}, A_i) - the same number of showers, calculated under the assumption that the mass number of the nucleus is equal to A_i, and P (A_i) - the probability that a storm of energy E_0, is formed by the primary particle A_i. Then

$$N_T(n) = \sum_{i=1}^{n} P(A_i) \cdot N_T(X_{max}, A_i) \tag{4}$$

Analysis of the shape of experimental and calculated distribution of X_{max} showed that at optimal value of χ^2 obtained result does not conflict with the following relationships for 5 nuclei components:

1). $\bar{E_0} = 5 \cdot 10^{17}$ eV - p : (39 ± 11) % , α: (31 ± 13) % , M : (18 ± 10) % , H : (7 ± 6) % , Fe : (5 ± 4) % ;

2). $\bar{E_0} = 1 \cdot 10^{18}$ eV - p : (41 ± 8) % , α : (32 ± 11) % , M : (16 ± 9) % , H : (6 ± 4) % , Fe : (5 ± 3) % ;

3). $\bar{E_0} = 5 \cdot 10^{18}$ eV - p : (60 ± 14) % , α : (21 ± 13) % , M : (10 ± 8) % , H : (5 ± 4) % , Fe : (3 ± 3) %.

Thus, in the framework of the QGSJET 01 the indication was obtained: mass of the PCR in the transition from the energy (5 ± 30)· 10^{17} eV to the energy (3 ± 10)·10^{18} eV changes. For $E_0 \cdot 3 \cdot 10^{18}$ eV primary cosmic radiation consists on ~70% of protons and helium nuclei, the proportion of other nuclei does not exceed ~30%.

2.4. Multicomponent analysis
In order to interpret experimental data of Yakutsk array we used CORSIKA code (v. 6.0. QGSJET model) to generate database of X_{max} and $\rho_e(600)$ [4]. Simulation were done for five primaries (P, He, C, Si, Fe) and three energies 10^{17}, 10^{18}, 10^{19} eV. For each energy we simulated 100 showers in the standard atmosphere. In this paper, we used two-dimensional probability density F(X_{max}, ρ_e (600)), preliminary standardized experimental data of the entire array data (X_{max}, ρ_e (600)) for a given energy. At numerical implementation of this method instead of (X_{max}, ρ_e (600)) and used variables τ and ρ_e:

$$\tau = (X_{max}/\sigma_x) - < (X_{max}/\sigma_x) > \tag{5}$$

$$\rho = (lg\rho(600)/\sigma_{lg}\rho) - < (lg\rho(600)/\sigma_{lg}\rho) > \tag{6}$$

Where σ standart deviation of the value.

Standardization performed on pooled data (X_{max}, ρ_e (600)) for all groups of nuclei and each energy 10^{17}, 10^{18}, 10^{19} eV. Distribution on X_{max} and ρ_e (600) separately and joint distributions for τ and ρ are described respectively by dimensional F (X_{max}) F (ρ(600)) and two-dimensional

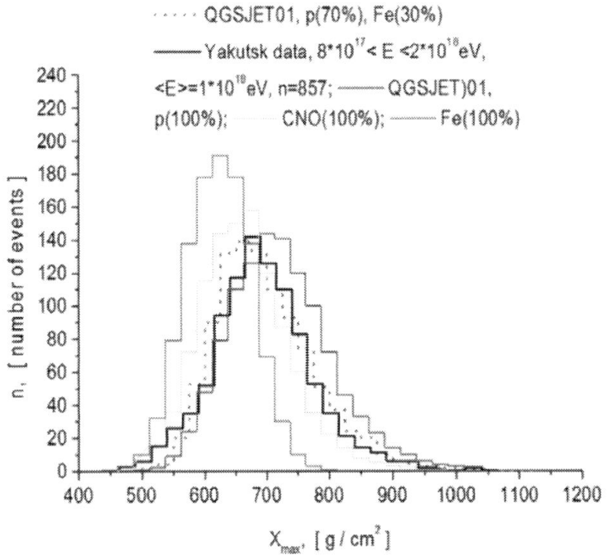

Figure 1. Distribution of X_{max} at fixed energy. Point result - test provided p (70%) and Fe (30%)

f(τ, ρ) logarithmically normal distribution. For each given energy and different types of primary nuclei, including for nuclei, combined in groups P + He, C, Si + Fe, were plotted probability distribution density f(τ, ρ). The intersection of f(τ, ρ) layers gives lines m1 and m2, which optimally separates nuclei into 3 groups: (P + He), C and (Si + Fe) respectively.

Fig. 2 shows the result of a multi-component analysis of the data binding (X_{max} ρ_e (600)) of Yakutsk array. A cloud of points in such a representation reflects standardized values, and lines represent areas that are directly associated with the mass number of the primary particle. In this case, the line m1 and lines m2 are optimally separates nuclei into groups (P + He), and C (Si + Fe).

Analysis has shown that the proportion of nuclei (p + He) increases from 50% to 53%, and the proportion of carbon nuclei from 23% to 31%. At the same time the proportion of nuclei of heavy chemical elements decreases from 27% to 16% while increasing energy from $2.4 \cdot 10^{17}$ to $4.8 \cdot 10^{18}$ eV.

2.5. Proportion of muons analysis method depending on the length of the track of the particles in the atmosphere

In paper [5] considered the dependence ρ_μ / ρ_s the length of the track of the particles after the maximum of EAS $\Delta\lambda = X_0 / \cos\theta - X_{max}$ where $X_0 = 1020$ g / cm^2 for Yakutsk. Here X_{max} determined from measurements of the Cherenkov light and ρ_μ and ρ_s by measuring a large EAS. Next, the experiment was compared with model calculations of QGSJETII 03 and EPOS. It is known that X_{max} of showers greatly differs depending on primary nucleus, therefore, this fact can be used to analyze the mass composition of cosmic rays, for example, by fixing the parameter $\Delta\lambda$ and analyzing fluctuations in the ratio of ρ_μ / ρ_s. This method is somewhat similar to the method of Christiansen, proposed in 1981 [9]. With sufficient precision of measurements of each parameter (better than 5%) in the distribution allocated single peaks from different nuclei.

Comparison of distribution of muon proportion with calculation results indicates to mixed composition at energies above 10^{18} eV. Large fluctuations do not allow allocating separate groups of nuclei with a good precision and evaluating percentage of each group. But, the use of pure response of muon detectors leads to conclusion that MC at energies 10^{18} 10^{19} eV is light [10].

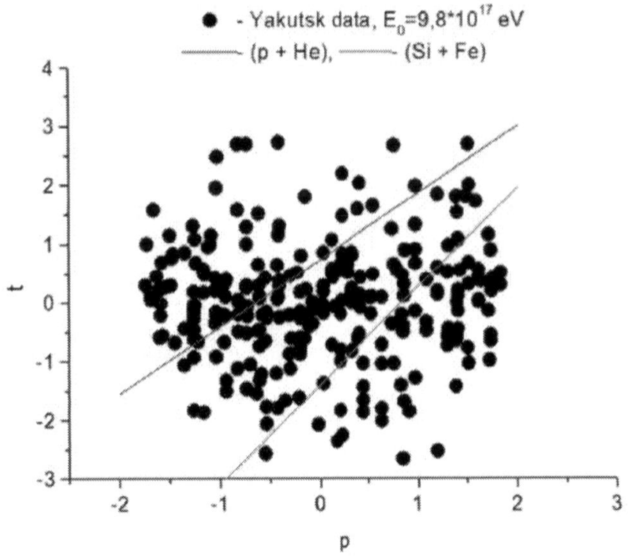

Figure 2. Standardized experimental data for X_{max} and ρ_e (600) at different energies. Line m1 divides nuclei (P + He) and C, line m2 divides C and (Si + Fe)

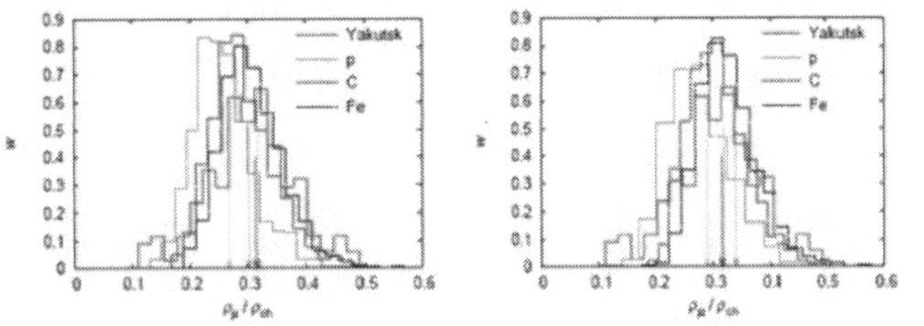

Figure 3. A distribution of the ρ_μ / ρ_s relation normalized to the track length 500 g / cm^2. On the left according to models QGSJet II(FLUKA), on the right according to EPOS(UrQMD)

2.6. Evaluation of the mass composition at average depth of maximum of EAS development. Interpolation method

In papers [11, 12], dependence of X_{max} from energy at range ~10^{15} to $5 \cdot 10^{19}$ eV was considered. MC of PCR evaluated by this formula:

$$< lnA >\equiv \sum a_i \cdot lnA_i \tag{7}$$

Where a_i the relative proportion of nuclei with mass number A_i.

In each case experimental data compared with QGSJET 03 calculation made for proton and iron in the frame of superposition model:

$$< lnA >= ((P^{exp.} - P^p))/(P^{Fe} - P^p)) \cdot lnA_{Fe} \tag{8}$$

Where P_i parameter, that characterize longitudinal development of air showers X_{max}.

In Fig. 4 shows dependence of X_{max} from energy (dots) derived from experiment and from simulation (lines) of this characteristics calculated by QGSJET 03 and SIBYLL model for proton and iron nuclei. Fig. 5 shows Yakutsk array results of MC PCR derived by method described above. Data obtained in the frame of QGSJETII 03 model and dual component MC (proton iron). Value ¡ln(A)¿ in each case was determined by interpolation method.

Fig. 5 shows that nature of dependence of value ¡lnA¿ with increase of energy changes reaching maximum at energy range (5 - 30)$\cdot 10^{16}$ eV. This means, that MC of CR changes after first kink in the spectrum at ~$3 \cdot 10^{15}$ eV, reaching heavier particles (3 - 30)$\cdot 10^{16}$ eV and then, starting at energies $3 \cdot 10^{17}$ eV becomes much lighter.

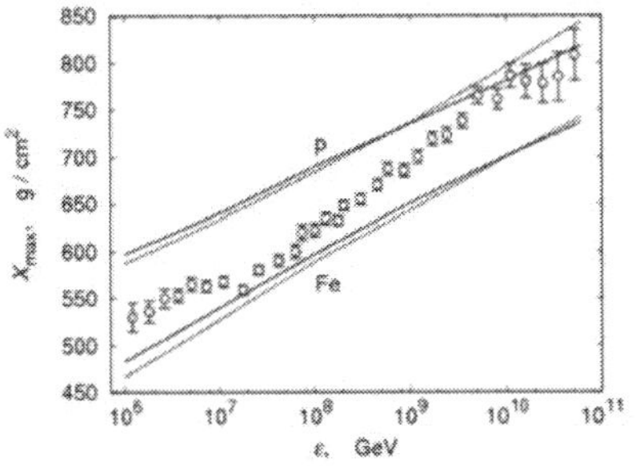

Figure 4. Dependence of X_{max} from energy. Lines are calculated values for proton and iron nuclei.

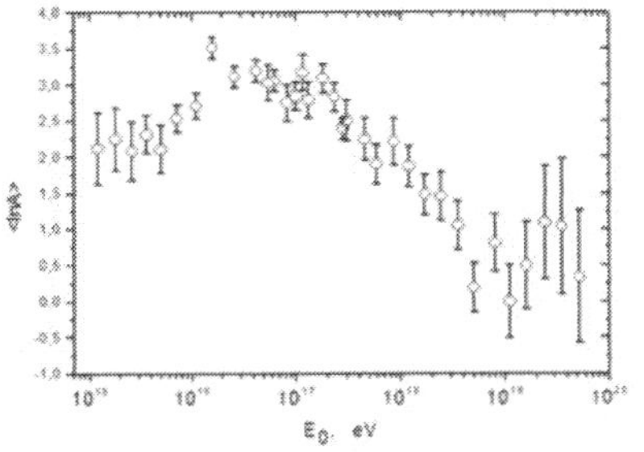

Figure 5. Mass composition of Cosmic rays highest energy are obtained at Yakutsk. Model QGSJETII-03

3. Conclusion

a) For more than 40 years Yakutsk array continuously records air showers with ultrahigh energies. We obtain information about all main components of the shower: electrons, photons, hadrons and muons. All these data at different times were used to estimate the mass composition of cosmic rays involving different methods. This follows from the numerous publications in journals and proceedings of scientific conferences. According to the data shown in Figure 6 [13]. The mass composition is not uniform over a wide energy range, and has a peak at (0.8 - 2) 10^{17} eV

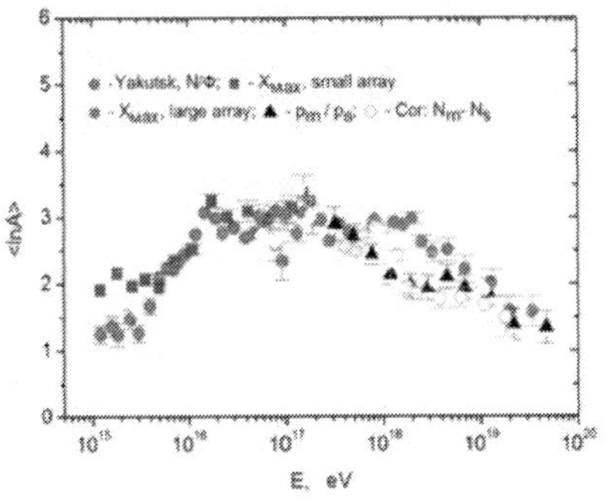

Figure 6. Estimation of MC CR with different method by use of different characteristics of air showers

Figure 7 shows the latest results on MC obtained using the interpolation method (see. Section 2.6) and the model QGSJET 04. Also the figure shows results obtained at compact and the large arrays. From Figure 7 it follows that MC is undergoing a change in the energy range (8 - 20) $\cdot 10^{16}$ eV and (8 - 20) $\cdot 10^{18}$eV. And most likely this is due to the nature of the formation of cosmic rays in the sources and their distribution in the galactic and intergalactic space.

b) At the Yakutsk we measured energy spectrum of CR (see. [14, 15]) and evaluated MC over a wide range in energy [11]. If we compare the energy scale with studied spectrum of CR and obtained results of cosmic rays of MC, than we observe matching of energy intervals, where the change in the shape of the spectrum and the change in the value of A = <lnA> are the

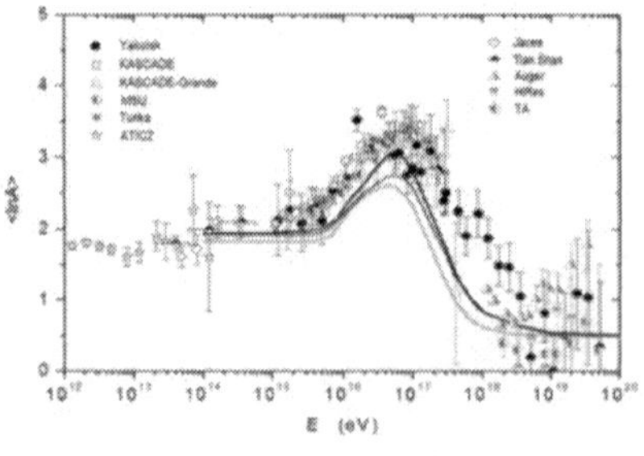

Figure 7. Mass composition of CRs from measurements of different EAS arrays. Lines show MC calculated from paper [16] in the case of near SNR

same. Most likely, these two results are related and caused by the same astrophysical processes.

c) On the boundary of the transition from galactic to metagalactic cosmic rays. Recently developed nonlinear kinetic theory of CR acceleration in supernova remnants has allowed not only to achieve agreement in shape of CR spectrum up to energies $\sim 10^{17}$ with experimental data, but also to choose a class of SNR, which responsible for MC of particles similar to those observed in the satellite, balloon and ground experiments [11]. This is confirmed by the results of the calculation of work [16], which are shown in Figure 7 (lines). Figure 7 shows a comparison of MC obtained at different arrays, with MC generated in the sources, which are remnants of supernovas. There is not only a satisfactory agreement of experimental data with calculation in the energy range 10^{15} - 10^{19} eV, but also indicates that the sharp change in the MC at an energy $\sim 2{\cdot}10^{17}$ eV may be associated with the boundary of the transition from galactic CR to metagalactic CR. In this case mass composition of CRs at energies above $\sim 2{\cdot}10^{17}$ eV should be presented primarily by protons, which is consistent with mass composition obtained at the Yakutsk EAS.

Acknowledgments

The research was supported by "Scientific and Educational Foundation for Young Scientists of Republic of Sakha (Yakutia)"(N 201302010098), Presidium SB RAS (integration project "Modernization of Yakutsk array"), RFBR (grant 13-02-12036 ofi_m)

References

[1] Dyakonov M N, Egorova V P, Ivanov A A, Knurenko S P et al 1987 *Proc. 20th ICRC.* **6** 147
[2] Dyakonov M N, Egorova V P, Ivanov A A, Knurenko S P et al 1987 *JETP Let.* **50** 408
[3] Knurenko S P, Ivanov A A, Kolosov V A et al 2005 *Int. Jour. of Modern Phys.* **20** 6894
[4] Knurenko S P, Ivanov A A, Pravdin M I et al. 2008 *Nucl. Phys. B* 201 408
[5] Knurenko S P, Makarov A K, Pravdin M I, Sabourov A V 2011 *Bull. of the RAoS.* **75** 320
[6] Landau L D 1953 *Izv. AS USSR.* **17**
[7] Kaidalov A V et al. 1986 *Nuclear Physics* **43** 1282
[8] Kaidalov A V et al 1986 *Izv AS USSR* **50** 2087
[9] Atrashkevich V B, Kalmykov M N, Christiansen G B 1981 *JETP Let.* **33** 236
[10] Knurenko S P, Makarov I T, Pravdin M I, Sabourov A V 2010 *Proc. XVI Int. Symp.*
[11] Berezhko E G, Knurenko S P, Ksenofontof L T 2012 *Astrop. Phys.* **36** 31
[12] Knurenko S P, Sabourov A V 2011 *Astrophys. Space Sci. Trans.* **7** 251
[13] Knurenko S P, Ivanov A A, Sabourov A V 2007 *JETP Let.* **86** 709
[14] Ivanov A A, Knurenko S P, Sleptsov I Y 2009 *New J Phys.* **11**
[15] Knurenko S P et al. 2013 *Proc. 33 ICRC*
[16] Kotera K and Lemoine M 2008 *arXiv:0706.1891v2 [astro-Ph]*

Optic detectors calibration for measuring ultra-high energy extensive air showers Cherenkov radiation by 532 nm laser

Stanislav Knurenko, Igor Petrov and Yuri Egorov

Yu. G. Shafer Institute of Cosmophysical Research and Aeronomy SB RAS, Lenin ave. 31, Yakutsk, Russia

E-mail: knurenko@ikfia.sbras.ru, igor.petrov@ikfia.sbras.ru

Abstract. Calibration of a PMT matrix is crucial for the treatment of the data obtained with Cherenkov tracking detector. Furthermore, due to high variability of the aerosol abundance in the atmosphere depending on season, weather etc. A constant monitoring of the atmospheric transparency is required during the measurements. For this purpose, besides traditional methods, a station for laser atmospheric probing is used.

1. Introduction

At the Yakutsk array, atmosphere laser sensing station is deployed. Laser radiation of remote atmosphere sensing attenuates when passing through molecules of gases and aerosols of the atmosphere. Part of radiation scatters off aerosols particles and falls through a narrow gap on tiled PMT. Amplitude of the signal determines by physics and properties of atmosphere to scatter photons of Cherenkov light. In our case, analysis of these data allows to clarify some air showers parameters, which are determined from Cherenkov radiation, for example, estimation of shower energy [1], determined as all ionization losses of electrons and muons in the atmosphere. In addition, LIDAR allows to determined spectral transparency of atmosphere directly [2], with respect to previously used in [3] relative method.

At Yakutsk array, measurement of Cherenkov radiation characteristics carried out by large number of integral and track detectors [4]. Cherenkov detectors operate only during moonless and clear nights. Track detector consists of chamber with narrow straight and long gap under which located perpendicularly to gap direction tiles of PMT (Fig. 1). Each PMT of track detector registers Cherenkov photons throughout dense part of the atmosphere. Such detector can trace the entire longitudinal development of the shower particles and allows us to study the physics of the process. Since these measurements are precise nature, the accuracy of the cascade curve is crucial for this technique, LIDAR data can significantly improve the accuracy of the experiment.

2. Weather condition requirements for Cherenkov Observations

Cherenkov radiation observation conducted mainly in the wintertime (for Yakutia, October - April), when the atmosphere is practically frozen out and the sky partly cloudy or clear [2]. In the absence of man-made and the Moon lights during this period has a minimum background

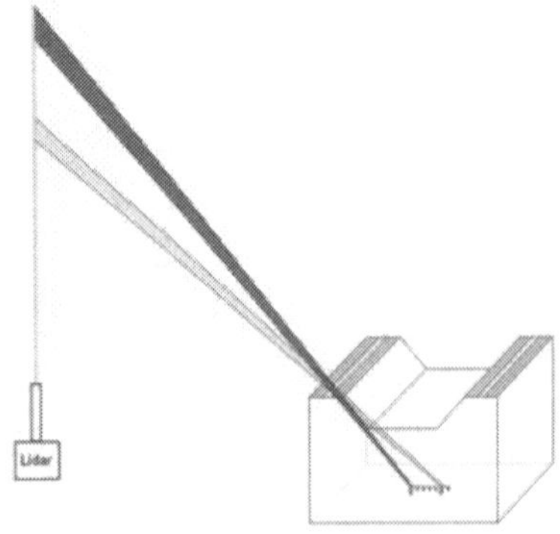

Figure 1. Scheme of calibration of Cherenkov track detector with 532 nm laser

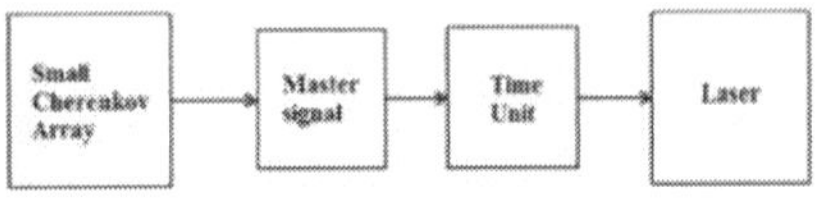

Figure 2. Laser launch control flowchart for calibration of Cherenkov detectors.

light of the night sky in the wavelength range from 300 nm to 800 nm, which is the ideal condition for optical observations. However, as long-term observation implemented various weather conditions, the regular measurements of atmospheric transparency needed. In our case, they do contribute to clarifying estimates of various characteristics of EAS in the calculations by using the correction factors, taking into account the transparency of the atmosphere [6].

Figure 3. LIDAR beam at Yakutsk array

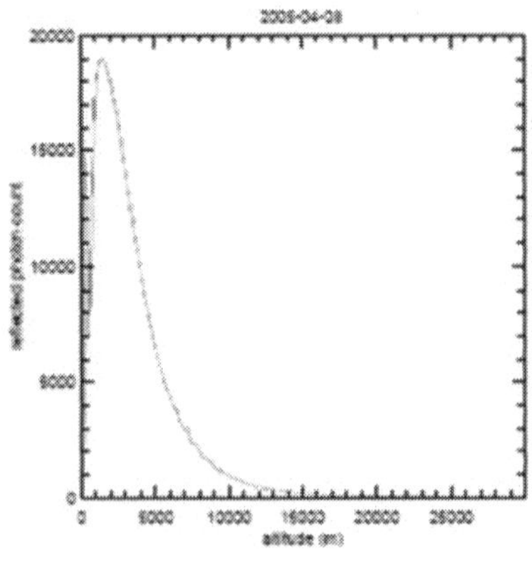

Figure 4. LIDAR measurements of transparency of the atmosphere

Cherenkov light observations requires an information about transmission of the atmosphere at the time of observation. Therefore, an essential part of these observations is to organize regular measurements of the spectral or integrated transparency of the atmosphere. There are many ways of measuring the transparency of the atmosphere [1, 2, 3, 4, 5], but the optimal method is based on measurement of the same Cherenkov radiation [4, 5]. The LIDAR measurements is also suitable as monitor of the state of the atmosphere, especially when used for this purpose tunable to different wavelength. The Yakutsk array uses laser at a wavelength of 532 nm, which is close to the maximum of the spectral characteristics of the PMT-49B, which is used in the Cherenkov detector as a receiver of EAS Cherenkov light.

Profile measurements of the transmission function of the atmosphere conducted in the area of Yakutsk array showed that in periods of observation implemented a number of scenarios of

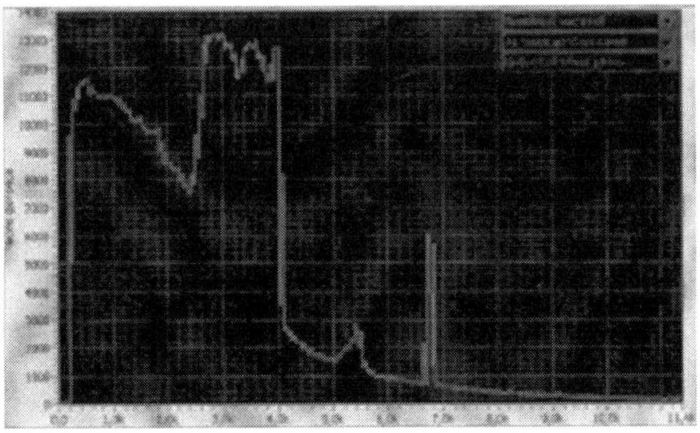

Figure 5. LIDAR measurements of transparency of the atmosphere (low height)

atmospheric conditions, from Rayleigh scattering, when practically we have the clear atmosphere (Fig. 4), to almost full or partial absorption of laser radiation in a thick continuous layer of clouds or to separate layers, as shown in the Figure 5.

3. Calibration algorithm Cherenkov detectors using laser

Since the spectral characteristics of Cherenkov detectors is within wavelength range 300-800 nm, selected monochromatic laser $\lambda = 532$ nm with a sufficiently high output power is able to create a flash of light up to stratospheric heights. Which is more than enough to overlap the maximum luminescence of air shower Cherenkov light (maximum is located at a height of 3 - 5 km above sea level), which allow us to observe photons from both initial stage of shower development and from lower layers of the atmosphere near sea level. The latter allow us to control the sensitivity of the detector near threshold level and use the signals as a test measurement.

The response time of LIDAR is constant and has an exponential shape with the only Rayleigh scattering and may be used for calibration, because area under the curve reflects total number of photons collected from all heights. Comparison of LIDAR response and PMT response at the same solid angles by comparing areas under the curves and allows one to calibrate the track detector.

For differential detectors, which overlook certain height, this fact makes it possible to directly attribute to the response of the PMT, i.e. pulse area known number of photons generated by the laser (see Fig. 1).

Table 1. Laser radiation heights visible by track detector PMT

N PMT	1	2	3	4	5	6	7	8	9	10
Min Height	1134	972	852	756	681	618	567	525	1047	546
Max Height	1455	1197	1020	885	885	702	639	582	1314	609

We restored height of each monitored PMT (Table) by using geometric method and taking into account the aperture of Cherenkov detectors and pulse parameters including pulse area are assigned for these heights. In this case, the pulse reflects the number of photons that come from specified heights and with certain intensity. Changes of intensity and hence the pulse form is associated with only optical signal propagation conditions in the atmosphere and therefore with good weather conditions (only in the case of Rayleigh scattering), this signal can be considered as calibration signal.

4. Method of calibration and measurements testing of EAS Cherenkov light

Figure 2 shows a flowchart of process control Shooting laser, permanently installed in the center of Yakutsk array and shoot vertically to the plane of device. The control was carried out by Cherenkov trigger of Small Cherenkov array after the laser power ups at predetermined wavelength λ=532 nm.

Besides from being used as calibration for Cherenkov detectors, the laser used as a test of the optical part of the Yakutsk array and for measurement of atmosphere transparency. The shooting session occurred after each event of air showers or every 15 minutes. As the station of LIDAR measurements is located at the center of the array (Fig. 6), then we can calibrate Cherenkov detectors without change of the output characteristics of the laser.

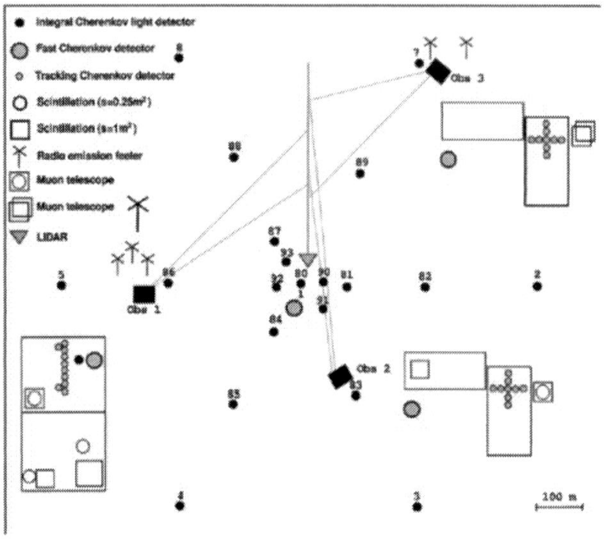

Figure 6. Small Cherenkov array

5. Conclusion

Laser testing in 2013 season showed a sufficiently high stability of the signal reproduction of track detectors up to 20 Hz. Registered pulses had a constant amplitude and a constant shape that is necessary for calibration.

Statistical analysis showed that the accuracy of counting the number of photons for each PMT in the case of vertical sounding and great weather conditions does not go beyond 15 $\pm$ 3%. This is significantly higher than if the Cherenkov receivers calibrated using a reference block of methyl methacrylate [5].

Acknowledgments

The research was supported by "Scientific and Educational Foundation for Young Scientists of Republic of Sakha (Yakutia)"(N 201302010098), Presidium SB RAS (integration project "Modernization of Yakutsk array"), RFBR (grant 13-02-12036 ofi_m)

References

[1] Knurenko S P, Ivanov A A, Sleptsov I Ye and Sabourov A V 2006 *JETP Let.* **83** 563
[2] Knurenko S P, Nikolashkin S V, Saburov A V and Sleptsov I Ye 2006 *Proc. Of SPIE* **6522**
[3] Dyakonov M N, Knurenko S P, Kolosov V A and Sleptsov I Ye 1991 *Optics of atmosphere* **4** 868
[4] Knurenko S P and Sabourov A V 2011 *Astrophys. Space Sci. Trans* **7** 251
[5] Ivanov A A, Knurenko S P and Sleptsov I Ye 2009 *New J. Phys.* 6 065008
[6] Dyakonov M N, Knurenko S P, Kolosov V A and Sleptsov I Ye 1999 *Optic of atmosphere and ocean* **12** 315

Journal of Physics: Conference Series **632** (2015) 012100 doi:10.1088/1742-6596/632/1/012100

Radio signal correlation at 32 MHz with extensive air showers parameters

Stanislav Knurenko and Igor Petrov

Yu. G. Shafer Institute of Cosmophysical Research and Aeronomy SB RAS, Lenin ave. 31, Yakutsk, Russia

E-mail: knurenko@ikfia.sbras.ru, igor.petrov@ikfia.sbras.ru

Abstract. The paper presents correlations of radio signals measured at the Yakutsk array with air shower parameters: the shower energy E_0 and the depth of maximum X_{max}. It is shown that from radio emission measurements of air showers one can obtain individual shower parameters and hence, the mass composition of cosmic rays. In addition, we also derived a generalized formula for calculating the primary energy of the air showers.

1. Introduction

The study of cosmic-ray properties by measuring the radio emission from charged particles in extensive air showers might be an alternative to traditional methods for large arrays [1, 2, 3, 4]. Such arrays consist of hundreds and thousands of scintillation detectors for the registration of charged particles, or of detectors, recording the optical emission from the relativistic particles in the EAS. Such arrays are very costly because of the large amount of detectors and the required complex technical equipment. On the other hand, the radio method is much cheaper and easier to operate with a nearly 100% duty cycle. It is sufficient to have an antenna and a simple radio receiver tuned to a given frequency. One of the main problems is to choose a noise free frequency range [5, 6, 7, 8]. The air shower registration is conducted at a frequency of 32 MHz, free from industrial noise. Yakutsk Radio Array operates since 2009. Data obtained during those several seasons includes showers with energy above 10^{19} eV

2. Theoretical Background

The radio emission mechanisms are meanwhile well known and calculations [9] show that the radio emission of EAS depends on the development of the electromagnetic cascade in the atmosphere and the magnetic field near the level of observation. That means, the radio emission is directly associated with the physics of the air shower development: production of electrons and positrons in the atmosphere (longitudinal development of air showers) and the total number of charged particles (EAS energy).

The electric field induced on the antennas of the radio arrays, according to [10] , can be represented as follows:

$$\varepsilon_\nu = \sqrt{\frac{4\pi\nu^2\mu_0}{G_{(\theta,\phi,\nu)}C} \cdot \frac{1}{A_{ele(\nu)}R_{ADC}} \frac{V_{ADC}}{\Delta\nu}} \tag{1}$$

Here ν is the observing frequency, $G_{(\theta,\phi,\nu)}$ is the gain of the antenna, $A_{ele(\nu)}$ is the total gain of the electronics, V_{ADC} is the voltage measured by the ADC, and R_{ADC} is the input impedance of the ADC.

It was also shown that the amplitude A_{max} of the radio pulse is proportional to the energy of the electromagnetic component of the EAS with $E_{em} = \mathrm{c} \cdot \mathrm{A}_{max}$ (μV^{-1} m GeV), where c - is a shower geometry dependent parameter [9].

Combining our radio measurements and calculations of [9], we can determine the shower energy, regardless of the registration method, and then compare the two energy estimations, namely that by Cherenkov detectors data [11] and that by radio emission data.

In [9] calculations are performed, which show that the slope of the lateral distribution function (LDF) of the radio emission depends on the depth of the shower maximum X_{max}. This slope can be determined as the ratio of the measured amplitudes of the radio pulse at the flat and the steep part of the LDF. In our case the distances of 175 m and 725 m from the shower axis were used. Analytically, this relationship can be expressed by the formula [9]:

$$X_{max} = a\left[ln\left(b\frac{A_{175m}}{A_{725m}}\right)\right]^c \quad (2)$$

Here a, b, c, are constants coefficients, $a = 856.1$, $b = 0.3149$, $c = 0.4340$. They are adapted to the latitude of the Yakutsk and coincide with calculation of [9] within experimental errors.

3. Short Description of Yakutsk Radio Array

First experiments at the Yakutsk array to register radio pulses from air showers have been performed in 1987-1989.

Radio signals from 6250 showers with energies above 10^{17} eV were registered, including some events with $E_0 \geq 10^{19}$ eV. In 2009, 6 antennas (half-wave dipole) for radio emission registration were installed at the Yakutsk array. The antennas are located at 300, 350 and 500 m distance from the array center. For an optimal frequency choice, the background frequency spectrum from 1 to 200 MHz was analyzed [6]. At frequencies up to 20 MHz due to the presence of large natural radio noise (primarily storm origin), it is not possible to distinguish air showers pulses with sufficient efficiency. Therefore, it is reasonable to select frequencies above 20 MHz. Over this frequency range, the amplitude of galactic noises decreases much slower with the frequency than storm noises. At 32 MHz it is 1.0 - 2.0 μV$\cdot$m$^{-1}\cdot$MHz^{-1}. Thermal noise of the antenna is much smaller than the galactic noise at frequencies up to 100 MHz and has almost no influence on our measurements. Therefore, the most favorable frequency range for the measurements at the Yakutsk array is 30 - 40 MHz. We installed an array tuned to a frequency of about 32 MHz with a bandwidth of 8 MHz. It consists of antennas, amplifiers and registration device with data storage [8]. We used a half-wave dipole installed at $\lambda/4$ above ground (Fig. 1). Crossed dipoles are oriented along East-West (on the magnetic latitude), and North-South (along the magnetic meridian). The bandwidth is $\pm$4 MHz, sensitivity $\sim$10μV (2 μV$\cdot$m^{-1} $\cdot$ MHz^{-1}), the dynamic range is 50 dB.

Receiving channels are based on the principle of direct signal amplification and subsequent detection. Antenna amplifiers are placed into special containers and are located directly at the antenna. The main paths are based on the cascade amplification circuit with mismatched contours.

The array currently consists two set of antennas, the distance between them is 500 m. The first set consists of 8, the second of 4 antennas. Each set is connected to an industrial computer. Directly under the antennas are electronic devices and matching amplifiers and grounded shielding grid. Testing and calibration of measurement including the antennas is done once in a measurement season (typically the period from September to May) using a broadband high-frequency generator and a control PC. Registration of radio emission is triggered by one of

Figure 1. Design of the receiving antenna

two event triggers (masters) from the Yakutsk array. The location of the antenna sets is shown in Fig. 2.

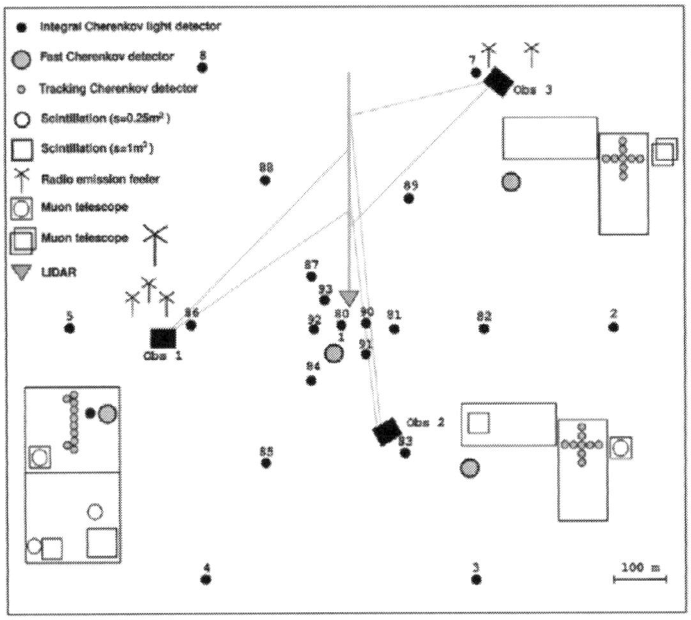

Figure 2. Arrangement of observation stations on the Small Cherenkov array

The first of the two possible triggers is the main Yakutsk array trigger, which registers in an area of 12 km^2 showers with energy more than 10^{17} eV. The Small Cherenkov array registers in an area of 1 km^2 showers with energy 10^{15} - $5{\cdot}10^{17}$ eV. The data collection system is based on an industrial computer and is able to recording signal from multiple antennas simultaneously. We use a fast 8-bit ADC, which records the prehistory (before the trigger arrival) for 25 μs and history (after the arrival of the trigger) for 15 μs (Fig. 3).

Fig. 3 shows the ADC values of the radio signals from 4 antennas, the X axis spans 40 μs. Vertical line marks the arrival time of the trigger signal.

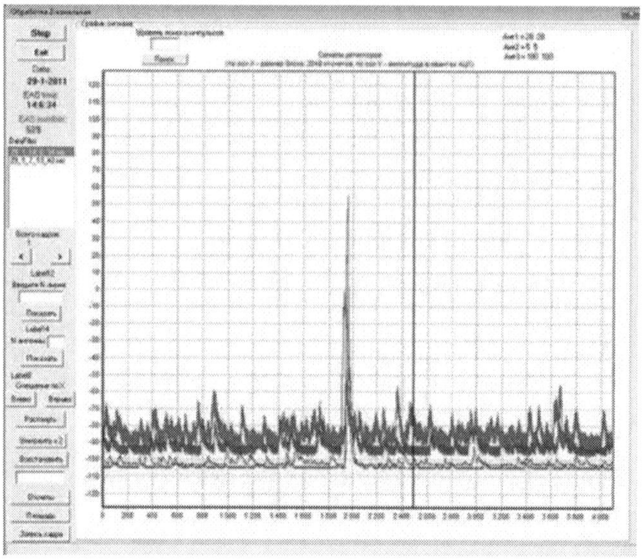

Figure 3. Radio emission pulses from 4 antennas.

4. Results

The Radio Array [8] benefits from the co-location with a particle detector experiment, which provides values for the fundamental parameters defining an air shower: primary energy, depth of shower maximum, shower core and arrival direction [12, 13]. The goal of this work is to find a correlation: a) between signal amplitude and air shower energy (the energy is estimated from the Cherenkov light flux [11, 14]); b) between LDF and longitudinal development (depth of shower maximum as determined by the Cherenkov data [15, 16]). If such a correlation is experimentally confirmed, then using theory and calculations [9] it is possible to determine energy and depth of the maximum by radio emission data and to use these parameters to study high energy cosmic rays. In this section we present the recent results obtained at Yakutsk.

As seen in Fig. 4, the average slope of the LDF varies with the distance. At small distances (50-200 m) the slope is flat, at larger distance (~300 m) it becomes steeper [17]. For example in the showers with energy $\sim 10^{17}$, slope at a distance 100 meters is $\sim 0.71 \pm 0.1$, at the distance of 100-700 m $\sim 0.97 \pm 0.1$.

As it is shown in Fig. 4 the average slope of the LDF varies with distance. At large distances from the shower axis, the radio signal attenuates slowly.

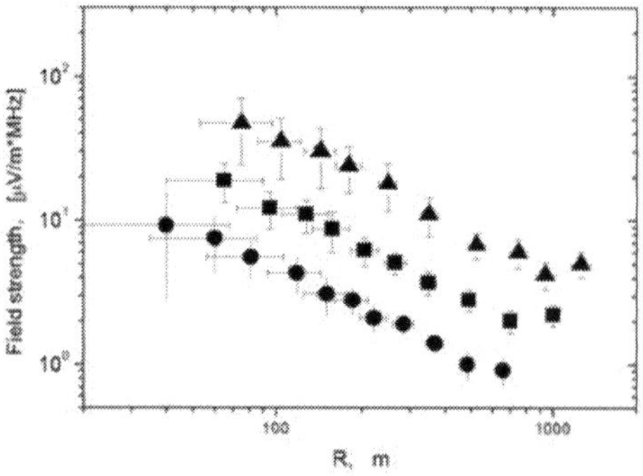

Figure 4. Average LDF of radio emission at 32 MHz in showers with energies $1.73 \cdot 10^{17}$ eV, $4.38 \cdot 10^{17}$ eV and $1.32 \cdot 10^{18}$ eV.

Using equations (1) and (2) and the measured EAS radio signals we tried to find correlations between a) the amplitude of the radio signal and the shower energy E_{em} (derived from the

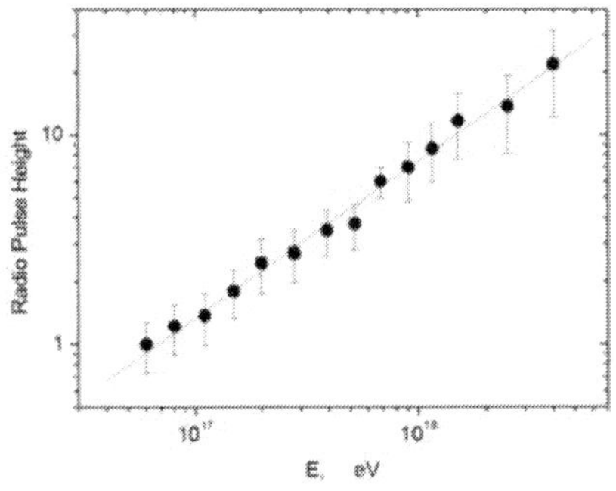

Figure 5. Dependence of the amplitude of the radio emission pulse from the shower energy

Cherenkov light signal) and b) between the ratio of the radio signals at 175 m and 725 m and the depth of the shower maximum X_{max} [15, 16].

For this analysis, we selected 421 events for which simultaneously have been measured charged particles, muons and Cherenkov light. In Fig. 5 the correlation between radio signal amplitude and shower energy is shown. Data can be fitted by formula (3).

$$\varepsilon_{EW} = (1.3 \pm 0.3) \left(\frac{E_0}{10^{17}}\right)^{(0.99 \pm 0.04)} \tag{3}$$

Here ε_{EW} denotes the estimated radio amplitude and E_0 the energy of the primary particle. The generalized formula (4) allows to determine the energy of individual shower coming with zenith angle θ, the amplitude of the radio signal registered at a distance of 350 m from the shower axis:

$$\varepsilon = (15 \pm 1)(1 - \cos\theta)^{1.16 \pm 0.05} \cdot \exp\left(-\frac{R}{350 \pm 25.41}\right) \cdot \left(\frac{E_p}{10^{17}}\right)^{0.99 \pm 0.04} \tag{4}$$

where, θ - zenith angle, R - the distance to the antenna, E_p - energy of the primary particle. As seen from calculations [9], the slope of the LDF is correlated with the longitudinal development of the EAS. Using the amplitude of the radio signal at different distances from the shower axis, with formula (2) we can determine X_{max} of the shower. We can estimate X_{max} using the average LDF data (Fig. 4) with correcting for the influence of noise as it proposed in [18]. The estimation of X_{max} from radio data is compared in Table 1 with the X_{max} obtained from Cherenkov data [16].

Table 1. Depth of maximum by radio data and Cherenkov light data

E_0, eV	A_{175m}	A_{725m}	A_{175m}/A_{725m}	X^{rad}_{max},g/cm^2	X^{Cher}_{max},g/cm^2
$1.7\cdot10^{17}$	2.71	0.54	5.02	610±25	644±18
$4.3\cdot10^{17}$	8.52	1.67	5.11	621±25	675±15
$1.3\cdot10^{18}$	29.71	4.95	6.01	704±25	710±15

One sees that the values agree within experimental uncertainty. This confirms the possibility of using radio emission for the study of the longitudinal development of air showers. In future we plan to estimate the mass of the primary cosmic rays with larger statistics like in [19].

The calculations in [19] show that the X_{max} distribution at fixed energy represents the contribution of different groups of nuclei and can be used to analyze the mass composition of cosmic rays.

5. Conclusion

The radio data of the Yakutsk array show that: a) the measured amplitude of the radio signal correlates with the shower energy; b) the depth of the shower maximum can be determined from radio measurements if a reliable adjustment of the noise contribution in the radio signal [18] can be performed.

Acknowledgments

The research was supported by "Scientific and Educational Foundation for Young Scientists of Republic of Sakha (Yakutia)"(N 201302010098), Presidium SB RAS (integration project "Modernization of Yakutsk array"), RFBR (grant 13-02-12036 ofi_m)

References

[1] Schröder F for the PA col. Radio detection of air showers with the Auger Engineering Radio Array, *ICRC 2013*, Brazil. http://www.cbpf.br/ icrc2013/papers/icrc2013-0899.pdf.
[2] Hiller R et al. 2013 *Proc. 33rd ICRC (Rio de Janeiro)* arXiv:1308.0917
[3] Huege T, Apel W D and Arteaga J C et al. 2010 *Nucl. Instr. and Meth. in Phys. Res.* **A 662** 72-79
[4] Apel W D et al. 2010 *Astropart. Phys.* **32** 294.
[5] Jasik H 1961 Antenna Engineering *Mcgraw-Hill.*
[6] Kozlov V I, Knurenko S P, Petrov Z E and Pravdin M I 2011 *Proc. XVII Intern. Symp. on Atmosp. and Ocean Opt.: Atmosp. Phys.* D1.
[7] Knurenko S P et al. 2010 *Proc. of 22 ECRS* 252.
[8] Petrov Z E, Borschevsky D S and Knurenko S P et al. 2011 *NEFU Bulletin* p 5.
[9] Huege T, Ulrich R and Engel R 2008 *Astropart. Phys.* **30** 96
[10] Horneffer A et al. 2008 *Proc. of the 30th ICRC (Merida)* Vol 4 p 83.
[11] Knurenko S P et al. 2006 *JETP Lett.* **83** 11 563.
[12] Knurenko S P, Kolosov V A and Petrov Z E et al. 1998 *Sci. and Educ.* **4** 46.
[13] Artamonov V P, Glushkov A V and Knurenko S P et al. 1994 *Bulletin of the Rus. Acad. of Sci.* **58** 12 92.
[14] Ivanov A A, Knurenko S P and Sleptsov I Ye. 2007 *JETP Lett.* **131** 6 1001.
[15] Knurenko S P, Kolosov V A and Petrov Z E et al. 2001 *Proc. 27 ICRC (Hamburg)*, **1** 157.
[16] Knurenko S P and Sabourov A V. 2011 *Astrophys. Space Sci. Trans.* **7** 251 doi:10.5194/astra-7-251-2011
[17] Knurenko S P et al. 2013 *Bulletin of the Rus. Acad. of Sci.* **77** 11 1559.
[18] Schroder F et al. 2012 *Nucl. Instr. and Meth.* **A 662** S238 doi:10.1016/j.nima.2010.11.009.
[19] Kalmykov N N, Konstantinov A A and Vedeneev O V. 2012 *J. Phys.: Conf. Series* **1** 409.

On the determination of source spectrum of ultra high energy cosmic rays

V.Ptuskin, S.Rogovaya and V.Zirakashvili
IZMIRAN, Troitsk, Moscow 142190 Russia
E-mail: vptuskin@izmiran.ru

Abstract. The calculations of source spectra of extragalactic cosmic rays with ultra high energies are made based on the solution of inverse problem for a system of transport equations that describes the propagation of protons and nuclei with energies 10^{18} to 10^{21} eV in the expanding Universe filled with the background radiation. It is assumed that protons and Iron nuclei are the most abundant species in the sources. The data from the Auger experiment is used to illustrate the method.

1. Introduction

The origin of cosmic rays with energies $E > 10^{18}$ eV observed at the Earth is associated with the most active extragalactic objects - active galactic nuclei, gamma-ray burst progenitors, newborn rapidly rotating pulsars, interactive galaxies etc., see review [1]. The present knowledge about the highest energy cosmic rays was mainly acquired from the High Resolution Fly's Eye Experiment (HiRes), Pierre Auger Observatory (Auger), Telescope Array experiment (TA), and from the Yakutsk EAS array. The transport of cosmic ray nuclei in the intergalactic medium is accompanied by the transformation of energy spectrum and elemental composition due to interactions with background electromagnetic radiation that fills the expanding Universe.

The solution of direct problem when the calculations follow cosmic ray propagation from the source to the observer is commonly used for the determination of source spectrum and composition, e. g. [2]. A power law source spectrum with some maximum energy is usually assumed. The source characteristics are determined by the trial-and-error method when the calculations can be repeated for the refined source spectrum after comparison with observations. In the present work, we inverse the procedure and determine the source properties starting from the spectrum observed at the Earth without ad hoc assumptions about the shape of source spectrum. A simple case of the source composition that includes only protons and Iron nuclei is considered and the analytical approximation of the data from the Auger experiment is used.

2. Calculations by the inverse problem method

We use the following transport equation for cosmic ray protons and nuclei interacting with the background electromagnetic radiation in the expanding Universe (see [3] for detail):

$$-H(z)(1+z)\frac{\partial}{\partial z}\left(\frac{F(A,\varepsilon,z)}{(1+z)^3}\right) -$$

$$-\frac{\partial}{\partial\varepsilon}\left(\varepsilon\left(\frac{H(z)}{(1+z)^3}+\frac{1}{\tau(A,\varepsilon,z)}\right)F(A,\varepsilon,z)\right)+\nu(A,\varepsilon,z)F(A,\varepsilon,z)$$
$$=\sum_{i=1,2...}\nu(A+i\to A,\varepsilon,z)F(A+i,\varepsilon,z)+q(A,\varepsilon)(1+z)^m. \quad (1)$$

The system of eqs.(1) for all kinds of nuclei with different mass numbers A from Iron to Hydrogen should be solved simultaneously. The energy per nucleon $\varepsilon = E/A$ is used here because it is approximately conserved in a process of nuclear photodisintegration, $F(A,\varepsilon,z)$ is the corresponding cosmic-ray distribution function, z is the redshift, $q(A,\varepsilon)$ is the density of cosmic-ray sources at the present epoch $z = 0$, m characterizes the source evolution (the evolution is absent for $m = 0$), $\tau(A,\varepsilon,z)$ is the characteristic time of energy loss by the production of e^-e^+ pairs and pions, $\nu(A,\varepsilon,z)$ is the frequency of nuclear photodisintegration, the sum in the right side of eq. (1) describes the contribution of secondary nuclei produced by the photodisintegration of heavier nuclei, $H(z) = H_0((1+z)^3\Omega_m + \Omega_\Lambda)^{1/2}$ is the Hubble parameter in a flat universe with the matter density $\Omega_m(= 0.3)$ and the Λ-term $\Omega_\Lambda(= 0.7)$.

The solution of eqs. (1) at the Earth can be presented as

$$F(A,\varepsilon,z=0)=\sum_{A'}\int d\varepsilon' G(A,\varepsilon;A',\varepsilon')q(A',\varepsilon'), \quad (2)$$

where the Green function G is introduced.

The spectrum on the total energy E at the Earth is given by the equation

$$N(A,E,z=0)=A^{-1}\sum_{A'}\int d\varepsilon' G(A,E/A;A',\varepsilon')q(A',\varepsilon'), \quad (3)$$

In a process of numerical solution of the inverse problem, eqs. (2,3) are presented in a matrix form and inverted to find the source functions $q(A,\varepsilon)$ for different types of nuclei. This procedure is described at length in [4]. The numerical solution of cosmic-ray transport equations follows the finite differences method. The variables are z and $\log(E/A)$.

The all-particle energy spectrum of cosmic rays $J(E)$ obtained from Auger observations [5] is used in our calculations. The following analytical approximation of the spectrum is accepted (see also Figure 1):

$$J(E)\propto E^{-3.23}, E < 5\times10^{18}\text{eV};$$
$$J(E)\propto E^{-2.63}\times[1+exp(log(E/10^{19.63}\text{eV})/0.15)]^{-1}\times$$
$$exp(-(E/(1.5\times10^{20}eV))^4), E > 5\times10^{18}\text{eV}. \quad (4)$$

This formula is similar to the equation suggested by the Auger team but contains $exp(-(E/1.5\times 10^{20}eV)^4)$ factor of cosmic ray flux suppression at energies $\geq 1.5\times10^{20}$ eV.

Using the Auger data on energy dependence of the mean logarithm of the atomic mass number $\langle lnA\rangle$ calculated in the EPOS-LHC model of particle interactions in the atmosphere [5], we accept the following approximation

$$\langle lnA\rangle = 0.5 + 4.2\times(E/10^{20}\text{eV})^{0.6}. \quad (5)$$

The calculated source spectra of extragalactic sources that reproduce the all-particle cosmic ray spectrum observed at the Earth (4) are illustrated in Figures 2, 4.

Figure 2 shows the spectra calculated on the assumption that accelerated protons and Iron nuclei have similar spectra on magnetic rigidity E/Z. Different curves correspond to the different

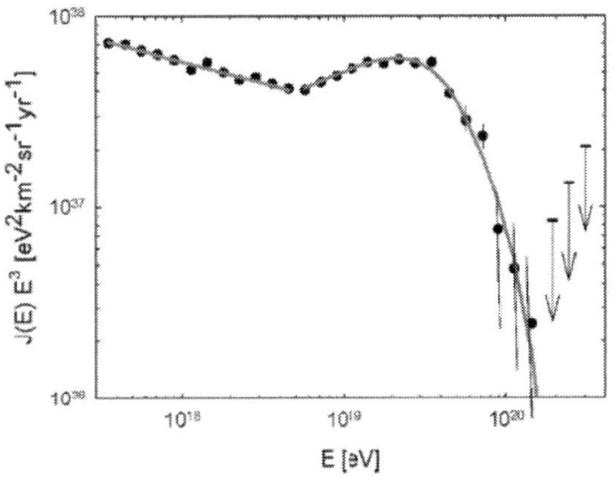

Figure 1. Analytical approximation (4) used in the present calculations to describe the Auger data are shown by solid line together with Auger data [5].

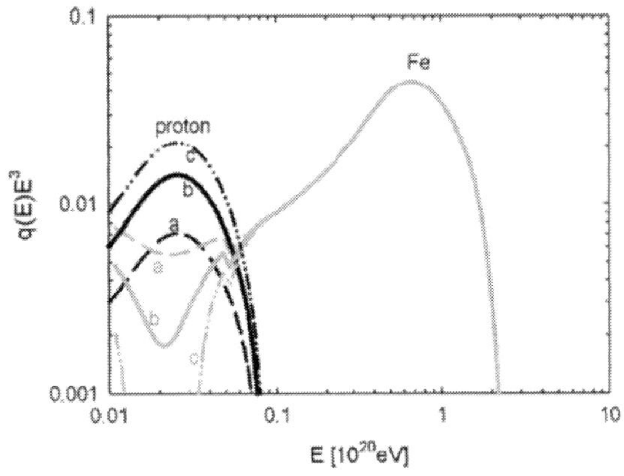

Figure 2. Calculated source spectra in arbitrary units based on the Auger data for different Iron-to-proton source ratios labelled (a) for $S_{Fe}/S_p = 2 \times 10^{-2}$, (b) for $S_{Fe}/S_p = 10^{-2}$, and (c) for $S_{Fe}/S_p = 6.7 \times 10^{-3}$. Black lines for proton source; gray lines for Iron source. Proton and Iron source spectra are assumed to have the same dependence on magnetic rigidity.

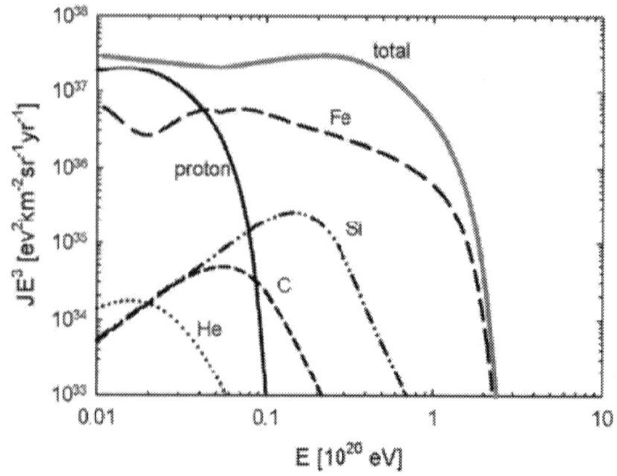

Figure 3. Calculated spectra of different types of nuclei for Iron-to-proton source ratio $S_{Fe}/S_p = 10^{-2}$ and the total observed Auger cosmic ray spectrum at the Earth.

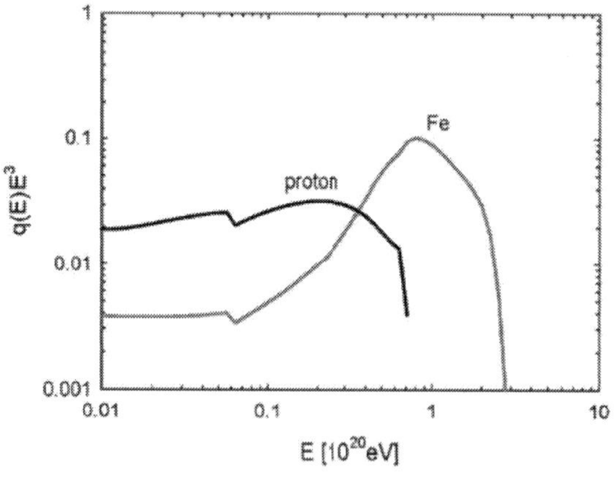

Figure 4. Calculated source spectra of protons and Iron based on Auger data on cosmic ray spectrum and $\langle lnA \rangle$ function.

assumptions on the proton and Iron abundance at the source. The calculated maximum energy which particles reach in a process of acceleration is approximately equal to $4 \times 10^{18} Z$ eV. The obtained shape of the source spectrum differs from the power-law. The calculated elemental composition of cosmic rays at the Earth for the case $S_{Fe}/S_p = 10^{-2}$ is presented in Figure 3.

Figure 4 present the proton and Iron source spectra calculated without assumption that they are similar functions of magnetic rigidity. Instead, the observed dependence of $\langle lnA \rangle$ on energy (5) is imposed. It is evident that the source energy spectra have different dependence on rigidity in this case and the Iron spectrum is not of the power-law form.

The results of our research and conclusions are presented in more detail in [4].

Acknowledgments

The work was supported by the Russian Foundation for Basic Research grant 13-02-00056 and by the Russian Federation Ministry of Science and Education contract 14.518.11.7046. Part of this work was fulfilled during VSP visit to the University of Maryland, USA, where it was supported by the NASA grant NNX13AC46G.

References

[1] Kotera K and Olinto A V 2011 *The astrophysics of ultrahigh energy cosmic rays*, *Annu. Rev. Astron. Astrophys.* **49** 119
[2] Allard D 2012 *Extragalactic propagation of ultrahigh energy cosmic rays*, *Astropart. Phys.* **39** 33
[3] Ptuskin V S, Rogovaya S I and Zirakashvili V N 2013 *On ultra-high energy cosmic rays: origin in AGN jets and transport in expanding Universe*, *Adv. Space Res.* **51** 315
[4] Ptuskin V S, Rogovaya S I and Zirakashvili V N 2014 *Inverse problem for extragalactic transport of ultra-high energy cosmic rays*, *eprint arXiv*: 1409.4654
[5] Letessier-Selvon A for the Pierre Auger Collaboration 2013 *Highlights from the Pierre Auger Observatory 33rd International Cosmic Ray Conference, Rio de Janeiro, 2013*, in press *eprint arXiv*:1310.2118v1

24th European Cosmic Ray Symposium (ECRS2014) IOP Publishing
Journal of Physics: Conference Series **632** (2015) 012102 doi:10.1088/1742-6596/632/1/012102

LOPES – Recent Results and Open Questions on the Radio Detection of Air Showers

F G Schröder[1], W D Apel[1], J C Arteaga-Velázquez[2], L Bähren[3], K Bekk[1], M Bertaina[4], P L Biermann[5,1], J Blümer[1,6], H Bozdog[1], I M Brancus[7], E Cantoni[4,8], A Chiavassa[4], K Daumiller[1], V de Souza[9], F Di Pierro[4], P Doll[1], R Engel[1], H Falcke[10,3,5], B Fuchs[6], H Gemmeke[11], C Grupen[12], A Haungs[1], D Heck[1], J R Hörandel[10], A Horneffer[5], D Huber[6], T Huege[1], P G Isar[13], K-H Kampert[14], D Kang[6], O Krömer[11], J Kuijpers[10], K Link[6], P Łuczak[15], M Ludwig[6], H J Mathes[1], M Melissas[6], C Morello[8], J Oehlschläger[1], N Palmieri[6], T Pierog[1], J Rautenberg[14], H Rebel[1], M Roth[1], C Rühle[11], A Saftoiu[7], H Schieler[1], A Schmidt[11], S Schoo[1], O Sima[16], G Toma[7], G C Trinchero[8], A Weindl[1], J Wochele[1], J Zabierowski[15] and J A Zensus[5]

[1] Institut für Kernphysik, Karlsruher Institut für Technologie (KIT), Germany
[2] Instituto de Física y Matemáticas, Universidad Michoacana, Morelia, Mexico
[3] ASTRON, Dwingeloo, The Netherlands
[4] Dipartimento di Fisica, Università degli Studi di Torino, Torino, Italy
[5] Max-Planck-Institut für Radioastronomie, Bonn, Germany
[6] Institut für Experimentelle Kernphysik, Karlsruher Institut für Technologie (KIT), Germany
[7] National Institute of Physics and Nuclear Engineering, Bucharest-Magurele, Romania
[8] Osservatorio Astrofisico di Torino, INAF Torino, Italy
[9] Universidade São Paulo, Instituto de Física de São Carlos, São Carlos, Brasil
[10] Department of Astrophysics, Radboud University Nijmegen, The Netherlands
[11] Institut für Prozessdatenverarbeitung und Elektronik, KIT, Germany
[12] Faculty of Natural Sciences and Engineering, Universität Siegen, Germany
[13] Institute for Space Sciences, Bucharest-Magurele, Romania
[14] Fachbereich C, Physik, Universität Wuppertal, Germany
[15] Department of Astrophysics, National Centre for Nuclear Research, Łódź, Poland
[16] Department of Physics, University of Bucharest, Bucharest, Romania

E-mail: `frank.schroeder@kit.edu`

Abstract. LOPES was a digital antenna array operating for approximately 10 years until spring 2013 at the Karlsruhe Institute of Technology (KIT). Triggered by the co-located KASCADE-Grande air-shower experiment, it measured the radio signal of around 1000 cosmic-ray air showers with energies $E \gtrsim 10^{17}\,$eV in an effective band of 43 – 74 MHz. Using the interferometric technique of cross-correlation beamforming, LOPES could reconstruct the shower direction with an accuracy $< 0.7°$, the shower energy with a precision $< 20\,\%$, and the atmospheric depth of the shower maximum, X_{max}, with a precision $< 95\,\mathrm{g/cm^2}$. In particular the reconstruction of the shower maximum suffers from significant measurement uncertainties due to the radio-loud environment of the site. This article summarizes our latest results on the reconstruction of the shower maximum, using two independent methods: the steepness of the hyperbolic radio wavefront and the slope of the lateral distribution of the radio amplitude. Moreover, we show vectorial measurements of the electric field with the tripole antennas of the latest LOPES setup. Finally, we discuss open questions as well as the potential impact of the lessons learned at LOPES for future antenna arrays.

1. Introduction

LOPES, the LOFAR prototype station, was a pathfinder experiment for digital detection of the radio emission by cosmic-ray air showers at MHz frequencies. Operating from 2003 until 2013 in different configurations, LOPES provided proof-of-principle for the capability to detect air showers with digital antenna arrays [1], and developed several technical methods, e.g., for amplitude and time calibration [2],[3]. Moreover, it demonstrated the feasibility for the offline reconstruction of air-shower properties from the digital radio signal, in particular, the direction [4],[5], the energy [6],[7], and the position of the shower maximum [8],[6],[5] which provides a statistical estimator for the mass composition of the primary particles. Finally, LOPES contributed to the understanding of the physics relevant for the radio emission by comparing the measured electric field with model predictions [9],[10], and by confronting Monte Carlo simulations with detailed measurements of the lateral distribution and the wavefront [11],[5].

All these results can be found in the various publications, and rather complete descriptions of the current understanding of the radio emission are available as well, see e.g., Ref. [12] for a recent summary. For this reason, we summarize just the latest results of LOPES, comparing for the first time our results on the shower maximum obtained with two different methods – namely the slope of the radio lateral distribution, and the steepness of the hyperbolic radio wavefront. Furthermore, the last section is devoted to open issues, which should be answered by future measurements and corresponding analysis methods with the new generation of radio arrays.

2. Experimental Setup and Measurements

LOPES was built as the radio extension of KASCADE-Grande [13] at the Karlsruhe Institute of Technology (KIT), Germany, where most of its antennas were located inside of the original KASCADE scintillator array [14]. As such LOPES was externally triggered by KASCADE-Grande which turned out to be a successful approach: by this all LOPES events are automatically hybrid measurements with the particle detector array, and we were able to study correlations with the air-shower properties measured by KASCADE-Grande. We also installed a few self-triggered antennas [15], but distinguishing air-shower pulses from background was close to impossible in the noisy environment of a research center. Even at areas which are more radio quiet, it is still questionable whether the additional effort required for self-triggering is justified, since hybrid measurements might be required anyway for most science goals, e.g., for determining the cosmic-ray composition with high accuracy.

In different periods LOPES consisted of 10 to 30 antennas placed in an area smaller than $200 \times 200\,\mathrm{m}^2$. Thus, LOPES can be considered a relatively dense array, since the distance between antennas is significantly smaller than both the radius of the Cherenkov ring and the typical decay distance of the lateral distribution, which both are in between 100 and 200 m. Consequently, we make two reasonable assumptions for the analysis of our measurements: First, if there is a significant radio signal detected by LOPES, then all antennas should have measured it, though, it could be lower than the noise level at some of the antennas. Second, the signal shape is approximately the same in all antennas, i.e., we neglect the known dependence of the pulse shape on the distance to the shower axis [16],[17].

Under these assumptions cross-correlation beamforming is a very powerful technique: Using the air-shower core and direction reconstructed by KASCADE-Grande as starting values, we digitally shift all recorded radio signals in time according to the calculated arrival time of the radio wavefront at each antenna. Then, we determine the cross-correlation and maximize it by slightly varying the direction and the shape of the wavefront using a simplex fit. In our latest analyses we assumed a hyperbolic shape for the wavefront and vary in particular the angle of the asymptotic cone of the hyperbola. By this interferometric technique we can reconstruct many events which have no significant signal in individual antennas, but still show a sufficiently high signal-to-noise ratio in the cross-correlation beam.

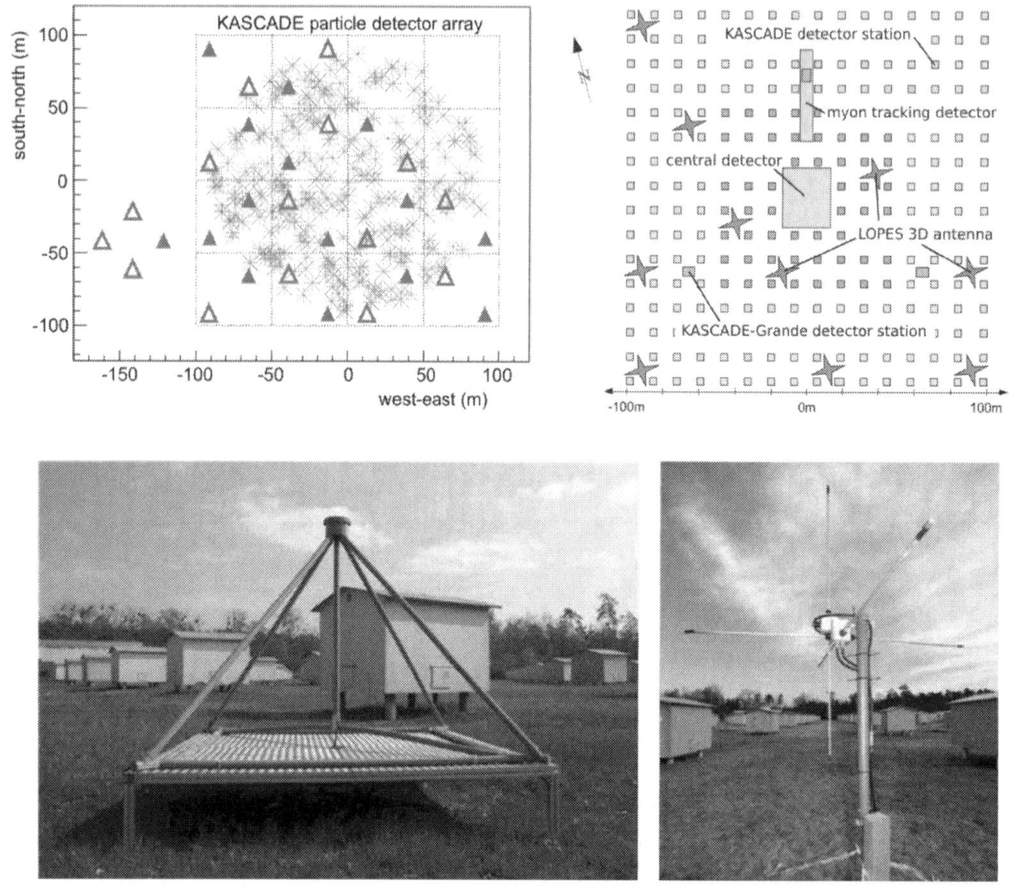

Figure 1. Left (top and bottom): LOPES setup from 2005-2009 and a photo of the inverted-V-shape dipole antenna used at this time. Filled triangles indicate the positions of 15 east-west aligned antennas present in the full data set; open triangles indicate 15 antennas present in about one third of the used data set. The shower cores of the events selected for a typical analysis are marked as stars (reconstructed from KASCADE data only). Right (top and bottom): LOPES-3D setup in operation from 2010 until spring 2013 consisting of 10 tripole antennas (photo) [18]. Apart from the antennas and preamplifiers, the analog and digital electronics was the same in both setups, providing an effective bandwidth of 43 – 74 MHz.

Using this technique, we detected more than 1000 events during the whole live time of LOPES. For our recent analyses, though, we used only approximately 300 events recorded with 15 – 30 east-west aligned antennas from 2005-2009, with energies $E > 10^{17}$ eV, zenith angles $\theta < 40°$, and shower core inside of the fiducial area of the KASCADE array. In earlier years the calibration was not yet finished. In later years we tested different types of antenna for the reconstruction of the electric field vectors [18]. These tripole antennas forming the LOPES-3D setup could have been beneficial in particular for inclined showers, but unfortunately we discovered that the background at the research center increases dramatically towards the horizon. Thus, after reasonable quality cuts, LOPES-3D would have increased the total LOPES statistics only marginally, and we decided to use only the former dipole antennas for our standard analyses.

Figure 1 shows photos of the two antenna types and maps of the different setups. The shower cores of a typical data selection are also plotted.

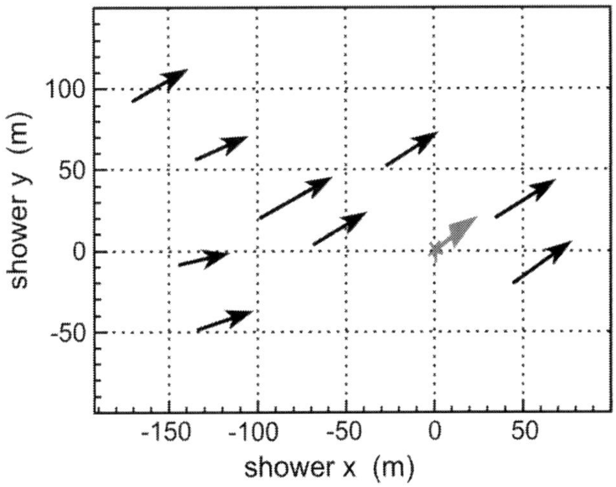

Figure 2. LOPES-3D event featuring a signal-to-noise ratio sufficient for the reconstruction of the electric field vector in the individual antennas. Each black arrow indicates the reconstructed direction and amplitude of the electric field vector at the position of one antenna, and the red arrow denotes the direction predicted for geomagnetic radio emission [10].

3. Results

Our most recent results are on the reconstruction of the shower maximum, and on the measurement of the electric field vector with the tripole antennas of the LOPES-3D setup. Despite the high background in the vertical polarization, LOPES-3D could confirm that a vertically aligned antenna increases the sensitivity for inclined showers [19],[10]. This could be important for future large-scale arrays, since inclined showers have very large footprints and make sparse and economic arrays feasible [20].

Moreover, we could confirm earlier results [9] that the measured polarization or, in this case, the reconstructed direction of the electric field vector, is compatible with the direction of the geomagnetically induced transverse currents in the air shower (see figure 2 for an example). This means that the vectorial measurements of LOPES-3D provide another evidence that the radio emission from air-showers is dominated by the geomagnetic effect. However, the sensitivity of LOPES-3D is insufficient to distinguish the contribution of the sub-dominant Askaryan effect. Nevertheless, indications for the Askaryan effect have been found at other experiments in regions with lower background [21],[22],[23].

While the high background is a problem for some analyses, for others it is not so relevant. Thanks to the interferometric beamforming, LOPES achieves a competitive accuracy for the air shower direction of better than 0.7°. Also the energy precision is competitive to other air shower techniques, namely at least as good as the precision of the KASCADE particle-detector array, i.e., better than 20 %. Since the radio technique provides a calorimetric measurement of the shower energy, both energy and arrival direction are first-order effects, and thus relatively easy to reconstruct: The arrival direction determines in first-order the arrival time of the radio signal at each antenna, and the shower energy determines in first-order the amplitude of the radio signal.

In contrast, the position of the shower maximum is a second-order effect on these observables: the distance from the radio array to the shower maximum determines by how much the arrival times deviate from a plane wavefront, and how fast the amplitude of the radio signal changes with distance to the shower axis. In other words: The closer the shower maximum, the steeper the

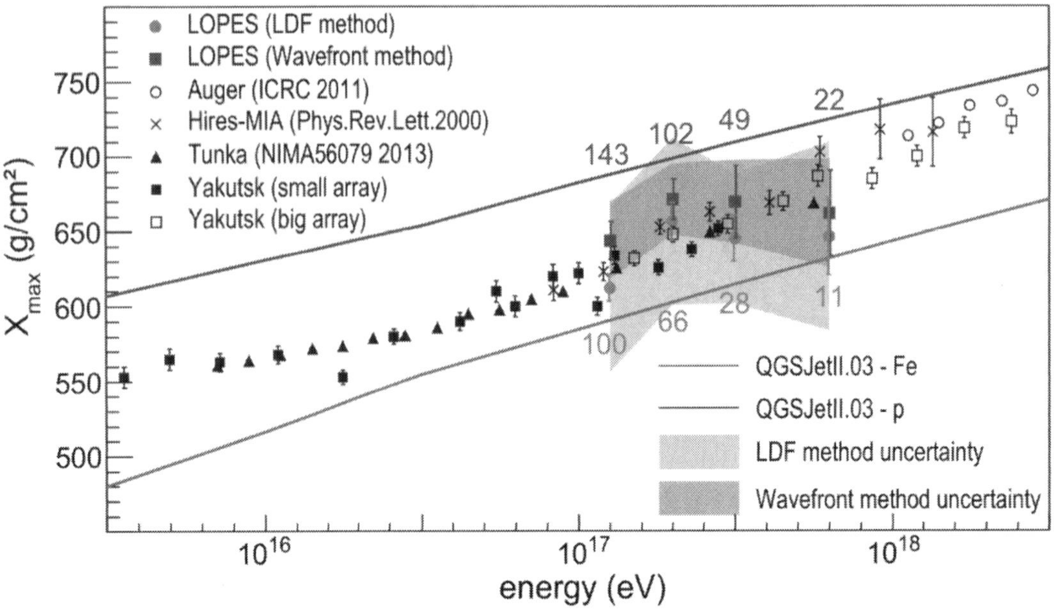

Figure 3. Measurements of different experiments of the mean atmospheric depth of the shower maximum, X_{max}, in different energy bins. The bands around the LOPES values are for both methods the average deviation between the true and reconstructed X_{max} values of the CoREAS simulations, but not the total systematic uncertainty, which is not known. The indicated numbers of events per bin deviate between both methods due to different quality cuts. For references to the data points see Refs. [6] and [5].

radio wavefront, and the steeper the lateral distribution of the radio amplitude. Consequently, the reconstruction of the atmospheric depth of the shower maximum, X_{max}, is more challenging than the reconstruction of the energy and the shower direction. This is because any measurement uncertainties generally have a larger effect on the second-order effects relevant for X_{max} than on the first-order effects relevant for energy and arrival direction. Still, we were able to provide a practical proof-of-principle for both methods, performing a reconstruction of X_{max} using high-quality LOPES events of the years 2005 to 2009 [5],[6].

First, we could show that the radio wavefront is approximately of hyperbolic shape. According to CoREAS simulations [24], the steepness of the asymptotic cone of the hyperbola is clearly correlated with the distance to the shower maximum [5]. CoREAS simulations predict that under ideal conditions, i.e., negligible noise and measurement uncertainties, a precision of better than $30\,\mathrm{g/cm^2}$ for X_{max} should be possible for a LOPES-like experiment. With LOPES measurements, however, we achieved a precision of $140\,\mathrm{g/cm^2}$. Nevertheless, we could demonstrate that the method in principle works, since we experimentally confirmed expected correlations of the wavefront steepness with other observables, e.g., the shower age reconstructed by the KASCADE array.

Second, we exploited the dependence of the slope of the lateral distribution on the shower development, for which we already had found experimental evidence [8]. Again, we used CoREAS simulations to study for different zenith angles how exactly the lateral slope depends on the shower maximum. Then, we used the input of the simulations to reconstruct X_{max} for measured events. Although the predicted precision for the ideal case without measurement uncertainties is only $50\,\mathrm{g/cm^2}$ and, thus, worse than for the wavefront method, the practically achieved precision at LOPES is better than for the wavefront method, namely better than $95\,\mathrm{g/cm^2}$ [6].

In figure 3 we compare the reconstructed mean $X_{\max}$ obtained with both methods to measurements of other experiments. The event selections for both methods are not identical, since both methods require slightly different quality cuts. Moreover, we have not studied the effect of a potential selection bias, and we do not yet know the total systematic uncertainty of the LOPES measurements (e.g., due to varying atmospheric conditions). In any case, the total uncertainty will be larger than the pure statistical uncertainties. Consequently, the results of both methods are compatible with each other and in agreement with the results of independent experiments using other methods, namely air-fluorescence and air-Cherenkov measurements.

4. Open questions

4.1. Absolute scale

The general features of the radio emission by air showers seem to be understood: The latest generation of several simulation codes and models for the radio emission of air showers, e.g., ZHAires [25], EVA [26], SELFAS2 [27], and CoREAS [28],[24], yield comparable results and agree in the features of the radio signal [17]. For simplicity and due to limited resources we have compared LOPES data only to CoREAS simulations and its predecessor REAS 3.11. Results from AERA [29] indicate that qualitatively similar results could be expected for other simulation codes.

The radio extension CoREAS of the CORSIKA simulation package for air showers incorporates the present state of knowledge, and is able to reproduce the experimentally observed features of the lateral distribution and the wavefront within the measurement accuracy. Earlier published results indicated that the LOPES measurements would be in better agreement with REAS 3.11 than with CoREAS, which predicted a lower amplitude scale [11]. Using new input provided by the company which produced our external calibration source, we started to re-analyze the absolute amplitude calibration of LOPES. The new calibration indeed leads to a lower amplitude scale, moving the measurements closer to the CoREAS simulations. At the present state of analysis we cannot yet say, in which way this will alter conclusions for the comparison of measurements and simulations.

4.2. Subtraction of external noise

The detection of the air-shower radio signal at MHz frequencies suffers significantly from external noise, which is mainly composed of galactic and anthropogenic noise in the frequency range relevant for LOPES. Already galactic noise exceeds thermal noise by an order of magnitude. Including rural areas, anthropogenic background can be even higher [30]. Thus, sophisticated techniques for the reduction of background could significantly improve the performance of the radio technique. The detection threshold could be lowered by almost an order of magnitude if external noise could be completely suppressed.

At LOPES we successfully demonstrated that the detection threshold can be lowered by cross-correlation beamforming, exploiting the correlation of the air-shower signal arriving in different antennas at different times. In principle, also the background should be correlated, but this is not yet exploited: Most of the background, be it galactic or anthropogenic, is the sum of the radio emission originating from many different individual sources. This means that the background recorded by individual antennas is composed of many signals arriving from different directions. In principle, each antenna records the same background contributions just at different times, depending on the position of the antenna and the direction of the individual background signals. Consequently, using background observations before and after the air-shower signal, it should be in principle possible to predict and calculate the background in each antenna from the background measured at other antennas. This requires that the array consists of a sufficient number of antennas to split the sky in a sufficient number of direction bins, or alternatively, when working in the Fourier domain, a sufficient number of different baselines. Once the background at one antenna is predictable from the measurements in other antennas (at least to a certain precision), it can be subtracted and the signal-to-noise ratio can be improved.

Such a method for rejection of directional background has not yet been tried. It corresponds to the solution of a multi-dimensional equation system for the time series recorded at the individual antennas, and requires sufficient computing power. Since all triggered events are stored on disk, this offers the opportunity to later increase the efficiency of the already terminated experiment, once the technique for noise reduction becomes available. Moreover, this method could be applied to radio astronomy whenever background cannot be reduced by integration, e.g., when observing fast transients.

4.3. Feasibility for large observatories

At the start of LOPES there was hope that the radio technique could be extended at reasonable cost to very large scale arrays of several 1000 km^2, which are necessary for the detection of ultra-high energy cosmic rays. Scaling the radio technique to huge areas would be economically feasible only when the array is much sparser than LOPES, or when the costs per station is reduced significantly. However, the methods and techniques developed for LOPES cannot be applied to an arbitrary large antenna spacing, for two reasons: First, we achieve the highest precision for the reconstructed shower energy using the radio amplitude at a distance to the shower axis of approximately 100 m. Second, interferometric beamforming requires several antennas with signal. This means that several antennas have to be placed within the radio footprint, which has a diameter of at most a few 100 m, except for very inclined showers.

Consequently, the LOPES technique requires an antenna spacing of not more than 100 to 200 m. At the present state of technology this might be too expensive for huge areas, in particular when a ns-timing precision is required for interferometry. This does not mean that radio detection would be completely unfeasible at the highest energies. However, certain modification compared to LOPES would be necessary when applying the radio technique to huge areas. Perhaps one could focus on inclined showers with large footprints or better exploit the information contained in the pulse shape and polarization measured at a single station. Balloons or satellites might be an alternative to ground arrays, which is currently under investigation by ANITA [31]. Finally, other target materials than the Earth's atmosphere could be observed, e.g, the moon [32].

5. Conclusion

With LOPES we could demonstrate that digital radio arrays are a suitable extension to particle detector arrays for air showers at energies $E \gtrsim 10^{17}$ eV. Regarding the precision of the shower direction and energy, radio measurements can compete with other detection techniques, even in the noisy environment of the LOPES site. Furthermore, the position of the shower maximum can be reconstructed from the radio measurements. But for $X_{\max}$, the precision of LOPES is not competitive and just sufficient to distinguish on average the extreme cases of protons and iron nuclei as primary particles. Nevertheless, LOFAR has demonstrated a competitive precision for the shower maximum with a denser array at a site with lower ambient background. Still, it is not yet clear, if only the background level, or also the antenna density is essential for the precision. This question will be answered by the currently running, sparser radio arrays AERA [33] and Tunka-Rex [34], which are located at radio quiet sites, but have a larger antenna spacing in the order of 200 m.

Even if for larger radio arrays only the arrival direction and the shower energy could be determined with decent precision, this could still make it worth to build such arrays. The combination of a particle detector array providing a measurement of the muonic component, and a complementary radio array providing a precise measurement of the energy of the electromagnetic component, together could bring a significant step in accuracy, e.g., for an energy spectrum distinguishing different mass groups. This would be an essential input for the open questions with respect to the transition from galactic to extra-galactic cosmic rays at energies larger than 10^{17} eV, i.e., exactly in the energy range where the radio technique becomes efficient.

Acknowledgments

Acknowledgement LOPES and KASCADE-Grande have been supported by the German Federal Ministry of Education and Research. KASCADE-Grande is partly supported by the MIUR and INAF of Italy, the Polish Ministry of Science and Higher Education and by the Romanian Authority for Scientific Research UEFISCDI (PNII-IDEI grant 271/2011). This research has been supported by grant number VH-NG-413 of the Helmholtz Association. The present study is supported by the 'Helmholtz Alliance for Astroparticle Physics - HAP' funded by the Initiative and Networking Fund of the Helmholtz Association, Germany.

References

[1] Falcke H *et al* - LOPES Coll 2005 *Nature* **435** 313–316
[2] Nehls S, Hakenjos A, Arts M J *et al.* 2008 *Nucl. Instr. and Meth. A* **589** 350 – 361
[3] Schröder F G, Asch T, Bähren L *et al.* 2010 *Nucl. Instr. and Meth. A* **615** 277 – 284
[4] Nigl A *et al* - LOPES Coll 2008 *Astronomy & Astrophysics* **487** 781–788
[5] Apel W D *et al* - LOPES Coll 2014 *J. Cosmol. Astropart. Phys.* JCAP09(2014)025
[6] Apel W D *et al* - LOPES Coll 2014 *Phys. Rev. D* **90** 062001
[7] Schröder F G *et al* - LOPES Coll 2013 *JPCS* **409** 012075
[8] Apel W D *et al* - LOPES Coll 2012 *Phys. Rev. D* **85** 071101(R)
[9] Isar P G *et al* - LOPES Coll 2009 *Nucl. Instr. and Meth. A* **604, Supp. 1** S81–S84
[10] Huber D 2014 PhD Thesis URL `digbib.ubka.uni-karlsruhe.de/volltexte/1000043289`
[11] Apel W D *et al* - LOPES Coll 2013 *Astroparticle Physics* **50-52** 76–91
[12] Huege T 2013 *Proc. of 33rd ICRC* #1294 arxiv.org:1310.6927
[13] Apel W D *et al* - KASCADE-Grande Coll 2010 *Nucl. Instr. and Meth. A* **620** 202–216
[14] Antoni T *et al* - KASCADE Coll 2003 *Nucl. Instr. and Meth. A* **513** 490–510
[15] Asch T, Forschungszentrum Karlsruhe 2009 *FZKA Report* **7459**
[16] Grebe S *et al* -Pierre Auger Coll 2013 *Proc. of 5th ARENA 2012, AIP Conf. Proc.* **1535** 73–77
[17] Huege T 2013 *Proc. of 5th ARENA 2012, AIP Conf. Proc.* **1535** 121–127
[18] Apel W D *et al* - LOPES Coll 2012 *Nucl. Instr. and Meth. A* **696** 100–109
[19] Huber D *et al* - LOPES Coll 2014 *Proc. of 6th ARENA 2014, submitted to AIP Conf. Proc.*
[20] Holt D 2013 Diploma Thesis, Karlsruhe Institute of Technology (KIT), Germany
[21] Marin V *et al* - CODALEMA Coll 2011 *Proc. of the 32nd ICRC, Beijing, China* vol 1 p #0942
[22] Pierre Auger Coll 2014 *Phys. Rev. D* **89** 052002
[23] Schellart P *et al* - LOFAR Coll 2014 *JCAP* **10** 014
[24] Huege T and Ludwig M 2013 *Proc. of 5th ARENA 2012, AIP Conf. Proc.* **1535** 128–132
[25] Alvarez-Muñiz J *et al* 2012 *Astroparticle Physics* **35** 325–341
[26] Werner K *et al* 2012 *Astroparticle Physics* **37** 733–741
[27] Marin V *et al* 2012 *Astroparticle Physics* **35** 733–741
[28] Ludwig M and Huege T 2011 *Astroparticle Physics* **34** 438–446
[29] Schröder F G *et al* - Pierre Auger Coll 2013 *Proc. of 33rd ICRC* #0899 arxiv.org:1307.5059
[30] CCIR (International Radio Consultative Committee), ITU, Geneva, Switzerland 1990 *CCIR Report* **670**
[31] Hoover S *et al* - ANITA Coll *Phys. Rev. Letters* **105** 151101
[32] James C W *et al* - LUNASKA 2010 *Phys. Rev. D* **81** 042003
[33] Pierre Auger Coll 2012 *Journal of Instr.* **7** P10011
[34] Kostunin D *et al* - Tunka-Rex Coll 2014 *Nucl. Instr. and Meth. A; Proc. of RICAP 2013* **742** 89–94

Journal of Physics: Conference Series **632** (2015) 012103 doi:10.1088/1742-6596/632/1/012103

Measurements of the muon content of air showers at the Pierre Auger Observatory

I. Valiño[1] for The Pierre Auger Collaboration[2]

[1] Dept. de Física de Partículas & IGFAE, Universidade de Santiago de Compostela, Spain.

[2] Av. San Martín Norte 304 (5613) Malargüe, Prov. de Mendonza, Argentina

(Full author list at http://www.auger.org/archive/authors_2014_12.html).

E-mail: inesvr@gmail.com

Abstract. The Pierre Auger Observatory offers a unique window to study cosmic rays and particle physics at energies above 3 EeV (corresponding to a centre-of-mass energy of 75 TeV in proton-proton collisions) inaccessible to accelerator experiments. We discuss the different methods of estimating the number of muons in showers recorded at the Pierre Auger Observatory, which is an observable sensitive to primary mass composition and to properties of the hadronic interactions in the shower. The muon content, derived from data with these methods, is presented and compared to predictions from the post-LHC hadronic interaction models for different primary composition. We find that models do not reproduce well the Auger observations, displaying a deficit of muons at the ground. In the light of these results, a better understanding of ultra-high energy extensive air showers and hadronic interactions is crucial to determine the composition of ultra-high energy cosmic rays. We report on the upgrade plans of the Pierre Auger Observatory to achieve this science goal.

1. Introduction

Understanding the mass composition of ultra-high energy cosmic rays (UHECR) is fundamental to unveiling their production and propagation mechanisms. The interpretations of observed anisotropies [1, 2] and of features of the flux, such as the ankle at an energy of about 5×10^{18} eV and the strong suppression above 4×10^{19} eV [3], rely on the assumed mass composition at Earth.

Ultra-high energy cosmic rays can only be observed indirectly through air showers initiated in the Earth's atmosphere. The mass composition can be inferred from certain shower observables, which are also dependent on the properties of the hadronic interactions in the shower. Inferring the mass composition from these measurements is subject to some level of uncertainty because it relies on the use of simulated showers generated by assuming a hadronic interaction model [4]. Such models are based on extrapolation from accelerator data over more than two orders of magnitude in energy in the centre-of-mass frame and in different phase space regions of interest. Disentangling the composition from models is of utmost importance and is one of the most compelling challenges in UHECR physics.

The number of muons in an air shower at ground-level is a powerful tracer of the cosmic-ray mass. It has been shown that the number of muons produced scales nearly linearly with the cosmic-ray energy, E, and increases with a small power of the cosmic-ray mass, A, as [5]:

$$N_\mu = A^{1-\beta}\left(\frac{E}{\xi_c^\pi}\right)^\beta, \tag{1}$$

where ξ_c^π is the critical energy at which particle production by charged pions ceases and β is predicted by simulations to be ≈ 0.9. Simulations also show further dependences on several properties of hadronic interactions, including the multiplicity, the charge ratio and the baryon antibaryon pair production [6, 7].

The Pierre Auger Observatory [8], conceived to characterise the properties of the UHECRs with energies above 10^{18} eV, is a hybrid instrument of air-shower detection. It combines an array of particle detectors, the Surface Detector array (SD) [9], to sample the air-shower front as it reaches the ground, and Fluorescence Detector (FD) telescopes to collect the ultraviolet light emitted by nitrogen as showers develop in the atmosphere [10]. The FD is used to monitor the atmosphere, on dark clear nights, above the 3 000 km^2 area over which the SD 1 660 water-Cherenkov stations (separated by 1.5 km) are laid out. The site is located near the town of Malargüe, Argentina, at an altitude of about 1 425 m above sea-level and at an average altitude of 35.2° S.

In this paper, we present the average muon content in showers measured with the Pierre Auger Observatory using different analysis techniques. We also compare these measurements to predictions of the post-LHC hadronic interaction models QGSJETII-04 [11] and EPOS LHC [12] for different primary masses. Finally, we discuss the potential upgrades of the Pierre Auger Observatory to enhance the capabilities of the SD array in order to achieve a clean measurement of the muon number. This is the key to estimating the primary composition on an event-by-event basis, and to study fundamental interactions at energies inaccessible to accelerator experiments.

2. Measurement with highly inclined showers

Highly inclined showers, those arriving at ground with zenith angle $\theta > 60°$, are characterised by the dominance of secondary energetic muons at the ground, since the electromagnetic (EM) component has been largely absorbed in the greatly enhanced atmospheric depth crossed by the shower before reaching ground. Such inclined showers provide an almost direct measurement of the muon number at ground-level [13, 14].

The approach to measuring the number of shower muons follows from developments that have been introduced to reconstruct inclined showers [15]. It is based on the observation that the shape of the muon distribution is universal for a given shower direction and that only the overall normalisation of the muon distribution depends on the shower energy and primary mass. It has also been shown that the lateral shape of the muon density is consistently reproduced by different hadronic interaction models and software packages used for air-shower simulations. These universal characteristics allow one to model the muon density $\rho_\mu(\vec{r})$ as:

$$\rho_\mu(\vec{r}) = N_{19}\, \rho_{\mu,19}(\vec{r}; \theta, \phi), \tag{2}$$

where N_{19} is defined as the scale factor of a particular event with respect to a reference muon distribution $\rho_{\mu,19}$, conventionally chosen to be the average muon density for primary protons of 10^{19} eV simulated with the hadronic interaction model QGSJETII-03 [16]. N_{19} at a given zenith angle is a relative measure of the number of muons produced N_μ (addressed in Eq. 1), and is inferred from fitting the measured signals at the SD stations to the expected muon patterns for a given reconstructed arrival direction.

Simulations show that N_{19} deviates, on average, around 5% from the true relative content of muons in a shower, R_μ^{MC} [13, 14, 15]. This ratio is computed for each event by counting the total number of muons N_μ at the ground in the simulation and dividing by the total number of muons $N_{\mu,19}$ in the reference muon distribution: $R_\mu^{\mathrm{MC}} = N_\mu/N_{\mu,19}$. Hence to get an unbiased estimator, the measured value N_{19} is corrected for this average bias, and renamed R_μ. By combining the uncertainty attributed to the hadronic interaction model assumption with that of the simulated muon response of the SD detectors at $\theta > 60°$ [17], the systematic uncertainty of R_μ is estimated to be 11%.

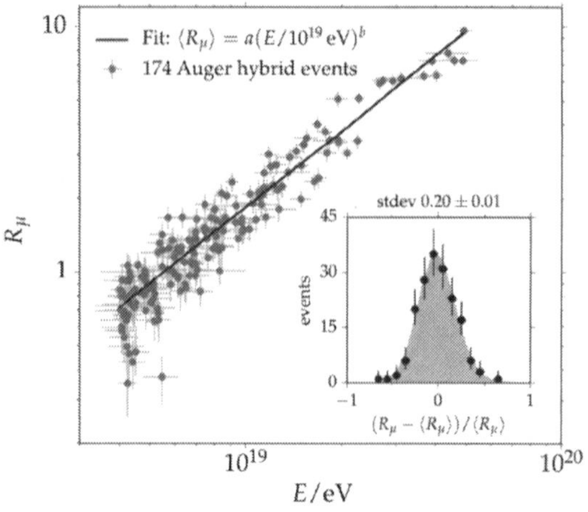

Figure 1. Fit of the calibration $\langle R_\mu \rangle = a\,(E/10^{19}\,\mathrm{eV})^b$ to the selected hybrid events. Only statistical uncertainties are shown.

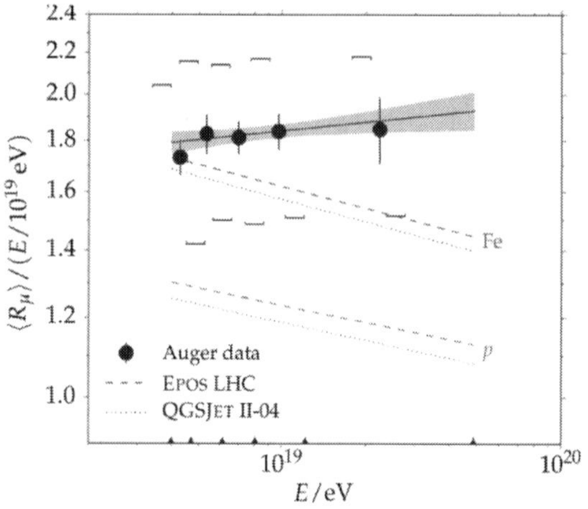

Figure 2. Average muon content $\langle R_\mu \rangle$ per shower energy $E/10^{19}\,\mathrm{eV}$ as a function of the shower energy E. The brackets indicate the systematic uncertainties and systematic shifts due to 14 % of the energy scale uncertainty [18]. Theoretical curves for proton and iron showers are shown for comparison.

The muon content R_μ of showers with zenith angles $62° < \theta < 80°$ is studied as a function of the calorimetric measurement of the cosmic-ray energy E from high-quality events measured simultaneously with the SD and the FD. Only events well-contained in the SD array with energies above $4\times10^{18}\,\mathrm{eV}$ (corresponding to 100% SD trigger probability) are accepted for the analysis. More quality selection criteria and fiducial cuts used in this analysis are explained in detail in [14]. For the period from 1 January 2004 to 31 December 2012, a total of 174 events were accepted.

The correlation between R_μ and E can be described by fitting the parametrisation

$$R_\mu = a\,(E/10^{19}\,\mathrm{eV})^b, \tag{3}$$

to the selected data. The a parameter represents the average muon content at $10^{19}\,\mathrm{eV}$, and the b parameter the logarithmic gain $\mathrm{d}\langle \ln R_\mu \rangle/\mathrm{d}\ln E \simeq \mathrm{d}\ln N_\mu/\mathrm{d}\ln E$ of muons with increasing energy. The data and results of the fit are shown in figure 1. The average muon content at $10^{19}\,\mathrm{eV}$ obtained is $(1.841 \pm 0.029 \pm 0.324\,(\mathrm{sys.}))$, which corresponds to $(2.68 \pm 0.04 \pm 0.48\,(\mathrm{sys.}))\times10^7$ muons with energies larger than 0.3 GeV that reach 1425 m altitude in an average $10^{19}\,\mathrm{eV}$ shower with a zenith angle of 67°.

A simple comparison between data and predictions for proton and iron-induced showers simulated at the mean zenith angle of 67° with the hadronic interaction models QGSJETII-04 and EPOS LHC is shown in figure 2. The absolute values of $\langle \ln R_\mu \rangle$ in data correspond to masses close to or even heavier than iron. This interpretation is in contradiction with recent studies based on the depth of shower maximum, X_{max}, that show an average logarithmic mass $\langle \ln A \rangle$ between proton and iron in this energy range [19]. To reach consistency between data and predictions based on $\langle \ln A \rangle$ data, the mean muon number around $10^{19}\,\mathrm{eV}$ in simulations would have to be increased by 30 % to 80 %, depending on the model.

The logarithmic gain $\mathrm{d}\langle \ln R_\mu \rangle/\mathrm{d}\ln E$ is also large compared to predictions for constant pure composition. The hypothesis of a constant proton (iron) composition is disfavoured at the level of $2.2\,\sigma$ ($2.6\,\sigma$). Our measurement favours a transition from lighter to heavier elements in the considered energy range. If all statistical and systematic uncertainties are added in quadrature, the deviation between the measured logarithmic gain $\mathrm{d}\langle\ln R_\mu\rangle/\mathrm{d}\ln E$ and predictions based on the energy evolution of $\langle\ln A\rangle$ estimated from the X_{max} measurements [19] is 1.3 to $1.4\,\sigma$.

3. Measurement from temporal structure of signals in surface detectors

The number of muons in less inclined air showers has also been explored, but the measurement in this case is complicated by the need to separate the electromagnetic and the muonic signals recorded in the SD stations.

Sophisticated methods have been developed to extract the muon information by analysing the time structure of the signal generated by shower particles in the detector [20, 21, 22]. These methods are based on two features of the signals which depend on the type of shower component: the amplitude distribution of the particle responses and the arrival time distributions. On one hand, the mean amplitude of a single EM particle is much smaller, and the number of EM particles are, on average, an order of magnitude larger than the number of muons. On the other hand, muons typically arrive earlier than EM particles. These two features make the muon signal short and spiky, and the EM signal long and smooth.

The two methods presented below [22], measure the mean fraction of the SD signal due to muons of air showers with energy $E = 10^{19}\,\mathrm{eV}$ at a reference distance of $r = 1000\,\mathrm{m}$ from the shower axis. Due to the similar energy scaling of the total and muon signals in the detectors at about 1000 m, the muon fraction is mostly insensitive to the systematic uncertainty of the reconstructed energy, which is 14 % [18]. The main source of uncertainty is due to high-energy photons ($\gtrsim 300\,\mathrm{MeV}$) that can produce a signal similar to that of a muon. Their contribution is estimated to be less than 10 % to 15 % for proton and iron primaries, respectively, in the considered energy and angular range.

Note that the muon signal, S_μ, is the pure muonic signal, so the electromagnetic halo produced by muon interactions and muon decays in the atmosphere [23] is not included in S_μ and is treated as an EM component. Therefore, the smooth part of the signal is assumed to be the sum of the primary EM component and the EM muon halo. For vertical showers, the contribution of the EM muon halo ranges from $\sim 8\,\%$ at $\theta < 40°$ up to $\sim 12\,\%$ at 60°.

3.1. Multivariate method

The basic idea of this method is to combine a large number of muon-content-sensitive observables of the FADC signal[1] to estimate the muon fraction, f_μ, using a multivariate regressor. The optimal estimator found is:

$$\hat{f} = a + b\,\hat{\theta} + c\,g_{0.5}^2 + d\,\hat{\theta}\,P_0 + e\,\hat{r} \tag{4}$$

where $\hat{\theta}$ is the reconstructed zenith angle of the shower and $\hat{r}$ is the distance of the detector from the reconstructed shower axis. The observable $g_{0.5}^2$ is the portion of the signal in FADC bins greater than 0.5 VEM. P_0 is the normalised zero-frequency component of the power spectrum. Both $g_{0.5}^2$ and P_0 are sensitive to large relative fluctuations in temporal structure of individual traces and short signal peaks, which are signatures of high muon content.

The fit parameters (a, b, c, d, e) are estimated using simulations as described in [22]. Simulations for different models and primary particles show that $\hat{f}$ deviates, on average, about 2 % from the true muon fraction, and that the average resolution is about 8 %.

[1] The Cherenkov photons produced by the shower particles in the detectors are sampled by three photomultipliers. The analogue signal is digitised with FADCs in 25 ns bins within a 10 bit dynamic range. The raw signal is calibrated such that the integrated signal of a typical vertical atmospheric muon traversing the detector through its centre is 1. The signal in each time bin is thus measured in units of "vertical equivalent muon" of VEM [24].

3.2. Smoothing method
The basic idea of this method is to apply a low-pass filter a few times on the signal to gradually separate the low-frequency smooth EM component from the high-frequency component which is attributed to muons. In each step, the muon estimate is subtracted from the output of the previous iteration and the signal is re-smoothed. The final muonic signal is the sum of the non-smooth positive differences between the original trace and the smoothed one at each step (more details can be found in [20, 22]). The muon fraction, $\hat{f}$, is then estimated by dividing the muonic signal by the total signal. Simulations show that $\hat{f}$ deviates, on average, about 5 % from the true muon fraction, and that the average resolution is about 8 %.

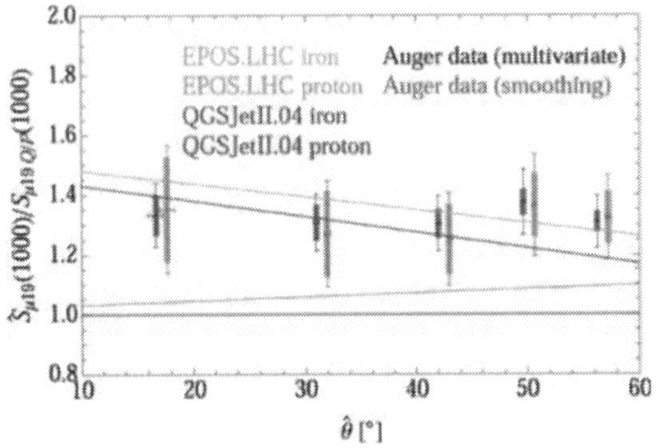

Figure 3. The measured muon signal rescaled to $E = 10^{19}$ eV and at 1000 m from the shower axis as a function of the shower zenith angle, with respect to the reference based on QGSJETII-04 proton simulations. The boxes indicate the systematic uncertainties and the error bars represent the total uncertainty, also including the statistical uncertainties.

3.3. Muon signal rescaling
Both methods are used to estimate the muon fraction $\hat{f}$ for every detector in the distance range $\hat{r} \in [950, 1050]$ m of SD events with zenith angle below 60° and reconstructed energy $\log_{10}(\hat{E}/\text{eV}) \in [18.98, 19.02]$. For the period from 1 January 2004 to 31 December 2012, a total of 521 SD signals were analysed.

The muon signal for every detector can be retrieved by multiplying the muon fraction by the total signal at 1000 m from the shower axis. The results with respect to the chosen reference based on proton showers simulated with QGSJETII-04 are shown in figure 3, with an average value of $(1.33 \pm 0.02 \pm 0.05\,(\text{sys.}))$ (multivariate) and $(1.31 \pm 0.02 \pm 0.09\,(\text{sys.}))$ (smoothing). The two analysis methods show a good agreement. The measurement is bracketed by model predictions for proton and iron primaries with QGSJETII-04 and EPOS LHC within the systematic uncertainties.

A simple comparison of these results with the measurement from highly inclined showers can be performed by accounting the scale factor of the muon number densities derived for QGSJETII-04 proton showers relative to those for QGSJETII-03, which is ~ 1.2 [15]. We found that vertical and inclined measurements are compatible within uncertainties.

4. Measurement with hybrid vertical events
The interplay between SD and FD data can be further explored using showers that have been well recorded by both detectors. For each observed event, showers are picked from a simulated library of pure proton or mixed composition so their FD longitudinal profiles show the best match

with the observed one. Then the ground signal of the selected simulated events is compared to the observed one in data [21, 25].

Since the Monte Carlo predictions in general do not match the observed signals, regardless of the hadronic model being used, two independent rescaling factors are introduced in order to achieve an agreement. These factors are R_E which acts as a rescaling of the primary energy (it rescales the total signal of the event uniformly), and R_μ which acts as a rescaling of the muon size[2] (it rescales only the contribution to the ground signal of inherently hadronic origin).

The data used for this study were narrowed down to the energy bin $10^{18.8} < E < 10^{19.2}$ eV, sufficient to have adequate statistics while being small enough that the primary cosmic-ray mass composition does not evolve significantly. For the period from 1 January 2004 to 31 December 2012, a total of 411 hybrid events were accepted after applying quality selection cuts [18, 26].

Each event of the dataset was compared with the results obtained from simulations using the hadronic interaction models QGSJETII-04 and EPOS LHC, both for a set of proton showers and for a set of mixed composition based on the $\langle \ln A \rangle$ estimated from the $X_{\max}$ measurements for this energy bin [19]. The values of the scaling parameters that best reproduce the data are shown in figure 4. The results suggest that while the predicted energy is compatible with the observed one (R_E is compatible with 1 within the systematics on the absolute energy scale), the muon rescaling parameter demands an increase of a factor 1.3 to 1.6 of the muon size in simulations, depending on the model.

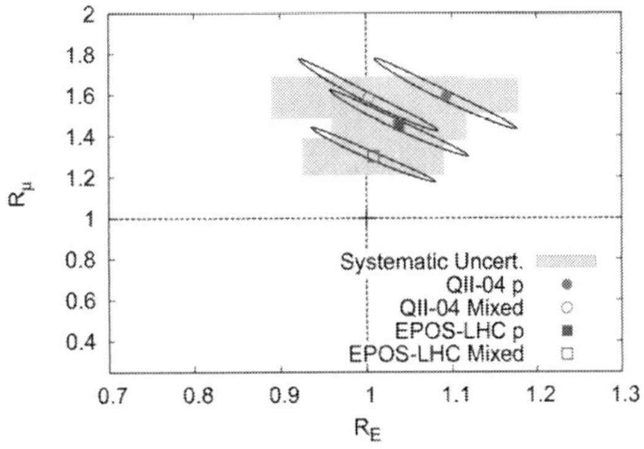

Figure 4. Value of the energy rescaling parameter R_E and of the hadronic rescaling parameter R_μ that best represent the hybrid data at 10^{19} eV.

5. Conclusions

The Pierre Auger Observatory has unique capabilities to measure several observables sensitive to both the mass composition of cosmic rays and hadronic interaction properties. Despite not being originally designed for such a purpose, the Pierre Auger Observatory is able to measure the muon content in showers at the ground under various conditions. These measurements provide relevant information to constraint the models that attempt to describe the hadronic interactions at extremely high energies inaccessible to accelerator experiments. The observed muon content in 10^{19} eV air showers ($E_{\rm CM} = 135$ TeV) suggests a muon deficit in simulations using the latest hadronic interaction models tuned to LHC and lower energy accelerator data, even for the best

[2] Note that the factor R_μ is not directly comparable to direct muon number determinations obtained from inclined showers [13, 14] or from the FADC traces of vertical showers [22], which have been reported in the previous sections.

case of EPOS LHC with mixed composition. In addition, the result from the analysis of hybrid vertical showers proves that none of the tested models calls for an overall rescaling of the energy.

Further progress is expected with more data, particularly thanks to the currently planned upgrade aimed at a more precise muon measurement by improving the discrimination between the electromagnetic and muonic components of the showers at the highest energies. A precise measurement of the number of muons is the key to estimating the primary mass on an event-by-event basis, to improve our sensitivity to photon/neutrino searches and to study fundamental interactions at centre-of-mass energies an order of magnitude above those presently attainable at the LHC. After testing in the field several prototypes based on different ideas to measure muons with the required precision, the solution agreed by the collaboration is to deploy scintillators on top of all the water-Cherenkov detectors to measure charged particles. The combination of two independent measurements will allow us to obtain an estimation of the two shower components. The upgrade is expected to be completed by 2018 with Observatory operation through 2023.

Acknowledgments

The successful installation, commissioning, and operation of the Pierre Auger Observatory would not have been possible without the strong commitment and effort from the technical and administrative staff in Malargüe. We are very grateful to all the national agencies and organisations for financial support. The author thanks the organisers for the nice scientific atmosphere and hospitality.

References

[1] Abraham J *et al.* (Pierre Auger Collaboration) 2007 *Science* **318** 938-943
[2] Abreu P *et al.* (Pierre Auger Collaboration) 2010 *Astropart. Phys.* **34** 314-326
[3] Abraham J *et al.* (Pierre Auger Collaboration) 2010 *Phys. Lett.* B **685** 239-246
[4] Abreu P *et al.* (Pierre Auger Collaboration) 2013 *JCAP* **02** 026
[5] Matthews J 2005 *Astropart. Phys.* **22** 387-397
[6] Pierog T and Werner K 2008 *Phys. Rev. Lett.* **101**, 171101
[7] Ulrich R *et al.* 2011 *Phys. Rev.* D **83** 054026
[8] Abraham J *et al.* (Pierre Auger Collaboration) 2004 *Nucl. Instrum. Meth.* A **523** 50-95
[9] Abraham J *et al.* (Pierre Auger Collaboration) 2010 *Nucl. Instrum. Meth.* A **613** 29-39
[10] Abraham J *et al.* (Pierre Auger Collaboration) 2010 *Nucl. Instrum. Meth.* A **620** 227-251
[11] Ostapchenko S S 2011 *Phys. Rev.* D **83** 014018
[12] Pierog T *et al.* 2013 EPOS LHC: test of collective hadronization with LHC data *Preprint* hep-ph/1306.0121
[13] Valiño I (Pierre Auger Collaboration) 2013 *Proc. 33rd ICRC* (Rio de Janeiro, Brazil) *Preprint* astro-ph/1307.5059
[14] Aab A *et al.* (Pierre Auger Collaboration) 2014 accepted for publication in *Phys. Rev.* D (*Preprint* astro-ph/1408.1421)
[15] Aab A *et al.* (Pierre Auger Collaboration) 2014 *JCAP* **08** 019
[16] Ostapchenko S S 2006 *Nucl. Phys.* B *Proc. Suppl.* **151** 143-146 (*Preprint* hep-ph/0412332)
[17] Ghia P (Pierre Auger Collaboration) 2007 *Proc. 30th ICRC* (Mérida, Mexico) **4** 315-318
[18] Verzi V (Pierre Auger Collaboration) 2013 *Proc. 33rd ICRC* (Rio de Janeiro, Brazil) *Preprint* astro-ph/1307.5059
[19] Aab A *et al.* (Pierre Auger Collaboration) 2014 accepted for publication in *Phys. Rev.* D (*Preprint* astro-ph/1409.4809)
[20] Castellina A (Pierre Auger Collaboration) 2009 *Proc. 31st ICRC* (Łódź, Poland) *Preprint* astro-ph/0906.2319
[21] Allen J (Pierre Auger Collaboration) 2011 *Proc. 32nd ICRC* (Beijing, China) *Preprint* astro-ph/1107.4804
[22] Kegl B (Pierre Auger Collaboration) 2013 *Proc. 33rd ICRC* (Rio de Janeiro, Brazil) *Preprint* astro-ph/1307.5059
[23] Valino I, Alvarez-Muniz J, Roth M, Vazquez R A and Zas E 2010 *Astropart. Phys.* **32** 304
[24] Bertou X *et al.* (Pierre Auger Collaboration) 2006 *Nucl. Instrum. Meth.* A **568** 839
[25] Farrar G (Pierre Auger Collaboration) 2013 *Proc. 33rd ICRC* (Rio de Janeiro, Brazil) *Preprint* astro-ph/1307.5059
[26] Pesce R (Pierre Auger Collaboration) 2011 *Proc. 32nd ICRC* (Beijing, China) *Preprint* astro-ph/1107.4809

Beyond the Galaxy: UHECR results from the Pierre Auger Observatory and the Telescope Array

Enrique Zas

Departamento de Física de Partículas,
and Instituto Galego de Física de Altas Enerxías.
Universidad de Santiago, 15782 Santiago de Compostela, Spain

E-mail: enrique.zas@usc.es

Abstract. The beginning of the 21st century has brought much progress into the field of Ultra High Energy Cosmic Rays particularly through the completion of the Pierre Auger Observatory first and the Telescope Array later on. The success of these detectors follows from their hybrid character combining the advantages of the fluorescence technique for energy determination with the larger exposure provided by an array of particle detectors. In this article we review the important contributions made by both experiments concerning the spectrum, anisotropy searches and composition studies. We discuss how these measurements have contributed to progress in the field, partially solving long standing puzzles, and how they have also led to new challenges that are likely to constitute the driving force of the field in the immediate future.

1. Introduction

The existence of ultra high energy cosmic rays reaching energies in the vicinity of 10^{20} eV (100 EeV) has been known for about half a century now [1]. These extremely energetic particles are uniquely interesting for multiple reasons both in astrophysics and in particle physics. It is difficult to envisage astrophysical objects capable of accelerating particles to such extreme energies or alternative mechanisms of production. Because of their large rigidity, they approximately keep the direction with which they are produced over paths greatly exceeding galactic dimensions and, if they are of extragalactic origin, which seems to be most plausible, they must have short ranges on cosmological scales due to interactions with the Cosmic Microwave Background (CMB) and with the intergalactic magnetic fields. Their collisions with the CMB must produce fluxes of neutrinos in the EeV range which could be detected, while their collisions with atmospheric nuclei have center of mass energies that exceed those achieved in particle accelerators by several orders of magnitude. Not surprisingly they have attracted much attention and many efforts to observe them with increasing statistics and precision since the early days. In spite of enormous progress there are still crucial issues which are unresolved. Finding answers to the nature and origin of these particles has a high priority since they are likely to forge the future of the field.

The two largest and most precise detectors to date are the Pierre Auger Observatory in Argentina (Mendoza) and the Telescope Array in the US (Utah). Both detectors are hybrid, combining an array of particle detectors to sample extensive air showers when they reach the ground and telescopes to collect the fluorescence light of the atmospheric nitrogen excited by billions of shower particles crossing it. This combination of techniques optimizes their performance and keeps the need to use simulations to a bare minimum.

The two experimental facilities have accumulated large exposures during years of operation and their data have provided measurements of the spectrum, the arrival directions and the shower development with unprecedented precision, leading to significant advances in the field such as the establishment of the suppression of the flux at the highest energies. The interpretation of these measurements is however far from settled, particularly concerning the arrival directions and the primary composition since both experiments arrive at different conclusions. I here review the progress made in the measurement of the spectrum, the study of anisotropies, and the primary mass composition of these enigmatic particles.

2. The success of the hybrid approach

2.1. Auger Observatory and Telescope Array

Both the Pierre Auger Observatory and the Telescope Array are based on similar detector concepts. They are both located at roughly similar latitudes respectively in the southern and northern hemispheres and at an average altitude of 1,400 m. The main characteristics of these detectors are summarized in table 1 and further details can be found elsewhere [2, 3, 4]. The most fundamental difference lies in their surface detectors (SD) which are based on different detector concepts. The particle detectors in the Auger SD are cylindrical tanks of 10 m^2 surface and 1.2 m height filled with purified water, with three photomultiplier tubes (PMT) to detect the Cherenkov light of particles in the shower front. In the TA they consist of two 3 m^2 slabs of plastic scintillator on top of one another which give light pulses also read by PMTs. The former are sensitive to the track of the charged particles and the latter count charged particles as they cross. The water tanks are relatively much more sensitive to shower muons which usually traverse the tank from wall to wall while the counts in the TA detectors are dominated by electrons and positrons that typically outnumber the muons in the shower front. The detectors of the Pierre Auger Observatory are well suited to detect highly inclined showers, increasing the exposure and the sky coverage. The analysis of inclined showers has also been used for neutrino searches and to establish the muon content of the showers.

The Auger Observatory extends over an area of 3,000 km^2 while the TA covers about 700 km^2. The fluorescence telescopes are located on the perimeter of the two observatories to monitor practically the whole atmospheric volume just above the surface arrays. They have photomultiplier cameras in the focal plane which measure the arrival time, direction and intensity of the emitted light as the shower develops. Some details like the telescope number, size, pixeling and viewing angles can be compared in the aforementioned table. The Pierre Auger Observatory contains a smaller area of 23.5 km^2 with stations separated by 750 m (750 m array) which can be combined with three additional telescopes pointing at higher elevations (HEAT) for lower energy measurements. Similarly the Telescope Array has two sub-arrays of 46 and 35 stations separated by about 600 m and 400 m over a surface of $\sim$20 km^2, together with ten telescopes covering from 31° to 59° of elevation (TALE). These detectors are designed to measure the spectrum for energies above 30 PeV [5]. The Pierre Auger Observatory is run by an international collaboration involving over 500 scientists from more than 100 institutions in 19 countries. The Telescope Array is run by about 150 scientists from 27 institutions in 5 countries.

The combination of two detection techniques which are well established has multiple advantages. The fluorescence detector (FD) allows a calorimetric measurement of the energy deposition in the atmosphere providing a reliable determination of the shower energy. In addition it also allows a measurement of the depth of shower maximum which gives a good handle on primary composition. On the other hand the surface detector (SD) has a very well defined exposure, related to the geometrical area of the surface array, and a duty cycle of nearly 100%, crucial to improve the statistical precision. Furthermore the redundancy introduced by the two

Table 1. Comparison of characteristics of the Pierre Auger Observatory and the Telescope Array. The low energy extensions for each observatory, HEAT and TALE, are not included.

			Auger	TA
SD	Average latitude		35.3° N	39.4° S
	Average altitude		1,400 m	1,400 m
	Surface area		3,000 km^2	700 km^2
	Lattice		1.5 km hexagon	1.2 km square
	Detector	Type	water-Cherenkov	Plastic scintillator
		Size	10 m^2 × 1.2 m	(2×) 3 m^2 × 1.2 cm
		Sampling	25 ns	20 ns
FD	Sites		4	3
	Telescopes	Number	24	36
		Size	13 m^2	6.8 m^2/3 m^2
		Field of view	28.5° × 30°	16° × 14°/18° × 15°
		Pixels	440	256

techniques opens the way to explore and test new approaches, such as constraining interaction models and composition through the time structure of the signals [6].

2.2. Reconstruction of events

2.2.1. Fluorescence Technique The reconstruction of FD events, pioneered by the Fly's Eye group [7], amounts to first establishing the arrival direction of the shower using timing information and then converting the measured light signal to the energy deposition in the atmosphere after accounting for absorption and dispersion. Sophisticated monitoring systems of the atmosphere are implemented in both observatories to minimize the uncertainties in the calculation of the emitted light. If the shower is viewed from a single site, the uncertainty in the arrival direction can be quite large. Both collaborations use the hybrid potential to get around this problem, combining timing information of at least one SD station with that of the signals in the pixels of the camera. The amount of emitted light is assumed to be proportional to energy deposition and the relation between particle numbers and emitted light is given by the fluorescence yield which has been established experimentally. Unfortunately both collaborations use different parameterizations of the fluorescence yield, which complicate comparisons between results.

2.2.2. Surface array for zenith angles below 60° The reconstruction of events with the SD is very similar in the two observatories. The first step consists on establishing the arrival direction using the arrival times of the signals in the particle detectors. Once the direction is known, the pattern of signals on the ground is fitted to obtain the position of the shower core and the shower size. For showers with zenith angles, θ, below 60° the signal pattern can be assumed to have approximate circular symmetry around the shower axis so that it is parameterized as a function of a single variable, r, the distance to shower axis. For a fixed mass composition and arrival direction the shower size scales with energy, and it is characterized by the value of the fitted signal, S, at a reference distance, chosen to minimize fluctuations. These parameters are $S(1000)$ at 1000 m for the Auger data with $\theta < 60°$, $S(800)$ at 800 m for the TA data, restricted to $\theta < 45°$, and $S(450)$ at 450 m for the data of the Auger 750 m array with $\theta < 55°$.

The TA Collaboration compares measured values of $S(800)$ to simulations to obtain a first estimate of shower energy. The conversion is made with a table obtained from detailed simulations of extensive air showers and the signals they produce on the SD. These simulations have been made assuming constant proton composition, an energy spectrum compatible to earlier measurements, and using the QGSJETII-03 [8] hadronic interaction model. The Auger Collaboration however first finds a quantity that scales with energy and which is independent of zenith angle to use it as an energy estimator. For the regular (750 m) array this quantity is chosen to be S_{38} (S_{35}), the value of $S(1000)$ ($S(450)$) that the shower would have had if it had the same energy but arrived at the median zenith angle of 38° (35°). Assuming isotropy of the arrival directions the conversion is made with the constant intensity cut method, avoiding the use of simulation[9].

2.2.3. Surface array for zenith angles exceeding 60° For high zenith angles the geomagnetic field breaks the cylindrical symmetry and a specific method has been devised to fit the expected patterns on the ground to the data [10]. For a given arrival direction the number density of muons has a two-dimensional pattern [11] approximately universal in shape, independently of shower energy, primary composition and hadronic model used [12].

For the Auger data the distributions used for reference correspond to the average values of 10 EeV-proton showers simulated with QGSJETII-03 which have been conveniently parameterized [13]. The fitted normalization, N_{19}, is a size parameter which is by construction independent of arrival direction and thus can be used as an energy estimator.

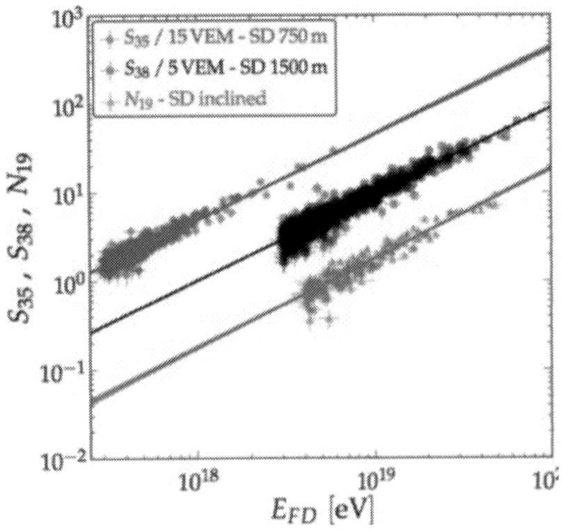

Figure 1. Calibration curves for SD measurements at the Pierre Auger Observatory. The energy estimators S_{38}, N_{19} and S_{35} (respectively for events with $\theta < 60°$, for events with $62° < \theta < 80°$ and for events obtained in the 750 m array with $\theta < 55°$) are represented on the y-axis and the FD energy on the x-axis. The lines indicate the fitted relation between energy and each estimator.

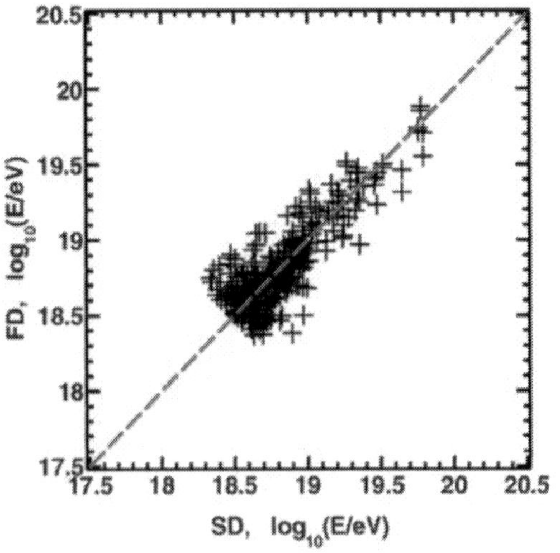

Figure 2. Correlation between the FD energy and SD energy for data of the Telescope Array which can be simultaneously reconstructed with the FD and SD techniques. The SD energy, obtained by directly comparing S(800) to simulations, has been rescaled by a constant factor of 1.27.

2.2.4. Energy calibration The showers detected with the surface detector of the two observatories have been calibrated with their fluorescence detectors. Both collaborations take advantage of the hybrid measurements using a sub-sample of events well reconstructed with both the SD and FD techniques. This allows them to translate the more precise measurement of the energy obtained with the FD, which is however limited in duty cycle to moonless clear nights, to the larger statistical sample obtained with the SD array. The Pierre Auger Collaboration calculates the correlation between the energy estimator and the FD energy [14]. A relation between FD energy and the value of each of the three estimators is obtained as illustrated in Fig.1. The TA has found that rescaling the derived SD energy with a constant factor of 1.27 ensures a good matching to the measured FD energy as illustrated in Fig. 2. The plot is rather similar to the calibration plots of the Pierre Auger Collaboration but we note that the TA procedure does not allow a deviation from linear scaling between the measured energy and the inferred energy from simulations. Such deviations should however be expected in the case that there was a significant change of composition in the explored energy range. Shower simulations show that when composition is changed for a given energy and arrival direction, the average position of shower maximum the average number of muons changes. Consequently the average signal at a given distance to shower axis differs depending on mass composition.

3. The shape of the UHECR energy spectrum

The spectrum is calculated selecting data that fall inside a given region of the detector, and dividing the energy distribution by the corresponding exposure. If the showers have sufficient energy, the trigger efficiency of the surface detector becomes close to 100% and the exposure is independent of energy and simply obtained by integrating over the surface area, time and solid angle. The SD of the Pierre Auger Observatory becomes fully efficient above 3 EeV for events with zenith angle, θ, below 60°. For inclined events with θ exceeding 60° the energy above which the efficiency is approximately 100% is 4 EeV. For events detected with the 750 m array full efficiency is reached at energies greater than 0.3 EeV. The TA Collaboration includes the efficiency in the calculation of the exposure and measures the spectrum for energies above 1.6 EeV. At the lower energies the detection efficiency is of order 10%, and uncertainties in the efficiency directly translate into the flux determination. Full efficiency is achieved at the TA for energies above about 10 EeV.

Three independent fluxes have been obtained with the Auger Observatory at the highest energies[15, 16]. The data collected up to 31st December 2012 with $\theta < 60°$ give the largest exposure corresponding to 82,318 events of energies above 3 EeV. The highly inclined data ($62° < \theta < 80°$) with about a third of the exposure correspond to 11,074 events of energy above 4 EeV and the hybrid data add up to 11,155 events above 1 EeV during the same period. The results of these three fluxes and the flux obtained at lower energies with the 750 m array are compared in Fig. 3. The Telescope Array with about a tenth of the total Auger SD exposure has collected 14,747 events above 1.6 EeV during the period between 11 May 2008 and 20 May 2012.

The results of the spectrum obtained with the TA data are compared in Fig. 4 to those obtained by the Pierre Auger Collaboration after combining the four spectra obtained with hybrid data and with each of the three sets of SD data ($\theta < 60°$, very inclined and 750 m array). They provide unambiguous evidence of flux suppression at the highest energies. In addition a break in the spectrum is clearly appreciated, the *ankle*, at about 5 EeV. Above this energy, the flux is harder. The agreement between the spectra of the two observatories for energies below 40 EeV is excellent, well within the systematic uncertainties. The ankle energy is found at 5.25 ± 0.27 EeV and 4.6 ± 0.3 EeV respectively for the Auger and TA data and the spectral indices at either side are fully compatible. The difference between the fluxes is of the same order as that obtained when the convention of the fluorescence yield from one experiment is changed to that used by the other.

At the highest energies, the differences between the two fluxes are much more significant leading to different interpretations. The energy at which the integrated flux is half of that

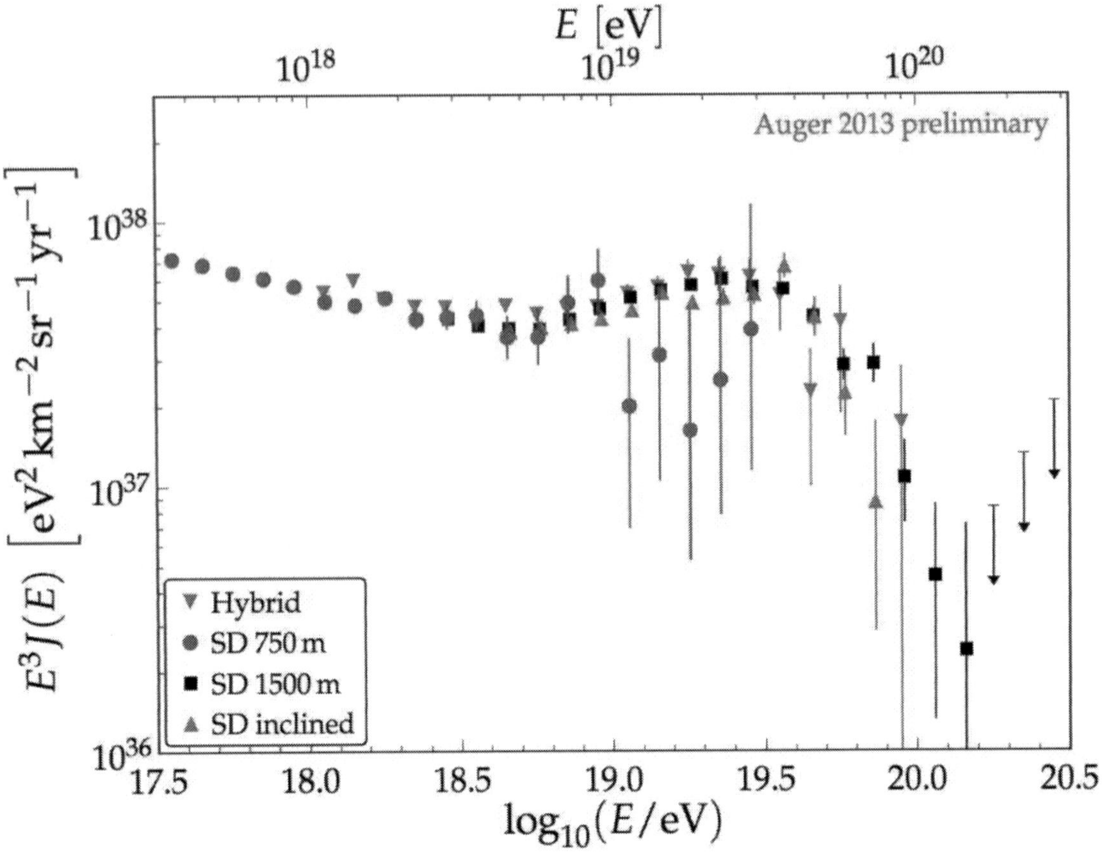

Figure 3. Comparison of the four spectra obtained with the Auger Observatory. The spectrum obtained with the SD for events with $\theta < 60°$ is shown with black squares the inclined spectrum ($62° < \theta < 80°$) is plotted with red upright triangles and that obtained in the 750 array is indicated with blue circles. The spectrum obtained with hybrid data is indicated with green triangles. Only statistical uncertainties are indicated with the error bars.

expected by extrapolating the spectral index found above the ankle is usually denoted as $E_{1/2}$. The Pierre Auger Collaboration finds $E_{1/2} = 43 \pm 1$ EeV to be compared with 60 EeV for the Telescope Array. Besides the fluorescence yield, we note differences in the treatment of the missing energy correction (energy that is not deposited in the atmosphere, mainly attributed to neutrinos and muons) in the procedure used for calibrating the energy to the fluorescence measurement and note that the two experiments are looking at opossite parts of the sky. It is possible that composition changes and/or anisotropies at these energies contribute to these differences.

The shape of the spectrum alone can give us information about possible sources and locations of UHECR. It is tempting to favor a pure proton scenario in which both the suppression and the ankle are features naturally resulting from interactions with the CMB, as the TA Collaboration claims. Models in which protons are produced by isotropically distributed sources emitting protons with free parameters such as the maximum accelerating energy, the spectral index of the emission at the sources and the cosmological evolution of the sources have been successfully fit to the spectral data [17]. The same exercise has been done by the Pierre Auger Collaboration which has also found good fits to the data. However these fits reveal that the maximum accelerating

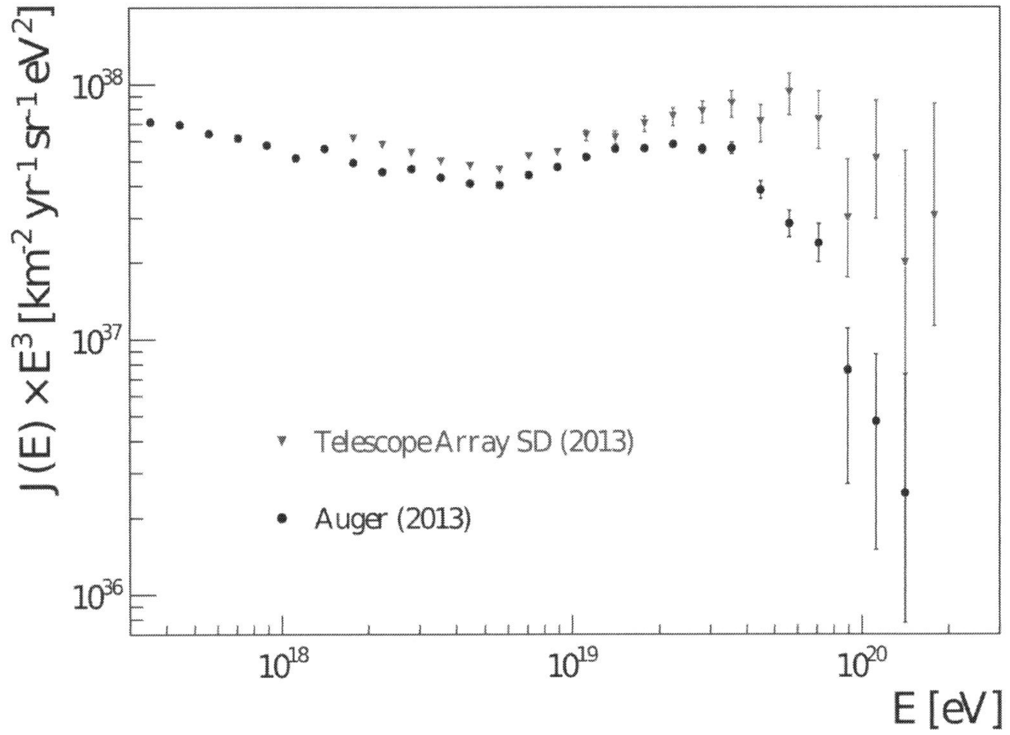

Figure 4. Comparison of the spectra obtained with the Pierre Auger Observatory (black dots) and the Telescope Array (blue triangles). Only statistical uncertainties are plotted as error bars.

energy of these sources must be about 100 EeV. This implies that the observed suppression at the highest energies is an intrinsic property of the sources and no longer the effect of the interactions with the CMB alone. A pure proton composition is not consistent with other data from the Auger Observatory as will be discussed below. Other scenarios with pure iron or mixed composition have also been fitted to the spectrum measured by the Pierre Auger Collaboration. In these scenarios the ankle is interpreted as the transition from galactic to extragalactic origin [18] and ceases to be a natural consequence of particle interactions unless the spectrum is dominated by protons at the highest energies.

4. Hints of anisotropies

The arrival directions provide an important handle on fundamental issues relating to the origin and nature of UHECR. Magnetic deviation can be sufficiently small if these particles have large rigidities and they are not produced too far away. This would be the case if the transition from galactic to extragalactic origin is responsible for the ankle feature. Large-scale anisotropies would be expected in the EeV range, signalling particles that escape from the galaxy. The expected patterns are however not easily predictable due to the uncertainty in magnetic fields and the unknown distribution of sources. Naturally, the patterns would be much stronger in case the composition is light, in agreement with most EeV data. Even in the case that these cosmic rays are extragalactic and are highly isotropic, our bulk motions relative to the rest mass frame of the cosmic rays would be expected to provide some large-scale anisotropy signal at the level of about 0.6% due to our relative motions with respect to the rest frame of the CMB radiation [19].

Above the 50 EeV region, interactions with the background radiation fields restrict the range of proton and iron particles to distances of order 200 Mpc or less and this becomes even lower for particles of intermediate charge [20]. Intergalactic magnetic fields are often assumed to be coherent over regions of ~1 Mpc with intensity below ~1 nG which leads to a few degrees deflection for protons over these distances [21]. As the matter distribution is not homogeneous in our vicinity at scales of order 100 Mpc it is again reasonable to expect anisotropies, particularly if the particles are protons. The attractive possibility of doing astronomy with charged particles requires light mass composition at the highest energies. Composition measurements are needed. Both the Pierre Auger and the TA Collaborations have extensive programs to search for and measure anisotropies of UHECR. These include auto-correlation functions, correlation with existing catalogs, searches for excess regions in the sky, for point sources, for large scale anisotropies, as well as correlations with other experiments.

4.1. Correlations

The first anisotropy signal was claimed by the Pierre Auger Collaboration from the analysis of correlations of arrival directions of UHECR with catalogues of nearby objects. A correlation was reported for events with energies above 57 EeV with objects of redshift below $z < 0.018$ in the Véron-Cetty and Véron catalogue of Active Galactic Nuclei (AGN), assuming a maximum of 3.1° directionality loss [22]. The signal was clear and strong at first but weakened to a $38^{+7}_{-6}\%$ correlation when the data increased by a factor of 3 [23] and more recently to $28.1^{+3.8}_{-3.6}\%$ [24] with over ten years of data, to be compared 21% expected from isotropy. The TA Collaboration has also searched for such correlations and found 11 events correlating when only 5.9 are expected from isotropy, what represents a chance probability of 2% [25]. The results of this search were recently reported to be also consistent with the signal reported by the Pierre Auger Collaboration, with a combined chance probability to be due to an isotropic distribution of $p = 10^{-3}$ [26]. Cross correlations with other flux limited catalogs of nearby extragalactic objects have also been scanned. The corresponding probabilities for the observed correlations to arise from isotropy are at the percent level. These probabilities are actually heavily penalized for the multiple scans in UHECR energy threshold, maximum object distance, random angular spreading of directions and catalogues.

Analysis for directions with excess number of events over background has been used to search for anisotropies. The search directions can be either arbitrary or preselected according to plausible scenarios or to claims of signals from other experiments. Neither the TA or the Pierre Auger Collaboration have found any significant excess coming from the galactic center [27, 28] as claimed by AGASA [29]. The Pierre Auger Collaboration has reported an excess of events in a broad circular region within 18° of a direction in the vicinity of Centaurus A for events above 57 EeV after a scan. Expectations from isotropy would lead to 3.2 events in this region while 13 events were found [23]. The latest update gives 14 events above 58 EeV in a region within 15° around the same region [24]. The chance probability of obtaining such an excess from an isotropic distribution has been calculated to be 1.4%, not low enough to claim a signal. It has been noted that this contributes significantly to autocorrelation measurements and to the correlation with different populations of nearby extragalactic objects. More recently, the TA has reported a significant excess of events above the same energy and within a 20° circle around a direction centered at R.A.(α)=146.7°, Dec.(δ)=43.2° in the northern hemisphere[30]. The chance probability first reported for five years of data was equivalent to a 3.4σ effect (post-trial). Including the sixth year increased the chance probability to 4.1σ [30].

4.2. Large scale anisotropies

The search for large-scale anisotropies has recently attracted a lot of attention. In the multipolar approach the flux is expanded in spherical harmonics, $Y_{lm}(\hat{\mathbf{n}})$ to obtain the multipole coefficients of the flux. For a direction-dependent exposure the data are weighted with the inverse of the corresponding relative exposure [31]. Clearly this procedure requires full sky coverage. Otherwise the multipole expansion can be found for the product of the flux times the exposure as measured by the detector. A relation can be found between this expansion and the standard expansion of the flux in spherical harmonics. If the multipole series of the flux alone is assumed to be bounded, the two sets of coefficients are related by a finite matrix multiplication. To obtain the flux coefficients the matrix must be inverted. Unfortunately the uncertainties involved become large when the number of multipoles is high [32].

Alternatively the rotation of the Earth allows one to search for anisotropies because the exposure is approximately invariant in right ascension. The flux is Fourier-expanded in this coordinate in Rayleigh analysis. The first terms can be also obtained taking the difference of fluxes in the East and West sectors. Since both sectors have identical instantaneous exposures, the difference eliminates to first order detector instabilities and weather effects, at the price of a reduced sensitivity. The Fourier coefficients in right ascension can be related to the standard multipole coefficients. When the odd multipoles in the flux expansion are assumed to be reduced to a single dipole (i.e. to $l = 1$) the first two Fourier terms are exactly a measurement of the dipole projected onto the equatorial plane [33]. In general, this relation holds if the dipole and quadrupole dominate the flux, but it is only approximate.

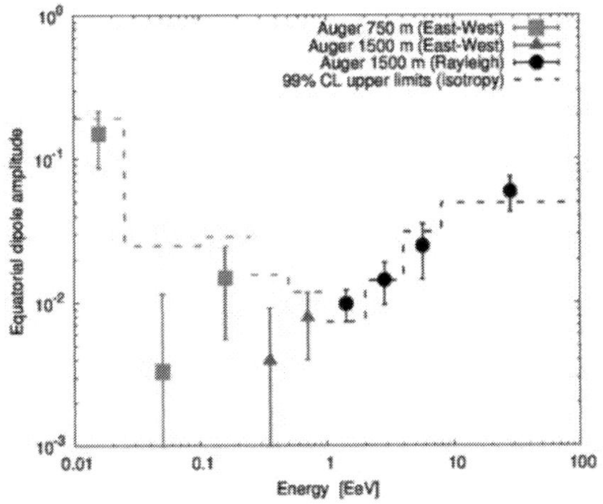

Figure 5. Amplitude of the first term in the Fourier expansion of the flux measured at the Auger Observatory in terms of R.A. as a function of energy. It can be related to the projection of a dipole amplitude onto the equatorial plane.

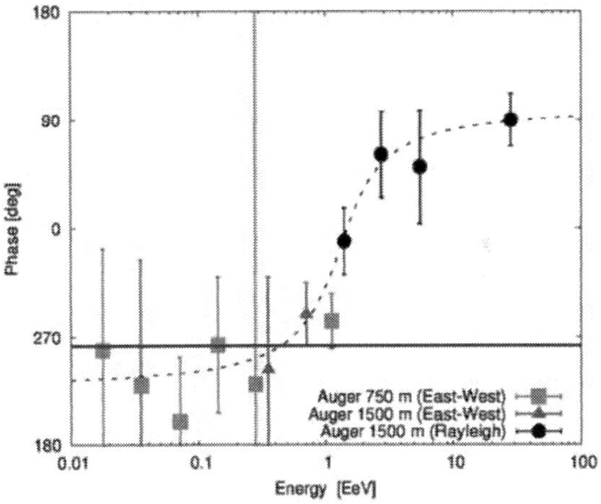

Figure 6. Phase of the first harmonic in R.A. as a function of energy using data from the Pierre Auger Observatory from January 1 2004 to December 31 2010 for the larger array, and from September 12 2007 to April 11 2011 for the infill.

A modified Rayleigh analysis has been applied to Auger showers of energy exceeding 1 EeV and zenith angles below 60° including corrections to compensate for small modulations from non-uniform exposure and weather effects. For lower energies, the differential East-West method has been chosen both on the regular and the 750 m arrays [34]. It has been remarked that the amplitudes of the first harmonic in the modulation of the right ascension at a few energy bins lie just beyond the region of 99% C.L. expected from isotropy [35]. Moreover, the dipole phases obtained in adjacent energy bins are consistent with a smooth transition from a dipole direction pointing close to the galactic center at low energies, to one which is rotated by about 180° at energies above a few EeV. It is expected that in case that some anisotropy exists the first signs would show up precisely in the measured phases that cease to be randomly distributed as expected from statistical noise. A prescription has been established to test these possible hints of anisotropy [34]. The results obtained of the dipole amplitude compared to expectations from isotropy and the dipole phases presented when the prescription was midterm [35], are shown in Figs. 5,6.

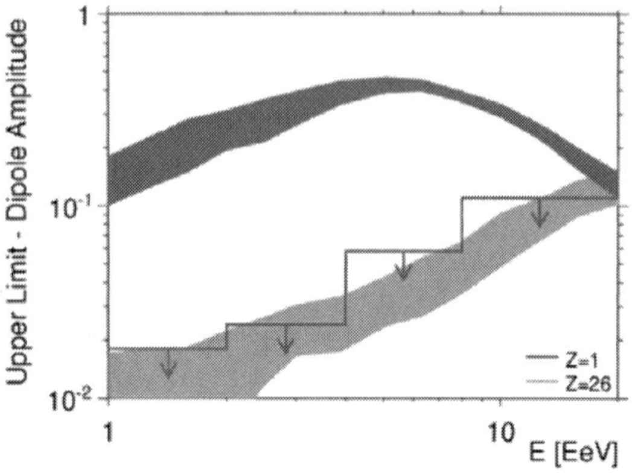

Figure 7. Bound on the dipole amplitude at 99% C.L. compared to expectations from galactic sources of proton and iron uniformly distributed in the galactic disk.

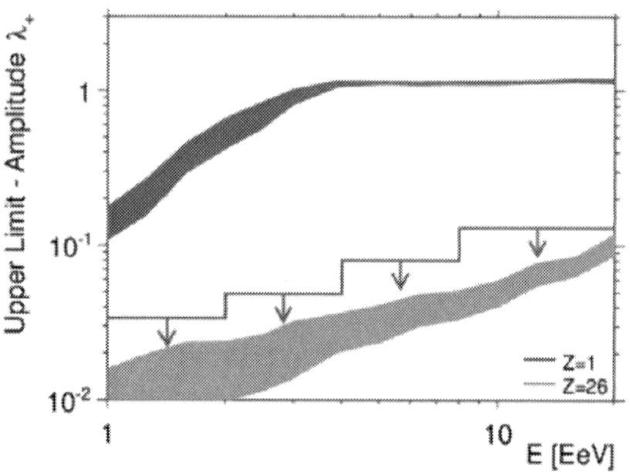

Figure 8. Same as Fig. 7 for one of the quadrupole amplitudes. The bands illustrate the RMS fluctuations due to the stochastic nature of the random component of the magnetic field assumed.

Multipole coefficients have also been obtained for energies above 1 EeV, truncating the expansion to $l \leq 2$, i.e. assuming that all the anisotropy is reduced to a single dipole or a dipole plus a quadrupole [37]. The 99% C.L. bounds obtained on the dipole and quadrupole amplitudes are at the 2-12 % level. They are illustrated in Figs. 7,8 compared to expectations from generic models in which EeV particles are proton or iron, produced in stationary galactic sources distributed in the galactic disk and emitting isotropically [38]. The regular component of the magnetic field is assumed to have Bisymmetric Spiral Structure with an antisymmetric halo with respect to the galactic plane [39]. Pure iron sources would be marginally allowed and pure protons excluded in this scenario. This is quite restrictive because it will be later shown that iron composition is not at all favored at these energies by either observatory. As a result galactic sources of UHECR are unlikely according to the data. It is not feasible to go further in l in the expansion because the uncertainty becomes too high [37]. This is the consequence of not having full sky coverage. In such case assumptions about the flux itself must be made to obtain the coefficients.

4.3. Recent developments

In the time period between the conference and the writing of this article a number of important developments have followed. The combination of the data from the two experiments provides full sky coverage facilitating the direct extraction of the harmonic coefficients for the first time [31]. The energy scales of the two experiments were compared using the declination band in which both detectors overlap as shown in Fig. 9. The results, compatible with isotropy, are limited due

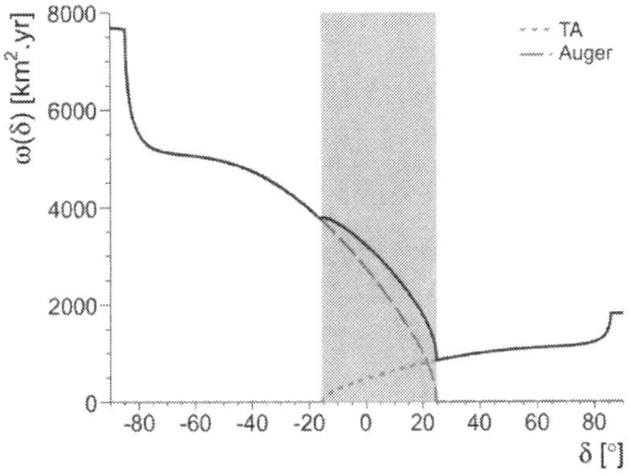

Figure 9. Directional exposure of the Pierre Auger Observatory and the Telescope Array for energies above 10 EeV as a function of declination.

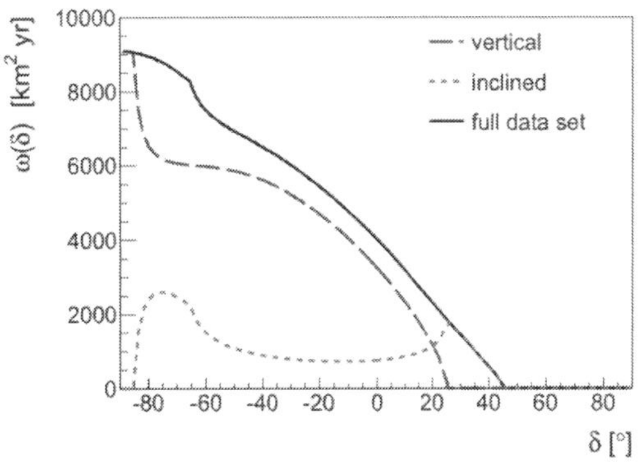

Figure 10. Directional exposures for events of $\theta < 60°$ and for events $60° < \theta < 80°$ at the Pierre Auger Observatory for energies exceeding 4 EeV as a function of declination.

to systematic uncertainties, particularly in the cross calibration of the energy scales. Much more recently, the highly inclined events with zenith angles between 60° and 80° and energies above 4 EeV detected with the Pierre Auger Observatory have been for the first time combined with those below 60° to update large-scale anisotropy searches [36]. Very inclined showers not only represent a further increase of statistics by nearly 30% but they also enhance the sky coverage to 85%, significantly contributing to reduced uncertainties. The exposures of the two sets of data are compared in Fig. 10. Results were obtained for first and second harmonic modulations in both right ascension and in azimuth as well as for multipole coefficients in the hypotheses of either a single dipole or a dipole plus a quadrupole.

While the results are compatible with isotropy for energies between 4 and 8 EeV, the raw significance assuming a single dipole is already at the 4σ level for events with $E > 8$ EeV, thus strengthening the hint of anisotropy. The multipole coefficients assuming a single dipole give a magnitude of 0.073 ± 0.015 in the direction $(\delta, \alpha) = (95° \pm 13°, -39° \pm 13°)$. Assuming a dipole plus a quadrupole gives results consistent with those reported above, but the quadrupole coefficients are not significant. Both the dipole strength and the direction are consistent with previous results. A map of the measured flux above 8 EeV smoothed in 45° windows is shown in Fig 11 resulting in a contrast (maximum relative difference between fluxes) of 21%.

5. Unravelling UHECR composition

Understanding composition is a very hard task since showers are stochastic processes. Showers of a given energy and direction tend to be rather similar regardless of composition since the shower development is a process by which the initial particle energy is converted to matter, mostly pions and kaons, in successive particle interactions. In addition, all our knowledge about the subtle differences between showers induced by different primaries must be deduced from extensive simulation programs that must extrapolate the properties of particle interactions to center-of-mass energies far exceeding those achieved in accelerator experiments. A number of simulation packets for air showers exist and they can use different models to extrapolate properties of particle interactions such as cross sections, inelasticity and multiplicity. Unfortunately the differences between these extrapolations are subtle and change the shower evolution at levels which are similar to those due to changing the primary composition. As a result, not only has it been notably difficult to extract properties about composition from air shower measurements,

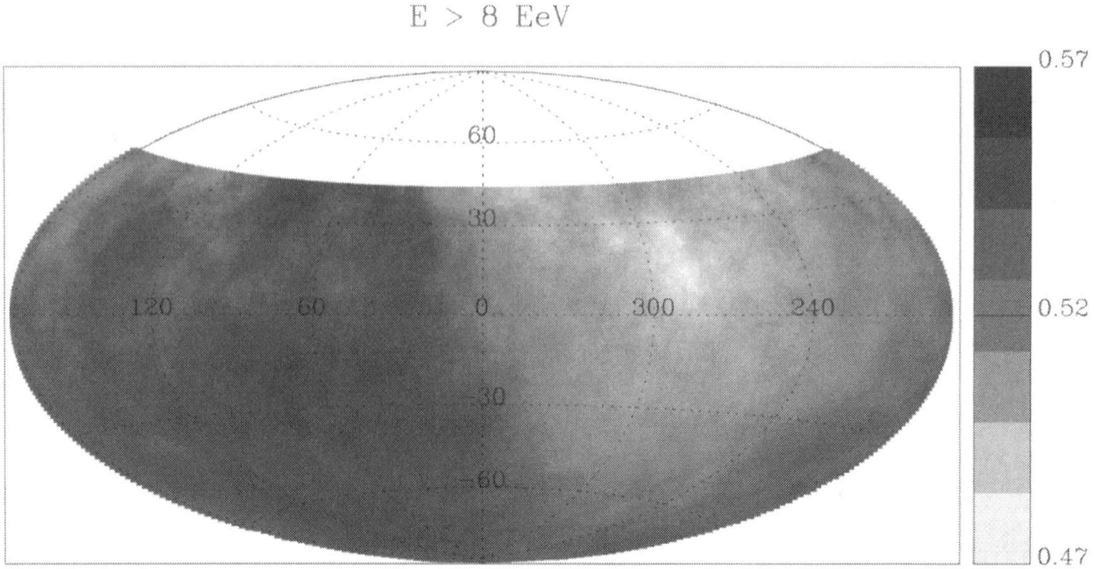

Figure 11. Sky map of the flux in equatorial coordinates expressed in $km^{-2}y^{-1}sr^{-1}$ for energies above 8 EeV smoothed in an angular window of 45°.

but the coupling between an unknown composition and uncertainties in the hadronic models severely limits the conclusions that can be obtained.

5.1. Measurements of the average depth of maximum

Composition can be addressed using the depth of the position of the maximum in the number of shower particles, abbreviated as shower maximum, X_{max}, which is measured by the FD. In the simple superposition model, air showers of energy E produced by elements of mass A can be approximated by the superposition of A proton showers of energy E/A. As the shower maximum scales with the logarithm of the traversed depth, showers produced by heavy nuclei are shallower. In addition the superposition of A showers of lower energy has an X_{max} distribution which is much narrower. In this picture the resulting X_{max} is some kind of average of the maxima of the A single proton showers that make it up, which naturally fluctuates less.

Both experiments have made precise measurements of the depth of shower maximum with their FDs that allow a partial reconstruction of shower development. The TA (Auger) Collaboration has reported X_{max} data for energies above 1.6 (0.63) EeV. While the results look at first rather similar, the interpretation made by each collaboration is qualitatively quite different. The TA has interpreted these measurements as compatible with a pure proton composition [40] while the Pierre Auger Collaboration claims that they indicate a composition that changes from being light at 1 EeV to becoming heavier [41].

Part of this discrepancy may lie in the way that the two collaborations address the bias that arises because of the limited field of view of the telescopes and the quality cuts that are required to accurately reconstruct X_{max}. Depending on the zenith angle and impact point of the shower, an FD views a different range of depths as illustrated in Fig. 12 for three shower geometries. A given shower will be selected if the maximum is well within the field of view (shaded region). In general, the distributions will be artificially cut by the field of view and the quality cuts, potentially biasing the result. If there is a mixture of primaries there will also be a composition bias in the selected data.

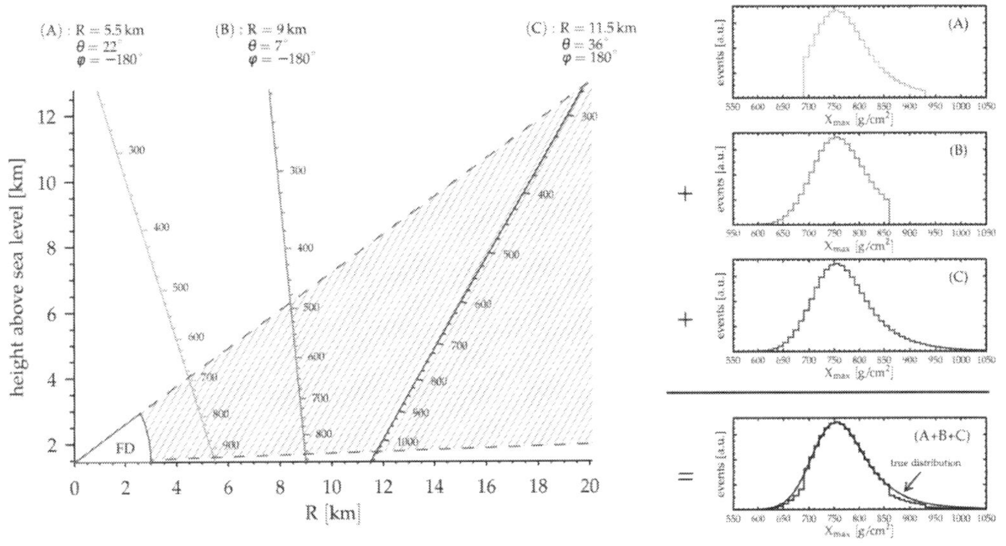

Figure 12. Left: Range of depths (gcm^{-2}) accessible to an FD for three different shower geometries labelled A, B and C. Only the shaded area is viewed by a fluorescence telescope. Right: The observed X_{max} distributions for showers arriving with the three geometries indicated on the left, illustrating the truncation of the observed distribution due to the field of view of the telescope.

The Telescope Array Collaboration measures an effective distribution of shower maximum and compares it to simulations which undergo identical selection as the data. Because of the selection bias, the measured average and variance of X_{max} will not match those of the true distribution. The comparison between data and simulation is valid provided the models used accurately describe the ranges of X_{max}. The Pierre Auger Collaboration attempts an unbiased measurement introducing fiducial cuts which reduce the statistics [42]. For each shower that passes the quality cuts the range of values of X_{max} that can be selected must contain that of the bulk of the X_{max} distribution for the corresponding energy characterized by its extremes. These extremes are chosen ensuring that, when they are changed, $< X_{max} >$ remains sufficiently stable.

The different approaches of the two collaborations means that they are not measuring the same quantity so that the values obtained cannot be directly compared to one another. The predictions for the measurements made by each collaboration for a given hadronic model and composition are also different. In spite of this, the two sets of measurements [43, 44] have been superposed in the right panel of Fig. 13. This figure must be interpreted with great care. The two sets of measurements are compared to predictions for protons for the true $< X_{max} >$ using EPOS-LHC [45], SYBILL2.1 [46] and QGSJETII-04 [47] (the three upper lines) and for the effective $< X_{max} >$ expected in the Telescope Array, using QGSJET-01 [48], SYBILL2.1 and QGSJETII-03 both for proton and iron primaries (respectively the middle and lower sets of lines). We note that the most recent hadronic models EPOS-LHC and QGSJETII-04, and SYBILL2.1 have systematically deeper X_{max} than earlier versions of QGSJET and EPOS. Predictions for protons for the two experiments with SYBILL2.1 illustrate the magnitude of the selection bias of the Telescope Array data. Clearly the Pierre Auger measurements of X_{max} are compatible with

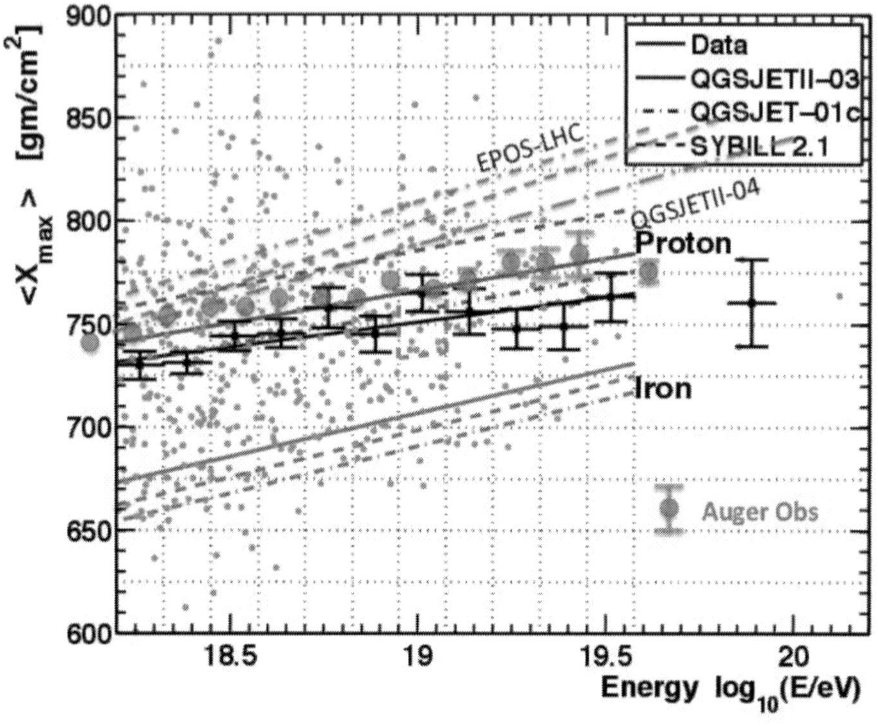

Figure 13. Effective $< X_{max} >$ measured at the Telescope Array (small black squares) and $< X_{max} >$ obtained from the Auger Observatory data with fiducial cuts (large brown circles) compared to expectations for the effective X_{max} measured by the TA for proton (middle set of lines; blue in on-line version) and iron (lower set of lines; red in on-line version), using QGSJET-01 (dot dashed), QGSJETII-03 (full) and SYBILL-2.1 (dashed). The upper set of lines (purple in on-line version) correspond to X_{max} for proton with EPOS-LHC (dot dashed), SYBILL-2.1 (dashed) and QGSJETII-04 (dot long dashes) from top to bottom (see text).

a proton like composition at energies of order 1 EeV. The rate at which the shower maximum changes with energy, or *elongation rate*, is much smaller than expectations for a pure composition suggesting that composition is becoming heavier. The Telescope Array results are, in contrast, compatible with expectations for protons using QGSJETII-03.

5.2. Combining the average and variance of the depth of maximum

Study of the X_{max} distributions has not helped to clarify this controversy. The variance measured by the Pierre Auger Collaboration has consistently shown a decrease with energy above 1.8 EeV which is suggestive of a composition becoming both heavier and less mixed. The TA results indicate distributions in effective X_{max} which are compatible with proton simulations. It is worth noting that the total number of points in the TA analysis with energy exceeding 1.6 EeV (438) is of the same order as those reported in the single bin between 5 and 6.3 EeV (413) in the last update by the Pierre Auger Collaboration [49].

The implications of these measurements depend on the interaction model. The mean value of the X_{max} distribution is indicative of the average mass composition and the variance can be related to the dispersion in composition. This can be exemplified using a simple generalization of the Heitler model for shower development [50]. In this simple model it can be shown that the average and variance of X_{max} are given by:

$$< X_{max} > \;=\; < X_{max} >_p + f_E < ln\ A > \quad (1)$$

$$\sigma^2(X_{max}) \;=\; < \sigma^2_{sh} > + f_E^2 \sigma^2_{ln\ A} \quad (2)$$

Here the sub-index p refers to a pure proton composition and f_E is an energy dependent term. The variance of X_{max} has a contribution from the variance of each primary, σ^2_{sh}, and another because each primary has a different $< X_{max} >$. Here σ_{sh} can be regarded as a function of mass number, A, and can be parameterized as:

$$\sigma^2_{sh} = \sigma^2_p \left[1 + a\ lnA + b\ (lnA)^2\right] \quad (3)$$

Taking the average of Eq. 3 it is straightforward to invert the above equations to give an explicit relation of the average and variance of lnA in terms of measurable quantities:

$$< ln\ A > \;=\; \frac{< X_{max} > - < X_{max} >_p}{f_E} \quad (4)$$

$$\sigma^2_{ln\ A} \;=\; \frac{\sigma^2(X_{max}) - \sigma^2_{sh}(< lnA >)}{b\ \sigma^2_p + f_E^2} \quad (5)$$

These relations depend on the hadronic model though $< X_{max} >_p$, f_E and $< \sigma_{sh} >$.

The results for three different models are shown in Fig. 14. The averages obtained of lnA indicate a clear transition from light to heavier composition as the energy rises above 2 EeV. The variance of lnA indicates the level of mixing and a value of zero corresponds to a pure composition no matter what. The negative variances obtained for QGSJETII-04 are unphysical, suggesting that either the model is not compatible with the data or that the measurements are unreliable [50]. These results have been recently updated and completed with an analysis of the full X_{max} distributions, which allows estimates of the relative abundances assuming a mixed composition of proton, helium, nitrogen and iron [49]. The conclusions are basically the same.

5.3. The distributions of depth of maximum

The analysis of the mean and variance does not fully exploit the quality and richness of the available data which can be obtained from the distributions of X_{max}. The Telescope Array has compared the effective X_{max} distributions obtained in different energy bins and has found them compatible with proton composition and QGSJETII-03 [40]. The distributions obtained by the Pierre Auger Collaboration have been fitted to obtain the fractions of primaries according to different hadronic models. Templates are obtained for distributions of a restricted number of pure primaries, namely proton, helium, nitrogen and iron. Best fit values of the relative fractions of the primary composition have been obtained in three simplified scenarios, namely restricting the number of different species to proton and iron only, considering proton, helium and iron and finally using a four component mix by also including nitrogen [41].

The first scenario with only proton and iron is shown to give poor quality fits but the situation improves as intermediate mass primaries are introduced. The results obtained with EPOS-LHC are satisfactory for all the energy range while QGSJETII-04 does not fit the distributions well. It can be argued that adding extra components will not improve the fit for this model. This is consistent with the unphysical results obtained for QGSJETII-04 when the first two moments (mean and variance) of the X_{max} distribution are interpreted in terms of the mean and variance of lnA as discussed earlier. The predictions of proton and iron fractions are remarkably similar

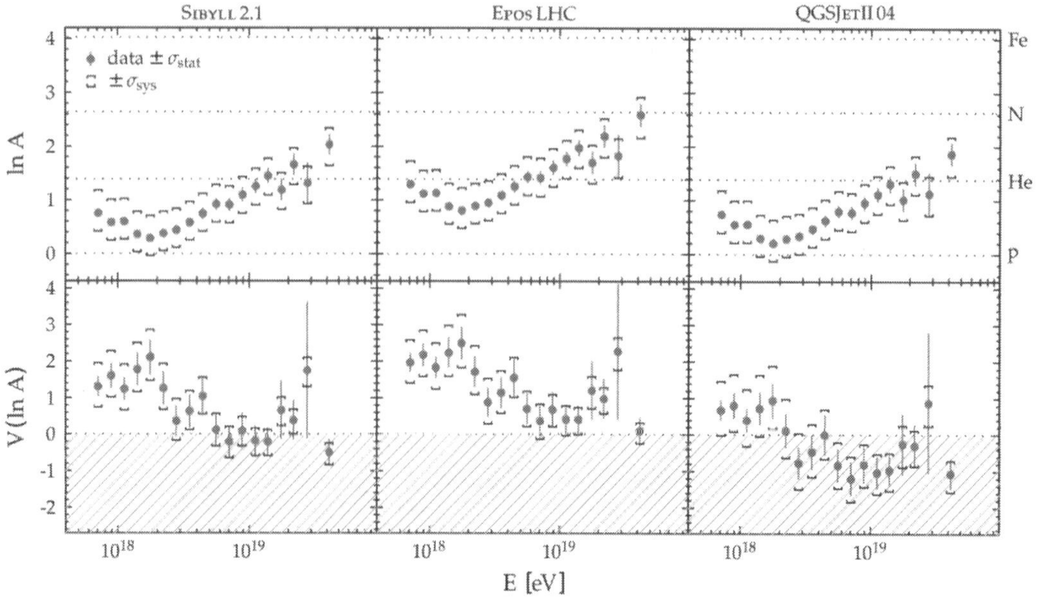

Figure 14. Results for $< ln\ A >$ (top row) and $\sigma^2_{ln\ A}$ (bottom row) inferred with the Auger X_{max} data for three different hadronic models.

for all models while they differ greatly on the composition of intermediate mass ranges. The proton fraction is energy dependent changing from values between 60% and 80% at energies around 2 EeV but drops to very low fractions as the energy increases to 10 EeV. The proton fraction in the EeV range seems to be high to explain the ankle feature with the transition to extragalactic sources unless there is a proton contribution at EeV energies from the galaxy too. But this is disfavored by anisotropy measurements as explained in the previous section. On the other hand a pure proton scenario would require substantial changes to the hadronic models.

5.4. Measurements of the muon content of showers

Highly inclined showers consist mainly of muons and thus provide an opportunity to measure the muon content of air showers [51]. Since N_{19} is the relative size with respect to a 10 EeV proton shower simulated with QGSJETII-03, it is also an approximate estimate of the relative number of muons with respect to the same reference. Detailed comparisons have been made between the number of muons relative to those of the reference and the fitted values of N_{19} in simulated showers using different composition and hadronic models. They reveal a small energy dependent bias and spread. A new variable R_μ is defined correcting N_{19} at the level of a few percent to compensate this bias and to minimize the dispersion from hadronic models [52]. The calibration data for inclined showers provides a measurement of R_μ and an independent measurement of the FD energy. The results are also consistent with a change to heavier composition in the EeV range [51]. There are more muons in the data that in the reference by about a factor of 1.8. This can be marginally accommodated using iron and the most recent hadronic models. However measurements of X_{max} favor a much lighter composition which is in tension with these results. This is illustrated in Fig. 15 that displays a two-dimensional plot of $< X_{max} >$ and $ln\ R_\mu$ at an energy of 10 EeV compared to expectations from different hadronic models and mass compositions.

Similar results have been observed when comparing the energy of events with $\theta < 60°$ obtained

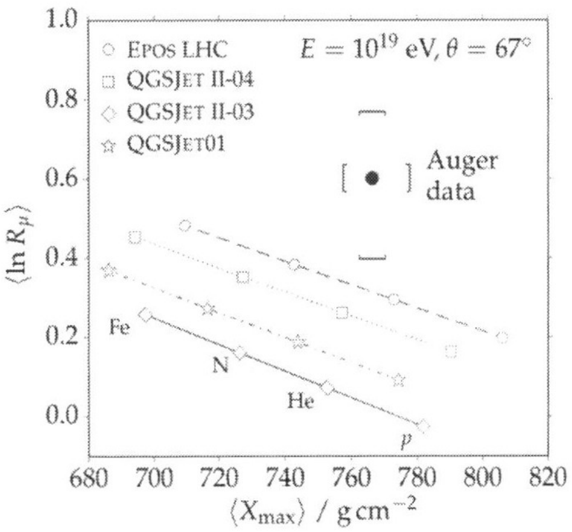

Figure 15. Two dimensional plot of $ln\ R_{\mu}$ obtained from inclined showers and X_{max} at the Pierre Auger Observatory compared to expectations form different models and different primaries.

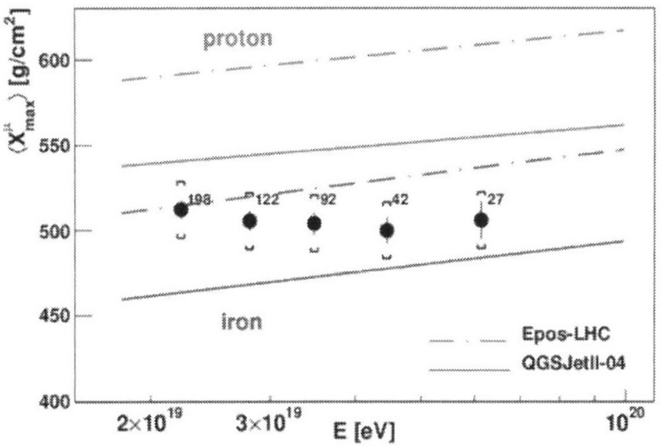

Figure 16. Measurements of the average value of X^{μ}_{max} as a function of energy using the Pierre Auger Observatory SD data. The results are compared to expectations for both proton and iron using the most recent hadronic models.

using simulations to those obtained with the calibration procedure and with other methods to count muons [53, 54]. This is not new, similar evidence has been found from experiments measuring horizontal muon bundles [55], from lower energy surface arrays such as KASCADE-Grande [57] and from the Large Hadron Collider (LHC) [56].

5.5. Measurements of the muon production depth

The hybrid approach has great potential to explore new SD observables in order to further constrain composition and hadronic models. The Pierre Auger Collaboration has explored observables related to the muons in the shower front because it is very sensitive to them. An interesting method has been developed by studying the time structure of the muon front [58]. The time delays of the muons with respect to a plane front can be shown to be dominated by geometry and a simple relation holds between the lateral displacement, the time delay and the distance at which the muon was produced. Under some conditions, this relation can be inverted to obtain the distribution of the depths at which the shower muons are produced, using the time structure of the data of the SD stations [6]. This variable has been designated as X^{μ}. The depth at which the maximum production is achieved, X^{μ}_{max}, depends on composition and hadronic models as could be expected, similarly to X_{max}.

The average values of X^{μ}_{max} measured by the Pierre Auger Observatory are displayed as a function of energy in Fig. 16 compared to expectations from several models. Interestingly, one of the required conditions to obtain it is that the number of sampled muons is high, so that the method is more sensitive at the highest energies where the hybrid data are scarce. The results are also consistent with the primary mass composition becoming heavier. However the values obtained are out of the bracketed values obtained with EPOS-LHC. The measured value of $< X^{\mu}_{max} >$ can be also converted to an average value of $< lnA >$ mimicking the procedure used for $< X_{max} >$. The results for the values of $< lnA >$ obtained using both X_{max} and X^{μ}_{max} are combined in Fig. 17 for QGSJETII-04 and EPOS-LHC. The two most recent hadronic models QGSJETII-04 and EPOS-LHC, updated using LHC data, seem to be inconsistent with X_{max}

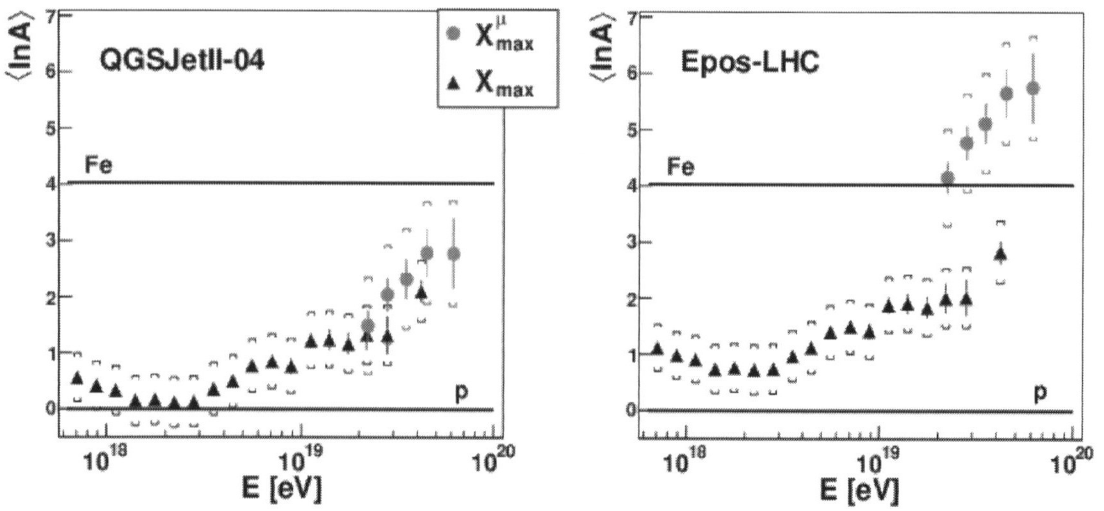

Figure 17. Results for $< ln\ A >$ obtained using the measurements of both X_{max} and X^{μ}_{max} data at the Pierre Auger Observatory using the most recent hadronic models.

and X^{μ}_{max} data taken together, pointing to the possibility that part of the puzzle lies in the extrapolations of accelerator data. Surely this is one of the paths that must be further explored in the future [59].

6. Summary and outlook

Recent results obtained by the TA and the Pierre Auger Collaborations have been reviewed. The spectrum measurements agree below 50 EeV and display the ankle at about 5 EeV. For energies greater than 50 EeV, both spectra clearly indicate a suppression but there are some important discrepancies. TA results are consistent with models in which protons are accelerated and they interact with the CMB producing the GZK cutoff. The Auger spectrum can only be made consistent with such scenario if the fluxes at the sources are suppressed at similar energies. But such a scenario is in fact in tension with X_{max} and X^{μ}_{max} data from the Auger Observatory.

Anisotropy searches with the Auger data have shown some evidence of large-scale anisotropy which is becoming stronger and both observatories have indications of possible hot spots in each hemisphere. The significance of the TA hot spot is stronger. Recent updates of these measurements enhancing the sky coverage with inclined showers and combining the data of the two observatories have given compatible results strengthening the case for large-scale anisotropies. It is in fact not ruled out that large scale anisotropies may have some impact in the flux discrepancies between the northern and southern hemisphere.

Composition measurements also indicate discrepancies between the two experiments. Unbiased X_{max} measurements can be related to primary mass if particle interactions are under control. Unfortunately this is not the case and hadronic models are evolving as new data appear. While the TA group measures an effective X_{max} because of selection effects, the Pierre Auger Collaboration has developed fiducial cuts to avoid bias. Results are not easy to compare since each group uses observables not strictly identical and compares to predictions using different hadronic models. The TA flux and X_{max} data are claimed to be consistent with a pure proton scenario. The average and variance of X_{max} Auger data indicate a transition from light to heavier elements starting at $E \sim 4$ EeV. We note that at roughly this same energy the spectrum displays the ankle and there are indications of a small anisotropy signal.

Auger data allow the measurement of X^{μ}_{max} and the numbers of muons in inclined showers (and with other methods). These data also suggest that the primary mass composition becomes heavier as the energy increases, however new difficulties arise when all the data are tried to be interpreted in terms of conventional hadronic models. The measured number of muons in the showers is difficult to reconcile with the deduced composition inferred from X_{max} measurements. On the other hand the interpretation of X^{μ}_{max} and X_{max} results in terms of composition is inconsistent for EPOS-LHC. The average and variance of the X_{max} distributions measured at the Auger Observatory disfavor QGSJETII-04, giving a negative variance of lnA. Older models do not improve the picture. Finally other interesting findings can be attributed to the two groups, such as negative searches for photons, neutrons and neutrinos that have led to relevant bounds.

Auger results related to mass composition using these independent methods suggest a mixed composition in which, qualitatively, nuclei of increasing A are dominating the composition as the energy increases. When these results are considered together with the energy spectrum, new classes of models have been proposed such as those supporting a heavy composition at the sources which gets degraded through interactions with the CMB, or those which assume that the sources have a maximum acceleration energy which increases with the nuclear charge, Z [20, 60, 61, 62, 63]. But these interpretations must be taken with caution because standard hadronic models fail to interpret all the data in a consistent way.

While much progress has been made in the study of UHECR there are still many fundamental issues to be solved. The different interpretations obtained by the Telescope Array and the Pierre Auger collaborations should be understood. There are several joint groups working together on the comparison of the data and combined analyses, notably for spectrum, composition and anisotropy studies. There are plans for new enhancements and extensions of the two observatories, to increase the size of the TA [64] and to add muon counters at the Auger Observatory. The results obtained with these enhanced detectors and the collaborative efforts of the two collaborations should clarify the picture in the next few years of what is likely to shape the future of the field.

Acknowledgments

I am grateful to the Telescope Array Collaboration for having facilitated much material to make this article possible and I am indebted to my colleagues from the Pierre Auger Collaboration with whom I have acquired my views in the field. I also thank R. Clay, C. Dobrigkeit and A.A. Watson for suggestions after carefully reading the manuscript. I thank the Ministerio de Economía (FPA2012-39489), the Consolider-Ingenio 2010 CPAN Programme (CSD2007-00042), the Xunta de Galicia (GRC2013-024), Feder Funds and Marie Curie-IRSES/ EPLANET (European Particle physics Latin American NETwork), 7th Framework Program (PIRSES- 2009-GA-246806). I am also grateful to CESGA for computing resources.

References

[1] Linsley J 1963 *Phys. Rev. Let.* **10** 146.
[2] Abraham J *et al.* [Pierre Auger Collab.] 2004. *Nucl. Instrum. Meth.* **A523** 50.
[3] Abreu P *et al.* [Pierre Auger Collab.] 2010 *Nucl. Instrum. Meth.* **A620** 227.
[4] Abu-Zayyad T *et al.* [TA Collab.] 2012 *Nucl. Instrum. Meth.* **A689** 87.
[5] Thomson G these proceedings.
[6] Cazón L *et al.* 2005 *Astropart. Phys.* **23** 393 and 2012 **36** 211.
[7] Baltrusaitis *et al.* [Fly's Eye Collab.] 1985 *Nucl. Instrum. Meth.* **A240** 410.
[8] Ostapcehnko S S 2006 *Phys. Rev. D* **74** 014026.
[9] Hersil J *et al.* 1961 *Phys. Rev. Lett.* **6** 22.
[10] Aab A *et al.* 2014 *JCAP* **8** 019.
[11] Ave M, Vazquez R A and Zas E 2000 *Astropart. Phys.* **14** 91.
[12] Ave M, Hinton J A, Vázquez R A, Watson A A and Zas E 2002 *Phys. Rev. D* **65** 063007.

[13] Dembinski H P, Billoir P *et al.* 2010 *Astropart. Phys.* **34** 128.
[14] Abraham J *et al.* [Pierre Auger Collab.] 2008 *Phys. Rev. Let.* **101** 061101.
[15] Schulz A for the Pierre Auger Collab. 2013 *Proc. of the 33rd Int. Cosmic Rray Conf.*, Rio de Janeiro, arXiv astro-ph:1307.5059.
[16] Lettesier-Selvon A for the Pierre Auger Collab. 2013 *Proc. of the 33rd ICRC*, Rio de Janeiro, (highlight talk) arXiv astro-ph:1310.4620.
[17] Kido E and Kalashev O E, for the TA Collab. 2013 2013 *Proc. of the 33rd ICRC*, Rio de Janeiro.
[18] Linsley J 1963 *Proc. of the 8th Int. Cosmic Ray Conf.*, Jaipur 1963, vol. 4 p. 77.
[19] Kachelriess M and Serpico D P 2006 *Phys. Lett.* **B640** 225.
[20] Allard D 2012 *Astropart. Phys.* **39-40** 33.
[21] Harari D 2014 *Proc. of Phys. Dark Univ.* **4** 23. DOI: 10.1016/j.dark.2014.04.003.
[22] Abreu P *et al.* [Pierre Auger Collab.] 2007 *Science* **318** 938.
[23] Abreu P *et al.* [Pierre Auger Collab.] 2010 *Astropart. Phys.* **34** 314.
[24] Aab A *et al.* [Pierre Auger Collab.] Submitted to *Astrophys. J.*; arXiv astro-ph:1411.6111.
[25] Abu-Zayyad T *et al.* [TA Collab.] 2012 *Astrophys. J.*; **5757** 26.
[26] Kampert K H and Tinyakov P 2014 *Proc. of TeV Particle Astrophysics* to be submitted to CRAS; arXiv astro-ph 1405.0575.
[27] Abraham J *et al.* [Pierre Auger Collab.] 2007 *Astropart. Phys.* **27** 244.
[28] Fukushima M *et al.* 2013 *Proc. of the 33rd ICRC*, Rio de Janeiro.
[29] Hayasida N *et al.* 1999 *Astropart. Phys.* **10** 303.
[30] Abbasi R U *et al.* [TA Collab.] 2014 *Astrophys. J. Let.* **790** L21.
[31] Aab A *et al.* [Pierre Auger Collab.] 2014 *Astrophys. J.* **749 (2)** 172.
[32] Billoir P and Deligny O 2008 *JCAP* **02** 009.
[33] Tinyakov P G and Urban F R; arXiv astro-ph:1411.2486.
[34] Abreu P *et al.* [Pierre Auger Collab.] 2011 *Astropart. Phys.* **34** 627.
[35] Sidelnik I for the Pierre Auger Collab. 2013 *Proc. of the 33rd ICRC*, Rio de Janeiro, arXiv astro-ph:1307.5059.
[36] Aab A *et al.* [Pierre Auger Collab.] Submitted to *Astrophys. J.* arXiv astro-ph:1411.6953.
[37] Abreu P *et al.* [Pierre Auger Collab.] 2012 *Astrophys. J. Sup.* **203** 34.
[38] Abreu P *et al.* [Pierre Auger Collab.] 2013 *Astrophys. J. Let.* **762** L13.
[39] Pshirkov M S *et al.* 2011 *Astrophys. J.* **738** 192.
[40] Abbasi R U *et al.* [TA Collab.] Submitted to *Astropart. Phys.* arXiv astro-ph:1408.1726.
[41] Aab A *et al.* [Pierre Auger Collab.] 2014 *Phys. Rev.* **D 90** 122006 .
[42] Abraham J *et al.* [Pierre Auger Collab.] 2010 *Phys. Rev. Letts.* **104** 091101
[43] Abbasi R U *et al.* [TA Collab.] 2015 *Astropart. Phys.* **64** 49
[44] de Souza V for the Pierre Auger Collab. 2013 *Proc. of the 33rd ICRC*, Rio de Janeiro, arXiv astro-ph:1307.5059.
[45] Pierog T *et al.* 2013 DESY-13-125, arXiv hep-ph:1306.0121.
[46] Engel R, Gaisser T K, Lipari P and Stanev T 1992 *Phys. Rev.* **D46** 5013.
[47] Ostapcehnko S S 2011 *Phys. Rev.* **D83** 014018.
[48] Kalmykov N N and Ostapchenko S S 1993 *Phys. Atom. Nucl.* **56** 346 [1993 *Yad. Fiz.* **56N3** 105]; Kalmykov N N, Ostapchenko S S and Pavlov A I 1994 *Bull. Russ. Acad. Sci. (Physics)* **58** 1966-69.
[49] Aab A *et al.* [Pierre Auger Collab.] 2014 *Phys. Rev.* **D 90** 122005.
[50] Abreu P *et al.* [Pierre Auger Collab.] 2013 *JCAP* **2** 26.
[51] Valiño I for the Pierre Auger Collab., these proceedings.
[52] Aab A *et al.* [Pierre Auger Collab.] Submitted to *Phys. Rev.* **D**; arXiv astro-ph 1408.1421.
[53] Farrar G for the Pierre Auger Collab. 2013 *Proc. of the 33rd ICRC*, Rio de Janeiro, arXiv astro-ph:1307.5059.
[54] Kegl B for the Pierre Auger Collab. 2013 *Proc. of the 33rd ICRC*, Rio de Janeiro, arXiv astro-ph:1307.5059.
[55] Kokoulin R 2012 *Proc. of ECRS 2012* Moscow Erlykin A D (ed.).
[56] Bongi M these proceedings.
[57] Arteaga-Velázquez J C these proceedings.
[58] Cazón L, Vázquez R A, Watson A A and Zas E 2004 *Astropart. Phys.* **21** 71.
[59] Ulrich R these proceedings.
[60] Allard D these proceedings.
[61] Taylor A 2014 *Astropart. Phys.* **54** 48.
[62] Aloisio R 2014 submitted to *JCAP*.
[63] Stanev T *et al.* 2014 *Nuc. Inst. Meth.* **A742** 42.
[64] Fukushima M these proceedings.

Institute of Physics
Dirac House, Temple Back
Bristol BS1 6BE UK

ISSN: 1742-6588
ISBN 978-1-5108-1313-7

9 781510 813137

CURRAN ASSOCIATES INC.
proceedings
.com